SIGNAL PROCESSING VI
THEORIES AND APPLICATIONS

SIGNAL PROCESSING VI
THEORIES AND APPLICATIONS

Proceedings of EUSIPCO-92
Sixth European Signal Processing Conference
Brussels, Belgium, August 24–27, 1992

Edited by

J. VANDEWALLE
ESAT Laboratory
Katholieke Universiteit Leuven
Heverlee, Belgium

R. BOITE
Département E.P.I.
Faculté Polytechnique de Mons
Mons, Belgium

M. MOONEN
A. OOSTERLINCK
ESAT Laboratory
Katholieke Universiteit Leuven
Heverlee, Belgium

VOLUME II

1992

ELSEVIER
AMSTERDAM • LONDON • NEW YORK • TOKYO

ELSEVIER SCIENCE PUBLISHERS B.V.
Sara Burgerhartstraat 25
P.O. Box 211, 1000 AE Amsterdam, The Netherlands

Library of Congress Cataloging-in-Publication Data

EUSIPCO-92 (1992 : Brussels, Belgium)
Signal processing VI : theories and applications : proceedings of EUSIPCO-92, Sixth European Signal Processing Conference, Brussels, Belgium, August 24-27, 1992 / edited by J. Vandewalle ... [et al.].
p. cm.
Includes bibliographical references and index.
Contents: v. 1. Tutorials, special sessions, speech processing, knowledge engineering, and signal processing -- v. 2. Theory of signals and systems, neural networks for signal processing -- v. 3. Image processing and multidimensional signal processing, implementations, applications.
ISBN 0-444-89587-6 (set : acid-free paper)
1. Signal processing--Congresses. I. Vandewalle, J. (Joost), 1948- . II. Title. III. Title: Signal processing 6. IV. Title: Signal processing six.
TK5102.5.E9 1992
621.382'2--dc20 92-24842
CIP

ISBN: 0 444 89587 6

This book is printed on acid-free paper.

Printed in The Netherlands.

FOREWORD

EUSIPCO-92, the European Signal Processing Conference, is the sixth in the sequence of the International Conferences promoted and organized by EURASIP, The European Association for Signal Processing. The Conference was held from 24 to 27 August 1992, in Brussels, the capital of Belgium, and the administrative heart of the European Community. These three volumes present the Proceedings of the Conference.

EUSIPCO-92 consisted of 48 separate sessions, organized in 6 parallel programs. The Program Committee reviewed over 640 submissions to select the 410 papers that were presented at the Conference. Each submission was assessed by at least two independent reviewers. In addition, 13 tutorials were given by well known experts in the field, and 7 special sessions were held, with 44 contributions on relevant actual topics.

The technical sessions were organized in 7 themes, containing the following topics :

1. Theory of Signals and Systems :
   a) Detection, b) Estimation, c) Filtering, d) Spectral Estimation, e) Adaptive Systems, f) Modeling, g) Digital Transforms, h) Digital Filtering.

2. Image Processing and Multidimensional Signal Processing :
   a) Coding, b) Enhancement, c) Restoration, d) Medical Image Processing.

3. Speech Processing :
   a) Coding, b) Synthesis, c) Recognition and Understanding, d) Enhancement.

4. Implementations :
   a) Hardware, b) Software, c) VLSI, d) Novel Architectures, e) Array Processing.

5. Knowledge Engineering and Signal Processing :
   a) Expert Systems, b) Pattern Recognition, c) Signal Interpretation, d) Image Understanding.

6. Neural Networks for Signal Processing :
   a) Theory, b) Speech, c) Vision, d) Implementations.

7. Applications :
   a) Radar, b) Sonar, c) Communications, d) Geophysics, e) Digital Audio, f) Biomedics, g) Sensing, h) Robotics, i) Astrophysics, j) Mechanics, k) other.

The volumes are organized as follows :

Volume I :
Tutorials, Special Sessions, Speech Processing, Knowledge Engineering and Signal Processing

Volume II :
Theory of Signals and Systems, Neural Networks for Signal Processing

Volume III :
Image Processing and Multidimensional Signal Processing, Implementations, Applications

We would like to thank all the EUSIPCO-92 participants and in particular those whose contributions made these proceedings so attractive. Our thanks are also due to all the co-sponsoring institutions for their financial support, and to Elsevier Science Publishers B.V. - North Holland for the realization of these Proceedings. We would be remiss not to thank the Conference Committee, the Scientific Committee, the Program Committee, the Session Chairmen, and the Tutorial Speakers. Last, but by no means least, we are most grateful to Ann Deforce, Rita De Wolf and Ingrid Tokka, for their outstanding clerical support.

Heverlee, May 1992

J. Vandewalle
R. Boite
M. Moonen
A. Oosterlinck

CONFERENCE COMMITTEE

General Chairman
Oosterlinck A.
Katholieke Universiteit Leuven

Vice-Chairman
Van Eck J.L.
Université Libre de Bruxelles

Program Chairmen
Boite R.
Faculté Polytechnique de Mons

Vandewalle J.
Katholieke Universiteit Leuven

Finance
Govaerts R.
Katholieke Universiteit Leuven

Local Arrangements
Grenez F.
Université Libre de Bruxelles

Exhibition
Van Compernolle D.
Katholieke Universiteit Leuven

Publications
Moonen M.
Katholieke Universiteit Leuven

PROGRAM COMMITTEE

Acheroy M.
Barbé A.
Boite R.
Bruylandt I.
Catthoor F.
David J.
Delogne P.
Deprettere E.
Destiné J.
Fawé A.
Fettweis A.
Gielen G.
Grenez F.
Hasler M.
Hoffman G.
Kunt M.
Laloux A.
Leich H.
Martens J.P.
Moonen M.
Sas P.
Schoentgen J.
Schoukens J.
Slock D.
Steyaert M.
Suetens P.
Trullemans Ch.
Vanbiesen L.
Van Compernolle D.
Vanderschoot J.
Vandewalle J.
Van Eycken L.
Van Huffel S.
Vanwormhoudt M.
Vanpoucke F.
Vloeberghs C.
Wambacq P.
Wellekens C.

SCIENTIFIC COMMITTEE

Almeida L.
Atal B.S.
Bellanger M.
Benveniste A.
Biemond J.
Butterweck H.J.
Capellini V.
Carayannis G.
Castanié F.
Claasen T.A.C.M.
Constantinides A.
Deliyannis Th.
Deprettere E.
Devijver P.
Dewilde P.
Durrani T.S.
Escudié B.
Fallside F.
Fettweis A.
Figueiras-Vidal A.R.
Fliege N.
Galand C.
Gersho A.
Granlund G.H.
Gueguen C.
Hasler M.
Haton J.P.
Herault J.
Hernandez-Gomez L.
Herrmann O.
Hess W.
Heute U.
Kalouptsidis N.
Kunt M.
Lacoume J.L.
Lacroix A.
Lagunas M.A.
Macchi O.
McWhirther J.G.
Mecklenbräuker W.
Mersereau R.M.
Mitra S.K.
Moschytz G.
Neuvo Y.
Niemann H.
Peek J.B.H.
Pellandini F.
Picinbono B.
Sankur B.
Schüssler H.W.
Scharf L.
Sicuranza G.L.
Vaidyanathan P.P.
Venetsanopoulos A.
Vetterli M.
Vich R.
Weinrichter H.
Wolf D.

ORGANIZATION

ORGANIZER :

European Association for Signal Processing
EURASIP

in co-operation with

IEEE Benelux Section

SPONSORS :

Belgian National Fund for Scientific Research (N.F.W.O.)
Egmontstraat 5
1050 Brussel
Belgium

CERA
Brusselsesteenweg 100
3000 Leuven
Belgium

LMS International
Interleuvenlaan 68
3001 Leuven
Belgium

EXHIBITORS :

Comdisco Systems, Inc.
First Floor, Aztec Centre
Aztec West, Almondsbury
Bristol BS12 4TD
England

Digital Equipment NV
Luchtschipstraat 1
1140 Evere
Belgium

DVC Digitalvideo Computing Gmbh
Seestrasse 7
D-8036 Herrsching
Deutschland

ELSEVIER Science Publishers B.V.
Sara Burgerhartstraat 25
1055 KV Amsterdam
The Netherlands

European Development Center NV
Abdijstraat 34
3001 Heverlee
Belgium

John Wiley & Sons Ltd
Baffins Lane
Chichester, Sussex PO19 IUD
United Kingdom

KLUWER Academic Publishers Group
P.O. Box 989
3300 AZ Dordrecht
The Netherlands

Loughborough Sound Images Ltd
The Technology Centre
Epinal way
Loughbourough
Leicestershire LE11 OQE
England

National Instruments Belgium
Leuvensesteenweg 613
1930 Zaventem
Belgium

OROS
13, Chemin des Prés
B.P. 26 Zirst
38241 Meylan Cedex
France

Prentice Hall
Simon & Schuster Int. Group
Campus 400
Maylands Avenue
Hemel Hempstead
Herts. HP2 7EZ
England

Sun Microsystems Belgium
Rue Colonel Bourg straat 127/129
1140 Bruxelles
Belgium

ARIEL Corporation
433 River Road
Highland Park
New Jersey 08904
U. S. A.

VOLUME I

OPENING SESSION

TUTORIALS

SPECIAL SESSIONS

Special Session 1

Imaging, Algorithms and Architectures

Special Session 2

Adaptive Filtering in Telecommunications

Special Session 3

Acoustic Articulatory Inversion

Special Session 4

Chaos in Signal Processing

Session S4.L

Speech Recognition

Session S5.L

Speech Coding

Session S6.L

Speech Coding

KNOWLEDGE ENGINEERING AND SIGNAL PROCESSING

Session K1.P

Knowledge Based Engineering

Session K2.L

Knowledge Based Engineering

VOLUME II

THEORY OF SIGNALS AND SYSTEMS

Session T1.L

Array Processing: Detection

Session T2.L

Estimation

Session T4.L

Spectral Estimation

Session T5.P

Estimation

Session T6.L

Detection and Filtering

Session T7.L

Modeling and Interpolation

Session T8.L

Digital Transforms

Session T11.P

Filtering and Transforms

Session T14.L

Filtering

NEURAL NETWORKS FOR SIGNAL PROCESSING

Session N1.L

Neural Networks

Session N2.P

Neural Networks

VOLUME III

IMAGE PROCESSING AND MULTIDIMENSIONAL SIGNAL PROCESSING

Session I1.L

Image Coding

Session I2.L

Image Coding

Session I3.L

Image Coding

Session I4.L

Image Coding

Session I5.L

Image Enhancement & Restoration

Session I6.L

Medical Image Processing

Session I7.P

Image Processing

IMPLEMENTATIONS

Session V1.P

Implementations

APPLICATIONS

Session A1.L

Spread Spectrum

Session A2.L

Echo Cancellation and Equalization

Session A3.L

Data Communications

Session A4.L

Data Communications

Index

SIGNAL PROCESSING VI: Theories and Applications
J. Vandewalle, R. Boite, M. Moonen, A. Oosterlinck (eds.)
1992 Elsevier Science Publishers B.V.

IMPROVED PERFORMANCE OF ROOT-MUSIC WITH SENSOR PERTURBATIONS

SHAMLA MATHUR

Department of Electronics and Communications Engineering,
Osmania University, Hyderabad - 500 007, India.

The performance of Root-Music, for estimating the direction of arrival (DOA) of plane waves in the presence of sensor gain and phase perturbations, in the case of a linear equispaced sensor array was studied. It is found that the performance of Root-Music with sensor perturbations can be improved by simple teoplitzation of the array covariance matrix (R) for DOA estimation. Also the teoplitzed R matrix tends to the true array covariance matrix with increase in number of sensors. Computer simulations are provided to substantiate the analysis.

1 INTRODUCTION

Most of the eigen decomposition based methods decompose the observed covariance matrix into two orthogonal spaces, commonly referred to as signal and noise subspaces, and estimate the DOA's from one of these spaces [1] - [4]. In Root- Music [5], the DOA's are determined form the roots of a polynomial formed from the noise subspace.

2. PROBLEM FORMULATION

Assume that D narrowband sources impinge on a linear array consisting of M equispaced omnidirectional sensors, and that the directions for arrival (DOAs) of these sources are $\{\theta_1, \theta_2, \ldots, \theta_D\}$. The output of the m-th sensor can be expressed as:

$$r_m(t) = \sum_{d=1}^{D} s_d(t)\, e^{-j\omega_o (m-1)\tau_d} + n_m(t) \qquad \ldots\ldots\ldots(1)$$

where $s_d(.)$ denotes the signal emitted by the d-th source as observed at sensor one, ω_o is the centre frequency of the sources and $n_m(.)$ is the additive noise, assumed to be a zero-mean stationary random process that is independent from sensor to sensor. The interelement path delay is given by $\tau_d = \Delta \sin\theta_d/c$, where Δ denotes spacing of the sensor, θ_d is the direction of arrival of the d-th source and c is the velocity of the planewaves. Let

$$r^T(t) = [r_1(t), r_2(t), \quad , r_M(t)] \qquad (2)$$

be the simultaneously sampled vector (snapshot) of array signals, s(.) and n(.) are similarly defined. The array signal vector can then be expressed as:

$$r(t) = As(t) + n(t) \qquad (3)$$

where A is a MxD direction matrix with the i-th column

$$a(\theta_i) = [1\ e^{-j\omega_o\tau_i} \ldots e^{-j\omega_o(M-1)\tau_i}]^T \qquad \ldots\ldots(4)$$

The array covariance matrix can then be expressed as:

$$R = E\,[r(t)\, r^+(t)] = ASA^+ + \sigma^2 I \qquad (5)$$

where $S=E[s(t)s^+(t)]$ denotes the source covariance matrix, σ^2 is the variance of the additive noice.

In Root-Music, the DOAs can be estimated accurately only when the gain and phase characteristics of the elements in the array are identical, which may be difficult to ensure in practice. Any deviation in gain and phase characteristics can cause errors in the estimation of the DOAs.

3 EFFECT OF SENSOR GAIN AND PHASE PERTURBATIONS

Let the deviation of gain and phase of the mth sensor be δ_{a_m} and δ_{p_m} respectively and assume that the array covariance matrix in the presence of gain and phase errors can be written as

$$\tilde{R} = GASA^{+}G^{+} + \sigma^{2}I \tag{6}$$

where G is a diagonal matrix of size MxM with elements

$$g_m = (1 + \delta_{am})\, e^{j\delta_{pm}} \tag{7}$$

Note that in equation (7), a nominal gain of unity and a nominal phase of zero for the sensors is assumed for the unperturbed array. We also assume that both δ_{a_m} and δ_{p_m} are small and zero - mean random variables with the assumption (7) can be simplified as;

$$g = (1 + \delta_{am} + j\delta_{pm}) \tag{8}$$

In the presence of sensor gain and phase errors, the noise subspace E_n is modified to $\tilde{E}_n$ where $\tilde{E}_n$ is the noise subspace obtained from $\tilde{R}$. Consequently, the polynomial obtained using $\tilde{E}_n$ will be perturbed and hence the DOAs. The degradation in the performance of the algorithm with sensor perturbations has been studied extensively in terms of bais and variance through computer simulations and is explained in the next section.

4 IMPROVED PERFORMANCE OF ROOT-MUSIC ALGORITHM IN THE PRESENCE OF SENSOR PERTURBATIONS

With sensor perturbations, the elements of the perturbed array covariance matrix $\tilde{R}$ can be written as:

$$\tilde{R}_{ij} = (g_i g_j^{*})\, \gamma_i \gamma_j^{*} \tag{9a}$$

$$\text{and } \tilde{R}_{ii} = |g_i|^{2}\, \gamma_i \gamma_i^{*} + \sigma^{2} \tag{9b}$$

$$\text{where } \gamma_m = \sum_{d=1}^{D} s_d(t)\, e^{-jw_o(m-1)\tau_d} \tag{10}$$

The teoplitzation of $\tilde{R}$ is done by obtaining the average estimate of each of the lth diagonal of $\tilde{R}$ and replacing that diagonal with this computed estimate. Let us denote the teoplitzed matrix by $\hat{R}$. To see how teoplitzation helps in improving the DOA estimates, for gain errors only consider the following.

The averaged estimate of the main diagonal of $\tilde{R}$ is

$$\hat{r}(0) = \frac{1}{M} \sum_{m=1}^{M} (g_m^{2}\, \gamma_m \gamma_m^{*}) + \sigma^{2} \tag{11}$$

$$\text{where } g_m^{2} = (1 + \delta_{am})^{2} \tag{12}$$

$$\hat{r}(0) = \frac{1}{M} \left[\sum_{m=1}^{M} (1 + 2\delta_{am} + \delta_{am}^{2})\gamma_m \gamma_m^{*} \right] + \sigma^{2} \tag{13}$$

Assuming a zero - mean, uniform random distribution, $\hat{\gamma}(0)$ can be approximated as

$$\hat{r}(0) = \left(1 + \frac{\epsilon_g^{2}}{3M}\right) \hat{\gamma}(0) \tag{14}$$

where ϵ_g is the upper limit of the uniform distribution

$$\hat{r}(0) = \frac{1}{M} \sum_{m=1}^{M} \gamma_m \gamma_m^{*} \tag{15}$$

As M increases, the contribution of the second term will decrease and hence M--> ∞, $\hat{r}(0)$ --> $\hat{\gamma}(0)$. Thereby reducing significantly the sensor gain perturbation effects.

Consider next the averaged estimate of the first sub - diagonal of $\tilde{R}$, which is given by,

$$\hat{r}(1) = \frac{1}{M-1} \sum_{m=1}^{M-1} (g_m\, g_{m+1})\, \gamma_m \gamma_{m+1}^{*} \tag{16}$$

$$\text{where } g_m\, g_{m+1} = (1 + \delta_{am})(1 + \delta_{a(m+1)}) \tag{17}$$

$$\hat{r}(1) = \frac{1}{M-1} \sum_{m=1}^{M-1} [1 + (\delta_{am} + \delta_{a(m+1)}) + \delta_{am}\, \delta_{a(m+1)}] \sum_{m=1}^{M-1} \gamma_m \gamma_{m+1}^{*} \tag{18}$$

$$\frac{1}{M-1} [(M-1) + (\delta_{a1} + \delta_{a2} + .. + \delta_{a(M-1)}) + (\delta_{a2} + \delta_{a3} + ... + \delta_{aM}) + (\delta_{a1}\delta_{a2} + \delta_{a2}\delta_{a3} + .. + \delta_{a(m-1)}\delta_{aM})]\, \hat{\gamma}(1) \tag{19}$$

$$\text{where } \hat{\gamma}(1) = \frac{1}{M-1} \sum_{m=1}^{M-1} \gamma_m \gamma^*_{m+1} \qquad (20)$$

with increase in M, $\hat{r}(1)$ --> $\hat{\gamma}(1)$, thereby reducing the effects of sensor gain perturbations.

Similarly, it can be shown that the averaged estimate of each of the lth sub - diagonal of $\tilde{R}$ can be reduced to $\hat{\gamma}(1)$. Similar mathematical analysis can be carried out in the case of gain and phase errors.

5 SIMULATION STUDIES

The following scenario is assumed for the purpose of simulation. Two sources from 20° and 30° with respect to broadside direction of displacement vector impinge on a linear array whose interelement spacing is $\lambda/2$. The source and sensor noise powers were assumed to be zero db each. The nominal gain and phase values for the sensors are unity and zero respectively. Since the analysis of the previous section was based on the asymptotic covariance matrices, simulations were carried out for the same.

The perturbed covariance matrix was formed using $\tilde{R}$.The DOAs were estimated using the teoplitzed matrix, $\hat{R}$. The bias and variance in the angle estimates thus obtained were computed.

The above simulations were carried out for various values of a_{max} and P_{max}. Fig. 1a and Fig. 1b the estimates and variance of the DOAs as a function of a_{max} , for M = 6 are plotted, while Fig. 2a and Fig. 2b show the plots for M= 8.The estimates and variance in the DOAs were also computed without teoplitzation ie. using $\tilde{R}$ and these plots are shown Figs. 1 & 2 by dased lines. From these plots, the significant improvement in the performace of Root-Music by teoplitzation is apparent.

The combined effect of gain and phase perturbations, some sample results are presented in table 1. With teoplitzation of the array covariance matrix, the results are presented in table 2. Comparing tables 1 and 2, we can see the improved performance of Root - Music by toeplitzation.

M = 8, [20° 30°], [0 0]db and N = 100					
ε_g	ε_p	$\hat{\theta}_1$	$\hat{\theta}_2$	ν_1	ν_2
0.01	10°	19.8255	30.2053	0.2385	0.2724
0.1	5°	19.9	30.1223	0.1698	0.2054
0.2	10°	19.5812	30.4198	0.6606	0.6717
0.3	15°	19.4119	30.7671	0.9162	1.3648
0.4	2°	19.482	30.5522	1.2375	1.7341

TABLE 1

6 CONCLUSIONS

In this paper we have proposed a simple method to improve the performace of Root - Music algorithm in the presence of sensor gain and phase perturbations. It is found that, with teoplitzation the bias and variance in the angle estimates reduces significantly with increase in the number of sensors.

References

[1] Schmidt, R.O.,"Multiple emitter location and signal parameter estimation". IEEE Trans. Antennas Propagat., Vol. AP. 84, pp 276 - 280, Mar. 1986.

[2] Roy, R., Paulraj, A.,and Kailath, T.,"Estimation of signal parameters via rotaional invariance techniques - ESPRIT". in Proc. 19th Asilomar Conf. Ciruits, Syst. Computing, Monterey, CA, Nov. 1985.

[3] Roy, R., and Kailath, T.,"Total least squares ESPRIT", in Proc. 21st Asilomar conf. Circuits, Syst. computing, Monterey, CA, Nov. 1987.

[4] Kumarasan, R., and Tufts, D.W.,"Estimating the angle of arrival of multiple plane waves", IEEE Trans, Aerosp. Electron. Syst., vol AES - 19, pp 134 - 139, Jan 1983.

[5] Rao, B.D., and Hari, K.V.S.,"Performance Analysis of Root - Music", IEEE Trnas. Acoustics, Speech and signal Processing, Vol. ASSP - 37, pp 1939 - 1949, Dec. 1989.

ε_g	ε_p	$\hat{\theta}_1$	$\hat{\theta}_2$	ν_1	ν_2
0.01	10°	20.0173	30.0139	0.1088	0.1289
0.1	5°	20.0081	30.0071	0.0276	0.0333
0.2	10°	20.0184	30.0118	0.1117	0.1356
0.3	15°	20.0311	30.0136	0.2558	0.3117
0.4	2°	20.0103	29.9948	0.009	0.0123

TABLE 2

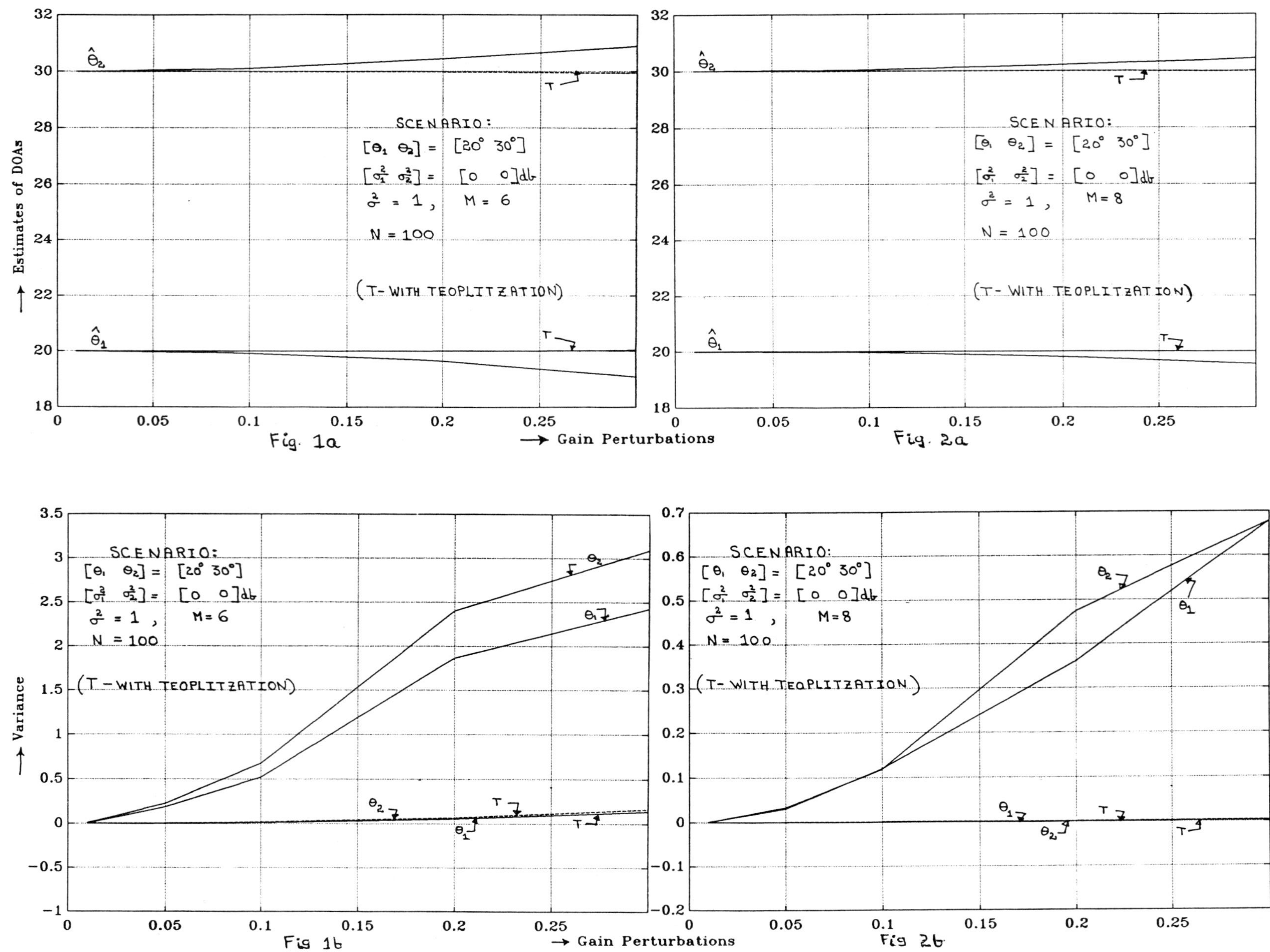

Fig. 1a

Fig. 2a

Fig 1b

Fig 2b

SIGNAL PROCESSING VI: Theories and Applications
J. Vandewalle, R. Boite, M. Moonen, A. Oosterlinck (eds.)

PERFORMANCE CHARACTERISTICS OF A GENERALIZED NEYMAN-PEARSON DETECTOR WITH FUZZY INFORMATION *

Jae Cheol Son, Iickho Song, Sangyoub Kim, Seong Ill Park
Jin Seon Yun, and Yoon Kyoo Jhee

Department of Electrical Engineering
Korea Advanced Institutue of Science and Technology (KAIST)
373-1 Guseong Dong, Yuseong Gu, Daejeon 305-701, Korea
Tel: +82-42-869-3445, 5445
Fax: +82-42-869-3410
e-mail: isong@Sejong.kaist.ac.kr

In this paper statistical characteristics of the locally optimum fuzzy detector test statistic for known signals are discussed. Specifically, the probability distribution of the locally optimum fuzzy detector test statistic is obtained from the exact and approximate approaches. Performance characteristics of the locally optimum fuzzy detector are also considered and compared to those of the locally optimum detector.

1. Introduction

In several studies it has been suggested that under certain circumstances the fuzzy set theoretic techniques can be applied to signal detection problems [*e.g.*, 3, 7]. For example, in [7, 8] it was pointed out that techniques of statistical hypotheses testing with fuzzy information can be used in detecting signals when the detection processor produces a self-noise.

As an application of the fuzzy set theory in signal detection, the quantizer-detector scheme was considered in [7, 8] assuming that the observations can be interpreted as fuzzy informations since it is customary to implement the nonlinearity of the locally optimum (LO) detector by a quantizer [*e.g.*, 5]. We will denote this concatenation of a quantizer and the LO detector by the LOQ detector. The rationale to consider the fuzzy information in the quantizer-detector structure can be explained as follows. In digital signal processing [*e.g.*, 6] it is usually assumed that the probability distribution of the quantization error process is uniform over the *range* of the *quantization error*, and this assumption is known to yield reasonable results in some situations. In other cases, however, this assumption does not give an accurate description of the behavior of the quantizer [*e.g.*, 9]. In addition, the assumption is not appropriate either when the input signals corrupted by a noise (*e.g.*, self-noise) are quantized. Based on this observation there have been some studies on statistical analysis of the quantized noisy signals [*e.g.*, 1]. However, it seems that to employ fuzzy testing of statistical hypotheses is more convenient and practical for signal detection since exact analysis of the quantization error is cumbersome in practice.

In this paper, we will investigate the statistical characteristics and performance of the locally optimum fuzzy (LOF) detector test statistic as a natural extension of our previous study [7]. Two approaches will be used to obtain the distribution of the LOF detector test statistic in Section 3. In addition, performance characteristics of the LOF detector will be investigated and an important observation will be obtained in Section 4.

2. Basics

Let us consider the frequently encountered signal-detection problem which can be expressed by the following two hypotheses H_0 and H_1:

$$H_0:\ Y_i = W_i\,, \tag{1}$$

versus

$$H_1:\ Y_i = \theta e_i + W_i\,, \tag{2}$$

for $i = 1, 2, \ldots, n$. In (1) and (2), Y_i is the observation, e_i is the known signal component, and W_i is the purely-additive noise (PAN) component at the i-th sampling instant. θ is the amplitude parameter which controls the signal strength. The PAN components W_i, $i = 1, 2, \ldots, n$, are assumed to be independent and identically distributed (i.i.d.) with the common continuous probability density function (pdf) f. It is also assumed that the pdf f is zero-mean and even.

In order to handle the observation Y_i as fuzzy information based on the descriptions in Section 1, let us introduce some definitions. Let (X^n, B_{X^n}, F) be a probability space where X^n is a Euclidean n-dimensional space, B_{X^n} is the Borel σ-field, and F is a probability measure over X^n. A *fuzzy information system* τ is a fuzzy partition of the real line X by means of fuzzy events [2]. An n tuple of elements in τ, $\bar{\kappa} = (\kappa_1, \kappa_2, \ldots, \kappa_n)$, representing the algebraic product of $\kappa_1, \kappa_2, \ldots, \kappa_n$, is called the *sample fuzzy information of size* n The set consisting of all algebraic products of $\kappa_1, \kappa_2, \ldots, \kappa_n$ is called the *fuzzy random sample of size* n and is denoted by $\tau^{(n)}$. The *probability distribution* of $\tau^{(n)}$ is given by

$$P(\bar{\kappa}) = \int_{X^n} \lambda_{\kappa_1}(y_1) \cdots \lambda_{\kappa_n}(y_n)\, dF(y_1, \ldots, y_n)\,, \tag{3}$$

where the integral is the Lebesgue-Stieltjes integral and $\lambda_{\kappa_i}(y)$ is called the *membership function* of κ_i. For notational convenience, we will denote by $P_0(\bar{\kappa}) = P(\bar{\kappa} \mid H_0)$ and $P_1(\bar{\kappa}) = P(\bar{\kappa} \mid H_1)$ the probability distributions of $\tau^{(n)}$ under H_0 and

* This research was supported by Korea Advanced Institute of Science and Technology (KAIST) and by Korea Science and Engineering Foundation (KOSEF), for which the authors would like to express their thanks.

H_1, respectively.

In addition, two types of detection systems, the midtread-quantizer (MTQ) detector system and the midrise-quantizer (MRQ) detector system [4], will be considered in this paper. More specifically, it is assumed in this paper that the inputs for both of the LOQ and LOF detection processor are $Q_i = Q(Y_i)$, where $Q(\cdot)$ is the quantizer characteristic. It is noteworthy that the LOQ detector makes a decision based on this output of the quantizer, while the LOF detector makes a decision regarding the quantizer output as fuzzy information. The input-output characteristics of the MRQ and MTQ are assumed to be odd-symmetric: a typical input-output characteristic of the MRQ is shown in Figure 1(a) where $\{\pm b_i\}_{i=1}^{i=m}$ and $\{\pm l_i\}_{i=1}^{i=m}$ are called the breakpoints and quantization levels of the quantizer, respectively.

From the above construction we are now able to define a specific fuzzy information. For example, a fuzzy information τ_2 obtained by the MRQ implies that the observed value approximately lies in $[b_1, b_2]$ (see Figure 1(b)). The shape of the membership function $\lambda_{\tau_i}(y)$ can arbitrarily be given provided that it satisfies the constraint of orthogonality [2]. As a simple but reasonable example, the trapezoidal membership functions for the MRQ detector are illustrated in Figure 1(b), where the incredibility Δ is a measure of the amount of fuzziness in the observed fuzzy information.

For use in the theorems given later in this paper, we define some more terms. The collection of all possible values of the random vector $(Q_1, \ldots, Q_n)$ is called the *sample space*. The collection of informations based on which the crisp detector makes a decision will be called the *crisp information space* and that based on which the fuzzy detector makes a decision will be called the *fuzzy information space*. The *ordered* crisp information space is the crisp information space with elements arranged in the decreasing order of corresponding value of the crisp detector test statistic. Similarly, the *ordered* fuzzy information space is the fuzzy information space with elements arranged in the decreasing order of corresponding value of the fuzzy detector test statistic.

In addition, Ψ_C is defined as the subspace of the ordered crisp information space in which a crisp detector accepts H_1 for a given false-alarm probability based on the classical definition of probability. Similarly, we will denote the subspace of the ordered fuzzy information space in which a fuzzy detector accepts H_1 for a given false-alarm probability based on the Zadeh's definition of probability (3) by Ψ_F.

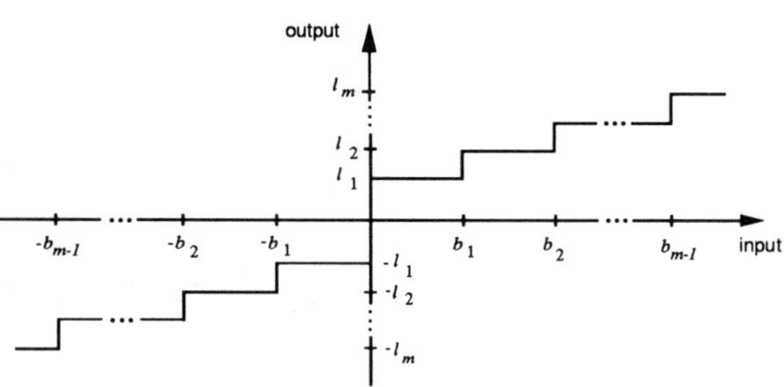

(a) input-output characteristic

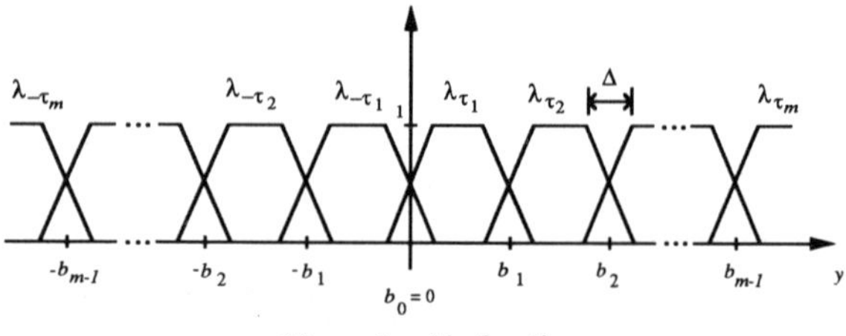

(b) membership functions

Figure 1. Input-output characteristic and membership functions of fuzzy informations for the midrise quantizer

3. Distributions of the Test Statistic

It was already shown in [7] that the LOF detector test statistic for known signals can be expressed as

$$T_{LOF}(\overline{\kappa}) = \sum_{i=1}^{n} e_i \, g_{LOF}(\kappa_i), \tag{4}$$

where

$$g_{LOF}(\kappa_i) = \frac{E\{\lambda_{\kappa_i}'(y)\}}{E\{\lambda_{\kappa_i}(y)\}} \tag{5}$$

with $E\{\cdot\}$ denoting the statistical expectation with respect to the pdf f.

The practical methods in most radar (or sonar) signal detection problems relies upon the Neyman-Pearson procedure. When we use the Neyman-Pearson procedure in detection systems, it is necessary to obtain the probability distribution of the test statistic under the null hypothesis to find the exact detection threshold. In this section, we will consider two approaches to obtain the distribution function of the LOF detector test statistic for known signals.

3.1. Exact Distribution

Let us consider here the $2m$-level MRQ only, since a similar approach can be applied to the MTQ. To obtain the distribution of the test statistic (4), let us first define $\{d_1, d_2, \ldots, d_{2m}\}$ as a set of nonnegative integers such that $\sum_{i=1}^{2m} d_i = n$. Then the probability of the set $\{A_1 \cap A_2 \cap \cdots \cap A_{2m}\}$, where A_i, $i = 1, \ldots, 2m$, are the events defined as $A_1 \triangleq \{$ the fuzzy information $-\tau_m$ appears exactly d_1 times in n observations $\}$, $A_2 \triangleq \{$ the fuzzy information $-\tau_{m-1}$ appears exactly d_2 times in n observations $\}, \ldots,$ and $A_{2m} \triangleq \{$ the fuzzy information τ_m appears exactly d_{2m} times in n observations $\}$, is easily obtained to be

$$\frac{n!}{d_1! \, d_2! \cdots d_{2m}!} p^{d_1}(-\tau_m) p^{d_2}(-\tau_{m-1}) \cdots p^{d_{2m}}(\tau_m), \tag{6}$$

where $0 \le d_i \le n$ and $p(\tau)$ is the probability that the fuzzy information τ is observed when a sample fuzzy information of size 1 is taken from the experiment.

The distribution of the test statistic can now be easily obtained since the probability that the test statistic takes a specific value is just one value or the sum of two or more values evaluated by (6). It should be emphasized, however, that the above procedure becomes unmanageable as the sample size grows. To obtain the probability distribution of the LOF detector test statistic, we should calculate the joint probability for all elements in $\tau^{(n)}$. If we assume that quantization level is 256 and the sample-size is 20, then the joint probability for 256^{20} ($\approx 1.46 \times 10^{48}$) cases should be obtained which seems to be too large to handle.

3.2. Approximate Approach

If we would like to find the exact threshold to satisfy a given false-alarm restriction, the procedure in the previous subsection is unavoidable. If the threshold is not required to be set exactly, however, the calculation burden due to the large sample size can be alleviated by exploiting the well-known central limit theorem. For brevity, let us assume that e_i, $i = 1, ..., n$, are all equal to 1, which is a common and reasonable assumption.

We first see that the mean and variance of the LOF detector test statistic under the null hypothesis are obtained to be

$$\mu_0 = \sum_{\kappa \in \tau} P_0(\kappa)\, g_{LOF}(\kappa), \tag{7}$$

and

$$\sigma_0^2 = \sum_{\kappa \in \tau} P_0(\kappa)\, \{ g_{LOF}(\kappa) - \mu_0 \}^2, \tag{8}$$

respectively. The following theorem describes how we can find the approximate value of the detection threshold for large sample-size.

Theorem 1. Assume that the mean and variance of $g_{LOF}(\kappa)$ under H_0 are finite. Then the fuzzy test accepting H_1 when $\overline{\kappa} \in \Xi$ and rejecting it when $\overline{\kappa} \in \Xi^c$ is asymptotically locally optimum for testing H_1 against H_0 at level α, where

$$\Xi = \{ \overline{\kappa} \mid \overline{\kappa} \in \tau^{(n)} \text{ with } T_n^* = \frac{T_{LOF}(\overline{\kappa}) - n\mu_0}{\sigma_0 \sqrt{n}} > t \}. \tag{9}$$

In (9) t is the $100(1-\alpha)th$ percentile of the standard normal distribution.

Proof: Denoting $g_{LOF}(\kappa_i)$ by G_i, we see that $G_1, G_2, \cdots$ are clearly i.i.d. nondegenerate random variables. In addition, the mean and the variance of G_i under H_0 are μ_0 and σ_0^2, respectively. Since both μ_0 and σ_0^2 are assumed to be finite, we see from the central limit theorem that T_n^* converges in law to T as $n \to \infty$, where T is a standard normal random variable. Then the statement of the theorem is immediately obtained. □

Strictly speaking, Theorem 1 can be applied when the sample size n approaches to infinity. If we consider a sufficiently large value of n, however, we see that Theorem 1 can be used to find the approximate value of the threshold. Since it is easy to see that $\mu_0 = 0$ because of the odd-symmetric property of the g_{LOF} [7], the approximate distribution of the LOF detector test statistic is normal with mean zero and variance $n \sum_{\kappa \in \tau} P_0(\kappa)\, g_{LOF}^2(\kappa)$. Thus the approximate value of the threshold for level α is the $100(1-\alpha)th$ percentile of this normal distribution.

Now we concisely discuss the usefulness of the approximate approach. As we have seen from the discussion in Subsection 3.1, the probability distribution of the test statistic is available only after the joint probabilities for all $(2m)^n$ cases are calculated. In the approximate approach, however, what is required to be known is the probabilities of $2m$ fuzzy informations, not the joint probabilities of $(2m)^n$ cases.

4. Performance Characteristics

Since intensive calculations are required when the quantization-level is large, we will investigate the performance characteristics of the LOF detector when the quantization level is 2 and 3.

4.1. 2-level MRQ Detector

We first consider the 2-level MRQ detector. Since it was found [7] that we can not detect weak known signals when $\Delta = \infty$, it is assumed here that $\Delta < \infty$. Note that the 2-level MRQ detector based on crisp information is the sign detector in the classical detection theory. Let us now consider the following theorem.

Theorem 2. If the pdf f is zero-mean and even, then the 2-level MRQ detector is a *nonparametric* detector.

Proof: Since f is zero-mean and even-symmetric, we have $P_0(-\tau_1) = P_0(\tau_1) = 0.5$ for any value of Δ. Hence we see that the probability for any element in $\tau^{(n)}$ under H_0 is always 2^{-n}. Therefore, p_{fa} does not depend on the noise pdf f and the theorem is proved. □

It is well-known that the classical sign detector is a nonparametric detector. We can therefore expect that the two detection schemes the 2-level LOF and sign detectors would have quite similar characteristics. This anticipation becomes more clear if we consider the following theorem.

Theorem 3. The 2-level LOQ detector based on crisp information and the 2-level LOF detector based on fuzzy information make the same decision for the same observation when the sizes of the two detectors are equal.

Proof: Note that a test problem on the fuzzy information space can be converted into a test problem on the crisp information space. Converting Ψ_F with a given value of size α into the corresponding subspace of the ordered crisp information space, we see that it is equal to Ψ_C with the same value of α, since the probability for any element of the crisp and information spaces under H_0 is 2^{-n} and the order of the ordered fuzzy information space is preserved after the conversion. Hence the theorem is proved. □

Theorem 3 implies that *it makes no difference* between employing fuzzy information and crisp information in making a decision when the level is two.

4.2. 3-level MTQ Detector

Let us now assume that the input signals are quantized by a 3-level MTQ which provides output values -1, 0, and 1. Suppose that the experimentation provides us with the following fuzzy information: $-\tau_1$ = "approximately less than -1", τ_0 = "approximately 0", and τ_1 = "approximately greater than 1", where the membership functions associated with the fuzzy informations $-\tau_1$, τ_0, and τ_1 are

$$\lambda_{-\tau_1}(y) = \begin{cases} 1 & y < -1 \\ -y & -1 \le y < 0 \\ 0 & otherwise, \end{cases}$$

$$\lambda_{\tau_0}(y) = \begin{cases} 1+y & -1 < y < 0 \\ 1-y & 0 \le y < 1 \\ 0 & otherwise, \end{cases} \tag{10}$$

and

$$\lambda_{\tau_1}(y) = \begin{cases} 1 & y > 1 \\ y & 0 < y \le 1 \\ 0 & otherwise, \end{cases}$$

respectively. In (10), it is easy to see that $\Delta = 2$.

We assume that a decision is made for every three observations, and that $\alpha = 0.03155$ and $W_i \tilde{} N(0,1)$, that is, each of the noise components is normally distributed with zero mean and unit variance. The false-alarm probabilities (P_{fa}) and detection probabilities (P_d) of the LOF and LOQ detectors in various cases are obtained and shown in Tables 1-4.

We now consider two situations. Let us first consider the ideal case when there is no self-noise. From Table 1, we see that the characteristic of the LOQ detector is almost the same as that predicted by theoretic values which were obtained by an exact calculation under the assumption that we have all the information about the whole 'noise' environment. On the contrary we see that the characteristic of the LOF detector deviates from that predicted by the theoretical values since there is no fuzziness in the observation.

In Tables 2-4, however, the errors associated with the LOF detector are reduced whereas those associated with the LOQ detector are increased, when the distributions of the self-noise are $N(0, 0.5)$, $N(0, 1)$, and $N(0, 3)$. We see that for a preassigned size α the LOF detector has a similar characteristic to that of the LOQ detector obtained with perfect information about the noise environment, while the LOQ detector without information about the self-noise sacrifices as the self-noise becomes important to the overall noise characteristics. When the variance of the self-noise is 0.5, it is hard to say whether the LOF detector is better than the LOQ detector. This is due to the fact that the incredibility is chosen to be so large relative to the fuzziness resulting from the self-noise. When the variance of the self-noise is 3, however, we may conclude that the incredibility is chosen appropriately. This shows that the choice of the incredibility and the shape of the membership function is an important factor in the design and performance analysis of fuzzy detectors.

Now note that the size of the LOQ detector in Tables 2-4 is greater than 0.03155 although the threshold was set to satisfy the constraint of size α, since the threshold was calculated without considering the self-noise. The reason for the low P_d of the LOF detector is, hence, due to this low P_{fa} compared to that of the LOQ detector.

As a final observation, let us consider the following. As it is known, when we compare the performance of two tests, it is required to set the false-alarm probabilities of the two tests equal in the Neyman-Pearson procedure. Let us now consider the generalization of Theorem 3.

Theorem 4. If we make the LOQ and LOF detectors have the same size α based on the classical probabilistic definition, the two detectors will make the same decision for the same observation provided that the order of the ordered fuzzy information space is preserved after the fuzzy test problem is converted into a test problem on the crisp information space.

Proof: If Ψ_F is converted into the corresponding subspace of the ordered crisp information space under the assumption that the order is preserved, then the only way to achieve the same size for the two detectors based on the classical probabilistic definition is to make the converted Ψ_F equal to Ψ_C. Then the two detectors make the same decision for the same observation. □

It is immediate from Theorem 4 that the two detectors have the same probability of detection when the order of the ordered fuzzy information space is preserved after the fuzzy test problem is converted into a test problem on the crisp information space. It should be remarked that preservation of the order is not a very strong condition and we can find many examples which satisfy the assumption.

5. Conclusion

In this paper, statistical characteristics of the locally optimum fuzzy detector for known signals were investigated, and the performance of the detector was analyzed. We showed that the locally optimum fuzzy detector provides performance similar to that of the locally optimum detector which uses crisp information for a preassigned false-alarm probability with perfect information about the whole noise environment: the crisp locally optimum detector without complete information about the self-noise, however, sacrifices as the self-noise becomes more important to the overall noise characteristics. In other words, the locally optimum fuzzy detector has a robustness property.

Type	θ	LOF	LOQ	Theoretic
p_{fa}	0.0	4.500E-3	3.160E-2	3.155E-2
p_d	0.5	2.988E-2	1.246E-1	1.246E-1
	1.0	1.254E-1	3.139E-1	3.161E-1
	2.0	5.968E-1	7.740E-1	7.739E-1

Table 1. Comparison of p_{fa} and p_d when there is no self-noise

Type	θ	LOF	LOQ	Theoretic
p_{fa}	0.0	9.380E-3	4.912E-2	3.155E-2
p_d	0.5	4.056E-2	1.408E-1	9.753E-2
	1.0	1.248E-1	3.052E-1	2.262E-1
	2.0	4.958E-1	7.001E-1	6.119E-1

Table 2. Comparison of p_{fa} and p_d when there is normal self-noise of variance 0.5

Type	θ	LOF	LOQ	Theoretic
p_{fa}	0.0	1.404E-2	6.178E-2	3.155E-2
p_d	0.5	4.798E-2	1.498E-1	8.575E-2
	1.0	1.231E-1	2.942E-1	1.876E-1
	2.0	4.393E-1	6.469E-1	5.209E-1

Table 3. Comparison of p_{fa} and p_d when there is normal self-noise of variance 1

Type	θ	LOF	LOQ	Theoretic
p_{fa}	0.0	2.960E-2	8.738E-2	3.155E-2
p_d	0.5	6.268E-2	1.577E-1	6.822E-2
	1.0	1.243E-1	2.633E-1	1.301E-1
	2.0	3.311E-1	5.158E-1	3.376E-1

Table 4. Comparison of p_{fa} and p_d when there is normal self-noise of variance 3

References

[1] M.K. Brown, "On quantization of noisy signals", *IEEE Trans. Signal Proc.*, vol. SP-39, April 1991, pp. 836-841.

[2] M.R. Casals, M.A. Gil, and P. Gil, "On the use of Zadeh's probabilistic definition for testing statistical hypotheses from fuzzy information", *Fuzzy Sets and Systems*, vol. 20, October 1986, pp. 175-190.

[3] A. Dziech and M.B. Gorzalczany, "Decision making in signal transmission problems with interval-valued fuzzy sets", *Fuzzy Sets and Systems*, vol. 23, August 1987, pp. 191-203.

[4] N.S. Jayant and P. Noll, *Digital Coding of Waveforms*, Prentice-Hall, Englewood Cliffs, NJ, U.S.A., 1984.

[5] S.A. Kassam, *Signal Detection in Non-Gaussian Noise*, Springer-Verlag, New York, NY, U.S.A., 1988.

[6] L.R. Rabiner and R.W. Schafer, *Digital Processing of Speech Signals*, Prentice-Hall, Englewood Cliffs, NJ, U.S.A., 1978.

[7] J.C. Son, S.Y. Kim, I. Song, and H.M. Kim, "A fuzzy set theoretic approach to weak-known signal detection", *Signal Processing*, vol. 28, August 1992 (to be published).

[8] J.C. Son, I. Song, and H.Y. Kim, "A fuzzy decision problem based on the generalized Neyman-Pearson criterion", *Fuzzy Sets and Systems*, vol. 48, June 1992 (to be published).

[9] P.W. Wong, "Quantization noise, fixed-point multiplicative roundoff noise, and dithering", *IEEE Trans. Acous. Speech, Signal Proc.*, vol. ASSP-38, February 1990, pp. 286-300.

SIGNAL PROCESSING VI: Theories and Applications
J. Vandewalle, R. Boite, M. Moonen, A. Oosterlinck (eds.)

Tests for Optimizing Sensor Positions Using Linear Regression

A.M. Zoubir, T. Grewing and J.F. Böhme

Department of Electrical Engineering, Ruhr University Bochum, D-4630 Bochum, FRG

In this study, we propose a method for finding optimum sensor positions in a group of vibration sensors for knock detection in spark ignition engines. It differs from other techniques in that only signal processing and statistical tests are used. Our method is based on linearly predicting a reference signal from the output of an arbitrary sensor group. We derive a linear regression model in the frequency domain and discuss rank tests for various hypotheses that are tested with respect to the model parameters. This leads us to a technique for testing sensor irrelevancy. Simulation and experimental results emphasize the usefulness of the method.

1. Introduction

The efficiency of a spark ignition engine can be improved by an increase of its compression ratio. However, for high compression ratio engines the angle of ignition for minimum fuel consumption lies in a region where knock occurs. Knock is known as an abnormal combustion of the gas mixture causing rapid rise of temperature and pressure. This may lead to engine damage, particularly when it occurs at high speed. Knock detection systems with structural vibration sensors are used to control high compression engines for better fuel economy while avoiding engine damage as well as noise annoyance to the vehicle occupants.

The use of vibration sensors on the surface of the engine is easy and economical but makes the detection of knock difficult. This is due to the poor signal to noise power ratio (SNR) of the vibration signals. It is obvious that knock detectability can be improved if the vibration sensors are placed optimally on the engine surface. Holographic techniques for imaging knock centres have been proposed, but their application is complicated and does not necessarily lead to a practical solution [9]. Up to now, the sensor positions have been chosen heuristically.

In this contribution, we present a method for finding optimum positions of sensors within a sensor group being positioned for example heuristically on the engine. The method is less complex than holographic techniques because only signal processing and statistical tests are used. Alternatively to [10], we propose nonparametric tests.

In section 2., we introduce the data model. In section 3., we propose rank based tests for identifying optimum sensor positions in a group of sensors. Section 4. concerns test results with simulated data. We discuss experimental results in section 5. before concluding.

2. Data Model

Let $S(t)$ be a reference signal and $Z_i(t)$, $i = 1,\ldots,q/2$ sensor signals, all zero-mean and stationary, $t = 0,\pm1,\pm2,\ldots$. The series $S(t)$ and $Z_i(t)$ model the cylinder pressure signal recorded via a pressure transducer during a combustion cycle for reference purposes and a vibration sensor signal, respectively. Consider the linear prediction model

$$S(t) = \sum_{i=1}^{q/2} \sum_{u=-\infty}^{\infty} h_i(u)\, Z_i(t-u) + \mathcal{E}(t)\,, \qquad (1)$$

where $h_i(\cdot)$ is the ith impulse response of the prediction filter and $\mathcal{E}(t)$ is the prediction error series. Suppose we are given $N/2$ data blocks of length T each, $\mathbf{S}(t) = (S^1(t),\ldots,S^{N/2}(t))'$, $\mathbf{Z}_i(t) = (Z_i^1(t),\ldots,Z_i^{N/2}(t))'$, $i = 1,\ldots q/2, t = 0,\ldots,T-1$. We compute the finite Fourier transforms of the time series data, and (1) becomes

$$\mathbf{d}_S(\omega) \approx \sum_{i=1}^{q/2} H_i(\omega)\, \mathbf{d}_{Z_i}(\omega) + \mathbf{d}_{\mathcal{E}}(\omega)\,, \qquad (2)$$

approximatively a complex regression at each frequency ω. For the sake of simplicity, we drop the argument variable ω. Assuming independence and the same distribution function for the real and imaginary part of $\mathbf{d}_{\mathcal{E}}$, we obtain the real valued linear regression

$$\mathbf{Y} = \mathbf{V}\,\mathbf{B} + \mathbf{E}\,. \qquad (3)$$

Herein, $\mathbf{Y} = \left(\mathsf{Re}\,\{\mathbf{d}_\mathbf{S}'\}, \mathsf{Im}\,\{\mathbf{d}_\mathbf{S}'\}\right)'$ and $\mathbf{B}$, $\mathbf{V}$ contain the suitable arranged real and imaginary parts of the prediction filter frequency responses and the Fourier transformed sensor signals, respectively.

The aim of this work is to present an approach to reduce the dimension of $\mathbf{Z}(t)$, leading to outputs of sensors that are optimally positioned for knock detection. For this purpose, we use a nonparametric test to first decide whether there is no sensor giving relevant information on the reference signal by testing the corresponding coefficient $\mathbf{B}$ to be zero. In the case of rejection, we then apply a test to decide the relevancy of additional sensors.

3. Quadratic Rank Tests

Parametric methods to test the hypothesis H: $\mathbf{B} = \mathbf{0}$ against the alternative K: $\mathbf{B} \neq \mathbf{0}$ assume the vector $\mathbf{E}$ to be normally distributed. Though, this assumption holds asymptotically for large T [3], it is not necessarily guaranted in our case. Therefore, we use the quadratic rank test proposed by Adichie [1]. Following [1], we calculate

$$M_N = \frac{(\mathbf{V}'\boldsymbol{\psi}_N(\boldsymbol{\mathcal{R}}_Y(\mathbf{Y})))'\,(\mathbf{V}'\mathbf{V})^{-1}\,(\mathbf{V}'\boldsymbol{\psi}_N(\boldsymbol{\mathcal{R}}_Y(\mathbf{Y})))}{A^2(\psi)}, \tag{4}$$

wherein $\boldsymbol{\psi}_N(\boldsymbol{\mathcal{R}}_Y(\mathbf{Y}))$ is a function of the ranks $\boldsymbol{\mathcal{R}}_Y(\mathbf{Y})$ of $\mathbf{Y}$ with $\psi_N(i) = \psi(i/(N+1))$ and $A^2(\psi) = \int \psi(u)^2\,du - \bar{\psi}^2$, $\bar{\psi} = \int \psi(u)\,du$. For the Wilcoxon scores generated by $\psi(u) = u$ that are used in this contribution, we have $A^2(\psi) = 1/12$ [4].

The exact distribution of M_N under the hypothesis H can be calculated by permutation technique for any given matrix $\mathbf{V}$. However, this requires $N!$ computations of the statistic. This can be inpracticable in some cases where N is large. Thus, we use the asymptotic distribution of M_N that has been proposed in [1]. Under H and some regularity conditions, the statistic in (4) has asymptotically for large N a chi-square distribution with q degrees of freedom [1].

For the decision, whether additional sensors improve detection, it is necessary to test only some components of $\mathbf{B}$. For this, it is more convenient to rewrite (3) in the form

$$\mathbf{Y} = \mathbf{V}_1\,\mathbf{B}_1 + \mathbf{V}_2\,\mathbf{B}_2 + \mathbf{E}\,, \tag{5}$$

wherein $\mathbf{B} = (\mathbf{B}_1', \mathbf{B}_2')'$ and $\mathbf{V} = (\mathbf{V}_1, \mathbf{V}_2)$. The vectors $\mathbf{Y}$ and $\mathbf{E}$ are as before N vector-valued. $\mathbf{B}_s$ is q_s vector-valued and the matrix $\mathbf{V}_s$ has dimension $N \times q_s$ for $s = 1, 2$. It is of interest to test H_2: $\mathbf{B}_2 = \mathbf{0}$, $\mathbf{B}_1$ unspecified, against the alternative K_2: $\mathbf{B}_2 \neq \mathbf{0}$. Herein, $\mathbf{B}_1$ corresponds to already chosen sensors and $\mathbf{B}_2$ to further ones. A test similar to the one given above has been proposed by Adichie [1] and is summarized below.

To eliminate the influence of the unknown vector $\mathbf{B}_1$, an estimate $\hat{\mathbf{B}}_1$ for $\mathbf{B}_1$ is calculated. In this contribution, we use the least squares estimate. Though, any other estimate that is consistent and translation invariant might be used. An estimate $\hat{\mathbf{B}}_1$ for $\mathbf{B}_1$ is translation invariant if for all $\mathbf{B}_1$, $\hat{\mathbf{B}}_1(\mathbf{Y} - \mathbf{V}_1\mathbf{B}_1) = \hat{\mathbf{B}}_1(\mathbf{Y}) - \mathbf{B}_1$, where $\hat{\mathbf{B}}_1(\mathbf{Y})$ denotes estimates computed from $\mathbf{Y}$ [1]. With this estimate, the cleared data

$$\widehat{\mathbf{Y}} = \mathbf{Y} - \mathbf{V}_1\,\hat{\mathbf{B}}_1$$

are computed. Let us now define the matrices

$$\begin{aligned} \mathbf{C}_1 &= \mathbf{V}_1\,(\mathbf{V}_1'\mathbf{V}_1)^{-1}\,\mathbf{V}_1'\,, \\ \widehat{\mathbf{T}}_2 &= \mathbf{V}_2'\,(\mathbf{I}_N - \mathbf{C}_1)\,\boldsymbol{\psi}_N(\boldsymbol{\mathcal{R}}_{\widehat{Y}}(\widehat{\mathbf{Y}}))\,, \\ \mathbf{C}_* &= \mathbf{V}_2'\,(\mathbf{I}_N - \mathbf{C}_1)\,\mathbf{V}_2\,, \end{aligned}$$

where $\mathbf{I}_N$ is the identity matrix of dimension $N \times N$. For testing the hypothesis H_2 against K_2, Adichie proposes the statistic

$$\check{M}_N = \frac{\widehat{\mathbf{T}}_2'\mathbf{C}_*{}^{-1}\widehat{\mathbf{T}}_2}{A^2(\psi)}\,. \tag{6}$$

Under the hypothesis H_2, $\check{M}_N$ has asymptotically for large N a chi-square distribution with q_2 degrees of freedom. This holds under some regularity conditions stated in [1,8].

To compare the suggested tests with classical likelihood ratio tests, we employ a measure of relative efficiency due to Pitman, cf. [7,1]. Given two tests of the same size of the same statistical hypothesis, the relative efficiency of the second test with respect to the first is given by the ratio n_1/n_2, where n_2 is the sample size of the second test required to achieve the same power for a given alternative as is achieved by the first test with respect to the same alternative when using a sample of size n_1. In general, the relative efficiency will depend on n_1 as well as on the particular alternative chosen. Thus, we use the relative asymptotic efficiency, taking a sequence of alternatives changing with n in such a way that as $n \to \infty$ the power of the corresponding sequence of tests converges to some number less than 1. Andrews [2] has shown that if under the same sequence of alternatives, two test statistics have noncentral chi-square limit distributions with the same degrees of freedom, then their relative asymptotic efficiency is given by the ratio of their noncentrality parameters. Considering the suggested test and the classical likelihood ratio test for the hypothesis H (H_2) and a sequence of alternatives K_N: $\mathbf{B} = \mathbf{b}/\sqrt{N}$ ($\mathbf{B}_2 = \mathbf{b}_2/\sqrt{N}$), $\|\mathbf{b}\| < c$ ($\|\mathbf{b}_2\| < c$), the asymptotic relative efficiency of $M_N(\check{M}_N)$ yields to

$$\varepsilon = \sigma^2(F)\,B^2(F)\,/A^2(\psi)\,. \tag{7}$$

Herein, F denotes the distribution function of the components of $\mathbf{Y}$ with finite variance $\sigma^2(F)$ and $B^2(F) = \int \psi'(F(u))\,dF(u)$. Expression (7) was shown never to fall beneath 0.864 by Hodges and Lehmann [6]. In the normal distribution case, we have $\varepsilon = 3/\pi \approx 0.955$. This result shows that use of quadratic rank tests can never entail a serious loss of efficiency.

In practice, the simultaneous consideration of some frequencies of interest often arises. Especially in knock detection, we are interested in a few number of resonance frequencies that are excited by knock [5]. Because of dampings due to the physical process, we also consider neighboring frequencies. Thus, we test

$$\mathrm{H}_0 := \bigcap_{p=1}^{P}\,\bigcap_{k=-m}^{m} \mathrm{H}_{p,k}\,, \quad \mathrm{H}_{p,k} : \mathbf{B}_{p,k} = \mathbf{0} \tag{8}$$

against the two-sided alternative K_0: $\mathbf{B}_{p,k} \neq \mathbf{0}$ for at least one pair (p,k). Herein, $\mathbf{B}_{p,k} = \mathbf{B}(\omega_{p,k})$, $k = -m, \ldots, m$, where $\omega_{p,0} = \omega_p$, $p = 1, \ldots, P$ denote resonance frequencies. Under the assumption that the M_N statistics at different frequencies are independent which holds for large T, we may test the irrelevancy of a sensor by considering $\mathcal{M}_N = \sum_{p=1}^{P} \sum_{k=-m}^{m} M_N(\omega_{p,k})$. Under the hypothesis *no regression possible*, i.e. the sensor is irrelevant, this sum is approximately chi-square distributed with $q(2m+1)P$ degrees of freedom. We reject H_0 if, for a given level of significance α, the statistic $\mathcal{M}_N$ is too large. The boundary of the rejection region can be determined from a chi-square table. Similarly we construct $\breve{\mathcal{M}}_N = \sum_{p=1}^{P} \sum_{k=-m}^{m} \breve{M}_N(\omega_{p,k})$ to test $\breve{H}_0 := \bigcap_{p=1}^{P} \bigcap_{k=-m}^{m} H_{2,p,k}$, $H_{2,p,k}$: $\mathbf{B}_{2,p,k} = \mathbf{0}$ against the two-sided alternative.

In order to find the optimally placed sensor group with dimension lower than $q/2$, we propose a stepwise procedure. First, we test whether a single sensor is suitable to predict the reference signal. For this purpose, we compute $\mathcal{M}_N$. Sensors for which the hypothesis H_0 is not rejected are removed from the group. Among the remaining sensors, the one concerning to the highest $\mathcal{M}_N$ is chosen. In the following steps, the gain reached by adding a further sensor to the already determined one is tested. Using $\breve{\mathcal{M}}_N$, we remove sensors for which the hypothesis $\breve{H}_0$ is not rejected and select the one with the highest $\breve{\mathcal{M}}_N$. This procedure is repeated until enough sensors are determined or no sensor is left, i. e. all sensors are either chosen or removed.

4. Simulations

To investigate the proposed selection procedure, we have simulated a pressure signal and various sensor signals. The model

$$S(t) = \sum_{p=1}^{P} A_p e^{-d_p t} \cos(\omega_p t + \Phi_p) + Z(t) \qquad (9)$$

is used for the reference signal. Herein, A_p and Φ_p are mutually independent random amplitudes and phases, respectively, d_p dampings, ω_p the cavity resonance frequencies for $p = 1, \ldots, P$ and $Z(t)$ is a white noise process. To simulate the pressure signal, we have fixed $P = 4$ and generated 200 records. We have then simulated eight sensor signals. For this, some of the bands centered at the four resonance frequencies have been filtered out. These are numerated in increasing order and given in Table 1. Noise has been added to generate a sensor signal. The estimated SNRs of the sensor signals are given in Table 1.

In the first step, the procedure yields to the expected sensor S1 with the highest SNR. In the next step, sensor S6 of which output signal contains the missing resonances 3 and 4 in the signal of sensor S1 is selected. The selection of the third sensor is free among the sensors of which signals contain two resonances. It can be seen from Table 1 that $\breve{\mathcal{M}}_N$ for the sensors S3 and S5 are nearly equal. Sensor S3 is chosen. Thus, in the next step, sensor S4 is expected to be chosen. Nevertheless, the fifth sensor has the highest $\breve{\mathcal{M}}_N$. Sensor S4 lies nearly 8% beneath it.

Simulation results								
Sensor	S1	S2	S3	S4	S5	S6	S7	S8
SNR [dB]	-23	-28	-28	-28	-28	-28	-28	-28
Reson.	1,2	1,3	1,4	2,3	2,4	3,4	4	3
1. Step	**238**	194	196	181	193	170	130	86
2. Step		143	157	138	152	**165**	127	78
3. Step		111	**130**	112	126		91	50
4. Step		97		108	**117**		75	50
Results	1.	–	3.	–	4.	2.	–	–

Table 1: $\mathcal{M}_N$ and $\breve{\mathcal{M}}_N$ of the simulated signals. Removed sensors appear slanted, selected ones boldfaced

5. Experimental Results

Experiments were performed using a 1.8 ℓ, 4 cylinder engine with production knock control system. The high-pass filtered pressure signals of the cylinders 1 and 3 (D1 and D3 in Figure 1) was recorded by an instrumentation tape recorder together with the output of eight acceleration sensors. Two of these sensors (sensor SD and S2 in Figure 1) had been mounted at the place of the standard sensor, the other ones at heuristically chosen positions on the engine wall and on the cylinder head (sensor SK1). The sensors SA2 and SA23 had been mounted at the outlet side of the engine block, all others at the inlet side. Figure 1 schematically shows these positions. With the exception of the standard sensor SD, all the sensors were of the same type.

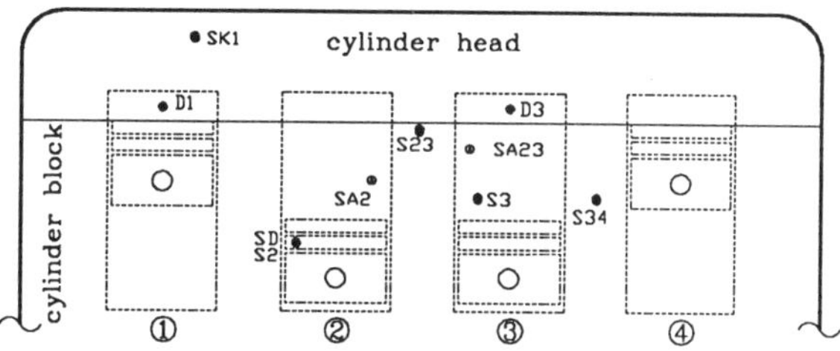

Figure 1: Sensor positions on the engine

The engine was running at 5250 rpm and at full load. Strong knock intensity was adjusted. Three thousand cycles were digitized with respect to cylinder 3 for our analyses. In order to identify knocking cycles, we compared the difference between successive samples of the cylinder pressure signal at an interval ranging from 10 to 40° crank angle with a suitably chosen threshold. This method exploits the fact that knock results in a steep pressure rise. The decision was verified by comparing the estimated pressure signal power of the selected cycles with that of the non-knocking cycles. In this way, 208 cycles were found.

Resonance frequencies were estimated by averaging periodograms of the knocking cycles of $S(t)$. Spectral peaks were found at 8.7, 13.1, 17.4 and 21.8 kHz. Two neighboring frequency bins to each resonance frequency have also been considered and the tests were performed as described above.

Experimental results								
	Sensor							
Step	S2	S23	S3	S34	SA2	SA23	SK1	SD
1.	141	201	127	122	170	**223**	*33*	160
2.	101	118	100	**134**	126		–	110
3.	45	68	37		**71**		–	48
4.	38	**46**	*32*				–	40
Ergebnis	–	4.	–	2.	3.	1.	–	–

Table 2: $\mathcal{M}_N$ and $\check{\mathcal{M}}_N$ of the real signals. Removed sensors appear slanted, selected ones boldfaced

The experimental results are given in Table 2. The proposed procedure yields to sensor SA23 of which signal gives the highest $\mathcal{M}_N$. The hypothesis H for testing sensor SK1 in the first step is not rejected. Thus, we remove this sensor from the group and proceed in the second step with the test of the hypothesis H_2 in order to add a further sensor to SA23. It should be noted that sensor SK1 was rather intended to detect noise generated by valve train than to detect knock. Sensor S34 which is placed between cylinder 3 and 4 is determined in the second step. Together with sensor SA23, it constitutes the optimally placed group of two sensors. This result is realistic because we have only considered cylinder 3 and the two sensors found are close to it. Nevertheless, Table 2 shows that sensor S3 performs poorly and is removed in the fourth step.

From this experiment, it is seen that the position proposed by the engine manufacturer (sensors SD and S2) is not optimum. However, it should be noted that this proposal was based on heuristical considerations. The approach presented here has also been applied to further cylinders and rotation speeds. In some cases where the engine speed is low and cylinder 1 is considered, good results have been registered for sensor S2 or SD. However, a selection procedure developed in [4] shows that both sensors perform poorly when considering simultaneously three rotation speeds and two cylinders.

6. Conclusions

In the precedings sections, we have presented a method to test the relevancy of sensors in an arbitrary array that is used to detect knock in spark ignition engines. Based on the linear prediction of a reference signal from vibration signals, we have derived a linear regression model in the frequency domain. We have then discussed quadratic rank based tests for hypotheses with respect to the model parameters. We have claimed that the relative asymptotic efficiency of the rank tests relative to the corresponding likelihood ratio tests is close to one. The latter tests assume normality of the regression error that is not necessarily guaranted in our case. We have proposed a stepwise procedure for identifying an optimally placed single sensor or a sensor group under consideration of multiple resonance frequencies. The method has been investigated by simulations. We have shown that in the very noisy case, the procedure leads to the expected sensors. Results of an experiment performed on a 4 cylinder engine suggest a sensor position that differs from the one proposed by the engine manufacturer.

REFERENCES

[1] Adichie, J.N.: Rank Tests in Linear Models, *Handbook of Statistics* **4**, P.R. Krishnaiah and P.K. Sen, eds., North-Holland, Amsterdam, 229 – 257, 1984.

[2] Andrews, F.C.: Asymptotic Behavior of some Rank Tests for Analysis of Variance, *Ann. Math. Statist.* **25**, 724 – 735, 1954.

[3] Brillinger D.R.: *Time Series: Data Analysis and Theory*, Expanded Ed., Holden-Day, San Fransisco, 1981.

[4] Grewing, T.: *Rangtests in linearen Modellen zur Bestimmung optimaler Sensorgruppierungen,* Diplomarbeit am Lehrstuhl für Signaltheorie, Ruhr-Universität Bochum, 1991. In German.

[5] Hickling R. et al.: Cavity Resonances in Engine Combustion Chambers and Some Applications, *J. Acoust. Soc. Am.* **73**(4), pp. 1170-1178, 1983.

[6] Hodges, J.L. and Lehmann, E.L.: The Efficiency of some Nonparametric Competitors of the t-Test, *Ann. Math. Statist.* **27**, 324 – 335, 1966.

[7] Noether, G.E.: On a Theorem of Pitman, *Ann. Math. Statist.* **26**, 64 – 68, 1955.

[8] Sen, P.K. und Puri, M.L.: Asymptotically Distribution-Free Aligned Rank Order Tests for Composite Hypotheses for General Multivariate Linear Models, *Zeit. Wahr. verw. Geb.* **39**, 175 – 186, 1977.

[9] Stoffregen, B.: Applications of Holography and Laser-Doppler Techniques for Vibration Analysis, *VDI Berichte*, **617**, 145 – 167, Düsseldorf, 1986. In German.

[10] Zoubir, A.M. and Böhme, J.F.: Simultaneous Tests for Optimizing Sensor Positions in Knock Detection, *Signal Processing V: Theories and Applications,* L.Torres et al., eds., North-Holland, Amsterdam, 677 – 680, 1990.

SIGNAL PROCESSING VI: Theories and Applications
J. Vandewalle, R. Boite, M. Moonen, A. Oosterlinck (eds.)

PATTERN RECONSTRUCTION USING PARAMETRIC TIME-FREQUENCY METHOD

J. MARS

CEPHAG- ENSIEG
Domaine Universitaire BP 46,
38402 Saint-Martin d'Hères Cedex, France

Non stationary signal processing methods are often applied to signals obtained from mechanics and rotating machines. This paper concerns the study of signals whose energetic time frequency representation exhibits several patterns. Firstly, non related Fourier time-frequency methods named Sliding Lagunas methods is used in order to characterize patterns. Then a new filtering applied to the time-frequency domain is proposed in a order to extract the individual objects of this representation. The validation of this processing is shown on synthetic data.

1. INTRODUCTION

Time-frequency signal analysis has recently becomes the focus research as more and more analysts are confronted with the widely reported insufficiencies of classical signal analysis tools, based either on time domain or frequency domain representations of the signal. In this paper, several non stationary methods are proposed : Short Time Fourier Transform (STFT), Sliding Lagunas method issued from Capon's method (Filter bank). Opposite to Short Time FT, Lagunas and Capon's methods do not need Fourier analysis, but they require a knowledge on the signal structure. The original method is issued from classical spectral analysis. A sliding memory of the filter along the total signal duration allows the formulation of an adaptive spectral density. A local stationarity of the signal is necessary. A time-frequency representation allows us to localize different patterns, and we apply a filtering in time-frequency on this result to select each pattern [3].

2. TIME-FREQUENCY REPRESENTATIONS

2.1. SHORT-TIME TRANSFORM FOURIER

Given a finite energy signal x(t) and a sliding window h(t), a classical linear time-frequency representation can be obtained by computing the Short Time Fourier Transform (STFT) [1] :

$$STFT(t,\nu) = \int_{-\infty}^{+\infty} x(k)\, h^*(k-t)\, e^{-2i\pi\nu k}\, dk$$

The spectrogram defined as the energy distribution associated to the STFT, ie $|STFT(t,\nu)|^2$, has been widely used for many signal processing tasks. The local stationarity is needed on successive windows. A classical time and frequency resolution trade off underlies the structure of the spectrogram : the choice of an analyzing window of short duration ensures a good time localization but, at expenses of a poor frequency resolution by Fourier duality.

2.2. CAPON'S METHOD IN TIME FREQUENCY REPRESENTATION

The objective of Capon's method [2];[4] is to build a band-pass filter around a frequency ν_k while ensuring the following conditions :

a) The spectrum at the frequency ν_k is not altered. The output power of the filter equals the input power at frequency ν_k.

b) The influence of interferences due to other frequencies is minimized.

For this, Capon uses a moving average filter (MA) whose input-output relation is given by :

(1) $$Y(n) = \underline{A}^T . \underline{S}(n)$$

where ^{T} denotes the transpose,

with $\underline{S}^T(n) = [S(n), ..., S(n-P)]$ the filter input,

$\underline{A}^T = [A_0, .. ,A_P]$ the impulse response of the filter,

Y(n) the filter output.

The first condition implies that, at frequency ν_k, the transfer function of the filter equals one. To satisfy the second condition, the interferences due to other frequencies present in the signal are minimized at each time. The two conditions lead to :

a) $\underline{A}^H \underline{Z}_k = 1$,

b) $\underline{A}^H \underline{\underline{R}}_s \underline{A}$ minimum,

with (^{H} is the conjugate transpose)

$$\underline{Z}_k^T = \left[1,..,\ z_k^{-P} \right] \text{ and } z_k = e^{2i\pi\nu_k\Delta t}$$

$\underline{\underline{R}}_s$ is the correlation matrix of the signal of dimension (P+1)(P+1).

The solution to this problem of minimization under constraint using Lagrange multipliers is:

$$(2)\quad \underline{A} = \frac{\underline{\underline{R}}_s^{-1}.\underline{Z}_k}{\underline{Z}_k^H.\underline{\underline{R}}_s^{-1}.\underline{Z}_k}$$

and gives the impulse response of the Capon filter. The output power of the filter is an estimator of the power of the signal at frequency ν_k. By generalizing to all frequencies, the power estimator can be written :

$$(3)\quad P_{cap} = \underline{A}^H\ \underline{\underline{R}}_s\ \underline{A} = \frac{1}{\underline{Z}^H.\underline{\underline{R}}_s^{-1}.\underline{Z}}$$

With this method, the estimated filter is adapted to the signal at each frequency, whereas, in the Fourier analysis, the estimated filter is independent of the signal. It is clear that Capon's method estimates the power of the signal from the estimation of the signal correlation matrix and its resolution is better than spectrogram resolution. In a nonstationary context, a sliding memory of the filter along the total signal duration allows the formulation of an adaptive spectral density [7].

2.3. PRINCIPLES OF SLIDING LAGUNAS'S METHOD

To obtain a power spectrum density estimator called Lag(n,ν) from Capon's method, we normalize the power P_{cap} by the effective bandwidth B_{eq} [5] :

$$P_{cap} = \int_{-B_{eq}/2}^{-B_{eq}/2} \text{Lag}(n,\nu)\ |A(\nu,\nu')|^2 d\nu'$$

If B_{eq} is very tight, we can considerate that the spectral density is flat over the band ($|A(\nu,\nu')|=1$) Then

$$P_{cap} = \text{Lag}(n,\nu) \int_{-B_{eq}/2}^{-B_{eq}/2} |A(\nu,\nu')|^2 d\nu'.$$

The final spectral density is :

$$\text{Lag}(n,\nu) = \frac{P_{cap}}{\underline{A}^T\underline{A}} = \frac{\underline{Z}^H.\underline{\underline{R}}_s^{-1}.\underline{Z}}{\underline{Z}^H.\underline{\underline{R}}_s^{-2}.\underline{Z}}$$

Spectral information is available at the output of the filter. This original method is issued from classical spectral analysis, and a sliding memory of the filter along the total signal duration permits the formulation of an adaptive spectral density.

3. PRINCIPLE OF FILTERING IN TIME-FREQUENCY DOMAIN

The result of time frequency methods on a nonstationary signal gives several patterns in time-frequency domain. Generally, direct separation of patterns in time or in frequency is not easy. By a filtering process in the time frequency domain, we propose to isolate each pattern.

Consider a signal $S(t) = S_1(t) + S_2(t-\tau)$, where $S_1(t)$ is a chirp. We choose a basis of projection as :

$$\begin{cases} M(t) = F_T(t-T/2)\ .\ \cos(\Phi(t)) \\ M_Q(t) = F_T(t-T/2)\ .\ \sin(\Phi(t)) \end{cases}$$

with $F_T(t) = 1$ if $t \geq 0$, and 0 if $t < 0$, and $\Phi(t)$ is the instantaneous phase terms.
In this basis, each component can be written as :

$S_i(t) = P_i(t)\ M(t)\ -\ Q_i(t)\ M_Q(t)$

and $S(t) = \big(P_1(t)M(t) - Q_1(t)M_Q(t)\big)$

$\qquad + \big(P_2(t-\tau)M(t-\tau) - Q_2(t-\tau)M_Q(t-\tau)\big).$

The principle of filtering is to separate the component $P_1(t)$ and $Q_1(t)$ of $P_2(t-\tau)$ and $Q_2(t-\tau)$. For this, we multiply S(t) by each model M(t) and $M_Q(t)$. In this product, we obtain an expression with low frequency and high frequency terms. The spectrum of $P_1(t)$ is centered on the 0 frequency with a bandwidth $\Delta\nu_1$ ($\Delta\nu_1$ is a maximal bandwidth around a pattern). With a low pass filter on low frequency term, it is possible to select $P_1(t)$ (figure 1).

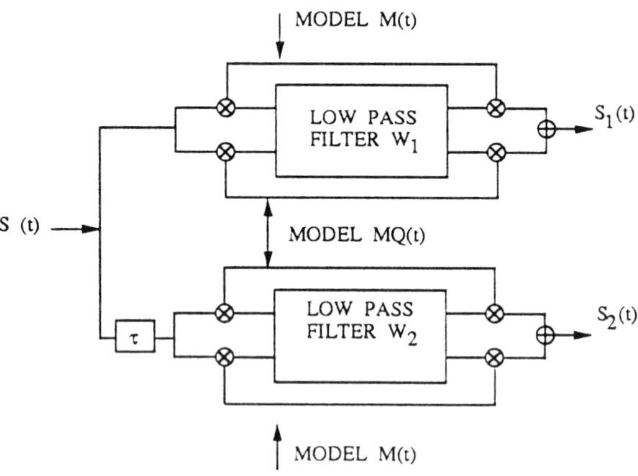

Figure 1 : Process Filter

The main difficulty is to construct the model. For this, we plot on the time-frequency representation, the points of the selected pattern in order to compute its instantaneous frequency $f_i(t)$, where $f_i(t) = \frac{1}{2\pi}\frac{d\Phi(t)}{dt}$

with $\Phi(t) = \text{Arctgt}\left(\frac{M(t)}{\tilde{M}(t)}\right)$

$\tilde{M}(t)$ is the Hilbert transform of M(t). Generally the instantaneous frequency presents several wrong points. In order to obtain a correct model without this points, it is necessary to apply a interpolation.
Finally, we can considerate that the filtering method is a variable frequency demodulation whose development made up of three steps.

- Elaboration of models.
- Multiplication by the models.
- Low-pass filter.

4. APPLICATION

A synthetic model composed by a summation of a sinusoidal frequency modulation (Signal A) and a parabolic frequency modulation (Signal B) was used to test these processes. The used parameters for the signal A are ν_{min} = 0.3 Hz; ν_{max} = 0.4 Hz; and for the signal B these parameters are : ν_{min} = 0.05 Hz; ν_{max} = 0.4 Hz. The sampling frequency is 1 Hz. The signal to noise ratio is 20 dB. Temporal representation of its signal recorded on 2048 sampling is given on figure 2.

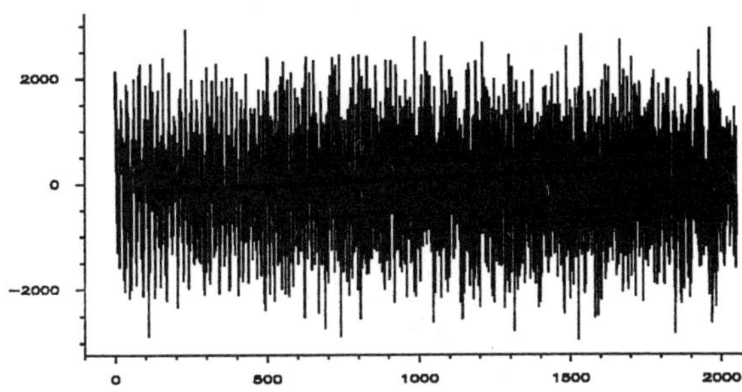

Figure 2 : Synthetic data

A time frequency representation F-t-ρ results from the STFT method with an analyzing window of 63 samples is presented on figure 3. The frequency F and the time t are respectively plotted on the abscissa and on the ordinates. The module ρ of each frequency represents the third dimension. We can observe an estimation of energy distribution which gives an information on the number of frequencies presents at each time on this signal (Two frequencies in this case).
With a sliding Lagunas method used with an order equal to 14 (7 real frequencies are sought : two for the signal and five for the noise) on a window length equal to 51 samples (memory), we can observe clearly each component in a time frequency representation (figure 4). In this case, the module ρ represents the four maxima of each spectrum. At each time, we have two frequencies associated to the signal and two others frequencies at random position which characterizes noise. Using this representation, we can constructs two models for filtering. For this application, only one model is plotted (figure 5).
Time frequency representation of sliding Lagunas's method applied on the result of the filtering with different bandwidth $\Delta\nu_1$ is presented. If the $\Delta\nu_1$ is short (0.01 Hz), the selection of the pattern is efficient (figure 6a), but if the bandwidth of the low pass filter increases, outline of the filter appears in the time frequency representation, where the module ρ represents the nth maxima of each spectrum. To have an information on the power of the frequencies obtained, it is necessary to applied Capon's method on the data (see also 2.2). Therefore, by associating the Lagunas's method and Capon's method, a hybrid method is obtained [7]. The result of this association on the signal filtered with a large bandwidth $\Delta\nu_1$ (0.1 Hz) shows the signal A and a part of signal B (figure 6b). The signal B is characterized by the difference between initial data and the result of filtering (Figure 7).

5. CONCLUSION

The use of parametric time frequency methods not based on Fourier analysis allows us to detect clearly dispersive patterns. Thus the construction of models is easier. The application of filtering which is a variable frequency demodulation by choosing models permits to characterize and separate different patterns. This technique which has been validated on simulation, has been efficient to separate two dispersives subsurfaces waves coming from a earthquake located in Aleoutian Islands [6].

6. REFERENCES

[1] ALLEN J., RABINER L.A., Unified Approach to Short Time Fourier Analysis and Synthesis. *Proc IEEE*, Vol 65 n°11, 1977, pp 1558-1564.

[2] CAPON J., High-resolution frequency wavenumber spectrum analysis, *Proc. IEEE*, Vol 57, 1969, pp 1048-1418.

[3] FARGETTON H., GLANGEAUD F., JOURDAIN G., Filtrage dans le plan Temps-Fréquence. Caractérisation de signaux UBF et du milieu magnétosphérique. *Extrait des annales des Télécommunications*. Tome 3, n°3-4, 1979.

[4] LACOSS R.T., Data adaptive spectral analysis methods. *Geophysics*, Vol 36, 1971, p. 661-675.

[5] LAGUNAS M.A., GASULLL-LAMPADAS A., An improved maximum likelihood method for power spectral density estimation. *Proc. IEEE, Trans. Acoustics, Speech and Signal Processing*, Vol ASSP-32, n°1, feb 1984.

[6] MARS J., PEDERSEN H.A., Mise en oeuvre et validation d'analyse temps-fréquence. GRETSI, Juan-les-pins, septembre 1991, pp 373-376 .

[7] MARS J.,MARTIN N., LACOUME J.L., DUBESSET M., Analysis of signal over short time windows. *Signal Processing*, Vol 26, n° 2, pp147-159.

Figure 3 : Short Time Fourier Transform on initial data

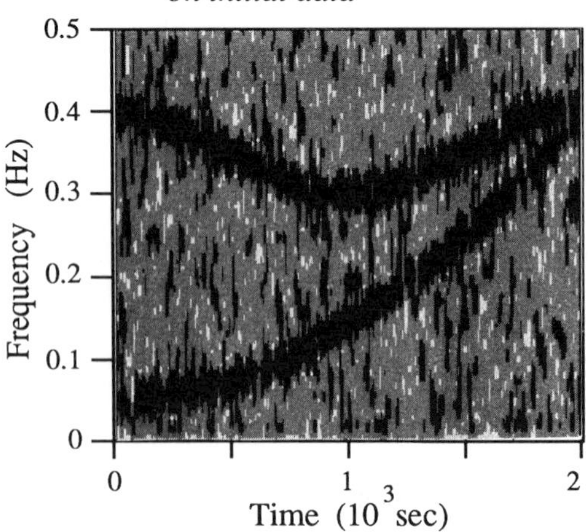

Figure 4 : Lagunas's method on initial data
Fe = 1 Hz ; Order = 14 ; Memory = 51

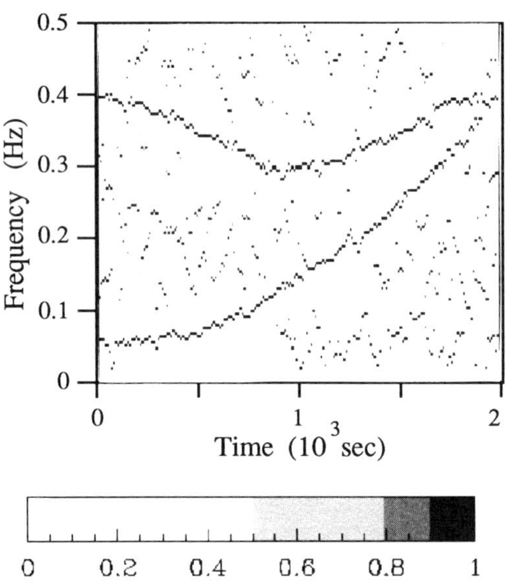

Figure 5 : Instantaneous frequency of model A

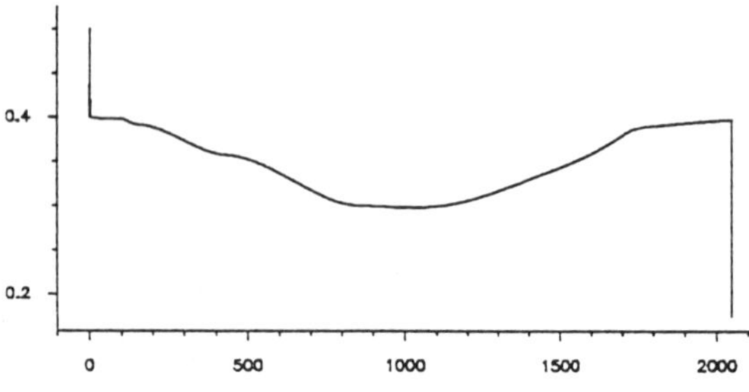

Figure 6 a : Result of Filtering (Lagunas)
Order = 6 ; Memory = 31 ; $\Delta\nu_1$ = 0.01 Hz

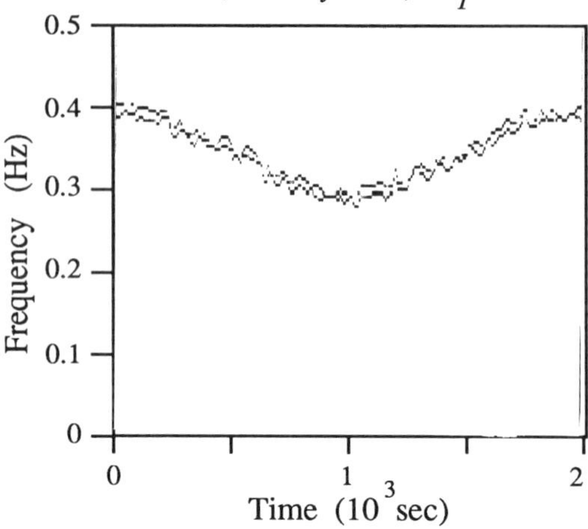

Figure 6 b : Result of Filtering (Hybrid)
Order = 10 ; Memory = 41 ; $\Delta\nu_1$ = 0.1 Hz

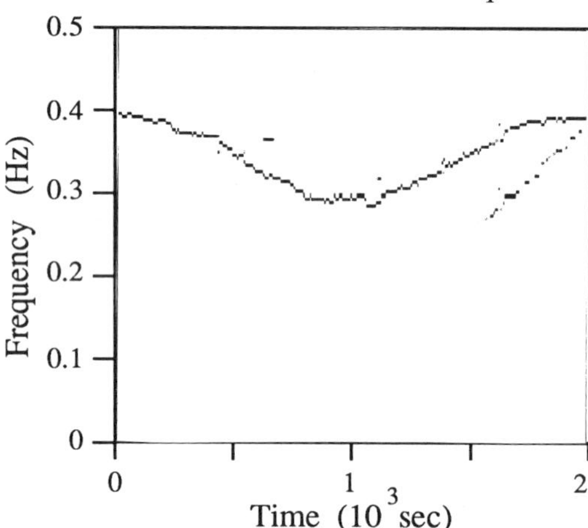

Figure 7 : Difference Initial data - Result of filter (Lagunas) Order = 8 ; Memory = 31

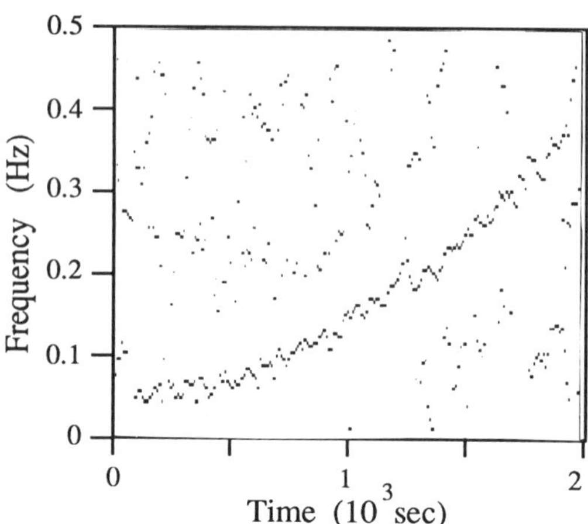

SIGNAL PROCESSING VI: Theories and Applications
J. Vandewalle, R. Boite, M. Moonen, A. Oosterlinck (eds.)

Maximum Likelihood Source Location Estimation via EM Algorithm

D. Kraus, D. Maiwald and J.F. Böhme

Department of Electrical Engineering, Ruhr University Bochum, 4630 Bochum, FRG

Approximate maximum likelihood source location estimation is investigated. An extended model of the spectral density matrix of the sensor array output for coherent sources (multipath propagation) is introduced. We show that the EM algorithm can be successfully appplied for this special case of practical importance. In contrast to [1], the EM iteration scheme is derived by exploiting that the distribution type of the finite Fourier transformed sensor outputs belongs to the exponential family. Finally, we investigate the loss of information due to the missing data.

1. Introduction

Approximate maximum likelihood estimates (AMLSEs) for source locating of wide band sources in the presence of partly unknown noise fields are developed. In comparison to other estimation methods, a significant disadvantage is the cumbersome computation of AMLEs. An efficient alternative to a brute-force algorithm, i.e. global search and a local optimization technique, is the use of the expectation maximization (EM) algorithm, cf. [5], [2], [3] and [1]. In continuation of [1], we introduce an extended model of the sensor array output for coherent sources (multipath propagation) and derive a modified EM iteration scheme by exploiting the distributional properties (exponential family) of the finite Fourier-transformed sensor output of successive data pieces, cf. [5]. This approach requires neither the additional sensor noise parameters $\nu_{0,i}$ used in [1] nor an alternating step by step optimization of spectral and source location pa-Fisher's sense and the increase in variance and mention how it can be reduced by spatial filtering if instead of the complete data only the incomplete data can be measured .

An outline of the paper follows. In section 2, the data model and the parameter structure are introduced. The EM algorithm for incomplete data from an exponential family is developed in section 3. In section 4, the EM algorithm is applied for AMLE of source location. In section 5, the informationloss is derived. We conclude with some remarks.

2. Data Model

A conventional propagation-reception model is used. Sources generate signals which are transmitted by wavefields. At an array of sensors, the signals are measured additively disturbed by noise. The outputs of the sensors $n = 1, \ldots, N$ are finite Fourier transformed with a smooth, normalized window of lenght T. For every frequency ω of interest, we get data $\underline{X}^k(\omega) = \left(X_1^k(\omega), \ldots, X_N^k(\omega)\right)^T$ of $k = 1, \ldots, K$ successive pieces of sensor outputs.

Suppose there are M' groups of uncorrelated signals. Each group consists of L_i $(i = 1, \ldots, M')$ correlated signals with $\sum_{i=1}^{M'} L_i = M$. Then, the $(N \times N)$ spectral density matrix of the array output can be expressed by

$$\mathbf{C}_{\underline{X}}(\omega) = \sum_{i=1}^{M'+I} \mathbf{C}_{\underline{Y}_i}(\omega)$$

where

$$\begin{aligned} \mathbf{C}_{\underline{Y}_i}(\omega) &= \mathbf{H}_i(\omega)\mathbf{C}_{\underline{S}_i}(\omega)\mathbf{H}_i(\omega)^* + \epsilon_i(\omega)\mathbf{I} \\ & \quad (i = 1, \ldots, M') \end{aligned} \tag{1}$$

and

$$\begin{aligned} \mathbf{C}_{\underline{Y}_{M'+i}}(\omega) &= \nu_i(\omega)\mathbf{J}_i \\ & \quad (i = 1, \ldots, I) \end{aligned}$$

with $\epsilon_i(\omega)$ known, $\epsilon_i(\omega) \geq 0$ and $\sum_{i=1}^{M'} \epsilon_i(\omega) << \nu_1(\omega)$. The columns of the $(N \times L_i)$ matrix $\mathbf{H}_i(\omega)$ are known as steering vectors $\underline{d}_{i,l}(\omega)$ $(l = 1, \ldots, L_i)$ that contain the unknown wave parameter vector $\underline{\xi}_i$ $(i = 1, \ldots, M')$, e.g. bearings $\beta_{i,l}$ and ranges $\rho_{i,l}$ $(l = 1, \ldots, L_i)$ of the signals. The signal spectral parameter vectors $\underline{\zeta}_i(\omega)$ $(i = 1, \ldots, M')$ are given by the entries of the $(L_i \times L_i)$ matrix $\mathbf{C}_{\underline{S}_i}(\omega)$ itself. The noise spectral parameters are collected in the vector $\underline{\nu}(\omega) = (\nu_1(\omega), \ldots, \nu_I(\omega))^T$ and the matrices $\mathbf{J}_i(\omega)$ $(i = 1, \ldots, I)$ are supposed to be known nonnegative hermitian matrices. For example, they describe angle bandlimited noise of different directions and widths.

Furthermore, let the groups of signals with $L_i > 1$ ($i = 1,\ldots,M'$) to be due to multipath propagation (perfect coherence), $\mathbf{C}_{\underline{Y}_i}(\omega)$ can be rewritten to

$$\mathbf{C}_{\underline{Y}_i}(\omega) = \sigma_i(\omega)\underline{d}_i(\omega)\underline{d}_i(\omega)^* + \epsilon_i(\omega)\mathbf{I} \quad (i = 1,\ldots,M') \qquad (2)$$

where $\underline{d}_i(\omega)$ denotes for $L_i = 1$ or $L_i > 1$ ($i = 1,\ldots,M'$) the usual or a generalized steering vector, respectively. For an arbitary array geometry the generalized steering vector is given by

$$\underline{d}_i(\omega) = \sum_{l=1}^{L_i} \kappa_{i,l}(\omega)\underline{d}_{i,l}(\omega),\ \kappa_{i,1} \equiv 1,\ \kappa_{i,l} \in \mathbb{C}, \qquad (3)$$

where $\underline{d}_{i,l}(\omega)$ denotes the usual steering vector for the l-th propagation path of the i-th signal group (source). If a line array is employed and plane wave propagation is assumed $\underline{d}_i(\omega)$ can be expressed by

$$\underline{d}_i(\omega) = \left(\sum_{l=1}^{L_i} \gamma_{i,l}(\omega)\mathbf{D}_{i,l}(\omega)\right)\underline{d}_{i,1}(\omega), \quad \gamma_{i,1} \equiv 1,\ \gamma_{i,l} \in \mathbb{R} \qquad (4)$$

with

$$\mathbf{D}_{i,l}(\omega) = \operatorname{diag}\left(1, e^{-j\omega\tau_{i,l}},\ldots,e^{-j\omega(N-1)\tau_{i,l}}\right), \quad \tau_{i,1} = 0,\ \tau_{i,l} \in \mathbb{R}.$$

The $\mathbf{C}_{\underline{Y}_i}(\omega)$ can be understood as the spectral density matrix corresponding to a virtual array output $\underline{Y}_i^k(\omega) = \left(Y_{i,1}^k(\omega),\ldots,Y_{i,N}^k(\omega)\right)^T$ ($k = 1,\ldots,K$) generated either by a group of correlated signals or by a noise component. The incomplete data $\underline{X}^k(\omega)$ are related to the complete one, $\underline{Y}^k(\omega) = (\underline{Y}_1^k(\omega)^T,\ldots,\underline{Y}_{M'+I}^k(\omega)^T)^T$ by the linear many-to-one mapping

$$\underline{X}^k(\omega) = (\mathbf{I},\ldots,\mathbf{I})\underline{Y}^k(\omega).$$

Under the conditions leading to (4), the number of unknown parameters using model (1), model (2) with (3) and model (2) with (4) are $P = L_i^2 + L_i$, $P = 3L_i - 1$, and $P = 2L_i$, respectively. The parameter vector is given by $\underline{\theta}(\omega) = (\underline{\xi}_1^T,\ldots,\underline{\xi}_{M'}^T,\underline{\zeta}_1(\omega)^T,\ldots,\underline{\zeta}_{M'}(\omega)^T,\underline{\nu}(\omega)^T)^T$, where in $\underline{\xi}_i$ the wave parameters and in $\underline{\zeta}_i(\omega)$ the signal parameters of the i-th group of signals corresponding to the chosen model are suitable collected. Generally, the numbers M' and L_i ($i = 1,\ldots,M'$) are unknown. In this paper we assume to know M' and L_i ($i = 1,\ldots,M'$) with $\sum_{i=1}^{M'} L_i = M < N$.

In order to define likelihood functions, the distributional properties of the finite Fourier transformed complete data $\underline{Y}^k(\omega)$ are required. Let us state the well known asymptotic properties of $\underline{Y}^k(\omega)$ if the window length T is large and $0 < \omega_1 < \ldots < \omega_J < \pi$:

i) $\underline{Y}^k(\omega_1),\ldots,\underline{Y}^k(\omega_J)$ are independent, complex normally distributed random vectors with zero mean and block diagonal covariance matrices $\mathbf{C}_{\underline{Y}}(\omega_j) = \operatorname{diag}\left(\mathbf{C}_{\underline{Y}_1}(\omega_j),\ldots,\mathbf{C}_{\underline{Y}_{M'+I}}(\omega_j)\right)$ ($j = 1,\ldots,J$), respectively.

ii) $\underline{Y}^1(\omega),\ldots,\underline{Y}^K(\omega)$ are independent and identically complex normally distributed random vectors with zero mean and block diagonal covariance matrix $\mathbf{C}_{\underline{Y}}(\omega) = \operatorname{diag}(\mathbf{C}_{\underline{Y}_1}(\omega),\ldots,\mathbf{C}_{\underline{Y}_{M'+I}}(\omega))$.

In the sequel, we shall use the nonparametric spectral density estimates

$$\hat{\mathbf{C}}_{\underline{Y}}(\omega) = \frac{1}{K}\sum_{k=1}^{K} \underline{Y}^k(\omega)\underline{Y}^k(\omega)^*,$$

$$\hat{\mathbf{C}}_{\underline{Y}_i}(\omega) = \frac{1}{K}\sum_{k=1}^{K} \underline{Y}_i^k(\omega)\underline{Y}_i^k(\omega)^*,$$

and

$$\hat{\mathbf{C}}_{\underline{X}}(\omega) = \frac{1}{K}\sum_{k=1}^{K} \underline{X}^k(\omega)\underline{X}^k(\omega)^*,$$

for notational convenience.

3. EM-Algorithm

In contrast to the application of the EM algorithm in [1], we exploit that the density function of the complete data belongs to the (curved) exponential family, cf. [5] and [2]. Hence, we can write

$$f_{\underline{Y}}(\underline{y};\underline{\theta}) = a(\underline{\alpha}(\underline{\theta}))^{-1}\exp\{\underline{\alpha}(\underline{\theta})^*\underline{t}(\underline{y})\}, \qquad (5)$$

with respect to a σ-finite measure $\mu_{\underline{Y}}$. After some calculations, when the notation

$$a_{\underline{X}}(\underline{\alpha}(\underline{\theta})) = \int \exp\{\underline{\alpha}(\underline{\theta})^*\underline{t}(\underline{y})d\mu_{\underline{Y}|\underline{X}}(\underline{y})$$

is introduced, the conditional density function

$$f_{\underline{Y}|\underline{X}}(\underline{y};\underline{\theta}) = a_{\underline{X}}(\underline{\alpha}(\underline{\theta}))^{-1}\exp\{\underline{\alpha}(\underline{\theta})^*\underline{t}(\underline{y})\}, \qquad (6)$$

and the density function

$$f_{\underline{X}}(\underline{x};\underline{\theta}) = \frac{a_{\underline{X}}(\underline{\alpha}(\underline{\theta}))}{a(\underline{\alpha}(\underline{\theta}))}$$

can be deduced, with respect to the measures

$$\mu_{\underline{Y}|\underline{X}} = \begin{cases} \mu_{\underline{Y}} & : \ \underline{X} = \underline{g}(\underline{Y}) \\ 0 & : \ \text{else} \end{cases}$$

and $\mu_{\underline{X}}$, respectively. The log-likelihood function of the incomplete data can be represented by

$$L_{\underline{X}}(\underline{\theta}) = \log f_{\underline{X}}(\underline{x};\underline{\theta}) = \log\frac{a_{\underline{X}}(\underline{\alpha}(\underline{\theta}))}{a(\underline{\alpha}(\underline{\theta}))}.$$

Using some results well known for the exponential family and $\underline{T} = \underline{t}(\underline{Y})$, the necessary conditions for an optimum $\underline{\theta}$ provide after some algebra

$$\left[\frac{\partial\underline{\alpha}(\underline{\theta})}{\partial\underline{\theta}}\right]^* (E_{\underline{\alpha}(\underline{\theta})}(\underline{T}) - E_{\underline{\alpha}(\underline{\theta})}(\underline{T}|\underline{X})) = 0. \quad (7)$$

This nonlinear equation system motivates the iterations of the EM algorithm that can be described as follows.

E-step: calculating $\underline{t}^n = E_{\underline{\alpha}(\underline{\theta}^n)}(\underline{T}|\underline{X})$,

M-step: solving $\left[\frac{\partial\underline{\alpha}(\underline{\theta})}{\partial\underline{\theta}}\right]^* (E_{\underline{\alpha}(\underline{\theta})}(\underline{T}) - \underline{t}^n) = 0 \Longrightarrow \underline{\theta}^{n+1}$.

Because this iteration scheme is a special case of that developed in [2], the same convergence properties as indicated in [6] and [7] can be claimed.

4. Approximate MLE

Let us assume that the sensor outputs can be sufficiently well described in the frequency domain by a number J of narrow bands or by a spectrum, varying slowly with frequency ω, such that it can be reconstructed satisfactorily by a finite number J of interpolation points. These assumptions in conjunction with the asymptotic distributional properties i) and ii) suggest the approximate likelihood functions for the complete data

$$\begin{aligned} f_{\underline{Y}}(\underline{y},\underline{\theta}) &= \prod_{j=1}^{J} \det(\mathbf{C}_{\underline{Y}}(\omega_j,\underline{\theta}(\omega_j))^{-1} \cdot \\ &\quad \exp\{-\operatorname{tr}[\mathbf{C}_{\underline{Y}}(\omega_j,\underline{\theta}(\omega_j))^{-1}\hat{\mathbf{C}}_{\underline{Y}}(\omega_j)]\}, \end{aligned}$$

where the $\underline{\theta}(\omega_j)$ $(j = 1,\ldots,J)$ are appropriately collected in $\underline{\theta}$. Exploiting the block diagonal structure of $\mathbf{C}_{\underline{Y}}(\omega)$, we obtain

$$\begin{aligned} f_{\underline{Y}}(\underline{y},\underline{\theta}) &= \prod_{j=1}^{J}\prod_{j=1}^{M'+I} \det(\mathbf{C}_{\underline{Y}_i}(\omega_j,\underline{\theta}_i(\omega_j))^{-1} \cdot \\ &\quad \exp\{-\operatorname{tr}[\mathbf{C}_{\underline{Y}_i}(\omega_j,\underline{\theta}_i(\omega_j))^{-1}\hat{\mathbf{C}}_{\underline{Y}_i}(\omega_j)]\} \quad (8) \end{aligned}$$

where $\underline{\theta}_i(\omega_j) = (\underline{\xi}_i^T, \underline{\zeta}_i(\omega_j)^T)^T$ $(i = 1,\ldots,M')$ and $\underline{\theta}_{M'+I}(\omega_j) = \nu_i(\omega_j)$ $(i = 1,\ldots,I)$, Comparing (5) with (8), the statistic $\underline{T}$ and the vector-valued function $\underline{\alpha}(\underline{\theta})$ can be identified by

$$\begin{aligned} \underline{T} &= (\operatorname{vec}(\hat{\mathbf{C}}_{\underline{Y}_1}(\omega_1))^T,\ldots,\operatorname{vec}(\hat{\mathbf{C}}_{\underline{Y}_1}(\omega_J))^T, \\ &\quad \ldots,\operatorname{vec}(\hat{\mathbf{C}}_{\underline{Y}_{M'+I}}(\omega_1))^T,\ldots,\operatorname{vec}(\hat{\mathbf{C}}_{\underline{Y}_{M'+I}}(\omega_J))^T)^T \end{aligned}$$

and

$$\begin{aligned} \underline{\alpha}(\underline{\theta}) &= -(\operatorname{vec}(\mathbf{C}_{\underline{Y}_1}(\omega_1,\underline{\theta}_1(\omega_1))^{-1})^T, \\ &\quad \ldots,\operatorname{vec}(\mathbf{C}_{\underline{Y}_1}(\omega_J,\underline{\theta}_1(\omega_J))^{-1})^T, \\ &\quad \ldots,\operatorname{vec}(\mathbf{C}_{\underline{Y}_{M'+I}}(\omega_1,\underline{\theta}_{M'+I}(\omega_1))^{-1})^T, \\ &\quad \ldots,\operatorname{vec}(\mathbf{C}_{\underline{Y}_{M'+I}}(\omega_J,\underline{\theta}_{M'+I}(\omega_J))^{-1})^T)^T. \end{aligned}$$

After rearranging the nonlinear equation system (7), where the nonlinear equation system can be decoupled in $(M' + I)$ systems of lower dimensions, and with

$$\begin{aligned} \mathbf{C}_{\underline{Y}_i}(\omega_j,\underline{\theta}_i(\omega_j)) &= \frac{1}{\epsilon_i(\omega_j)} \\ &\quad \left[\mathbf{I} - \frac{\sigma_i(\omega_j)}{N\sigma_i(\omega_j)+\epsilon_i(\omega_j)}\underline{d}_i(\omega_j)\underline{d}_i(\omega_j)^*\right] \end{aligned}$$

the EM algorithm provides the following iteration scheme

E-step: for $i = 1,\ldots,M'+I$, cf. [1]

$$\begin{aligned} &\bar{\mathbf{C}}_{\underline{Y}_i}(\omega_j,\underline{\theta}^n(\omega_j)) = E_{\underline{\theta}_j^n}(\hat{\mathbf{C}}_{\underline{Y}_i}(\omega_j)|\underline{X}(\omega_j)) = \\ &\mathbf{C}_{\underline{Y}_i}(\omega_j,\underline{\theta}_i^n(\omega_j)) - \\ &\mathbf{C}_{\underline{Y}_i}(\omega_j,\underline{\theta}_i^n(\omega_j))\mathbf{C}_{\underline{X}}(\omega_j,\underline{\theta}^n(\omega_j))^{-1}\mathbf{C}_{\underline{Y}_i}(\omega_j,\underline{\theta}_i^n(\omega_j)) + \\ &\mathbf{C}_{\underline{Y}_i}(\omega_j,\underline{\theta}_i^n(\omega_j))\mathbf{C}_{\underline{X}}(\omega_j,\underline{\theta}^n(\omega_j))^{-1} \\ &\hat{\mathbf{C}}_{\underline{X}}(\omega_j)\mathbf{C}_{\underline{X}}(\omega_j,\underline{\theta}^n(\omega_j))^{-1}\mathbf{C}_{\underline{Y}_i}(\omega_j,\underline{\theta}_i^n(\omega_j)), \end{aligned}$$

M-step: for $i = 1,\ldots,M$ and $k = 1,\ldots,I$, respectively

$$\begin{aligned} &\sigma_i^{n+1}(\omega_j) = \\ &\frac{1}{N}\left(\frac{\underline{d}(\omega_j,\underline{\xi}_i^{n+1})^*\bar{\mathbf{C}}_{\underline{Y}_i}(\omega_j,\underline{\theta}_i^n(\omega_j))\underline{d}(\omega_j,\underline{\xi}_i^{n+1})}{N} - \epsilon_i(\omega_j)\right), \\ &\nu_k^{n+1}(\omega_j) = \frac{1}{N}\operatorname{tr}\left[\mathbf{J}_k(\omega_j)^{-1}\bar{\mathbf{C}}_{\underline{Y}_{M'+k}}(\omega_j,\underline{\theta}_i^n(\omega_j))\right], \end{aligned}$$

and $\underline{\xi}_i^{n+1}$ is obtained by solving the nonlinear equation system

$$\begin{aligned} &\sum_{j=1}^{J}\frac{\sigma_i^{n+1}(\omega_j)/\epsilon_i(\omega_j)}{N\sigma_i^{n+1}(\omega_j)+\epsilon_i(\omega_j)} \cdot \\ &\operatorname{Re}\left\{\underline{d}_i(\omega_j)^*\bar{\mathbf{C}}_{\underline{Y}_i}(\omega_j,\underline{\theta}_i^n(\omega_j))\frac{\partial\underline{d}_i(\omega_j)}{\partial\underline{\xi}_i}\right\} = \underline{0}, \quad (9) \end{aligned}$$

Observing that (9) is the necessary condition of the weighted classical beamformer criterion obtained in [1], this approach and that in [1] seem to be equivalent. Nevertheless, there are two major distinctions.

In [1], the log-likelihood function could only be simplified to the weighted classical beamformer criterion if the spectral parameters would be kept fixed and became more complicated if the spectral parameters would be replaced by their explicit solutions. The nonlinear equation system (9) could be derived without such considerations. Moreover, for $J = 1$ or the $\epsilon_i(\omega_j)$ appropriately chosen such that

$$\frac{\sigma_i^{n+1}(\omega_1)/\epsilon_i(\omega_1)}{N\sigma_i^{n+1}(\omega_1)+\epsilon_i(\omega_1)} \cong \ldots \cong \frac{\sigma_i^{n+1}(\omega_J)/\epsilon_i(\omega_J)}{N\sigma_i^{n+1}(\omega_J)+\epsilon_i(\omega_J)}$$

the equations of the signal spectral parameters and the wave parameters can be decoupled completely. Herein, the sensor noise parameter and the ambient noise parameters are treated equivalently. Therefore, instead of the additional and unknown sensor noise parameters $\nu_{0,i}(\omega_j)$ used in [1], we have introduced the known auxillary parameters $\epsilon_i(\omega_j)$ to guarantee the positive definiteness of $\mathbf{C}_{\underline{Y}_i}(\omega_j)$.

5. Informationloss and Increase in Variance

The log-likelhood function for the complete data is given by

$$L_{\underline{Y}}(\underline{\theta}) = L_{\underline{Y}|\underline{X}}(\underline{\theta}) + L_{\underline{X}}(\underline{\theta}), \qquad (10)$$

where $L_{\underline{Y}|\underline{X}}(\underline{\theta})$ denotes the logarithm of (6). Taking the derivative of (10) with respect to $\underline{\theta}_p$ $(p = 1, \ldots, P)$ and then the conditional expectation given $\underline{X} = \underline{x}$, we obtain

$$E(\underline{T}_{\underline{Y},p}|\underline{X})) = T_{\underline{X},p}, \qquad (11)$$

where $E(\underline{T}_{\underline{Y}|\underline{X},p}|\underline{X})) = 0$ is exploited and $T_{\underline{Y},p}$, $T_{\underline{X},p}$ and $T_{\underline{Y}|\underline{X},p}$ denote $\partial L_{\underline{Y}}(\underline{\theta})/\partial\theta_p$, $\partial L_{\underline{X}}(\underline{\theta})/\partial\theta_p$ and $\partial L_{\underline{Y}|\underline{X}}(\underline{\theta})/\partial\theta_p$, respectively. Using the well known identity

$$\begin{aligned} Cov(T_{\underline{Y},p}, T_{\underline{Y},q}) &= Cov(E(T_{\underline{Y},p}|\underline{X}), E(T_{\underline{Y},q}|\underline{X})) \\ &+ E(Cov(T_{\underline{Y},p}, T_{\underline{Y},q}|\underline{X})) \end{aligned}$$

together with (11), we can derive

$$\begin{aligned} E(T_{\underline{Y},p}, T_{\underline{Y},q}) &= E(T_{\underline{X},p}(T_{\underline{X},q}) \\ &+ E(Cov(T_{\underline{Y},p}, T_{\underline{Y},q}|\underline{X})). \end{aligned}$$

Interpretation in terms of information matrices yields

$$I_{\underline{Y}}(\underline{\theta}) = I_{\underline{X}}(\underline{\theta}) + I_{\underline{Y}|\underline{X}}(\underline{\theta}),$$

where the last quantity on the right denotes the so-called lost information. Applying the equation

$$I_{\underline{Y}}(\underline{\theta})(I_{\underline{X}}(\underline{\theta})^{-1} - I_{\underline{Y}}(\underline{\theta})^{-1}) = I_{\underline{Y}|\underline{X}}(\underline{\theta})I_{\underline{X}}(\underline{\theta})^{-1},$$

the increase in variance and the informationloss can be expressed by

$$I_{\underline{X}}(\underline{\theta})^{-1} - I_{\underline{Y}}(\underline{\theta})^{-1} = I_{\underline{Y}}(\underline{\theta})^{-1} I_{\underline{Y}|\underline{X}}(\underline{\theta}) I_{\underline{X}}(\underline{\theta})^{-1}$$

and

$$I_{\underline{Y}|\underline{X}}(\underline{\theta}) = I_{\underline{Y}}(\underline{\theta})(I_{\underline{X}}(\underline{\theta})^{-1} - I_{\underline{Y}}(\underline{\theta})^{-1})I_{\underline{Y}}(\underline{\theta}),$$

respectively.

Therefore, in order to get more accurate estimates techniques recovering missing data should be employed. One possibility might be spatial filtering as e.g. spheroidal beamforming, cf. [4]. Another way seems to consist in the following two steps procedure

i) Computing the AMLEs via EM algorithm as described in section 4.

ii) The (initial) estimates obtained from i) are used to recover some missing data by removing certained signals from the data, alternatingly. Then, with less missing data the EM algorithm is applied again.

6. Concluding Remarks

We have introduced an extended model of the spectral density matrix of the sensor array output for sources in a multipath propagating environment. In comparison with [1], we have derived a modified EM iteration scheme and AMLEs. Furthermore, we have investigated the loss of information and the increase in variance due to the missing data, and mentioned two alternatives how it can be reduced.

Numerical experiments with the real data applying the two steps procedure indicated at the end of the last section are under work. The links between EM algorithm, the Kullback-Leibler divergence and and the maximum entropy methods are currently investigated.

REFERENCES

[1] Böhme J.F. and Kraus D.: *Parametric Methods for Source Location Estimation.* Proc. 9-th IFAC/IFORS Symposium on Identification and System Parameter Estimation, Budapest July 1991, pp. 1379–1364.

[2] Dempster et. al.: *Maximum Likelihood from Incomplete Data via the EM Algorithm.* J. Roy. Statist. Soc. Ser.39, 1977, pp. 1–38.

[3] Feder M. and Weinstein E.: *Parameter Estimation of Superimposed Signals Using the EM Algorithm.* IEEE Transactions on Acoustics, Speech, and Signal Processing, Vol.36, No.4, April 1988, pp. 477–489.

[4] Foster P., and Vezzosi G.: *Application of Spheroidal Sequences to Array Processing.* Proc. IEEE-ICASSP, Dallas, 1987, pp. 2268–2271.

[5] Sundberg R.: *Maximum Likelihood Theory for Incomplete Data from Exponential Family.* Scand. J. Statist., 1974, pp. 49–58.

[6] Sundberg R.: *An Iterative Method for Solution of the Likelihood Equitions for Incomplete Data from Exponential Families.* Comm. Statist.- Simula. Comput., B5(1), 1976, pp.55-64

[7] Wu C. F.: *On the Convergence Properties of the EM Algorithm.* Annals of Statistics, Vol. 11, No.1, 1983, pp. 95-103.

SIGNAL PROCESSING VI: Theories and Applications
J. Vandewalle, R. Boite, M. Moonen, A. Oosterlinck (eds.)

An Improved Algorithm for the Spatial Analysis of Correlated Sources

Dominic Grenier, Gilles Y. Delisle
Electrical Engineering Department
Laval University
Québec (QC), Canada G1K 7P4

tel: (418) 656-2806 (418) 656-2981
email: dgrenier@gel.ulaval.ca gdelisle@gel.ulaval.ca

Abstract

A high resolving power algorithm adapted to narrow band correlated sources detection without the use of a scanning analysis vector in the area of interest is presented.

Source correlation and computer time requirements are the major obstacles to a more important use of superresolutive methods. The proposed new algorithm used the translation and rotation invariances of the covariance matrix eigenvectors in both directions (forward and reverse) to generate the estimated sources subspace. The source subspace is therefore created again as it is the case for source vectors such as in DEESE. However, the required computations are limited since the vectorial multiplications associated with the scanning using paralled network techniques are cancelled and therefore allows a simple way of finding and saving the positions of the transmitting sources.

1 Introduction

Spatial analysis using second order signal processing techniques permits to achieve a better resolving power for an array of sensors. Even if this characteristic is desirable, the use of this technique is still to be generalized, even if second order signal processing is now more frequently used in specialized applications. This is due to some difficulties associated with the algorithm's hypothesis and additional considerations such as:

- second order characteristics that can be used on the received datas only for stationnary or quasi-stationnary system.
- zero or close to zero correlation between sources that does not apply to active systems or those with interferences.
- mathematical operations on complex matrices that generate large computing time.

Over the past few years, many research works have tried to avoid these difficulties. Methods using array or source properties have recently been used in particular situations. The majority of these methods addresses the case of correlated sources using spatial or frequency diversity since it is the most frequent cases [1-7] Some others methods are however looking at a greater tolerance with respect to the low number of samples [8-10] or simplify the computations or the sensitivity of the material [11-13] using the geometry.

Following this trend, the algorithm presented here combined in an advantageous way all the different manners to use the geometry of a linear array of identical equally spaced elements as well as the invariance in rotation and in translation. This algorithm resolved not only correlated sources but it also avoids the multiple vectorial products required for scanning with an analysis vector and obtains directly the positions of the sources with therefore simplifier the data storage problem.

The rank increase of the source matrix comes from the modified space diversity (translation invariance) to which the partition of the array in subarrays in both directions like in DEESE [6] is added. A change in the partionning direction is equivalent to a phase reserval of 180°about the center, i.e. the array's sensor numbering is interchanged and the signal's angles of arrival as well. The removal of scanning, follows algorithms such as ESPRIT [12] and TAM [11], namely the parallel way between two array or subarrays do not modify sources and noise subspaces in an independant sources environment (no interference figures).

2 Sources Subspace

An array of "N" identical, aligned and equally spaced (by a distance "d") elements is considered upon which "n" plane wavefronts ($n < N$) issued from narrow band sources in the directions $[\theta_1, \theta_2, \ldots, \theta_n]$ as in figure (1).

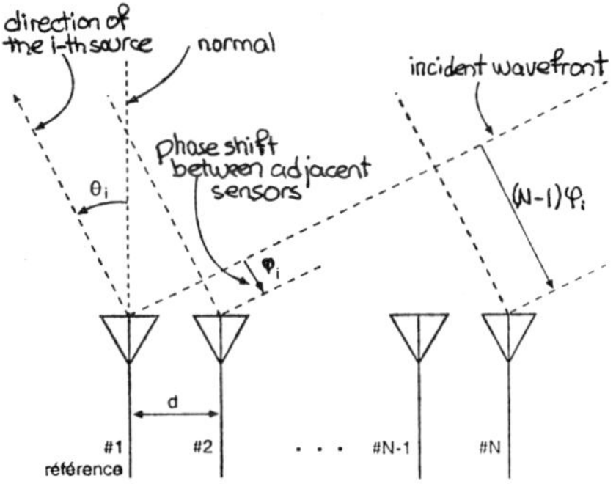

Figure (1): Array geometry.

The noise free received signals vector, at the central frequency, can be expressed as

$$s = U\,\alpha \tag{1}$$

where U is the sources matrix ($N \times n$) formed by the juxtaposition of the sources vectors u_i

$$u_i = \frac{1}{\sqrt{N}}\left[1,\, e^{j\phi_i},\, \ldots,\, e^{j(N-1)\phi_i}\right]^T \tag{2}$$

$$\phi_i = 2\pi \frac{d}{\lambda} \sin(\theta_i) \tag{3}$$

and α, the sources complex enveloppes vector α_i. The covariance matrix of the noiseless received signals R_s ($N \times N$) writes as

$$R_s = E\left\{ss^\dagger\right\} = U S U^\dagger \tag{4}$$

$$S = E\left\{\alpha\alpha^\dagger\right\} \tag{5}$$

when S is the sources ($n \times n$) covariance matrix. It can then be written as

$$R_s = V \Lambda V^\dagger \tag{6}$$

$$V = [v_1, v_2, \ldots, v_N], \tag{7}$$

$$\Lambda = \operatorname{diag}\{\lambda_1, \lambda_2, \ldots, \lambda_g, \underbrace{\lambda_{g+1}, \ldots, \lambda_N}_{0}\}. \tag{8}$$

where the λ_i are the eigenvalues ranked in decreasing order of R_s and v_i are the associated eigengectors.

If an additive noise is added to s, yielding x, a noise only covariance matrix R_b is adding to R_s, making all eigenvalues of R non-zero such as

$$R = R_s + R_b. \tag{9}$$

The problem created by the correlation between sources comes from the progressive ill-conditionning of the sources covariance matrix S and, consequently, of the covariance matrix R_s. Therefore, the image of R_s become a subspace with dimensions equal to the number of groups "g" and is not anymore equal to the number of sources "n". This subspace, noted the estimated source subspace is located within the real sources subspace, i.e. the subspace created by the source vectors. Then, the eigenvectors associated with the non-zero eigenvalues of R_s constitute now an orthonormal base of the estimated sources subspace. The source-vectors are not located inside this smaller subspace i.e. the subspace generated by the eigenvectors associated with the non-zero eigenvalues of R_s (these two subspaces are identical in the uncorrelated case). The source subspace must therefore be regenerated.

The spatial smoothing [1] is a known technique to increase the rank of R_s up to "n" in a coherent sources environment. Smoothing is a preprocessing issued from the split of the main array into subarrays and it uses translation invariance. As presented however, this solution does not sufficiently decorrelate the closely spaced sources and consequently maintaining the superresolution property, this neglecting that the effective number of elements is decreasing. The DEESE [6, 14] method offers a better alternative to the use of translation invariance by decomposing in "r" subvectors of "m" elements ($r = N - m + 1$) each vector of the estimated source subspace:

$$v_{i(k)} = \left[v_{i,k},\, v_{i,k+1},\, \ldots,\, v_{i,k+m-1}\right]^T \tag{10}$$

The property of rotation invariance is also used to decompose the vectors in the other direction, i.e. by inverting the elements. These new subvectors must however be conjugated since only the reference is reversed, not the source's position:

$$w_{i(k)} = \left[v^*_{i,N-k},\, v^*_{i,N-k-1},\, \ldots,\, v^*_{i,N-k-m+1}\right]^T \tag{11}$$

It can be shown that the $v_{i(k)}$ et $w_{i(k)}$ ($1 \le i \le g$, $1 \le k \le r$ subvectors generate an incoherent subspace which include all the source-vectors if

- $m > n$;
- $2r \ge p$

where "p" is the number of sources in the larger group. The coherent subspace converges toward the real sources subspace even under additive noise and can therefore be used as a good estimation of this subspace. However, since the eigenvectors do not constitutes an orthonormal base, the euclidian distance between the analysis vector and the incoherent subspace cannot be considered as a measure of the projection norm of this vector on the base. The solution envisaged makes use of eigenvectors a_j ($1 \le j \le n$) orthonormal to the noiseless image of $Q_b^{(r)}$ matrix, constituting the incoherent subspace,

$$Q_b^{(r)} = V_{(r)} W_{(r)} W^\dagger_{(r)} V^\dagger_{(r)}; \tag{12}$$

with

$$V^{(r)} = \left[v_{1(1)},\, v_{1(2)},\, \ldots,\, v_{i(k)},\, \ldots,\, v_{g(r)}\right] \tag{13}$$

$$W^{(r)} = \left[w_{1(1)},\, w_{1(2)},\, \ldots,\, w_{i(k)},\, \ldots,\, w_{g(r)}\right] \tag{14}$$

The advantages associated with the operation are

- uses of a full covariance matrix ($N \times N$)
- optimal use of the array geometry to increase the number of effective elements
- noise reduction by the exclusive selection of the eigenvectors associated, with a greater confidence, to the sources prior to the decomposition.

3 Rotation Matrix

The non-scanning techniques are using two or more parallel arrays [6, 11, 12] to obtain the rotation matrix Δ ($n \times n$). For a linear equispaces arrays, the translation invariance matrix is obtained in a similar way (also called rotation invariance by some authors because of the structure of the Δ matrix).

A structure of the U matrix and the operator $\uparrow$ considered in such a way that it produces a shift of one line up, namely

$$U = \left[r^\dagger, r^\dagger \Delta, \ldots, r^\dagger \Delta^{N-1}\right]^\top \tag{15}$$

with

$$r = \frac{1}{\sqrt{N}} [1, 1, \ldots, 1]^\top \tag{16}$$

$$\Delta = \mathrm{diag}\left\{e^{j\phi_1}, e^{j\phi_2}, \ldots, e^{j\phi_n}\right\} \tag{17}$$

wich yields:

$$U^\dagger = U\Delta. \tag{18}$$

The a_j vectors, similarly to the sources-vectors u_i form a base for the source-space and a linear dependance exists between these vectors express bythe regular matrix T:

$$UT = A \quad and \quad U^\dagger T = A^\dagger, \tag{19}$$

where A $(m \times n)$ is formed by the vectors a_j placed side-by-side. Using the pseudo-inverse of A, noted $A^{\sim 1}$, it is possible to find back the diagonal elements of Δ and the directives θ_i:

$$A^\dagger T^{-1} = AT^{-1}\Delta \tag{20}$$

$$A^{\sim 1} = \begin{cases} (A^\dagger A)^{-1} A^\dagger A^\dagger \\ T^{-1}\Delta T \end{cases}. \tag{21}$$

The eigenvalues of $A^{\sim 1}$, λ_i^A are the $e^{j\phi_i}$-type element looked for.

The closer to unity that the modules at the eigenvectors of $A^{\sim 1}$ are (the eigenvalues are complex since the matrix is not of hermitian symetry), the stronger will be the hypothesis that the eigenvalue associated to a source and his argument will correspond to the electrical phase shift that it produces between adjacent elements.

4 Simulation Results

The results presented here were obtained on a SUN-SPARC station using Monte-Carlo simulations. For each trial, 200 snapshots were obtained on a 8 elements $d = \lambda/2$ apart array ($N = 8$), using phase quadrature synchronous demodulation. The theorical Rayleigh limit is 15° (20° in practice). The results obtained are plotted in polar coordinates plane and give the position of the eigenvalues of the $A^{\sim 1}$ in the complex plane.

The first simulation is using 2 sources, one with a signal-to-noise ratio of $6dB$ located at 5° and the second one of $2dB$ at −2°. The correlation factor between the two sources as compared to the reference element is $1\angle 53°$. Figures (2) and (3) shows the results obtained after 100 trials when the number of sources was overtimated to 3 ($\tilde{n} = 3$) with the first version of DEESE and with the improved algorithm presented in this paper. It is clearly recognize that two clusters of points along the unit circle identifies the two sources (superresolution factor of 3) and the points spreaded in all direction within this circle correspond to the non-existing third source. It is also noted that the more concentrated cluster of points corresponds to the stronger source, showing a variance smaller the position estimation error, the results computed for each source are $\sigma^2_{(5°)} = 0.17$, $\mu_{(5°)} = 5.06$ and $\sigma^2_{(-2°)} = 0.35$, $\mu_{(-2°)} = -2.04$.

Finally, the same system has been simulated with an

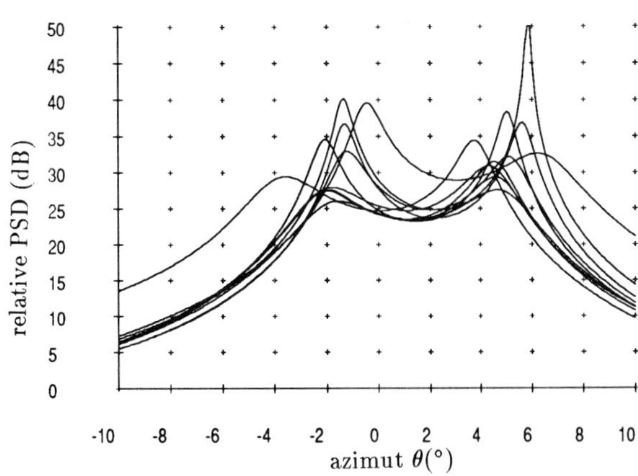

Figure (2): Spectra given by DEESE for simulations (10 trials) $p = 2, g = 1$.

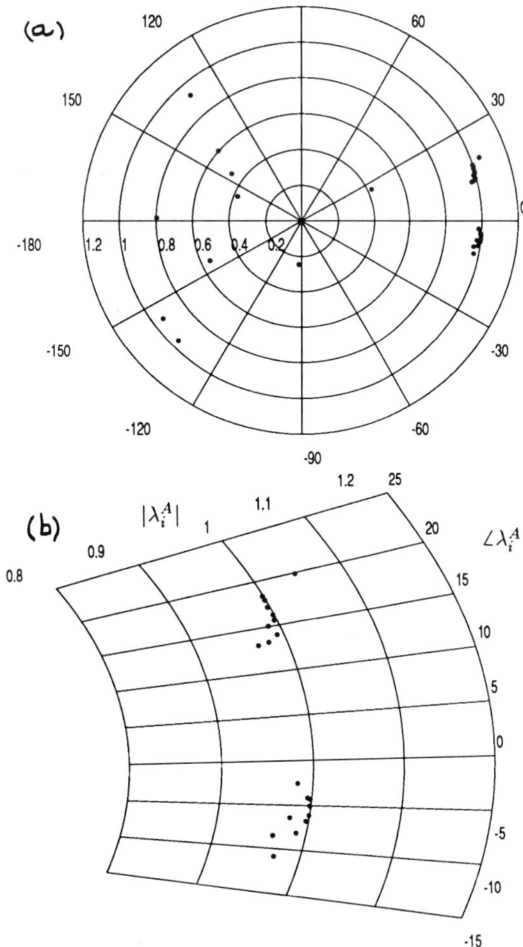

Figure (3): Pole positions by new algorithm for exactly the same simulations as (2).

independant source of $2dB$ located at $-11°$. This increase the number of group to two ($g = 2$) preserving however the size of the highest group to two ($p = 2$). Here, the correlation factor between the first and second sources at the reference element is taken equal to $1\angle 23°$ while this one between the first and third sources equal $1\angle\text{-}21°$ in figure (4) or $1\angle 31°$ in figure (5). It is noted that the separation of each source is less evident because of the interference created by this supplementary source on the others and also that the quality of separation depends of the phase relation between sources.

5 Conclusion

This paper reports initial results obtained up to now for correlated sources signal processing. The simulation and experimental results obtained so far show, however, the applicability and the efficiency of the method for strongly correlated sources. The absence of scanning decreases the required computer time and makes easier to store the position estimates since it is done using only the eigenvalues λ_i^A and the resultant the θ_i angles instead of a family of curves. Otherwise, the possible fonctions of the sources would have to be search for by an algorithm.

The proposed approach is promising; the problem still to be solved however is the interference between sources of a same group or from one group to the other, a problem that increases with the number of sources. Many solutions are envisaged for implementation; one of these would be the use of frequency diversity. It must be taken into account however that the computation time is already important and such a scheme should not increased if significantly.

References

[1] T.J. Shan, M. Max, and T. Kailath, "On Spatial Smoothing for Direction of Arrival Estimation of Coherent Signals", IEEE Trans., ASSP-33:806-811, August 1985.

[2] G. Su, M. Morf, "Modal Decomposition Signal Subspace Algorithms", IEEE Trans., ASSP- 34:585-602, June 1986.

[3] M. Zoltowski, F. Haber, "A Vector Space Approach to Direction Finding in a Coherent Multipath Environment", IEEE Trans., AP-34:1069-1079, Sept. 1986.

[4] J.A. Cadzow, "A High Resolution Direction of Arrival Algorithm for Narrow-Band Coherent and Incoherent Sources", IEEE Trans., ASSP-36:965-979, July 1988.

[5] J.A. Cadzow, Y.S. Kim, and D.C. Shiue, "General Direction-of-Arrival Estimation: a Signal Subspace Approach", IEEE Trans., AES-25: 31-46, Jan. 1989.

[6] D. Grenier, "Analysis of narrow-band correlated sources with signal processing techniques" (in French), Ph.D. thesis, Université Laval, Québec, February 1989.

[7] P. Stoica, K.C. Sharman, "Novel Eigenanalysis Method for Direction Estimation", IEEE Proc., 137, Pt.F:19-26, Feb. 1990.

[8] B.D. Van Veen, "Eigenstructure Based Partially Adaptive Array Design", IEEE Trans., AP-36:357-362, March 1988.

[9] M.D. Zoltowski, D. Stavrinides, "Sensor Array Signal Processing Via a Procrustes Rotations Based Eigenanalysis of the ESPRIT Data Pencil", IEEE Trans., ASSP-37:832-861, june 1989.

[10] B. Ottersen, T. Kailath, "Direction-of-Arrival Estimation for Wide-Band Signals Using the ESPRIT Algorithm", IEEE Trans., ASSP-38:317-327, Feb. 1990.

[11] R. Kamuresan, D.W. Tufts, "Estimating the Angles of Arrival of Multiple Plane Waves", IEEE Trans., AES-19:134-139, Jan. 1983.

[12] A. Paulraj, R. Roy, and T. Kailath, "Estimation of Signal Parameter Via Rotational Invariance Techniques - ESPRIT", in ASILOMAR Conf., pages 83-89, Pacific Grove (CA), 1985.

[13] R. Kumaresan, A.K. Shaw, "Superresolution by Structured Matrix Approximation", IEEE Trans., AP-36:34-44, Jan. 1988.

[14] D. Grenier, G.Y. Delisle, "Identification superrésolutive de sources correlées par décomposition de la base du sous-espace source estimé", submitted at the "Traitement du Signal" Journal, February 1992.

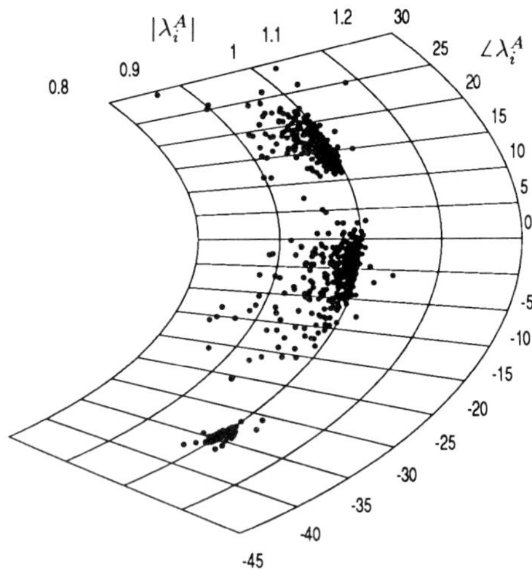

Figure (4): Pole positions by new algorithm for $p = g = 2$, case #1.

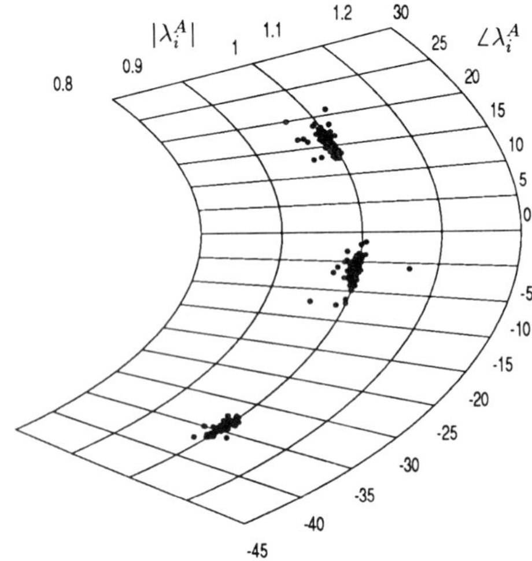

Figure (5): Pole positions by new algorithm for $p = g = 2$, case #2.

SIGNAL PROCESSING VI: Theories and Applications
J. Vandewalle, R. Boite, M. Moonen, A. Oosterlinck (eds.)

A Complex Adaptive Eigensubspace Algorithm for DOE or Frequency Estimation and Tracking

J.P. DELMAS

Dpt Electronique et Communications, Institut National des Télécommunications,
9 rue Charles Fourier, 91011 Evry cédex, France

Abstract

Many authors have proposed adaptive estimation methods of complete noise (or signal) subspace and have suggested different implementation methods. Among them, Regalia [R90] presented an adaptive unit norm filter based on planar rotators which offers interesting properties. We first reformulate this idea and extend it to complex processes, and then prove the convergence of the stochastic coupled algorithms with the help of the associated Ordinary Differential Equation.

1] *Introduction*

Signal-subspace techniques such as MUSIC [S79], minimum norm [KT83], root Music [B83] and reduced order MUSIC [O86] have became classical superresolution methods for the estimation of frequencies of sinusoids or the directions of arrival of plane waves by an array of sensors. However, implementations of these techniques have been mainly based on batch estimation of eigenvectors derived from the estimated covariance matrix, making them unsuitable for adaptive processing that is needed when tracking nonstationary signal parameters.

Owsley [078] was the first to introduce an adaptive procedure to estimate the eigenvectors. Thompson [T80] proposed a gradient based adaptive algorithm for the estimation of the eigenvector corresponding to the minimum eigenvalue. Reddy et al [REK82] developed a least squares algorithm for the same problem. Yang and Kaweh [TK88] studied several gradient based search algorithms and Sarkar and Yang [SY89] proposed a conjugate gradient algorithm for interactively finding the eigenvectors corresponding to complete noise or signal subspace estimation.

Recently, an adaptive unit norm filter was proposed by P.Regalia [R90]. It performs the Karhunen-Loève transformation of the input real signal by an orthogonal-based triangular array.

This paper is organized as follows. After introducing some notations and formulating the problem in Section 2, our solution, based on global parametrization of orthonormal eigenvectors of correlation matrix of complex signal, is described in Section 3. Coupled minimizations are solved by coupled stochastic gradient algorithms in Section 4 and finally the convergence of these coupled stochastic algorithms are proved using the ODE method in Section 5.

2] *Notation and formulation of the problem*

Let us introduce some notations: for an N element array and P incoherent sources with $P<N$, the algorithm studied adaptively estimates the eigenvectors corresponding to either the $N-P$ smallest or the P largest eigenvalues of a continuously updated sample covariance matrix of order N :

$\boldsymbol{R}_x=\mathrm{E}[\boldsymbol{x}^*(k)\boldsymbol{x}^t(k)]$ with $\boldsymbol{x}(k)=[x(1,k),..,x(N,k)]^t$. Let λ_i and $\boldsymbol{h}_i$ $i=1,..,N$ be these ordered eigenvalues ($\lambda_1\leq..\lambda_{N-P}<\lambda_{N-P+1}..\leq\lambda_N$) and corresponding normalized orthogonal eigenvectors. (*,t,+ stand respectively for conjugate, transpose and conjugate transpose).

We shall describe the estimation of the noise subspace, since the estimation of the signal subspace problem can be formulated similarly (e.g., replace the minimizations [1] by maximizations, L defined below by $L=P$ and the sign – by + in the gradient algorithms).

h_1 is the solution of the minimization problem

$$\min_{\|h\|=1} h^+ R_x h \qquad [1a]$$

and each h_i for $i=2,..,L=N-P$ is the solution of the minimization problem

$$\min_{\|h\|=1,\ h\perp sp\{h_1,..,h_{i-1}\}} h^+ R_x h \qquad [1b]$$

We note that the solutions of these minimizations are defined up to a multiplicative unit modulus complex constant.

3] *Global parametrization of $h_1,.,h_i,.,h_L$*

The orthonormal conditions on $h_1,.,h_i,.,h_L$ imply L^2 real constaints. Therefore these vectors must be parametrized by a set of $2NL-L^2$ real parameters.

One easily proves that the vectors $h_1,.,h_i,.,h_L$ defined up to a multiplicative unit modulus complex constant, may be parametrized by a unique set of $\frac{1}{2}[(2NL-L^2)-L]$ couples of rotation angles $(\theta_{i,j},\phi_{i,j})$ $i=1,..L$ and $j=1,..N-i$ satisfying $\phi_{i,j}\in]-\frac{\pi}{2},\frac{\pi}{2}]$ and $\theta_{i,j}\in]-\frac{\pi}{2},\frac{\pi}{2}]$ as follows. (The factor L comes from an extra condition on each of the L vectors that fixes the multiplicative unit modulus complex constant).

h_1 is the last column of a unitary matrix Q_1:

$$h_1=Q_1\begin{bmatrix}0\\1\end{bmatrix} \text{ and } h_2=Q_1\begin{bmatrix}Q_2\begin{bmatrix}0\\1\end{bmatrix}\\0\end{bmatrix},$$

$$.., h_L=Q_1\begin{bmatrix}Q_2\begin{bmatrix}\begin{bmatrix}Q_L\begin{bmatrix}0\\1\end{bmatrix}\\0\end{bmatrix}\\ \\0\end{bmatrix}\\0\end{bmatrix} \qquad [2]$$

with each Q_i a unitary matrix of order $N-i+1$:

$$Q_i=U_{i,1}..U_{i,j}..U_{i,N-i}$$

$$\text{with } U_{i,j}=\begin{bmatrix}I_{j-1}&0&0&0\\0&-\sin\theta_{i,j}&\cos\theta_{i,j}&0\\0&e^{i\phi_{i,j}}\cos\theta_{i,j}&e^{i\phi_{i,j}}\sin\theta_{i,j}&0\\0&0&0&I_{N-i-j}\end{bmatrix} \qquad [3]$$

Proof: Suppose that $h_{1,1}$ is real nonnegative. Since h_1^t is the bottom row Q_1^t (with Q_1 defined by [3]), the equation:

$$h_1=[h_{1,1},..,h_{1,N}]^t=Q_1\begin{bmatrix}0\\1\end{bmatrix}$$

is equivalent to:

$$h_{1,1}=\prod_{i=1}^{N-1}\cos\theta_{1,i}$$

$$h_{1,k}=[\prod_{i=k}^{N-1}\cos\theta_{1,i}]\sin\theta_{1,k-1}e^{i\phi_{1,k-1}} \quad \text{for } k=2,..N-1$$

$$h_{1,N}=\sin\theta_{1,N-1}e^{i\phi_{1,N-1}}$$

which in turn, is equivalent to:

$$\frac{h_{1,2}}{h_{1,1}}=\mathrm{tg}\theta_{1,1}e^{i\phi_{1,1}} \quad (\text{if } h_{1,1}=0,\ \theta_{1,1}=\frac{\pi}{2})$$

$$\frac{h_{1,k+1}}{h_{1,k}}=\mathrm{tg}\theta_{1,k}e^{i\phi_{1,k}}\left(\frac{e^{-i\phi_{1,k-1}}}{\sin\theta_{1,k-1}}\right)$$

(if $h_{1,k}=0$, $\theta_{1,k}=\frac{\pi}{2}$) for $k=2,..N-1$

$$h_{1,1}=\prod_{i=1}^{N-1}\cos\theta_{1,i}$$

Step by step, starting from $i=1$, the couples $(\theta_{1,i},\phi_{1,i})\in]-\frac{\pi}{2},+\frac{\pi}{2}]^2$ are uniquely determined and the relation $h_{1,1}=\prod_{i=1}^{N-1}\cos\theta_{1,i}$ is automatically satisfied from the $N-1$ preceding relations since $h_{1,1}$ is assumed real nonnegative and $\|h_1\|=1$.

Then $h_2\perp h_1 \Leftrightarrow h_2=Q_1\begin{bmatrix}h\\0\end{bmatrix}$ with $h\in\mathbb{C}^{N-1}$ and $\|h\|=1$. Using the same argument as for h_1,

$$h=Q_2\begin{bmatrix}0\\1\end{bmatrix}.$$

With similar arguments we prove iteratively the existence and uniqueness of [3] if no component of h_1 nor of h is zero. We note that if a componant of h_1 or of h is zero, existence is proved but uniqueness is no longer guaranteed.

4] *Coupled minimizations*

The minimization [1a] can be performed with the help of a stochastic gradient algorithm, in which the parameter is:

$$p=[\theta_{1,1},\phi_{1,1},..,\theta_{1,N-1},\phi_{1,N-1}]^t.$$

With $y_1(k)=h_1^t x(k)=[0^t,1]Q_1^t x(k)=[0^t,1]Q_1^t z_0(k)$ with $z_0(k)\triangleq x(k)$, we must minimize $E\{V_1[p,x(k)]\}$ for p. (Here we denote $V_1(p,x(k))\triangleq|y_1(k)|^2=h^+x^*(k)x^t(k)h)$.

Since the minimization [1b] in h contains parameters $h_1,..,h_{i-1}$, it can also be performed according to a stochastic gradient algorithm in which the parameters $p_n=[\theta_{n,1},\phi_{n,1},..,\theta_{n,N-n},\phi_{n,N-n}]^t$, for $n=1,..i-1$, are injected from the $i-1$ previous algorithms.

With $y_i(k)=\boldsymbol{h}_i^t\boldsymbol{x}(k)=[\boldsymbol{0}^t,1]\boldsymbol{Q}_i^t z_{i-1}(k)$, $z_i(k)$ defined by $\begin{bmatrix} z_1(k) \\ y_1(k) \end{bmatrix} \triangleq \boldsymbol{Q}_1^t\boldsymbol{x}(k)$ and $\begin{bmatrix} z_i(k) \\ y_i(k) \end{bmatrix} \triangleq \boldsymbol{Q}_i^t z_{i-1}(k)$, we must minimize $\mathrm{E}\{V_i[\boldsymbol{p}_1,..,\boldsymbol{p}_{i-1},\boldsymbol{p},\boldsymbol{z}(k)]\}$ with respect to $\boldsymbol{p}$. (Here we denote $V_i[\boldsymbol{p}_1,..,\boldsymbol{p}_{i-1},\boldsymbol{p}_i,\boldsymbol{x}(k)] = |y_i(k)|^2 = \boldsymbol{h}_i^+\boldsymbol{x}^*(k)\boldsymbol{x}^t(k)\boldsymbol{h}_i$).

Thus we obtain L coupled stochastic gradient algorithms, derived from the stochastic approximation algorithms associated with the L minimizations of [1].

Now we can show, using the following figure, that for $i=1,..L$, the gradient of $|y_i(k)|^2$ with respect to the parameters $\boldsymbol{p}_i$ is given by :
(we denote $z_i(k) \triangleq [z_{i,1}(k),..,z_{i,N-i}(k)]^t$)

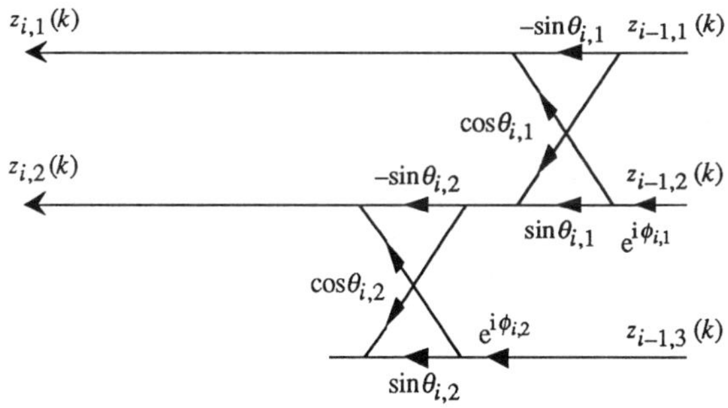

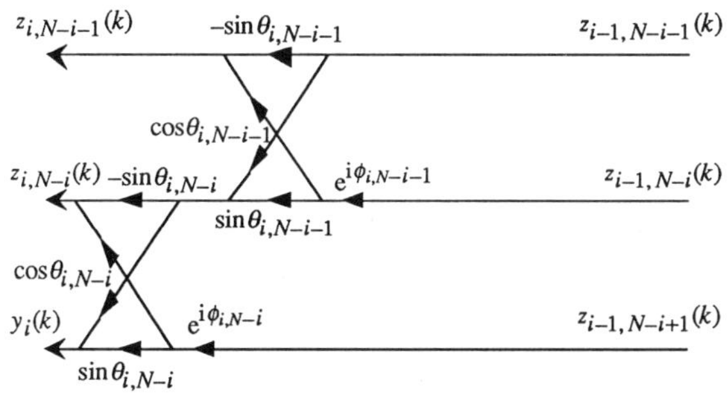

$$\frac{\partial y_i(k)}{\partial \theta_{i,N-i}} = z_{i,N-i}(k)$$

$$\frac{\partial y_i(k)}{\partial \theta_{i,j}} = z_{i,j}(k) \prod_{l=j+1}^{N-i} \cos\theta_{i,l} \quad \text{for } j=1,..N-i-1$$

$$\frac{\partial y_i(k)}{\partial \phi_{i,N-i}} = \mathrm{i}e^{\mathrm{i}\phi_{i,N-i}} \sin\theta_{i,N-i}\, z_{i-1,N-i+1}(k)$$

$$\frac{\partial y_i(k)}{\partial \phi_{i,j}} = \mathrm{i}e^{\mathrm{i}\phi_{i,j}} \sin\theta_{i,j}\, [\prod_{l=j+1}^{N-i} \cos\theta_{i,l}]\, z_{i-1,j+1}(k) \quad \text{for}$$

$j=1,..N-i-1$

The $L(2N-L-1)$ rotation angles are updated according to :

$$\theta_{i,j}(k) = \theta_{i,j}(k-1) - \gamma\, \mathrm{Re}[\frac{\partial y_i(k)}{\partial \theta_{i,j}} y_i^*(k)]$$

$$\phi_{i,j}(k) = \phi_{i,j}(k-1) - \gamma\, \mathrm{Re}[\frac{\partial y_i(k)}{\partial \phi_{i,j}} y_i^*(k)]$$

$$i=1,..L \text{ and } j=1,..N-i \qquad [4]$$

This leads to a truncated triangular array comprising $\frac{1}{2}L(2N-L-1)$ rotator sections, as in [R90], but using now complex rotation cells.

For example, in case $N=5$ and $L=2$, we obtain the following flowgraph:

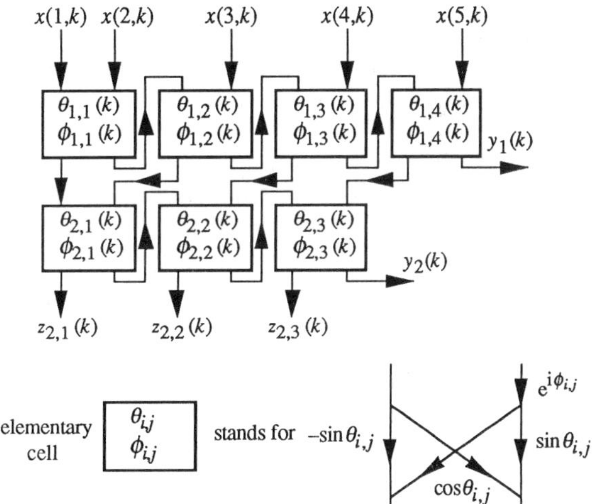

5] Convergence of the stochastic coupled algorithms with the help of the Ordinary Differential Equation.

We note that in the algorithms [4], $\boldsymbol{p}_n(k)=[\theta_{n,1}(k),\phi_{n,1}(k),..,\theta_{n,N-n}(k),\phi_{n,N-n}(k)]^t$, for $n=1,..i-1$, are injected from the $i-1$ previous algorithms. The problem of the convergence of the L coupled stochastic gradient algorithms described previously is stated. In the stationary context, it can be studied by the **ODE** method [LS83].

The stochastic gradient algorithms can be globally written as

$$\begin{bmatrix} \boldsymbol{p}_1 \\ \cdot \\ \boldsymbol{p}_L \end{bmatrix}_{k+1} = \begin{bmatrix} \boldsymbol{p}_1 \\ \cdot \\ \boldsymbol{p}_L \end{bmatrix}_k - \gamma_k \begin{bmatrix} f_1(\boldsymbol{p}_1(k),\boldsymbol{x}(k)) \\ \cdot \\ f_L(\boldsymbol{p}_1(k),..,\boldsymbol{p}_L(k),\boldsymbol{x}(k)) \end{bmatrix} \qquad [5]$$

with $f_i(\boldsymbol{p}_1(k),..,\boldsymbol{p}_i(k),\boldsymbol{x}(k)) \triangleq \mathbf{grad}_{/\boldsymbol{p}_i} V_i[\boldsymbol{p}_1(k),..,\boldsymbol{p}_i(k),\boldsymbol{x}(k)]$

If the gain sequence γ_k satisfies the conditions $\sum_{k=1}^{+\infty} \gamma_k = +\infty$ and $\lim_{k\to\infty} \gamma_k = 0$ (e.g., $\gamma_k = \frac{\gamma}{k}$), the associate ODE is:

$$\frac{dp_1(t)}{dt} = -\mathrm{E}\Big[\,\mathbf{grad}_{/p_1} V_1[p_1(t),x(t)]\Big] \qquad [6a]$$

....

$$\frac{dp_L(t)}{dt} = -\mathrm{E}\Big[\,\mathbf{grad}_{/p_L} V_L[p_1(t),..,p_L(t),x(t)]\Big] \qquad [6b]$$

Studying the convergence of [5] is thus equivalent to studying the stability of the system [6]. We show that the parameters that minimize the expressions [1] are globally asymptotically stable for the equations [6].

Proof:

Thanks to the uniqueness of the parametrization of $h_1,..,h_i,..,h_L$, the coupled estimation problem has a unique solution in $(\theta_{i,j},\phi_{i,j})$ $i=1,..L$ and $j=1,..N-i$ satisfying $\phi_{i,j}\in\,]-\frac{\pi}{2}, \frac{\pi}{2}]$ and $\theta_{i,j}\in\,]-\frac{\pi}{2}, \frac{\pi}{2}]$. We shall show that this solution is globally asymptotically stable for the equations [6]. Then the convergence of the coupled stochastic algorithms is ensured.

We suppose that $L=2$, as the extension to $L>2$ is straightforward.

Since $\mathbf{grad}_{/p_1} V_1[p_1(t),x(t)]$ is the derivative of a positive gradient field, the stationary point (denoted $p_1(+\infty)$) of [6a] is globally asymptotically stable for that equation.

Similarly for $\mathbf{grad}_{/p_2} V_2[p_1^0,p_2(t)]$, the stationary point $p_{2_{p_1^0}}(+\infty)$ of [6b] (in which $p_1(t)$ is replaced by p_1^0) is globally asymptotically stable for that equation.

Since $\mathrm{E}[\mathbf{grad}_{/p_2} V_2[p_1^0,p_2(t)]]$ is an analytic function in p_1^0 and p_2, the stationary point $p_{2_{p_1^0}}(+\infty)$ is a continuous function of p_1^0.

The solution $p_{2_{p_1^0}}(t)$ of the first differential equation

$$\frac{dp_2(t)}{dt} = -\mathrm{E}[\mathbf{grad}_{/p_2} V_2[p_1^0,p_2(t)]]$$

converges uniformly with respect to p_1^0 to $p_{2_{p_1^0}}(+\infty)$ when $t\rightarrow+\infty$, provided we choose $t>T$ such that $\|p_1^0 - p_1(+\infty)\|\leq\varepsilon$.

Then, since

$$\|p_2(t)-p_{2_{p_1^0(+\infty)}}(+\infty)\| \leq \|p_2(t)-p_{2_{p_1^0(t)}}(+\infty)\| + \|p_{2_{p_1^0(t)}}(+\infty)-p_{2_{p_1^0(+\infty)}}(+\infty)\|,$$

the equations [6] have an asymptotically stable solution in which the stationary point is $[p_1(+\infty),p_{2_{p_1^0(+\infty)}}(+\infty)]$.

Consequently, the convergence of the 2 coupled stochastic algorithms is ensured.

References

[B83] A.J.Barabell, *Improving the resolution performance of eigenstructure-based direction-finding algorithms*, Proc.ICASSP, pp.336-339, Boston 1983.

[KT83] R.Kumaresan and D.W.Tufts, *Estimation the angles of arrival of multiple plane waves*, IEEE Trans. Aerospace and Elect.Syst.vol.19, no.1 pp.134-139, Jan.1983.

[LS83] L.Ljung and T.Söderström, *Theory and practice of recursive identification*, MIT Press 1983.

[O86] S.Orfanidis, *A reduced MUSIC algorithm*, Proc.Third ASSP Workshop on Spectrum Estimation and Modeling, Boston, 1986.

[O78] N.L.Owsley, *Adaptive data orthogonalization*, Proceedings ICASSP 1978.

[REK82] V.Reddy, B.Egardt and T.Kailath, *Least square type algorithm for adaptive implementation of Pisarenko harmonic retrieval method*, IEEE Trans. ASSP vol.30, no.3 pp.399-405 June 1982.

[R90] P.A.Regalia, *An adaptative unit norm filter with applications to signal analysis and Karhunen-Loeve transformation*, IEEE Trans. Circuits and Systems, vol.37, no.5 pp.646-649 May 1990.

[SY89] T.K.Sarkar and X.Yang, *Application of conjugate gradient and steepest descent for computing the eigenvalue of an operator*, Signal Processing, vol.17 pp.31-38, May 1989.

[S79] R.O.Schmidt, *Multiple emitter location and signal parameter estimation*, In Proc.RADC Spectrum Estimation Workshop New York 1979.

[T80] P.A.Thompson, *Adaptive spectral analysis technique for unbiased frequency estimation in the presence of white noise*, In Proc.13th Asilomar Conf.Circ.Syst.Comp., Pacific Grove, CA 1980.

[YK88] J.P.Yang and M.Kaveh, *Adaptive eigensubspace algorithms for direction or frequency estimation and tracking*, IEEE Trans. ASSP vol.36, no.2, Feb.1988.

SIGNAL PROCESSING VI: Theories and Applications
J. Vandewalle, R. Boite, M. Moonen, A. Oosterlinck (eds.)

EXPLOITING MULTIVARIATE AUTOREGRESSIVE MODEL IN THE SEPARATION OF SOURCES

Seppo Rantala and Risto Suoranta

Machine Automation Laboratory, Technical Research Centre of Finland
P.O. Box 192, SF-33101 Tampere, Finland, Tel +358-31-163616, Fax +358-31-163694,
Internet: Seppo.Rantala@kau.vtt.fi, Risto.Suoranta@kau.vtt.fi

In this paper we propose a method to separate independent sources from the signals obtained by using an array of sensors. The method is based on a novel interpretation of results derived from a *multivariate autoregressive model, MAR-model.* Exploiting the model expressions related to source separation are derived. Due to the fact that *MAR-model* is black-box model and it can describe systems with feedback-loops, it is very general and useful tool to analyze linear systems. Particularly its strength is in cases, where there is no exact a priori knowledge of internal structures of the system.

1. INTRODUCTION

The problem of extracting independent sources based on measurements obtained from an array of sensors is an essential problem in signal processing. The signal received by a sensor is the sum of elementary contributions that can be called as sources. Sources as well as their mixtures are generally unknown. In this case this task is called *blind separation of sources* [2]. For instance, antenna or array of acoustic sensors receives a superimposed signal from all the sources which are in its receptive field.

Numerous approaches have been developed to solve the problem of separating different sources. Most of the works related to this problem use the information included in the received signals together with other information. Therefore the usage of these methods is limited to special cases. For instance, one possibility is to make an assumption that one of the sources produces deterministic signal.

In this work we construct a model of the system under study. Using the model it is possible to obtain insight into dynamics and internal relations of the system. If the chosen model is appropriate we are able to reveal the sources by utilizing the model.

The proposed method exploits *multivariate autoregressive model* (MAR-model) as a system analysis tool. The MAR-model describes the closed loop black-box system with internal feedback loops. Consequently it is an appropriate model to describe complicated systems without a need to do strict assumptions about the system.

In this method we identify the MAR-model and based on the model the multichannel spectral matrix is decomposed. Spectral decomposition offers a new possibility to analyze signals measured from the array of sensors. Utilizing the spectral decomposition we can efficiently estimate which signals can be regarded as independent variables (sources) and which signals are linear combinations of the other signals. Also the sources and their propagation paths inside the observed system can be analyzed applying proposed approach.

2. MULTIVARIATE AR-MODEL

MAR-model is a model for multivariate stochastic process where relationships between internal processes are described. MAR-model can be expressed using matrix notation; that is,

$$\sum_{i=0}^{p} \mathbf{a}(i)\mathbf{X}(k-i) = \mathcal{E}(k), \tag{1}$$

where $\mathbf{X}(k)$ is $m \times 1$ column vector containing m univariate stochastic processes included in the multivariate process. $\mathbf{a}(i)$, $(i = 1 \dots p)$ are $m \times m$ MAR-coefficient matrices and $\mathcal{E}(k)$ is a $m \times 1$ column vector containing m zero mean white noise processes $e(k)$; p is the order of AR-model.

As we can see from the Eq. 1, the MAR-model describes closed loop system with linear feedback connections between all signals. The MAR-model has no predetermined inputs and outputs. The driving force of the system is a set of noise sources $e(k)$ added to each node in the system. In Fig. 1 the structure of a bivariate AR-model is shown; N1 and N2 are internal noise sources as well as X1 and X2 are measured signals.

The existence of the straightforward recursive estimation algorithm for MAR-model parameters, presented by Wiggins and Robinson, makes the MAR-model method applicable and attractive to the analysis of multichannel data in practice. The fast algorithm can be found e.g. in [3, pp. 400–402].

Spectral matrix for multivariate stochastic process can be derived from the Eq. 1. The spectral matrix

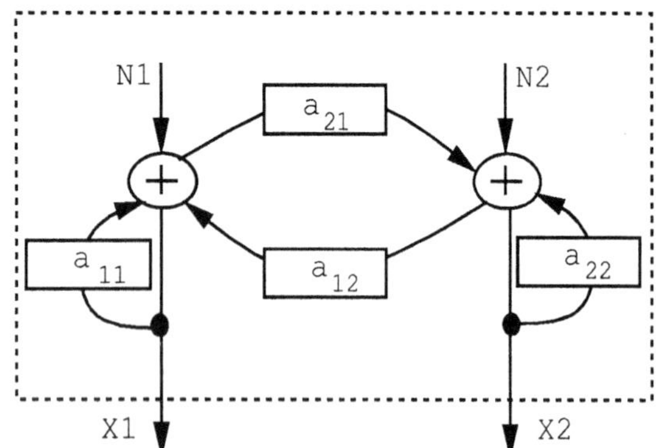

Figure 1. The structure of bivariate MAR-model

for multivariate process $\mathbf{X}$ is expressed as

$$\mathbf{P}_x(\omega) = (\mathbf{A}(e^{j\omega})^{-1})\ \mathbf{\Sigma}_e\ (\mathbf{A}(e^{j\omega})^*)^{-1}. \quad (2)$$

$\mathbf{A}(e^{j\omega})$ is the Fourier transform of MAR-coefficient matrices $\mathbf{a}(i)$ and $\mathbf{\Sigma}_e$ is the covariance matrix of white noise processes $e_i(k)$, $(i = 1, \ldots, m)$.

3. NOISE SOURCE ANALYSIS

Noise processes $e_i(k)$ are the only excitations of the modelled system. To understand the dynamics of the system it would be useful to know the properties of each noise source and how the effects of each noise source spread in the system. To analyze the effects of noise sources we present a concept of a *noise conditioned spectral matrix* (NCSM).

The *noise conditioned spectral matrix* (NCSM) is a spectral matrix, where the effects of selected noise sources are removed. The notation for the residual spectral matrix is $\mathbf{P}_{x|e_1,e_2}(\omega)$, where e_1 and e_2 are the noise sources to be excluded. An alternative notation for NCSM is $\mathbf{P}_{x:e_1,e_2}(\omega)$, where e_1 and e_2 are noise sources to be included in. In other words "|" denotes exclusion of variables and accordingly ":" denotes inclusion of variables.

To remove the effects of the noise process $e_i(k)$ we modify $\mathbf{\Sigma}_e$ by resetting those values in the covariance matrix to zero which correspond to the removed noise processes. This means that the removed noise process is regarded as zero and consequently the variance and corresponding cross-covariances will also be zero. The modification of the noise covariance matrix can be done by multiplying $\mathbf{\Sigma}_e$ with matrix $\mathbf{M}$, which is derived from the identity matrix $\mathbf{I}$ by zeroing those 1's in diagonal which correspond to the noise sources to be removed. Using modification matrix $\mathbf{M}$ we define *noise conditioned spectral matrix, NCSM*,

$$\mathbf{P}_{x|\epsilon}(\omega) = (\mathbf{A}(e^{j\omega}))^{-1}\ \mathbf{M}\ \mathbf{\Sigma}_e\ \mathbf{M}\ (\mathbf{A}(e^{j\omega})^*)^{-1}, \quad (3)$$

where ϵ is a list of excluded noise sources. If noise sources in $\mathcal{E}(k)$ are independent of each other we can write,

$$\mathbf{P}_x(\omega) = \sum_{i=1}^{m} \mathbf{P}_{x:e_i}(\omega). \quad (4)$$

By exploiting the noise conditioned spectral matrix we can define a multiple coherence function, which is called *noise conditioned multiple coherence, (NCMCH)*. The NCMCH is a measure of linear dependence between one variable and a set of selected noise sources. The NCMCH between variable $X_j(k)$ and single noise source $e_i(k)$ is defined as

$$\gamma^2_{j:e_i}(\omega) = \frac{P_{x:e_i(jj)}(\omega)}{P_{x(jj)}(\omega)}. \quad (5)$$

A subscript (ij) denotes the element in row i and column j in corresponding matrix. If the noise sources in $\mathcal{E}(k)$ are independent of each other, then

$$\sum_{i=1}^{m} \gamma^2_{j:e_i}(\omega) = 1. \quad (6)$$

The NCMCH describes the linear relationship between noise sources and variable in one frequency ω. In many cases it would be valuable to know what is the noise power contribution of the noise source to the variables of the system in the selected frequency band. By utilizing the concept of NCSM we can build up a new measure that describes how much power in selected frequency band is originated from the noise source $e_i(k)$. New *noise contribution ratio (NCR)* is defined as

$$q_{ji}(\omega_1, \omega_2) = \frac{\int_{\omega_1}^{\omega_2} P_{x:e_i(jj)}(\omega)\ d\omega}{\int_{\omega_1}^{\omega_2} P_{x(jj)}(\omega)\ d\omega}, \quad (7)$$

where ω_1 is the lower bound and ω_2 is the higher bound of the defined frequency band. By combining all values $q_{ji}(\omega_1, \omega_2)$ to one $m \times m$ matrix we obtain a *system noise contribution matrix, (SNCM)*, that is

$$\mathbf{Q}(\omega_1, \omega_2) = \begin{bmatrix} q_{11} & q_{12} & \cdots & q_{1m} \\ q_{21} & q_{22} & & \\ \vdots & & \ddots & \vdots \\ q_{1m} & & \cdots & q_{mm} \end{bmatrix}. \quad (8)$$

The *system noise contribution matrix* describes the dynamics of the whole system. If the $\mathbf{Q}(0, 0.5)$ matrix is a identity matrix, it means that there are no cross-effects between variables, i.e., all signals are described only by their own noise sources. By studying the matrix $\mathbf{Q}$ we can resolve which signals are sources of the system and which are dependent signals.

4. NUMERICAL EXAMPLE

To illustrate possibilities of the new method we have simulated data obtained from the system depicted in Fig. 2. System includes three variables (sensors) and two sources.

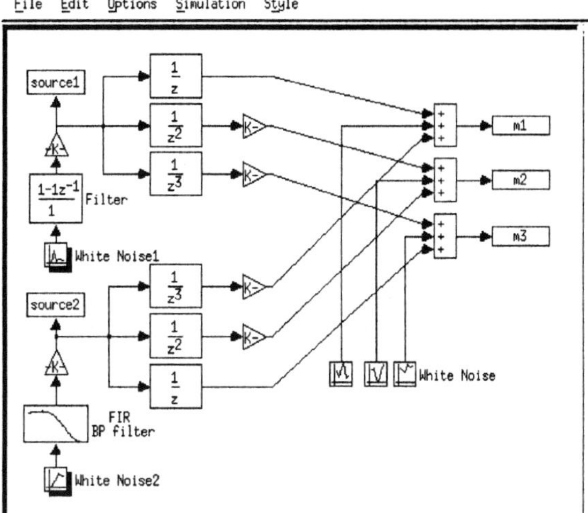

Figure 2. Simulated array of three sensors

There is two sources connected to the system (source1 and source2) and we have measured three signals m_1, m_2 and m_3. In Fig. 3 the spectra of sources are shown. The system has a marked correspondence to acoustic system, where transfer functions consists of time delays and dampings.

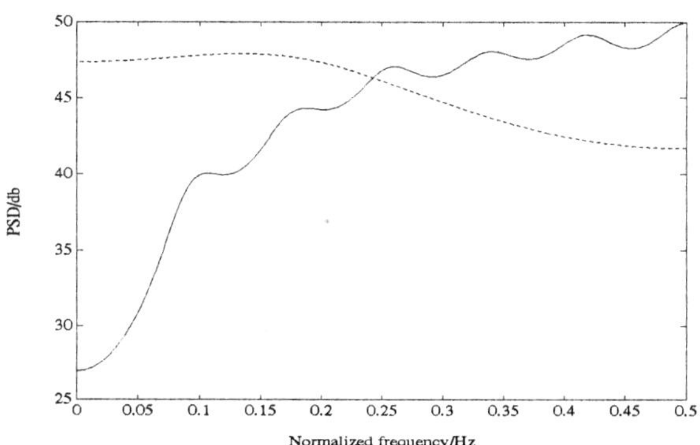

Figure 3. Spectra of sources: source1 (solid line) and source2 (dashed line)

In this case 4000 samples were simulated from each channel of the system. The reason for the length was the comparison to the classical Fourier spectral analysis, which needs longer data records.

Due to the cross effect between sources measured signals are all very similar and near white noise. Classical multivariate Fourier spectral analysis [1] offers mainly coherence based estimates to the problem of separating the sources. In this work we have tried separate sources by using traditional ordinary, conditional(partial) and multiple coherence, but the sources were impossible to determine. As an example ordinary and conditional coherence function estimates are shown in Fig. 4.

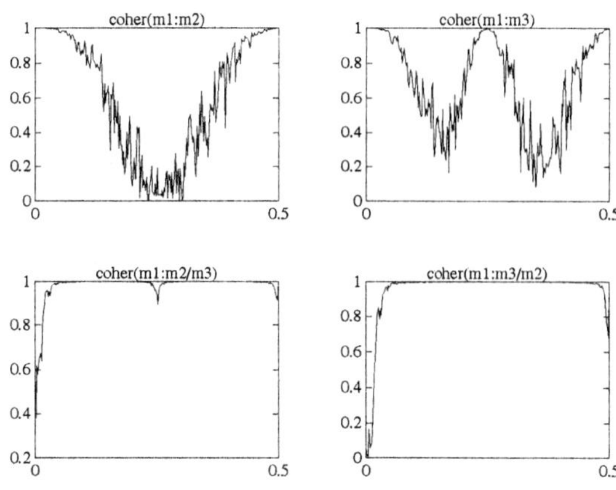

Figure 4. Coherence estimates between m_1m_2 and m_1m_3

First step in applying MAR-model is to identify the model. The model order p arouses some problems: *Akaike information criterion*(AIC) suggests order 29, but it must be too high, because the simulation system has maximum delay of 3 samples and used filters are low order. Too high model order may cause artifacts to the spectral quantities. If the AIC criterion and its difference are examined further (Fig. 5), the better model order can be chosen from the point, where the convergence of the difference is clear. Therefore the model order 10 was selected.

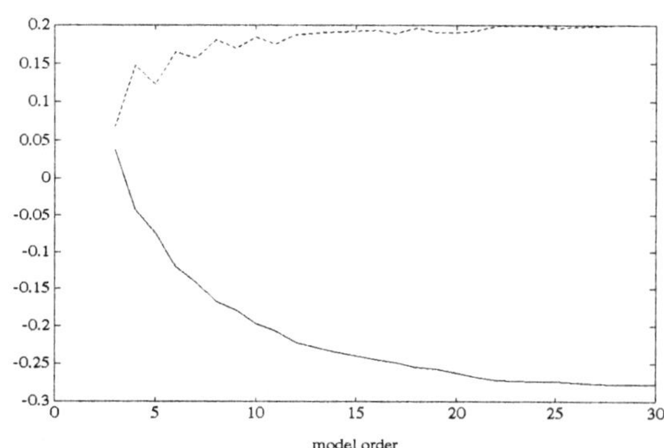

Figure 5. Model order by AIC(solid line) and its difference (dashed line)

Utilizing the identified MAR-model the *system noise contribution matrix* is computed. In Fig. 6 the $\mathbf{Q}-$*matrix* computed over the whole frequency band can been seen. The $\mathbf{Q}$ − *matrix* is visualized with bars. The maximum height of the bar is 1, and the sum of bars along one row corresponding to one variable is 1. From the matrix it can be remarked that the system has two sources: N1 represents the noise source of the variable m_1 (X1) and correspondingly N3 noise source of the variable m_3 (X3). There is no energy coming to the system from the noise source N2, which corresponds to the variable m_2 (X2). This can be seen from the column of N2: there is no visible bars at all.

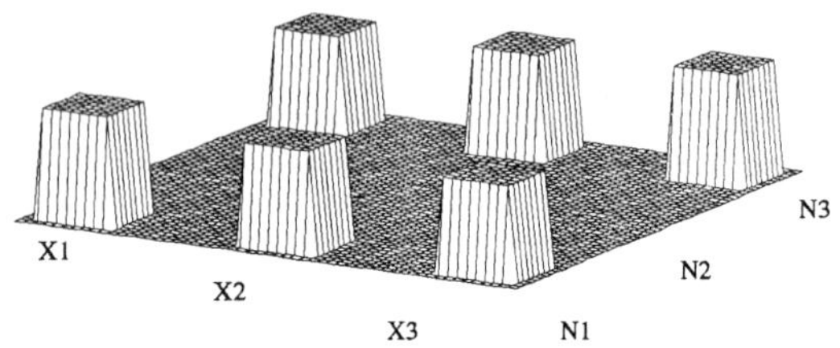

Figure 6. System noise contribution plot

Although N1 has effect to all variables, it has most influence to the variable m_1. So we can draw a conclusion that the first source is near sensor m_1. And correspondingly, since the bar of N3 is highest on row X3, we can conclude that the second source is near sensor m_3.

Using the model we have possibility to examine the spectrum of estimated source and compare it to the original source and measurement (Fig. 7 and Fig. 8). From the plots we can see that the spectra of estimated sources are extremely good: in both cases the shape of estimated spectrum corresponds rather well to the true one.

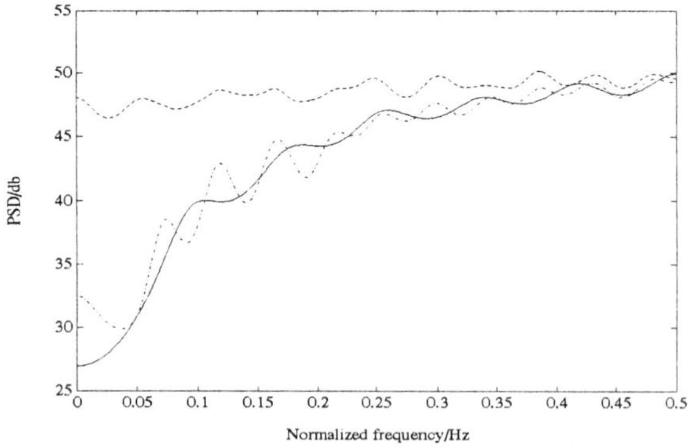

Figure 7. Spectra of source1 (solid line), measured signal (dashed line) and estimated source (dashdotted line)

5. SUMMARY

In this paper a novel method for the separation of sources is presented. In presented method *multivariate autoregressive model* is exploited to construct conditioned spectral matrix NCSM. Conditioned spectral matrix is utilized to construct new measures which are efficient to extract independent sources.

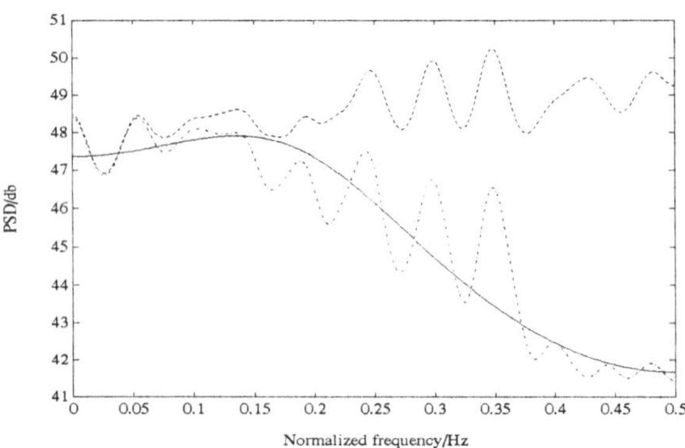

Figure 8. Spectra of source2 (solid line), measured signal (dashed line) and estimated source (dashdotted line)

REFERENCES

[1] Julius S. Bendat and Allan G. Piersol. *Engineering Applications of Correlation and Spectral Analysis.* John Wiley & Sons, New York, NY, USA, 1980.

[2] Christian Jutten and Jeanny Herault. Blind separation of sources, part 1: An adaptive algorithm based on neurominetic architecture. *Signal Processing*, 24:1–10, 1991.

[3] S. Lawrence Marple Jr. *Digital Spectral Analysis with Applications.* Prentice-Hall, Englewood Cliffs, NJ, USA, 1987.

[4] Risto Suoranta. Analysis of linear systems applying multivariate autoregressive model. Master's thesis, Tampere University of Technology, Measurement Engineering Laboratory, Tampere, Finland, 1990.

[5] Risto Suoranta and Seppo Rantala. Multivariate autoregressive model applied to conditioned spectral analysis of complex systems. In *Proc. IEEE Int. Conf. on Systems Engineering*, pages 190–193, Pittsburgh, PA, USA, Aug 1990. IEEE.

[6] Risto Suoranta and Seppo Rantala. Utilizing multivariate autoregressive model to reveal internal dependencies in multichannel measurement data. In *Proc. of the IEEE Instrumentation and Measurement Technology Conference*, Atlanta, GA, USA, 1991. IEEE.

SIGNAL PROCESSING VI: Theories and Applications
J. Vandewalle, R. Boite, M. Moonen, A. Oosterlinck (eds.)

A Data-Based Enumeration Technique Of Coherent Signals*

H. Krim and J. G. Proakis‡
Laboratoire des Signaux et Systèmes,
Plateau du Moulon, 91192, Gif-Sur-Yvette, France,
‡Communications and Digital Signal Processing Center
Northeastern University, Boston, MA 02115

The combination of a smoothing transformation on an array covariance matrix with signal enumeration criteria is shown to be insufficient when dealing with fully correlated signals impinging on a Uniform Linear Array (ULA). We account for the "vestigial" correlation following a smoothing transformation of a matrix appropriately constructed with the received data, and derive a Minimum Description Length (MDL)-based detection criterion. The additional constraint which is introduced, is shown to achieve a significant performance improvement, particularly with short data records and closely spaced signals.

1 Introduction

Eigenstructure-based parameter estimation techniques have been the focus of a great research activity because of their wide applications in a number of fields.

In a Direction Of Arrival estimation problem, for example, determining the number of signals impinging on an array, is the first step critical to the success of any of these techniques, selected for the solution. Many algorithms for order estimation of an observed process which have appeared in the literature [2, 8, 10, 9, 11] most often simplify to a test of equality of the smallest eigenvalues of the array covariance matrix. The consistency‡of these detection or enumeration techniques has been very useful as a comparison measure among them. It is in fact, on account of this asymptotic behavior, that a method based on the MDL principle, proposed by Wax *et al.*[9, 10], has become very popular. In the presence of fully correlated signals, it requires, however, a Multidimensional (MD) search, since the signal coherence causes a failure of all the classical methods.

A computationally more attractive solution consists in first recovering the rank through a smoothing transformation of the covariance matrix [4, 8] and then applying a Smoothed Rank Profile (SRP)§test. This consists in tracking the increase and eventual stabilization of the rank as the rank of the smoothing matrix varies. The stable rank corresponds to the number of signals present in the observed process with probability 1 (w.p.1) [4]. These techniques, however, share a common performance threshold problem which is due to the remaining correlation among the averaged diagonal submatrices after the smoothing takes place. We will show here that a significant performance gain is achieved if this "vestigial" correlation is accounted for.

We shall assume throughout, a uniform linear array composed of L identical, omnidirectional sensors with equal spacing $d = \lambda/2$ where λ is the wavelength. We assume that $M < L$ narrowband planewaves (centered about the known frequency ω_0) impinge from (distinct) directions θ_i, $i = 1, 2, \cdots, M$ and that $M - (L' - 1)$ of them are fully correlated. The effective rank¶of $\mathbf{R} = E\{\mathbf{x}\mathbf{x}^H\}$, is thus denoted by L'. The received data vector is denoted by $\mathbf{x}(t) = (x_1(t), \ldots, x_L(t))^{\mathrm{T}}$, where $x_i(t)$ is the data received by the ith sensor and where "T" denotes transposition and "H" conjugate transposition. We also assume, that the array outputs and the corresponding noise, $\mathbf{x}(t)$ and $\mathbf{n}(t)$, respectively, are stationary and ergodic complex-valued random processes having zero mean; and that the noise is uncorrelated with the signals; the noise terms are also mutually uncorrelated with *unknown* but identical variances, σ^2.

The problem is to directly estimate the number of signals impinging on the above array when given a short data record of length $N < 2L$. We recall that some or all of these signals may be coherent. For this purpose, we choose to use a data matrix $\mathbf{X}$ shown below. The ith column of $\mathbf{X}$ corresponds to the data vector recorded at the $(L - i + 1)^{th}$ sensor:

$$\mathbf{X} = \begin{pmatrix} x_L(1) & x_{L-1}(1) & \ldots & x_1(1) \\ \vdots & \ldots & \ldots & \vdots \\ x_L(N) & x_{L-1}(N) & \ldots & x_1(N) \end{pmatrix}. \tag{1}$$

2 A New Data-Domain Variation

It is well known that full correlation among two or more signals folds the subspace spanned by the eigenvectors with the largest eigenvalues [6]. It was also shown in [6] that an increase in a column dimension of a smoothing transformation applied to the data matrix in Eq. (1), induced a rank increase of the resulting matrix (this is equivalent to decreasing the subarray size). The smoothing was shown to be achieved by first defining the following windowing matrix $\mathbf{F} = [\mathbf{F}_{1p}|\mathbf{F}_{2p}|\cdots|\mathbf{F}_{n_sp}]$, where $\mathbf{F}_{ip} = [\mathbf{0}_{p\times(i-1)}|\mathbf{I}_p|\mathbf{0}_{p\times(L-i-p+1)}]$, $i = 1, 2, \cdots, L-p$, and then carrying out the following formal compact

*This work was supported in part by MITRE Corp., (Bedford, MA.) and by NSF grant MIP-8904109.

product

$$\widetilde{\mathbf{X}} = [\mathbf{I}_{n_s} \otimes \mathbf{X}]\mathbf{F}^T, \quad (2)$$

where $\otimes$ denotes the Kronecker product, and $\mathbf{I}_{n_s}$ is the $n_s \times n_s$ identity matrix, with $n_s = L - p + 1$.

Realizing the key role noise plays in model identification, we use Eq. (2) to fit a linear model to the observed process and thus to unravel the underlying signal structure. It is thus reasonable to assume that errors will be incurred when a linear model is fitted to the observed data,

$$\widetilde{\mathbf{X}}'\mathbf{d} + \mathbf{e} = \tilde{\mathbf{x}}_p, \quad (3)$$

where $\mathbf{d}$ is the $p \times 1$ linear prediction vector, and $\mathbf{e} = [\mathbf{e}(p), \mathbf{e}(p+1), \cdots, \mathbf{e}(L)]^T$ is the $(L-p+1)N \times 1$ error vector. We shall assume that $(L - p + 1)N$ is large, so that the errors can be taken to be normally distributed with a covariance matrix $\mathbf{\Gamma} = \mathrm{E}[\mathbf{e}\mathbf{e}^H]$, and have minimal end effects. The components of an error subvector (e.g., $\mathbf{e}(p)$) are clearly uncorrelated, since time samples are assumed independent, and thus, uncorrelated. There exists, however, some correlation between the error subvectors since a subblock of data of any block subarray is used in subsequent ones as can be seen from Eqs. (2) and (3). To simplify the density needed to estimate the model, we whiten the model error. This, in turn requires explicit knowledge of $\mathbf{\Gamma}$. For that purpose, Eq. (3) is first rewritten as

$$\mathbf{d}'^T\widetilde{\mathbf{X}}^T = \mathbf{e}^T, \quad (4)$$

where $\mathbf{d}' = [1, \mathbf{d}^T]^T$. The Vec operator with its properties implies the following,

$$\mathrm{Vec}(\mathbf{d}'^T\widetilde{\mathbf{X}}^T) = (\mathbf{I} \otimes \mathbf{d}'^T)\,\mathrm{Vec}(\widetilde{\mathbf{X}}^T).$$

Using the latter expression in $\mathbf{\Gamma}$ and after a simple manipulation, one can write,

$$\mathbf{\Gamma} = (\mathbf{I} \otimes \mathbf{d}'^T)\mathrm{E}[\mathrm{Vec}(\widetilde{\mathbf{X}}^T)\mathrm{Vec}(\widetilde{\mathbf{X}}^T)^H] \times (\mathbf{I} \otimes \mathbf{d}'^T)^H.$$

Recall that each data point has a signal component contaminated with noise (which is not to be confused with the signal subspace component). Denoting the noise components of $\widetilde{\mathbf{X}}$ by $\widetilde{\mathbf{N}}$ implies that $\mathrm{Vec}(\widetilde{\mathbf{X}}^T) = \mathrm{Vec}(\widetilde{\mathbf{X}}_{sc}^T) + \mathrm{Vec}(\widetilde{\mathbf{N}}^T)$, from which

$$\mathbf{e} \approx 0 + (\mathbf{I} \otimes \mathbf{d}'^T)\,\mathrm{Vec}(\widetilde{\mathbf{N}}^T), \quad (5)$$

since the signal components (cisoids) theoretically fit the linear model perfectly. The expression for $\mathbf{\Gamma}$ is now simply deduced as

$$\begin{aligned}\mathbf{\Gamma} &= (\mathbf{I} \otimes \mathbf{d}'^T)\,\mathrm{E}[\mathrm{Vec}(\widetilde{\mathbf{N}}^T)\mathrm{Vec}(\widetilde{\mathbf{N}}^T)^H] \times \\ &\quad (\mathbf{I} \otimes \mathbf{d}'^T)^H \\ &= \sigma^2(\mathbf{I} \otimes \mathbf{d}'^T)\tilde{\mathbf{I}}(\mathbf{I} \otimes \mathbf{d}'^T)^H = \sigma^2\tilde{\mathbf{\Gamma}} \quad (6)\end{aligned}$$

where $\tilde{\mathbf{I}}$ is an indicator matrix with 1 and 0 components which result from the covariance of the noise components, and σ^2 is the noise variance. By premultiplying Eq. (4) by $\mathbf{\Gamma}^{-\frac{1}{2}}$, we obtain,

$$\underset{\sim}{\mathbf{X}}'\,\mathbf{d} + \varepsilon = \underset{\sim}{\mathbf{x}}_p, \quad (7)$$

where $\mathrm{E}[\varepsilon\varepsilon^H] = \sigma^2\mathbf{I}$.

Using the fact that ε is Gaussian, we obtain the conditional probability density function of $\underset{\sim}{\mathbf{x}}_p$, and derive an MDL-based criterion after obtaining the Maximum Likelihood (ML) parameter vector $\widehat{\beta}$ (by maximizing the log-likelihood function $\mathcal{L}(\beta) = -\log f(\underset{\sim}{\mathbf{x}}_p|\,\mathbf{d}^T, \sigma^2)$.

We first obtain,

$$\begin{aligned}\widehat{\sigma^2} &= \frac{1}{\alpha}\,\|\underset{\sim}{\mathbf{x}}_p - \underset{\sim}{\mathbf{X}}'\,\mathbf{d}\,\|^2, \text{ and} \\ \hat{\mathbf{d}} &= (\underset{\sim}{\mathbf{X}}'^H\underset{\sim}{\mathbf{X}}')^{\#}\,\underset{\sim}{\mathbf{X}}'^H\underset{\sim}{\mathbf{x}}_p, \quad (8)\end{aligned}$$

where # is a symbol used for the pseudo-inverse. Using $\widehat{\beta} = [\hat{\mathbf{d}}^T, \widehat{\sigma^2}]$, we can rewrite $\mathcal{L}(\cdot)$

$$\begin{aligned}\mathcal{L}(\widehat{\beta}) &= \frac{\alpha}{2}\log 2\pi + \frac{\alpha}{2} \\ &+ \tfrac{\alpha}{2}\log\left(\tfrac{1}{\alpha}\,\|\underset{\sim}{\mathbf{x}}_p - \underset{\sim}{\mathbf{X}}'\,(\underset{\sim}{\mathbf{X}}'^H\underset{\sim}{\mathbf{X}}')^{\#}\,\underset{\sim}{\mathbf{X}}'^H\underset{\sim}{\mathbf{x}}_p\|^2\right) \quad (9)\end{aligned}$$

where $\alpha = N \times n_s$. Further simplification results when the Singular Value Decomposition (SVD) of $\underset{\sim}{\mathbf{X}}' = \underset{\sim}{\mathbf{V}}\underset{\sim}{\mathbf{\Sigma}}\underset{\sim}{\mathbf{U}}^H$ is used to obtain its rank-k approximating matrix. This results in, $\widehat{\underset{\sim}{\mathbf{X}}}' = \underset{\sim}{\mathbf{V}}\begin{bmatrix}\underset{\sim}{\mathbf{\Sigma}}_k & \mathbf{0} \\ \mathbf{0} & \mathbf{0}\end{bmatrix}\underset{\sim}{\mathbf{U}}^H$, where the vectors $\underset{\sim}{\mathbf{V}}$ and $\underset{\sim}{\mathbf{U}}$ are respectively the left and the right singular vectors of $\underset{\sim}{\mathbf{X}}'$, and $\underset{\sim}{\mathbf{\Sigma}}_k$ is a $(k \times k)$ diagonal matrix of the largest singular values (assuming that k signals impinge on the receiving array).[||]Using $\widehat{\underset{\sim}{\mathbf{X}}}'$, the argument of the second term in Eq. (9) simplifies to,

$$\begin{aligned}\varepsilon &= \left\{\underset{\sim}{\mathbf{x}}_p - \underset{\sim}{\mathbf{X}}'\,(\underset{\sim}{\mathbf{X}}'^H\underset{\sim}{\mathbf{X}}')^{\#}\,\underset{\sim}{\mathbf{X}}'^H\underset{\sim}{\mathbf{x}}_p\right\} \\ &= \underset{\sim}{\mathbf{x}}_p - \underset{\sim}{\mathbf{V}}\begin{bmatrix}\mathbf{I}_k & \mathbf{0} \\ \mathbf{0} & \mathbf{0}\end{bmatrix}\underset{\sim}{\mathbf{V}}^H\,\underset{\sim}{\mathbf{x}}_p\,. \quad (10)\end{aligned}$$

The invariance of $\|\,\varepsilon\,\|$ under a unitary transformation $\underset{\sim}{\mathbf{V}}^H$ further simplifies [3] as,

$$\|\,\varepsilon^k\,\|^2 = \sum_{i=k+1}^{(L-p+1)N} \|\,\xi_i\,\|^2, \quad (11)$$

where $\xi = \underset{\sim}{\mathbf{V}}^H\,\underset{\sim}{\mathbf{x}}_p$ and the superscript on ε^k indicates the assumed effective rank.

For an assumed number k of impinging signals, the optimal linear prediction vector $\mathbf{d}$ is achieved by minimizing $\|\,\varepsilon\,\|^2$. Since only k entries of the vector ε are used in the minimization, and, given that complex signals are considered herein, the number of adjustable parameters are the $2k$ entries of ε. It is now possible to write down an MDL-based test which, when minimized over all possible k, will provide the model order for the chosen subarray size p:

$$\begin{aligned}\hat{k} &= \arg\left\{\min_{k\in\{1,\cdots,p-1\}} \mathrm{MDL}(k)\right\}, \text{ where} \\ \mathrm{MDL}(k) &= \left\{\frac{\alpha}{2} + \frac{\alpha}{2}\log 2\pi + \right. \\ &\quad \frac{\alpha}{2}\log\{(\frac{1}{\alpha})\,\|\,\varepsilon^k\,\|^2\} \\ &\quad \left. + \frac{1}{2}(2k+1)\log\left(N(L-p+1)\right)\right\}, \quad (12)\end{aligned}$$

where α as previously defined, represents the length of the vector $\tilde{\mathbf{x}}_p$ used in estimating the model. An order $\hat{k}$ (one which minimizes the MDL(k)) is obtained, for several subarray sizes p. If the number of degrees of freedom is sufficient, this order $\hat{k}$ (effective rank of $\widetilde{\mathbf{X}}$) will stabilize w.p.1 [4]. This value of $\hat{k}$ for which the stabilization takes place, is as described next to be a consistent estimate of the number of signals [6].

The algorithm can now be summarized below:

Summary of Algorithm

1. For a subarray size p (usually chosen to be $(L-1)$), obtain an estimate of the linear prediction vector $\mathbf{d}$ as in equation (3);
2. Obtain an estimate of the error covariance matrix $\mathbf{\Gamma}$ as in (6)
3. Whiten the noise via the inverse hermitian square root of $\mathbf{\Gamma}$ as described in equation (7);
4. Use the norm of ξ for different model orders as they are tested to minimize the MDL in equation (12) to obtain a model order estimate (or effective rank estimate of $\widetilde{\mathbf{X}}$) $\hat{k}$;
5. Increase (or decrease) p and repeat steps 2-4;
6. If $\hat{k}$ stabilizes for two or more consecutive subarray sizes (as p varies), it should be picked as the number of signals; if the rank does not stabilize the number of degrees of freedom is not sufficient to obtain a solution (and there is no solution).

2.1 Consistency of the Detection Scheme

The power of a detection procedure is evaluated by the bias in its estimate of the model order (or number of signals), as the length of the data vector used for prediction grows without bound (or is very large for practical purposes).

Lemma 1 *The order estimate $\hat{k}$ obtained via the above algorithm, is a consistent estimate of the number of signals incident on a uniform linear array, i.e.,*

$$\arg\{\lim_{\alpha\to\infty} \mathrm{MDL}(\hat{k})\} = M \quad a.s. (almost\ surely). \quad (13)$$

Proof: We sketch the proof for the above lemma. We first use the criterion developed by Cozzens, *et al.* [4] to say that as p varies, the rank will stabilize w.p.1 to the number of signals if the number of degrees of freedom (number of sensors) is sufficient. It is then necessary to prove that the effective rank (of $\widetilde{\mathbf{X}}$) determined with the above procedure is asymptotically consistent. We thus want to prove that the length MDL(k) is always greater than the length given by the optimal or true order M (or MDL(M)). By using the following two result [2],

$$\text{if, } \lim_{N\to\infty} C_N/N \to 0, \text{ and } \lim_{N\to\infty} C_N \to \infty,$$

then,

$$\text{then, } \log\frac{\| \varepsilon^k \|}{\| \varepsilon^M \|} \geq -(k-M)C_N/N, \text{when } k > M \ ;$$

$$\log\frac{\| \varepsilon^k \|}{\| \varepsilon^M \|} > (M-k)C_N/N, \text{when } k < M (14)$$

one then proceeds to prove that,
for $k > M$

$$(\mathrm{MDL}(k) - \mathrm{MDL}(M)) \geq 0 \quad (15)$$

and for $k < M$

$$(\mathrm{MDL}(k) - \mathrm{MDL}(M)) \geq 0, \quad (16)$$

proving the consistency since any length is greater than the one given by the true number M. ■

3 Simulations

Example: A 10 element array with equal spacing of $\lambda/2$ is considered. Two coherent narrowband signals with a common normalized frequency of $f_{1(2)} = .25$ Hz are assumed to impinge on the array from distinct directions ($\theta_{1(2)} = \pm 2.5$ deg). The data record length is N=15 snapshots. A set of 100 random trials is used for each SNR to evaluate the detection performance of the proposed method.

When we perform additional processing to whiten the model error, the detection performance threshold as encountered in the previously described methods is improved. The performance of this DDDET method, as noted earlier, will be commensurate with the goodness of estimation of the noise covariance matrix and the data record length. The performance improvement of the DDDET is illustrated in Fig. 1 and confirms that the vestigial correlation between the overlapping subarrays affects the detection criterion. We should mention that the rank stabilization criterion described earlier is applied throughout this evaluation. The overall detection performance is summarized in Table 3.1. A performance comparison with some of the previously described methods favors the present approach [6].

4 Conclusion

It was shown that the vestigial correlation of the signals following a smoothing transformation needs to be accounted for if a unidimensional signal enumeration technique is to be used. The performance improvement was shown to be significant as a consequence, with a modest computational requirement.

Footnotes

‡A detection algorithm is said to be consistent if the estimated number of signals tends to the true number w.p.1, as the number of snapshots N (data record length) grows without bounds.

§The rank of a sequence of smoothed covariance matrices is referred to as a smoothed rank profile.

¶The effective rank is the rank coinciding with the number of the largest eigenvalues of the covariance matrix. It is different from the mathematical rank which is full.

‖The approximation for $\underset{\sim}{\mathbf{X}}'$ proceeds by first assuming that the process model order is k, and then by selecting the k largest singular values of $\underset{\sim}{\mathbf{\Sigma}}$ to form $\underset{\sim}{\mathbf{\Sigma}}_k$.

Acknowledgements

One of us wishes to thank Dr. J. H. Cozzens for helpful comments and suggestions.

References

[1] H. Akaike, A New Look at the Statistical Model. Identification. *IEEE Trans. on Automatic Control, Vol. AC-19, No. 6, Dec. 1974, pp.716-723.*

[2] Z. D. Bai, K. Subramanyam and L. C. Zhao. On Determination of the Order of an Autoregressive Model. *Multivariate Statistics and Probability, Vol. 27, 1988, pp. 40-52, .*

[3] L. S. Biradar and V. U. Reddy. SVD Based Information criteria for Detection of the number of sinusoids. *Proc. of the ICASSP'90, pp. 2519-2522.*

[4] J. H. Cozzens, M. J. Sousa and R. Dipietro. Enumeration of Fully Correlated Signals *IEEE Trans. on S.P.* to appear.

[5] H. Krim, P. Forster and J. G. Proakis. Performance Analysis of Smoothed Subspace-Based Methods. *Proc. ICASSP'91, Toronto, May 1991.*

[6] H. Krim. Detection and Parameter Estimation of Correlated Signals in Noise Ph.D. dissertation, Northeastern University, June 1990.

[7] S. Prasad and B. Chandna. An Aungmented Smoothed Rank Algorithm for Determination of Source Coherence Structure. *"IEEE Trans. on ASSP, Vol. 37, No. 7, pp. 1144-1146, 1989".*

[8] T. J. Shan, A. Paulraj and T. Kailath. On Smoothed Rank Profile Testing in Eigenstructure Methods for Direction Of Arrival Estimation. *IEEE Trans. on ASSP, Vol. 35, No. 10, pp. 1377-1385, 1985.*

[9] M. Wax and T. Kailath. Detection of Signals by Information Theoretic Criteria. *IEEE Trans. on ASSP, Vol. 33, No. 2, pp. 387-392, 1985.*

[10] M. Wax and I. Ziskind. Detection of the Number of Coherent Signals By the MDL Principle. *IEEE Tran. on ASSP, Vol. 37, No. 8, Aug. 1989.*

[11] L. C. Zhao, P. R. Krishnaiah and Z. D. Bai. On Detection of the Number of Signals in the Presence of White Noise. *J. Multivariate Anals., Vol. 20, pp. 1-20, 1986.*

DDDET					
SNR (dB)	p=4	p=5	p=6	p=7	p=8
20	100	100	100	100	100
17.5	100	100	100	100	98
15	100	100	100	98	66
12.5	97	99	95	70	17
10	52	66	53	18	0

Table 1: Detection performance of modified DDDET

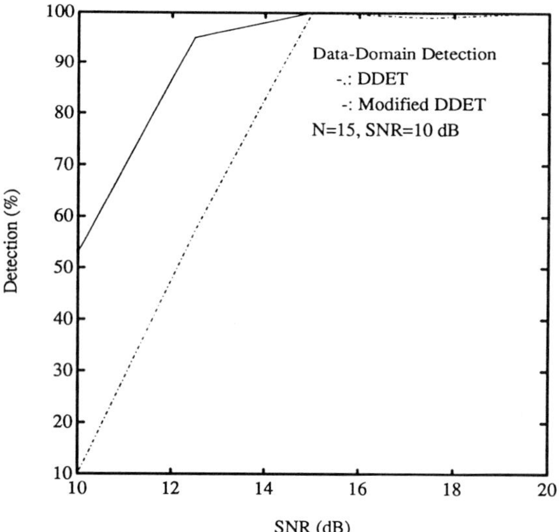

Figure 1: Performance improvement of modified DDDET

SIGNAL PROCESSING VI: Theories and Applications
J. Vandewalle, R. Boite, M. Moonen, A. Oosterlinck (eds.)

On the equivalence of SVD and TLS based Yule–Walker approaches to frequency estimation

Petre Stoica
Department of Automatic Control
Polytechnic Institute of Bucharest
Splaiul Independentei 313
R–77206 Bucharest, Roumania

Torsten Söderström
Automatic Control and
Systems Analysis Group
Department of Technology,
Uppsala University, Box 27
S–751 03 Uppsala, Sweden

Sabine Van Huffel
ESAT Laboratory
Department of Electrical Engineering
Katholieke Universiteit Leuven
Kardinaal Mercierlaan 94
B-3030 Heverlee, Belgium

The total least–squares (TLS) technique for solving noise–corrupted linear systems of equations is receiving a considerable attention in the signal processing literature. Here, we study the use of TLS within a high–order Yule–Walker (HOYW) procedure for estimating the frequencies of sinusoidal signals, and show that the TLS–based HOYW method has the same large–sample properties as the commonly used singular value decomposition (SVD)–based HOYW procedure.

1 Introduction

The TLS technique formally introduced in [1] is currently receiving a significant attention in the signal processing and system identification literature. See [2] for a recent and comprehensive overview. Applications of TLS to signal processing problems include the TLS–ESPRIT and TLS–Linear Prediction (LP) methods for estimating the frequencies of sinusoidal signals or the bearings of plane waves, from noisy measurements. See [3], [21], and [4], respectively. The two aforementioned estimation problems are closely related and can be treated similarly. However, for the sake of clarity, we focus here on the sinusoidal frequency estimation application.

TLS–ESPRIT has been analysed in [5] (see also [21]), where it was shown to have the same large–sample properties as the computationally–simpler LS–ESPRIT. This result of [5] can be obtained as a special case of a more general result proved in the following, as explained in Section 4.

The TLS–LP method [4] has been studied recently in [6] where it was shown that in the high signal–to–noise ratio (SNR) case, the TLS-LP has the same statistical properties as the commonly–used SVD–based LP technique of [7]. Concerning the TLS–LP, it is interesting to note that this technique is identical to the Min–Norm method of [8]. This fact, overlooked in [4], essentially follows from the original work [8]. It has been stated explicitly in [9], [17],and recently rediscovered in [10].

Here, we study the use of TLS in the HOYW procedure of sinusoidal frequency estimation, and show that the TLS–based HOYW frequency estimates have the same large–sample properties as the commonly employed SVD–based HOYW estimates ([11]–[13]). This type of first–order equivalence result is similar to the previously quoted result of [6]. Note, however, that the two results concern different estimation methods, and are derived by different analysis techniques. Furthermore, the result in [6] holds under the assumption of high SNR, whereas our results require that the number of data samples is large (which appears to be a more natural assumption).

2 SVD–based and TLS–based HOYW methods

Consider the noisy sinusoidal signal

$$y(t) = \sum_{k=1}^{n} \alpha_k e^{i(\omega_k t + \varphi_k)} + \varepsilon(t) \qquad (2.1)$$

where $\alpha_k > 0$, the frequencies $\{\omega_k\}$ are distinct, the initial phases $\{\varphi_k\}$ are independently and uniformly distributed over $[0, 2\pi)$, and $\varepsilon(t)$ is complex white noise with zero mean and the following variances

$$E\varepsilon(t)\varepsilon^*(s) = \sigma^2 \delta_{t,s} \quad E\varepsilon(t)\varepsilon(s) = 0 \qquad (2.2)$$

Here E is the expectation operator, the superscript " $*$ " denotes the conjugate transpose, and $\delta_{t,s}$ is the Kronecker delta. It is also assumed that $\{\varepsilon(t)\}$ and $\{\varphi_k\}$ are independent random variables. The problem is to estimate the sinusoidal frequencies $\{\omega_k\}$. Once estimates of $\{\omega_k\}$ are available, the estimation of the other signal parameters in (2.1) reduces to a simple LS fit (see, for example, [18], [22]). The number of sine waves is assumed to be known (given or estimated).

Define

$$R = E\begin{pmatrix} y(t-L-1) \\ \vdots \\ y(t-L-M) \end{pmatrix} (y^*(t-1)\dots y^*(t-L)) \tag{2.3}$$

$$r = E\begin{pmatrix} y(t-L-1) \\ \vdots \\ y(t-L-M) \end{pmatrix} y^*(t) \tag{2.4}$$

where $L \geq n, M \geq n$. Also, define

$$\Theta(z) = z^L + \theta_1 z^{L-1} + \dots + \theta_L \tag{2.5}$$

where

$$\theta = (\theta_1 \dots \theta_L)^T \tag{2.6}$$

is the minimum norm solution to the following system of so-called Yule-Walker equations, (see, for example, [11]-[13], [22]),

$$R\theta = -r \tag{2.7}$$

It is well known that n zeros of the polynomial $\Theta(z)$ are located on the unit circle at $\{e^{i\omega_k}\}_{k=1}^n$, and the remaining $(L-n)$ zeros are situated strictly inside the unit circle [11]-[14]. This property of $\Theta(z)$ can be used to obtain consistent estimates of the frequencies $\{\omega_k\}$, in the following steps.

(i) Compute the sample covariances from the available samples $\{y(1), \dots, y(N)\}$ and form $\hat{R}$ and $\hat{r}$.

(ii) Determine a consistent estimate $\hat{\theta}$ of θ from the (sample) HOYW system of equations,

$$\hat{R}\hat{\theta} \simeq -\hat{r} \tag{2.8}$$

(the symbol "$\simeq$" used above indicates that (2.8) will in general have no exact solution).

(iiia) (Root form). Estimate the frequencies as the angular positions $\{\hat{\omega}_k\}$ of the n largest-modulus zeros of $\hat{\Theta}(z)$ (the polynomial $\hat{\Theta}(z)$ is made from $\hat{\theta}$ determined in step (ii)); or, alternatively,

(iiib) (Spectral form). Estimate the frequencies as the locations $\{\hat{\omega}_k\}$ of the n largest peaks of the "pseudo-spectrum" $1/|\hat{\Theta}(e^{i\omega})|^2$, $\omega \in (0, \pi)$.

Steps (i) and (iii) above are standard. The only critical problem is performing Step (ii). Some possibilities for approaching the problem of Step (ii) are briefly reviewed in the following.

A. LS-based HOYW

The LS solution to the linear system of (approximate) equations (2.8) is given by

$$\hat{\theta}_{LS} = -\hat{R}^\dagger \hat{r} \tag{2.9}$$

where $\hat{R}^\dagger$ denotes the Moore-Penrose pseudoinverse of $\hat{R}$ (see, for example, [2], [15]). Now, if $L > n$, then $\hat{\theta}_{LS}$ in (2.9) is a very poor estimate of θ. This is so since rank $(R) = n$ (see [11]-[14]), whereas $\hat{R}$ has full rank (w p 1) , which implies that $\hat{R}^\dagger$ does not approach $R^\dagger$ in such a case, even when N increases without bound. The a priori knowledge that rank $(R) = n$ is used by the ensuing approaches B and C, to refine the LS-based estimate (which is never used in the form of (2.9)).

B. SVD-based HOYW

Let $\hat{R}_n$ denote the best (in the Frobenius norm sense) rank-n approximation of $\hat{R}$, see [2], [19]. Replace $\hat{R}$ in (2.8) by $\hat{R}_n$ to obtain the following modified system of equations

$$\hat{R}_n\hat{\theta} \simeq -\hat{r} \tag{2.10}$$

Note that, in general, (2.10) has no exact solution. The minimum norm LS solution of (2.10) is given by

$$\hat{\theta}_{\text{SVD}} = -(\hat{R}_n)^\dagger \hat{r} \tag{2.11}$$

The HOYW method which estimates θ as in (2.11), is said to be SVD-based since the computation of $\hat{R}_n$ from $\hat{R}$ requires a SVD step (as is well known). This method has been found to possess excellent statistical performance, especially for reasonably large values of M and L (see [11]-[14]).

C. TLS-based HOYW

The SVD-based approach only corrects $\hat{R}$ in (2.8) using the a priori information that rank$(R) = n$, and the corrected system remains (generically) incompatible. A more natural way to proceed is to make the system of equations (2.8) compatible by correcting both $\hat{R}$ and $\hat{r}$ using the information that

$$\text{rank}[r \quad R] = n \tag{2.12}$$

This is the basic idea of the TLS-based approach.

Let $[\hat{\gamma} \quad \hat{\Lambda}] \triangleq [\hat{r} \quad \hat{R}]_n$ denote the best rank-n approximation of the (generically full-rank) matrix $[\hat{r} \quad \hat{R}]$. Then, the TLS-estimate of θ is given by the solution to the following system of equations

$$[\hat{\gamma} \quad \hat{\Lambda}] \begin{pmatrix} 1 \\ \phi \end{pmatrix} = 0 \tag{2.13}$$

Any solution to $\hat{\Lambda}\theta = -\hat{\gamma}$ is a TLS solution. The minimum norm solution, which we call the TLS estimate of θ, is given by

$$\hat{\theta}_{\text{TLS}} = -\hat{\Lambda}^\dagger \hat{\gamma} \tag{2.14}$$

3 Theoretical analysis

The TLS-based estimate of $\{\omega_k\}$ is more appealing, from an intuitive standpoint, than the SVD-based estimate (as discussed in the previous section). However, these two estimates can be shown to be asymptotically (i.e. for $N \gg 1$) equivalent.

Theorem 3.1. The SVD-HOYW estimate of $\{\omega_k\}$ is asymptotically equivalent to the TLS-HOYW estimate in the sense that the two estimates coincide (to within a first-order approximation), for sufficiently large values of N. This is true, regardless the form ("root" or "spectral") used in Step (iii).

Proof. See [23].

The above theorem implies that the SVD–based and TLS–based HOYW estimates of $\{\omega_k\}$ have the same asymptotic distribution. This distribution has been shown, in [12], [14], to be multivariate Gaussian. An explicit formula for the covariance matrix of the estimation errors has also been derived in these references. The theoretical variances shown in the numerical simulations in the next section were computed using the quoted formula of [14].

4 Numerical study

This section provides numerical evidence lending support to the fact that the asymptotic equivalence shown in the previous section may hold in practice for reasonably–sized sample lengths. See also [23].

Consider a signal consisting of two sine waves with

$$\begin{aligned} \alpha_1 &= 1 \quad \alpha_2 = 1 \quad \sigma^2 = 1 \quad (\text{hence SNR} = 0 \text{ dB}) \\ \omega_1 &= 0.2 \times 2\pi \quad \omega_2 = 0.23 \times 2\pi \end{aligned}$$

Two sample lengths have been considered (as also shown in Figure 1). In each case, 50 independent realizations of the noisy signal have been generated. The root–form SVD–based and TLS–based HOYW frequency estimates were then computed using the procedures described in [14] and [2], [9], respectively. The values of L and M considered are shown in Table 1, along with an attached code used in order to ease the plotting.

L	2	2	5	5	8	8	15	15	25
M	2	5	5	8	8	15	15	25	25
Code	1	2	3	4	5	6	7	8	9

Table 1: The considered combination of L and M.

It should be noted from Figure 1 that the SVD–based and TLS–based procedures behave quite similarly in the cases considered. For small estimation errors, this behaviour was predicted by the theory developed in the previous sections. However, it is interesting that the two procedures have similar statistical performance also in the case of fairly large estimation errors. This behaviour, which is not explained by the previously developed theory, can be observed for small values of (L, M), and has also been noted in several other simulations not shown here. Some comments on it can be found in the next section.

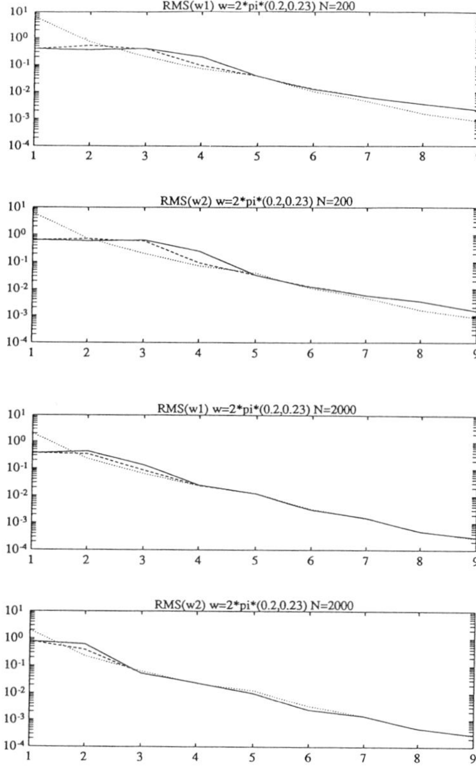

Figure 1: Empirical and theoretical RMS values versus (L, M) (see Table 1). Solid line = empirical RMS for SVD–based HOYW; Dashed line = empirical RMS for TLS–based HOYW; Dotted line = theoretical (asymptotical) RMS for both SVD–based and TLS–based HOYW; Upper two curves $N = 200$, lower two curves $N = 2000$.

5 Some extensions

The previous analysis has also a more general implication discussed in the following.

Consider a *general* system of equations of the form (2.7) and its noise–corrupted version in (2.8). The matrix $[r \;\; R]$ is assumed to satisfy the rank constraint (2.12), but otherwise it may be arbitrary (i.e. R and r are not necessarily covariance matrices as in (2.3), (2.4) etc). Then, under the assumption of small errors $(\hat{R} - R)$ and $(\hat{r} - r)$, it follows from the proof of Theorem 3.1 in [23] that

$$R(\hat{\theta}_{\text{SVD}} - \hat{\theta}_{\text{TLS}}) \approx 0 \qquad (5.1)$$

(to within a first–order approximation).

If $L = n$ and R has full (column) rank, then (5.1) implies $\hat{\theta}_{\text{SVD}} \approx \hat{\theta}_{\text{TLS}}$, which shows that our results encompass the result in [5] on the equivalence of LS–based and TLS–based ESPRIT procedures.

If $L > n$, then (5.1) no longer implies $\hat{\theta}_{\text{SVD}} \approx \hat{\theta}_{\text{TLS}}$. However, (5.1) is still useful in establishing the equivalence of some sinusoidal frequency or bearing estimation procedures, others than HOYW, based on SVD or TLS steps. For instance, the result in [6] on the first–order equivalence, in high SNR scenarios, of SVD–based and TLS–based LP methods can be derived at once from (5.1).

It should be noted that the first-order equality $\hat{\theta}_{\rm SVD} \approx \hat{\theta}_{\rm TLS}$ holds in the case of $L > n$ as well. This follows from the more elaborated (but also more involved) analysis in [20], where it was shown that (in the notation of this paper)

$$\hat{\theta}_{\rm SVD} - \hat{\theta}_{\rm TLS} = {\rm const}_1 \times \parallel [\hat{r} \quad \hat{R}] - [r \quad R] \parallel^2 \qquad (5.2)$$

whereas it is well known that

$$\tilde{\theta} = \hat{\theta} - \theta = {\rm const}_2 \times \parallel [\hat{r} \quad \hat{R}] - [r \quad R] \parallel \qquad (5.3)$$

for both the SVD-based and TLS-based approaches. Note from (5.2) and (5.3) that the difference $\hat{\theta}_{\rm SVD} - \hat{\theta}_{\rm TLS}$ may be small even if neither $\tilde{\theta}_{\rm SVD}$ nor $\tilde{\theta}_{\rm TLS}$ is so. This shows that the behaviour observed in the numerical example, reported in Section 4, is indeed possible. A theoretical study of this behaviour would, however, require a detailed analysis of const_1 and const_2, which is yet to be done.

References

[1] G H Golub and C Van Loan (1989)
Matrix Computations. 2nd ed, The John Hopkins University Press, Baltimore, MD.

[2] S Van Huffel and J Vandewalle (1991)
The Total Least Squares Problem: Computational Aspects and Analysis. *Frontiers in Applied Mathematics Series,* vol 9, SIAM, Philadelphia.

[3] R Roy and T Kailath (1989)
ESPRIT – estimation of signal parameters via rotational invariance technique. *IEEE Transactions on Acoustics, Speech and Signal Processing,* vol ASSP-37, pp 984–995.

[4] M A Rahman and K B Yu (1987)
Total least squares approach for frequency estimation using linear prediction. *IEEE Transactions on Acoustics, Speech and Signal Processing,* vol ASSP-35, pp 1440–1454.

[5] B D Rao and K V S Hari (1989)
Performance analysis of ESPRIT and TAM in determining the direction of arrival of plane waves in noise. *IEEE Transactions on Acoustics, Speech and Signal Processing,* vol ASSP-37, pp 1990–1995.

[6] Y Hua and T K Sarkar (1990)
A perturbation property of the TLS-LP method. *IEEE Transactions on Acoustics, Speech and Signal Processing,* vol ASSP-38, pp 2004–2005.

[7] D W Tufts and R Kumaresan (1982)
Estimation of frequencies of multiple sinusoids: making linear prediction perform like maximum likelihood. *Proc IEEE,* vol 70, pp 975–989.

[8] R Kumaresan and D W Tufts (1983)
Estimating the angle of arrival of multiple plane waves. *IEEE Transactions on Aerospace Electronic Systems,* vol AES-19, pp 134–139.

[9] Y Hua and T K Sarkar (1990)
On the total least squares linear prediction method for frequency estimation. *IEEE Transactions on Acoustics, Speech and Signal Processing,* vol ASSP-38, pp 2186–2189.

[10] E M Dowling and R D DeGroat (1991)
The equivalence of the total least squares and minimum norm methods. *IEEE Transactions on Signal Processing,* vol 39, p 1891–1892.

[11] Y T Chan and R P Langford (1982)
Spectral estimation via the high-order Yule-Walker equations. *IEEE Transactions on Acoustics, Speech and Signal Processing,* vol ASSP-30, pp 689–698.

[12] P Stoica, T Söderström and F-N Ti (1989)
Asymptotic properties of the high-order Yule-Walker estimates of sinusoidal frequencies. *IEEE Transactions on Acoustics, Speech and Signal Processing,* vol ASSP-37, pp 1721–1734.

[13] J A Cadzow (1982)
Spectral estimation: an overdetermined rational model equation approach. *Proc IEEE,* vol 70, pp 907–939.

[14] T Söderström and P Stoica (1991)
On the accuracy of high-order Yule-Walker methods for cisoids. *IEEE International Conference on Acoustics, Speech and Signal Processing,* (ICASSP 91), Toronto, Canada, May 14–17, 1991.

[15] T Söderström and P Stoica (1989)
System Identification. Prentice-Hall International, Hemel Hempstead, UK.

[16] P Stoica and T Söderström (1991)
On spectral and root forms of sinusoidal frequency estimators. *Signal Processing,* vol 24, pp 91–103.

[17] P Stoica (1990)
On the equivalence of Min-Norm and TLS approaches to frequency estimation. Unpublished notes, Uppsala University, Uppsala, Sweden, July 1990.

[18] P Stoica, R Moses, B Friedlander and T Söderström (1989)
Maximum likelihood estimation of the parameters of multiple sinusoids from noisy measurements. *IEEE Transactions on Acoustics, Speech and Signal Processing,* vol ASSP-37, pp 378–392.

[19] G W Stewart (1973)
Introduction to Matrix Computations. Academic Press, New York.

[20] A van der Sluis and G W Veltkamp (1979)
Restoring rank and consistency by orthogonal projection. *Linear Algebra and Its Applications,* vol 28, pp 257–278.

[21] B Ottersten, M Viberg and T Kailath (1991)
Performance analysis of the total least squares ESPRIT algorithm. *IEEE Transactions on Signal Processing,* vol 39, pp 1122–1135.

[22] A A (Louis) Beex and L L Scharf (1981)
Covariance sequence approximation for parametric spectrum modeling. *IEEE Transactions on Acoustics, Speech and Signal Processing,* vol ASSP-29, pp 1042–1052, October 1981.

[23] P Stoica, T Söderström and S Van Huffel (1992)
On SVD-based and TLS-based high-order Yule-Walker methods of frequency estimation. Submitted to *Signal Processing.*

SIGNAL PROCESSING VI: Theories and Applications
J. Vandewalle, R. Boite, M. Moonen, A. Oosterlinck (eds.)

Computationally Efficient Multiple Parameter Estimation of Composite Signals in Noise with Periodic Components

A. Mertins, N. Fliege
Telecommunications Group, Technical University of Hamburg, FRG

The problem of estimating amplitudes and arrival times of composite signals in additive noise is addressed. The noise process considered is a composition of Gaussian noise and periodic components with known frequency. The proposed new method is based on a simultaneous estimation of the required signal parameters and some parameters that describe the periodic components. The advantage is that the true amplitudes of the periodic components will have no influence on the estimates of the required signal parameters. It will be shown that for an increasing number of basic signals the variances of the estimates decrease to a minimum which is given by the signal to noise ratio (without periodic components). *A decomposition of the estimation algorithm that requires only a linearly increasing computation effort according to the number of basic signals will be presented.* This allows to handle large data records and use their advantages while keeping the computation effort relatively low.

1 Introduction

Composite signals appear in many important applications such as data communication systems or measurement problems where estimates of signal amplitudes and time delays with high precision are required.

In practice, the receiver input signals are often not corrupted by additive Gaussian noise only. An offset and further periodic components may occur due to several physical phenomena. The noise process considered is the sum of a white Gaussian noise process and some periodic components (including offset). The assumption of white noise is no general restriction, because the noise process can be prewhitened [3]. Another solution to the coloured-noise-problem can be found by introducing the covariance matrix of the noise process to the optimization criterion [4], but this method would not allow the decomposition of the problem to a very low number of operations.

Standard prefiltering techniques such as highpass or notch filtering cannot always help to eliminate the influence of the periodic components on the estimates. For example, eliminating an offset by highpass filtering is only possible if the signal itself has zero mean value.

The signals considered are linear combinations of basic signals of the form

$$r(kT) = n(kT) + \sum_{i=1}^{I} a_i \cdot s(kT - (i-1)\tau - D). \quad (1)$$

Throughout this paper the basic signal $s(t)$ and the delay τ are assumed to be known, while the amplitudes $a_1, \ldots, a_I$ and the delay D are unknown parameters that have to be estimated in real time. In applications like the analysis of mass spectrometer signals the delay τ is also not known a priori, but it can be estimated via a least-squares error minimization in a "warming-up phase".

For the estimation algorithm presented in this paper, the delay τ does not have to be an integer multiple of the sampling rate T, so that *we do not require any sampling rate synchronization.*

We can write (1) in vector notation as

$$\mathbf{r}(kT) = \mathbf{S}(kT - D) \cdot \mathbf{a} + \mathbf{n}(kT) \quad (2)$$

where $\mathbf{r}(kT)$ includes m samples of the received signal, $\mathbf{a}$ contains the amplitudes, and the $m \times I$- dimensional signal matrix $\mathbf{S}$ contains the samples of the delayed versions of the signal $s(t)$.

In order to estimate the delay D from the sampled input signal we use the decomposition

$$D = \Delta + \delta, \ \ |\delta| \leq T/2, \quad (3)$$

where Δ is an integer multiple of T. For $kT = \Delta$ we can rewrite (2) as

$$\mathbf{r}(\Delta) = \mathbf{S}(\delta) \cdot \mathbf{a} + \mathbf{n}(\Delta). \quad (4)$$

The noise process can be written as

$$\mathbf{n}(kT) = \mathbf{n}_W(kT) + \mathbf{n}_P(kT) = \mathbf{n}_W(kT) + \mathbf{N} \cdot \mathbf{c}(kT) \quad (5)$$

where $\mathbf{n}_W(kT)$ and $\mathbf{n}_P(kT)$ denote the white Gaussian noise process and the periodic components respectively. The matrix $\mathbf{N}$ contains samples of the periodic base functions and the vector $\mathbf{c}$ represents the periodic components. By combining the matrices and vectors as follows

$$\begin{aligned} \mathbf{B} &= [\mathbf{S}, \mathbf{N}], & (6) \\ \mathbf{b}^T &= [\mathbf{a}^T, \mathbf{c}^T], & (7) \end{aligned}$$

we can rewrite (4) in the short form

$$\mathbf{r}(\Delta) = \mathbf{B}(\delta)\cdot\mathbf{b} + \mathbf{n}_W(\Delta). \tag{8}$$

In order to get short and distinct formulae, the time dependence of vectors and matrices will not be written explicitly in the following sections.

2 Maximum Likelihood Estimation

The maximum likelihood estimates $\hat{\mathbf{b}}, \hat{D}$ in additive white Gaussian noise [2], [4], [5] have to be deduced from

$$\left.\|\mathbf{r}-\mathbf{Bb}\|^2\right|_{\mathbf{b}=\hat{\mathbf{b}},\,D=\hat{D}} \stackrel{!}{=} \min, \tag{9}$$

where $\|\ \|$ denotes the Euclidean norm ($\|\mathbf{x}\|^2 = \mathbf{x}^T\mathbf{x}$). A straightforward derivation of the corresponding likelihood equations

$$\left.\frac{\partial}{\partial\mathbf{b}}\|\mathbf{r}-\mathbf{Bb}\|^2\right|_{\mathbf{b}=\hat{\mathbf{b}},\,D=\hat{D}} = \mathbf{0} \tag{10}$$

$$\left.\frac{\partial}{\partial\delta}\|\mathbf{r}-\mathbf{Bb}\|^2\right|_{\mathbf{b}=\hat{\mathbf{b}},\,D=\hat{D}} = 0 \tag{11}$$

leads to the orthogonality relations

$$\left.\mathbf{B}^T\cdot[\mathbf{r}-\mathbf{Bb}]\right|_{\mathbf{b}=\hat{\mathbf{b}},\,D=\hat{D}} = \mathbf{0}, \tag{12}$$

$$\left.\mathbf{b}^T\dot{\mathbf{B}}^T\cdot[\mathbf{r}-\mathbf{Bb}]\right|_{\mathbf{b}=\hat{\mathbf{b}},\,D=\hat{D}} = 0. \tag{13}$$

Equations (12) and (13) say that the estimated noise $\mathbf{r}-\mathbf{B}\hat{\mathbf{b}}$ has to be orthogonal to all signals in $\mathbf{B}$ and to $\dot{\mathbf{B}}\hat{\mathbf{b}}$. For building up the matrix $\dot{\mathbf{B}}$ see Appendix .

It is clear that we can only find unbiased amplitude estimates, if $\mathbf{B}$ contains linear independent rows. Assuming an existing matrix $\left[\mathbf{B}^T\mathbf{B}\right]^{-1}$, the solution of (12) is given by

$$\boxed{\left.\hat{\mathbf{b}} = \left[\mathbf{B}^T\mathbf{B}\right]^{-1}\mathbf{B}^T\mathbf{r}\right|_{D=\hat{D}}} \tag{14}$$

By substituting (14) into (13), we find the scalar target function

$$\boxed{\left.Q(D) = \hat{\mathbf{b}}^T\dot{\mathbf{B}}^T\mathbf{r} - \hat{\mathbf{b}}^T\dot{\mathbf{B}}^T\mathbf{B}\hat{\mathbf{b}}\right|_{D=\hat{D}} = 0} \tag{15}$$

for the time delay estimation problem. This target function is a nonlinear function of the delay $D = \Delta + \delta$ **and** of the input samples. For only one basic signal it can be shown that a target function corresponding to (15) can be produced at the output of a linear filter (see Appendix).

Decomposition

We will now present decompositions of the equations (14) and (15) which require only a low number of operations. Since $\mathbf{N}$ is independent of δ, the matrix $\dot{\mathbf{B}}$ has the structure (see Appendix)

$$\dot{\mathbf{B}} = \left[\dot{\mathbf{S}}, \mathbf{0}\right]. \tag{16}$$

By resubstitution of (6), (7) and using (16) we can rewrite (15) as

$$\boxed{Q(D) = \hat{\mathbf{a}}^T\dot{\mathbf{S}}^T\mathbf{r} - \hat{\mathbf{a}}^T\dot{\mathbf{S}}^T\mathbf{N}\hat{\mathbf{c}}.} \tag{17}$$

The method of calculating the amplitude estimates is most important for the computation effort. Resubstitution of (6) and (7) into (14), partitioned matrix inversion and some straightforward numerical operations lead to

$$\boxed{\hat{\mathbf{c}} = \mathbf{Cr}},\qquad \boxed{\hat{\mathbf{a}} = \mathbf{Mr} - \mathbf{MN}\hat{\mathbf{c}}}, \tag{18}$$

where $\mathbf{M}$ and $\mathbf{C}$ are given by

$$\mathbf{C} = \left[\mathbf{N}^T\mathbf{N} - \mathbf{N}^T\mathbf{S}\left[\mathbf{S}^T\mathbf{S}\right]^{-1}\mathbf{S}^T\mathbf{N}\right]^{-1} \cdot \mathbf{N}^T\left[\mathbf{I} - \mathbf{S}\left[\mathbf{S}^T\mathbf{S}\right]^{-1}\mathbf{S}^T\right], \tag{19}$$

$$\mathbf{M} = \left[\mathbf{S}^T\mathbf{S}\right]^{-1}\mathbf{S}^T. \tag{20}$$

It should be pointed out that $\mathbf{C}$ has only a few rows and that $\mathbf{M}$ contains only a low number of non-zero elements if the basic signals are mutually orthogonal, especially if they are not overlapping. For a matrix $\mathbf{S}$ that contains orthonormal rows, the pseudo-inverse is given by $\mathbf{M} = \mathbf{S}^T$. In general, $\mathbf{M}$ should be calculated from a singulary value decomposition (SVD) of $\mathbf{S}$ for achieving a high numerical stability. The matrix $\mathbf{C}$ can be found to be the pseudo-inverse of $(\mathbf{I} - \mathbf{S}\left[\mathbf{S}^T\mathbf{S}\right]^{-1}\mathbf{S}^T)\mathbf{N}$ and can also be calculated from a SVD.

3 Application

A typical application is the estimation of the concentration of mass from mass spectrometer signals. Fig. 1 shows a typical signal where the concentration of mass is proportional to the amplitudes of the impulses. In this application, a typical number of basic signals is $I = 100$. For parameters $T = 0.1msec$, $\tau = 10\cdot T$ and ten non-zero samples of each signal, we receive a total number of $m = 10\cdot I = 1000$ samples. An offset and a sinusoidal disturbance with the frequency $\omega_1 = 2\pi 50Hz$

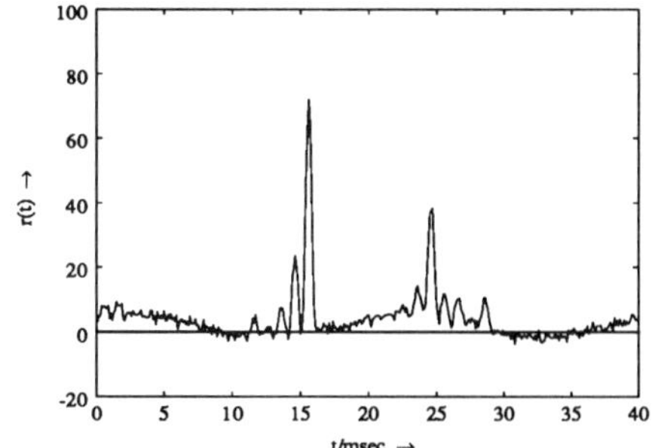

Fig. 1: Typical mass spectrometer signal

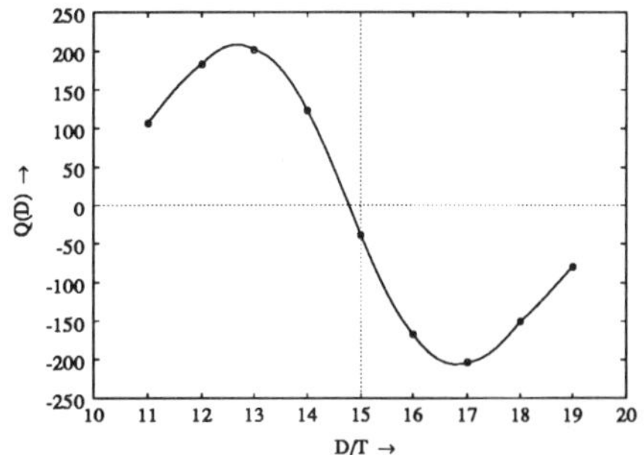

Fig. 2: Example for the target function

are typical, so that the matrix $\mathbf{N}$ and the vector $\mathbf{c}$ can be written as

$$\mathbf{N} = \begin{bmatrix} 1 & cos(0T\omega_1) & sin(0T\omega_1) \\ \vdots & \vdots & \vdots \\ 1 & cos([m-1]T\omega_1) & sin([m-1]T\omega_1) \end{bmatrix}, \tag{21}$$

$$\mathbf{c} = [c_0, c_1, c_2]^T. \tag{22}$$

It is clear that an extension to further periodic components can simply be made.

4 Computation Effort

In order to minimize the computation effort we can calculate all matrices a priori for a number of $p+1$ equally spaced interpolation times

$$\delta^{(i)} = -T/2 + i \cdot T/p, \qquad i = 0, 1, \ldots, p \tag{23}$$

so that we only have to choose the best matrices in a real time application. The number p depends on the desired accuracy.

The time delay has to be found from the zero of $Q(D)$. An example is shown in figure 2, where the true delay is $D = 15T$, and its estimate is $\hat{D} = 14.8T$.

Close to its zero the target function $Q(D)$ is relatively linear. Therefore we can estimate $\hat{\delta}$ without great error by calculating the zero of the linear interpolation of $Q(D-\delta)$ and $Q(D+T-\delta)$. For higher precision nonlinear search methods like a Newton-Raphson search or a modified Newton-Raphson search can be used [1].

We will compare the computation effort of the different solutions while using the example of analysing mass spectrometer signals with the parameters from section 3.

Both equations (14) and (18), respectively, for amplitude estimation can be interpreted as *different implementations of the same multidimensional linear filter.* We want to show the computation effort for non-overlapping orthonormal base signals in $\mathbf{S}$, where each basic signal in $\mathbf{S}$ has ten non-zero samples. This means: $\mathbf{M}$ is given by $\mathbf{M} = \mathbf{S}^T$, contains $m = 10 \cdot I$ non-zero samples, and $\mathbf{S}^T\mathbf{r}$ requires $10 \cdot I$ multiplications. Under these conditions, we need $10 \cdot I^2$ multiplications for the direct solution $\left([\mathbf{B}^T\mathbf{B}]^{-1}\mathbf{B}^T\right) \cdot \mathbf{r}$, or $40 \cdot I + I^2$ multiplications if we first compute $\mathbf{B}^T \cdot \mathbf{r}$ and multiply the result with $[\mathbf{B}^T\mathbf{B}]^{-1}$.

The implementation (18) only leads to $43 \cdot I$ multiplications (see Table 1). The important point is that the effort of (18) increases linearly with the number I of basic signals. This means that the relative effort for analysing one basic signal is independent of the total number I.

Tab. 1: Computation Effort for eq. (18)

$\hat{\mathbf{c}} = \mathbf{Cr}$	$\mathbf{Mr}$	$(\mathbf{MN})\mathbf{r}$	$\sum$
$30 \cdot I$	$10 \cdot I$	$3 \cdot I$	$43 \cdot I$

As is shown in table 2, the computation of the nonlinear target function $Q(D)$ requires a linearly increasing effort if it is computed from (17).

Tab. 2: Computation Effort for $Q(D)$

$\mathbf{r}^T\dot{\mathbf{S}}$	$[\mathbf{r}^T\dot{\mathbf{S}}]\hat{\mathbf{a}}$	$[\mathbf{N}^T\dot{\mathbf{S}}]\hat{\mathbf{a}}$	$\hat{\mathbf{c}}^T[\mathbf{N}^T\dot{\mathbf{S}}\hat{\mathbf{a}}]$	$\sum$
$10 \cdot I$	I	$3 \cdot I$	3	$14 \cdot I + 3$

5 Computer Simulation Results

Simulations have been made while using the parameters from section 3. The energy of the basic signals is one. The variance of the white noise process is $\sigma^2_{nW} = 1$, and the expected energy of the signal amplitudes is

$$E\left\{a_i^2\right\} = 10, \qquad i = 1, \ldots, I. \tag{24}$$

Figure 3 shows the variances $\sigma^2_{\hat{D}}$ and $\sigma^2_{\hat{a}}$ in relation to the number of basic signals. The actual value of the sinusoidal disturbances has no influence on the estimates because they are estimated simultaneously. It can be seen that both variances decrease rapidly if I increases. For a high number I we can reach a value for $\sigma^2_{\hat{a}}$ which is close to the minimum $\sigma^2_{\hat{a}_{min}} = 1$ given by the signal-to-noise-ratio in the white noise case $\mathbf{r} = \mathbf{Sa} + \mathbf{n}_W$. There are two reasons for the reduction of variances: 1) the periodic noise-to-white noise-ratio increases if I increases, 2) under condition (24), we can find $\sigma^2_{\hat{D}} \sim 1/I$, [6].

6 Conclusion

A new method for estimating amplitudes and arrival times of composite signals in additive noise with periodic components has been presented. The method is based on a simultaneous estimation of the required signal parameters and some parameters that describe the periodic components, so that the true amplitudes of the periodic components will have no influence on the estimates of the wanted signal parameters. The variances of the estimates decrease to a minimum which is given by the signal to noise ratio for an increasing number of basic signals. A new decomposition of the estimation algorithm that requires only a *linearly increasing* computation effort according to the number of basic signals has been presented. This allows to handle large data records and use their advantages without having an effectively increased computation effort.

A Appendix

In order to find unbiased estimates, the matrices $\mathbf{B}, \dot{\mathbf{B}}$ should achieve the condition

$$\dot{\mathbf{B}}^T(\delta) \cdot \mathbf{B}(\delta) = \mathbf{0} \tag{25}$$

for any δ.

We can build up a matrix $\dot{\mathbf{S}}$, analogous to $\mathbf{S}$, from the samples of $\frac{\partial}{\partial \delta} s(t - [i-1]\tau - \Delta - \delta)$ and use an orthogonal projection to achieve (25)

$$\dot{\mathbf{B}} = \left[\mathbf{I} - \mathbf{B}\left[\mathbf{B}^T\mathbf{B}\right]^{-1}\mathbf{B}^T\right]\left[\dot{\mathbf{S}}, \mathbf{0}\right]. \tag{26}$$

The disadvantage of (26) is that $\dot{\mathbf{B}}$ will contain a large number of non-zero elements. Since the orthogonal projection from (26) is included in the target function (15) which can be written as

$$Q(D) = \hat{\mathbf{b}}^T\dot{\mathbf{B}}^T\left[\mathbf{I} - \mathbf{B}\left[\mathbf{B}^T\mathbf{B}\right]^{-1}\mathbf{B}^T\right] \cdot \mathbf{r}, \tag{27}$$

we can use (16) directly.

By using (18) and (19) the target function (17) can be written as

$$Q(D) = \hat{\mathbf{a}}^T\Big(\dot{\mathbf{S}}^T - \hat{\mathbf{a}}^T\dot{\mathbf{S}}^T\mathbf{N}\left[\mathbf{N}^T\mathbf{N} - \mathbf{N}^T\mathbf{S}\left[\mathbf{S}^T\mathbf{S}\right]^{-1}\mathbf{S}^T\mathbf{N}\right]^{-1} \cdot \mathbf{N}^T\left[\mathbf{I} - \mathbf{S}\left[\mathbf{S}^T\mathbf{S}\right]^{-1}\mathbf{S}^T\right]\Big)\mathbf{r} \tag{28}$$

If we only have one basic signal, the expression in round brackets in (28) can be interpreted as an impulse response of a linear filter, and by ignoring the amplitude estimate, we could produce a corresponding target function at the output of a linear filter.

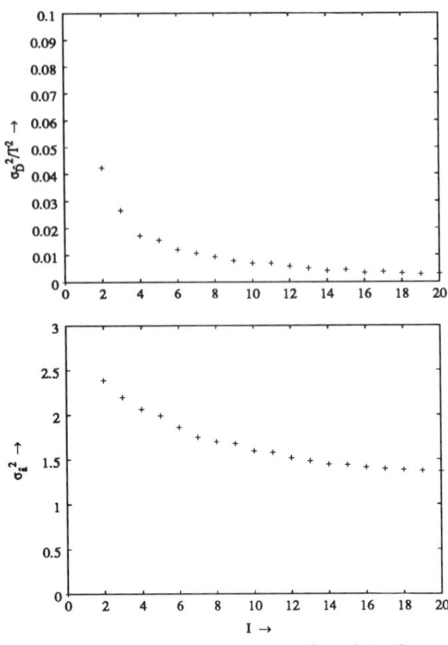

Fig. 3: Variances $\sigma_{\hat{a}}^2$, $\sigma_{\hat{D}}^2/T^2$

References

[1] Burkhardt, H.; Moll, H.: A Modified Newton-Raphson-Search for the Model-Adaptive Identification of Delays. In: Isermann, R. (ed.): Identification and System Parameter Estimation, pp. 1279 - 1286. Pergamon 1979.

[2] Goldberger, A.S.: Econometric Theory. New York: John Wiley & Sons, Inc. 1964.

[3] Kammeyer, K.D.; Kroschel, K.: Digitale Signalverarbeitung. Stuttgart: B. G. Teubner 1989.

[4] Sage, A.P.; Melsa, J.L.: Estimation Theory with Applications to Communications and Control. New York: Mc Graw Hill 1971.

[5] Van Trees, H.L.: Detection, Estimation, and Modulation Theory, Part I. New York: Wiley 1968.

[6] Mertins, A.; Fliege, N.: Multiple Parameter Estimation of Composite Signals in Uncharacterized Impulsive Noise. Proc. IEEE Int. Conf. Acoust., Speech, Signal Processing, March 1992.

SIGNAL PROCESSING VI: Theories and Applications
J. Vandewalle, R. Boite, M. Moonen, A. Oosterlinck (eds.)

THRESHOLD-FREE BAYESIAN ESTIMATION USING CENSORED MARGINAL INFERENCE

Anthony P. Quinn

Signal Processing and Communications Laboratory, Department of Engineering,
University of Cambridge, Trumpington Street, Cambridge, CB2 1PZ, England.

We contrast the strong Bayesian policies of joint and marginal inference for the deterministic parameters of a general signal model. The *marginal inference* uniquely trades data support for the model against the *Ockham Parameter Inference (OPI)* which seeks a reduction in model complexity. In stressful regimes, the latter dominates, offering an objective criterion for *inference rejection.* We present the new *Censored Marginal (CMaAP) Estimator*, which is threshold-free and optimal. High accuracy estimates of the resolution frequency between two cisoids are obtained in regimes where the joint and ML estimators fail. The CMaAP estimator embodies an *objective rule* for rejection of the Estimation Problem when ill-posed.

1 Strong and Weak Bayesian Inference

The Orthodox School rejects the extended rôle of Probability Theory as the logic of belief in hypotheses, and admits probabilities only on a *random* variable Θ, for which a (conceptual) frequency distribution of outcomes may be postulated. Orthodox and Bayesian inference for *random* parameters are equivalent since, in the former case, a frequency-based prior is proposed on the space $\overline{\Theta}$ of Θ, which modifies the likelihood function $l(\theta \mid \mathbf{d}, \mathcal{I})$, via Bayes' Theorem, to yield the *a posteriori* (AP) inference:

$$p(\theta \mid \mathbf{d}, \mathcal{I}) \propto l(\theta \mid \mathbf{d}, \mathcal{I})\, p(\theta \mid \mathcal{I}) \tag{1}$$

Probabilistic random parameter inference is 'weakly' Bayesian [1] in the sense that it does not require an acceptance of the belief basis of Probability Theory. The distinction between Bayesian and orthodox estimation has been clouded by repeated assertions that Bayesian and 'random' parameter estimation are synonymous [2]. We stress that the inference policies which are being developed here are both Bayesian *and* deterministic. We classify this inference as 'strongly' Bayesian [1] since it *requires* the admission of a prior distribution on the deterministic parameter space $\overline{\Theta}$, something which is excluded by the orthodox philosophy. The currency of the term 'random variable', to refer to the arguments of a probability function, prejudices the Bayesian philosophy and is misleading. We shall prefer the term 'probabilistic parameter' (p.p.) [1] which embraces the diverse nature of such parameters, both fixed and random.

2 Joint (JAP) and Marginal (MaAP) Bayesian Point Inference

Since the prior, $p(\theta \mid \mathcal{I})$, for deterministic θ, is excluded by the orthodox school, the likelihood function remains as the frequentist 'inference machine'. This is a *point* function [3] whose mode is chosen heuristically as an 'appropriate' estimate of θ, being the one for which the observed data are most probable. In contrast, the AP inference (1) is a *measure* function [3] over the space $\overline{\Theta}$ of Θ, from which a variety of point inferences may be drawn in order to estimate Θ, using decision theoretic arguments (e.g. see [1] [4]). The Maximum *a Posteriori* (MAP) estimator is revealed as the minimum risk estimator under a uniform loss hypothesis.

Consider the regression of N data samples d_i onto the General Non-Linear Signal (GNLS) model $g(t)$ [1] [4], formed as the linear combination of m basis functions, $G_k(\cdot)$, each parameterized non-linearly in unknown vector $\boldsymbol{\omega}$:

$$d_i = \sum_{k=1}^{m} b_k G_k(t_i, \boldsymbol{\omega}) + e_i \tag{2}$$

where e_i is an additive noise process. In vector-matrix form, $\mathbf{d} = \mathbf{G}\mathbf{b} + \mathbf{e}$, which is the parametric decomposition of the data set into systematic and random components. In the Estimation Problem, the model order m and the basis function type, $G_k(\cdot)$, are known *a priori*. The deterministic parameters $\boldsymbol{\omega}$ and $\mathbf{b}$ are unknown, and are therefore interpreted as p.p.s in the Bayesian Paradigm. The Maximum Entropy Method (MEM), with second order moment constraints, assigns conservative Gaussian priors on the random p.p. vector $\mathbf{e}$ and on the deterministic p.p. vectors $\boldsymbol{\omega}$ and $\mathbf{b}$. Assuming complex data:

$$\begin{array}{lll} p(\mathbf{e} \mid \boldsymbol{\Sigma}, \mathcal{I}) & \propto \exp\left(-\mathbf{e}^H \boldsymbol{\Sigma}^{-1} \mathbf{e}\right) & \mathbf{e} \in \overline{\mathbb{C}}^N \\ p(\mathbf{b} \mid \mathbf{B}, \mathcal{I}) & \propto \exp(-\mathbf{b}^H \mathbf{B}^{-1} \mathbf{b}) & \mathbf{b} \in \overline{\mathbf{B}} \\ p(\boldsymbol{\omega} \mid \mathbf{W}, \mathcal{I}) & \propto \exp(-\frac{1}{2}\boldsymbol{\omega}^T \mathbf{W}^{-1} \boldsymbol{\omega}) & \boldsymbol{\omega} \in \overline{\mathbf{W}} \end{array} \tag{3}$$

The hyperparameters, $\boldsymbol{\Sigma}$, $\mathbf{B}$ and $\mathbf{W}$, are testable information in the MEM, and are assumed known. These densities may be employed in Bayes' Theorem to reveal the AP inference $p(\boldsymbol{\omega}, \mathbf{b} \mid \mathbf{d}, \boldsymbol{\Sigma}, \mathbf{B}, \mathbf{W}, \mathcal{I})$ for the parameters of $g(t)$, as described in [1] [4].

2.1 JAP Estimation of $\boldsymbol{\omega}$ and $\mathbf{b}$

If estimates of *all* $\boldsymbol{\theta} = (\boldsymbol{\omega}^T, \mathbf{b}^T)^T$ are required, then the complete inference must be maximized *jointly* in all its arguments.

The JAP sufficient functions for estimating ω and $\mathbf{b}$ are revealed as [4]

$$\mathcal{S}_{\mathrm{JAP}}(\omega) = \mathbf{d}^H \Sigma^{-1} \mathbf{G} \mathcal{B}_{\mathrm{JAP}}(\omega) - \frac{1}{2}\omega^T \mathbf{W}^{-1}\omega \quad (4)$$

$$\mathcal{B}_{\mathrm{JAP}}(\omega) = (\mathbf{X}^H\mathbf{X})^{-1}\mathbf{G}^H \Sigma^{-1}\mathbf{d} \quad (5)$$

where $\mathbf{X}^H\mathbf{X} = \mathbf{G}^H \Sigma^{-1}\mathbf{G} + \mathbf{B}^{-1}$. (4) is maximized at ω_{JAP} and the accompanying JAP estimate of $\mathbf{b}$ is found by substituting ω_{JAP} into $\mathcal{B}_{\mathrm{JAP}}(\cdot)$.

2.2 MaAP Estimation of ω

Since $p(\omega, \mathbf{b} \mid \mathbf{d}, \Sigma, \mathbf{B}, \mathbf{W}, \mathcal{I})$ is a measure function, Probability Theory permits any of its parameters to be marginalized out by integration. Performing m integrations over $\overline{\mathrm{B}}$, the sufficient function for marginal estimation of ω is found to be [4]

$$\begin{aligned}\mathcal{S}_{\mathrm{MaAP}}(\omega) &= \mathcal{S}_{\mathrm{JAP}}(\omega) - \ln|\mathbf{G}^H\Sigma^{-1}\mathbf{G} + \mathbf{B}^{-1}| \\ &= \mathcal{S}_{\mathrm{JAP}}(\omega) + \mathcal{D}(\omega) \quad (6)\end{aligned}$$

where $\mathcal{S}_{\mathrm{MaAP}}(\omega)$ is also known as the marginal *support function* for ω [1]. The term $\mathcal{D}(\omega)$ has been ignored in the literature on Bayesian parameter estimation, e.g. [5], principally because the performance of the marginal estimator has only been investigated in unstressful regimes, where $\mathcal{D}(\omega)$ is inactive. We shall now show that it is dominant in stressful regimes and confers *robustness* on the point inference endeavour.

3 Ockham Parameter Inference (OPI)

If the data $\mathbf{d}$ are dominantly random, then the fitting of a systematic model, $\mathbf{s} = \mathbf{Gb}$, violates Ockham's Desideratum of Simplicity. This requires that observed variations be treated as random unless there is support for the systematic hypothesis [1]. Consider the null hypothesis $\mathcal{I}_0 \equiv$ '$\mathbf{Gb} = \mathbf{e}$' which asserts that the model has been fitted erroneously to noise. The inference for ω, conditioned on $\mathcal{I}_0$, is given by [1]

$$\ln P(\omega \mid \Sigma, \mathbf{B}, \mathcal{I}_0) = -\ln|\mathbf{G}^H\Sigma^{-1}\mathbf{G} + \mathbf{B}^{-1}| = \mathcal{D}(\omega)$$

assuming a diffuse prior for ω. Thus, $\mathcal{D}(\omega)$ measures the hypothesis that we have fitted randomness with determinism in accepting the model (2), and, as such, it is a direct metric of the degree to which the assumed model contravenes Ockham's Rule. We refer to it as the *Ockham Parameter Inference (OPI)*. If $\mathrm{rank}(\mathbf{G}) = q < m$ at some $\omega = \omega_0$, then linear dependence has set in among $m - q$ of the basis functions, and the complexity of the model has diminished. Since $\mathcal{D}(\omega) \to \infty$ in this case (assuming a dominant likelihood [4]), we conclude that *the OPI overwhelmingly supports a reduction in model complexity.*

3.1 An Absolute Criterion for Inference Rejection

If $\omega_0 \in \overline{\mathrm{W}}$, the space of ω, then the OPI dominates $\mathcal{S}_{\mathrm{JAP}}(\omega)$, and ω_0—being the redundancy inference—will be returned in all data regimes. However, *it is inconsistent to test for the redundancy of an accepted prior model hypothesis $\mathcal{I}$.* A regularizing parameter, $\omega_l > \mathbf{0}$, which expresses our belief in the closest approach to redundancy, must be proposed as a hyperparameter of the inference. The following features of (6) establish the Ockham-based behaviour of the marginal inference policy for $\omega_l > \mathbf{0}$:

(i) The data enter the inference only via $\mathcal{S}_{\mathrm{JAP}}(\omega)$;

(ii) $\mathcal{S}_{\mathrm{JAP}}(\omega)$ dominates the inference when the data strongly support the model, and estimates of ω may therefore be inferred;

(iii) In the absence of data support, $\mathcal{D}(\omega)$ dominates, returning an inference of model redundancy. The estimation objective is rejected.

Data support is strong when (i) the basis functions are well resolved in the data and (ii) the SNR is large. The marginal inference inherently 'calibrates' this support against the OPI. Estimation is only acceptable when $\mathcal{S}_{\mathrm{JAP}}(\omega)$ dominates $\mathcal{D}(\omega)$. This is a quantitative statement of Ockham's Rule, and furnishes an *absolute* criterion for rejection of parameter inference from $\mathbf{d}$, conditioned on $\mathcal{I}$. The criterion is regularized by the hyperparameters Σ, $\mathbf{B}$, $\mathbf{W}$ and ω_l of the hypothesis. Attempts to inspect $\mathcal{S}_{\mathrm{JAP}}(\omega)$ in isolation, in order to determine model redundancy, are invariably heuristic.

3.2 The Necessity of Marginalization

JAP inference is conditioned upon an absolute acceptance of the model $\mathcal{I}$, and may therefore incur the overfitting of the data. Thus, *JAP (and ML) estimation are ill-posed in stressful regimes.* They suffer a threshold effect with increasing noise, due to the chaotic proliferation of viable point estimates, as $\mathbf{d}$ recedes from the signal subspace $\overline{\mathrm{G}}$ [1].

Marginalization over $\overline{\mathrm{B}}$ (6) measures our belief, not in particular points of the parameter space, but in the *entire* subspace spanned by the model basis functions at a particular ω. Consequently, $p(\omega \mid \mathbf{d}, \Sigma, \mathbf{B}, \mathbf{W}, \mathcal{I})$ employs a *volume displacement metric* of the signal subspace $\overline{\mathrm{G}}$ from $\mathbf{d}$ [1]. This metric is functional on (i) the closest approach of $\overline{\mathrm{G}}$ to $\mathbf{d}$ (measured by $\mathcal{S}_{\mathrm{JAP}}(\omega)$) and (ii) the volume element entrapped by the basis functions of $\mathbf{G}$ (measured by $\mathcal{D}(\omega)$). Smaller volume elements imply increased basis function redundancy and receive greater support via (6). Thus, $\mathcal{D}(\omega)$ behaves as an "intra-hypothesis Ockham factor" [4].

We conclude that joint inference is inadequate as a policy for admitting the Simplicity Desideratum, which should guide all attempts at model-based inference. The Desideratum should *always* be active in stressful regimes, *necessitating the adoption of the marginal policy.* We reject a number of fallacies which often accompany the interpretation of marginalization in the Bayesian Paradigm [1]:

(i) It does not 'get rid of' $\mathbf{b}$, since $p(\omega \mid \mathbf{d}, \Sigma, \mathbf{B}, \mathbf{W}, \mathcal{I})$ remains conditioned on $\mathcal{I}$, which implies both $\mathbf{b}$ and ω, via (2);

(ii) The space in which numerical optimization is required is not reduced by marginalizing over $\overline{\mathrm{B}}$. In contrast, the numerical load is *increased* because of the requirement to calculate $\mathcal{D}(\omega)$ in (6);

(iii) Marginal estimates cannot be combined to form a joint estimate of the complete parameter set. A policy which is consistent with the inference question should be adopted.

4 Bayesian Superresolution of Two Complex Tones

Consider the problem of estimating the resolution frequency ω (scalar) in the following model:

$$g(t) = e^{j\omega_1 t}\left[b_1 + b_2 e^{j\omega t}\right] \quad (7)$$

Assuming ω_1 known, and adopting the uncorrelated priors $\boldsymbol{\Sigma} = \sigma^2\mathbf{I}_N$ and $\mathbf{B} = \delta^2\mathbf{I}_m$, and a diffuse prior for ω, then, from (3), (4) and (5):

$$\mathcal{S}_{\mathrm{JAP}}(\omega) = \mathbf{f}^H \boldsymbol{\mathcal{B}}_{\mathrm{JAP}}(\omega) \quad (8)$$

where $\omega \geq \omega_l > 0$, $\boldsymbol{\mathcal{B}}_{\mathrm{JAP}}(\cdot) = \left(\mathbf{G}^H\mathbf{G} + \rho\mathbf{I}_2\right)^{-1}\mathbf{f}$ and $\rho = \frac{\sigma^2}{\delta^2}$. $\mathbf{f} = \{\ F(j\omega_1),\ \ F[j(\omega_1+\omega)]\ \}^T$, being samples of the DFT of $\mathbf{d}$ at candidate frequencies ω_1 and $\omega_1 + \omega$ [1] [4]. The OPI is given by

$$\mathcal{D}(\omega) = -\sigma^2 \ln\left[(N+\rho)^2 - \frac{\sin^2\left(\frac{N\omega T}{2}\right)}{\sin^2\left(\frac{\omega T}{2}\right)}\right] \quad (9)$$

which depends only on the difference frequency ω, as is appropriate for a function which measures the 'degree of redundancy' in the model. It is plotted in Fig. 1 for $\omega \geq \frac{1}{6}$ DFT bins, where a DFT bin-width is defined to be $\frac{\omega_s}{N}$, ω_s the sampling frequency. The OPI exhibits the characteristic redundancy peak below the first DFT bin, manifesting the Ockham support for the coalescence of the cisoidal basis functions in (7), which occurs at $\omega = 0$. The OPI is minimized 'on-bin', where the cisoidal basis functions are orthogonal.

5 Robust Sub-Threshold Inference using the Censored Marginal (CMaAP) Estimator

A point inference, ω_{MaAP}, drawn from $\mathcal{S}_{\mathrm{MaAP}}(\omega)$, is classified as follows:

(i) If the data-based support exceeds the support for redundancy, ω_{MaAP} should be used to estimate ω;

(ii) If the OPI support for redundancy exceeds the data support, ω_{MaAP} should be used to reject the model.

In a worst case scenario, a point inference, ω_{MaAP}, is obtained when

$$\mathcal{S}_{\mathrm{MaAP}}(\omega_{\mathrm{MaAP}}) = [\mathcal{S}_{\mathrm{MaAP}}(\omega_l)]^+ \quad (10)$$

i.e. the overall support for ω_{MaAP} just exceeds the support for redundancy at ω_l. We define the *censoring threshold*, ω_c, for estimation, to be that frequency whose redundancy support is equal to its data support in this worst case scenario:

$$\mathcal{S}_{\mathrm{JAP}}(\omega_c) - \mathcal{S}_{\mathrm{JAP}}(\omega_l) = \mathcal{D}(\omega_c) - \mathcal{D}\left(\frac{\omega_s}{N}\right) \quad (11)$$

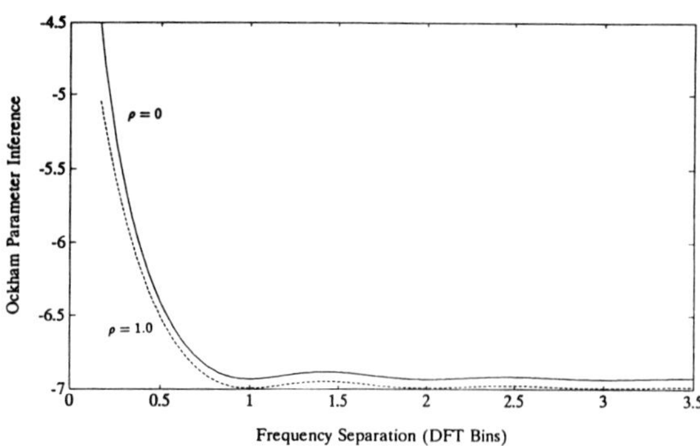

Figure 1: The 2-Cis OPI ($\sigma^2 = 1$; $N = 32$).

Eliminating $\mathcal{S}_{\mathrm{JAP}}(\cdot)$ from (6), (10) and (11), we obtain

$$\mathcal{D}(\omega_c) = \frac{1}{2}\left[\mathcal{D}(\omega_l) + \mathcal{D}\left(\frac{\omega_s}{N}\right)\right] \quad (12)$$

In the CMaAP estimation policy, we reject an estimate, and the model $\mathcal{I}$, if $\omega_{\mathrm{MaAP}} < \omega_c$. An estimate $\omega_{\mathrm{MaAP}} \geq \omega_c$ is accepted, being data-dominated.

(i) ω_c depends on the *fixed* parameters of the hypothesis $\mathcal{I}$, namely m, N, σ^2, ρ, ω_s and ω_l, but *not* on the parameters of $\mathbf{d}$, i.e. ω and SNR. It therefore constitutes an *objective* criterion for censoring an estimate;

(ii) Since $\omega_{\mathrm{CMaAP}} \not< \omega_c$, then ω_c is the smallest estimable frequency resolution using this policy;

(iii) The rule is conservative. If $\mathcal{S}_{\mathrm{MaAP}}(\omega_{\mathrm{MaAP}}) > \mathcal{S}_{\mathrm{MaAP}}(\omega_l)$ for $\omega_{\mathrm{MaAP}} < \omega_c$, then its data support may still be dominant. The 'hard' threshold ensures that *all* CMaAP estimates are data-dominated.

5.1 Performance of the CMaAP Estimator

We shall take $N = 32$, $f_s = 32$ Hz, $\sigma^2 = 1$ and $\omega_1 = 2.5$ DFT bins. The lower limit on ω is chosen to be $\omega_l = \frac{1}{600}$ DFT bins. We take $\rho = 0$, so that JAP and ML estimation are synonymous. The censoring threshold ω_c is determined by substituting (9) into (12) yielding the following relation:

$$\frac{\sin^2(\pi\omega_c)}{\sin^2\left(\frac{\pi\omega_c}{N}\right)} = N^2\left[1 - \sqrt{1 - \frac{\sin^2\left(\frac{\pi}{600}\right)}{N^2\sin^2\left(\frac{\pi}{600N}\right)}}\right] \quad (13)$$

The solution is found at $\omega_c = \frac{19}{600}$ DFT bins. 2000 Monte Carlo trials were undertaken at each tested value of SNR and ω. We note the following (Fig. 2):

(i) The JAP/ML and MaAP policies exhibit thresholds. The Threshold SNRs increase as ω diminishes;

(ii) The deleterious consequences of ignoring Ockham's Rule are manifested in the proliferation of outlier estimates, and failure of the JAP/ML policy, below threshold;

(iii) The MaAP performances bifurcate from the JAP/ML curves, and the MaAP bias falls to -100%, as the OPI becomes more dominant with decreasing SNR. The variance exhibits a turning point and decreases in very noisy regimes;

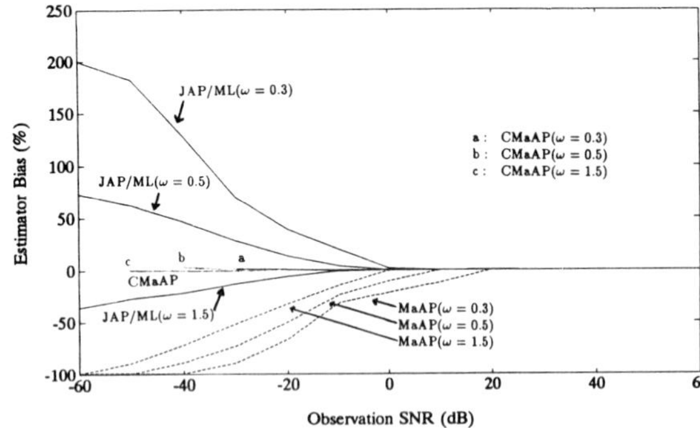

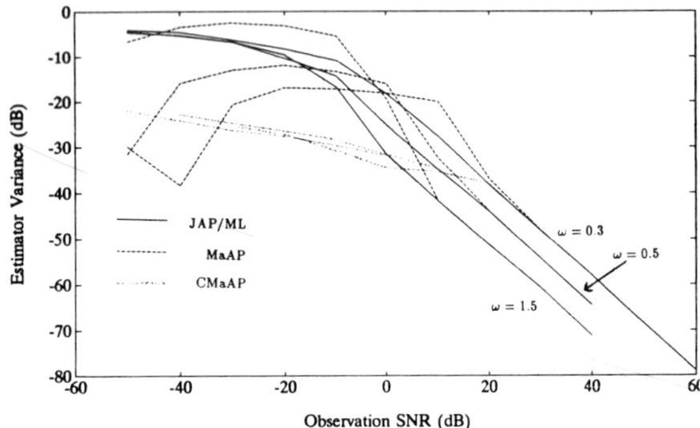

Figure 2: Performance Characteristics for the JAP/ML, MaAP and CMaAP Bayesian Estimators.

(iv) The CMaAP estimator behaves as an 'outlier detector' [1], and is threshold-free. It is virtually unbiased below the JAP/ML threshold, and the rate of increase of the CMaAP variance with falling SNR is retarded;

(v) The danger—which attends the JAP/ML policy—of estimation below the noise threshold is eliminated in the CMaAP policy. Estimates which exceed ω_c are highly confident;

(vi) As the SNR decreases, so does the probability of being able to make an estimate under the CMaAP policy. The probability also diminishes with decreasing ω, as indicated in Fig. 3. There were no data-dominant estimates of $\omega = 0.3$ for SNR ≤ -40 dB over the 2000 trials. This sets an upper bound on the noise level in the data for which estimation is possible at $\omega = 0.3$. The bound increases as ω increases;

(vii) The odds in favour of being able to make a CMaAP estimate below the JAP/ML threshold are good: e.g. there is a 50% chance of being able to estimate $\omega = 0.3$ bins at -15 dB, with zero bias and an estimator variance of -28 dB. This stressful regime incurs a JAP/ML bias of $+34\%$ with a variance of -9 dB.

6 Asking the Right Question

The inference policy which we adopt must be consistent with the nature of the question being asked. Questions about the

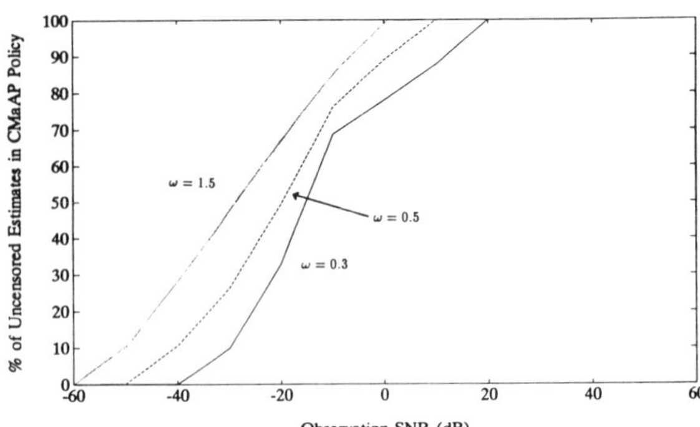

Figure 3: Probability of Estimation in the CMaAP Policy.

values of the global parameter set are ill-posed in stressful regimes and must be avoided. This encourages the adoption of marginal inference questions which admit Ockham's Rule. The admission of Ockham within the ambit of Estimation has not received attention before. It is known to be operative within the evidence framework for Model *Selection* [5]. However, we have shown [1] that it will be circumvented, once again, if the inference question does not induce an appropriate partition of $\mathcal{I}$ for which model complexity is measured, a fact which is overlooked. Techniques for simultaneous model selection *with* estimation have been developed [1] and will be reported shortly.

Marginalization over deterministic parameter spaces is the exclusive territory of the strong Bayesian. Marginal inference uniquely engrains the stresses inherent in model-based signal processing, trading the belief in the model as 'truth' against the belief in the model as 'error'. The embodiment of the Simplicity Desideratum, and the ability to undertake robust estimation below threshold, are powerful incentives for the adoption of the strong Bayesian philosophy. Marginalization deserves a pre-eminence in Inference, since it confers a reductive aptitude on the inductive task.

References

[1] A. P. Quinn. *Bayesian Point Inference in Signal Processing.* PhD thesis, Cambridge University Engineering Department, 1992.

[2] S. M. Kay. *Modern Spectral Estimation: Theory and Application.* Prentice Hall Signal Processing Series, 1988.

[3] B. de Finetti. *Theory of Probability,* volume 2. Wiley, 1975.

[4] A. P. Quinn. The bias and variance of Bayesian estimators in the superresolution of signal parameters. *IEEE Int. Conf. on Acoust., Sp. and Sig. Proc. (ICASSP),* San Francisco, 1992.

[5] G. L. Bretthorst. *Bayesian Spectrum Analysis and Parameter Estimation.* Springer-Verlag, 1989.

SIGNAL PROCESSING VI: Theories and Applications
J. Vandewalle, R. Boite, M. Moonen, A. Oosterlinck (eds.)

EFFICIENT RECURSIVE "ML" BEARING ESTIMATION IN NON STATIONNARY CONTEXT

P.LARZABAL* and H.CLERGEOT†

* THOMSON-CSF, Division RGS, Service traitement du signal, 66 rue du fossé blanc 92231 Gennevilliers FRANCE---------Tel (1)-46-13-20-00---Fax 46-13-25-55
†LESIR,URA CNRS n° 1375, Ecole Normale Supérieure, 61 avenue du président Wilson, 94235 CACHAN CEDEX, FRANCE--Tel (1)-47-40-21-19---Fax 47-40-20-74

ML or approximate ML may be considered as the upper state of the art in high resolution methods, but they suffer from an extensive multidimensionnal search. Starting from a crude initialization with a robust low resolution method, the "recursive" ML presented in [5] avoids this extensive multidimensionnal search and provides an opportunity for extension to moving sources traking. Emphasis in this paper is on the non stationnary case.

1. INTRODUCTION

1.1 The problem addressed

Maximum likelihood is presently recognized as the most promising approach in array processing for high resolution bearing estimation, including such important topics as array perturbations. In a stationnary context, three very similar algorithms, Maximum likelihood, Approximate Maximum Likelihood or Weighted Subspace Fitting methods provide very similar practical implementations of the same ML principle [1,2,3]. We have recently proposed in [4] a recursive formulation with moving sources, assuming a known, fixed number of sources and a good initialization. In [5] our interest was focused on these two last remaining fondamental topics (sources detection and initialization), but it was restricted to stationnary sources. We propose here to extend this concept to the case of maneuvering targets which can appear or disappear .

1. 2 Existing solutions

Presently existing algorithms amount to application of stationnary methods on adjacent intervals short enough to neglect the sources deplacement. There is a compromise to find between an interval long enough to resolve closely spaced sources but short enough to minimize non stationnarity effects.

1.3 Main features of our algorithm

We propose a time recursive algorithm with an update on a small number of snapshots (even with one snapshot). In [4] the source's number was known, here it is not, so that the algorithm is conceived for simultaneous estimation and detection with the ability to detect a new source and update M at each iteration. This enables the use of a low resolution method for initialization with a value $M<M_0$. We present an implementation for moving sources with occasional on /off power switching.

2 METHOD DESCRIPTION IN A STATIONNARY CONTEXT

2.1. Bearing estimation algorithm

2.1.1. Notations and Model

No assumption is made about array's geometry. We suppose an N dimensionnal observation vector in the form:

$$\underline{X}(t) = \underline{Y}(t) + \underline{N}(t) \qquad (1)$$

$\underline{N}$ is an additive noise, which for simplicity, is assumed to be a circular complex white noise with covariance $\sigma^2.I$. The steering vector associated to a source with azimut θ is $\underline{S}(\theta)$. Let's denote $\underline{G}(t)$ the deterministic unknown sources amplitude vector. Expression (1) becomes:

$$\underline{X}(t) = S.\,\underline{G}(t) + \underline{N}(t) \qquad (2)$$

where

$$S = [S(\theta_1), S(\theta_2), \ldots, S(\theta_M)] \qquad (3)$$

2.1.2.Maximum Likelihood Method

The loglikelihood for this model can be easily calculated from a set of T independant snapshots:

$$L\left(\overset{\Delta}{\theta}, \overset{\Delta}{G}\right) = -N.T.\log(\pi.\sigma^2) - \frac{1}{\sigma^2}.\sum_{t=1}^{T} |X(t) - S.G(t)|^2 \qquad (4)$$

where $\hat{\theta}, \hat{G}$ are the candidate values for vectors $\underline{\theta}$ and $\underline{G}$. Separate maximisation with respect to the amplitude vectors $\underline{G}(t)$ yields:

$$L_0\left(\hat{\theta}\right) = \mathrm{Tr}\left[\hat{\Pi}_N \hat{R}_X\right] \tag{5}$$

with $\hat{R}_X$ the estimated covariance matrix and $\hat{\Pi}_N$ the projector onto noise subspace. If **P** is the sample covariance matrix of the amplitudes, we have:

$$\begin{aligned} E\left[\hat{R}_X\right] &= S.P.S^{\dagger} + \sigma^2.I \\ &= R_Y + \sigma^2.I \end{aligned} \tag{6}$$

where R_Y stands for the noiseless covariance. Note that

$$R_Y = S.P.S^{\dagger} = \Pi_S.E\left[\hat{R}_X - \sigma^2.I\right].\Pi_S \tag{7}$$

2.1.3. Subspace Fitting Methods

According to standard EVD technics, the signal subspace is identified as the space spanned by the eigenvectors associated to the largest eigenvalues of R_X.

Let $\hat{\Pi}_S = \hat{E}_S.\hat{E}_S^{\dagger}$ be the corresponding signal projector. Then R_Y may be estimated by:

$$\hat{R}_Y = \hat{\Pi}_S.\left[\hat{R}_X - \sigma^2.I\right].\hat{\Pi}_S \tag{8}$$

But for the statistical deviation of the noise eigenvalues from σ^2,we obtain:

$$\hat{R}_X \cong \hat{R}_Y + \sigma^2.I \tag{9}$$

Substitution in (5) results in a first loglikelihood approximation

$$L_0\left(\hat{\theta}\right) = \mathrm{Tr}\left[\hat{\Pi}_N.\hat{R}_Y\right] + (N-M).\sigma^2 \tag{10}$$

This expression fits an unifying presentation of bearing estimators recently proposed by Viberg[3], by minimization of $\mathrm{Tr}\left[\hat{\Pi}_N.\hat{E}_S.W.\hat{E}_S^{\dagger}\right]$, where, in (10), $W = \hat{\Lambda}_S - \sigma^2$. For high SNR, this is very close to the "optimal" weighting matrix derived by Viberg:

$$W_{opt} = \left(\hat{\Lambda}_S - \sigma^2\right).\left(1 - \sigma^2.\hat{\Lambda}_S^{-1}\right) \tag{11}$$

2.1.4 Approximate Maximum Likelihood Method (AMLM)

In [2], we demonstrate that (10) could again be locally approximated by:

$$\mathrm{Tr}\left[\hat{\Pi}_N.\hat{R}_Y\right] = \mathrm{Tr}\left[\hat{\Pi}_N.\hat{R}_Y\right] = \mathrm{Tr}\left[\hat{\Pi}_N.\hat{S}.P.\hat{S}^{\dagger}\right] \tag{12}$$

In this equation **P** will be replaced by an estimate obtained recursively from the last $\hat{S}$ and from observations,according to:

$$\hat{P} = \left(S^{\dagger}.S\right)^{-1}.S^{\dagger}.\hat{R}_Y.S.\left(S^{\dagger}.S\right)^{-1} \tag{13}$$

2.1.5 Corresponding Time Recursive Algorithms

The three previous methods are asymptotically efficient, with the Hessian at the convergence point equal to the Fisher matrix. From simulations it appears that algorithms taking advantage of EVD may be more robust at low SNR that those without EVD. They are very similar concerning gradient and Hessien computation.

We address the problem of real time localization with sources position update every T snapshots. In a stationnary context, one could update the value of R_X integrated over all the past and perform the corresponding ML estimation using a Newton Gauss algorithm with starting positions provided by the previous estimation. This was done in [5]. We have also proposed a recursive update, the advantage of which is to be extensible to the moving sources case [4,6]. It may be interpreted as a Kalman filter generalisation [6].

The remaining problem is in initialization and sources number selection.

2.2. First bearing initialization

Roughly speaking sources should be initialized within one beamwidth from there true location. As pointed out in the introduction they don't need being resolved at this step. For decorrelated sources, a good initialization point could be obtained by MUSIC. In our simulations, we have prefered conventional beamforming, which proves to be very robust, even for correlated sources and for a very small snapshots number.

We also need an initial guess for the bearings covariance: We use a diagonal matrix with standard deviation equal to about one quarter beamwidth. This value is not critical.

2.3. Recursive sources detection

2.3.1. Principle

Along the same idea than in [7], when M sources have been identified with there estimated position $\underline{\theta}$, we construct the corresponding projector $\Pi_N(\underline{\theta})$ on the noise subspace, so that the projection

$$\underline{Z}(t) = \Pi_N(\underline{\theta}).\underline{X}(t) \tag{14}$$

should be residual noise only. The first point is then to

perform a "sphericity test "on $\underline{Z}$, based on $\Pi_N(\underline{\theta}) \cdot R_X \cdot \Pi_N(\underline{\theta})$ eigenvalues.If the noise only hypothesis is false , M is incremented.

When a new source is detected, the second point will be to provide a correct initialization for the sources (position and covariance): this is obtained by a modified beamforming method in the noise subspace.

2.3.2. Projection Sphericity Test

This is reduced to a test of "equality"of the non zero eigenvalues of $\Pi_N(\underline{\theta}) \cdot R_X \cdot \Pi_N(\underline{\theta})$. This is a very classical topic. We use a new criterium, giving a better insigth to the problem.

Due to statistical fluctuations, even under the noise only hypothesis the eigenvalues are not equal . When sorted in order of non increasing values, they approximatively fit an exponential law, with a decay rate which can be evaluated, using the first and second order of the distribution [8]. The next figure illustrates this point. The natural eigenvalues logarithm is reported on the graph. Of course, one source can be detected in the upper graph.

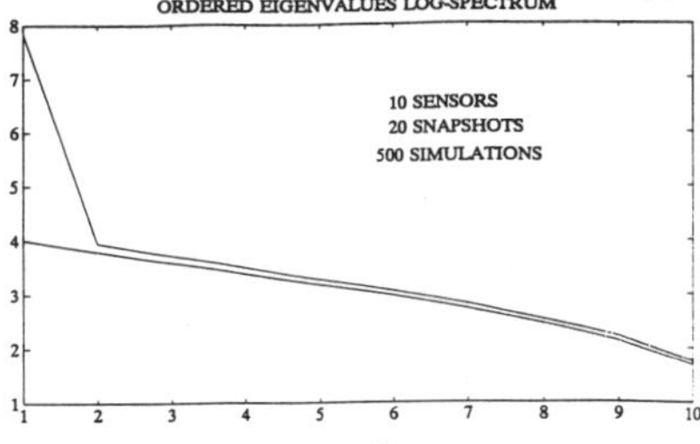

Our test is based on detection of any decay rate departure to the theoretical law[8]. When detection occurs, we also record the eigenvectors U_S corresponding to the maximum eigenvalue of R_Z, which will be used for source bearing initialization.

2.3.3 New Source Initialization

Let $\underline{U}(\theta) = \Pi_N.S(\theta)/\mathrm{norm}(\Pi_N.S(\theta))$ be our modifed steering vector. The new source bearing is located by a "modified beamforming" method as the maximum of $f(\theta) = |\underline{U}(\theta)^{\dagger}.\underline{U}_S|^2$. See [5] for more detail. The recursive estimation performance is now illustraded .The following simulations show the good behavior of our algorithm in difficult cases which, to our knowledge can't be resolved by existing methods. An uniform linear array of five elements with half wavelength is used. The beamwidth is 25°. In fig (2), 3 non equipowered signals with 100% correlation arrive from -30°,5°,10° relative to the array broadside. At each iteration only 5 snapshots are used.

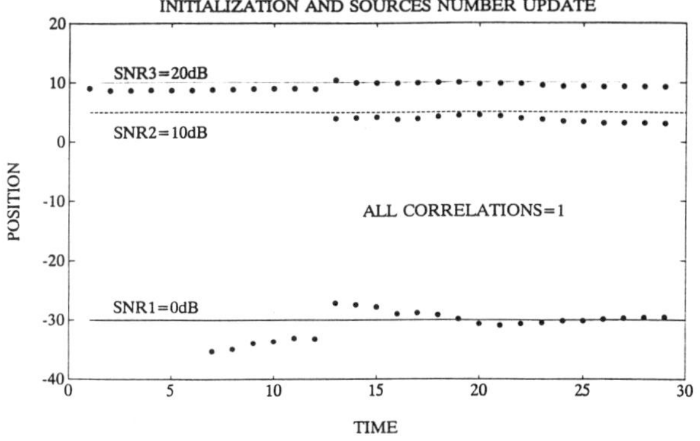

3 TRACKING OF MANOEUVERING TARGETS

3.1 Introduction

In many practical problems of tracking a maneuvering target, a simple kinematic model can fairly accurately describe the target dynamic characteristics for a wide class of maneuvers. However, since the target can exhibit a wide range of dynamics caracteristics, no fixed Kalman filter can be matched to estimate, to the required accuracy, the states of the target for every specific maneuver. For a complete solution of the problem it is necessary to detect the instant of initiation or a change of the maneuver and estimate the proper value of the Kalman filter process noise covariance matrix. A simplified target tracking scheme based on an adaptive Kalman filter is presented, it includes an estimation on line of sources speed and an original adaptive estimation of the modelling noise. The scheme is suitable for tracking after targets which may have substantial different dynamic characteristics. The effectivness of the proposed filter is demonstrated by extensive computer simulation.

Dealing with manoeuvering targets implies a Kalman like recursive algorithm with adaptive estimation of the targets dynamics (see 3.2). Possible source disappearing will be detected by estimated power monitoring (see 3.3). Contrary to stationnary case, new sources detection will use residual noise correlation smoothing on a finite sliding time intervall and not on all the past

3.2 Recursive algorithm

Assume that evolution model is given by:

$$\underline{\theta}_{t+1} = \underline{\theta}_t + \underline{u}_t + \underline{\varepsilon}_t \qquad (15)$$

$\underline{\varepsilon}_t$ account for "fast" decorrelated position increments (random walk) and $\underline{u}_t$ for long term average motion (see [5] for more details).

$\underline{u}_t$ will be estimated by prediction from past observations by smothing of $\widehat{\underline{\theta}}_{t'} - \widehat{\underline{\theta}}_{t'-1}$ .The estimation of $R=E\{\varepsilon.\varepsilon^{\dagger}\}$ needs some particular attention as we'll see now.

The loglikelihood is approximatively quadratic in $\underline{\theta}$

$$L(\underline{\theta}) = \frac{1}{2}.(\underline{\theta} - \underline{\theta}_{ML})^{\dagger}.\ H\ .(\underline{\theta} - \underline{\theta}_{ML}) + C^{te} \qquad (16)$$

H is the Hessien. The gradient in $\underline{\theta}$ is given by :

$$\nabla = H.(\underline{\theta} - \underline{\theta}_{ML}) \qquad (17)$$

At iteration t, the a priori estimate is $\underline{\theta}_t$.

$$
\begin{aligned}
cov(H^{-1}.\nabla_t) &= cov(\underline{\theta}_t - \underline{\theta}_{ML}) \\
&= \left(cov(\tilde{\underline{\theta}}_t) + cov(\underline{\theta}_{ML})\right) \\
&= \left(R_t + \Gamma_t + H^{-1}\right)
\end{aligned}
\qquad (18)
$$

R can easily be obtained from this last equation. However, due to statistical fluctuation a smoothed estimation will give better result:

$$\widehat{R}_t = \widehat{R}_{t-1} + \mu . H^{\neq}\left\{\nabla_t . \nabla_t^{\dagger} - H(\Gamma_{t-1} + R_{t-1})H - H\right\}H^{\neq} \quad (19)$$

The next simulation confirms good results obtained with this estimator. Two completely correlated sources have stochastic walk of 1.7°. Fig(3) shows the standard. deviation given by the estimated R diagonal elements.

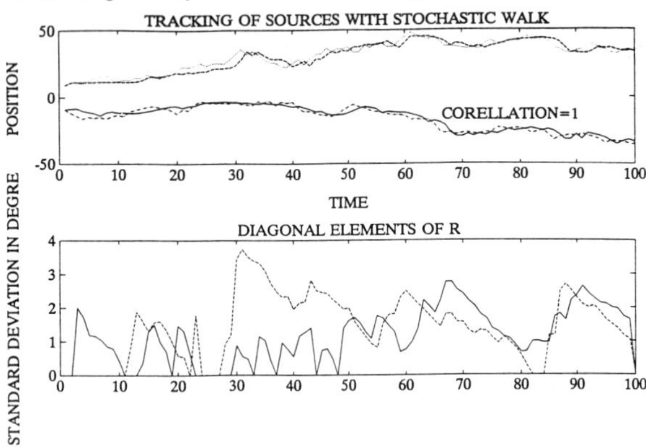

3.3 Source disappearing

There are detected by monitoring of the sources power matrix according to (13), (see simulations).

3.4 Modification of new sources detection

Due to non stationnarity, only finite averaging memory can be used. Detection test will work with an average of the residual noise correlation matrix R_Z:

$$\widehat{R}_Z(t) = \widehat{R}_Z(t-1) + \mu\left\{\Pi_N(t).R_X(t).\Pi_N(t) - \widehat{R}_Z(t-1)\right\} \quad (20)$$

By adjusting the coefficient μ , the sliding intervall length will be regulated.

4 SIMULATIONS RESULTS

In the next simulation, there are 2 sources with 100% correlation and constant positions at -2° and 2° relative to the array broadside but with occasionnal on / off switching of the power. The figure (4) shows the real positions (in dotted lines) and the estimated positions. On figure (5), we can see the corresponding power's monitoring which enable us to detect a disappearance.
In the figure (6) we have two 100% correlated targets which move parallel to linear antenna and cross them on array broadside. The second source really appears at iteration number 11 and disappears at number 21. The adaptive Kalman filter provides excellent results.

fig(4)

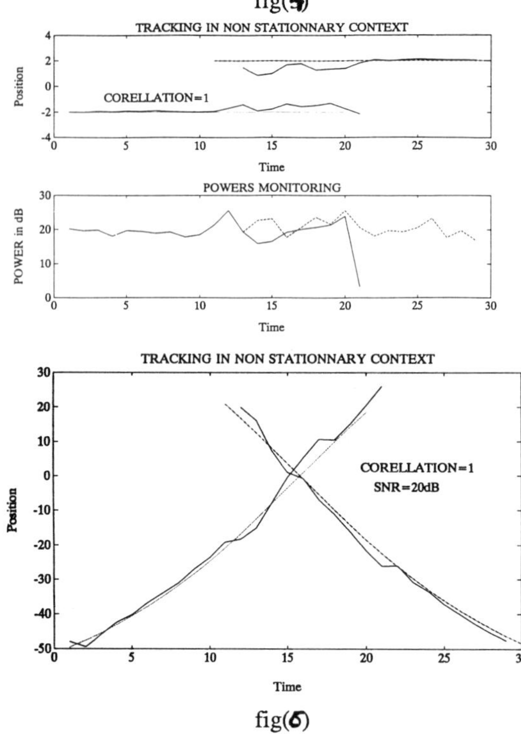

fig(6)

5 CONCLUSIONS

We have developped a new detection / bearing estimation algorithm which gives good results in difficult non-stationnary contexts. Sources can exhibit a wide range of dynamic caracteristics.

REFERENCES

[1] P.Stoica, A.Nehorai," MUSIC, Maximum likelihood and Cramer Rao bounds" , IEEE Trans ASSP 37 n° 5 May 1989, pp 720_741.

[2] H.Clergeot and al "A new maximum likelihood method for estimation of correlated sources-Comparison with existing methods " Proc of Digital Signal Processing, pp71-75; Florence 87

[3] M.Viberg, V.Ottersten "Sensor array processing based on subspace fitting" IEEE trans ASSP-39, May 1991.

[4] O.Michel and H.Clergeot "Multiple sources tracking using a high resolution method" Proc of ICASSP 91, Toronto, pp 1277-1280

[5] P.Larzabal and H.Clergeot "Recursive ML bearing estimation: Initialization and sources number update" ICASSP 92 San Fransisco

[6] H.Clergeot, O.Michel "Généralisation du filtrage de Kalman au cas de la localisation de sources mobiles par une méthode haute résolution" Proc of GRETSI 91 ,Juan les pins

[7] R.H.Roy "ESPRIT, Estimation of signal parameters via rotational invariance techniques" PhD thesis, Stanford university, August 1987

[8] O.Michel,P.Larzabal,H.Clergeot "Critères de détection du nombre de sources coréllées pour les méthodes "HR" en traitement d'antenne" Proc of GRETSI 91, Juan les pins

SIGNAL PROCESSING VI: Theories and Applications
J. Vandewalle, R. Boite, M. Moonen, A. Oosterlinck (eds.)

EFFICIENT IMPLEMENTATION OF NONLINEAR VOLTERRA SYSTEMS VIA LAGUERRE ORTHOGONAL FUNCTIONS

S. A. ALshebeili[1], B. G. Mertzios, and A. N. Venetsanopoulos

[1]Department of Electrical Engineering, King Saud University, Riyadh, Saudi Arabia
Department of Electrical Engineering, University of Toronto, Toronto, Canada M5S 1A4

ABSTRACT: In this paper we present an efficient implementation structure for nonlinear Volterra-type systems. The implementation structure that we propose is based on orthogonal expansion of kernels of the nonlinear system using discrete Laguerre orthogonal functions. The resulting expansion permits the use of current implementation structures available in the literature, with a substantial reduction in the dimensionality of kernels. To illustrate the idea two implementation structures are discussed, namely the direct form implementation and the matrix decomposition based implementation.

I. Introduction

In recognizing the important role played by the discrete nonlinear Volterra-type systems in solving many problems of digital signal processing and the pervasive advances in microelectronic technology, several authors [1-7] proposed different realizations/implementations for the discrete Volterra series taking into account the great opportunities offered by VLSI technology.

The appearance of VLSI has resulted in enormous possibilities for the implementation of sophisticated algorithms of high complexity, in a variety of important applications, by using low-cost, very efficient, high-speed special purpose hardware. The emphasis on minimizing the number of dynamic elements (registers, adders, multipliers, etc.) of a single device has been replaced by a new set of criteria based on considering modularity and regularity of structures and reducing the data throughput delay.

In this paper we study the implementation problem of nonlinear systems represented by Volterra-type models. The implementation structure that we propose is based on orthogonal expansion of the kernels of the nonlinear system using discrete Laguerre orthogonal functions (DLOF) given by Gottlien [8]. This type of expansion, however, resembles Wiener's expansion [9] proposed in the continuous case.

The main features motivated the use of DLOF over other orthogonal functions are their time and frequency domain properties. In the time domain the p-order Laguerre function $l_p(n)$ is zero for $n < 0$, and hence it is possible to construct a causal linear time-invariant (LTI) system with unit impulse response equal to $l_p(n)$. In the frequency domain the transfer function of $l_p(n)$ takes the form of a cascade connection of a first-order low-pass system with p identical first-order all-pass systems [10]. The impulse responses produced at the outputs of the cascade chain are simply the set of DOLF $l_0(n), l_1(n), \ldots, l_p(n)$.

In the next section we introduce Laguerre representation of Volterra kernels, and in section III we use this representation to model an ith-order Volterra operator. The direct and matrix decomposition implementation structures of the expanded operator are also presented and discussed.

II. Laguerre Representation of Volterra Kernels

The input-output relationship of a discrete kth-order Volterra system is given by

$$\begin{aligned} y(n) &= h_0 + \sum_i^k \sum_{\tau_1,\tau_2,\ldots,\tau_i} h_i(\tau_1,\ldots,\tau_i)x(n-\tau_1)\cdots \\ &\qquad \cdots x(n-\tau_i) \\ &= h_0 + \sum_i H_i[x(n)] \qquad (1) \end{aligned}$$

where

$$y_i(n) = H_i[x(n)] = \sum_{\tau_1,\tau_2,\ldots,\tau_i} h_i(\tau_1,\ldots,\tau_i)x(n-\tau_1)\cdots x(n-\tau_i) \quad (2)$$

The symbol H_i is called an ith-order Volterra operator, while the functions $h_i(\tau_1, \tau_2, \ldots, \tau_i)$ are referred to as the Volterra kernels of the system. The kernels are bounded and discrete in each τ_i, symmetric functions of their arguments, and, for causal systems, $h_i(\tau_1, \tau_2, \ldots, \tau_i) = 0$ for any $\tau_i < 0$.

In this section we introduce Laguerre representation of Volterra kernels. Generally, any arbitrary kernel of finite energy can be expanded into an infinite series of orthogonal functions defined over a suitable domain. The coefficients of this expansion then become the information bearing quantities. Since we are considering discrete nonlinear systems with finite energy kernels, we expand the kernels of the system in a series of DLOF.

The discrete Laguerre function of order p is given by [8,11]

$$l_p(n) = (-1)^p \left(\frac{1-\beta}{\beta^{p-n}}\right)^{1/2} \phi_p(n) \qquad (3)$$

In (3) $\phi_p(n), p = 0, 1, 2, \ldots$ are the discrete orthogonal Laguerre polynomials (DOLP) defined by

$$\phi_p(n) = \beta^p \sum_{r=0}^{n} \left(1-\beta\right)^r \binom{p}{r} \binom{n}{r} \qquad (4)$$

where $\beta = \exp(-\mu)$ (μ is a positive real number), and $\binom{p}{r}$ is the binomial coefficient.

The functions $l_p(n)$ are orthogonal and satisfy the relation

$$\sum_{n=0}^{\infty} l_i(n) l_j(n) = \delta(i, j) \qquad (5)$$

where δ is the Kronecker delta. The parameter $\beta \in (0, 1)$ is called the discount factor. The orthonormal functions (3) form a complete set of basis functions in a space which has an infinite number of dimensions. The elements $a(p_1, p_2, \ldots, p_i)$, which constitute the 'Laguerre spectrum' are obtained by the Laguerre series as follows

$$\begin{aligned} a(p_1, p_2, \ldots, p_i) &= \sum_{\tau_1, \tau_2, \ldots, \tau_i = 0}^{\infty} h(\tau_1, \ldots, \tau_i) l_{p_1}(\tau_1). \\ &\quad l_{p_2}(\tau_2) \cdots l_{p_i}(\tau_i) \qquad (6) \end{aligned}$$

The element of a given causal kernel $\{h(\tau_1, \tau_2, \ldots, \tau_p)\}$ can in turn be expressed in terms of the discrete functions $l_p(n)$ by way of inverse Laguerre series

$$h(\tau_1, \tau_2, \ldots, \tau_i) = \sum_{p_1, p_2, \ldots, p_i = 0}^{\infty} a(p_1, \ldots, p_i) l_{p_1}(\tau_1) \cdots \cdots l_{p_i}(\tau_i) \qquad (7)$$

By analogy with the classical Fourier transform (6) and (7) can be viewed as a transform pair.

In practice the first few terms of the series of orthogonal functions often suffice to yield a very good approximation to the desired response. That is,

$$h(\tau_1, \tau_2, \ldots, \tau_i) = \sum_{p_1, \ldots, p_i = 0}^{N-1} a(p_1, p_2, \ldots, p_i) l_{p_1}(\tau_1) \cdots \cdots l_{p_i}(\tau_i) \qquad (8)$$

where we have truncated the infinite sum of (7) to the first 'N' terms. The error introduced by the above truncation can be reduced by optimally selecting the value of the free parameter 'β' of the Laguerre function.

III. Laguerre Representation of Volterra Operators

In this section we introduce Laguerre representation of Volterra operators. Based on the results obtained in the previous section we can express the output $y_i(n)$ of an ith-order Volterra operator $H_i[x(n)]$ in terms of the DOLF as follows

$$\begin{aligned} y_i(n) &= H_i[x(n)] \\ &= \sum_{\tau_1, \tau_2, \ldots, \tau_i} \sum_{p_1, p_2, \ldots, p_i = 0}^{N-1} a(p_1, p_2, \ldots, p_i) l_{p_1}(\tau_1) \cdots \\ &\qquad \cdots l_{p_i}(\tau_i) x(n-\tau_1) \cdots x(n-\tau_i) \\ &= \sum_{p_1, p_2, \ldots, p_i = 0}^{N-1} a(p_1, p_2, \ldots, p_i)(l_{p_1}(n) * x(n)) \cdots \\ &\qquad \cdots (l_{p_i}(n) * x(n)) \qquad (9) \end{aligned}$$

where '*' denotes convolution operation. If we set $s_k(n) = l_k(n) * x(n)$, (9) can be written in the following form

$$y_i(n) = \sum_{p_1, p_2, \ldots, p_i = 0}^{N-1} a(p_1, \ldots, p_i) s_{p_1}(n) \cdots s_{p_i}(n) \qquad (10)$$

Remark: Equation (10) resembles (2). Therefore any technique used for the implementation of (2) can be also used for the implementation of (10). The main advantage of using (10) over (2) is that by expanding the Volterra kernels using orthogonal functions the dimensionality of kernels is considerably reduced.

Two different implementation structures which are based on Laguerre series expansion of the Volterra kernels are now in order.

Direct Form Implementation

Based on the above remark we expect that the direct implementation of (10) will result in a substantial saving in the number of multipliers compared to the direct implementation of unexpanded kernels, (2). Figure I shows the direct implementation structure of (10). We have used the fact that the z-transform $L_p(z)$ of the Laguerre function $l_p(n)$ is given by

$$\begin{aligned} L_p(z) &= \sqrt{1-\beta}\left(\frac{z}{z-\sqrt{\beta}}\right)\left(\frac{1-\sqrt{\beta}z}{z-\sqrt{\beta}}\right)^p \\ &= LP(z)\Big[AP(z)\Big]^p \end{aligned} \quad (11)$$

Note that the function $l_p(n)$ takes the form of a cascade connection of a first-order low-pass system with p identical first-order all-pass systems. When the input to the system is $x(n)$ the responses produced at the outputs of the cascade chain are simply the set: $s_0(n), s_1(n), \ldots, s_p(n)$.

Matrix Decomposition Based Implementation

For a second-order Volterra operator (10) can be written in a matrix form as shown below

$$y_2(n) = \mathbf{s}^T(n)\mathbf{A}\mathbf{s}(n) \quad (12)$$

where $\mathbf{s}^T = (s_0(n)s_1(n)\cdots s_{N-1}(n))$, and $\mathbf{A} = \{a(p_1,p_2)\}$ is $N \times N$ symmetric matrix. Applying the SVD to the symmetric matrix $\mathbf{A}$ [1,7,12] may be decomposed exactly into a finite sum of rank one matrices as follows

$$\mathbf{A} = \sum_{j=1}^{q} \lambda_j \mathbf{R}_j \mathbf{R}_j^T \quad (13)$$

where q is the rank of the matrix, λ_j are real scalars, and $\mathbf{R}_j = (r_j(0)\cdots r_j(N-1))^T$ are $M \times 1$ vectors. Substituting (13) into (12) one can easily obtain

$$\begin{aligned} y_2(n) &= \sum_{j=1}^{q} \lambda_j[\mathbf{s}^T(n)\mathbf{R}_j][\mathbf{R}_j^T\mathbf{s}(n)] \\ &= \sum_{j=1}^{q} \lambda_j d_j^2(n) \end{aligned} \quad (14)$$

where

$$\begin{aligned} d_j(n) &= \mathbf{s}^T(n)\mathbf{R}_j \\ &= \sum_{k=0}^{N-1} r_j(k)s_k(n), \quad j = 1,2,\ldots,q \end{aligned} \quad (15)$$

Equation (15) describes q independent, linear, time-invarient, parallel digital systems, and (14) represents a set of q parallel square-in-add-out type of operations. The implementation structure of (14) is shown in Figure II.

References

[1] B. G. Mertzios and A. N. Venetsanopoulos," Implementation of Quadratic Digital Filters via VLSI Array Processors," AEU, vol. 43, pp. 153-157, 1989.

[2] B. G. Mertzios, G. L. Sicuranza, and A. N. Venetsanopoulos," Efficient Realization of Two-dimensional Quadratic Digital Filters," *IEEE Trans. on Acoust., Speech, Signal Processing,* vol. ASSP-37, pp. 765-768, 1989.

[3] B. G. Mertzios, G. L. Sicuranza, and A. N. Venetsanopoulos," Efficient Structures for Two-dimensional Quadratic Filters," *Photogrammetria,* vol. 37, pp. 157-106, 1989.

[4] G. L. Sicuranza, "Nonlinear Digital Realization by Distributed Arithmetic," *IEEE Trans. on Acoust., Speech, Signal Processing,* vol. ASSP-33, pp. 939-945, 1985.

[5] E. Biglieri,"Theory of Volterra Processors and Some Applications,"*in Proc. ICASSP-82, Paris, France,* pp. 294, May 1982.

[6] B. G. Mertzios and A. N. Venetsanopoulos," Fast Implementation of Nonlinear Volterra Digital Filters via Systolic Arrays," *Proc. of the Communication, Control, and Signal Processing Conf.*, pp. 1135-1141, Ankara, Turkey, July 1990.

[7] H. H. Chang, C. L. Nikias, and A. N. Venetsanopoulos," Efficient Implementations of Quadratic Digital Filters," *IEEE Trans. on Acoust., Speech, Signal Processing,* vol. ASSP-34, pp. 1511-1528, 1986.

[8] M. J. Gottlieb,"Polynomials Orthogonal on a Finite or Enumerable Set of Points," *J. Math.*, 60, pp. 453-458, 1938.

[9] N. Wiener, Nonlinear Problems in Random Theory, New York: Technology Press M.I.T. and Wiley, 1958.

[10] M. Schetzen, The Volterra and Wiener Theories of Nonlinear Systems, New York: Wiley, 1980.

[11] R. E. King and P. N. Paraskevopoulos,"Digital Laguerre Filters," *Circuit Theory and Applications,* Vol. 5, pp. 81-91, 1977.

[12] B. G. Mertzios and A. N. Venetsanopoulos," A Decomposition Theorem and its Implementation to the Design and Realization of Two-dimensional Filters," *IEEE Trans. on Acoust., Speech, Signal Processing,* vol. ASSP-33, pp. 1502-1575, 1985.

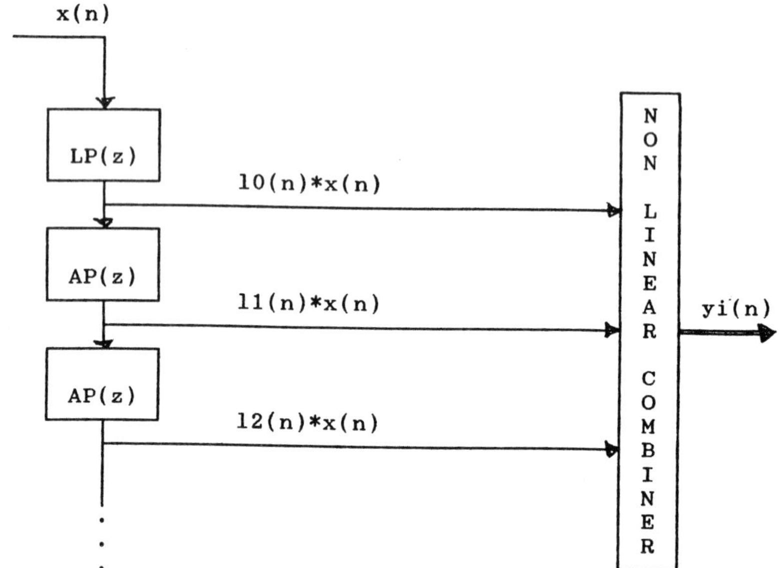

Figure I.

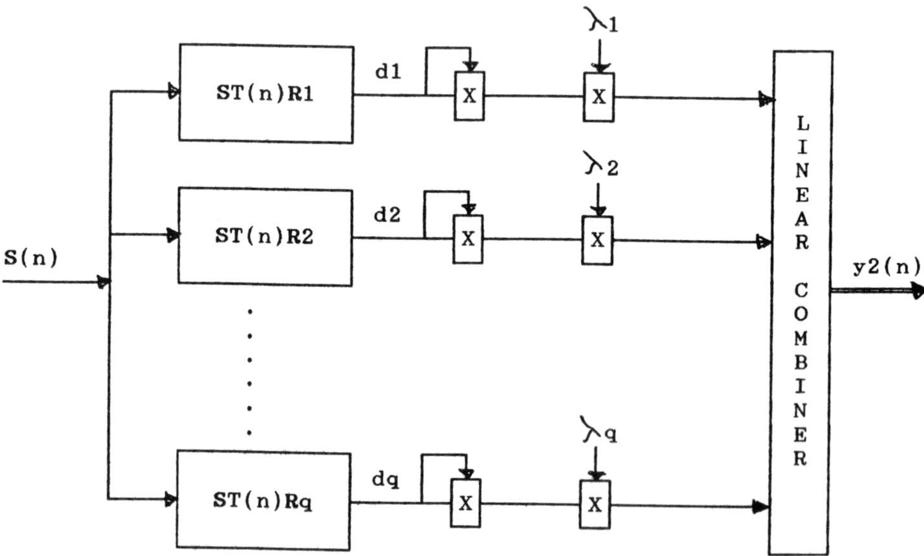

Figure II.

The "Z-slice": an Identification Method for Noisy MA Systems using Higher Order Statistics

A. G. Constantinides, Dept. of Elec. Eng., Imperial College, Exhibition Road, London, UK
E. Fleuet, Sponsored by Electricité de France, DER/TAI, 6 quai Watier, 78400 Chatou, France

In this paper, we present a new approach for the blind identification of MA systems. The algorithm is based on slices of the Z-transform of the cumulants, and it uses all the information provided by the cumulants. The extraction of the signal samples, at some stage of the algorithm, relies on the solution of a set of non linear equations, and this is achieved through the use of some novel techniques. The results appear superior to other methods.

Introduction

The identification of systems only from output measurements is an important task in many areas. And it is normally the case that the unknown systems are considered as AR, MA or ARMA models. The parameters of these models are almost invariably determined by employing some estimate of the covariance matrix of the output signal. In [1], Mendel stressed that the estimation of an ARMA model could be divided in two steps: an AR estimation and a MA estimation. Thus the subset of methods for blind estimation of MA models covers a broad area of applications. Several methods using Higher Order Statistics (HOS) have already been proposed to solve the issue. Some techniques in [1] [2] [3] are based on recursion, but their main drawback is the propagation of errors. Others such as those in [4] [5] [6] are based on linear algebra, and provide more robust solutions. However they depend on the estimation of 2nd order moments which are of course sensitive to additive noise.

This paper presents a new approach, "the Z Slice Method", to determine the parameters of a MA-system from the observed noise degraded signal. The algorithm makes use of HOS in order to benefit from the following properties: 1/ to preserve the true phase of the signal spectrum, 2/ to reduce the influence of additive Gaussian noise. From these, a good estimate of the MA parameters is obtained. In this paper the problem and its resolution are presented and compared with previous methods.

1 Presentation of the Z-slice method

In this paper third order statistics are used but the method can be easily generalised to higher orders. The signals are generated according to figure (fig. 1). The output y of a finite-dimensional MA model satisfies the equation (1) where u is assumed to be non-Gaussian, i.i.d., zero-mean with $E(u^2(k)) = \sigma^2$, $E(u^3(k)) = \beta$. The orders p (causal), q (anticausal) are assumed to be known.

$$y(n) = \sum_{k=q}^{p} h_k u(n-k) \quad , \quad v(n) = y(n) + w(n) \qquad (1)$$

The signal y is contaminated by additive Gaussian noise w. Our aim is to find the parameters h_k of the MA-system from the observed zero-mean process v.

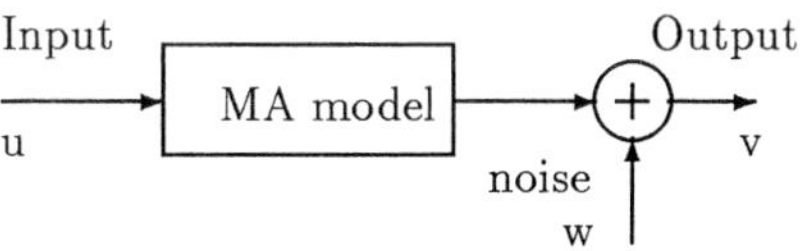

Figure 1: General schema of the signal generation

The Gaussian noise is removed or, at any rate, its influence reduced by taking the third order cumulant $cum_3(k,l)$ of v, in which case, in theory, $cum_3(k,l)$ are also the cumulants of y. Reference [7] gives all the appropriate definitions of the cumulants, in our case the third order cumulants of v is $cum_3(k,l) = E(v(n)v(n+l)v(n+k))$ [8]. But the cumulants can also be expressed by using the impulse response (or the parameters) of MA models such as in (2) [1]. From the parameters point of view the cumulants contain considerable redundancies.

$$cum_3(k,l) = \sum_{n=N}^{M} h_n h_{n+k} h_{n+l} \qquad (2)$$

$$N = Max(q, q-k, q-l) \; ; \; M = Min(p, p-k, p-l)$$

That is the reason why earlier methods to identify MA model with HOS require only the computation of the cumulant along specific lines. In so doing, the Swami and Mendel recursion (SWA) in [2] or the C(k,q) method in [1] are not exploiting the redundancy of information in third order cumulant plane. The GM algorithm [4] and its adaptations T1 [5], T2 [6] partly incorporate redundant information for increased robustness, but they still employ autocorrelation values.

Our aim is to build an approach which guarantees

good robustness to the estimation error, and which also makes good use of all the cumulant values. To achieve that, we take the Z-transform $\mathcal{C}_3$ of $cum_3(k,l)$ which can be expressed in terms of the Z-transform H of the impulse response of the MA process as [1] (the Z-space version of equation (2)).

$$\mathcal{C}_3(z_1,z_2) = \beta H(z_1)H(z_2)H(z_1^{-1}z_2^{-1}) \quad (3)$$

Consider the slice $\mathcal{S}_3$ of $\mathcal{C}_3$ along $z = z_1 = z_2$. Then equation (3) leads to (4).

$$\mathcal{S}_3(z) = \sum_{j=-2(p-q)}^{2(p-q)} P_j(\vec{h})\, z^{-j} = \sum_{j=-2(p-q)}^{2(p-q)} g_j\, z^{-j} \quad (4)$$

$P_j()$ are (p-q+1)-variable polynomials of order 3, and g_j correspond to the summation of the cumulants along the lines $k+l=j$. By a simple process of coefficient identification between each of $P_j()$ and g_j, we obtain 4(p-q)+1 non-linear equations. Now the blind identification issue consists of finding an accurate solution for h_k from the set of these non linear equations.

2 New Recursive Solution

Despite the lack of accuracy of recursive solutions due to the error propagation, our first method uses alternatively equations (5) and (6) below to obtain a recursive solution $\vec{h}_R$. The recursion can be initialized by assuming the value h_q to be known a priori or by calculating h_q and h_p from (5) and (6) with $n=0$. Equations (5) and (6) are obtained by multiplying $\mathcal{S}_3(z)$ and $\mathcal{S}_3(z^{-1})$ by $z^{2(p-q)}$, and taking the nth derivative at $z=0$.

$$2h_q h_p h_{q+n} = g_{2(q-p)+n} - \sum_{i+j+2k=n} h_{q+i}h_{q+j}h_{p-k} \quad (5)$$

$$2h_p h_q h_{p-n} = g_{2(p-q)-n} - \sum_{i+j+2k=n} h_{p-i}h_{p-j}h_{q+k} \quad (6)$$

$$i,j \geq 0 \;\; k > 0$$

3 Optimisation Solutions

A second approach to the solution is to obtain a solution for $\vec{h}_M$ as that vector that minimises $\psi(\vec{h}) = \sum(P_j(\vec{h}) - g_j)^2$. This can conceivably be achieved by using classical mathematical tools. But in view of the third order non linearity one cannot guarantee, in this case, that $\vec{h}_M$ provides the global minimum. This is so, even if we start the solution with an initial guess from the recursive solution $\vec{h}_R$ to make the procedure more robust. As there is still an uncertainity about the confidence one can attach to the solution $\vec{h}_M$, an alternative method has been deployed relying on "The Lagrange Programming Neural Network" (LPNN) [9]. The problem is now formulated as "to minimize $f(\vec{h})$ under some constraints $P_i(\vec{h}) - g_i = 0$". $f()$ could be any appropriate function but for our problem here, we have taken $f(\vec{h}) = \sum(P_i(\vec{h}) - g_i)^2$. Thus, the Lagrange function $\mathcal{L}$ and the set of dynamic equations below are obtained

$$\mathcal{L}(\vec{h},\vec{\lambda}) = f(\vec{h}) + \sum_i \lambda_i (P_i(\vec{h}) - g_i) \quad (7)$$

$$\frac{d\vec{h}}{dt} = -\nabla_h \mathcal{L}(\vec{h},\vec{\lambda}) \;\; and \;\; \frac{d\vec{\lambda}}{dt} = \nabla_\lambda \mathcal{L}(\vec{h},\vec{\lambda}) \quad (8)$$

One can distribute $(P_i(\vec{h}) - g_i)$ either in $f(\vec{h})$, in the set of constraints or both. According to [9], if we consider the Lagrange function (7), the equilibrium of (8) is reached for the global solution $\vec{h}_L$ when the Hessian of L is always positive definite.

4 Results

All the presented tests have been performed for length of signal of 4096 (32*128) samples. The cumulants have been computed according to the algorithm given by Nikias in [10]. Two example models are used model 1 (mixed phased) $1.0 - 0.8z^{-1} + 0.6z^{-2} - 0.4z^{-3} + 0.2z^{-4}$ and model 2 (minimum phase) $1.0 - 0.85z^{-1} + 0.175z^{-2}$. The phase rather than the amplitude has been plotted on figure 2 because its variations are more visible.

If fast computation is required, the recursive solution $\vec{h}_R$ offers a good estimate of $\vec{h}$ when the orders (p,q) of the model are small or the length of the data is large. At any rate, $\vec{h}_R$ is a good initial guess for the optimisation based methods which then obtain better estimates of $\vec{h}_M$. $\vec{h}_M$ is computed by a Quasi-Newton Method. It gives accurate results as those in figure 2 and in tables 1, 2. Moreover, one has several ways for evaluating the accuracy of $\vec{h}_M$. As h_0 is not fixed, its closeness to 1.0 (this value is assumed by others methods) indicates the goodness of $\vec{h}_M$. The value of $\psi(\vec{h}_M)$ and the cumulant error (the distance between the estimated cumulants and the reconstructed cumulants from $\vec{h}_M$) also indicate the confidence one can attach $\vec{h}_M$.

In practice, (8) is computed by the Runge-Kutta algorithm. In the LPNN method, the step-size of this algorithm slows down considerably the computation of $\vec{h}_L$. The differences between $\vec{h}_M$ and $\vec{h}_L$ are due to the introduction of constraints in the LPNN. It has an adverse effect on $\vec{h}_L$ in table 2 but a good one in figure 2. It depends mainly on the goodness of the g_i involved in the constraints.

C(k,q), SWA and h_R are fast solutions to our problem. But they need an accurate estimation of the cumulants to achieve useful results. GM, SWA are sensitive to additive Gaussian noise. GM, T1, T2 are based on different set of linear equations involving specific cumulants. When these cumulants are badly estimated $\vec{h}_M$, $\vec{h}_L$ always give better results specially with additive noise.

h	1.000	-0.85	0.175
h_R	1.000	-0.755	0.224
3dB	1.000	-0.842	0.255
OdB	1.000	-0.907	0.275
h_M	1.032	-0.892	0.191
3dB	1.050	-0.879	0.215
OdB	1.061	-0.850	0.234
h_L	1.020	-0.891	0.174
3dB	1.046	-0.881	0.214
OdB	1.058	-0.852	0.234
GM	1.000	-0.755	0.042
3dB	1.000	-0.881	0.214
OdB	1.000	-0.907	-0.160
T1	1.000	-0.662	0.136
3dB	1.000	-0.669	0.108
OdB	1.000	-0.690	0.095
T2	1.000	-0.883	0.241
3dB	1.000	-1.149	0.349
OdB	1.000	-1.306	0.395

Table 1: *Sensitivity to additive Gaussian noise of Impulse response identified by different approaches*

h	1.000	-0.800	0.600	-0.400	0.200
h_R	1.000	-0.811	0.598	-407	0.187
var	—	0.138	0.415	0.130	0.032
SWA	1.000	-1.138	0.547	-0.189	0.098
var	—	1.129	0.118	0.498	0.252
C(k,q)	1.000	-0.811	0.547	-0.407	0.098
var	—	0.092	0.118	0.130	0.251
h_M	0.998	-0.812	0.587	-0.395	0.182
var	0.023	0.029	0.041	0.036	0.031
h_L	1.005	-0.812	0.570	-0.397	0.152
var	0.018	0.032	0.037	0.025	0.028
GM	1.000	-0.837	0.571	-0.375	0.209
var	—	0.058	0.074	0.072	0.056
T1	1.000	-0.766	0.577	-0.397	0.170
var	—	0.149	0.174	0.087	0.089
T2	1.000	-0.803	0.582	-0.424	0.198
var	—	0.118	0.104	0.103	0.066

Table 2: *Impulse response identified by different approaches for MA model 1 (mean and variance over 20 random set of data)*

Conclusion

Methods are presented for the identification of MA systems through Higher Order Cumulants using the new approach named "The Z slice". They provide a good solution for the estimation of the parameters of an MA model even with small data lengths, and are robust in the presence of additive gaussian noise. The entire set of cumulants of a specific order are involved in the computation of the solution, which permits the use of all the information known on the model and the solution is therefore less sensitive to estimation errors.

References

[1] Mendel J. M. Tutorial on higher-order statistics (spectra) in signal processing and system theory: Theoretical results and some applications. *Proceedings of the IEEE*, Vol. 79(No 3):pp. 278–304, March 1991.

[2] Swami A and Mendel J. M. Closed-form recursive estimation of ma coefficients using autocorrelations and third-order cumulants. *IEEE Trans. on Acou. Speech and Sig. Proc.*, Vol. 37(No 11):pp. 1794–1795, November 1989.

[3] Giannakis G.B. Cumulants: A powerful tool in signal processing. *Proc. IEEE*, Vol. 75:pp.1333–1334, 1987.

[4] Giannakis G. B. and Mendel J. M. Identification of nonminimum phase systems using higher order statistics. *IEEE Trans. on Acou. Speech and Sig. Proc.*, Vol. 37(No 3):pp. 360–377, March 1989.

[5] Tugnait J. K. Approaches to fir system identification with noisy data using higher order statistics. *IEEE Trans. on Acou. Speech and Sig. Proc.*, Vol. 38(No 5):pp. 1307–1317, July 1990.

[6] Tugnait J.K. New results on fir system identification using hos. *IEEE Trans. on Sig. Proc.*, Vol. 39(No 10), October 1991.

[7] Brillinger D. R. An introduction to polyspectra. *Ann. Math. Statist.*, Vol. 36:pp. 1351–1374, 1965.

[8] Nikias C. L. and Raghuveer M. R. Bispectrum estimation: A digital signal processing framework. *Proceedings of the IEEE*, Vol. 75(No 7):pp. 869–889, 1987.

[9] Zhang S. W. and Constantinides A. G. Lagrange programming neural networks. *Submitted paper*, 1991.

[10] Nikias C. L. Arma bispectrum approach to nonminimum phase system identification. *IEEE Trans. on Acou. Speech and Sig. Proc.*, Vol. 36(No 4):pp.513–524, April 1988.

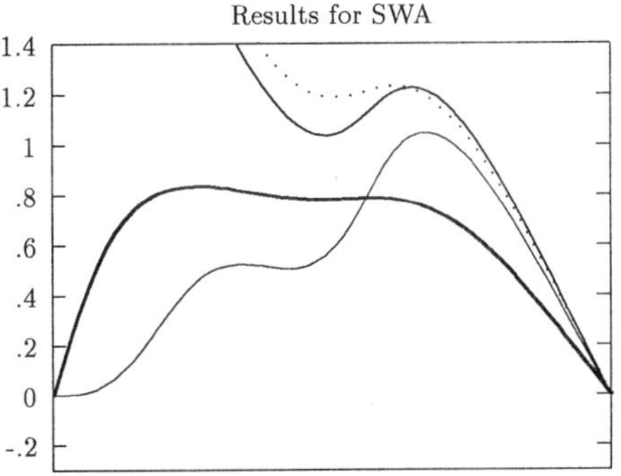

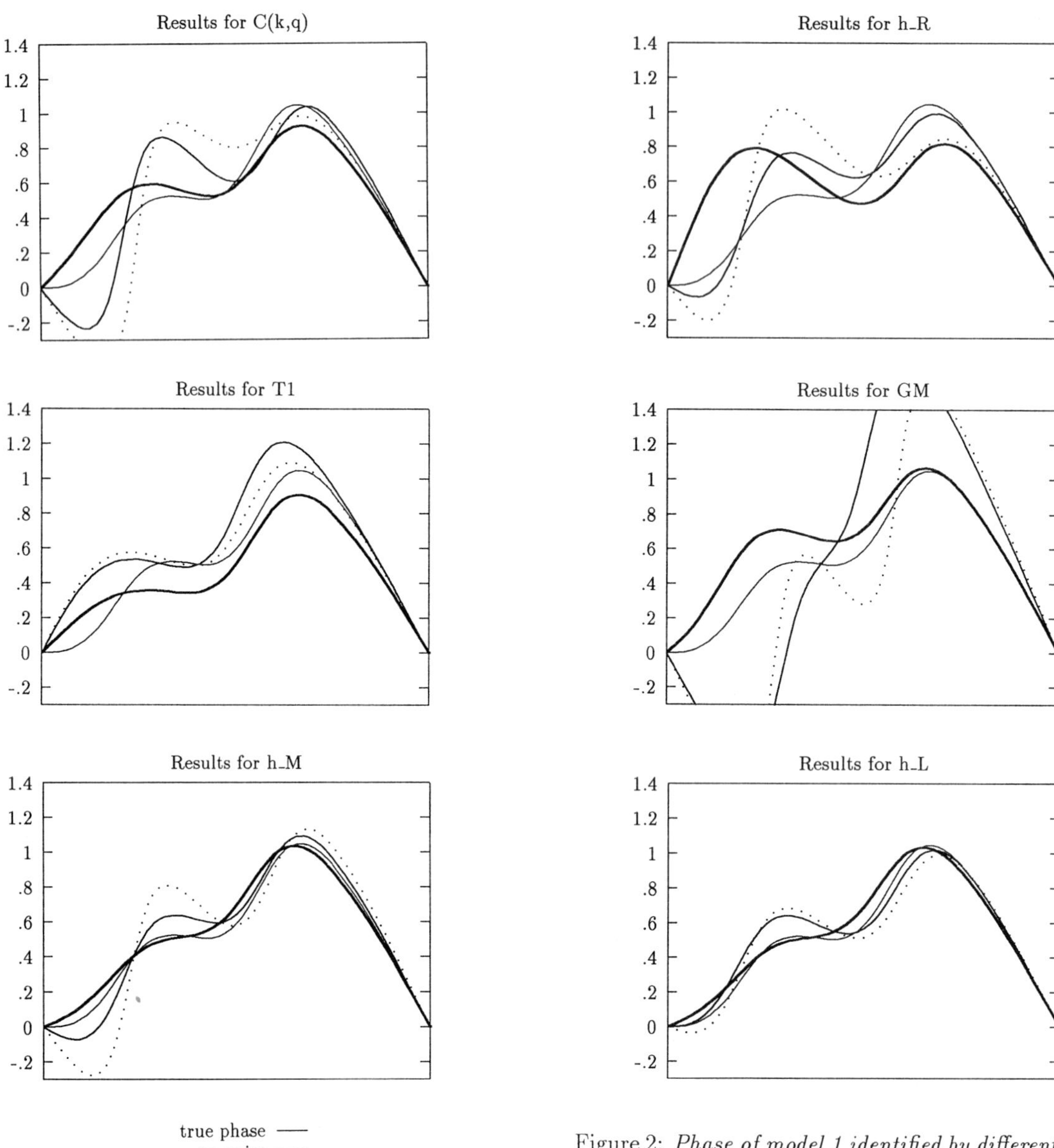

Figure 2: *Phase of model 1 identified by different methods without noise and with SNR = 3dB, 0dB and from the same set of 4096 samples.*

SIGNAL PROCESSING VI: Theories and Applications
J. Vandewalle, R. Boite, M. Moonen, A. Oosterlinck (eds.)

THE DISCRETE-TIME RUNNING CROSS-CORRELATION FUNCTION ESTIMATED FROM THE LAGUERRE SPECTRA

A. C. DEN BRINKER

EE-Department, Eindhoven University of Technology, Eindhoven
P.O. Box 513, NL-5600 MB Eindhoven, The Netherlands

We propose a simple and computationally efficient mechanism to calculate a *running* or *local cross-correlation function* of two discrete-time signals. In order to obtain a running cross-correlation function, the signals must be windowed. It is argued that an appropriate window for a local cross-correlation is an exponential sequence. To obtain a computationally efficient mechanism, the windowed functions are decomposed in discrete-time Laguerre series. Cross-correlation of the windowed functions is identical to cross-correlation of the Laguerre coefficients and subsequent multiplication by cross-correlation pattern functions.

1. INTRODUCTION

In many applications of signal processing, running or local measures of certain characteristics of the signal are required, instead of global measures. We consider the problem of constructing a simple mechanism to calculate a *running* or *local cross-correlation function.* Such a mechanism can be used in a large number of applications, for instance in adaptive filtering or system identification.

First, we will consider which window is appropriate to construct a running cross-correlation function. Next, we represent the windowed signals as a series of orthogonal functions. In our case this will be a Laguerre series. The coefficients of this series are obtained by filtering the input signals. The local cross-correlation function is subsequently expressed in terms of the orthogonal set of functions. We show that the running cross-correlation function is a weighted summation of cross-correlation pattern functions where the weights are determined as the cross-correlation of the Laguerre coefficients. This method is computationally efficient if the number of coefficients of the Laguerre series that is required to represent the signal accurately is small. To this end the window function and the set of orthonormal functions are matched.

2. THE WINDOW FUNCTION

In the case of time-domain signals, taking a running or local measure as a meaningful quantity is based on the following two assumptions. First, running measures prohibit the use of future signal values, i.e., we only know the signal at the current moment n *and* its past. Second, the signal in the distant past of n should not, or at most only weakly, determine our running measure at n. Thus, in order to obtain a local cross-correlation, we have to window the signals, i.e., we multiply the signals, say $x(m)$ and $y(m)$, by $w(n-m)$. The function $w(m)$ is called the *window function.*

Since we may not use information of future signal values, we set $w(m) = 0$ for $m < 0$. Furthermore, we require that the *shape* of the windowed function $x(m)w(n-m)$ is independent of the 'current moment' n. This is a necessary requirement since cross-correlation involves a comparison of the shape of the signals. The only window satisfying this requirement of constancy of the shape of the windowed function is given by an exponential sequence: $w(m) = w(0)\xi^m$. Without loss of generality we can take $w(0) = 1$. Since we want to consider real-valued windows and to take past events less into account we take a real scaling parameter and $0 < \xi < 1$. To summarize, we use the window function $w(m)$ given by $w(m) = \xi^m \varepsilon(m)$, where $\varepsilon(m)$ is the unit step function.

The suitable range of the scaling parameter ξ involves a compromise. On the one hand we want to take ξ as close as possible to unity in order to get an accurate estimate of the cross-correlation function, while on the other hand for $\xi \to 1$ the effect of windowing is lost.

Consider now the windowed function $\hat{x}$ according to $\hat{x}(n;m) = x(m)w(n-m)$. The local cross-correlation function $R_{xy}(n;l)$ is simply defined as the cross-correlation of the windowed signals $\hat{x}$ and $\hat{y}$

$$R_{xy}(n;l) = \sum_{m=-\infty}^{\infty} \hat{x}(n;m)\hat{y}(n;m+l). \tag{1}$$

3. THE LAGUERRE TRANSFORM

Instead of directly calculating the cross-correlation of the windowed signals, these signals are transformed first to the Laguerre domain. Next, we will show that then the cross-correlation results in a simple expression with respect to the Laguerre coefficients. Our approach links up with the renewed interest in the possibilities of signal descriptions by orthogonal polynomials in both system identification (e.g., [1]-[6]) and coding [7].

We write a windowed function as a discrete-time Laguerre series:

$$\hat{u}(n;m) = \sum_{k=0}^{\infty} g_{uk}(n)\phi_k(n-m). \tag{2}$$

u stands for both x and y, g_{uk} are the Laguerre coefficients or the Laguerre spectrum, and $\phi_k(m)$ are the *pattern functions.* One may consider this as a transform: $\hat{u}(n;m) \rightarrow g_{uk}(n)$ which is called a *forward polynomial transform.* The Laguerre spectrum depends on n and thus constitutes a running decomposition. Since the transform involves Laguerre polynomials, we speak about a *Laguerre transform.* The pattern functions are defined by

$$\phi_k(m) = (-1)^k \nu_k \gamma_k(m)\theta^{m/2}, \tag{3}$$

where $\nu_k = \sqrt{(1-\theta)/\theta^k}$ and $\gamma_k(m)$ is the discrete Laguerre polynomial [8] defined by

$$\gamma_k(m) = \theta^{-m}\Delta^k \left[\begin{pmatrix} m \\ k \end{pmatrix} \theta^m \right] \varepsilon(m), \tag{4}$$

with k the order of the polynomial, Δ the forward difference operator, and $0 < \theta < 1$. The pattern functions $\phi_k(m)$ constitute a complete orthonormal set of functions on the interval $[0,\infty)$

$$\sum_{m=0}^{\infty} \phi_k(m)\phi_{k'}(m) = \delta_{kk'}, \tag{5}$$

where $\delta_{kk'}$ is the Kronecker delta. Assuming that $u(n)$ is a bounded signal, the windowed signal $u(n;m)$ is square-summable and the proposed decomposition (2) exists. The first three pattern functions for $\theta = 0.9$ are shown in Fig. 1. The number of zero-crossings in the pattern functions is equal to the index k, and so, higher-order pattern functions are increasingly oscillatory.

Taking $\theta = \xi^2$ amounts to matching the signal decomposition on the window function. The Laguerre coefficients can be easily derived from the original signal $u(n)$ via convolution $g_{uk}(n) = u(n) * d_k(n)$, where $d_k(n)$ are the impulse responses of the *decomposition filters* according to $d_k(n) = \phi_k(n)w(n)$.

We denote the Z-transforms of decomposition filters and pattern functions by $\Phi_k(z)$ and $D_k(z)$, respectively. We find

$$\Phi_k(z) = \sqrt{1-\xi^2}\frac{z}{z-\xi}\left(\frac{1-\xi z}{z-\xi}\right)^k, \tag{6}$$

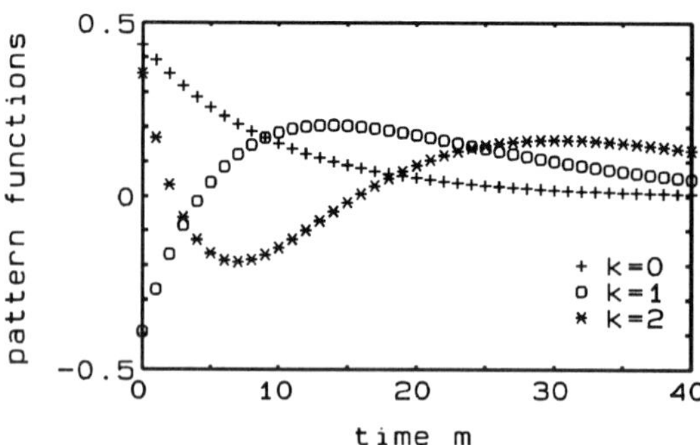

Figure 1: The pattern functions $\phi_k(m)$ for $\theta = 0.9$ and $k = 0, 1, 2$ (as indicated in the plot).

$$D_k(z) = \sqrt{1-\theta}\frac{z}{z-\theta}\left(\frac{\sqrt{\theta}(1-z)}{z-\theta}\right)^k. \tag{7}$$

In Fig. 2 an efficient filtering scheme to obtain the coefficients g_{uk} from the original signal is shown. The set of decomposition filters encompasses the effect of the windowing operation as well as the Laguerre transform.

$u(n)$ → $\frac{\sqrt{1-\theta}z}{z-\theta}$ → $\frac{\sqrt{\theta}(1-z)}{z-\theta}$ → $\frac{\sqrt{\theta}(1-z)}{z-\theta}$ → ...

$g_{u0}(n)$ $g_{u1}(n)$ $g_{u2}(n)$

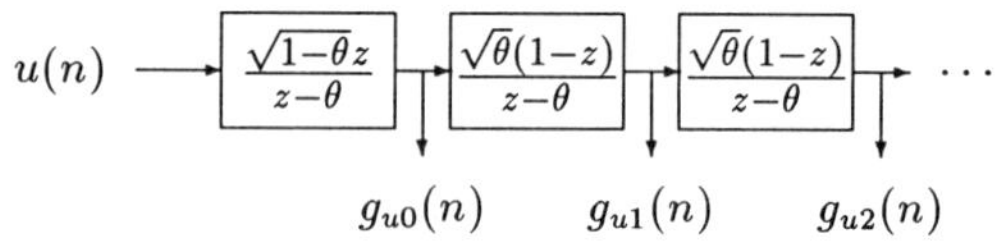

Figure 2: Filter bank to derive the running Laguerre coefficients $g_{uk}(n)$ from the original signal $u(n)$.

4. THE LOCAL CROSS-CORRELATION FUNCTION

We now proceed to express the local cross-correlation function $R_{xy}(n;m)$ in terms of the Laguerre coefficients of the windowed signals. By substituting the Laguerre series for the windowed signals in (1) we obtain

$$R_{xy}(n;l) = \sum_{k=0}^{\infty}\sum_{m=0}^{\infty} g_{xk}(n)g_{ym}(n)\tilde{q}_{k,m}(l), \tag{8}$$

where $\tilde{q}_{k,m}(l) = \mathrm{Z}^{-1}\{\tilde{Q}_{k,m}(z)\}$ (Z^{-1} denotes the inverse Z-transform) and $\tilde{Q}_{k,m}(z) = \Phi_k(z)\Phi_m(z^{-1})$. Substituting the Z-transforms (6), we find that the function $\tilde{Q}_{k,m}$ depends only on the difference $k-m$

$$\begin{aligned}\tilde{Q}_{k,m}(z) &= Q_{k-m}(z) \\ &= \frac{1-\theta}{(z-\xi)(z^{-1}-\xi)}\left(\frac{1-\xi z}{z-\xi}\right)^{k-m}.\end{aligned} \tag{9}$$

We define $q_k(l)$ as the inverse Z-transform of $Q_k(z)$. The local cross-correlation function can now be expressed as

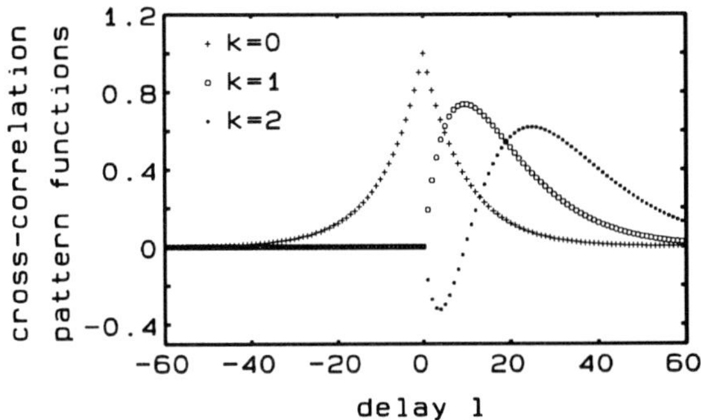

Figure 3: The cross-correlation pattern functions $q_k(l)$ for $\theta = 0.9$ and $k = 0, 1, 2$ (as indicated).

a summation of weighted *local cross-correlation pattern functions* q_k according to

$$R_{xy}(n;l) = \sum_{k=-\infty}^{\infty} c_k(n) q_k(l). \tag{10}$$

The weighting factors c_k which are called the *cross-correlation coefficients* are given by

$$c_k(n) = \sum_{m=-\infty}^{\infty} g_{x,k+m}(n) g_{ym}(n), \tag{11}$$

where the Laguerre spectra g_{xm}, g_{ym} for negative m are defined as $g_{xm} = g_{ym} = 0$. We find that the local cross-correlation function is a weighted summation of cross-correlation pattern functions (10). The weights are calculated by cross-correlating the Laguerre spectra of the windowed signals (11). The advantage of calculating $R_{xy}(n;l)$ by (10) instead of a direct cross-correlation of the windowed functions, is its computational simplicity in the case the Laguerre series has a fast convergence. To that end, we already matched the scale parameter of the window function and that of the Laguerre series.

From the Z-transforms of the cross-correlation pattern functions we see that $Q_k(z) = Q_{-k}(1/z)$ and thus that $q_k(l) = q_{-k}(-l)$. The function $q_0(l)$ has a simple explicit expression by $q_0(l) = \xi^{|l|}$, and for positive k the cross-correlation pattern function can be expressed as the sum of two pattern functions ϕ:

$$q_k(l) = \frac{1}{\sqrt{1-\xi^2}} \left[\phi_k(l) + \xi \phi_{k-1}(l) \right]. \tag{12}$$

Three cross-correlation pattern functions are shown in Fig. 3 for $\theta = 0.9$ and $k = 0, 1, 2$.

5. DISCUSSION

We have given a formal definition of the discrete-time local cross-correlation function. We argued that the best choice for a window function is an exponential window, since cross-correlation involves a comparison of the shape of two signals and the shape of a windowed signal is independent of its location in the case of an exponential window. Furthermore, we have shown that the local cross-correlation function can be simply computed from the Laguerre coefficients of the windowed signals.

The definition of the local cross-correlation function and its method of calculation immediately raises two questions. First, in which cases is the proposed local cross-correlation function a meaningful quantity? Second, when is the proposed method of calculation efficient?

The local character ensures that the given definition is meaningful not only for stationary processes, but also for quasi-stationary processes. However, in that case the time scale associated with changes in the stochastics must be much larger than that associated with the signals themselves. In practical situations this will not be a very restrictive requirement.

The proposed method of calculation is efficient if the number of Laguerre coefficients needed to represent the signals accurately is small. This is the case if the time scale $1/(1-\xi)$ associated with the parameter ξ is matched to that of the signal content of x and y (or more precisely, to the parts of x and y that are correlated). The choice for ξ is in essence the question on which time scale we want to view the signals, i.e., which scale provides the best representation of the characteristics that are of interest. At the same time, it is a question of computational simplicity: the best scale is that which provides a minimum of significant coefficients of the Laguerre transform.

A restriction that follows from a specific choice of ξ is that the proposed local cross-correlation function is only meaningful if the time scale of the cross-correlation $E[x(n)y(n-l)]$ is also within the scale $1/(1-\xi)$. This is obvious from the fact that all cross-correlation pattern functions are on this scale. For instance, if x and y are delayed replicas with delay L and if $\xi^L \ll 1$, we will not be able to estimate this delay reliably from the local cross-correlation function.

To summarize, there are three time scales, associated with the change in stochastics, the signals x and y, and the cross-correlation function, respectively. On the other hand we have only one free parameter (ξ) in the definition of the local cross-correlation function. The proposed definition and its method of calculation is meaningful and efficient under the following three requirements. The time scale associated with the change in stochastics is much larger than that associated with the signals. The time scale associated with the cross-correlation function is smaller than or equal to that associated with the signals. The parameter ξ is matched to the time scale of the signals.

We feel that in many areas a formal definition of a local cross-correlation function and an easy assessment thereof may be a useful tool. In particular we are thinking of applications in the field of system identification and adaptive filtering [9].

REFERENCES

[1] King, R.E. and Paraskevopoulos, P.M., "Parametric identification of discrete-time SISO systems," Int. J. Contr. 30 (1979) pp. 1023-1029.

[2] Paraskevopoulos, P.N., "System analysis and synthesis via orthogonal polynomial series and Fourier series," Mathematics and Computers in Simulation 27 (1985) pp. 453-469.

[3] Zervos, C., Bélanger, P.R. and Dumont, G.A., "On PID controller tuning using orthonormal series identification," Automatica 24 (1985) pp. 165-175.

[4] Mäkilä, P.M., "Approximation of stable systems by Laguerre filters," Automatica 26 (1990) pp. 333-345.

[5] Heuberger, P., On approximate system identification with system based orthonormal functions. (PhD Dissertation Delft University of Technology, Delft, 1991).

[6] Wahlberg, B., "System identification using Laguerre models," IEEE Trans. Automat. Contr. 36 (1991) pp. 551-562.

[7] Martens, J.B., "The Hermite transform - theory," IEEE Trans. Acoust. Speech Signal Processing ASSP-38 (1990) pp. 1595-1606.

[8] Gottlieb, M.J., "Concerning some polynomials orthogonal on a finite or enumerable set of points," Am. J. Math. 60 (1938) pp. 453-458.

[9] Brinker, A.C. den, "Laguerre-domain adaptive filters," in preparation.

SIGNAL PROCESSING VI: Theories and Applications
J. Vandewalle, R. Boite, M. Moonen, A. Oosterlinck (eds.)

Some Robust Stability Results Concerning Sparse Predictors

D. Docampo*, F. Pérez,* J.R. Fernández** and C. Abdallah†

* BSP Group, ** Chronobiology Group
Departamento de Tecnologiás de las Comunicaciones
ETSI Telecomunicación, Universidad de Vigo, 36200-Vigo, Spain

† BSP Group, Department of Electrical and Computer Engineering
University of New Mexico, Albuquerque, NM 87131, USA

This paper applies new extreme-point robust stability results, to the study of the stationarity (stability) of sparse AR processes. Point estimates of the coefficients of those sparse predictors are uncertain if the inaccuracies introduced by the variance of the estimators are considered. Making use of bootstrap techniques, non-parametric confidence intervals are obtained around the nominal coefficients (point estimators) and then robust stability properties of the All-Pole Filters associated with the sparse AR models are discussed.

1 Introduction

Since the seminal work of Kharitonov [1] concerning the stability of continuous-time uncertain systems, the stability of polynomials with structured perturbations has become a very active area of research. In recent years, the discrete-time counterpart has also commended a great deal of research effort. It is now known that a strictly strong Kharitonov-like result does not exist in discrete time. However, weak or partial results are available in some special cases, see for example [2] and [3]. Such cases allow us to handle perturbations affecting only half of the coefficients and the case of symmetric or antisymmetric polynomials.

The first objective of this paper is to present the results obtained previously by some of the authors, see [4] amd [5] in the discrete-time domain. By expressing a discrete polynomial using its Barycentric coordinates (BC) and using standard results, we were able to generalize some existing discrete-time stability results but more importantly, to introduce and prove more general results. In fact, we were able to show that a Kharitonov-like result holds even though the upper-half of the polynomial coefficients are uncertain. This result will simplify the stability test of a family of uncertain polynomials to that of some combinations of the extremes of the family. Moreover, when the uncertainty under consideration is symmetric or antisymmetric (see section 2), all the coefficients are allowed to vary independently while an extreme-point check for stability is still valid. This implies that a generalized linear-phase type of perturbations may be analyzed with the so called *weak-Kharitonov tests*.

Once those results have been reviewed, we show how sparse predictors (namely, those having a number of non-zero coefficients lower than the actual length of the predictor) fit into the cases of uncertain filters as previously studied. Examples of such predictors can be AR processes [6] where only a short number of coefficients are non-zero, or long-term predictors used in Linear Predictive Speech Coding. We will use this last example, where assuring stability for the predictor is required, as a showcase of our results. With our robust stability test available, only a small number of polynomials need to be considered in order to study the stationarity of all the possible filters obtained by combining all the possible parameter values falling in some precalculated confidence intervals.

2 Robust Stability Results

An uncertain polynomial may be described as a member of a set of possible polynomials. Then, the stability of the polynomial may be deduced from the stability of the underlying set. The original problem of Kharitonov deals with uncertainties in the continuous-time case. Specifically, consider the family of polynomials $\mathcal{P}$, where each polynomial can be written as

$$p(s,q) = q_0 + q_1 s + q_2 s^2 + \cdots + q_n s^n, \quad (1)$$

where each coefficient q_i takes values in the range $[q_i^-, q_i^+]$. The variation in each coefficient is assumed independent from all others. The question answered by Kharitonov [1] is to find necessary and sufficient conditions such that all polynomials in $\mathcal{P}$ are strictly Hurwitz (i.e., all the roots lie in the strict left half plane). His theorem solves this problem with the following simple test. Given the four polynomials

$$\begin{aligned} P_1(s) &= q_0^- + q_1^- s + q_2^+ s^2 + q_3^+ s^3 \\ &+ q_4^- s^4 + q_5^- s^5 + \cdots \quad (2) \\ P_2(s) &= q_0^+ + q_1^+ s + q_2^- s^2 + q_3^- s^3 \\ &+ q_4^+ s^4 + q_5^+ s^5 + \cdots \quad (3) \\ P_3(s) &= q_0^+ + q_1^- s + q_2^- s^2 \\ &+ q_3^+ s^3 + q_4^+ s^4 + q_5^- s^5 + \cdots \quad (4) \\ P_4(s) &= q_0^- + q_1^+ s + q_2^+ s^2 \\ &+ q_3^- s^3 + q_4^- s^4 + q_5^+ s^5 + \cdots \quad (5) \end{aligned}$$

then the stability of these four polynomials is necessary and sufficient to guarantee the stability of the family $\mathcal{P}$. Moreover, when the degree of the polynomials involved is less than 6, then a reduced stability test is possible, where fewer than four polynomials are needed. Therefore, Kharitonov's theorem provides necessary and sufficient conditions for the analysis of the stability of a family of continuous time polynomials subject to independent perturbations. Note the important property of the test, that the number of polynomials to check does not depend on the degree of the polynomials in $\mathcal{P}$.

After the results of Kharitonov became widely known, a great effort was expanded trying to adapt them to the discrete-time case. Unfortunately, the extension of Kharitonov's work to discrete-time systems is not straightforward. Actually, it is known that a four-polynomials test does not hold for the discrete-time case, and neither does a test in which the stability of all extreme polynomials (not just a special set of four) is checked. Clearly, this means that for a n-th order interval polynomial family, the 2^n combinations of corners can be stable, but some member of the family is not. This fact was established in some well-known counterexamples; e.g., see [7]. Some partial results are however available for discrete-time polynomials, some of wich are discussed below.

In the following, we will deal with discrete-time polynomials with real coefficients. Such polynomials can be written as

$$P(z) = a_0 z^n + a_1 z^{n-1} + ... + a_n. \quad (6)$$

We will consider first the case pointed out in [2] where only the upper half of the coefficients are allowed to vary. Concretely, we have that

$$a_i \in [a_i^- \, a_i^+] \, , \; i = \lfloor \frac{n+1}{2} \rfloor, \cdots n. \quad (7)$$

where $\lfloor \frac{n+1}{2} \rfloor$ denotes the integer part of $(n+1)/2$. Then, the stability of the entire family is equivalent to that of the extreme polynomials, obtained by combining all the possible extremes of the coefficients defined before. This result is, of course, weaker than Kharitonov's theorem but, as we will see later, it will be useful in studying the stability of some specific kinds of uncertainties.

Another interesting partial result applies to symmetric or antisymmetric perturbations [3]. In this case, the variation in the coefficients is such that

$$a_i - a_i^- = a_{n-i} - a_{n-i}^- \quad (8)$$

or

$$a_i^+ - a_i = a_{n-i} - a_{n-i}^- \quad (9)$$

Again, the complete family of polynomials is stable if and only if all the combinations of extreme-polynomials are stable. It should be noted that this result contains two important constraints:

1. The coefficients must be coupled as $(a_i, \; a_{n-i})$. This means that different types of coupling are not allowed nor is a single coefficient allowed to vary on its own.

2. The edges allowed in the pairwise variation defined in 1) have $\pi/4$ and $3\pi/4$ slopes in the parameter space.

In [4] and [5], we showed that the previous results can be generalized by removing those restrictions. This was done by using the barycentric coordinates i.e. the bilinear transformation, and exploiting some symmetries found in the transformation matrix. The following theorem was then obtained:

Theorem 1 *Consider a polytope in the coefficient space where each pair $(a_i, \; a_k)$, for $0 \le i \le n$ and $n - i \le k \le n$ is varying inside a polygon with edges sloped in the closed interval $[\pi/4, \; 3\pi/4]$ and where the pairwise variations $(a_i, \; a_l)$, $k \ne l$ are not allowed simultaneously. Then, every polynomial in the polytope will be stable if and only if all the polynomials obtained by combining all the polygon corners are stable.*

Clearly, if $i = k$ for $i = \lfloor \frac{n+1}{2} \rfloor, \cdots n$ we obtain the result of [2]. When $k = n - i$ for $i = \lfloor \frac{n+1}{2} \rfloor, \cdots n$ and the slope is set to $\pi/4$ and $3\pi/4$ the result of [3] is recovered.

3 Stationarity in AR Processes

In a number of applications of Digital Signal Processing we are interested in fitting an all-pole model to a signal, usually based on short data records. If stationarity for the time series (stability for the all-pole model) is desired, we have to deal not only with the nominal coefficients of the model (obtained through point estimations taken from the data), but also with all the possible values for the parameters that lie in the confidence intervals that can be placed around the actual values of the estimators.

Once those intervals are obtained, we are in a position to study the stability of the whole family of possible values of the parameters (robust stability), where now the perturbations of the nominal coefficients are precisely the ranges of variation inside the confidence intervals. Due to the lack of *a priori* information about the underlying probabilistic properties of the estimators, it is not possible to get theoretical bounds for those intervals. Nevertheless, that problem can be overcome using non-parametric bootstrap techniques [8] as it is usual in fitting linear regression models [9].

In order to summarize the boostrap analysis, we assume the validity of a model structure used to fit the data, and find residuals around the nominal model over the actual data record; then, resampling those residuals we can generate artificial data that provides new estimations for the parameters of the model. Repeating the process as many times as necessary the bootstrap distribution of the parameters is obtained; such an empirical distribution can then be taken as an approximation of the true distribution of the parameters estimates. Finally, using the bootstrap distribution, empirical confidence intervals for the parameters can be computed, and a value for the error in the estimation can also be obtained. The great advantage of this technique resides in its ability to handle problems where no statistical information is available, and no assumptions are made about the underlying statistics of the model, improving the classical theory of confidence intervals construction.

In the next section, those intervals are taken as uncertainties (or structured perturbations) of the parameters, and a case is made for the study of the robust stability properties of such models.

4 Numerical Example

As an example for testing the results and techniques developed in the paper, we have made simulations with a zero-mean AR process having only three non-zero coefficients, specifically the last three. This is actually a good example of a long term predictor taken from those we find in Linear Predictive Speech Coders, as the CELP or Multipulse among others.

Bootstrap confidence intervals were computed from a data record containing 256 samples. A modified version (assuming that only the last three coefficients are different from zero) of the classical LMS algorithm was used to estimate the parameters of the predictor. The results are shown in Table 1, where values for the point-estimations and confidence intervals can be found. It should be pointed out that these three intervals do not contain zero, so that we can assure that the true parameters of the model are not null.

The predictor under consideration can then be written as

$$A(z) = 1 + a_{24}z^{-24} + a_{25}z^{-25} + a_{26}z^{-26}. \qquad (10)$$

Therefore, since the variation is only on a small part of the upper half of the coefficients, i.e. in $a_i\,; i \geq 13$, our results can be applied to the study of the stationarity of this AR process. As we have three free coefficients, we will need to check all the possible combinations of extremes, that is, 2^3 according to Section 2. In Table 2 we present the eight extreme combinations and the maximum value of the modulus of the roots for each corresponding polynomial. Since this maximum is less than 1 for all the extremes, it can be concluded that for every possible value of the parameters inside the confidence intervals, sationarity is guaranteed.

Coefficient	Point Estimates	Confidence Intervals
24	-0.19	-0.31 $\leq a_{24} \leq$ -0.039
25	0.23	0.15 $\leq a_{25} \leq$ 0.27
26	-0.23	-0.38 $\leq a_{26} \leq$ -0.065

Table 1: Estimates and Confidence Intervals

a_{24}	a_{25}	a_{26}	Max. Modulus
-0.33	0.15	-0.38	0.99399
-0.039	0.15	-0.38	0.97821
-0.33	0.27	-0.38	0.98919
-0.039	0.27	-0.38	0.98549
-0.33	0.15	-0.065	0.97754
-0.039	0.15	-0.065	0.94689
-0.33	0.27	-0.065	0.98355
-0.039	0.27	-0.065	0.96513

Table 2: Extreme Polynomials and Roots

5 Conclusions

The stationarity of sparse AR processes has been analyzed when the uncertainties introduced by the point estimators are considered. The main contribution of the paper resides in the use of techniques from robust stability and bootstrap analysis to study classical signal processing problems. Our future research will study the stability of generalized linear-phase filters under perturbations in the filter coeeficients and the study of the complex filters case as it applies to multipath communications problems.

References

[1] V. Kharitonov, "Asymptotic stability of an equilibrium position of a family of systems of linear differential equations," *Differential Equations*, vol. 14, pp. 1483–1485, 1979.

[2] C. Hollot and A. Bartlett, "Some discrete-time counterparts to kharitonov's stability theorem for uncertain systems," *IEEE Trans. Auto. Control*, vol. AC-31, pp. 355–356, 1986.

[3] F. Kraus, M. Mansour, and B. Anderson, "Robust schur polynomial stability and kharitonov's theorem," in *Proceedings 26th IEEE CDC*, (Los Angeles, CA), pp. 2088–2095, 1987.

[4] F. Pérez, D. Docampo, and C. Abdallah, "New extreme-point robust stability results for discrete-time polynomials," in *Proceedings of the IEEE American Control Conference*, (Chicago, IL), 1992.

[5] F. Pérez, D. Docampo, and C. Abdallah, "Simple proofs of existing and new extreme-point robust stability results for discrete-time polynomials," in *Proceedings of the Symposium on Fundamentals of Discrete-Time Systems*, (Chicago, IL), 1992.

[6] A. Figueiras, D. Docampo, J. Casar, and A. Artés, "On using the adaptive delay filter in sparse identification and extensions to sparse deconvolution and sinusoidal detection," in *Proc. IEEE Conf. on Acoustics, Speech and Signal Processing*, (Albuquerque, NM), pp. 2503–2506, 1990.

[7] J. Cieslik, "On possibilities of the extension of kharitonov's stability test for interval polynomials to the discrete-time case," *IEEE Trans. Auto. Control*, vol. AC-32, pp. 237–138, 1987.

[8] B. Efron, "The jacknife, the bootstrap and other resampling plans," in *SIAM, Vol. 38*, 1982.

[9] D. Freedman and S. Peters, "Bootstrapping a regression equation: Some empirical results," *Journal of the American Statistical Association*, vol. 79, pp. 97–106, 1984.

SIGNAL PROCESSING VI: Theories and Applications
J. Vandewalle, R. Boite, M. Moonen, A. Oosterlinck (eds.)

The Scale-Angle Representation in Image Analysis with 2D Wavelet Transform

Jean-Pierre ANTOINE, Pierre CARRETTE and Romain MURENZI

Institut de Physique Théorique, Université Catholique de Louvain
B - 1348 Louvain-la-Neuve, Belgium

Wavelet analysis [1,2] is by now recognized as a powerful tool for the analysis and reconstruction of signals. Since it is a scale-space decomposition, it is particularly efficient for detecting features which are well-localized both in scale and in space. In 1 dimension, this means basically discontinuities in the signal or one of its derivatives: sharp peaks (spectral lines, e.g. in NMR spectroscopy), onset of a signal (e.g. in speech or music analysis), etc. In 2 dimensions, i.e. in images, typical features are edges, tips, sharp luminance changes, coherent structures in 2D turbulent flows, etc. In this contribution we will focus on the 2D case. The use of 2D wavelets in image processing has been advocated by Mallat [3], using a method based on multiresolution analysis. This is an instance of a *discrete* wavelet transform (WT), and it is essentially confined to Cartesian geometry: the 2D analysis is obtained by combining (tensor product) 1D analyses for the x- and y-directions separately. Thus the method is not well-adapted to the detection of *directional* features of an image. For that purpose, one has to use the *continuous* WT, that we now briefly describe [4]. A 2D signal is represented by a bounded nonnegative function s, square-integrable over the plane $\mathbb{R}^2$. An *analyzing wavelet* is a complex-valued, square-integrable function $\psi \in L^2(\mathbb{R}^2, d^2\vec{x})$, satisfying an admissibility condition, which, for ψ regular enough, amounts to a zero mean condition :

$$\int d^2\vec{x}\,\psi(\vec{x}) = 0. \qquad (1)$$

The WT of the signal s, with respect to the wavelet ψ (suitably normalized) is the function:

$$\begin{aligned} S(a,\theta,\vec{b}) &= \langle\psi_{a,\theta,\vec{b}}|s\rangle \qquad (2)\\ &= a^{-1}\int d^2\vec{x}\,\bar{\psi}(a^{-1}r_{-\theta}(\vec{x}-\vec{b}))s(\vec{x}). \end{aligned}$$

Here $\langle\cdot|\cdot\rangle$ is the usual scalar product in $L^2(\mathbb{R}^2, d^2\vec{x})$, and $\psi_{a,\theta,\vec{b}}$ is obtained from ψ through dilation by $a > 0$, translation by $\vec{b} \in \mathbb{R}^2$ and rotation by $\theta \in [0, 2\pi)$:

$$\begin{aligned} \psi_{a,\theta,\vec{b}}(\vec{x}) &= a^{-1}\,\psi(a^{-1}r_{-\theta}(\vec{x}-\vec{b})),\\ r_\theta(\vec{x}) &= (x\,cos\theta - y\sin\theta, x\sin\theta + y\cos\theta),\\ & \qquad 0 \le \theta < 2\pi. \qquad (3) \end{aligned}$$

These transformations generate the 2D Euclidean group with dilations, and indeed the whole WT formalism stems from a suitable representation of that group [4]. The linear transform $W_\psi : s \mapsto S$ can be inverted on its range by the transposed map, and thus one gets integral reconstruction formulas, such as :

$$s(\vec{x}) = \iiint a^{-3}da\,d\theta\,d^2\vec{b}\;\;\psi_{a,\theta,\vec{b}}(\vec{x})\;S(a,\theta,\vec{b}). \qquad (4)$$

Of course, for practical purposes, both the direct WT (2) and the inverse WT (4) have to be discretized, which poses the question of determining the optimal sampling rate in all variables (see below). Because of the mean zero condition (1), the wavelet transform (2) acts as a very efficient filter in position ($\vec{b}$), in scale (a) and in direction (θ) : the WT $S(a,\theta,\vec{b})$ is appreciable only in those regions of parameter space $a,\theta,\vec{b}$ where the signal is appreciable, that is, where the wavelet $\psi_{a,\theta,\vec{b}}$ 'matches' the features of the signal s. This explains why the WT may be interpreted as a 'mathematical microscope', with optics ψ, global magnification $1/a$ and orientation tuning parameter θ. Now the WT $S(a,\theta,\vec{b})$ is a function of 4 variables : $a > 0, \theta \in [0,2\pi), \vec{b} \in \mathbb{R}^2$. Clearly, in order to obtain a manageable tool, one must limit oneself to special sections of the parameter space $\{a,\theta,\vec{b}\}$. It turns out that two such sections, both 2-dimensional, are very natural : the *position* representation, where (a,θ) are fixed and $S(a,\theta,\vec{b}) \equiv S_{a,\theta}(\vec{b})$ is considered as a function of (b_x, b_y), and the *scale-angle* representation, where (b_x, b_y) are fixed and $S(a,\theta,\vec{b}) \equiv S_{\vec{b}}(a,\theta)$ is considered as a function of (a,θ). Actually the pair (a^{-1},θ) plays the role of spatial frequency, expressed in polar coordinates; this interpretation is supported both by mathematical analysis (phase space geometry) and by some physiological evidence [6].

The position representation is standard and has been used in all applications so far [2-5]. The aim of this contribution is to demonstrate the usefulness of the scale-angle representation. Here the WT is examined through a key-hole located at $\vec{b}$, and examined at all scales and in all directions :

$$S_{\vec{b}}(a,\theta) = a^{-1} \iint dx dy\, \bar{\psi}(a^{-1} r_{-\theta}(\vec{x})) s_{\vec{b}}(x,y), \tag{5}$$

where $s_{\vec{b}}(x,y) = s(x+b_x, y+b_y)$ represents the signal seen in a Cartesian frame located in $\vec{b}$. The expression (5) is evaluated with the FFT algorithm, after transcription in polar coordinates and use of a Mellin transform.

Whereas the position representation is used for the general purposes of image processing (detection of specific features, such as contours, edges, etc., noise filtering, data compression), the scale-angle representation is useful for several applications [2,4,5]:

- determination of the sampling grid in the variables (a,θ) for the numerical evaluation of the reconstruction integral (4), using the reproducing kernel of the continuous WT;
- (multi) fractal recognition, since the signal is analyzed at all scales ;
- determination of the local regularity of the function s at the point $\vec{b}$;
- disentangling of a superposition of plane waves.

Let us sketch this last application. The basic problem is to decompose a complex signal into single components which have a simple form, more or less known a priori, and to isolate their individual contribution to the WT; this is the key for noise filtering or elimination of unwanted components. A typical example in 1D is the estimation of spectral lines (e.g. NMR spectroscopy, geophysical time series). The problem is to identify with great precision the position of the peaks (local maxima) and to measure their parameters [7]. Another 1D example is the analysis of refracted waves in underwater acoustics [8], although this particular example is in fact a 3D situation. We shall consider here the 2D version of the problem. The signal is a superposition of plane waves, damped in a given direction, and we want to measure the various parameters of each component. We shall take the signal in the following form :

$$f(\vec{x}) = \sum_{n=1}^{N} c_n\, e^{i\vec{k}_n\cdot\vec{x}}\, e^{-\vec{l}_n\cdot\vec{x}}, \tag{6}$$

where $c_n = A_n e^{i\phi_n}$, with A_n and ϕ_n respectively the modulus and phase of the nth component at the origin; $\vec{k}_n$ is the wave vector and the vector $\vec{l}_n$ in a (oriented) damping factor.

The choice of an exponential damping factor $\exp(-\vec{l}_n\cdot\vec{x})$ makes of course the computation especially simple, but it is by no means essential. For instance, similar results would be obtained with a damping factor acting in one direction only, such as $\vartheta(\vec{l}_n\cdot\vec{x})\exp(-\vec{l}_n\cdot\vec{x})$, where ϑ is the jump function or a smoothed version thereof. The important point is that the perturbation be local: the method still allows the determination of the various parameters sufficiently far away from the perturbation (as in 1D).

A straightforward evaluation, using a Morlet wavelet (Gabor function) with isotropic modulus yields the WT of the signal (6) :

$$F(a,\theta,\vec{b}) = 2\pi e^{\frac{1}{2}k_o^2} \sum_{n=1}^{N} c_n\, e^{i\vec{k}_n\cdot\vec{b}}\, e^{-\vec{l}_n\cdot\vec{b}} F_{\vec{b},n}(a,\theta). \tag{7}$$

Of course, the linearity of the WT implies that the result is the linear superposition of the contributions of the various components. In a first step, we fix $\vec{b}$ and analyze the (a, θ) dependence of $F(a, \theta, \vec{b})$. More precisely, writing

$$F_{\vec{b}}(a,\theta) = \sum_{n=1}^{N} c_{\vec{b},n}\, F_{\vec{b},n}(a,\theta), \qquad (8)$$

we observe that $a^{-1}|F_{\vec{b},n}(a,\theta)|$ has a *unique* maximum localized at $(a_n^{max}, \theta_n^{max})$. Now, in the full transform (7), each term has its own local maximum, but these need not be well separated : one maximum may hide another one, totally or partially. This masking effect will happen, for instance, when:

- one component has a much bigger amplitude, $|c_{\vec{b},n}| >> |c_{\vec{b},m}|$, for all m not equal to n (total masking);
- two wave vectors are close to each other, $\vec{k}_n \simeq \vec{k}_n$, but with different amplitudes, $|c_{\vec{b},n}| > |c_{\vec{b},m}|$ (partial masking).

In both cases, the two waves can be separated, by increasing the selectivity of the wavelet (for instance, using a Morlet wavelet with a more anisotropic modulus [5]). If the two waves have close wave vectors ($\vec{k}_n \simeq \vec{k}_n$) with similar amplitudes ($|c_{\vec{b},n}| \simeq |c_{\vec{b},m}|$), but different damping vectors ($\vec{l}_n \neq \vec{l}_m$), then they can still be separated, by changing the observation point $\vec{b}$. Otherwise the method will fail, the two waves interfere inextricably, none of them dominates the other one.

When the masking effect is not too important, the maxima will be sufficiently prominent that the interferences between the different components will become negligible (in the modulus) and one may write

$$a^{-1}|F_{\vec{b},n}(a,\theta)| \simeq \sum_{n=1}^{N} |c_{\vec{b},n}|\, a^{-1}\, |F_{\vec{b},n}(a,\theta)|. \quad (9)$$

This is the formula used in practice. Now we come back to the full transform (7) and evaluate it successively at every local maximum $(a_n^{max}, \theta_n^{max})$, thus reverting to the position representation. This amounts to filter out all components except the nth :

$$F_{a_n^{max},\theta_n^{max}}(b_x, b_y) \simeq d_n\, e^{i\vec{k}_n\cdot\vec{b}}\, e^{-\vec{l}_n\cdot\vec{b}}. \qquad (10)$$

This expression, which we call $F(\vec{b})$ for short, represents a single damped plane wave. For evaluating the wave parameters $\vec{k}_n = k_n(\cos\alpha_n, \sin\alpha_n), \vec{l}_n = l_n(\cos\beta_n, \sin\beta_n)$, it suffices now to consider the lines of constant modulus and constant phase of $F(\vec{b})$: their orientation yields the angles α_n and β_n, their spacing the norms k_n and l_n, respectively. To get rid of noise, one may average $\log|F(\vec{b})|$ and $\arg F(\vec{b})$ on the corresponding line of constancy passing through the point $\vec{b}$ considered. Finally the original coefficients $c_n = A_n e^{i\phi_n}$ are evaluated by measuring $F(\vec{b})$ itself at a particular point $\vec{b}_o$, typically 0.

In conclusion, the feasibility of this procedure rests, as in 1D, on two key features:

(i) the linearity of the CWT, which allows to treat each component separately ;

(ii) the filtering effect of the CWT, which essentially eliminates all components except one.

We may remark that, in case of a masking, one of the components hides the other one. In that case, one must evaluate first the dominant component and then subtract it from the full transform. This is the 2D analogue of the subtraction of unwanted spectral lines in 1D, one of the spectacular achievements in the analysis of NMR spectra [7]. The comparison with the 1D case also confirms that the parameters (a^{-1}, θ) together play the role of a wave vector : localizing the peaks (local maxima) in the (a, θ) plot corresponds to the first step in the iterative procedure of localization of spectral lines.

The method has been tested on an explicit example, consisting of a superposition of four waves, with the parameters chosen so as to exhibit a partial masking ($\alpha_1 \simeq \alpha_4$). This signal is analyzed by the Morlet wavelet $\psi(x, y) = \exp(i6y)\exp[-\frac{1}{2}(\epsilon^{-1}x^2 + y^2)]$. Fig.1 shows the level curves of the renormalized modulus of the WT, seen from the origin, $a^{-1}|F_{0,0}(a,\theta)|$. Only the prominent features survive, the function is negligible except in the neighborhood of the local maxima. Fig.1(a) is obtained with the isotropic wavelet ($\epsilon = 1$) ; it clearly shows the presence of four waves (four maxima), but those numbered 1 and 4 are not well separated. Thus the analysis is redone with $\epsilon = 5$, and the result plotted in Fig.1(b) clearly shows the improved selectivity, the two maxima are now clearly identifiable. The position of the four local maxima is measured by a standard algorithm, then the full WT is evaluated at each of the maxima, which allows to measure the parameters of each plane wave.

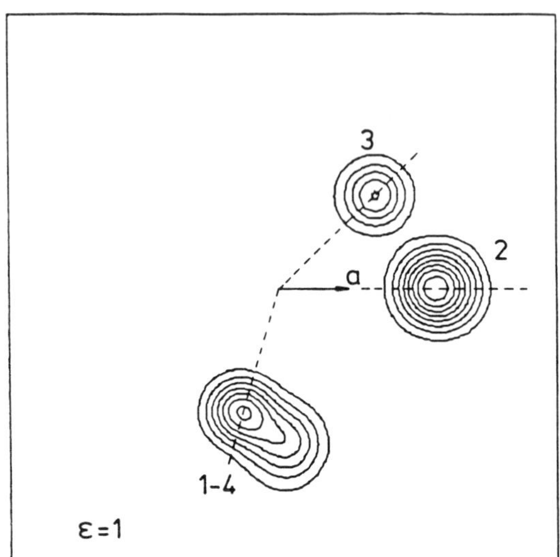

Fig.1(a).- Analysis of a four wave train: (a, θ) representation of $a^{-1}|F_{0,0}(a, \theta)|$, for a Morlet wavelet with $\epsilon = 1$. Notice the partial masking of wave 4 by wave 1.

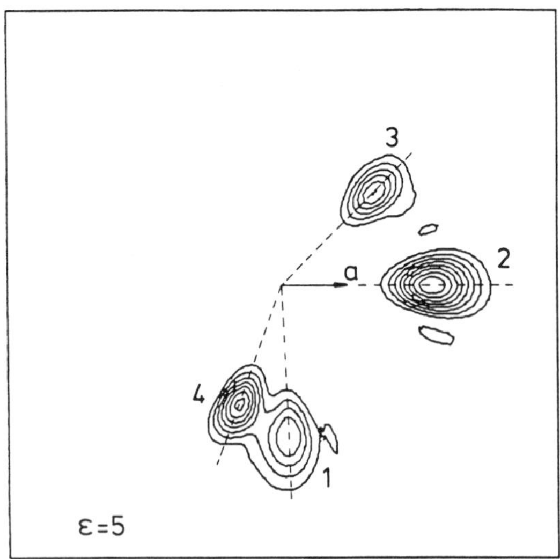

Fig.1(b).- The same with $\epsilon = 5$. The waves 1 and 4 are now well resolved.

The result is that all parameters are recovered with an excellent precision (relative error of the order of 10^{-3}), except for the amplitude A_4. This is, of course, a consequence of the masking of wave 4 by wave 1. If one plots the phase of each individual wave, as given by (10), each plot indeed represents the phase of a pure plane wave, except for side effects and the interference between waves 1 and 4, mostly visible in the plot of wave 4. For improving the result, we could subtract the (reconstructed) wave 1 from the signal and repeat the analysis. Also a better precision could be obtained by a finer discretization of the signals (the grid used throughout is 128 × 128 points).

References

[1] *Wavelets, Time-Frequency Methods and Phase Space (Proc. Marseille 1987)*, J.-M. Combes, A. Grossmann, Ph. Tchamitchian (eds.), Berlin: Springer-Verlag, 1989; 2d ed. 1990.

[2] *Wavelets and their Applications (Proc. Marseille 1989)*, Y. Meyer (ed.), Berlin: Springer-Verlag, and Paris: Masson, 1991.

[3] S.G. Mallat, Multifrequency channel decompositions of images and wavelet models, *IEEE Trans. ASSP*, **37** (1989) 2091-2110.

[4] J.-P. Antoine, P. Carrette, R. Murenzi, B. Piette, Image analysis with two-dimensional continuous wavelet transform, report UCL-IPT-91-01, Louvain-la-Neuve.

[5] J.-P. Antoine, Wavelet analysis in image processing, this volume.

[6] R. De Valois, K. De Valois, *Spatial Vision*, New York: Oxford Univ. Press, 1988 (see in particular Chap.4, pp. 137-143).

[7] P. Guillemain, R. Kronland-Martinet, B. Martens, Estimation of spectral lines with the help of the wavelet transform, applications in NMR spectroscopy, in [2], pp. 38-60.

[8] G. Saracco, A. Grossmann, Ph. Tchamitchian, Use of wavelet transforms in the study of propagation of transient acoustic signals acroos a plane interface between two homogeneous media, in [1], pp. 139-146.

Lowering the Threshold SNR of Fourier Signal Subspace

Jyrki Joutsensalo and Juha Karhunen

Helsinki University of Technology,
Laboratory of Computer and Information Sciences,
Rakentajanaukio 2 C, SF-02150 Espoo, Finland
Fax: +358-0-4513277

Abstract

In sinusoidal frequency estimation and array processing subspace type methods are currently studied extensively. In these methods (e.g. MUSIC and ESPRIT) the signal subspace which defines the unknown frequencies or directions of arrival is approximated by computing the eigenvectors of the sample correlation matrix. We have earlier proposed two methods which use the Discrete Fourier Transform (DFT) for estimating the signal subspace instead of eigenvectors. These methods can make use of prior information provided by classical spectral estimators. They are also more robust against noise and overestimation of the number of signals. In this work we present two new methods in which the signal subspace is first estimated crudely using DFT. This estimate is then postprocessed, leading to a lowered threshold SNR and improved performance compared to our earlier methods. We consider sinusoidal frequency estimation, but with small restrictions and modifications the methods are applicable to array processing, too.

1 Data model

The data are assumed to consist of M complex sinusoids in additive white noise. The N available data samples $x(0)$,$x(1)$,...,$x(N-1)$ are thus modeled as

$$x(n) = \sum_{m=1}^{M} a_m e^{j(\omega_m n + \theta_m)} + w(n), \qquad (1)$$

where the amplitudes a_m, angular frequencies ω_m, and phases θ_m are unknown constants. The complex white noise term $w(n)$ has zero mean and variance σ^2. In practical estimation a data matrix

$$\mathbf{X} = \begin{bmatrix} x(0) & x(1) & \cdots & x(N-L) \\ x(1) & x(2) & \cdots & x(N-L+1) \\ \vdots & \vdots & \ddots & \vdots \\ x(L-1) & x(L) & \cdots & x(N-1) \end{bmatrix} \qquad (2)$$

$$= \mathbf{X}_s + \mathbf{W} = \mathbf{E}_{LM}\mathbf{A}(\mathbf{E}_{KM})^T + \mathbf{W} \qquad (3)$$

is used, where

$$\mathbf{E}_{LM} = \begin{bmatrix} \mathbf{e}_L(\omega_1) & \mathbf{e}_L(\omega_2) & \cdots \mathbf{e}_L(\omega_M) \end{bmatrix},$$

$$\mathbf{e}_L(\omega_m) = \begin{bmatrix} 1 & e^{j\omega_m} & \cdots & e^{j(L-1)\omega_m} \end{bmatrix}^T,$$

$$\mathbf{A} = \mathrm{diag}\left\{a_1 e^{j\theta_1}, a_2 e^{j\theta_2}, \ldots, a_M e^{j\theta_M}\right\},$$

and $\mathbf{W}$ is a noise matrix having similar structure as (2). Superscripts T and H denote transpose and complex conjugate transpose, respectively. $\mathbf{E}_{LM}$ is an $L \times M$ matrix and it consists of *frequency information* or *direction* vectors $\mathbf{e}_L(\omega_m)$. $\mathbf{E}_{KM}$ has the same structure as $\mathbf{E}_{LM}$, but with the size $K \times M$, where $K = N - L + 1$ is the number of columns (data vectors) in $\mathbf{X}$. We use this subscript convention also later to denote the size of a matrix. The first term $\mathbf{X}_s$ in Eq. (3) is deterministic signal part, and its rank is M. The number M of signals is assumed to be known, at least roughly.

2 Estimation of signal subspace using DFT

In the subspace type methods a vector-valued DFT can be used for estimating the signal subspace [2,3]. We have used two different approaches for forming the DFT operator matrix. The first approach is to use the periodogram; the second approach is to use the Bartlett estimate. Before presenting the new methods we define these operators.

In the periodogram approach, the vectorial DFT operator matrix is constructed using frequency estimates given by the periodogram. The form of the DFT matrix depends on the form of the matrix to be transformed. In the case of $\mathbf{X}$ one tries to get the DFT matrix to be as close as possible to the complex conjugate of $(\mathbf{E}_{KM})^T$ in Eq. (3). Then the DFT matrix is

$$\hat{\mathbf{E}}_{KM} = \begin{bmatrix} \mathbf{e}_K(-\hat{\omega}_1) & \mathbf{e}_K(-\hat{\omega}_2) & \cdots & \mathbf{e}_K(-\hat{\omega}_M) \end{bmatrix}, \quad (4)$$

where $\hat{\omega}_m$, $m = 1, 2, \ldots, M$ yield the M highest peaks of the periodogram of data sequence which is assumed to be of the form (1). The reason to this type of choice is that the Frobenius norm of the signal part is improved with respect to the Frobenius norm of the noise part:

$$\frac{\| \mathbf{X}_s \hat{\mathbf{E}}_{KM} \|_F}{\| \mathbf{W}\hat{\mathbf{E}}_{KM} \|_F} > \frac{\| \mathbf{X}_s \|_F}{\| \mathbf{W} \|_F}.$$

In general there may exist other frequency bins than those closest to the correct ones yielding high peaks, too. In the new methods we take this into account by choosing H ($H > M$) frequencies in the DFT matrix: $\hat{\mathbf{E}}_{KM}$ in Eq. (4) is replaced by $\hat{\mathbf{E}}_{KH}$. Multiplying the data matrix $\mathbf{X}$ by $\hat{\mathbf{E}}_{KH}$ corresponds to a filtering where one tries to retain all the relevant signal information and remove as much noise as possible.

When the data matrix $\mathbf{X}$ is Fourier transformed – e.g. in DFT-MUSIC – one can alternatively use the Bartlett estimate. In this case one first constructs K vectors $\hat{\mathbf{q}}_k$ corresponding to the uniformly spaced frequencies ζ_k:

$$\hat{\mathbf{q}}_k = \mathbf{X}\mathbf{e}_K(\zeta_k), \qquad (5)$$

$$\zeta_k = \frac{2\pi(k-1)}{K}, \quad k = 1, 2, \ldots, K.$$

Then one computes the squared norms

$$\begin{aligned} \| \hat{\mathbf{q}}_k \|^2 &= \hat{\mathbf{q}}_k^H \hat{\mathbf{q}}_k = \mathbf{e}_K^H(\zeta_k)\mathbf{X}^H\mathbf{X}\mathbf{e}_K(\zeta_k) \\ &= \mathbf{e}_K^H(\zeta_k)\mathbf{R}\mathbf{e}_K(\zeta_k), \quad k = 1, 2, \ldots, K, \qquad (6) \end{aligned}$$

and chooses the vectors $\hat{\mathbf{q}}_k$ corresponding to the H largest norms. These H vectors are taken as the columns of the matrix $\hat{\mathbf{Q}}$. Eq. (6) is clearly the Bartlett estimate.

3 New methods

From the fact that the rank of $\mathbf{X}_s$ is M, we know that the data vectors lie in the M-dimensional signal subspace if there exist no noise. The idea in the new methods is to first prefilter the data by forming H vectors $\hat{\mathbf{q}}_k$, $k = 1, 2, \ldots, H$ using either of the methods explained in Section 2. After this one estimates the true signal subspace by searching for a smaller M-dimensional subspace in which this vector set approximately lies. The search is done by applying again the vectorial DFT to the correlation matrix constructed of the $\hat{\mathbf{q}}_k$-vectors. The eigenvector approach can be used, alternatively. The algorithms are as follows:

1. (Only in the periodogram version) Compute the periodogram of the samples $x(0), x(1), \ldots, x(N-1)$. Select the negative counterparts $-\hat{\omega}_m$ of the frequencies $\hat{\omega}_m$, $m = 1, 2, \ldots, H$, $H > M$ corresponding to the H highest peaks of the periodogram. Using these frequencies, construct $\hat{\mathbf{E}}_{KH}$ as in Eq. (4).

2. Construct a data matrix $\mathbf{X}$ from the samples using Eq. (2).

3. (Periodogram version) Take the DFT of $\mathbf{X}$ at selected frequencies:

$$\hat{\mathbf{Q}} = \mathbf{X}\hat{\mathbf{E}}_{KH}. \qquad (7)$$

(Bartlett estimate version) Take the DFT of $\mathbf{X}$ in K frequency bins using Eq. (5). Select the H vectors with largest norms from the set $\hat{\mathbf{q}}_k$, $k = 1, 2, \ldots, K$ (Eq. (6)); form the matrix $\hat{\mathbf{Q}}$ from these vectors.

4. Algorithm 1: Construct a new DFT operator $\hat{\mathbf{E}}_{LM}$ corresponding to the M highest peaks $\hat{\omega}_k$ or ζ_k:s yielding M largest norms and take the Fourier transform of $\hat{\mathbf{Q}}\hat{\mathbf{Q}}^H$:

$$\hat{\mathbf{V}} = \hat{\mathbf{Q}}\hat{\mathbf{Q}}^H\hat{\mathbf{E}}_{LM}. \qquad (8)$$

Algorithm 2: Compute the M principal eigenvectors of $\hat{\mathbf{Q}}\hat{\mathbf{Q}}^H$ and form the matrix $\hat{\mathbf{V}}$ from these vectors.

5. Orthonormalize $\hat{\mathbf{V}}$ and insert the result to the MUSIC or ESPRIT estimate [1].

In these methods one can use the FB estimates: matrix product $\hat{\mathbf{Q}}\hat{\mathbf{Q}}^H$ is replaced by the product $\hat{\mathbf{Q}}\hat{\mathbf{Q}}^H + \hat{\mathbf{S}}\hat{\mathbf{S}}^H$ in which the first row of $\hat{\mathbf{S}}$ is the last row of $\hat{\mathbf{Q}}$ complex conjugated, the second row of $\hat{\mathbf{S}}$ is the second last row of $\mathbf{Q}$ complex conjugated, and so on. Both of these methods are computationally less complex than the basic version of MUSIC.

Illustrative example: Consider the simple model

$$x(n) = e^{j2\pi \cdot 0.1n} + e^{j2\pi \cdot 0.2n} + w(n), \quad n = 0, \ldots, 24.$$

Let us construct 10×16 matrix $\mathbf{X}$. By taking the periodogram of 25 point sequence with FFT we get 32 values corresponding to 32 frequency bins. Assume that the four ($= H$) highest peaks in decreasing order are obtained at frequency bins 3/32, 6/32, 4/32, and 7/32. Then 3/32 and 6/32 yield the two ($= M$) highest peaks, and $\hat{\mathbf{E}}_{16,4}$ in (7) and $\hat{\mathbf{E}}_{10,2}$ in (8) are

$$\hat{\mathbf{E}}_{16,4} = \left[\; \mathbf{e}_{16}(\tfrac{-2\pi\cdot 3}{32}) \;\; \mathbf{e}_{16}(\tfrac{-2\pi\cdot 4}{32}) \;\; \mathbf{e}_{16}(\tfrac{-2\pi\cdot 6}{32}) \;\; \mathbf{e}_{16}(\tfrac{-2\pi\cdot 7}{32}) \;\right],$$

$$\hat{\mathbf{E}}_{10,2} = \left[\; \mathbf{e}_{10}(\tfrac{2\pi\cdot 3}{32}) \;\; \mathbf{e}_{10}(\tfrac{2\pi\cdot 6}{32}) \;\right].$$

4 Analysis of the algorithms

For simplicity, we analyze only the periodogram and backward version. Our goal is to show that the methods yield the matrices spanning the same subspace as $\mathbf{E}_{LM}$. With this condition MUSIC and ESPRIT can extract the frequencies. In Step 4 one forms a matrix $\hat{\mathbf{P}}$ with

$$\mathbf{P} = \mathcal{E}\{\hat{\mathbf{P}}\} = \mathcal{E}\{\hat{\mathbf{Q}}\hat{\mathbf{Q}}^H\}, \qquad (9)$$

where $\mathcal{E}\{\cdot\}$ denotes the expectation operator. From Eq. (3) and Step 3 we get

$$\hat{\mathbf{Q}} = \mathbf{X}\hat{\mathbf{E}}_{KH} = \mathbf{E}_{LM}\mathbf{K} + \mathbf{W}\hat{\mathbf{E}}_{KH}, \qquad (10)$$

where $\mathbf{K} = \mathbf{A}(\mathbf{E}_{KM})^T\hat{\mathbf{E}}_{KH}$. Using Eqs. (9), (10), and the facts that the absolute values of the elements of the DFT matrices are 1, and that the noise is white with zero mean, we get

$$\begin{aligned}\mathbf{P} &= \mathbf{E}_{LM}\mathbf{K}\mathbf{K}^H(\mathbf{E}_{LM})^H + \mathbf{E}_{LM}\mathbf{K}(\hat{\mathbf{E}}_{KH})^H\mathcal{E}\{\mathbf{W}^H\} \\ &+ \mathcal{E}\{\mathbf{W}\}\hat{\mathbf{E}}_{KH}\mathbf{K}^H(\mathbf{E}_{LM})^H + \mathcal{E}\{\mathbf{W}\hat{\mathbf{E}}_{KH}(\hat{\mathbf{E}}_{KH})^H\mathbf{W}^H\} \\ &= \mathbf{E}_{LM}\mathbf{B}(\mathbf{E}_{LM})^H + K\sigma^2\mathbf{I}, \qquad (11)\end{aligned}$$

$$\mathbf{B} = \mathbf{K}\mathbf{K}^H = \mathbf{A}(\mathbf{E}_{KM})^T\hat{\mathbf{E}}_{KH}(\hat{\mathbf{E}}_{KH})^H(\mathbf{E}_{KM})^*\mathbf{A}^H,$$

where $*$ denotes the complex conjugate. The rank of $\mathbf{B}$ is M due to the $\mathbf{E}_{KM}$:s Vandermonde property so that the signal part of the matrix $\mathbf{P}$ is of full rank M. This is a *necessary condition* for subspace type methods. In algorithm 2 one computes the M principal eigenvectors of $\mathbf{P}$ *which span theoretically the same subspace as the columns of* $\mathbf{E}_{LM}$, as we see from Eq. (11). This is a *sufficient condition*; so the algorithm 2 works theoretically. In algorithm 1 one takes the DFT in Step 4:

$$\mathbf{V} = \mathbf{P}\hat{\mathbf{E}}_{LM} = \mathbf{E}_{LM}\mathbf{H} + K\sigma^2\hat{\mathbf{E}}_{LM}, \qquad (12)$$

$$\mathbf{H} = \mathbf{B}(\mathbf{E}_{LM})^H\hat{\mathbf{E}}_{LM}.$$

Denoting the m:th row and the k:th column element of $\mathbf{H}$ by h_{mk}, we get

$$\begin{aligned}h_{mk} &= \tilde{a}_m(0)\sum_{n=1}^{M}\tilde{a}_n^*(0)\sum_{l=1}^{H}g_K(-\omega_m+\hat{\omega}_l)g_K(\omega_n-\hat{\omega}_l) \\ &\times g_L(\omega_n-\hat{\omega}_k), \quad m,k=1,2,\ldots,M,\end{aligned}$$

$$g_K(\omega) = \frac{\sin(K\omega/2)}{\sin(\omega/2)}e^{j(K-1)\omega/2}.$$

$g_K(\omega)$ is a sinc function. The signal part $\mathbf{E}_{LM}\mathbf{H}$ in Eq. (12) *spans the same subspace as* $\mathbf{E}_{LM}$. In general $\hat{\omega}_m \approx \omega_m$, and there exists (usually negligible) bias in the MUSIC spectrum due to the matrix $K\sigma^2\hat{\mathbf{E}}_{LM}$.

For studying the ability of Fourier transform in improving the SNR we must make certain simplifications. Assume that the estimates corresponding to the highest peaks of periodogram are exactly the same as the true frequencies: $\hat{\omega}_m = \omega_m$, $m = 1,2,\ldots,M$. Then $\mathbf{V}$ in Eq. (12) *spans exactly the desired subspace.* Assume also that $H = 2M$, and that there exists to each frequency a pair close to it: $\hat{\omega}_m \approx \hat{\omega}_{m+M}$. To make more simplifications, let the frequencies be orthogonal to each other: $\mathbf{e}_L^H(\omega_i)\mathbf{e}_L(\omega_j) = L\delta_{ij}$, $\mathbf{e}_K^H(\omega_i)\mathbf{e}_K(\omega_j) = K\delta_{ij}$, where δ_{ij} is Kronecker delta. Then

$$(\mathbf{E}_{KM})^T\hat{\mathbf{E}}_{KH} \approx K\begin{bmatrix}\mathbf{I} & \mathbf{I}\end{bmatrix}, \quad \mathbf{E}_{LM}^H\hat{\mathbf{E}}_{LM} = L\mathbf{I},$$

and an approximation to $\mathbf{V}$ is:

$$\mathbf{V} \approx 2K^2L\mathbf{E}_{LM}\mathbf{A}\mathbf{A}^H + K\sigma^2\mathbf{E}_{LM}.$$

Magnification factor $2K^2L/K = 2KL$ and $\mathbf{A}\mathbf{A}^H$ (diagonal matrix containing the powers of amplitudes in its elements) imply that SNR *is clearly improved.*

5 Generalization

The new algorithms, as well as the methods introduced in [2,3], belong to a more general category. Matrices $\mathbf{X}$ and $\hat{\mathbf{P}}$ are of to special type: Fourier transform enhances the energy of a signal part with respect to the noise part. We define a general matrix $\hat{\mathbf{M}}$ having this property. It has the form

$$\hat{\mathbf{M}} = \hat{\mathbf{M}}_s + \hat{\mathbf{M}}_w, \qquad (13)$$

$$\mathcal{E}\{\hat{\mathbf{M}}\} = \mathbf{M},$$

$$\mathcal{E}\{\hat{\mathbf{M}}_s\} = \mathbf{E}_{LM}\mathbf{C}(\mathbf{G}_{KM})^H,$$

$$\mathcal{E}\{\hat{\mathbf{M}}_w\} = \alpha\mathbf{J},$$

$$\Rightarrow \quad \mathbf{M} = \mathbf{E}_{LM}\mathbf{C}(\mathbf{G}_{KM})^H + \alpha\mathbf{J}, \qquad (14)$$

$$\mathbf{G}_{KM} = \begin{bmatrix}\mathbf{e}_K(\nu_1) & \mathbf{e}_K(\nu_2) & \cdots & \mathbf{e}_K(\nu_M)\end{bmatrix}, \qquad (15)$$

$$\mathbf{J} = \begin{bmatrix}\mathbf{I} & \mathbf{0}\end{bmatrix}, \qquad (16)$$

$$\begin{cases}\alpha \in \mathcal{C}, & \text{when } \omega_m = \nu_m, \quad m = 1,2,\ldots,M \\ \alpha = 0, & \text{otherwise}\end{cases}. \qquad (17)$$

Here $\mathbf{C}$ is arbitrary full rank $M \times M$ matrix. $\mathcal{C}$ is the complex number set, and the parameters ω_m and ν_m, $m = 1,2,\ldots,M$ are unknown. The ω_m:s are the frequencies or directions of arrival (DOAs) that are to be estimated. The ν_m:s are frequencies which are used in the prefiltering. The DFT operations corresponding to ν_m:s yield large norm linear combinations of $\mathbf{e}_L(\omega_m)$-vectors:

$$\begin{aligned}\mathbf{V} &= \mathbf{M}\hat{\mathbf{G}}_{KM} = \mathbf{E}_{LM}\mathbf{C}(\mathbf{G}_{KM})^H\hat{\mathbf{G}}_{KM} + \alpha\mathbf{J}\hat{\mathbf{G}}_{KM} \\ &= \mathbf{E}_{LM}\mathbf{C}(\mathbf{G}_{KM})^H\hat{\mathbf{G}}_{KM} + \alpha\hat{\mathbf{G}}_{LM}. \qquad (18)\end{aligned}$$

The condition (17) on parameter α is explained as follows. If $\alpha \neq 0$, then $\alpha\hat{\mathbf{G}}_{LM}$ in Eq. (18) may cause peaks in the locations ν_m. If $\omega_m \neq \nu_m$, they are undesired spurious peaks; ortherwise they are near the correct frequency locations.

There are several ways of constructing the matrix $\hat{\mathbf{M}}$ fulfilling the conditions (13)–(17). For example, with mild assumptions, correlation and crosscorrelation matrices belong to this category. The matrix $\hat{\mathbf{M}}$ arises in its nearly most general form when two sensor arrays are used:

$$\hat{\mathbf{M}} = \frac{1}{N}\sum_{n=0}^{N-1}\mathbf{x}(n)\mathbf{y}^H(n),$$

where $\mathbf{x}(n)$:s are snapshots of the first array, and $\mathbf{y}(n)$:s are snapshots of the second array. If the arrays are not in parallel direction, then they "see" different DOAs: ν_m:s and ω_m:s are different. In the data matrix $\mathbf{X}$ they also are different: $\nu_m = -\omega_m$. Then it must hold: $\alpha = 0$. This holds when noise is zero mean. The advantage of the generalized matrix is that it is possible to resolve the frequencies or DOAs from the crosscorrelation matrix that suppresses noise or in general from a matrix that is indefinite so that eigenvector analysis is not possible. We must only know the frequencies ν_m yielding the highest peaks of periodogram or largest norm vectors of the Bartlett estimate; then we can estimate the signal subspace.

In the array signal processing applications, the theory can be extended from linear arrays to deal with arbitrary array geometries

$$\mathbf{x}(n) = \mathbf{D}_{LM}\mathbf{a}(n) + \mathbf{w}(n), \quad n = 0, \ldots, N-1.$$

Here the general matrix $\mathbf{D}_{LM}$ has taken the place of $\mathbf{E}_{LM}$, and it describes the structure of the array. $\mathbf{a}(n)$ contains the amplitudes of the signals, and $\mathbf{x}(n)$ and $\mathbf{w}(n)$ are L-dimensional data and noise vectors, respectively. The estimated matrix is $\mathbf{D}_{LM}$, and the linear matrix transform is no more of Fourier type, but reflects the array structure.

6 Experiments

In the experiments we compare the new methods to forward-backward MUSIC. MUSIC estimate

$$\hat{P}(\omega) = \frac{1}{L - \mathbf{e}_L^H(\omega)\hat{\mathbf{V}}\hat{\mathbf{V}}^H\mathbf{e}_L(\omega)}$$

was evaluated at 2048 frequencies. The compared properties are probability of resolution (Table 1), and the number of spurious peaks when the dimensionality of the signal subspace was overestimated by one (Table 2). The SNR varied from −4 dB to 16 dB (first row in the tables). 100 simulations were made at every SNR. Noise process, amplitudes (a), frequencies ($f = \omega/2\pi$), and phases (θ) were chosen randomly in each realization so that they satisfied the conditions

$$a_1, a_2 \in (1,4), \quad |f_1 - f_2| = 0.02, \quad \theta_1, \theta_2 \in [0, 2\pi).$$

Each realization consisted of $N = 40$ samples so that $|f_1 - f_2| < 1/N$. $L = 17$. The frequencies were considered to be resolved if

$$\sqrt{\frac{1}{2}[(f_1 - \hat{f}_1)^2 + (f_2 - \hat{f}_2)^2]} \leq 0.02.$$

In determining the average number of peaks, peaks at most 30 dB below the maximum value of MUSIC spectrum were counted. The results in Tables 1 and 2 show that the algorithms 1 and 2 are clearly more reliable than MUSIC in noisy circumstances, and at high SNRs they yield spurious peaks rather seldom.

7 Conclusions

In this paper we presented new methods for estimating sinusoidal frequencies. They are combinations of classical Fourier methods and modern subspace type methods. Experiments show that these algorithms are clearly more robust than MUSIC, and they perform somewhat better than those proposed in [2,3]. A heuristic explanation is that the use of DFT makes the estimated signal subspace more robust, while the subspace approach yields high resolution. In this way one can combine beneficially classical Fourier-based spectral estimation with the high resolution subspace techniques.

References

1. S.J. Orfanidis, Optimum signal processing: an introduction. Macmillan, 1988.
2. J. Karhunen and J. Joutsensalo, "Sinusoidal frequency estimation by signal subspace approximation," to be published in IEEE Trans. Signal Processing, December 1992.
3. J. Karhunen and J. Joutsensalo, "Robust MUSIC based on direct signal subspace estimation", Proc. ICASSP-91, Toronto, May 1991, pp. 3357-3360.

Algorithm SNR →	-4	-2	0	2	4	6	8	10	12	14	16
MUSIC	4	8	13	19	37	48	67	80	89	95	**100**
Alg. 1, Periodogram ($H = 3$)	30	43	54	62	68	85	**93**	94	**98**	**99**	**100**
Alg. 1, Norm ($H = 3$)	10	13	18	22	28	48	64	76	88	92	98
Alg. 2, Periodogram ($H = 3$)	33	**48**	**64**	**71**	**76**	**89**	**93**	**96**	**98**	**99**	**100**
Alg. 2, Norm ($H = 3$)	6	12	20	31	42	62	74	86	93	97	99

Table 1: Probability of resolution.

Algorithm SNR →	-4	-2	0	2	4	6	8	10	12	14	16
MUSIC	2.89	2.86	2.83	2.77	2.75	2.67	2.56	2.60	2.57	2.59	2.49
Alg. 1, Per ($H = 4$)	**2.77**	**2.68**	**2.55**	2.39	2.15	2.07	**1.99**	**1.96**	**2.01**	**2.02**	**2.01**
Alg. 1, Norm ($H = 4$)	**2.77**	2.74	2.68	2.64	2.59	2.39	2.24	2.20	2.14	2.05	2.02
Alg. 2, Per ($H = 4$)	2.82	2.73	2.56	**2.28**	**2.12**	**2.05**	**1.99**	1.95	**2.01**	**2.02**	2.02
Alg. 2, Norm ($H = 4$)	2.91	2.91	2.91	2.84	2.76	2.68	2.50	2.43	2.39	2.29	2.25

Table 2: Average number of peaks when the dimensionality of the signal subspace was overestimated by one ($M = 3$).

SIGNAL PROCESSING VI: Theories and Applications
J. Vandewalle, R. Boite, M. Moonen, A. Oosterlinck (eds.)

ESTIMATION OF THE FUNDAMENTAL FREQUENCY OF A NOISY SUM OF CISOIDS WITH HARMONIC RELATED FREQUENCIES - APPLICATION TO SONAR MULTIPATHS DELAY ESTIMATION *

A. Ferrari, G. Alengrin and C. Theys

LASSY Université de Nice-Sophia Antipolis, Département de l'I3S
41 Bd Napoléon III, 06041 CEDEX FRANCE
tel. : +33-93217953, e-mail : ferrari@mimosa.unice.fr

We tackle in this paper the problem of the estimation of the fundamental frequency of a noisy complex sinusoidal signal with p harmonic related frequencies. This problem is studied considering an extension of the MUSIC algorithm based on the minimisation of a function of the p principal angles between two p-dimensional subspaces. This one-dimension minimisation is achieved via a quasi-Newton algorithm. The method is then applied to the multipaths delay estimation problem. Results on a SONAR signal are presented.

1. INTRODUCTION

This paper deals with the problem of estimating the fundamental frequency ω_0 of a complex sinusoidal signal with p harmonic related frequencies $k.\omega_0$, $k = 1, \ldots, p$, corrupted by a white Gaussian noise. We propose in this paper to take into account the constraints between the frequencies to reduce the problem to the estimation of ω_0.

2. MATHEMATICAL DEVELOPMENT

2.1. Signal subspace approach

We will derive in this section the principle of the algorithm in the general non-harmonic case. A vector $\mathbf{z}$ of M successive noisy observations of a complex sinusoidal signal with frequencies ω_k^0, $k = 1, \ldots, p$, can be modeled in the classical way:

$$\mathbf{z} = W(\Omega^0).\mathbf{x}_0 + \mathbf{n} \qquad (1)$$

where, the $M \times p$ Vandermonde matrix $W(\Omega^0)$ is defined as:

$$W(\Omega^0) = (\mathbf{d}(\omega_1^0) \mid \cdots \mid \mathbf{d}(\omega_p^0)) \qquad (2)$$

with the notations:

$$\mathbf{d}(\omega) = (1 \ \exp j\omega \ \cdots \ \exp j(M-1)\omega)^t \qquad (3)$$

$$\Omega^0 = (\omega_1^0 \cdots \omega_p^0)^t \qquad (4)$$

According to the classical result on the eigen-decomposition of the autocorrelation matrix of the process $\mathbf{z}$, we have:

$$\mathcal{R}W(\Omega^0) = \mathcal{R}E_s \qquad (5)$$

where:
– E_s denotes the $M \times p$ matrix associated to the signal subspace
– $\mathcal{R}Q$ denotes the range of the matrix Q
So, when Ω varies, $\mathcal{R}W(\Omega)$ is a continuum that has an intersection with $\mathcal{R}E_s$ for $\Omega = \Omega^0$. Hence,

* Work partly supported by DCN St. Tropez.

Ω^0 is the global minimizer of the distance between $\mathcal{R}W(\Omega)$ and $\mathcal{R}E_s$. In the general case, this approach leads to a p-dimensional minimization problem. This complex problem is solved in a sub-optimal way in the MUSIC algorithm replacing the global p-dimensional minimization by the 1-dimensional minimization of the distance between $\mathcal{R}E_s$ and a particular vector of $\mathcal{R}W(\Omega)$:

$$W(\Omega).(1\,0\,\cdots\,0)^t = \mathbf{d}(\omega) \tag{6}$$

We will in the sequel investigate the global optimization problem. This method is related to the minimization of the distance $\mathcal{D}$ between two linear spaces of dimension p. This distance is related to the notion of principal angles [2]. If we note θ_k, $k = 1, \ldots, p$ the p principal angles between two linear spaces A and B, $(0 \leq \theta_1 \leq \cdots \leq \theta_p \leq \pi/2)$, we have :

$$\mathcal{D}(A, B) = \sin\theta_p. \tag{7}$$

The angles between A and B are usually computed using the singular value decomposition of the matrix $Q_A^H.Q_B$ where Q_A is a matrix having as columns an orthogonal basis of A. The singular values of the above matrix are $\cos\theta_k$, $k = 1, \ldots, p$. This method has a very high computational cost, in fact it requires for every Ω, an orthogonalization of $W(\Omega)$ and the computation of the singular value decomposition of an order p matrix.

A less expensive method has then been investigated. It is based on the property that the matrix E_s being orthonormal, the matrix $T(\Omega)$ defined as:

$$T(\Omega) = E_s^H.W(\Omega).(W(\Omega)^H W(\Omega))^{-1}.W(\Omega)^H.E_s \tag{8}$$

has $\cos^2\theta_k$, $k = 1, \ldots, p$ as eigenvalues, [1]. Even if the computational cost has been considerably reduced, this method still requires the maximisation of the largest eigenvalue of the $p \times p$ matrix $T(\Omega)$.

To cope with this problem, we will slightly change the estimation criterion. We will replace the minimisation of θ_p by the minimisation of a convex function of all the principal angles. Moreover, we will impose the constraints:

– The function is convex on the domain $[0, \pi/2]^p$.

– The global minimum of the function is at the origin.

– The function is symmetric over its p variables.

This choice is also motivated by the fact that, according to the inequality between the θ_k, $\mathcal{R}W(\omega) = \mathcal{R}E_s$ is equivalent to $\theta_1 = \cdots = \theta_p = 0$. So, in the ideal case, the two criteria are identical. According to these remarks, we decide to minimize the new cost function $v(\Omega)$ defined as:

$$v(\Omega) = -\prod_{k=1}^{p}\cos^2\theta_k \tag{9}$$

$$= -|T(\Omega)| \tag{10}$$

$$= -\frac{\mathrm{abs}|W(\Omega)^H.E_S|^2}{|W(\Omega)^H.W(\Omega)|} \tag{11}$$

We can note that the general approach described above includes the Weighted Subspace Fitting criterion of M. Viber and al., [4]. In fact, this last criterion corresponds to the maximization of the function $\sum_{k=1}^{p}\cos^2\theta_k$.

2.2. Harmonic related frequencies case

In this particular case, the cost function $v(\Omega)$ is a one dimensional function of the fundamental frequency ω. We hence have to perform a one dimensional optimization of $v(\omega)$. Moreover a fondamental property of $W(\omega)$ is that, due to the particular relation between the frequencies, this matrix has a double Vandermonde structure, vertical and horizontal. So, $W(\omega).W(\omega)^H$ and $W(\omega)^H.W(\omega)$ have both a Tœplitz structure. This leads to the efficient formula for the criterion $v(\omega)$:

$$v(\omega) = -\frac{\text{abs}|W(\omega)^H.E_s|^2}{\prod_{k=1}^{p}\epsilon_k^2(\omega)} \tag{12}$$

where $\epsilon_k^2(\omega)$, $k = 1,\ldots,p$ are the variances of prediction errors obtained applying the Levinson recursion to $W(\omega)^H.W(\omega)$.

2.3. Optimization algorithm

The cost function $v(\omega)$ is minimized using a Newton method [3]. Using the expression of the differential of the determinant of a matrix, we obtain the following expression of the first derivative for $v(\omega)$:

$$v'(\omega) = 2.v(\omega).\Re\,\text{tr}(W'(\omega)^H.E_s.(W(\omega)^H.E_s)^{-1} - W'(\omega)^H.W(\omega).(W(\omega)^H.W(\omega))^{-1}) \tag{13}$$

where $\Re\,\text{tr}Q$ is the real part of the trace of Q. This expression can be efficiently evaluated using the structure of $W(\omega)^H.W(\omega)$. The formula of the second derivative of $v(\omega)$ being rather complicated, this quantity is computed using finite difference (quasi-Newton). An initialisation scheme coresponding to the studied problem is presented in the simulation section.

3. MULTIPATHS DELAY ESTIMATION

This methcd will be applied to the multipaths delay estimation in SONAR signals.

3.1. Problem formulation

The SONAR signal underscope is composed of a direct path returned from reflection on the target additionned with multipaths returned from

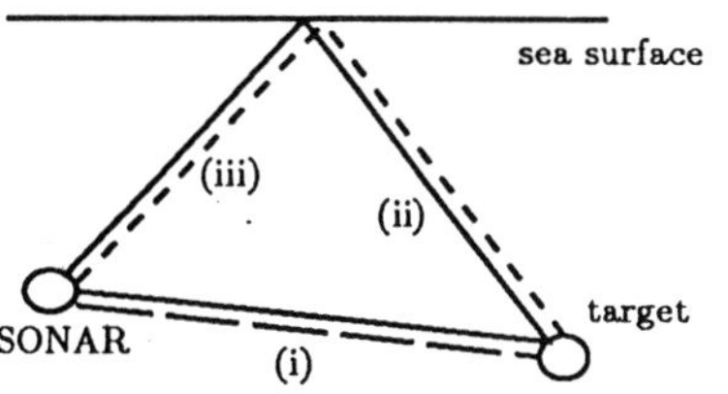

Fig. 1. Multipaths on sea surface

reflection on sea surface. When the sea surface is smooth, the intensity of sound reflected from the sea is nearly equal to that incident upon it, and so this reflection can falsify and even prevent detection and location of the target. Consequently we have to separate the instants of reflection on sea surface from the instant of reflection on the target, we hence have to estimate the multipath delays. Figure 1 represents the first three successive reflections .

(i) is the direct path: SONAR-Target-SONAR.

(ii) is the first reflection on the sea surface: SONAR-Target-Sea Surface-SONAR.

(iii) is the second reflection on the sea surface: SONAR-Sea Surface-Target-Sea Surface-SONAR.

If we note the delays of these successive reflections respectively t_d, t_1 and t_2, we notice that:

$$t_2 - t_1 = t_1 - t_d \tag{14}$$

$$t_2 - t_d = 2.(t_1 - t_d) \tag{15}$$

and more generally:

$$t_k - td = k.(t1 - td) \tag{16}$$

where t_k denotes the kth instant of arrival.

Taking into account the fact that the delays of the reflections are regularly spaced, (16), this signal can be modeled as:

$$x(t) = \sum_{k=1}^{p} A_k.s(t - k.\tau) \tag{17}$$

where $s(t)$ is the direct received signal, A_k and $k.\tau$ are respectively the amplitude and the delay of the kth reflected path, $(\tau = t_1 - td)$. We assume there is no noise and the propagation affects linearly the amplitude of the emitted signal. If we note $X(f)$ the Fourier transform of $x(t)$ and $S(f)$ the Fourier transform of $s(t)$, we obtain:

$$Y(f) = \frac{X(f)}{S(f)} = \sum_{k=1}^{p} A_k.\exp(-2\pi.j.f.k.\tau) \quad (18)$$

If we consider now the N points signal Y_k obtained sampling uniformely $Y(f)$ in the frequency domain where $S(f) \neq 0$, we are reduced to the estimation of p harmonic related frequencies.

3.2. Simulation results

Figure (2) represents 512 data points of a band-pass filtered SONAR signal with multipaths. The signal s_n is estimated from the 100 first samples of this signal. The signal Y_k is obtained using a Fast Fourier Transform of s_n and x_n. Figure (3) represents a Fast Fourier Transform of Y_k. The attenuation due to the successive reflections are clearly visible on this plot. The optimization algorithm is initialized with the frequency associated to the most powerfull spectrum line (the first reflection). Moreover, we have chosen $p = 5$ and $M = 8$. The estimated fundamental frequencies of the seven first iterations given by the algorithm and converted in time samples are: 500, 745, 526, 523, 521, 522, 522. The first and last one of these values are represented on the figure (2) by dashed and solid lines. The corresponding value calculated according to the geometry of the system measured during the experiment is 530 samples. In regard to these results, the proposed method is a good solution to the multipaths delay estimation problem.

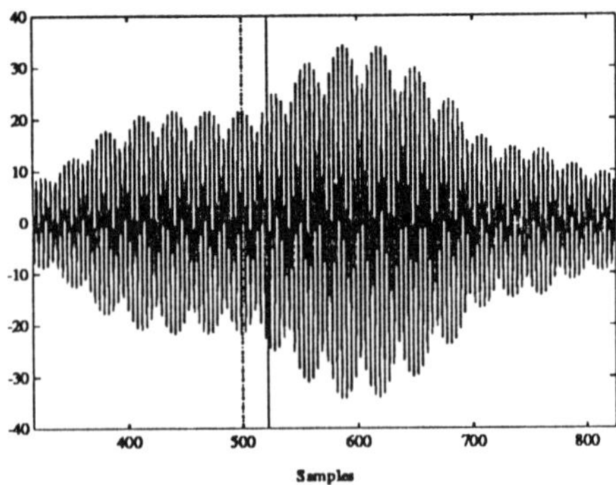

Fig. 2. SONAR signal

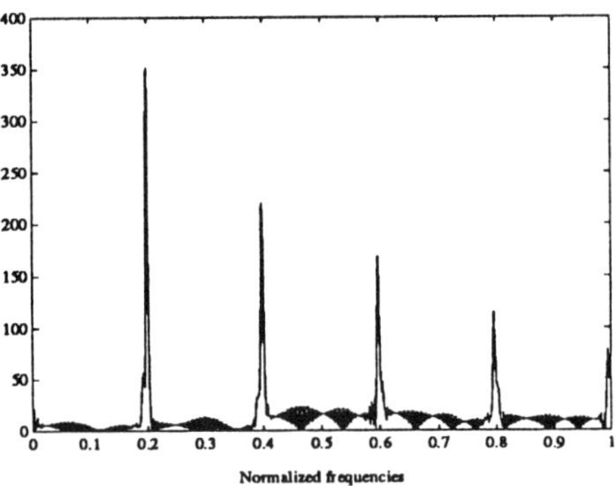

Fig. 3. Fast Fourier Transform of Y_k

References

[1] A. Björck and G. H. Golub. Numerical Methods for Computing Angles Between Linear Subspaces. *Mathematics of Computation*, 27(123):579–594, July 1973.

[2] G. H. Golub and C. F. Van Loan. *Matrix Computations*. Series in Mathematical Sciences. The Johns Hopkins University Press, 1989.

[3] D. Kahaner, C. Moler, and S. Nash. *Numerical Methods and Software*. Prentice-Hall, 1989.

[4] M. Viberg and B. Ottersten. Sensor Array Procesing Based on Subspace Fitting. *IEEE Transactions Signal Processing*, 39(5):1110–1121, May 1991.

SIGNAL PROCESSING VI: Theories and Applications
J. Vandewalle, R. Boite, M. Moonen, A. Oosterlinck (eds.)

WHAT IS A GOOD ESTIMATE?

P. L. Combettes[†] and *W. W. Edmonson*[‡]
[†]Department of Electrical Engineering, City College and Graduate School,
City University of New York, New York, NY 10031, USA.
[‡]Computational Science and Engineering Research Center,
Howard University, Washington, DC 20059, USA.

Numerous point and set theoretic estimation procedures have been used in signal processing. One can however question the adequacy of standard estimation objectives and their subsequent performance evaluation criteria in practical applications. In this paper, we address these questions and analyze the important factors that should be involved in the rational quest for meaningful estimates.

1. INTRODUCTION

The goal of most signal processing and systems theory problems is to produce estimates of objects relevant to specific analysis or synthesis purposes. Thus, in system identification, one estimates the parameters involved in the model of a system; in signal recovery, the goal is to estimate the original form of a signal; in spectral estimation, the objective is to estimate the spectral distribution of a stochastic process.

One tends to select an estimation scheme on grounds which are seldom related to rational and practical goals reflecting the specificities of the problem at hand. In general, one favors standard estimation procedures that provide, without too much effort, a solution. This point is illustrated by the widespread use of the maximum likelihood estimator, whose practical virtues have yet to be rationally established but which yields a solution in a fairly straightforward manner. In most cases, little or no attention is paid to critical factors such as the incorporation of *a priori* information and the relevance of the estimation objective. Thus, procedures which yield "optimal" solutions have been very popular despite the fact that the optimality criterion is often arbitrary and that the resulting optimization problem is not easily modifiable to incorporate whatever *a priori* information may be available about the problem. An alternative approach, which has proven effective in a steadily increasing number of systems theory and signal processing problems, is that of set theoretic estimation. In that framework, more emphasis is placed on the uncertainty that surrounds the problem. The requirement for a single estimate is relaxed and the solution is a set of equally valid solutions.

Various definitions for "good" estimates have been proposed. Thus, some will argue that a maximum likelihood estimate is desirable while others will discount it on account of its many pathologies. In fact, such pathologies exist for almost every type of estimation procedure and have given rise to comparable controversies [6], [7], [8], [9], [14].

The purpose of this paper is not to propose a formal and universal definition of what a good estimate should be but, rather, to discuss the various factors that should be involved in the selection of an estimation method in order to obtain physically meaningful estimates in an objective manner. In the next section, we briefly describe the most common estimation procedures. In Section 3, we proceed to analyze the tools available to evaluate the performance of these procedures. We discuss various estimation procedures in Section 4 and conclude the paper in Section 5.

2. ESTIMATION OBJECTIVES

2.1. General Model

Two key elements in an estimation procedure are the solution space and the observation space. The solution space Ξ is the space to which the true object h (the state of nature) belongs. Depending on how h is modeled, this space can take various forms, e.g., a field of scalars, a matrix space, a functional or distributional space. On the other hand, the observation space Δ is the space to which the observed data x belong. An estimation procedure is a mapping $\mathcal{T}$ which assigns to the observed data x a subset $\mathcal{T}(x)$ of Ξ, that is,

$$\begin{aligned} \mathcal{T} : \Delta &\to 2^{\Xi} \\ x &\mapsto \mathcal{T}(x). \end{aligned} \tag{1}$$

The set of points $\mathcal{T}(x)$ represents our guess of the value of the true state of nature h given the data x and some *a priori* knowledge.

In statistics, the data model is probabilistic. If $(\Omega, \mathcal{F}, \mathsf{P})$ denotes the underlying probability space, the observed data are regarded as a realization $x = X(\omega)$ of a stochastic process $X : \Omega \to \Delta$ and a set theoretic estimator is a measurable map $\mathcal{M} = \mathcal{T} \circ X : \Omega \to \Sigma$, where Σ is a σ-algebra of subsets of Ξ [1].

2.2. Point Estimates

The objective of a point estimation procedure is to provide a single estimate, i.e., $\mathcal{T}(x) = \{a(x)\}$. The estimate $a(x)$ is usually obtained by optimizing an objective function over Ξ. Various optimality criteria have been proposed for point estimates, e.g., maximum likelihood, minimax, maximum entropy, maximum *a posteriori* and other Bayesian criteria [1], [10], [12]. We describe the main approaches here.

Let $(\Omega, \mathcal{F}, (\mathsf{P}_h)_{h\in\Xi})$ be the underlying statistical model, assumed to be dominated by P. Then a maximum likelihood estimator is a point $a(x)$ which maximizes the likelihood function $\mathcal{L}(a) = d\mathsf{P}_a/d\mathsf{P}$ over Ξ. In the Bayesian approach, the true object h is regarded as a realization of a random variable whose distribution (the prior) is known. Moreover, one selects a function L, where $L(a, h)$ is the loss associated with estimating h by a. The associated Bayesian estimate $a(x)$ is the point which minimizes the posterior expected loss, i.e., the expected loss conditioned on the data. The posterior distribution involved in the computation of a Bayesian estimate comprises a prior distribution, which reflects the *a priori* information on h, and a likelihood function, which reflects sample information. In the minimax approach, one seeks to minimize the maximum expected loss, which is is usually regarded as a conservative procedure.

2.3. Set Estimates

The result of a set theoretic estimation procedure is not a single point but a subset $\mathcal{T}(x)$ of Ξ, which usually does not reduce to a singleton. Several approaches have been proposed to define $\mathcal{T}(x)$.

In statistics, confidence regions constitute a well established set theoretic method of estimation [1], [12]. Let ω be the elementary event giving rise to the observation $X(\omega) = x$ of the data process. A confidence region is a subset $\mathcal{T}(x) = \mathcal{T}(X(\omega))$ of Ξ such that

$$(\forall h \in \Xi)\ \ \mathsf{P}_h\{\omega \in \Omega \mid h \in \mathcal{T}(X(\omega))\} \geq \alpha, \qquad (2)$$

where $(\Omega, \mathcal{F}, (\mathsf{P}_h)_{h\in\Xi})$ is the underlying statistical model.

Another set theoretic approach is the set-valued Bayesian framework proposed in [5] and further discussed in [13]. In Bayesian point estimation, the specification of a prior distribution is critical for the estimator often lacks robustness, particularly if the data record is not extensive. One is required to select a single prior distribution that will properly model *a priori* knowledge and, at the same time, yield a tractable optimization problem. A way to relax this requirement while remaining faithful to the Bayesian philosophy is to consider a set of prior distributions, each of which is an equally acceptable candidate to model the uncertainty surrounding the true state of nature. An estimate is obtained for each prior distribution and $\mathcal{T}(x)$ is therefore a set.

In [4], a general set theoretic estimation framework is presented in which the emphasis is placed on consistency with respect to all the information available about the problem, i.e., the *a priori* knowledge and the data. Let $\Psi_\iota : \Xi \to [0, 1]$ be the fuzzy logic proposition associated with a particular piece of information. If $(\Psi_\iota)_{\iota\in I}$ is the family of all such propositions, given a real number ψ in $]0, 1]$, a family $(S_\iota)_{\iota\in I}$ of property sets in the solution space Ξ can be constructed as follows

$$(\forall \iota \in I)\ S_\iota = \{a \in \Xi \mid \Psi_\iota(a) \geq \psi\}. \qquad (3)$$

S_ι is the set of all estimates which are consistent with the information carried by Ψ_ι at level ψ. The pair $(\Xi, (S_\iota)_{\iota\in I})$ is called a ψ-level set theoretic formulation. The solution is the feasibility set $\mathcal{T}(x) = \bigcap_{\iota\in I} S_\iota$. Any point in S is called a set theoretic estimate. Although it is nonstatistical, this set theoretic framework encompasses the two above set theoretic procedures if appropriate statistical information is present. Indeed, the sets produced by these procedures can be incorporated to the set theoretic formulation as specific property sets, i.e., $S_\iota = \mathcal{T}(x)$.

3. EVALUATION TOOLS

It is generally accepted that an estimation procedure should be accompanied by an estimate of its accuracy. However, as noted by Savage [12], there are conceptual problems associated with evaluation procedures. First of all, it creates an endless regression since an estimate of accuracy should be accompanied by an estimate of its own accuracy, etc. Second, one can question the purpose of an evaluation procedure. Except for satisfying our curiosity, there is little practical use to know how accurate the estimate is once the estimation process is completed. One could argue that evaluation tools may help in deciding that a given estimate is better than another, but such a decision is subjective since it is based on the choice of the accuracy criterion.

3.1. Point Estimates

There exist a plethora of tools for evaluating point estimates and we mention only the most important here.

A popular index to evaluate the accuracy of a point estimate is the mean-square error, $\mathsf{E}|a - h|^2$, which reduces to the variance in the unbiased case. In connection with this index, the so-called Cramér-Rao[1] bound (CRB) has become the most often used criterion in determining the accuracy of an estimator. The CRB represents the minimum variance that can be achieved by an unbiased estimator. Its popularity is mainly due to the fact that it is usually easy to compute.

Another way to evaluate the accuracy of an estimation procedure is to compute the divergence between the actual probability structure and the probability structure resulting from the estimate. These criteria fall into three basic

[1]Historical note: the name of the bound refers to the work of Rao (1945) and Cramér (1946) but is not justified. Indeed, the bound was derived by Fréchet in 1943 and, for the multi-parameter case, by Darmois in 1945 [12].

categories: the generalized Kolmogorov variational distance, the f-divergence, and the Chernoff distance (e.g., see [2], [11] and the references therein). The Kullback-Leibler and Bhattacharyya distances are two of the more popular criteria of this class.

The CRB and the information and distance criteria are based on two assumptions, infinite sample and known probability distributions, which prove to be unrealistic in practice. The number of samples is always finite where these criteria are assuming that we have infinite data. In most instances, knowledge of the probability distributions is either incomplete or nonexistent. Under these conditions, estimates of the probability distributions should be used, but they do not perform as well as predicted from theory. Finally, let us stress that the CRB is useful only for unbiased estimates. This is a severe restriction since there is no sound theoretical or practical reason to prefer unbiased estimates to biased ones.

Utility [12] is a statistical concept that has received little attention in the evaluation of estimators. Utility is used along with the above criteria. It is a function of the intended use of the estimate and of the risk and benefits associated with the criterion. As an example, suppose that we are concerned with estimating the spectral content of a time series only within a specific frequency band. One estimator is more accurate over the complete spectrum while the second is more accurate within the desired frequency band. Then the second estimator has more utility than the first.

Finally, let us mention that a straightforward way to evaluate the accuracy of an estimation procedure is to run Monte Carlo computer simulations to obtain a qualitative indication of the dispersion of the estimates over realizations. This brute-force approach does not rely on a particular analytical criterion but provides a global qualitative description of the accuracy of the estimation procedure.

3.2. Set Estimates

The accuracy of a set theoretic estimation procedure can be assessed by the dispersion of the estimates over the solution set. In [3], various monotone set functions, called mensurations, were proposed for this purpose. Assuming that the solution space Ξ is a metric space with distance d, examples of mensurations of a solution set S are are the diameter of S; the thickness of S (i.e., the diameter of the largest open ball contained in S); $\mu(S)$, $\int_{\Xi} \delta_S d\mu$, and μ-ess $\sup\{\delta_S(a) \,|\, a \in \Xi\}$, where μ is a measure and δ_S a proper deviation function.

4. DISCUSSION

We have seen that two main types of estimation procedures exist. One, in which there is a single "good" estimate which is optimal in some predetermined sense and one in which there is a set of equally "good" estimates, usually determined in terms of some feasibility argument.

Following Zadeh [14], we can can question the rationality of the quest for optimal estimates when different users, through their own interpretation of the optimal way to solve the problem, may obtain different solutions. In fact, our insistence on optimal solutions often leads to arbitrary decisions because the selection of a criterion of performance is inherently subjective and solving the problem may require oversimplifications in its formulation. For some particular point estimation procedures, more specific reservations can be expressed. For instance, the maximum likelihood (ML) estimator has enjoyed widespread use on account of its "desirable" properties such as normality and efficiency. However, the practical validity of these properties is quite debatable and, moreover, they are only asymptotic. Theoretically, the ML estimator may not be defined or lead to irrelevant results. Finally, computational tractability generally imposes standard assumptions, such as normality of the random processes involved, and leaves almost no room for the incorporation of *a priori* knowledge. On that last point, the Bayesian framework is better suited. However, it requires to model the object h to be estimated as the realization of a random variable to which a distribution function is assigned (in most situations, there is however nothing random about h, which may for instance be a fixed but unknown physical constant). In addition, not all *a priori* information can be conveniently described in probabilistic terms and the resulting prior distribution is usually too complex to yield a tractable minimization of the resulting conditional expectation. At all events, a Bayesian technique is fundamentally subjective since the specification of the prior distribution and of the loss function is left to the user. In the extreme, the user could reach any conclusion he desires through biased choices of these specifications.

The central principle governing set theoretic estimation is that of feasibility. An acceptable solution is any object which is consistent with all the information available about the problem, namely the data and the *a priori* knowledge. By recasting this concept in a solution space, the estimation problem is to find a point in the intersection of property sets, each of which represents the class of objects consistent with a particular piece of information. It is clear that set theoretic estimation departs radically from the conventional approach which relies on optimal decision theory and in which feasibility is either ignored or of secondary importance. In this regard, it has been charged that the validity of set theoretic methods could be questioned on the grounds that they do not yield a unique solution but a set of solutions and are therefore suboptimal. Let us first remark that, although it may be gratifying to have obtained the "best" solution, optimality claims were seen to have little practical value. At best, their result can be regarded as a qualitative selection of a feasible solution. Moreover, from a conceptual standpoint, demanding that only one solution be acceptable for problems which are notoriously affected by uncontrollable factors (e.g. noise, uncertain data formation model) seems somewhat unwise. Let us also note that methods which yield unique solutions are usually iterative and their solution depends on the stopping rule. Certainly, such a criterion does not define a unique solution.

The basic objective of the set theoretic approach is to provide a flexible framework for the incorporation of a wide range of information. This is contingent on the availability of methods to generate feasible solutions, i.e., methods to find a point in the intersection of a collection of property sets. If all the sets are closed and convex and Ξ is hilbertian, several methods are available to solve this problem; in the nonconvex case, methods have been developed but they do not apply to all cases [4]. Hence, the lack of algorithm for the nonconvex case constitutes the main limitation of the set theoretic framework as there are many problems for which all the information cannot be associated with closed and convex subsets of a Hilbert solution space.

The selection of set theoretic or conventional estimation theory to solve a problem depends on several factors. First of all, it should be noted that even if it is at the expense of rationally posing the estimation problem, an advantage of some conventional estimation methods is to yield simple problem formulations and, in some cases, closed-form solutions. Indeed, one can always have recourse to standard cost functions and assumptions to simplify the problem. This approach has at least the merit of leading to a solution. On the other hand, although there are methods for computing set theoretic estimates for a wide class of problems, setting up a tractable set theoretic estimation formulation may not always be possible. Even if it is, the use of whatever conventional estimation method seems appropriate should not be precluded, especially if a solution can be computed efficiently. This solution can then be tested for feasibility with respect to available information. If it is feasible, it must be accepted; if not, one should expect a set theoretic solution to bring improvement. Of course, an ideal estimation procedure would be one in which a fully relevant objective function is rationally selected and minimized over the feasibility set. Unfortunately, given the state-of-the-art in constrained optimization theory, this approach seldom viable.

5. EPILOGUE

The title of this paper is similar to Alice in Wonderland asking the Cheshire Cat, "Would you tell me, please, which way I ought to go from here?" - "That depends a good deal on where you want to go," said the Cat. Then the Cat began to vanish, first the tail and then the body of conventional statistical methods. The last visible part was the set theorist's grin, "which remained some time after the rest of it had gone". But that, too, disappeared.

REFERENCES

[1] J. R. Barra, *Notions Fondamentales de Statistique Mathématique*. Paris: Dunod, 1971.

[2] C. H. Chen, "On Information and Distance Measures, Error Bounds, and Feature Selection," *Information Sciences,* vol. 10, pp. 159-173, 1976.

[3] P. L. Combettes and T. J. Chaussalet, "Critères de Qualité en Estimation Ensembliste," *Actes du Treizième Colloque GRETSI,* pp. 249-252. Juan-les-Pins, France, September 16-20, 1991.

[4] P. L. Combettes, "The Foundations of Set Theoretic Estimation," *Proceedings of the IEEE,* in review.

[5] A. P. Dempster, "A Generalization of Bayesian Inference," *Journal of the Royal Statistical Society,* vol. B30, pp. 205-247, 1968.

[6] B. Efron, "Controversies in the Foundations of Statistics" *The American Mathematical Monthly,* vol. 85, no. 4, pp. 231-246, April 1978.

[7] B. Efron, "Why Isn't Everyone a Bayesian?" *The American Statistician,* vol. 40, no. 1, pp. 1-5, February 1986.

[8] S. S. Gupta and J. O. Berger (Editors), *Statistical Decision Theory and Related Topics IV,* vol. 1. New York: Springer-Verlag, 1988.

[9] G. T. Herman, "Mathematical Optimization versus Practical Performance: A Case Study Based on the Maximum Entropy Criterion in Image Reconstruction," *Mathematical Programming Study,* vol. 20, pp. 96-112, October 1982.

[10] J. N. Kapur, "Twenty-Five Years of Maximum-Entropy Principle," *Journal of Mathematical and Physical Sciences,* vol. 17, no. 2, pp. 103-156, April 1983.

[11] Y. Lee, "Information-Theoretic Distortion Measures for Speech Recognition," *IEEE Transactions on Signal Processing,* vol. 39, no. 2, pp. 330-335, February 1991.

[12] L. J. Savage, *The Foundations of Statistics,* second edition. New York: Dover, 1972.

[13] W. C. Stirling and D. R. Morrell, "Convex Bayes Decision Theory" *IEEE Transactions on Systems, Man, and Cybernetics,* vol. 21, no. 1, pp. 173-183, January 1991.

[14] L. A. Zadeh, "What Is Optimal?," *IRE Transactions on Information Theory,* vol. IT-4, no. 1, p. 3, March 1958.

SIGNAL PROCESSING VI: Theories and Applications
J. Vandewalle, R. Boite, M. Moonen, A. Oosterlinck (eds.)

OPTIMAL RECONSTRUCTION OF TRANSIENT SIGNALS USING DECONVOLUTION ALGORITHMS[1]

T. Dabóczi and I. Kollár

Department of Measurement and Instrument Engineering, Technical University of Budapest
H-1521 Budapest, Hungary, Phone, fax: + 36 1 166-4938
e-mail: h4427dab@ella.uucp

Estimation of the transient input signal of a system from its measured output is an often arising problem, because a dynamical system usually significantly distort the waveform. There are several inverse filtering methods for this purpose. However, the optimization method is generally heuristic, and dedicated for a certain inverse filtering algorithm.

In this paper a new approach is proposed to calculate the error energy in the estimation of the input signal as a tool to find the optimal reconstruction in least squares sense. The proposed algorithm requires a rough approximation only of the absolute value of the input spectrum and noise variance. It is shown that the quality of the optimal reconstruction is robust with respect to slight errors in input spectrum and noise variance. The proposed method can be applied for any inverse filtering algorithm to get the optimal reconstruction.

Keywords: inverse filtering, deconvolution, signal reconstruction, optimal filtering.

1. Introduction

The measurement of fast transient signals is usually a difficult task, since the bandwidth of the measuring instrument is often lower than that of the measured signal, that is, the measuring instrument distorts the waveform. When this distortion is not acceptable, the effect of the system should be compensated for. Considering a linear time invariant measuring system, the relation between the input and output signals can be described by convolution:

$$y(t) = x(t) * h(t) \quad , \tag{1}$$

where $x(t)$ is the input signal, $y(t)$ is the output signal, and $h(t)$ is the impulse response of the system. The exact output waveform is usually not known, because the measured signal is always corrupted by noise (electromagnetic interferences, quantization noise etc.):

$$y_n(t) = x(t) * h(t) + n(t), \tag{2}$$

where $y_n(t)$ is the observed (noisy) signal and $n(t)$ is the noise component. In the frequency domain this convolution corresponds to multiplication:

$$Y_n(f) = X(f)H(f) + N(f), \tag{3}$$

where the capital letters correspond to the frequency domain signals. For the input signal the following estimator eliminates the effect of the measuring system:

$$X_{est}(f) = \frac{Y_n(f)}{H(f)} = X(f) + \frac{N(f)}{H(f)}, \tag{4}$$

where X_{est} is the estimated input spectrum. From Eq. (4) it can be seen that the noise will be amplified at the frequencies where $H(f)$ is small, usually in the high frequency band. The effect of the noise increases rapidly with the order of the transfer function of the system, and makes the estimation result unusable. It will be called *rough deconvolution*. Therefore, the noise should be suppressed to get a smooth result. Since the noise and the useful signal component usually cannot be separated, the noise reduction will lead to a bias in the estimation. A compromise has to be found between variance and input signal distortion.

There are several factors which influence the performance of the deconvolution:

1. input signal shape (or, equivalently, its spectrum),
2. transfer function of the measurement system,
3. distribution and level of the measurement noise,
4. errors in the identification of the measurement system,
5. deconvolution algorithm,
6. degree of noise suppression vs. bias.

Throughout this paper the first three factors will be supposed to be given.

---

[1] This research was sponsored by the Hungarian Fund for Scientific Research (OTKA), under contract no. 780/91.

In the deconvolution process the transfer function of the system is assumed to be exactly given. In practice this is not true, since non-parametric identification is also a deconvolution problem, and the noisy measurement causes similar problems as above. Fortunately, in many cases the input signal estimation is not too sensitive with respect to slight identification errors.

The deconvolution algorithm can be chosen by the user. There are several methods described in the literature. The different error criteria lead to different deconvolution algorithms [1]-[7]. The obtained results are close to each other, though they are not the same.

The degree of noise reduction vs. bias can be adjusted by a certain parameter in every algorithm, which can be either manually or automatically chosen. However, in the case of transient deconvolution the optimal setting of deconvolution parameters is often heuristic. A systematic approach, deduced from a mathematical model, is usually lacking.

In this paper the tradeoff between bias and noise variance is studied. A new method is proposed to calculate the approximate error energy of the input signal estimation as a tool to find the optimal deconvolution in least squares (LS) sense. Since energy estimation is approximated, the given result is suboptimal, but it will be shown that the obtained optimum is very close to the exact LS solution.

2. MODEL FOR THE APPROXIMATION OF THE ERROR ENERGY

The error energy of the input signal estimation can be calculated for a simplified model of the measurement setup. The model is shown in Fig. 1.

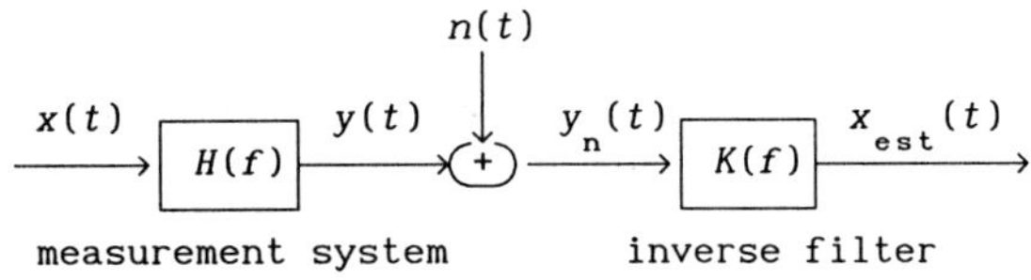

Fig. 1. Model of the measurement setup

In the frequency domain the noise can be reduced to the input:

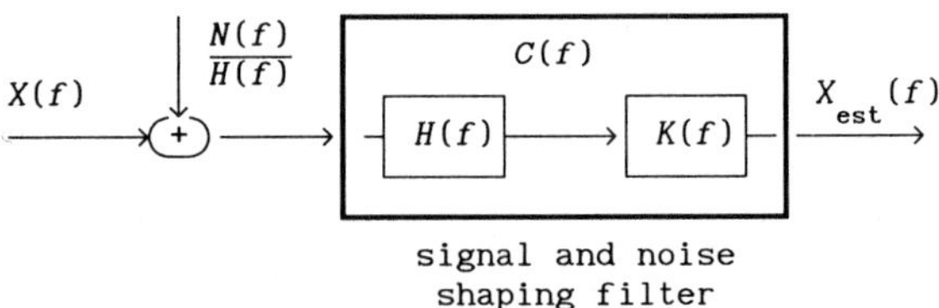

Fig. 2. Model of the measurement setup
The noise is reduced to the input

The error energy is then:

$$EE = T_s \sum_{n=-\infty}^{\infty} \left(x(n) - x_{est}(n)\right)^2, \tag{5}$$

where EE is the error energy, T_s is the sampling period, $x(n)$ is the measured input signal at time instant $n*T_s$ and x_{est} is the estimated input signal. The energy can be computed in the frequency domain using Eq. (4) and Parseval's theorem:

$$EE = \frac{1}{2\pi}\int_{-\pi}^{\pi}\left| X(e^{j\Omega}) - X_{est}(e^{j\Omega})\right|^2 d\Omega$$

$$= \frac{1}{2\pi}\int_{-\pi}^{\pi}\left| X(e^{j\Omega}) - \left(X(e^{j\Omega}) + \frac{N(e^{j\Omega})}{H(e^{j\Omega})}\right) C(e^{j\Omega})\right|^2 d\Omega$$

$$= \frac{1}{2\pi}\int_{-\pi}^{\pi}\left| X(e^{j\Omega})\left(1 - C(e^{j\Omega})\right) - \frac{N(e^{j\Omega})}{H(e^{j\Omega})} C(e^{j\Omega})\right|^2 d\Omega$$

$$= \frac{1}{2\pi}\int_{-\pi}^{\pi}\left| X(e^{j\Omega})\left(1-C(e^{j\Omega})\right)\right|^2 d\Omega + \frac{1}{2\pi}\int_{-\pi}^{\pi}\left| \frac{N(e^{j\Omega})}{H(e^{j\Omega})} C(e^{j\Omega})\right|^2 d\Omega$$

$$+ \frac{1}{2\pi}\int_{-\pi}^{\pi} 2\left| X(e^{j\Omega})\left(1-C(e^{j\Omega})\right)\right| \left| \frac{N(e^{j\Omega})}{H(e^{j\Omega})} C(e^{j\Omega})\right| \cos(\varphi) d\Omega \tag{6}$$

$$EE = EE_{bias} + EE_{noise} + EE_{bias,noise}, \tag{7}$$

where the capital letters correspond to the Fourier transforms of the signal sequences, $C(e^{j\Omega})$ is the transfer function of the discrete signal and noise shaping filter, and φ is the angle of the two absolute valued terms in the last integral. The error energy is split into three terms, where $EE_{bias,noise}$ is due to the cross connection of the bias and noise term. The theoretically correct error energy can be calculated from Eq. (6). In practice, however some approximations have to be used because of the following reasons:

1. The signal sequences can only be measured in a finite time interval,
2. The noise and input signal spectra are not exactly known.

The finite time record causes edge effects, but these can be neglected if the measurement time is long enough, compared to the transient duration.

Another problem is that the noise is a stochastic signal, and the longer is the measurement time, it has the more energy. In order to avoid this uncertainty, the record length must be standardized.

For the second problem an approximate solution is proposed. In many cases the observation noise can be assumed to have a white spectrum, its variance can be more or less accurately measured. With the above assumptions the output noise spectrum can be approximated by a constant.

The input spectrum can be approximated on the basis of *rough deconvolution* (Eq. (4)). Considering a lowpass system, the noise will be significantly amplified only at high frequencies, the low frequency band remains useful. The absolute value of the input spectrum can be approximated by fitting an appropriate signal model to the low frequency part of the spectrum of the rough deconvolution. For the signal model a differential equation of appropriate order can be given, which can be well fitted in frequency domain by adjusting the coefficients.

In the term $EE_{bias,noise}$ the angle of two terms should be given. Since the phase of the spectra of the noise and the input signal is not known, an upper limit can be given by $\cos(\varphi)=1$. From the simulation it was found that usually the first two terms dominate in the energy expression (7), and this last term can be neglected. In the following sections only the bias and noise term will be calculated.

In the next sections the proposed method will be verified by simulations and experimentally.

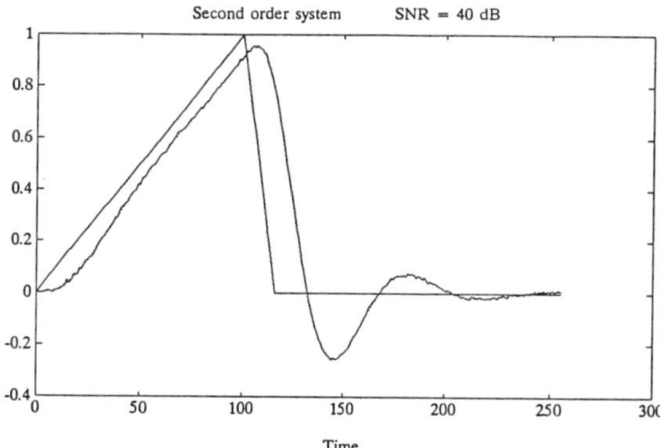

Fig. 3. Simulated input and output signals

3. STUDY OF THE PERFORMANCE LIMITATION

A standard test signal [1] was used to investigate the input error energy and to verify the accuracy of the proposed approximation. The simulated signals are shown in Fig. 3. The two ramps form a good model for front chopped high voltage impulses. A second order system was taken into consideration by applying the backward difference transform to $H(f)$. The output signal was corrupted by white Gaussian noise sequence with 40 dB signal-to-noise ratio, where the SNR is defined by:

$$\mathrm{SNR} = 10\,\log_{10}\left(\frac{\frac{1}{T}\int_0^T y(t)^2\,dt}{\sigma^2}\right), \qquad (8)$$

where σ^2 is the noise variance. For the inverse filter the constrained least squares solution was used [2], [4]:

$$K(f) = \frac{|H(f)|^2}{|H(f)|^2 + \lambda}, \qquad (9)$$

where λ is the parameter which has to be optimized to find the optimal compromise between bias and variance. (The measurement setup and deconvolution algorithm given here are only examples for the problem of the inverse filtering.)

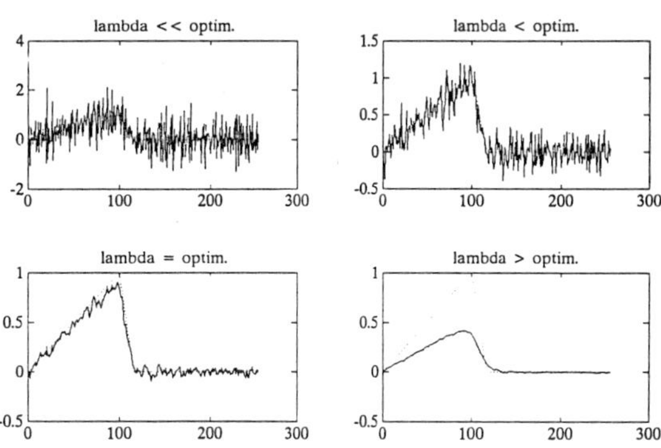

Fig. 4. Deconvolution results with different settings of the optimization parameter

In Fig. 4. the deconvolution results can be seen for different λ values. If λ is very small, i.e. there is no noise reduction, the estimated signal oscillates by an amplitude, higher than the input signal peak by a factor of 30. Increasing λ, the noise will be suppressed while the useful signal becomes biased. The bias can be well seen by large values of λ.

The proposed energy estimation requires the approximation of only the absolute vale of the spectra. The output noise variance is assumed to be known, its energy density spectrum will be approximated by a constant. For the input signal a rough approximation was given. A second order model was fitted to the low frequency part of the absolute value of the spectrum of the rough deconvolution. The input spectra are to be seen in Fig. 5. The model approximates well the absolute value of the input spectrum. The error energy of the deconvolution process was computed for different λ values, substituting the approximated input spectrum into the first two terms of Eq. (6). The true and the approximated error energies are shown in Fig. 6. The estimation is close to the true one. The relative error is no more than 40%. The most important aspect is the precision of the minimum detection since the main goal is the optimal reconstruction. The approximated minimum was found by a 30% precision. Fortunately, the constrained LS algorithm is not too sensitive with respect to the λ value, the input estimation with approximated λ_{min} differs only at few points from the LS solution (Fig. 7), what means that the result obtained by the approximated parameter can be considered to be optimal.

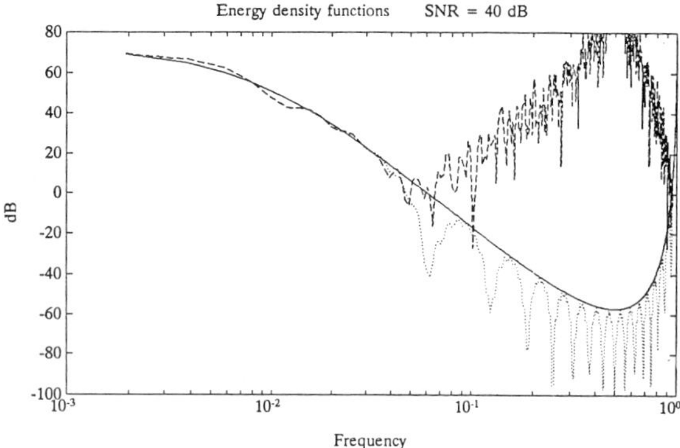

Fig. 5. The input (. . .), noisy input (- -) and approximated input spectra (——)

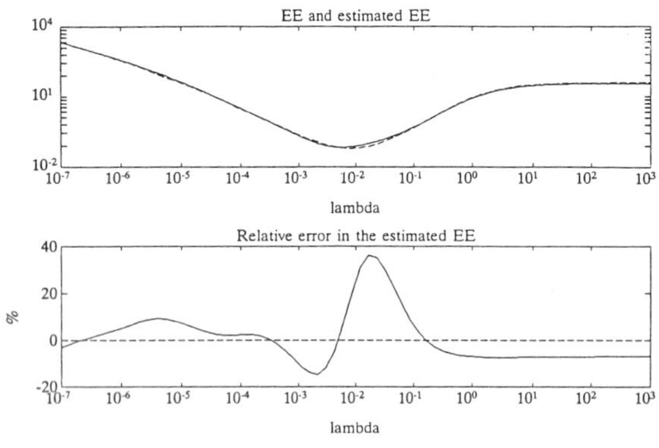

Fig. 6. Error energy (- - -), energy estimation (——)

The precision of the energy estimation depends on the approximations of the input spectrum and noise variance. In the next sections we are going to investigate, how robust is the proposed approximation with respect to errors in the input signal model and the output noise level.

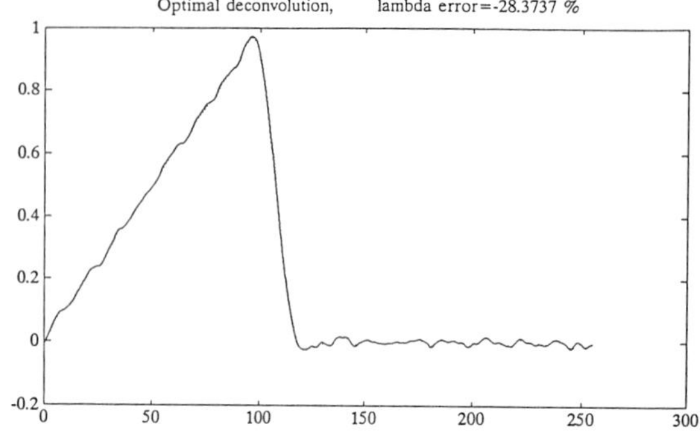

Fig. 7. Optimal deconvolution in LS sense (. . .), and the approximated optimum (——)

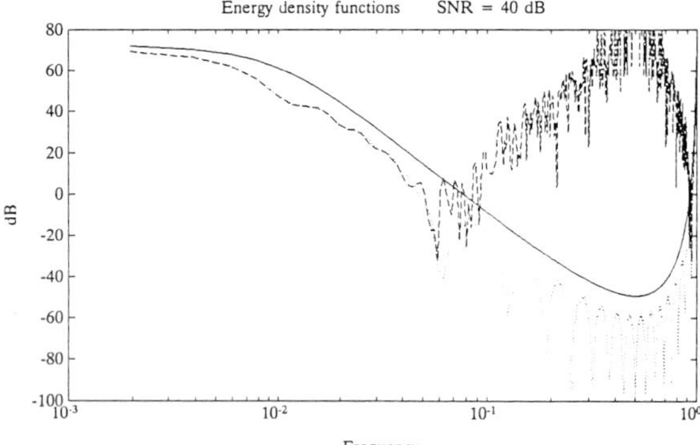

Fig. 8. Badly modeled input signal

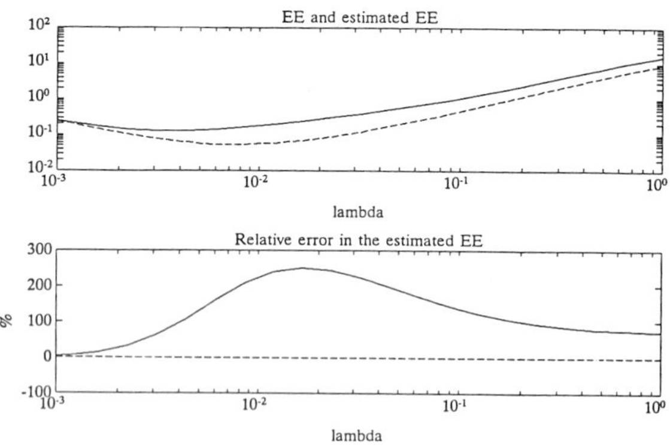

Fig. 9. Estimation of the error energy using a bad input signal model

3.1. Bad input signal model

The time constants of the spectrum estimation were drastically modified, by a factor of about 2. The badly modeled spectrum is drawn in Fig. 8. The error in the EE estimation was increased to 200% at larger λ values, where the bias term dominates (Fig. 9), but the minimum was obtained with about 60% error only. The LS and the obtained optimal deconvolution are close to each other (Fig. 10). The rising slope is nearly the same, while the peak is about 2% higher than that of the LS solution. Such a large error can usually be avoided at the input spectrum, so the precision of the minimum detection is robust enough with respect to the input signal model.

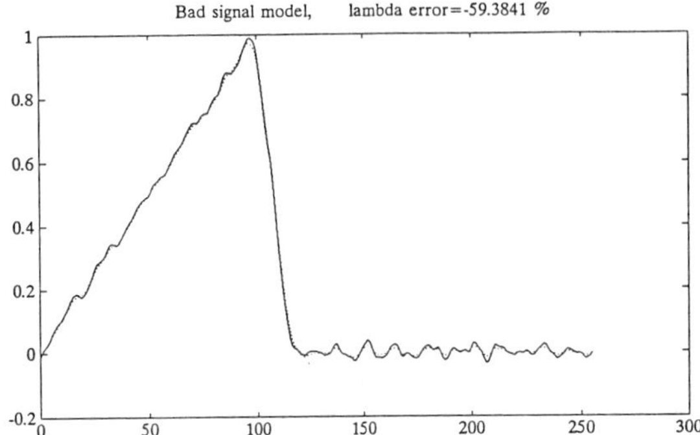

Fig. 10. Deconvolution result using a bad signal model.

3.2. Erroneous noise level estimation

The variance estimation of the measurement noise was increased by a factor of 2. For EE estimation we obtained 100% error at low λ values, where the noise term dominates. The result is obvious, since in Eq. (6) the noise variance is a multiplicative factor assuming white spectrum. At the region of optimal deconvolution the error is only 10-20%. The optimal λ was found by about 20% error, which caused only a slight amplification or suppression, without significant waveform distortion. From the simulation it was found that the optimum detection is not too sensitive with respect to noise level estimation, only the estimated EE error will be significantly modified at very small λ values.

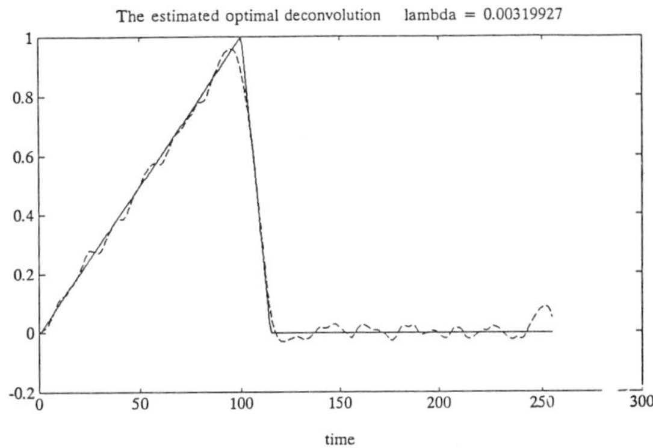

Fig. 11. Approximated optimum for measurement

3.3. Experimental verification

A second order Sallen-Key lowpass filter was excited with the input signal shown in Fig. 3. The time constants of the filter are the same as in the simulations. A Gaussian noise was added to the output of the filter by a noise generator (SNR = 39 dB) and the noisy response was measured. For the input signal a smooth optimum was obtained, depicted in Fig. 11.

4. CONCLUSIONS

A new approximate method was proposed to estimate the error energy in deconvolution, which makes possible to find the optimal solution in LS sense. Since the error energy is an approximation, the obtained optimum is suboptimal, but it was shown that the solution is very close to the true LS one. The calculation of the error energy requires the estimation of the noise level and a rough approximation about the absolute value of the input spectrum. It was shown that the optimum is robust with respect to errors in estimation of the noise level and input spectrum, respectively. Furthermore, the optimal reconstruction can be found for any deconvolution algorithms, it requires only the knowledge of the signal and noise shaping filter characteristics.

References:

[1] K. Schon and W. Gitt, "Reconstruction of High Impulse Voltages Considering the Step Response of the Measuring System", IEEE Trans. on PAS, Vol. PAS-101, No. 10, 1982, pp. 4147-4155.

[2] N. S. Nahman and M. E. Guillaume, "Deconvolution of Time Domain Waveforms in the Presence of Noise", NBS Tech. Note 1047, NBS, Boulder, CO, 1981.

[3] J. V. Candy and J. E. Zicker, "Deconvolution of Noisy Transient Signals: A Kalman Filtering Application", IEEE Conf. on Decision and Control, Orlando, FL, 1982, pp. 211-216.

[4] B. Parruck, S.M. Riad, "An optimization criterion for iterative deconvolution", IEEE Tr. on IM, Vol. IM-32, No 10, 1983, pp. 137-140.

[5] I. Kollár, P. Osváth and W. Zaengl, "Numerical Correction and Deconvolution of Noisy HV Impulses by Means of Kalman Filtering", 1988 IEEE International Symposium on Electrical Insulation, Boston, MA, June 5-8, 1988. Conference Record, CH2594-0/88, pp. 359-363.

[6] C. D. Taylor, D. P. Nicolas, "An adaptive data-smoothing routine", Computer in Physics, Mar/Apr 1989, pp. 63-64.

[7] C. Narduzzi, G. Zingales: "An Inverse Filtering Approach to Reconstruction of HV Impulses", 6th Int. Symp. on High Voltage Eng., New Orleans, LA, USA, 1989. Paper 47.09

SIGNAL PROCESSING VI: Theories and Applications
J. Vandewalle, R. Boite, M. Moonen, A. Oosterlinck (eds.)

UNIFICATION OF SEVERAL APPROACHES TO MEAN FREQUENCY ESTIMATION FOR DOPPLER DETERMINATION

G. D. Cain and A. Yardim

School of Electronic & Manufacturing Systems Engineering
Polytechnic of Central London
115 New Cavendish Street
London W1M 8JS, England

1. ABSTRACT

Doppler tracking may be viewed as a progression of short-term evaluations of the mean of the energy density spectrum of moving-windowed fragments of the time signal. This is particularly appropriate when the spectrum is broadband, failing to exhibit the near-monochromaticity that so significantly aids some frequency estimation techniques. Several methods employing differing mixes of correlation, differentiation and Hilbert transformation to achieve such spectral "centroid tracking" frequently crop up in frequency estimation studies. Each appears in isolation and, since there is no guidance given on alternatives, readers are faced with a number of dissimilar-appearing formulations all aiming at a common measurement goal. These individual competitors are shown to be variants of the same basic calculation, and prime candidates for a unified treatment. A "square-root" filter is shown to be a very cost-effective replacement for most of the variants examined.

2. BROADBAND EXPRESSIONS

The measurement of frequency is an instrumentation problem elemental in nature and of seemingly never-ending interest to engineers. Scarcely a month goes by without a leading technical journal carrying a paper examining this issue from yet another perspective.

The wealth of literature which has built up can crudely be partitioned into two segments: that dealing with pure sinusoidal models contaminated by noise and that aimed at broadband spectral characterization resulting in descriptors of "a frequency" somehow representative of the gross shape of the spectrum. Reference [1] gives an authoritative background relating to the first category, while papers in the second category - most often concerned with Doppler frequency determination - are typified by the classic [2]. There have been relatively few attempts to draw together the disparate estimation approaches into quantified comparisons; among these, [3]-[7] attacked subsets (for digital estimation) of that problem. Though the zero-crossing method is simple and extremely popular, a second approach - that of spectral moments - is also very prominent. This paper focuses on deterministic calculation of spectral means, and is cast in continuous-time terms prior to arriving at a digital filter-based formulation of a proposed estimator structure.

We start by assuming a real-valued signal s(t) possessing an energy density spectrum $|S(f)|^2$ and a (time) autocorrelation function $R_s(\tau)$. Our interest centres on determining the centroid of the positive-frequency portion of the (even-symmetric) energy density spectrum:

$$f_m = \frac{\int_0^\infty f|S(f)|^2 \, df}{\frac{1}{2}R_s(0)} \tag{1}$$

where normalization by half the signal's energy, $R_s(0)$, is - as in analogous moment calculations for force distributions - required for correct calibration of the energy centroid location, f_m. Our efforts are directed at evading explicit evaluation of the integral in (1), relying instead upon various arrangements of simple transform properties to permit our processing - though equivalent to (1) - to be wholly couched in the time domain.

The key is to trade for a double-sided integral so that we tackle

$$\int_{-\infty}^{\infty} |f| . |S(f)|^2 \, df = R_s(0) f_m \tag{2}$$

The possibilities which open up for processing strategies become obvious when we make just one of the possible re-castings of a term that lies at the heart of the integrand:

$$|f| = (j2\pi f)(-jsgnf)(\tfrac{1}{2\pi}) \tag{3}$$

since this can be readily recognized as the spectral effect imparted by the (scaled) cascading of a differentiator and a Hilbert transform filter. This immediately allows us to write

$$2\pi\, R_s(0) f_m = \int_{-\infty}^{\infty} (j2\pi f)\, S^*(f)[-jsgnf\, S(f)]\, df \tag{4}$$

or

$$f_m = \dot{R}_{s\hat{s}}(0) / 2\pi R_s(0) \tag{5}$$

where $R_{s\hat{s}}(\tau)$ is our nomenclature for the cross-correlation of the original signal and its Hilbert transform, $\hat{s}(t)$, and $\dot{R}(\tau)$ signifies the τ-derivative of a correlation function.

Of course (5) suggests a form of hardware that could be employed for a Doppler-determining system. However, this is but one of a multitude of formulations that can be obtained by rearrangement of the groups of ingredients in our fundamental equation (4). About the simplest change from (5) is to associate the differentiation with the $S^*(f)$ term in (4) so that

$$f_m = -R_{\dot{s}\hat{s}}(0) / 2\pi R_s(0) \tag{6}$$

Or, we could pull both operations out so that only the autocorrelation function is worked upon:

$$f_m = \hat{\dot{R}}_s(0) / 2\pi R_s(0) \tag{7}$$

This type of arrangement is alluded to in [8] and made very explicit in [9]. Clearly there are numerous variations possible that involve mixes of operations on the raw signal and correlation functions derived from it.

3. BASEBAND AND INSTANTANEOUS FREQUENCY ALTERNATIVES

We can move away from the formulation where processing takes place at the original spectral location of s(t) (which we call the "broadband" processing situation) and down-modulate to a baseband scenario. One particularly popular embodiment involves derivatives of in-phase and quadrature components (e.g. [10] and [11]).

If we take f_c as a convenient "carrier frequency" to be nearby f_m, then we need to find the offset frequency f_o that corrects to give our mean:

$$f_m = f_c + f_o \tag{8}$$

We can do this by a complex envelope description of our given s(t):

$$s(t) = \Re e[z(t) e^{j2\pi f_c t}] = \Re e\{[x(t) + jy(t)] e^{j2\pi f_c t}\} \tag{9}$$

where x(t) is the in-phase component of s(t) and y(t) is the associated quadrature component (for the given f_c value assumed). Subjecting z(t) to the same kind of energy density centroid calculation as was used in (1), we immediately obtain

$$f_o = \frac{j\dot{R}_{zz}(0)}{2\pi R_z(0)} = \frac{\int_{-\infty}^{\infty} [x(t)\dot{y}(t) - \dot{x}(t) y(t)]\, dt}{2\pi \int_{-\infty}^{\infty} [x^2(t) + y^2(t)]\, dt} \tag{10}$$

where over-dots denote time derivatives. This baseband formulation not only enables us to treat low-frequency signals, but also permits an analytically handy doubling-up of integration that would not have been meaningful in the context of (1) (due to the symmetry of $|S(f)|^2$ for the real signal s(t)). Of course, the price is the double processing load associated with a complex envelope signal z(t).

It is no accident that (10) bears a strong resemblance to z(t)'s "instantaneous frequency" definable [12] as

$$f_i(t) = \frac{d}{dt}\{\tan^{-1}[\tfrac{y(t)}{x(t)}]\} = \frac{x(t)\dot{y}(t) - \dot{x}(t) y(t)}{x^2(t) + y^2(t)} \tag{11}$$

and, as pointed out by Papoulis [13, p. 321], a

mean frequency based on statistical autocorrelation is a weighted average which has the same form as (10), where the counterpart of time integration is expectation.

Obviously we can again proceed to move the differentation operation about to modify (10); Angelsen and Kristoffersen prefer

$$f_o = \frac{-j\dot{R}_z(0)}{2\pi R_z(0)} \tag{12}$$

in their series of useful papers on this topic [4].

4. MORE DIRECT DISCRETE-TIME BROAD-BAND APPROACHES: A VEE FILTER AND A SQUARE-ROOT FILTER

There are five papers ([8], [9], [14] - [16]) known to us which set out the case for time-domain broadband mean frequency calculation in its very essence, without the trappings of differentiation and Hilbert transformation which are unnecessary in the context of this particular problem. Curiously enough, all four authoring teams have done their work in connection with biomedical measurements.

In [8], [9] and [16] what we call the "vee digital filter" seems to have been independently derived. The idea is to view the discrete-time counterpart of (2) (where ν is normalized frequency and k is the discrete-time index),

$$\nu_m R_s(0) = \int_{-0.5}^{0.5} |\nu|.|S(\nu)|^2\,d\nu = \sum_{k=-\infty}^{\infty} h_\nu(k)R_s(k) \tag{13}$$

as the initial value of a convolution in which a filter impulse response sequence $h_\nu(k)$ fully engages all samples of the $R_s(k)$ autorcorrelation sequence. Of course practical realization restricts $h_\nu(k)$ to causality and forces departure from the idealized vee-shaped transfer function $|\nu|$ for $\nu \in (-0.5, 0.5)$. Closed-form expressions for a simple FIR approximation to $h_\nu(k)$ are given in [8], [9] and [16].

We much prefer to avoid autocorrelation calculation altogether and to concentrate instead on direct filtering of the raw signal s(k). Of course (2) suggests that this can be done by choosing a digital filter with transfer function $\sqrt{|\nu|}$ for $\nu \in (-0.5, 0.5)$. It seems that Gerzberg and Meindl were first to report this approach (e.g. [14]), though (apparently independently) [8] and [15] report the same approach. In [8] Strube points out the advantage of an FIR realization of this "square-root filter" to permit moving-window tracking of a nonstationary spectral mean. The elegance and economy of this approach, which we advocate as the preferred solution for spectral mean calculation, can be seen from the complete processing system's block diagram (where α is the linear-phase square-root filter's group delay):

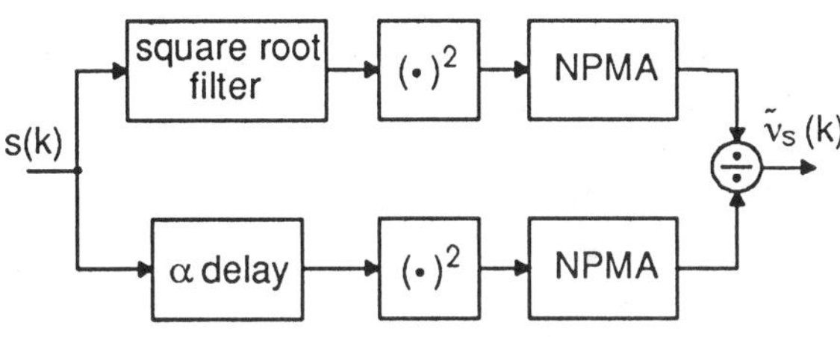

Figure 1

and where "NPMA" signifies an N-point moving window averaging filter and $\tilde{\nu}_s(k)$ is a time-varying estimate of the centroid of the spectrum of s(k).

The square-root filter should, in our view, be FIR and may well need to have a different number of coefficients from the NPMA smoothing filter. Aside from a single closed-form expression in [8], we have found no report of how to design the square-root filter apart from our own work in [17] and [18]. We have found this design target to be an interesting challenge, particularly as regards low frequency performance, and are extending our investigation of practical realizations. We have one 512-point square-root filter already in routine real-time use for a vehicular speed measuring testbed, yielding excellent tracking results.

5. REFERENCES

[1] Rife, D.C., "Digital tone parameter estimation in the presence of Gaussian noise", Ph.D. dissertation, Polytech. Inst. Brooklyn, Brooklyn, N.Y., (University Microfilms 73-24, 783), June 1973.

[2] Berger, F.B., "The design of airborne Doppler velocity measuring systems", *IRE Trans. Aeronautical and Navigational Electronics*, vol. ANE-4, pp. 157-175, December 1957.

[3] Sirmans, D. and B. Bumgarner, "Numerical comparison of five mean frequency estimators", *Journal of Applied Meteorology*, vol. 14, pp. 991-1003, September 1975.

[4] Kristoffersen, K. and B.A.J. Angelsen, "A comparison between mean frequency estimators for multigated Doppler systems with serial signal processing", vol. BME-22, no. 9, pp. 645-657, September 1985.

[5] Barrett, R.F. and D.R.A. McMahon, "Comparison of frequency estimators for underwater acoustic data", *J. Acoustic Soc. Am.*, vol. 79, no. 5, pp. 1461-1471, May 1986.

[6] Cain, G.D. and A. Yardim, "Digital estimation of Doppler frequency in wideband signalling applications", presented at the Fifth *International Conference on Digital Processing of Signals in Communications*, University of Loughborough, 19-23 September 1988.

[7] Boashash, B. et al, "Statistical/computational comparison of some estimators for instantaneous frequency", *Proc. ICASSP*, vol. 5, pp. 3193-3196, July 1991.

[8] Strube, H.W., "Spectral moments from the ACF or directly from the signal", *Signal Processing*, vol. 9, no. 1, pp. 51-55, August 1985.

[9] Sjoberg, S., "Computationally efficient estimation of the mean frequency for real-valued signals", *Proc. ICASSP*, pp. 38.3.1-38.3.4, March 1984.

[10] Denenberg, J.N., "The power mean frequency estimator: another approach to the FM detector", *IEEE Trans. Broadcast and Television Receivers*, vol. BTR-20, no. 3, pp. 201-204, August 1974.

[11] Barber, W.D. et al, "A new time domain technique for velocity measurements using Doppler ultrasound", *IEEE Trans. Biomed. Engr.*, BME-32, no. 3, ppl. 213-229, March 1985.

[12] Berthomier, C., "Instantaneous frequency and energy distribution of a signal", *Signal Processing*, vol. 5, no. 1, pp. 31-45, January 1983.

[13] Papoulis, A., *Probability, Random Variables and Stochastic Processes*, Second Edition, New York: McGraw-Hill, 1984.

[14] Gerzberg, L. and J.D. Meindl, "Power-spectrum centroid detection for Doppler systems applications", *Ultrasonic Imaging 2*, pp. 232-261, 1980.

[15] Pritchard, T.B. et al, "A microprocessor-based signal processing system for measurement of vascular and urethral parameters", *IEEE Trans. Biomed. Engr.*, vol. BME-33, no. 12, pp. 1197-1203, December 1986.

[16] Saltzberg, B., "An efficient formula for estimating the generalized moments of the power spectral density (PSD) without computing the Fourier transform", *IEEE Trans. Biomed. Engr.* BME-33, no. 12, pp. 1134-1136, December 1986.

[17] Cain, G.D., A. Yardim and E. Katsaros, "Performance of an FIR filter-based centroid tracker for Doppler determination", *Proc. 1991 International Symposium on Circuits and Systems,* Singapore, vol. 5, pp. 2455-2458, June 1991.

[18] Marchesi, M., G.D. Cain and A. Yardim, "Design of a square-root filter for use in FIR spectral centroid tracking", submitted for publication.

SIGNAL PROCESSING VI: Theories and Applications
J. Vandewalle, R. Boite, M. Moonen, A. Oosterlinck (eds.)

OPTIMAL SVD-BASED DETERMINATION OF NUMBER OF SINUSOIDS

H. Tsuji† and A. Sano‡

†Communications Research Laboratory, M. P. T., 4-2-1 Nukuikitamachi, Koganei 184, Japan
‡Department of Electrical Engineering, Keio University, 3-14-1 Hiyoshi, Kohoku-ku, Yokohama 223, Japan

By introducing multiple adjustable free parameters into the singular value decomposition (SVD) and determining their optimal values, the optimal truncation scheme for the singular values is derived. The number of retained singular values gives the estimated number of sinusoids in white noise. Three criteria are presented in this paper to decide the free parameters; one is the minimum MSE-like criterion, the second one is based on the generalized cross validation method, and the third one is based on the Bayesian information criterion. The new algorithms using the obtained optimal free parameters are given and compared to the AIC and MDL in numerical simulations.

1. INTRODUCTION

The eigenstructure approach has recently become a powerful tool in wide area of digital signal processing [1]. The singular value decomposition (SVD) can attain the immunity of noise effects in spectral estimation by truncating small singular values of a signal data matrix adequately [2]. The number of the retained singular values corresponds to the dimension of the signal subspace, which indicates the number of harmonics embedded in white noise. In actual cases, however, large and small singular values do not separate neatly in the presence of large noises. Therefore theoretical criteria are required to determine the number of principal singular values, which will be useful in the SVD-based signal processing.

In this paper, we introduce multiple free parameters into the SVD of the signal data matrix and determine them so as to minimize various types of criterion. We first clarify the existence of the optimal values in the sense of the minimum mean squares error (MSE). The truncation scheme of the singular values can be given by comparing the optimal free parameters with the corresponding singular values. We propose three types of criterion for deciding the free parameters by using only accessible signal data. The first is an MSE-like criterion which has been investigated by the authors [3] in ill-conditioned parameter estimation. The second one is based on the generalized cross validation (GCV) approach proposed originally in [4]. The third one is based on the Bayesian statistical approach [5]. The proposed algorithms for deciding the number of sinusoids are compared to other criteria such as the AIC and MDL in numerical simulations.

2. PARAMETRIZED INVERSE IN USE OF SVD

Let the observed signal $\{y_k\}$ be a complex autoregressive (AR) process described by

$$y_k = \sum_{i=1}^{L} x_i y_{k-i} + e_k \tag{1}$$

where e_k is a Gaussian white noise with zero-mean and the variance σ_e^2. If we adopt a least squares fitting of the AR model to the given data, (1) is used as the linear prediction model as

$$\hat{y}_k = \sum_{i=1}^{L} \hat{x}_i y_{k-i} \tag{2}$$

where e_k can be regarded as a prediction error defined by

$$e_k \equiv y_k - \hat{y}_k$$

Then (1) can be expressed in the LP method as

$$\begin{bmatrix} y_{L+1} \\ y_{L+2} \\ \vdots \\ y_N \end{bmatrix} = \begin{bmatrix} y_L & y_{L-1} & \cdots & y_1 \\ y_{L+1} & y_L & \cdots & y_2 \\ \vdots & \vdots & \ddots & \vdots \\ y_{N-1} & y_{N-2} & \cdots & y_{N-L} \end{bmatrix} \begin{bmatrix} x_1 \\ x_2 \\ \vdots \\ x_L \end{bmatrix} + \begin{bmatrix} e_{L+1} \\ e_{L+2} \\ \vdots \\ e_N \end{bmatrix}$$

or

$$\boldsymbol{y} = \boldsymbol{A}\boldsymbol{x} + \boldsymbol{e} \tag{3}$$

where L is the order of prediction, N the data number and x_i are the LP coefficients. The least squares estimate in (2) can be expressed by

$$\hat{\boldsymbol{x}} = \sum_{i=1}^{L} \frac{1}{\sigma_i} \boldsymbol{v}_i \boldsymbol{u}_i^H \boldsymbol{y} \tag{4}$$

where σ_i is singular value of $\boldsymbol{A}$, $\boldsymbol{v}_i$ and $\boldsymbol{u}_i$ are the singular vectors. The accuracy of the solution (4) is degraded in a case of low SN ratio. Tuft and Kumaresan [2] suggested a truncated SVD in which the number of retained singular values is chosen the number of sinusoids if it is known.

In this paper, in order to treat with the case when the number of sinusoids is unknown, we introduce multiple adjustable free parameters ρ_i into the expression (4) to give a parametrized estimate of the LP coefficients as

$$\hat{\boldsymbol{x}}(\{\rho_i\}) = \sum_{i=1}^{L} \frac{\sigma_i}{\sigma_i^2 + \rho_i} \boldsymbol{v}_i \boldsymbol{u}_i^H \boldsymbol{y} \tag{5}$$

If ρ_i=0 for i=1,...,K and $\rho_i = \infty$ for i=K+1,...,L, then (5) is reduced to the truncated SVD-based estimate, where K is referred to as the truncation number. Generally, the problem is how to determine the optimal values of $\{\rho_i\}$.

3. OPTIMAL FREE PARAMETERS IN SENSE OF MINIMUM MEAN SQUARES ERROR

As a criterion for determining the free parameters $\{\rho_i\}$ in (5), first we consider the mean squares error (MSE) criterion, which can be evaluated approximately, for sufficiently large N, as

$$MSE(\{\rho_i\}) \equiv E\left\{\left\|\boldsymbol{x} - \hat{\boldsymbol{x}}(\{\rho_i\})\right\|^2\right\}$$
$$\approx \sum_{i=1}^{L} \frac{\rho_i^2}{(\sigma_i^2 + \rho_i)^2} \left|\boldsymbol{v}_i^H \boldsymbol{x}\right|^2 + \sum_{i=1}^{L} \frac{\sigma_e^2 \sigma_i^2}{(\sigma_i^2 + \rho_i)^2} \tag{6}$$

The optimal values $\{\rho_i{}^*\}$ that minimize the RHS of (6) can be analytically given by

$$\rho_i^* = \frac{\sigma_e^2}{\left|\boldsymbol{v}_i^H \boldsymbol{x}\right|^2}, \qquad \text{for } i = 1,\cdots,L \tag{7}$$

The aim is to determine the truncation number K, which also indicates the dimension of the signal subspace. The optimal truncation problem can be formulated as the minimization of (6) subject to that $\{\rho_i\}$ takes either 0 or ∞. The result is given as [3]:

$$\rho_i^{(T)} = \begin{cases} 0, & \text{if} \quad \sigma_i^2 \geq \rho_i^* \\ \infty, & \text{if} \quad \sigma_i^2 < \rho_i^* \end{cases} \tag{8}$$

It is seen from (8) that $\rho_i{}^*$ works as threshold values for deciding whether the singular value σ_i should be retained or discarded. Therefore, if $\rho_i{}^{(T)}$ given by (8) satisfies that

$$\rho_i^{(T)} = \begin{cases} 0, & \text{for} \quad i = 1,\cdots,K \\ \infty, & \text{for} \quad i = K+1,\cdots,L \end{cases} \tag{9}$$

(9) gives the optimal truncation law for deciding K.

However, the optimal values $\{\rho_i{}^*\}$ include the unknown true values of $\boldsymbol{x}$ and $\sigma_e{}^2$. In order to overcome this difficulty, we can take a modified MSE criterion using only accessible data [3], as

$$EMSE(\{\rho_i\}) \equiv \sum_{i=1}^{L} \frac{\sigma_i^2 \rho_i^2}{(\sigma_i^2 + \rho_i)^4} \left|\boldsymbol{u}_i^H \boldsymbol{y}\right|^2 + \sum_{i=1}^{L} \frac{\hat{\sigma}_e^2(\{\rho_i\}) \sigma_i^2}{(\sigma_i^2 + \rho_i)^2}$$
$$\hat{\sigma}_e^2(\{\rho_i\}) = \frac{1}{N-L}\left(\|\boldsymbol{y}\|^2 - \sum_{i=1}^{L} \frac{\sigma_i^2}{\sigma_i^2 + \rho_i} \left|\boldsymbol{u}_i^H \boldsymbol{y}\right|^2\right) \tag{10}$$

By using (8) and (9), we can obtain the truncation number K by replacing $\rho_i{}^*$ with ρ_i which minimize the EMSE in (10).

4. GCV CRITERION FOR FREE PARAMETERS

We present the generalized cross-validation (GCV) criterion without using the unknown $\boldsymbol{x}$ and $\sigma_e{}^2$ to determine the multiple free parameters. The criterion is derived on a basis of a rotation-invariant version of the ordinary cross-validation [4] as

$$P(\{\rho_i\}) \equiv \frac{1}{n} \sum_{k=L+1}^{N} \left(\left[\boldsymbol{A}\boldsymbol{x}^{(k)}(\{\rho_i\})\right]_k - y_k\right)^2 w_k(\{\rho_i\})$$
$$= \frac{1}{n}\left\|(\boldsymbol{I} - \boldsymbol{Q}(\{\rho_i\}))\boldsymbol{y}\right\|^2 \left[\frac{1}{n}\mathrm{Trace}(\boldsymbol{I} - \boldsymbol{Q}(\{\rho_i\}))\right]^{-2} \tag{11}$$

where n=N–L and $\boldsymbol{x}^{(k)}(\{\rho_i\})$ represents the least squares estimate of $\boldsymbol{x}$ with the k-th data point y_k omitted. $[\ ]_k$ denotes the k-th element of the vector. $w_k(\{\rho_i\})$ is the weighted function as

$$w_k(\{\rho_i\}) = \frac{1 - q_{kk}(\{\rho_i\})}{1 - \frac{1}{n}\mathrm{Trace}\,\boldsymbol{Q}(\{\rho_i\})}$$

$$\boldsymbol{Q}(\{\rho_i\}) = \boldsymbol{A}\left(\boldsymbol{A}^H \boldsymbol{A} + \boldsymbol{G}\right)^{-1} \boldsymbol{A}^H, \quad \boldsymbol{G} = \sum_{i=1}^{L} \rho_i \boldsymbol{v}_i \boldsymbol{v}_i^H \tag{12}$$

where $q_{kk}(\{\rho_i\})$ is the (k,k)-th entry of $\boldsymbol{Q}(\{\rho_i\})$. From the SVD point of view, we can modify the GCV criterion in terms of $\{\rho_i\}$ explicitly as

$$P(\{\rho_i\}) = n\left[\sum_{i=1}^{n}\left(\frac{\rho_i}{\lambda_i + \rho_i}\right)^2 \left|\boldsymbol{u}_i^H \boldsymbol{y}\right|^2\right] \cdot \left[\sum_{i=1}^{L} \frac{\rho_i}{\lambda_i + \rho_i} + n - L\right]^{-2} \tag{13}$$

where λ_i, for i=1,...,n, are the eigenvalues of $\boldsymbol{AA}^H$ and λ_i=$\sigma_i{}^2$ for i=1,...,L, and λ_i=0 for i=L+1,...,n. We can find out the optimal free parameters $\{\rho_i\}$ through minimization of (13) by applying nonlinear optimization techniques and determine the number of harmonics on a basis of (8) and (9).

5. BAYESIAN INFORMATION CRITERION FOR FREE PARAMETERS

Finally we take the Bayesian statistical approach to determine the optimal value of the adjustable free parameters in (5) as well as the optimal truncation number K by using only accessible observed data.

We first assume the prior distribution of $\boldsymbol{x}$ as

$$P_a\left(\boldsymbol{x}\middle|\sigma_e^2,\{\rho_i\}\right) = \frac{\det(\boldsymbol{G})}{\pi^L \sigma_e^{2L}} \exp\left(-\frac{1}{\sigma_e^2}\boldsymbol{x}^H \boldsymbol{G}\boldsymbol{x}\right) \tag{14}$$

where $\boldsymbol{G}$ is already defined in (12) which includes the free parameters $\{\rho_i\}$. Since $\boldsymbol{e}$ is a zero mean white Gaussian random vector, the conditional probability density of $\boldsymbol{y}$ is given by

$$P_r\left(\boldsymbol{y}\middle|\boldsymbol{x},\sigma_e^2\right) = \frac{1}{\pi^{N-L}\sigma_e^{2(N-L)}} \exp\left(-\frac{1}{\sigma_e^2}\|\boldsymbol{y} - \boldsymbol{A}\boldsymbol{x}\|^2\right) \tag{15}$$

Using (14) and (15), we can specify the posterior distribution, when the observed signal $\boldsymbol{y}$ is given, as

$$P_p\left(\boldsymbol{x}\middle|\boldsymbol{y},\sigma_e^2,\{\rho_i\}\right)\propto P_r\left(\boldsymbol{y}\middle|\boldsymbol{x},\sigma_e^2\right)P_a\left(\boldsymbol{x}\middle|\sigma_e^2,\{\rho_i\}\right)$$
$$=\frac{\det(\boldsymbol{G})}{\pi^N\sigma_e^{2N}}\exp\left(-\frac{1}{\sigma_e^2}\left\{\|\boldsymbol{y}-\boldsymbol{A}\boldsymbol{x}\|^2+\boldsymbol{x}^H\boldsymbol{G}\boldsymbol{x}\right\}\right)$$
$$\equiv L\left(\boldsymbol{x},\sigma_e^2,\{\rho_i\}\middle|\boldsymbol{y}\right) \qquad (16)$$

where $L(.\,|\,.)$ is called the Bayesian likelihood. We take the Bayesian statistical approach [5] to give the ABIC for determining $\{\rho_i\}$ and σ_e^2. The Bayesian information criterion (ABIC) is finally given by integrating (16) with respect to $\boldsymbol{x}$.

Then we finally obtain the form of ABIC [7] as

$$ABIC(\{\rho_i\})=\sum_{i=1}^{L}\log\left(\frac{\sigma_e^2}{\rho_i}+1\right)$$
$$+n\log\left(\|\boldsymbol{y}\|^2-\sum_{i=1}^{L}\frac{1}{\sigma_i^2+\rho_i}\left|\boldsymbol{v}_i^H\boldsymbol{A}^H\boldsymbol{y}\right|^2\right) \qquad (17)$$

By calculating the optimal values of $\{\rho_i\}$ so as to minimize (17), we can estimate K from (8) and (9) by replacing $\{\rho_i{*}\}$ with $\{\rho_i\}$ which minimize (17). However, it should be remarked that the ABIC is different from the EMSE in (10) and the GCV in (13) in that the free parameters ρ_i cannot take zero. Therefore, it is not simple to give the ABIC criterion in terms of K explicitly by putting zero and infinity on ρ_i. Then by using small positive δ, we can rewrite the ABIC (17) as:

$$ABIC(K)=\sum_{i=1}^{K}\log\left(\frac{\sigma_e^2}{\delta}+1\right)$$
$$+n\log\left(\|\boldsymbol{y}\|^2-\sum_{i=1}^{K}\frac{1}{\sigma_i^2+\delta}\left|\boldsymbol{v}_i^H\boldsymbol{A}^H\boldsymbol{y}\right|^2\right) \qquad (18)$$

where δ should be sufficiently small positive value. Practically we may set up δ as $\varepsilon\sigma_K^2$, where σ_K is the smallest singular value in the adopted ones and ε is chosen small such as 0.01 so that the value of $\varepsilon\sigma_K^2$ may not effect the calculation of estimate (5). Therefore we use the following form:

$$ABIC(K)=\sum_{i=1}^{K}\log\left(\frac{\sigma_e^2}{\varepsilon\sigma_K^2}+1\right)$$
$$+n\log\left(\|\boldsymbol{y}\|^2-\sum_{i=1}^{K}\frac{1}{\sigma_i^2+\varepsilon\sigma_K^2}\left|\boldsymbol{v}_i^H\boldsymbol{A}^H\boldsymbol{y}\right|^2\right) \qquad (19)$$

6. DETECTION OF NUMBER OF SINUSOIDS

The observed signal consists of p complex sinusoids and white Gaussian noise as

$$y_k=\sum_{i=1}^{p}a_i\exp(j2\pi f_i k)+w_k, \quad k=1,\cdots N \qquad (20)$$

where a_i is the complex amplitude with unknown magnitude and phase. f_i is the unknown frequency and p denotes the number of sinusoids. w_k is the complex white Gaussian noise with zero mean and the covariance σ_w^2. The aim of this section is to give an algorithm for detecting the number of the sinusoids in (20) by the optimal truncation of the singular values.

The outline of the algorithm is given as follows:

(1) Calculate the singular values $\{\sigma_i\}$ and the singular vectors $\boldsymbol{u}_i$, $\boldsymbol{v}_i$ of $\boldsymbol{A}$.
(2) Calculate the multiple free parameters $\{\rho_i\}$ so as to minimize (10), (13) or (17) by using the quasi-Newton method.
(3) Find the truncation number K satisfying that $\sigma_K^2>\rho_K$ and $\sigma_{K+1}^2<\rho_{K+1}$ using (8) and (9).

We consider the first example in which the observed signal consists of three complex sinusoids in white noise with f_1=0.30, f_2=0.33, f_3=0.35, a_1=a_2=a_3=1, N= 64, L=30 for various SNR's. Fig. 1 shows the profiles of the singular values σ_i^2, $\rho_i{*}$ in (7) and the calculated parameters ρ_i minimizing (13) and (17) for i=1,...,6, in the case of SNR=–3dB. The plotted curves of $\{\rho_i\}$ intersect the profile of σ_i^2 at between i=3 and i=4, then it is decided that the number of sinusoids is three. In cases of low SNR, it is difficult to determine the optimal truncation by watching the profile of the singular values σ_i^2. The proposed scheme can give the distinctive threshold for the decision by calculating the optimal ρ_i minimizing (10), (13) or (17).

Table 1 shows the distribution of the detected number of sinusoids for 100 trials using the EMSE, GCV, ABIC, AIC and MDL criterion. The MDL criterion has the following form [6] as

$$MDL(K)=-2\log\left[\frac{\prod_{i=K+1}^{L}\lambda_i^{1/(L-K)}}{\frac{1}{L-K}\sum_{i=K+1}^{L}\lambda_i}\right]^{n(L-K)}$$
$$+K(2L-K)\log n \qquad (21)$$

where λ_i is the eigenvalue of the sample covariance matrix. In the case of the AIC, the second term in (21) is replaced by $2K(2L–K)$. The MDL and AIC criteria depend on only the singular values. In the case of high SNR, the GCV, ABIC and MDL have similar performance, while in the case of low SNR ABIC($\{\rho_i\}$) can detect most precisely compared to the MDL and other criteria. The AIC tends to overestimate the number of harmonics and the MDL tends to underestimate it. The proposed schemes can detect the number of sinusoids more accurately than the AIC or MDL criterion.

Next, we consider the second example with f_1=0.30, f_2=0.34, a_1=a_2=1, N= 16, L=8. The results are summarized in Table 2, which indicates the proposed method based on the ABIC($\{\rho_i\}$) performs best even in cases of small number of data and low SNR.

7. CONCLUSIONS

By choosing the adjustable free parameters induced to the SVD, we can give the optimal truncation of the singular values and the algorithm for deciding the number of sinusoids embedded in white noise. Three types of criterion for deciding the free parameters have been presented.

REFERENCES

[1] Deprettere, F., ed., SVD and Signal Processing, (North-Holland, Amsterdam, 1988).

[2] Tuft, D. W. and Kumaresan, R., Estimation of frequencies of multiple sinusoids: Making linear prediction perform like maximum likelihood, Proc. IEEE, (70-9, 1982) pp.975-989.

[3] Sano, A., Tsuji, H. and Ohmori, H., Optimized Singular Value Decomposition with Applications to Signal Extrapolation and Detection of Sinusoidal Number, Proc. IFAC World Congress, (1990) pp.123-128.

[4] Golub, G. H., Heath, M. and Wahba, G., Gene- ralized cross-validation as a method for choosing a good ridge parameter, Technometrics, 21-2 (1979) pp.215-223.

[5] Akaike, H., Likelihood and the Bayesian Procedure, in Bayesian Statistics eds. J. M. Berbardo et al, (Univ.Press, 1980) pp.143-166.

[6] Wax , M. and Kailath, T., Detection of signals by information theoretic criteria, IEEE Trans. Acoust. Speech, Signal Processing, ASSP-33-2, (1985) pp.387-392

[7] Sano, A, Tsuji, H. and Nagasawa, K., Bayesian Information Theoretic Criterion for Detection of Number of Signals in Array Processing, Proc. Fifth IEEE ASSP Workshop on Spectrum Estimation and Modeling, (Rochester, NewYork, 1990) pp.129-133.

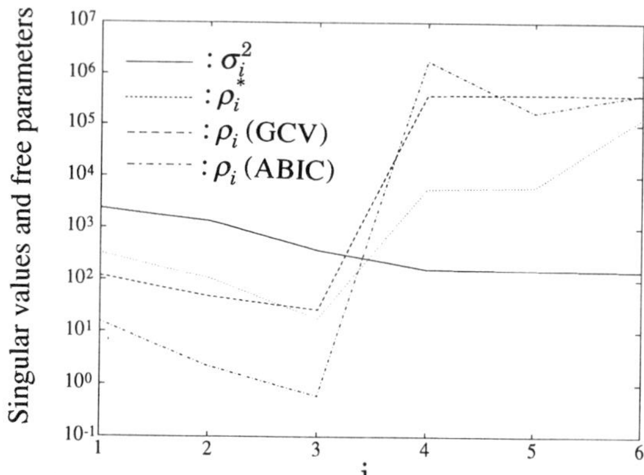

Fig. 1. Plots of the calculated free parameters and singular values

Table 1. Comparison of the proposed schems with other methods minimizing AIC and MDL

SNR	3dB						0dB						−3dB						−6dB					
K	1	2	3	4	5	6	1	2	3	4	5	6	1	2	3	4	5	6	1	2	3	4	5	6
EMSE($\{\rho_i\}$)	0	0	89	9	2	0	0	0	89	8	3	0	0	0	80	14	5	1	1	8	59	18	11	3
GCV($\{\rho_i\}$)	0	0	100	0	0	0	0	0	99	1	0	0	1	8	80	10	1	0	11	27	41	16	5	0
ABIC($\{\rho_i\}$)	0	0	100	0	0	0	0	0	100	0	0	0	0	0	100	0	0	0	0	2	90	7	1	0
ABIC(K)	0	0	100	0	0	0	0	0	100	0	0	0	0	1	96	3	0	0	1	15	61	19	3	1
AIC(K)	0	0	71	18	8	3	0	0	71	18	8	3	0	2	70	17	8	3	2	44	42	7	4	1
MDL(K)	0	0	98	2	0	0	0	0	98	2	0	0	0	23	76	1	0	0	19	73	8	0	0	0

Table 2. The distribution of detected number of sinusoids in example 2

SNR	10dB						6dB						3dB					
K	0	1	2	3	4	5	0	1	2	3	4	5	0	1	2	3	4	5
EMSE($\{\rho_i\}$)	0	0	90	10	0	0	10	7	76	7	0	0	15	13	60	8	4	0
GCV($\{\rho_i\}$)	0	3	97	0	0	0	0	12	88	0	0	0	0	10	65	24	1	0
ABIC($\{\rho_i\}$)	0	0	100	0	0	0	0	0	96	4	0	0	1	6	82	11	0	0
ABIC(K)	0	0	94	6	0	0	0	8	80	11	0	1	20	17	51	12	0	0
AIC(K)	0	0	91	7	1	1	16	15	61	6	1	1	36	31	28	3	1	1
MDL(K)	0	0	91	7	1	1	16	16	60	6	1	1	42	26	27	3	1	1

SIGNAL PROCESSING VI: Theories and Applications
J. Vandewalle, R. Boite, M. Moonen, A. Oosterlinck (eds.)
1992 Elsevier Science Publishers B.V.

APPLICATION OF THE BOOTSTRAPPED TOTAL LEAST SQUARES (BTLS) ESTIMATOR IN LINEAR SYSTEM IDENTIFICATION.

H. VAN HAMME and R. PINTELON*

Vrije Universiteit Brussel, Dienst ELEC, Pleinlaan 2, B-1050 Brussel, Belgium.
Tel. : 32-2-641 2944; Fax. : 32-2-641 2850; e-mail : hvhamme@vnet3.vub.ac.be

Abstract - The estimation of the parameters of linear time-invariant single-input-single-output continuous-time or discrete-time systems with an errors-in-variables model is examined. The Bootstrapped Total Least Squares Estimator (BTLS; pronounce BeaTLeS) is shown to produce consistent estimates in this setting. The near-maximum likelihood performance of BTLS is shown through asymptotic analysis and verified through simulation. The BTLS estimates have a smaller mean-square estimation error than the generalized total least squares (GTLS) estimates.

1. Introduction - modeling linear systems.

We consider the problem of estimating the parameters of a linear time-invariant continuous-time or discrete-time single-input-single-output system from noise-corrupted input/output samples. Let t represent time (integer values for discrete-time systems) and let u(t) and y(t) denote the input and output signal respectively. Using the notation $^{(k)}$ for the k-th order derivative, the continuous-time system obeys the *differential* equation:

$$a_n y^{(n)}(t) + a_{n-1} y^{(n-1)}(t) + \dots + a_0 y^{(0)}(t) = b_m u^{(m)}(t) + b_{n-1} u^{(m-1)}(t) + \dots + b_0 u^{(0)}(t). \qquad (1.1a)$$

The discrete system obeys the *difference* equation:

$$a_n y(t-n) + a_{n-1} y(t-n+1) + \dots + a_0 y(t) = b_m u(t-m) + b_{n-1} u(t-m+1) + \dots + b_0 u(t) \qquad (1.1b)$$

Equation (1.1b) can be seen as an algebraic relation among the *system parameters* $a_n, \dots, a_0, b_m, \dots, b_0$. This relation can be evaluated at different values of t. The same is valid for (1.1a), except that it requires the derivatives of the input and the output, which are not known if only an input/output record is available. However, (1.1a) can be transformed into an algebraic relation [1]:

$$a_n y_n(t) + a_{n-1} y_{n-1}(t) + \dots + a_0 y_0(t) = b_m u_m(t) + b_{n-1} u_{m-1}(t) + \dots + b_0 u_0(t) \qquad (1.1c)$$

where $u_k(t)$ is obtained by passing u(t) through a hybrid linear system consisting of an analog anti-alias filter followed by a numeric filter. The whole chain has a frequency response:

$$(j\omega)^k (1 + \delta(j\omega))^k G(j\omega) \qquad (1.2)$$

where $G(j\omega)$ is some complex weighting function and $|\delta(j\omega)\, G(j\omega)|$ is less than some prescribed accuracy level for all ω. The $y_k(t)$ are obtained likewise from y(t). The transformation of (1.1a) to (1.1c) can be made arbitrarily accurate *without* increasing the sampling frequency [1].

When (1.1a), (1.1b) or (1.1c) is evaluated at t = 0 ... K-1 (the sampling frequency is 1 Hz without loss of generality), it can be written as

$$\mathbf{h}_t' \mathbf{p} = 0 \qquad t = 0 \dots K-1 \qquad (1.1d)$$

Here, prime denotes matrix transpose, the unknown system parameters are placed in the vector $\mathbf{p} \in \mathbb{R}^{L\times 1}$ with $\mathbf{p}' = (a_n, \dots, a_0, b_m, \dots, b_0)$, L = n + m + 2, and $\mathbf{h}_t \in \mathbb{R}^{L\times 1}$ are the coefficients of the system parameters in (1.1a) - (1.1c). In matrix notation:

$$\mathbf{H}\,\mathbf{p} = \mathbf{0} \qquad (1.1e)$$

where $\mathbf{H} \in \mathbb{R}^{K\times L}$ is a matrix whose t-th row is $\mathbf{h}_t'$. We exclude the trivial solution $\mathbf{p} = \mathbf{0}$ by fixing the first entry of $\mathbf{p}$ at 1 (any entry could have been chosen). This is written symbolically as $\mathbf{p}'\,\mathbf{e}_1 = 1$, where $\mathbf{e}_1$ is the first unit vector.

In the next sections, we consider the least squares, the generalized total least squares, the (approximate) maximum likelihood and the bootstrapped total least squares estimation method for estimation of the parameters in model (1.1e). Due to space limitations, the exposition will be based on heuristics, rather than on hard mathematical proof. The reader is referred to [1] for more details.

2. Parameter Estimation.

When the input/output record is contaminated by additive zero-mean noise of any joint distribution, all $\mathbf{h}_t$ will be perturbed by noise $\mathbf{n}_t \in \mathbb{R}^{L\times 1}$, yielding the observations $\mathbf{g}_t$:

* H. Van hamme is a research assistant and R. Pintelon is a research associate, both of the Belgian National Fund for Scientific Research (NFWO).

$$g_t = h_t + n_t \qquad t = 0 \ldots K\text{-}1 \tag{2.1}$$

Equivalently, $\mathbf{H}$ will be perturbed by some noise matrix $\mathbf{N} \in \mathbb{R}^{KxL}$: $\mathbf{G} = \mathbf{H} + \mathbf{N}$ and generally $\mathbf{G}\,\mathbf{p}$ will not be zero for any value of $\mathbf{p}$. We then need to find a 'best' solution of an overdetermined set of linear equations, where all columns of $\mathbf{G}$ are subject to noise. One obvious approach is to solve $\mathbf{G}\,\mathbf{p} = \mathbf{0}$ in a least squares sense:

$$\mathbf{p}_{LS} = \underset{\mathbf{p}\text{ with }\mathbf{p}'\mathbf{e}_1 = 1}{\arg\min}\ \mathbf{p}'\,\mathbf{G}'\,\mathbf{G}\,\mathbf{p} = \begin{pmatrix} 1 \\ (-\mathbf{A}'\,\mathbf{A})^{-1}\,\mathbf{A}'\,\mathbf{b} \end{pmatrix} \tag{2.2}$$

where $\mathbf{b}$ is the first column of $\mathbf{G}$ and $\mathbf{A}$ is the matrix containing the remaining columns: $\mathbf{G} = (\mathbf{b} \quad \mathbf{A})$. The least squares (LS) method will generally produce asymptotically biased results.

To obtain consistent estimates, the generalized total least squares (GTLS) solution can be used:

$$\mathbf{p}_{GTLS} = \underset{\mathbf{p}\text{ with }\mathbf{p}'\mathbf{e}_1 = 1}{\arg\min}\ \frac{\mathbf{p}'\,\mathbf{G}'\,\mathbf{G}\,\mathbf{p}}{\mathbf{p}'\,\mathbf{C}_0\,\mathbf{p}} = \frac{\mathbf{v}}{\mathbf{v}'\,\mathbf{e}_1} \tag{2.3}$$

where $\mathbf{v}$ is the right-singular vector corresponding to the smallest singular value of $\mathbf{G}\,\mathbf{T}$ and $\mathbf{C}_0 = (\mathbf{T}\,\mathbf{T}')^{-1} = E\{\mathbf{N}'\,\mathbf{N}\}$ (see Van Huffel and Vandewalle, 1991).

To obtain the maximum likelihood estimates (MLE) for Gaussian observation noise, we define $\mathbf{B}_\mathbf{p} = \mathbf{I}_K \otimes \mathbf{p}'$ ($\otimes$ denotes Kronecker product), $\mathbf{n}' = (\mathbf{n}_1'\ \mathbf{n}_2' \ldots \mathbf{n}_K')$, i.e. $\mathbf{n} \in \mathbb{R}^{KLx1}$ is obtained by stacking the $\mathbf{n}_t$ on top of each other, and $\mathbf{C}^{(\mathbf{n})} = E\{\mathbf{n}\,\mathbf{n}'\}$. Then the MLE for Gaussian $\mathbf{n}$ is given by:

$$\mathbf{p}_{MLE} = \underset{\mathbf{p}\text{ with }\mathbf{p}'\mathbf{e}_1 = 1}{\arg\min}\ \mathbf{p}'\,\mathbf{G}'\,(\mathbf{B}_\mathbf{p}\,\mathbf{C}^{(\mathbf{n})}\,\mathbf{B}_\mathbf{p}')^{-1}\,\mathbf{G}\,\mathbf{p}. \tag{2.4a}$$

In contrast to (2.2) and (2.3), an explicit solution of the minimization problem is *not* given here, because at present, no closed form solution is known to the authors. The minimizer in (2.4a) is calculated using iterative minimization techniques. It requires an initial guess, which can be provided by e.g. GTLS.

Alternatively, an *approximate* maximum likelihood estimator can be defined, where $\mathbf{C}^{(\mathbf{n})}$ is replaced by $\hat{\mathbf{D}}$

$$\mathbf{p}_{AMLE} = \underset{\mathbf{p}\text{ with }\mathbf{p}'\mathbf{e}_1 = 1}{\arg\min}\ \mathbf{p}'\,\mathbf{G}'\,(\mathbf{B}_\mathbf{p}\,\hat{\mathbf{D}}\,\mathbf{B}_\mathbf{p}')^{-1}\,\mathbf{G}\,\mathbf{p}. \tag{2.4b}$$

The block-Toeplitz $\hat{\mathbf{D}}$ is related to $\mathbf{C}^{(\mathbf{n})}$ in a way that will be specified below.

All estimators above are written as the minimizer of a cost function $L_K(\mathbf{p}, \mathbf{G})$:

$$L_K(\mathbf{p}, \mathbf{G}) = \frac{1}{2K}\,\mathbf{p}'\,\mathbf{G}'\,(\mathbf{B}_\mathbf{p}\,\mathbf{D}\,\mathbf{B}_\mathbf{p}')^{-1}\,\mathbf{G}\,\mathbf{p} \tag{2.5}$$

where the index K denotes the number of rows in $\mathbf{G}$ and where $\mathbf{D}$ is a weighting matrix that depends on the estimator. Comparing (2.2), (2.3) and (2.4) with (2.5), we find for the specific estimators:

LS: $$\mathbf{D} = (\mathbf{e}_1\,\mathbf{e}_1') \otimes \mathbf{I}_K \tag{2.6a}$$

GTLS: $$\mathbf{D} = \mathbf{C}_0 \otimes \mathbf{I}_K \tag{2.6b}$$

MLE: $$\mathbf{D} = \mathbf{C}^{(\mathbf{n})} \tag{2.6c}$$

AMLE: $$\mathbf{D} = \hat{\mathbf{D}}.$$

The cost in (2.5) was scaled by 2K for later convenience, which doesn't affect its minimizer. Also, $L_K(\mathbf{p}, \mathbf{G})$ is invariant for scaling of all entries of $\mathbf{p}$ by the same constant. Hence, the constraint $\mathbf{p}'\,\mathbf{e}_1 = 1$ can be omitted.

3. Asymptotic Analysis.

Assume that the system excitation is quasi-stationary and that the noise on the input/output record is wide-sense stationary and statistically independent of the excitation, i.e.

$$\mathbf{C}_n = E\{\mathbf{n}_k\,\mathbf{n}_{n+k}'\} \in \mathbb{R}^{LxL} \tag{3.1a}$$

is independent of k for any n and

$$\mathbf{S}_n = \lim_{K \to \infty} \frac{1}{K} \sum_{k=\max(1,1-k)}^{\min(K,K-k)} E\{\mathbf{h}_k\,\mathbf{h}_{n+k}'\} \tag{3.1b}$$

exists. One can define the *noise and signal spectra* $\Phi_\mathbf{C}(\omega)$ and $d\Psi_\mathbf{S}(\omega)$ via the Fourier relations:

$$\mathbf{C}_n = \frac{1}{2\pi} \int_{-\pi}^{\pi} \Phi_\mathbf{C}(\omega)\,e^{-j\omega n}\,d\omega \tag{3.2a}$$

$$\text{and } \mathbf{S}_n = \frac{1}{2\pi} \int_{-\pi}^{\pi} e^{-j\omega k}\,d\Psi_\mathbf{S}(\omega) \tag{3.2b}$$

where (3.2b) is a Stieltjes integral. Likewise, since $\mathbf{D}$ is a block-Toeplitz matrix with LxL blocks for LS, GTLS, MLE and AMLE, it is possible to define a *weighting spectrum* $\Phi_\mathbf{D}(\omega)$ via

$$\mathbf{D}_n = \frac{1}{2\pi} \int_{-\pi}^{\pi} \Phi_\mathbf{D}(\omega)\,e^{-j\omega n}\,d\omega \in \mathbb{R}^{LxL},\ n = -\infty \ldots \infty \tag{3.2c}$$

where $\mathbf{D}_{l-k}$ is the k,l-th LxL block of $\mathbf{D}$. With (2.6):

LS: $$\Phi_\mathbf{D}(\omega) = \mathbf{e}_1\,\mathbf{e}_1' \tag{3.3a}$$

GTLS: $$\Phi_\mathbf{D}(\omega) = \mathbf{C}_0 \tag{3.3b}$$

MLE: $$\Phi_\mathbf{D}(\omega) = \Phi_\mathbf{C}(\omega). \tag{3.3c}$$

In the AMLE approach, we choose:

AMLE: $$\Phi_\mathbf{D}(\omega) = f(\omega)\,\Phi_\mathbf{C}(\omega). \tag{3.3d}$$

where $f(\omega)$ is a real-valued positive weighting function. The matrix $\hat{\mathbf{D}}$ in (2.3b) is defined via (3.2c) with (3.3d). For $f(\omega) = 1$, the AMLE reduces to MLE.

It was shown in [1] that under fairly weak restrictions, the cost function (2.5) converges uniformly in $\mathbf{p}$ to:

$$L(\mathbf{p}) = \frac{1}{4\pi} \int_{-\pi}^{\pi} \frac{\mathbf{p}'\,d\Psi_\mathbf{S}(\omega)\,\mathbf{p}}{\mathbf{p}'\,\Phi_\mathbf{D}(\omega)\,\mathbf{p}} + \frac{1}{4\pi} \int_{-\pi}^{\pi} \frac{\mathbf{p}'\,\Phi_\mathbf{C}(\omega)\,\mathbf{p}}{\mathbf{p}'\,\Phi_\mathbf{D}(\omega)\,\mathbf{p}}\,d\omega \tag{3.4}$$

as K tends to infinity. Due to the uniform convergence of the cost, the asymptotic estimates are given by the minimizer of $L(\mathbf{p})$. The two *positive* terms in (3.3) represent the contribution to the asymptotic cost of the signal and the noise respectively.

The quantity $\mathbf{p}'\,d\Psi_\mathbf{S}(\omega)\,\mathbf{p}$ is the contribution of the excitation to the power spectrum of the residuals $\mathbf{G\,p}$ and $\mathbf{p}'\,\Phi_C(\omega)\,\mathbf{p}$ is the contribution of the noise to this residual power spectrum. In (3.4), these spectra are both weighted by the estimator-dependent function $\mathbf{p}'\,\Phi_D(\omega)\,\mathbf{p}$. For the MLE (3.3c), this function equals $\mathbf{p}'\,\Phi_C(\omega)\,\mathbf{p}$, i.e. the residuals are weighted *according to their noise spectrum.* AMLE offers an extra freedom in the choice of $f(\omega)$, which can be used to emphasize some frequencies in (3.4). For LS and GTLS, $\mathbf{p}'\,\Phi_D(\omega)\,\mathbf{p}$ is independent of frequency, which makes them more susceptible to noise perturbation. This will now be illustrated in the context of system identification, when the input (output) signals are perturbed by mutually uncorrelated noise with a power spectral density $\sigma_u^2(\omega)$ ($\sigma_y^2(\omega)$). For discrete-time systems:

$$\mathbf{p}'\,\Phi_C(\omega)\,\mathbf{p} = \sigma_u^2(\omega)\,|B(e^{j\omega})|^2 + \sigma_y^2(\omega)\,|A(e^{j\omega})|^2, \qquad (3.5a)$$

where $B(e^{j\omega})/A(e^{j\omega})$ is the system frequency response associated with the parameter vector $\mathbf{p}$. (3.5a) is obtained by computing the joint power spectrum of the noise sequence $\{\mathbf{n}_t\}$, keeping the relations (1.1b) and (2.1) in mind. For continuous-time systems, we obtain with (1.2):

$$\mathbf{p}'\,\Phi_C(\omega)\,\mathbf{p} \approx |G(j\omega)|^2\,(\sigma_u^2(\omega)\,|B(j\omega)|^2 + \sigma_y^2(\omega)\,|A(j\omega)|^2), \qquad (3.5b)$$

where $B(j\omega)/A(j\omega)$ is the frequency response associated with the parameter vector $\mathbf{p}$. The functions (3.5) tend to increase for higher frequencies. Hence, since this dependence on the frequency is missing for LS and GTLS, these estimators over-emphasize high frequencies. Deviations at low frequencies will hardly be penalized, resulting in poor asymptotic estimates for the transfer function at these frequencies.

Now suppose that there exists a *true* parameter vector $\mathbf{p_e}$ that satisfies (1.1) exactly, i.e. $\mathbf{H\,p_e} = \mathbf{0}$. In view of (3.1b), the positive form $\mathbf{p}'\,d\Psi_S(\omega)\,\mathbf{p}$ will be zero for $\mathbf{p} = \mathbf{p_e}$. Under the persistence condition that $\mathrm{rank}(\mathbf{S}_0) \geq L\text{-}1$ (where $\mathbf{S}_0$ is defined in (3.2b)), the first term will *only* be zero when $\mathbf{p}$ is proportional to $\mathbf{p_e}$. Since this term is positive, it is minimal at $\mathbf{p_e}$. With (3.2) and (3.3), the second term equals :

$$\text{LS:}\quad \frac{1}{4\pi}\int_{-\pi}^{\pi}\frac{\mathbf{p}'\,\Phi_C(\omega)\,\mathbf{p}}{(\mathbf{p}'\,\mathbf{e}_1)^2}\,d\omega = \frac{\mathbf{p}'\,\mathbf{C}_0\,\mathbf{p}}{(\mathbf{p}'\,\mathbf{e}_1)^2} \qquad (3.6a)$$

$$\text{GTLS:}\quad \frac{1}{2\pi\,\mathbf{p}'\,\mathbf{C}_0\,\mathbf{p}}\int_{-\pi}^{\pi}\mathbf{p}'\,\Phi_C(\omega)\,\mathbf{p}\,d\omega = \frac{1}{2} \qquad (3.6b)$$

$$\text{MLE:}\quad \frac{1}{4\pi}\int_{-\pi}^{\pi}\frac{\mathbf{p}'\,\Phi_C(\omega)\,\mathbf{p}}{\mathbf{p}'\,\Phi_C(\omega)\,\mathbf{p}}\,d\omega = \frac{1}{2} \qquad (3.6c)$$

$$\text{AMLE:}\quad \frac{1}{4\pi}\int_{-\pi}^{\pi}\frac{\mathbf{p}'\,\Phi_C(\omega)\,\mathbf{p}}{f(\omega)\,\mathbf{p}'\,\Phi_C(\omega)\,\mathbf{p}}\,d\omega = \frac{1}{4\pi}\int_{-\pi}^{\pi}\frac{1}{f(\omega)}\,d\omega \qquad (3.6d)$$

For GTLS, MLE and AMLE, the second term in (3.4) is a constant, so the asymptotic estimates will be given by the minimizer of the first term, which is $\mathbf{p_e}$ for all estimators considered. Hence, these estimators will be consistent. For LS, the (3.6a) is *not* independent of $\mathbf{p}$, and the minimizer of (3.4) will *not* coincide with the true parameters and the estimates are inconsistent. LS will therefore be omitted from further discussion.

4. The Bootstrapped Total Least Squares (BTLS) Estimator.

The GTLS estimator is attractive because it yields consistent estimates that can be calculated quite efficiently [2]. Its poor frequency-domain weighting can be improved by premultipling the observation matrix $\mathbf{G}$ with $(\mathbf{B_{pe}}\,\hat{\mathbf{D}}\,\mathbf{B}'_{\mathbf{pe}})^{-1/2}$ to yield $\tilde{\mathbf{G}}$ (likewise, $\tilde{\mathbf{N}} = (\mathbf{B_{pe}}\,\hat{\mathbf{D}}\,\mathbf{B}'_{\mathbf{pe}})^{-1/2}\,\mathbf{N}$), and then applying the GTLS technique to the set of equations $\tilde{\mathbf{G}}\,\mathbf{p} \approx \mathbf{0}$. The purpose of incorporating this weight is to approximate the AMLE cost as close as possible, while retaining the computational advantage of GTLS. The transformed GTLS problem is solved by minimizing:

$$\frac{\mathbf{p}'\,\mathbf{G}'\,(\mathbf{B_{pe}}\,\hat{\mathbf{D}}\,\mathbf{B}'_{\mathbf{pe}})^{-1}\,\mathbf{G\,p}}{\mathbf{p}'\,\mathbf{D}_{\mathbf{pe},btls}\,\mathbf{p}} \qquad (4.1)$$

where $\mathbf{D}_{\mathbf{pe},btls} = E\{\tilde{\mathbf{N}}'\,\tilde{\mathbf{N}}\} = E\{\mathbf{N}'\,(\mathbf{B_{pe}}\,\hat{\mathbf{D}}\,\mathbf{B}'_{\mathbf{pe}})^{-1}\,\mathbf{N}\}$.
The cost (4.1) is minimized by the right-singular vector corresponding to the smallest singular value of $\tilde{\mathbf{G}}\,\mathbf{T}$, where $\mathbf{D}_{\mathbf{pe},btls} = (\mathbf{T}\,\mathbf{T}')^{-1}$. Comparing (2.5) and (4.1) yields the weighting matrix:

$$\mathbf{D} = (\mathbf{B_{pe}}\,\hat{\mathbf{D}}\,\mathbf{B}'_{\mathbf{pe}}) \otimes \mathbf{D}_{\mathbf{pe},btls} \qquad (4.2a)$$

from which the weighting spectrum is found:

$$\Phi_D(\omega) = f(\omega)\,(\mathbf{p_e}'\,\Phi_C(\omega)\,\mathbf{p_e})\,\mathbf{D}_{\mathbf{pe},btls}. \qquad (4.2b)$$

With (3.4) and knowing that $\mathbf{D}_{\mathbf{pe},btls}$ converges to [1]:

$$\frac{1}{2\pi}\int_{-\pi}^{\pi}\frac{\Phi_C(\omega)}{f(\omega)\,\mathbf{p_e}'\,\Phi_C(\omega)\,\mathbf{p_e}}\,d\omega, \qquad (4.3)$$

the asymptotic cost is found:

$$\frac{1}{4\pi\,\mathbf{p}'\mathbf{D}_{\mathbf{pe},btls}\,\mathbf{p}}\int_{-\pi}^{\pi}\frac{\mathbf{p}'\,d\Psi_S(\omega)\,\mathbf{p}}{f(\omega)\,\mathbf{p_e}'\,\Phi_C(\omega)\,\mathbf{p_e}} + \frac{1}{4\pi}\int_{-\pi}^{\pi}\frac{1}{f(\omega)}\,d\omega \qquad (4.4)$$

In (4.4), the contribution of the noise is a constant. As for GTLS, MLE and AMLE, this ensures the *consistency* of the estimates. Also, when minimizing (4.1), the de-emphasis at high frequencies is present in the asymptotic cost (4.4). However, the construction of the weights $\mathbf{B_{pe}}\,\mathbf{C}^{(\mathbf{n})}\,\mathbf{B}'_{\mathbf{pe}}$ requires the knowledge of the true parameters $\mathbf{p_e}$. Therefore, $\mathbf{p_e}$ is replaced by an estimate $\mathbf{p}_i$, which is obtained from a previous step, hence the name of *bootstrapped* total least squares. The BTLS algorithm is then:

1. choose a $\mathbf{p}_0$; set $i = 0$.
2. find $\mathbf{p}_{i+1}$, the minimizer of (4.1) where $\mathbf{p_e}$ is replaced by $\mathbf{p}_i$. This is obtained by computing a right-singular vector as explained above.
3. if the stopping criterion is not met, increment i and go to 2, else, the final estimate is $\mathbf{p}_{i+1}$.

The initial value $\mathbf{p}_0$ is not critical. Possible choices are a vector of ones, or a parameter vector that corresponds to B = 0 and A = 1. In the latter case, the weights are those used by the GTLS estimator when the observation noise is white. Reasonable stopping criteria are that i exceeds a prescribed value (e.g. 5) or that the parameter change is less than some bound.

An important property of the BTLS algorithm is that every step yields consistent estimates and that all $\mathbf{p}_i$ for $i \geq 2$ have the same asymptotic distribution (see [1]). Therefore, limiting the number of iterations is sensible. For Gaussian observation noise $\mathbf{N}$, its asymptotic covariance differs from the asymptotic covariance of the AMLE only by a small positive definite contribution. In the calculation of the covariance for $\mathbf{p}_i$, it is essential that $\mathbf{p}_{i-1}$ is a consistent estimate of $\mathbf{p}_e$. Simulation studies show that the difference is small, even at high noise levels. For instance, for a continuous-time 5th order plant measured with noise excitation and with observation noise on the input and the output samples with a standard deviation that is nearly 50 % of the excitation noise standard deviation, the standard deviation on the tranfer function estimate increases by no more than 10 %. Since this deterioration is proportional to the observation noise level, even better results are obtained at lower noise levels.

The implementation of (4.1) can be accomplished by solving a Toeplitz set of equations. This can be done efficiently using the methods described in [3] and its references, or using filtering [1].

5. Simulations.

The consequences of the bad weighting in GTLS are illustrated by simulating a 5th order continuous-time Chebychev filter with 1 dB pass band ripple and a cut-off frequency of 0.15 Hz , sampled at 1 Hz. The estimation is based on (1.1c) and (1.2), where $|G(j\omega)| \approx 1$ below 0.25 Hz and $|G(j\omega)| \approx 0$ above that frequency with a sharp transition. This weighting function is required to suppress the systematical errors in (1.1c) which would otherwise be significant for frequencies above 0.25 Hz. To avoid that the estimator boosts these errors up again, $f(\omega)$ is chosen equal to $|G(j\omega)|^{-2}$ in AMLE and BTLS. The excitation is noise with a nearly flat spectrum up to the measurement frequency of 0.25 Hz. The observation noise is also white in this frequency range, but it is about 10 times smaller. The experiment is repeated for 100 independent observation noise realizations, where each time K = 150 rows of $\mathbf{G}$ are gathered. The mean square error (MSE) of the transfer function estimate is found and compared with the Cramér-Rao lower bound (CRB) on the variance of the transfer function. The ratio of these quantities is depicted in Fig. 1 in dB (= $10 \log_{10}(\text{MSE/CRB})$). We notice that BTLS and MLE outperform GTLS and that they both nearly attain the CRB.

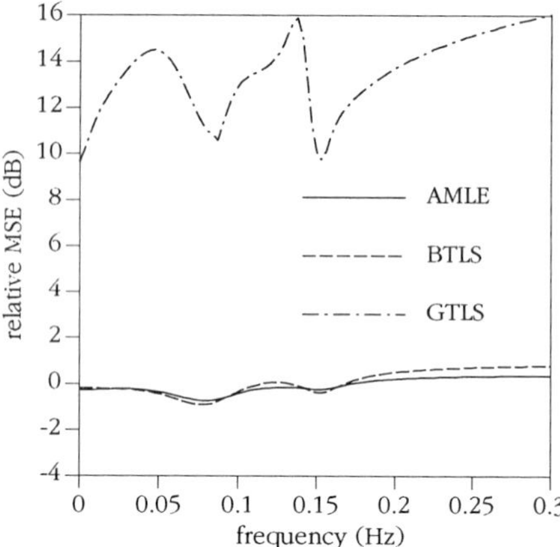

Figure 1 : ratio of the mean square transfer function estimation error and the Cramér-Rao bound.

It must be stressed that the profit of using BTLS instead of GTLS depends on $\mathbf{p}_e{}'\Phi_D(\omega)\,\mathbf{p}_e$ and hence on the plant. In [1], examples can be found where BTLS and MLE exhibit a standard deviation that is about 100 times smaller than that of GTLS. In other simulations, where $\mathbf{p}_e{}'\Phi_D(\omega)\,\mathbf{p}_e$ is nearly constant in the measurement frequency range, the increased computational load of BTLS is not justified by its reduced variance.

6. Conclusion.

We have presented a new estimator that has MLE performance but with reduced computational load. It is based on the GTLS estimator but weights the rows of the observation matrix to improve statistical efficiency. The idea can be applied to other problems that can be formulated as a regression problem.

References.

[1] Van hamme H. *Identification of Linear Systems from Time- or Frequency-Domain Measurements,* PhD thesis, April 1992.

[2] Van Huffel S. and J. Vandewalle. *The Total Least Squares Problem: Computational Aspects and Analysis, Volume 9 of Frontiers in Applied Mathematics,* SIAM, Philadelphia, 1991.

[3] Ammar G. S. and W. B. Gragg. Superfast Solution of Real Positive Definite Toeplitz Systems. *SIAM J. Matrix Anal. Appl.,* Vol. 9, No. 1, 1988, pp. 61-76.

SIGNAL PROCESSING VI: Theories and Applications
J. Vandewalle, R. Boite, M. Moonen, A. Oosterlinck (eds.)

GAUSSIAN NOISE INSENSITIVE FIR SYSTEM IDENTIFICATION

Gaetano Scarano, Fausto Corina**, Alessandro Neri** and Giovanni Jacovitti***

* Istituto di Elettronica, Università di Perugia, S.Lucia Canetola, I-06143 Perugia, Italy
** Dip. INFOCOM, University of Rome "La Sapienza", Via Eudossiana 18, I-00184 Roma, Italy

ABSTRACT

A contribution to the problem of FIR system identification based on higher order statistics of the output only is presented. As all equation error methods the parameters estimates are obtained through least-squares solution of a (overdetermined) linear set of equations, in which, in the proposed method, only higher than two cumulants are involved. This fact implies that, in principle, the solution is insentitive to any additive, possibly coloured, Gaussian noise contaminating the FIR system output.

1. INTRODUCTION

In this contribution we address the problem of estimating the parameters of a moving average (MA) model from higher-order statistics of the noisy observations of its output. To be specific, let us consider a FIR system of order q

$$x(n) = \sum_{i=0}^{q} b(i) \cdot w(n-i) \tag{1}$$

where the driving noise $w(n)$ is a zero mean, independent and identically distributed (i.i.d.) *non-Gaussian* series. In many application of interest, such as digital communication, geophysical signal processing, astronomical signal processing, *etc.*, the parameters $b(i)$'s characterize a nonminimum phase system, which must be identified from N consecutive samples of the system output (1) contaminated by additive Gaussian noise

$$y(n) = x(n) + v(n) \quad ; \qquad n = 0, 1, \cdots, N-1.$$

It is well known that second-order (spectral) analysis of the output enables to measure the magnitude of the transfer function of the system, but it does not provide any information about the phase of the transformation. Higher-order analysis (higher-order moments, cumulants and associate polyspectra) allows, in principle, for unambiguous identification.

In this paper, we are concerned with the problem of recovering the coefficients $b(i)$'s from the cumulant statistics taken from the available observations. This problem has been addressed by several authors (for instance see [1,2,3,4,5,6,7,8]) and a very interesting tutorial has been recently published [9]. Other applications of higher-order statistics and polyspectra in more general parameter estimation problems can be found in [10,11,12].

Approaches to the FIR system identification are usually classified in three broad classes [4]:

1. **Inverse filter error:** this class includes algorithms based on minimization of some criterion involving functionals computed from the output of an *inverse filter* of the FIR transformation, designed to obtain an estimate $\hat{w}(n)$ of the input $w(n)$;
2. **Fitting error:** these solutions are obtained by matching the estimated (measured) statistics to the model (*a priori*) statistics;
3. **Equation error:** generally, in this class are collected algorithms where $b(i)$'s are directly obtained by solving linear equations ideally satisfied by ensemble moments.

In the first two classes we find algorithms which result into non-linear optimization of suitably defined functionals, whereas in the last class the MA parameters are non-optimally determined by a linear set of equations. For this reason, equation error algorithms can be also used somewhere as coarse estimates for setting the starting point for (non-linear) estimation procedures belonging to the first two classes.

Almost all equation error methods involve both second-order statistics as well as higher-order statistics. For this reason the observation noise $v(n)$ can be accomodated by modeling it as a MA process of order $\overline{q}$ significantly lower than the order q of the MA process of interest. In this way, by eliminating those equations which involve the non-zero noise autocorrelation lags (see [4]), noise insensitive algorithms are obtained. This elimination allows to include also second-order (autocorrelation) statistics which are more statistically stable than higher-order statistics. Clearly, when the noise cannot be modelled as a MA process, these methods give biased estimates.

Here, we propose a MA parameter estimation equation error based method which makes use of higher-order cumulant statistics only, in order to obtain coloured Gaussian noise insensitive estimates without resorting to any equation elimination. Simulation results show that the proposed technique generally performs better than autocorrelation-including methods for medium signal-to-noise ratio values, while resulting in insignificant performance losses for high SNR values.

2. FIR SYSTEM IDENTIFICATION

Let us briefly recall the basic steps of equation error FIR system identification techniques based on relationships between output cumulants and MA parameters. For simplicity, let us refer to the third order cumulant of the noise contaminated observation, defined as

$$c_{3y}(k,l) = \mathrm{E}\{y(n)y(n+k)y(n+l)\}. \quad (2)$$

The assumed independence of the additive noise $v(n)$ allows to decompose (2) in a sum of cumulants

$$c_{3y}(k,l) = c_{3x}(k,l) + c_{3v}(k,l). \quad (3)$$

If the noise is Gaussian, all cumulants of order greater than two vanish and (3) becomes

$$c_{3y}(k,l) = c_{3x}(k,l). \quad (4)$$

Now, (4) allows to simply relate the noisy output cumulants to the unknown MA parameters. In fact we have

$$c_{3x}(k,l) = \gamma_{3w} \sum_{i=0}^{q} b(i)\cdot b(i+k)\cdot b(i+l) \quad (5)$$

where $\gamma_{3w} = \mathrm{E}\{w^3(n)\}$ is the third order cumulant of the i.i.d. driving noise $w(n)$. In the following we suppose that the system order q is perfectly known. This means that we must resort to system order estimation before MA parameters estimation: this can be done by classical techniques as specified also in [4]. Here, we focus on the MA parameters estimation only.

From the basic equation (5), we obtain the following sets of equations setting the lag $l = k$ and $l = q$, respectively:

$$\begin{aligned} c_{3y}(k,k) &= \gamma_{3w}\cdot\sum_{i=0}^{q} b^2(i+k)\cdot b(i) \quad ; \quad |k|\le q \\ c_{3y}(k,q) &= \gamma_{3w}\cdot b(0)b(q)\cdot b(k) \quad ; \quad 0\le k\le q \end{aligned}$$

Taking $\mathcal{Z}$-transforms of both sides of the previous equations yields

$$\begin{aligned} \mathcal{Z}\{c_{3y}(k,k)\} &= \gamma_{3w}\cdot B(z^{-1})\,[B(z)*B(z)] \quad (6) \\ \mathcal{Z}\{c_{3y}(k,q)\} &= \gamma_{3w}\cdot b(0)b(q)\cdot B(z) \quad (7) \end{aligned}$$

Along with these equations, classical methods makes use of the power spectral density also, expressed by

$$\mathcal{Z}\{c_{2y}(k)\} = \sigma_w^2\cdot B(z)\cdot B(z^{-1}) + \mathcal{Z}\{c_{2v}(k)\} \quad (8)$$

where $c_{2y}(k) = \mathrm{E}\{y(n+k)y(n)\}$ is the noisy output series autocorrelation, $c_{2v}(k) = \mathrm{E}\{v(n+k)v(n)\}$ is the observation noise autocorrelation and σ_w^2 is the variance of the input $w(n)$. Elimination of $B(z)$ from (7) and (8) and inverse transformation yields the following sets of $2q+1$ equations [4]

$$c_{2y}(k) = \frac{\sigma_w^2}{\gamma_{3w}b(0)b(q)}\sum_{i=0}^{q} b(i)c_{3y}(k+i,q) + c_{2v}(k) \qquad -q\le k\le q. \quad (9)$$

Note that these equations contain the unknown correlation of the unobservable noise $v(n)$. If this noise is white, by eliminating the equation for $k=0$ we supress any noise correlation contribution in (9) (and consequently, any biasing effects on MA parameters estimates). We recall that the equation error estimation method of [2] makes use of (6) and (8), solving equations by by elimination of $B(z^{-1})$. This approach is affected by the noise correlation in different way with respect to (9) and it is not possible to suppress white noise contribution by eliminating suitable equations (as done in (9)). Moreover, the consistency of the estimates is not assured (see [4]).

For non-white noise, it is suggested in [4] that, by modelling it as an MA series of order $\overline{q}\le\frac{q}{2}$, it is possible to write $2q+1-2\overline{q}$ equations obtained from (9) for $|k| = \overline{q}+1,\cdots,q$

$$\sum_{i=1}^{q} b(i)c_{3y}(k+i,q) - \epsilon'\cdot c_{2y}(k) = -c_{3y}(k,q) \qquad |k| = \overline{q}+1,\cdots,q \quad (10)$$

Here $\epsilon' = \frac{\gamma_{3w}\cdot b(q)}{\sigma_w^2}$ and $b(0)$ has been assumed equal to 1 without loss of generality.

Now, (10) do not include unestimable quantities but they suffer the fact that the MA observation noise order must be known *a priori*. Moreover, the number of available equations has been reduced and, as we will see later, this fact impairs the statistical stability of the MA parameters estimates. This statistical unstability depends also on the fact that large cumulant lags are involved.

To overcome these problems we have derived a different set of equations, which results in completely noise contribution free system. To this purpose let us take the (complex) convolution of (7) with itself

$$\mathcal{Z}\{c_{3y}(k,q)\} * \mathcal{Z}\{c_{3y}(k,q)\} = [\gamma_{3w}\cdot b(q)]^2\cdot[B(z)*B(z)]. \quad (11)$$

By eliminating $[B(z)*B(z)]$ from (6) and (11), we obtain, after inverse transformation, the following equations relating the MA parameters to the cumulant statistics of the output

$$\sum_{i=1}^{q} b(i) \cdot c_{3y}^2(i+k,q) - \epsilon'' \cdot c_{3y}(k,k) = -c_{3y}^2(k,q) \quad (12)$$

$$|k| \leq q$$

where $\epsilon'' = \gamma_{3w} b(q)^2$.

Here, the blind noise rejection relies on the Gaussianity of the noise, and further assumptions on its model has not been made.

Both (10) and (12) are solved in a least squares sense considering ϵ' and ϵ'' as unknowns independent of the MA parameters $b(1), b(2), \cdots, b(q)$. Estimates are obtained by substituting expectations with sample moments. The proof of the consistency of the estimation follows the same guidelines presented in [4]. Note that any additive Gaussian noise affects the performance of the estimation only, not the derivation of the solving equations (12) (see (2)). Extension to fourth-order cumulant statistics is straightforward.

3. SIMULATION RESULTS

To test the effectiveness of the presented additive Gaussian noise insensitive FIR identification method (12) against the method (10), computer simulations have been carried out for non-zero skewness driving noise $w(n)$ using third-order cumulants, and for zero skewness, non-zero kurtosis driving noise using fourth-order cumulants. A second-order FIR system ($b(1) = -2.05$, $b(2) = 1$), already reported in literature, has been considered.

The bias and the root mean square error (RMSE) of the MA parameters estimates obtained using third order cumulants are shown in Fig. 1 for $b(1) = -2.05$ and Fig. 2 for $b(2) = 1$. Here the driving noise is exponential distributed and the additive coloured noise is a first order MA series obtained filtering white noise through the FIR filter with transfer function $H(z) = 1+0.9z^{-1}$. The estimates are taken over 2000 observed samples and 100 independent MonteCarlo runs.

Similarly, in Figs. 3 and 4, the driving noise is an independent binary series with symbols ± 1 having the same probability while the additive coloured noise has a Gaussian shaped autocorrelation $c_{2v}(k) = 0.9^{-k^2}$. The estimates are taken over 4000 observed samples and 100 independent MonteCarlo runs.

In both cases we observe that the proposed methos reduces the RMSE essentially by reducing the bias of the estimates induced by the additive coloured noise, with a slight increase of the variance. Note that the first order MA model of the additive noise of Fig.1 cannot be accomodated by (10), resulting this latter in a underdetermined set of equations. To overcome this fact we have used (10) by considering the additive noise as it were white, *i.e.* by eliminating the equation for $k = 0$, as indicated in [4].

Simulations on other classical scenarios considered in literature show that the proposed method generally outperforms existing methods, even though, in some cases, have been observed moderate performance losses. General conclusion should be made on a theoretical basis following the guidelines of [5] and it will be matter of a forthcoming paper.

In conclusion, the proposed Gaussian noise insensitive FIR system identification can be usefully employed in *on-line* applications, where one-shot solutions methods are required, as well as in initial (rough) estimation, where it provides good starting point for more computationally intensive (iterative) non-linear optimum estimation solutions.

REFERENCES

[1] G.B. Giannakis, "Cumulants: A powerful tool in signal processing", *Proc. IEEE*, vol.75, pp.1333-1334, Sept. 1987.

[2] G.B Giannakis, J.M. Mendel, "Identification of nonminimum phase system using higher order statistics", *IEEE Trans. on ASSP*, vol.ASSP-37, pp.360-377, March 1989.

[3] A. Swami, J.M. Mendel, "Closed-form, recursive estimation of MA coefficients using autocorrelation and third-order cumulants" *IEEE Trans. on ASSP*, vol.ASSP-37, pp.1794-1795, Nov. 1989.

[4] J.K. Tugnait, "Approaches to FIR system identification with noisy data using higher-order statistics", *IEEE Trans. on ASSP*, vol.ASSP-38, no.7, pp.1307-317, July 1990.

[5] B. Porat and B.Friedlander, "Performance analysis of parameter estimation algorithms based on higher-order moments", *Int. Journal of Adap. Contr. and Sig. Proc.*, vol.3, pp.191-229, 1989.

[6] B. Porat and B.Friedlander, "Performance analysis of parameter estimation algorithms based on higher-order moments", *IEEE Trans. on AC*, vol.AC-35, pp.27-35, January 1990.

[7] K.S. Lii and M. Rosenblatt, "Deconvolution and estimation of transfer function phase and coefficients for non-Gaussian linear processes" *Ann. Stat.*, vol.10, pp.1195-1208, 1982.

[8] C.L. Nikias, "ARMA bispectrum approach to nonminimum phase system identification", *IEEE Trans. on ASSP*, vol.ASSP-36, pp.513-524, April 1988.

[9] J.M. Mendel, "Tutorial on higher-order statistics (spectra) in signal processing and system theory: theoretical results and some applications", *Proc. IEEE*, vol.79, March 1991.

[10] Special section on *Higher-Order Spectral Analysis*, C.L. Nikias and J.M. Mendel editors, *IEEE Trans. on ASSP*, vol.ASSP-38, no.7, July 1990.

[11] *Proceedings of the Workshop on Higher-Order Spectral Analysis*, Vail (Co), U.S.A., June 1989.

[12] *Proceedings of the Int. Sig. Proc. Workshop on Higher Order Statistics*, Chamrousse, France, July 1991.

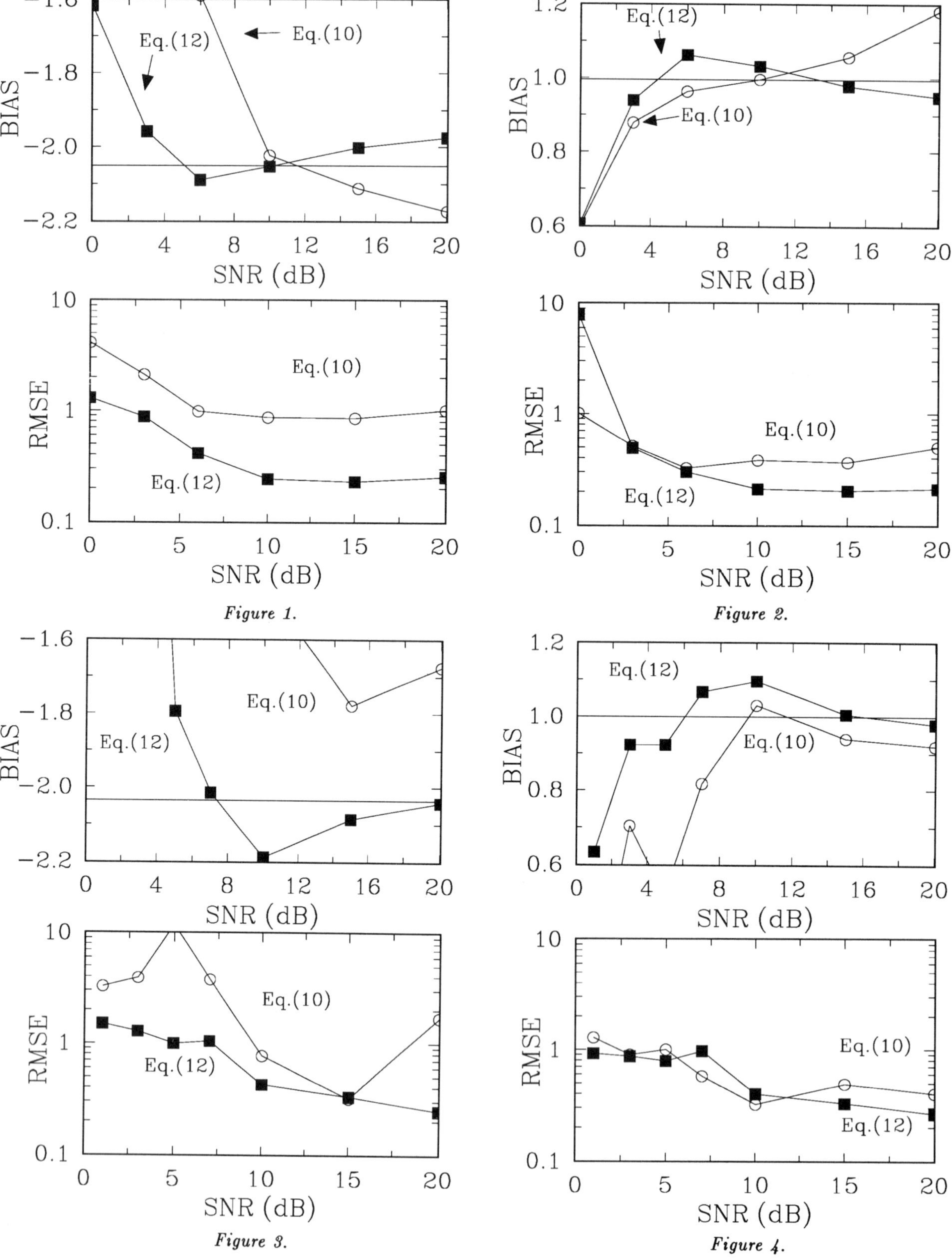

Figure 1.

Figure 2.

Figure 3.

Figure 4.

SIGNAL PROCESSING VI: Theories and Applications
J. Vandewalle, R. Boite, M. Moonen, A. Oosterlinck (eds.)

ITERATIVE TECHNIQUES FOR BLIND SOURCE SEPARATION USING ONLY FOURTH-ORDER CUMULANTS

Jean-François Cardoso

Télécom Paris. Dept Signal, 46 rue Barrault, 75634 Paris CEDEX 13, FRANCE.
Email : cardoso@sig.enst.fr

Abstract. "Blind source separation" is an array processing problem without a priori information (no array manifold). This model can be identified resorting to 4th-order cumulants only via the concept of *4th-order signal subspace* (FOSS) which is defined as a matrix space. This idea leads to a "Blind MUSIC" approach where identification is achieved by looking for the (approximate) intersections between the FOSS and the manifold of 1D projection matrices. Pratical implementations of these ideas are discussed and illustrated with computer simulations.

1. INTRODUCTION

This paper adresses the problem of *blind source separation* or *independent component analysis* in the complex case where it can be seen as a narrow-band array processing problem: the output, denoted $\boldsymbol{x}(t)$, of an array of m sensors listening at n discrete sources takes the form:

$$\boldsymbol{x}(t) = \sum_{p=1,n} s_p(t)\, \boldsymbol{a}_p + \boldsymbol{n}(t) \tag{1}$$

where $s_p(t)$ denotes the complex signal emmited by the p-th source; where $\boldsymbol{a}_p$ is a fixed (deterministic) vector called the p-th *source signature*; and where $\boldsymbol{n}(t)$ is an independent additive noise assumed to be normally distributed with arbitrary covariance.

In "standard" array processing, the same data model is used but the source signatures depend on very few location parameters and this dependence is assumed to be known via the *array manifold*. In contrast, we adress here the **blind** problem where **no** a priori information is available about source signatures. Hence *blind* qualifies any processing based on the sole obervations. "Blindness" is compensated by exploiting the hypothesized assumption of source independence. In the following it is assumed that:

- source signals are statistically independent,
- source signals have non vanishing kurtosis,
- source signatures are linearly independent.

Blind source separation is understood as estimation of the source signals $s_p(t)$, while blind identification refers to the estimation of the source signatures $\boldsymbol{a}_p$. In the following, we focus on identfication since separation can be based on signature estimates.

The blind problem is interesting because its solution allows to process narrow band array data without explicit knowledge about array geometry and without assumptions about wavefront shapes. Various solutions relying on the use of **both** 2nd-order and 4th-order cumulant statistics of the array output have already been reported [1-4]. These approaches make use of 2nd-order information to whiten the data, and 4th-order information is then used to process the resulting orthogonalized problem. The additive noise is assumed to be normally distributed : it has no effect on the (exact) 4th-order cumulants but 2nd-order noise cumulants do **not** vanish so that the spatial structure of the noise covariance has to be known, modelled or estimated in order to achieve consistent 2nd-order prewhitening. These limitations can be overcome by giving up the idea of 2nd-order whitening and resorting to 4th-order cumulants only [5-6]. It is the purpose of this communication to show how the concept of *fourth-order signal subspace* yields simple implementations of 4th-order only blind identification.

2. BLIND IDENTIFICATION

2.1 On identifiability. The blind context does not lead to full identifiability of the model (1) because any complex factor can be exchanged between s_p and $\boldsymbol{a}_p$ without modifying the observation. Hence if no a priori information is available, each signature can be identified only up to a scale factor. We take advantage of this to assume, without any loss of generality, that each signature $\boldsymbol{a}_p$ has unit norm. With this constraint, an unidentifiable phase term is still present in $\boldsymbol{a}_p$. For each source p, we denote as Π_p the orthogonal projector onto the 1D space where the pth component lives. It is the space spanned by the signature $\boldsymbol{a}_p$ and the projector onto it is

$$\Pi_p \triangleq \boldsymbol{a}_p\, \boldsymbol{a}_p^* \tag{2}$$

This hermitian matrix is unaffected by any phase term in $\boldsymbol{a}_p$ and conversely determines $\boldsymbol{a}_p$ up to a phase term. It follows that the projectors Π_p are the algebraic quantities that can be, at best, identified in the blind context. It is easily seen that knowing these projectors is sufficient to perform **blind separation** of the source signals $s_p(t)$ because they allow to construct for each source the linear filter that zeroes all the components but the one specified. We then define **blind identification** as the problem of estimating the projectors Π_p from sample statistics only.

2.2 Blind MUSIC. The approach to blind identification presented in this contribution can be seen as a blind 4th-order version of the celebrated MUSIC algorithm. The MUSIC technique is based on the concept of *signal subspace* which is the vector space spanned by the steering vectors. It can be summarized as i) Estimate the signal subspace using the covariance. ii) Search for the steering vectors which are the closest to the signal subspace. The search is, of course, across the array manifold: this is how MUSIC exploits the a priori information contained in the parameterization of the steering vectors.

The *fourth-order signal subspace* (FOSS) is defined as the real span of the projectors Π_p i.e. as the linear **matrix** space made of all possible linear combinations with real coefficients of the projectors Π_p :

(3) $$\text{FOSS} = \{ M \mid M = \sum_{p=1,n} \gamma_p \, \boldsymbol{a}_p \boldsymbol{a}_p^* \, , \gamma_p \in \mathbf{R} \}$$

Let us now consider the following idea for blind identification i.e. estimation of the Π_p: i) Estimate the FOSS using 4th-order cumulants. ii) Search for the orthogonal 1D projectors which are the closest to the FOSS. The closest projectors to the FOSS are taken as estimates of the source projectors Π_p. Such an idea could be termed "blind MUSIC" because, in spite of its strong analogy with the classical MUSIC, no signature parameterization is assumed here: the search is across the so called *rank one manifold* (ROM) which is defined as the set of all rank-one unit-norm hermitian matrices i.e. across all the 1D orthogonal projectors.

The reason why Blind MUSIC works is that a matrix space with structure as in eq. (3) is shown, under mild conditions, to contain no other 1D projectors than the ones used in its construction, i.e. the Π_p. This is obviously true as soon as the signatures $\boldsymbol{a}_p$ are linearly independent since in that case, a matrix M as in eq. (3) has a rank equal to the number of non-zero coefficients γ_p. If M is a 1D projector, it has rank one, hence all the coefficients γ_p but one are zero and M is then necessarily one of the Π_p.

We now have to discuss i) FOSS estimation from 4th-order cumulants ii) practical implementations of the blind MUSIC search.

3. FOURTH-ORDER SIGNAL SUBSPACE

3.1 Quadricovariance. We find convenient to make temporary use of indexed notations to express the cumulants of the vector process $\boldsymbol{x}$. Let us denote by x_i the i-th coordinate of vector $\boldsymbol{x}$ and by x^i the i-th coordinate of its dual $\boldsymbol{x}^*$. Of course, only orthonormal basis are used: x^i is just the complex conjugate of x_i. The covariance classically is the matrix R whose (i,j)-coordinate denoted r_i^j is the 2nd-order cumulant of x_i and x^j :

(4) $$r_i^j \triangleq Cum\,(\, x_i \, , x^j \,) \qquad 1 \le i,j \le m$$

Similarly, we define the *quadricovariance* of $\boldsymbol{x}$ as the set of m^4 complex scalars, q_{il}^{jk}, $1 \le i,j,k,l \le m$:

(5) $$q_{il}^{jk} \triangleq Cum\,(\, x_i \, , x^j \, , x^k \, , x_l \,)$$

Our approach to process 4th-order information is to consider the quadricovariance as a matrix mapping denoted Q, which to any matrix M with coordinates m_i^j associates the matrix $N = Q(M)$ with coordinates n_i^j according to:

(6) $$n_i^j = \sum_{1 \le k,l \le m} q_{il}^{jk} \, m_k^l$$

The quadricovariance has the two following properties: i) it maps any hermitian matrix to another hermitian matrix. ii) it is itself an hermitian operator in the (usual) sense that for any matrices M and N we have $\langle N \mid Q(M) \rangle^* = \langle M \mid Q(N) \rangle$ with the Euclidian scalar product $\langle M \mid N \rangle \triangleq Tr(NM^H)$. These are trivial consequences of cumulant symmetries. It follows [3] that the quadricovariance admits m^2 real eigenvalues, denoted μ_i , $i=1,m^2$ and m^2 corresponding orthonormal hermitian eigen-matrices, denoted E_i , $i=1,m^2$, verifying:

$$\forall i=1,m^2 \quad Q(E_i) = \mu_i \, E_i \qquad \text{with}$$
$$E_i = E_i^H \, , \quad \mu_i \in \mathbf{R} \, , \quad \mathrm{Tr}(E_i E_j) = \delta(i,j)$$

As a simple consequence [3] of cumulant additivity and multilinearity, the quadricovariance of a linear mixture (1) of independent components takes the special form:

(7) $$Q(M) = \sum_{p=1,n} k_p \; \boldsymbol{a}_p^* M \boldsymbol{a}_p \; \boldsymbol{a}_p \boldsymbol{a}_p^*$$

with no contribution from the additive noise (since it has been assumed Gaussian and independent of the signals) and where the *kurtosis* of the p-th source is denoted by k_p :

(8) $$k_p \triangleq \mathrm{Cum}\,(\, s_p \, , s_p^* \, , s_p^* \, , s_p \,)$$

Equation (7) evidences that the image space of the quadricovariance Q is spanned by the projectors $\Pi_p = \boldsymbol{a}_p \boldsymbol{a}_p^*$ (hence the name "FOSS"). It has exactly rank n if no kurtosis k_p is zero and if the projectors Π_p are linearly independent. This last condition is fulfilled whenever the signatures $\boldsymbol{a}_p$ are themselves independent. It follows that quadricovariance eigen-decomposition shows only n non-zero eigenvalues. Let us assume that they are numbered in such a way that the corresponding n eigen-matrices are $(E_i \mid i=1,n)$. These eigen-matrices form an hermitian orthonormal basis of the FOSS.

3.2 FOSS estimation. When a strongly consistent estimate of the signal covariance is used, the 2nd-order signal subspace estimate obtained via eigen-decomposition also is strongly consistent. The same can be shown to hold for the FOSS estimates obtained from an eigen-decomposition of the sample quadricovariance into eigen-matrices. This should be the preferred FOSS estimation method for small arrays but eigen-decomposition of the quadricovariance may be too expensive with large arrays.

Note however that only a small number of eigen-matrices need to be estimated (n and not m^2) and this fact can lead to large computational savings (see [7]). Even in that case, the whole set of 4th-order cumulants is needed, and quadricovariance estimation cost may be prohibitive. Fortunately, the FOSS can be estimated in a simpler manner. We demonstrate this in the (rather common) case where signals are circurlarly distributed. Cumulant expression in terms of the moments then reduces to:

$$(9) \qquad q_{il}^{jk} = E\{ \; x_i \, x^j \, x^k \, x_l \; \} \; - r_i^j \, r_l^k - r_l^j \, r_i^k$$

and it is readily checked that the quadricovariance image of any matrix M accordingly reduces to:

$$(10) \quad Q(M) \; = \; E\{ \; (x^* M x) \; x x^* \} \; - RMR - R\mathrm{Tr}(MR)$$

This expression admits an obvious sample counterpart showing that $Q(M)$ can be estimated at a cost similar to the covariance. This suggests to choose a priori a set of n hermitian matrices M_i and to estimate $Q(M_i)$ according to (10). The result will be a set of n almost surely independent matrices of the FOSS. They can then be orthonormalized (by a Gram-Schmidt procedure for instance) into an orthonormal hermitian basis of the FOSS. Such a procedure obviously yields FOSS estimates with higher variance than those obtained by eigen-decomposition of the whole set of 4th-order cumulants. Since it is not possible to ensure in advance that the $Q(M_i)$ actually are independent, a safer solution would be to use a number of M_i larger than n.

4. BLIND MUSIC IMPLEMENTATION

From now on, we assume that a FOSS estimate is available in the form of a set of n hermitian orthonormal matrices: $(M_i \mid i=1,n)$ forming a basis of the estimated FOSS. The following search implementations do not depend on the particular FOSS estimation technique.

Blind MUSIC can be implemented as searching through the FOSS the closest ROM matrix (see the *PQN* technique below) or, alternatively, as searching through the ROM the closest FOSS matrix (see $\Pi V3$) In both cases, the suggested techniques do not implement the search via a gradient (or similar) approach but expresses the Blind MUSIC estimates as **fixed points** of an appropriate mapping. In our simulations, we have found that these fixed points were the only stable points: the blind MUSIC search can then be implemented as the iteration of these mappings with arbitrary starting points.

4.1 ΠV^3 Search. The natural approach to blind MUSIC is to maximize the norm of the projection onto the FOSS of a matrix A under the constraint that it is a 1D projector. Using an orthonormal hermitian basis, the squared norm of this projection, denoted d, is the sum of the squared projections onto each basis matrix.

$$(11) \qquad d \; = \; \sum_{i=1,n} \mathrm{Tr}^2 \, (A \, M_i \,)$$

Blind MUSIC estimates are obtained as the maximizers of d under the constraint that $A = vv^*$ with $v^* v = 1$. Since $\mathrm{Tr}(A \, M_i) = \; v^* \, M_i \, v$, the variation with respect to v of a Lagrange function $L=1/2d - \lambda v^* v$ associated to this constrained optimization problem is:

$$\delta L = \sum_{i=1,n} (v^* \, M_i \, v)(v^* M_i \delta v + \delta v^* M_i v) - \lambda \, (v^* \delta v + v^* \delta v)$$

Defining the cubic vector mapping $v \rightarrow \phi(v)$ as :

$$(13) \qquad \phi(v) \triangleq \Sigma_{i=1,n} \, (\, v^* \, M_i \, v \,) \, M_i v$$

the Lagrange function variation is rewritten in:

$$(14) \qquad \delta L = (\phi - \lambda v)^* \, \delta v \; + \; \delta v^* (\phi - \lambda v)$$

which is zero for any δv iff $\phi(v) = \lambda \, v$. This is equivalent to v being a fixed point of the mapping $v \rightarrow \Phi(v)$ where:

$$(15) \qquad \Phi(v) \triangleq \phi(v) \, / \, | \, \phi(v) \, |$$

The $\Pi V3$ search (where $V3$ is a reminder for the cubic dependence on the iterated vector) starts with a random vector and then iteratively computes its image through Φ.

4.2 PQN Search. An alternate approach is to search for matrices of the FOSS that are as close as possible to the ROM. The basic idea is to start with an arbitrary matrix of the FOSS and to repeatedly project it onto the ROM and back onto the FOSS. Projection onto the ROM is equivalent to truncating the matrix to its first principal eigen-component, which requires an eigen-decomposition at each step. On the other hand, repeatedly squaring a matrix has the effect of enhancing the dominant eigenvalue. In our experiments, we have found that the projection onto the ROM, being included in the iteration loop, could be replaced by a simple matrix squaring followed by renormalization The PQN algorithm is just cycling through the three steps of projection, quadration, and normalization, hence the acronym "PQN". After convergence, the dominating eigen-vector is extracted, providing an estimate of one of the source signatures.

Quadration and projection can be efficiently implemented in a single step by representing the iterated matrix, say A, by its (real) coordinates a_i, $i=1,n$ in the FOSS basis: $A = \sum a_i \, M_i$. The squared matrix then is $A^2 = \sum a_i \, a_j \, M_i M_j$. Since an orthonormal basis is used, the projection of A^2 onto the FOSS has coordinates a_k' given by:

$$(16) \quad a_k' = \Sigma_{i,j} \, t_{ijk} a_i a_j \quad \text{with} \qquad t_{ijk} \triangleq \mathrm{Tr} \, (\, M_i \, M_j \, M_k \,)$$

Hence, by pre-computing the table t_{ijk}, each quadration-projection is computed in the single step (16), involving only n^3 real multiplications.

4.3 Simulation results The following simulation results are for a uniform linear half-wavelength array of 4 sensors, two independent PSK modulated sources of unit variance located respectively at 0 and 20 degrees (i.e. under the same lobe). The signal is corrupted by additive Gaussian white noise with covariance σI. We performed 20 Monte-Carlo runs using the PQN algorithm for data lengths of 50, 100, 200, 500, 1000 and for noise levels σ= -10, 0, 10, 20 dB. For each run, we plot a performance index ρ defined as

$$\rho = \frac{1}{n} \sum_{p=1,n} 1 - \mathrm{Tr}\,(\hat{\Pi}_p\, \Pi_p\,) \qquad (17)$$

where each $\hat{\Pi}_p$ is the estimated p-th projector. This index also is the squared sine of the angle between each signature and its estimate (averaged on the sources).

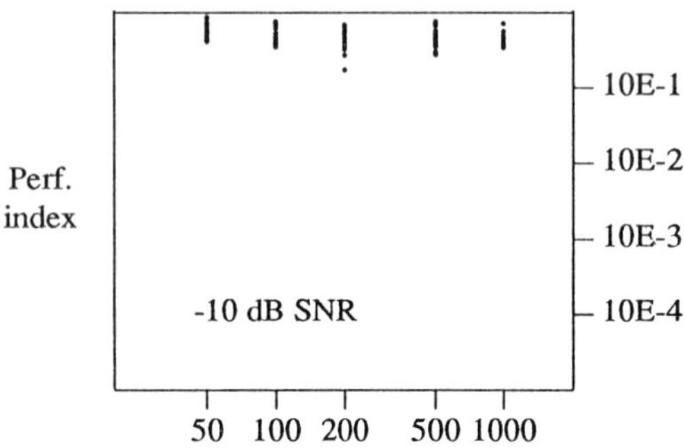

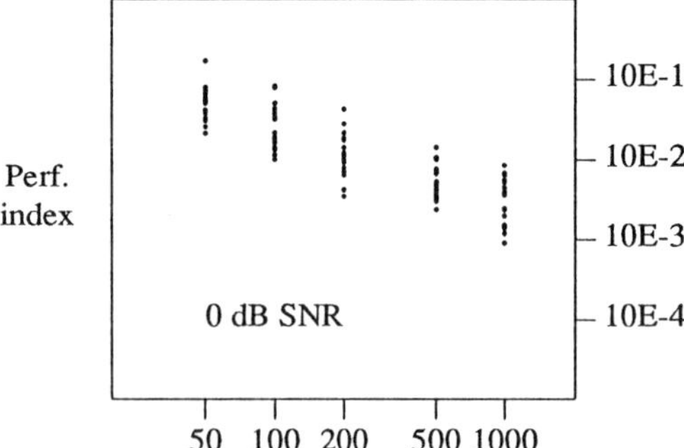

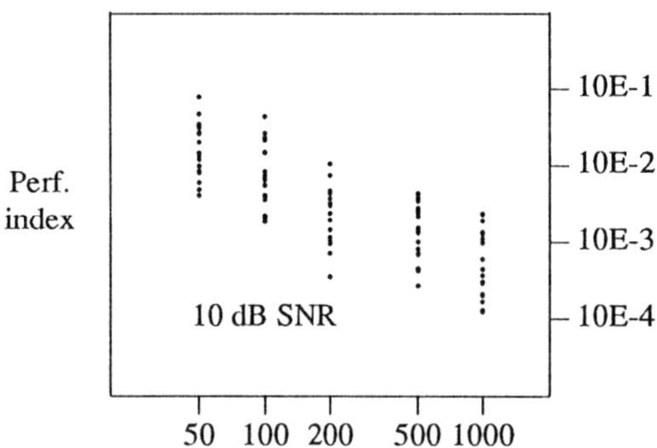

For negative SNR, the data lengths used here do not allow any correct estimation: the performance index is close to 0.5, indicating "random estimation" ! But as the SNR gets positive meaningful estimates are obtained with relatively short data lengths. Also note that there is no significant improvement when the SNR goes from 10 dB to 20 dB. This is a general feature of 4th-order-only blind techniques: when the noise level is low enough, the performance is dominated by the sample size. This could be contrasted with the standard parametric MUSIC (2nd-order or 4th-order) where, at low noise levels, the variance

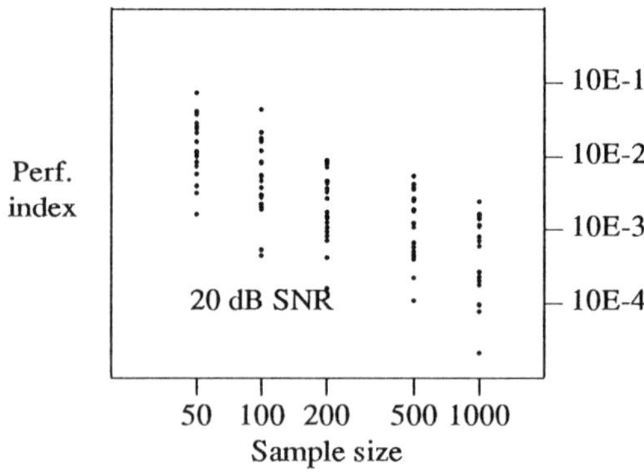

of the estimates is proportional to the noise power. Another remark is that these simulations are for equipowered sources: the performance would degrade for a source when its power gets weaker. This effect shows at any order but is naturally more severe at fourth-order than at 2nd-order.

Just as in the MUSIC case, asymptotic performance can be obtained in closed form but room is definitely lacking for exposition of the results. They will be presented at the conference.

CONCLUSION. The notion of *fourth-order signal subspace* (FOSS) has been introduced. This matrix space, function of the 4th-order cumulants, is the natural 4th-order counterpart of the classical 2nd-order signal subspace. Exploited in a "blind MUSIC" fashion, it allows for blind identification without resorting to 2nd-order information. This can be done with one of the low cost fixed point techniques presented here.

REFERENCES

[1] P. Comon, "Independent Component Analysis", Proc. Int. Workshop on Higher-Order Stat., Chamrousse, France, Jul. 91, pp. 111-120.

[2] M. Gaeta, J.L. Lacoume, "Source Separation Without A Priori Knowledge: the Maximum Likelihood Solution", Proc. EUSIPCO, Barcelona, Spain, Sept. 90, pp. 621-624.

[3] J.F. Cardoso, "Eigen-structure of the fourth-order cumulant tensor with application to the blind source separation problem", Proc. ICASSP'90, pp. 2655-2658, Albuquerque, 1990.

[4] V. C. Soon, L.Tong, Y. F. Huang, R. Liu, "An extended fourth order blind identification algorithm in spatially correlated noise", Proc. ICASSP'90, pp. 1365-1368, Albuquerque, 1990.

[5] J.F. Cardoso, "Super-Symmetric Decomposition of the Fourth-Order Cumulant Tensor. Blind Identification of More Sources than Sensors", Proc. ICASSP'91, pp. 3109-3112, Toronto, 1991.

[6] G. Giannakis, S. Shamsunder, "Modelling of non Gaussian array data using cumulants : DOA estimation with less sensors than sources", Proc. of conf. on Info. Sci. and Syst., Baltimore, MD, 1991.

[7] J.F. Cardoso, P. Comon, "Tensor-based Independent Component Analysis", Proc. EUSIPCO, Barcelona, Spain, Sept. 90, pp. 673-676.

SIGNAL PROCESSING VI: Theories and Applications
J. Vandewalle, R. Boite, M. Moonen, A. Oosterlinck (eds.)

PHASE RECONSTRUCTION OF AUTOREGRESSIVE PROCESSES USING THIRD ORDER CUMULANTS

Asoke K NANDI[1)] and Ralf MEHLAN[2)]

[1)]Department of Electronic and Electrical Engineering, University of Strathclyde, Royal College, 204 George Street, Glasgow, G1 1XW, UK

[2)]Lehrstuhl für Elektrische Regelungstechnik, RWTH Aachen, Templergraben 55, 5100 Aachen, Germany

The knowledge of the phase is very important in many situations. It is well known [1-2] that phase information is, in principle, available from the third order cumulants (except when third order cumulants happen to be zero as in the Gaussian case). Methods for reconstructing phases in autoregressive (AR) processes, using estimated third order cumulant sequences from the time-series data are presented in this paper and they are shown to have the consistency of performance.

1. INTRODUCTION

In many situations the knowledge of the phase is very useful and sometimes extremely important. While the second order statistics (SOS) is widely used and is very useful, it has its limitations. It is well known [1-2] that the phase information is preserved in higher order statistics (HOS) - that is, for order greater than two. In principle then, the phase information is available from the third order cumulants (except when third order cumulants happen to be zero as in the Gaussian case).

The bispectrum, $B(\omega_1, \omega_2)$, of a time-series, $x(n)$, is in general a complex quantity and it can therefore be written as $B(\omega_1, \omega_2) = |B(\omega_1, \omega_2)| \exp(j\psi(\omega_1, \omega_2))$, where the $|B(\omega_1, \omega_2)|$ is referred to as the bispectral magnitude and the $\psi(\omega_1, \omega_2)$ is called the bispectral phase. If the time-series can be well modelled as the output of an AR filter, with the system function $H(z)$, driven by ideal, white noise the bispectrum can be written as $B(\omega_1, \omega_2) = \beta . H(\omega_1) . H(\omega_2) . H^*(\omega_1, \omega_2)$, where the constant β is often called the skewness or $C_3(0,0)$ of the ideal, white noise. The phase of the bispectrum is related to the phase of the filter by $\psi(\omega_1, \omega_2) = \phi(\omega_1) + \phi(\omega_2) - \phi(\omega_1 + \omega_2)$, where $\phi(\omega)$ is the phase of the model filter and is also referred to as the phase of the observed signal, $x(n)$. Non-parametric phase reconstruction algorithms [3-6] try to recover the phase samples $\phi(\omega)$ from the estimates of the bispectrum phase, $\psi(\omega_1, \omega_2)$.

Parametric phase reconstruction methods are based on the assumption that the observed time-series is the output from a specific type of process, say obeying an AR model. Once such identification of the process is made the transfer function of the filter, $H(\omega)$, becomes a known function with unknown model parameters, a_i's, which are the parameters of the model. Now if these model parameters can be obtained, the $H(\omega)$ can be numerically computed and subsequently the phase, $\phi(\omega)$, can be estimated. Thus the problem in parametric phase reconstruction methods can be reduced to how to best estimate the model parameters (and, if need be, the model order) from the observed time-series, $x(n)$. Usual (dis)advantages of parametric methods over non-parametric methods as in the SOS apply here.

2. TOR EQUATIONS

For AR models a set of linear equations in terms of the model parameters and third order cumulants can be derived [7] much along the lines of Yule-Walker equations for second order cumulants [8]. In [7] Nikias and Raghuveer derived the following set of linear equations for AR models

$$C_3(-m,-n) + \sum_{i=1}^{p} a_i . C_3(i-m,i-n) = \beta . \delta(m,n) \quad (2.1)$$

where p is the order of the AR model, a_i's are the model parameters, and $C_3(m,n)$ represents the third order cumulant of lags m and n of the observed time-series. These were termed the third order recursion (TOR) equations.

By chosing the diagonal slice, namely setting m = n and choosing m = 0, 1, ..., p, a set of (p+1) linear equations are obtained with (p+1) unknown parameters of the AR model. In matrix notation the set of equations can be written as

$$\mathfrak{R} . \mathcal{A} = \mathfrak{B}, \quad (2.2)$$

where

$$\mathcal{R} = \begin{bmatrix} C_3(0,0) & C_3(1,1) & \cdot\cdot & C_3(p,p) \\ C_3(-1,-1) & C_3(0,0) & \cdot\cdot & C_3(p-1,p-1) \\ \cdot & \cdot & \cdot\cdot & \cdot \\ \cdot & \cdot & \cdot\cdot & \cdot \\ C_3(-p,-p) & \cdot & \cdot\cdot & C_3(0,0) \end{bmatrix}$$

$\mathcal{A} = [a_0\ a_1\ .\ .\ a_p]^T$, and $\mathcal{B} = [\beta\ 0\ .\ .\ 0]^T$. The matrix $\mathcal{R}$ is a non-symmetric Toeplitz matrix and therefore equation (2.2) can be solved with a fast and efficient algorithm.

Various applications of the TOR equations have been proposed [9-11]. As the exact third order cumulants of the observed process are not available but only their estimates the derived parameters can only be estimated and now the accuracy of any model based on cumulants estimates becomes a very important consideration in its choice for reconstructing the phase.

3. PROPOSED METHOD

The TOR equations (2.2), as given above[7], express a specific relationship between AR parameters and the third order cumulants of the observed AR process. This algorithm does not use all available information on third order cumulants but only those along a diagonal on the cumulant indices plane. These represent a 1-D slice and this can only be a full rank slice if the matrix representing the set of linear equations is full rank. However it has been shown that such a full rank slice may not always exist [13]; therefore the linear equations may not be solved and the AR model parameters cannot always be estimated in this way.

Within the TOR context though, there are no theoretical constraint as to cumulants of what lags should be used, there are, of course, practical constraints of a general nature. For example, it is perhaps better to avoid cumulants of high lags as they are difficult to estimate well from a small number of samples. In this paper the above idea is modified to guarantee the robust estimation of AR parameters from a suitable but a priori choice of a overdetermined set of linear equations.

3.1. PROPOSED ALGORITHM

Step 1: Form $N \geq (p+1)$ TOR equations from (2.1) by chosing the lags m, n in a well defined way;

Step 2: Form an overdetermined set of equations and set up the matrix equation $\mathcal{G}.\mathcal{A} = \mathcal{B}$ (3.1)

with $\mathcal{A} = [a_0\ a_1\ .\ .\ a_p]^T$ of dimension $(p+1)\text{x}1$,

$\mathcal{B} = [\beta\ 0\ .\ .\ 0]^T$ of dimension Nx1, $\mathcal{G}$ is a Nx(p+1) dimensional matrix;

Step 3: Obtain the least squares solution of (3.1) from

$$\mathcal{A} = (\mathcal{G}^T.\mathcal{G})^{-1}.\mathcal{G}^T.\mathcal{B} \qquad (3.2)$$

In this algorithm, it is yet to be specified how many TOR equations should be used (i.e. the value of N) and which specific third order cumulants (values of m and n) should be included in this overdetermined set of equations.

3.2. SELECTION OF TOR EQUATIONS

The aim of this section is to outline how the investigation is carried out to discover the set of indices (m,n) that generate TOR equations, from (2.1), in the form of (3.1) such that the matrix $\mathcal{G}$ is of rank (p+1). Three strategies are adopted with respect to two criteria - 1) the matrix $\mathcal{G}$ should be of rank (p+1) with the minimum of TOR equations, and 2) the accuracy of phase estimates from (3.2) as measured in terms of the mean square error with respect to the exact phase must be less than that obtained from (2.2).

The following three strategies are used to determine the selection of TOR equations -

A) Here the TOR equations come from a diagonal slice in the (m,n) index-plane but differently from (2.2). First the (p+1) equations are chosen corresponding to m=n=0, 1, . ., p. Then additional equations are chosen along the diagonal of the (m.n) index-plane to satisfy the above two criteria.

B) Now the TOR equations come from a growing triangle in (m,n) index plane with column by column growth. It starts with the following (m,n) values in the order they are stated - (0,0),{(1,0), (1,1)}, {(2,0), (2,1), (2,2)}, and so on.

C) In this strategy diagonal as well as non-diagonal indices are used, as in B) but in the following order of (m,n) values - first the full triangle {(0,0), (1,1), . ., (p,p)}, {(1,0), (2,1), . ., (p,p-1)}, {(2,0), (3,1), . ., (p,p-2)}, . ., (p,0), and then the columns {(p+1,p+1), (p+1,p), . ., (p+1,0)}, {(p+2,p+2), (p+2,p+1), . ., (p+2,0)}, and so on.

Simulations have been carried out to answer which of above three strategies is the best and how many TOR equations are needed.

4. SIMULATION

Twenty different AR models of order upto seven are chosen; some of these have been used in [7,14]. For each of these models the exact third order cumulants were distorted with the following error model to represent the 'estimated' cumulants

$$\hat{C}_3(m,n) = C_3(m,n).[1 + \epsilon]$$

where $\epsilon \in [-\nu, +\nu]$ and is uniformly distributed. For example, ν=0.1 means that no estimated cumulant deviates from the exact value by more than 10%. For each model, ten different values of ν are chosen. Thus for each of the twenty AR models and each of the ten values of ν, a set of $\hat{C}_3(m,n)$ for various values of m,n (i.e. there are 200 such sets) is obtained.

Now in each of these 200 sets:

1) samples of the phase are estimated via (2.2) and the mean square error (MSE) is computed as

$$MSE(2.2) = \sum_{i=0}^{P-1}[\phi(i) - \hat{\phi}(i)]^2/P$$

where $\phi(i)$ and $\hat{\phi}(i)$ represent the exact phase and the estimated phase via (2.2), and P is the number of phase samples;

2) samples of the phase are estimated via (3.2) according strategies A, B, and C but for various values of N, the number of TOR equations. For each strategy, the MSE depends on N and is computed from

$$MSE(A,N) = \sum_{i=0}^{P-1}[\phi(i) - \hat{\phi}_{A,N}(i)]^2/P$$

where $\phi(i)$ represent the exact phase, while $\hat{\phi}_{A,N}(i)$ represent the estimated phase via (3.2), and P is the number of phase samples; correspondingly MSE(B,N) and MSE(C,N) can be defined;

3) for each values of N, calculate the following normalised mean square errors

NMSE(A,N) = MSE(A,N)/MSE(2.2),
NMSE(B,N) = MSE(B,N)/MSE(2.2),
NMSE(C,N) = MSE(C,N)/MSE(2.2).

Note that if NMSE(C,N) is less than 1.0, the proposed method with strategy C with N TOR equations is able to reconstruct the phase better than using (2.2) which uses cumulants on the specific diagonal slice.

4.1. RESULTS - PART I

The following general points are observed:
1) Selecting the TOR equations following strategy A appears to have large fluctuations in NMSE(A,N) around the value of 1 as N changes. No clear advantage in the strategy A over (2.2) emerges.

2) The NMSE(B,N) is a decreasing function of N although the rate of decrease slows down for large N values. When N is small (the matrix $\mathcal{I}$ is not of rank (p+1)), NMSE(B,N) is larger than 1; while NMSE(B,N) is smaller than 1 for larger values of N.

3) In the cases where TOR equations follow the strategy C, the NMSE(C,N) decreasingly fluctuates for smaller values of N and settles down to small values, much less than 1, as N increases.

Based on the above observations, which are independent of the values of ν and the AR process studied, it is concluded that the strategy A should be discarded and that the strategy C appears to be the best choice. With the strategy C and for an AR(p) process, the value of N = (p+1)(p+2)/2 tends to be the point beyond which the NMSE(C,N) settles down to the 'lowest error level'. It is suggested that a total of $N^*=(p+2)(p+3)/2$ TOR equations following the

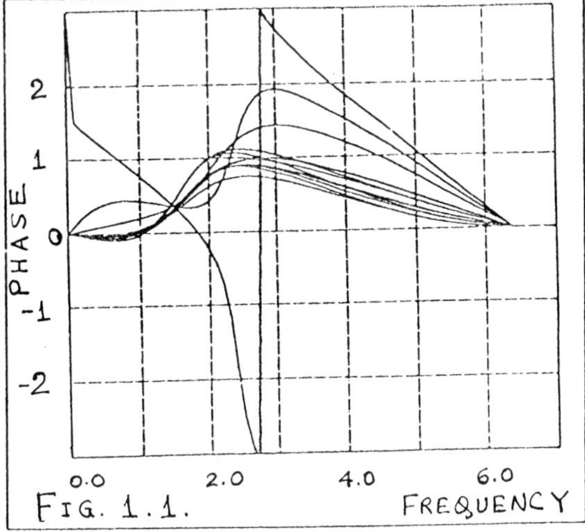

FIG. 1.1.

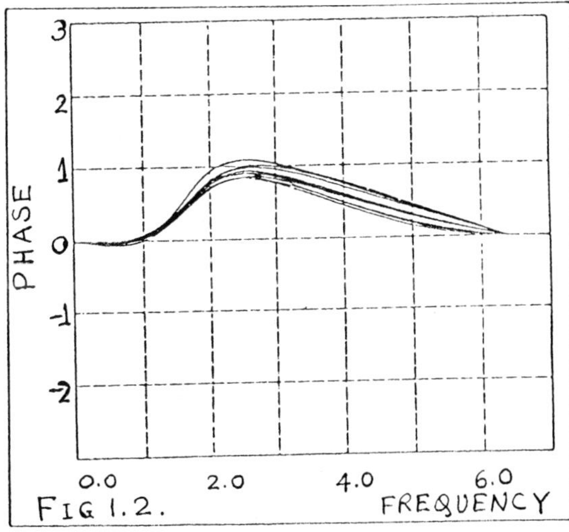

FIG 1.2.

strategy C is sufficient to gurrantee a considerable reduction in the error of the phase estimates of an AR(p) process.

4.2. RESULTS - PART II

As an example, an AR(3) process is chosen with parameters $a_0=1$, $a_1=-0.7$, $a_2=0.2$, and $a_3=0.15$. The value of ν is set to 0.2 and ten different sets of the third order cumulants, $C_3(m,n)$, are obtained. For each of these ten sets of cumulants, phases are estimated using the diagonal slice with (2.2). The resulting ten sets of phase estimates are displayed in fig. 1.1. In the figures here, estimated phases (in radians) of the AR process under study are plotted versus the 'frequency' (in radians). Using the same cumulants, phase estimates are obtained using (3.2) with N^* TOR equations following strategy C. These new sets of phase estimates are plotted in fig. 1.2. It is clear that estimates in fig. 1.2 are lot less variable than the ones in fig. 1.1.

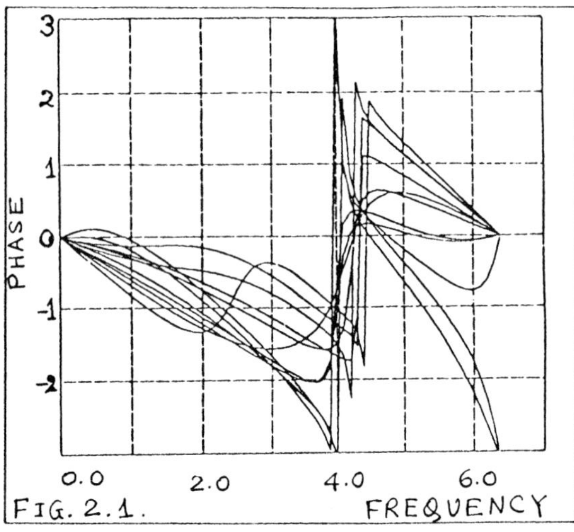

FIG. 2.1.

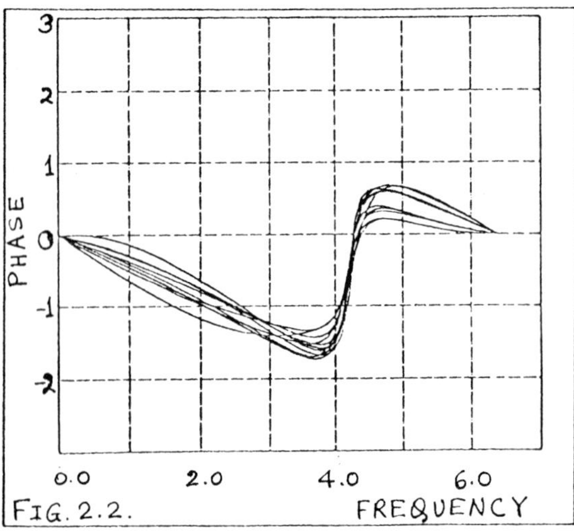

FIG. 2.2.

As a second example, an AR(4) process is chosen with parameters $a_0=1$, $a_1=0.8$, $a_2=0.7$, $a_3=-0.2$, and $a_4=-0.1$. Setting the maximum error in any cumulant to be 20% (i.e. $\nu=0.2$), ten different sets of the third order cumulants, $C_3(m,n)$, are obtained. For each of these ten sets of cumulants, phases are estimated using the diagonal slice with (2.2). The resulting ten sets of phase estimates are displayed in fig. 2.1. For each of the same sets of cumulants, phase estimates are obtained from (3.2) with 21 TOR equations following strategy C. These new sets of phase estimates are plotted in fig. 2.2. It is clear that estimates in fig. 2.2 (from the proposed method) are lot less variable than the ones in fig. 2.1.

5. SUMMARY

In this paper a new algorithm based on (3.2) has been proposed to estimate the phase of an AR filter. Although the algorithm based on (2.2) is computationally efficient the proposed algorithm in this paper provides much less variable and more consistent phase estimates than the algorithm based on (2.2).

REFERENCES

[1] Nikias, C. L. and Raghuveer, M. R., Proc. of the IEEE, vol. 75, pp. 869 (1987).
[2] Mendel, J. M., Proc IEEE, vol. 79, pp. 278 (1991).
[3] Brillinger, D. R., Biometrica, vol. 65, pp. 509 (1977).
[4] Lii, K. S. and Rosenblatt, M., Annals of Statistics, vol. 10, pp. 1195 (1982).
[5] Matsuoka, T. and Ulrych, T. J., Proc. of the IEEE, vol. 72, pp. 1403 (1984).
[6] Nandi, A. K., Internal Report, Imperial College, London, UK.
[7] Raghuveer, M. R. and Nikias, C. L., IEEE Transc., ASSP-33, pp. 1213 (1985).
[8] Kay, S. M., Modern Spectral Estimation (Prentice Hall, Englewood Cliffs, New Jersey 07632, 1988).
[9] Nikias, C. L. and Chiang, H., IEEE Transc. on ASSP, ASSP-36, pp. 1911 (1988).
[10] Nikias, C. L., IEEE Transc. on ASSP, ASSP-36, pp. 521 (1988).
[11] Nikias, C. L. and Pan, R., IEEE Transc. on ASSP, ASSP-35, pp. 895 (1987).
[12] Giannikis, G. B. and Mendel, J. M., IEEE Transc., ASSP-37, pp. 360 (1989).
[13] Swami, A. and Mendel, J. M., Workshop on Higher Order Spectral Analysis, pp. 1 (1989).
[14] Raghuveer, M. R. and Nikias, C. L., Signal Processing, vol. 10, pp. 35 (1986).

SIGNAL PROCESSING VI: Theories and Applications
J. Vandewalle, R. Boite, M. Moonen, A. Oosterlinck (eds.)
1992 Elsevier Science Publishers B.V.

DIGITAL CALCULATION OF THE FUNDAMENTAL SINEWAVE OF SIGNALS USING METHODS CONSTRAINED IN FREQUENCY DOMAIN

Tadeusz ŁOBOS Piotr RUCZEWSKI Jarosław M. SZYMAŃDA

Technical University of Wrocław, PL-50370 Wrocław, pl.Grunwaldzki 13, Poland *

Real-time determination of parameters of voltages and currents is very important for measurement, control and protection tasks in electrical power systems. The paper presents a method which enables us to develop fast filter algorithms with auxiliary constraints. The problem of designing is converted to an optimization problem with constraints using standard least-square criterion. The minimization problem is mapped to a simpler problem of solving a set of linear equations, by using the Lagrange multipliers. Using the method we can develop filter algorithms for estimating the phasor of fundamental waveforms of signals optimizing the algorithms for concrete application.

1. INTRODUCTION

Serious problems for the control and protection of electrical power systems arise in estimating the fundamental waveforms of voltages and currents from measured signals with high noise levels. Various algorithms have been suggested to estimate the fundamental waveforms burried in noise and distorted by higher harmonics [1-4]. The methods presented in the literature optimize the estimation in the time domain.

The waveforms of voltages and currents in electrical power systems may include harmonics of basic waveforms due to nonlinear elements and random noise due to faults and other disturbances. In some case, e.g. in circuits with thyristor converters, it is possible to predict the frequencies of the higher harmonics. To estimate the phasor of the fundamental waveform in such situations we have developed and investigated filter algorithms which enable us to suppress specific higher harmonics effectively.

To estimate the real part of the phasor the argument of the transfer function of the algorithm for the fundamental frequency should be equal to zero, and for imaginary part equal to $-\pi/2$. In the both cases the magnitude of the transfer function for the fundamental frequency should be equal to one. In the developed method we can prescribe the values of the magnitude and argument, the values of the derivatives of the magnitude and argument at any frequency, adjusting the filter characteristics to special requirements of the application.

The proposed methods are illustrated by some numerical examples.

*This work was mainly supported by the State Committee for Scientific Research (Poland), grant 306619101

2. CONSTRAINTS OF THE FREQUENCY RESPONSE

The frequency characteristic of a digital nonrecursive filter should approximate the periodic function $G(j\omega)$

$$G(j\omega)=\sum_{-\infty}^{+\infty} g_k e^{-jk\omega T} \tag{1}$$

where

$$g_k=\frac{T}{2\pi}\int_{-\pi/T}^{\pi/T} G(j\omega)e^{jk\omega T}d\omega \tag{2}$$

T - sampling interval,
ω - angular frequency,
g_k - coefficients of the filter being the discrete values of the impulse response of an ideal filter.

The function (1) describes the characteristic of a noncausal ideal filter. We are looking for a causal filter with a finite impulse response, which frequency characteristic is possibly closed to the ideal. For a given characteristic $G(j\omega)$ and a given filter order N we have to calculate the coefficients h_k of the real filter by minimizing the objective function

$$Q=\frac{\pi}{4T}\int_{-\pi/T}^{\pi/T} \left|G(j\omega)-H(j\omega)\right|^2 d\omega \tag{3}$$

where

$$H(j\omega)=\sum_{k=0}^{N-1} h_k e^{-jk\omega T}$$

It is required that the partial derivatives of the function with respect to h_k be zero:

$$\frac{\partial Q}{\partial h_k} = 0 \qquad \text{for } k=0,1,\ldots,N-1 \tag{4}$$

In this way we obtain the desired coefficients h_k, which are the coefficients of a truncated Fourier series ($h_k=\bar{g}_k$).
In the proposed method we will prescribe the values of the magnitude and argument, the values of the derivatives of the magnitude and argument at any points, chosen according to the desired application. In this way we obtain a set of linear equations which can be solve for h_k.

Introducing the notation

$$H(j\omega)=H(\omega)e^{j\varphi(\omega)} \tag{5}$$

and using the eqn. 3 we obtain

$$H(\omega)=\sum_{k=0}^{N-1} h_k e^{-j[k\omega T+\varphi(\omega)]} \tag{6}$$

The complex eqn. 6 can be expressed in a form of two real equations

$$\sum_{k=0}^{N-1} h_k \cos[\varphi(\omega)+k\omega T]=H(\omega) \tag{7}$$

$$\sum_{k=0}^{N-1} h_k \sin[\varphi(\omega)+k\omega T]=0 \tag{8}$$

Using (7) and (8) we can also introduce the non--zero constraints.
If the magnitude of a transfer function for assumed frequency should be equal to zero, the argument at the point is not determinated and eqns. 7 and 8 take the form

$$\sum_{k=0}^{N-1} h_k \cos[k\omega T]=0 \; ; \; \sum_{k=0}^{N-1} h_k \sin[k\omega T]=0 \tag{9; 10}$$

Now we can obtain equations representing the conditions for derivatives by differentiation of the eqn. 6 with respect to the frequency to the frequency ω.
In this way we can determine the number of equations we need in order to take into account the assumed constraints. The filter order is equal to the number of equations. As solution of the set of equations we obtain the coefficients h_k ($k=1,0,\ldots,N-1$) of a nonrecursive filter algorithm.

The set of equations can be described as

$$\sum_{k=0}^{N-1} C_{ik} h_k = f_i \qquad \text{for } i=1,2,\ldots,D \tag{11}$$

where

C_{ik}, f_i - coefficients depending on the constraints and the sampling frequency,
D=N - number of equations equal to the filter order.

3. APPROXIMATION OF FREQUENCY CHARACTERISTICS WITH CONSTRAINTS

In this chapter we will study design methods which can be classified as optimization methods with constraints. When the filter order N can be higher than the number of design equations (11) we can achieve improved frequency characteristic, adjusting it to requirements of the application. The set of equations cannot be solved explicitly to give the filter coefficients h_k. We apply a mathematical optimization procedure that minimizes an error function, subject to the characteristic of an ideal filter and to the constraints. Using the Lagrange multipliers [5] we obtain the objective function.

$$Q_D=Q+\sum_{i=0}^{D} u_i \left[\sum_{k=0}^{N-1} C_{ik} h_k - f_i\right] \tag{12}$$

where

u_i - Lagrange multipliers

The optimization procedure requires that the partial derivatives of the function Q_D with respect to coefficients h_k and Lagrange multipliers u_i be zero

$$\frac{\partial Q_D}{\partial u_i} = 0 \qquad \text{and} \qquad \frac{\partial Q_D}{\partial h_k} = 0 \tag{13}$$

Applying the coefficients of a truncated Fourier series $\bar{g}_k$ (as in section 2) describing the approximated characteristic of an assumed filter, we obtain the set of linear equations

$$\sum_{k=0}^{N-1} C_{ik} h = f_i \; ; \; h_k + \sum_{i=0}^{D} C_{ik} u_i = \bar{g}_k \tag{14; 15}$$

which can be written in a compact matrix notation

$$\mathbb{C}\,\mathbb{H} = \mathbb{F} \; ; \; \mathbb{H} + \mathbb{C}^T \mathbb{U} = \mathbb{G} \tag{16; 17}$$

where

$\mathbb{F}=(f_1,f_2,\ldots,f_D)^T$ - coefficients depending on the constraints and the sampling frequency,

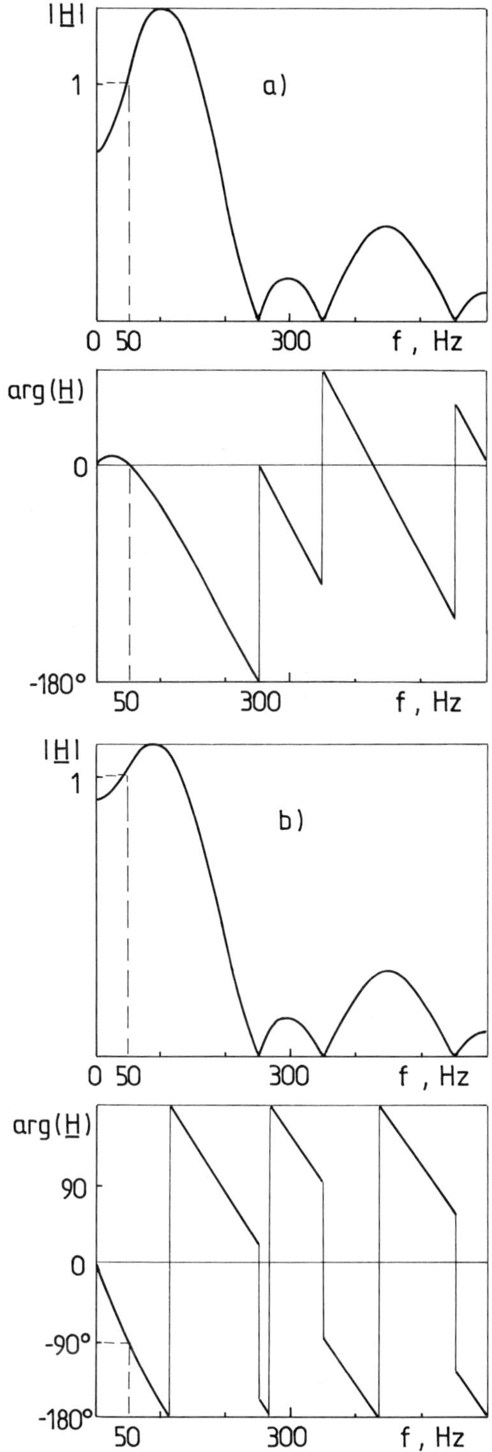

Fig. 1. Magnitude and phase spectrum of transfer function of algorithm for a) real part (basic sinewave), b) imaginary part (orthogonal function) of a phasor. The zeros of transfer function at the 5th, 7th and 11th harmonics, f_s=1200 Hz, N=8.

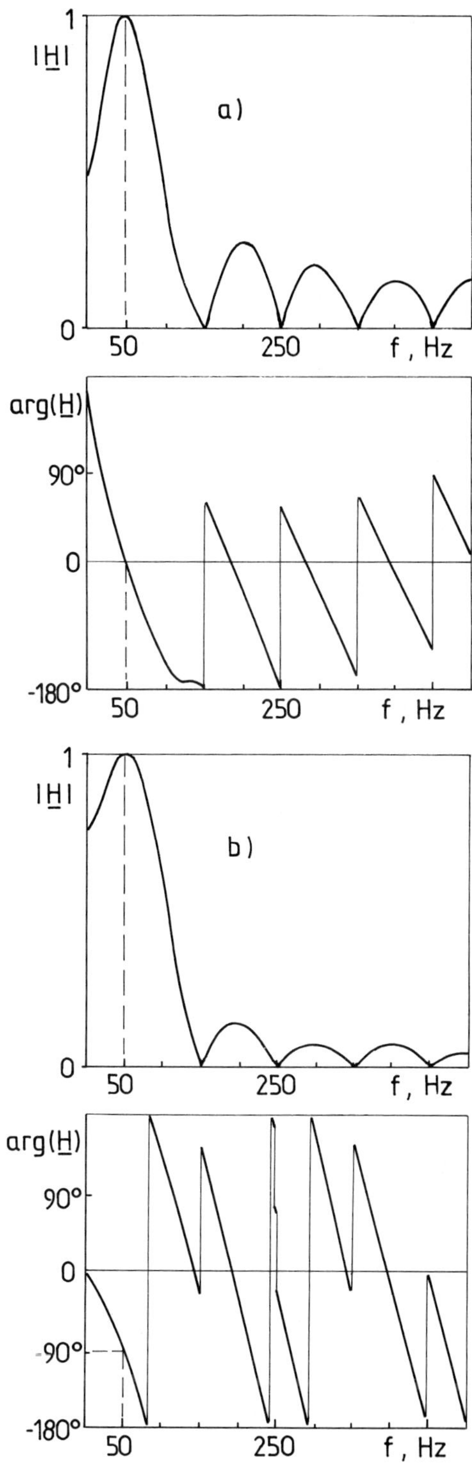

Fig. 2. Magnitude and phase spectrum of transfer function of algorithm for a) real part (basic sinewave), b) imaginary part (orthogonal function) of a phasor. Maximum of transfer function at 50Hz (basic waveform) Zeros at all odd harmonics, f_s=1000Hz, N=15.

$\mathbb{G}=(\bar{g}_0,\bar{g}_1,\ldots,\bar{g}_{N-1})^T$ - coefficients of a truncated Fourier series,

$\mathbb{H}=(h_0,h_1,\ldots,h_{N-1})^T$ - the desired coefficients of a filter algorithm,

$\mathbb{U}=(u_1,u_2,\ldots,u_D)^T$ - constraints of a filter characteristic,

$\mathbb{C}=[C_{ik}]_{(D\times N)}$ - coefficients depending on the constraints and the sampling frequency.

Hence it follows the solution of the problem

$$\mathbb{H} = \mathbb{G} + \mathbb{C}^T\left[\mathbb{C}\,\mathbb{C}^T\right]^{-1}\left[\mathbb{F} - \mathbb{C}\,\mathbb{G}\right] \quad (18)$$

The special cases are

1. $\mathbb{H} = \mathbb{G}$ - truncated Fourier series (D=0)
2. $\mathbb{H} = \mathbb{C}^{-1}\mathbb{F}$ - where D=N

4. FILTERS DESIGNING

A computer program for designing of the filter algorithms has been elaborated and frequency characteristics of developed algorithms have been investigated.

Fig.1. shows frequency magnitude and phase response of algorithms for estimating the phasor of the fundamental sinewave of currents and voltages in three-phase circuits with thyristor converters. As constraints the magnitude of the frequency response of the algorithms is zero at the 5th, 7th and 11th harmonics. The phase response of the algorithm for estimation of the real part of the phasor is, at the fundamental frequency (50 Hz), equal to zero, and for the imaginary part equal to $-\pi/2$. Substituting into eqns.7 and 8 $\omega=2\pi 50, \varphi(\omega_n)=0$ and $H(\omega)=1$ we obtain two equations, and substituting into eq.9 and 10 successive $2\pi 250, 2\pi 350$ and $2\pi 550$ we obtain the next six equations.

The set of 8 equations can be solved for h_k (N=8), which are coefficients of the algorithm for determination of the real part of the phasor. To calculate the coefficients of the algorithm for the imaginary part of the phasor, we substitute into eqns.7 and 8 $\varphi(\omega_n)=-\pi/2$. The other six equations remain the same as for algorithm for the real part of the phasor.

By the sampling frequency of f_s=1200 Hz the filter order is equal to N=8. In this way we have minimized the sampling window and the filter order. The transient response time is equal to one third period of the main sinewave. When estimating the phasor of the fundamental sinewave, it is important that the values of the magnitude of transfer function do not change considerably by varying the network frequency in a small range. This can be achieved when the maximum of the frequency response is exactly at the fundamental frequency. Thus, the first derivative of the magnitude spectrum at the frequency should be equal to zero. The widely used Fourier algorithms [3] do not have a maximum at the fundamental frequency if the sampling window is not exactly equal to one period. As an example Fig.2 shows the frequency characteristics of algorithms developed according to optimization method with constraints.

In this case we have assumed zeros of the transfer function at all odd harmonics and the first derivative of the function at the fundamental harmonic equal to zero. The optimization procedure minimizes the error function, subject to the characteristic of an ideal lowpass filter with a cutoff frequency of f_c = 100 Hz, taking into account the constraints.

5. CONCLUSIONS

The widely used algorithms for online estimation of the fundamental sinewave of signals in electrical power systems are based on the Fourier technique or on the Kalman filter theory. They optimize the estimation in the time domain approximating the distorted signals by sinusoidal signals. The proposed method enables us to develop filter algorithms for estimation the phasor of the main component, which frequency responses have specific properties. It is possible to prescribe the values of the transfer function and its derivatives at any frequency. The optimization procedure makes it possible to minimize an error function subject to the characteristic of an ideal filter, taking into account the additive constraints. Using this method we can optimize the selection of algorithms for concrete application.

REFERENCES

[1] Thorp, J.S., Phadke, A.G. and Karimi, K.J., IEEE Trans., PAS-104, (1985) pp.3098-3106.

[2] Łobos, T. and Schneider, E., New digital algorithms for online determination of symmetrical components in power systems, in:Signal Processing IV; Theory and Applications, Lacoume, J.L., Chehikian, A., Martin, N. and Malbos, J. (eds.), Elsevier S.P.B.V., EURASIP, (1988) pp.1321-1324.

[3] Łobos, T., IEE Proc. Pt.C vol. 136, (1989) pp.347-351.

[4] Eichhorn, K.F. and Łobos, T.,IEE Proc. Pt.C vol. 138, (1991) pp.469-470.

[5] Steffen, P.,AEÜ, Electronics and Communication, Bd 38, (1984) pp.363-367.

SIGNAL PROCESSING VI: Theories and Applications
J. Vandewalle, R. Boite, M. Moonen, A. Oosterlinck (eds.)

Generalized Wigner-Ville Distibutions: Examples and Properties

Boualem Boashash and Branko Ristic

Centre for Signal Processing Research
Queensland University of Technology
GPO Box 2434, Brisbane, Q 4001, Australia

This paper presents the concept of reduced time-varying higher order spectra which is based on generalized Wigner-Ville distributions. The relationship with the higher-order Wigner-Ville distributions and the general properties in the context of the time-frequency distributions are studied. Two members of the class are addressed and their applications illustrated.

1 Introduction

Time-varying higher order spectra have become a very attractive research area recently, due to their potential applications. The earlier work by Gerr [9], defined the third-order Wigner distribution (i.e. Wigner bispectrum) as the two-dimensional Fourier transform of the third order time-varying moment $m_3(t, \tau_1, \tau_2)$:

$$W_3(t, f_1, f_2) = \underset{\tau_1 \to f_1}{\mathcal{F}} \; \underset{\tau_2 \to f_2}{\mathcal{F}} \{ m_3(t, \tau_1, \tau_2) \} \tag{1}$$

Subsequently, the work by Gerr was extended and generalized in [13], [8] and [10]. Swami, [13], studied the properties of the Wigner bispectrum and proposed its extension to higher-order time-frequency distributions. In [8], Wigner higher-order spectra are defined as a slightly modified extension of eq.(1). Finally, in [10], a cumulant based higher-order Wigner-Ville distribution (WVD) was proposed, again following the method of Gerr [9].

Definition of the higher-order WVD. Consider the k-th order 'moment' of a deterministic signal $x(t)$:

$$m_k(t, \underline{\tau}) = x(t + a_1)\, x(t + a_2) \ldots x(t + a_k) \tag{2}$$

where $\underline{\tau} = [\tau_1 \ldots \tau_{k-1}]$ and the lags $a_1, a_2, \ldots, a_k$ are required to satisfy:

$$a_1 + a_2 + \ldots + a_k = 0 \quad \text{(lag centering)} \tag{3}$$

$$a_2 - a_1 = \tau_1; a_3 - a_1 = \tau_2; \ldots; a_k - a_1 = \tau_{k-1} \tag{4}$$

The k-th order WVD is defined as:

$$W_k(t, \underline{f}) = \underbrace{\int_{-\infty}^{\infty} \cdots \int_{-\infty}^{\infty}}_{(k-1)\,\text{times}} m_k(t, \underline{\tau}) e^{-j2\pi \underline{f}\, \underline{\tau}^T} \, d\tau_1 \ldots d\tau_{k-1} \tag{5}$$

where $\underline{f} = [f_1 \ldots f_{k-1}]$ and T denotes a transpose.

If $x(t)$ is a random signal, then $m_k(t, \underline{\tau})$ is defined with the expectation value operator $\boldsymbol{E}$. In [8], eq.(2) is slightly modified, having $x^*(t + a_1)$ instead of $x(t + a_1)$.

The constraints (3) and (4) are common to [13], [8] and [10]. They ensure the k-th order WVD of a stationary zero mean random signal to be equal to the k-th order (time-invariant) moment spectra of the signal. The entire concept of the higher-order WVDs can be expressed in terms of cumulants rather than moments [10].

In parallel with the work on the higher-order WVDs, a generalized WVD (GWVD) was introduced and studied in [5], [4], [6] and [3].

Definition of the generalized WVD. For a deterministic signal $x(t)$, the generalized WVD is defined as a Fourier transform of a k-th order time-varying moment reduced to one lag variable, and expressed in a symmetric form as:

$$m_k^g(t, \tau) = \prod_{i=1}^{q} x^{b_i}(t + c_i \tau)[x^*(t + c_{-i}\tau)]^{b_{-i}} \tag{6}$$

that is:

$$W_k^g(t, f) = \int_{-\infty}^{\infty} m_k^g(t, \tau) e^{-j2\pi f \tau} d\tau \tag{7}$$

where $k = \sum_{i=1}^{q}(|b_i| + |b_{-i}|)$. For zero-mean random signals, the expectation value is used in definition (6).

The choice of the coefficients q, b_i and c_i depends on the particular problem/application, as it will be illustrated in section 2. Comparing the two different concepts, the higher-order WVDs and the generalized WVDs, we observe the following:

1. $W_k^g(t,f)$ would be a two-dimensional slice of the k-dimensional $W_k(t,\underline{f})$, if the appropriate complex-conjugate operations were used in (2).

2. In the generalized WVD, lag constraints are imposed so that the GWVD of a complex sinusoid $x(t) = e^{j2\pi f_o t}$ is:

$$W^g_{k\{x(t)\}}(t,f) = \delta(f - f_0) \qquad (\forall k \in N) \qquad (8)$$

Satisfying the above condition, the GWVD automatically fulfills constraints (3) and (4), that is a k-th order GWVD of a stationary signal equals the one-dimensional slice of its k-th order moment spectra.

3. The physical interpretation of the generalized WVD for deterministic or random signals is streithforward for two reasons: (i) it has a single frequency axis; (ii) the energy of a sinusoid appears as a spike at the frequency f_0, eq.(8). The interpretation of the higher-order WVDs is very difficult.

2 Two members of the class of the GWVD

In this section we illustrate two members of the class of the generalized WVDs. Each of them is dedicated to a specific problem (application) that can not be optimally solved using conventional time-frequency distributions. Notice that the WVD is a member of the class of GWVDs with parameters: $q = 1$, $b_1 = b_{-1} = 1$, $(k = 2)$, and $c_1 = -c_{-1} = 0.5$.

2.1 A tool for analysis of polynomial phase signals

Consider a polynomial phase signal in noise:

$$x(t) = a(t)\, e^{j\phi(t)} + n(t) \qquad (9)$$

where

$$\phi(t) = 2\pi \sum_{m=0}^{p} b_m t^m \qquad (10)$$

If $p \leq 2$ (linear FM) and $a(t)$ non-random, the conventional WVD would produce the optimal time-frequency representation of the signal $x(t)$. Moreover, the estimate of the instantaneous frequency from the peak of the WVD, if noise $n(t)$ is white and Gaussian, would be unbiased and its variance would meet the CR bound [2]. For case $p > 2$, we could either use the dechirping approach proposed in [11], or the generalized WVD in order to obtain an optimal time-frequency signal representation. For example, if $p = 3$ or $p = 4$, the 6-th order reduced time-varying moment has to be used [4], [6], with parameters:

$$\begin{array}{lll} q = 2 & & \\ b_1 = b_{-1} = 2 & b_2 = b_{-2} = 1 & (k = 6) \\ c_1 = -c_{-1} = 0.675 & c_2 = -c_{-2} = -0.85 & \end{array} \qquad (11)$$

Figures 1(a) and 1(b) illustrate the conventional WVD and the generalized WVD with parameters (11), of the same quadratic FM ($p = 3$) signal. The superior behaviour of the later is indicated by the sharpness of the peaks in the Fig.1(b). From the peaks of the GWVD, the quadratic law can be recovered easily as opposed to the smeared WVD showing oscillations without any physical meaning.

2.2 A tool for analysis of FM signals in multiplicative noise

Consider again the signal given by (9) and (10) and assume that $a(t)$ is a real white noise. By extending the work by Dwyer [7] and Abeysekera [1], it can be shown that a 4-th order reduced spectra is a very effective tool for analysis of such signals. This member of the class of the generalized WVD is studied in detail in [12] and is referred to as a reduced Wigner trispectrum (RWT). Its parameters are:

$$\begin{array}{ll} q = 1 & \\ b_1 = b_{-1} = 2 & (k = 4) \\ c_1 = -c_{-1} = 0.25 & \end{array}$$

Figure 2(a) and 2(b) show the WVD and the RWT of the same linear FM signal. Notice that the WVD is completely blind in this example, unable to detect the presence of a deterministic component in the noisy signal. The RWT clearly shows the existence of a linear FM.

3 Properties of the generalized WVDs

The generalized WVD has preserved and generalized a great deal of the properties that characterize the WVD. In derivation of the following properties we assumed: $b_i = b_{-i}$ and $c_i = -c_{-i}$. The properties are valid for $\forall k \in N$.

P-1. The GWVD is real for any signal $x(t)$:

$$W^g_{k\{x(t)\}}(t,f) = \left[W^g_{k\{x(t)\}}(t,f)\right]^* \qquad (12)$$

P-2. The GWVD is an even function of frequency if $x(t)$ is real:

$$W^g_{k\{x^*(t)\}}(t,-f) = W^g_{k\{x(t)\}}(t,f) \qquad (13)$$

P-3. A shift in time by T and in frequency by F (i.e. modulation by $e^{j2\pi Ft}$) of signal $x(t)$ results in the same shift in time and frequency of the GWVD:

$$W^g_{k\{x(t-T)e^{j2\pi Ft}\}}(t,f) = W^g_{k\{x(t)\}}(t-T, f-F) \qquad (14)$$

$(\forall T, F \in R)$

P-4. The GWVD preserves the convolution and modulation. If $x(t) = w(t) * z(t)$ and $y(t) = w(t)z(t)$ then:

$$W^g_{k\{x(t)\}}(t,f) = W^g_{k\{w(t)\}}(t,f) *_t W^g_{k\{z(t)\}}(t,f) \qquad (15)$$

$$W^g_{k\{x(t)\}}(t,f) = W^g_{k\{w(t)\}}(t,f) *_f W^g_{k\{z(t)\}}(t,f) \qquad (16)$$

P-5. Projection of the $W^g_{k\{x(t)\}}(t,f)$ to the time axis:

$$\int_{-\infty}^{\infty} W^g_{k\{x(t)\}}(t,f)df = |x(t)|^k \qquad (17)$$

P-6. The local moment of the GWVD in the frequency gives the instantaneous frequency of the signal $x(t)$:

$$\frac{\int_{-\infty}^{\infty} f W^g_{k\{x(t)\}}(t,f)df}{\int_{-\infty}^{\infty} W^g_{k\{x(t)\}}(t,f)df} = \frac{1}{2\pi}\phi'(t) \qquad (18)$$

P-7. Moyal's formula:

$$\int_{-\infty}^{\infty}\int_{-\infty}^{\infty} \left[W^g_{k\{x(t)\}}(t,f)\right]^2 dtdf = \left|\int_{-\infty}^{\infty} x^k(t)dt\right|^2 \qquad (19)$$

P-8. Projection of $W^g_{k\{x(t)\}}(t,f)$ to the frequency axis:

$$\int_{-\infty}^{\infty} W^g_{k\{x(t)\}}(t,f)dt =$$

$$|\underbrace{X(\frac{f}{2c_1}) *_f \ldots *_f X(\frac{f}{2c_1})}_{b_1 times} *_f \underbrace{X(\frac{f}{2c_2}) \ldots *_f}_{b_2 times} \ldots|^2 \qquad (20)$$

P-9. Total volume under the generalized WVD:

$$\int_{-\infty}^{\infty}\int_{-\infty}^{\infty} W^g_{k\{x(t)\}}(t,f)dfdt = \int_{-\infty}^{\infty} |x(t)|^k dt \qquad (21)$$

P-10. Suppose $x(t) = e^{j2\pi f_1 t} + e^{j2\pi f_2 t}$. From P-8. it follows that RWT is:

$$W^g_{4\{x(t)\}}(t,f)dt = \delta(f-f_1) + \delta(f-f_2) + 4\,\delta(f - \frac{f_1+f_2}{2}) \qquad (22)$$

The last term in (22) is a cross-term which is indistinguishable from the two auto-terms. This property of the GWVDs (as well as the higher order spectra in general) makes the analysis of multicomponent signals very difficult.

4 Conclusion

This paper compared two different approaches to time-varying higher-order spectra, the higher-order WVDs and the generalized WVDs. Two members of the class of the GWVDs are shown to have practical meaning since they can optimally solve problems that conventional signal processing tools can not. Finally, the properties of such GWVDs are presented. These properties are either identical or generalized versions of the properties of the WVD.

References

[1] R. M. S. S. Abeysekera and B. Boashash. DSTO progress report: Detection of radar signals in the presence of non-Gaussian clutter. Technical Report 1, CSPR, QUT, Dec, 1991.

[2] B. Boashash. Interpreting and estimating the instantaneous frequency of a signal - Part I: Fundamentals, Part II: Algorithms. *Proceedings of the IEEE*, 1992. to appear.

[3] B. Boashash and G. Frazer. Time varying higher order spectra, generalized Wigner-ville distributions and the analysis of underwater acoustic data. In *Proceedings of the Intern. Conf. Acoustic, Speech and Signal Proc. (ICASSP)*, San Francisco, March, 1992.

[4] B. Boashash and P. O'Shea. Time-varying higher order spectra. In Franklin T. Luk, editor, *Advanced Signal Processing Algorithms, Architectures and Implementations*, San Diego, USA, July 1991. Proceedings of SPIE.

[5] B. Boashash and P. J. O'Shea. Time-varying higher order spectra. In *Proceedings of 9th Kobe International Conference on Electronics and Information Sciences*, Kobe, Japan, June 1991.

[6] B. Boashash and B. Ristic. Time varying higher order spectra. In *Proceedings of 25th Asilomar Conference*, Pacific Grove, California, Nov. 1991.

[7] R. F. Dwyer. Fourth-order spectra of Gaussian amplitude-modulated sinusoids. *J. Acoust. Soc. Am.*, pages 919–926, August 1991.

[8] J. R. Fonollosa and C. L. Nikias. Wigner polyspectra: HOS spectra in TV signal processing. In *Proc. Int. Conf. Acoustic, Speech and Signal Processing (ICASSP)*, pages 3085–8, Toronto, Canada, May 1991.

[9] N. L. Gerr. Introducing a third order Wigner distribution. *Proceedings of the IEEE*, 76:290–292, March 1988.

[10] G. B. Giannakis and A. Dandawate. Polyspectral analysis of non-stationary signals: bases, consistency and HOS-WV. In *Proc. Int. Signal Processing Workshop on Higher-Order Statistics*, pages 167–170, Chamrousse, France, July 1991.

[11] B. Ristic and B. Boashash. Instantaneous frequency estimation using generalized bilinear representations. In *IREECON International 91*, Sydney, Australia, September 1991.

[12] B. Ristic and S. Abysekera nd B. Boashash. Reduced Wigner trispectrum. 1992. (submitted to IEEE Trans. SP).

[13] A. Swami. Third-order Wigner distributions: definitions and properties. In *Proc. Int. Conf Acoustic, Speech and Signal Processing (ICASSP)*, pages 3081–4, Toronto, Canada, May, 1991.

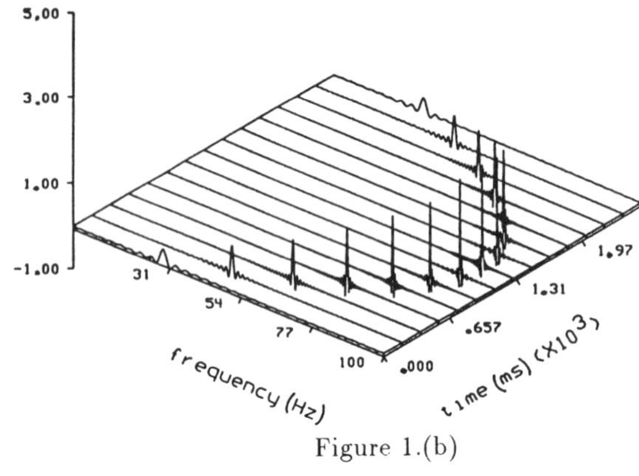

Figure 1.(b)

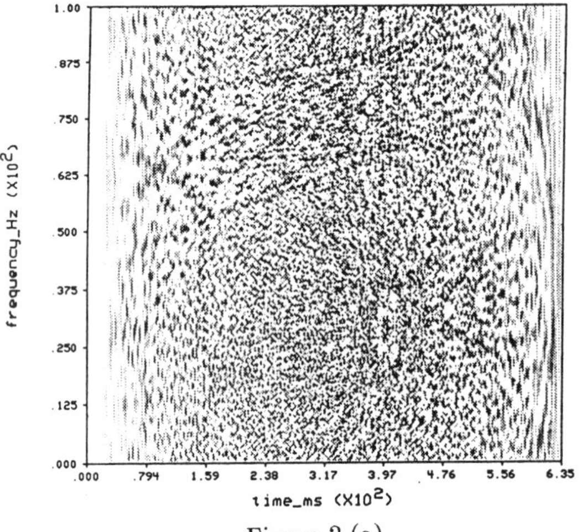

Figure 2.(a)

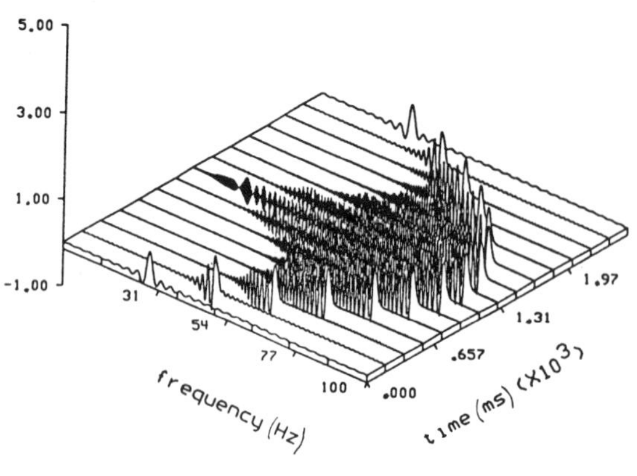

Figure 1. (a)

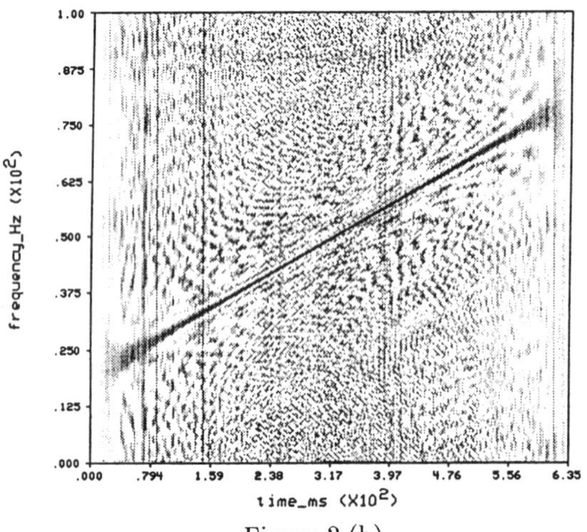

Figure 2.(b)

LINEAR PREDICTION, MAXIMUM FLATNESS, MAXIMUM HIGHER-ORDER ENTROPY ON AR POLYSPECTRAL ESTIMATION

Chong-Yung Chi

Department of Electrical Engineering, National Tsing Hua University, Hsinchu, Taiwan, ROC.

Abstract: It is well-known that linear prediction (LP) spectral estimator is equivalent to maximum entropy spectral estimator [1,2] and they are also equivalent to maximum spectral flatness spectral estimator [3,4] for AR processes of known order. In this paper, we propose a new higher-order statistics (HOS) based linear prediction error (LPE) filter. We also present LP polyspectral estimator, maximum polyspectral flatness polyspectral estimator, maximum higher-order entropy polyspectral estimator and equivalencies on these polyspectral estimators. The results presented in this paper provide some theoretical foundations on the polyspectral estimation and modeling of non-Gaussian AR processes.

1. Introduction

It is well–known on correlation (second–order statistics) based autoregressive (AR) spectral estimation that linear prediction (LP) spectral estimator is equivalent to maximum entropy spectral estimator [1–2] as well as maximum spectral flatness spectral estimator [3,4] as the order of AR processes is known a priori, although they are based on different theoretical foundations. Among the existing cumulant (higher order statistics (HOS)) based AR parameter estimators, many of them ([6] and references therein) are developed based on fitting a set of linear equations by least–squares method without resort to the minimization of a cost function of the prediction error as LP spectral estimator is based on minimum mean–squared prediction error, because, to the author's knowledge, a well–defined linear prediction error filter (LPE) based on HOS is never reported in the open literature.

In this paper, in order to provide some theoretical foundations on the polyspectral estimation and modeling of non–Gaussian AR processes, we begin with a brief review of non–Gaussian AR processes in Section 2, and then present a new HOS based LPE filter, maximum polyspectral flatness measure and maximum higher–order entropy criterion for AR polyspectral estimation, in Sections 3, 4 and 5, respectively. We also present equivalencies on these polyspectral estimators. Finally, we provide a discussion and some conclusions.

2. Non-Gaussian AR Processes

Assume that x(k) is the output of a causal autoregressive (AR) model of order p as follows:

$$x(k) = -\sum_{i=1}^{p} a(i)\, x(k-i) + u(k) \qquad (1)$$

where u(k) is real, zero–mean, independent identically distributed (i.i.d.), non–Gaussian with Mth–order cumulant γ_M. The transfer function 1/A(z) where

$$A(z) = 1 + \sum_{i=1}^{p} a(i)\, z^{-i} \qquad (2)$$

is assumed to be minimum–phase for x(k) to be stationary. Let $C_{M,x}(\underline{k})$ where $\underline{k} = (k_1,k_2, ...,k_{M-1})$ denote the Mth–order cumulant function of x(k). The Mth–order polyspectrum, denoted $S_{M,x}(\underline{f})$ where $\underline{f} = (f_1, f_2, ..., f_{M-1})$, of x(k), which is the (M–1)–dimensional Fourier transform of $C_{M,x}(\underline{k})$, is given by

$$S_{M,x}(\underline{f}) = \gamma_M \frac{1}{A(f_1)} \cdots \frac{1}{A(f_{M-1})} \cdot \frac{1}{A^*(f_1+\cdots+f_{M-1})} \qquad (3)$$

where $A(f)=A(z=e^{j2\pi f})$. Next, let us proceed with the new LPE filter based on the HOS of x(k).

3. Prediction Error Filter Based on HOS

Let $\hat{A}(z)$ be the transfer function of the LPE filter of order p with $\hat{a}(0)=1$. The output, denoted e(k), of the LPE filter with the input x(k) is then

$$e(k) = x(k) + \sum_{i=1}^{p} \hat{a}(i)\, x(k-i) \ . \qquad (4)$$

The new HOS based LPE filter $\hat{A}(z)$ is the one which minimizes the sum of absolute squares of Mth–order cumulant function of e(k) as follows:

$$J(\hat{A}(z)) = \sum_{k_1=-\infty}^{\infty} \cdots \sum_{k_{M-1}=-\infty}^{\infty} \left| C_{M,e}(\underline{k}) \right|^2 . \qquad (5)$$

The associated LP polyspectrum can be computed by (3) with A(f) replaced by the HOS based LPE filter

This work was supported by the National Science Council under Grant NSC81-0404-E-007-001 Telecommunication Laboratories under Grant TL-NSC-81-5201.

$\hat{A}(f)$ associated with minimum $J(\hat{A}(z))$.

It is well–known that the conventional LPE filter based on the minimization of $E[e^2(k)]$ is minimum–phase [4]. We show, in Appendix A, the following result.

(R1) The HOS based LPE filter $\hat{A}(z)$ is minimum–phase.

Remark that for M=2, $J(\hat{A}(z))$ reduces to

$$J(\hat{A}(z)) = \sum_{k=-\infty}^{\infty} \left| r_{ee}(k) \right|^2 \qquad (6)$$

where $r_{ee}(k)$ is the autocorrelation function of e(k). It is sufficient that $J(\hat{A}(z))$ given by (6) is minimum if $|r_{ee}(k)|$ is minimum for all k. A set of sufficient conditions for this to occur is that $r_{ee}(0) = E[e^2(k)] \geq 0$ is minimum and $r_{ee}(k) = 0$ for $k \neq 0$ in the meantime. In other words, minimizing $J(\hat{A}(z))$ given by (6) is equivalent to minimizing $E[e^2(k)]$ because the latter leads to the well–known fact that e(k) is white when x(k) is an AR process of order p. Therefore, we have shown the following result.

(R2) The HOS based LPE filter for M=2 is identical with the conventional LPE filter.

4. Maximum Flatness of Polyspectra

Let ξ_e denote the flatness of Mth–order polyspectrum of e(k) defined as

$$\xi_e = \frac{\exp\{ \int_0^1 \cdots \int_0^1 \ln [\, |S_{M,e}(\underline{f})|^L] \, d\underline{f} \, \}}{\int_0^1 \cdots \int_0^1 |S_{M,e}(\underline{f})|^L \, d\underline{f}} \qquad (7)$$

where L is a positive integer and $d\underline{f}$ denotes $df_1 \cdot df_2 \cdots df_{M-1}$. The maximum polyspectral flatness (MPF) polyspectrum is obtained by (3) with A(z) replaced by the optimum $\hat{A}(z)$ associated with maximum ξ_e. Note that ξ_e is the geometric mean of $|S_{M,e}(\underline{f})|^L$ divided by its arithmetic mean and therefore can be shown to satisfy $0 \leq \xi_e \leq 1$. Note that $\xi_e = 1$ if $|S_{M,e}(\underline{f})|$ equals a constant for all $\underline{f}$. On the other hand, $\xi_e \approx 0$ if $|S_{M,e}(\underline{f})|$ is very peaky. Remark that for L=1 and M=2, ξ_e is exactly the same as the well–known Gray and Markel's spectral flatness measure [3,4].

The following results associated with the MPF polyspectrum can be proven.

(R3) The optimum $\hat{A}(z)$ associated with maximum ξ_e and the HOS based LPE filter are identical as long as x(k) is a stationary non–Gaussian linear process.

(R4) The optimum $\hat{A}(z)$ associated with maximum ξ_e is equal to A(z). Hence, the LP polyspectral estimator is equivalent to the MPF polyspectral estimator, and both of them offer the same polyspectrum as $S_{M,x}(\underline{f})$.

The proof of (R3) and the proof of (R4) are given in Appendices B and C, respectively. By (R3) and (R4), we conclude that the HOS based LPE filter is an "Mth–order whitening" filter which suggests that γ_M can be estimated as the sample cumulant $\hat{C}_{M,e}(\underline{k}=\underline{0})$ when data are finite.

5. Maximum Higher-Order Entropy

We define the Mth–order entropy, denoted $\Gamma(\hat{S}_{M,x}(\underline{f}))$, of the polyspectrum $\hat{S}_{M,x}(\underline{f})$ of non–Gaussian linear process x(k), as

$$\Gamma(\hat{S}_{M,x}(\underline{f})) = \frac{1}{2} \int_0^1 \cdots \int_0^1 \ln \left| \hat{S}_{M,x}(\underline{f}) \right|^2 d\underline{f}. \qquad (8)$$

The maximum Mth–order entropy (MME) polyspectrum of x(k) is the one which maximizes $\Gamma(\hat{S}_{M,x}(\underline{f}))$ subject to the constraint

$$\hat{C}_{M,x}(\underline{k}) = C_{M,x}(\underline{k}) \text{ for all } \underline{k} \in R(M,p) \qquad (9)$$

where R(M,p) denotes the finite domain of support associated with the Mth–order cumulant function of any non–Gaussian moving average (MA) process of order p, and $C_{M,x}(\underline{k})$ for $\underline{k} \in R(M,p)$ are the known cumulant samples associated with $\hat{S}_{M,x}(\underline{f})$. In other words, the MME polyspectral estimator attempts to extrapolate the Mth–order cumulant function for all $\underline{k} \notin R(M,p)$ from the known Mth–order cumulants $C_{M,x}(\underline{k})$ for all $\underline{k} \in R(M,p)$ such that $\Gamma(\hat{S}_{M,x}(\underline{f}))$ is maximum. Remark that for M=2, $\Gamma(\hat{S}_{M,x}(\underline{f}))$ reduces to

$$\Gamma(\hat{S}_{xx}(f)) = \int_0^1 \ln \hat{S}_{xx}(f) \, df \qquad (10)$$

which is proportional to the entropy rate associated with the well–known Burg's maximum entropy spectral estimator, where $\hat{S}_{xx}(f)$ denotes power spectrum estimate of x(k).

We show the following two results associated with the MME polyspectrum in Appendix D.

(R5) The MME polyspectrum is equivalent to the Mth–order polyspectrum of a stationary non–Guassian AR process of order p.

(R6) $S_{M,x}(\underline{f})$ given by (3) is an MME polyspectrum.

Furthermore, we have the following result by (R4) and (R6).

(R7) Both the LP polyspectrum and the MPF polyspectrum are an MME polyspectrum.

Remark that if the optimum $\hat{A}(z)$ (minimum–phase) associated with $\Gamma(\hat{S}_{M,x}(\underline{f}))$ described in (R5) is unique, then $\hat{A}(z)=A(z)$ and the MME polyspectrum is the same as $S_{M,x}(\underline{f})$ due to (R6) and (R7). Unfortunately, it is still unknown whether the solution $\hat{A}(z)=A(z)$ under the constraint given by (9) is unique or not although this is true for the case of M=2.

6. Discussion and Conclusions

The previous results described in (R1) through (R7) are based on the assumption that x(k) is a non–Gaussian AR process of known order p. However, when the order p of the AR process of interest is not known in advance, the proposed HOS based LPE filter $\hat{A}(z)$ is then no longer identical with the true A(z) when the order, $\hat{p}$, of $\hat{A}(z)$ is less than p. On the other hand, if $\hat{p} \geq p$, $\hat{a}(i) = a(i)$ for $1 \leq i \leq p$ and $\hat{a}(i)=0$ for i>p.

Assume that y(k) is the output of an allpass filter with input x(k). It can be easily shown, by Parseval's theorem, that $J(\hat{A}(z))$ associated with the signal x(k) is the same as that associated with the signal y(k). Due to this fact as well as (R3) we have the following result.

(R8) The LP polyspectral estimator is allpass factor blind, so is the MPF polyspectral estimator.

Corresponding to (R8) for M=2 is the well–known fact that all power spectral estimators are allpass factor blind.

The results we have presented contribute to theoretical foundations on the polyspectral estimation and modeling of non–Gaussian AR processes.

The performance of the proposed HOS based PEF filter with finite data will be reported in the future.

Appendix A. Proof of (R1)

Assume that $\hat{A}(z)$ is not minimum–phase with a zero z_i outside the unit circle (i.e. $|z_i|>1$). We can then express $\hat{A}(z)$ as

$$\hat{A}(z) = (1 - z_i z^{-1}) A'(z) \qquad (A1)$$

where A'(z) is a (p–1)th–order polynomial of z^{-1}. We need the following inequality in the proof.

$$|1-z_i \exp\{-j2\pi f\}| = |z_i| \cdot \left| 1 - (1/z_i^*) \exp\{-j2\pi f\} \right| > \left| 1 - (1/z_i^*) \exp\{-j2\pi f\} \right|. \qquad (A2)$$

Let

$$\hat{B}(z) = [1 - (1/z_i^*) z^{-1}] A'(z). \qquad (A3)$$

One can easily see, from (A1), (A2) and (A3), that

$$|\hat{A}(f)| > |\hat{B}(f)|. \qquad (A4)$$

Next, we infer, from (5) and (A4), that

$$J(\hat{A}(z)) = \int_0^1 \cdots \int_0^1 \left|S_{M,e}(\underline{f})\right|^2 d\underline{f}$$

$$= \int_0^1 \cdots \int_0^1 \left|S_{M,x}(\underline{f})\right|^2 \cdot \left|\hat{A}(f_1)\cdot\hat{A}(f_2) \cdots \cdot\hat{A}(f_{M-1})\cdot\hat{A}(f_1+\ldots+f_{M-1})\right|^2 d\underline{f}$$

$$> \int_0^1 \cdots \int_0^1 \left|S_{M,x}(\underline{f})\right|^2 \cdot \left|\hat{B}(f_1)\, \hat{B}(f_2) \cdots \cdot\hat{B}(f_{M-1}) \cdot\hat{B}(f_1+\ldots+f_{M-1})\right|^2 d\underline{f} = J(\hat{B}(z)) \quad \text{(by (A4))} \qquad (A5)$$

which implies that $J(\hat{A}(z))$ can never be minimum unless $\hat{A}(z)$ is minimum–phase.

Appendix B. Proof of (R3)

If $\hat{A}(z)$ is minimum–phase [5] then

$$\int_0^1 \ln \left|\hat{A}(f)\right|^2 df = 0. \qquad (B1)$$

Then we have

$$\int_0^1 \cdots \int_0^1 \ln\left|S_{M,e}(\underline{f})\right|^L d\underline{f} = \frac{L}{2}\int_0^1 \cdots \int_0^1 \ln\left|S_{M,e}(\underline{f})\right|^2 d\underline{f}$$

$$= \frac{L}{2}\int_0^1 \cdots \int_0^1 \ln\left|S_{M,x}(\underline{f})\right|^2 d\underline{f}$$

$$+ \frac{L}{2}\int_0^1 \cdots \int_0^1 \left\{ \ln\left|\hat{A}(f_1)\right|^2 + \ldots + \ln\left|\hat{A}(f_{M-1})\right|^2 + \ln\left|\hat{A}(f_1+\ldots+f_{M-1})\right|^2 \right\} d\underline{f}$$

$$= \int_0^1 \cdots \int_0^1 \ln\left|S_{M,x}(\underline{f})\right|^L d\underline{f} \quad \text{(by (B1))} \qquad (B2)$$

which is not a function of $\hat{A}(z)$. Substituting (B2) into (7) provides

$$\xi_e = \frac{\exp\{ \int_0^1 \cdots \int_0^1 \ln[|S_{M,x}(\underline{f})|^L]\, d\underline{f} \}}{J'(\hat{A}(z))} \qquad (B3)$$

where

$$J'(\hat{A}(z)) = \int_0^1 \cdots \int_0^1 \left|S_{M,e}(\underline{f})\right|^L d\underline{f}. \qquad (B4)$$

Thus, maximizing ξ_e is equivalent to minimizing $J'(\hat{A}(z))$. Letting L=2 in $J'(\hat{A}(z))$ yields

$$J'(\hat{A}(z)) = \int_0^1 \cdots \int_0^1 \left|S_{M,e}(\underline{f})\right|^2 d\underline{f}$$

$$= \sum_{k_1=-\infty}^{\infty} \cdots \sum_{k_{M-1}=-\infty}^{\infty} \left|c_{M,e}(\underline{k})\right|^2 = J(\hat{A}(z)). \qquad (B5)$$

Hence, minimizing ξ_e is equivalent to minimizing $J(\hat{A}(z))$ which implies R(3). The assumption of a minimum–phase $\hat{A}(z)$ is not restrictive due to (R1) [4].

Appendix C. Proof of (R4)

Let us solve for the optimum $\hat{A}(z)$ by letting $\xi_e=1$ (maximum value of ξ_e) which requires

$$\left|S_{M,e}(\underline{f})\right| = \left|S_{M,x}(\underline{f})\right| \cdot \left|\hat{A}(f_1)\right| \cdot \left|\hat{A}(f_2)\right| \cdots \cdot \left|\hat{A}(f_{M-1})\right| \cdot \left|\hat{A}(f_1+f_2+\ldots+f_{M-1})\right|$$

$$= |\gamma_M| \cdot \left|\frac{\hat{A}(f_1)}{A(f_1)}\right| \cdot \left|\frac{\hat{A}(f_2)}{A(f_2)}\right| \cdots \left|\frac{\hat{A}(f_{M-1})}{A(f_{M-1})}\right| \cdot$$

$$\left|\frac{\hat{A}(f_1+f_2+\ldots+f_{M-1})}{A(f_1+f_2+\ldots+f_{M-1})}\right| = K, \text{ for all } \underline{f} \qquad (C1)$$

where K is a positive constant. Note that A(z) is minimum–phase by assumption and that $\hat{A}(z)$ is also minimum–phase due to (R1) and (R3). Therefore, the pth–order rational function $\hat{A}(z)/A(z)$, which is also minimum–phase, never forms a pth–order allpass filter which is maximum–phase. Thus, we conclude, from (C1), that

$$\hat{A}(z)/A(z) = K' \qquad (C2)$$

where K' is also a constant. Since $\hat{a}(0)=a(0)=1$, the constant K' can only take the value of unity. In other words, $\hat{A}(z) = A(z)$.

Appendix D. Proof of (R5) and (R6)

Proof: Let $\mathfrak{L}$ be

$$\mathfrak{L} = \Gamma(\hat{S}_{M,x}(\underline{f})) - \sum_{\underline{k}\in R(M,p)} \lambda(\underline{k}) \left\{ \hat{C}_{M,x}(\underline{k}) - C_{M,x}(\underline{k}) \right\} \qquad (D1)$$

where $\lambda(\underline{k})$ is the Lagrange multiplier. Note that $\lambda(\underline{k})$ is real since $\hat{C}_{M,x}(\underline{k})$ is real. Taking partial derivative of $\mathfrak{L}$ with respect to $\hat{C}_{M,x}(\underline{k})$ and setting the result to zero yield

$$\frac{1}{2}\int_0^1 \cdots \int_0^1 \left\{ \frac{e^{-j2\pi(\underline{f}'\underline{k})}}{\hat{S}_{M,x}(\underline{f})} + \frac{e^{+j2\pi(\underline{f}'\underline{k})}}{\hat{S}^*_{M,x}(\underline{f})} \right\} d\underline{f} = \begin{cases} \lambda(\underline{k}), & \underline{k}\in R(M,p) \\ 0, & \text{otherwise.} \end{cases} \qquad (D2)$$

Let us define $\Lambda(\underline{k})$ to be

$$\Lambda(\underline{k}) = \begin{cases} \lambda(-\underline{k}), & \underline{k}\in R(M,p) \\ 0, & \text{otherwise.} \end{cases} \qquad (D3)$$

It can be easily shown, from (D2) and (D3), that $1/\hat{S}_{M,x}(\underline{f})$ and $\Lambda(\underline{k})$ form an (M–1)–dimensional Fourier transform pair, i.e.,

$$\frac{1}{\hat{S}_{M,x}(\underline{f})} = \sum_{\underline{k}\in R(M,p)} \Lambda(\underline{k}) \exp\{-j2\pi\underline{f}'\underline{k}\} \qquad (D4)$$

where $\Lambda(\underline{k})$ also satisfies all the inherent symmetry properties of the Mth–order cumulant function since $\hat{S}_{M,x}(\underline{f})$ is a polyspectrum of a non–Gaussian linear process by assumption. Moreover, one can see, from (D4), that $1/\hat{S}_{M,x}(\underline{f})$ can be viewed as the Mth–order polyspectrum of a non–Gaussian MA process of order p and therefore can be factored as the product

$$\frac{1}{\hat{S}_{M,x}(\underline{f})} = \frac{1}{\hat{\gamma}_M} \hat{A}(f_1) \cdots \hat{A}(f_{M-1}) \cdot \hat{A}^*(f_1+\ldots+f_{M-1}) \qquad (D5)$$

where $\hat{A}(f)=\hat{A}(z=\exp\{j2\pi f\})$ and $\hat{A}(z)$ is a pth–order polynomial of z^{-1}, or

$$\hat{S}_{M,x}(\underline{f}) = \hat{\gamma}_M \frac{1}{\hat{A}(f_1)} \cdots \frac{1}{\hat{A}(f_{M-1})} \cdot \frac{1}{\hat{A}^*(f_1+\ldots+f_{M-1})}. \qquad (D6)$$

Substituting (D6) back into (8) gives rise to

$$\Gamma(\hat{S}_{M,x}(\underline{f})) = \ln|\hat{\gamma}_M| - \frac{1}{2}\int_0^1 \cdots \int_0^1 \ln\left|\hat{A}(f_1)\hat{A}(f_2)\cdots\hat{A}(f_{M-1})\cdot\hat{A}(f_1+\ldots+f_{M-1})\right|^2 d\underline{f}. \qquad (D7)$$

Next, let us show that $\hat{A}(z)$ associated with the maximum of $\Gamma(\hat{S}_{M,x}(\underline{f}))$ given by (D7) is minimum–phase.

Assume that $\hat{A}(z)$ is expressed as (A1) where $|z_i|>1$. We can easily infer, from (A4) and (D7), that the value of $\Gamma(\hat{S}_{M,x}(\underline{f}))$ is larger if $\hat{A}(z)$ is replaced by $\hat{B}(z)$ given by (A3). Therefore, the optimum $\hat{A}(z)$ has to be minimum–phase, and the MME polyspectrum $\hat{S}_{M,x}(\underline{f})$ given by (D6) can be thought of as the Mth–order polyspectrum of a stationary non–Gaussian AR process of order p. Thus, we have shown (R5).

By (R5), solving for the MME polyspectrum is equivalent to solving for the minimum–phase $\hat{A}(z)$ from $\hat{C}_{M,x}(\underline{k}) = C_{M,x}(\underline{k})$ for all $\underline{k}\in R(M,p)$. Obviously, $\hat{A}(z)=A(z)$ is a solution. Hence, the associated AR polyspectrm $S_{M,x}(\underline{f})$ given by (3) is an MME polyspectrum. Thus we have proven (R6).

References

[1] J. Burg, "Maximum entropy spectral analysis," Ph.D. dissertation, Stanford University, 1975.

[2] S. S. Haykin and S. Kesler, Prediction error filtering and maximum–entropy spectral estimation, Chap. 2 in Nonlinear Methods of Spectral Analysis, 2nd ed., S. Haykin, ed., Springer–Verlag, 1983.

[3] A. H. Gray, Jr. and J. D. Markel, "A spectral flatness measure for studying the autocorrelation method of linear prediction of speech," IEEE Trans. Acoustics, Speech, and Signal Processing, vol. ASSP–22, pp. 207–217, June 1974.

[4] Steven M. Kay, Modern spectral estimation, Prentice–Hall, 1988.

[5] J. D. Markel and A. H. Gray, Jr., Linear prediction of speech, Springer–Verlag, 1976.

[6] J. M. Mendel, "Tutorial on higher–order statistics (spectra) in signal processing and system theory: theoretical results and some applications," Proc. IEEE, vol. 79, no. 3, pp. 278–305, March 1991.

SIGNAL PROCESSING VI: Theories and Applications
J. Vandewalle, R. Boite, M. Moonen, A. Oosterlinck (eds.)

AN ADAPTIVE ARMA LATTICE FILTER FOR SPECTRAL ESTIMATION WITH FREQUENCY WEIGHTING

Miki HASEYAMA, Nobuo NAGAI, and Nobuhiro MIKI

Research Institute for Electronic Science, Hokkaido University, Sapporo 060, JAPAN

This paper proposes a method of spectral estimation with frequency weighting (SEFW) using an adaptive ARMA lattice filter. By this SEFW lattice filter algorithm, we can obtain accurate spectral estimation in the frequency band focused. The SEFW lattice filter algorithm is constructed with two algorithms:

Algorithm(1): Realization algorithm of an ARMA lattice filter with frequency-weighting.

Algorithm(2): An algorithm which computes ARMA parameters from the coefficients of the lattice filter realized by *Algorithm(1)*.

In this paper, we present the above two algorithms.

1. INTRODUCTION

Lattice filters for digital signal processing became popular in the mid-1970's [2] [5].

In this paper, we propose a method of spectral estimation with frequency weighting (SEFW) using an adaptive ARMA lattice filter. It is well-known that spectral estimation is significant for digital signal processing is significant. Using this SEFW lattice filter algorithm, we can obtain accurate spectral estimation in the frequency band focused. In the SEFW, the band in which the spectrum is estimated more accurately is set by 'frequency-weighting function' which can be designed with an ARMA model arbitrarily [6] [4]. Actually, in spectral estimation of speech, since important information is not equally presented throughout all region of the half of the sampling frequency, the spectrum has often been estimated with frequency weighting, such as the Mel spectral weighting method. It has not been derived such an SEFW lattice-filter algorithm by using other ARMA lattice filters.

The SEFW lattice filter algorithm proposed in this paper is constructed with the following two algorithms:

Algorithm(1): Realization algorithm of an ARMA lattice filer with frequency-weighting.

Algorithm(2): Algorithm which computes ARMA parameters from the coefficients of the lattice filter realized by *Algorithm(1)*.

In this paper, in Sec.2, we present structure of an ARMA lattice filter which is used in the SEFW lattice algorithm. And in Sec.3, we propose *Algorithm(1)* and *Algorithm(2)*.

2. STRUCTURE OF AN ARMA LATTICE FILTER

The ARMA lattice filter used in the SEFW lattice algorithm is constructed with the following prediction errors:

$\hat{\epsilon}^f_{s,t}(i|k)$ and $\bar{\epsilon}^f_{s,t}(i|k)$: forward prediction errors for an input signal $x(i)$ with the ARMA order (s,t)

$$\hat{\epsilon}^f_{s,t}(i|k) \triangleq x(i) - \boldsymbol{h}^T_{s+1,t}(i|i-1)\boldsymbol{R}^{-1}_{s+1,t}(k|k-1) \cdot \boldsymbol{H}^T_{s+1,t}(k|k-1)\boldsymbol{\Lambda x}(k) \quad (1)$$

$$\bar{\epsilon}^f_{s,t}(i|k) \triangleq x(i) - \boldsymbol{h}^T_{s,t}(i-1|i-1)\boldsymbol{R}^{-1}_{s,t}(k-1|k-1) \cdot \boldsymbol{H}^T_{s,t}(k-1|k-1)\boldsymbol{\Lambda x}(k) \quad (2)$$

$\bar{\nu}^f_{s,t}(i|k)$ and $\hat{\nu}^f_{s,t}(i|k)$: forward prediction errors for the output signal $y(i)$ with the ARMA order (s,t)

$$\hat{\nu}^f_{s,t}(i|k) \triangleq y(i) - \boldsymbol{h}^T_{s,t+1}(i-1|i)\boldsymbol{R}^{-1}_{s,t+1}(k-1|k) \cdot \boldsymbol{H}^T_{s,t+1}(k-1|k)\boldsymbol{\Lambda x}(k) \quad (3)$$

$$\bar{\nu}^f_{s,t}(i|k) \triangleq y(i) - \boldsymbol{h}^T_{s,t}(i-1|i-1)\boldsymbol{R}^{-1}_{s,t}(k-1|k-1) \cdot \boldsymbol{H}^T_{s,t}(k-1|k-1)\boldsymbol{\Lambda x}(k) \quad (4)$$

$\hat{\epsilon}^b_{s,t}(i|k)$ and $\bar{\epsilon}^b_{s,t}(i|k)$: backward prediction errors for the input signal $x(i-t)$ with the ARMA order (s,t)

$$\hat{\epsilon}^b_{s,t}(i|k) \triangleq x(i-t) - \boldsymbol{h}^T_{s+1,t}(i|i)\boldsymbol{R}^{-1}_{s+1,t}(k|k) \cdot \boldsymbol{H}^T_{s+1,t}(k|k)\boldsymbol{\Lambda x}(k-t) \quad (5)$$

$$\bar{\epsilon}^b_{s,t}(i|k) \triangleq x(i-t) - \boldsymbol{h}^T_{s,t}(i|i)\boldsymbol{R}^{-1}_{s,t}(k|k) \cdot \boldsymbol{H}^T_{s,t}(k|k)\boldsymbol{\Lambda x}(k-t) \quad (6)$$

$\bar{\nu}^b_{s,t}(i|k)$ and $\hat{\nu}^b_{s,t}(i|k)$: backward prediction errors for the output signal $y(i-s)$ with the ARMA order (s,t)

$$\hat{\nu}^b_{s,t}(i|k) \triangleq y(i-s) - \boldsymbol{h}^T_{s,t+1}(i|i)\boldsymbol{R}^{-1}_{s,t+1}(k|k) \cdot \boldsymbol{H}^T_{s,t+1}(k|k)\boldsymbol{\Lambda y}(k-s) \quad (7)$$

$$\bar{\nu}^b_{s,t}(i|k) \triangleq y(i-s) - \boldsymbol{h}^T_{s,t}(i|i)\boldsymbol{R}^{-1}_{s,t}(k|k) \cdot \boldsymbol{H}^T_{s,t}(k|k)\boldsymbol{\Lambda y}(k-s) \quad (8)$$

where,

$$\boldsymbol{x}(k-l) \triangleq [x(k-l)\cdots x(k-l-w+1)]^T$$
$$\boldsymbol{y}(k-m) \triangleq [y(k-m)\cdots y(k-m-w+1)]^T$$
$$(m=0,\ldots,s)(l=0,\ldots,t)$$
$$\boldsymbol{h}_{s,t}(i|j)^T \triangleq [y(i)\cdots y(i-t+1)x(j)\cdots x(j-s+1)]^T$$
$$\boldsymbol{H}^T_{s,t}(k|k) \triangleq [\boldsymbol{y}(k)\cdots \boldsymbol{y}(k-t+1)\boldsymbol{x}(k)\cdots \boldsymbol{x}(k-s+1)]$$
$$\boldsymbol{\Lambda} \triangleq \mathrm{diag}\left[\prod_{j=i}^{w-1}\lambda(j)\cdots\prod_{j=i}^{1}\lambda(j) \quad 1\right]$$

$$\boldsymbol{R}_{s,t}(k|k) \triangleq \boldsymbol{H}^T_{s,t}(k|k)\boldsymbol{\Lambda}\boldsymbol{H}_{s,t}(k|k) \quad (9)$$

The arguments for each prediction error are written in the form $(i|k)$, where the first index i is the time index of the quantity predicted, and the second index k is the time index of the newest measurement used in making the prediction.

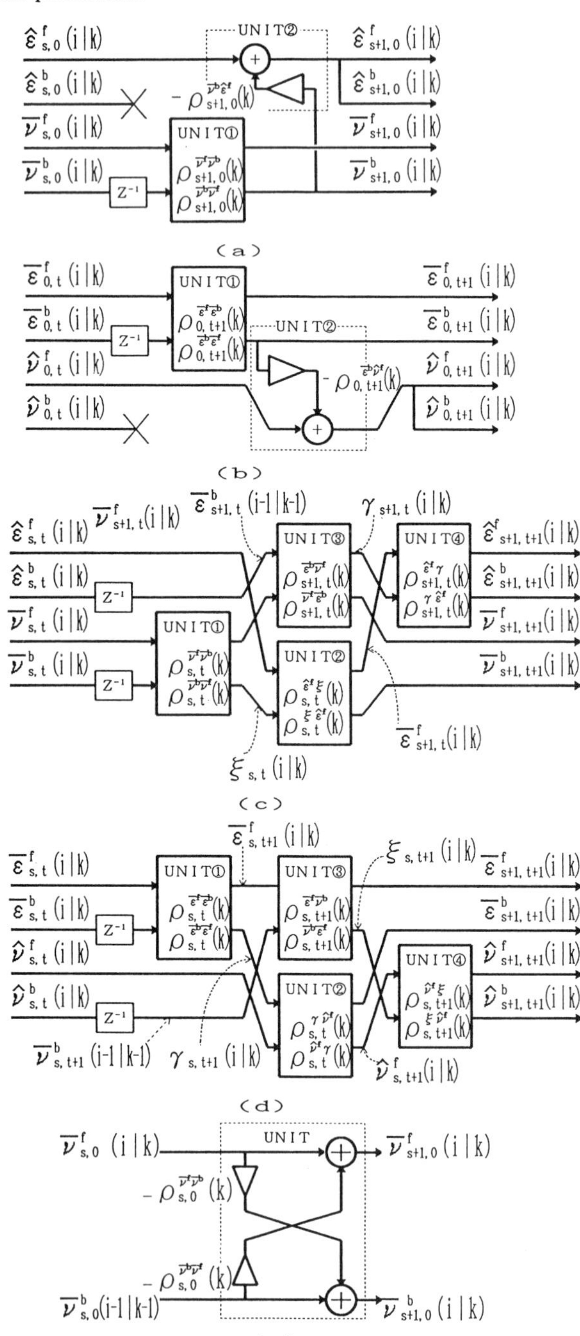

Figure 1: ARMA lattice filter elementary sections
(a) AR0 section (b) MA0 section (c) ARMA-1 section
(d) ARMA-2 section (e) The structure of UNIT

Further, by using the prediction errors defined in Eqs.(1) $\sim$ (8), we can derived four types of order-update recursive formulas, AR0, MA0, ARMA-1 and ARMA-2 types. The AR0 order-update recursive formulas increase the AR order of the prediction errors by one order when their MA orders are zero. And the MA0 order-update recursive formulas increase the MA order by one when the AR orders are zero. Using the ARMA-1 and ARMA-2 order-update recursive formulas, we can increase both the AR order and the MA order by one when AR order $\geq$ MA order or AR order $<$ MA order. Based on the order-update recursive formulas, the ARMA lattice filter elementary sections are realized as in Fig. 1 (a) $\sim$ (d). And the structure of 'UNIT' used in Fig. 1 is illustrated in Fig. 1 (e), where

$$V^{\bar{\nu}^f\bar{\nu}^b}_{s,t}(k) \triangleq \sum_{i=k-w+1}^{k} \lambda^{k-i}\bar{\nu}^f_{s,t}(i|k)\bar{\nu}^b_{s,t}(i-1|k-1)$$

$$V^{\bar{\nu}^f}_{s,t}(k) \triangleq \sum_{i=k-w+1}^{k} \lambda^{k-i}\left\{\bar{\nu}^f_{s,t}(i|k)\right\}^2$$

$$\rho^{\bar{\nu}^f\bar{\nu}^b}_{s,t}(k) \triangleq \frac{V^{\bar{\nu}^f\bar{\nu}^b}_{s,t}(k)}{V^{\bar{\nu}^f}_{s,t}(k)}. \quad (10)$$

In the above equations, w is a sliding window length [3] and λ is a forgetting factor [1], therefore, our algorithm can perform the SEFW with the exponentially weighted sliding window [5].

In order to realize the ARMA four-line lattice filter as adaptive form, we need to time-update the filter coefficients shown in Fig. 1. Since each elementary sections are constructed with some units shown in Fig. 1 (e), the filter coefficients can be time-updated by time-updating the coefficients of the unit.

For example, let us consider the time-update recursive formula for the coefficients of the first unit (UNIT ①) of the AR0 type elementary section. The coefficients of the unit are calculated with the correlations $V^{\bar{\nu}^f\bar{\nu}^b}_{s,t}(k)$, $V^{\bar{\nu}^f}_{s,t}(k)$, and $V^{\bar{\nu}^b}_{s,t}(k)$. The values $V^{\bar{\nu}^f}_{s,t}(k)$ and $V^{\bar{\nu}^b}_{s,t}(k)$ can be obtain from the lattice coefficients and correlations of the previous sections. The time-update recursive formulas of $V^{\bar{\nu}^f\bar{\nu}^b}_{s,t}(k)$ is shown hereunder:

$$\begin{aligned} V^{\bar{\nu}^f\bar{\nu}^b}_{s,t}(k) &= \lambda\left\{V^{\bar{\nu}^f\bar{\nu}^b}_{s,t}(k-1)\right. \\ &\quad \left.+A_1\bar{\nu}^f_{s,t}(k|k-1)\bar{\nu}^b_{s,t}(k-1|k-2)\right\} \\ &\quad -A_2\left\{\bar{\nu}^f_{s,t}(k-w|k-1)-A_3\bar{\nu}^f_{s,t}(k|k-1)\right\} \\ &\quad \cdot\left\{\bar{\nu}^b_{s,t}(k-w|k-1)-A_3\bar{\nu}^b_{s,t}(k|k-1)\right\} \end{aligned} \quad (11)$$

where

$$A_1 \triangleq (\lambda+\alpha_{s,t}(k-1))^{-1}$$

$$A_2 \triangleq \lambda^w\{1-\lambda^w\alpha_{s,t}(k-w)\}^{-1}$$

$$A_3 \triangleq \alpha_{s,t}(k-w-1|k-1)A_1$$

$$\alpha_{s,t}(k-i|k-j) \triangleq \boldsymbol{h}^T_{s,t}(i-1|i-1)\boldsymbol{R}^{-1}_{s,t}(k-2|k-2)\,\boldsymbol{h}_{s,t}(j-1|j-1)$$

(i, j = k-w, k-1 and if i=j, $\alpha_{s,t}(k-i|k-j)=\alpha_{s,t}(k-i)$.) (12)

Furthermore, in our algorithm, order-update recursive formulas of $\alpha_{s,t}(k-i|k-j)$ defined in Eq.(12). are needed. These formulas are described as follows:

i) $\alpha_{s,t}(k-i|k-j)$ which is used for time-updating coefficients of AR0 section's ①UNIT, ARMA-1 section's ①UNIT, and ARMA-2 section's ③UNIT

$$\alpha_{s+1,t}(k-i|k-j) = \alpha_{s,t}(k-i|k-j) + \frac{\bar{\nu}^b_{s,t}(k-i|k-1)\bar{\nu}^b_{s,t}(k-j|k-1)}{V^{\bar{\nu}^b}_{s,t}(k-1)} \quad (13)$$

ii) $\alpha_{s,t}(k-i|k-j)$ which is used for time-updating coefficients of MA0 section's ①UNIT, ARMA-1 section's ③UNIT, and ARMA-2 section's ①UNIT

$$\alpha_{s,t+1}(k-i|k-j) = \alpha_{s,t}(k-i|k-j) + \frac{\bar{\epsilon}^b_{s,t}(k-i|k-1)\bar{\epsilon}^b_{s,t}(k-j|k-1)}{V^{\bar{\epsilon}^b}_{s,t}(k-1)} \quad (14)$$

iii) $\alpha'_{s,t}(k-i|k-j)$ which is used for time-updating coefficients of AR0 section's ②UNIT, ARMA-1 section's ②UNIT, and ARMA-1 section's ④UNIT

$$\alpha'_{s,t}(k-i|k-j) = \alpha_{s,t}(k-i|k-j) + \frac{\bar{\nu}^f_{s,t}(k-i+1|k)\bar{\nu}^f_{s,t}(k-j+1|k)}{V^{\bar{\nu}^f}_{s,t}(k)} \quad (15)$$

iv) $\alpha'_{s,t}(k-i|k-j)$ which is used for time-updating coefficients of MA0 section's ②UNIT, ARMA-2 section's ②UNIT, and ARMA-2 section's ④UNIT

$$\alpha'_{s,t}(k-i|k-j) = \alpha_{s,t}(k-i|k-j) + \frac{\bar{\epsilon}^f_{s,t}(k-i+1|k)\bar{\epsilon}^f_{s,t}(k-j+1|k)}{V^{\bar{\epsilon}^f}_{s,t}(k)} \quad (16)$$

where

$$\alpha_{0,0}(k-i|k-j) = 0. \quad (17)$$

The *Algorithm (1)* shown in Sec. 3 has two parts; One is formed based on order-update recursive formulas and the other is formed based on time-update recursive formulas of the elementary-section coefficients.

3. ALGORITHM(1) AND ALGORITHM(2)

The proposed lattice-filter structure is similar to that shown in Ref. [5]. Comparing Ref. [5] with Fig. 1, we can see that there is no significant differences between the number of their lattice-filter coefficients. However, the proposed algorithm is different from Ref. [5] not only on the ability of the SEFW, but also on the following points.

(1) In the case of AR order $\geq$ MA order, an input estimation procedure can be enbedded in *Algorithm(1)* without increasing calculation costs. (The input estimation problem often occurs in the case of ARMA spectral estimation.)

(2) Using *Algorithm(2)*, we can obtain the ARMA parameters. (The spectrum can be easily obtained from the ARMA parameters.)

3.1. Frequency-Weighting Method

The block diagram shown in Fig. 2 (a) expresses an error for the least-square (LS) estimation problem without frequency-weighting, where, in Fig. 2, $u(i)$ is a white Gaussian noise. Minimizing the value

$$\sum_{i=k-w+1}^{k} \lambda^{k-i} e^2_{s,t}(i|k), \quad (18)$$

We can obtain the optimal estimated parameters of $\hat{A}(q^{-1})$ and $\hat{B}(q^{-1})$; the parameters can be estimated by the lattice filter denoted in Sec. **2**.

Further, using a transfer function $W(q^{-1})$, where

$$\frac{1}{2\pi j}\oint_{|z|=1} \left|W(z^{-1})\right|^2 \frac{dz}{z} = 1, \quad (19)$$

we try to modify Fig. 2 (a) frequency-weighting LS estimation problem. Comparing Fig. 2 (a) and (b), we can find frequency-weighting model identification is achieved by the lattice filter derived in Sec.2.

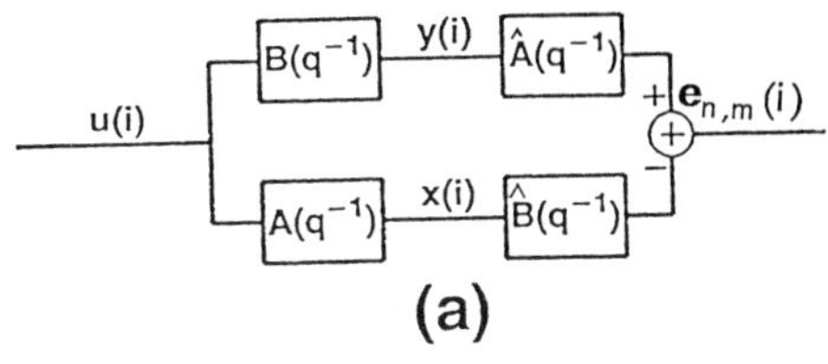

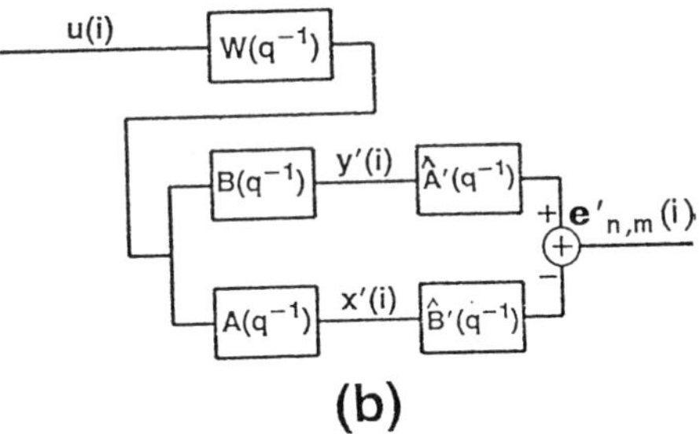

Figure 2: Diagrams for LS and frequency-weighting LS

3.2. Algorithm(1)

Algorithm(1) achieves the realization of ARMA lattice filter with the frequency weighting and the input estimation. In the case of AR order $\geq$ MA order, the procedures of *Algorithm(1)* are shown in Table I. *Algorithm(1)* of case of AR order $<$ MA order can be easily obtained.

Table I: Algorithm(1)

begin algorithm
initialize V, $\hat{\epsilon}^f$, $\hat{\epsilon}^b$, $\bar{\nu}^f$, $\bar{\nu}^b$, and α.
n:=AR order; m:=MA order; w:=the rectangular window length; λ:=the forgetting factor
k:= 1
while ($y(k)$ exists)
    read $y(k)$ {$y(k)$ is the observed signal.}
    $y'(k) = W(q^{-1})y(k)$ {If frequency-weighting is not needed, $W(q^{-1}) = 1$.}
    $\bar{\nu}^f_{0,0}(k|k-1) = y'(k)$
    s:=$n-m$
    for j:=1 to s step by 1
        compute $\bar{\nu}^f_{j,0}(k|k-1)$
        {by order-update recursive formula}
    end of for j

13. for l:=0 to $m-1$ step by 1
14. compute $\bar{\nu}^{f}_{s+l+1,t+l+1}(k|k-1)$
15. end of for l
{input estimator}
16. $\hat{u}(k) = \frac{\lambda(k-1)}{\lambda(k-1)+\alpha_{n,m}(k-1)} \bar{\nu}^{f}_{n,m}(k|k-1)$
17. $x'(k) = W(q^{-1})\hat{u}(k)$ {If the frequency-weighting is not needed, $W(q^{-1}) = 1$.}
{the estimated input and output signals are fed into the lattice filter.}
18. $\hat{\epsilon}^{f}_{0,0}(k|k-1) = x'(k) - \frac{r_{xy}}{r_y} y'(k)$
$\hat{\epsilon}^{b}_{0,0}(k|k-1) = \hat{\epsilon}^{f}_{0,0}(k|k-1)$
$\bar{\nu}^{b}_{0,0}(k|k-1) = y'(k)$
{the order-update for the other prediction errors and α}
19. for i:=1 to s step by 1
20. order-update $\hat{\epsilon}^{f}_{i,0}(k|k-1)$, $\hat{\epsilon}^{b}_{i,0}(k|k-1)$, and $\bar{\nu}^{b}_{i,0}(k|k-1)$.
21. compute $\alpha_{i,0}(k)$ {by Eq.(13)}
22. end of for j
23. for l:=1 to m step by 1
24. compute $\hat{\epsilon}^{f}_{j+l,l}(k|k-1)$, $\hat{\epsilon}^{b}_{j+l,l}(k|k-1)$, and $\bar{\nu}^{b}_{j+l,l}(k|k-1)$
25. compute $\alpha_{j+l,l}(k)$ and $\alpha'_{j+l,l}(k)$
{by Eqs.(13) ~(16)}
26. end of for j
27. time-update all of the lattice parameters by Eq.(11)
28. $r_y = \lambda r_y + y'^2(k)$
$r_{xy} = \lambda r_{xy} + x'(k)y'(k)$
29. end of while
30. end of algorithm

3.3. Algorithm(2)

Algorithm(2) computes the estimated parameters of ARMA transfer function from the parameters of the ARMA lattice filter realized by *Algorithm(1)*. The procedures of *Algorithm(2)* are shown in Table II. In Table II, the procedures on the case of AR order $\geq$ MA order are shown. Similarly, the procedure of the case of AR order $<$ MA order can be obtained.

Table II: Algorithm(2)

1. begin algorithm
2. define $\boldsymbol{A}$; (3,2) or (4,2) matrix
define $\boldsymbol{B}$; (3,3) or (4,4) matrix
{ each element of the matrices $\boldsymbol{A}$ and $\boldsymbol{B}$ is a polynomial of z^{-1}; $\boldsymbol{A} = [a_{i,j}(z^{-1})]$ }
3. k:=the time index; $\boldsymbol{B}$:=$\boldsymbol{O}$; $\boldsymbol{A}$:=$\boldsymbol{I}$
4. n:= AR order; m:= MA order; m-n:= t
5. $\boldsymbol{A} := \begin{bmatrix} 1 & 0 \\ 1 & 0 \\ -\frac{r_{xy}}{r_x} & 1 \end{bmatrix}$
6. for j:=0 to $t-1$ step by 1
7. $\boldsymbol{B} := \begin{bmatrix} 1 & -\rho^{\bar{\epsilon}^b \bar{\epsilon}^f}_{0,j}(k) z^{-1} & 0 \\ -\rho^{\bar{\epsilon}^f \bar{\epsilon}^b}_{0,j}(k) & z^{-1} & 0 \\ 0 & 0 & 1 \end{bmatrix}$
8. $\boldsymbol{A} = \boldsymbol{B}^* \boldsymbol{A}$
9. $\boldsymbol{B} := \begin{bmatrix} 1 & 0 & 0 \\ 0 & 1 & 0 \\ 0 & -\rho^{\bar{\epsilon}^b \hat{\nu}^f}_{0,j}(k) & 1 \end{bmatrix}$
10. $\boldsymbol{A} = \boldsymbol{B}^* \boldsymbol{A}$
11. end of for j
12. $\boldsymbol{A} = \begin{bmatrix} 1 & 0 & 0 \\ 0 & 1 & 0 \\ 0 & 0 & 1 \\ 0 & 0 & 1 \end{bmatrix} {}^* \boldsymbol{A}$
13. for l:=0 to $n-1$ step by 1
14. $\boldsymbol{B} := \begin{bmatrix} 1 & -\rho^{\bar{\epsilon}^b \bar{\epsilon}^f}_{l,t+l}(k) z^{-1} & 0 & 0 \\ -\rho^{\bar{\epsilon}^f \bar{\epsilon}^b}_{l,t+l}(k) & z^{-1} & 0 & 0 \\ 0 & 0 & 1 & 0 \\ 0 & 0 & 0 & 1 \end{bmatrix}$
15. $\boldsymbol{A} = \boldsymbol{B}^* \boldsymbol{A}$
16. $\boldsymbol{B} := \begin{bmatrix} 1 & 0 & 0 & 0 \\ 0 & 1 & -\rho^{\hat{\nu}^f \gamma}_{l,t+l}(k) & 0 \\ 0 & -\rho^{\gamma \hat{\nu}^f}_{l,t+l}(k) & 1 & 0 \\ 0 & 0 & 0 & 1 \end{bmatrix}$
17. $\boldsymbol{A} = \boldsymbol{B}^* \boldsymbol{A}$
18. $\boldsymbol{B} :=$
$\begin{bmatrix} 1 & 0 & 0 & -\rho^{\bar{\nu}^b \bar{\epsilon}^f}_{l,t+l+1}(k) z^{-1} \\ 0 & 1 & 0 & 0 \\ 0 & 0 & 1 & 0 \\ -\rho^{\bar{\epsilon}^f \bar{\nu}^b}_{l,t+l+1}(k) z^{-1} & 0 & 0 & z^{-1} \end{bmatrix}$
19. $\boldsymbol{A} = \boldsymbol{B}^* \boldsymbol{A}$
20. $\boldsymbol{B} := \begin{bmatrix} 1 & 0 & 0 & 0 \\ 0 & 1 & 0 & 0 \\ 0 & 0 & 1 & -\rho^{\xi \hat{\nu}^f}_{l,t+l+1}(k) \\ 0 & 0 & -\rho^{\hat{\nu}^f \xi}_{l,t+l+1}(k) & 1 \end{bmatrix}$
21. $\boldsymbol{A} = \boldsymbol{B}^* \boldsymbol{A}$
22. end of for l
23. { AR(z^{-1}) and MA(z^{-1}) are polynomials of z^{-1}
$\frac{\mathrm{MA}(z^{-1})}{\mathrm{AR}(z^{-1})}$: the estimated ARMA transfer function}
24. AR (z^{-1}) = - $a_{3,2}(z^{-1})$
25. MA(z^{-1}) = $a_{3,1}(z^{-1})$
26. end of algorithm

4. CONCLUSIONS

We present the SEFW lattice filter algorithm which is constructed with two algorithms: *Algorithm(1)* and *Algorithm(2)*. Using this SEFW lattice filter algorithm, we can obtain the accurate spectral estimation in the frequency band focused. Therefore, this method is suitable for the realization of a reduced order model.

REFERENCES

[1] P.Eykhoff:" Trends and progress in system identification," Pergamon Press, 1981.

[2] M. R. Gevers and V. J. Wertz: "A D-step predictor in lattice and ladder form," IEEE Trans. Automatic Control, **AC-28**, 4, pp.465-476, (April 1983)

[3] H. L. Honig and D. G. Messerschmitt: " Adaptive filters structures, algorithms, and applications," Kluwer Academic Publishers (1984)

[4] Lennart Ljung: "System identification Theory for the user," Prentice Hall, 1987.

[5] E.Karlsson and M.H.Hayes: " Least squares ARMA modeling of linear time-varying systems: lattice filter structures and fast RLS algorithms," IEEE Trans. Acoust., Speech & Signal Process., **ASSP-35**, 7, pp.994-1014, (July 1987)

[6] H.Watanabe, Y.Miyanaga, N.Miki and N.Nagai: "ARMA spectral approximation method with weighting functions," Electronics and Communications in Japan, Part 1, vol.68, No.4, 1985, Scripta Technica, Inc.

Application of SVD to AR Spectral Estimation

Poul M. Rands Jensen* Kjeld Hermansen

Department of Communication Technology, Institute of Electronic Systems, Aalborg University, DK-9220 Aalborg Ø, Denmark

Abstract

During recent years much interest has been given to the application of Singular Value Decomposition in association with extended-order and overdetermined evaluation in the finite parametric spectral estimation domain. Such approaches have been shown to perform superior to other methods for the harmonic retrieval problem.

We apply a similar SVD approach to wide-banded AR processes embedded in white noise which has not been issued previously. This situation typically arises from feature extraction of speech signals as the pre-processing part of a speech recognition.

It is found that the extraneous poles of the lower rank solution influence the spectral estimate. A novel approach, the Direction Weighted Total Least Squares solution, which enforces the extraneous poles to be located at the origin while maintaining the good properties of the aforementioned approach, is therefore introduced.

Computer simulation experiments performed so fare indicate this approach to be superior to existing overdetermined and extended- or parsimonious-order methods on noisy data.

1 Introduction

Since the introduction of parametric modeling techniques in spectral estimation like sinusoidal and rational models, these techniques have become widely used due to their higher resolution and spectral fidelity compared to the periodogram.

It has been shown by Cadzow [2] that upon using more than the minimal number of equations, a reduction in data-induced model parameter hyper sensitivity is obtained and an improvement in estimation performance is obtained. This is especially true for noisy environments. Further increased performance can often be obtained by using the **T***otal* **L***east* **S***quares* formula, cf. [3, 11]. Moreover, improvements can be obtained using an *extended-order* model and apply SVD to reduce the rank of the obtained matrix, cf. [2, 9]. In the spectral estimation domain most interest has been given to the harmonic retrieval problem, cf. [1, 2, 5, 10], in which the detection of the number of sinusoids in presence of white noise is issued.

We consider the application of SVD to spectral estimation of wide-banded AR processes embedded in white noise, like voiced speech in noisy environments. The aim of this approach is to remove noise influence in the LPC coefficients to achieve improved spectral estimates. This *continuous frequency* application presents a number of new problems compared to the *discontinuous frequency* harmonic retrieval problem associated with the influence of the spectral estimate by the *extraneous* poles, caused by the extended-order.

We therefore propose a new approach called *the* **D***irection* **W***eighted* **T***otal* **L***east* **S***quares solution*, which utilizes overdetermination, extended-order, TLS and rank reduction via SVD, but by weighting the direction of the solution, enforces the extraneous poles to be located at the origin while maintaining the good properties of the aforementioned approaches.

2 AR Spectral Estimation

For estimating voiced speech and alike signals, the most often applied parametric model is the autoregressive model:

$$y(k) = b_0 \cdot u(k) - \sum_{i=1}^{p} \alpha_i \cdot y(k-i), \tag{2.1}$$

*Currently supported by a grant from the Danish Technical Research Council, Grant No. 16-4895

where $s(k) = y(k) + \nu(k)$ is the *observable* signal, e.g. speech, $u(k)$ is an unknown white noise input and $\nu(k)$ is white noise perturbation. The estimate, $\widehat{s}(k)$, of $s(k)$ is then given as:

$$\widehat{s}(k) = -\sum_{i=1}^{p} a_i \cdot s(k-i), \tag{2.2}$$

where $\{a_i\}_1^p$ are the parameters to be determined. Expanding (2.2) we can write the estimation problem as a LS problem, $\mathbf{A}\underline{a} = \underline{b} + \underline{r}$, using the *deterministic autocovariance approach* [4]:

$$\begin{bmatrix} s(k+N-1) & \cdots & s(k) \\ s(k+N) & \cdots & s(k+1) \\ \vdots & & \vdots \\ s(k+M+N-1) & \cdots & s(k+M) \end{bmatrix} \begin{bmatrix} a_1 \\ a_2 \\ \vdots \\ a_N \end{bmatrix} =$$

$$\begin{bmatrix} -s(k+N) \\ -s(k+N+1) \\ \vdots \\ -s(k+M+N) \end{bmatrix} + \begin{bmatrix} e(k+N) \\ e(k+N+1) \\ \vdots \\ e(k+M+N) \end{bmatrix} \tag{2.3}$$

where $\{e(k)\}$ is an *unobservable* random variable of zero-mean, introduced into the model to account for its inaccuracy and which obviously is to be minimized. Alternatively we could have used the autocorrelation method or even have used the autocorrelation matrix, $\mathbf{R}$, since the things issued in below do apply in these cases too.

(2.3) is readily solved as $\underline{a} = (\mathbf{A}^T\mathbf{A})^{-1}\mathbf{A}^T\underline{b}$. To improve performance, however, we may use an extended-order model, i.e. $N > p$, and apply SVD to obtain a **T**runcated SVD solution of rank p, $\underline{x}_p$. This is given as:

$$\underline{x}_p = \sum_{i=1}^{p} \underline{v}_i \frac{\underline{u}_i^T}{\sigma_i} \underline{b}, \tag{2.4}$$

where $\{\underline{u}_i\}_1^M$, $\{\underline{v}_i\}_1^N$, $\{\sigma_i\}_1^N$ are the left and right *singular vectors* and *singular values* of $\mathbf{A}$, respectively. SVD allows for *removal of noise influence in* $\underline{a}_p$ since (2.4) *expresses a partitioning of the domain of* $\mathbf{A}$ *into a p dimensional* **signal-** *and a* $(N-p)$ *dimensional* **noise** *space*, and hence, an *improved spectral estimate* can be gained.

When $\{\nu(k)\}$ is non-zero, $\mathbf{A}$ as well as $\underline{b}$ are perturbed which makes a TLS solution more feasible [5, 11]. Forming $\mathbf{B} = [\underline{b}\ \mathbf{A}]$ and computing its SVD,

$$\mathbf{B} = \mathbf{U} \cdot \mathbf{\Sigma} \cdot \begin{bmatrix} \mathbf{V}_{11} & \mathbf{V}_{12} \\ \mathbf{V}_{21} & \mathbf{V}_{22} \end{bmatrix}^T, \tag{2.5}$$

the **T***runcated* TLS solution, $\underline{a}_{TLS,p}$, is given as:

$$\underline{a}_{TLS,p} = \mathbf{V}_{22} \frac{\mathbf{V}_{12}^T}{\|\mathbf{V}_{12}\|^2} \tag{2.6}$$

with $\mathbf{V}_{11} \in \mathsf{R}^{1\times p}$, $\mathbf{V}_{12} \in \mathsf{R}^{1\times N+1-p}$, $\mathbf{V}_{21} \in \mathsf{R}^{N\times p}$ and $\mathbf{V}_{22} \in \mathsf{R}^{N\times N+1-p}$.

3 The Direction Weighted Total Least Squares solution

Revealing a proper value for p is non-trivial since the problem is *severely ill-conditioned*, cf. [1, 2, 7]. This problem has been issued in e.g. [1, 8] and is not within the scope of this paper. Hence, we consider the computation of the parameters, $\underline{a}$.

This may readily be carried out using (2.4) for a TSVD solution or (2.6) for a TTLS solution. The extraneous poles that obviously are contained in these solutions will ideally be uniformly distributed inside the unit circle [6]. However, for a perturbed system the extraneous poles do fall closer to, on or even outside the unit circle and are not likely to be uniformly distributed.

Unless the signals are very narrow-banded like sinusoids, the extraneous poles will influence the spectrum. Hence, for speech signals considered here and other such AR processes, this will lead to wrong spectral estimates. We therefore propose a novel solution, the DW TLS solution, *characterized by only computing a p-elements vector for a rank p solution.*

Reconsider the TLS problem resulting from $[\underline{b}\mathbf{A}]\underline{x} \simeq \underline{0}$, and assume $rank([\underline{b}\ \mathbf{A}]) = p$ where $p < N$. This implies that we only need p columns of $\mathbf{A}$, and hence a $\{a_i\}_1^p$ to form a linear combination equal to $\underline{b}$.

The DW TLS solution enforces the extraneous poles to be located at the origin. That is, only $\{a_i\}_1^p$ are allowed to differ from zero, while $\{a_i\}_{p+1}^N$ are enforced to equal zero.

First, consider the matrix $\mathbf{B} \in \mathsf{R}^{M\times(N+1)}$ formed as above, i.e. $\mathbf{B} = [\underline{b}\ \mathbf{A}]$. Next, assume $\mathbf{B}$ has a SVD, $\mathbf{B} = \mathbf{U} \cdot \mathbf{\Sigma} \cdot \mathbf{V}^T$, as in (2.5). We then know that the (minimal norm) TTLS solution, $\underline{x}_{TLS,p}$, is given by (2.6) which, of course, is unique. However, any $\underline{x}$ given as a linear combination of $\{\underline{v}_i\}_{p+1}^{N+1}$ and with its first element equal to 1 is a valid solution.

Using the á priori knowledge that we only need $\{a_i\}_1^p$, we search for a solution of the form $\underline{x} = [1\ a_1 \cdots a_p\ 0 \ldots 0]^T$:

$$\begin{bmatrix} v_{1,p+1} & \cdots & v_{1,N+1} \\ v_{2,p+1} & \cdots & v_{2,N+1} \\ \vdots & \cdots & \vdots \\ v_{p+1,p+1} & \cdots & v_{p+1,N+1} \\ v_{p+2,p+1} & \cdots & v_{p+2,N+1} \\ \vdots & & \vdots \\ v_{N+1,p+1} & \cdots & v_{N+1,N+1} \end{bmatrix} \begin{bmatrix} \xi_{p+1} \\ \xi_{p+2} \\ \vdots \\ \xi_{N+1} \end{bmatrix} = \begin{bmatrix} 1 \\ a_1 \\ \vdots \\ a_p \\ 0 \\ \vdots \\ 0 \end{bmatrix}$$

DW TLS versus parsimonious-order QR solution						
	n	ω_c	Filter Q	$\overline{\Delta H}(\omega)$	$\Delta H(\omega)_{max}$	$SNR\,[dB]$
DW TLS	990	$2.21 \cdot 10^{-2}$	7.95	$3.35 \cdot 10^{-2}$	2.19	∞
QRD 2	980	$2.63 \cdot 10^{-2}$	4.82	$7.79 \cdot 10^{-2}$	2.37	
DW TLS	980	$2.21 \cdot 10^{-2}$	7.64	$4.75 \cdot 10^{-2}$	2.93	20
QRD 6	983	$2.29 \cdot 10^{-2}$	5.66	6.90	$3.04 \cdot 10^{2}$	
DW TLS	743	$1.95 \cdot 10^{-2}$	4.58	$8.27 \cdot 10^{-2}$	3.49	10
QRD 6	469	$1.84 \cdot 10^{-2}$	1.25	4.63	$5.77 \cdot 10^{1}$	
DW TLS	513	$2.15 \cdot 10^{-2}$	4.70	$1.01 \cdot 10^{-1}$	3.14	6
QRD 6	142	$2.00 \cdot 10^{-2}$	4.28	4.14	$4.03 \cdot 10^{1}$	
DW TLS	353	$3.35 \cdot 10^{-2}$	$1.35 \cdot 10^{1}$	$1.80 \cdot 10^{-1}$	6.04	3
QRD 6	35	$3.66 \cdot 10^{-2}$	$1.48 \cdot 10^{1}$	3.76	$3.21 \cdot 10^{1}$	
DW TLS	231	$7.47 \cdot 10^{-2}$	$4.02 \cdot 10^{1}$	$4.09 \cdot 10^{-1}$	6.75	0
QRD 6	15	$2.67 \cdot 10^{-1}$	$6.83 \cdot 10^{1}$	2.38	8.76	

Table 1: *'n' equals the number of trials for which a complex pole pair was obtained, and only these are included in the other quantities. QRD x means a model order of x.*

This leads to the requirement:

$$\begin{bmatrix} v_{1,p+1} & \cdots & v_{1,N+1} \\ v_{p+2,p+1} & \cdots & v_{p+2,N+1} \\ \vdots & & \vdots \\ v_{N+1,p+1} & \cdots & v_{N+1,N+1} \end{bmatrix} \begin{bmatrix} \xi_{p+1} \\ \xi_{p+2} \\ \vdots \\ \xi_{N+1} \end{bmatrix} = \begin{bmatrix} 1 \\ 0 \\ \vdots \\ 0 \end{bmatrix} \tag{3.1}$$

The DW TLS solution, $\underline{a}_{DW,p}$ with elements $\{a_i\}_1^p$, is finally given as

$$a_i = \sum_{j=p+1}^{N+1} \xi_j \cdot v_{i+1,j}, \qquad i \in [1;p] \tag{3.2}$$

Also refer to [7].

4 Results

We have performed a number of computer simulations on known AR processes comparing extended- and parsimonious-order standard solutions with extended-order LS and TLS solutions using SVD and the DW TLS solution.

E.g. we have made a comparison between DW TLS and parsimonious-order standard solution, in terms of QRD for a known 2nd order process embedded in white noise. The process is characterized by centerfrequency, $\omega = 2.20 \cdot 10^{-2}$, and filter $Q = 5.63$. We have used a model order of 12 and $p = 2$ for DW TLS in all cases while the parsimonious-order for the QRD solution has been slightly increased for decreased SNR. For each SNR 1000 Monte Carlo trials each of 92 samples have been carried out.

In table 1 are summarized the results of this comparison showing the superior performance of DW TLS, where the evaluation quantities are given by (4.1).

$$\begin{aligned} \omega_c &= \left\{\omega \middle| \max\left(\widehat{H}(\omega)\right)\right\} \\ Q &= \tfrac{\omega_c}{\omega_{3dB}} \\ \overline{\Delta H}(\omega) &= \mathbf{E}\left[\left|H(\omega) - \widehat{H}(\omega)\right|\right] \\ \Delta H(\omega)_{max} &= \max_\omega \left|H(\omega) - \widehat{H}(\omega)\right| \end{aligned} \tag{4.1}$$

5 Conclusion

In this contribution we have considered the application of the SVD to finite parametric AR spectral estimation. It has been found that the extraneous poles of the TSVD solutions spoil the spectral estimate. The presented DW TLS solution enforces these extraneous poles to be located at the origin while maintaining the good properties of the SVD approaches.

Computer simulation experiments carried out indicate that this approach in the case considered is superior performance of the novel approach compared to existing parsimonious-order methods on noisy data provided knowledge of the parsimonious-order. More work is needed, especially on real data such as voiced speech, which is an issue of our current research [8].

References

[1] S. Bakamidis, M. Dendrinos, and G. Carayannis. SVD Analysis by Synthesis of Harmonic Signals. *IEEE Transactions on Signal Processing*, 39(2), 1991.

[2] J. A. Cadzow. Spectral Estimation: An Overdetermined Rational Model Approach. *Proceedings of the IEEE*, 70(9), 1982.

[3] G. H. Golub and C. F. Van Loan. An Analysis of the Total Least Squares problem. *SIAM Journal on Numerical Analysis*, 17(6), 1980.

[4] S. S. Haykin. *Adaptive Filter Theory*. Prentice–Hall Inc., Englewood Cliffs, 1986.

[5] Y. Hua and T. K. Sarkar. On the Total Least Squares Linear Prediction Method for Frequency Estimation. *IEEE Transactions on Acoustics, Speech and Signal Processing*, 38(12), 1990.

[6] R. Kumaresan. On the Zeroes of the Linear Prediction–Error Filter for Deterministic Signals. *IEEE Transactions on Acoustics, Speech and Signal Processing*, 31(1), 1983.

[7] P. M. Rands Jensen. Singular Value Decomposition and Its Application to AutoRegressive Parametric Spectral Estimation. Technical report R. 91–33, Department of Communication Technology, Institute of Electronic Systems, Aalborg University, Aalborg, Denmark, August 1991.

[8] P. M. Rands Jensen and K. Hermansen. A SVD based approach for noise-robust speech signal feature extraction. to be submitted: IEEE Transactions on Signal Processing, 1992.

[9] L. L. Scharf. The SVD and Reduced-Rank Signal Processing. In R. Vaccaro, editor, *SVD and Signal Processing II*, pages 3–31. Elseviers Science Publishers, 1991.

[10] D. W. Tufts and R. Kumaresan. Estimation of Frequencies of Multiple Sinusoids: Making Linear Prediction Perform Like Maximum Likelihood. *Proceedings of the IEEE*, 70(9), 1982.

[11] S. Van Huffel and J. Vandewalle. The Total Least Squares Technique: Computation, Properties and Applications. In E. F. Deprettere, editor, *SVD and Signal Processing*, pages 189–207. Elseviers Science Publishers, 1988.

SIGNAL PROCESSING VI: Theories and Applications
J. Vandewalle, R. Boite, M. Moonen, A. Oosterlinck (eds.)

A NEW APPROACH TO ESTIMATION OF THE FREQUENCY OF A NOISY CISOID *

A. Ferrari, G. Alengrin and J. Ménez

LASSY Université de Nice-Sophia Antipolis, Département de l'I3S
41 Bd Napoléon III, 06041 CEDEX FRANCE
tel. : +33–93217953, e-mail : ferrari@mimosa.unice.fr

We present in this paper a method to estimate the frequency of a cisoid from a small number of noisy samples based on the Constrained Total Least Squares. The estimated frequency is the global minimizer of a function consisting in the sum of the spectral density of the signal weighted by a bank of windows. The principle of this estimation scheme is studied. This leads to a new frame of estimators based on the presented scheme and using window banks associated to different criteria. The windows associated to the maximum likelihood estimator are given.

1. INTRODUCTION

The estimation of the frequency of a cisoid from a small number of noisy samples is a problem that arises in many fields (i.e. Doppler estimation in RADAR signals). Many algorithms relying on the well known property that the maximum likelihood estimator (MLE) is given by the global maximizer of the periodogram are present in the litterature. In this paper, we address the problem using the Constrained Total Least Squares of Abatzoglou and Mendel, [1].

2. CTLS estimation

The Total Least Squares (TLS) is used to solve an overdetermined system of linear equations where both right and left hand sides are noisy. The CTLS assumes that the noise on each column of these matrices is a linear transformation of a unique white random vector. This formulation can be mathematically stated as a minimisation under constraints. In [1], the authors have shown that the above problem is equivalent to the unconstrained minimisation of a highly nonlinear function Λ_{CTLS}. We refer to [2] for a complete summary on this topic.

Let assume now that the observed signal vector $\mathbf{x}$ is constituted by N samples of a complex noisy sine wave with frequency ω_0 corrupted by a white noise sequence. The CTLS can then be applied to the $N-1 \times 2$ forward linear prediction matrix associated to $\mathbf{x}$. We have demonstrated in [4] that in this particular case, the function Λ_{CTLS} can be rewritten as:

$$\Lambda_{CTLS}(\omega) = \mathbf{s}_\omega^H.X^H.P_{CTLS}^{-1}.X.\mathbf{s}_\omega \tag{1}$$

where the $N-1 \times 1$ vector $\mathbf{s}_\omega$ is:

$$\mathbf{s}_\omega = (1 \exp j\omega \ldots \exp j(N-1)\omega)^H \tag{2}$$

the $N-1 \times N$ rectangular matrix X is:

* Work partly supported by CESDA/DCN Toulon.

$$X = \begin{pmatrix} x_1 & -x_2 & 0 & \dots & 0 \\ 0 & x_2 & -x_3 & \ddots & \vdots \\ \vdots & \ddots & \ddots & \ddots & 0 \\ 0 & \dots & 0 & x_{N-1} & -x_N \end{pmatrix} \quad (3)$$

and the $N-1 \times N-1$ matrix P_{CTLS} is:

$$P_{CTLS} = \begin{pmatrix} 3 & -1 & 0 & \dots & 0 \\ -1 & 2 & \ddots & \ddots & \vdots \\ 0 & \ddots & \ddots & \ddots & 0 \\ \vdots & \ddots & \ddots & 2 & -1 \\ 0 & \dots & 0 & -1 & 3 \end{pmatrix} \quad (4)$$

(H denotes the conjugate transpose operator.)

This expression can be simplified using the eigendecomposition of the matrix P_{CTLS}. If we denote $(\sigma_i^{CTLS}, \mathbf{v}_i^{CTLS})$, $i = 1, \dots, N-1$ the eigenvalue-eigenvector pairs of P_{CTLS} and if we define the normalized eigenvectors $\mathbf{V}_i^{CTLS}$ as:

$$\mathbf{V}_i^{CTLS} = \mathbf{v}_i^{CTLS} / \sqrt{\sigma_i^{CTLS}} \quad (5)$$

the previous expression of $\Lambda_{CTLS}(\omega)$ can be rewritten :

$$\Lambda_{CTLS}(\omega) = \sum_{i=1}^{N-1} |(\mathbf{V}_i^{CTLS})^H . X . \mathbf{s}_\omega|^2 \quad (6)$$

If we now define the $N-1$ weighting vectors of order N $\mathbf{w}_i^{CTLS}$ $i = 1, \dots, N-1$ as :

$$\mathbf{w}_i^{CTLS} = \begin{pmatrix} \mathbf{V}_i^{CTLS} \\ 0 \end{pmatrix} - \begin{pmatrix} 0 \\ \mathbf{V}_i^{CTLS} \end{pmatrix} \quad (7)$$

we finally obtain the expression :

$$\Lambda_{CTLS}(\omega) = \sum_{i=1}^{N-1} |\mathcal{F}(\mathbf{w}_i^{CTLS} \times \mathbf{x})|^2 \quad (8)$$

($\mathcal{F}(\mathbf{z})$ denotes the discrete Fourier transform of $\mathbf{z}$ at the frequency ω and $\mathbf{w} \times \mathbf{z}$ represents the signal $\mathbf{w}$ windowed by z.)

According to this expression, we can make the following remarks:

– The above expression of $\Lambda_{CTLS}(\omega)$ has a highly parallel implementation. It consists in $N-1$ parallel windowed Fourier transform of the signal $\mathbf{x}$.

– In the preceeding expression of $\Lambda_{CTLS}(\omega)$, the criterion of estimation is relegated to the choice of the $N-1$ time windows $\mathbf{w}_i^{CTLS}$, $i = 1, \dots, N-1$. We can easily imagine to generalize this scheme of estimation using windows based on other criteria. This point will be investigated in the next section

– It has been demonstrated in [4] that in the noiseless case, the function $\Lambda_{CTLS}(\omega)$ does not present as much oscillations as the periodogram. This property is an important advantage of the method in fact, due to the well known oscillations produced by the Dirichlet kernel, [5], the computation of the global maximum of the periodogram with a classical iterative Newton algorithm is strongly dependent of the initial point, [3].

3. Principle of the estimation scheme

We will derive in this section the principle of the estimation scheme and show how it can be generalized to other criteria.

Let first consider the ith term of the sum (8). This term can be written:

$$\mathcal{F}(\mathbf{w}_i^{CTLS} \times \mathbf{x}) = \mathcal{F}(\mathbf{w}_i^{CTLS}) \otimes \mathcal{F}(\mathbf{x}) \quad (9)$$

where $\otimes$ denotes the convolution operator. The Fourier transform of the window $\mathbf{w}_i^{CTLS}$ is given by:

$$\mathcal{F}(\mathbf{w}_i^{CTLS}) = (1 - \exp j\omega) . \mathcal{F}(\mathbf{V}_i^{CTLS}) \quad (10)$$

We see that $\mathcal{F}(\mathbf{V}_i^{CTLS})$ is the key for the interpretation of the estimation scheme. According to the appendix, the $N-1$ components of the eigenvector $\mathbf{V}_i^{CTLS}$ are the samples of a sinewave at the frequency $i.\pi/(N-1)$. The Fourier transform of this vector is then the sum of two Dirichlet kernels $K_D(\omega)$ centered at frequencies $i.\pi/(N-1)$ and $-i.\pi/(N-1)$:

$$\mathcal{F}(V_i^{CTLS}) = K_D(\omega - \frac{i.\pi}{N-1}) + K_D(\omega + \frac{i.\pi}{N-1}) \quad (11)$$

More insight in the method can then be obtained considering the ideal case were $\mathcal{F}(\mathbf{x})$ is a pure spectral line. With no loss of generality, we can assume that this line is located at frequency zero:

$$\mathcal{F}(\mathbf{x}) = A.\delta_0 \quad (12)$$

The term corresponding to equation (9) is then given by:

$$\mathcal{F}(w_i^{CTLS} \times x) = A.(1 - \exp j\omega) .(K_D(\omega - \frac{i.\pi}{N-1}) + K_D(\omega + \frac{i.\pi}{N-1})) \quad (13)$$

and finally, the function $\Lambda_{CTLS}(\omega)$ will represent the sum of the contribution of all these kernels centered at frequencies regularly separated by $\pi/(N-1)$ with the exception of the central-frequency zero. This result is illustrated by figure (1).

Moreover, we can make the following remarks:
– The residual contributions of all the kernels at the frequency zero will be cancelled by the effect of $1 - \exp j\omega$. We can note that if the windows are constructed according to relation (7), this property will always hold.
– The eigenvectors being orthogonal, their Fourier transform will also be orthogonal according to Parseval theorem. The consequence of this property is a "regularization" of the effect of each window at the non zero frequencies.
– The effect of the normalization of each eigenvector by the square root of its associated eigenvalue is fundamental to obtain a function $\Lambda_{CTLS}(\omega)$ as flat as possible for the non zero frequencies.
– The squared modulus in (8) is fundamental for the same reason. However, it introduces a non linearity in the method that leads to the impossibility of the application of this scheme in the case where more than one cisoid is present in the signal.
– The overall process can be viewed as the construction of a filter rejecting the frequency present in the signal.
– If the effect of truncation on the observation of the cisoid is taken into account in the expression of its Fourier transform $\mathcal{F}(\mathbf{x})$, and if the noise spectrum is assumed flat, equations leading to a similar interpretation of the method are obtained.
– In the real case, the main difference is the non flatness of the noise spectrum. The windows w_i^{CTLS} are designed to minimize the influence of this effect according to a particular criterium.

It is clear that windows associated to other criteria can be used. From the above remarks, the principal conditions they must verify is to be orthogonal and to have a cumulative spectrum as constant as possible. We can mention for example a bank of exponential kernels with $\Delta\omega/\omega$ constant.

4. Maximum likelihood estimation windows

Using [1] and developping the same calculus as in [4], it can be shown that the MLE estimator obeys the same scheme as the one above replacing the matrix P_{CTLS} by the matrix P_{MLE} defined by:

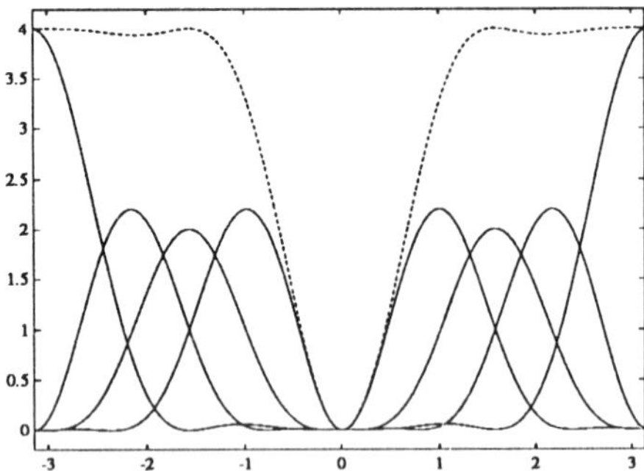

Fig. 1. Squared modulus of the Fourier transform of the windows w_i^{CTLS}, $i = 1, \ldots, 4$ (solid lines) and their sum (dashed lines).

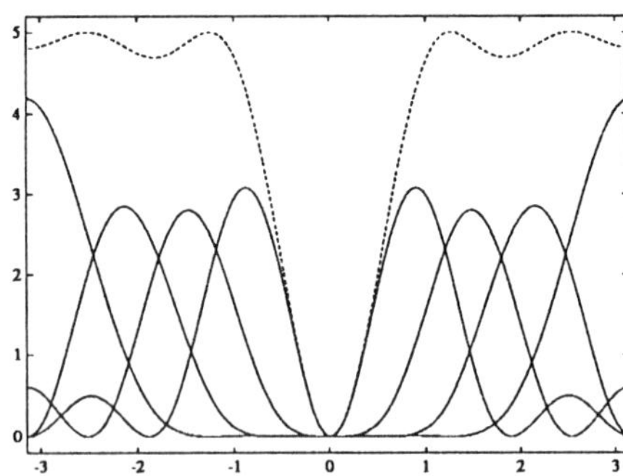

Fig. 2. Squared modulus of the Fourier transform of the windows w_i^{MLE}, $i = 1, \ldots, 4$ (solid lines) and their sum (dashed lines).

$$P_{MLE} = \begin{pmatrix} 2 & -1 & 0 & \ldots & 0 \\ -1 & 2 & \ddots & \ddots & \vdots \\ 0 & \ddots & \ddots & \ddots & 0 \\ \vdots & \ddots & \ddots & 2 & -1 \\ 0 & \ldots & 0 & -1 & 2 \end{pmatrix} \tag{14}$$

According to the Appendix, the same reasoning as the one derived above can be developped for this estimator. The figure (2) illustrates this result.

Appendix A

The eigendecomposition of the $N-1 \times N-1$ matrices P_{CTLS} and P_{MLE} can be obtained analiticaly. The eigenvalues of P_{CTLS} are given by:

$$\sigma_i^{CTLS} = 4.\sin^2\frac{i.\pi}{2.(N-1)} \tag{A.1}$$

and the eigenvalues of P_{MLE} by:

$$\sigma_i^{MLE} = 4.\sin^2\frac{i.\pi}{2.N} \tag{A.2}$$

The coordinate k of the eigenvectors of P_{CTLS} are given by:

$$v_i^{CTLS}(k) = \sqrt{\frac{c_i}{N-1}}.\sin\frac{(2k-1).i.\pi}{N-1} \tag{A.3}$$

where $c_i = 2$ for $i < N-1$ and $c_{N-1} = 1$, and the coordinate k of the eigenvectors of P_{MLE} by:

$$v_i^{MLE}(k) = \sqrt{\frac{2}{N}}.\sin\frac{k.i.\pi}{N} \tag{A.4}$$

References

[1] T. J. Abatzoglou and J. M. Mendel. Constrained Total Least Squares. In *International Conference on Acoustics, Speech and Signal Processing*, pages 1485–1488, 1987.

[2] T. J. Abatzoglou, J. M. Mendel, and G. A. Harada. The Constrained Total Least Squares Techniques and its Applications to Harmonic Superresolution. *IEEE Transactions on Acoustics, Speech and Signal Processing*, 39(5):1070–1087, May 1991.

[3] Y. T. Chan and R. Israelson. Estimation of the Parameters of a Sinusoid and its Detection. In *International Conference on Acoustics, Speech and Signal Processing*, pages 113–117, 1983.

[4] A. Ferrari, J. Ménez, and G. Alengrin. A single frequency efficient estimator. *Electronics letters*, 27(21):1939–1941, October 1991.

[5] F. J. Harris. On the Use of Windows for Harmonic Analysis with the Discrete Fourier Transform. *proceedings of the IEEE*, 66(1):12–83, January 1978.

SIGNAL PROCESSING VI: Theories and Applications
J. Vandewalle, R. Boite, M. Moonen, A. Oosterlinck (eds.)

SEPARATION OF A MIXTURE OF INDEPENDENT SOURCES THROUGH A MAXIMUM LIKELIHOOD APPROACH

Dinh Tuan PHAM *, Philippe GARAT * and Christian JUTTEN **

*Laboratory LMC/IMAG, C.N.R.S., BP 53X, 38041 Grenoble Cedex, France
** Laboratory TIRF, INPG 44 Avenue Félix-Vialet 38031 Grenoble Cedex, France.

The problem of separation of a mixture of independent (non Gaussian) sources without a priori knowledge is approached from the maximum likelihood principle, leading to the uses of nonlinear functions of observations. These functions are unknown as they depend on the densities of sources and the problem is how to choose them. To this end we investigate the performance of the method in term of the chosen functions. We also derive a simple way to determine the optimal choice from the data when the above functions are restricted to given vector space. Finally an algorithm for constructing the estimator of the mixture weights is proposed and tested by simulation.

1. INTRODUCTION

This paper deals with the problem of separation of a mixture of independent sources without any precise assumption on the structure of them. This problem has important applications in signal processing (e.g. speech analysis, radar and sonar processing, ...). Many methods have been proposed in the literature ([1], [2], [5]...), most of them are based on the use of higher order moments. Our approach is based on the maximum likelihood principle, but we do not adopt a model of sources. Instead of higher order moments, we are led to consider other statistics which are nonlinear functions of the observations (such functions arize naturally as we work with non Gaussian signals). Since no model of source is assumed, such seperating function have to be chosen arbitrarily. However, from the performance study of the estimator, we can derive a method for choosing them optimally in a given vector space of functions. Further, we propose an adaptive method in order to follow changes of the mixture weights. A similar algorithm, derived in an *ad-hoc* manner, has appeared in [4]. The maximum likelihood principle is also used in [3], but the source density is approximated by a Gram-Charlier expansion based on cumulants up to fourth order, leading to a different solution.

2. FORMULATION OF THE PROBLEM AND THE PROPOSED PROCEDURE

The problem can be formalized as follows: one observes T random vectors $\mathbf{X}(t)$, $t = 1, \ldots, T$, of dimension K', each component $X_i(t)$ of $\mathbf{X}(t)$ being a linear combination of K independent sources, i.e. $\mathbf{X}(t) = \mathbf{AS}(t)$ where $\mathbf{A}$ is some unknown matrix and $\mathbf{S}(t)$, $t = 1, \ldots, T$, are random K-vectors with independent components $S_1(t), \ldots, S_K(t)$. The aim is to reconstruct the "sources" $S_i(t)$ (up to a permutation and a scale factor) from the observations $\mathbf{X}(1), \ldots, \mathbf{X}(T)$. To simplify, we suppose here that the number of sources K is the same as the number of sensors K'. Suppose further that the $\mathbf{S}(t)$ at different times t are independently identically distributed (this independence assumption can be relaxed). Then, in the case where the probability density distribution of the i-th source $S_i(t)$ is known up to a scale factor, i.e. of the form $f_i(s/\sigma_i)/\sigma_i$ with known f_i, the log likelihood of the data can be written as:

$$L_T = T\{\sum_{i=1}^{K} \hat{\mathrm{E}} \ln[f_i(\mathbf{e}_i^T\mathbf{A}^{-1}\mathbf{X}/\sigma_i)/\sigma_i] - \ln|\det \mathbf{A}|\},$$

where $\hat{\mathrm{E}}$ denotes the time averaging operator: $\hat{\mathrm{E}}\, g(\mathbf{X}) = \{g[\mathbf{X}(1)] + \ldots + g[\mathbf{X}(T)]\}/T$, $\mathbf{e}_i$ denotes the i-th column of the identity matrix (of order K) and ^{T} denotes the transpose. The above function is to be maximized with respect to $\mathbf{A}$ and $\sigma_1, \ldots, \sigma_K$. To this end, one may equate its partial derivatives to zero. But is is more convenient to work with its differential instead. From $d\mathbf{A}^{-1} = -\mathbf{A}^{-1}d\mathbf{A}\mathbf{A}^{-1}$, $d \ln|\det \mathbf{A}| = \mathrm{tr}(\mathbf{A}^{-1}d\mathbf{A})$, (tr denoting the trace), the differential of $T^{-1}L_T$ can be seen to be

$$\sum_{i=1}^{K} \hat{\mathrm{E}}[\psi_i(\mathbf{e}_i^T\mathbf{A}^{-1}\mathbf{X}/\sigma_i)\mathbf{e}_i^T\mathbf{A}^{-1}d\mathbf{A}\mathbf{A}^{-1}\mathbf{X}/\sigma_i - \mathrm{tr}(\mathbf{A}^{-1}d\mathbf{A}) +$$

$$+\sum_{i=1}^{K} \{\hat{\mathrm{E}}[\psi_i(\mathbf{e}_i^T\mathbf{A}^{-1}\mathbf{X}/\sigma_i)\mathbf{e}_i^T\mathbf{A}^{-1}\mathbf{X}/\sigma_i^2] - 1/\sigma_i\}d\sigma_i,$$

where $\psi_i = -(\ln f_i)'$, ' denoting derivative. Note that the first two terms of the above differential can be written as

$$\sum_{i=1}^{K}\sum_{j=1}^{K} \widehat{E}[\psi_i(e_i^T A^{-1}X)/\sigma_i)e_j^T A^{-1}X/\sigma_i]\partial_{ij} - \sum_{i=1}^{K} \partial_{ii},$$

where ∂_{ij} denotes the general element of $A^{-1}dA$. At the maximum likelihood estimators $\widehat{A}$, $\widehat{\sigma}_i$ of A, σ_i, the above diferential must be zero for all infinitesimal increments $d\sigma_i$ and ∂_{ij}, yielding the estimating equations:

$$\widehat{\sigma}_i = \widehat{E}[\psi_i(e_i^T\widehat{A}^{-1}X/\widehat{\sigma}_i)e_i^T\widehat{A}^{-1}X], \qquad i = 1, ..., K,$$

$$\widehat{E}[\psi_i(e_i^T\widehat{A}^{-1}X/\widehat{\sigma}_i)e_j^T\widehat{A}^{-1}X] = 0, \qquad i \neq j = 1, ..., K.$$

In practice, one does not know f_i, but it can be seen that the above second system of equations still define a consistent estimator of A, when one uses *arbitrary* functions ψ_i, provided that the sources are *independent* and of *zero mean*. Note that the true sources satisfy $\widehat{E}[\psi_i(S_i/\sigma_i)S_j] \to 0$ as $T \to \infty$ by ergodicity. Of course the choice of ψ_i can greatly affect the performance of the method and this question will be studied later. By absorbing the scale factor σ_i in ψ_i, our separation procedure simply consists in solving (with respect to $\widehat{A}$)

$$\text{(2.1)} \qquad \widehat{E}[\psi_i(e_i^T\widehat{A}^{-1}X)e_i^T\widehat{A}^{-1}X] = 0, \qquad i \neq j = 1, ..., K.$$

3. ASYMPTOTIC PROPERTIES OF THE ESTIMATOR

To derive the asymptotic properties of the estimator, we shall assume that $\widehat{A}$ is close to the *true* mixture matrix, now denoted by A, so that one can make a Taylor expansion of (2.1) around A (it is assumed here that A is scaled in the same way as $\widehat{A}$ and its column may be permuted). Thus, one can write $\widehat{A} = A(I + \delta)$ where δ is a small matrix. The right hand side of (2.1) can then be written approximatively (for large T) as

$$\widehat{E}[\psi_i(S_i)S_j] - \sum_{k=1}^{K} \{\widehat{E}[\psi_i'(S_i)S_jS_k]\delta_{ik} + \widehat{E}[\psi_i(S_i)S_k]\delta_{jk}\}$$

where $S_i = e_i^T A^{-1}X$ and δ_{ij} denotes the general element of δ. We now appproximate $\widehat{E}[\psi_i'(S_i)S_jS_k]$ and $\widehat{E}[\psi_i(S_i)S_k]$ by the expectation of $\psi_i'(S_i)S_jS_k$ and $\psi_i(S_i)S_k$, which vanish unless $k = j$ or $k = i$, respectively. Thus, for $i \neq j = 1, ..., K$,

$$E[\psi_i'(S_i)S_j^2]\delta_{ij} + E[\psi_i(S_i)S_i]\delta_{ji} \approx \widehat{E}[\psi_i(S_i)S_j],$$

where E is the expectation operator. Write δ in the form of a K(K−1)-vector $\Delta = (\delta_{12}\ \ \delta_{21}\ \ ...)^T$ and let $\Psi = (\Psi_{12}\ \ \Psi_{21}\ \ ...)^T$ where $\Psi_{ij} = \psi_i(S_i)S_j$, one gets: $\Delta \approx H^{-1}\widehat{E}(\Psi)$, H being the block diagonal matrix with blocks

$$H_{(ij)} = \begin{bmatrix} E[\psi_i'(S_i)]E(S_j^2) & E[\psi_i(S_i)S_i] \\ E[\psi_j(S_j)S_j] & E[\psi_j'(S_j)]E(S_i^2) \end{bmatrix}$$

indexed by the pair (i, j), $i \neq j = 1, ..., K$. By the central limit Theorem, $\widehat{E}(\Psi)$ is asymptotically Gaussian with mean zero and covariance matrix G/T where $G = E(\Psi\Psi^T)$, *if the* $S(t)$ *are temporally independent* (this assumption is made to simplify the computation). Thus, Δ is asymptotically Gaussian with covariance matrix $H^{-1}G(H^{-1})^T/T$. Note that G is also block diagonal provided that $E[\psi_i(S_i)] = 0$ (which is satisfied e.g. if S_i has a symmetric distribution and ψ_i is an odd function). In this case, the random vectors $(\delta_{ij}\ \ \delta_{ji})^T$, for different pairs (i, j), are asymptotically independent with covariance matrix $H_{(ij)}^{-1}G_{(ij)}(H_{(ij)}^{-1})^T/T$ where

$$G_{(ij)} = \begin{bmatrix} E[\psi_i^2(S_i)]E(S_j^2) & E[\psi_i(S_i)S_i]E[\psi_j(S_j)S_j] \\ E[\psi_i(S_i)S_i]E[\psi_j(S_j)S_j] & E[\psi_j^2(S_j)]E(S_i^2) \end{bmatrix}.$$

We now show that the above covariance matrix is minimum when $\psi_i = \psi_i^* = -(\log f_i^*)'$, f_i^* being the *true* density of the i-th source. We first note that by integration by parts, one has, for any differentiable function g:

$$\text{(3.1)} \qquad E[g(S_i)\psi_i^*(S_i)] = E[g'(S_i)],$$

in particular $E[\psi_i^*(S_i)S_i] = 1$ and $E[\psi_i^*(S_i)] = 0$. Thus $H = E(\Psi\Psi^{*T})$ where Ψ^* is defined similarly to Ψ with ψ_i^* in place of ψ_i. Now, the partitionned matrix

$$E\left(\begin{bmatrix} \Psi \\ \Psi^* \end{bmatrix}[\Psi^T\ \ \Psi^{*T}]\right) = \begin{bmatrix} G & H \\ H^T & J \end{bmatrix}$$

where $J = E[\Psi^*\Psi^{*T}]$, is non negative, yielding $G - HJ^{-1}H^T \geq 0$. Thus $H^{-1}G(H^{-1})^T/T \geq J^{-1}/T$, and it can be easily seen that equality is attained when ψ_i is proportional to ψ_i^*. Since $E[\psi_i^*(S_i)] = 0$ by (3.1), the matrix J is also block diagonal as H, meaning that there is a *decoupling* of the separation into pairs of channels.

4. CONTRUCTION OF THE ESTIMATOR

An iterative algorithm for finding the solution of (2.1), based on the Newton method is now given. By expanding the equation (2.1) around an initial estimate $\overline{A}$ of A, using the same computation as in section 3, the next step estimate, denoted by $\overline{A}(I + \widetilde{\delta})$, can be constructed as the solution of

$$\sum_{k=1}^{K} \{\widehat{\mathrm{E}}[\psi_i'(\widetilde{S}_i)\widetilde{S}_j\widetilde{S}_k]\widetilde{\delta}_{ik} + \widehat{\mathrm{E}}[\psi_i(\widetilde{S}_i)\widetilde{S}_k]\widetilde{\delta}_{jk}\} = \widehat{\mathrm{E}}[\psi_i(\widetilde{S}_i)\widetilde{S}_j]$$

where $\widetilde{S}_i = \mathbf{e}_i^T\widetilde{\mathbf{A}}^{-1}\mathbf{X}$ and $\widetilde{\delta}_{ij}$ denote the general element of $\widetilde{\delta}$. As before, $\widehat{\mathrm{E}}[\psi_i'(\widetilde{S}_i)\widetilde{S}_j\widetilde{S}_k] \approx \mathrm{E}[\psi_i'(S_i)S_jS_k] = 0$ if $k \neq j$ and $\widehat{\mathrm{E}}[\psi_i(\widetilde{S}_i)\widetilde{S}_k] \approx \mathrm{E}[\psi_i(S_i)S_k] = 0$ if $k \neq i$. Thus the above equation may be replaced by

$$\widehat{\mathrm{E}}[\psi_i'(\widetilde{S}_i)\widetilde{S}_j^2]\widetilde{\delta}_{ij} + \widehat{\mathrm{E}}[\psi_i(\widetilde{S}_i)\widetilde{S}_i]\widetilde{\delta}_{ji} = \widehat{\mathrm{E}}[\psi_i(\widetilde{S}_i)\widetilde{S}_j],$$

$i \neq j = 1, \ldots, K$. These equations alows to calculate $\widetilde{\delta}$ and one can iterate the procedure until convergence. Since the sources are defined up to a scale factor, it is recommended to rescale each column of the estimate at each step, according to a chosen rule.

The above algorithm can be easily modified for use in an adaptative context, in which the estimate is updated each time a new observation is available. Let $\widehat{\mathbf{A}}(t-1)$ be an estimate of $\mathbf{A}$ at time $t-1$. The application of the above iteration leads to a new estimate at time t: $\widehat{\mathbf{A}}(t) = \widehat{\mathbf{A}}(t-1)[\mathbf{I} + \widehat{\delta}(t)]$ where $\widehat{\delta}(t)$ is the matrix with general element $\widehat{\delta}_{ij}(t)$, solution of
4

$$\{\sum_{s=1}^{t} \psi_i'[\widehat{S}_i(s)]\widehat{S}_j^2(s)\}\widehat{\delta}_{ij}(t) + \{\sum_{s=1}^{t} \psi_i[\widehat{S}_i(s)]\widehat{S}_i(s)\}\widehat{\delta}_{ji}(t) = \sum_{s=1}^{t} \psi_i[\widehat{S}_i(s)]\widehat{S}_j(s), \quad i \neq j = 1, \ldots, K.$$

Note that $\widehat{S}_i$ is now computed from $\widehat{\mathbf{A}}(t-1)$ and hence $\sum_{s=1}^{t-1} \psi_i[\widehat{S}_i(s)]\widehat{S}_j(s)$ is close to 0. Thus we simply replace above the right hand side by $\psi_i[\widehat{S}_i(t)]\widehat{S}_j(t)$, yielding the algorithm ($i = 1, \ldots, K, j \neq i = 1, \ldots, K$)

$$\widehat{S}_i(t) = \mathbf{e}_i^T\widehat{\mathbf{A}}(t-1)^{-1}\mathbf{X}(t),$$
$$\Omega_{ii}(t) = \lambda\Omega_{ii}(t-1) + \psi_i[\widehat{S}_i(t)]\widehat{S}_i(t)$$
$$\Omega_{ij}(t) = \lambda\Omega_{ij}(t-1) + \psi_i'[\widehat{S}_i(t)]\widehat{S}_j^2(t)$$
$$\Omega_{ij}(t)\widehat{\delta}_{ij}(t) + \Omega_{ii}(t)\widehat{\delta}_{ji}(t) = \psi_i[\widehat{S}_i(t)]\widehat{S}_j(t),$$

where $\lambda \in]0, 1]$ is a forgetting factor introduced to track the possible evolution of $\mathbf{A}$.

5. CHOICE OF SEPARATING FUNCTIONS

Since the optimal choice ψ_i^* for ψ_i depends on the unkown source distribution and is not easily estimated, we shall restrict the choice of ψ_i to a vector space of functions spanned by a basis $\phi_1, \ldots, \phi_N$, say, *with* $\phi_1(s) = s$. Thus $\psi_i = \sum_{n=1}^{N} c_{in}\,\phi_n$ for some constants c_{in}. The goal is to minimize the covariance matrix of the estimator, or equivalently to maximize $\mathbf{H}_{(ij)}^T\mathbf{G}_{(ij)}^{-1}\mathbf{H}_{(ij)}$ where $\mathbf{H}_{(ij)}$ and $\mathbf{G}_{(ij)}$ are defined in section 3. Note that $\mathbf{H}_{(ij)} = \mathrm{E}[\Psi_{(ij)}\Psi_{(ij)}^{*T}]$, $\mathbf{G}_{(ij)} = \mathrm{E}[\Psi_{(ij)}\Psi_{(ij)}^T]$ where $\Psi_{(ij)} = [\psi_i(S_i)S_j \;\; \psi_j(S_j)S_i]^T$ and $\Psi_{(ij)}^*$ is defined similarly, with ψ_i^* in place of ψ_i. We show here that the optimal choice of ψ_i is given by the projection of ψ_i^* onto the vector space spanned by the ϕ_n, i.e. by taking c_{in} equal to c_{in}^*, the solution of

$$(5.1) \qquad \sum_{n=1}^{N} c_{in}^*\,\mathrm{E}[\phi_n(S_i)\phi_m(S_i)] = \mathrm{E}[\psi_i^*(S_i)\phi_m(S_i)], \qquad m = 1, \ldots, N.$$

We note that $\Psi_{(ij)}$ may be written as $\mathbf{C}_{(ij)}\Phi_{(ij)}$ where

$$\mathbf{C}_{(ij)} = \begin{bmatrix} c_{i1} & c_{i2}\cdots & c_{iN} & 0\cdots & 0 \\ c_{j1} & 0\cdots & 0 & c_{j2}\cdots & c_{jN} \end{bmatrix}.$$

and $\Phi_{(ij)}$ is the vector with components $\phi_1(S_i)S_j, \ldots, \phi_N(S_i)S_j, \ldots, \phi_2(S_j)S_i, \ldots, \phi_N(S_j)S_i$. Let $\widehat{\Psi}_{(ij)}^*$ be the projection of $\Psi_{(ij)}^*$ on the $\Phi_{(ij)}$. Then

$$\mathrm{E}\left(\begin{bmatrix} \Psi_{(ij)} \\ \widehat{\Psi}_{(ij)}^* \end{bmatrix} [\Psi_{(ij)}^T \; \widehat{\Psi}_{(ij)}^{*T}]\right) = \begin{bmatrix} \mathbf{G}_{(ij)} & \mathbf{H}_{(ij)} \\ \mathbf{H}_{(ij)}^T & \mathrm{E}[\widehat{\Psi}_{(ij)}^*\widehat{\Psi}_{(ij)}^{*T}] \end{bmatrix}.$$

and from the non negativity of this matrix, $\mathbf{H}_{(ij)}^T\mathbf{G}_{(ij)}^{-1}\mathbf{H}_{(ij)} \leq \mathrm{E}[\widehat{\Psi}_{(ij)}^*\widehat{\Psi}_{(ij)}^{*T}]$ with equality if $\Psi_{(ij)}^* = \widehat{\Psi}_{(ij)}^*$. Thus, it suffices to show that for $\mathbf{C}_{(ij)}^*$ of the same form as $\mathbf{C}_{(ij)}$ with c_{in}^* in place of c_{in}, $\mathbf{C}_{(ij)}^*\mathrm{E}[\Phi_{(ij)}\Phi_{(ij)}^T] = \mathrm{E}[\Psi_{(ij)}^*\Phi_{(ij)}^T]$, since this implies $\widehat{\Psi}_{(ij)}^* = \mathbf{C}_{(ij)}^*\Phi_{(ij)}$ and hence by chosing $\mathbf{C}_{(ij)} = \mathbf{C}_{(ij)}^*$, optimality is achieved. Now, the first row of $\mathbf{C}_{(ij)}^*\mathrm{E}[\Phi_{(ij)}\Phi_{(ij)}^T]$ contains the elements

$$\sum_{n=1}^{N} c_{in}^*\,\mathrm{E}[\phi_n(S_i)S_j^2\phi_m(S_i)] \text{ and } \sum_{n=1}^{N} c_{in}^*\,\mathrm{E}[\phi_n(S_i)S_j\phi_m(S_j)S_i],$$

$m = 1, \ldots, N$ (they are repeated if $m = 1$), which by (5.1) are equal to $\mathrm{E}[\psi_i^*(S_i)\phi_m(S_i)]\mathrm{E}(S_j^2)$ and $\mathrm{E}[\psi_i^*(S_i)S_i]\mathrm{E}[\phi_m(S_j)S_j]$. Hence the first row of $\mathbf{C}_{(ij)}^*\mathrm{E}[\Phi_{(ij)}\Phi_{(ij)}^T]$ equals that of $\mathrm{E}[\Psi_{(ij)}^*\Phi_{(ij)}^T]$. The equality between the second rows of these matrices can be proved in the same way.

The important point is that one can easily estimate the c_{in}^*. Indeed, by (3.1), the right hand sides of (5.1) can be replaced by $\mathrm{E}[\phi_m'(S_i)]$. Then, replacing the expectation E operator by the time averaging operator $\widehat{\mathrm{E}}$ one gets a system of equations for estimating the coefficients c_{in}^*.

6. SOME EXPERIMENTAL STUDIES

Figure 1 illustrates the results for one typical simulation experiment testing the iterative (non adaptive) method of section 4. Two sources from the same uniform distribution are mixed and our algorithm, with $\psi_i(s) = s^3$, is applied on 256 sample points. We normalise the rows of $\mathbf{A}^{-1}$ to have unit Euclidian norm, and represent its off diagonal elements as coordinates of a point in the graph. The two solid curves correspond to the two equations (2.1). Their intersections yield $\widehat{\mathbf{A}}^{-1}$, which unfortunately, is not unique. Actually, ignoring permutation, there is a value of $\widehat{\mathbf{A}}^{-1}$, lying quite close to $\mathbf{A}^{-1}$ (+ in the graph) and another spurious one. The iteration may converge to the "right" (1st and 2nd iterations) or to the "wrong" solution (3rd iteration), depending on the initial value (it converges well and fast unless one starts at a point on the circle for which the corresponding matrix is singular). To resolve the above ambiguity, one may compute some index of dependence (e.g. sum of squares of the correlations) between the reconstructed sources and retain the solution for which they are least dependent. Also, by starting at some good initial estimate, one is more likely to attain the "right" solution.

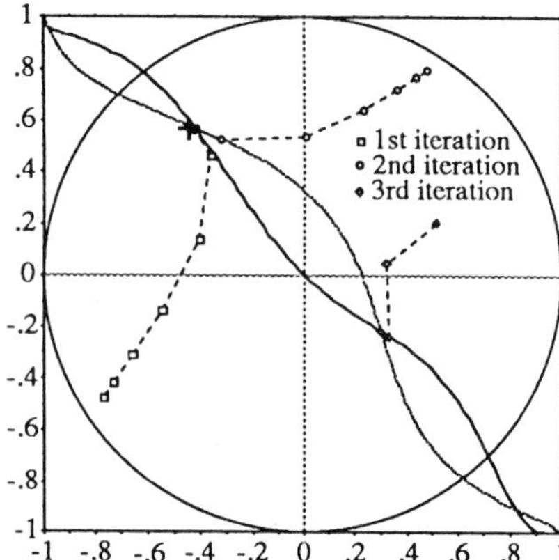

Figure 1: Examples of iterative computation of the estimator.

To study the effect of the functions ψ_i, we consider the case where the sources have a same distribution from the family of densities $f_p(s) = K_p\exp(-|s|^p/p)$, p being a shape parameter and K_p being the normalization constant. We take ψ_i to be a same function ψ. The asymptotic variance is smallest when $\psi = \varphi_p$ where $\varphi_p(s) = \text{sign}(s)|s|^{p-1}$. The ratios of this variance over the actual asymptotic variances of the estimator are called efficiencies and plotted in figure 2. As choices of ψ, the functions φ_4, $\varphi_{1.5}$, $\varphi_4 + c\varphi_{1.5}$ (with c chosen by the method of projection of section 5 and optimally by trial) and $\varphi_4 + c_1\varphi_2 + c_2\varphi_{1.5}$ (obtained by the method of projection, which is optimal) are considered. It is seen that the efficiency is quite good for φ_4 for $p > 2$ (unless p is very large) but poor for $p < 2$. The efficiency for $\varphi_{1.5}$, on the other hand, is good for $p < 2$ but poor for $p > 2$. By using the linear combination of φ_4 and $\varphi_{1.5}$ with optimal coefficient, good efficiency can be achieved overall, whereas the linear combination from the method of projection (not optimal since the function s is missing) can lead to catastrophic result for p near 2 (see figure 2). Note that the efficiency curves (except for $\varphi_4 + c_{\text{PRJ}}\varphi_{1.5}$) seem to be smooth at $p = 2$ even though the asymptotic variance diverges to infinity as $p \to 2$. This divergence is a consequence of the results of section 3 and means that Gaussian sources cannot be separated.

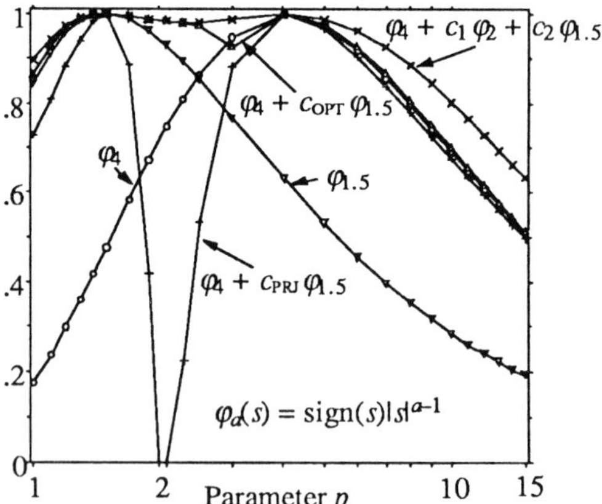

Figure 2: Asymptotic efficiencies of the estimator.

7. CONCLUSION

We have shown that the proposed method work well and yield highly efficient estimators, although some care needs to be taken with regard to the ambiguity problem.

REFERENCES

[1] COMON, P. Separation of sources using higher-order cumulants. Proc. SPIE: "Advanced Algorithms and Architecture for Signal Processing", San Diego, August 1989, 170-181.

[2] DUVAUT, P. Principle des méthodes de séparation de sources fondées sur les moments d'ordre supérieur. Traitement du Signal, Vol. 7 (1990), 5, 407-418.

[3] GAETA, M., LACOUME, J. L. Source separation without a priori knowledge: the maximum likelihood approach. In *Signal Processing* V, Proceedings of EUSIPCO 90, L.Tores, E. MasGrau and M. A. Lagunas eds. (1990), 621-624.

[4] JUTTEN, C. HERAULT, J. Blind separation of sources, Part I: an adaptative algorithm based on neuromimictic structure. Signal Processing, Vol. **24** (1991), 1-10.

[5] LACOUME, J. L., RUIZ, P. Sources identification: a solution based on the cumulants. IEEE Workshop on spectrum analysis and modelling. Mineapolis, 1988.

SIGNAL PROCESSING VI: Theories and Applications
J. Vandewalle, R. Boite, M. Moonen, A. Oosterlinck (eds.)

A FRAMEWORK FOR AUTOREGRESSIVE THEORY

P.M.T. Broersen
H.E. Wensink

Faculty of Applied Physics, Delft University of Technology
P.O. Box 5046, 2600 GA Delft, The Netherlands
phone: +31 (0)15-786419 fax: +31 (0)15-786081

Summary: Three levels of approximation can be distinguished in Autoregressive theory: Probability Limits, Asymptotic Theory and Finite Sample Theory. The Finite Sample Theory for autoregression takes the real data as a starting point, rather than relying on strict mathematical derivations. It provides formulae, which describe the behaviour of quantities like residual variance and prediction error in the practical situation of AR model estimation. Not only does the Finite Sample Theory lead to a more accurate order selection criterion, FSC, but it also evaluates the existing criteria, indicating ways of improvement. Some formulae are derived for the variance of reflection coefficients and parameters. They provide the necessary addition for an accurate and precise interpretation of the finite sample estimation results in AR model estimation.

1. INTRODUCTION

Suppose that an autoregressive series is generated by an AR(K) process:

$$x(n)+\sum_{i=1}^{K}\alpha(i)x(n-i)=\varepsilon(n), \tag{1}$$

where K is the process order. The driving process $\varepsilon(n)$ is gaussian white noise, with zero mean and variance σ_ε^2.

An AR model of order p can be fitted to a realization x(n), n=1,...,N. The entries a(i) of the parameter vector $\underline{a}_p$ are estimated and the order p has to be selected. Typical quantities in the estimation process are the residual variance $S^2(p)$ and the Prediction Error PE(p). The residual variance, which is defined as the fit of the estimated model to the series x(n) from which the parameters have been estimated, is as follows:

$$S^2(p)=\frac{1}{N-p}\sum_{n=p+1}^{N}\Big[x(n)+\sum_{i=1}^{p}a(i)x(n-i)\Big]^2. \tag{2}$$

The prediction error is defined as the average fit of the estimated model to an infinitely long realization y of the same stochastic process:

$$PE(p)=\overline{\Big[y(n)+\sum_{i=1}^{p}a(i)y(n-i)\Big]^2}. \tag{3}$$

This result is equal to the expectation of the one step ahead prediction error. The latter being defined as the error in predicting x(n+1) from a series, which is known up to x(n).

In general the search will be for the model with lowest *PE(p)*. The prediction error is not solely a measure for the forecasting capacities of the model. Small *PE(p)* indicates the model will provide the best parametric description of the data. Furthermore it can be shown via *NATFE(p)* that such a model has optimal spectral properties [1,2]. The definition in (3) refers to a theoretical quantity, which can only be computed in simulation experiments. In practice estimation from the data x(n) is unavoidable.

2. AR THEORY

Lately a theoretical framework for autoregressive estimation has been suggested by the authors [3]. Three levels of approximation to autoregressive theory are distinguished: probability limits, asymptotic theory and finite sample theory.

The following remarks concerning the table can be made:
- The probability limits were used by Mann and Wald [4] to apply results of classical linear regression to AR estimation. As such it constitutes the theoretical background to Autoregression. In the probability limits estimates assume their statistical expectations. The probability density functions of residuals and predictions become one and the same; $S^2(p)$ and $PE(p)$ converge to σ_ε^2 for $N\to\infty$. This accounts for the fact, that everything is known at this level. So process and model order are equal, all parameters being zero for p>K.
- At the asymptotic level the number of observations *N* is a given, finite, number. As to modeling no benefit can be taken from considerations with plim. To get an accurate description of the data, order selection becomes necessary. At this level an accuracy of order O(1/N) is needed, because essential differences between competing model orders are of this magnitude order [5,6,7]. This level is a source of confusion in literature, because several heuristic reasonings will fortuitously give some partially correct results [8]. Problems arise from the fact that, although stochastic, the autocovariance function of the data is taken to be exact. The influence of the fact that the autocovariance is actually stochastic is easily evaluated by means of a second order Taylor expansion [8]. More generally all bias terms of O(1/N) can be computed with this expansion.

For order selection purposes the Final Prediction Error, *FPE(p)*[5], has been developed. Selecting the model with lowest *FPE(p)*, it provides an asymptotically based estimate for the *PE(p)* of the estimated model. Later alternatives to *FPE(p)* were investigated, using concepts from information theory. Resulting criteria can together be described as a

Table: **Three levels of approximation to AR theory**

probability limits	asymptotic theory	finite sample theory
ML theory	Taylor approximation	real data
$N \to \infty$	O(1/N)	v(i,.)
$\underset{N\to\infty}{plim}\, S^2(p) = \sigma_\varepsilon^2 \quad p \geq K$	$\mathscr{E}[S^2(p)] = \sigma_\varepsilon^2 (1 - p/N) \quad p \geq K$	$\mathscr{E}[S^2(p)] = \sigma_\varepsilon^2 \prod_{i=0}^{p} [1 - v(i,.)] \quad p \geq K$
$\underset{N\to\infty}{plim}\, PE(p) = \sigma_\varepsilon^2 \quad p \geq K$	$\mathscr{E}[PE(p)] = \sigma_\varepsilon^2 (1 + p/N) \quad p \geq K$	$\mathscr{E}[PE(p)] = \sigma_\varepsilon^2 \prod_{i=0}^{p} [1 + v(i,.)] \quad p \geq K$
no order selection	$FPE(p) = S^2(p) \dfrac{N+p}{N-p}$	$FSC(p) = S^2(p) \prod_{i=0}^{p} \dfrac{1+v(i,.)}{1-v(i,.)}$
....	....	....
	$GIC(p,\alpha) = \ln[S^2(p)] + \alpha\, p/N$	$FIC(p,\alpha) = \ln[S^2(p)] + \alpha \sum_{i=0}^{p} v(i,.)$
Mann and Wald (1943)	*Akaike (1970, 1974)* *Bhansali (1986), etc.*	*Broersen (1985,1990)* *Broersen and Wensink (1991)*

Generalized Information Criterion *GIC(p,α)*. By substituting for α quantities like 2 [6], ln(N) [9,10] and 2lnln(N) [11], the *AIC(p)* and its consistent variants can be described.

- At finite sample level the characteristic behaviour of the data forms the starting point. The influence of the estimation method becomes important. It might be called a physical theory as it gives an accurate description of numerous simulation experiments. It is shown, that the variance of the last estimated parameter depends on the model order i currently under estimation. Moreover, each estimation method has its particular influence on the estimation results. This is the basis for the finite sample variance coefficients *v(i,.)*, which are defined as the variance of parameters above the true model order. They are plotted in Figure 1, for an AR(3) process with reflection coefficients $\kappa_1=0.8$, $\kappa_2=0.3$ and $\kappa_3= 0.5$. For their determination four estimation methods have been used: the Yule-Walker method (*YW*), the method of Burg (*Burg*), the least squares method minimizing both forward and backward residuals (*LSFB*) and the least squares method with one-sided residuals only (*LSF*). To the experimental results empirical approximations have been made, leading to the following expressions [1]:

$v(i,YW)=(N-i)/N(N+2)$
$v(i,Burg)=1/(N+1-i)$
$v(i,LSFB)=1/(N+1.5-1.5i)$
$v(i,LSF)=1/(N+2-2i)$

For all methods $v(0,.) = 0$. The *v(i,.)* are motivated rather by some simple theoretical background, than by the most optimal fit to the simulation results. With *v(i,LSF)* the asymptotical fact is reflected, that i parameters are estimated for N-i residuals, leaving N-2i degrees of freedom. The constant 2 in the denominator is used to obtain 1/N for first order, which is the best approximation. The same holds for *Burg* and *LSFB*. The coefficients for *Burg* reflect the N-i contributions to the estimated reflection coefficient $\hat{k}(i)$ in the lattice structure.

Thus the finite sample theory not only shows the deviation of practical results from the asymptotic variance 1/N, but it also explains the difference between the Yule-Walker method of estimation and all other methods: only in the Yule-Walker method the *v(i,.)* are smaller than 1/N.

Order selection is done making use of the Finite Sample Criterion for order selection, *FSC(p)*. The criteria based on information theory can be improved by means of finite sample concepts. This gives rise to the Finite sample Information Criterion *FIC(p,α)*, which is the finite sample improvement for *GIC(p,α)*, using the same α.

3. ORDER SELECTION

The Framework can now be used to tackle the problem of selecting a model order with low prediction error. Therefore first of all the *PE(p)* will be evaluated. This has been done in 500,000 Monte Carlo simulation runs with 30 observations each, generated by an AR(3) process. The results are plotted in Figure 2. The *PE(p)* is shown to have a minimum at the optimal model order. Moreover, the behaviour clearly depends on the model order and the on estimation method used for parameter estimation. It can be described for p≥K, because the Finite Sample Theory provides the expression for its statistical expectation:

$$\mathscr{E}[PE(p)] = \sigma_\varepsilon^2 \prod_{i=0}^{p} [1 + v(i,.)]. \tag{4}$$

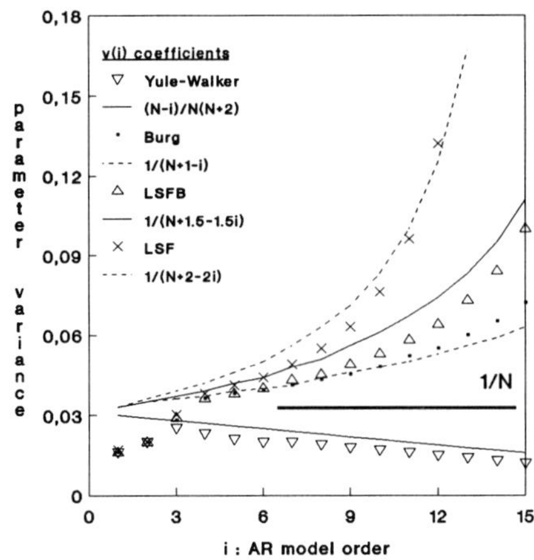

Fig.1 Variance of i-th parameter of AR(i) model from 30 AR(3) observations

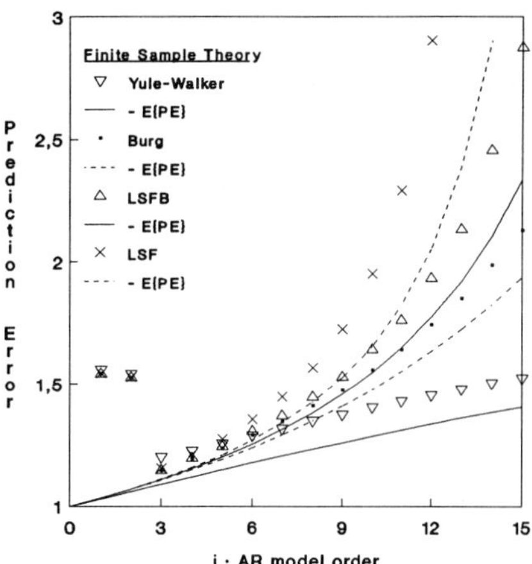

Fig.2 PE and E[PE] for different estimates from 30 AR(3) observations

In practical situations the *PE(p)* is not known. In the quest for the optimal model the *PE(p)* must be estimated from the same data that have been used to estimate the parameters. As a counterpart of the asymptotical *FPE(p)*, the Finite Sample Criterion *FSC(p)* has been developed for the finite sample case. Not only is it expected to select models with low *PE(p)*, but also does the criterion value provide an estimate of the actual prediction error of the finally estimated model.

Like Akaike's *FPE(p)*, the *FSC(p)* is based on the insight that each extra parameter above the best model order will increase *PE(p)* and decrease $S^2(p)$. The accuracy of *FSC(p)* as an estimator of the prediction error can be determined in simulation experiments, where an independent realization of the same process can be used to compute *PE(p)*. Figure 3 shows the average observed *PE(p)*, the theoretical expectation *E{PE(p)}* and estimates provided by the asymptotical *FPE(p)* and the finite sample *FSC(p)*. The results of *Burg* estimates are presented, but one-sided residuals *LSF* and least squares method with forward and backward residuals *LSFB* give similar results. The figure shows *FSC(p)* and *FPE(p)* to be quite alike for low model orders; the *v(i,.)* are still close to the asymptotical value of 1/N. The values of the two criteria separate for higher orders and *FPE(p)* will eventually have a wrong high order minimum.

4. VARIANCE OF PARAMETERS

The computation of the theoretical variance of finite sample parameters will now be treated. The covariance matrix $\boldsymbol{C}(\underline{a}_K)$ will be treated for the three levels of approximation to AR theory. For an accurate computation at finite sample level the bias of the parameters has to be taken into account. Therefore the Levinson recursion is used to compute the parameter variance from the finite sample covariance matrix of the reflection coefficients.

In the probability limit the variance of √N times the estimated parameters $\boldsymbol{C}^{plim}(\sqrt{N}\underline{a}_K)$ is given by [4]:

$$\boldsymbol{C}^{plim}(\sqrt{N}\underline{a}_K) = \sigma_\varepsilon^2\ \boldsymbol{R}^{-1}. \qquad (5)$$

The matrix $\boldsymbol{R}$ is the theoretical symmetric Toeplitz KxK auto-covariance matrix of the series x(n), which has expectation $\mathcal{E}\{x(n)x(n+i)\}$, for i=0,...,K-1.

At asymptotic level the variance of the parameters is given with accuracy of O(1/N):

$$\boldsymbol{C}^{ass}(\underline{a}_K) = \sigma_\varepsilon^2/N\ \boldsymbol{R}^{-1}. \qquad (6)$$

The asymptotic expression of the variance of estimated reflection coefficients k is given by [12]:

$$\boldsymbol{C}^{ass}(\underline{k}_K) = \sigma_\varepsilon^2/N\ \boldsymbol{D}\ \boldsymbol{R}^{-1}\ \boldsymbol{D}, \qquad (7)$$

where $\boldsymbol{D}$ is the matrix with entries $d(i,j) = \partial\kappa_i(\underline{a}_K)/\partial a_j$; κ_i denotes the theoretical reflection coefficient indexed i. This is the usual first-order Taylor approximation for non-linear functions of a stochastic variable.

At finite sample level the covariance matrix is defined as:

$$\boldsymbol{C}^{fst}(\underline{k}_K) = \sigma_\varepsilon^2\ \boldsymbol{V}\ \boldsymbol{D}\ \boldsymbol{R}^{-1}\ \boldsymbol{D}^T\ \boldsymbol{V}, \qquad (8)$$

where the matrix $\boldsymbol{V}$ is the diagonal matrix with elements $\sqrt{[v(i,.)]}$. For white noise $\boldsymbol{C}^{fst}(\underline{k}_K)$ becomes the diagonal matrix with elements *v(i,.)*, because $\boldsymbol{D} = \boldsymbol{I}$ and $\boldsymbol{R} = \sigma_\varepsilon^2\ \boldsymbol{I}$, where $\boldsymbol{I}$ is the diagonal unity matrix. For non-white processes, K>0, the matrix $\boldsymbol{C}^{fst}(\underline{k}_K)$ contains a KxK sub-matrix in the upper left corner with the covariance structure of the reflection coefficients of the AR(K) process. Its diagonal elements are 0.5298v(1,.), 0.5460v(2,.) and 0.7500v(3,.) in the example of Figure 1 (see Fig. 1).

The parameter variance is now build up applying essentially the Levinson recursion. The variable γ_p is introduced as $\gamma_p = \kappa_p + B(\kappa_p)$; the theoretical reflection coefficient corrected by its bias B. The finite sample bias is taken to be equal to the asymptotic bias, which follows from a second order Taylor expansion of k written as a function of $\hat{R}$. Thus is accounted for the bias and the recursion becomes as follows:

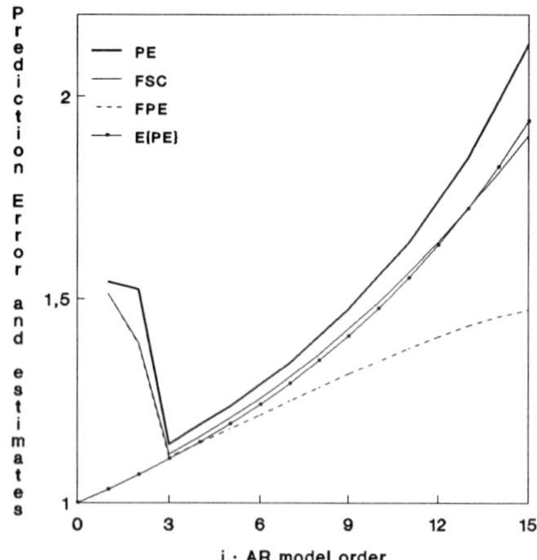

Fig.3 PE, FSC, FPE and E[PE] for Burg estimates from 30 AR(3) observations

$$a_p(i) = a_{p-1}(i) + a_{p-1}(p-i)\ \gamma_p, \tag{9}$$

where $a_p(i)$ denotes the theoretical construction of an estimated parameter i in a model of order p, for a given estimation method.
Defining, that

$$\mathrm{var}\{\gamma_p\} = \mathrm{var}\{\kappa_p\},$$

the recursion starts with

$$a_1(1) = \kappa_1 + B(\kappa_1)$$

and

$$\mathrm{cov}\{a_1(1), k_{p+j}\} = \mathrm{cov}\{k_1, k_{p+j}\}.$$

which is calculated using (8).

The covariance of parameters and reflection coefficients is for $p \geq 2$ and $i<p$ calculated as:

$$cov\{a_p(i),k_{p+j}\} = cov\{a_{p-1}(i),k_{p+j}\} + \gamma_p\ cov\{a_{p-1}(p-i),k_{p+j}\} + a_{p-1}(p-i)\ cov\{k_p,k_{p+j}\} \tag{10}$$

and using,

$$\mathrm{cov}\{a_p(p), k_{p+j}\} = \mathrm{cov}\{k_p, k_{p+j}\},\ \text{for } 0 \leq j \leq K\text{-}p.$$

The entries of the finite sample covariance matrix of the parameters $C^{fst}(\underline{a}_p)$ can now be found, for all orders p and $i,j<p$, with the following recursive expression:

$$\begin{aligned} cov\{a_p(i),a_p(j)\} &= cov\{a_{p-1}(i),a_{p-1}(j)\} + \\ &\gamma_p\ cov\{a_{p-1}(i),\ a_{p-1}(p-j)\} + a_{p-1}(p-j)\ cov\{a_{p-1}(i),k_p\} + \\ &\gamma_p\ cov\{a_{p-1}(p-i),\ a_{p-1}(j)\} + a_{p-1}(p-i)\ cov\{k_p,a_{p-1}(j)\} + \gamma_p^2 \\ &cov\{a_{p-1}(p-i),a_{p-1}(p-j)\} + \gamma_p\ a_{p-1}(p-j)\ cov\{k_p,a_{p-1}(p-i)\} + \\ &\gamma_p\ a_{p-1}(p-i)\ cov\{k_p,a_{p-1}(p-j)\} + a_{p-1}(p-i)\ a_{p-1}(p-j)\ var\{k_p\} + \\ &var\{k_p\}\ cov\{a_{p-1}(p-i),a_{p-1}(p-j)\} + \\ &cov\{k_p,a_{p-1}(p-j)\}\ cov\{k_p,a_{p-1}(p-i)\}, \end{aligned} \tag{11}$$

Simulation experiments have been carried out to evaluate the theoretical description provided by the formula. It turns out that expression (10) together with (11) provides an accurate description. For *LSFB* and *LSF* it rather sketches the general behaviour, still being much more accurate than the asymptotic results.

5. CONCLUSIONS

Three different levels of approximation to AR theory are described by the Framework for Autoregressive Theory. The levels differ from one another in theoretical description and practical applicability. The Finite Sample Theory converges to asymptotic theory and further to plim results when $v(i,.)$ is replaced by 1/N and $N \to \infty$. Its basis, however, does not result from theoretical derivations, but is founded on real data results. Therefore it may be called a physical theory. Its formula provide descriptions of the results, which actually occur in estimating AR models. Concluding it is stated, that the Finite Sample Theory gives immediate practical applicable results in AR model estimation from finite data sets.

REFERENCES

[1] Broersen, P.M.T., Selecting the Order of Autoregressive Models from Small Samples, IEEE Trans. on Acoust., Speech and Signal Proc., vol. ASSP-33 (1985) pp. 874-879.

[2] Parzen, E., Some Recent Advances in Time Series Modeling., IEEE Trans. on Automat. Contr., vol. AC-19, (1974) pp.723-730.

[3] Broersen, P.M.T. and H.E. Wensink., Finite Sample Theory for Autoregressive Model Estimation, Proc. 13-ième coll. GRETSI, (1991) pp. 297-300.

[4] Mann, H.B. and A. Wald., On the Statistical Treatment of Linear Stochastic Difference Equations, Econometrica, vol. 11, (1943) pp. 173-220.

[5] Akaike, H., Statistical Predictor Identification, Ann. Inst. Stat. Math., vol. 22, (1970) pp. 203-217.

[6] Akaike, H., A new Look at Statistical Model Identification, IEEE Trans. on Automat. Contr., vol AC-19, (1974) pp. 716-723.

[7] Bhansali, R.J., A Derivation of the Information Criteria for Selecting Autoregressive Models, Adv. Appl. Prob., vol.18, (1986) pp. 360-387.

[8] Broersen, P.M.T. and H.E. Wensink, On Asymptotic Autoregressive Theory, Offered for publication, (1992).

[9] Rissanen, J., Modelling by Shortest Data Description, Automatica, vol.14, (1978) pp. 465-471.

[10] Schwarz, G., Estimating the Dimension of a Model, The Annals of Statistics, vol. 6, (1978) pp.461-464.

[11] Hannan, E.J. and B.G. Quin., The Determination of the Order of an Autoregression, J. R. Stat. Soc. Ser. B, vol.B-41, (1979) pp. 190-195.

[12] Kay S. and J. Makhoul., On the Statistics of the Estimated Reflection Coefficients of an Autoregressive Process, IEEE Trans. on Acoust., Speech and Signal Proc., vol. ASSP-31, (1983) pp. 1447-1455.

A NEW IDENTIFICATION ALGORITHM FOR ALLPASS SYSTEMS BY HIGHER-ORDER STATISTICS

Chong-Yung Chi and Jung-Yuan Kung

Department of Electrical Engineering, National Tsing Hua University, Hsinchu, Taiwan, ROC.

Abstract: In this paper, based on higher-order statistics (HOS) , we propose a new allpass system identification algorithm with only non-Gaussian output measurements. Besides, not only the order determination but also the determination of whether the system of interest is an allpass system are also included in the proposed identification algorithm. It is applicable in channel equalization for the case of phase distortion channel [2]. It can be used to remove the remaining phase distortion of nonminimum-phase source wavelet in the predictive deconvolved data. On the other hand, the identification of an ARMA system H(z) can be converted into the identification of an allpass system [3,4] when minimum-phase (MP) - allpass (AP) decomposition based methods are used, and then it can be applied even when H(z) includes allpass factors whereas Chi-Kung's allpass system classification algorithm [5] is not applicable in this case. Some simulation results are provided to support that the proposed algorithm works well.

1. Introduction

Cumulant (higher–order statistics) based identification of nonminimum–phase linear time–invariant (LTI) systems with only non–Gaussian output measurements has drawn extensive attention in various signal processing applications. A general parametric model for the LTI system is known as autoregressive moving average (ARMA) model, denoted H(z). One can preprocess the output of H(z) by the inverse filter $1/H_{MP}(z)$ where $H_{MP}(z)$ is the spectrally equivalent minimum–phase system obtained by correlation based spectral estimators such that the identification of H(z) is equivalent to the identification of the allpass system $H_{AP}(z) = H(z)/H_{MP}(z)$ [3,4]. When H(z) does not include any allpass factors, the poles of each possible candidate $H_{AP}(z)$ are also zeros of $H_{MP}(z)$. Chi and Kung [5] proposed an allpass system classification algorithm by a single cumulant sample for determining the $H_{AP}(z)$ for the case that $H_{MP}(z)$ is given in advance. On the other hand, in some applications, the LTI system H(z) is known to be an allpass system without any prior information about pole locations such as the phase distortion channel (allpass system) in channel equalization [2] where the channel input is often a sequence of non–Gaussian M–ary symbols. Another interesting instance is that the source wavelet in the well–known predictive deconvolution [6] is assumed to be minimum–phase but it might be nonminimum–phase in practice. Therefore, the deconvolved results can be viewed as the output of a phase distortion channel whose input is a non–Gaussian sparse reflectivity sequence, and removing the remaining phase distortion of nonminimum–phase source wavelet is needed in order to improve the quality of the deconvolved results.

In this paper, we propose a new cumulant based allpass system identification algorithm without any prior knowledge about pole locations and order of the system.

2. The Allpass System Identification Algorithm

Assume that data x(k), k=0, 1, ..., N–1, are noisy output measurements of the allpass system h(k) to be identified as follows:

$$x(k) = u(k) * h(k) + w(k) \qquad (1)$$

under the following assumptions about h(k), u(k) and w(k).

(A1) The allpass system h(k) is a real causal and exponentially stable pth–order allpass system with the transfer function H(z)

$$H(z) = \frac{B(z)}{A(z)} \qquad (2)$$

where the minimum–phase A(z) is given by

$$A(z) = A_p(z) = 1 + a_1 z^{-1} + \cdots + a_p z^{-p} \qquad (3)$$

and B(z) (maximum–phase) is given by

$$B(z) = B_p(z) = A_p(z^{-1})\cdot z^{-p}. \qquad (4)$$

(A2) The input u(k) is real, zero–mean, independent identically distributed (i.i.d.), non–Gaussian with nth–order cumulant γ_n. Moreover, $|\gamma_n|<\infty$ for $2\leq n\leq 2M$.

(A3) The measurement noise w(k) is real Gaussian which can be white or colored with unknown statistics, and is statistically independent of u(k).

This work was supported by the National Science Council under Grant NSC81-0404-E-007-001 and the Telecommunication Laboratories under Grant TL-NSC-81-5201.

For ease of later use, let $\underline{k}=(k_1, \ldots, k_{M-1})$ and $S_{AP}(L)$ denote the set of all Lth-order anticausal stable allpass filters as follows:

$$\hat{H}_{AP}(z) = A_L(z) / B_L(z) \qquad (5)$$

where $A_L(z)$ and $B_L(z)$ have been defined in (3) and (4), respectively. Note that $|\hat{H}_{AP}(f)| = |\hat{H}_{AP}(z=e^{j2\pi f})| = 1$ for all f, and that $|H(f)| = 1$ for all f and $1/H(z) \in S_{AP}(p)$. The new identification algorithm is based on Theorem 1 and Fact 1 below.

Theorem 1. Assume that x(k) was generated from (1) under the assumptions (A1), (A2) and (A3). Let y(k) be the output of a pth-order allpass filter $\hat{H}_{AP}(z) \in S_{AP}(p)$ with the input x(k). Then the absolute Mth-order ($M \geq 3$) cumulant $|C_{M,y}(\underline{k}=\underline{0})|$ of y(k) is maximum if and only if $\hat{H}_{AP}(z)=1/H(z)$. Furthermore, $\max\{|C_{M,y}(\underline{k}=\underline{0})|\} = |\gamma_M|$.

Fact 1. Assume that x(k) was generated from (1) under the assumptions (A1), (A2) and (A3). Let x(k) be the input of an arbitrary allpass system with amplitude response equal to unity and y(k) be the corresponding output. Then

$$\sum_{k_1=-\infty}^{\infty} \cdots \sum_{k_{M-1}=-\infty}^{\infty} C_{M,y}^2(\underline{k}) = \sum_{k_1=-\infty}^{\infty} \cdots \sum_{k_{M-1}=-\infty}^{\infty} C_{M,x}^2(\underline{k}) = \gamma_M^2 . \qquad (6)$$

The new identification algorithm is basically to maximize the following objective function

$$J(L) = \hat{C}_{M,y}^2(\underline{k}=\underline{0}) \qquad (7)$$

where $\hat{C}_{M,y}(\underline{k}=\underline{0})$ is the sample cumulant of the output y(k) of an Lth-order allpass system $\hat{H}_{AP}(z) \in S_{AP}(L)$ with input x(k). Note that y(k) must be computed from x(k) recursively backwards since $\hat{H}_{AP}(z)$ is an anticausal stable IIR filter and that J(L) is a nonlinear function of the coefficients of $\hat{H}_{AP}(z)$. We, next, illuminate the identification procedure.

Parameter Estimation: (The order is known a priori.)

(s0) Set L=p.

(s1) Search for the $J_{max}(L)$ (maximum of J(L)) and associated $\hat{H}_{AP}(z) \in S_{AP}(L)$ by a numerical optimization algorithm such as Newton-Raphson type algorithm.

Thus the optimum $\hat{H}(z) = 1/\hat{H}_{AP}(z)$ and the optimum estimate $\hat{\gamma}_M = \hat{C}_{M,y}(\underline{k}=\underline{0})$.

Order Determination and Parameter Estimation:

(s2) Set L = 0 and compute $J_{max}(0) = \hat{C}_{M,x}^2(\underline{k}=\underline{0})$.

(s3) Set L = L + 1.

(s4) Execute **(s1)**.

(s5) If $[J_{max}(L) - J_{max}(L-1)]/J_{max}(L-1) < \xi$ where ξ is a preassigned positive constant, then compute the ratio

$$R(L) = J_{max}(L) / D(L) \qquad (8)$$

where

$$D(L) = \sum_{k_1=-\infty}^{\infty} \cdots \sum_{k_{M-1}=-\infty}^{\infty} \hat{C}_{M,y}^2(\underline{k}), \qquad (9)$$

otherwise go to **(s3)**.

If $R(L) \approx 1$, $H(z) = 1/\hat{H}_{AP}(z)$ and the optimum estimate $\hat{\gamma}_M = \hat{C}_{M,y}(\underline{k}=\underline{0})$; if $R(L) << 1$, $\hat{H}(z)$ is not an allpass system. Remark that we calculate D(L) in **(s5)** over F(q) for a preassigned value of q where

$$F(q) = \left\{ \begin{array}{l} \text{Domain of support of the} \\ \text{Mth-order cumulant function} \\ \text{of non-Gaussian MA(q) process} \end{array} \right\} \qquad (10)$$

The proposed algorithm works well due to the following characteristics:

(C1) The proposed algorithm is guaranteed for convergence no matter whether H(z) is an allpass system or not.

(C2) If H(z) is indeed an allpass system, R(L)=1 as well as $\hat{H}_{AP}(z) = 1/\hat{H}(z) \in S_{AP}(p)$ (by both Theorem 1 and Fact 1) for all $L \geq p$ as N is large. On the other hand, if $R(L) << 1$, one can infer that H(z) is not an allpass system.

(C3) With no doubt, various cumulant based least-squares (LS) methods such as [1] can be applied to estimating the AR parameters of H(z) by treating H(z) as a general ARMA model, while the MA parameters of H(z) are automatically determined (see (4)). The proposed algorithm outperforms cumulant based LS estimators simply due to more accurate model (allpass model) used by the former.

(C4) The proposed allpass system identification algorithm is a consistent estimator.

(C5) The proposed identification algorithm can be used for any $M \geq 3$ as long as $\gamma_M \neq 0$.

3. Simulation Results

Two simulation examples are to be presented to justify the good performance of the proposed allpass system identification algorithm. We calculated the quantity D(L) over F(q=15) (see (9) and (10)) in our simulation.

Example 1.

The driving input u(k) used was exponentially distributed with zero mean and skewness $\gamma_3=2$. Hence the cumulant order M=3 was used. A second–order allpass model (taken from [1]) was used for the case of w(k) being colored Gaussian noise which was generated from a highpass second–order FIR filter $B(z) = 1 - 1.2z^{-1} + 0.32z^{-2}$ driven by a zero–mean white Gaussian noise sequence. The total data used in each run was N=1024.

First of all, let us consider the case that the order of H(z) is known a priori. The simulation results are shown in Table 1, from which one can observe that AR parameter estimates show smaller bias and smaller standard deviation as SNR is larger, and that the values of bias and standard deviation for SNR=1 (a quite low SNR) are also small. These simulation results support the good performance of the proposed allpass system identification algorithm.

Next, let us consider the case that the order of H(z) is unknown. We only show the simulation results for SNR=10 since those for other SNR's are similar in every respect. Thirty $J_{max}(L)$'s and associated R(L)'s are shown in Fig. 1. In the top part of Fig. 1, different kinds of lines are used to make each single $J_{max}(L)$ discernible from other $J_{max}(L)$'s. The bottom part of Fig. 1 depicts thirty R(L)'s in the same manner. Note, from Fig. 1, that each $J_{max}(L)$ increases monotonically with L and that each $J_{max}(L)$ converges for all $L \geq p=2$ (true order), and that all R(L)'s are larger than 0.85 (close to unity). These simulation results also justify (C1) and support (C2).

Example 2.

Assume that h(k) is not an allpass system but a third–order nonminimum–phase source wavelet as follows:

$$H(z) = \frac{1 + 0.1z^{-1} - 3.2725z^{-2} + 1.41125z^{-3}}{1 - 1.9z^{-1} + 1.1525z^{-2} - 0.1625z^{-3}}$$

whose zeros are located at −2.0415, 1.4719 and 0.4696. The input u(k) was a sparse spike sequence with random amplitudes generated by use of the well–known Bernoulli–Gaussian (B–G) model [7] which has been used in seismic deconvolution and ultrasonic biomedical imaging. The cumulant order M=4 was used since $\gamma_4 \neq 0$ but $\gamma_3=0$ for this case.

The synthetic data x(k) for SNR=100 shown in Fig. 2(a) were generated for the case of w(k) being white Gaussian. We preprocessed x(k) by the minimum–phase prediction error filter (PEF) of order equal to 40, which was obtained from x(k) by Burg's algorithm [8], to get the deconvolved data e(k) which is shown in the top part of Fig. 2(b), from which, one can see that each spike in u(k) (solid line) is associated with a wavelet in e(k) (dotted line) which begins with two opposite peaks and gradually decays due to the remaining phase distortion of source wavelet. Then we processed e(k) with the proposed algorithm. The output y(k) (dotted line) of the optimum $\hat{H}_{AP}(z)$ is shown in the bottom part of Fig. 2(b), from which, one can see that y(k) approximates u(k) very well except for a scale factor. In other words, the phase distortion in e(k) has been considerably removed by the proposed algorithm. Hence, the previous simulation results justify that our allpass system identification algorithm can remove the remaining phase distortion of nonminimum–phase source wavelet and improve the quality of the predictive deconvolved results.

Table 1
Simulation results for parameter estimation associated with Example 1

$a_1 = -0.3$, $a_2 = -0.4$, $\gamma_3 = 2$, N=1024, 30 independent runs.			
	Estimated values (mean ± one standard deviation)		
SNR	$\hat{a}_1$	$\hat{a}_2$	$\hat{\gamma}_3$
∞	-0.3021±0.0147	-0.4001±0.0162	1.9892±0.0638
100	-0.3032±0.0182	-0.3974±0.0187	1.9765±0.0823
10	-0.3048±0.0246	-0.3952±0.0393	2.0287±0.1303
1	-0.3276±0.0715	-0.3723±0.0658	2.0474±0.2918

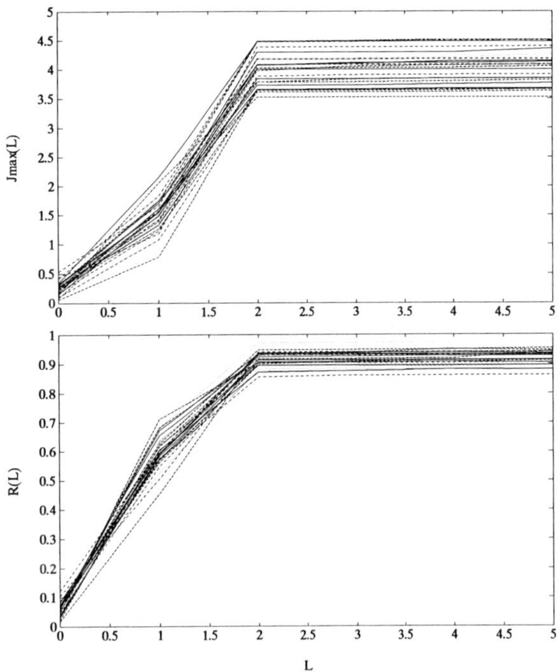

Fig. 1. Simulation results for order determination associated with Example 1. The top part includes thirty $J_{max}(L)$'s; the bottom includes the associated R(L)'s for SNR=10.

4. Conclusions

In this paper, we have presented a new cumulant based allpass system identification algorithm with only non–Gaussian output measurements. Besides, not only the order determination but also the determination of whether the system of interest is an allpass system are also included. The proposed algorithm possesses five nice characteristics described in (C1) through (C5) in Section 2. It is applicable for any LTI phase distortion systems such as phase–distortion communication channels and phase distortion of source wavelet in predictive deconvolved seismic signals. We also showed two simulation examples to support the proposed algorithm. As mentioned in Section 1, the identification of an ARMA(p,q) system H(z) can be converted into the identification of an allpass system when MP–AP decomposition based methods are used, and our allpass system identification algorithm can be applied no matter whether H(z) includes allpass factors, whereas Chi–Kung's [5] algorithm is not applicable when H(z) includes allpass factors.

References

[1] G. B. Giannakis and J. M. Mendel, "Identification of nonminimum phase systems using higher order statistics," IEEE Trans. Acoustics, Speech, and Signal Processing, vol. 37, pp. 360–377, March 1989.

[2] A. P. Clark, Equalizers for digital modems, Wiley, New York, 1985.

[3] G. B. Giannakis and J. M. Mendel, "Stochastic realization of non–minimum phase systems," Proc. 1986 Amer. Control Conf., vol. 2, pp. 1254–1259, Seattle, WA.

[4] G. B. Giannakis and J. M. Mendel, "Approximate realization and model reduction of nonminimum phase stochastic systems," Proc. 25th IEEE Conf. on Decision and Control, pp. 1079–1084, Athens, Greece, 1986.

[5] C.–Y. Chi and J.–Y. Kung, "A fast phase determination method by a single cumulant sample," Proc. International Signal Processing Workshop on Higher Order Statistics, Chamrousse, France, pp. 83–86, July 10–12, 1991.

[6] M. T. Silvia and E. A. Robinson, Geophysical signal analysis, Prentice–Hall, 1980.

[7] J. M. Mendel, Maximum–likelihood deconvolution: A journey into model–based signal processing, Springer–Verlag, 1990.

[8] J. Burg, "Maximum entropy spectral analysis," Ph.D. dissertation, Stanford University, 1975.

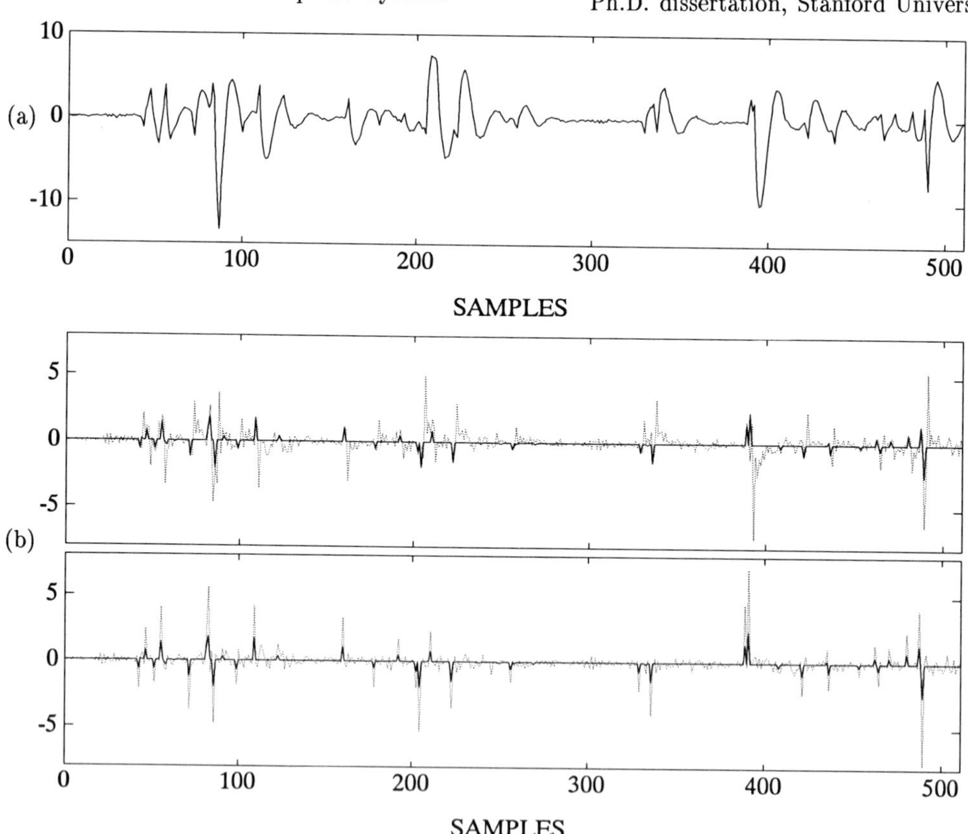

Fig. 2. Simulation results associated with Example 2. (a) Synthetic noisy data for SNR=100. (b) The top part includes true input signal u(k) (solid line) and the predictive deconvolved data e(k) (dotted line); the bottom includes true input signal u(k) (solid line) and the output y(k) (dotted line) of the optimum $\hat{H}_{AP}(z)$ of order equal to 2.

SIGNAL PROCESSING VI: Theories and Applications
J. Vandewalle, R. Boite, M. Moonen, A. Oosterlinck (eds.)

OPTIMAL BAYESIAN DETECTION OF SPECTRAL CHANGES IN RANDOM SIGNALS

E. LE CARPENTIER - C. DONCARLI - B. VOZEL

Laboratoire d'Automatique de Nantes, E.C.N. / C.N.R.S.
1 rue de la Noë, 44072 NANTES CEDEX, FRANCE

The purpose of this paper is the presentation of a new sequential parametric procedure to detect and estimate simultaneously abrupt spectral changes in random signals. After a recall of the problem position, a classification of practically used methods is then considered. The application of the attractive optimum bayesian technique to the detection and estimation problem, inspired by a previous work of the authors, dealing with supervised classification of finite length random signals, is afterwards developed and the corresponding decision criteria are then derived. Finally, a simulation study is presented to light the improvement of the results owing to this new approach before concluding.

1 INTRODUCTION

The early detection and diagnosis of abnormal behaviors in the dynamic properties of random signals are realistic problems which are encountered in many fields of application. It is a natural way to improve performances of very numerous and applied processes (mechanical devices, medical monitoring,...), for which we might be interested in taking action after detecting the change instant,with minimum time delay.
Moreover, tracking of time-varying phenomena leads also to improve performances of adaptive filters, which from a theoritical viewpoint can only deal with slow changes.
In this paper, we mainly focuse our attention on the on-line parametric approach to detect spectral abrupt changes.
After the definition of the studied signals and changes (§2), the state of the art is briefly described (§3). Then, a new sequential procedure, based on a specific distance, is introduced (§4) and a comparison study with classical methods is performed on a very large set of signals (§5). Finally, some concluding comments are given.

2 PROBLEM STATEMENT

The signals are assumed to be described by a linear parametric model (AR or ARMA), which can be written in the following form:

$$y_k = \theta^T \Phi_k + e_k$$

with (e_k) zero mean i.i.d. gaussian sequence, $\mathrm{var}(e_k)=\sigma_e^2$

$$\theta^T = (a_1, a_2,\ldots, a_p)$$

$$\Phi_k^T = (-y_{k-1}, -y_{k-2},\ldots, -y_{k-p}) \text{ for an AR (p) model}$$

$$\theta^T = (a_1,\ldots, a_p, c_1,\ldots, c_q)$$

$$\Phi_k^T = (-y_{k-1},\ldots,-y_{k-p}, e_{k-1},\ldots,e_{k-q}) \text{ for ARMA (p,q)}$$

Then, spectral changes are either change in the variance or in the dynamic parameter vector of the signal model.
The detection of a change can be stated as follows; given:
- samples $(y_t)_{1\le t\le n}$ of a stochastic process (Y_t) with conditional density $p_\theta(y_t/Y_1^{t-1})$ where $(Y_1^{t-1})^T=(y_1,\ldots, y_{t-1})$,
- a nominal time invariant parameter vector $\Theta^T = (\theta^T, \sigma)^T$ possibly identified, we test on-line if the observations are generated by the nominal parameter vector or by a different set of parameters. We assume that no change occurs in the model order or in the statistical distribution either.

Formally, observing the sample $y_0, y_1,\ldots, y_t$ up to time n, we have to decide between the two hypotheses:

H_0: $\Theta = \Theta_0$
H_1: there exists an instant r ($1\le r\le t$) such that:
$\Theta = \Theta_0$ $1\le k\le r-1$
$\Theta = \Theta_1$ $r\le k\le t$

If (H_1) is decided, then the estimation of the change time r and possibly Θ_1 are required.
According to the assumption that r is a deterministic unknown value, the standard optimality criteria [1] of interest in this paper (on-line detection theory) consists in minimizing the mean detection delay for a fixed false alarm rate, namely the mean time between false alarms.
Let t_a be the alarm time at which a detection occurs, the delay to be minimized is:

$$\bar{\tau}= \sup_{r\ge1}\ \operatorname{ess}\sup_{\theta_1\ne\theta_0}\ \sup_{\theta_1} E_{\theta_1} \{t_a-r+1 / Y_1^{r-1}, t_a \ge r\}$$

while the mean time between false alarms satisfies:

$$\bar{T}= E_{\theta_0} (t_a) \ge T$$

But the factors which allow us to estimate the advantages and drawbacks of the different procedures (τ, false alarms rate) are obviously non linear and complicated functions of the random variables. Moreover, the difficulty to evaluate the efficiency of the various methods is strongly related to the difference between Θ_0 and Θ_1 and the value of the change time r. As a result, a precise analytical assessment of their efficiency becomes quickly difficult, and a comparison based on extensive simulations is often considered.[2],[3],[4]

3 STATE OF THE ART

Roughly speaking, two classes of sequential procedures for detecting and estimating a change in the signal parametric model can be distinguished and briefly recalled:

3.1 First Class

In the first group, we find algorithms which test the statistical properties of a prediction error computed via a suitable filter: LPE (Linear Prediction Error)[5], EKF (Extended Kalman Filter)[6], RML (Recursive Maximum

Likelihood)[7].

According to the method used to design the filters, the prediction error is studied under both hypotheses H_0 and H_1. For any of these methods, we can consider the prediction error ε_k as the output of a linear system having e_k as input. As a result, the sequence ε_k is centered under both the hypotheses but two types of behavior may be expected in case of change: a change in θ makes the prediction error ε_k becoming correlated under H_1 and probably more energetic while independent under H_0. A change in σ_e^2 makes only σ_ε^2 suddenly increasing after the change time.

Thus, each method implements a specific test to detect the expected behavior of the prediction error when a change occurs.

3.1.1 LPE

The linear prediction error is defined by:

$$\varepsilon_k = y_k + \sum_{i=1}^{p} \hat{a}_i\, y_{k-i}$$

where $(\hat{a}_i)\ 1 \le i \le p$ denotes an identified AR model of the signal (before change)

Then, for each type of changes (in θ or in σ_e^2), the test consists in finding the characteristic number of times L , ε_k overreaches a threshold d [5]. The threshold can be computed a priori by fixing an upper false alarms bound T and approximating the false alarm rate in the following form:

$$P_{FA} \# \text{Prob}\,[\,(|\varepsilon_k| > d)^L\,] \le T$$

where, under the gaussian assumption

$$\text{Prob}\,[\,|\varepsilon_k| > d\,] = \text{Erfc}\,(\frac{d}{\sigma_\varepsilon \sqrt{2}})$$

If $\varepsilon_k > d$ at least L consecutives samples, H_1 has to be decided, H_0 otherwise.

3.1.2 EKF

The prediction error (pseudo-innovation process) is now defined by:

$$\varepsilon_k = y_k + \sum_{i=1}^{p} \hat{a}_i\, y_{k-i} - \sum_{i=1}^{q} \hat{c}_i\, \varepsilon_{k-i}$$

where $((\hat{a}_i)\ 1 \le i \le p\ ;\ (\hat{c}_i)\ 1 \le i \le q)$,denote the coefficients of an identified ARMA model of the signal by EKF.

The test consists in computing an ergodic estimation of the correlation between ε_k and ε_{k-1}:

$$\rho_k = \alpha\, \rho_{k-1} + (1-\alpha)\, \varepsilon_k\, \varepsilon_{k-1}$$

A change is detected when $|\,\rho_k\,|$ suddenly increases.

3.2 Second class

The second class musters algorithms which are based on Maximum Likelihood Estimation of the time and magnitude of the jumps, in order to get statistical performance near to optimum.

The basis statistical approach is the GLR method [8].

3.2.1 GLR

Considering an AR or ARMA modelisation of the signal, and assuming the following model for the joint density of the observations Y_1^k:

$$P(Y_1^{k-1}/\,k{\ge}r) = p_{\Theta_0}(y_1) \prod_{i=2}^{r-1} p_{\Theta_0}(y_i/Y_1^{i-1}) \prod_{i=r}^{k} p_{\Theta_1}(y_i/\,Y_1^{i-1})$$

The log-likelihood ratio between the two hypotheses for a n-size sample is:

$$S_1^n\,(H_0/\,H_1) = \sum_{k=1}^{n} \text{Log}\frac{p_{\Theta_1}(y_k\,/\,Y_1^{k-1})}{p_{\Theta_0}(y_k\,/\,Y_1^{k-1})}$$

The change instant r and the parameter vector θ_1 which are unknown are then replacing by their maximum likelihood estimates:

$$\hat{r} = \underset{r}{\text{argmax}}\ S_1^n\,(H_0/\,H_1)$$

$$\hat{\Theta}_1 = \underset{\Theta_1}{\text{argmax}}\ S_1^n\,(H_0/\,H_1)$$

Then, the change detector, conditionned by a threshold λ is given by:

$$g_n = \max_{1\le r\le n}\ \max_{\Theta_1}\ S_1^n\,(H_0/\,H_1) \underset{H_0}{\overset{H_1}{\gtrless}} \lambda$$

When g_n exceeds λ, H_1 has to be decided, H_0 otherwise.

But the double maximisation in the previous sum is complex and time consuming and so complicates the recursive structure of the algorithm.Thus, the practically used methods (Brandt's GLR [9],the divergence test [4]) consist in an approximated on-line implementation of the GLR. Roughly speaking, they can be described as follows: At time t, two or eventually three observation windows of different lengths are sequentially considered (according to the selected method): a reference large one which takes into account all or almost all the past of the signal, and a test short one which only contains the L past observations. Within these observation windows, an AR or ARMA parametric model is identified. When a change occurs, the large windowed data model can be supposed to remain constant ; on the other hand, the short windowed data one, obtained from a finite length data set quickly changes.Then, the detection process consists in comparing the two previous models and to decide if they are different by the mean of a suitable distance.

3.2.2 Brandt's GLR

Here, the GLR test consists first in detecting a change in the signal model variance σ_e^2 by computing an estimation of a signal AR model within three observation windows defined as follows:

```
1        W1        k-1  k     W2     t
|-------------------|   |------------|
          W0
|------------------------------------|
```

where W_1 is a growing reference window starting from the last boundary detected; W_2 a sliding test window of constant length (L), and W_0 the pooled window. Then, the estimation of the change instant is performed. Assuming the change is detected at $t = t_d$ (end of W_2), a new GLR test is applied between the hypotheses:

H_0': there exists an instant r $(1\le r\le t_d\text{-}L)$ such that:

$\Theta = \Theta_0 \quad 1\le k\le r\text{-}1$

$\Theta = \Theta_0' \quad r< k \le t_d$

H_1: $\Theta = \Theta_1^0 \quad 1\le k\le t_d\text{-}L$

$\Theta = \Theta_2 \quad t_d\text{-}L<k\le t_d$

This test is applied from $n = t_d$ to t_d+L and the change instant is given by the last value taken by r.

3.2.3 The Divergence Test

A cumulative distance measure $W_t=\sum_{k=1}^{t} w_k$ between two AR models Θ_0 and Θ_1 as shown in the next figure (cross entropy between the conditional distributions of the two models) is computed.

This cumulative sum can be shown to have a zero conditional drift under H_0 and a negative one under H_1. A procedure of Page-Hinkley [10],[11] is then applied to detect such a behavior of the statistics W_t $1 \leq t \leq n$.

$$\overset{1}{\vdash}\!\!\!\!\!\!-\!\!-\!\!-\Theta_0-\!\!-\!\!-\!\!\overset{t}{\dashv}$$
$$\overset{t-L+1}{\vdash}\!\!-\!\!-\!\!-\Theta_1-\!\!-\!\!-\!\!\dashv$$

Many other distance measurements, which depend on the choice of the representation space (model or cepstral coefficients, spectral distribution,..) can be used in a same way.
However, both the previous algorithms, used with any distance assume the models to be compared are exactly known. Unfortunately, even if the large windowed data model can be reasonabily supposed very closed to the true model (before change), the short windowed data one must be considered as an estimation obtained from just a finite length data set and thus, must be processed as a random variable.
The original part of this paper consists then in introducing a sequential procedure for detecting and estimating parameter changes via the use of a specific distance, where the random character of the short windowed data model is taken into account.This method is derived from a previous one dealing with supervised classification of finite length random signals [12], where the random character of the estimated models is introduced into the decision laws.

4 A NEW SEQUENTIAL PROCEDURE

4.1 Introduction

In a previous work [12], the authors pointed out that, to classify ARMA models, one must take into account that considered models are just estimated from finite length data sequences. Then, the uncertainty of knowledge must be included in the decision process.

Let us consider a N-points data sequence $\{y_k\}_{1 \leq k < N}$ of a random signal driven by an unknown ARMA(p,q) model $\theta = [a_1,\ldots,a_p,c_1,\ldots,c_q]$: $y_k = \Phi_k^T \theta + e_k$
with $\Phi_k^T = [-y_{k-1},\ldots,y_{k-p},e_{k-1},\ldots e_{k-q}]$
The Maximum Likelihood (ML) estimator x of θ is:

$$x = \arg\min_\theta \sum_{k=1}^{N} \varepsilon_k^2(\theta)$$

where ε_k is recursively defined by:

$$\varepsilon_k(\theta) = y_k - \phi_k^T(\theta).\theta$$
$$\phi_k^T(\theta) = [-y_{k-1},\ldots,-y_{k-p},\varepsilon_{k-1}(\theta),\ldots,\varepsilon_{k-q}(\theta)]$$

Then $\sqrt{N}(x-\theta)$ is approximately gaussian with zero mean and covariance matrix V defined by:

$$V = [\sum_{k=1}^{N} \varepsilon_k^2(x)]\,[\sum_{k=1}^{N} \psi_k(x).\psi_k^T(x)]^{-1}$$

where $\psi_k^T(\theta) = -\dfrac{d\varepsilon_k}{d\theta}(\theta)$

Then, a bayesian viewpoint led the authors to propose in [12] the following distance measure between two ARMA models x_1 and x_2 identified by ML method with data sequences of length N_1 and N_2, which corresponds to covariance matrices $P_1 = V_1/N_1$ and $P_2 = V_2/N_2$:

$$d = \text{Log Det}\,(P_1 + P_2) + (x_2 - x_1)^T (P_1 + P_2)^{-1} (x_2 - x_1)$$

4.2 Application to segmentation

We propose to use the two models approach, as in the divergence test, where the short-time model and the long-time one are identified by the Recursive Maximum Likelihood (RML) method.
According to [7], we use the following recursive algorithm, with initializations x_0, P_0, ϕ_1, ψ_1:

$$\varepsilon_k = y_k - \phi_k^T x_k$$
$$\mu_k = \mu_{k-1} + \gamma_k (\varepsilon_k^2 - \mu_{k-1})$$
$$K_k = P_{k-1} \psi_k (\psi_k^T P_{k-1} \psi_k + \lambda_k \mu_k)^{-1}$$
$$x_k = x_k + K_k e_k$$
$$P_k = [P_{k-1} - K_k \psi_k^T P_{k-1}] \lambda_k^{-1}$$

see [7] to compute recursively ϕ_k and ψ_k.
γ_k and λ_k vary according to:

$$\gamma_0 = 1;\ \gamma_k = \frac{1}{1 + \lambda_k \gamma_{k-1}^{-1}}$$
$$\lambda_k = \alpha\,\lambda_{k-1} + (1-\alpha)\,\lambda_\infty$$

with $0 << \lambda_1 \leq 1$, $0 << \lambda_\infty \leq 1$, $0 \leq \alpha < 1$

if $\lambda_k \equiv 1$ then $\gamma_k \leq 1$
if $\lambda_k \equiv \lambda < 1$ then $\gamma_k = (1-\lambda)\,(1-\lambda^k)^{-1}$, $\gamma_\infty = 1 - \lambda$: the filter is adaptive and λ sets the adaptation speed.
Using $\lambda_k = \alpha\,\lambda_{k-1} + (1-\alpha)\,\lambda_\infty$, with $\lambda_1 < \lambda_\infty \leq 1$, then the filter is adaptive in transient case (to quickly forget the initializations), and λ_∞ sets the adaptation speed in permanent case.
Then, $\gamma_k^{-1} P_k$ is nothing but the covariance matrix V defined above.
Here, we use the following values for the long-time model and the short-time one:
$\lambda_{1,long} = 0.95$, $\lambda_{\infty,long} = 1$, $\alpha_{long} = 0.95$
$\lambda_{1,short} = 0.95$, $\lambda_{\infty,short} = 0.99$, $\alpha_{short} = 0.95$
Then, in transient case, the 2 filters have the same behaviour, and (omitting the time-index k):

$$P_{short} + P_{long} \cong (\gamma_{short} + \gamma_{long})\,V = \frac{\gamma_{short} + \gamma_{long}}{\gamma_{short}} P_{short}$$

In permanent case, $\gamma_{long} \cong 0$, $P_{long} \cong 0$, and

$$P_{short} + P_{long} \cong \frac{\gamma_{short} + \gamma_{long}}{\gamma_{short}} P_{short}$$

Then, according to 4.1., the distance measure between the 2 models is defined by:

$$d = \text{Log Det} \frac{\gamma_{short}+\gamma_{long}}{\gamma_{short}} P_{short} + (x_{short} - x_{long})^T \left(\frac{\gamma_{short}+\gamma_{long}}{\gamma_{short}} P_{short}\right)^{-1} (x_{short} - x_{long})$$

Since the first term in d just depends on the short-time model, we can pick it out. Then the distance used for the segmentation is:

$$\Delta = (x_{short} - x_{long})^T \left(\frac{\gamma_{short}+\gamma_{long}}{\gamma_{short}} P_{short}\right)^{-1} (x_{short} - x_{long})$$

5 SIMULATIONS

5.1 Method

We simulated an AR or ARMA process, with a jump at time 1000. For three different distances (namely cepstral distance, Kullback divergence between joint laws, proposed distance), and for each values of a varying threshold, we measured the false alarm percentage between times 1 and 1000, and the detection delay. Then, representing the false alarm rate versus the detection delay, we obtain a curve from point (0,100) (nil threshold) to (+∞,0) (infinite threshold). Then, we can conclude that the best distance is the one which gives the curve which is the closest from the point (0,0).

5.2 Results

We made two trials
- AR (3) models (fig. 1):
before jump: [0.65, -0.33, -0.05]
poles: -0.944, 0.42, -0.126
after jump: [0.5, -0.55, -0.1]
poles: -0.963, 0.628, 0.165
- ARMA (2,2) models (fig. 2):
before jump: [-1.79, 0.836, 0.146, 0.141]
poles: 0.895 ± 0.187j (norm 0.914)
zeros: -0.073 ± 0.368j (norm 0.375)
after jump: [-1.78, 0.829, 0.391, 0.244]
poles: 0.89 ± 0.192j (norm 0.91)
zeros: -0.196 ± 0.454j (norm 0.494)
We obtain the following curves.

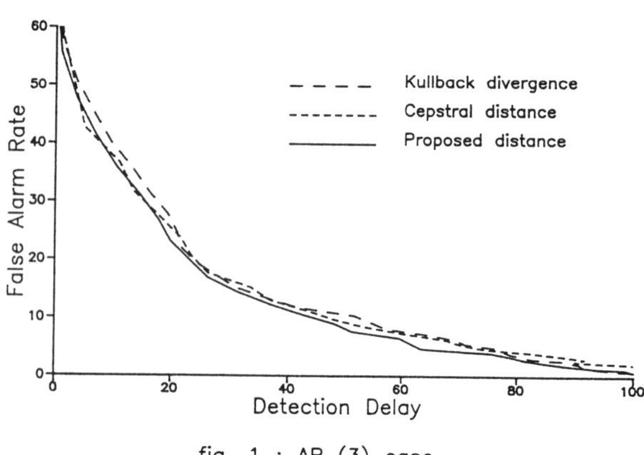

fig. 1 : AR (3) case

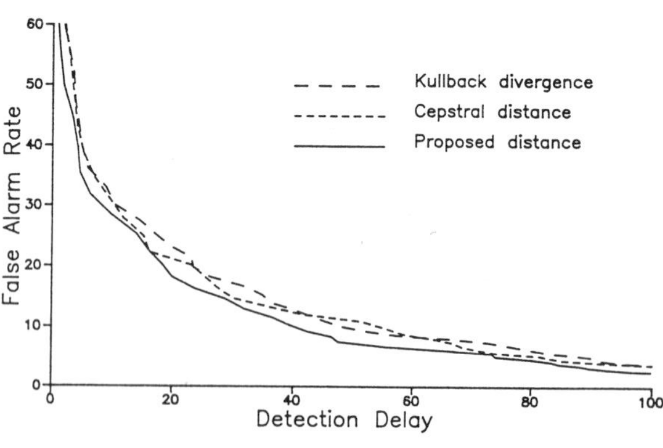

fig. 2 : ARMA (2,2) case

6 CONCLUSION

The two previous trials show a significant improvement with the proposed method, which can be explained because we explicitely take into account the random nature of the identified models.
Therefore, we restricted ourselves to compare the jump in the distance between the two models of the two-models approach. We must point out the problems of methodology to compare objectively two different segmentation methods, because of the numerous parameters which are involved (choice of model, distance, window lengths, identification method...).

REFERENCES

[1] G. Lorden (1971).Procedures for reacting to a change in distribution. Annals Math. Statistics, vol.42, pp1897-1908

[2] R. André-Obrecht (1988). A new statistical approach for the automatic segmentation of continuous speech signals. IEEE Trans. Acoustics, Speech Signal Proc, vol. ASSP-36, n°1,pp.29-40

[3] M. Basseville (1986). The two models approach for the on line detection of changes in AR processes.In Detection of abrupt changes in signals and dynamical systems (M.Basseville, A. Benveniste,eds.) chap.6.Lecture Notes in Control and Information Sciences, LNCIS77, eds(1986), Spinger-Verlag, New-York

[4] U.A. Appel and A.V. Brandt (1984).Performance comparison of two segmentation algorithms using growing reference windows.Proc. 6th Int. Conf. on Analysis and Optimisation of Systems, pp 156-169, Nice, June 19-22

[5] F. Castanie and P. Soule (1983). Méthode de détection de non stationnarités par analyse des propriétés de franchissement d'un seuil par l'erreur de prédiction linéaire. Actes du 9ème colloque GRETSI, pp105-109, Nice 1983

[6] C. Doncarli (1983). Un algorithme séquentiel de détection des ruptures d'un modèle ARMA. 9e colloque GRETSI, pp 1035-1039, Nice 1983

[7] L. Ljung and T.Soderstrom (1983). Theory and practice of recursive identification. Mit Press, Cambridge, Massachusetts

[8] A.S. Willsky and H.L. Jones (1976). A generalized likelihood ratio approach to the detection and estimation of jumps in linear systems.IEEE Trans. Aut. Control, 21, pp108-112

[9] A.V. Brandt (1983). An efficient GLR algorithm for adaptive segmentation of non stationary signals. Proc. of EUSIPCO, pp763-766, Erlangen (RFA)

[10] E.S. Page (1954). Continuous inspection schemes. Biometrika, 41, pp 100-115

[11] D.V. Hinkley (1971). Inference about the change-point from cumulative sum tests. Biometrika,58,pp509-523

[12] C. Doncarli and E. Le Carpentier (1991). An optimal approach for random signals classifications. IEEE Transact. P.A.M.I., vol 13,n°19

SIGNAL PROCESSING VI: Theories and Applications
J. Vandewalle, R. Boite, M. Moonen, A. Oosterlinck (eds.)

LOWERING THE SNR THRESHOLD IN SVD-BASED FREQUENCY ESTIMATION ALGORITHMS

J. I. Portillo-García, J. R. Casar-Corredera

Dpto. Señales, Sistemas y Radiocomunicaciones
E. T. S. Ing. de Telecomunicacion - Universidad Politécnica de Madrid
Ciudad Universitaria s/n
28040 MADRID (SPAIN)
Tfn: 34-1-5495700 x 206

Some high resolution methods for frequency estimation obtain a vector in the noise subspace of the estimated covariance matrix. The frequencies are estimated from the phases of the complex roots of the polynomial associated to that vector. Those methods need a number of components of the noise subspace estimated vector reasonably greater than the number of frequencies to be estimated. However, this makes difficult to distinguish between the signal roots and the spurious roots originated by the higher degree of the polynomial. Minimum-norm method overcomes this difficulty moving the spurious roots inside the unit circle, which allows to pick the roots nearest to the unit circumference. We propose in this communication a new method to perform the correct signal-root selection. The procedure improves the performance of the minimun-norm method and allows to employ other vectors in the noise subspace. As a side result, a sub-optimum root selection method improves the performance of FBLP, and avoids the necessity to compute the SVD of the data covariance matrix.

1. INTRODUCTION

Frequency estimation of complex exponentials in noise is a very important subject in applications such as: radar, sonar, radioastronomy, speech and music processing, etc. There have been proposed a lot of solutions to this problem: some of them are based on classical spectral analysis, some other are based on maximum-likelihood approaches, or on linear prediction methods (such as FBLP) [1] [2]. FBLP method [3] [4] have been widely used, due to its good behaviour in moderate-high SNR. However, when SNR lowers, the performance gets worse very quickly.

Lately, a family of high resolution methods have been applied to this problem, with good performance at low SNR. These methods are based on computing the SVD of the data covariance matrix [5]. Some of them, such as minimum-norm [6], obtain a vector in the noise subspace of the data covariance matrix. Using that vector, an associated polynomial is formed:

$$G(z)=\sum_{i=0}^{L} g_i z^{L-i} \tag{1}$$

where g_i is the i-th component of the minmum-norm vector. Supposing we know we have p complex exponentials, the p roots which are nearest to the unit circumference are selected. The p frequencies are then estimated as:

$$f_i=\tan^{-1}\frac{\Im(z_i)}{\Re(z_i)} \tag{2}$$

where z_i is one of the p roots selected. Minimum-norm property in the noise subspace vector selected is very desirable in the sense that it tends to locate p signal roots on the unit circumference, with the adequate phases, and to move the spurious $(L-p)$ roots inside the unit circle, far away from the p signal roots. It is then easy to select the correct signal roots.

However, when SNR is too low, the spurious roots show an erratic behaviour, and some of them move towards the unit circumference.

Besides, some signal roots may move away from the unit circumference. If one uses the root selection procedure previously commented, it will result in a wrong root selection, which in turn causes an increment in the mean squared value of the frequency error. The final effect is a threshold effect: at some value of SNR, the frequency errors grow very fast, due to the effect of the noise spurious roots.

2. PROPOSED METHOD

We propose in this contribution a new method to select the correct roots of the polynomial defined in (1). Let us define the steering vector at frequency f as:

$$\mathbf{d}(f)=\begin{bmatrix}1 & e^{(j2\pi f)} & \dots & e^{(j2\pi Lf)}\end{bmatrix}^T \quad (3)$$

Theorically, the steering vectors corresponding to the signal frequencies will be in the signal subspace, and the steering vectors corresponding to the spurious frequencies will not be in that subspace. Therefore, we may expect the norm of the projection of steering vectors corresponding to signal frequencies onto the signal subspace to be larger than the projection of steering vectors corresponding to spurious frequencies. We then propose to select the p frequencies which maximize the norm of the associated steering vector onto the signal subspace:

$$P(f)=\mathbf{d}^H(f)\mathbf{P}_S\mathbf{d}(f) \quad (4)$$

where $\mathbf{P}_S$ is the signal subspace projection matrix, defined as:

$$\mathbf{P}_S=\sum_{i=0}^{p-1}\mathbf{u}_i\mathbf{u}_i^H \quad (5)$$

and $\mathbf{u}_i$ are the eigenvectors of the data correlation matrix which span the signal subspace.

The employment of this root selection procedure has two main advantages:

1) It improves the performance of the minimum-norm method. Since the method is not directly based on computing distances to the unit circumference, it is more robust against root movements in complex plane due to the low SNR. The threshold lowers, as can be seen in an example below.

2) It allows the employment of some other vectors in the noise subspace. Minimum-norm is a well-behaved method if one uses a root selection criterion based on the position of the roots with respect to the unit circumference. If the root selection criterion changes, some other noise subspace vectors may be used, even if they locate the spuriuos roots near of the unit circumference.

As a side result, we will examine an improvement of the FBLP method. We use p columns of the data covariance matrix to estimate a sub-optimum signal subspace projection matrix:

$$\mathbf{P}'_S=\begin{bmatrix}\mathbf{r}_{i_1} & \mathbf{r}_{i_2} & \dots & \mathbf{r}_{i_p}\end{bmatrix}\begin{bmatrix}\mathbf{r}_{i_1} & \mathbf{r}_{i_2} & \dots & \mathbf{r}_{i_p}\end{bmatrix}^H \quad (6)$$

where $\mathbf{r}_{i_k}$ is the i_k-th column of the data covariance matrix. We select p roots, using the function defined in (4), as explained before, using the sub-optimum projection matrix $\mathbf{P}'_S$. Note that this procedure avoids SVD computing. Therefore, its computational cost will be lower than SVD-based methods, with good results, as it is shown below.

3. SOME RESULTS

In this section, we show some results that illustrate the improvement of the proposed root selection method. 25 points of the following signal have been generated:

$$x[n]=e^{(j2\pi 0.1n)}+e^{(j2\pi 0.12n+\pi/4)}+w[n] \quad (7)$$

where $w[n]$ is a Gaussian, white noise. SNR is varied from -2 dB to 25 dB in steps of 1 dB. 500 data registers have been generated at each SNR.

Figure 1 shows the improvement in the MSE of the normalized frequency 0.1, when we select the roots in the minimum-norm method using the proposed procedure. Note the lower threshold obtained.

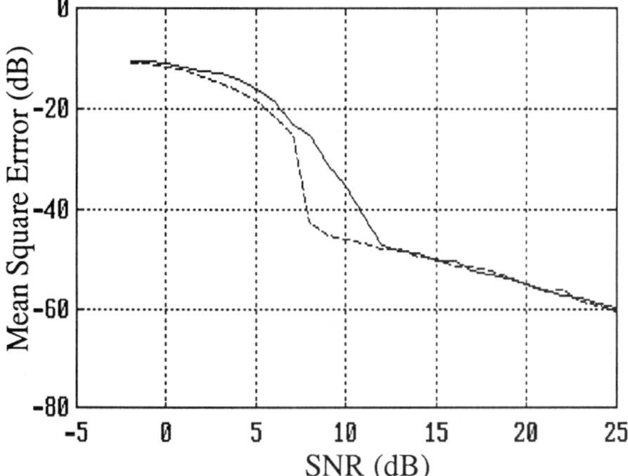

Figure 1. Results for the proposed root selection method. Minimum–norm vector. f = 0.1.
—— Classical root selection.
- - - - Proposed root selection.

As we stated before, the proposed procedure allows the employment of some other noise subspace vectors. Figure 2 shows the improvement in the MSE of the normalized frequency 0.1, when we use a noise subspace proposed vector, $\mathbf{g}_{fbN}$: the projection onto the noise subspace of the FBLP prediction error filter $\mathbf{g}_{fb}$:

$$\mathbf{g}_{fbN} = \mathbf{P}_N \mathbf{g}_{fb} \tag{8}$$

where $\mathbf{P}_N$ is the projection matrix onto the noise subspace. Note the improvement obtained using the proposed root selection procedure. Without it, this vector would not be useful.

Figure 3 is a comparison between minimum-norm and the proposed vector, both cases using the proposed root selection method. We can appreciate the improvement in the threshold using the proposed noise subspace vector. Furthermore, this vector, as we will show in a future contribution, is more robust when there are errors in the estimation of the model order (p).

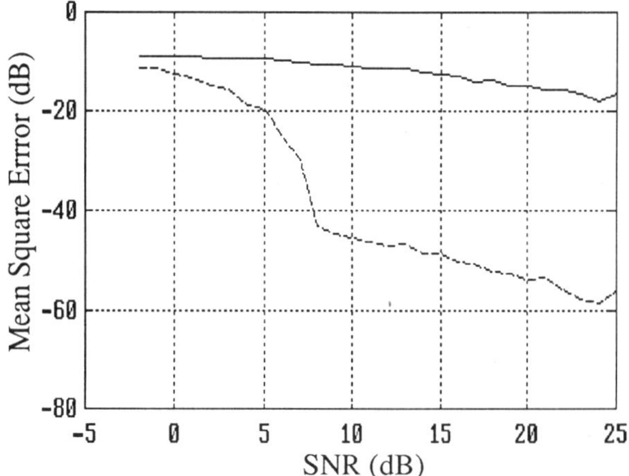

Figure 2. Results for the proposed noise subspace vector. f = 0.1.
—— Classical root selection.
- - - - Proposed root selection.

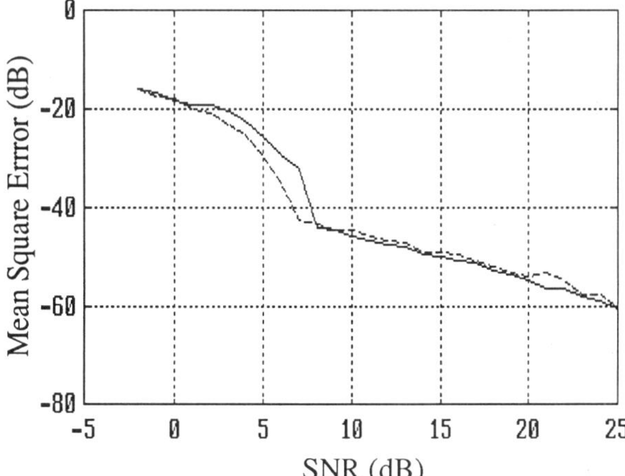

Figure 3. Comparison of minimum–norm vector results and the proposed noise subspace vector results. f = 0.1.
—— Minimum–norm.
- - - Proposed vector.

Finally, figure 4 shows the performance of FBLP method using the suboptimal root selection procedure, with the matrix $\mathbf{P}'_S$, defined in (6). Note the improvement obtained, compared with the use of FBLP and selection of the p roots closest to the unit circumference.

4. CONCLUSIONS

In this paper, we propose a different method for the selection of the signal roots of polynomials associated to vectors in the noise subspace of the

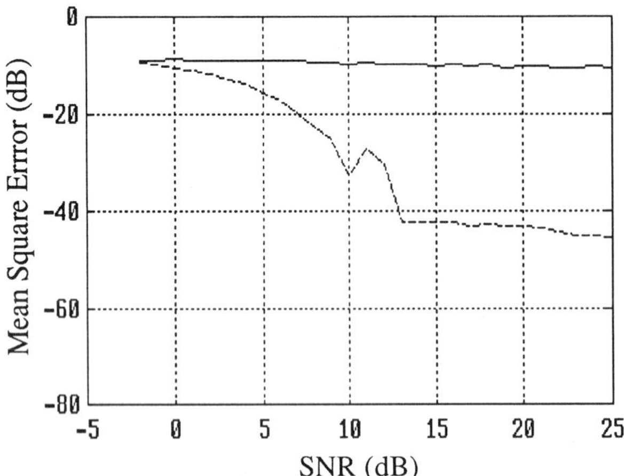

Figure 4. Improvement in FBLP results using the suboptimum root selection procedure. f = 0.1.

— Classical root selection.

- - - Proposed root selection.

data covariance matrix. The method is based on constructing a steering vector for each root, and selecting the p roots for which the projection of that steering vector onto the signal subspace is maximum. Therefore, the method is not based on computing any distance to the unit circumference, and offers better results when it is applied to minimum-norm vector, as it can be seen in simulation examples. Besides, this root selection procedure is general, and may applied to some other vectors in the noise subspace. We propose the use of a noise subspace vector obtained as the projection of the FBLP error prediction filter vector onto the noise subspace. Again, simulations presen examples of the better performance of this vector. However, the roots of the polynomial associated to this vector are very bad behaved, and, without the use of the root selection procedure proposed, this vector could not be employed. As a side result, we apply the root selection procedure to FBLP method, using a suboptimal signal subspace projection matrix, with improved results.

5. REFERENCES

[1] Kay, S., M., "Modern Spectral Estimation-Theory and Application", Prentice-Hall, 1988.

[2] Marple Jr, S., L., "Digital Spectral Analysis with Applications", Prentice-Hall, 1987.

[3] Nuttall, A., H., "Spectral Analysis of an Univariate Process with Bad Data Points, via Maximum Entropy and Linear Prediction Techniques", Naval Underwater Systems Center, Technical Report TR-5303, New London, 1976.

[4] Ulrych, T., J., Clayton, R., W., "TIme Series Modelling and Maximum Entropy", Phys Earth Planetary Interiors, vol. 12.

[5] Bouvet, M., Clergeot, H., "Eigen and Singular Value Decomposition Techniques for the Solution of Harmonic Retrieval Problems" in "SVD and Signal Processing", pp. 93-114, Ed. F. Deprettere, North-Holland 1988.

[6] Kumaresan, R., Tufts, D., W., "Estimating the Angles of Arrival of Multiple Plane Waves", IEEE Trans. Aerospace and Electronic Systems, vol. AES-19, no. 1, Jun 1983.

SIGNAL PROCESSING VI: Theories and Applications
J. Vandewalle, R. Boite, M. Moonen, A. Oosterlinck (eds.)

DEGENERACY IN NON-LINEAR PREDICTORS

B. Mulgrew, K. Nisbet & S. McLaughlin
The Univerity of Edinburgh

Abstract This paper describes an initial attempt to reduce the complexity of Volterra series and radial basis function nonlinear predictors. The algorithm used exploits signal subspace concepts to reduce the number of nonlinear terms while attempting to maintain the same MSE performance.

1. Introduction

At present two of the major choices in non-linear signal prediction are the radial basis function (RBF) network and the Volterra series combiner (VSC). They share a common architecture - a non-linear expansion followed by a linear combiner. The resultant linear least squares (LS) optimisation problem is usually overdetermined and resort is often made to a minimum norm solution [1].

The recent work reported in [2] and [3] have proposed exploiting the overdetermined nature of the problem in a positive way to select centres [2] or to remove polynomial terms [3]. If a system of linear equations is overdetermined, then the amount by which it is overdetermined tell us how many parameters or elements in the weight vector can arbitrarily be set to zero without degrading the performance of the predictor. Setting a weight to zero effectively removes an RBF centre or a polynomial term from a VSC. These methods, which we might call state reduction techniques, tell us how many elements we can remove but do not tell us which ones would be the best choice.

In this paper we present a geometrical interpretation of the problem in terms of the signal subspace. The geometrical interpretation indicates that some choices of reduced state subspace may be better than others with respect to numerical conditioning. Further it is suggested that the reduced state subspace which is "close" to the original subspace is the best choice. Initial results indicate that the proximity of the reduced state subspace to the original signal subspace leads to solutions which tends to preserve the good numerical properties of the minimum norm solution.

2. Background

If for example we have an L-element signal vector $\underline{x}(k)$ which contains the last L signal samples. This vector is used to predict the next signal sample $x(k+1)$. A fixed non-linearity such as a Volterra series expansion or a radial basis function expansion is applied to each vector $\underline{x}(k)$ to produce an N-vector $\underline{z}(k)$ [4]. Usually N is greater than L. The sequence of $\underline{z}(k)$ vectors has associated with it an $(N \times N)$ autocorrelation matrix ϕ_{zz}. The matrix has rank M which is usually less than N. In order facilitate the development of a geometrical interpretation consider the simple case where $N = 3$ and $M = 2$. Since the autocorrelation matrix has rank 2 the sequence of 3-vectors $\underline{z}(k)$ spans a subspace of order 2. This plane is illustrated in Figure 1 and is labelled the "signal subspace". The axes set $\hat{\underline{z}}_1$, $\hat{\underline{z}}_2$ and $\hat{\underline{z}}_3$ define orthonormal directions for outputs from individual non-linear functions in the Volterra or RBF expansions. For example:

$$\underline{z}(k) = \begin{bmatrix} 3 \\ 2 \\ 1 \end{bmatrix} = 3\,\hat{\underline{z}}_1 + 2\,\hat{\underline{z}}_2 + 1\,\hat{\underline{z}}_3$$

The signal subspace is spanned by the eigenvectors $\underline{v}_2$ and $\underline{v}_3$ which are associated with the non-zero eigenvalues λ_2 and λ_3. The remaining eigenvector $\underline{v}_1$ is associated with the zero eigenvalue λ_1 and is by definition orthogonal to the signal subspace.

3. The Nearest Subspace

In reducing the number of terms in a Volterra expansion or selecting centres for an RBF, we essentially wish to throw away some of the axes vectors while maintaining the quality of the prediction. One possible way of doing this is to chose a subset of axes vectors which are close to the signal subspace. Thus we have to define a measure of distance between a unit axis vector (e.g. $\hat{\underline{z}}_1$) and the signal subspace. Such a measure could be obtained by calculating the length of the projection of a unit axis vector onto the signal subspace. The greater the length the "closer" the axes vector to the subspace. The projection can be calculated using the eigenvectors which define the signal subspace. The algorithm can be summarised as follows: (i) calculate eigenspectrum and hence rank M; (ii) measure the distance between all axis vectors and the signal subspace; (iii) choose the M axis vectors which are closest to the subspace; (iv) recalculate prediction coefficients and compare prediction quality.

4. Other Possible Solutions

The minimum norm solution $\underline{h}_{\min}$ lies in the signal subspace. Other equally good solutions (in a mean square error (MSE) sense) are defined by the minimum norm solution plus a weighted sum of the eigenvectors which are orthogonal to the signal subspace. In the simple example illustrated in Figure 1 this would be:

$$\underline{h} = \underline{h}_{\min} + \alpha\, \underline{v}_1$$

where α is a constant. A more general way of looking at this effect is facilitated if a rank 1 autocorrelation matrix is considered. This situation is illustrated in Figure 2. The signal subspace is now a line defined by the eigenvector $\underline{v}_3$. The minimum norm solution lies in this subspace at the point $\underline{h}_{\min}$. All other solutions lie in a subspace which is orthogonal to $\underline{v}_3$ and intersects it at $\underline{h}_{\min}$. Thus the "plane of other possible solutions" is defined:

$$\underline{h} = \underline{h}_{\min} + \alpha_1\, \underline{v}_1 + \alpha_2\, \underline{v}_2$$

In [3] this property was utilised to arbitrarily set elements of $\underline{h}$ to zero. In the example of Figure 2 we could set the components along the $\hat{z}_1$ and $\hat{z}_3$ axes to zero in which case the solution for $\underline{h}$ would be found at the point where the "plane of other possible solutions" intersects the axis $\hat{z}_2$. However it would be equally valid to set the components along $\hat{z}_1$ and $\hat{z}_2$ to zero in which case the solution for $\underline{h}$ would be found at the point where the "plane of other possible solutions" intersects the axis $\hat{z}_3$. The most worrying possibility would be to choose a solution along $\hat{z}_1$ since the "plane" is almost parallel to that axis and hence they will intersect at a very large value of $\underline{h}$. This is why the minimum norm solution is often chosen as the solution to a set of overdetermined linear equations - because it is numerically well-conditioned.

Step (iv) of the nearest subspace method appears at first to be an approximation. It is indeed an approximation to the minimum norm solution. A more important question to ask is whether it can find one of the other possible solutions which is close to the minimum norm solution? Consider Figure 2 once more. The axis $\hat{z}_2$ is clearly the nearest axis to the signal subspace defined by $\underline{v}_3$. We would suggest that the solution which lies on $\hat{z}_2$ defines a "best solution" to the problem of removing degeneracy from a nonlinear predictor in that it is both close to the minimum norm solution and sets 2 components to zero?

5. Algorithm

We can summarise the steps towards reducing the number of centres or non-linear elements in a systematic way as follows:

(i) calculate eigenvectors and eigenvalues of the matrix $\sum_k \underline{z}(k)\underline{z}^T(k)$ which defines the LS problem

$$\sum_k \underline{z}(k)\underline{z}^T(k)\, \underline{h} = \sum_k \underline{z}(k)\, x(k+1) \tag{1}$$

and hence deduce the rank M; the eigenvectors are collected into a matrix $\underline{V} = [\underline{v}_1\ \underline{v}_2\ \ldots.\ \underline{v}_N]$;

(ii) project each of the axis vectors $\hat{z}_1, \hat{z}_2, \cdots \hat{z}_M$ onto the signal subspace; in general the length of this projection has the form $d_j = \hat{z}_j^T\, \underline{V}_s\, \underline{V}_s^T\, \hat{z}_j$ where $\underline{V}_s$ is the matrix of eigenvectors which span the signal subspace. The indices or addresses of the M largest elements of $\underline{d}$ are the indices or addresses of the centres or nonlinearities which are selected. The remaining indices indicate the weights which will effectively be set to zero.

(iii) Other possible solutions *with the same MSE* are given by: $\underline{h} = \underline{h}_{\min} + \underline{V}_n\, \underline{b}$ where $\underline{V}_n$ is the matrix of eigenvectors orthogonal to the signal subspace and $\underline{b}$ is a column vector with $N-M$ elements. By suitable manipulation the appropriate elements of $\underline{h}$, identified in (ii), can be set to zero and hence the remaining non-zero elements of $\underline{h}$ can be calculated. Alternatively appropriate rows and columns of (1) can be forced to zero and the reduced state LS problem can be solved directly.

6. Results

To demonstrate the use of the technique we consider Duffing's equation [5] in the chaotic region as the signal generation mechanism i.e.

$$\frac{d^2x}{dt^2} + 0.25\,\frac{dx}{dt} - x + x^3 = 0.3\cos(\,t\,)$$

This is solved using fourth order Runga-Kutta with a step size $\Delta t = 0.1$. An appropriate choice for the embedding dimension is made - in this case $L = 6$. The predictions are initially constructed using a 3-term (cubic) Volterra series with a full compliment of $N = 84$ polynomial terms. The performance of this "full compliment" predictor sets the lower bound on performance in this example. Reduced state predictors are also formed from the 84 possible polynomial terms using the method described in section 5 with $M = 69$ and $M = 42$

Figure 3 illustrates the normalised MSE performance of these predictors against iterate or sample number n for a sequence of 10000 data samples. The normalised MSE is defined as:

$$MSE(n) = 10\log_{10}\left(\frac{\sum_{k=1}^{n} e^2(k)}{\sum_{k=1}^{n} x^2(k)}\right)$$

where $e(k)$ is the one-step prediction error and $x(k)$ is the signal at sample k. The performance of the reduced state predictor with $M = 69$ polynomial terms (Figure 3(b)) is within 2 dB of the performance of the full predictor (Figure

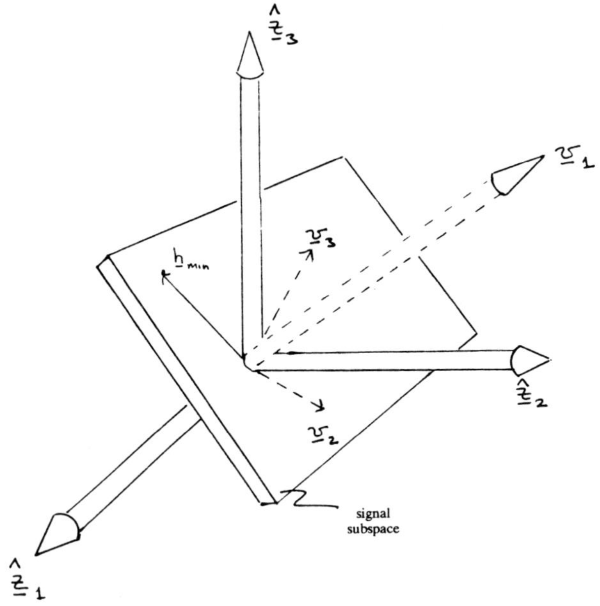

FIGURE 1 Rank-2 Signal Subspace and Associated Eigenvectors.

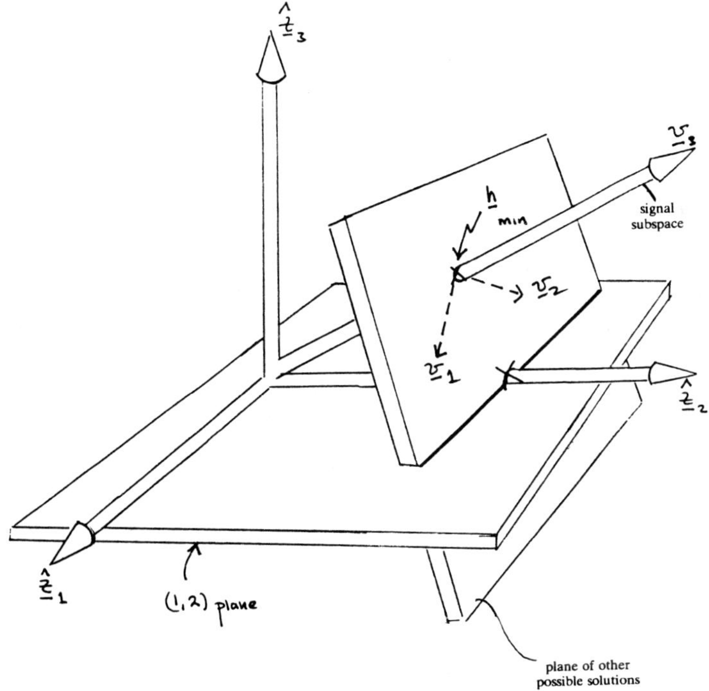

FIGURE 2 Other Possible Solutions (rank-1 signal subspace).

3(a)) with $M = 84$ terms. Reducing the number of term to $M = 42$ degrades the performance of the predictor by a further 15dB to approximately -43 dB (Figure 3(c)).

7. Discussion

In this paper some intial results are presented which demonstrate how the concept of a "nearest subspace" can be used to systematically reduce the number of polynomial terms in a Voterra series predictor used on a chaotic process. In principal the technique can also be used in conjunction with RBF predictors to remove centres in a similar manner to the technique described in [2]. In the particular example illustrated in Figure 3 it should be noted that 15 of the 84 polynomial terms can be removed without significantly degrading the quality of the predictions.

8. Acknowledgement

The authors wish to acknowledge the support of the UK Science and Engineering Research Council.

References

1. D. S. Broomhead and D. Lowe, "Multivariable Functional Interpolation and Adaptive Networks," *Complex Systems*, **Vol. 2**, pp. 321-355, (1988).

2. S. Chen, C.F.N. Cowan, and P.M. Grant, "Orthogonal Least Squares Learning Algorithm for Radial Basis function Networks," *IEEE Trans Neural Networks*, **Vol. 2**, pp. 302-309, (1991).

3. M. R. Lynch, P. J. Rayner, and S. B. Holden, "Removal of Degeneracy in Adaptive Volterra Networks by Dynamic Structuring," *Proceedings of IEEE ICASSP 91*, pp. 2069-2072, (1991).

4. K. Nisbet, B. Mulgrew, and S. McLaughlin, "Nonlinear Prediction and the Wiener Process," *Proceeding S.P.I.E.*, (1991).

5. J. Guckenheimer and P. Holmes, *Nonlinear Oscillations, Dynamical Systems and Bifurcations of Vector Fields*, Springer-Verlag, (1983).

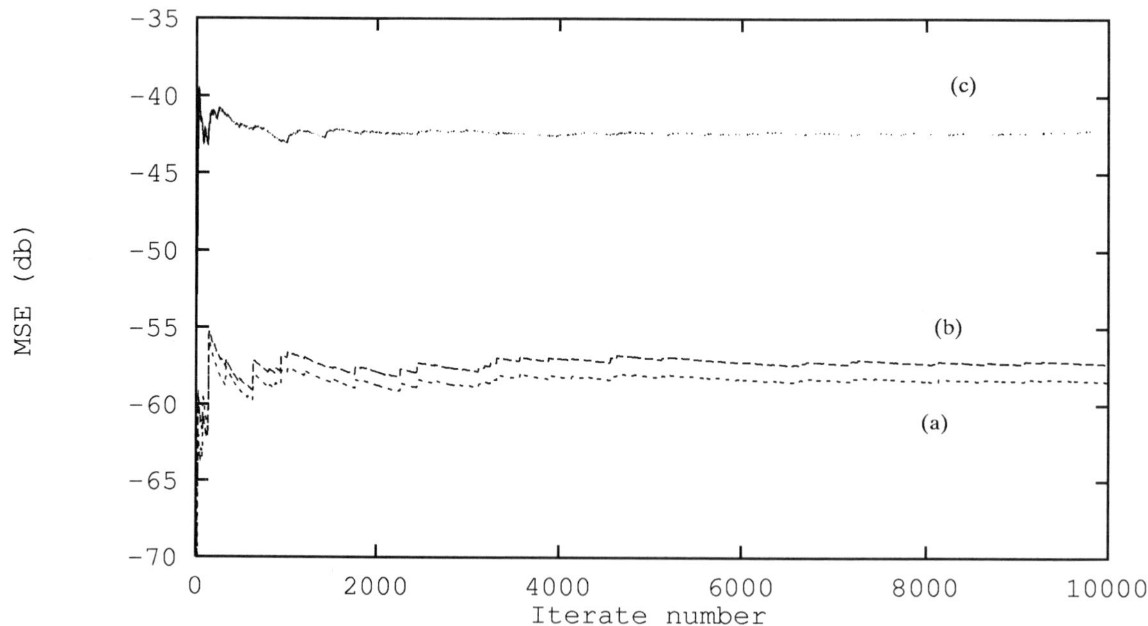

FIGURE 3 Normalised MSE performance of cubic Volterra one-step predictor on chaotic Duffing's equation: (a) full set of 84 polynomial terms; (b) reduced state with 69 polynomial terms; (c) reduced state with 42 polynomial terms.

SIGNAL PROCESSING VI: Theories and Applications
J. Vandewalle, R. Boite, M. Moonen, A. Oosterlinck (eds.)

POLYNOMIAL PHASE SIGNAL ESTIMATION

Z.FARAJ and F.CASTANIE (Member EURASIP)

GAPSE-ENSEEIHT, 2 rue Camichel, 31071 TOULOUSE, FRANCE

ABSTRACT : To overcome the problem of estimating the parameters of complex polynomial FM signals, a new algorithm, based on a combination of two new transforms, is proposed. The polynomial autocorrelation function and the modified Fourier transform, which can be considered as a generalization of the standard autocorrelation function and the standard Fourier transform, are adapted for polynomial phase signals. Numerical examples are given in order to illustrate the performances of this algorithm and to show the good agreement with the theory.

I. INTRODUCTION

Signals having constant amplitude and polynomial phase are common in radar and communication applications [1]. The estimation of the coefficients of the phase polynomial is important in various applications such as are found in deep space telemetry systems [2].

In a recent correspondence [1], a new transform was introduced for estimating the phase of complex signals that have constant amplitude and polynomial phase. Let $g(t)$ be a complex valued function of a real variable t, q an integer, τ a positive real number and M a positive integer. The Polynomial Transform $PT_M[g,\omega,\tau]$ is defined as the Fourier transform of $P_M[g(t),\tau]$ over the interval $[0,T]$, i.e.,

$$PT_M[g,\omega,\tau]=\int_0^T P_M[g(t),\tau]\exp\{-j\omega t\}dt$$

where :

$$P_M[g(t),\tau]=\prod_{q=0}^{M-1}[g^{\$q}(t-q\tau)]^{\binom{M-1}{q}}$$

$g^{\$q}(t)=g(t)$ if q is even.

$g^{\$q}(t)=g^*(t)$ if q is odd.

$(.)^*$ denotes complex conjugation.

The following special cases illustrate this definition :

$$PT_1[g,\omega,\tau]=\int_0^T g(t)\exp\{-j\omega t\}dt$$

$$PT_2[g,\omega,\tau]=\int_0^T g(t)g^*(t-\tau)\exp\{-j\omega t\}dt$$

PT_1 is the standard Fourier transform. PT_2 is the ambiguity function. PT_3 is new and so are the higher order transforms. One of the basic properties of PT_M is that $|PT_M[g,\omega,\tau]|$ has a global maximum at $\omega=M!\tau^{M-1}a_M$ if $g(t)$ is a complex function with constant amplitude and polynomial phase on the interval $[0,T]$, i.e.,

$$g(t)=b_0\exp\left\{j\sum_{m=0}^{M}a_m t^m\right\}\quad 0\le t\le T$$

The choice of τ is arbitrary, but in practice it will affect the accuracy of the estimated coefficients.

In this paper, **two new transforms** are proposed for **estimating the phase of complex signals of this type :**

- **The Polynomial Autocorrelation Function.**
- **The Modified Fourier Transform.**

II.POLYNOMIAL AUTOCORRELATION FUNCTION

The Polynomial Autocorrelation Function **(P.A.F)** $C_M[g(t),\tau]$ is defined by :

$$C_M[g(t),\tau]=E\{P_{M+1}[g(t),\tau]\}=E\left\{\prod_{q=0}^{M}[g^{\$q}(t-q\tau)]^{\binom{M}{q}}\right\}$$

$E\{.\}$ denotes the expectation operator.

The following special cases illustrate this definition :

$$C_1[g(t),\tau]=E\{g(t)g^*(t-\tau)\}$$

$$C_2[g(t),\tau]=E\{g(t)[g^*(t-\tau)]^2 g(t-2\tau)\}$$

C_1 is the standard autocorrelation function. C_2 is new, and so are the higher order functions.

The basic properties of the Polynomial Autocorrelation Function are given in the following two new theorems :

Theorem 1 : Let $g(t)$ be a complex function with constant amplitude and polynomial phase, i.e.,

$$g(t)=b_0\exp\left\{j\sum_{m=0}^{M}a_m t^m\right\}$$

Then :

$$C_M[g(t),\tau]=C_M[\tau]=b_0^{(2^M)}\exp\{jM!a_M\tau^M\}$$

Proof :

$$P_{M+1}[g(t),\tau]=\prod_{q=0}^{M}[g^{*q}(t-q\tau)]^{\binom{M}{q}}$$

$$P_{M+1}[g(t),\tau]=\prod_{q=0}^{M}\left[b_0\exp\left\{j(-1)^q\sum_{m=0}^{M}a_m(t-q\tau)^m\right\}\right]^{\binom{M}{q}}$$

$$P_{M+1}[g(t),\tau]=\prod_{q=0}^{M}b_0^{\binom{M}{q}}\prod_{q=0}^{M}\exp\left\{j(-1)^q\binom{M}{q}\sum_{m=0}^{M}a_m(t-q\tau)^m\right\}$$

$$P_{M+1}[g(t),\tau]=b_0^{\sum_{q=0}^{M}\binom{M}{q}}\exp\left\{j\sum_{q=0}^{M}(-1)^q\binom{M}{q}\sum_{m=0}^{M}a_m(t-q\tau)^m\right\}$$

$$P_{M+1}[g(t),\tau]=b_0^{(2^M)}\exp\left\{j\sum_{m=0}^{M}a_m\sum_{q=0}^{M}(-1)^q\binom{M}{q}(t-q\tau)^m\right\}$$

By Lemmas A.1 and A.2 [1], we have :

$$P_{M+1}[g(t),\tau]=b_0^{(2^M)}\exp\{jM!a_M\tau^M\}$$

Therefore,

$$C_M[g(t),\tau]=E\{P_{M+1}[g(t),\tau]\}=b_0^{(2^M)}\exp\{jM!a_M\tau^M\}=C_M[\tau]$$

Lemma A.1 : $\sum_{q=0}^{M}(-1)^q\binom{M}{q}(t-q\tau)^M=M!\tau^M \quad M\geq 0$

Lemma A.2 : $\sum_{q=0}^{M}(-1)^q\binom{M}{q}(t-q\tau)^m=0 \quad 0\leq m\leq M-1$

Theorem 2 : Let $g(t)$ be a complex function having constant amplitude and polynomial phase with an additive noise component, i.e.,

$$g(t)=b_0\exp\left\{j\sum_{m=0}^{M}a_m t^m\right\}+B(t)$$

where $B(t)$ is a complex white Gaussian noise with zero mean and variance σ_B^2. Then :

$$C_M[g(t),\tau]=C_M[\tau]=b_0^{(2^M)}\exp\{jM!a_M\tau^M\}+P_{BM}\delta(\tau)$$

$$P_{BM}=\sum_{m=1}^{L}\sum_{\substack{n=0\\ n\ \text{even}}}^{m}\binom{L}{m}\binom{m}{n}\binom{n}{\frac{n}{2}}b_0^{2L-2m+n}\left(m-\frac{n}{2}\right)!\sigma_B^{2m-n} \quad L=2^{M-1}$$

The proof of this theorem do not appear in this paper for lack of space. It can be found in the forth-coming thesis [4].

Suppose the signal is sampled at interval T_e, and let k be a positive integer. Then the **unbiased Polynomial Autocorrelation Function estimator** is defined by :

$$\hat{C}_M[g(n),k]=\frac{1}{N-Mk}\sum_{n=Mk+1}^{N}\prod_{q=0}^{M}[g^{*q}(n-qk)]^{\binom{M}{q}}$$

$$0\leq k\leq N_M=\frac{N-1}{M}$$

where :

$$g(n)=b_0\exp\left\{j\sum_{m=0}^{M}a_m T_e^m n^m\right\}+B(n) \quad 1\leq n\leq N$$

III. THE MODIFIED FOURIER TRANSFORM

The Modified Fourier Transform (**M.F.T.**) is introduced for estimating the highest order coefficient a_M. Let $x(\tau)$ be a complex valued function of a real variable τ over the interval $[0,T]$, the Modified Fourier Transform is defined by :

$$FT_M[x(\tau),\omega]=\int_0^T\tau^{M-1}x(\tau)\exp\{-j\omega\tau^M\}d\tau$$

The following special cases illustrate this definition :

$$FT_1[x(\tau),\omega]=\int_0^T x(\tau)\exp\{-j\omega\tau\}d\tau$$

$$FT_2[x(\tau),\omega]=\int_0^T \tau x(\tau)\exp\{-j\omega\tau^2\}d\tau$$

FT_1 is the standard Fourier transform. FT_2 is new, and so are the higher order transforms.

One of the basic properties of the Modified Fourier Transform is given in the following theorem :

> *Theorem 3* : Let $x(\tau)$ be a complex function with constant amplitude and polynomial phase where only the highest order coefficient is nonzero on the interval $[0, T]$, i.e.,
>
> $$x(\tau) = A\exp\{j\omega_M \tau^M\} \quad 0 \le \tau \le T$$
>
> Then $|FT_M[x(\tau), \omega]|$ has a global maximum at $\omega = \omega_M$, i.e.,
>
> $$argmax\,|FT_M[x(\tau), \omega]| = \omega_M$$

Proof :

$$x(\tau) = Aexp\{j\omega_M \tau^M\} \quad 0 \le \tau \le T$$

$$FT_M[x(\tau), \omega] = \int_0^T \tau^{M-1} x(\tau)\exp\{-j\omega\tau^M\}d\tau$$

$$FT_M[x(\tau), \omega] = \int_0^T A\tau^{M-1}\exp\{j[\omega_M - \omega]\tau^M\}d\tau$$

Put : $t = \tau^M$, then : $dt = M\tau^{M-1}d\tau$. This gives,

$$FT_M[x(\tau), \omega] = \int_0^{T^M} \frac{A}{M}\exp\{j[\omega_M - \omega]t\}dt$$

$$FT_M[x(\tau), \omega] = \frac{A}{M}T^M \exp\left\{j[\omega_M - \omega]\frac{T^M}{2}\right\} sinc\left\{[\omega_M - \omega]\frac{T^M}{2}\right\}$$

Hence,

$$argmax\,|FT_M[x(\tau), \omega]| = argmax\,\left|sinc\left\{[\omega_M - \omega]\frac{T^M}{2}\right\}\right| = \omega_M$$

Suppose the signal is sampled at interval T_e. Then the **Discrete Modified Fourier Transform** (D.M.F.T.) is defined by :

$$DFT_M[x(k), \omega] = \sum_{k=1}^{N_M} k^{M-1}x(k)\exp\{-j\omega T_e^M k^M\}$$

where :

$$x(k) = A\exp\{j\omega_M T_e^M k^M\} \quad 1 \le k \le N_M$$

IV. POLYNOMIAL PHASE SIGNAL ESTIMATION

Under the assumptions of theorems 1, 2 and 3, the highest order coefficient a_M of a polynomial phase signal $g(t)$ is given by :

$$a_M = \frac{1}{M!}argmax\,|FT_M[C_M[g(t), \tau], \omega]|$$

We can also estimate the coefficients $\{a_m\}$ sequentially, starting at the highest order coefficient a_M. At each step, the effect of the phase term of the higher order is removed :

> 1) Substitute $m = M$ and $g_m(n) = g(n) \quad 1 \le n \le N$
>
> 2) Estimate the P.A.F. by :
>
> $$\hat{C}_m[g_m(n), k] = \frac{1}{N - mk}\sum_{n=mk+1}^{N}\prod_{q=0}^{m}[g_m^{*q}(n - qk)]^{\binom{m}{q}}$$
>
> $$1 \le k \le N_m = \frac{N-1}{m}$$
>
> 3) Compute the D.M.F.T. by :
>
> $$DFT_m[\hat{C}_m[g_m(n), k], \omega] = \sum_{k=1}^{N_m} k^{m-1}\hat{C}_m[g_m(n), k]\exp\{-j\omega T_e^m k^m\}$$
>
> 4) Estimate a_m by :
>
> $$\hat{a}_m = \frac{1}{m!}argmax\,|DFT_m[\hat{C}_m[g_m(n), k], \omega]|$$
>
> 5) Let : $g_{m-1}(n) = g_m(n)\exp\{-j\hat{a}_m T_e^m n^m\} \quad 1 \le n \le N$ and set $m \leftarrow m - 1$
>
> 6) if $m \ge 1$, go back to Step 2). Else go to Step 7).
>
> 7) Estimate a_0 from $g_0(n)$ using the formula :
>
> $$\hat{a}_0 = \Im\left\{\mathrm{Log}\left[\sum_{n=1}^{N} g_0(n)\right]\right\}$$

The phase polynomial in the following example has degree 2. The instantaneous frequency is given by : $f(n) = f_0 + f_1 T_e n$. The signal parameters are :

$f_0 = 10.34Hz \quad f_1 = 5.24Hz/S \quad T_e = 0.01S \quad N = 201$

The Signal to Noise Ratio which is defined by : $SNR = b_0^2/\sigma_B^2$ is equal to 10 db

Figures (1.a) and (1.b) show, respectively, the real and the imaginary part of the P.A.F. : $\hat{C}_2[g_2(n), k]$. Figure (1.c) shows the magnitude of the D.M.F.T. : $DFT_2[\hat{C}_2[g_2(n), k], \omega]$.

Figures (2.a) and (2.b) show, respectively, the real and the imaginary part of the P.A.F. : $\hat{C}_1[g_1(n), k]$ after removing estimated quadratic phase term. Figure (2.c) shows the magnitude of the D.M.F.T. : $DFT_1[\hat{C}_1[g_1(n), k], \omega]$.

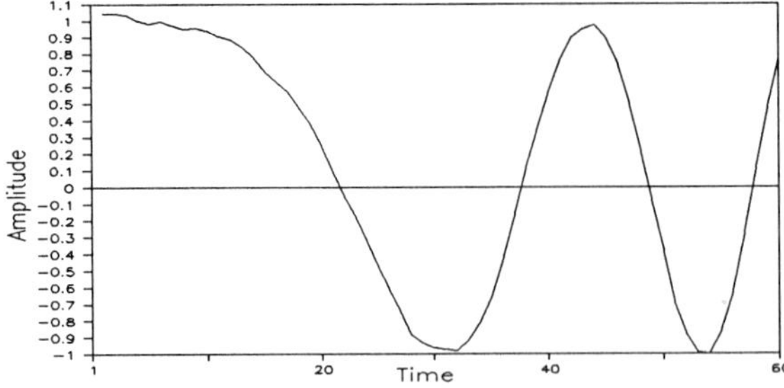

Fig 1.a : $\Re\{\hat{C}_2[g_2(n),k]\}$

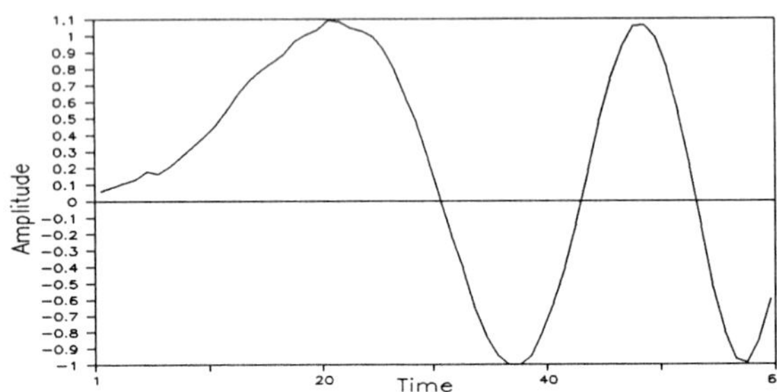

Fig 1.b : $\Im\{\hat{C}_2[g_2(n),k]\}$

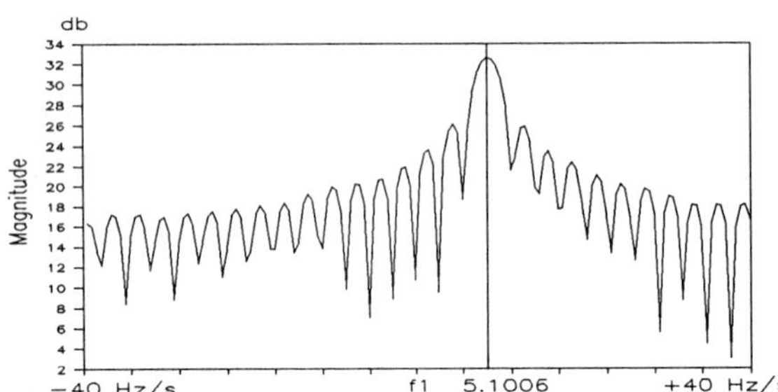

Fig 1.c : $|DFT_2[\hat{C}_2[g_2(n),k],\omega]|$

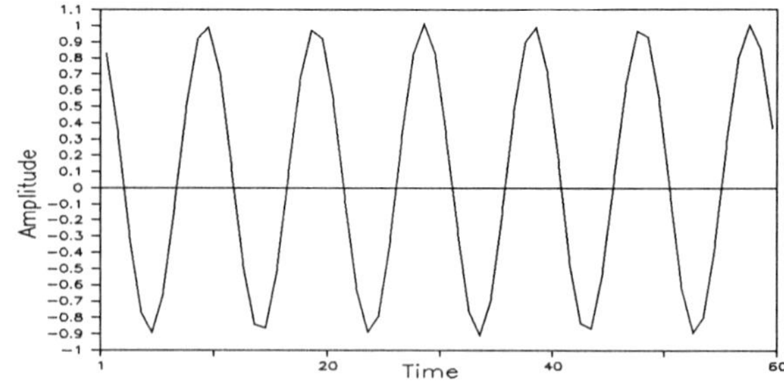

Fig 2.a : $\Re\{\hat{C}_1[g_1(n),k]\}$

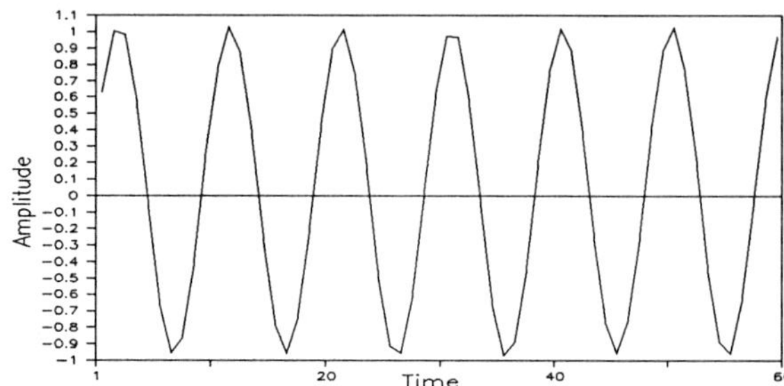

Fig 2.b : $\Im\{\hat{C}_1[g_1(n),k]\}$

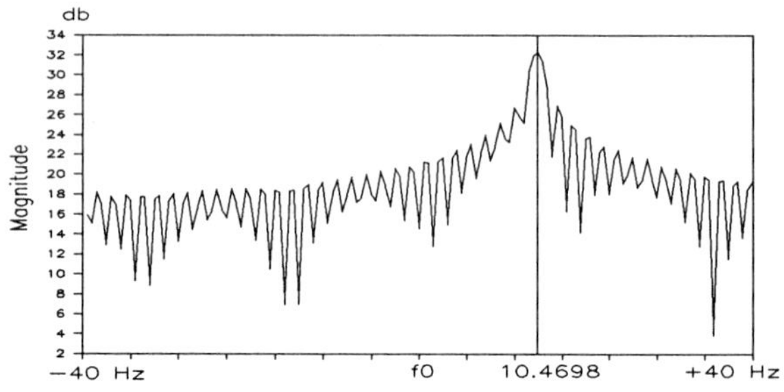

Fig 2.c : $|DFT_1[\hat{C}_1[g_1(n),k],\omega]|$

V. CONCLUSION

In this paper, two new transforms are presented for estimating the phase of complex signals that have constant amplitude and polynomial phase : the polynomial autocorrelation function and the modified Fourier transform.

To overcome the problem of estimating the parameters of complex polynomial FM signals , a new algorithm, based on a combination of the two new transforms, is proposed and numerical examples are given in order to illustrate the performances of this algorithm and to show the good agreement with the theory.

This work is a part of Z.Faraj's thesis and has been supported by the French Space Agency (CNES) under contract N° 874 CNES 89579200.

REFERENCES

[1] Shimon Peleg and Boaz Porat, "Estimation and Classification of Polynomial Phase Signals", IEEE Trans. on Information Theory, Vol.37, N°.2, March 1991.

[2] Z.Faraj, "Traitement du Signal des Sondes Lointaines", Contract Report 874, CNES 89579200, French Space Agency, May 1990.

[3] Shimon Peleg and Boaz Porat, "The Cramer-Rao Lower Bound for signals with Constant Amplitude and Polynomial Phase", IEEE Trans. on Signal Processing, Vol.39, N°.3, March 1991.

[4] Z.Faraj, "Traitement du Signal des Sondes Lointaines", Thèse de Doctorat de l'INP de Toulouse, France.

THE MV SPECTRAL CONVERGENCE APPLIED TO RANDOM FIELDS

(1) D. E. LYON, (2) P. J. SHERMAN, (3) A. E. FRAZHO

(1) University of New Brunswick, Fredericton, N.B., Canada E3B 5A3
(2) Iowa State University, Ames, IA 50011, U.S.A.
(3) Purdue University, West Lafayette, IN 47907, U.S.A.

ABSTRACT

This work addresses the problem of identifying a multichannel harmonic signal field that has been corrupted by an unknown homogeneous noise field. Because recursive spectral-based algorithms such as the Levinson algorithm do not exist in the d-dimensional setting (for d > 1), the computational expense associated with multichannel spectral analysis on a d-dimensional field can play a major role in the types of models that can be considered and in the resolution of the d-dimensional lattice on which it is computed. This is particularly true of a recently developed method of Frazho and Sherman [1] which extends the multichannel convergence-based method of [2] to the d-dimensional setting. As noted [2], in the random process setting (d = 1) the MV(n) spectra can be computed in a recursive fashion. This work presents a method for utilizing this recursive nature for the d > 1 setting. A simple approach would be to apply the MV(n) method to each dimension of the field. This has the disadvantage of not utilizing the joint information contained in other dimensions of the field. The approach presented here allows this simple approach as a special case. More generally however, it allows information from the other dimensions to be utilized in a compact way.

1. BACKGROUND

The problem of identifying the periodic components of a multichannel random field arises in a variety of areas, including radar and sonar processing, seismic analysis, active noise control, and acoustic holography, to name a few. The most common approach to solving this problem is to identify the peaks in the autospectra of the estimated spectral matrix. This matrix can be obtained by a number of methods. The most popular method is the discrete Fourier transform (DFT) approach. Improved algorithms for computation of the multidimensional DFT have been proposed by [3], [4], while a real time implementation using an FFT chip has recently been proposed by [5]. The autoregressive (AR) method has been popular in settings involving multichannel random processes, and single-channel 2-D random fields (see [6] for references), but has not been considered to any degree for the more general multichannel random field setting. One obvious reason for this is no doubt related to the complexities involved in extensions to the random field setting from the scalar setting (e.g. [7]). The minimum-variance (MV) method, otherwise known as Capon's maximum likelihood method, was originally proposed by Capon [8] in the single-channel 3-D random field setting, and was perhaps the first alternative to the DFT in this setting. Actually, Capon first computed the DFT of the time domain signal, and then applied the MV method to the 2-D spatial dimensions of this transformed data at each time frequency. It has been a popular alternative to the DFT approach in sonar processing, but only as a fixed order estimate. In the d-dimensional setting the MV(n) spectral estimate has the general form [1]:

$$MV_n(z) = [E(z)\hat{R}_n^{-1}E(z)^*]^{-1}$$

where $E(z) = [I, z^{\sigma_1}I, \ldots, z^{\sigma_n}I]$. Here I is the identity on $\mathbb{C}^p$ (p equals the number of channels), $z \in \mathbb{C}^d$ has elements of modulus one and takes the form $e^{i\omega}$ for $d = 1$, $\{\sigma_j\}_{j=1}^{n}$ is an ordering of the d-dimensional lattice, and $z^{\sigma_j} = \prod_{k=1}^{d} z_k^{\sigma_j^{(k)}}$. The $p \times p$ block covariance matrix $\hat{R}_n$ is in general not Toeplitz. Thus, for each order, n, the computation of the MV(n) spectrum will entail $O(np)^4$ operations for each value of z using conventional methods.

Even though the results in [1] provide the theoretical elements for utilizing the convergence properties of the MV(n) spectra in the d-dimensional setting, to implement these results in the fashion addressed in [9] would entail a notable effort. For example, the rate of convergence, so central to that work, will clearly depend on the order in which one proceeds along the d-dimensional lattice (i.e. the ordering $\{\sigma_j\}_1^{\infty}$). For this reason, the method of [1] will not be addressed in this work. Instead, we will present an approach for using the MV(n) spectral approach in [2] to identify harmonic signals in the random field setting. There are instances wherein the random field spectral information can be obtained by considering only a subset of the random field data. For example, if the field is known to be isotropic [e.g. the correlation function $R(\mathbf{x}) = R(\|\mathbf{x}\|)$] and $\mathbf{x}$ is in $\mathbb{C}^d$, then all of the spectral information is contained in the correlation sequence $R_j(k)$, corresponding to the random process associated with any j^{th} component of the random field. Theoretically then, one could obtain the desired spectral information by utilizing spectral methods for random processes. Similarly, if the random field consists of harmonic plane waves then theoretically one could recover their wavenumbers from information contained in the component correlation sequences. Even though this approach would significantly reduce the computational load, it is suboptimal in the sense that it does not utilize all of the information contained in the random field. A compromise between these two extremes would be to somehow include all of this information in the component correlation sequences. The purpose of this work is to present a method for doing just that, in order to identify harmonic signals in arbitrary un known homogeneous noise fields.

2. WAVENUMBER IDENTIFICATION

Consider a p-channel random field

$$\mathbf{Y}(\mathbf{x}) = \mathbf{S}(\mathbf{x}) + \mathbf{N}(\mathbf{x}), \quad \mathbf{x} \in \mathbb{C}^d \tag{1}$$

where the noise field $\mathbf{N}(\mathbf{x})$ is homogeneous and purely nondeterministic, and the signal field is purely deterministic and has the form

$$\mathbf{S}(\mathbf{x}) = \sum_{n=1}^{q} \mathbf{a}_n [e^{(i[<\mathbf{K}_n,\mathbf{x}>+\phi_{1n}])}, \ldots, e^{(i[<\mathbf{K}_n,\mathbf{x}>+\phi_{pn}])}]^t \tag{2}$$

Here $<\mathbf{x},\mathbf{y}> = \mathbf{y}^*\mathbf{x}$ denotes the usual inner product. The purpose of this work is to present a new technique which uses the multichannel random process signal recovery method of [2] to identify the signal wavenumbers $\{\mathbf{K}_n\}_1^q$ and amplitude matrices $\{\mathbf{a}_n\}_1^q$. Under the assumption that the signal phases $\{\phi_{jn};\ n = 1,\ldots,q \text{ and } j = 1,\ldots,p\}$ are mutually independent and identically distributed uniformly over $[0,2\pi]$, the correlation function corresponding to (1) is given by

$$\mathbf{R}_{\mathbf{Y}}(\mathbf{x}) = \mathbf{R}_{\mathbf{S}}(\mathbf{x}) + \mathbf{R}_{\mathbf{N}}(\mathbf{x})$$

$$= \sum_{n=1}^{q} \mathbf{a}_n \mathbf{a}_n^* e^{i<\mathbf{K}_n,\mathbf{x}>} + \mathbf{R}_{\mathbf{N}}(\mathbf{x}), \quad \mathbf{x} \in \mathbb{C}^d . \tag{3}$$

Note, when $d = 1$ then (1) is termed a p-channel random process, and the signal information can be identified using the method of [2].

Our proposed method for identification of the wavenumbers $\{\mathbf{K}_n\}_1^q$ proceeds in two stages. Denote $\mathbf{K}_n = [k_{n1},k_{n2},\ldots,k_{nd}]^t$. The first stage successively identifies the collections $\{k_{nj}\}_{n=1}^q$ for $j = 1,2,\ldots,d$; that is, the collection of first components of each wavenumber are estimated initially, then the second, and so forth. The second stage determines the appro-priate combination of d components, one from each collection, to form the nth wavenumber $\mathbf{K}_n$ for each $n = 1,2,\ldots,q$.
We begin by identifying $\{k_{n1}\}_{n=1}^q$ in no particular order. The same procedure can be applied repeatedly to recover the remaining collections of components. Denote

$$\mathbf{K}_n = [k_{n1},k_{n2},\ldots,k_{nd}]^t \triangleq [k_{n1},\tilde{\mathbf{K}}_n]^t$$

and in (3) let $\mathbf{x} = [x,\tilde{\mathbf{x}}]^t$ where x member $\mathbb{C}$ and $\tilde{\mathbf{x}} \in \mathbb{C}^{d-1}$. The information associated with $\tilde{\mathbf{x}}$ will be condensed by selecting a collection of functions $\{\mathbf{f}_k\}_1^p$ with values in $\mathbb{C}^p$, from the space

$$L^2(\mu,\mathbb{C}^{d-1}) \triangleq \{\mathbf{f}(\tilde{\mathbf{x}}) \mid \int_{\mathbb{C}^{d-1}} \| \mathbf{f}(\tilde{\mathbf{x}}) \|^2 \, d\mu(\tilde{\mathbf{x}}) < \infty\} \tag{4a}$$

where the inner product in this space is denoted

$$<\mathbf{f},\mathbf{g}>_\mu \triangleq \int_{\mathbb{C}^{d-1}} \mathbf{g}^*(\tilde{\mathbf{x}})\mathbf{f}(\tilde{\mathbf{x}})d\mu(\tilde{\mathbf{x}}) \tag{4b}$$

and $\mu(\tilde{\mathbf{x}})$ is a specified measure. The following p-channel random process can now be formed:

$$\mathbf{y}(x) = [y_1(x),y_2(x),\ldots,y_p(x)]^t \tag{5}$$

where

$$y_j(x) = <\mathbf{Y}([x,\tilde{\mathbf{x}}]^t,\mathbf{f}_j(\tilde{\mathbf{x}})>_\mu .$$

From (1) and (2) it follows that

$$y_j(x) = \sum_{n=1}^{q} \mathbf{A}_{jn} [e^{i(k_{n1}x+\phi_{1n})}, \ldots, e^{i(k_{n1}x+\phi_{pn})}]^t$$

$$+ n_j(x), \quad j = 1,2,\ldots,p \tag{6a}$$

where $\mathbf{A}_{jn}$ is a row-vector of the form

$$\mathbf{A}_{jn} = \int_{\mathbb{C}^{d-1}} \mathbf{f}_j^*(\tilde{\mathbf{x}}) e^{i<\tilde{\mathbf{K}}_n,\tilde{\mathbf{x}}>} d\mu(\tilde{\mathbf{x}}) \mathbf{a}_n \tag{6b}$$

and the noise random process is

$$n_j(x) = <\mathbf{N}([x,\tilde{\mathbf{x}}]^t,\mathbf{f}_j(\tilde{\mathbf{x}})>_\mu . \tag{6c}$$

In vector form (6) becomes

$$\mathbf{y}(x) = \sum_{n=1}^{q} \mathbf{A}_n [e^{i(k_{n1}x+\phi_{1n})}, \ldots, e^{i(k_{n1}x+\phi_{pn})}]^t$$

$$+ \mathbf{n}(x) \triangleq \mathbf{s}(x) + \mathbf{n}(x) \tag{7}$$

where $\mathbf{A}_n = [\mathbf{A}_{1n}^t,\mathbf{A}_{2n}^t,\ldots,\mathbf{A}_{pn}^t]^t$. The matrix-alued correlation function $\mathbf{r}_{\mathbf{y}}(x) = E[\mathbf{y}(x+w)\mathbf{y}^*(w)]$ corresponding to the p-channel random process (7) has elements computed from (5) as

$$[\mathbf{r}_{\mathbf{y}}(x)]_{jk} = <\Gamma_x^{\mathbf{Y}}\mathbf{f}_k(\tilde{\mathbf{x}}),\mathbf{f}_j(\tilde{\mathbf{x}})>_\mu \quad j,k = 1,\ldots,p \tag{8a}$$

where the operator

$$\Gamma_x^{\mathbf{Y}}\mathbf{f}_k(\tilde{\mathbf{x}}) = \int_{\mathbb{C}^{d-1}} \mathbf{R}_{\mathbf{Y}}([x,\tilde{\mathbf{x}} - \tilde{\mathbf{w}}]^t)\mathbf{f}_k(\tilde{\mathbf{w}})d\mu(\tilde{\mathbf{w}}) \tag{8b}$$

is on $L^2(\mathbb{C}^{d-1},\mu)$. Notice that (8) admits utilization of all of the information in $\mathbf{R}_{\mathbf{Y}}(x,\tilde{\mathbf{x}})$. The correlation information associated with the first component of $\mathbf{x} = [x,\tilde{\mathbf{x}}]$ is retained, while the remainder of the information is condensed, and depends on the selected functions $\{\mathbf{f}_j\}_1^p$, as well as the measure μ. The form (7) reduced (8) to

$$r_Y(x) = \sum_{n=1}^{q} A_n A_n^* e^{ik_{n1}x} + r_n(x) \triangleq r_s(x) + r_n(x) \ . \quad (9)$$

where $[r_n(x)]_{jk}$ is identical to (8) upon replacing the observed field $\mathbf{Y}$ replaced by the noise field $\mathbf{N}$. The results of [2] for the d = 1 setting are now applicable. In particular, they will identify the collection $\{k_{n1}\}_{n=1}^{q}$ as well as $\{A_n A_n^*\}_1^q$.

Recovery of $\{k_{n2}\}_{n=1}^{q}$ proceeds in the same fashion as above. In this case, one obtains another set of amplitudes which generally will differ from the above set. Also, one might desire to choose a different set $\{f_k\}_1^p$ and a different measure μ. Repeated application will ultimately result in $\{k_{nj}\}_{n=1}^{q}$ and $\{A_n^{(j)} A_n^{(j)*}\}_{n=1}^{q}$ for j = 1,2,...,d. The signal spectral matrices $\{a_n a_n^*\}_1^q$ can then be recovered from this information. Prior knowledge concerning the signal wavenumbers can be incorporated into the selection of the $\{f_k^{(j)}\}_{k=1}^{q}$ as well as $\mu^{(j)}$ for j = 1,...,d to enhance the rate of convergence of the MV(n) spectra to the point spectrum matrix, as demonstrated in the next section.

To identify the wavenumber vectors from the collections $\{k_{nj}\}_{n=1}^{q}$ for j = 1,...,d, without loss of generality, assume that the collection $\{k_{n1}\}_1^q$ has the most distinct components ($\leq$ q). To recover k_{n2} corresponding to k_{n1} we use (3) to compute:

$$\frac{1}{T}\int_{-T/2}^{T/2} \mathbf{R_Y}\left(\begin{bmatrix} x_1 \\ x_2 \\ 0 \end{bmatrix}\right) e^{-ik_{n1}x_1} dx_1 \cong a_n a_n^* e^{ik_{n2}x_2} \ . \quad (10)$$

The approximate equality in (10) will hold for sufficiently large T. A subsequent Fourier transformation of any $(j,k)^{th}$ component of the matrix-valued function of (10) will yield a peak at location k_{n2} with a magnitude approximately equal to the value of $(a_n a_n^*)_{j,k}$. The choice of the (j,k) element in $R_Y(\mathbf{x})$ admits verification. Thus, identification of k_{n2} proceeds by noting the location where the maximum occurs in the Fourier transform of the left hand side of (10). The element in $\{k_{j2}\}_{j=1}^{q}$ closest to this peak is identified as k_{n2}. Alternatively, one can apply the multichannel algorithm of [2] to (10). This would yield an estimate of k_{n2}, as well as $a_n a_n^*$. We remark that the approximate equality in (10) is not necessary. All that is required is that the DFT of (10) yield its largest peak nearer to k_{n2} than to any of the other second components. A large value of T increases the chances of this, and admits an estimate of $a_n a_n^*$.

If k_{n1} is repeated then two such peaks will be present. In the case of a unique k_{n1} we can next recover k_{n3} by computing

$$\frac{1}{T}\int_{-T/2}^{T/2} \mathbf{R_Y}\left(\begin{bmatrix} x_1 \\ 0 \\ x_3 \\ 0 \end{bmatrix}\right) e^{-ik_{n1}x_1} dx_1 \cong a_n a_n^* e^{ik_{n3}x_3} \ . \quad (11)$$

and taking a subsequent Fourier transform. If k_{n1} has identified two second components k_{n2_1} we compute

$$\frac{1}{T}\int_{-T/2}^{T/2} \mathbf{R_Y}\left(\begin{bmatrix} x_1 \\ x_2 \\ x_3 \\ 0 \end{bmatrix}\right) e^{-i(k_{n1}x_1 + k_{n2}x_2)} dx_1 dx_2$$

$$\cong a_n a_n^* e^{ik_{n3}x_3} \ . \quad (12)$$

Example. Consider a 3-channel 2-dimensional random field setting. The noise field, N(**x**), has a correlation function given by

$$\mathbf{R_N}(\mathbf{x}) = e^{-0.5|x_1|} e^{-0.5|x_2|} \mathbf{I} \ . \quad (13)$$

This type of noise field is found in image processing. The signal field includes two components with amplitude matrices and wavenumbers in Table 1.

Table 1. Signal wavenumber and amplitude information.

$K_1/2\pi$	a_1			$K_2/2\pi$	a_2		
0.129	10	0	0	0.254	10	4	0
0.000	0	13	0	0.160	0	8	0
	0	0	16		0	0	0.

Our measure is chosen to be $\mu(x) = 1$ for $|x| < 64$, and our 3-channel functions are $f_1(x) = [1,0,0]^t$, $f_2(x) = [0,1,0]^t$ and $f_3(x) = [0,0,1]^t$ for all $x \in \mathbb{C}$.

The collection {0.129,0.254} of first components, normalized by 2π, are easily identified by the locations of convergence on the minimum eigenvalue plot of Figure 1(a). That 0.125 is present on all three channels, while 0.254 is present on only two is suggested by Figures 1(b-c). Identification of the collection {0,0.160} of normalized second components and their presence on the appropriate number of channels proceeds in exactly the same way using the information in Figure 2. To determine which second component corresponds to the first component, k_{11}= 0.129, the Fourier transform of the (1,1) entry of (10) was computed. The peak correctly identified k_{12}= 0.0. The value of this peak also gave a good estimate of the (1,1) component of $a_n a_n^*$. In particular, (6b) yielded

$$A_n A_n^* = [\sin(Tk_{n2})/k_{n2}]^2 a_n a_n^* \ . \quad (14)$$

Using the estimate of $A_n A_n^*$ obtained from the MV(160) spectrum resulted in the power estimates given in Table 2.

Table 2. Signal spectral power estimates obtained from (14). [True values are given in parentheses.]

$a_1 a_1^*$			$a_2 a_2^*$		
100(100)	0(0)	0(0)	131(116)	34(32)	0(0)
0(0)	169(169)	0(0)	34(32)	65(64)	0(0)
0(0)	0(0)	256(256)	0(0)	0(0)	0(0)

The estimate of $\mathbf{a}_1\mathbf{a}_1^*$ is perfect, while that of $\mathbf{a}_2\mathbf{a}_2^*$ is slightly high. Notice that because one can only recover $\mathbf{a}_n$ up to a right unitary transformation, the information in $\mathbf{a}_n\mathbf{a}_n^*$ does not readily yield the relative source strengths measured at a given receiver. Its diagonals do give the total power measured at that channel. The diagonal structure of $\mathbf{a}_1\mathbf{a}_1^*$ in Table 2 does reflect the fact that each receiver is related to a distinct source.

REFERENCES

[1] Frazho, A.E. and Sherman, P.J., "A Geometric Approach to Multichannel, Multidimensional Harmonic Signal Analysis in Nonstationary Noise", Proc. 28th Annual Allerton Conference on Communications, Control and Computing, Oct. 3-5, 1990, University of Illinois, Champaign, Urbana, IL.

[2] Foias, C., Frazho, A.E. and Sherman, P.J., "A new approach for determining the spectral data of multichannel harmonic signals in noise", *Mathematics of Control, Signals and Systems*, **3**(1), 31-43, January 1990.

[3] Auslander, L., Feig, E. and Winograd, S., "New Algorithms for the Multidimensional Discrete Fourier Transform", IEEE Trans. on ASSP, Vol. ASSP-1, No. 2, pp. 388-403, April 1983.

[4] Shamesh, M. and Gertner, I., "A New Parallel Algorithm for the Multidimensional Fourier Transform Processing", ICASSP, Vol. 4, pp. 1993-1996, Albuquerque, NM, 1990.

[5] Magotra, N., Bitto, J. and McCoy, J.W., "Real Time Implementation of Multidimensional Fast Fourier Transforms", ICASSP, Vol. 3, pp. 1767-1770, Albuquerque, NM, 1990.

[6] Dudgeon, D.E. and Mersereau, R.M., "*Multidimensional Digital Signal Processing*", Prentice Hall, 1984.

[7] Rudin, W., "The extension problem for positive-definite functions", Ill. Jnl. Math., Vol. 7, pp. 523-539, 1963.

[8] Capon, J., "High-resolution frequency-wavenumber spectrum analysis", *Proceedings of the IEEE*, **57**, 1408-1418, 1969.

[9] Lyon, D.E., Sherman, P.J. and Frazho, A.E., "On the performance of the family of multichannel ML spectra for recovery of point spectrum", *5TH ASSP Workshop on Spectrum Estimation and Modeling*, Rochester, NY, October 10-12, 1990.

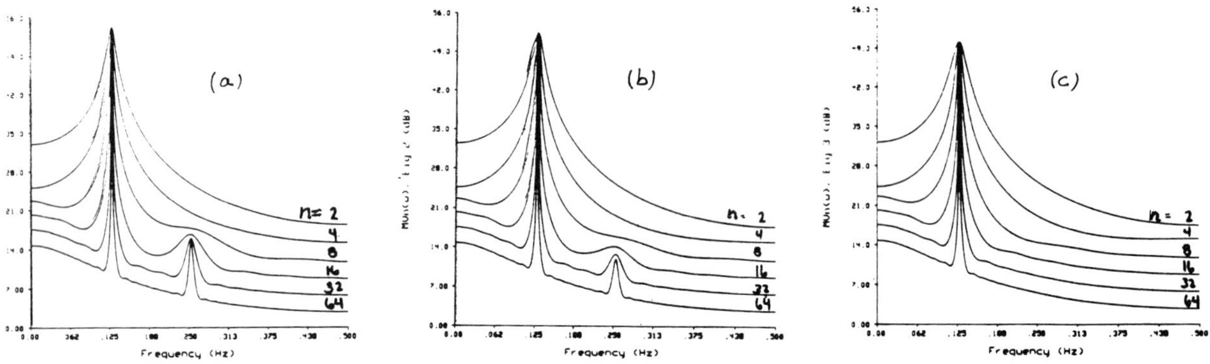

Figure 1. MV(k) spectra corresponding to the collection of first wavenumber components in Example 3. The eigenvalues $\lambda_1(k) \le \lambda_2(k) \le \lambda_3(k)$ corresponding to MV(k) are given in (a)-(c), respectively.

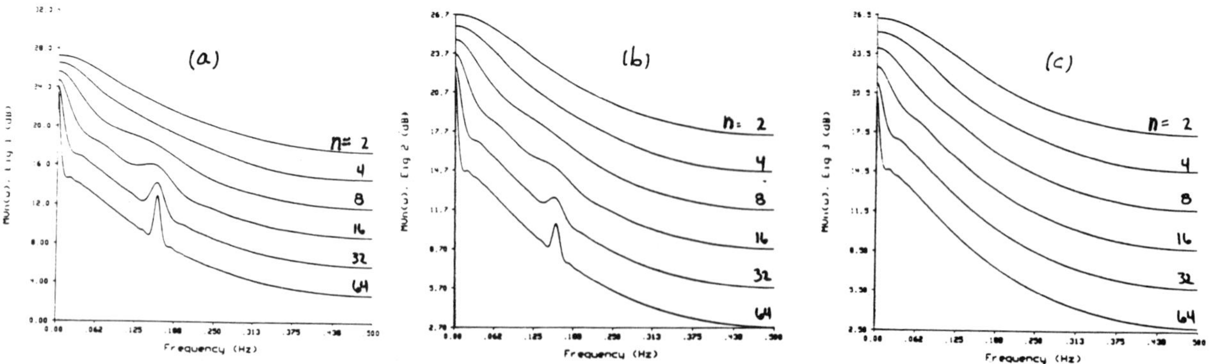

Figure 2. MV(k) spectra corresponding to the collection of second wavenumber components in Example 3. The eigenvalues $\lambda_1(k) \le \lambda_2(k) \le \lambda_3(k)$ corresponding to MV(k) are given in (a)-(c), respectively.

SIGNAL PROCESSING VI: Theories and Applications
J. Vandewalle, R. Boite, M. Moonen, A. Oosterlinck (eds.)

A RECURSIVE LEAST SQUARE APPROACH WITH EXPONENTIAL WEIGHTING FACTOR TO SHORT TIME FOURIER ANALYSIS

P.Gruber

C-R&D Landis & Gyr Betriebs AG
CH-6301 Zug, Switzerland

A recursive filter is derived which is able to track the amplitudes and phases of frequency components with known frequencies that are part of an arbitrary measured signal. The filter is based on the recursive least square method with exponential weighting factor. In the case of estimating the phase and amplitude of one frequency only, the complete parametrized filter is solved analytically. The resulting filter is closely related to the Kalman filter solution to that estimation problem based on a random walk model. Its structure is also similar to a discrete quadrature/correlator structure.

1. INTRODUCTION

In many application one is interested in estimating in real time the parameters $a_i(k)$, $i=0,..,M$ and $b_i(k)$, $i=1,..,M$ of a signal consisting of a sum of sinusoidal functions with time varying amplitudes and phases but with known and constant frequencies, which is part of an arbitrary measured signal y(k). y(k) can thus be represented as

$$y(k) = a_0(k) + \sum_{i=1}^{M}[A_i(k)\cos(\omega_i kT-\psi_i(k))] + w(k)$$

$$= a_0(k) + \sum_{i=1}^{M}[a_i(k)\cos\omega_i kT + b_i(k)\sin\omega_i kT] + w(k) \quad (1)$$

$a_i(k) = A_i(k)\cos\psi_i(k) \quad i=1.,,.M$
$b_i(k) = A_i(k)\sin\psi_i(k)$
T: sampling period
w(k): measurement noise,disturbances

The frequencies can be chosen arbitrarily except for the condition that the time variation of the signal parameters $a_i(k)$ and $b_i(k)$ are much slower than the frequencies ω_i. If, for instance,

i) the frequencies ω_i are related by
$\omega_i = i\omega_1 \quad i=1,..,M \quad \omega_1 = 2\pi/\tau$
$\tau = NT \quad N = 2M$

ii) the measurement noise w(k) is equal to zero for all k and

iii) the parameters are constant,

then the right hand side of (1) is the Fourier representation of the signal y(k). For the estimation of the time varying parameters various methods can be used. The classical short time Fourier analysis using either block transforms or discrete running Fourier series methods is strongly noise dependent and does not allow an arbitrary selection of frequencies ω_i. That's why in [1] a random walk model was chosen for the variation of the parameters and then the Kalman filter theory has been applied to solve the estimation problem, resulting in an optimal filter with periodic coefficients. Another possibility is to assume an underlying oscillator model leading to a different Kalman filter structure [2]. Comparisons of the two above methods and the Fourier transform method are given in [3]. Here a recursive least square method with a constant exponential weighting factor λ is used for the estimation of the time varying parameters. Fig.1 shows the signal processing scheme for the estimation of the amplitudes and phases. In a first stage the coefficients $\underline{a}(k)$ and $\underline{b}(k)$ are estimated by a linear estimator. In a second stage the estimates of the amplitudes $\underline{A}(k)$ and the phases $\underline{\psi}(k)$ are determined by a nonlinear transformation:

$$\hat{A}_i(k) = \sqrt{\hat{a}_i(k)^2+\hat{b}_i(k)^2}$$
$$\hat{\psi}_i(k) = \mathrm{arctg}(\hat{b}_i(k)/\hat{a}_i(k))$$

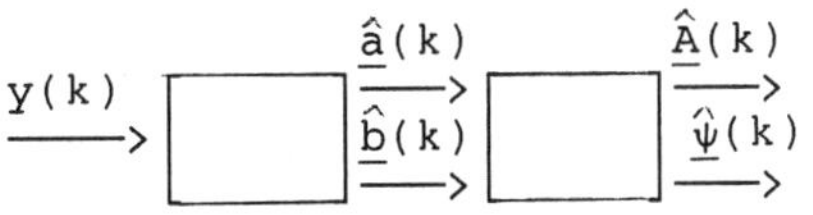

y(k) ──> [ ] ──$\hat{\underline{a}}(k)$, $\hat{\underline{b}}(k)$──> [ ] ──$\hat{\underline{A}}(k)$, $\hat{\underline{\psi}}(k)$──>

Fig.1: Estimation scheme

2. DERIVATION OF THE FILTER EQUATIONS

The performance index to be minimized is given by [4]

$$J(k,\lambda,\underline{x}(k)) = \sum_{i=1}^{k}\lambda^{k-i}e(i)^2 + \lambda^k J_0$$

$$= \sum_{i=1}^{k}\lambda^{k-i}(y(i)-\underline{c}'(i)\underline{x}(i))^2 + \lambda^k J_0$$

J_0: contribution of the initial values
$\underline{c}'(k) = [1,\cos(\omega_1 kT),\sin(\omega_1 kT),......]$
$\underline{x}(k) = [a_0(k),a_1(k),b_1(k),......]$

To simplify the expressions we concentrate us here on one frequency only and set $x_1(k) = a_1(k) \qquad x_2(k) = b_1(k)$

Then $J(k,\lambda,\underline{x}(k))$ is given by

$$J(k,\lambda,\underline{x}(k))=\sum_{i=1}^{k}\lambda^{k-i}\{y(i)-x_1(i)\cos(\omega_1 iT)-x_2(i)\sin(\omega_1 iT)\}^2+\lambda^k J_0 \quad (2)$$

The recursive solution for the estimates $\hat{\underline{x}}(k)$ minimizing (2) can be written as

$$\hat{\underline{x}}(k)=\hat{\underline{x}}(k-1)+\underline{K}(k)(y(k)-\underline{c}'(k)\hat{\underline{x}}(k-1)) \quad (3)$$

$$\underline{K}(k)=P(k-1)\underline{c}(k)[r+\underline{c}'(k)P(k-1)\underline{c}(k)]^{-1} \quad (4)$$

$$P(k)=\lambda^{-1}(P(k-1)-\underline{K}(k)\underline{c}'(k)P(k-1)) \quad (5)$$

$\hat{\underline{x}}(0)=\hat{\underline{x}}_0$ $\quad P(0)=P_0$ positive definit
$0<\lambda<1$ $\quad r>0$ normally 1

If the least square method is formulated in a stochastic way, $P(k)$ corresponds to the symmetric error covariance matrix $P(k)=\mathrm{Var}[\underline{x}(k)-\hat{\underline{x}}(k)]$

$$P(k)=\begin{bmatrix}p_1(k) & p_3(k)\\ p_3(k) & p_2(k)\end{bmatrix} \qquad \phi=2\pi/N$$

$$\underline{c}'(k)=[\cos k\phi \quad \sin k\phi]=[c_1(k)\ \ c_2(k)]$$

$\underline{K}(k)$ is a time-varying gain that weighs the error between the incoming signal $y(k)$ and its one step ahead estimate $\hat{y}(k)=\underline{c}'(k)\hat{\underline{x}}(k-1)$.

3. DIFFERENCE EQUATIONS FOR P(k)

By introducing the new variables

$$Z(k)=p_1(k)c_1(k+1)^2+p_2(k)c_2(k+1)^2+2p_3(k)c_1(k+1)c_2(k+1)+r \quad (6)$$

$$D(k)=p_1(k)p_2(k)-p_3(k)^2 \quad (7)$$

one can rewrite equations (4) and (5) in the following way:

$$p_1(k+1)=\lambda^{-1}[D(k)c_2(k+1)^2+rp_1(k)]/Z(k) \quad (8)$$

$$p_2(k+1)=\lambda^{-1}[D(k)c_1(k+1)^2+rp_2(k)]/Z(k) \quad (9)$$

$$p_3(k+1)=[rp_3(k)-D(k)c_1(k+1)c_2(k+1)/\lambda Z(k) \quad (10)$$

This represents a set of <u>three nonlinear time varying difference equations</u>. The solutions can be determined independent of the data $y(k)$. The new variable $D(k)$ is the determinant of the the positiv definit matrix $P(k)$ and is therefore >0, and $Z(k)$, the denominator of equation (5) is a quadratic function of $\underline{c}(k)$ and thus also >0 for $r>0$. Fig.2 shows the behaviour of the elements $p_1(k),p_2(k)$ and $p_3(k)$ of $P(k)$ for the case $N=20$, $\lambda=0.9$ and $p_1(0)=p_2(0)=0.1$ and $p_3(0)=0$.

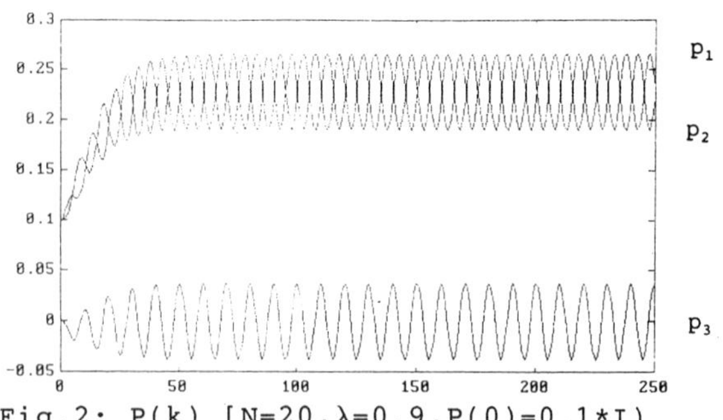

Fig.2: $P(k)$ $[N=20,\lambda=0.9,P(0)=0.1*I)$

Due to the periodicity in $\underline{c}(k)$, periodic solutions are obtained for $P(k)$ in the steady state. For λ values close to 1 (e.g $\lambda=0.99$), the steady state is much slower reached than for smaller λ values (e.g. $\lambda=0.9$). In the limiting case $\lambda=1$ all $p_i(k)$ values tend to zero for $k\to\infty$, which means that the gain $\underline{K}(k)$ tends towards zero too. In that case no adaptation would be possible. The periodic states of $P(k)$ are sinusoidal with an offset for $p_1(k)$ and $p_2(k)$ and frequency $2\omega_1$ (double frequency of $\underline{c}(k)$ frequency content) and with a 180° phase shift bet ween $p_1(k)$ and $p_2(k)$ and a $\underline{+}90°$ phase shift between $p_1(k)$ [or $p_2(k)$] and $p_3(k)$ Beside $Z(k)$ and $D(k)$ the following two functions of $P(k)$ are of interest as can be seen later

$$T(k)=p_1(k)+p_2(k)=\text{trace of } P(k) \quad (11)$$

$$S(k)=[c_1(k+1))^2-s_1(k+1)^2]p_3(k)+c_1(k+1)c_2(k+1)[p_2(k)-p_1(k)] \quad (12)$$

The steady state of the variables $Z(k)$ $D(k)$, $T(k)$ and $S(k)$ are constant and denoted by Z^0, D^0, T^0 and S^0. The values of Table 1 and 2 were found by simulations. They are a function of N and λ. It will be shown in part 4 that these steady state values can also be found by solving a system of algebraic equations. The maximum of p_1 and p_2 is a function of N and λ too. It can easily be computed from the steady state values D^0 and T^0. From equations (7) and (11) follows

$$p_1(k)=T^0/2\pm\sqrt{(T^0/2)^2-D^0-p_3(k)^2} \quad (13)$$

Equation (13) clearly has a maximum if $p_3(k)=0$, that means

$$p_{1max}=T^0/2+\sqrt{(T^0/2)^2-D^0} \quad (14)$$

Table 1 shows that for a given N the denominator Z^0 decreases with increasing λ with a limit $Z^0=1$ for $\lambda=1$, wheras the other variables T^0, D^0 and S^0 decrease towards zero. Table 2 shows the dependence of the steady states on the number of samples N per period. It is interes-

λ	Z^0	T^0	D^0	S^0	p_{1max} (or p_{2max})
0.5	4.	9.236	9.236	3.078	8.095
0.8	1.562	1.131	0.283	0.192	0.758
0.9	1.235	0.457	0.051	0.038	0.267
0.95	1.108	0.212	0.011	0.009	0.112
0.99	1.022	0.043	0.0005	0.0003	0.026
1	1	0	0	0	0

Table 1: Steady state values of Z,T,D and S and p_{11max} for different λ values and fixed N (N=20) and r (r=1)

N	Z^0	T^0	D^0	S^0	p_{1max}
4	1.235	0.446	0.050	0.005	0.236
8	1.235	0.447	0.050	0.012	0.239
16	1.235	0.453	0.050	0.030	0.258
32	1.235	0.477	0.053	0.062	0.301
64	1.235	0.573	0.064	0.125	0.422
128	1.235	0.957	0.106	0.251	0.829
256	1.235	2.494	0.277	0.503	2.378
512	1.235	8.643	0.960	1.006	3.530
1024	1.235	33.236	3.693	2.012	33.124

Table 2: Steady state values of Z,T,D and S and p_{1max} for different N values and fixed λ (λ=0.9) and r (r=1)

ting to see that Z^0 is independent on N, while T^0, D^0 and S^0 are increasing with increasing N.

4. DIFFERENCE EQUATIONS FOR D(k),T(k), Z(k) AND S(k)

Writing down the difference equations for the new variables T(k),D(k),S(k) and Z(k) yields after some tedious manipulations with $\phi=2\pi/N$

$$T(k+1)=\lambda^{-1}[D(k)+rT(k)]/Z(k) \tag{15}$$

$$D(k+1)=\lambda^{-2}rD(k)/Z(k) \tag{16}$$

$$Z(k+1)=\lambda^{-1}[\{D(k)+rT(k)\}\sin^2\phi+r(Z(k)-r)\cos 2\phi+rS(k)\sin 2\phi+r\lambda Z(k)]/Z(k) \tag{17}$$

$$S(k+1)=\lambda^{-1}[0.5\{D(k)+rT(k)-2r(Z(k)-r)\}\sin 2\phi+rS(k)\cos 2\phi]/Z(k) \tag{18}$$

It is interesting to note that the above system of nonlinear difference equations is <u>time invariant</u>. The steady state of this system correspond to the constant values Z^0,T^0,D^0 and S^0. These values can be found by solving the following algebraic equations:

$$T^0=\lambda^{-1}[D^0+rT^0]/Z^0 \tag{19}$$

$$D^0=\lambda^{-2}rD^0/Z^0 \tag{20}$$

$$Z^0=\lambda^{-1}[(D^0+rT^0)\sin^2\phi+r(Z^0-r)\cos 2\phi+rS^0\sin 2\phi+r\lambda Z^0]/Z^0 \tag{21}$$

$$S^0=\lambda^{-1}[0.5\{D^0+rT^0-2r(Z^0-r)\}\sin 2\phi+rS^0\cos 2\phi]/Z^0 \tag{22}$$

Although the equations are nonlinear they can be solved analytically. If a Kalman filter is used for the estimation the corresponding system of nonlinear equations (19)-(22) cannot be solved analytically. From equation (20) follows

$$Z^0 = r/\lambda^2 \tag{23}$$

which is indeed independent on N or ϕ. Equation (19) yields with (23)

$$D^0 = (1-\lambda)rT^0/\lambda \tag{24}$$

If D^0 and Z^0 are plugged into equations (21) and (22) one obtains the following system of linear equations:

$$A\begin{bmatrix}S^0\\ T^0\end{bmatrix} = \underline{b};\quad A = \begin{bmatrix}\lambda\sin 2\phi & \sin^2\phi\\ (1-\lambda\cos 2\phi) & -0.5\sin 2\phi\end{bmatrix}$$

$$\underline{b} = \begin{bmatrix}r\{(\lambda^{-2}-1-\lambda^{-1}(1-\lambda^2)\}\cos 2\phi\\ -r\lambda^{-1}(1-\lambda^2)\sin 2\phi\end{bmatrix}$$

This can be solved leading to

$$S^0=r(1-\lambda^{-1})^2\mathrm{ctg}\phi \tag{25}$$

$$T^0=r(1-\lambda)(1+\lambda^{-2}-2\lambda^{-1}\cos 2\phi)/\sin^2\phi \tag{26}$$

Equations (23)-(26) represent a complete set of solutions. All steady states are linearly dependent on r.

5. THE RESULTING FILTER

The definition equations of S(k),T(k) and Z(k) can be put into the following form:

$$\begin{bmatrix}1 & 1 & 0\\ -\alpha & \alpha & \beta-\alpha\\ \beta & \gamma & 2\alpha\end{bmatrix}\begin{bmatrix}p_1(k)\\ p_2(k)\\ p_3(k)\end{bmatrix} = \begin{bmatrix}T(k)\\ S(k)\\ Z(k)-r\end{bmatrix}$$

with $\alpha=c_1(k)c_2(k)=0.5\sin 2k\phi$
$\beta=c_1(k)^2=\cos^2 k\phi$; $\gamma=c_2(k)^2=\sin^2 k\phi$
From this we get the element of P(k):

$$\begin{bmatrix}p_1(k)\\ p_2(k)\\ p_3(k)\end{bmatrix} = \begin{bmatrix}\gamma & -2\alpha & \beta-\gamma\\ \beta & 2\alpha & \gamma-\beta\\ -\alpha & \beta-\gamma & 2\alpha\end{bmatrix}\begin{bmatrix}T(k)\\ S(k)\\ Z(k)-r\end{bmatrix} \tag{27}$$

The gain vector $\underline{K}(k)$ for the correction of the estimates $\underline{x}(k)$ are obtained from P(k) leading to time varying gains. The gains can be computed offline. They are given with equations (4) and (27) by

$$K_1(k)=-\sin k\phi(S(k)/Z(k))+\cos k\phi(1-r/Z(k))$$

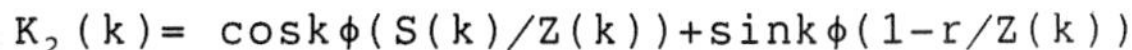

$K_2(k) = \cos k\phi(S(k)/Z(k)) + \sin k\phi(1 - r/Z(k))$

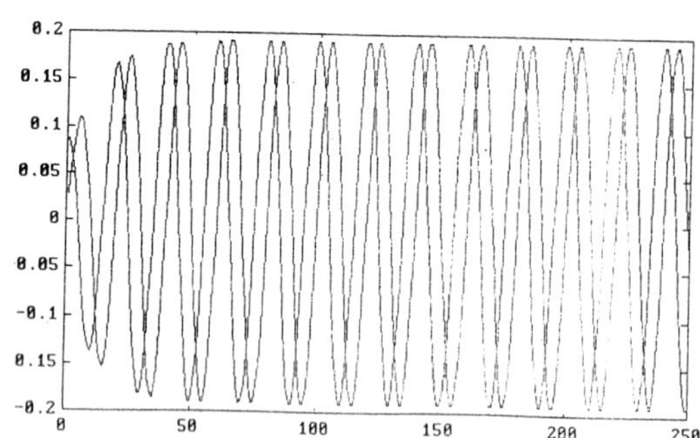

Fig.3: Gain vector $\underline{K}(k)$ $[N=20, \lambda=1]$

Fig.3 shows the gains for the example mentioned above. In the steady state the gains reduce to the simple form:

$$K_1(k) = -\sin k\phi(1-\lambda)^2 \operatorname{ctg}\phi + \cos k\phi(1-\lambda^2)$$

$$K_2(k) = \cos k\phi(1-\lambda)^2 \operatorname{ctg}\phi + \sin k\phi(1-\lambda^2)$$

It is interesting to note that the steady state filter is independent on r. Fig 4 shows the resulting steady state filter structure for the fundamental harmonic component. The gain is split into a harmonic part (I) and a time-varying part (II) whose coefficients converge to the above constant values in the steady state.

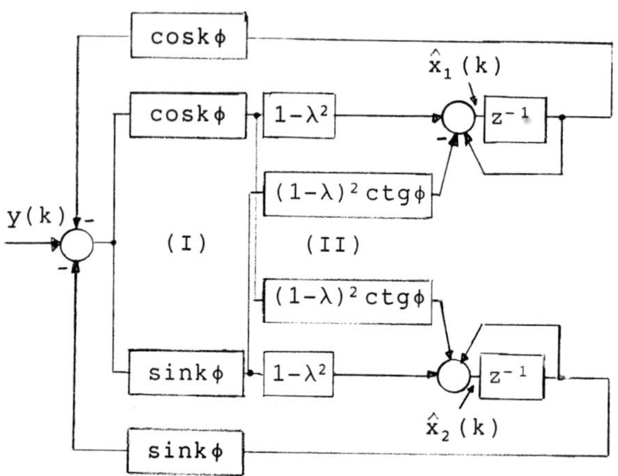

Fig.4: Filter structure of the recursive least square estimator with exponential weighting factor

The resulting estimator structure is similiar to the one that is obtained if a Kalman filter estimator based on a random walk model is used (see [1]). It is also interesting that the structure is related to the quadrature demodulation scheme of Fig.5.

The difference of the quadrature demodulator are:

1) the missing time varying feedback

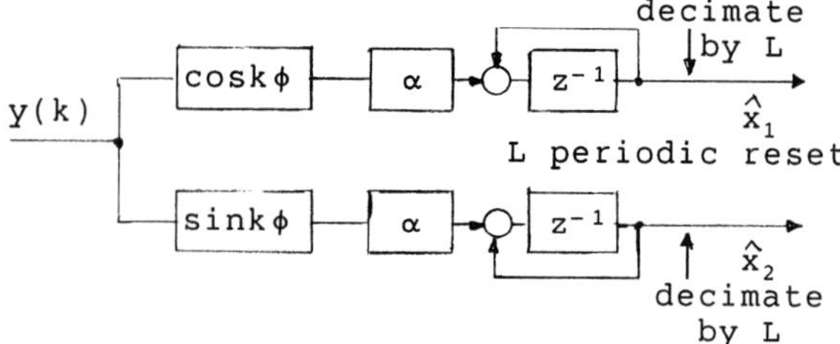

Fig.5: discrete quadrature demodulator front end

loop generating the error signal $y(k) - \hat{y}(k)$

2) the links in the forward path between the sin and cos path
3) summation over a finite interval and reset for the determination of the estimates.

6. CONCLUSION

An estimator has been derived that allows to estimate on line the time varying amplitude and phase of a sinusoid of given frequency in the presence of other sinusoidal disturbances with known frequencies and noise. It is based on the recursive least square method with exponential weighting factor. The method can be extended to the determination of a higher order filter for the estimation of time-varying amplitudes and phases of several sinusoids of known frequencies. Analytical results however are not yet available for that case. The structure of the optimal filter has some interesting properties related to Kalman filters and quadrature demodulators.

7. ACKNOWLEDGEMENT

The author likes to thank J.Tödtli for his valuable discussions.

[1] P.Gruber, J.Tödtli: A recursive estimator for the determination of unknown constant or slowly varying Fourier coefficients, EUSIPCO 88, p.1433-1436, Grenoble

[2] R.R.Bitmead, A.C.Tsoi, P.J.Parker: A Kalman filter approach to short time Fourier analysis, IEEE TR-ASSP, Vol 34, no 6, Dec 86.

[3] P.Gruber,J.Tödtli: Estimation of quasiperiodic signal parameters by means of dynamic signal models, Internal report 89, C-R & D, Landis & Gyr Betriebs AG, Zug, Switzerland, submitted to IEEE -ASSP Transactions

[4] J.Tödtli: Rekursive Parameterschätzverfahren nach der Methode der kleinsten Quadrate: zugehörige Optimierungsprobleme und direkte Lösungen, Scientia Electrica Vol 29, no.4, 1983, S.1-46

SIGNAL PROCESSING VI: Theories and Applications
J. Vandewalle, R. Boite, M. Moonen, A. Oosterlinck (eds.)

Doubly Normalized Schur RLS Lattice Filter and CORDIC Array Architecture

Peter Strobach

Siemens AG, Zentralabteilung Forschung und Entwicklung
ZFE ST SN 61, Otto-Hahn-Ring 6, W-8000 München 83, Germany

Abstract - This paper presents the following novel results in Schur RLS adaptive filtering: (1) It is shown that both boundary and internal cells of the square-root Schur RLS adaptive filter array can be realized with the same elementary CORDIC cell. (2) This CORDIC cell can be used additionally in a second order moment estimation section based on recursive Gray-Markel lattice filters. (3) The structure of the square-root Schur RLS adaptive filter is invariant with respect to an additional input power normalization. (4) All internal variables of the doubly (input and residual) normalized algorithm are scaled to a value of less or equal one in magnitude and fully utilize this interval in successive stages due to the autoscaling feature of hyperbolic rotations. These important properties make the presented CORDIC array of the doubly normalized Schur RLS adaptive filter an ideal candidate for a systolic VLSI implementation.

I. Introduction

Schur algorithms have been a subject of widespread interest in signal processing [1-8]. Their square-root forms have the key feature that they provide a Cholesky factorization of an underlying covariance matrix by local hyperbolic rotations and hence are the natural counterparts of QR-based algorithms which provide a Cholesky factorization by local circular rotations. Recent investigations [6-8] have shown that the square-root Schur approach extends to the recursive least squares (RLS) adaptive filtering case. The resulting square-root Schur RLS adaptive filter [7] has outstanding numerical and structural properties. It provides an efficient Cholesky factorization of the (non-Toeplitz) data covariance matrix and facilitates an easy incorporation of smooth finite or infinite forgetting functions in the RLS time updating process. Thus Schur RLS adaptive filters can outperform other types of RLS adaptive filters which are restricted on the application of simple exponential or rectangular forgetting functions, or simple modifications thereof.

It is the purpose of this paper to note some further properties of the square-root Schur RLS adaptive filter presented in [7]. First, we want to show that even in the true nonstationary case, the boundary cell of the square-root Schur RLS adaptive filter array of [7] can be computed with the same hyperbolic CORDIC cell as is used for internal cells. The resulting pure CORDIC structure suggests that the second order moments (inputs of the triangular array) may also be computed with a CORDIC based structure. We show how this can be accomplished on the basis of recursive Gray-Markel lattice filters. Finally, we disvcovered the particularly useful property that the square-root Schur RLS adaptive filter is invariant with respect to input power normalization. In case of input power normalization, all internal variables of the algorithm remain bounded in the unit interval and fully utilize it in successive stages due to a nice autoscaling feature of the involved hyperbolic rotations. These properties have a significant impact on the VLSI realization of the algorithm. It is conjectured that the presented structures can become a useful part of future high-speed adaptive equalizer/demodulator circuits.

II. The Square-Root Schur RLS Adaptive Filter

The square-root Schur RLS adaptive filter [7] is based on the feed-forward square-root lattice form, a cascaded structure of hyperbolic lattice rotations,

$$\begin{bmatrix} \mathbf{v}_m(t) \\ \mathbf{w}_m(t) \end{bmatrix} = \boldsymbol{\Theta}_m(t) \begin{bmatrix} \mathbf{v}_{m-1}(t) \\ \mathbf{w}_{m-1}(t-1) \end{bmatrix} , \tag{1a}$$

where $\boldsymbol{\Theta}_m(t)$ is the square-root lattice rotor,

$$\boldsymbol{\Theta}_m(t) = \left[1 - \rho_m^2(t)\right]^{-1/2} \begin{bmatrix} 1 & \rho_m(t) \\ \rho_m(t) & 1 \end{bmatrix} , \tag{1b}$$

and

$$\mathbf{v}_m(t) = \mathbf{e}_m(t) E_m^{-1/2}(t) , \tag{2a}$$

$$\mathbf{w}_m(t) = \mathbf{r}_m(t) R_m^{-1/2}(t) , \tag{2b}$$

are normalized residual vectors with initial condition $\mathbf{e}_0(t) = \mathbf{r}_0(t) = \mathbf{x}(t)$, where

$$\mathbf{x}(t) = \left[x(t), x(t-1), x(t-2), \ldots\right] \tag{3}$$

is a sliding or growing window observation of $\{x\}$. Note that $E_m(t)$ and $R_m(t)$ are forward/backward residual energies:

$$E_m(t) = \mathbf{e}_m^T(t)\mathbf{e}_m(t) , \tag{4a}$$

$$R_m(t) = \mathbf{r}_m^T(t)\mathbf{r}_m(t) . \tag{4b}$$

Thus, by definition, $\mathbf{v}_m^T(t)\, \mathbf{v}_m(t) = 1$ and the same holds for $\mathbf{w}_m(t)$. The interesting point with $\boldsymbol{\Theta}_m(t)$ is that this rotor is J-unitary, i.e., $\boldsymbol{\Theta}_m(t)$ represents a standard hyperbolic rotation and thus can be implemented easily by using standard vector rotation hardware, such as Coordinate Rotation Digital Computer (CORDIC) processors [9].

The square-root lattice rotor is uniquely determined by the square-root lattice parameter $\rho_m(t)$,

$$\rho_{m+1}(t) = - \mathbf{v}_m^T(t)\mathbf{w}_m(t-1) , \tag{5}$$

which has also the physical meaning of a reflection coefficient in scattering theory. This physical interpretation suggests that we must have $|\rho_m(t)| \leq 1$. An order recursion of $\rho_m(t)$ can be defined on the basis of square-root generalized covariances (inner products of shifted data and normalized residual vectors):

$$D_{m,j}^f(t) = \mathbf{x}^T(t-j)\mathbf{v}_m(t) , \tag{6a}$$

$$D_{m,j}^b(t) = \mathbf{x}^T(t-j)\mathbf{w}_m(t) . \tag{6b}$$

The key result in square-root Schur RLS adaptive filtering is that an order recursion of the square-root generalized covariances exists and this recursion has exactly the structure of the square-root lattice filter:

$$\begin{bmatrix} D^f_{m,j}(t) \\ D^b_{m,j}(t) \end{bmatrix} = \boldsymbol{\Theta}_m(t) \begin{bmatrix} D^f_{m-1,j}(t) \\ D^b_{m-1,j-1}(t-1) \end{bmatrix} \quad . \tag{7}$$

Next note that from (2), (4), (5) and (6a) it follows that the reflection coefficient can be expressed as the product of the normalized generalized forward covariance of degree m+1 and the inverse square-root backward residual energy:

$$\rho_{m+1}(t) = -\mathbf{v}^T_m(t)\mathbf{w}_m(t-1) = -D^f_{m,m+1}(t)\,R^{-1/2}_m(t-1) \quad . \tag{8}$$

In order to find a recursion for $R_m(t)$ one exploits a result from linear prediction theory [8], namely,

$$R_m(t) = \mathbf{x}^T(t-m)\mathbf{r}_m(t) = \mathbf{x}^T(t-m)\,\mathbf{w}_m(t)\,R^{1/2}_m(t) \quad . \tag{9}$$

Thus it is easy to check [comparison with (6b)] that

$$R^{1/2}_m(t) = \mathbf{x}^T(t-m)\,\mathbf{w}_m(t) = D^b_{m,m}(t) \quad . \tag{10}$$

Strictly speaking, expression (10) reveals that the square-root of the backward residual energy is identical with the normalized generalized backward covariance of order m and degree m. Thus we can ultimately use rotation (7) for a computation of $R^{1/2}_m(t)$, or equivalently, $D^b_{m,m}(t)$ as follows:

$$\begin{bmatrix} D^f_{m,m}(t) \\ D^b_{m,m}(t) \end{bmatrix} = \boldsymbol{\Theta}_m(t) \begin{bmatrix} D^f_{m-1,m}(t) \\ D^b_{m-1,m-1}(t-1) \end{bmatrix} \quad . \tag{11}$$

Even more notable is the role of the quantity $D^f_{m,m}(t)$, which appears as a by-product in the computation of $D^b_{m,m}(t)$ via (11). From (10) and (8) it follows that

$$\rho_m(t) = -D^f_{m-1,m}(t) \,/\, D^b_{m-1,m-1}(t-1) \quad . \tag{12}$$

Thus an evaluation of the upper equation in (11) gives

$$D^f_{m,m}(t) = 0 \tag{13}$$

for all practical values of $\rho_m(t)$. Hence one may employ a single CORDIC processor for annihilation of the top component $D^f_{m,m}(t)$ by iterative hyperbolic rotation (vectoring mode) according to

$$\begin{bmatrix} D^f_{m,m}(t) \longrightarrow 0 \\ D^b_{m,m}(t) \end{bmatrix} = \boldsymbol{\Theta}_m(t) \begin{bmatrix} D^f_{m-1,m}(t) \\ D^b_{m-1,m-1}(t-1) \end{bmatrix} \quad . \tag{14}$$

As soon as $D^f_{m,m}(t)$ is sufficiently close to zero one obtains the desired value of $D^b_{m,m}(t)$ at the second output of the CORDIC processor. Most notable in this context is the fact that this operation additionally produces the reflection coefficient $\rho_m(t)$ implicitly as a "by-product" in terms of the resulting "angular variable" $\varphi_m(t)$ because

$$\sinh \varphi_m(t) = \rho_m(t)\left[1 - \rho^2_m(t)\right]^{-1/2} \quad , \tag{15a}$$

$$\cosh \varphi_m(t) = \left[1 - \rho^2_m(t)\right]^{-1/2} \quad . \tag{15b}$$

The recursions of the square-root Schur RLS adaptive filter can be summarized as follows:

1. Boundary Cell: Compute $\rho_m(t)$ and $D^b_{m,m}(t)$ via (14) [CORDIC processor in hyperbolic vectoring mode. See Fig. 1].

2. Internal Cell: Compute order recursion of $D^f_{m,j}(t)$ and $D^b_{m,j}(t)$ via (7) [CORDIC processor in hyperbolic rotation mode. See Fig. 1].

3. Filter Cell: Compute adaptive square-root lattice filter via (1a,b) [CORDIC processor in hyperbolic rotation mode. See Fig. 1].

III. Input Normalization

The recursions of the square-root Schur RLS adaptive filter are initialized from the square-root normalized second-order information of an observed process. This information has a compressed dynamic range, yet it is not bounded in the unit interval. In this section, we demonstrate that power normalization can be applied on the shifted data vectors $\mathbf{x}(t)$, $\mathbf{x}(t-1)$, . . . , $\mathbf{x}(t-m)$. In a geometrical interpretation the power normalized second-order information is just the set of cosines of the angles between the shifted data vectors and is hence bounded in the unit interval. It is shown that the recursions of the square-root Schur RLS adaptive filter are invariant to input power normalization. Further it is shown that all internal variables of the square-root Schur RLS adaptive filter remain bounded in the unit interval when initialized from the power normalized second-order information.

Consider the power normalized data vectors

$$\tilde{\mathbf{x}}(t-j) = \mathbf{x}(t-j)\,c_o^{-1/2}(t-j) \quad ; \qquad 0 \le j \le N \quad . \tag{16}$$

Clearly, $\tilde{\mathbf{x}}^T(t-j)\,\tilde{\mathbf{x}}(t-j) = 1$ and

$$\tilde{c}_j(t) = \tilde{\mathbf{x}}^T(t)\tilde{\mathbf{x}}(t-j) = c_j(t)\,c_o^{-1/2}(t)\,c_o^{-1/2}(t-j) \quad . \tag{17}$$

A quick inspection of (17) shows that the power normalized second-order information (short term autocorrelation coefficients) $\tilde{c}_j(t)$ are just the cosines of the angles between shifted data vectors. Moreover, one can show that with this type of normalization, on obtains normalized residual vectors of the type

$$\tilde{\mathbf{e}}_m(t) = \mathbf{e}_m(t)\,c_o^{-1/2}(t) \quad , \tag{18a}$$

$$\tilde{\mathbf{r}}_m(t) = \mathbf{r}_m(t)\,c_o^{-1/2}(t-m) \quad . \tag{18b}$$

Thus:

$$\tilde{E}_m(t) = E_m(t)\,c_o^{-1}(t) \quad , \tag{19a}$$

$$\tilde{R}_m(t) = R_m(t)\,c_o^{-1}(t-m) \quad . \tag{19b}$$

With these normalized quantities one finds that:

$$\tilde{\mathbf{v}}_m(t) = \tilde{\mathbf{e}}_m(t)\,\tilde{E}^{-1/2}_m(t) = \mathbf{v}_m(t) \quad , \tag{20a}$$

$$\tilde{\mathbf{w}}_m(t) = \tilde{\mathbf{r}}_m(t)\,\tilde{R}^{-1/2}_m(t) = \mathbf{w}_m(t) \quad . \tag{20b}$$

Strictly speaking, the square-root normalized residual vectors remain unaffected by additional power normalization of the shifted data vectors. Moreover,

$$\tilde{\rho}_{m+1}(t) = -\tilde{\mathbf{v}}^T_m(t)\,\tilde{\mathbf{w}}_m(t-1) = -\mathbf{v}^T_m(t)\,\mathbf{w}_m(t-1) = \rho_{m+1}(t) \quad , \tag{21}$$

which means that the same statement holds for the reflection coefficients.

Finally it remains to determine the dynamic range requirements of the square-root normalized generalized covariances when initialized from the time varying cosines (17). A normalization of the generalized covariance in consideration of (20a,b) gives:

$$\tilde{D}^f_{m,j}(t) = \tilde{\mathbf{x}}^T(t-j)\,\tilde{\mathbf{v}}_m(t) = \tilde{\mathbf{x}}^T(t-j)\,\mathbf{v}_m(t) \quad , \tag{22a}$$

$$\tilde{D}^b_{m,j}(t) = \tilde{\mathbf{x}}^T(t-j)\,\tilde{\mathbf{w}}_m(t) = \tilde{\mathbf{x}}^T(t-j)\,\mathbf{w}_m(t) \quad . \tag{22b}$$

Again inner products of unit energy vectors are always less or equal one in magnitude and hence:

$$\left|\tilde{D}^f_{m,j}(t)\right| \le 1 \;, \quad \text{(23a)} \quad ; \qquad \left|\tilde{D}^b_{m,j}(t)\right| \le 1 \quad . \quad \text{(23b)}$$

Thus we have shown that all internal variables of the square-root Schur RLS adaptive filter remain bounded in the unit interval when only the square-root generalized covariances are initialized from the power normalized second-order information (time varying cosines) of the observed process:

$$\tilde{D}^f_{0,j}(t) = \tilde{D}^b_{0,j}(t) = \tilde{c}_j(t) \quad , \qquad 1 \le j \le N \; . \qquad (24)$$

IV. Estimation of Second-Order Information

The second-order information $\{c_j(t),\ 0 \le j \le N\}$ can be expressed as a weighted inner product of shifted data vectors:

$$c_j(t) = \mathbf{x}^T(t)\mathbf{W}\mathbf{x}(t-j) \quad ; \quad \mathbf{W} = \mathrm{diag}\left[w_0,\ w_1,\ w_2,\ \dots\ \right] . \qquad (25)$$

Usually this weighted inner product is updated recursively. Such recursions typically exhibit the structure of recursive low-pass filters with the sample products $x(t)x(t-j)$ as input and the "cumulative" second-order information as output. Particularly interesting cases arise with recursive filters with transfer function

$$H(z) = \text{const.}\ \frac{z^n}{(z-\delta)^n} \quad ; \qquad 0 < \delta < 1 \quad . \qquad (26)$$

Such systems have n-fold poles at $z = \delta$ on the real axis in the z-plane. They can be realized by the following order n cascaded Gray-Markel feed-back lattice filter [10], with instantaneous autocorrelation coefficients $x(t)x(t-j) = p^{(j)}_n(t)$ as input and the estimated second-order information $c_j(t) = p^{(j)}_0(t) = q^{(j)}_0(t)$ as output:

$$\begin{bmatrix} p^{(j)}_{m-1}(t) \\ q^{(j)}_m(t) \end{bmatrix} = \begin{bmatrix} \left[1-K^2_m\right]^{1/2} & -K_m \\ K_m & \left[1-K^2_m\right]^{1/2} \end{bmatrix} \begin{bmatrix} p^{(j)}_m(t) \\ q^{(j)}_{m-1}(t-1) \end{bmatrix} , \qquad (27)$$

where $\{K_m,\ 1 \le m \le n\}$ are the fixed parameters of the Gray-Markel lattice filter of order n which are determined such that the overall filter has the desired transfer function (26). Clearly, the classical exponential forgetting function drops out as the special case $n = 1$ and the second-order (or modified Barnwell) window [6] is just the special case of an order $n = 2$ filter.

The most interesting point with this type of recursive forgetting is in fact the observation that the Gray-Markel lattice recursion (27) is just a circular rotation. Thus this type of a recursive lattice filter can be implemented by a cascade of CORDIC processors performing circular rotations as described in [11]. This CORDIC-based estimator of the second-order information fits very well into the overall CORDIC array structure of the square-root Schur RLS adaptive filter.

V. Systolic Array

Fig. 1 illustrates the described hyperbolic rotation and hyperbolic vectoring modes of the CORDIC elementary cell which is used in the adaptation and in the filter sections of the square-root Schur RLS adaptive filter. Fig. 2 shows the CORDIC array architecture of the parameter adaptation and adaptive filter sections. The triangular part of the array (parameter adaptation section) computes the angles $\varphi_1(t)$ - $\varphi_4(t)$ from the second order information (time varying cosines) provided at the inputs. The angles $\varphi_1(t)$ - $\varphi_4(t)$ are finally transmitted to the adaptive filter section at the bottom of the triangular array. In the case of a time invariant second-order information, the proposed algorithm and structures reduce to the Toeplitz system solver of Jou, Hu and Feng [12].

It becomes apparent that in an all-CORDIC realization of the square-root Schur RLS adaptive filter, one prefers to operate with the angular variable $\varphi_m(t)$ rather than with the explicit reflection coefficient $\rho_m(t)$. Once the angular variable has been determined by a CORDIC processor in vectoring mode, it is fed through all rotational processors in column m of the array and the angular variable input of these processors is preset by $\varphi_m(t)$ to obtain a prescribed rotation by iterative annihilation of the angular variable (rotational mode) [9].

VI. Conclusions

A doubly normalized Schur RLS adaptive filter and its fully systolic structure have been presented. The algorithm is very amenable to a VLSI realization because it consist of a mere locally connected CORDIC cells. Moreover, it has been shown that the algorithm has a particularly useful autoscaling feature in that the hyperbolic rotors always fully utilize a given dynamic range. Thus it is conjectured that the proposed structure can become a useful technique in high-speed data equalization and adaptive receiver applications. But we may not tie ourselves beforehand to a pure digital implementation of this array. Since all elementary cells perform passive or J-passive operations the computations could probably been mapped on even more exciting structures with light or other electromagnetic waves, depending on the medium chosen for implementation (e.g. fiber optics, surface acoustic wave (SAW) devices [13], etc.).

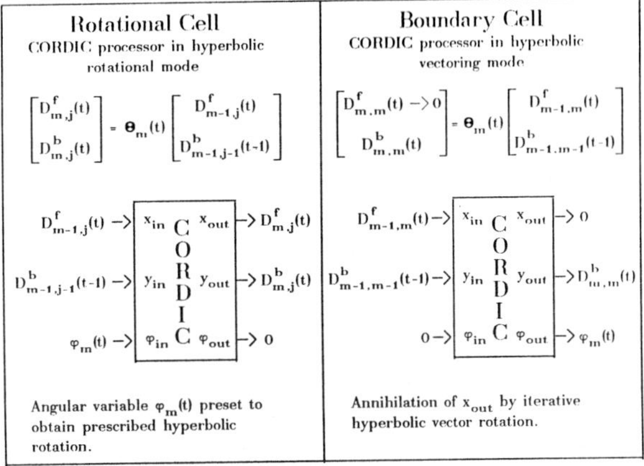

Figure 1: CORDIC processor configurations for rotational and boundary cell operations of the doubly normalized Schur RLS adaptive filter.

References

[1] J. Schur, "Über Potenzreihen die im Innern des Einheitskreises beschränkt sind", J. Reine Angew. Math., Vol. 147, pp. 205-232, 1917.

[2] S.C. Lie and E. Deprettere, "On the computational complexity of lattice-recursive algorithms", in Proc. Int. Conf. on DSP, Florence, Italy, Sept. 1981.

[3] E. Deprettere, "Mixed form time variant lattice recursions", in *Outils et modeles math. pour l'automatique*, pp. 545-562, CNRS, Paris 1982.

[4] S. Theodoridis, "Pipeline architecture for block adaptive LS FIR filtering and prediction", IEEE Trans. on Signal Proc., Vol. 38, pp. 81-90, Jan. 1990.

[5] F. Desbouvries and C Gueguen, "A Gohberg-Semencul formula for linear time-varying systems", in Proc. EUSIPCO-90, Barcelona, Spain, 1990.

[6] P. Strobach, "New forms of Levinson and Schur algorithms", IEEE Signal Processing Magazine, Jan. 1991.

[7] P. Strobach, "The square-root Schur RLS adaptive filter", in Proc. ICASSP-91, pp. 1845-1848, Toronto, May 1991.

[8] P. Strobach, *Linear Prediction Theory: A Mathematical Basis for Adaptive Systems,* Springer Series in Information Sciences, Vol. 21, Springer, New York, Jan. 1990.

[9] J.F. Böhme, D. Timmermann, H. Hahn, and B.J. Hosticka, "CORDIC processor architectures", in Proc. SPIE Int. Conf. Advanced Signal Proc., San Diego, CA, July 1991.

[10] A.H. Gray and J.D. Markel, "A normalized digital filter structure", IEEE Trans. on ASSP, Vol. 23, pp. 268-277, 1975.

[11] B.J. Hosticka, J. Büddefeld and U. Kleine, "Power-wave digital filters using CORDIC adaptors", AEÜ, Vol. 39, pp. 242-244, 1985.

[12] I.C. Jou, Y.H. Hu and W.S. Feng, "A novel implementation of pipelined Toeplitz system solver", Proc. IEEE, Vol. 74, pp. 1463-1464, 1986.

[13] R. Männer, "Semidigital multiplication by discrete convolution using surface acoustic wave (SAW) devices", IEEE Trans. on Signal Proc., Vol. 39, pp. 935-938, 1991.

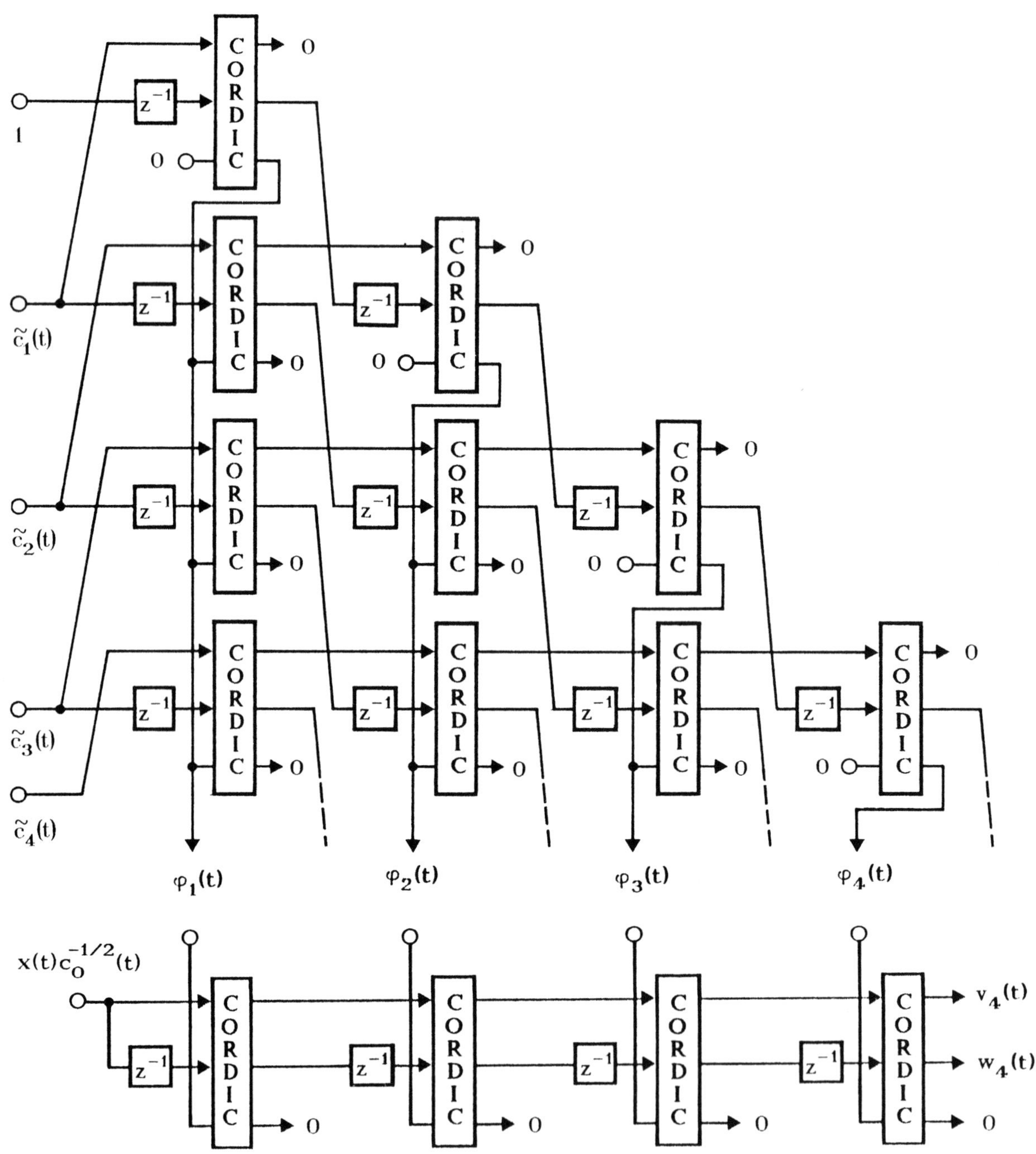

Figure 2: Division-free systolic all-CORDIC array architecture of the doubly normalized Schur RLS adaptive filter (order N=4 example).

SIGNAL PROCESSING VI: Theories and Applications
J. Vandewalle, R. Boite, M. Moonen, A. Oosterlinck (eds.)

A RLS with an adaptive forgetting factor

Alison Carter, Pascal Ruiz

Techniphone Développement, BP 44, F-13610 PUY STE REPARADE
Phone Number : 42338733 ; Fax : 42618961

We present here a Fast RLS algorithm with an adaptive forgetting factor. The performances are compared with a classical Fast RLS algorithm, on both simulations and real situations. The achievement of the former algorithm is then showed, whereas the calculation cost is still "reasonable".

1. Introduction

Adaptive signal processing is highly conditioned by the performances of the algorithms which minimize an energy function [1,2]. The challenge has always been the compromise between the speed convergence of the algorithm, and if whether or not the solution obtained is the optimum one. Moreover, real time applications has involved the birth of fast algorithms, with other problems of convergence. Nowadays, the LMS and Fast RLS algorithms have been largely used, because of their properties of stability. Therefore, several parameters must be initialized, as the forgetting factor, or the adaptation step. For the LMS, the adaptation step μ will influence the convergence speed and the bias of the solution. A weak value of μ will be usefull near the minimum of the performance surface, whereas a larger value at the beginning of the minimization will accelerate the descent towards the minimum. A solution for improving the algorithm would consist in the calculation of an adaptive μ. Such a procedure, from a calculation point of view, implies a greater but still reasonable cost since it remains in O(N) per sample, and not in O(N^2) (N is the order of the FIR filter to estimate). P. Bragard [3] has proposed an adaptive optimisation of μ in the LMS algorithm. The accuracy of such an algorithm has been proved. Since the RLS algorithm has better performances than the LMS, we propose here an adaptive optimisation of the RLS algorithm. Part 2 will detail the algorithm, and Part 3 will show some results.

2. The adaptive RLS

We present here how we have defined the adaptive RLS algorithm with a gradient procedure, in minimizing the mean square error defined versus the filter coefficients, but also versus the forgetting factor w.

2.1 The RLS algorithm [4,5,6]

Let x_n be the nth sample of the observed signal (primary input), and $\underline{X}_n$ the N-dimension observation vector. The purpose of the RLS algorithm is to minimize

$$J_d(n) = \sum_{p=0}^{n} (e_p)^2, \text{ where } e_p = s_p - H^t_{p-1}X_{p-1}.$$

The RLS algorithm estimates H_p in a recursive way :

$$H_p = H_{p-1} + R^{-1}(p)X_p(s_p - H^t_{p-1}X_p),$$

where R is the estimated covariance matrix. In a non stationary case, one can apply a forgetting factor w<1, and minimize now :

$$J_d(n) = \sum_{p=0}^{n} w^{n-p}(e_p)^2$$

2.2 The Fast RLS (FRLS) algorithm

The FRLS consists in estimating $R^{-1}(p)$ versus $R^{-1}(p-1)$, among the determination of the backward and forward prediction coefficients $\underline{B}_n^t$ and $\underline{F}_n^t$ which minimize a function of $E_f(n)$ and $E_b(n)$, where :

$$E_f(n) = \sum_{p=1}^{n} w^{n-p} \left| x_p - \underline{F}_n^t \underline{X}_{p-1} \right|^2 \text{ and}$$

$$E_b(n) = \sum_{p=1}^{n} w^{n-p} \left| x_{p-N} - \underline{B}_n^t \underline{X}_p \right|^2$$

Our problem is now reduced in estimating the backward and forward prediction coefficients $\underline{B}_n^t$ and $\underline{F}_n^t$ which minimize a function $J(E_f;E_b)$ of $E_f(n)$ and $E_b(n)$.
Noticing that $J(E_f;E_b)$ is also a function of w, our purpose is here to make w be dependent of n, and to update it in a recursive way.

2.3 The adaptive FRLS

In that algorithm, the forgetting factor w will be dependent of n : the way of updating it would be :

$$w_{n+1} = w_n - \alpha \frac{\partial J(E_f;E_b)}{\partial w_n}, \text{ where } J(E_f;E_b) \text{ is a}$$

function of the energies (we take here $J(E_f;E_b) = E_f + E_b$), and where α is the second adaptive step. From the previous formula we obtain :

$$w_{n+1} = w_n - \alpha \left(\sum_{p=1}^{n} (n-p)\ w^{n-p-1} \left(\left| x_p - \underline{F}_n^t \underline{X}_{p-1} \right|^2 + \left| x_{p-N} - \underline{B}_n^t \underline{X}_p \right|^2 \right) \right)$$

2.3.1. The algorithm

The complete FRLS with an adaptive forgetting factor will be :

$e^f_{n+1} = x_{n+1} - \underline{F}_n^t \underline{X}_n$; forward prediction error

$\underline{F}_{n+1} = \underline{F}_n + \underline{G}_n e^f_{n+1}$; forward prediction coefficients

$\varepsilon^f_{n+1} = x_{n+1} - \underline{F}_{n+1}^t \underline{X}_n$; forward prediction error (a posteriori)

$w_{n+1} = w_n - \alpha \sum_{p=1}^{n} (n-p)\ w_n^{n-p-1} \left(\left| x_p - \underline{F}_n^t \underline{X}_{p-1} \right|^2 + \left| x_{p-N} - \underline{B}_n^t \underline{X}_p \right|^2 \right)$; forgetting factor update

$E^f_{n+1} = w_{n+1}E^f_n + e^f_n \varepsilon^f_n + c$; cumulated forward prediction error update

$$\begin{bmatrix} \underline{M}_{n+1} \\ m_{n+1} \end{bmatrix} = \begin{bmatrix} 0 \\ \underline{G}_n \end{bmatrix} + r\left(\frac{1}{w_{n+1}}\right) \begin{bmatrix} 1 \\ -\underline{F}_{n+1} \end{bmatrix}$$

$e^b_{n+1} = x_{n+1-N} - \underline{B}_n^t \underline{X}_{n+1}$; backward prediction error

$\underline{G}_{n+1} = \dfrac{\underline{M}_{n+1} + m_{n+1} \underline{B}_n}{1 - m_{n+1} e^b_{n+1}}$; Kalman gain

$\underline{B}_{n+1} = \underline{B}_n + \underline{G}_{n+1} e^b_{n+1}$; backward prediction coefficients update

$\hat{s}_{n+1} = y_{n+1} - \underline{H}_n^t \underline{X}_{n+1}$; error signal

$\underline{H}_{n+1} = \underline{H}_n + \underline{G}_{n+1} \hat{s}_{n+1}$; filter coefficients

2.3.2 Remarks

• Because the forgetting factor cannot be negative, a positivity constraint is applied :

$$w_{n+1} = \frac{|w_{n+1}| + w_{n+1}}{2}$$

• The choice of α would be such that w_n converges between two transient signal, with slow fluctuations. That is quite an empirical choice. We chose it here around 0,001.

• The calculation cost will be around 11 N multiplications/divisions and 10 N additions/substractions, where N is the order of the filter.

3. Results

3.1. Results on simulated signals

We compare the performances between the FRLS and the FRLS algorithm with a variable forgetting factor. The reference input x(n) is a sine with a period of 32 points. The primary input is y(n) = h(n)*x(n) + b(n), where h(n) is a F.I.R. filter of order 10 which variates slowly, where * is the convolution operator, and where b(n) is an additive noise uncorrelated with x(n). The signal to noise ratio between x(n) and b(n) is approximatively 20 dB. In figure 1 as in figure 2 we represented the ratio

$$\frac{E((y(n)-\hat{h}(n)*x(n))^2)}{E(y^2(n))}$$

for the LMS algorithms (μ = 0,02), the FRLS algorithm (w = 0,98), and the FRLS with a variable forgetting factor (w_0 = 0,98 and $\alpha = 10^{-4}$). The figures show the accuracy of the FRLS with a variable forgetting factor, compared with the FRLS.

3.2 Results on real situations

We compare here the performances between the FRLS, LMS, and our FRLS with a variable forgetting factor. The algorithms are applied on real signals : the reference input (figure3) is a motor noise, the primary input is the motor noise (figure 4) inside a cabin, plus an uncorrelated desired response. In figure 5 we represent $\sum_{i=1}^{n} y^2(i)$ input versus n, and the residual output $\sum_{i=1}^{n} (y(i)-\hat{h}(i)*x(i))^2$ for the different algorithms. The performances of the adaptive FRLS are still better than the other algorithms.

References :

[1] B. Widrow, S. Stearns : Adaptive Signal Processing, Prentice Hall, 1986.

[2] A. Benveniste, R. Metivier, P. Priouret : Algorithmes adaptatifs et approximations stochastiques. Théorie et applications, Masson, 1987.

[3] P. Bragard : Egalisation adaptative de données transmises dans le canal acoustique sous-marin en contexte non-stationnaire, PhD, Grenoble, June 1990.

[4] D. Baudois, C. Serviere, A. Silvent : Soustraction de bruit, Cephag Report n° 12/88, 1988.

[5] B. Toplis, S.P. Supathy : Tracking improvements in fast-RLS algorithms using a variable forgetting factor, IEEE on ASSP, pp 206-227, 1988.

[6] O. Macchi, Bershad : "Adaptive recovery of a chirped sinusoid in noise. Part 1 : performance of the RLS algorithm". IEEE on ASSP, 1990.

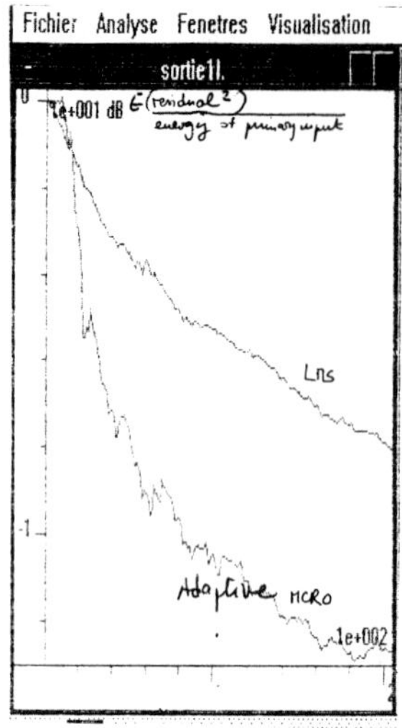

Figure 1.

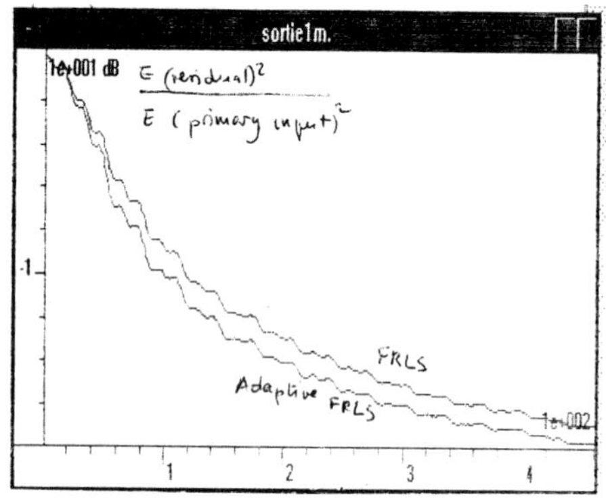

Figure 2.

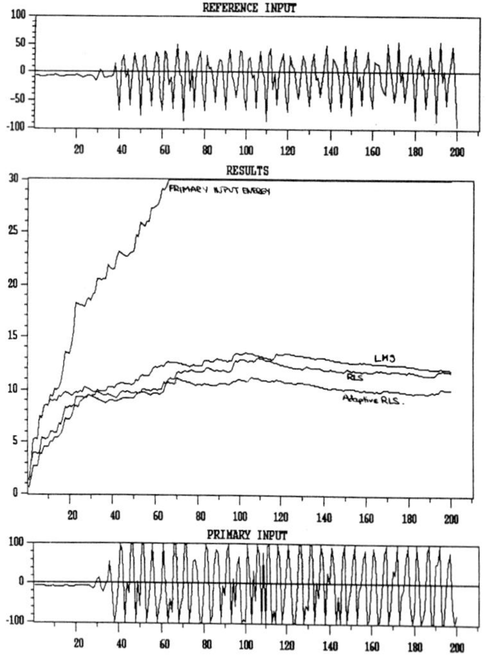

Figures 3, 4, 5.

SIGNAL PROCESSING VI: Theories and Applications
J. Vandewalle, R. Boite, M. Moonen, A. Oosterlinck (eds.)

A FOURTH ORDER CUMULANT BASED ADAPTIVE ALGORITHM FOR THE IDENTIFICATION OF NONLINEAR HAMMERSTEIN MODELS

Deergha Rao Korrai and Devender C Reddy

R & T Unit for Navigational Electronics, Osmania University
Hyderabad-500007, India.

An adaptive algorithm known as ORIV and based on HOS has been proposed [1] for the identification of IIR systems. Although powerful, yet nevertheless the method has the shortcoming that a proper choice of instrument variables be made, which at times is difficult. A new on-line technique that uses ORLS in conjunction with RLS is now proposed for the identification of non-linear Hammerstein models driven by Gaussian input (this is equivalent to the identification of linear ARMA models driven by non-Gaussian inputs) which is relatively easy to apply. However, due to paucity of time simulation studies in respect of an MA process only using ORLS was attempted. The results obtained were found to be quite satisfactory.

1. INTRODUCTION

Both the theory and practice of adaptive signal processing are almost exclusively based on the second order statistics of the underlying process. This is because of the implicit assumption that the process is Gaussian. But non-Gaussian processes are quite common [1]. The tendency in contemporary Signal Processing by and large is to ignore the non-Gaussian nature of underlying signal. However, substantial benefits may accrue by making use of the non-Gaussian nature of these signals. This has been made possible by the use of higher order statistics (HOS). Since HOS provide an operational calculus for studying non-Gaussian processes and nonlinear systems, vanish for coloured Gaussian noise and convey additional (phase) information about the underlying process they have found direct application in nonlinear filter identification, parameter estimation, speech and signal processing, phase estimation of linear processes, blind deconvolution, tests for linearity, measuring association, detection and classification [2].

This being the case the search for HOS-based algorithms with improved accuracy, continues.

1.1. HOS in the Identification of Non-linear System

As has been mentioned above, HOS seem to be natural candidates to use when one is confronted with non-Gaussian models. A case in point is the nonlinear Hammerstein model which is composed of a zero memory nonlinear part N(.), a linear ARMA(p,q) part and driven by Gaussian white noise (see fig. 1). Observe that although the input to the system is Gaussian nevertheless the input to the ARMA part of the system is non-Gaussian.

One of the important problems in system identification is the determination of the system parameters of such a model. A noniterative procedure, using HOS of the observed data, has been suggested [3] for the estimation of parameters of the linear MA(q) part of the Hammerstein model (fig. 1). However, literature seems to be lacking in adaptive techniques that are based on HOS for the identification of nonlinear, ARMA Hammerstein models. The object of this article, therefore, is to suggest an adaptive algorithm for the identification of nonlinear Hammerstein models driven by Gaussian white noise, using fourth order cumulants. Fourth order cumulants are to be used when the random process is symmetrically distributed [6] for this the third order cumulants equal zero.

We shall therefore consider such a system in which the third order cumulants are zero but the fourth order cumulants are non zero and develop an algorithm for parameter identification.Towards this end, as a first step we shall consider a MA process whose parameters are to be estimated.

2. THE OVERDETERMINED RECURSIVE LEAST SQUARES ALGORITHM FOR MA PROCESSES

Let the output of the observed process be defined by

the eq:

$$y_t = u_t + \sum_{k=1}^{q} b_k u_{t-k} \tag{1}$$

where

$\{u_t\}$ is a zero mean, non-Gaussian white noise sequence and b_k's are constant coefficients.

For this process, it has been shown [4] that the covariances,r(m), and the diagonal fourth order cumulants,c(m) satisfy the following relationship:

$$r(m) + \sum_{k=1}^{q} b_k^3 r(m-k) = \epsilon \left[c(m) + \sum_{k=1}^{q} b_k c(m-k)\right] \tag{2}$$

$$-q \leq m \leq 2q$$

where

$r(m) = E[y_t y_{t+m}]$; $c(m) = E[y_t y_{t+m}^3] - 3r(o)r(m)$

and

$\epsilon = \sigma^2/\gamma_4; \sigma^2 = E[u_t^2]$;

$\gamma_4 = E[u_t^4] - 3\sigma^4$; $\gamma_4 \neq 0$;

Eq.(2) may also be written in the more compact matrix form:

$$3q+1\left\{\left[\begin{array}{cccc|ccc} c(-q) & 0 & \dots & 0 & 0 & \dots & 0 \\ c(-q+1) & c(-q) & \dots & 0 & -r(q) & \dots & 0 \\ \vdots & \vdots & \vdots & \vdots & \vdots & \vdots & \vdots \\ c(q) & c(q-1) & \dots & c(0) & -r(q-1) & \dots & r(0) \\ 0 & c(q) & \dots & 0 & 0 & \dots & 0 \\ \vdots & \vdots & \vdots & \vdots & \vdots & \vdots & \vdots \\ 0 & 0 & \dots & c(q) & 0 & \dots & -r(q) \end{array}\right]\right. \begin{bmatrix} \epsilon \\ \epsilon b_1 \\ \vdots \\ \epsilon b_q \\ b_1^3 \\ \vdots \\ b_q^3 \end{bmatrix} = \begin{bmatrix} r(q) \\ r(q-1) \\ \vdots \\ r(q) \\ 0 \\ \vdots \\ 0 \end{bmatrix} \tag{3}$$

(first block of columns: $q+1$; second block: q)

$$Y\theta = \underline{r} \tag{4}$$

where the definition of Y, θ and $\underline{r}$ are obvious from eq.(3).

Eq.(3) represents a set of overdetermined set of equations in the parameter vector θ.

The least squares solution of this overdetermined system of equations is given by the eq.:

$$\theta_{LS} = (Y^T Y)^{-1} Y^T \underline{r} \tag{5}$$

Observe that eq.(5) is not an iterative solution. Friedlander and Porat [1] has proposed an adaptive algorithm, known as ORIV, for solving equation (3). However, this method requires the choice of instrument variables which at times may not be too obvious. We therefore propose a different on-line technique, known as ORLS (overdetermined recursive least squares) for this purpose.

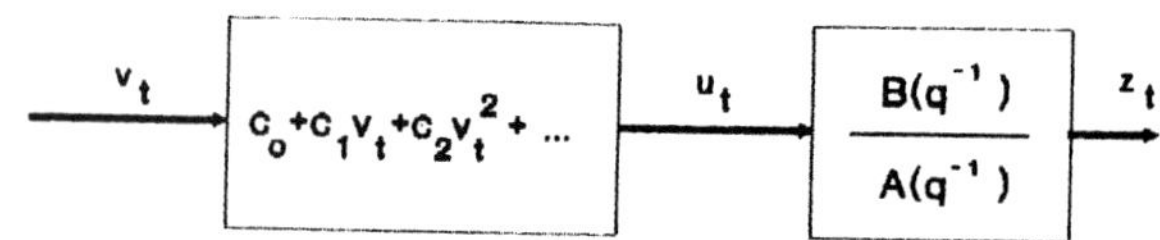

Fig.1: Simple Hammerstein model
q^{-1} : delay operator

2.1. Derivation of the ORLS Algorithm

Consider the eq :

$$\theta(N) = \left[\sum_{t=1}^{N} Y^T(t) \sum_{t=1}^{N} Y(t)\right]^{-1} \left[\sum_{t=1}^{N} Y^T(t) \sum_{t=1}^{N} \underline{r}(t)\right] \tag{6}$$

where

$\underline{r}(t) = Y_1(t) y_{t-q}$

$Y(t) = Y_1(t) Y_2^T(t)$

$Y_1(t) = [y_t, \dots\dots, y_{t-3q}]^T$

$$Y_2(t) = [[y_{t-q}^3, \dots\dots, y_{t-2q}^3 \vdots\ y_{t-1-q}, \dots\dots, y_{t-2q}]$$
$$-3\hat{r}_t(0)[y_{t-q}, \dots\dots, y_{t-2q} \vdots\ 0, \dots\dots, 0]]^T$$

where

$$\hat{r}_t(0) = \frac{1}{t+1} \sum_{k=0}^{t} y_k^2$$

Now it is desired to obtain a recursive solution to eq.(6). Towards this end let

Denote

$$R(t) = \sum_{k=1}^{t} Y(k) = \sum_{k=1}^{t} Y_1(k) Y_2^T(k) \tag{7}$$

Then, from eq(6), it follows that

$$\sum_{k=1}^{t-1} Y^T(k) \sum_{k=1}^{t-1} \underline{r}(k) = \sum_{k=1}^{t-1} Y_1^T(k) Y_2(k) \sum_{k=1}^{t-1} \underline{r}(k)$$
$$= \left[R^T(t-1) R(t-1)\right] \hat{\theta}(t-1) \tag{8}$$

From the definition of R(t) it follows that

$$R(t-1) = R(t) - Y_1(t) Y_2^T(t) \tag{9}$$

Hence

$$\hat{\theta}(t) = \left[R^T(t)R(t)\right]^{-1} R^T(t) \sum_{k=1}^{t} \underline{r}(k) \tag{10}$$

An exponential forgetting factor $0 \leq \alpha \leq 1$ which enables the algorithm to track time varying parameters can be introduced by premultiplying eq.(10) by diag $\{\alpha^t, \alpha^{t-1}, \cdots\cdots\cdots\cdots\cdots, \alpha, 1\}$. In this derivation, the forgetting factor α is assumed to be unity.

Next let

$$P_t = \left[R^T(t)R(t)\right]^{-1} \tag{11}$$

Hence

$$P_t^{-1} = R^T(t)R(t) \tag{12}$$

Also it follows that

$$\begin{aligned} P_t^{-1} &= \left[R^T(t-1) + Y_2(t)Y_1^T(t)\right] [R(t-1) \\ &\qquad + Y_1(t)Y_2^T(t)] \\ &= R^T(t-1)R(t-1) + Y_2(t)Y_1^T(t)R(t-1) \\ &\quad + R^T(t-1)Y_1(t)Y_2^T(t) + Y_2(t)Y_1^T(t)Y_1(t)Y_2^T(t) \\ &= P_{t-1}^{-1} + W(t)Y_2^T(t) + Y_2(t)W^T(t) \\ &\qquad + Y_2(t)K_b(t)Y_2^T(t) \end{aligned} \tag{13}$$

where

$$W(t)_{n\times 1} = R^T(t-1)Y_1(t) \tag{14}$$

$$K_b(t)_{1\times 1} = Y_1^T(t)Y_1(t) \tag{15}$$

This can be written more compactly as

$$P_t^{-1} = P_{t-1}^{-1} + X(t)\begin{bmatrix} 0 & 1 \\ 1 & K_b(t) \end{bmatrix} X^T(t) \tag{16}$$

where
$X(t)_{n\times 2} = \left[W(t) \vdots Y_2(t)\right]$
Inversion of eq(13) yields

$$\begin{aligned} P(t) &= P(t-1) - P(t-1)X(t)\left[X^T(t)P(t-1)\right. \\ &\quad X(t) - \begin{bmatrix} K_b(t) & -1 \\ -1 & 0 \end{bmatrix}]^{-1} X^T(t)P(t-1) \end{aligned} \tag{17}$$

From eq.(10), we note that

$$\begin{aligned} \hat{\theta}(t) &= P(t)\left(R^T(t-1) + Y_2(t)Y_1^T(t)\right) \\ &\quad (\underline{r}(t-1) + Y_1(t)y_{t-q}) \\ &= P(t)R^T(t-1)\underline{r}(t-1) \\ &\quad + P(t)\left(Y_2(t)Y_1^T(t)\underline{r}(t-1)\right. \\ &\quad + R^T(t-1)Y_1(t)y_{t-q} \\ &\quad + Y_2(t)K_b(t)y_{t-q}) \\ &= P(t)P^{-1}(t-1)\hat{\theta}(t-1) + P(t) \\ &\quad \left(Y_2(t)Y_1^T(t)\underline{r}(t-1)\right. \\ &\quad + R^T(t-1)Y_1(t)y_{t-q} \\ &\quad + Y_2(t)K_b(t)y_{t-q}) \end{aligned} \tag{18}$$

From eq.(13), it follows that :

$$\begin{aligned} P(t)P^{-1}(t-1) &= I - P(t)\left(W(t)Y_2^T(t)\right. \\ &\quad + Y_2(t)W^T(t) \\ &\quad \left. + Y_2(t)K_b(t)Y_2^T(t)\right) \end{aligned} \tag{19}$$

Substituting eq.(19) in eq.(18), we obtain

$$\begin{aligned} \hat{\theta}(t) &= \hat{\theta}(t-1) - P(t) \\ &\quad \left(W(t)Y_2^T(t) + Y_2(t)W^T(t)\right. \\ &\quad \left. + Y_2(t)K_b(t)Y_2^T(t)\right)\hat{\theta}(t-1) \\ &\quad + P(t)\left(Y_2(t)Y_1^T(t)\underline{r}(t-1)\right. \\ &\quad + R^T(t-1)Y_1(t)y_{t-q} \\ &\quad + Y_2(t)K_b(t)y_{t-q}) \end{aligned} \tag{20}$$

Eq.(20) can also be written as :

$$\begin{aligned} \hat{\theta}(t) &= \hat{\theta}(t-1) + P(t)X(t)\begin{bmatrix} 0 & 1 \\ 1 & K_b(t) \end{bmatrix} \\ & S(t) \quad X^T(t)\hat{\theta}(t \quad 1) \end{aligned} \tag{21}$$

where

$$S(t)_{2\times 1} = \begin{bmatrix} Y_1^T(t)\underline{r}(t-1) \\ y_{t-q} \end{bmatrix} \tag{22}$$

Now we shall use the above result for developing an algorithm for the estimation of ARMA process parameters.

3. A COMBINED ADAPTIVE ALGORITHM FOR ESTIMATION OF ARMA PROCESS PARAMETERS

Let the ARMA process be represented by the eq:

$$z_t = -\sum_{k=1}^{p} a_k z_{t-k} + u_t + \sum_{k=1}^{q} b_k u_{t-k} \tag{23}$$

where
$\{u_t\}$ is a zero mean, non-Gaussian white noise sequence. Now assuming that $\{b_k(1 \leq k \leq q)\}$ and $\{u_s, s < t\}$ are known exactly, eq.(24) may be written as

$$y_t = z_t + \sum_{k=1}^{p} \hat{a}_k z_{t-k} \tag{24}$$

where
$\hat{a}_k$'s are the estimates of $\{a_k\}$ Next define the error :

$$e_t = z_t + \sum_{k=1}^{p} \hat{a}_k z_{t-k} - \sum_{k=1}^{q} b_k u_{t-k} \tag{25}$$

The parameters $\hat{a}_k$ can then be estimated by minimizing the total square error $\sum e_t^2$ using the RLS algorithm [5]. Since in practice, the parameters $\{b_k\}$ and $\{u_k\}$ are unknown, replacing $\{b_k\}$ by its estimate, $\{\hat{b}_k\}$ we generate an approximate $\hat{e}_t$ by recursively computing it as :

$$\hat{e}_t = z_t + \sum_{k=1}^{p} \hat{a}_k z_{t-k} - \sum_{k=1}^{q} \hat{b}_k e_{t-k} = y_t - \sum_{k=1}^{q} \hat{b}_k e_{t-k} \quad (26)$$

This algorithm is as shown in fig.(2).

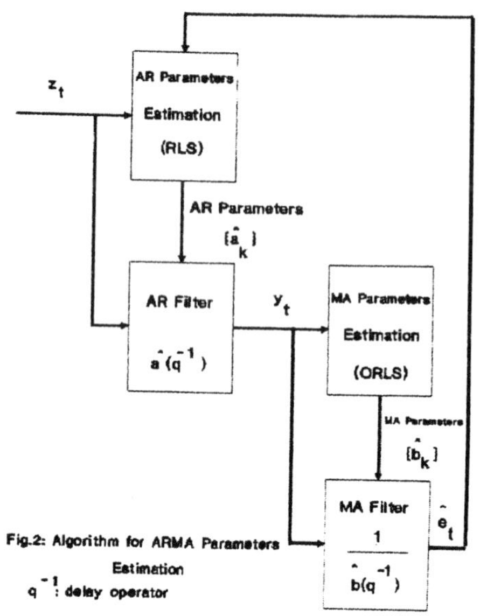

Fig.2: Algorithm for ARMA Parameters Estimation
q^{-1}: delay operator

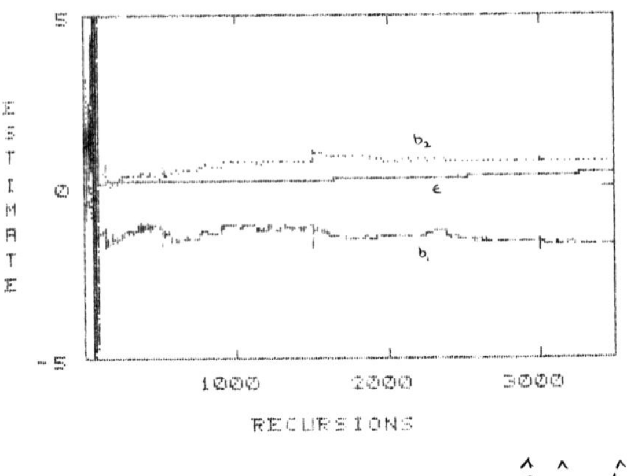

FIG. 3: TIME HISTORIES OF $\hat{\varepsilon}, \hat{b}_1, \hat{b}_2$

4. SIMULATION STUDIES

The performance of the adaptive ORLS algorithm for an MA process was tested through a computer simulation of the Hammerstein model given by the following equation :

$$y_t = u_t - 1.663u_{t-1} + 0.542u_{t-2} \quad (27)$$

$$u_t = v_t^3 \quad (28)$$

where v_t is zero mean ,Gaussian white noise sequence of variance 0.2045. The time histories of $\hat{\epsilon}$, $\hat{b}_1$ and $\hat{b}_2$ up to 3500 recursions are as given in fig-3 where it may be observed that the parameters to be identified converged to the values -1.6631 and 0.6922. Off line simulation of this process notably yielded -1.6115 and 0.5347. Here the sequence length was 10240 which was divided into 80 records of length 128 samples.

5. CONCLUSIONS

A new on-line technique, that uses ORLS in conjunction with RLS is proposed for the identification of non-linear Hammerstein models driven by Gaussian input (this is equivalent to the identification of linear ARMA models driven by non-Gaussian inputs). Simulation studies are carried out in respect of an MA process using ORLS only. The results obtained are found to be quite satisfactory. Work relating to the use of ORLS in conjunction with RLS is currently in progress.

REFERENCES

[1] B.Friedlander and B.Porat, 'Adaptive I I R algorithms based on High-order Statistics', I E E E Trans. Acoust., Speech, Signal Processing, Vol. ASSP-37, pp.485-495,1989.

[2] D.R.Brillinger, 'Basic Aspects of Higher-order Spectra and Some of Their Uses', Proceedings of International Signal Processing Workshop on Higher Order Statistics, Chamrousse, France, July, 1991.

[3] L.Pinxing and M.Shiyi, 'On the Modelling of non-gaussian signal Identification of Hammerstein Models and Higher Order Statistics', Proceedings of 1990 Bilkent International Conference on New trends in Communication, Control and Signal Processing, Ankara, Turkey, July, 1990

[4] G.B.Giannakis and J.M.Mendel, 'Identification of Non-minimum Phase Systems using Higher Order Statistics', IEEE Trans. Acoust., Speech, Signal Processing, Vol. ASSP-37, No. 3, March, 1989.

[5] L. Ljung and T. Soderstrom, Theory and Practice of Recursive Identification, Cambridge, MA: M.I.T. Press, 1986.

[6] J.M.Mendel,'Tutorial on Higher-Order Statistics (Spec tra) in Signal Processing and System Theory: Theoretical Results and Some Applications', Proceedings of the IEEE, Vol.79,No.3,March,!991.

SIGNAL PROCESSING VI: Theories and Applications
J. Vandewalle, R. Boite, M. Moonen, A. Oosterlinck (eds.)

BOOTSTRAPPED SPATIAL SEPARATION OF WIDEBAND SUPERIMPOSED SIGNALS [1]

Hagit Messer[2] and Yeheskel Bar-Ness
Dept. of ECE
New Jersey Institute of Technology
University heights
Newark, NJ 07102
U.S.A

Abstract

Bootstrapped algorithms were developed and used for separation of two signals, when two versions of their weighted sum is given. In this paper we apply the bootstrap principle to the separation of N signals transmitted from point sources at different, unknown locations, when received by an array of M sensors. We present a general structure of the separation scheme which consists of delay elements and summation only. Its input is the M-sensor output signals and its output is the estimates of the N source signals. We show that if the source locations are known, this system provides a least squares estimate of the source signals. If not, it can adaptively converge to the least squares solution, provided that some prior information about the source signals is available. In particular, we present a detailed study of the bootstrapped algorithm for the separation of two sources received by two sensors. A simplified version of the algorithm is presented and the idea of adaptively controlling the unknown delays is discussed. We show that the proposed algorithm is a powerful tool for the decomposition of spatially mixed wideband signals.

I. Introduction and Background

The paper deals with the scenario in which N point sources are received by M omni-directional sensors. The received signal at the output of each of the M sensors can be modeled by:

$$\begin{aligned} z_m(t) &= \sum_{n=1}^{N} s_n(t-\tau_m(\theta_n)) + e_m(t) \\ m &= 1,...,M \quad ; \quad |t| \leq T/2 \end{aligned} \tag{1}$$

where: $s_n(t)$ is the signal radiated from the n-th source; θ_n represents the coordinates (location) of the n-th source; and $e_m(t)$ is the additive noise at the m-th sensor. In the special case of a two dimensional array and far-field sources, θ_n is the source bearing. $\tau_m(\theta_n)$, the travel time of the n-th source from the array origin to the m-th sensor, is given by:

$$\tau_m(\theta_n) = \frac{1}{c}[x_m sin\theta_n + y_m cos\theta_n] \tag{2}$$

where c is the propagation velocity of the signal wavefront and (x_m, y_m) are the Cartesian coordinates of the m-th sensor. If the array is an equally spaced linear array (ESLA) then y_m =0 and x_m=$(m-1)d$, m=1,...,M where d is the separation between successive sensors and the plane origin is assumed at the coordinates of the first sensor. Thus, (2) becomes:

$$\tau_m(\theta_n) = (m-1)\frac{d}{c}sin\theta_n \tag{3}$$

In general there are MxN (or (M-1)xN) delays which are a function of the N source locations $\theta_1,...\theta_N$. For an ESLA there are only N different delays which carry all the spatial information.
The above model can match many practical applications in different fields. In passive sonar the source signals are wideband, noise-like random processes and the unknown source location vector, $\underline{\theta} = (\theta_1, .., ...\theta_N)^T$, has to be estimated. In the active case the source signals are basically known, but the aim is the same - estimation of the source locations. However, in communication systems one is usually interested in the source signals themselves and not in their locations, which are sometimes known.

We assume broadband signals, which is more general than the usually assumed narrowband case, where a delay can be regarded as a phase shift. The received sensor data (1) can be represented in the frequency domain:

[1]This work was partially supported by grant from ROME (AFSC, Griffiss Air Force Base, NY) under contract F 30602-88-D-0025, Task C-0-2456.

[2]On sabbatical leave from the Dept. of Electrical Engineering - Systems, Tel Aviv University,Tel Aviv 69978, Israel.

$$Z_m(\omega_k) = \frac{1}{T}\int_{-T/2}^{T/2} z_m(t)e^{-j\omega_k t}dt$$
$$m = 1,...,M \;;\; k = 1,...L \qquad (4)$$

$Z_m(\omega_k)$ is the Fourier coefficient of the output from the m-th sensor at frequency $(2\pi/T)k$. The processing bandwidth is, therefore, $B = L/T$. At each frequency, the array output is given by the m-dimensional vector:

$$\underline{Z}(\omega_k) = (Z_1(\omega_k), ..., Z_M(\omega_k))^T \qquad (5)$$

Following the model of (1), $\underline{Z}(\omega_k)$ can be written as

$$\begin{aligned}\underline{Z}(\omega_k) &= \sum_{n=1}^{N} S_n(\omega_k)\underline{a}(\omega_k, \theta_n) + \underline{E}(\omega_k)\\ &= A(\omega_k\underline{\theta})\underline{S}(\omega_k) + \underline{E}(\omega_k) \qquad (6)\end{aligned}$$

where $\underline{S}(\omega_k) = (S_1(\omega_k), ..., S_N(\omega_k))^T$ and $\underline{E}(\omega_k = (E_1(\omega_k), ..., E_M(\omega_k))^T$. $S_n(\omega_k)$ and $E_m(\omega_k)$ are the Fourier coefficients of the n-th source signal and the noise in the m-th sensor, respectively. Also,

$$A(\omega_k, \underline{\theta}) = [\underline{a}_1(\omega_k) : ... : \underline{a}_N(\omega_k)] \qquad (7)$$

where

$$\underline{a}(\omega_k, \theta_n) = \underline{a}_n(\omega_k) = (e^{j\omega_k\tau_1(\theta_n)}, ..., e^{j\omega_k\tau_M(\theta_n)})^T \qquad (8)$$

If $M \geq N$ the least-squares estimate of the frequency domain vector of source signals, $S(\omega)$, given the data vector $\underline{Z}(\omega)$, is [1]:

$$\underline{\hat{S}}(\omega_k) = [A^*(\omega_k, \underline{\theta})A(\omega_k, \underline{\theta})]^{-1}A^*(\omega_k, \underline{\theta})\underline{Z}(\omega_k) \qquad (9)$$

For the special case of $M = N = 2$ it easily can be verified that

$$[A^*(\omega_k, \underline{\theta})A(\omega_k, \underline{\theta})]^{-1}A^*(\omega_k, \underline{\theta})\underline{Z}(\omega_k) = \frac{1}{\sin^2\omega_k\Delta}\cdot \begin{pmatrix} e^{j\omega_k D_1} - \cos\omega_k\Delta e^{j\omega_k D_2} & e^{-j\omega_k D_1} - \cos\omega_k\Delta e^{-j\omega_k D_2} \\ e^{j\omega_k D_2} - \cos\omega_k\Delta e^{j\omega_k D_1} & e^{-j\omega_k D_2} - \cos\omega_k\Delta e^{-j\omega_k D_1} \end{pmatrix} \qquad (10)$$

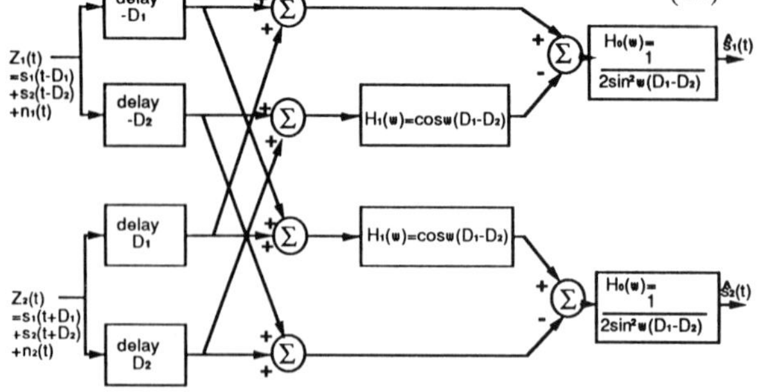

Fig. 1: Direct implementation of the least-squares operator

Where $\Delta = D_1 - D_2$. In (8) we also assume that the array origin reference is the mid location between the elements, so that $\tau_1(\theta_1) = -\tau_2(\theta_1) = D_1$, $\tau_1(\theta_2) = -\tau_2(\theta_2) = D_2$. We see that, even if D_1 and D_2 are known, the implementation of the least squares solution of (9) requires filtering the array outputs $z_i(t)$ using filters having transfer functions of the form: $\cos\omega\Delta$ or $1/\sin^2\omega\Delta$, as well as pure delays (see Fig. 1). The implementation of the trigonometric filters is difficult, especially when δ is unknown and is to be estimated adaptively. One possible approach to deal with this implementation problem is to *approximate* the trigonometric filters by FIR (or IIR) filters. In the sequel, we show that applying the bootstrap principle to this problem results in *excat* implementation of (9) which uses only delay elements and summations, in a feedback configuration.

II. The Bootstrapped Algorithm

Bootstrapped systems are multi-input multi-output feed-back systems in which each output "helps" to improve the other outputs. The idea is to assume that a system output is indeed the desired response and to use it, via feedback, to get other outputs. This approach has been successfully applied in satellite communication to improve separation of cross-pol signals [2]. In our problem, given any N-1 source signals one can get a good estimate of the remaining signal, say $s_n(t)$ from any of the sensor outputs using:

$$\begin{aligned}\hat{s}_{nm}(t) &= \hat{s}_n(t - \tau_m(\theta_n))\\ &= z_m(t) - \sum_{(i=1)_{i\neq n}}^{N} s_i(t - \tau_m(\theta_i))) \qquad (11)\end{aligned}$$

If no noise exists, then the left hand side of (11) is indeed a delayed version of $s_n(t)$. However, since noise is never zero, averaging over the estimated version of a certain signal from all sensor outputs will improve SNR output. Therefore, we have:

$$\hat{s}_n(t) = \frac{1}{M}\sum_{m=1}^{M} \hat{s}_{nm}(t + \tau_m(\theta_i)) \qquad (12)$$

The outputs of the proposed algorithm are N signals, $y_1(t), ..., y_N(t)$ which are desired to be the best possible estimates of the N source signals $s_1(t), ..., s_N(t)$. Following the bootstrapped approach, we replace the known source signals in (11) by their estimates $\{y_i\}$. Therefore, the proposed scheme is described by the NxM equations:

$$\begin{aligned} y_{nm}(t) &= y_n(t-\tau_m(\theta_n)) \\ &= z_m(t) - \sum_{(i=1)_{i\neq n}}^{N} y_i(t-\tau_m(\theta_i)) \\ m &= 1,...,M \;;\; n=1,...,N \end{aligned} \quad (13)$$

$$y_n(t) = \frac{1}{M}\sum_{m=1}^{M} y_{nm}(t+\tau_m(\theta_i)) \quad (14)$$

By transforming (13) to the frequency domain we get

$$\begin{aligned} Y_n(\omega) &= \frac{1}{M}\sum_{m=1}^{M} e^{j\omega\tau_m(\theta_n)}[Z_m(\omega) \quad (15) \\ &- \sum_{(i=1)_{i\neq n}}^{N} Y_i(\omega)e^{-j\omega\tau_i(\theta_n)}] \\ m &= 1,...,M \;;\; n=1,...,N \quad (16) \end{aligned}$$

In a matrix form this equation is equivalent to

$$A^*(\omega,\underline{\theta})A(\omega,\underline{\theta})\underline{Y}(\omega) = A^*(\omega,\underline{\theta})\underline{Z}(\omega) \quad (17)$$

where $A(\omega,\underline{\theta})$ is given by (7) and (8). Therefore, the frequency domain representation of the output vector $\underline{y}(t) = (y_1(t),...,y_N(t))^T$ is exactly the same as $\hat{S}(\omega_k)$ of (9). i.e., the system described by (12) is a realization of the least squares estimator of the source signals. In Fig. 2 we present a block diagram of this system for the special case of $N = M = 2$.

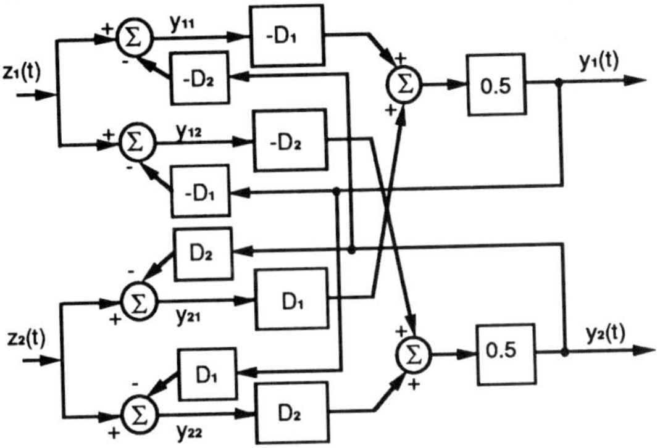

Fig. 2: Bootstrapped implementation of the least-squares separator

III. Bootstrapped Separation of Sources at Unknown Locations

If the sources location is unknown, then $\underline{\theta} = (\theta_1,...,\theta_N)^T$ is unknown, and $\tau_m(\theta_i)$ in (9) is replaced by its estimate, $\hat{\tau}_{mi} = \tau_m(\hat{\theta}_i)$, $m = 1,...,M$, $i = 1,...,N$. In that case, the frequency domain representation of the output vector is:

$$\begin{aligned} \underline{Y}(\omega_k) &= [A^*(\omega_k,\hat{\underline{\theta}})A(\omega_k,\hat{\underline{\theta}})]^{-1}A^*(\omega_k,\hat{\underline{\theta}})\underline{Z}(\omega_k) \\ &= [\hat{A}^*\hat{A}]^{-1}\hat{A}^*\underline{Z} \end{aligned} \quad (18)$$

This is no-longer the least square estimate of $\underline{S}(\omega_k)$, but only an approximation. In the sequel, we propose an adaptive algorithm by which an estimate of the delays D_i will be found. We show that, in case of no additive noise, the algorithm converges to the least squares solution of (9). We demonstrate our results for the special case where $N = M = 2$. However, generalization of the algorithm for any $M \geq N \geq 2$ is straight forward. Consider the system of Fig. 3. It can be shown that for $\tau_1(\theta_1) = -\tau_2(\theta_1) = D_1$, $\tau_1(\theta_2) = -\tau_2(\theta_2) = D_2$, the 2x2 transfer function matrix, $H(\omega)$, which relates the two outputs, $y_1(t)$ and $y_2(t)$ to the two inputs $z_1(t)$ and $z_2(t)$ is exactly the same as those of the system of Figs. 1 and 2. That is, the system of Fig. 3 is another, alternative implementation of a least squares separator with $H(\omega)$ given by (10). If initially $\tau_1 \neq D_1$ and/or $\tau_2 \neq D_2$ then we intend to adapt τ_1 and τ_2 so that in the steady state they reach these optimal values. We notice that the scheme of Fig. 3 is similar to the bootstrapped "power-power" separator of [2-7] which is applied to separate weighted sum of two uncorrelated signals. There, the delays τ_1 and τ_2 are replaced by complex weights, say W_1 and

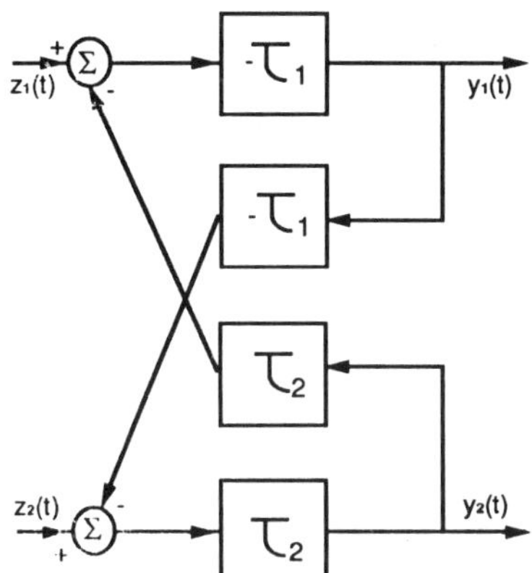

Fig. 3: Alternative implementation of the bootstrapped least-squares separator.

W_2. In the frequency domain our unknown delays are represented by $e^{j\omega\tau_1}$ and $e^{-j\omega\tau_2}$ and the input signals are weighted sums of the uncorrelated frequency domain signals $S_1(\omega)$ and $S_2(\omega)$.

Therefore, the frequency domain representation of our problem is equivalent to the time domain separation problem of [2-10] and a similar approach can be considered. It was shown there that without noise ($n_1 = n_2 = 0$), the power of the two output signals is minimal if and only if the controlled weights are equal to the unknown model parameters. It was also shown that for this configuration, the optimization criterion of minimum power is equivalent to the criterion of zero correlation between the two outputs [3-4]. By analogy, in our problem the equivalent optimization criterion should be minimum power (or zero correlation) in the frequency domain. However, since power and correlation are preserved when transforming from time to frequency and vice versa (Parseval), this criterion can be applied to our problem either in the frequency domain, or in the time domain. Inspirated by a possible hardware implementation using voltage controlled delay lines, we prefer the time domain approach. That is, we suggest to control the unknown delays τ_1 and τ_2 by an adaptive algorithm which seeks for the minimum power of both outputs, simultaneously (or, for the minimum power of their cross correlation signal). Notice, however, that this control procedure cannot be employed unless some information that distinguishes the signals to be separated is available. Mathematically, it can be shown that all possible optimization criteria (zero correlation[1], minimum power) yield the same, or linearly dependent, control equations. Notice that to control both τ_1 and τ_2 one needs *two* independent equations. This difficulty can also be predicted by considering the separation problem as a multi-input multi-output identification problem, where it is well known that weighted sums cannot be separated if nothing is known about their components. In our application, it is also well known that the resolution capacity of an array of M sensors is bounded by the number of sensors (i.e., $M \geq N$) if absolutely nothing is known about the signals. Conversely, if prior information is available, it can be used to discriminate between the two signals in the control loops. As shown in [3,4], such dependency problem can be handled by a "discriminator" which

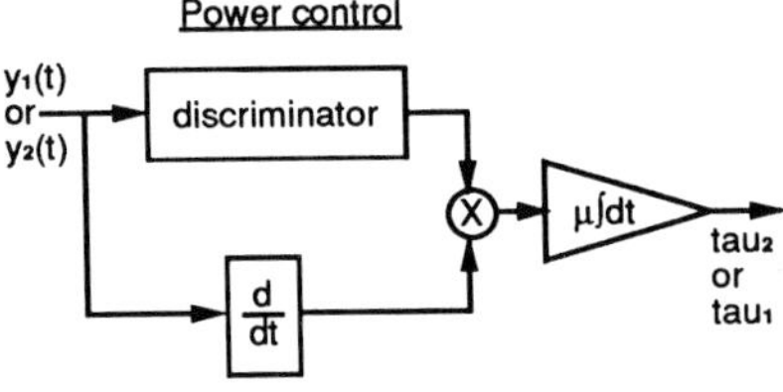

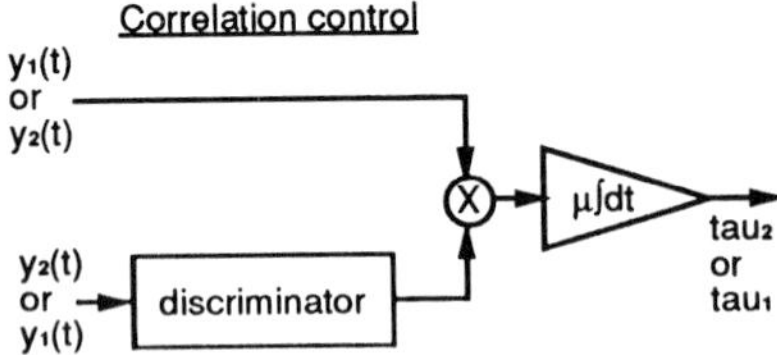

Fig. 4: Block diagrams of the delay control algorithms

uses the distinguishing information to emphasize $s_1(t)$ in one of the control loops and another one which emphasizes $s_2(t)$ in the other loop. This procedure yields two independent control equations, which guarantees convergence of the adaptive algorithm to the desired solution $\tau_1 = D_1$ and $\tau_2 = D_2$. The possible control loops are depicted in Fig. 4.

IV. Conclusions

For separation of signals radiated from point sources, we propose the system of Fig. 3, where the delays are controlled by any of the algorithms of Fig. 4 (there are 4 different combinations, for two unknown delays). We have shown that if the delays are adapted to the unknown model parameters exactly, then the outputs of the proposed bootstrapped system are the least-squares estimates of the source signals. In a further study we investigate the adaptive algorithm with more details, we suggest alternative implementations of the separation configuration and we study the effect of the additive noise on the performance of the proposed separator.

References

[1] J.M. Mendel, *Lessons in Digital Estimation Theory*, Prentice Hall, 1987.
[2] Y. Bar-Ness, J. Rokach: "Cross-Coupled Bootstrapped Interference Canceler", *Proc. of the 1981 SP-S International Symposium*, pp. 292-295.
[3] Y. Bar-Ness, J.W. Carlin, M.L. Steinberger: "Bootstrapping Adaptive Cross-Pol Canceller for Satellite Communication", *Proc. of the Int. Conf. on Comm.*, paper 4F.S, June 1982.
[4] Y. Bar-Ness, J.W. Carlin, M.L. Steinberger: "Bootstrapping Adaptive Interference Cancelers: Some Practical Limitations", *Proc. of the Globecom,* Paper F3.7, 1982.
[5] J. Carlin, Y. Bar-Ness et. al.: "An IF Cross-Pol Canceler for Microwave radio", *IEEE Trans. SAC-S*, pp. 502-514, April 1987.
[6] Y. Bar-Ness, A. Dinc: "Performance Comparison of LMS, Diagonalizer and Bootstrapped Adaptive Cross-Pol Canceler Over Non-Dispersive Channel", MILCOM '1990, paper 3.7.
[7] A. Dinc, Y. Bar-Ness: "Error Probability of Bootstrapped Blind Adaptive Cross-Pol Cancelers for Many GAM Signals Over Non=Dispersive Fading Channels", ICASSP-92, March 1992.
[8] C. Jutten, J. Herault: "Blind Separation of Sources, Part I: An Adaptive Algorithm Based on Neuromimetic Architecture", *Signal Processing*, Vol. 24, No. 1, pp.1-10, July 1991.
[9] P. Common, C. Jutten, J. Herault: "Blind Separation of Sources, Part II: Problem Statement", *Signal Processing,* Vol. 24, No. 1, pp. 11-20, July 1991.
[10] E. Sorouchyati, C. Jutten, J. Herault: "Blind Separation of Sources, Part III: Stability Analysis", *Signal Processing,* Vol. 24, No. 1, pp.21-29, July 1991.
[11] E. Weinstein, M. Feder, O.V. Oppenhiem: "Multi-Channel Signal Separation Based on Decorrelation", submitted for publication in the IEEE Trans. on Signal Processing.

[1]Lately, the decorrelation approach is also used in [11]

SIGNAL PROCESSING VI: Theories and Applications
J. Vandewalle, R. Boite, M. Moonen, A. Oosterlinck (eds.)

Hybrid Mean/Median Wiener Filters

James L. Lansford
Department of Electrical and Computer Engineering
University of Colorado at Colorado Springs
1420 Austin Bluffs Parkway
Colorado Springs, CO 80903

Abstract

Wiener filters are used as whitening filters to remove correlation in the noise. Such filters, because they are based on a least squares model, exhibit poor performance in environments where noise exhibits burst or impulsive characteristics. The traditional formulation of the Wiener filter can be modified to allow a non-least squares criterion; this adaptive approach is more robust to transient events like burst noise.

Introduction

The traditional Wiener filter is based on an autoregressive model which can be described by the difference equation:

$$y(n)=b_0x(n)+a_1y(n-1)+a_2y(n-2)+\cdots a_{N-1}y(n-N+1)+a_Ny(n-N) \quad (1)$$

If written in matrix form starting at n=N and assuming that M samples {0...M-1} of the sequence y are available, this results in the overdetermined system [1]:

$$\begin{bmatrix} y(N-1) & y(N-2)\cdots & y(0) \\ y(N) & y(N-1)\cdots & y(1) \\ \vdots & \vdots & \vdots \\ y(M-2) & y(M-3)\cdots & y(M-N-1) \end{bmatrix} \begin{bmatrix} a_1 \\ a_2 \\ \vdots \\ a_N \end{bmatrix} = \begin{bmatrix} y(N) \\ y(N+1) \\ \vdots \\ y(M-1) \end{bmatrix} \quad (2)$$

of M-N equations in N unknowns, which could be written in matrix form as $\mathbf{G}a=f$. Recovering the a_i from a given sequence of outputs in this way is blind deconvolution with an assumed x(n) that is spectrally flat, such as an impulse or white noise. Even if this is not the case, effects of the source will be incorporated into the model [2].

A solution to the equations given in (2) can be developed by manipulating the matrices to obtain:

$$a=\left[\mathbf{G}^t\mathbf{G}\right]^{-1}\mathbf{G}^t f \quad (3)$$

which is the least squares solution. This solution is not robust to impulses, bursts, or other transients because of the nature of the least squares solution [4]; furthermore, the model is maximum likelihood only if it can be shown that the residual sequence $e=f-\mathbf{G}a$ is Gaussian [5].

In a communication system, the whitening process involves subtracting samples of the predicted values **G**a from the input sequence f to yield the residual sequence e; Figure 1 shows this process in block diagram form.

In terms of frequency response, the spectral estimate derived from the autoregressive model given by (2) is:

$$H(z) = \frac{b_0}{1 - \sum_{i=1}^{N} a_i z^{-i}} \tag{4}$$

Clearly, the matrix expression **G**a is an FIR filter of the form:

$$H_{inv}(z) = 1 - \sum_{i=1}^{N} a_i z^{-i} \tag{5}$$

which is, of course, the inverse filter for the process y(n). In the literature, little emphasis has been placed on the effect of the solution method for (2) on the spectral estimate.

Robust Wiener Filter

The least squares solution given by (3) is only one of an infinity of solutions to the overdetermined system in (2). In general, a p-normed error criterion can be defined:

$$e = \sum_{i=1}^{N} |f_i - (\mathbf{G}a)_i|^p \tag{6}$$

where minimization of e under the constraint **G**a=f when p=2 is the least squares solution. This generalization allows a "robust" solution, which is also known as the least absolute value (p=1) solution. In terms of maximum likelihood estimators, this formulation is quite comprehensive, encompassing Gaussian (p=2), Laplacian (p=1), and uniform ($p \to \infty$) densities; the family of probability density functions is called the generalized Gaussian or p-Gaussian densities. The functional form is:

$$p(x)= \frac{1}{\phi\Gamma(1+1/p)2^{1+1/p}} \exp\left(-\frac{1}{2}\left|\frac{x-\theta}{\phi}\right|^p\right) \qquad -\infty < x < \infty \tag{7}$$

Solution to (2) with the objective function given by (6) is more complicated than the Cholesky decomposition typically used for the least squares formulation. As suggested by Yarlagadda et al, a modified residual steepest descent (RSD) method works well in this application. This algorithm implies a computational complexity of about 4 times that of the Cholesky decomposition, since the least squares problem must be solved in each iteration of the algorithm, and an average of around four iterations are required for convergence in typical situations.

Robust Filtering of Communications Signals

Using the approach described above, the technique was evaluated with a continuous sinusoid of the form:

$$y(n)=\sin(\omega_0 n)+w(n)+\delta(n-k) \tag{8}$$

where w(n) is a white Gaussian noise process whose variance dictates the SNR, and $\delta(n-k)$ is a delta function that occurs sometime during the analysis interval. If the p-normed algorithms are applied to this signal and the spectral estimate is computed, the sequence of spectra shown in Figure 2 results. It is interesting to note that if the noise had been purely impulsive, the spectrum for p=1 would have been maximum likelihood, while Gaussian noise alone would have had p=2 as its maximum likelihood estimator. The fact that the noise is a blend of the two types is reflected in the peak in the spectral estimate for a value of p between one and two.

The effects of impulsive noise on the spectral estimate can be further observed to cause a bias in the spectral estimate. Figure 3 shows the L_1 and L_2 estimates for the case of $\omega_0=1.5$. As this figure shows, the least squares estimate is biased away from the "true" value because of the impulse, while the least absolute value estimator more closely represented the spectrum.

Finally, the output of the inverse filter is an indicator of how well the signal can be smoothed; Figure 4 shows the signal y(n) and the smoothed estimate **Ga** with an L_1 estimator. This figure clearly shows that the sinusoid's frequency is preserved (but not its phase) and that the impulse has been excised.

Summary

This paper has shown a generalized Wiener filter that can blend the properties of a median filter and an averager to generate a filter (and a spectral estimate) that is maximum likelihood for a wide variety of distributions, including Laplacian and Gaussian.

References

1. 1. Kay, S.M., *Modern Spectral Estimation: Theory and Practice*, Prentice Hall, 1988, p. 239.
2. Gonin, R., and A.H. Money, *Nonlinear Lp-Norm Estimation*, Marcel Dekker, 1989, p. 222.
3. Gonin and Money, *Ibid*, p. 222.
4. Lansford, J. L., "Lp Models in Speech Coding and Markov Models in Speech Recognition", Ph. D. Dissertation, Oklahoma State University, Stillwater, OK, 1988.
5. Bednar, J. Bee, R. Yarlagadda, and T. L. Watt, "L1 Deconvolution and its application to Seismic Signal Processing", *IEEE Trans. ASSP*, ASSP-34, Dec. 1986, pp. 1655-1658.
6. Lansford, J. L., "Optimal Lp Approach to Speech Coding", ICASSP-88 Proceedings, pp. 335-338.
7. Kriel, C., "Sonar Signal Processing Using the L_p Norm",SPIE -88 Proceedings
8. Schroeder, J. E, Lansford, J. L., et al, "Lp Normed Spectral Estimation of Mechanical Signals", Twenty-Fourth Asilomar Conference on Signals, Systems, and Computers, 1989.
9. Yarlagadda, R., et al, "Fast Algorithms for Lp Deconvolution", *IEEE Trans. ASSP*, ASSP-33, Feb 1985, pp. 174-182.

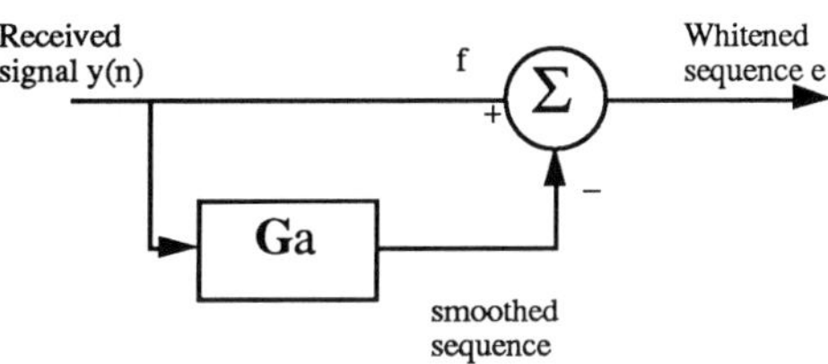

Figure 1: Wiener filtering model

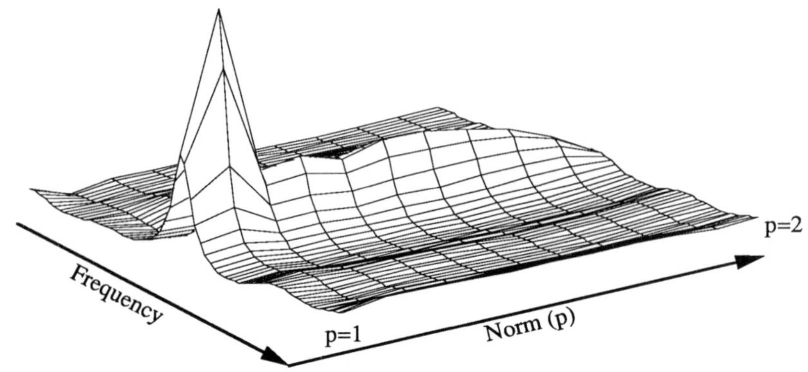

Figure 2: Spectral estimate as a function of p

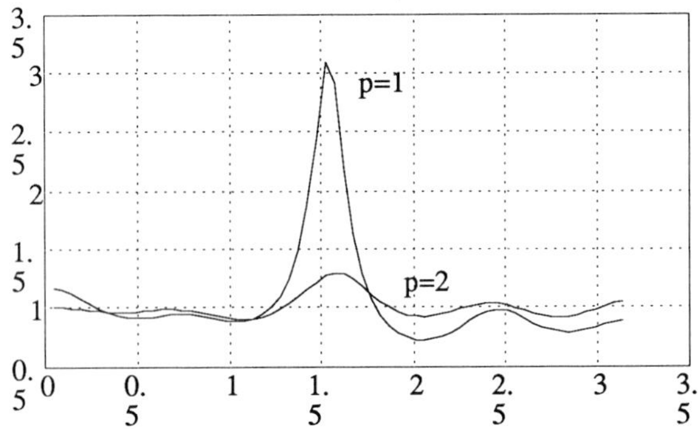

Figure 3: Comparison of spectra for p=1 and p=2

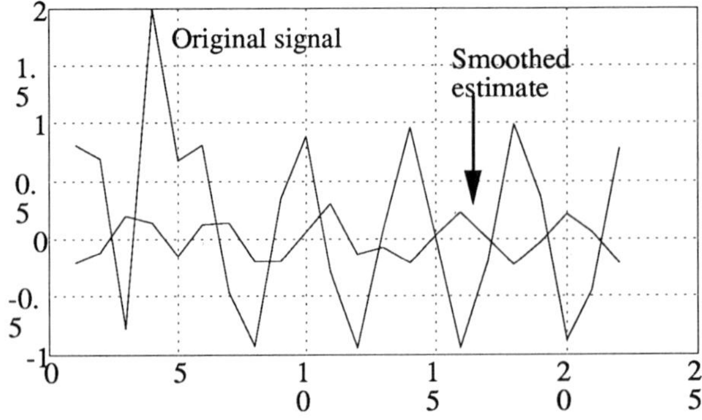

Figure 4: Smoothing filter output

SIGNAL PROCESSING VI: Theories and Applications
J. Vandewalle, R. Boite, M. Moonen, A. Oosterlinck (eds.)

STUDY OF MA MODEL IDENTIFICATION USING ORDER STATISTICS IN THE PRESENCE OF NOISE

J.Y TOURNERET and B LACAZE

ENSEEIHT / GAPSE, 2 Rue Camichel, 31 071 Toulouse, FRANCE

A new estimator of Moving Average (MA) parameters was presented in the Twenty-Sixth Annual Conference on Information Sciences and Systems [1]. This new estimator was studied for a model driven by a bounded input having a non zero probability of reaching its maximum. Such input laws are obtained in the case of binary, discrete or clipped inputs. The most frequently used MA parameter estimators are approximations of the Maximum Likelihood Method. **These Maximum Likelihood Estimators (M.L.E.) have a zero error asymptotically.** Our estimator is based on Order Statistics and will be called **O.S. Estimator** in the following. In absence of noise, **its main property is to give a zero error for a finite signal point number with a high known probability.** The aim of this paper is to study **O.S** estimator performances in the case of a model embedded in an additive noise. The study is done in the case of a binary input but can be extended to discrete or clipped inputs. In the first section, our new method of MA identification using Order Statistics is briefly described. The second section deals with the behaviour of the **O.S.** estimator in the presence of noise.

1 INTRODUCTION

Most commonly used Moving Average (MA) parameter estimators are approximations to the Maximum Likelihood Estimator (M.L.E.) [2],[3]. The use of the M.L.E. for parameter estimation can be justified for large data records on the basis of its asymptotic unbiased and minimum variance properties. In particular, the M.L.E. has a zero error asymptotically. A new estimator of MA parameters based on Order Statistics, which is denoted by **O.S. estimator**, was presented in the 26th C.I.S.S. [1]. The main property of this estimator is to give, in absence of noise, a zero error for a finite signal point number with a high known probability. In the first section, MA parameter identification using Order Statistics is briefly described. The second section deals with the estimator performances in the case of a MA model embedded with an additive gaussian noise. The study is carried out for a binary input but can be extended to discrete or clipped inputs.

2 MA IDENTIFICATION

Let *e(k)* be the input noise of an order *q* MA model, the parameters of which are denoted by $b_0, b_1, \ldots, b_q$. Let us denote *x(k)* the model output, *n(k)* the additive noise and *z(k)* =*x(k)*+*n(k)*. The model is then defined by the following recursive equations :

$$k = 0, 1, \ldots, K-1 \qquad z(k) = n(k) + \sum_{i=0}^{q} b_i e(k-i) \quad (1)$$

The noise *n(k)* is assumed to be Independent Identically Distributed (i.i.d.) and statistically independent of *x(k)*. In what follows, we propose a new method to estimate the parameters $\{b_i\}$ for a special class of input white noise : *e(n)* is bounded by its maximum $E = \mathrm{Max}\{e(k)\}$ such that :

$$Pmax \overset{\Delta}{=} P[e(k) = E] = P[e(k) = -E] > 0 \qquad (2)$$

Let us cascade at the output *z(k)* a set of *q+1* order *1* MA models with respective parameters c_j , for *j=0,1,...,q*, which will be referred to as **"Analysis Filters"**. The Analysis filter outputs, shown in *Fig 1* and denoted by $y_j(k)$ for *j=0,1,...,q* are given by :

$$y_j(k) = b_0 e(k) + \sum_{i=1}^{q} (b_i + c_j b_{i-1}) e(k-i)$$
$$+ c_j b_q e(k-q-1) + n(k) + c_j n(k-1) \qquad (3)$$

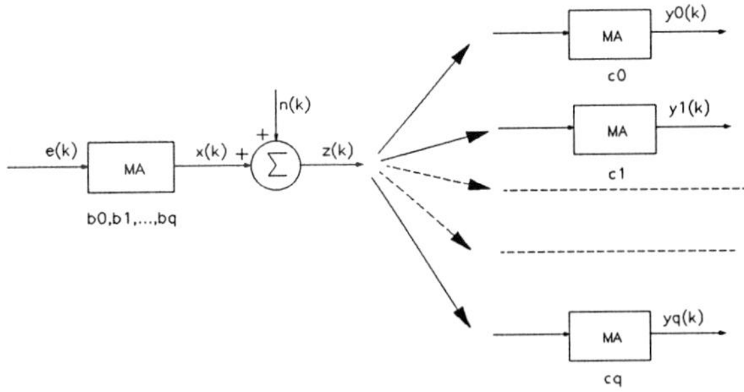

Fig 1 : Single-channel system with Analysis Filters

When there is no additive noise ($n(k)=0.$), the maximum of $y_j(k)$ for $k \in N$, denoted by $\mathrm{Max}\{y_j(k)\}$, is given by :

$$\mathrm{Max}\{y_j(k)\} = E.\left\{ |b_0| + \sum_{i=1}^{q} |b_i + c_j b_{i-1}| + |c_j b_q| \right\} \quad (4)$$

By denoting $S_j(i)$ for $i=1,2,...,q$, the sign of $(b_i + c_j b_{i-1})$ and respectively $S(0), S_j(q+1)$ the signs of b_0 and $c_j b_q$, *Eq.(4)* can be written in the following way :

$$\mathrm{Max}\{y_j(k)\}/E = b_0 S_j(0) + \sum_{i=1}^{q} (b_i + c_j b_{i-1}) S_j(i)$$

$$+ c_j b_q S_j(q+1) = \sum_{i=0}^{q} b_i [S_j(i) + c_j S_j(i+1)] \quad (5)$$

The concatenation of *Eq.(5)* for $j=0,1,...,q$ leads to a system of $q+1$ equations with $q+1$ unknowns which can be written :

$$\underline{M} = A\underline{b} \quad (6)$$

with :

$$\underline{M} = \frac{1}{E}[\mathrm{Max}\{y_0(k)\}, \ldots, \mathrm{Max}\{y_q(k)\}]^T$$

$$\underline{b} = [b_0, \ldots, b_q]^T \text{ and } A = (a_{ji}) = [S_j(i) + c_j S_j(i+1)]$$

The matrix equation *(6)* will yield unique MA parameters provided that the $(q+1)\times(q+1)$ matrix A is of full rank $q+1$. In [1], a method for choosing the parameters c_j, in order to have a full rank matrix A is studied.

For practical implementation, $\mathrm{Max}\{y_j(k)\}$ values are estimated with K Analysis Filter outputs $y_j(0), y_j(1), \ldots, y_j(K-1)$ and will be then denoted by $\mathrm{Max}_K\{y_j(k)\}$. The different $S_j(i)$ sign estimations, obtained with rough estimations of MA parameters, are supposed to be rigourously exact. The matrix system *(6)* leads to :

$$A\underline{\hat{b}} = \underline{\hat{M}}_K \quad (7)$$

with

$$\underline{\hat{b}} = [\hat{b}_0, \hat{b}_1, \ldots, \hat{b}_q]^T$$

$$\underline{\hat{M}}_K = \frac{1}{E}[\mathrm{Max}_K\{y_0(k)\}, \ldots, \mathrm{Max}_K\{y_q(k)\}]^T \quad (8)$$

The solution for this system gives MA parameter estimators $\hat{b}_0, \hat{b}_1, \ldots, \hat{b}_q$. **When the q+1 absolute maxima $\mathrm{Max}\{y_j(k)\}_{j=0,1,\ldots,q}$ are reached, that is to say $\mathrm{Max}_K\{y_j(k)\} = \mathrm{Max}\{y_j(k)\}$ for j=0,1,...,q, these MA parameter estimators are exactly equal to $b_0, b_1, \ldots, b_q$ and then have a zero error.** The aim of this paper is to study the performances of the **O.S.** estimator in the case of a model embedded with an additive gaussian noise.

3 NOISE EFFECTS

In this section, the model output is embedded with an additive noise which is assumed to be continuous. The probability density of $y_j(k)$ (see *Eq.(3)*) and then of $\mathrm{Max}_K\{y_j(k)\}$ depends on that of the noise. The effects of noise are then studied in the case of a binary input $e(n)$, taking on equally likely values $+1$ and -1, but this study could be extended to discrete or clipped inputs. Then the variable $y_j(k)$ is the sum of a discrete variable

$$e_j(k) = b_0 e(k) + \sum_{i=1}^{q} (b_i + c_j b_{i-1}) e(k-i) + c_j b_q e(k-q-1)$$

taking on 2^{q+2} different values denoted by u_i for $i = 1, 2, \ldots, 2^{q+2}$ and of a continuous variable $n_j(k) = n(k) + c_j n(k-1)$ (see *Eq.(3)*). The probability density function (p.d.f) of $n_j(k)$ is denoted by $f_j(t)$. With these notations, the p.d.f. of $y_j(k)$ is given by :

$$f_{y_j}(t) = \frac{1}{2^{q+2}} \sum_{i=1}^{2^{q+2}} f_j(t-u_i) \quad (9)$$

$n(k)$ is assumed to be a white gaussian noise with zero mean and variance σ^2. Then $n_j(k) = n(k) + c_j n(k-1)$ is gaussian with zero mean and variance $\sigma_j^2 = (1 + c_j^2)\sigma^2$. The $y_j(k)$ p.d.f is then given by :

$$f_{y_j}(t) = \frac{1}{2^{q+2}} \sum_{i=1}^{2^{q+2}} \frac{1}{\sqrt{2\pi\sigma_j^2}} \exp\left[-\frac{(t-u_i)^2}{2\sigma_j^2} \right] \quad (10)$$

The determination of $\mathrm{Max}_K\{y_j(k)\}$ distribution is very difficult because the variables are not independent. A lot of work has been done in the extreme value field. Gumbel [4] has shown the existence of three asymptotic distributions of largest values. Each assumes a specific behaviour for absolute large values of the variable. From *(10)*, the asymptotic $\mathrm{Max}_K\{y_j(k)\}$ distribution can be determined with the help of the following :

1. The distribution of the largest normal values converges towards the Gumbel first asymptote [4].
2. Given N p.d.f. $f_i(t)$ (For $i=1,...,N$) of N variables whose largest value distributions converge towards the Gumbel first asymptote, it can be shown that the variable whose p.d.f. is $\frac{1}{N}\sum_{i=1}^{N} f_i(t)$ has a largest value distribution converging towards the same asymptote.

Then, for a high signal point number K, the $\mathrm{Max}_K\{y_j(k)\}$ distribution is given by the first asymptote :

$$f(t) = \alpha e^{-\alpha(t-u)} \exp[-e^{-\alpha(x-u)}] \text{ with } \alpha > 0 \quad .$$

The parameters α and u have then to be estimated. Numerous estimation methods of these parameters are available in the literature [4]. They lead to an estimated $\mathrm{Max}_K\{y_j(k)\}$ probability density function which has been compared with the $\mathrm{Max}_K\{y_j(k)\}$ histogram. For a *K=10,000* signal point number, simulations show the good agreement between the two estimations. The results that follow deal with an order *2* MA Model with parameters $b_0 = 1, b_1 = 0.2$ and $b_2 = 0.5$:

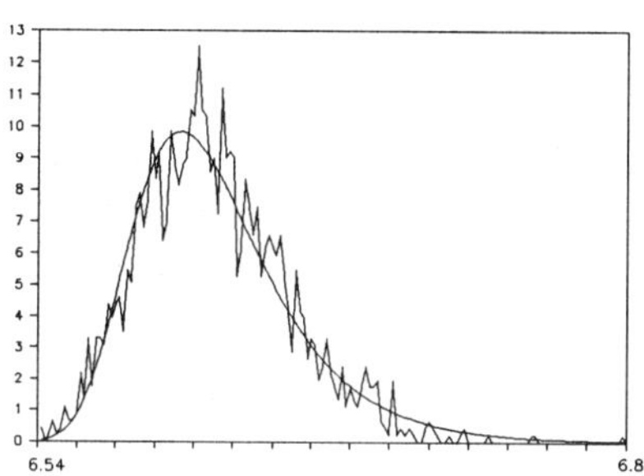

Fig 2a : $\mathrm{Max}_K\{y_1(k)\}$ *p.d.f.*

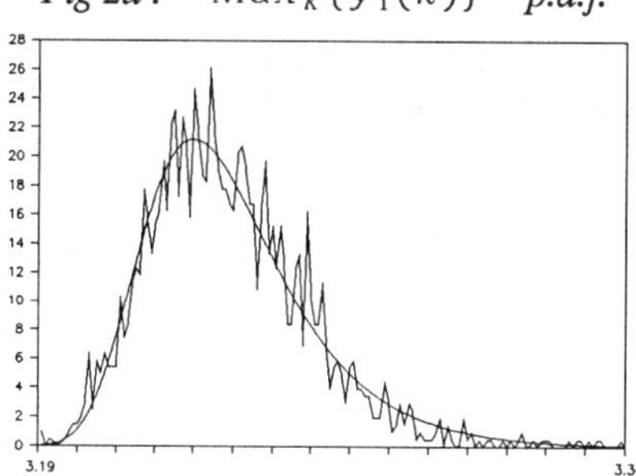

Fig 2b : $\mathrm{Max}_K\{y_2(k)\}$ *p.d.f.*

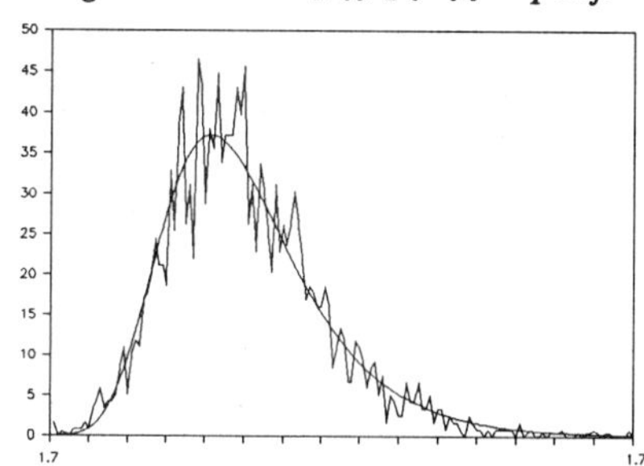

Fig 2c : $\mathrm{Max}_K\{y_3(k)\}$ *p.d.f.*

Fig 2a,2b,2c : Comparison between theoretical and experimental $\hat{\mathrm{Max}}_K\{y_j(n)\}$ *p.d.f.*
For an order 2 MA model
with parameters $b_0 = 1, b_1 = 0.2$ *and* $b_2 = 0.5$

The $\hat{\mathrm{Max}}_K\{y_j(n)\}$ asymptotic distribution moments allow us to estimate **the bias of the O.S. estimator**. We thus obtain :

$$E[\underline{\hat{b}}] - \underline{b} = A^{-1}(E[\underline{\hat{M}}_K] - \underline{M})$$

with $E[\hat{\mathrm{Max}}_K\{y_j(k)\}] = u + \frac{\gamma}{\alpha}$, γ being the Euler constant ($\gamma = 0.577$) *[4]*.

In order to determine **the O.S. estimator covariance matrix**, that of $\underline{\hat{M}}_K$ has to be computed. But this is a very difficult task because the variables $\hat{\mathrm{Max}}_K\{y_j(n)\}$ are not independent. For the sake of simplicity, let us consider a simpler model :

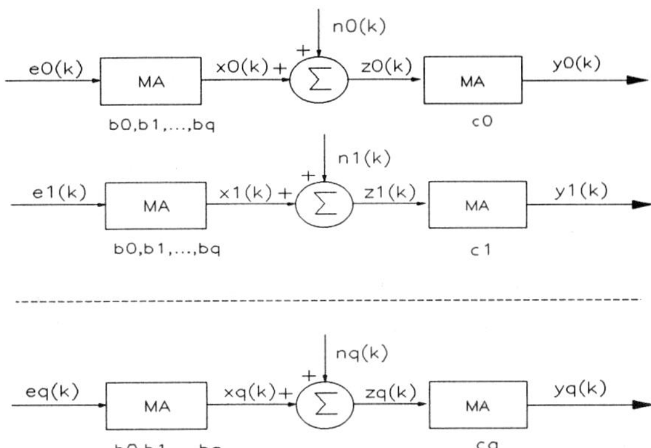

Fig 3 : q+1 single-channel model.

The variables $\hat{\mathrm{Max}}_K\{y_j(k)\}$ are then independent and the corresponding covariance matrix, denoted by $C_{\hat{M}}$, is diagonal. The $C_{\hat{M}}$ diagonal terms are equal to the $\hat{\mathrm{Max}}_K\{y_j(k)\}$ variances that is to say to the first asymptote order *2* moment ($\frac{\pi^2}{6\alpha^2}$ *[4]*). The **O.S.** estimator covariance matrix is then given by :

$$E[(\underline{\hat{b}} - \underline{b})(\underline{\hat{b}} - \underline{b})^T] = A^{-1}.C_{\hat{M}}.(A^{-1})^T$$

The independence assumption is clearly not valid for the application to an actual MA model. But we show here-under that, although this theoretical approach is not realistic, it yields experimental results in satisfactory agreement with the model of Fig.*3*. then, the covariance matrices of $\hat{M}_{\underline{K}}$ and $\underline{\hat{b}}$ have been computed as the by-products of numerous simulations (*2,000* estimations of $\hat{\mathrm{Max}}_K\{y_j(k)\}$ and of $\underline{\hat{b}}$). For the previous order *2* MA model, with a Signal to Noise Ratio (SNR) equal to *20dB*, the following results are obtained :

$$\begin{pmatrix} 1.8e-3 & 7.3e-6 & 2.3e-6 \\ 7.3e-6 & 4.4e-4 & -1.6e-6 \\ 2.3e-6 & -1.6e-6 & 1.3e-4 \end{pmatrix}$$

$\underline{\hat{M}}$ *simulated covariance matrix*

The $M_{\underline{K}}$ covariance matrix extra-diagonal terms are negligible. The variables $\hat{\mathrm{Max}}_K\{y_j(n)\}$ can then be assumed to be independent and the theoretical $\underline{\hat{b}}$

covariance matrix can be determined. We then obtain :

$$\begin{pmatrix} 5.2e-4 & -3.2e-5 & -5.5e-4 \\ -3.2e-5 & 5.5e-5 & 3.9e-5 \\ -5.5e-4 & 3.8e-5 & 6.2e-4 \end{pmatrix}$$

$\underline{b}$ *theoretical covariance matrix*

which can be compared with the experimental covariance matrix :

$$\begin{pmatrix} 4.3e-4 & -2.9e-5 & -4.5e-4 \\ -2.9e-5 & 4.8e-5 & 3.4e-5 \\ -4.5e-4 & 3.4e-5 & 5.1e-4 \end{pmatrix}$$

$\underline{\hat{b}}$ *experimental covariance matrix*

The matrix norm chosen for the comparison with the simulated and theoretical covariance matrix is the **Trace** of the difference of these two matrices. The relative error between the two covariance matrices is then approximately equal to *18%*, which is a quite acceptable result. This shows the good agreement between simulations and theory.

As a matter of illustration, the MA parameter **O.S.** estimator variance is compared, for different Signal to Noise ratio, to that obtained using the conventional Durbin algorithm *[2],[3]*. The choice of the signal point number K in terms of the MA model order is studied in *[1]*. For our simulations, it has been fixed to $K=10{,}000$. The following results deal with an order *2* MA Model with parameters $b_0 = 1$, $b_1 = 0.2$ and $b_2 = 0.5$. *M.L.E.* and *O.S.E.* stand respectively for Maximum Likelihood Estimator and Order Statistic Estimator.

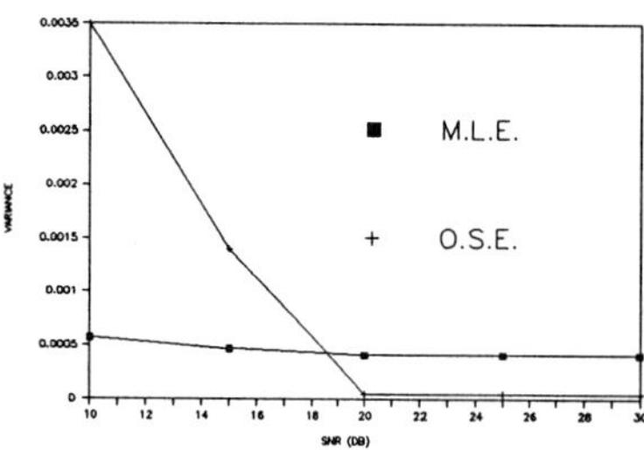

Fig 4a : $\hat{b}_1$ *variance*

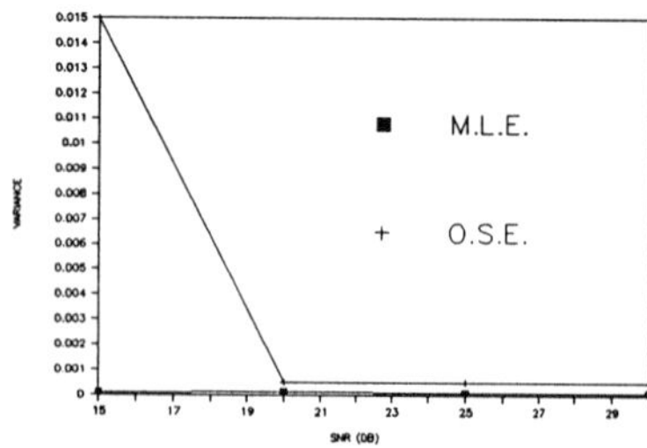

Fig 4b : $\hat{b}_2$ *variance*

Fig 4a,4b : Comparison between M.L.E. and *O.S.E. variances*

For Signal to Noise Ratio greater than *15* dB, the *M.L.E.* and *O.S.E.* variances have the same order of magnitude but for *S.N.R.* lower than *15dB*, the Maximum Likelihood Estimator performs better.

The main drawback of the O.S. Estimator, **in presence of gaussian noise** (more generally unbounded noise), lies in its bias. Namely, when *n(k)* is not bounded, the following property is obviously verified :

$$\lim_{K\to+\infty} \mathrm{Max}_K[n(k)] = \lim_{K\to+\infty} \mathrm{Max}_K[y_j(k)] = +\infty$$

In order to solve this problem, an unbiased Order Statistic Estimator will be presented in a forthcoming publication.

4 CONCLUSION

The behaviour of a new estimator of Moving Average Model parameters, based on Order Statistics, has been studied in presence of an a additive gaussian noise. An expression of the theoretical covariance matrix of this estimator has been computed under a simplifying assumption and compared with an experimental one. Then, M.L.E. And O.S.E. variances have been compared for various Signal to Noise Ratio. For high Signal to Noise Ratios, the O.S. estimator performs well, that is to say the MA parameter variances have the same order of magnitude as those obtained with the Maximum Likelihood Estimator. For lower Signal to Noise ratio, the classical Maximum Likelihood estimator will be preferred. On the other hand, the main effect of additive gaussian noise on the Order Statistic Estimator is its bias. For the case of an amplitude bounded noise with an approximately gaussian shape, an unbiased version of this estimator has been designed and will be published in a forthcoming publication.

5 REFERENCES

[1] J.Y. Tourneret and B. Lacaze, "A new estimator of Moving Average Model parameters based on Order Statistics" in proc. of the 26th Annual Conf. on Information Sciences and Systems, Computers

[2] J. Durbin, "Efficient estimation of parameters in Moving-Average Models," Biometrica, Vol. 46, pp. 306-316, 1959.

[3] S. Kay, Modern Spectral Estimation : theory and application. In Prentice Hall, 1988.

[4] E. J. Gumbel, Statistics of Extremes. New York : Columbia University Press, 1958.

SIGNAL PROCESSING VI: Theories and Applications
J. Vandewalle, R. Boite, M. Moonen, A. Oosterlinck (eds.)
1992 Elsevier Science Publishers B.V.

A NEURAL NETWORK FOR ONLINE ESTIMATION OF PARAMETERS OF NOISY SINUSOIDAL SIGNALS

Andrzej CICHOCKI

Technical University of Warsaw
PL-00661 Warszawa, Poland
ul. Koszykowa 75

Tadeusz ŁOBOS

Technical University of Wrocław
PL-50370 Wrocław, Poland
pl. Grunwaldzki 13

New massively parallel algorithms for estimation of parameters of sinewave contaminated by noise are proposed. The approach can be viewed as extension and generalization of the standard least-squares minimization method. The implementation of the algorithms by an appropriate neural network is also given. Illustrative computer simulations results confirm validity and high performance of the proposed solution.

1. INTRODUCTION

The problem of estimating the amplitudes, phase angles and frequencies of sinusoidal signals from noisy and distorted data have received considerable attention recently. For control and protection of electrical power systems it is desired to estimate in real-time parameters of basic waveform (fundamental harmonic) of voltages and currents. For this purpose different numerical algorithms have been developed e.g. based on the Fourier and Kalman filtering [1].
In this paper we shall present new algorithms associated with system of differential or difference equations which can be implemented (simulated) by analog VLSI circuits employing some neural network principles [2,3,4,5]. Neural networks may provide a new kind of solution for many standard signal processing problems, e.g. recovering of analog signal corrupted by noise. The main purpose of this paper is to present a new algorithm and associated new architectures of analog artificial neuron-like processors for online estimation of parameters of sinusoidal signal distorted by higher harmonics and corrupted by noise. The problem of estimation is formulated as an optimization problem [4].

2. STATEMENT OF THE PROBLEM

Consider the following sinusoidal signal

$$x(t) = X_a \sin(\omega_x t) + X_b \cos(\omega_x t) \qquad (1)$$

in which X_a, X_b are unknown amplitudes,

$\omega_x = 2\pi f_x$ is unknown angular frequency.

Let $y(t)$ denote the noise corrupted measurement of $x(t)$ i.e.

$$y(t) = x(t) + \varepsilon(t) \qquad (2)$$

where $\varepsilon(t)$ is unknown error. This error includes random noise and distortion caused e.g. by measurement instruments, or higher harmonics. Consider the practical case where the signal of interest $y(t)$ is observed (measured) during a finite time and only N samples of this signal are available i.e. the discrete-time signal can be specified as

$$y(t)\big|_{t=mT} = y_m = X_a \sin(m\omega_x T) + X_b \cos(m\omega_x T) + e(mT) \qquad (3)$$

for $m = 1,2,\dots,N$,

where T is the sampling interval.

Hence, the error at the moment $t = mT$ can be expressed as

$$e_m = y_m - x_m \qquad (4)$$

There exist a need for an online algorithm which can satisfy two simultaneous requirements:
a) directly providing estimates of the parameters (X_a, X_b, ω_x) on basis of data samples y_m obtained from noisy measurement, and b) providing automatically new estimates as each data sample is received or providing improved estimate as updated samples are received.

To formulate the above problem in terms of artificial neural networks the key step is to construct an appropriate energy or cost function

$E(\mathcal{X})$ where $\mathcal{X} = \left[X_a, X_b, \omega_x\right]^T$, so the lowest energy state will correspond to desired solution $\mathcal{X}^*$. Mathematically the problem can be formulated as follows: *find a vector* $\mathcal{X}$ *which minimizes the scalar energy function*

$$E(\mathcal{X}) = \left[\sum_{m=1}^{N} |e_m|^p\right]^{1/p} = \|\mathbf{e}\|_p \tag{5}$$

with $1 \le p \le \infty$, where $\mathbf{e} = [e_1, e_2, \ldots, e_N]^T$

For the above problem there are three important for practice special cases:

1) for p=1 the estimation problem is referred as least-absolute-deviations (L_1-norm) signal model fitting;
2) for p=2 the problem is standard (L_2-norm) least-squares model fitting;
3) for p=∞ the problem is minimax, Chebyshev or L_∞-norm model fitting.

The proper choice of the norm used depends on the distribution of the noise error in the sampled data. For normal distribution the best choice will be the least-squares norm. However, if there are outliers and wild noise the L_1-norm is preferable. If the noise have uniform distribution the L_∞-norm may be the most suitable choice [4,6,8].

3. ESTIMATION OF PARAMETERS USING CHEBYSHEV (L_∞-) NORM CRITERION

The problem can be formulated as the minimax (Chebyshev norm) optimization problem:
find the parameters $\mathcal{X}$ *which minimize a scalar cost (error) function*

$$E_\infty(X_a, X_b, \omega_x) = \max_{1 \le m \le N} \{|e_m(X_a, X_b, \omega_x)|\}. \tag{6}$$

which can be written shortly as

$$\min_{\mathcal{X} \in R^3} \max_{1 \le m \le N} \{|e_m|\} \tag{7}$$

In order to solve the problem we use dynamic gradient system [4,5], i.e. we apply the steepest descent continuous-time optimization algorithm which leads to set of nonlinear equations

$$\frac{dX_a}{dt} = \frac{1}{\tau_1} \sum_{m=1}^{N} S_m \operatorname{sign}(e_m) \sin(m\omega_x T) \tag{8}$$

$$\frac{dX_b}{dt} = \frac{1}{\tau_2} \sum_{m=1}^{N} S_m \operatorname{sign}(e_m) \cos(m\omega_x T) \tag{9}$$

$$\frac{d\omega_x}{dt} = \frac{1}{\tau_3} \sum_{m=1}^{N} S_m\, mT \operatorname{sign}(e_m)[X_a \cos(m\omega_x T) + - X_b \sin(m\omega_x T)] \tag{10}$$

where τ_1, τ_2, τ_3 are the positive coefficients representing time constant of integrators and

$$S_m = \begin{cases} 1 & \text{if } |e_m| = \max_{1 \le i \le N} \{|e_i|\} \\ 0 & \text{otherwise.} \end{cases}$$

Fig.1 shows a functional block diagram of an artificial neuron-like network which enables for parallel implementation of the above set of equations, and therefore estimation of the parameters in real-time. The network of Fig.1 consist of integrators, summers, multipliers, signum (hard limiter) activation functions and trigonometric (sin, cos) functions generators. The switches S are controlled by special subnetwork called Winner-Take-All (WTA) circuit. The function of the WTA is to select the strongest, i.e. the largest in absolute value instantaneous error [7]. The sign of selected error is transmitted for further processing while the other *weaker* error signals are completely inhibited by opening corresponding switches. The WTA circuit usually consist of mutually inhibitory links between processing units. Each units inhibits all others. A collective dynamic competition take place; each unit compares the maximum of the incoming signals to its own *strength.* As a result of this competition the largest signal is selected. In the network shown in Fig.1 we have used a simple WTA circuit which employs an integrator acting as a maximum finder, i.e. its output signal tracks continuously in time the signal

$$E(t) := \max_{1 \le m \le N} \{|e_m|\} \tag{11}$$

This signal is compared with all actual error signals in voltage comparators. The switches are controlled by voltage comparators [5].
Extensive computer simulation experiments have confirmed that the neural network shown in Fig.1 allows to estimate in real-time desired parameters. Furthermore, it has been found that the network exhibits notable robustness since its functionality is not affected by parameter variations in a wide range. For example signum ac-

tivation functions can be replaced by sigmoid (tangent hyperbolic) or linear function without affecting the functionality of the network and its final accuracy.

4. ESTIMATION BASED ON OTHER CRITERIA

It is interesting to note that the neural network shown in Fig.1 can easily be modified to perform estimation of the parameters according to least-absolute values and least-squares criteria. For example by closing all the switches or simply by removing them and the associated WTA circuit the network will act according to the least-absolute-values criterion.
Analogously, in order to estimate parameters according to the least-squares criterion all signum activation functions must be replaced by linear functions or simply removed and all switches must be closed.
Summerizing, the proposed neural network is able to estimate desired parameters of the observed noisy sinusoidal signal according to L_1-, L_2- or **L_∞-norm criterion. The choice of the criterion depends on distribution of noise in the measured signals.**

5. COMPUTER SIMULATION EXPERIMENTS

In order to check validity and performance of the proposed algorithms the associated network has been extensively simulated on computer. The simulations fully confirmed correctness of the presented approach and good agreement with theoretical considerations has been obtained. Due to limit of space we present here only one illustrative example.

Example

Let us consider sinusoidal signal contaminated by white Gaussian and impulsive noise.
The observed signal has the form

$$y(t) = -120\sin(2\pi ft) + 40\cos(2\pi ft) + \varepsilon(t)$$

where $f = 50$ Hz and $\varepsilon(t)$ is a noise.

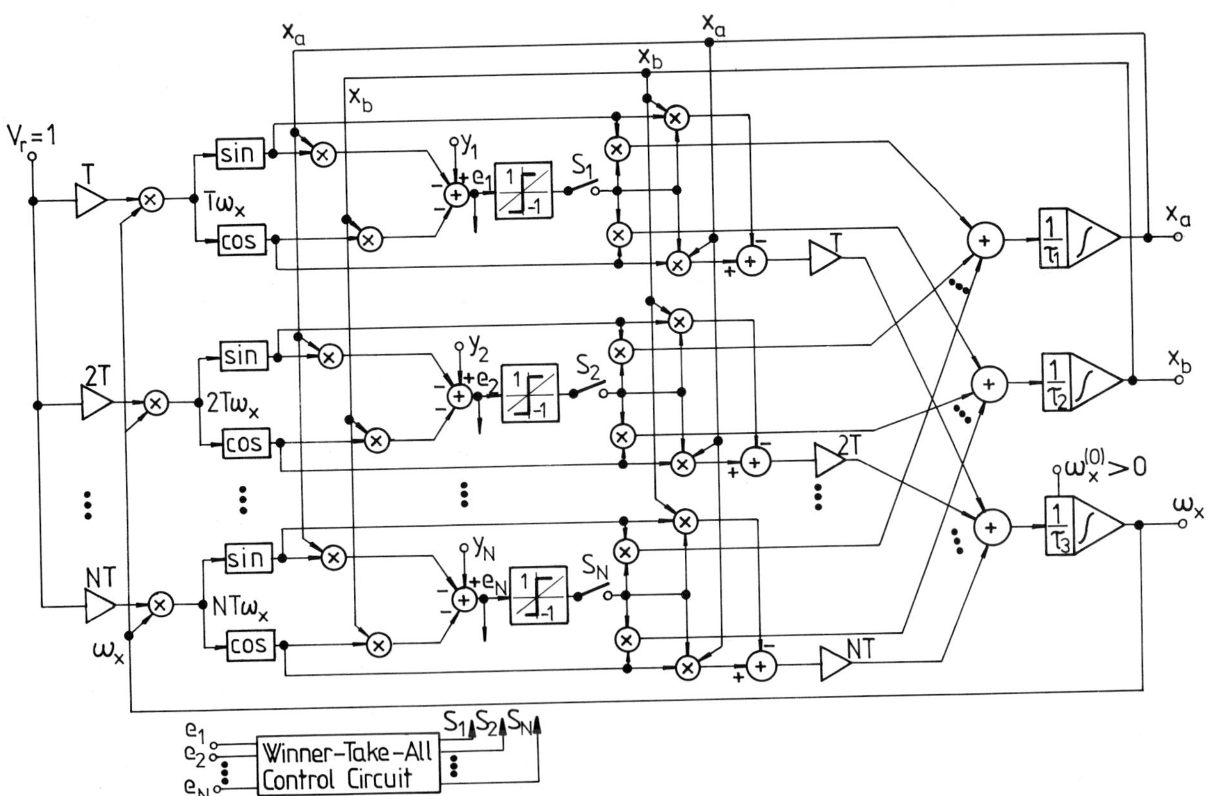

Fig. 1 Functional block diagram of neural network for estimating parameters of observed signal

It is required to find the parameters on basis of N=20 samples of the signal with sampling interval $0.8\cdot10^{-3}$ seconds. The network was able to find in time less than 50 nanoseconds the following parameters:

$$X_a = -119.73 , \quad X_b = 39.81 , \quad \omega_x = 314.12$$

which are in good agreement with actual values of parameters of the measured signal. Exemplary trajectories of estimated parameters for an other case (Fig.2) are shown in Fig. 3.

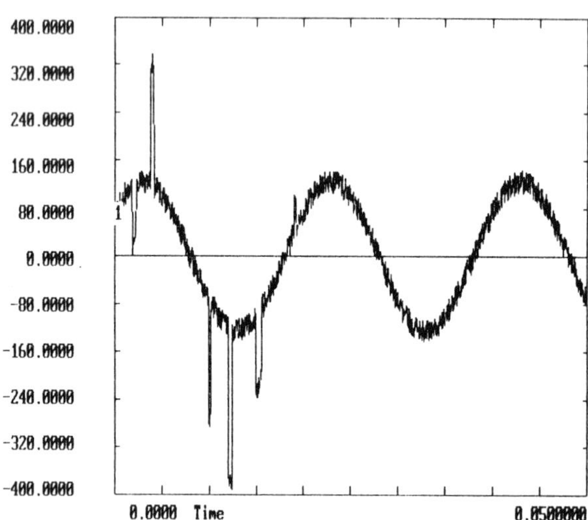

Fig. 2 Observed noisy signal
$y(t) = 100\sin(\omega t) + 75\cos(\omega t) + \varepsilon(t)$
$\omega = 314.159$ rd/s.

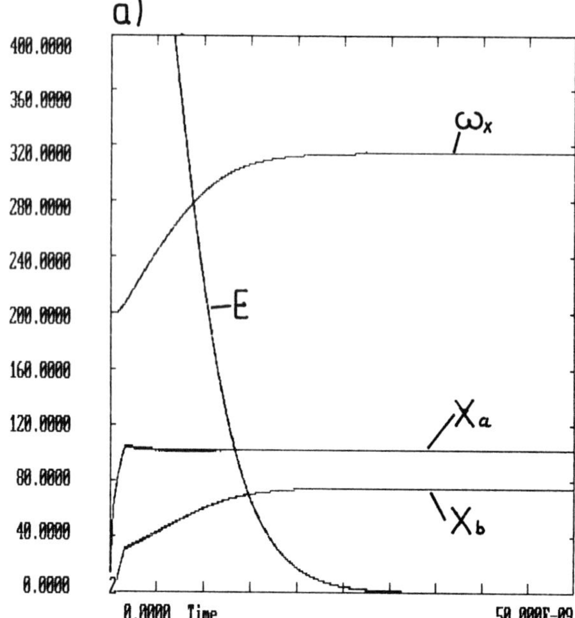

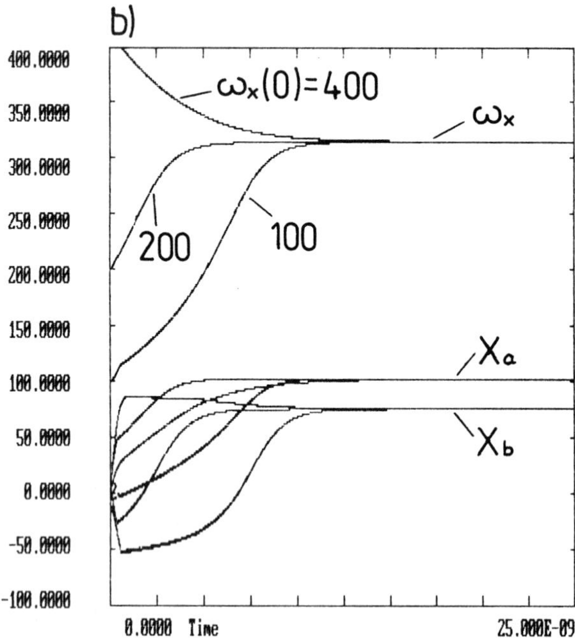

Fig. 3 (a) Computer simulated trajectories of estimated parameters and error function, (b) trajectories for several initial conditions for ω_x and zero initial conditions for X_a and X_b

REFERENCES

[1] Łobos, T., IEE Proc. Pt.C (1989) pp.347-351.

[2] Tank, D.W. and Hopfield, J., IEEE Trans. CAS-33 (1986) pp. 533-541.

[3] Osowski, S., Electronic Letters (1990) pp. 689-691.

[4] Cichocki, A. and Unbehauen, R., Neural Networks for Optimization, Computing and Signal Processing (Teubner-J.Wiley, Stuttgart, 1992)-in print.

[5] Cichocki, A. and Unbehauen, R., Electronic Letters (1991) pp. 2026-2028.

[6] Van Den Bos, A., Automatica (1988) pp.803-808.

[7] Cichocki, A. and Unbehauen, R., Int. Journ. Circuit Theory and Appl. (1991) pp.161-187.

[8] Barrodale, I. and Zala, C.A., L_1 and L_∞ Curve Fitting and Linear Programming Algorithms and Applications, in: Mohamed, J.L. and Walsh, J.E. (eds.), Numerical Algorithms (Oxford University Press, 1986) pp. 220-238.

SIGNAL PROCESSING VI: Theories and Applications
J. Vandewalle, R. Boite, M. Moonen, A. Oosterlinck (eds.)

BLIND IDENTIFICATION IN PRESENCE OF NOISE

Pierre COMON

THOMSON SINTRA, BP 138, F-06561 Valbonne Cedex, France

The problem of blind identification of noisy linear models of general form is addressed. Several aspects of the problem can be studied. The first concerns general statements that can be done independently of any algorithms, such as the definition of optimization criteria, or the determination of conditions of identifiability. The second aspect is more practical and includes the deflation of sufficient assumptions, and the design of numerical algorithms. Lastly, some properties can only be accessed via simulations, namely the robustness of an algorithm in presence of non-gaussian unknown noise, or the effect of short data lengths. These questions are analyzed in the paper.

1. INTRODUCTION

In this framework, the following linear statistical model is assumed:

$$\mathbf{y} = \mathrm{M}\,\mathbf{x} + \mathbf{v}, \qquad (1\text{-}1)$$

where **x**, **y** and **v** are random vectors with values in $\mathbb{R}$ or $\mathbb{C}$ and with zero mean and finite covariance, and M is a rectangular matrix with at most as many columns as rows (if M had more columns than rows, then some components x_i should be incorporated in the noise part, **v**).

Given realizations of **y**, it is desired to estimate both M and the corresponding realizations of **x**, under the assumption that vector **x** has statistically independent components. However, because of the presence of the noise **v**, it is in general impossible to recover exactly **x**, especially if **v** is non-gaussian. Thus with the goal of delivering a "best estimate" F of M, we define a "contrast" function that serves as optimization criterion.

The qualifiers "blind" or "myopic" are often used when only the outputs of the system considered are observed, as it is the case in (1-1). Quantities to be estimated are either M or **x**, or both. This blind identification problem is referred to as "sources separation" or "Independent Component Analysis" (ICA). When the matrix M in model (1-1) is Toeplitz triangular, it is actually dealt with a "Blind Deconvolution" (BD) problem, where vector **x** contains successive samples of a scalar white stochastic process. If in principle ICA contains BD as a particular case, general-purpose algorithms computing ICA are unmatched to BD precisely because the structure of the mixing matrix is not imposed. We shall be concerned in this paper only by the unconstrained ICA problem.

Since the early works on the subject drawing back to 1986 [1] [2] [3], see [12], more attention has been retained by ICA during the last years. It has been utilized for instance for localization problems in narrow-band antenna array processing [4] [5]. ICA-based localization techniques are of High-Resolution (HR) nature, but can be compared to the so-called Esprit algorithm where the mixing matrix is first estimated without taking advantage of the antenna manifold, the directions of arrival being estimated afterwards in a second stage. For wide-band signals, it is probably more appropriate to resort to blind identification techniques in time domain [6] [7] [8], especially for short data lengths.

2. GENERAL STATEMENTS

DEFINITION (2-1)

The ICA of a random vector **y** of size N with finite covariance V_y is a pair $\{F, \Delta\}$ of matrices such that

(a) the covariance factorizes into $V_y = F\,\Delta^2\,F^*$, where Δ is $\rho\times\rho$ diagonal real positive and F is $\rho\times N$ full rank;

(b) Δ^2 is the covariance of a $\rho\times 1$ random vector, **z**, whose components are "the most independent possible", in the sense of the maximization of a given "contrast function".

DEFINITION (2-2)

A contrast is a mapping Ψ from a space of random vectors, $\mathbb{E}$, to $\mathbb{R}$ satisfying the 3 requirements:

- $\Psi(\mathbf{x})$ depends only on $p_{\mathbf{x}}$, $\forall\, \mathbf{x} \in \mathbb{E}$, each component x_i playing the same role.
- Ψ is invariant by scale change, that is: $\Psi(\Lambda\mathbf{x}) = \Psi(\mathbf{x})$, $\forall\, \Lambda$ diagonal regular.
- if **x** has independent components, then: $\Psi(A\,\mathbf{x}) \le \Psi(\mathbf{x})$, $\forall$ A regular.

In order for a contrast to be of practical interest, it must also be discriminating:

DEFINITION (2-3)

A contrast will be said discriminating over a set $\mathbb{E}$ if, for any random vector $\mathbf{x} \in \mathbb{E}$ having independent components, the equality $\Psi(A\,\mathbf{x}) = \Psi(\mathbf{x})$ holds only when A is of the form Λ P, where Λ is diagonal and P a permutation.

If a contrast is discriminating over a set $\mathbb{E}$, then the condition $\mathbf{x} \in \mathbb{E}$ determines identifiability conditions in the noiseless case [10] [12]. Identifiability conditions in the noisy case can hardly be investigated.

Now remark that if a pair $\{F, \Delta\}$ is an ICA of a random vector, $\mathbf{y}$, then so is the pair $\{F', \Delta'\}$with

$$F' = F \, \Lambda \, D \, P, \text{ and } \Delta' = P^* \, \Lambda^{-1} \, \Delta \, P, \qquad (2\text{-}4)$$

where Λ is a scaling $\rho \mathrm{x} \rho$ regular diagonal real positive matrix, D is a $\rho \mathrm{x} \rho$ diagonal matrix with entries of unit modulus, and P is a $\rho \mathrm{x} \rho$ permutation. So (2-1) defines an equivalence class of decompositions rather than a single one. It is possible to define a unique representative by adding the following assumptions:

(c) the columns of F have unit norm;
(d) the entries of Δ are sorted in decreasing order;
(e) the entry of largest modulus in each column of F is positive real. (2-5)

Each of the constraint above removes one of the degree of freedom exhibited in (2-4). More precisely, (c), (d) and (e) determine Λ, P, and D, respectively.

Now for any zero-mean random vector $\mathbf{y}$ with finite variance V_y, we define the standardized variable $\tilde{\mathbf{y}}$ as follows. Denote

$$V_y = U \, S^2 \, U^* \qquad (2\text{-}6)$$

the eigenvalue decomposition of V_y, where U is $N\mathrm{x}\rho$ and S is $\rho \mathrm{x} \rho$ full rank. Then $\tilde{\mathbf{y}}$ is the ρ-dimensional random variable with unit variance:

$$\tilde{\mathbf{y}} = S^{-1} U^* \mathbf{y}. \qquad (2\text{-}7)$$

It may be shown [10] [12] that the opposite of the mutual information is a contrast, discriminating over the set of non-deterministic standardized random variables having at most one gaussian component:

$$\Psi(z) = -\int p_{\tilde{z}}(u) \log \frac{p_{\tilde{z}}(u)}{\prod_i p_{\tilde{z}_i}(u_i)} \, du. \qquad (2\text{-}8)$$

3. PRACTICAL SUFFICIENT ASSUMPTIONS

In practice, the mutual information cannot be used because the probability density of observations $\mathbf{y}$ is generally unknown. Therefore, it is proposed to resort to contrasts based on moments of y, that we can easily estimate. We look for a matrix F and a random vector $\mathbf{z}$ such that $\Psi(p_z)$ is maximized with $\mathbf{y} = F\mathbf{z}$. With S and U defined in (2-7), we can set $F = USQ^*$ and look for the unitary part Q, bearing in mind that $\tilde{\mathbf{y}} = Q^* \tilde{\mathbf{z}}$. This can be done by searching for the unitary matrix Q that maximizes the criterion $\psi(Q) = \Psi(p_{\tilde{z}})$ [10] [12].

Denote $\Gamma_{ij...r}$ and $K_{ij...r}$ the joint cumulants of $\{\tilde{y}_i, \tilde{y}_j, \ldots, \tilde{y}_r\}$ and $\{\tilde{z}_i, \tilde{z}_j, \ldots, \tilde{z}_r\}$, respectively, and assume $\tilde{\mathbf{z}} = Q\tilde{\mathbf{y}}$, where Q is a $\rho \mathrm{x} \rho$ unitary matrix. Then because of well-known properties [11] they are related by a multilinear relation. For instance, at order 4 we have:

$$K_{ijkl} = \sum_{mnpq} Q_{im} \, Q_{jn} \, Q_{kp} \, Q_{lq} \, \Gamma_{mnpq}. \qquad (3\text{-}1)$$

A key result is given by the theorem below:

THEOREM (3-2)

The sum of squares of marginal standardized cumulants of order r, denoted

$$\psi(Q) = \sum_{i=1}^{\rho} K^2_{ii...i}, \qquad (3\text{-}3)$$

is a contrast for any $r \geq 2$. Moreover this contrast is discriminating over the set of random vectors having at most one null marginal cumulant of order r, for any $r > 2$.

Proof

Let Q be $\rho \mathrm{x} \rho$ unitary. Denote $\overline{Q}$ the matrix defined by $\overline{Q}_{ij} = |Q_{ij}|$, and ${}^r\overline{Q}$ the matrix defined by ${}^r\overline{Q}_{ij} = \overline{Q}^r_{ij}$. Since Q is unitary, the matrix ${}^2\overline{Q}$ is doubly stochastic, i.e. the sum of its entries in any row or column is equal to 1. Yet from Birkhoff theorem, the set of doubly stochastic matrices is a convex polyhedron whose vertices are permutations. In other words ${}^2\overline{Q}$ can be decomposed as

$${}^2\overline{Q} = \sum_{s=1}^{\rho!} \alpha_s \, P_s, \; \alpha_s \geq 0, \text{ with } \sum_s \alpha_s = 1, \qquad (3\text{-}4)$$

where P_s are permutation matrices. Thus for any vector $\mathbf{u}$, if we denote $\overline{\mathbf{u}}$ the vector with components $|u_i|$, we have the inequality

$$\| {}^2Q \, \mathbf{u} \| \leq \| {}^2\overline{Q} \, \overline{\mathbf{u}} \| \leq \sum_s \alpha_s \| P_s \overline{\mathbf{u}} \| = \| \mathbf{u} \|. \qquad (3\text{-}5)$$

On the other hand, since $|Q_{ij}| \leq 1$, we also have for $r \geq 2$:

$${}^r\overline{Q}_{ij} \leq {}^2\overline{Q}_{ij}. \qquad (3\text{-}6)$$

Applying the triangular inequality yields, using (3-6):

$$\sum_{i,j,k} Q^r_{ki} Q^r_{kj} u_i u_j \leq \sum_{i,j,k} {}^r\overline{Q}_{ki} {}^r\overline{Q}_{kj} \bar{u}_i \bar{u}_j \leq \sum_{i,j,k} {}^2\overline{Q}_{ki} {}^2\overline{Q}_{kj} \bar{u}_i \bar{u}_j,$$

so that we can write because of (3-5)

$$\| {}^rQ \, \mathbf{u} \|^2 \leq \| {}^r\overline{Q} \, \overline{\mathbf{u}} \|^2 \leq \| {}^2\overline{Q} \, \overline{\mathbf{u}} \|^2 \leq \| \overline{\mathbf{u}} \|^2 = \| \mathbf{u} \|^2. \qquad (3\text{-}7)$$

Now consider a standardized random vector $\tilde{\mathbf{y}}$ having zero cross-cumulants of order r and denote u_i its marginal cumulants of order r. Because of the multilinearity of cumulants (3-1), the very left hand side of (3-7) is nothing else than $\Psi(Q\,\tilde{\mathbf{y}})$. So we have just proved with (3-7) that $\Psi(Q\,\tilde{\mathbf{y}}) \leq \Psi(\tilde{\mathbf{y}})$ for any unitary matrix, Q. This shows that (3-3) is indeed a contrast, as soon as $r \geq 2$.

Note that for r=2, the equality $\| {}^2\overline{Q} \, \overline{\mathbf{u}} \| = \| \mathbf{u} \|$ always holds, so that the contrast is actually degenerated. It remains to prove that (3-3) is discriminating for any r>2.

Assume equality in (3-7). Then in particular

$$\| {}^2\overline{Q} \, \overline{\mathbf{u}} \|^2 - \| {}^r\overline{Q} \, \overline{\mathbf{u}} \|^2 = 0, \qquad (3\text{-}8)$$

for some vector $\mathbf{u}$ having at most one null component. But because of (3-6), this implies that all the differences involved in (3-8) individually vanish:

$$({}^2\overline{Q} \, \overline{\mathbf{u}} \,)^2_i - ({}^r\overline{Q} \, \overline{\mathbf{u}} \,)^2_i = 0, \; \forall i.$$

Again, because $\bar{u}_j$, ${}^r\overline{Q}_{ij}$, and ${}^2\overline{Q}_{ij}$ are positive, this yields

$$({}^2\overline{Q}\,\overline{u})_i - ({}^r\overline{Q}\,\overline{u})_i = 0, \forall i, \quad (3\text{-}9)$$

and next $$({}^2\overline{Q}_{ij} - {}^r\overline{Q}_{ij})\overline{u}_j = 0, \forall i,j. \quad (3\text{-}10)$$

Consequently, for all values of i, and for at least ρ–1 values of j, we have $\overline{Q}_{ij}^2 = \overline{Q}_{ij}^r$. Since $0 \le \overline{Q}_{ij} \le 1$, ${}^2\overline{Q}_{ij}$ is necessarily either zero or one for those pairs (i,j). Now, a ρ by ρ doubly stochastic matrix ${}^2\overline{Q}$ containing only zeros or ones in a ρ by ρ–1 submatrix is a permutation. The matrix Q is thus equal to a permutation up to multiplication of complex numbers of unit modulus. This may be written as Q = D P, where D needs only be diagonal with unit modulus entries. This proves the second part of the theorem. □

It can be shown [10] [12] that the practical contrast (3-3) is actually an approximation of (2-8) in the Edgeworth sense.

4. ALGORITHM

The following algorithm is proposed for computing the ICA of a NxT data matrix, Y:

ICA algorithm (4-1)

1) Compute the SVD of the data matrix as $Y = V\,\Sigma\,U^*$, where V is Nxρ, Σ is ρxρ, and U^* is ρxT, and ρ denotes the rank of Y. Therefore, use preferably Chan's algorithm where the QR factorization is first computed, since T is often much larger than N. Then $Z = \sqrt{T}\,U^*$, and $L = V\,\Sigma / \sqrt{T}$.
2) Initialize F = L.
3) Begin a loop on the sweeps: $1 \le k \le 1 + \sqrt{\rho}$.
4) Sweep the ρ(ρ–1)/2 pairs (i,j), according to a fixed ordering. For each pair:
   - **a.** estimate the 4th-order cumulants of $(Z_{i:}, Z_{j:})$.
   - **b.** find the angle α maximizing $\psi(Q^{(i,j)})$, where $Q^{(i,j)}$ is the Givens rotation of angle α, α∈]–π/4,π/4].
   - **c.** accumulate $F := F\,Q^{(i,j)*}$.
   - **d.** update $Z := Q^{(i,j)}\,Z$.
5) End of the loop on k if $k > 1+\sqrt{\rho}$, or if all estimated angles are very small compared to 1/T.
6) Compute the norm of the columns of F: $\Delta_{ii} = \| F_{:i} \|$.
7) Sort the entries of Δ in decreasing order: $\Delta := P\,\Delta\,P^*$ and $F := F\,P^*$.
8) Normalize F by the transform $F := F\,\Delta^{-1}$.
9) Fix the phase (sign) of each column of F according to (2-5e). This yields $F := F\,\Lambda$.

Computational complexity

This algorithm can be shown [12] to require a total of

$$6\,K\,N^2\,T + O(K\,N^2) + O(N^3)$$

floating point operations, where K denotes the maximum number of sweeps, if real data are processed. In practice, T must be much larger than N and K is smaller than $\sqrt{N}$. A reasonable order of magnitude is obtained by taking for instance $T = O(N^{5/2})$, which yields an amount of $O(N^5)$ flops.

Note that step 4**b**, which requires to find the absolute maximum of a rational function, can be performed by simply rooting a polynomial [10] [12].

5. EFFECT OF NON GAUSSIAN NOISE

In this section, matrix M is a Toeplitz circulant matrix with as first row

$$\mathbf{m} = [3 \quad 0 \quad 2 \quad 1 \quad -1 \quad 1 \quad 0 \quad 1 \quad -1 \quad -2].$$

In order to observe the effect of additive noise, consider the following observation model:

$$\mathbf{y} = (1-\mu)\,M\,\mathbf{x} + \mu\,\|\,\mathbf{m}\,\|\,\mathbf{w}, \quad (5\text{-}1)$$

where μ is the noise rate, $0 \le \mu \le 1$. In order to facilitate the definition of signal to noise ratio, it is assumed that signal and noise kurtosis are equal in modulus. Thanks to the scaling factor $\|\mathbf{m}\|$, a signal to noise ratio of 0 dB is thus obtained for $\mu = 0.5$.

There are several ways to access performances of ICA. It has been chosen to use a distance measure between M and F that is invariant by postmultiplication by a factor ΛP, where Λ is diagonal regular and P a permutation. This measure has been called a "gap" because it is not a distance in the strict sense, and is defined as follows:

$$\varepsilon(A,\hat{A}) = \sum_i \left|\sum_j |D_{ij}| - 1\right|^2 + \sum_j \left|\sum_i |D_{ij}| - 1\right|^2 + \sum_i \left|\sum_j |D_{ij}|^2 - 1\right| + \sum_j \left|\sum_i |D_{ij}|^2 - 1\right|, \quad (5\text{-}2)$$

where

$$\underline{A} = A\,\Delta^{-1},\ \underline{\hat{A}} = \hat{A}\,\hat{\Delta}^{-1}, \text{ with } \Delta_{kk} = \|A_{:k}\|,\ \hat{\Delta}_{kk} = \|\hat{A}_{:k}\|,$$

and

$$D = \underline{A}^{-1}\,\underline{\hat{A}}.$$

This gap vanishes if and only if A and $\hat{A}$ are linked by a relation [10] [12] of the form $\hat{A} = A\,\Lambda\,P$; in other words, (5-2) defines a distance between equivalence classes.

In section 6, the algorithm (4-1) is utilized to compute an expectation of ε(M,F) for various values of μ and data length T. But in the present section, asymptotic performances are analyzed (i.e. for infinite T). For this purpose, it is not necessary to generate real data, and simulations can be carried out directly from cumulants. This procedure is referred to as "validation algorithm" in this paper. Let the matrix

$$A_\mu = [\,(1-\mu)\,M \quad \mu\,\|\mathbf{m}\|\,I\,]. \quad (5\text{-}3)$$

Denote Γ_x and Γ_w the vectors containing the marginal standardized cumulants of **x** and **w**, respectively, and denote Γ the vector $\Gamma = [\Gamma_x\ \Gamma_w]$. Then the validation algorithm can be designed [12] by replacing steps 1 and 4a of the algorithm (4-1) described in section 4 by the following:

Validation algorithm (5-4)

1) Compute the SVD of A_μ as $A_\mu = V\,S\,U^*$, where V is Nxρ, S is ρxρ, and U is 2Nxρ. Set $A := U^*$ and L=VS.
4) **a.** Calculate the required cumulants using the multilinearity relation (3-1).

Figure 1 represents the gap ε(M,F) obtained when running the validation algorithm for 2 values of Γ_x:

$\Gamma_x = -1.2$ [1 1 1 1 1 –1 –1 –1 –1 –1]: solid line;

$\Gamma_x = -1.2$ [1 1 1 1 1 1 1 1 1 1]: dashed line;

$\Gamma_w = -1.2$ [1 1 1 1 1 1 1 1 1 1]: in both cases.

It can be noticed that the variation around 0 dB is very steep, showing that ICA is very robust against non-gaussian noise.

Figure 2 shows initial and final values of the contrast for the same two simulations. The values obtained can be compared to the contrast of sources alone, $\|\Gamma_x\|^2$=14.4, or noise alone, $\|\Gamma_w\|^2$=14.4. It can be checked out that these values are indeed reached for $\mu = 0$ and $\mu = 1$.

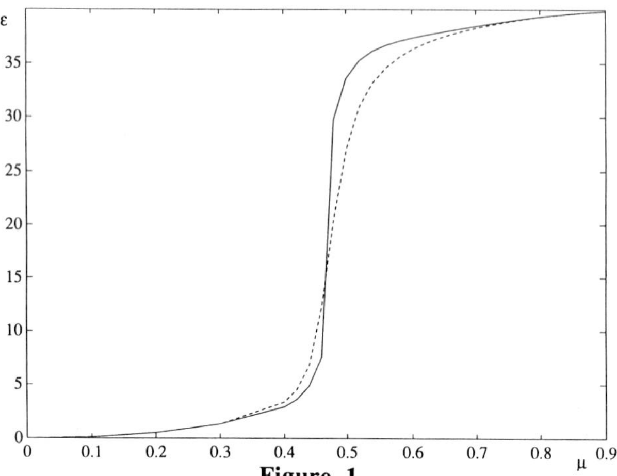

Figure 1

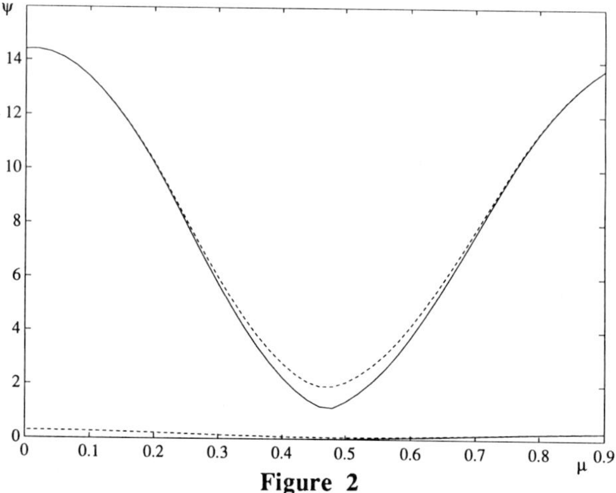

Figure 2

6. ADDITIONAL EFFECT OF DATA LENGTH

When cumulants are estimated from data of limited size, estimation errors enter in the process exactly as an extraneous noise. The shorter the data length, the larger the extraneous noise. This is observed in figure 3, where the effect of data length becomes very sensitive below T=300. For instance, the computation of the ICA in the noiseless case (curve μ=0) for T = 100 is equivalent to the computation of the ICA with T=300 with an additive uniform noise of 40% (curve μ=0.4).

For convenience, simulations have been run for N=2 in order to limit the computational burden. In fact, many trials are necessary for statistical averaging purposes (the product between T and the number of averages has been chosen to be 24000 in this simulation). The mixing matrix was

$$M = \begin{pmatrix} 1 & 3 \\ -3 & 1 \end{pmatrix}, \text{ with } \mathbf{y} = (1-\mu)M\mathbf{x} + \mu\mathbf{w}.$$

Sources x_i and noises w_i were independent zero-mean standardized random variables uniformly distributed in $[-\sqrt{3},\sqrt{3}]$; they have thus all a kurtosis of −1.2.

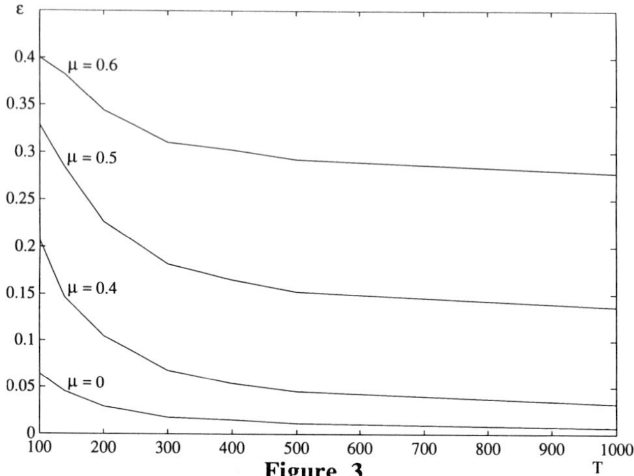

Figure 3

REFERENCES

[1] C.JUTTEN and J.HERAULT, "Independent Component Analysis versus Principal Component Analysis", *EUSIPCO 88*, Grenoble, 643-646.

[2] J.L.LACOUME and P.RUIZ, "Extraction of Independent Components from Correlated Inputs", *Proc Workshop Higher-Order Spectral Analysis,* june 1989, Vail, Colorado, 146-151.

[3] C.JUTTEN and J.HERAULT, "Blind Separation of Sources", *Signal Processing,* vol.24, n°1, 1991, 1-30.

[4] J.F.CARDOSO, "Sources Separation using High Order Moments", *Conference ICASSP,* 23-26 may 1989, Glasgow, 2109-2112.

[5] P.COMON, "Higher Order Separation", *Conf. EUSIPCO 90,* sept 18-21, 1990, Barcelona, 277-280.

[6] G.GIANNAKIS, Y.INOUYE and J.M.MENDEL, "Cumulant-based Identification of MA Models", *IEEE AC,* vol.34, july 1989, 783-787.

[7] P.COMON, "MA Identification using Fourth-Order Cumulants", *Signal Processing,* vol.26, n°3, 1992.

[8] A.SWAMI and J.M.MENDEL, "ARMA Parameter Estimation using Only Output Cumulants", *IEEE ASSP,* vol.38, n°7, july 1990, 1257-1265.

[9] J.F.CARDOSO and P.COMON, "Tensor-Based Independent Component Analysis", *Conf. EUSIPCO 90,* sept 18-21, 1990, Barcelona, 673-676.

[10] P.COMON, "Independent Component Analysis" *Int. Sig. Proc. Workshop on High-Order Statistics,* Chamrousse, France, july 10-12, 1991, 111-120.

[11] BRILLINGER, *Time Series Data Analysis and Theory,* Holden Day, 1981.

[12] P.COMON, "Independent Component Analysis, a New Concept ?", submitted to *Signal processing,* march 1992.

1: This work has been supported in part by the D.R.E.T, Paris.

SIGNAL PROCESSING VI: Theories and Applications
J. Vandewalle, R. Boite, M. Moonen, A. Oosterlinck (eds.)

THE L^∞ WIENER PROBLEM: AN APPLICATION OF THE L^∞ EXTENSION PROBLEM*

Cristian J. Fajre and Clifford T. Mullis

Department of Electrical and Computer Engineering, University of Colorado at Boulder, Boulder, CO 80309-0425, U.S.A.

This work presents a solution to the L^∞ Wiener problem, $\min_H \|F - GH\|_\infty$, where $F(z)$, $G(z)$ and $H(z)$ are rational transfer functions. The norm used is $\|F\|_\infty = \sup_\theta |F(e^{j\theta})|$. The solution is computed by reducing this problem to an L^∞ Extension Problem, which can be solved by applying the properties of Hankel operators and their Schmidt pairs. A closed form solution is computed, using efficient techniques for the spectral factorization of polynomials.

1. INTRODUCTION

The problem of minimizing $\|F - GH\|$ when the norm is the 2-norm is a typical equalization problem, when we want to minimize the energy of the error. The approach we now present corresponds to the minimization of the maximum absolute value of the error in our approximation.

Introducing some properties of All-Pass functions, we can express the minimization equation as

$$\min_H \|F - GH\|_\infty = \min_{H_2} \|\widetilde{F_2} + H_2\|_\infty,$$

where $\widetilde{F_2}(z)$ is a known anticausal rational function and $H_2(z)$ is a causal rational function. The left-hand side of the previous equation corresponds to the family of L^∞ Extension Problems, as stated in [1]. This problem has been solved in [2] using the state variable description of the system, and it is considered as an approximation of anticausal functions by causal functions. Throughout this work, $F(z)$ is defined as in [3],

$$F(z) = \sum_{k=-\infty}^{\infty} h_k z^{-k}.$$

2. THE APPROXIMATION PROBLEM

The statement of our approximation problem is,

$$\begin{array}{rll} \textbf{given:} & F(z), G(z), & \\ \textbf{find:} & H(z), & \text{causal}, \\ \textbf{such that:} & \|F - GH\|_\infty & \text{is minimized,} \end{array} \quad (1)$$

where F and G are given as general transfer functions. In order to solve this problem, consider the factorization of $G(z)$ as

$$G(z) = G_0(z)E(z), \quad (2)$$

where G_0 is completely causal and minimum phase, and E is an all-pass transfer function, such that $E(z)E(z^{-1}) \equiv 1$.

Now set

$$H(z) = \frac{H_0(z)}{G_0(z)}, \quad (3)$$

with H_0 causal, hence H will also be causal. Replacing these two equations back into the norm equation, we have

$$\|F - GH\|_\infty = \|F\widetilde{E} - H_0\|_\infty,$$

where $\widetilde{E}(z) = E(z^{-1})$ and then $E\widetilde{E} = 1$. The function $F\widetilde{E}$ can be represented as:

$$F(z)\widetilde{E}(z) = F_1(z) + \widetilde{F}_2(z),$$

where F_1 is causal and $\widetilde{F}_2$ is anticausal. If we now set $H_0(z) = F_1(z) - H_2(z)$, with H_2 a causal function, we will have

$$\|F - GH\|_\infty = \|\widetilde{F}_2 + H_2\|_\infty. \quad (4)$$

In order to solve the causal approximation problem, we have to find a causal extension function H_2 to the anticausal tail $\widetilde{F}_2$. Thus, the solution for the approximation problem stated in (1) is

$$H(z) = \frac{F_1(z) - H_2(z)}{G_0(z)}. \quad (5)$$

We have to consider two cases for eqn. (4). We will talk about the **Causal Extension** when F_2 is an FIR, and about the **Semi-infinite Interval Extension** when F_2 is a rational IIR.

3. THE CAUSAL EXTENSION

In this case we have $\widetilde{F}_2(z) = \sum_{k=-n}^{-1} f_k z^{-k}$, which means that we are looking for a function $H(z)$ that satisfies

$$H(z) = \widetilde{F}_2(z) + H_2(z) = \sum_{k=-n}^{-1} f_k z^{-k} + \sum_{k=0}^{\infty} h_k z^{-k}, \quad (6)$$

yielding the minimum ∞-norm. This can be accomplished if we consider a theorem by Carathéodory and Fejér, which states

*This research was supported by the Army Research Office, under Contract DAAL03-88-K-0058, and by the National Science Foundation, under Contract MIP-8717714.

Theorem 3..1 *Given the sequence* $\{f_0, \ldots, f_m\}$, *construct the matrix* $\mathbf{H}(f)$ *as*

$$\mathbf{H}(f) = \begin{bmatrix} f_m & \cdots & f_0 \\ \vdots & \cdot^{\cdot^{\cdot}} & \\ f_0 & & \mathbf{0} \end{bmatrix}. \tag{7}$$

Now let μ_0^2 *be the largest eigenvalue of* $\mathbf{HH}^*$, *then, every* $F(z) = \sum_{k=0}^{\infty} f_k z^{-k}$ *satisfies* $\|F\|_\infty \geq \mu_0$. *Equality holds if, and only if,*

$$F(z) = \mu_0 e^{i\gamma} \prod_{k=1}^{r} \frac{z^{-1} - w_k}{1 - \bar{w}_k z^{-1}},$$

for γ *a real constant,* $r \leq m$, *and* $|w_k| < 1$.

This minimizing function can be constructed as

$$F(z) = \mu_0 \frac{u_m + \cdots + u_0 z^{-m}}{u_0 + \cdots + u_m z^{-m}} = \mu_0 \frac{\widetilde{U}(z)}{U(z)}, \tag{8}$$

where $[u_0, \ldots, u_m]^T$ is an eigenvector corresponding to μ_0. Clearly, the solution function is an all-pass transfer function, since it has a constant value over the unit circle. This result and the proof of the theorem can be found in [1, 4, 5, 6].

With this in mind, we can see that the solution for the extension problem in eqn. (6), is $H(z) = z^n F(z)$, using the sequence $\{f_{-n}, \ldots, f_{-1}\}$ to construct $\mathbf{H}$ and $F(z)$, as shown in equations (7) and (8).

4. SEMI-INFINITE INTERVAL EXTENSION

Suppose now that $\widetilde{F}_2$ is a rational function, with denominator degree greater than zero. Repeating the analysis done for eqn. (6), we are now looking for a function H that satisfies

$$H(z) = \widetilde{F}_2(z) + H_2(z) = \sum_{k=-\infty}^{-1} f_k z^{-k} + \sum_{k=0}^{\infty} h_k z^{-k}, \tag{9}$$

and yields minimum ∞-norm. This shows clearly that in order to use the same procedure as in Section 3., we would need the eigenvalue decomposition of the infinite Hankel matrix constructed from the sequence $\{\ldots, f_{-2}, f_{-1}\}$. Instead, we can make use of the theory of Hankel Operators presented in [7, 8], and consider Nehari's theorem, which shows that the solution of this problem is of the form

$$H(z) = \mu_0 \frac{z\, V(z^{-1})}{U(z)}, \tag{10}$$

where the functions V and U are constructed from the pair of infinite dimensional vectors $(\mathbf{u}, \mathbf{v})$, which constitute the Schmidt Pair corresponding to the largest singular value of the Hankel operator, μ_0, as defined in [7, 8].

If we now apply this to eqn. (9), and considering real coefficients for the Hankel operator, we have that

$$\widehat{H}(z) = \widetilde{F}_2(z) + H_2(z) = \mu_0 \frac{z\, A(z^{-1})}{A(z)}, \tag{11}$$

with $A(z) = \sum_{k=0}^{\infty} a_k z^{-k}$. Recalling that $\widetilde{F}_2(z) = F_2(z^{-1})$, and that $H_2(z)$ is completely causal, we get that

$$zF_2(z)A(z^{-1}) = \mu_0\, A(z), \text{ in } [0, \infty). \tag{12}$$

If we now set

$$F_2(z) = \frac{\widehat{\beta}(z)}{\widehat{\alpha}(z)}, \quad A(z) = \frac{\widehat{\gamma}(z)}{\widehat{\alpha}(z)}, \quad H_2(z) = \frac{\widehat{\delta}(z)}{\widehat{\gamma}(z)}, \tag{13}$$

where $\widehat{\alpha}(z)$ is of degree n, and $\widehat{\gamma}(z)$ and $\widehat{\delta}(z)$ are of degree $n-1$. Equation (12) becomes

$$\mu_0 \frac{\widetilde{\gamma}(z)}{\widehat{\alpha}(z)} = \frac{\widetilde{\gamma}(z)\widehat{\beta}(z)}{\widehat{\alpha}(z)\widetilde{\alpha}(z)} = \widetilde{\gamma}(z) Y(z), \text{ in } [0, \infty), \tag{14}$$

where $\widetilde{\alpha}(z) = z^{-n}\widehat{\alpha}(z^{-1})$, and $\widetilde{\gamma}(z) = z^{-n+1}\widehat{\gamma}(z^{-1})$. This yields the matrix equation

$$\mu_0\, \underline{\gamma} = \mathbf{AYJ}\underline{\gamma} \ = \ \mathbf{M}\underline{\gamma}, \tag{15}$$

where

$$\underline{\gamma} = \begin{bmatrix} \gamma_0 \\ \vdots \\ \gamma_{n-1} \end{bmatrix}, \quad \mathbf{A} = \begin{bmatrix} 1 & & \mathbf{0} \\ \vdots & \ddots & \\ \alpha_{n-1} & \cdots & 1 \end{bmatrix}, \quad \mathbf{Y} = \begin{bmatrix} y_0 & \cdots & y_{-(n-1)} \\ \vdots & \ddots & \vdots \\ y_{n-1} & \cdots & y_0 \end{bmatrix}, \text{ and} \quad \mathbf{J} = \begin{bmatrix} \mathbf{0} & & 1 \\ & \cdot^{\cdot^{\cdot}} & \\ 1 & & \mathbf{0} \end{bmatrix}.$$

Equation (15) asks for the solution of an eigenvalue problem, but this time the matrices involved are of finite dimension. Once we have calculated μ_0 and $\underline{\gamma}$, we get

$$\begin{aligned} \widehat{H}(z) &= \mu_0 \frac{zA(z^{-1})}{A(z)} = \mu_0 \frac{\widetilde{\gamma}(z)\widehat{\alpha}(z)}{\widehat{\gamma}(z)\widetilde{\alpha}(z)} \text{ and} \\ H_2(z) &= \widehat{H}(z) - \widetilde{F}_2(z), \end{aligned} \tag{16}$$

which is our solution.

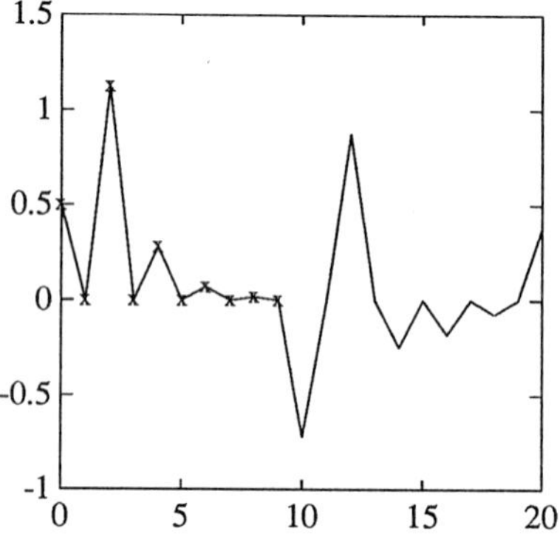

Figure 1: Impulse Response of $H(z)$. The given points are marked in the picture with a $(\times)$.

5. RESULTS

5.1. Causal Extension

In this case we take a sequence of values and compute the rational function with impulse response starting with that sequence, as shown in Figure 1. This is done by computing the eigenvalue with maximum absolute value and its corresponding eigenvector. This case is interesting in order to observe what happens when the eigenvalue has multiplicity greater than one. As shown in Figure 2, the polynomials cor-

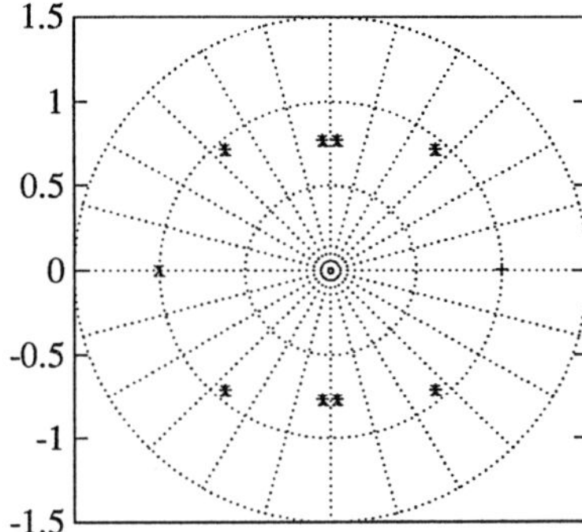

Figure 2: Roots of $U_0(z)$: ($\times$), and $U_1(z)$: (+).

responding to these eigenvalues only differ on roots located on the unit circle, as presented in [9, 1]. This property shows that the final order of the resulting all-pass function will be $n - p + 1$, when the length of the sequence is $n + 1$ and the multiplicity of the eigenvalue is p.

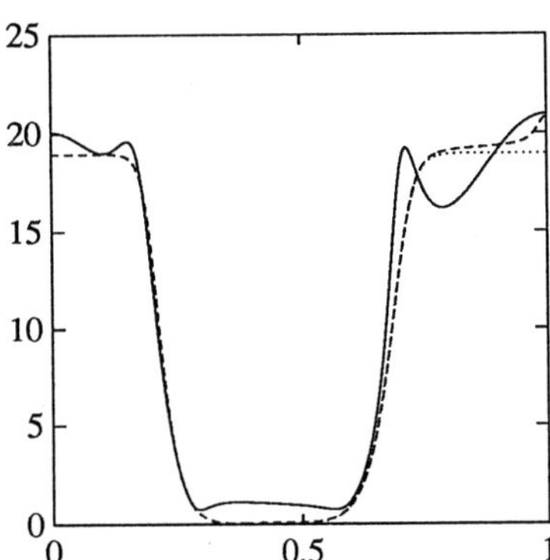

Figure 3: Amplitude response of $F(z)$: (—), $H_{inf}(z)$: (– –),and $H_W(z)$: ($\cdots$).

5.2. Semi-Infinite Interval Extension

In order to actually compute the coefficients of the rational solution we apply the techniques for factorization of polynomials presented in [10]. Based on this, we could implement a computational toolkit oriented to operatiions involving rational functions. This toolkit is also used to compute the L^2 approximation, which is shown in Figure 3, along with the L^∞ result and the desired function. In this case we took $G(z) \equiv 1$.

The characterization of the best approximation in L^∞ is that the error curve is perfectly circular, as stated in [11, 12, 13], and this can be oberved in Figure 4.

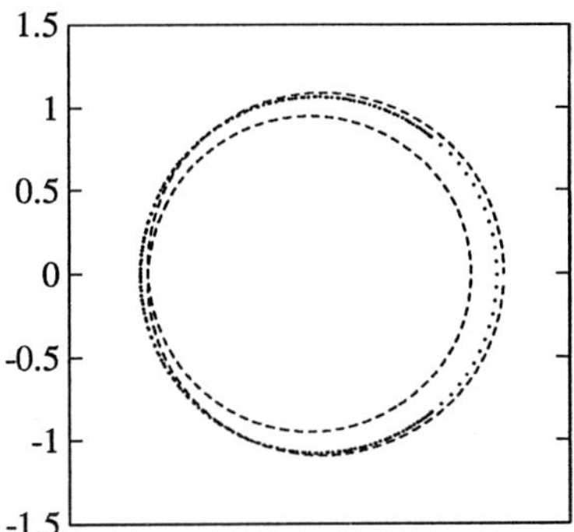

Figure 4: Error Curves for $H_{inf}(z)$: ($\cdots$), and $H_W(z)$: (– –).

REFERENCES

[1] C. J. Fajre. The application of the spectral decomposition of Hankel matrices to L^∞ approximation problems. Master's thesis, University of Colorado at Boulder, 1991.

[2] K. Glover. All optimal Hankel-norm approximations of linear multivariable systems and their L^∞-error bounds. *Int. J. Control*, 39(6):1115–1193, 1984.

[3] R. A. Roberts and C. T. Mullis. *Digital Signal Processing.* Addison-Wesley Publishing Company, Reading, Mass., 1987.

[4] U. Grenander and G. Szegö. *Toeplitz Forms and their applications.* Univ. of California Press, Berkeley, Calif., 1958.

[5] D. N. Clark. On the spectra of bounded, Hermitian, Hankel matrices. *Amer. J. Math. Soc.*, 90:625–656, 1968.

[6] H. W. Strube. How to make an All-Pass filter with a desired impulse response. *IEEE Trans. on ASSP*, ASSP-30(2):336–337, 1982.

[7] V. M. Adamjan, D. Z. Arov, and M. G. Krein. Analytic properties of Schmidt pairs for a Hankel operator and the generalized Schur-Takagi problem. *Math. USSR Sbornik*, 15(1):31–73, 1971.

[8] E. F. Deprettere, editor. *SVD and Signal Processing.* North-Holland, New York, N.Y., 1988.

[9] D. N. Clark. Hankel forms, Toeplitz forms and meromorphic functions. *Trans. Amer. Math. Soc.*, 134:109–116, 1968.

[10] C. J. Demeure and C. T. Mullis. A Newton-Raphson method for Moving-Average Spectral Factorization using the Euclid Algorithm. *IEEE Trans. on ASSP*, ASSP-38(10):1697–1709, 1990.

[11] M. H. Gutknecht, J. O. Smith, and L. N. Trefethen. The Carathéodory-Fejér method for recursive digital filter design. *IEEE Trans. on ASSP*, ASSP-31(6):1417–1426, 1983.

[12] L. N. Trefethen. Rational Chebyshev approximation on the unit disk. *Numer. Math.*, 37:297–320, 1981.

[13] L. N. Trefethen. Circularity of the error curve and sharpness of the CF method in complex Chebyshev approximation. *SIAM J. of Numer. Anal.*, 20(6):1258–1263, 1983.

SIGNAL PROCESSING VI: Theories and Applications
J. Vandewalle, R. Boite, M. Moonen, A. Oosterlinck (eds.)

A Non-Parametric Approach for Detecting Changes in the Autocorrelation Structure

Michail K. Tsatsanis and Georgios B. Giannakis

Department of Electrical Engineering
University of Virginia
Charlottesville, VA 22903-2442

Abstract — In this paper we derive a new test for detecting a jump in the autocorrelation function of a signal. The new test is simple, non-parametric, and does not assume knowledge of the signal's p.d.f. We show consistency and asymptotic normality of the statistic and we derive expressions for the asymptotic variance. In this way we are able to assess the performance of the test and set the threshold in an optimal way.

I. INTRODUCTION

The detection of an abrupt change in a time series is an important engineering problem and has attracted the interest of many researchers over the past 30 years. It has numerous applications in quality control procedures and failure detection, (e.g., in ECG and econometric signals). It is also used for segmentation of non-stationary signals, (e.g. speech and seismic signals). Finally, in image processing, it can be used for texture segmentation, which is a basic step in every image understanding procedure.

Early work on this problem has focused on detecting the jumps in the mean of i.i.d. Gaussian observations (e.g. see [1] for a review). This formulation is rather restrictive and does not address the more interesting problem of changes in the spectral characteristics of a time series. Parametric techniques have been employed to solve this more general problem. Changes in the parameters of ARMA or state-space models can be detected using tests on the innovations of the associated whitening filter. If the alternative model is *a-priori* known, then a dynamic programming algorithm can be used to determine the transition times, [2]. If not, local tests must be used considering small jumps, [4]. These parametric tests are rather involved, assume Gaussian data, lack the generality of the non-parametric ones and require knowledge of the underlying model.

Non-parametric tests have been derived in [5] using evolutionary spectra. Unfortunately, these tests can not be used to detect the exact transition time, due to the windowing inherent in the estimation of the evolutionary spectra.

In this paper we propose time-domain, non parametric tests for detecting jumps in the autocorrelation function of a time series. These tests can be easily generalized to detect changes in higher-order statistics of the process. They require no knowledge of the signal's p.d.f. and are computationally simple.

In the next section we will state the problem and derive the test from a mean square criterion. We will show consistency of the test as the data length goes to infinity. In Section III we will consider the threshold selection problem and we will discuss the performance of the test. Finally, in Section IV we will show some preliminary simulations and in Section V we will draw the conclusions.

II. Problem Statement-Results

Let $\{x(t)\}_{t=0}^{T}$ be a given record of $T+1$ samples from the process $x(t)$ with time-varying autocorrelation $r(t;\tau) = E\{x(t)\,x(t+\tau)\}$. More specifically we wish to discriminate between the following hypotheses

$$H_0 \quad : \quad r(t;\tau) = c \ , \ t = 0,\ldots,T-\tau \tag{1}$$

$$H_s \quad : \quad r(t;\tau) = \begin{cases} c & t = 0,\ldots,s-\tau-1 \\ c+\nu & t = s,\ldots,T-\tau \end{cases} \tag{2}$$

so that H_0 corresponds to a constant autocorrelation, while H_s corresponds to a jump in $r(t;\tau)$ at time point s.

Notice that this setup considers only one lag of the autocorrelation function. As we will see later though, the extension to include more lags is rather straightforward. Also, notice that we have not specified the value of $r(t;\tau)$ for $t = s-\tau,\ldots,s-1$, (the transition period). There is no general rule about how $r(t;\tau)$ goes from c to $c+\nu$ as this depends on the specific application. A linear slope, or even a constant value between c and $c+\nu$ may, in general, provide a good approximation. To simplify computations, we will assume $r(t;\tau) = c$, $t = s-\tau,\ldots,s-1$ in the sequel. The error introduced is anyway small, since both ν and τ are small numbers.

In order to estimate the time varying autocorrelation $r(t;\tau)$ and test the hypotheses we consider the mean square criterion

$$J(t,s) = E\{[x(t)x(t+\tau) - \theta_s(t;\tau)]^2\} \quad (3)$$

depending on the parameter $\theta_s(t;\tau)$. Since we are not interested on the precise value c of the autocorrelation but rather on the jump, we modify (3) to a mean compensated criterion

$$J(t,s) = E\{[x(t)x(t+\tau) - \bar{r}(\tau) - \theta_s(t;\tau)]^2\} \ , \quad (4)$$

where

$$\bar{r}(\tau) = \frac{1}{T-\tau+1}\sum_{t=0}^{T-\tau} r(t;\tau) \ . \quad (5)$$

Further, we will try to minimize

$$J_T(s) = \frac{1}{T-\tau+1}\sum_{t=0}^{T-\tau} J(t,s) \ . \quad (6)$$

Assume for the time being that c and ν are known and only s is to be determined. After substituting (4) to (6), expanding the square and dropping terms which do not depend on s we get

$$\begin{aligned} J_T(s) &= \frac{1}{T-\tau+1}\sum_{t=0}^{T-\tau}[\theta_s(t;\tau)]^2 \\ &- \frac{2}{T-\tau+1}\sum_{t=0}^{T-\tau}(r(t;\tau)-\bar{r})\theta_s(t;\tau) \ . \end{aligned} \quad (7)$$

(7) represents a matched filter based detector in the autocorrelation domain. It can be further simplified by scaling the $\theta_s(t;\tau)$ to have the same energy for all s, and omitting the first term. Also $\bar{r}$ does not really have to be subtracted, since now $\theta_s(t;\tau)$ has zero-mean.

After these simplifications the criterion to be maximized is

$$J_T(s) = \frac{1}{T-\tau+1}\sum_{t=0}^{T-\tau} r(t;\tau)\theta_s(t;\tau) \ , \quad (8)$$

where

$$\theta_s(t;\tau) = \begin{cases} -A_s & t = 0,\ldots,s-1 \\ B_s & t = s,\ldots,T-\tau \end{cases} \ , \quad (9)$$

and $A_s, B_s > 0$ are such that $\frac{1}{(T-\tau+1)}\sum_{t=0}^{T-\tau}\theta_s(t;\tau) = 0$, $\frac{1}{(T-\tau+1)}\sum_{t=0}^{T-\tau}[\theta_s(t;\tau)]^2 = 1$, i.e.,

$$A_s = \sqrt{\frac{(T-s-\tau+1)}{s}}, \quad B_s = \sqrt{\frac{s}{(T-s-\tau+1)}} \quad (10)$$

Notice that under H_0, $J_T(s) = 0$, for all s, so that (8) can indicate the presence of a change.

We estimate $J_T(s)$ by

$$\hat{J}_T(s) = \frac{1}{T-\tau+1}\sum_{t=0}^{T-\tau} x(t)x(t+\tau)\theta_s(t;\tau) \ , \quad (11)$$

and select $\hat{s}$ as

$$\hat{s} = \underset{0<s<T-\tau}{argmax}\ \hat{J}_T(s) \quad (12)$$

IIIA. Consistency

Next we show that $\hat{J}_T(s)$ is a meaningful estimator of $J_T(s)$ which converges as $T \to \infty$. We follow the approach taken in [3] to prove ergodicity.

Proposition 1: If $x(t)$ is a general non-linear process of the Volterra type, with absolutely summable cumulants of all orders, then

$$|\hat{J}_T(s) - J_T(s)| \xrightarrow{w.p.1} 0 \ , \quad (13)$$

as $T \to \infty$ and $s/T \to \alpha\epsilon(0,1)$. □

The proof follows the steps of [3] although in a somehow different setting. We skip the proof here.

IV. Performance Analysis

Up to this point we have seen that $J_T(s) = 0$ under H_0, while it peaks at the transition time under H_s. This suggests that we should compare $\hat{J}_T(s)$ to some threshold λ, to decide in favor of one hypothesis. In this section we will discuss the threshold selection as well as the performance of the test.

The threshold can be determined using a Neuman-Pearson approach, (for a given probability of false alarms), if the distribution of the statistic under H_0 is known. Unfortunately, in this case the distribution of $x(t)$ is unknown and hence, only asymptotic analysis seems to be feasible. Our analysis is based on the following result, which establishes asymptotic normality of the estimate $\hat{J}_T(s)$.

Proposition 2: Under the assumptions of proposition 1,

$$\sqrt{T-\tau+1}[\hat{J}_T(s) - J_T(s)] \xrightarrow{\mathcal{D}} \mathcal{N}(0,\sigma^2) \ , \quad (14)$$

as $T \to \infty$, where $\xrightarrow{\mathcal{D}}$ denotes convergence in distribution. Furthermore, if $x(t)$ is finite dependent the variance σ^2 can be computed as

$$\sigma^2 = lim_{T\to\infty}E\{|\sqrt{T-\tau+1}[\hat{J}_T(s) - J_T(s)]|^2\} \ . \ \Box \quad (15)$$

The proof is again along the lines of [3] and is omitted here.

Thus, we need to derive expressions for the mean and the variance of $\hat{J}_T(s)$ under the different hypotheses to conclude our discussion on the performance analysis. We again present only the final results.

i) Under H_0, $E\{\hat{J}_T(s)\} = 0$ and

$$E\{[\hat{J}_T(s)]^2\} = \left(\frac{A_s}{T-\tau+1}\right)^2 \sum_{t=-(s-1)}^{s-1}(s-|t|)\mu_4(\tau,t,t+\tau) \quad (16)$$

$$+ \left(\frac{B_s}{T-\tau+1}\right)^2 \sum_{t=-(T-\tau-s)}^{T-\tau-s} (T-\tau-s-|t|)\mu_4(\tau,t,t+\tau)$$

$$- \frac{2A_sB_s}{(T-\tau+1)^2}\left[\sum_{t=1}^{T-\tau-s} t\mu_4(\tau,t,t+\tau)\right.$$

$$+ (T-\tau-s-1)\sum_{t=T-\tau-s+1}^{s} \mu_4(\tau,t,t+\tau)$$

$$\left. + \sum_{t=s+1}^{T-\tau}(T-\tau-t-1)\mu_4(\tau,t,t+\tau)\right],$$

for $s > (T-\tau)/2$, with a similar expression for $s < (T-\tau)/2$. Here $\mu_4(\tau_1,\tau_2,\tau_3) = E\{x(t)x(t+\tau_1)x(t+\tau_2)x(t+\tau_3)\}$. By using (16) to compute the variance and exploiting the asymptotic normality of $\hat{J}_T(s)$ we can set the threshold, given a false alarm rate, from the tables of the normal distribution.

To compute the power of the test, we need the mean and variance of the statistic under H_s.

ii) Under H_s,

$$E\{\hat{J}_T(s)\} = \nu(\sqrt{s(T-\tau-s+1)}/(T-\tau+1) \quad (17)$$

and

$$E\{[\hat{J}_T(s)]^2\} = \left(\frac{A_s}{T-\tau+1}\right)^2 \sum_{t=-(s-1)}^{s-1} (s-|t|)\mu_4^{(1)}(\tau,t,t+\tau) \quad (18)$$

$$+ \left(\frac{B_s}{T-\tau+1}\right)^2 \sum_{t=-(T-\tau-s)}^{T-\tau-s} (T-\tau-s-|t|)\mu_4^{(2)}(\tau,t,t+\tau)$$

$$- \frac{2A_sB_s}{(T-\tau+1)^2} s(T-s)c(c+\nu) \ .$$

The superscript (1) above refers to the statistics of the process before the jump while (2) refers to the process after the jump. The expression in (18) is derived assuming that the processes before and after the jump are independent.

IVA. Extensions

The test can be easily extended to check for change in more than one autocorrelation lag. Let $\hat{\mathbf{J}}(s)$ be a vector containing p statistics of the form (11) for p different autocorrelation lags. Let $\hat{\Sigma} \triangleq E\{\hat{\mathbf{J}}(s)\hat{\mathbf{J}}(s)^T\}$. Then the quadratic form $\hat{q}(s) = \hat{\mathbf{J}}(s)^T\hat{\Sigma}\hat{\mathbf{J}}(s)$ is a chi-square r.v. with p degrees of freedom under H_0, while it is a non-central chi-square under H_s. Hence, probabilities of detection and false alarms can be established.

Also, the criterion of (8) can clearly be extended to test jumps in higher-order statistics of the process. For example, for third-order cumulants we should use

$$J_T^{(3)}(s) = \frac{1}{T-\tau+1}\sum_{t=0}^{T-\tau} E\{x(t)x(t+\tau_1)x(t+\tau_2)\}\theta_s(t;\tau) \quad (19)$$

where $\tau_1, \tau_2 > 0$, $\tau = max(\tau_1,\tau_2)$. Other extensions may include deriving a similar test in the frequency domain.

V. Simulations

We present two examples to illustrate the implementation issues. The first is a synthetic example with computer generated data while the second shows an application on real ECG data.

In Fig. 1 we show a realization of an MA(2) Gaussian process generated by the non-minimum phase filter $B_1(z) = 1 - z^{-1} + 2z^{-2}$ driven by standard normal i.i.d. r.v.'s. In Fig. 2 we show a realization of the same process for $t = 0,\ldots,500$, and of a different MA process with $B_2(z) = 1 - z^{-1} + 3z^{-2}$ for $t > 500$.

In Fig. 3 we see the statistic of (11) for $\tau = 0$, computed from the signal of Fig. 1. The statistic is close to zero, since the process is stationary. In the dashed line, we see the threshold that we should use, for a false alarm rate of 0.01. This threshold is computed using the variance expression (16). In Fig. 4 we see the statistic for the non stationary signal of Fig. 2, which peaks at the jump point. We also show the threshold.

In the second example we apply the non-parametric test on ECG data to monitor the variability of the heart beat rate. It is well known in the medical community that, if an infection is developing in the body, the heart beat rate becomes fairly constant, compared to the (more erratic) healthy one. This is very important in on-line monitoring of the patient's condition, especially in neo-natal intensive care units, where infants cannot express their symptoms, yet early diagnosis is crucial for their survival.

The heart rate variability is measured by the covariance of the beat-to-beat time interval (or R-R interval). In Fig. 5 we see the ECG signal of an (initially) healthy baby, that becomes sick. In Fig. 6, we plot the R-R intervals (extracted from the signal of Fig. 5) as a time-series. The test was applied to the time-series using a moving window of length 200 samples. Fig. 7 shows the statistic's value that peaks at the jump point.

V. Conclusions

We have derived a new test for detecting a jump in the autocorrelation function of a time series and we have analyzed its performance. We have proved consistency and asymptotic normality of the statistic. We have also shown its use through simulations.

References

[1] M. Basseville, 'On-line detection of jumps in the mean', in *Detection of abrupt changes in signals and systems*, ed. M. Basseville and A. Benveniste, Springer-Verlag, 1986.

[2] N. Burobin, V. Mottl' and I. Muchnik, 'An algorithm for detection of the instance of multiple change of properties of a random process based on the dynamic programming method', in *Detection of change in random processes*, ed. L. Telksnys, trans. from Russian, Optimization Software inc., 1986

[3] A. Dandawate, and G. B. Giannakis, 'Ergodic results for non-stationary processes', *Conf. on Info. Sciences and Systems (CISS'91)*, Mar. 1991.

[4] I. V. Nikiforov, 'Sequential detection of changes in stochastic systems', in *Detection of abrupt changes in signals and systems*, ed. M. Basseville and A. Benveniste, Springer-Verlag, 1986.

[5] T. Subba Rao, 'A cumulative sum test for detecting change in time series', *Int. J. Control*, vol. 34, No 2, pp. 285-293, 1981.

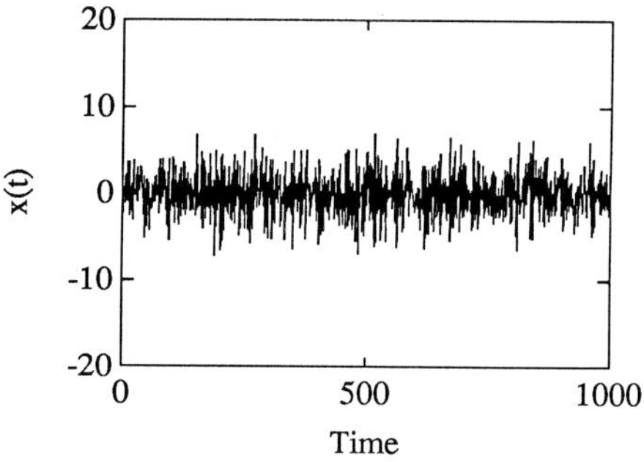

Figure 1: A stationary signal, (no jump)

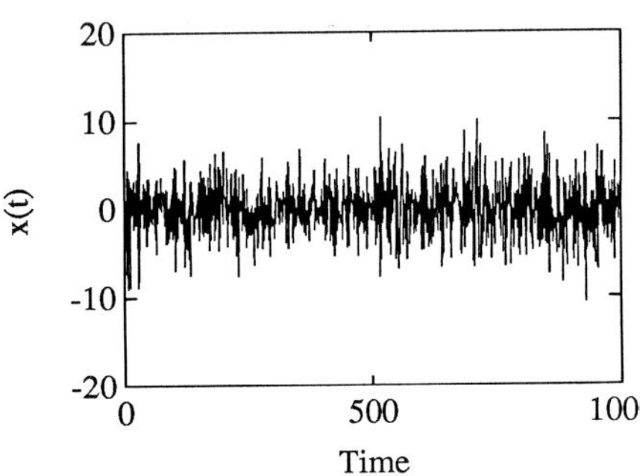

Figure 2: A non-stationary signal, (jump at $t = 500$)

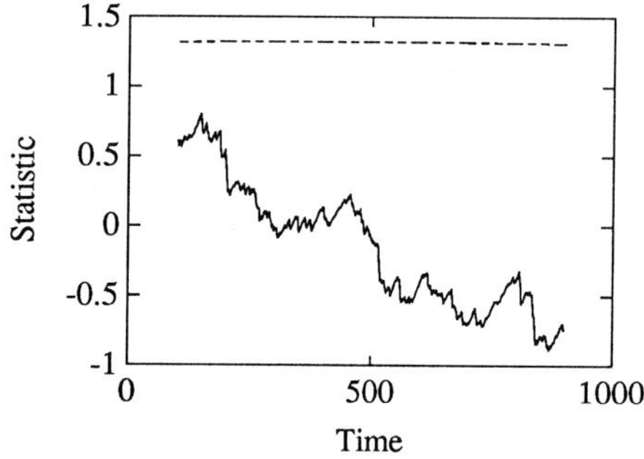

Figure 3: $J_T(s)$ for the signal in Fig. 1

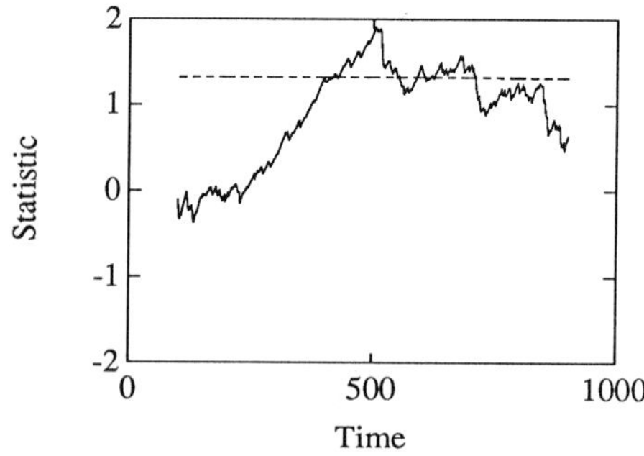

Figure 4: $J_T(s)$ for the signal in Fig. 2

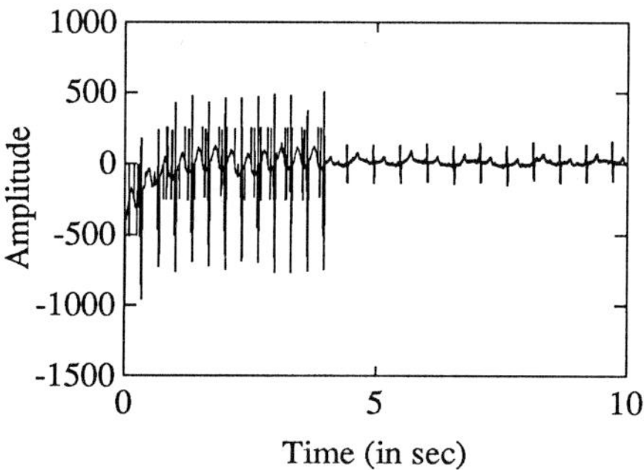

Figure 5: A baby's ECG signal

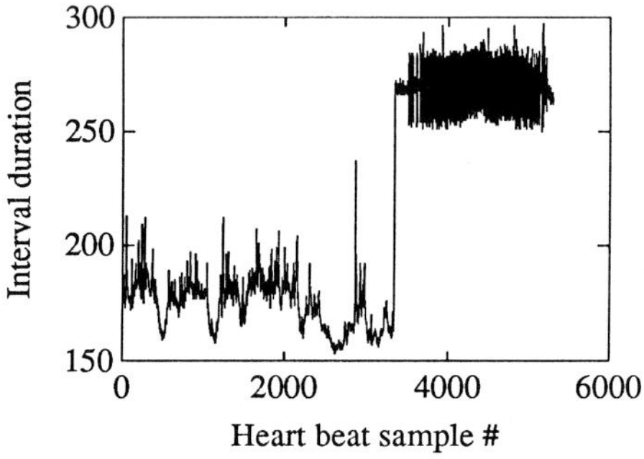

Figure 6: R-R intervals as a time series

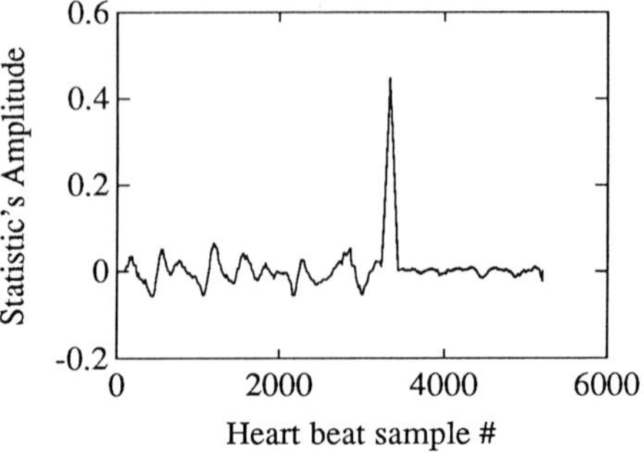

Figure 7: The test statistic for $\tau = 0$

SIGNAL PROCESSING VI: Theories and Applications
J. Vandewalle, R. Boite, M. Moonen, A. Oosterlinck (eds.)

Signal Representation using Operators

Klas Nordberg Hans Knutsson Gösta Granlund
Computer Vision Laboratory, Department of Electrical Engineering
Linköping University, S-581 83 LINKÖPING, SWEDEN
Phone: +46-13-28 16 34, Telefax: +46-13-13 85 26

Abstract

This paper presents a new way of representing a signal as well as an algorithm for extracting parameters of a specific class of features. The signal representation is based on the observation that we are often able to associate features with operators transforming the signal. The novel results presented in this paper are derived by restricting the operators to continuous one-parameter unitary operators. Under this restriction, it is shown that there exists a simple algorithm for extracting the parameters of the operators. The representation presented in this paper is intended for hierarchical processing structures.

1 Introduction

There are various ways of representing signals and the representation chosen is a reflection of assumptions and intentions regarding the signal. One common representation is the *ordered representation*, which implies that the signal elements can be ordered relative to each other. Take a sampled audio signal as an example. The elements of this signal correspond to a sequence of samples and each element is ordered according to when it was sampled. In image processing, the signals are often image neighbourhoods and it is therefore natural to assign spatial coordinates to the elements of the signal. Here, each element is ordered according to what spatial position is has relative to the others, which implies a two dimensional ordering. The ordered representation allows us to use Fourier analysis as a way of describing features of the signal. In the case of image processing, features such as local orientation and spatial frequency, curvature, etc, can be defined this way.

An other way of representing a signal is to consider it as a vector in a vector space, usually $\mathbf{R}^n$. The signal is then expressed as a linear combination of a set of basis vectors. We will refer to this type of representation as a *linear representation*. The choice of basis vectors is of course the main issue for this type of representation and at least two strategies for the choice deserve to be mentioned. The first is to minimize the number of basis vectors for the representation, which will allow us to describe a large part of the signal with just a few parameters, i.e. the coordinates of the signal relative to the basis. The second approach results in us being able to interpret the coordinates of the signal in some specific way, for example as an indication of class membership. Template matching is an example of this strategy. Each basis vector will correspond to a templet and if the signals coordinate relative to one of the basis vectors is large in magnitude, it is taken as an indication of the similarity between the signal and the corresponding templet.

Both ordered and linear representations are used in image processing as well as in other areas of signal processing. However, neither of them have proved fruitful tools in the context of hierarchical processing structures. In such a processing structure, features are represented and parameters extracted at each level and as we ascend through the hierarchy, the features become more and more abstract. In the higher parts of this hierarchy it is reasonable to expect the features detected to be relations between objects and perhaps also descriptions of possible responses the processing system can give for a certain event. As an example we might have to detect and represent the event "there is an obstacle on the road in front of the car, push the brake pedal and steer to the left of the obstacle". For signals describing such events, it will be difficult or even impossible to order its elements. The reason is that the features represented here simply do not correspond to spatial events like straight lines, curves, corners, etc. This may be one reason why it has proved difficult to define higher order features in most hierarchical processing systems.

The linear representation also seems to be a dead end. One reason is that even quite simple signals might have to be represented with a large number of basis vectors. Take as an example a cyclic image sequence showing a man waving his arms. We can construct this sequence so that each pixel of the image will be a continuous function of time. Let us consider the entire image as a signal vector in a vector space whose dimensionality is the same as the number of pixels in the image. As the image sequence is cyclic, the signal vector will move along a closed curve in the signal space. The signal is simple in the sense that we can characterize it using only one parameter which describes where it is on the curve. However, if we use a linear representation for this signal we would most certainly need a large number of basis vectors since the curve is not embedded in a linear subspace of small dimensionality. From this we infer that the simplicity of a general signal, reflected in the number of parameters needed to characterize it in the signal space, does not necessarily correspond to a simple linear representation of it, i.e. by few coordinates. In the worst case a linear representation for a signal might have to include the same number of basis vectors as the dimensionality of the signal space. Furthermore, this representation exhibits a somewhat more subtle flaw. Even if the signal can be linearly represented with a reasonably small basis, it will in general not be the case that the coordinates of the signal will correspond to our intuitive concept of a feature. Almost all feature extracting algorithms use various non-linear combinations of the coordinates to represent the feature, and there is no general approach to how this should be done.

We can summarize the problems of hierarchical information processing in the following way. First, it may not be possible to define an ordering of the signal elements. Therefore signal analysis based on Fourier methods can not be used as a general approach for feature extraction. Secondly, there must be a uniform way of defining signal features at all levels of the hierarchy. This implies that the features must be encoded in the signal in a very simple way in order to be detected by the processing units. For a thorough discussion of this subject, see [Granlund and Knutsson, 1983]. In the next sections we will present a signal representation which enables a definition of features for the signal in an intuitive way and when mathematically restricted yields a simple algorithm for extraction of the feature parameters.

2 An operator representation

As a tool for developing a new signal model we will use as a signal an image neighbourhood. Since suggested in [Granlund, 1978], it has become a widely accepted hypothesis that a sufficiently small neighbourhood only contains linear structures. Such a structure is characterized by its orientation, frequency and phase. Let us consider a specific instance of this signal. If we change the orientation of the linear structure corresponding to the signal we will generate a second instance of the signal. Changing the frequency will result in a third, etc. Thus, we are able to generate the complete set of signals, observed in the neighbourhood, by continuously changing the three parameters determining the orientation, frequency and phase of the corresponding linear structure. Any change in one of these parameters can be regarded as an operator transforming the signal, i.e. if we want to change the value of the orientation parameter by some amount, one and the same operator can be applied to the signal, regardless of its initial orientation. These operators are functions only of how much we change the parameters and independent or invariant of the signal itself.

Let us call the fixed instance of the signal $\mathbf{v}_0$. According to the previous discussion we can write any signal in the neighbourhood $\mathbf{v}(\theta,\omega,\varphi)$, with orientation θ, frequency ω and phase φ, as

$$\mathbf{v}(\theta,\omega,\varphi) = \mathbf{T}_\theta(\theta)\mathbf{T}_\omega(\omega)\mathbf{T}_\varphi(\varphi)\mathbf{v}_0,$$

where $\mathbf{T}_\theta$, $\mathbf{T}_\omega$ and $\mathbf{T}_\varphi$ are operator changing the values of the orientation, frequency and phase respectively. It is natural to set

$$\mathbf{T}_\theta(0) = \mathbf{T}_\omega(0) = \mathbf{T}_\varphi(0) = \mathbf{I},$$

where $\mathbf{I}$ is the identity operator. Hereby, the values of the orientation, frequency and phase of the signal $\mathbf{v}(\theta,\omega,\varphi)$ will be defined relative to the corresponding values of $\mathbf{v}_0$.

We will now generalize the ideas presented so far. We represent the signal in terms of operators. Each operator corresponds to a feature of the signal. This allows us to see an operator not merely as something changing the value of the signal elements, but rather as something changing the value of a parameter measuring a feature. For a signal which has n features, which can all be described with operators and each feature being quantified by a real variable x_k, we can then write

$$\mathbf{v}(x_1,x_2,\ldots,x_n) = \mathbf{T}_1(x_1)\mathbf{T}_2(x_2)\ldots\mathbf{T}_n(x_n)\mathbf{v}_0, \quad (1)$$

where $\mathbf{T}_k$ is an operator changing the value of the k-th feature. We will refer to the signal representation, defined by equation (1), as an *operator representation.* So far nothing new has been presented. Ideas along these lines have been presented in [Wilson and Knutsson, 1988], [Wilson and Spann, 1988], [Kanatani, 1990] and [Lenz, 1990]. In the following sections we will see what will happen if we restrict the operators to a specific type. To be able to formulate the operator representation in a convenient way, we briefly present a few results from matrix algebra.

3 Matrix algebra

Many of the results presented in this section are without any proof, for a more thorough presentation see [Nordberg, 1992]. We begin with some properties regarding the eigenvectors and eigenvalues of anti-Hermitian matrices, i.e. matrices $\mathbf{H}$ for which $\mathbf{H}^\star = -\mathbf{H}$. Here, the $\star$ sign indicates transponation and complex conjugation.

- The eigenvalues of an anti-Hermitian matrix are imaginary, i.e. are of the type $i\lambda$, $\lambda \in \mathbf{R}$.
- For an anti-Hermitian matrix $\mathbf{H}$ it is possible to find a set of orthogonal eigenvectors of $\mathbf{H}$, represented by the unitary matrix $\mathbf{E}$, containing the eigenvectors in its columns, such that

$$\mathbf{H} = \mathbf{E}\mathbf{D}\mathbf{E}^\star,$$

where $\mathbf{D}$ is a diagonal matrix containing the eigenvalues corresponding to the eigenvectors in $\mathbf{E}$.

Here, a matrix $\mathbf{E}$ is called unitary whenever $\mathbf{E}\mathbf{E}^{\star} = \mathbf{I}$, regardless of whether $\mathbf{E}$ is real or complex. Note also that the matrix $\mathbf{E}$ might not be unique, permutations of its columns disregared. The matrix $\mathbf{D}$ however, is unique, permutations disregarded.

Next we will define an exponential function for matrices. For our purposes it is enough to define this function only for anti-Hermitian matrices. For an anti-Hermitian matrix $\mathbf{H} = \mathbf{E}\mathbf{D}\mathbf{E}^{\star}$, according to the above, we define

$$e^{\mathbf{H}} = \mathbf{E}e^{\mathbf{D}}\mathbf{E}^{\star}.$$

As was mentioned above, $\mathbf{E}$ might not be unique. However, all matrices $\mathbf{E}$ and $\mathbf{D}$ satisfying $\mathbf{H} = \mathbf{E}\mathbf{D}\mathbf{E}^{\star}$ will produce the same value of $e^{\mathbf{H}}$.

If we denote the diagonal elements of $\mathbf{D}$ as $i\lambda_k$, then $e^{\mathbf{D}}$ is also a diagonal matrix and its diagonal elements are $e^{i\lambda_k}$. From this definition of the exponential function we immediately see that $e^{\mathbf{H}}$ must be unitary as it can be factorized into three unitary matrices. Two more properties are now at hand.

- If $\mathbf{H}$ is anti-Hermitian, then $e^{\mathbf{H}}$ is unitary.
- If $\mathbf{H}$ is real and anti-Hermitian, then $e^{\mathbf{H}}$ is real and orthogonal

If $\mathbf{H}$ is anti-Hermitian, then so is $x\mathbf{H}$ for any real x. Hence, we can define a continuous mapping from a real parameter x to unitary matrices $\mathbf{T}(x)$ in the following way

$$\mathbf{T}(x) = e^{x\mathbf{H}}.$$

In general, matrices, including unitary matrices, do not commute. For two arbitrary unitary matrices $\mathbf{T}_1$ and $\mathbf{T}_2$, it is thus in general the case that

$$\mathbf{T}_1\mathbf{T}_2 \neq \mathbf{T}_2\mathbf{T}_1.$$

However, if we can write $\mathbf{T}_1 = e^{\mathbf{H}_1}$ and $\mathbf{T}_2 = e^{\mathbf{H}_2}$, we have the following proposition.

- $\mathbf{T}_1\mathbf{T}_2 = \mathbf{T}_2\mathbf{T}_1$ if and only if $\mathbf{H}_1\mathbf{H}_2 = \mathbf{H}_2\mathbf{H}_1$. If this condition is met, then

$$e^{\mathbf{H}_1}e^{\mathbf{H}_2} = e^{\mathbf{H}_1+\mathbf{H}_2}. \tag{2}$$

4 A restricted operator representation

We will now restrict the operators used in the operator representation of a signal according to equation (1). Two restrictions will be made without any discussion of their validity. For a thorough presentation of this subject, see [Nordberg, 1992]. The two restrictions imposed on the operator are

- The operators are continuous one-parameter unitary operators. To each operator $\mathbf{T}_k(x_k)$ we associate an anti-Hermitian matrix $\mathbf{H}_k$, such that

$$\mathbf{T}_k(x_k) = e^{x_k\mathbf{H}_k}. \tag{3}$$

- All operators commute.

The first restriction provides us with a neat mathematical formulation of the operators. We have already stated that features of the signal corresponds to operators which changes the values of the parameters quantifying the features. Both a feature and an operator are quite general and abstract concepts. Now we can associate a mathematical object, an anti-Hermitian matrix, to each feature. Properties of the feature will then be reflected in properties of the matrix. We can combine equations (1) and (3), giving

$$\mathbf{v}(x_1, x_2, \ldots, x_n) = e^{x_1\mathbf{H}_1}e^{x_2\mathbf{H}_2}\ldots e^{x_n\mathbf{H}_n}\mathbf{v}_0. \tag{4}$$

The second restriction imposed on the operators allows us to use equation (2) to rewrite equation (4) as

$$\mathbf{v}(x_1, x_2, \ldots, x_n) = e^{x_1\mathbf{H}_1+x_2\mathbf{H}_2+\ldots x_n\mathbf{H}_n}\mathbf{v}_0. \tag{5}$$

According to equation (5) we have thus arrived to a linear representation of the signal. However, it is a linear representation of the operators transforming the signal, not of the signal itself. The parameters x_k can therefore be interpreted as *feature coordinates* of the signal. They define a point in the linear space of anti-Hermitian matrices, which in turn corresponds to a unitary operator, transforming $\mathbf{v}_0$ to $\mathbf{v}(x_1, x_2, \ldots, x_n)$.

5 Extracting the features

This section will describe an algorithm for the extraction of feature parameters. It is a generalization of the algorithm for estimation of local orientation as defined by [Knutsson, 1982], see [Nordberg, 1992]. We assume that the signal can be represented according to equation (5). We will consider only a signal which can be described with one operator corresponding to one feature of the signal. However, the algorithm presented here can be extended to multiple features, see [Nordberg, 1992]. Thus, we express the signal as

$$\mathbf{v}(x) = e^{x\mathbf{H}}\mathbf{v}_0. \tag{6}$$

Our goal now is to find a way of computing the real parameter x. There is of course no unique way to achieve this task and we will concentrate on one method. It is based on the assumption that we can find a matrix $\mathbf{X}$, such that

$$\mathbf{v}^{\star}(x)\mathbf{X}\mathbf{v}(x) = re^{i\kappa x}, \tag{7}$$

where $r > 0$ and κ are not yet determined constants. Inserting equation (6) in (7) yields

$$\mathbf{v}_0^\star e^{-x\mathbf{H}}\mathbf{X}e^{x\mathbf{H}}\mathbf{v}_0 = re^{i\kappa x}. \tag{8}$$

Differentiating the left and right hand side of equation (8) with respect to x results in

$$\mathbf{v}_0^\star e^{-x\mathbf{H}}(\mathbf{XH} - \mathbf{HX})e^{x\mathbf{H}}\mathbf{v}_0 = ri\kappa e^{i\kappa x},$$

which according to equation (8) is the same as

$$\mathbf{v}_0^\star e^{-x\mathbf{H}}(\mathbf{XH} - \mathbf{HX} - i\kappa\mathbf{X})e^{x\mathbf{H}}\mathbf{v}_0 = 0. \tag{9}$$

We see that equation (9) is satisfied if

$$\mathbf{XH} - \mathbf{HX} = i\kappa\mathbf{X}. \tag{10}$$

Note that equation (10) does not depend on the signal, only on the operator changing the signal, represented by the anti-Hermitian matrix $\mathbf{H}$. If we can find an $\mathbf{X}$ solving equation (10), we have proved the assumption equation (7) to be valid. Let $\mathbf{e}_m$ and $\mathbf{e}_n$ be eigenvectors of $\mathbf{H}$ with eigenvalues $i\lambda_m$ and $i\lambda_n$ such that $\lambda_n - \lambda_m = \kappa$. If we set $\mathbf{X} = \mathbf{e}_m\mathbf{e}_n^\star$, and insert this $\mathbf{X}$ in the left hand side of equation (10), we get

$$\mathbf{e}_m\mathbf{e}_n^\star\mathbf{H} - \mathbf{H}\mathbf{e}_m\mathbf{e}_n^\star = -\mathbf{e}_m(\mathbf{H}\mathbf{e}_n)^\star - (\mathbf{H}\mathbf{e}_m)\mathbf{e}_n^\star =$$

$$-\mathbf{e}_m(i\lambda_n\mathbf{e}_n)^\star - (i\lambda_m\mathbf{e}_m)\mathbf{e}_n^\star =$$

$$i\lambda_n\mathbf{e}_m\mathbf{e}_n^\star - i\lambda_m\mathbf{e}_m\mathbf{e}_n^\star = i(\lambda_n - \lambda_m)\mathbf{e}_m\mathbf{e}_n^\star = i\kappa\mathbf{X}.$$

We have thus proved $\mathbf{X} = \mathbf{e}_m\mathbf{e}_n^\star$ to be a solution of equation (10) and therefore the assumption made in equation (7) is valid. It can be proved, see [Nordberg, 1992], that we can only choose $i\kappa$ to be a difference between eigenvalues of $\mathbf{H}$, as indicated in the derivation above. There may of course be many pairs of eigenvalues having the same difference which leads us to the general solution of equation (10),

$$\mathbf{X} = \sum_{i=1}^n \sum_{j=1}^n c_{ij}\mathbf{e}_i\mathbf{e}_j^\star,$$

where c_{ij} is an arbitrary constant if $\lambda_j - \lambda_i = \kappa$ and zero otherwise. These constants determines the value of r according to

$$r = \sum_{i=1}^n \sum_{j=1}^n c_{ij}.$$

6 Summary

This paper has introduced a new way of representing signals. The representation makes a connection between features of the signal and operators acting on the signal so as to change the feature parameters. By restricting the operators to continuous one-parameter unitary operators it is possible to write a signal as

$$\mathbf{v}(x_1, x_2, \ldots, x_n) = e^{x_1\mathbf{H}_1 + x_2\mathbf{H}_2 + \ldots x_n\mathbf{H}_n}\mathbf{v}_0.$$

Each term in the exponent is an anti-Hermitian matrix and corresponds to a feature of the signal and each real x_k corresponds to a feature coordinate. For a signal which can be expressed as

$$\mathbf{v}(x) = e^{x\mathbf{H}}\mathbf{v}_0,$$

we have shown that it is possible to find a matrix $\mathbf{X}$, such that

$$\mathbf{v}^T(x)\mathbf{X}\mathbf{v}(x) = e^{i\kappa x},$$

for certain values of κ. This allows us to extract the feature coordinate x from the signal.

It is worth pointing out that both the representation and feature extraction algorithm presented in this paper is applicable to real as well as complex signals. In the case of complex signals we would have to use complex operators, defined by complex anti-Hermitian matrices.

7 References

Granlund, G. H. (1978). "In Search of a General Picture Processing Operator". Computer Graphics and Image Processing 8, p 155-173.

Granlund, G. H. and Knutsson, H. (1983). "Contrast of Structured and Homogenous Representations". Physical and Biological Processing of Images, vol 11, p 282-303. Editors 0. J. Braddick and A. C. Sleigh. Springer Verlag.

Kanatani, K. (1990). "Group Theoretical Methods in Image Understanding". Springer Series in Information Sciences, no 20. Springer-Verlag.

Knutsson, H. (1982). "Filtering and Reconstruction in Image Processing". PhD Thesis, Linköping University, Sweden. Dissertation No. 88.

Lenz, R. (1990). "Group Theoretical Methods in Image Processing". Lecture Notes in Computer Science, no 413. Springer-Verlag.

Nordberg, K. (1992) Report. Department of Electrical Engineering, Linköping University, Sweden. In preparation.

Wilson, R. and Knutsson, H. (1988). "Uncertainty and Interference in the Visual System". Systems, Man, and Cybernetics, Vol 18, No 2, March / April.

Wilson, R. and Spann, M. (1988). "Image Segmentation and Uncertainty", chapter 2. Letchworth, Research Studies Press.

SIGNAL PROCESSING VI: Theories and Applications
J. Vandewalle, R. Boite, M. Moonen, A. Oosterlinck (eds.)
1992 Elsevier Science Publishers B.V.

INSTANTANEOUS FREQUENCY AND ITS MEASUREMENT USING DIGITAL FILTERS

Václav V. ČÍŽEK

Institute of Radio Engineering and Electronics Czechoslovak Academy of Sciences, 182 51 Praha 8, Chaberská 57, Czechoslovakia

Evaluation of instantaneous frequency defined by an analytic signal requires determination of its Hilbert transform, its derivative and the derivative of the Hilbert transform. These auxiliary signals can be obtained with the limited accuracy only. In the paper the effect of inaccurately determined auxiliary signals on determination of instantaneous frequency for a harmonic signal, sum of two harmonic signals and for an amplitude modulated signal, is analyzed. It turns out that the relative error of the instantaneous frequency is in the case of filters with linear phase about of the same order as the relative deviation in amplitude characteristics of the filters. In the case of harmonic signals the filter inaccuracies cause ocurrence of periodic components in the envelope and in the instantaneous frequency. Results of the theoretical analysis are compared with the results obtained by computer simulation.

1. INTRODUCTION

Instantaneous frequency and envelope are finding their practical use in a number of applications (biomedicine, astronomy, speech processing, acoustics). Their definition based on an analytic signal requires calculation of Hilbert transform. In the literature, e.g. [1], [2], some devices for instantaneous frequency measurement have been described. As a rule, little requirements were placed on the accuracy of the instantaneous frequency evaluation, neither was the accuracy analyzed. In this contribution a method based on application of digital filters will be analyzed from the accuracy viewpoint. The often used method for instantaneous frequency determination based on calculation of the discrete Fourier transform is highly accurate, but only for periodic signals. In cases where instantaneous frequency is to be evaluated in real time or when we deal with nonperiodic signals, evaluation of instantaneous frequency by digital filtering is the only option we have. This also allows additional processing of instantaneous frequency signal by smoothing or filtering.

2. PRINCIPLE OF THE METHOD EMPLOYING DIGITAL FILTERS

The instantaneous frequency evaluation employing digital filters is based on expressing the instantaneous frequency of a signal $x(n)$ utilizing quadrature signal $y(n)$ and the derivatives $\dot{x}(n)$ and $\dot{y}(n)$. It can be expressed as

$$\omega(n)=(x(n)\dot{y}(n)-\dot{x}(n)y(n))A^{-2}(n) \tag{1}$$

where

$$A^2(n)=x^2(n)+y^2(n) \tag{2}$$

is the square of the signal envelope. In Fig.1 we denote by HF a Hilbert filter, by DF a differentiating filter, by DHF a differentiating Hilbert filter performing differentiation of the Hilbert transform and arithmetic operations by AO. The scheme with DHF is substantially more advantageous than realization of $\dot{y}(n)$ by a cascade connection of HF and DF as was shown in [3].

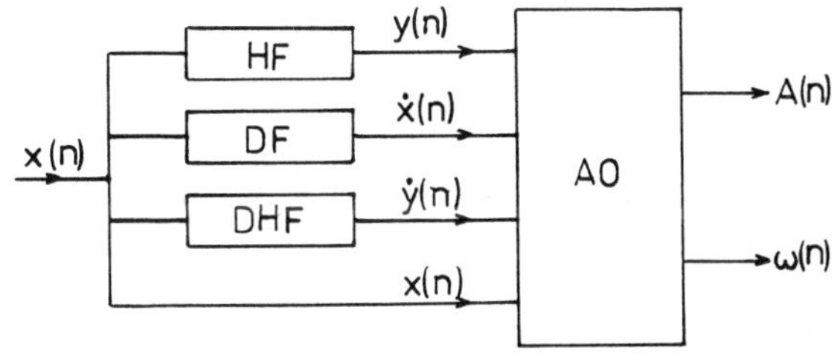

Figure 1

From the accuracy poing of view it is expedient to use filters with finite

impulse response and linear phase which introduce into the measurement only errors arising from approximation of amplitude characteristics of the ideal filters HF, DF and DHF. For error analysis it proved very advantageous to use filters with equi-ripple approximation in which a constant approximation deviation is known. Since HF, DF and DHF designed to have approximately equal deviation have not the same degree (lenght), they have to be completed by adding some delay elements.
The filter method makes the evaluation possible in a limited range of frequencies only, which is given by the approximation interval used during the filter design.
The ideal frequency characteristics of these filters are shown in Fig.2 where $H(e^{j\omega})$ is the frequency characteristic of filter HF, $D(e^{j\omega})$ is the frequency characteristic of filter DF and $G(e^{j\omega})$ is the frequency characteristic of filter DHF.

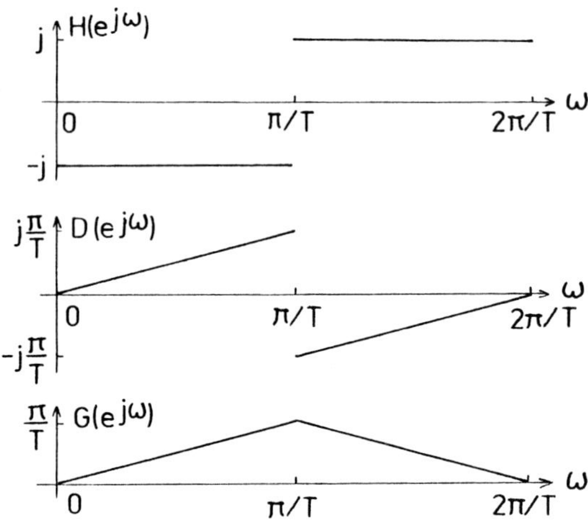

Figure 2

The scheme in Fig.1 was realized by computer simulation and at present its realization with the aid of a digital signal processor is being prepared.

3. EVALUATION ACCURACY

3.1. General solution

As a result of using filter with non-ideal characteristics, instead of signals $y(n)$, $\dot{x}(n)$ and $\dot{y}(n)$ we only get their approximate values

$$\tilde{y}(n) = y(n)+e_1(n) ,$$
$$\tilde{\dot{x}}(n) = \dot{x}(n)+e_2(n) ,$$
$$\tilde{\dot{y}}(n) = \dot{y}(n)+e_3(n) .$$

In this case, the signal envelope ist

$$\tilde{A}^2(n)=A^2(n)+2e_1(n)y(n)+e_1^2(n) \quad (3)$$

and its instantaneous frequency

$$\tilde{\omega}(n)=\omega(n)\frac{A^2(n)}{\tilde{A}^2(n)}+\frac{e_0(n)}{\tilde{A}^2(n)}, \quad (4)$$

where

$$e_0(n)=x(n)e_3(n)-\dot{x}(n)e_1(n)-y(n)e_2(n)-$$
$$-e_1(n)e_2(n).$$

3.2. Harmonic signal

The distorted signals corresponding to the harmonic signal $x(n)=\cos\omega_0 n$ are

$$\tilde{y}(n) = (1+d_1)\sin\omega_0 n ,$$
$$\tilde{\dot{x}}(n) =-(1+d_2)\omega_0\sin\omega_0 n ,$$
$$\tilde{\dot{y}}(n) = (1+d_3)\omega_0\cos\omega_0 n .$$

Assuming $d_i \ll 1$ and after some manipulation of the above relations we get for the instantaneous frequency and envelope the following expressions

$$\tilde{\omega}(n)\doteq\omega_0(1+d_1(n)), \quad (5)$$

$$d_1(n)=\frac{1}{2}(d_3+d_2-d_1)+\frac{1}{2}(d_3-d_2+d_1)\cos 2\omega_0 n$$

$$\tilde{A}(n)\doteq 1+\frac{1}{2}d_1-\frac{1}{2}d_1\cos 2\omega_0 n. \quad (6)$$

In the special case when $d_1=d_2=d_3=d$ compensation of errors takes place and it holds

$$\omega(n)=\omega_0(1+\frac{1}{2}d + \frac{1}{2}d \cos 2\omega_0 n).$$

From relations (5) and (6) it follows that the error caused by inaccuracy of the auxiliary signals $y(n)$, $x(n)$ and $y(n)$ manifests itself under the given conditions, distorting the envelope and frequency. The relative error of the envelope (6) and frequency (5) has a constant and a periodic component whose frequency is equal to the double of the signal frequency. Determination of the harmonic signal frequency with the aid of the filter method results in interpreting this signal as a signal with "amplitude and frequency modulation" by frequency $2\omega_0$. The maximum values of these "modulations" are given by deviations of the filters from ideal characteristics. One has to bear in mind that the quantities d_1, d_2 and d_3 are different for different frequencies and, as a result, the compensation of error in $\omega(n)$ can take place for some frequencies only. As a rule, the quantities d_1, d_2 and d_3 representing the filter inaccuracies are unknown. When the filters HF, DF and DHF have linear phase and have been designed with the aid of program [4] assuming equi-ripple approximation, we also obtain the values of the maximum relative approximation deviation. This program has to be slightly modified to be suitable for designing the DHF, as well. If we denote the maximum approximation deviations as D_1, D_2 and D_3, the following relations hold

$$|d_1| \leq D_1 \ , \ |d_2| \leq D_2 \text{ and } |d_3| \leq D_3$$

and the relative error of the envelope

$$\left|\frac{\tilde{A}(n) - A}{A}\right| \leq D_1 \tag{7}$$

and of instantaneous frequency

$$\left|\frac{\tilde{\omega}(n) - \omega_0}{\omega_0}\right| \leq D_1+D_2+D_3 \tag{8}$$

can be estimated.

The filter method according to Fig.1 was simulated on a computer for various filter orders. For smaller accuracy the following set of filters was simulated, where f_L and f_H is the lower and higher cut-off frequency, respectively.

	Filter	Degree	f_L	f_H	D
1.	HF	39	500	4500	$6,79.10^{-4}$
2.	DF	39	500	4500	$2,88.10^{-3}$
3.	DHF	27	0	4500	$2,14.10^{-4}$

From calculation of the relative errors the range of instantaneous frequency $1245.28\text{Hz} \leq f \leq 1254.71\text{Hz}$ follows for the signal frequency f_0=1250 Hz and the sampling frequency $f = 10^4$Hz. By simulation the range was found to be 1250.20Hz $\leq f \leq$ 1252.80Hz.
Another set of filters was simulated for high accuracy:

	Filter	Degree	f_L	f_H	D
1.	HF	127	410	7780	$1,16.10^{-5}$
2.	DF	127	410	7780	$5,6 \ .10^{-6}$
3.	DHF	63	0	7780	$5,6 \ .10^{-6}$

The estimates give the range of instantaneous frequency 1023.8 Hz $\leq f \leq$ 1024.1 Hz for f_0=1024Hz. From simulation the range 1023.92Hz $\leq f \leq$ 1024.06Hz was obtained. A simulation for different values of f_0 in the passband provided frequencies in the range 1024Hz $\pm$ 0.1Hz.
In Fig.3 the variation of instantaneous frequency for the first set of filters is depicted showing a good agreement with theoretical results.

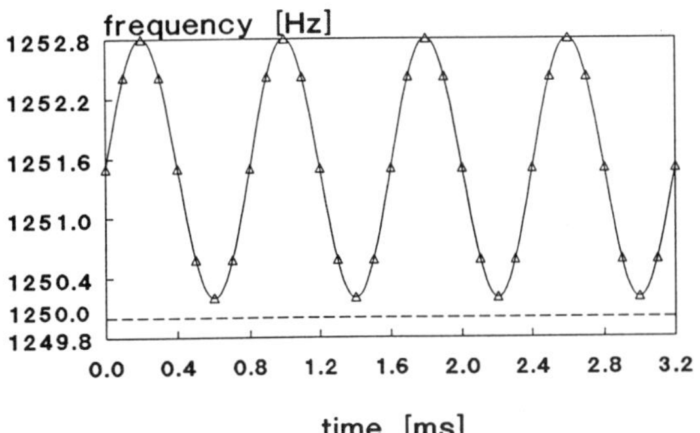

Figure 3

3.3. Two-frequency signal

The instantaneous envelope and frequency of a two-frequency signal

$x(t)=\cos\omega_1 t+K\cos(\omega_2 t+\varphi)$, $K>0$ is [5]

$$A^2(n)=1+K^2+2K\cos u, \quad u=(\omega_2-\omega_1)t+\varphi \ , \tag{9}$$

$$\omega(n)=(\omega_1+K^2\omega_2+K(\omega_1+\omega_2)\cos u).A^{-2}(n). \tag{10}$$

The instantaneous frequency then lie in the interval

$$\frac{K\omega_2 - \omega_1}{K - 1} \ , \ \frac{K\omega_2 + \omega_1}{K + 1}. \tag{11}$$

A general analysis of the effects of inaccuracies would in this case involve 6 quantities characterizing the filter inaccuracies at frequencies ω_1 and ω_2.
We therefore introduce a simplifying assumption that the distortion takes place only at frequency ω_2.

For $d_1=d_2=d_3=d$, $K \gg 1$ but $dK^2 \ll 1$ we get for instantaneous frequency an estimation

$$\tilde{\omega}(n) \doteq \omega(n)+Kd \ \frac{\omega_1+2K\omega_2}{A^2(n)} \ .$$

In the special case when K=1, the following estimation can be derived

$$\tilde{\omega}(n) \leq \frac{\omega_1+\omega_2}{2} \pm d(\omega_1+3\omega_2). \tag{12}$$

4. A SIMPLIFIED MODEL FOR FILTER INACCURACY EFFECTS

4.1. Two-frequency signal

Since we assume that the filter inaccuracy is caused by deviations of amplitude characteristics of the filters, we try to approximate the effect of these deviations in the case of a two-frequency signal by changing the amplitude K so that we shall investigate a signal

$$x(t)=\cos\omega_1 t+K(1+d_0)\cos(\omega_2 t+\varphi). \qquad (13)$$

From relations (9) and (10) we then obtain

$$\tilde{A}^2(n)=A^2(n)+2d_0K(K+\cos u)+d^2K^2 , \qquad (14)$$

$$\tilde{O}(n)=O(n)+2d_0K\omega_1\cos u+\omega_2(K+\cos u)+d_0^2K^2\omega_2 ,$$

$$O(n)=\omega_1+K^2\omega_2+K(\omega_1+\omega_2)\cos u, \omega(n)=\tilde{O}(n)\tilde{A}^{2}(n)$$

Under the assumption that $K\gg 1$ but $d^2K^2\ll 1$ we obtain the following relation for instantaneous frequency

$$\tilde{\omega}(n)=\omega(n)+2d_0K(\omega_1\cos u+\omega_2K).A^{-2}(n). \qquad (15)$$

It follows from relation (11) that the instantaneous frequency would take on values lying in an interval

$$\frac{K(1+d_0)\omega_2-\omega_1}{K(1+d_0)-1} , \frac{K(1+d_0)\omega_2+\omega_1}{K(1+d_0)+1} . \qquad (16)$$

For the case K=1 this estimation is true

$$\tilde{\omega}(n)\leq\frac{\omega_1+\omega_2}{2} \pm 2d_0(\omega_1+\omega_2). \qquad (17)$$

From comparison with section 3.3. it follows that this simplified model is an usable approximation.

4.2. Amplitude modulated signal

In the case of an amplitude modulation we evaluate the influence of the change in amplitude of one sideband of the signal

$$x(n)=a\cos(\omega_0-\Omega)n+\cos\omega_0 n+(a+e)\cos(\omega_0+\Omega)n$$

on the instantaneous frequency. Then, we get

$$\tilde{\omega}(n)=\omega_0+\Omega\frac{e(2a+\cos\omega_0 n)+e^2}{(1+(2a+e)\cos\omega_0 n)^2+e^2\sin^2\omega_0 n} .$$

The inaccuracy of filters manifests itself as frequency modulation and for $e\ll a$ the instantaneous frequency stays within the interval

$$\omega_0 - \frac{\Omega e}{1-2a} , \quad {}_0 + \frac{\Omega e}{1+2a} . \qquad (18)$$

5. CONCLUSIONS

The analysis has shown that approximation errors of the filter frequency characteristics result in approximately the same relative error in instantaneous frequency. The magnitude of this error depends on the order of the filters used in the procedure. The results obtained by simulation show that by using digital filters instantaneous frequency of the signals, whose spectrum lies within the frequency bandwidth of the filters, can be evaluated with sufficient accuracy.

6. REFERENCES

[1] Gabison, A. and Gendrin,R.,Appareil analogique destiné à la mesure en temps réel de l'amplitude, de la fréquence et de la phase instantanées de signaux variant dans le temps Ann. des télécom. 35 (1979) 3-4, 158-165.

[2] Ktonas, P.Y. and Papp,N., Instantaneous envelope and phase extraction from real signals: theory, implementation, and an application to EEG analysis, Signal processing 2 (1980) 373-385.

[3] Čížek, V., Differentiating FIR Hilbert transformer, in: Proc.1989 URSI Int. Symp. on Signals, Systems and Electronics - ISSSE'89,697-700.

[4] IEEE Digital Signal Processing Committee (ed.): Programs for Digital Signal Processing, IEEE Press, New York (1979).

[5] Čížek, V., Kinematische Theorie der Momentanfrequenz und Einhüllenden, Archiv für Elektronik und Übertragungstechnik 43 (1989) 5, 288-291.

SIGNAL PROCESSING VI: Theories and Applications
J. Vandewalle, R. Boite, M. Moonen, A. Oosterlinck (eds.)

Multichannel Multirate Sampling of Low-pass Stochastic Signals

Takuro KIDA and *Somsak SA-NGUANKOTCHAKORN*

Department of Information Processing, Graduate School at Nagatsuta, Tokyo Institute of Technology,

4259 Nagatsuta, Midori-ku, Yokohama-shi 227, Japan Tel.045-922-1111 Ext.2522
e-mail : somsak@ip.titech.ac.jp

Abstract

The extended form of the multichannel sampling theorem of a class of band-limited stochastic input signals is presented. The sample values used in this paper are nonuniform sample points repeating periodically. They are chosen from the output signals of a parallel bank of a finite number of linear shift invariant filters with the random process noise interfering with each channel prior to sampling. Firstly, we indicate that the interpolation functions can be derived by using the concept of the generating function for sampling theorem[4]-[6]. Next, under the condition that the signal treated here is not necessarily stationary, we establish a number of relations between the interpolation functions and the noises in the system by showing the covariance of the output signal and various kinds of power spectral density functions in terms of the interpolation functions.

1 Introduction

The sampling theorem for band-limited signal has been extended in a several directions, such as nonuniform sampling theorem and derivative sampling theorem[1]-[6]. In the papers of Papoulis[2] and Brown[3], for instances, they interpreted lucidly these sampling theorems and extended the domain of applicability to embrace multichannel sampling which resembles to the framework called filter banks. It was stated that one can restore the original signal by using the simultaneous sample points of the outputs of 2N linear time invariant filters having the input as a deterministic signal $f(t)$ band limited to $|\omega| \leq W$ with a uniform sampling rate of $W/\pi(2N)$ samples per second. Basically, the recovery of a signal from generalized samples is essentially a problem of designing appropriate linear filters called reconstruction (or interpolation) filters.

In this paper, the concept of multichannel sampling theorem is extended to the class of stochastic band-limited signal. We model the system as the multichannel sampling framework whose input is contained in the class of the signals mentioned above. We assume that the random process noises $n_k(\nu,t)$ with a random variable ν, are added to each of the corresponding output signals of the analysis filters prior to sampling. Further, the sample points have the form of $(nT + a_k)$ $(0 \leq a_1 < a_2 < .. < T \leq 2N\pi/W)$ $(n = 0, \pm1, \pm2, ..)$[5]. Hence, these are nonuniform sample points in general. Using this model, firstly we establish the general expression and derive the necessary and sufficient condition for the interpolation formula, by utilizing the concept of the generating function for sampling theorem [4]-[6], in order that the output signal always converges, in the statistical mean sense, to the input signal. In this discussion, all the interpolation functions can be derived directly by the Fourier series expansion of the generating function with respect to the angular frequency ω.

In addition, we also consider the covariances of the restored output signal and derive various kinds of power spectral density functions. It is important to note that the signal treated here is not necessarily wide sense stationary.

2 Notations and terminology

(a) The integers are expressed by the symbols N, k, m, n, p, q and i. The real numbers are indicated by W, T and a_k. Throughout this paper, $k = 1 \sim 2N$.

(b) Let $\Omega_m (m = 0, \pm1, \pm2, ..)$ be the closed intervals on the ω-axis given by

$$\Omega_m = [(N-m)W/N, (N-m+1)W/N] \tag{1}$$

The fact that ω belongs to Ω_m is written as $\omega \in \Omega_m$.

(c) The notation $f(t) \longrightarrow F(\omega)$ indicates that $f(t)$ and $F(\omega)$ form a Fourier transform pair satisfying

$$F(\omega) = \int_{-\infty}^{\infty} f(t) e^{-j\omega t} dt \tag{2}$$

$$f(t) = \frac{1}{2\pi} \int_{-\infty}^{\infty} F(\omega) e^{j\omega t} d\omega \tag{3}$$

(d) Bold-faced quantities denote matrices and vectors as in $\boldsymbol{A}_m(\omega), \boldsymbol{\ell}_m$ etc. The quantity z^* denotes the conjugate of z.

(e) Let $f(\nu,t)$ be a random process with a random variable ν. The statistical mean or expected value of a random process is defined by

$$f(t) = E[f(\nu,t)] \tag{4}$$

where $E[*]$ is called the expected value operator. The variance of $f(\nu,t)$ is defined as

$$\sigma_f^2 = E[| f(\nu,t) - f(t) |^2] \tag{5}$$

A random process $f(\nu,t)$ is said to be wide-sense stationary if two of its statistics, its mean and autocorrelation function, do not vary with a shift in time origin. That is, the random processes $f_k(\nu,t)$ satisfy

$$E[f_k(\nu,t)] = a_k \quad (constant) \tag{6}$$
$$E[f_k(\nu,t)f_m(\nu,s)] = \gamma_{km}(t-s) \tag{7}$$

where $\gamma_{km}(t-s)$ is the autocorrelation function depends only on $t-s$.

(f) In the following discussion, we consider $2N$ linear time invariant analysis filters whose transfer functions are denoted by $H_k(\omega)$ and the time domain impulse responses are written as $h_k(t)$. The random processes corresponding to external noise or observation error being added to each channel of the outputs of each analysis filters are written as $n_k(\nu,t)$. They are assumed to be independent of $h_k(t)*f(\nu,t)$ and satisfying $E[n_k(\nu,t)]=0$.

3 Interpolation of band-limited stochastic signal

In this section, the input signal $f(\nu,t)$, assumed to be band-limited stochastic signal (in the region $|\omega|\le W$) with random variable ν, is taken as the common input to each of the linear shift invariant filters $H_k(\omega)$. The corresponding signals are denoted by

$$g_k(\nu,t) = h_k(t)*f(\nu,t)+n_k(\nu,t) \tag{8}$$

where $h_k(t) \longleftrightarrow H_k(\omega)$. We assume additionally that $H_k(\omega)$ are square-integrable and bounded on the region $|\omega|\le W$.

Now, the output signal can be expressed as follows.

$$y(\nu,t)=\sum_{k=1}^{2N}\sum_{n=-\infty}^{\infty} g_k(\nu,nT+a_K)\psi_k(t-nT-a_k) \tag{9}$$

where the quantities T and a_k are the positive numbers satisfying

$$0\le a_1<a_2<\ldots<a_{2N}<T=\frac{2N\pi}{W} \tag{10}$$

and $\psi_k(t)$ are the deterministic time functions, called the interpolation functions, whose Fourier transforms $\Phi_k(\omega)$ are piece-wise continuous.We assume that the interpolation functions satisfy the next equation.

$$\sum_{k=1}^{2N}\sum_{n=-\infty}^{\infty}|\psi_k(t-nT-a_k)|^2<+\infty \tag{11}$$

In this discussion, it is assumed that the Fourier transform of $f(t)(=E[f(\nu,t)])$ satisfies

$$F(\omega)=0 \qquad (|\omega|>W) \tag{12}$$

Firstly, we will discuss about the necessary and sufficient condition that the output signal always converges to the input signal in the sense of statistical mean.

As a preliminaries, we provide the following lemma.

Lemma 1 *Let*

$$y(t)\longleftrightarrow Y(\omega) \tag{13}$$

where $y(t)=E[y(\nu,t)]$. Then, $Y(\omega)$ is expressed in Ω_m as

$$Y(\omega) = \frac{W}{2N\pi}\sum_{n=-m+1}^{-m+2N}\sum_{k=1}^{2N} e^{jnWa_k/N}H_k(\omega-nW/N) \Phi_{km}(\omega)F(\omega-nW/N) \tag{14}$$

where $m=0,\pm1,\pm2,..$

Next theorem and its corollary show how to choose $\Phi_k(\omega)$ in order that $y(t)$ always converges to $f(t)$.

Theorem 1 *The necessary and sufficient condition that $y(t)$ always converges to $f(t)$ is*
(a) in Ω_m $(m=1\sim 2N)$

$$\boldsymbol{A}_m(\omega)\boldsymbol{\Phi}_m(\omega)=(2\pi N/W)\boldsymbol{\ell}_m \tag{15}$$

(b) in Ω_m $(m<1$ or $m>2N)$

$$\boldsymbol{A}_m(\omega)\boldsymbol{\Phi}_m(\omega)=0 \tag{16}$$

where *(t:transpose)*

$$\boldsymbol{\ell}^t_m = [\overset{(1)}{0},..0,\overset{(m)}{1},0,..,\overset{(2N)}{0}] \tag{17}$$
$$\boldsymbol{\Phi}^t_m(\omega) = [\Phi_{1,m}(\omega),\Phi_{2,m}(\omega),..\Phi_{2N,m}(\omega)] \tag{18}$$
$$\begin{aligned}\boldsymbol{A}_m(\omega) &= [\alpha^m_{pq}] \qquad (p,q=1\sim 2N)\\ &= [e^{j(m-p)Wa_q/N}H_q(\omega+(m-p)W/N)]\end{aligned} \tag{19}$$

and $H_k(\omega)$ are the transfer functions of the analysis filters.

This theorem can be proved easily by using the concept of the generating function for sampling theorem which will be mentioned in the next section.

As is evident in theorem 1, if $\Phi_{k,m}(\omega)$ $(k,m=1\sim 2N)$ are chosen so as to satisfy the conditions (a) and (b) in the theorem, the next identity holds.

$$f(t)=\sum_{k=1}^{2N}\sum_{n=-\infty}^{\infty} g_k(nT+a_k)\psi_k(t-nT-a_k) \tag{20}$$

Eq.(20) provides a series of sampling theorems if appropriate sets of a_k and $H_k(\omega)$ are selected.

Let us now determine the closed form expressions of $\Phi_{k,m}(\omega)(k,m=1\sim 2N)$ under the condition that only $\boldsymbol{A}_1(j\omega)$ is nonsingular.

Corollary 1 *Suppose that the matrix*

$$\begin{aligned}\boldsymbol{A}_1(\omega) &= [a^1_{pq}] \quad (p,q=1\sim 2N)\\ &= [e^{-j(p-1)Wa_q/N}H_q(\omega-(p-1)W/N)]\end{aligned} \tag{21}$$

is nonsingular for all ω. Further, let

$$\boldsymbol{A}_1(\omega)^{-1}=[\Theta_{mn}(\omega)] \quad (m,n=1\sim 2N) \tag{22}$$

Then the solutions of Eq.(15) and Eq.(16) are given by
(a) in Ω_m $(m=1\sim 2N)$

$$\Phi_{k,m}(\omega)=\frac{2\pi N}{W}e^{-j(m-1)Wa_k/N}\Theta_{km}(\omega+(m-1)W/N) \tag{23}$$
$$(k,m=1\sim 2N)$$

(b) in Ω_m $(m<1$ or $m>2N)$

$$\Phi_{k,m}(\omega)=0 \quad (k=1\sim 2N; m<1 \text{ or } m>2N) \tag{24}$$

Corollary 1 can be proved easily if we notice the relation

$$A_m(\omega) = A_1(\omega + (m-1)W/N)\Delta[e^{j(m-1)Wa_n/N}] \quad (25)$$

where $\Delta[\bullet] = \Delta[\beta_n]$ means the diagonal matrix with the n-th diagonal element of which is $\beta_n (n = 1 \sim 2N)$.

The significant aspect of Corollary 1 is that all the $\Phi_{k,m}(\omega)$ can be obtained by evaluating only one matrix $A_1(\omega)^{-1}$.

As a direct consequence of Corollary 1, we have

Theorem 2 *If (a) and (b) in Corollary 1 are satisfied, then the interpolation functions $\psi_k(t)$ in Eq.(20) are given by*

$$\psi_k(t) = \frac{N}{W}\sum_{m=1}^{2N} e^{-j(m-1)(t+a_k)W/N}\int_{(N-1)W/N}^{W}\Theta_{km}(\omega)e^{j\omega t}d\omega \quad (26)$$

where $\psi_k(t) \longrightarrow \Phi_k(\omega)$ and $[\Theta_{km}(\omega)] = A_1(\omega)^{-1}$.

The proof is omitted.

4 Calculation of interpolation function by generating function

In this section, we show how to use the concept of the generating function for sampling theorem[4]-[6] to derive the interpolation functions.

Here, we assume again that the interpolation functions $\psi_k(t)$ satisfy Ineq.(11).

Now, we consider the 2-variable functions $\Psi_k(\omega, t)$ $(k = 1 \sim 2N)$ which satisfy the next 3 conditions at the same time.

(a) $\Psi_k(\omega, t)$ are periodic functions with period W/N.

$$\Psi_k(\omega, t) = \Psi_k(\omega - pW/N, t) \qquad (p = 0, 1, 2, ..) \quad (27)$$

(b) For any finite t, the following inequality holds.

$$\int_{\Omega_m} |\Psi_k(\omega, t)|^2 d\omega < +\infty \qquad ; \omega \in \Omega_m \quad (28)$$

(c) There exist the solutions $\Psi_k(\omega, t)$ of the next linear equations.

$$\sum_{k=1}^{2N} e^{j(\omega - pW/N)a_k} H_k(\omega - pW/N)\Psi_k(\omega, t) = e^{j(\omega - pW/N)t} \quad (29)$$

where the 2-variable functions $\Psi_k(\omega, t)$ here are called the generating functions. Then the next theorem holds.

Theorem 3 *The necessary and sufficient condition that $y(t)$ always converges to $f(t)$ is that there exist the functions $\Psi_k(\omega, t)$ satisfying conditions (a)$\sim$(c) simultaneously.*

In proving the necessity of the theorem, Riesz-Fischer's theorem is necessary. Hence, it is necessary to assume that Ineq.(11) is valid. Therefore, there exist $\Psi_k(\omega, t)$ $(k = 1 \sim 2N)$ satisfying

$$\Psi_k(\omega, t) = \sum_{n=-\infty}^{\infty} \psi_{kn}(t)e^{j\omega nT} \quad (30)$$

By Parseval's formula, the conditions (a) and (b) hold obviously. With respect to the necessity of the condition (c), see [6] for justification. we will show here only the guideline of the sufficiency of the proof. The similar proof in detail was shown in [6] of the case of uniform sampling.

Now, let the error between the statistical mean of input and output signal be defined as

$$\begin{aligned} e(t) &= |f(t) - y(t)| \quad (31) \\ &= \left| f(t) - \sum_{k=1}^{2N}\sum_{n=-\infty}^{\infty} g_k(nT + a_k)\psi_k(t - nT - a_k) \right| \\ &= \frac{1}{2\pi}\left| \int_{-W}^{W} F(\omega)\left\{ e^{j\omega t} - \sum_{k=1}^{2N}\sum_{n=-\infty}^{\infty} H_k(\omega) \right.\right. \\ &\quad \left.\left. \psi_k(t - nT - a_k)e^{j\omega(nT+a_k)}\right\} d\omega \right| \quad (32) \end{aligned}$$

Next, the generating functions $\Psi_k(\omega, t)$ are defined by

$$e^{j\omega t} = \sum_{k=1}^{2N} e^{j\omega a_k} H_k(\omega)\Psi_k(\omega, t) \quad (33)$$

Substituting Eq.(33) into Eq.(32), we have

$$\begin{aligned} e(t) &= \frac{1}{2\pi}\left| \int_{-W}^{W} F(\omega)\sum_{k=1}^{2N} e^{j\omega a_k} H_k(\omega)\left\{ \Psi_k(\omega, t) \right.\right. \\ &\quad \left.\left. - \sum_{n=-\infty}^{\infty} \psi_k(t - nT - a_k)e^{j\omega nT}\right\} d\omega \right| \quad (34) \end{aligned}$$

From the assumption of the sufficiency, we can use the conditions (a) and (b). Hence, Eq.(30) is obvious. Further, although the detail discussion is omitted, if the average rate of the sample points contains no redundancy, we can prove that

$$\psi_{kn}(t) = \psi_k(t - nT - a_k) \quad (35)$$

Substituting Eq.(30) and Eq.(35) into Eq.(34), we can easily prove that $e(t) = 0$.

Replacing ω in Eq.(33) by $\omega - pW/N$ $(p = 0, 1, 2, .., 2N - 1)$, we get

$$e^{j(\omega - pW/N)t} = \sum_{k=1}^{2N} e^{j(\omega - pW/N)a_k} H_k(\omega - pW/N)\Psi_k(\omega, t) \quad (36)$$

$$(p = 0, 1, 2, .., 2N - 1)$$

That is, the condition (c) holds obviously. Theorem 1 can be proved easily from Eq.(30), Eq.(35) and Eq.(36). Actually, the concept of the generating function is usable even in the case of oversampling. However, in this case, Eq.(35) is not always valid.

5 Covariance and variance of output signal $y(\nu, t)$

In this section, we wish to determine the covariance and the variance of the output signal $y(\nu, t)$.

Now, the error of the output signal is defined as

$$\begin{aligned} Er(\nu, t) &= y(\nu, t) - y(t) \\ &= \sum_{k=1}^{2N}\sum_{n=-\infty}^{\infty} \varepsilon_k(\nu, nT + a_k)\psi(t - nT - a_k) \quad (37) \end{aligned}$$

where $\varepsilon_k(\nu, t) = g_k(\nu, t) - g_k(t)$ and $y(t) = E[y(\nu, t)]$ is the statistical mean of the output signal.

Here, we assume that $\varepsilon_k(\nu,t)$ $(k=1\sim 2N)$ satisfy

$$\left.\begin{array}{l} E[\varepsilon_k(\nu,t)]=0 \\ E[\varepsilon_k(\nu,t)\varepsilon_m(\nu,s)]=\gamma_{km}(t-s)r_k(t)r_m(s) \end{array}\right\} \quad (38)$$

where $r_i(t) \longleftrightarrow R_i(\omega)$ $(i=1\sim 2N)$ are deterministic functions of t, for example, the output signals of $H_i(\omega)$ for a typical deterministic input signal $f(t)$. We suppose that $|R_i(\omega)|=0$ $(|\omega|>W)$.

In this case, the covariance of $y(\nu,t)$ is given by

$$\begin{aligned} c(t,\tau) &= E[Er(\nu,t+\tau)Er(\nu,t)] \\ &= \sum_{k=1}^{2N}\sum_{m=1}^{2N}\sum_{n=-\infty}^{\infty}\sum_{p=-\infty}^{\infty}\gamma_{km}((n-p)T+a_k-a_m) \\ &\quad r_k(nT+a_k)r_m(pT+a_m) \\ &\quad \phi_k(t+\tau-nT-a_k)\phi_m(t-pT-a_m) \end{aligned} \quad (39)$$

Further, let the power spectrum density function be written as $K(\omega,t)$, where $c(t,\tau) \longleftrightarrow K(\omega,t)$. Then the following theorem is obtained directly.

Theorem 4 *Let $\gamma_{km}(t) \longleftrightarrow \Gamma_{km}(\omega)$ .Further, let*

$$\hat{\Gamma}_{km}(\omega) = \sum_{n=-\infty}^{\infty}\Gamma_{km}(\omega-nW/N)e^{-jn(a_k-a_m)W/N} \quad (40)$$

$$\hat{R}_k(\omega) = \sum_{p=-\infty}^{\infty}R_k(\omega+pW/N)e^{jpa_kW/N} \quad (41)$$

Then, $K(\omega,t)$ can be expressed as

$$\begin{aligned} K(\omega,t) &= \left(\frac{W}{4\pi^2N}\right)^2\sum_{k=1}^{2N}\sum_{m=1}^{2N}\Phi_k(\omega)\int_{-\infty}^{\infty}\int_{-\infty}^{\infty}\hat{\Gamma}_{km}(w-u) \\ &\quad \hat{R}_k(u)R_m(v)^*\Phi_m(\omega-u+v)^*e^{j(u-v)t}dudv \end{aligned} \quad (42)$$

The proof is omitted.

If $\varepsilon_k(\nu,t)$ are wide sense stationary signals, that is

$$\left.\begin{array}{l} E[\varepsilon_k(\nu,t)]=0 \\ E[\varepsilon_k(\nu,t)\varepsilon_m(\nu,s)]=\gamma_{km}(t-s) \end{array}\right\} \quad (43)$$

holds, then the following theorem is immediate.

Theorem 5 *Under the condition that Eq.(43) holds, we have*

$$\begin{aligned} K(\omega,t) &= \frac{W^2}{4\pi^2N^2}\sum_{k=1}^{2N}\sum_{m=1}^{2N}\sum_{p=-\infty}^{\infty}\hat{\Gamma}_{km}(\omega)\Phi_k(\omega) \\ &\quad \Phi_m(\omega-pW/N)^*e^{jp(t-a_m)W/N} \end{aligned} \quad (44)$$

Substituting $R_k(\omega)=2\pi\delta(\omega)$ into Eq.(42), we may easily derive Eq.(44)

Further, if $\varepsilon_k(\nu,t)$ and $\varepsilon_m(\nu,t)$ are independent of each other, we may obtain the next corollary.

Corollary 1 *If $E[\varepsilon_k(\nu,t)\varepsilon_m(\nu,s)]=0$ $(k\neq m; k,m=1\sim 2N)$*

$$\begin{aligned} K(\omega,t) &= \left(\frac{W}{4\pi^2N}\right)^2\sum_{k=1}^{2N}\Phi_k(\omega)\int_{-\infty}^{\infty}\int_{-\infty}^{\infty}\Gamma_k(\omega-u) \\ &\quad \hat{R}_k(u)R_k(v)^*\Phi_k(\omega-u+v)^*e^{j(u-v)t}dudv \end{aligned} \quad (45)$$

where $\hat{R}_k(\omega)=$ Eq.(41) and

$$\Gamma_k(\omega)=\sum_{n=-\infty}^{\infty}\Gamma_{kk}(\omega-nW/N) \quad (46)$$

Moreover, if $\varepsilon_k(\nu,t)$ are wide sense stationary signals,

$$\begin{aligned} K(\omega,t) &= \frac{W^2}{4\pi^2N^2}\sum_{k=1}^{2N}\sum_{p=-\infty}^{\infty}\Gamma_k(\omega)\Phi_k(\omega) \\ &\quad \Phi_k(\omega-pW/N)^*e^{jpW(t-a_m)/N} \end{aligned} \quad (47)$$

$$\frac{W}{2\pi N}\int_0^{2\pi N/W}K(\omega,t)dt=\frac{W^2}{4\pi^2N^2}\sum_{k=1}^{2N}\Gamma_k(\omega)|\Phi_k(\omega)|^2 \quad (48)$$

Using this $K(\omega,t)$ function as a measure, we may achieve the optimum arrangement of the sample points, by the ordinary numerical optimization procedure, that minimize the error of output signal.

Conclusion

The extended form of the multichannel sampling of the class of stochastic band-limited signal with nonuniform sampling points has been investigated. Our discussion covered many kinds of errors, such as quantization error, round-off error, jitter etc, which occur in the physical applications of this transmission system. We study the effect of interpolation functions on such these kinds of errors by deriving various power spectral density functions.

Acknowledgement

The authors wish to thank Prof.A.Fettweis of Ruhr Universitat Bochum, Prof.Kenneth Jenkins of Coordinated Science Laboratory, University of Illinois, and my college Mr.Hiroshi Mochizuki of Tokyo Institute of Technology for helpful comments.

References

[1] A.Papoulis *Probability,Random Variables and Stochastic Processes*, McGrawhill (1984)

[2] A.Papoulis *Generalized Sampling Expansion*, IEEE,Trans.on Circuits and Systems,vol.cas-24,No.11, pp.652-654 (Nov.1977)

[3] John L. Brown,Jr. *Multi-Channel Sampling of Low-Pass Signals*, IEEE Trans. on Circuits and Systems,vol.cas-28,No.2, pp.101-106 (Feb.1981)

[4] T.Kida *Generating Functions for Sampling Theorem*, Trans.IEICE.,Japan,65-A,1, pp.46-53 (Jan.1982)

[5] T.Kida *On Interpolation Functions for classes of Sampling Theorems with Non-Uniform Sampling Points*, Trans.IEICE.,Japan,65-A,9, pp.919-926 (Sept.1982)

[6] T.Kida, S. Sa-nguankotchakorn *Study of Generating Function of Sampling Theorem and Optimum Interpolation Functions Minimizing Some Kinds of Measures* Trans.IEICE.,Japan,Vol.j74-A,No.8, pp.1332-1345 (Aug.1991)

SIGNAL PROCESSING VI: Theories and Applications
J. Vandewalle, R. Boite, M. Moonen, A. Oosterlinck (eds.)

OPTIMIZATION OF MULTIRESOLUTION ANALYSIS FOR FINITE SEQUENCES

Jean-Christophe PESQUET

Laboratoire des Signaux et Systèmes, CNRS/UPS, Ecole Supérieure d'Electricité,
Plateau de Moulon, 91192 Gif sur Yvette Cédex, France, pesquet@frese51.bitnet

At first, this paper presents a matrix formulation of the periodic wavelet transform. Then, the problem of the optimal choice of orthonormal wavelets is considered. A statistical approach is adopted in order to obtain a compact representation of a given set of input signals. Moreover, a parametrization of wavelets is introduced in the frequency domain and a multiscale method is proposed to estimate the frequency responses of the filters used at each resolution level.

1 INTRODUCTION

In the use of wavelet transforms [1] [2], a crucial step is to choose the wavelet which is best matched to the signal to be processed. The problem can also be solved in trying to find the set of wavelets which best approximates the Karhunen-Loève basis of the signal. A possible approach is to use a "dictionary" of wavelets and to look for the one which minimizes an entropy criterion [3]. However, the computational cost of this search is high, when the dictionary is large. Moreover, this method can only be successful if the dictionary contains a "good" wavelet. Another way to approach the problem, which is adopted here, is to design the wavelet through an optimization procedure. Because of the strong connections existing between wavelet transforms and octave-band filter banks [4] [5], this is equivalent to synthesizing optimal QMF filters. Such an approach has been used in [6]. The impulse responses of the filters characterizing the wavelet decomposition are then found by minimizing an upper bound of the distance between a deterministic signal and its approximation up to a given scale.

In this paper, we consider the multiresolution analyses of 2^K-periodic sequences [7]. More generally, the corresponding transform may be applied to finite sequences of 2^K samples to avoid the boundary effects occurring with the usual discrete wavelets [8]. We focus attention on the frequency domain implementation of this transform which leads to fast algorithms. At first, the principles of the discrete periodic wavelet transform are recalled and a matrix formulation of the transform is given. Then, a multiscale method is proposed to find the optimal transform.

2 WAVELET TRANSFORM FOR PERIODIC SEQUENCES

2.1 Reminders and notations

A multiresolution analysis of the space $\ell^2(\mathbb{Z}/2^K\mathbb{Z})$ of 2^K-periodic sequences is defined by vector subspaces $(V_k)_{1\le k\le K}$ and $(O_k)_{1\le k\le K}$ such that $\dim V_k = 2^{K-k}$, $V_k \oplus O_k = V_{k-1}$ (with $V_0 = \ell^2(\mathbb{Z}/2^K\mathbb{Z})$) and O_k is orthogonal to V_k. The spaces V_k and O_k are spanned by orthonormal bases, denoted respectively by $[\phi_{k,m}(n)]_{0\le m<2^{K-k}}$ and $[\psi_{k,m}(n)]_{0\le m<2^{K-k}}$ which satisfy the following translation property:

$$\phi_{k,m}(n) = \phi_{k0}(n-2^k m), \quad \psi_{k,m}(n) = \psi_{k0}(n-2^k m), \quad (1)$$

for $m \in \{0,\dots,2^{K-k}-1\}$. This definition implies that there exist sequences $[h_k(m)]_{0\le m<2^{K-k}}$ and $[g_k(m)]_{0\le m<2^{K-k}}$ such that

$$\phi_{k+1,0}(n) = \sum_{m=0}^{2^{K-k}-1} h_k(m)\,\phi_{k,0}(n-2^k m), \quad (2)$$

$$\psi_{k+1,0}(n) = \sum_{m=0}^{2^{K-k}-1} g_k(m)\,\phi_{k,0}(n-2^k m). \quad (3)$$

Furthermore, the wavelets $\psi_{k,m}(n)$ with $k \in \{1,\dots,K\}$ and $m \in \{0,\dots,2^{K-k}-1\}$ and the approximation sequence $\phi_{K0}(n)$ at the lowest resolution level form an orthonormal basis of $\ell^2(\mathbb{Z}/2^K\mathbb{Z})$.

Note that we do not impose any relations between the sequences $h_k(m)$ or $g_k(m)$ at different scales. Indeed, the above definition does not introduce any kind of "dilatation" property as usually required in the construction of wavelets. It must be emphasized that this constraint would restrict the set of possible bases and, at the same time, make their optimal choice more difficult.

The approximation coefficients $c_k(m)$ and the wavelet coefficients $d_k(m)$ of a sequence $x(n)$ of $\ell^2(\mathbb{Z}/2^K\mathbb{Z})$ are

$$c_k(m) = \sum_{n=0}^{2^K-1} x(n)\,\phi_{k,m}(n)^*, \quad (4)$$

$$d_k(m) = \sum_{n=0}^{2^K-1} x(n)\,\psi_{k,m}(n)^*. \quad (5)$$

A classical result is that $c_{k+1}(n)$ and $d_{k+1}(n)$ may be computed from $c_k(n)$ (with $c_0(n) = x(n)$) by the following recursions:

$$c_{k+1}(n) = \sum_{m=0}^{2^K-1} h_k(m)^*\, c_k(2n+m), \quad (6)$$

$$d_{k+1}(n) = \sum_{m=0}^{2^K-1} g_k(m)^*\, c_k(2n+m). \quad (7)$$

Subsequently, the normalized DFTs of $[c_k(n)]_{0\le n<2^{K-k}}$ and $[d_k(n)]_{0\le n<2^{K-k}}$ will be denoted respectively by $[C_k(p)]_{0\le p<2^{K-k}}$ and $[D_k(p)]_{0\le p<2^{K-k}}$ and the DFTs of $[\sqrt{2}h_k(m)]_{0\le m<2^{K-k}}$ and $[\sqrt{2}g_k(m)]_{0\le m<2^{K-k}}$ will be denoted respectively by $[H_k(p)]_{0\le p<2^{K-k}}$ and $[G_k(p)]_{0\le p<2^{K-k}}$.

2.2 Matrix formulation

For $k \in \{1,\dots,K\}$, let us define the vectors

$$\mathbf{c}_k = [c_k(0),\dots,c_k(2^{K-k}-1)]^t, \quad (8)$$
$$\mathbf{d}_k = [d_k(0),\dots,d_k(2^{K-k}-1)]^t. \quad (9)$$

A matrix form of equations (6) and (7) may be given:

$$\mathbf{C}_k = \mathbf{T}_k\mathbf{c}_k, \quad (10)$$

$$\begin{bmatrix} \mathbf{C}_{k+1} \\ \mathbf{D}_{k+1} \end{bmatrix} = \begin{bmatrix} \mathbf{H}_k^\dagger \\ \mathbf{G}_k^\dagger \end{bmatrix} \mathbf{C}_k, \quad (11)$$

$$\mathbf{c}_{k+1} = \mathbf{T}_{k+1}^\dagger \mathbf{C}_{k+1}, \quad \mathbf{d}_{k+1} = \mathbf{T}_{k+1}^\dagger \mathbf{D}_{k+1}, \quad (12)$$

where † represents the Hermitian transpose operation, $\mathbf{T}_k$ denotes the square matrix corresponding to the normalized DFT of length 2^{K-k} and $\mathbf{H}_k$ is the following diagonal block matrix,

$$\mathbf{H}_k = \begin{bmatrix} \mathbf{DH}_k(0, 2^{K-k-1}-1) \\ \mathbf{DH}_k(2^{K-k-1}, 2^{K-k}-1) \end{bmatrix}, \quad (13)$$

$$\mathbf{DH}_k(n,m) = \mathrm{Diag}(H_k(n),\dots,H_k(m)), \quad n \le m. \quad (14)$$

The matrix $\mathbf{G}_k$ is expressed from $[G_k(p)]_{0\le p<2^{K-k}}$, in a similar way. Let us note that the transform between $\mathbf{c}_k^t$ and $[\mathbf{c}_{k+1}^t \mathbf{d}_{k+1}^t]$ is unitary iff the matrices

$$\mathbf{M}_k(p) = \begin{bmatrix} H_k(p) & G_k(p) \\ H_k(2^{K-k-1}+p) & G_k(2^{K-k-1}+p) \end{bmatrix} \quad (15)$$

are unitary, for all $p \in \{0,\dots,2^{K-k-1}-1\}$. This result is known as the orthonormality or perfect reconstruction condition.

In the expression of the whole wavelet decomposition, a further simplification occurs as all the DFTs, except the first, and half the inverse DFTs may be omitted:

$$\begin{bmatrix} \mathbf{c}_K \\ \mathbf{d}_K \\ \vdots \\ \mathbf{d}_1 \end{bmatrix} = \mathbf{P} \begin{bmatrix} x(0) \\ x(1) \\ \vdots \\ x(2^K-1) \end{bmatrix}, \quad (16)$$

$$\mathbf{P} = \begin{bmatrix} \mathbf{H}_{K-1}^\dagger \cdots \mathbf{H}_0^\dagger \\ \mathbf{G}_{K-1}^\dagger \mathbf{H}_{K-2}^\dagger \cdots \mathbf{H}_0^\dagger \\ \mathbf{T}_{K-2}^\dagger \mathbf{G}_{K-2}^\dagger \mathbf{H}_{K-3}^\dagger \cdots \mathbf{H}_0^\dagger \\ \vdots \\ \mathbf{T}_2^\dagger \mathbf{G}_1^\dagger \mathbf{H}_0^\dagger \\ \mathbf{T}_1^\dagger \mathbf{G}_0^\dagger \end{bmatrix} \mathbf{T}_0. \quad (17)$$

2.3 Zeros of the basis sequences

Equations (2) and (3) show that the DFTs $\Phi_{k0}(p)$ of $\phi_{k0}(n)$ and $\Psi_{k0}(p)$ of $\psi_{k0}(n)$ are such that, for $p \in \{0,\dots,2^K-1\}$,

$$\Phi_{k0}(p) = 2^{\frac{k}{2}} \prod_{l=0}^{k-1} H_l(p), \quad (18)$$

$$\Psi_{k0}(p) = 2^{\frac{k}{2}} G_{k-1}(p) \prod_{l=0}^{k-2} H_l(p), \quad k > 1, \quad (19)$$

$$\Psi_{10}(p) = 2^{\frac{1}{2}} G_0(p), \quad (20)$$

where the sequences $H_k(p)$ and $G_k(p)$ now designate periodic extensions with period 2^{K-k}. Generally, $H_k(p)$ is the frequency response of a low-pass filter such that

$$H_k(2^{K-k-1}) = 0, \quad G_k(0) = 0. \quad (21)$$

So, according to (18), (19) and (20), $\Phi_{k0}(p)$ has at least $2^k - 1$ zeros on the interval $\{0,\dots,2^K-1\}$, for

$$p = 2^{K-k}q, \quad q \in \{1,\dots,2^k-1\}, \quad (22)$$

and $\Psi_{k0}(p)$ has at least 2^{k-1} zeros, for

$$p = 2^{K-k+1}q, \quad q \in \{0,\dots,2^{k-1}-1\}. \quad (23)$$

This implies, in particular, that $\phi_{K,0}(n) = 2^{-\frac{K}{2}}$ so that $c_k(0)/2^{\frac{K}{2}}$ is the mean value of the signal $x(n)$.

3 OPTIMIZATION OF THE WAVELET TRANSFORM

We will now try to choose the coefficients $H_k(p)$ and $G_k(p)$ in order to get the "best" orthonormal wavelet basis. To solve this problem, we will try to optimally concentrate the energy of the components of the transformed vector. The optimization procedure takes advantage of the structure of multiresolution analysis.

3.1 Parametrization of orthonormal wavelets

For orthonormal wavelets, the matrix $M_k(p)$ defined by (15) is unitary. So, it is straightforward to show that we have the following parametrization, for $k \in \{0,\ldots,K-1\}$ and $p \in \{0,\ldots,2^{K-k-1}-1\}$,

$$H_k(p) = \bar{H}'_k(p)\exp[j\alpha'_k(p)], \quad (24)$$
$$H_k(2^{K-k-1}+p) = \bar{H}'_k(2^{K-k-1}+p)\exp[j\alpha'_k(p)], \quad (25)$$
$$G_k(p) = \bar{G}'_k(p)\exp[j\beta'_k(p)], \quad (26)$$
$$G_k(2^{K-k-1}+p) = \bar{G}'_k(2^{K-k-1}+p)\exp[j\beta'_k(p)], \quad (27)$$

where

$$\bar{H}'_k(p) = \cos[\varphi_k(p)]\exp[j\Delta'_k(p)], \quad (28)$$
$$\bar{H}'_k(2^{K-k-1}+p) = \sin[\varphi_k(p)]\exp[-j\Delta'_k(p)], \quad (29)$$
$$\bar{G}'_k(p) = -\sin[\varphi_k(p)]\exp[j\Delta'_k(p)], \quad (30)$$
$$\bar{G}'_k(2^{K-k-1}+p) = \cos[\varphi_k(p)]\exp[-j\Delta'_k(p)], \quad (31)$$

and $(\alpha'_k(p), \beta'_k(p), \varphi_k(p), \Delta'_k(p)) \in \mathbb{R}^4$.

Then, by using equations (18) and (19), it can be proved that the wavelets may be parametrized as follows:

$$\Phi_{K0}(p) = 2^{\frac{K}{2}}\exp[j\alpha_{K-1}(0)]\prod_{l=0}^{K-1}\bar{H}_l(p), \quad (32)$$
$$\Psi_{k0}(p) = 2^{\frac{k}{2}}\exp[j\beta_{k-1}(p)]\bar{G}_{k-1}(p)\prod_{l=0}^{k-2}\bar{H}_l(p), \quad (33)$$

where $\alpha_{K-1}(0) \in \mathbb{R}$, $\beta_k(p) \in \ell^2(\mathbb{Z}/2^{K-k-1}\mathbb{Z})$ and $(\bar{H}_k(p), \bar{G}_k(p))(\in [\ell^2(\mathbb{Z}/2^{K-k}\mathbb{Z})]^2)$ is linked to $\varphi_k(p)$ and some variable $\Delta_k(p)$ by relations similar to (28)-(31). Therefore, the wavelet transform may be computed by replacing $[H_k(p)]_{k<K-1}$ by $[\bar{H}_k(p)]_{k<K-1}$, $H_{K-1}(p)$ by $\exp[j\alpha_{K-1}(0)]\bar{H}_{K-1}(p)$ and $[G_k(p)]_{k<K}$ by $[\exp[j\beta_k(p)]\bar{G}_k(p)]_{k<K}$. Moreover, a unique set of parameters may be found by taking

$$0 \le \varphi_k(p) \le \tfrac{\pi}{2}, \quad (34)$$
$$0 \le \Delta_k(p) < \pi \quad \text{if } 0 < \varphi_k(p) < \tfrac{\pi}{2}, \quad (35)$$
$$\Delta_k(p) = 0 \quad \text{if } \varphi_k(p) \in \{0, \tfrac{\pi}{2}\}, \quad (36)$$
$$0 \le \alpha_{K-1}(0) < 2\pi, \quad 0 \le \beta_k(p) < 2\pi. \quad (37)$$

3.2 One-scale optimization

Let us assume that $c_k(p)$ and $d_k(p)$ are known and let us optimize the choice of $(\beta_k(p), \varphi_k(p), \Delta_k(p))$. To do so, it may seem natural to maximize $\eta_k = \max\{\| \mathbf{c}_{k+1} \|, \| \mathbf{d}_{k+1} \|\}$ in order to obtain a concentration of the energy. As a result of relations (12), η_k may also be written as $\max\{\| \mathbf{C}_{k+1} \|, \| \mathbf{D}_{k+1} \|\}$.

To find $\varphi_k(p)$ and $\Delta_k(p)$, it is convenient to classify sequences $C_k(p)$ into two categories: a class $\mathcal{E}_k^+$ such that $\| \mathbf{C}_{k+1} \| > \| \mathbf{D}_{k+1} \|$ and a class $\mathcal{E}_k^-$ such that $\| \mathbf{C}_{k+1} \| < \| \mathbf{D}_{k+1} \|$. It is then straightforward to show that, if $\mathbf{C}_k$ belongs to $\mathcal{E}_k^+$,

$$\cos[\varphi_k(p)] = \frac{| C_k(p) |}{\sqrt{| C_k(p) |^2 + | C_k(2^{K-k-1}+p) |^2}}, \quad (38)$$
$$\Delta_k(p) = \frac{1}{2}\arg[C_k(p)C_k(2^{K-k-1}+p)^*][\pi], \quad (39)$$

whereas, if $\mathbf{C}_k$ belongs to $\mathcal{E}_k^-$,

$$\cos[\varphi_k(p)] = \frac{| C_k(2^{K-k-1}+p) |}{\sqrt{| C_k(p) |^2 + | C_k(2^{K-k-1}+p) |^2}}, \quad (40)$$
$$\Delta_k(p) = \frac{1}{2}\arg[C_k(p)C_k(2^{K-k-1}+p)^*] + \frac{\pi}{2}[\pi]. \quad (41)$$

To eliminate this ambiguity, it may be assumed that

$$\sum_{p=0}^{2^{K-k-2}-1}\cos[\varphi_k(p)]^2 + \sum_{p=2^{K-k-2}}^{2^{K-k-1}-1}\sin[\varphi_k(p)]^2 > \sigma_k, \quad (42)$$

with $\sigma_k = 2^{K-k-2}$. This assumption allows us to specify $H_k(p)$ as the frequency response of a low-pass filter. Then, the classification is simply achieved by checking the above condition. Note that formula (38) and (40) are not valid when $C_k(p) = C_k(2^{K-k-1}+p) = 0$. The corresponding values of p must be discarded (and σ_k must be decreased). According to section 2.3, this case may occur when the input signal is a wavelet. In a similar way, formula (39) and (41) cannot be used when $C_k(p) = 0$ or $C_k(2^{K-k-1}+p) = 0$. However, if $C_k(p) = 0$ and $C_k(2^{K-k-1}+p) \neq 0$ or if $C_k(p) \neq 0$ and $C_k(2^{K-k-1}+p) = 0$, (36) yields $\Delta_k(p) = 0$.

By defining

$$\bar{C}_{k+1}(p) = \bar{H}_k(p)^*C_k(p) + \bar{H}_k(2^{K-k-1}+p)^*C_k(2^{K-k-1}+p), \quad (43)$$
$$\bar{D}_{k+1}(p) = \bar{G}_k(p)^*C_k(p) + \bar{G}_k(2^{K-k-1}+p)^*C_k(2^{K-k-1}+p), \quad (44)$$

we have

$$D_{k+1}(p) = \exp[-j\beta_k(p)]\bar{D}_{k+1}(p). \quad (45)$$

An optimal decomposition is obtained by making $[d_{k+1}(n)]_n$ as close as possible to a Kronecker sequence, which is equivalent to minimize $| D_{k+1}(p) - \exp[-j(2\pi\frac{m_k p}{2^{K-k-1}} + \theta_k)] |$, where $\theta_k \in [0, 2\pi[$ and $m_k \in \{0,\ldots,2^{K-k-1}-1\}$. In this approach, the choice of m_k and θ_k is free and, by taking, $m_k = \theta_k = 0$,

$$\beta_k(p) = \arg[\bar{D}_{k+1}(p)]. \quad (46)$$

This formula can only be applied if $\bar{D}_{k+1}(p) \neq 0$, which is not satisfied when $\mathbf{C}_k$ belongs to $\mathcal{E}_k^+$. To determine $\alpha_K(0)$, the same approach may be used for the low-pass output, at the lowest resolution level.

3.3 Statistical estimations

In the previous method, the values of the parameters are undetermined in several cases. To solve this problem, several realizations of $\mathbf{C}_k$ must be considered. Moreover, this situation is even more interesting from a practical point of view.

To estimate $\varphi_k(p)$ and $\Delta_k(p)$, we propose to maximize the criterion

$$\eta'_k(p) = E_+\{| \bar{C}_{k+1}(p) |^2\}P_+ + E_-\{| \bar{D}_{k+1}(p) |^2\}P_-, \tag{47}$$

where $\bar{C}_{k+1}(p)$ and $\bar{D}_{k+1}(p)$ are given by relations (43) and (44), $E_+\{.\}$ (resp. $E_-\{.\}$) denotes the averaging operator when $\mathbf{C}_k \in \mathcal{E}_k^+$ (resp. $\mathcal{E}_k^-$) and P_+ (resp. P_-) is the probability measure of $\mathcal{E}_k^+$ (resp. $\mathcal{E}_k^-$). After some tractations, the expressions of the estimates may be found:

$$\varphi_k(p) = \frac{1}{2}\arctan\left[\frac{2 | A_k(p) |}{B_k(p)}\right][\frac{\pi}{2}], \tag{48}$$

$$\Delta_k(p) = \frac{1}{2}\arg[A_k(p)][\pi], \tag{49}$$

with

$$A_k(p) = E_+\{C_k(p)C_k(2^{K-k-1}+p)^*\}P_+ \\ -E_-\{C_k(p)C_k(2^{K-k-1}+p)^*\}P_-, \tag{50}$$

$$B_k(p) = E_+\{| C_k(p) |^2 - | C_k(2^{K-k-1}+p) |^2\}P_+ \\ -E_-\{| C_k(p) |^2 - | C_k(2^{K-k-1}+p) |^2\}P_-. \tag{51}$$

Estimating $\beta_k(p)$ is a more difficult problem. Indeed, the different realizations of $\mathbf{C}_k$ may lead to sequences $[d_{k+1}(n)]_n$ localized at different times. So, a null value can no longer be taken for the variables m_k and θ_k which have been defined in the previous section. An iterative method may be used which requires i) an *a priori* estimation of $\beta_k(p)$, ii) the determination of m_k and θ_k, for each realization of $[C_k(p)]_p$, and iii) an *a posteriori* statistical estimation of $\beta_k(p)$.

3.4 Multiscale optimization

To obtain the entire set $\varphi_k(p)$, $\Delta_k(p)$ and $\beta_k(p)$ when $k \in \{0,\ldots,K-1\}$ and $p \in \{0,\ldots,2^{K-k-1}-1\}$, the procedure is repeated recursively for each resolution level by starting from the highest resolution ($k=0$).

Since this method has been introduced in a rather intuitive way, one may wonder about its reliability. In fact, we have the following result which shows the validity of the algorithm: *the parameters $\varphi_k(p)$, $\Delta_k(p)$, $\beta_k(p)$ and $\alpha_{K-1}(0)$ characterizing a wavelet transform are correctly identified by the method if the sequences $\psi_{10}(n),\ldots,\psi_{K0}(n)$ and $\phi_{K0}(n)$ are taken as realizations of the input signal.* This result is still valid if the sequences $[\psi_{km}(n)]_{0\le k\le K, 0\le m<2^k}$ and $\phi_{K0}(n)$ are chosen as input sequences. However, $\beta_k(p)$ is identified up to a linear phase, which corresponds to reindexing $[\psi_{km}(n)]_m$.

4 CONCLUSION

In this paper, some methods have been proposed to optimize the choice of the coefficients characterizing the periodic discrete wavelet transforms. Our approach relies upon a frequency domain parametrization and uses some estimates of the parameters.

Several points need further studies. At first, it could be interesting to introduce further constraints in the optimization procedure to impose some regularity conditions. Secondly, more elaborate statistical estimations might be performed.

AKNOWLEDGEMENTS

The author would like to thank Dr. H. Krim for many interesting discussions.

References

[1] Y. Meyer, *Ondelettes, Ondelettes et opérateurs.* Tome 1, Hermann, Paris, 1990.

[2] O. Rioul and M. Vetterli, "Wavelets and signal processing", *IEEE SP Magazine*, Vol. 8, pp. 14-38, Oct. 1991.

[3] M.V. Wickerhauser, "INRIA lectures on wavelet packet algorithms", *Ondelettes et paquets d'ondelettes*, Roquencourt, Jun. 17-21, 1991, pp. 31-99.

[4] S.G. Mallat, "A theory for multiresolution signal decomposition: the wavelet representation", *IEEE Trans. PAMI*, Vol. 11, pp. 674-691, Jul. 1989.

[5] I. Daubechies, "Orthonormal bases of compactly supported wavelets", *Com. on Pure and Applied Math.*, Vol. XLI, pp. 909-996, 1988.

[6] A.H. Tewfik and P.E. Jorgensen, "On the choice of a wavelet for signal coding and processing", *Proc. ICASSP*, Toronto, May 14-17, pp. 2025-2028, 1991.

[7] P. Bonnet et D. Remond, "Une transformée rapide en ondelette", *Trait. Sig.*, Vol. 8, pp. 195-207, 1991.

[8] J.C. Pesquet "Orthonormal wavelets for finite sequences", *Proc. ICASSP*, San Francisco, Mar. 23-26, 1992.

SIGNAL PROCESSING VI: Theories and Applications
J. Vandewalle, R. Boite, M. Moonen, A. Oosterlinck (eds.)

NOISE CANCELLATION WITH UNCORRELATED REFERENCE – A NON LINEAR FILTER APPROACH

Min CAI* John BALL**

* Electricité De France, DER/SDM, 6, quai Watier 78401 Chatou, GdR, CNRS France

** Georgia Tech Lorraine, 2 et 3 rue Marconi, 57070 Metz, France

This paper investigates the situations of noise cancellation in which the useful and the reference signals are not necessarily correlated. We propose a method based upon a nonlinear structure which requires only that the signals be dependent, not necessarily correlated. We give the proof and the simulation results which confirme the previous proposition.

1. INTRODUCTION

The Wiener filter and the linear adaptive filter have been largely used in noise cancellation where it is assumed that a reference signal is correlated with the useful signal but uncorrelated with this noise. In other words, the reference and the useful signals must have overlapping spectral components. In many applications, this condition is not satisfied; one example is shock characterization in nuclear power process monitoring. The useful signal is a shock wave received by an accelerometer far from an incidence of collision and the reference is the same wave received by a nearer sensor. These sensor signals propagate through a metallic structure and are therefore filtered by a resonance response. They generally have separated spectral bands, thus classical methods do not work.

We propose a new method of noise cancellation which is based on a nonlinear structure. This method requires only that the signals are dependent, and thus operates well even when the reference and the useful signal are uncorrelated.

After giving the definition of the method, we show how the proposed method resolves the non-overlapping spectra problem. We detail the explanation in the simple sinusoid case. The result will be extended to the general case. Some simulation results will also be presented.

2. PRINCIPLE OF THE METHOD

Noise Cancellation

We use a sensor signal u which contains a useful signal s and an additive noise w:

$$u = s + w.$$

We have also a reference signal x of the useful signal . It is assumed that x is independent of w, but dependent on s. The noise cancellation problem intends to estimate s by filtering x.

When x is correlated to s, one solution is to use Wiener linear filter h which can be obtained by minimizing the mean separated error (difference between u and the filter output $y = h * x$):

$$\mathrm{Min}_h \; E|h * x|^2.$$

In this paper, we consider the case where x is not necessarily correlated to s (classical second order sense). In other words, components of their spectra do not necessarily overlap:

$$|X(f)\; S(f)|^2 = 0.$$

Non-Linear Filtering

We propose to use the following non-linear filtering defined by:

$$y = h_N * x^N + h_{N-1} * x^{N-1},$$

where x^k is the kth power of x and h_k a coefficient sequence. h_k is called kth order impulse response. Its memory length is denoted by M_k.

The previous non-linear filter is just a reduced form of the Volterra filter: all cross terms $x_{n-i}x_{n-j}$ and all terms of order lower than N-1 are omitted. Such a filter can be decomposed into two parts:

1) a fixed non-linear transformation from x to x^N, x^{N-1};
2) a coefficient dependent linear filtering.

Although the filter is non-linear, the relation between the output and the filter coefficients is linear. As in the linear filter case, s can be estimated by minimizing the mean square error:

$$\mathrm{Min}_{h_N, h_{N-1}} \; E|u - h_N * x^N - h_{N-1} * x^{N-1}|^2.$$

Algorithms such as LMS and RLS can therefore be used to calculate adaptively the optimal filter.

For the Nonoverlapping Spectra

If the non-linear part is missing, the band of a linear filter output B_y stays in the band of the input B_x. When the band of the useful signal B_s is separated from B_x: $B_x \cap B_s = \phi$, it is also separated from B_y. Thus, no linear filter can generate an output spectrum to estimate the useful spectrum.

Why can this problem be solved by using non-linearity? The basic idea is to use the harmonics of B_x to approach B_s. It is easy to see in the previous decomposition of the filter that

1) By the nonlinear transformation, we generate in x^N and x^{N-1} the harmonics of the frequencies of B_x. When N is suitably chosen, the harmonic bands will contain B_s;
2) A linear filtering on x^N and x^{N-1} only weights the harmonic spectrum. The error minimization generates a filter whose output spectrum approaches the useful one in B_s, and cancels the harmonics outside B_s. The mathematical formulation of this investigation will be given in the following sections.

3. SINUSOID CASE

Consider at first the following simple case where x and s are both sinusoids:

$$x_n = \alpha \sin(2\pi f_1 n + \phi)$$
$$s_n = \alpha' \sin(2\pi f_2 n + \phi').$$

We shall prove the existency of a reduced Volterra filter whose output y is exactly equal to s.

Harmonic Generation

When the input of the filter (the reference x) is a sinusoid of the frequency f_1, what is the output like? By using a simple calculation (see Appendix 1), one obtains the Fourier transform of x^k written as:

$$FT(x^k)=\sum\nolimits^{k}_{m=0}(\alpha/2j)^k C_k^m e^{j(2m-k)\phi}\delta_{f-(k-2m)f_1}.$$

If N is even, x^N contains all even harmonics $2mf_1$ from $-Nf_1$ to Nf_1, so as with x^{N-1} the odd harmonics $(2m+1)f_1$. The situation is analogous if N is odd. In any case, all harmonics of f_1 from $-Nf_1$ to Nf_1 are generated by x^N and x^{N-1}. That is why we use only the two highest order terms.

With the same parities as x^N and x^{N-1}, the linear filters h_N and h_{N-1} control the gain of the even or odd harmonics. The Fourier transform of the output $Y(f)$ depends on the filter transfer function $H_N(f)$, $H_N(f)$ by:

$$Y(f)=\sum\nolimits^{N}_{n=-N}Al(nf_1)e^{-jn\phi}\delta_{f-nf_1}$$

$$A_n(nf_1)=\begin{cases}(\alpha/2j)^N C^m_N H_N(nf_1) & \text{if } n=N-2m\\ (\alpha/2j)^{N-1}C^m_{N-1}H_{N-1}(nf_1) & \text{if } n=N-1-2m\end{cases}$$

Frequency equivalence

We know that the spectrum of a digital signal is periodic of period 1: $X(f)=X(f+1)$. For spectrum behavior; two frequencies f, f' are equivalent, if they are equivalent modulo 1: $f\equiv f'$ [1] (or $f-f'\in\mathbf{N}$). In the following, we define all spectra in the interval [0,1[. By aliasing, the harmonics nf_1 fold back into this interval, if it exceeds 1. Their equivalent frequencies in [0,1[are called basic frequencies.

Basic Frequencies of the Harmonics

We define the mapping function g from **N** to [0,1[which corresponds the harmonic number to the basic frequency:

$$g(n) = nf_1 - [nf_1],$$

where $[x]$ is the integer part of x.

We can see that $Image(g)$ has a finite number of elements, iff f_1 is a rational number.

Now, let's consider the case of the rational frequency. Denote $f_1=p/q$ its irreducible form. It is easy to verify that

1) g is periodic of period q: $g(n+q)=g(n)$;
2) $g(n)$ is a rational number with the same denominator: $g(n)=r/q$, $r\in\mathbf{N}$.

So we can reduce the definition domain into:

$\{1,2,\ldots q\} \longrightarrow \{0,1/q,\ldots(q-1)/q\}$.

It follows from the Appendix 2 that g defined such, is a bijection. So there are q basic frequencies of the harmonics and they are uniformly distributed from 0 to 1. $Image(g)=\{0,1/q,\ldots(q-1)/q\}$.

Estimation of the useful signal

Assume that the useful signal s is also at a rational frequency: $f_2=p'/q'$ and q' is a divisor of q: $q'/q=s \in \mathbf{N}$.

In this case, f_2 is in $Imag(g)$. There exists an integer N for which $g(n)=f_2$. If we use a N order filter such that

$$\begin{cases} h_{N-1} = 0 \\ h_N(nf_1) = 0, \quad \text{for } n \in \{1,2,,\ldots N-1\} \\ h_N(Nf_1) = (2j/\alpha)^N \alpha' e^{j(N\phi-\phi')}, \end{cases}$$

all harmonics will be cancelled, except Nf_1 ($\equiv f_2$). The output will be:

$$Y(f)=\alpha'(e^{j\phi'}\delta_{f-f_2} + e^{-j\phi'}\delta_{f+f_2}),$$

and equal to the useful signal: $y_n = s_n$.

4. NON-ZERO BANDWIDTH CASE

When the signal is not a pure sinusoid but a non-zero bandwidth signal, it can be written as a sinusoid modulated by a lower frequency base band signal:

$$x_n = a_n \sin(2\pi f_1 n)$$
$$s_n = b_n \sin(2\pi f_2 n)$$

with a_n, b_n the base band signals. The kth power of x can be seen as a_n^k modulated by the kth power of the sinusoid:

$$x^n_k = a_n^k \sin^k(2\pi f_1 n).$$

It is easy to prove that a_n^k is also a base band signal, but the bandwidth is larger than that of a_n. Since the power of the sinusoid generates the harmonics at kf_1, x^k is composed of harmonic bands which are centered at kf_1. The useful signal band B_s, centered at f_2 will be included in one of the harmonic bands. By using a suitable filtering on these "power" signals, we can preserve the band B_s and attenuate all others. We then obtain a filtered signal y_n quite near to the useful signal s_n.

5. SIMULATION RESULTS

Simulation results are shown in figures 1,2,3, and 4. The case of pure sinusoids is shown in figure 1, where an error convergence curve is given for an input signal at frequency $f_1 = 19/32$, and a useful signal at $f_2 = 25/32$. The filter of third order was calculated using LMS and was of memory $M_k = 5$. The filter converged after 1000 samples with an error residue on the order of 30 dB.

The results for the general case are shown in figures 2,3 and 4. The signals involved are random, generated by a low pass filtered gaussian noise which is then sinusoid modulated to frequencies $f_1 = 29/32$ and $f_2 = 26/32$. The second order filter was again calculated by LMS with a memory length $M_k = 10$. The error residue was on the order of 20 dB after 1000 samples. Figures 2 and 3 are the spectra of the reference and useful signals respectively showing clearly that they do not overlap. Figure 4 shows the spectrum of the filter output and one sees clearly that the spectrum of the filter output is a good approximation of the useful signal spectrum. It is seen in both cases that the reduced order Volterra filters produce useful results, justifing the theory developed earlier.

CONCLUSION

We have investigated the situations of noise cancellation in which the useful and referece signals are not necessarily correlated (or having nonoverlapping spectral components). A nonlinear method has been proposed and requires only that the signals be dependent, not necessarily correlated. The mathematical investigation and the simulation results have confirmed the previous proposition.

REFERENCES

[1] P. Chevalier, P. Duvaut and B. Picinbono "Le filtrage de Volterra transverse réel et complexe en traitement du signal" Traitement du signal, vol 7, No 5.

[2] B. Picinbono and P. Duvaut "Optimal Linear-Quadratic Systems for Detection and Estimation" IEEE Trans. on Inform. Theory, March 1988.

[3] M.J. Coker and D.N. Simkins "A nonlinear adaptive noise canceller" Proc. ICASSP 1980

APPENDIX 1

We try to calculate the output of the reduced Volterra filter defined as:

$$y = h_N * x^N + h_{N-1} * x^{N-1},$$

when the input signal is a sinusoid at the frequency f_1: $x_n = \alpha \sin(2\pi f_1 n + \phi)$.

$$x_n = \alpha/2j \left(e^{j(2\pi f_1 n+\phi)} - e^{-j(2\pi f_1 n+\phi)}\right)$$

$$x_n^k = (\alpha/2j)^k \sum_{m=0}^{k} C_k^m e^{j(2\pi f_1 n+\phi)m} e^{-j(2\pi f_1 n+\phi)(k-m)}$$
$$= \sum_{m=0}^{k} [(\alpha/2j)^k C_k^m e^{j(2m-k)\phi}]\, e^{-j2\pi(k-2m)f_1 n}$$

$$FT(x^k) = {}^{k} \quad (\alpha/2j)^k C\, {}^m e^{j(2m-k)\phi} \delta_{f-(k-2m)f} \cdot$$

Denote $Y(f)$, $H_N(f)$ and $H_{N-1}(f)$ the Fourier transform. So,

$$Y(f) = H_N(f)\sum_{m=0}^{N}(\alpha/2j)^N C_N^m e^{j(2m-N)\phi}\,\delta_{f-(N-2m)f_1} + H_{N-1}(f)\sum_{m=0}^{N-1}(\alpha/2j)^{N-1}C_{N-1}^m e^{j(2m-N+1)\phi}\,\delta_{f-(N-1-2m)f_1}$$

$$= \sum_{n=-N}^{N} A_n(nf_1)\, e^{-jn\phi}\,\delta_{f-nf_1},$$

with

$$A_n(nf_1) = \begin{cases} (\alpha/2j)^N C^m_N H_N(nf_1) & \text{if } n=N-2m \\ (\alpha/2j)^{N-1} C^m_{N-1} H_{N-1}(nf_1) & \text{if } n=N-1-2m \end{cases}$$

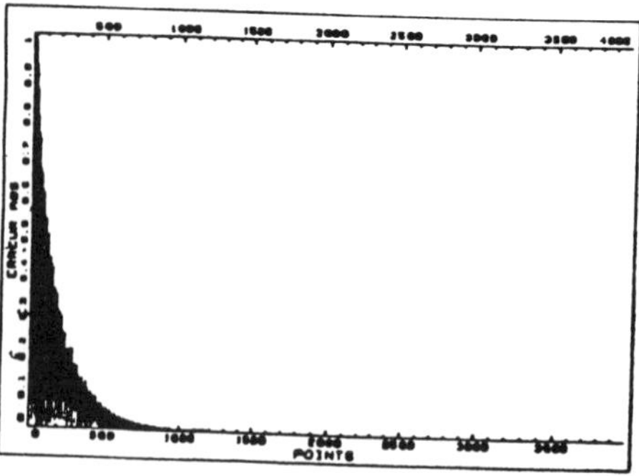

figure 1

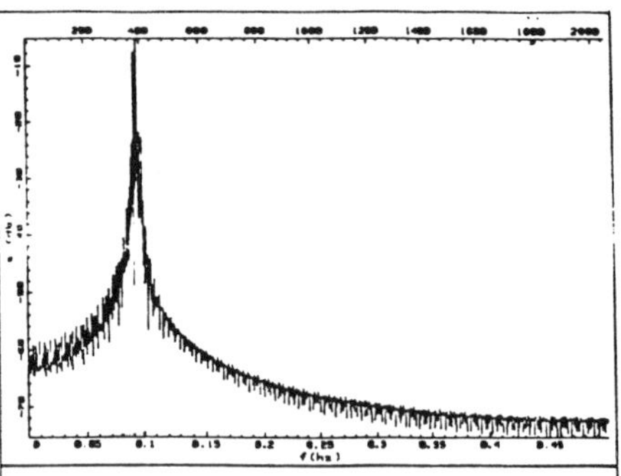

figure 2

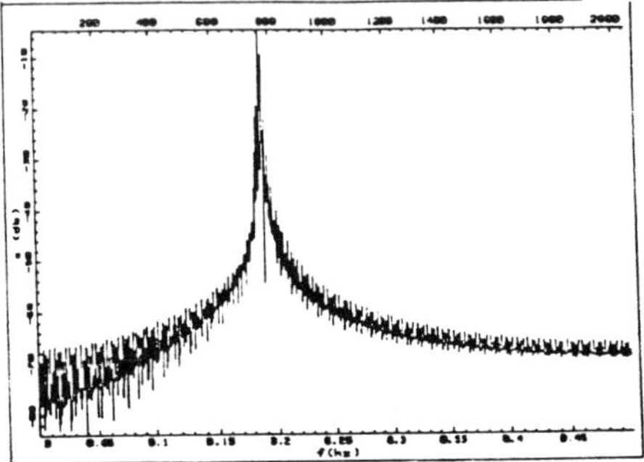

figure 3

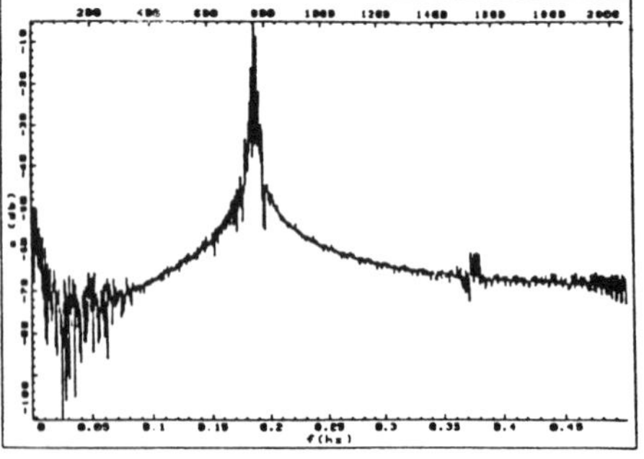

figure 4

APPENDIX 2

We want to prove the following proposition:

If f_1 is a rational number and p/q is its irreducible form, the function

$$g(n) = nf_1 - [nf_1],$$

with $[x]$ the integer part of x, is a bijection from $\{1,2,\ldots q\}$ to $\{0,1/q,\ldots(q-1)/q\}$.

a) $Image(g) \in \{0,1/q,\ldots(q-1)/q\}$:

$$g(n) = nf_1 - [nf_1] = np/q - [np/q] = (np - [np/q]q)/q;$$

and $[np/q] \le np/q < [np/q]+1$

$\Longrightarrow 0 \le np-[np/q]q < q$

b) $g(n)=g(n') \Longrightarrow n=n'$:

If $n>n'$, $g(n) = g(n')$

$\Longrightarrow nf_1-[nf_1] = n'f_1-[n'f_1]$

$\Longrightarrow f_1 = ([nf_1]-[n'f_1]) / (n-n')$.

Since p/q is the irreducible form of f_1, $(n-n')$ has to be divisible by q. But,

$$n\le q,\ n'\ge 1 \Longrightarrow n-n'\le q-1,$$

$(n-n')$ can not be divisible by q. That is the contradiction.

It follows from b) that Imag(g) has the same number of elements as the definition domain: $Card(Image(g))=q$. This implies with a) that: $Image(g) = \{0,1/q,\ldots(q-1)/q\}$, and g is a bijection.

SIGNAL PROCESSING VI: Theories and Applications
J. Vandewalle, R. Boite, M. Moonen, A. Oosterlinck (eds.)

DETECTION OF DETERMINISTIC TRANSIENT SIGNALS

A.MAGUER

SACLANT Undersea Research Centre - 19138 - La Spezia - ITALY

Abstract : An overview of receivers (matched filter, matched subspace filter, CFAR matched subspace filter, energy detector and the adaptive receiver proposed by Porat and Friedlander) for detection of deterministic transient signals is performed. The adaptive receiver is implemented and its empirical performance (computed by Monte-Carlo simulations) is compared to the other theoretical receiver ones. Its performance is shown to be lower than the CFAR matched subspace filter ones (to which it theoretically approximately tends) but in most cases well over than the energy detector ones.

1. Introduction

Several methods have been used for the detection of transients [1,2,3]. In this paper we deal with classical receivers (matched filter, energy detector, matched subspace filter, CFAR matched subspace filter) and the adaptive receiver (proposed in [4] by Porat and Friedlander) for detection of deterministic transient signals. First, theoretical computation of each receiver is performed and visualized on ROC curves. It is shown (as in [4]) that the performance of the adaptive receiver tends asymptotically towards those of the matched subspace filter for some appropriate choice of parameter values, which is a very interesting result. Then, empirical computation of the adaptive receiver ROC curve is performed for a specific exponentially damped sinusoid in order to verify the approximations made by Porat and Friedlander for the computation of the adaptive receiver. Several duration and several values of the constraint parameter of the adaptive receiver have been tested. The last part of this paper is the comparison of the theoretical ROC curves of the classical receivers with the empirical ROC curve obtained with the adaptive receiver.

2. Computation of the receivers and their performance

The signal r observed at the receiver is the sum of signal s and noise n under H_1 and equal to noise n under H_0.

$$\underline{r} : \begin{cases} \underline{n}, & \text{under } H_0 \\ \underline{s} + \underline{n}, & \text{under } H_1 \end{cases} \tag{1}$$

where $\underline{s}$ and $\underline{n}$ are N-dimensional vectors. $\underline{n}$ is gaussian, centered, decorrelated, with variance σ^2,and $E_s = \sum_{i=1}^{N} s_i^2$ is the energy of the signal s.

2.1 Matched filter

The matched filter corresponds to a situation where the signal waveform is known and represents an upper bound on the performance of any receiver. It has the structure:

$$\mathbf{r}_{\mathbf{MF}}(r) = \underline{s}'\underline{r} \underset{H_0}{\overset{H_1}{\gtrless}} \eta \tag{2}$$

Under H_0, $\mathbf{r}_{\mathbf{MF}}$ is gaussian with mean and variance equal to 0 and $\sigma^2.E_s$. Under H_1, $\mathbf{r}_{\mathbf{MF}}$ is gaussian with mean and variance equal to E_s and $\sigma^2.E_s$.

2.2 Energy detector

The energy detector makes no assumptions about the signal. Hence, it represents a lower bound on the performance for any receiver. It is given by :

$$\mathbf{r}_{\mathbf{ED}}(r) = \frac{1}{\sigma^2}\underline{r}'\underline{r} \underset{H_0}{\overset{H_1}{\gtrless}} \eta \tag{3}$$

Under H_0,$\mathbf{r}_{\mathbf{ED}}(r)$ is distributed as a χ^2 law with N degrees of freedom. Under H_1,$\mathbf{r}_{\mathbf{ED}}(r)$ is distributed as a decentered χ^2 law with N degrees of freedom. The decentralized parameter is equal to $\frac{E_s}{\sigma^2}$.

2.3 Receivers for partially unknown signals

In this part, the signal to be detected is supposed to be characterized by the impulse response of a rational transfer function. So, we have :

$$s_k = -\sum_{i=1}^{p} a_i.s_{k-i} + \sum_{j=1}^{q} b_j.\delta(k-j) \tag{4}$$

where $\delta(i)$ is the Dirac function. a_i and b_i are the coefficients of the transfer function (respectively poles and zeros).

In matrix form, we have :

$$A.\underline{s} = \underline{b} \qquad or \qquad \underline{s} = A^{-1}.\underline{b} \tag{5}$$

Then, by computing the singular value decomposition SVD of the matrix $X_{(N,p)}$ deduced from matrix A^{-1} (by conserving the first q columns) we obtain from (5) :

$$\underline{s} = U_s \Lambda V'\underline{b} = U_s \underline{\theta} \tag{6}$$

where Λ is a diagonal matrix composed by the eigenvalues of $X_{(N,p)}$;U_s and V are orthogonal matrices.

2.3.1 Matched subspace filter : $a_i(\omega, \alpha)$ known , b_i unknown. σ known

For this receiver σ^2 and the a_i are supposed to be known. So, the problem in this case is the detection of signals with known modes but with unknown initial conditions.

The generalized likelihood ratio is equal to :

$$G(r) = \frac{\underset{\theta}{Sup}\, p[\underline{r}/\theta, H_1]}{p[\underline{r}/H_0]} \tag{7}$$

Setting $\underline{t} = U'.\underline{r}$ ($\underline{t}$ is gaussian), where $U = [U_s V]$ is an orthogonal matrix ($U'U = I_N$) with V an $(N, N-p)$ arbitrary matrix, the matched subspace filter can be written :

$$\mathbf{r_{MSF}}(r) = \frac{1}{\sigma^2}.\underline{r}'.P_s.\underline{r} = \sum_{i=1}^{p}\left(\frac{t_i}{\sigma}\right)^2 \underset{H_0}{\overset{H_1}{\gtrless}} \eta \tag{8}$$

$P_S = U_S.U_S'$ is the orthogonal projector on the signal subspace.

Under H_0, $\mathbf{r_{MSF}}(r)$ is distributed as a centered χ^2 law with p degrees of freedom. Under H_1, $\mathbf{r_{MSF}}(r)$ is distributed as a decentered χ^2 law with p degrees of freedom. The decentralized parameter is the same as before.

2.3.2 CFAR matched subspace filter : $a_i(\omega, \alpha)$ known, b_i unknown. σ unknown

For this receiver the parameters a_i are supposed to be known as for the previous receiver. However, σ^2 is unknown.

The generalized likelihood ratio is equal to :

$$G(r) = \frac{\operatorname{Sup}_{\theta,\sigma} p[\underline{r}/\theta, \sigma, H_1]}{\operatorname{Sup}_{\sigma} p[\underline{r}/\sigma, H_0]} \tag{9}$$

After easy computations, the CFAR matched subspace filter defined as the function $(G(r))^{\frac{2}{N}} - 1$ can be expressed by :

$$\mathbf{r_{CMSF}}(r) = \frac{\|U_s'.\underline{r}\|^2}{\underline{r}'.\underline{r} - \|U_s'.\underline{r}\|^2} = \frac{\sum_{i=1}^{p} t_i^2}{\sum_{i=p+1}^{N} t_i^2} \underset{H_0}{\overset{H_1}{\gtrless}} \eta \tag{10}$$

Then, under H_0, $\frac{(N-p)}{p}\mathbf{r_{CMSF}}(r)$ is distributed as a centered F law with p and $N-p$ degrees of freedom. Under H_1, $\frac{(N-p)}{p}\mathbf{r_{CMSF}}(r)$ is distributed as a decentered F law with p and $N-p$ degrees of freedom. The decentralized parameter is the same as for the previous receiver.

2.3.3 Adaptive receiver : a_i and b_i unknown

For this receiver, neither the a_i or the b_i are known. So, this is a more general case than the two previous receivers.

In [4], Porat and Friedlander have proposed the replacement of the maximum likelihood estimator $\hat{\theta}$ by a new estimator $\hat{\theta}_F$ which minimizes :

$$C_{PF}(\theta) = \frac{1}{\sigma^2}.(\underline{r} - \underline{s}(\theta))'.(\underline{r} - \underline{s}(\theta)) + \mu^2.\underline{\theta}'.\underline{\theta} \tag{11}$$

where μ^2 is a constraint parameter which allows to regularize the Jacobian matrix $J(\theta)$ (rank m) and to avoid numerical problems when only noise is present (no signal).

Then, Porat and Friedlander have set :

$$\hat{\underline{\theta}}_F = \underline{\theta} + \tilde{\theta} \tag{12}$$

and have also make the following hypotheses for performing approximate analysis of their receiver:

$$\underline{s}(\hat{\underline{\theta}}_F) \simeq \underline{s}(\underline{\theta}) + J(\underline{\theta}).\tilde{\theta} \quad \textit{and} \quad J(\hat{\underline{\theta}}_F) \simeq J(\underline{\theta}) \tag{13}$$

Here, σ^2 is supposed to be known (remark : In [4], Porat and Friedlander have also analyzed the case where σ is unknown). The main results are summarized below.

Setting $\underline{z} = U_J'.\underline{n}$, the adaptive receiver is given by :

$$\mathbf{r_{AA}}(r)_{H_0} = \sum_{k=1}^{m} \frac{(\lambda_k^2 + 2.\sigma^2\mu^2).\lambda_k^2}{(\lambda_k^2 + \sigma^2\mu^2)^2}.\left(\frac{z_k}{\sigma}\right)^2 \tag{14}$$

$$\mathbf{r_{AA}}(r)_{H_1} = \beta_k.\sum_{k=1}^{m}\left(\frac{z_k}{\sigma} + \frac{1}{\sigma.\beta_k}.(q_k - \alpha_k.(V_J'.\theta)_k)\right)^2 + \gamma \tag{15}$$

with,

$$\underline{q} = U_J'.\underline{s} \quad , \quad J(\underline{\theta}) = U_J.\begin{pmatrix} \Lambda & 0 \\ 0 & 0 \end{pmatrix}.V_J' \tag{16}$$

$$\beta_k = \frac{(\lambda_k^2 + 2.\sigma^2\mu^2).\lambda_k^2}{(\lambda_k^2 + \sigma^2\mu^2)} \quad , \quad \alpha_k = \frac{(\sigma^2\mu^2)^2.\lambda_k}{(\lambda_k^2 + \sigma^2\mu^2)^2} \tag{17}$$

$$\gamma = \frac{1}{\sigma^2}.(\underline{q}'.\underline{q} - \underline{d}'.\underline{d}) - \sum_{k=1}^{m}\frac{1}{\sigma^2\beta_k}.(q_k - \alpha_k.(V_J'.\theta)_k)^2 \tag{18}$$

For rational signal all the λ_k are equal to 1. The rank m is exactly equal to half the number of unknown parameters.

particular case : $\sigma\mu << 1$ (that means, as σ is known, we must choose μ which verifies the previous conditions).

Then, $\beta_k \simeq 1$, $\alpha_k \simeq 0$, and $\gamma \simeq 0$. So, we deduce:

$$\mathbf{r_{AA}}(r)_{H_0} = \sum_{k=1}^{m}\left(\frac{z_k}{\sigma}\right)^2 \tag{19}$$

$$\mathbf{r_{AA}}(r)_{H_1} = \sum_{k=1}^{m}\left(\frac{z_k}{\sigma} + \frac{q_k}{\sigma}\right)^2 \tag{20}$$

So, under H_0, $\mathbf{r_{AA}}(r)$ is distributed, as a central χ^2 law with m degrees of freedom and under H_1, as a non central χ^2 law with m degrees of freedom. $\frac{q'q}{\sigma^2}$ is the decentralized parameter. As $\underline{q}'.\underline{q} = \underline{s}'.\underline{s}$ the decentralized parameter is the same as for the matched subspace filter.

Finally, when $\sigma\mu << 1$ the performance of the adaptive receiver (for σ known) tends to the matched subspace filter ones. So, the adaptive receiver will have the same performances as if the modes of the signal to be detected were perfectly known, which is a very interesting result and of course, the main interest of this detector.

3. Theoretical ROC curves

The theoretical ROC curves have been determined, for different given probability of false alarm P_{fa}, for different values of signal-to-noise ratio (from -10 dB to 20 dB). Some of the results are shown in figures 1 to 4, corresponding to 2 values of length N of the signal to be detected (N=10 or N=60) and two probability of false alarm P_{fa} (0.1 and 0.001).

On each figure, 5 curves are plotted. Four of them correspond to the theoretical ROC curves of the energy detector, the matched subspace filter (and also asymptotically the adaptive receiver), the CFAR matched subspace filter and finally the matched filter. The fifth one is the empirical ROC curve of the adaptive receiver which will be discussed later in the paper.

The performance of the matched subspace filter (and so theoretically of the adaptive receiver) are always well upper than those of the energy detector. Sometimes, they are not very far from the matched filter performance, specially for long duration signal and low P_{fa}. So, it encourages the use of the adpative receiver for the detection of deterministic transients.

4. Empirical ROC curves of the adaptive receiver

In order to validate the approximate results obtained by Porat and Friedlander, the adaptive receiver has been implemented by determining $\underline{\hat{\theta}}_F$ which minimizes $C_{PF}(\theta)$. This is equivalent to a non linear least squares problem for which a modified Gauss-Newton algorithm has been used.

4.1. Performance under H_0

The empirical cumulative distribution of $\mathbf{r_{AA}}(r)$ under H_0 has been determined (1000 Monte-Carlo simulations were run) and compared to the theoretical one. Several values of the product $\sigma\mu$ and also different values of signal duration N have been tested. Some of the results are visualized on figures 5 to 6 for values of the product $\sigma\mu$ equal respectively to 0.000001 and 5. In any case this distribution does not match any known statistical law and the differences between the theoretical and the empirical distribution are always great except for $\sigma\mu$ equal to 5 for which the two distributions are almost equal.

4.2. Performance under H_1

The computation of the probability of detection P_d under H_1 has been computed with Monte-Carlo simulation (1000 runs) for an exponentially damped sinusoid with 0.05 as damping coefficient and 0.35 as normalized frequency. The phase ϕ of the signal is fixed to 0. The amplitude of the signal has been chosen to generate, according a given white noise of known variance σ^2, a variation of signal-to-noise ratio from -10 dB to 20 dB. The best results in term of detection are always achieved for the product $\sigma\mu$ equal to 1, and this whatever the duration of the signal, and the probability of false alarm P_{fa}. Some of the results are shown on figures 1 to 4 (see adaptive receiver curve).

5. Comparison theoretical-empirical

The last part of this paper is the comparison of the empirical ROC curve of the adaptive receiver (for $\sigma\mu = 1$ which gives the best results) with the theoretical one of the energy detector, the matched subspace and CFAR matched subspace filters and finally the matched filter (see figures 1 to 4). For any value of N, differences between empirical distribution of the adaptive receiver and distribution of the matched subspace filter (which should be equal asymptotically) exist and are greater for small values of N. Moreover, the adaptive receiver is always greater than the energy detector, for great values of N. However, for small values of N (see figure 1. N=10.$P_{fa} = 0.1$), the differences are not very great and even null and then can limit the interest of this detector for medium band or large band signals.

6. Conclusions

In this paper performance of classical receivers and adaptive receivers have been compared for the detection of deterministic rational transient signals. From a theoretical point of view the performance of the adaptive receiver has been shown to tend asymptotically towards the performance of the matched subspace filter which is very interesting. However, the empirical performance of the adaptive receiver developed by Porat and Friedlander has been shown, depending on the signal you try to detect and on the P_{fa} you want, to be lower than the theoretical ones. However, in passive sonar it will probably always be difficult, even if the frequency (but never the damping coefficient) of the signal to be detected is precisely known, to set one of the sonar frequency bins accurately to the particular frequency of the signal. Therefore, loss of detection performance (due to mismatch) will be observed with the matched subspace and the CFAR matched subspace filters, which will decrease their potential superiority against the adaptive receiver. Moreover, the adaptive receiver has been shown to be well better than the energy detector for narrow band signals. So, the adaptive receiver seems to be a good detector to use for detection of some deterministic transients.

7. References

[1] Invariant detection of transient ARMA signals with unknown initial conditions.
S.KAY-L.L SCHARF.38.4.ICASSP 1984

[2] Detection of transient signals by Gabor representation.
B.FRIEDLANDER-B.PORAT.ASSP 37.2.1989

[3] Maximum likelihood detection of transient signals using sequenced short-time power spectra.
J.J WOLCIN. NUSC report TM831138.1983

[4] Adaptive detection of transient signals.
B.PORAT-B.FRIEDLANDER.ASSP 34,6,1986

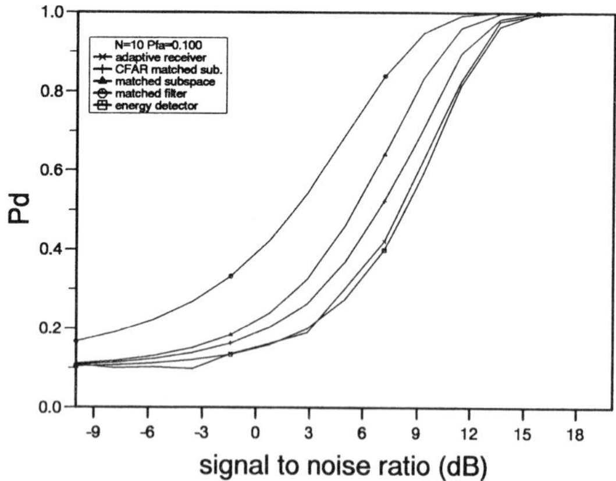

figure 1 : Performance curves. N=10 and $P_{fa} = 0.1$

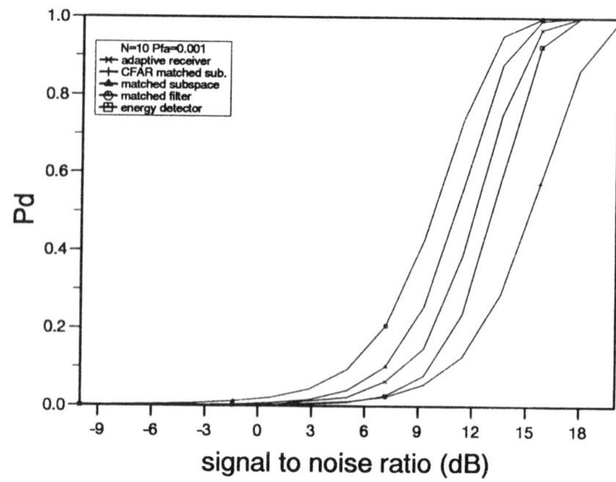

figure 2 : Performance curves. N=10 and $P_{fa} = 0.001$

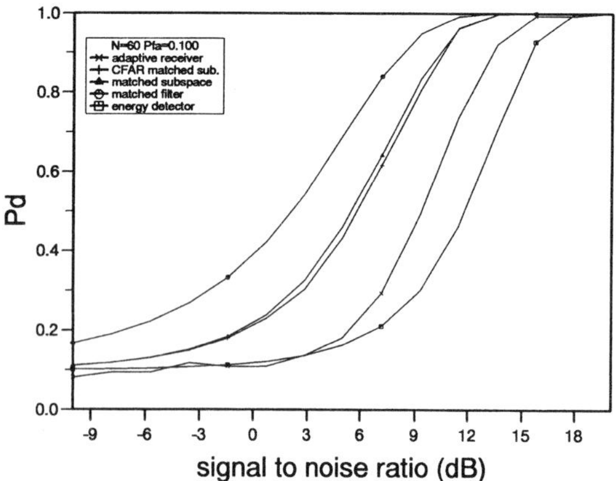

figure 3 : Performance curves. N=60 and $P_{fa} = 0.1$

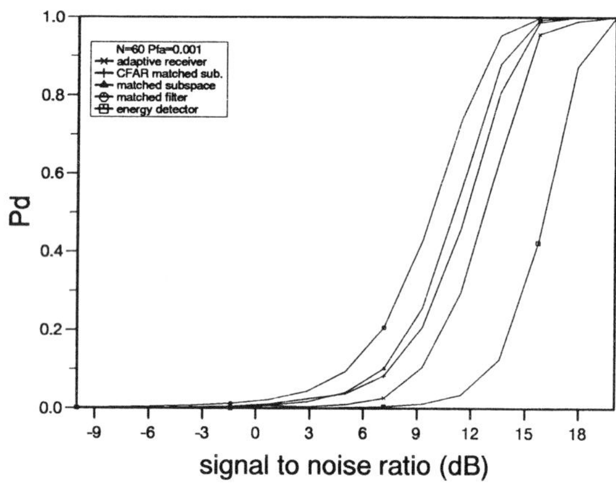

figure 4 : Performance curves. N=60 and $P_{fa} = 0.001$

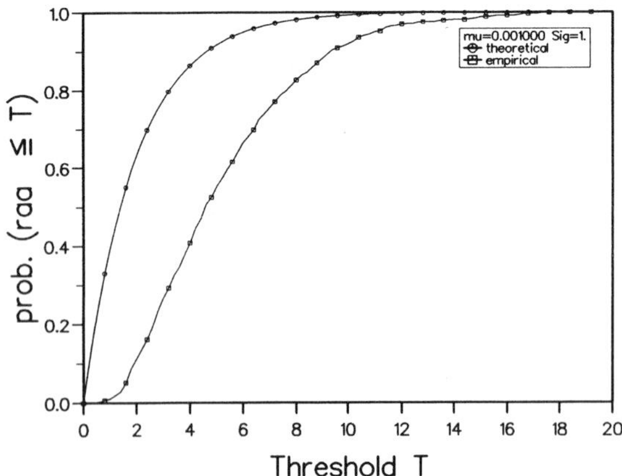

figure 5 : Cumulative distribution under $H_0.\sigma\mu = 0.000001$

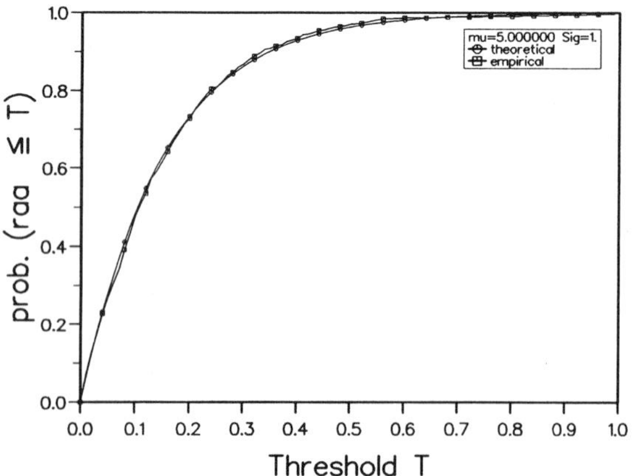

figure 6 : Cumulative distribution under $H_0.\sigma\mu = 5.$

SIGNAL PROCESSING VI: Theories and Applications
J. Vandewalle, R. Boite, M. Moonen, A. Oosterlinck (eds.)

A Class of Recursive Newton-Type Polynomial Extrapolation Filters

Seppo J. Ovaska

KONE Elevators
Research Center
P. O. Box 6, SF-05801 Hyvinkää, Finland

Olli Vainio

Tampere University of Technology
Signal Processing Laboratory
P. O. Box 553, SF-33101 Tampere, Finland

Modifications and extensions of Newton's extrapolation formula are introduced for digital signal processing purposes. The proposed class of computationally efficient polynomial extrapolation filters is called Recursive Linear Smoothed Newton (RLSN) predictors. Unlike the basic Newton algorithm, the RLSN predictors behave well also with noisy input signals. The characteristics of RLSN predictors compare favorably with other low-order polynomial prediction approaches.

1. INTRODUCTION

Extrapolation of limited data sets is a subject that dates back to antiquity, and comes up frequently in all application areas of digital signal processing, e.g., in instrumentation and measurement systems. The common problem is to forecast future samples of a primary signal corrupted by additive white Gaussian noise or uniformly distributed noise [1], [2]. An essential ingredient of the extrapolation problem is the postulation of a physical model for the underlying process [1]. In this paper, we consider polynomial extrapolation that is a special case of forward prediction. The primary signal $u(p)$ is assumed to be a polynomial of degree M, i.e.,

$$u(p) = \lambda_0 + \lambda_1 p + \ldots + \lambda_M p^M \qquad (1)$$

The polynomial coefficients $\lambda_i, i = 0, \ldots, M$, are unknown real constants. The noisy input sequence, available to the predictor, is

$$i(p) = u(p) + e(p) \qquad (2)$$

where $e(p)$ is the disturbing additive noise component. The difference equation of a general n-step-ahead predictor can be written as

$$\hat{u}(p+n) = \sum_{j=1}^{V} \alpha_j \hat{u}(p+n-j) + \sum_{k=0}^{W} \beta_k i(p-k) \qquad (3)$$

The coefficients α_j and β_k are real constants. In the case of FIR predictors, all the feedback coefficients α_j are equal to zero. In a noise-free situation we require that the steady-state prediction error is zero, i.e., $\hat{u}(p+n) = u(p+n)$.

The Newton's prediction algorithm has been used by mathematicians to extrapolate tables of polynomials and transcendental functions. Newton predictors are attractive for extrapolation of polynomials because of their low computational complexity and the straightforward design process. A general n-step-ahead Newton predictor of an Mth-order polynomial can be expressed as the z-domain transfer function [3]

$$H_M^n(z) = \sum_{k=0}^{M} (1 - z^{-n})^k \qquad (4)$$

Successful DSP applications of Newton predictors have been reported in the literature. Athani applied them to sample trend monitoring that was used for process failure detection [4]. Ovaska proposed Newton predictors for transmission error correction in a measuring oriented communication protocol [5], [6]. However, the predictors based on this computationally efficient algorithm have considerable gain at the higher frequencies because of the high-pass nature, and upper-band amplification of the difference operators. This property reduces the applicability of the basic Newton predictors in practical signal processing where the narrow-band primary signal is often corrupted by additive wide-band noise.

In this paper, we develop modifications and extensions of the original Newton predictors, aiming at considerably lowered overall noise gain while preserving the prediction property. The resulting new class of polynomial predictors, called the Recursive Linear Smoothed Newton (RLSN) predictors, is presented in the next Section. In Section 3, we quantitatively compare the characteristics of the RLSN predictors with some other predictor approaches, such as the optimal FIR predictors and IIR predictors designed in the frequency domain. Additional practical considerations and implementation issues are discussed in Section 4.

2. THE RLSN PREDICTORS

In a recent paper [7], two smoothed versions of the original Newton predictor were proposed. Those predictors were called the Linear Smoothed Newton (LSN) and the Median Smoothed Newton (MSN) predictors. The LSN predictor is recommended to be used with white Gaussian and uniformly distributed additive noises, and the MSN counterpart with impulsive or doubly exponentially distributed noise. In both cases, the smoothing operation is performed solely for the highest-order difference $(1 - z^{-n})^M$ of the Newton predictor of Eq. (4). The highest-order difference has the largest noise gain, and it approximates the highest-order non-zero derivative that attains a constant value if the polynomial is noiseless. All the lower-order derivatives are time varying, and all the higher-order derivatives are equal to zero. The general z-domain transfer function of LSN n-step-ahead predictors can be written as

$$G_M^n(z) = S(z)(1 - z^{-n})^M + \sum_{k=0}^{M-1} (1 - z^{-n})^k \qquad (5)$$

where $S(z)$ is the transfer function of some lowpass filter, preferably a moving averager,

$$S(z) = \frac{1}{N}\sum_{i=0}^{N-1} z^{-i}, \tag{6}$$

that smoothes the noisy Mth-order difference sequence (N is a positive integer). Insertion of a linear smoother into the original algorithm reduces the highpass nature and noise gain of the Newton predictors without adding any steady-state prediction error to the result. The MSN predictor is obtained by replacing the linear filter $S(z)$ with a median filter.

In order to further enhance the applicability of the LSN and MSN predictors, we first propose a recursive extension. It is desirable, however, to maintain the simple principal structure of the original LSN predictor. In this new approach, we feed back the predicted estimate $\hat{u}(p)$ of the current input sample $u(p)$, add it with weighting $(1-a)$ to the weighted true input $ai(p)$, and further add this sum to the sum of the outputs of the difference operators, which get their inputs $i(p)$ directly without prescaling. This kind of manipulation does not cause any steady-state error into the predicted output. The modified z-domain transfer function can be expressed as

$$P_M^n(z) = \frac{a + S(z)(1-z^{-n})^M + \sum_{k=1}^{M-1}(1-z^{-n})^k}{1-(1-a)z^{-n}} \tag{7}$$

This transfer function has only one additional multiplication and addition as compared to the nonrecursive LSN counterpart, and with appropriate values of the parameters a and N it has the desired lowpass nature. The width of the passband is mainly controlled by N, and the stopband attenuation by a.

For example, in case of the one-step-ahead ramp predictor, where $S(z)$ is an N-length moving averager of Eq. (6), the transfer function (7) can be reduced to the form

$$P_1^1(z) = \frac{(a+1/N) - z^{-N}/N}{1-(1-a)z^{-1}} \tag{8}$$

Delayed cascading of the basic sections (8) offers a viable way to further increase the stopband attenuation. It is also possible to obtain ramp predictors with a complex conjugate pole pair by cascading two basic sections with $a = a_{re} + ia_{im}$ and $a = a_{re} - ia_{im}$, respectively. The resulting transfer function can be expressed as

$$L_1^1(z) = \frac{c_0 + c_N z^{-N} + c_{2N} z^{-2N}}{1-(-2a_{re}+2)z^{-1} - (-a_{re}^2 + 2a_{re} - a_{im} - 1)z^{-2}} z^{-1} \tag{9}$$

where $c_0 = a_{re}^2 + 2a_{re}/N + a_{im} + 1/N^2$, $c_N = -2a_{re}/N - 2/N^2$, and $c_{2N} = 1/N^2$.

Although the ramp predictor of the form (8) may have the lowpass nature, the predictors (7) of higher-order polynomials have still gain greater than or equal to unity at the higher frequencies. The reason for this is that only the primary input signal and the highest-order successive difference are smoothed; the primary signal by an 'exponential' averager, and its highest-order non-zero derivative by a 'uniform' averager. All the other difference paths have disturbing gain at the higher frequencies. Smoothing of the approximations of time-varying derivatives cannot be carried out by a conventional lowpass filter or moving averager, because they all introduce lag into the primary signal and thus deteriorate the prediction function [7]. To overcome this problem, we propose an enhanced version of the basic recursive transfer function of Eq. (7). In the new transfer function, the lower-order successive differences, corresponding to the time-varying derivatives, are smoothed by polynomial filters, i.e., zero-step-ahead predictors, and so the polynomial nature of the signal is fully exploited. First, we need an additional ramp predictor to obtain a predictor for second-order polynomials; then we use both the ramp predictor and the second-order polynomial predictor to obtain the predictor for third-order polynomials, etc. Thus the general transfer function can be written in a form that contains a recursion

$$K_M^n(z) = \frac{a + S(z)(1-z^{-n})^M + z^{-1}\sum_{k=1}^{M-1} K_{M-k}^1(1-z^{-n})^k}{1-(1-a)z^{-n}}. \tag{10}$$

We call this new class of polynomial extrapolation filters Recursive Linear Smoothed Newton (RLSN). The RLSN extrapolator, depicted in Fig. 1, offers great flexibility to the user to optimize its characteristics for different signal shapes, noise conditions, and computational environments.

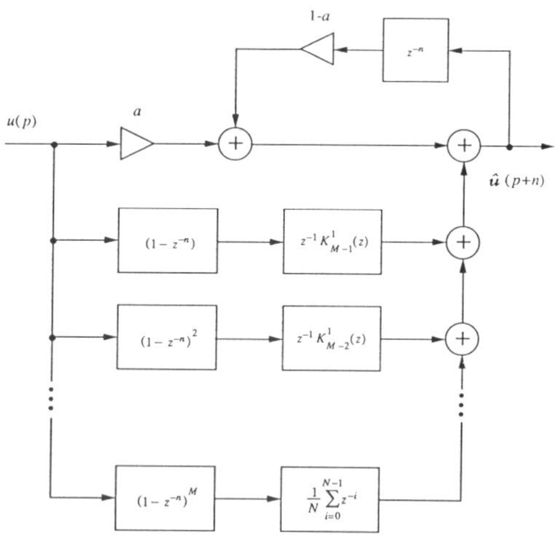

Fig. 1. Data flow diagram of the RLSN predictor.

Typical frequency response characteristics can be seen, for example, in the case of the second-order RLSN predictor. The transfer function can be reduced to:

$$K_2^1(z) = \frac{\frac{1}{N}\left((1+aN) + (a^2N + a - 1)z^{-1} - (aN+a)z^{-2}\right)}{(1-(1-a)z^{-1})^2} + \frac{\frac{1}{N}\left(-z^{-N} + (1-a)z^{-(N+1)} + az^{-(N+2)}\right)}{(1-(1-a)z^{-1})^2} \tag{11}$$

It has a double pole at $z = 1-a$, and $N+2$ zeros. The frequency response is shown in Fig. 2 as a 3-D plot for $a = 0\ldots1$ and $N = 16$. For higher-order transfer functions, the multiplicity of the pole causes practical upper limits for the parameter a, if undesirable bands of elevated gain are to be avoided. The practical lower limit of a is about $0.05\ldots0.1$.

3. NUMERICAL COMPARISONS

In this Section, we compare the computational complexities and noise gains of different polynomial predictors, taking the second- and fourth-order cases as examples. Assuming that the additive zero-mean input noise has a flat power spectrum, i.e., the noise samples are mutually uncorrelated, the noise gain (NG) can be computed either in the time domain or in the frequency domain as [9]

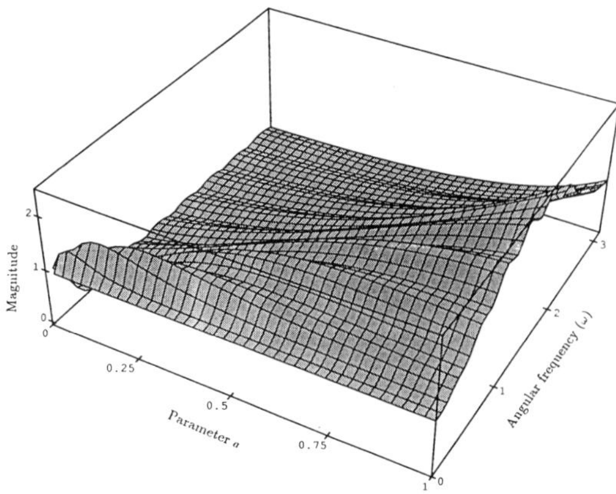

Fig. 2. Frequency response of the second-order RLSN predictor as function of the parameter a, for $N = 16$.

$$NG = \sum_{k=-\infty}^{\infty} |h(k)|^2 = \frac{1}{2\pi}\int_{-\pi}^{\pi} |H(e^{j\omega})|^2 d\omega \qquad (12)$$

where $h(k)$ is the impulse response of the prediction filter and $H(e^{j\omega})$ is the corresponding frequency response.

3.1. Other Approaches to Polynomial Prediction

The coefficients of a direct form FIR predictor can be optimized to minimize the noise power at the predictor output for a given polynomial degree [2]. The drawback of this approach is the number of arithmetic operations required, although recently a computationally efficient implementation has been introduced [8].

Besides the time-domain approaches discussed above, it is possible to design predictors in the frequency domain. For this purpose, one can try to approximate the ideal predictor characteristics by requiring that the magnitude $|H(e^{j\omega})| \approx 1$ and the group delay $grd[H(e^{j\omega})] \approx -1$. Such an approximation is only possible on a relatively narrow band $\omega = 0 \ldots \omega_p$, and the resulting predictors are not strictly polynomial predictors. For a second-order IIR section, optimized using the method of steepest descent, the zero locations are at $r_z = 0.552$, $\theta_z = \pm 0.851$, and the poles at $r_p = 0.939$, $\theta_p = \pm 0.803$. For a fourth-order IIR predictor, optimization places the zeros at $r_{z1} = r_{z2} = 0.616$, $\theta_{z1} = \theta_{z2} = \pm 0.870$, and the poles at $r_{p1} = r_{p2} = 0.819$, $\theta_{p1} = \theta_{p2} = \pm 0.846$. The prediction bandwidth ω_p is about 0.16π which is slightly wider than that of the RLSN predictors.

3.2. Results of the Comparison

In Tables I and II, we compare different types of one-step-ahead predictors for the second- and fourth-order cases, respectively. The following predictors are included in the comparison: the optimal FIR predictor [2], [8], the IIR predictor, the basic Newton predictor (Eq. (4)), and the three types of smoothed Newton predictors discussed in this paper (Eqs. (5), (7), and (10)). Note that for the IIR predictor, 'order' refers to the filter order rather than the order of the polynomial to be predicted.

It is clearly visible that the RLSN predictor offers the lowest noise gain with low computational complexity. The RLSN is also the only predictor that has true noise attenuation in fourth-order prediction - all the other predictors are amplifying the input noise. The frequency-domain approach does not give useful results for polynomial prediction.

Finally, let us equalize the noise gains of the optimal FIR predictor and the RLSN predictor, and compare their computational complexities. Choosing, for example, the second-order case and requiring that $NG = 0.5$, we get $k = 22$ for the FIR predictor, and $a = 0.27$ with $N = 16$ for the RLSN predictor. The number of arithmetic operations (MUL + ADD) in the FIR case is 18 while in the RLSN case it is only 11 ($\approx 40\%$ less). The number of unit delays is almost equal in both cases. We can conclude that the RLSN predictors have excellent noise attenuation characteristics and low computational complexity when compared to other commonly used low-order polynomial predictors.

4. IMPLEMENTATION CONSIDERATIONS

After expanding the z-domain transfer functions of some low-order one-step-ahead RLSN predictors, we can see that in their final substituted numerical forms the numerators have sparse impulse responses with several coefficients equal to zero. The relative sparsity increases as N increases. For the first- and second-order predictors, the most efficient implementation is based on the data flow graph of Fig. 1 directly. The flow graph implementation makes it possible to easily adjust the parameters a and N dynamically. For higher-order predictors with fixed parameters, the most practical approach is to expand the recursive transfer function (10) into a direct-form rational transfer function of the form

$$K_M^n(z) = \frac{\sum_{k=0}^{N+2(M-1)} b_k z^{-k}}{N(1-(1-a)z^{-1})^M} \qquad (13)$$

This form of the transfer function leads to the canonic direct form implementation. For example, the non-zero numerator coefficients of the third-order predictor are:

$b_0 = 1 + aN$ $\quad$ $b_1 = 2a^2N + 2a - 2$
$b_2 = a^3N + a^2 - 3aN - 4a + 1$ $\quad$ $b_3 = 2a + 2aN - 2a^2 - 3a^2N$
$b_4 = a^2N + a^2$ $\quad$ $b_N = -1$
$b_{N+1} = 2 - 2a$ $\quad$ $b_{N+2} = 4a - a^2 - 1$
$b_{N+3} = 2a^2 - 2a$ $\quad$ $b_{N+4} = -a^2$

The drawback of this approach is that the expanded coefficients must be recomputed every time the parameters are changed, or the polynomial order M is increased. This would be difficult to manage efficiently in a fixed-point signal processor implementation because of the large range of possible values.

The basic data flow graph of Fig. 1 can be easily divided into compact computational modules. Thus the efficient signal processor implementation is based on using subroutines or macros and constructing the program in a modular way. The parameters a and N are updated in memory locations. The recursive structure of the RLSN predictors is managed by allocating dedicated memory areas to the data at each level of the hierarchy.

In integrated circuit (ASIC) implementations, the wordlength can be adjusted according to the worst-case requirements. On the other hand, the chip area must be kept as small as possible, which limits the allocation of hardware resources. The microprogrammable VLSI architecture of Fig. 3 could provide the desired flexibility in a relatively compact structure. The on-chip RAM stores the data samples and internal delay lines as ring buffer structures. The multiplier-adder configuration allows computations using integer or fractional coefficients, or their combinations (multiple clock cycles are needed to implement coefficients outside of the range $-2 \ldots 2$). The accumulator accumulates intermediate and final results, and stores those into the RAM. The operations are sequentially activated by the micropro-

Table I. Comparison of different second-order predictors. (Notation: MUL = number of multiplications, ADD = number of additions/subtractions, DEL = number of unit delays.)

Predictor	Param.	NG	MUL	ADD	DEL
Optimal FIR	$k = 16$	0.73	5	13	21[(1)]
IIR	-	2.77	4	4	2
Newton (4)	-	19.00	-	4	2
LSN (5)	$N = 16$	5.39	-	4	17
LSN-FB (7)[(2)]	$N = 16$	1.36	2	5	18
RLSN (10)[(2)]	$N = 16$	0.23	4	7	20

[(1)] Campbell-Neuvo implementation [8]
[(2)] with $a = 0.1$.

Table II. Comparison of different fourth-order predictors

Predictor	Param.	NG	MUL	ADD	DEL
Optimal FIR	$k = 16$	3.66	9	15	15[(1)]
IIR	-	2.10	8	8	4
Newton (4)	-	251.00	2	4	4
LSN (5)	$N = 16$	73.53	-	8	19
LSN-FB (7)[(2)]	$N = 16$	22.58	2	9	20
RLSN (10)[(2)]	$N = 16$	0.50	18	17	22

[(1)] Direct form implementation
[(2)] with $a = 0.1$.

gram controller. In order to avoid the control logic becoming overly complicated, only a limited number of parameter combinations are supported.

The speed of this architecture is limited by the access time of the RAM, which is typically about 30 ns for contemporary CMOS technologies. Sampling rates exceeding 1 MHz could be achieved for low-order polynomial predictors.

5. CONCLUSIONS

The new Recursive Linear Smoothed Newton predictor offers good noise attenuation characteristics with low-order polynomials. The computational complexity of the RLSN predictors is usually low because the numerator of the transfer function has a sparse impulse response, i.e., most of the coefficients of the tapped delay line are equal to zero, and the order of the denominator is typically no more than three.

Several applications can be found for the RLSN predictor. These include use as a stand-alone predictor and a real-time curve fitter. Velocity prefiltering, commonly needed in motion control systems, is one potential application area of RLSN predictors. An appropriately delayed RLSN predictor can be used as a 'delayless' prefilter, or an estimator of a noisy tachometer signal if the underlying primary velocity pattern approximates some polynomial form.

Our future research will focus on developing adaptive extensions to the basic LSN- and RLSN-structures, and performing corresponding studies with the Median Smoothed Newton predictor. Although we have analyzed solely the properties of RLSN predictors in the one-dimensional case, the proposed approach can be applied to two-dimensional signal processing with similar benefits.

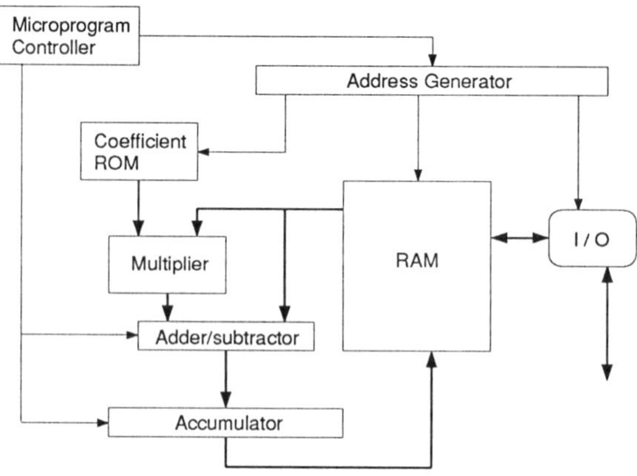

Fig. 3. A generic VLSI circuit architecture for implementation of the RLSN predictors. Thick arrows indicate data flow, and thin lines denote control flow.

REFERENCES

[1] D. R. Morgan and A. A. Aridgides, "Interpolation and extrapolation of an ideal band - limited random process," *IEEE Trans. Acoust., Speech, Signal Processing*, vol. 35, pp. 43-47, Jan. 1987.

[2] P. Heinonen and Y. Neuvo, "FIR-median hybrid filters with predictive FIR substructures," *IEEE Trans. Acoust., Speech, Signal Processing*, vol. 36, pp. 892-899, June 1988.

[3] S. J. Ovaska, "Newton-type predictors - a signal processing oriented viewpoint," *Signal Processing*, vol. 25, pp. 251-257, Nov. 1991.

[4] V. V. Athani, "Microprocessor based data acquisition system," *Microprocessors and Microsystems*, vol. 3, pp. 359-364, Oct. 1979.

[5] S. J. Ovaska, "Distributed architecture of an adaptive positioning servo," *IEEE Trans. Instrum. Meas.*, vol. 39, pp. 403-408, Apr. 1990.

[6] S. J. Ovaska, "Lost-sample predictor to replace a conventional transmission error correction method in instrumentation and measurement applications," in *Proc. 7th IMEKO Int. Symp. Techn. Diagn.*, Helsinki, Finland, Sept. 1990, pp. 384-391.

[7] S. J. Ovaska, "FIR prediction using Newton's backward interpolation algorithm with smoothed successive differences," *IEEE Trans. Instrum. Meas.*, vol. 40, pp. 811-815, Oct. 1991.

[8] T. G. Campbell and Y. Neuvo, "Predictive FIR filters with low computational complexity," *IEEE Trans. Circuits Syst.*, vol. 38, pp. 1067-1071, Sept. 1991.

[9] A. V. Oppenheim and R. W. Schafer, *Discrete-Time Signal Processing*. Englewood Cliffs, NJ: Prentice-Hall, 1989.

SIGNAL PROCESSING VI: Theories and Applications
J. Vandewalle, R. Boite, M. Moonen, A. Oosterlinck (eds.)

Gibbs random field texture model

Witold Czarnecki

Institute of Telecommunications, Warsaw Technical University
ul.Nowowiejska 15/19, 00-665 Warszawa,POLAND

The Gibbs random field model of a texture having a given marginal probability distribution and predefined correlation coefficients is proposed. An algorithm of computer generation of such textures based on the model is described and some results obtained in generation experiments are presented.

1. Introduction

Textures play a very important role in the image processing. They are *"portions of natural scenes that are devoid of significant details over large area and exhibit a repetitive structure"* [1]. The textures can be the visual representations of spatially homogeneous structures both natural and man-made in general sense. The characteristic features of textures can be found also in signals which do not form images in ordinary meaning of the word. It is useful to apply image processing techniques to "non image" signals too. The textures represent often noisy backgroud on which elements with information content occur. The texture modelling is a particular case of noise signal modelling. The visual representation of noise signals have the structure of textures i.e. have no details and are quasi-repetitive.

In numerous applications such 2-d signals are very important and it is desired to have computer programmes for their fast generation. If they are not white noises then their modelling and computer simulation is not easy [2]. It is mainly due to the difficulty of independent shaping of the probability distributions and the correlation. The noise signals are modelled as stochastic processes. The complexity of an adopted model depends on such properties as stationarity, normality, correlation and others. A model is valuable if it makes possible the definition of a simple algorithm suitable for computer simulation of the signal. It is not always easy to meet this requirement especially when a stationary noise signal is non-gaussian and has non-white spectrum That is why the creation of a model of any particular class of noise signals is interesting if it assures the effective computer simulation.

The proposed 2-d signal model is of Gibbs random field (GRF) class. It makes possible the formulation of a simple generation algorithm to be used in computer programmes for simulation studies of image processing problems. In many applications the simulation studies are inevitable and fast generation of the 2-d random signals becomes necessary.

2. Problem formulation

Let X be a discrete 2-d stationary and homogeneous random signal defined over a rectangular lattice L

$$\boldsymbol{X}=\{X_{i,j};\ (i,j)\in\boldsymbol{L}\} \tag{1}$$

$$\boldsymbol{L}=\{(i,j);\ i=1,2,\ldots,I;\ j=1,2,\ldots,J\}$$

The elements $X_{i,j}$ form a structure of I rows and J columns. For probabilistic description of X it is necessary to know the probability $P(X_1=x_1)$ where a subset X_1 of X is defined over an arbitrary subset L_1 of L. It is often supposed that marginal probability distribution and the moments of the second order probability distribution is sufficient.

The problem is to define a model of 2-d signal X of marginal distribution p(x) and correlation coefficients between the neighbouring elements of rows and columns ρ_h and ρ_v respectively. It is desired that the model of X enable to define a procedure of sequential generation of its elements $X_{i,j}$.

3. Gibbs Random Field Model

If a GRF model for a discrete, 2-d random signal **X** is defined over L then the neighbourhood system μ associated with it is a collection of the following subsets of L

$$\mu=\{\mu_{i,j};(i,j)\in L\} \qquad (2)$$

where $(i,j)\notin\mu_{i,j}$ and if $(k,l)\in\mu_{i,j}$ then $(i,j)\in\mu_{k,l}$ for any $(i,j)\in L$. The elements $c\in L$ called cliques such that

- c is a single element of L,
- if c is not a single element of L then a pair of elements $(i,j)\neq(k,l)$ belongs to c if $(i,j)\in\mu_{k,l}$

form the set $C(L,\mu)$. The cliques composed of one or two elements of L wil be called pointcliques or paircliques respectively. The signal **X** is a GRF with respect to the neighbourhood system μ if

$$P(\mathbf{X}=\mathbf{x})=P_{\mathbf{X}}(\mathbf{x})=\frac{1}{Z}e^{-U(\mathbf{X}=\mathbf{x})}=\frac{1}{Z}e^{-U_{\mathbf{X}}(\mathbf{x})} \qquad (3)$$

$$V_c(\mathbf{x}), \qquad U_{\mathbf{X}}(\mathbf{x})=\sum_{c\in C}V_c(\mathbf{x}), \qquad Z=\sum_{\mathbf{x}}e^{-U_{\mathbf{X}}(\mathbf{x})}$$

$V_c(x)$, $U_X(x)$ and Z are the potential associated with the clique c, the energy function of the realization **x** and the partition function respectively. To define a GRF model of **X** it is necessary to determine the neighbourhood system μ, the clique potentials $V_c(x)$ and the energy function $U_X(x)$.

4. Properties of GRF

The probability $P_X(x)$ is large when the energy $U_X(x)$ is small. The energy depends on the set of cliques **C** and the potentials of each clique $V_c(x)$. So they define the nature of the signal **X** and in consequence the correlations. Signal processing which modifies in a desired way its correlation can rely on forming such a permutation of signal el-ements which minimizes the given energy function. In such a case the desired modification of the correlation will not be followed by any change of the marginal distribution.

A realization **x** of the GRF **X** defined by μ, $V_c(x)$ and $U_X(x)$ can be a permutation of elements of some auxiliary signal $\mathbf{X}_{aux}$ having the same marginal probability as **X**. The permutation which minimizes the energy function $U_X(x)$ is the most probable realization of **X** composed of the elements of $\mathbf{X}_{aux}$. The GRF model enables to compare particular realization probability. It is natural then to permute the elements of a simple in a sense auxiliary signal $\mathbf{X}_{aux}$ (a white noise for example) in order to find the most probable realization of **X**.

To generate a GRF correlated signal **X** it is necessary to repeat many times, (like in the Metropolis algorithm), the following sequence of operations:

- choose at random a pair of elements $X_{i,j}$ and $X_{m,n}$,
- exchange their values if it takes the signal to the lower energy state.

The repetition of these operations will result in an enhancement of signal features defined by the energy function $U_X(x)$. In some cases its form could be expressed by horizontal and vertical correlation coefficients ρ_h, ρ_v.

The main inconveniance of the above procedure is its "globality". It needs prior generation of $\mathbf{X}_{aux}$. For great values of I and J it could be cumbersome. It is possible to modify the procedure to make it "local". If the lattice **L** is divided in the following parts: $L^{i,j}$ where the elements form already the generated signal **X**, $L-L^{i,j}-(i,j)$ where the elements form the auxiliary signal $\mathbf{X}_{aux}$ yet and (i,j) between them then it is possible to generate an elements $X_{i,j}$ as follows

- for each (i,j) taken in deterministic order of generation choose K different elements $X_{mk,nk}$; k=1,2,...,K taken at random from among $L-L^{i,j}$ positions,
- choose one out of K pairs $X_{i,j}$, $X_{mk,nk}$ for which the exchange of values takes the signal in $L^{i,j}+(i,j)$ part to the lowest energy state,
- exchange the values of the elements of the chosen pair $X_{i,j}$, $X_{mkopt,nkopt}$.

The sequence of K random elements $X_{mk,nk}$ of $\mathbf{X}_{aux}$ can be arranged in a set **R** not necessarily of 2-d nature. When the auxiliary signal $\mathbf{X}_{aux}$ is a white noise the set **R** is composed of K independent random numbers having the probability distribution p(x). One element from among them is chosen as $X_{i,j}$ provided it minimizes the energy function $U_X(x)$ over the region $L^{i,i}+(i,j)$.

5. Energy function

The correlations of the signal **X** depends on the energy function $U_X(x)$. The first order neighbourhood system μ_1 defines only two types of 2-element cliques: horizontal (both elements of the same row) and vertical (both elements of the same column). Such a neighbourhood system was adopted in all numerical experiments. It is easy to foresee that the correlations will depend on the

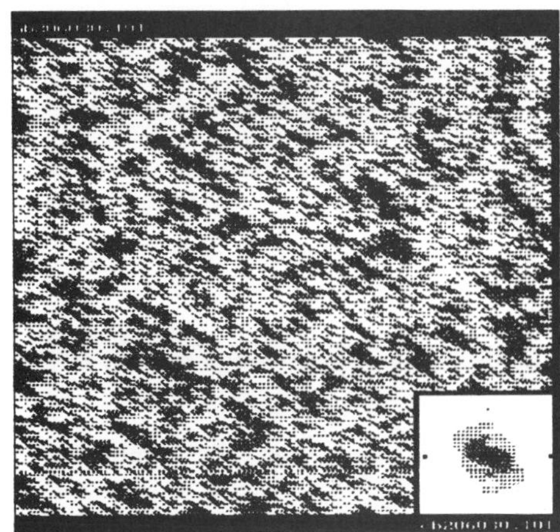

fig.1. Rayleigh texture and its correlation function (lexicographic order of generation, K=20, ρ_h=0.6, ρ_v=0.3.

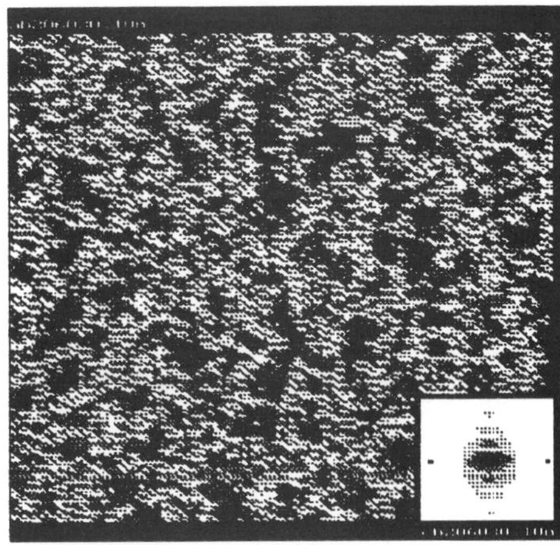

fig.2. Rayleigh texture and its correlation function (meanderlike order of generation, K=20, ρ_h=0.6, ρv=0.3.

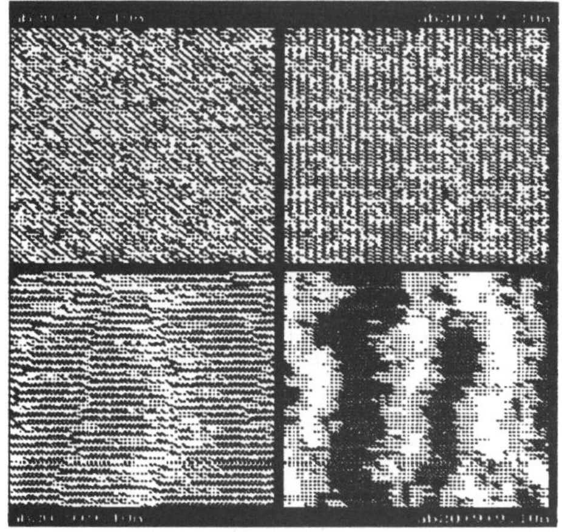

fig.3. Rayleigh texture realizations ρ_h=-0.9; ρ_v=±0.9 (upper row) and ρ_h=0.9, ρ_v=±0.9 (lower row).

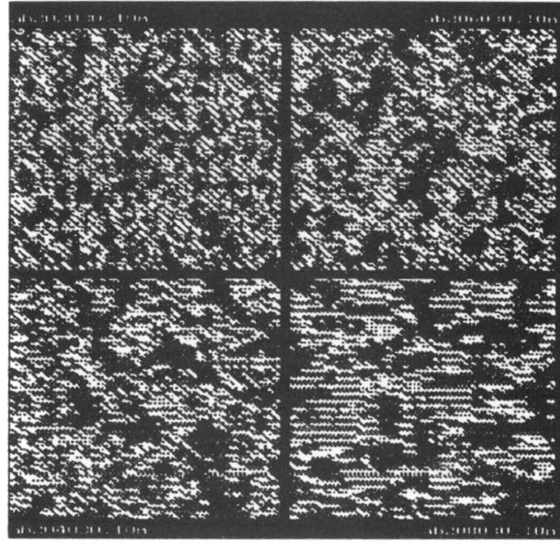

fig.4. Rayleigh texture realizations for ρ_v=0.3 and from upper left quadrant in clockwise manner: ρ_h=0.2, 0.4, 0.8, 0.6.

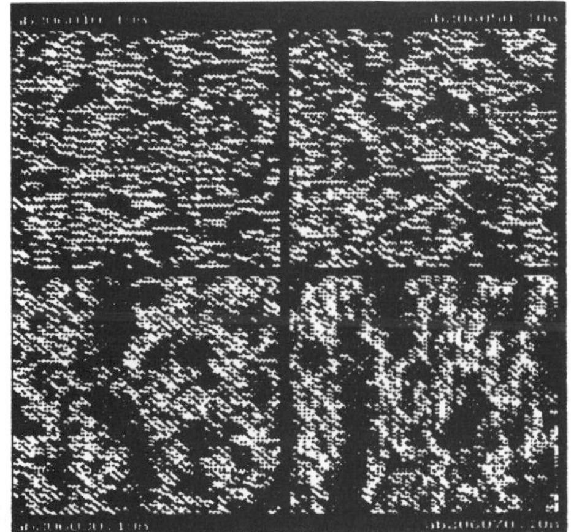

fig.5. Rayleigh texture realizations for ρ_h=0.6 and from upper left quadrant in clockwise manner: ρ_v=0.1, 0.3, 0.7, 0.5.

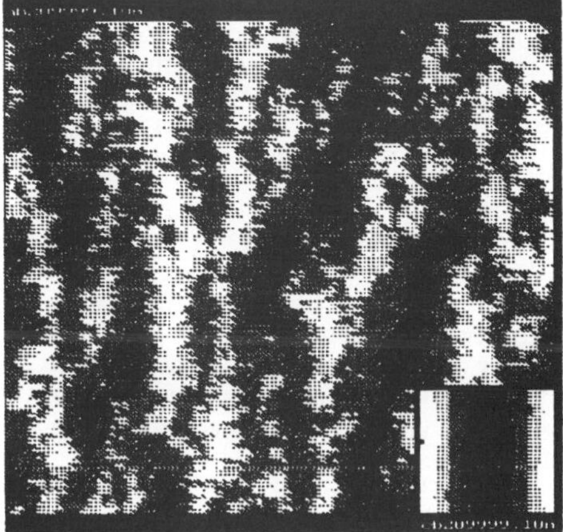

fig.6. Rayleigh texture and its correlation function (lexicographic order of generation, K=20, ρ_h=0.99, ρ_v=0.99.

paircliques potentials. The following energy function was proposed

$$U_X(x) = \left|\sum_{c\in C_h} V_c(x)\right| + \left|\sum_{c\in C_v} V_c(x)\right| \qquad (4)$$

with pairclique potentials (5)

$$V_{(i,j),(m,n)}(x) = (x_{i,j}-x_{m,n})^2 - 2(1-\rho_c)\sigma_x^2$$

In the preceding formulae the sets of horizontal and vertical paircliques of the neighbourhood system μ are denoted by C_h and C_v respectively. The marginal probability p(x) is of mean value m_x and variance σ_x^2. The correlation coefficient between the elements of pairclique c is ρ_c. It is easy to show that the simultaneous minimization of the two positive terms of $U_X(x)$ (4) will guarantee the desired correlation properties of **X**.

6. Generation algorithm

The GRF model of the texture **X** enables to define the following generation algorithm. An element $X_{i,j}$ can be taken as one out of K independent random numbers R_k of **R** which fits to the set $X^{i,j}$ of all previously generated elements

$$x^{i,j} = \{X_{i,j};\ (i,j)\in L^{i,j}\} \qquad (6)$$

If the elements of the signal are generated in a row after row manner and always from left to right (lexicographic order) then $L^{i,j}$ has the following form

$$L^{i,j} = \{(k,l);\ (1\le k<i \wedge 1\le l\le J) \vee (k=i \wedge 1\le l<j)\}$$

The elements can also be generated in a row after row manner but from left to right in the odd numbered rows and from right to left in the even numbered rows (meanderlike order) [6]. In this case the set of previously generated elements for (i,j) from odd numbered rows takes the form as follows

$$L^{i,j} = \{(k,l);\ (1\le k<i \wedge 1\le l\le J) \vee (k=i \wedge j<l\le J)\}$$

The criterion of fitness is the value of the energy function of $U_X(x^{ij}+x_{i,j})$. After each generation the chosen number R_k in **R** is replaced by a new random number. The generation algorithm takes the following form:

- for the given p(x), ρ_h and ρ_v define the potentials $V_c(x)$ as in (5),
- for the fixed value of K create the set **R** of K random numbers R_k
- take for $X_{1,1}$ a random number of probability distribution p(x),

$$X_{1,1} = random_{p(x)}$$

- in a row after row manner generate the successive elements of the signal taking for $X_{i,j}$ the value of R_{opt} which minimizes the energy function $U_X(x^{i,j}+x_{i,j})$; replace R_{opt} in **R** by a new random number,

$$U_X(x^{i,j}+R_{opt}) \le U_X(x^{i,j}+R_k)\ ;\ k=1,2,\ldots K$$

$$X_{i,j} = R_{opt};\ R_{opt} = random_{p(x)}$$

7. Simulation experiments

The computer simulations were done for the Rayleigh textures. It was found that for K≥10 the goodness of fit hypothesis for the marginal distribution of the generated signal was successfully verified and the relative errors of the correlation coefficients did not exceed 10%. The results of the simulation experiments (all for K=20) are presented on the fig.1-6. The figures represent the generated Rayleigh textures in the form of black and white images. In 3 cases the estimated correlation function are presented too.

References

1. Pratt W.K., Digital Image Processing, John Wiley & Sons, New York, 1976,
2. Liu B.,Munson D.C., Generation of Random Sequences Having a Jointly Specified Marginal Distribution and Autocovariance, IEEE Trans. Acoust., Speech, Signal Processing, Dec.1982, vol. ASSP- 30, pp 973-983,
3. Communications and Networks edited by J.F.Blake and H.K.Poor, Springer-Verlag, 1986,
4. G.R.Cross,A.K.Jain, Markov Random Field Texture Models, IEEE Trans. Pattern Anal. Machine Intell. vol. PAMI-5 No.1, Jan. 1983,
5. Devijver P.A., Modeling of Digital Image Using Hiden Markov Mesh Random Fields, Signal Processing IV, Proc. EUSIPCO 88, Grenoble, France, 1988
6. Czarnecki W., Modelling of a 2-d Discrete Stationary Random Signal Having Specified Probabilistic Properties, Signal Processing V, Proc.EUSIPCO 90, Barcelona, Spain 1990.

SIGNAL PROCESSING VI: Theories and Applications
J. Vandewalle, R. Boite, M. Moonen, A. Oosterlinck (eds.)

NOVEL SCHUR TYPE ALGORITHM FOR NEAR TO TOEPLITZ LINEAR SYSTEMS WITH MULTICHANNEL ENTRIES

A. Liavas & S. Theodoridis, Dept. of Computer Eng., University of Patras, Patras 26500, Greece

An efficient Schur type algorithm for block ρ-Toeplitz system solution is developed. The algorithm avoids matrix operations since it exploits the near to Toeplitz property of the coefficient matrix on a scalar level. As a consequence it can be implemented on a linear array processor consisted of scalar operators locally interconnected.

1. INTRODUCTION

Multichannel system identification, multirate filtering and seismic signal enhancement are some typical cases which result in the solution of a well structured block linear system of equations (block Toeplitz or block near to Toeplitz).

The notion of ρ-Toeplitz was introduced in [1], with ρ serving as a measure of how near to Toeplitz the matrix is.

All the algorithms for block ρ–Toeplitz system solution derived so far are straightforward generalizations of the single channel case [1, 2]. Matrix variables of order k and block vectors with entries $(k \times k)$ matrices appear in the place of scalar and vector variables respectively, complicating the resulting architectures associated with the Schur type algorithms, which offer enhanced parallelism compared to their Levinson type counterparts. Recently motivated by [3], we proposed a stairwise Levinson type algorithm for the solution of a block ρ-Toeplitz system which exploits the near to Toeplitz property of the coefficient matrix on a scalar level avoiding order k matrix operations [4]. In this paper starting from the stairwise Levinson type algorithm we will develop a Schur type algorithm which can be mapped in a straightforward way on a linear array processor, involving scalar operations and simplified localized interprocessor communications.

2. THE ALGORITHM

Suppose that we want to efficiently solve

$$R\underline{c} = -\underline{d} \tag{1}$$

where R is a $(p \times p)$ block near-to-Toeplitz matrix with entries $(k \times k)$ matrices and $\underline{d}$ is a $(p \times 1)$ block vector with entries $(k \times 1)$ vectors.

We define R_m $(1 \leq m \leq p)$ the block matrix that results from R after deleting its last $(p-m)$ block rows and columns and S_m $(1 \leq m \leq p-1)$ the block matrix that results from R_{m+1} after deleting its first block row and column. Thus

$$R_{m+1} = \begin{pmatrix} R_m & * \\ * & * \end{pmatrix} = \begin{pmatrix} * & * \\ * & S_m \end{pmatrix} \tag{2}$$

where * denotes block elements of no current interest.

Since R is a block near to Toeplitz matrix then [2]:

$$S_{p-1} - R_{p-1} = \sum_{i=1}^{\rho} (\pm) \underline{n}_i \underline{n}_i^{\prime t} \tag{3}$$

where $\underline{n}_i$ and $\underline{n}_i^{'}$ are $((p-1) \times 1)$ block vectors with entries $(k \times 1)$ vectors. ρ is assumed to be small with respect to p. If there is no representation in (3) with smaller number of terms, ρ is called the displacement rank. It can easily be seen that the low rank property of the difference of the two submatrices S_{p-1}, R_{p-1} is carried out for all principal submatrices of R. Thus if one defines

$$H = (\underline{n}_1 \quad \dots \quad \underline{n}_\rho) \qquad H^{'} = \begin{pmatrix} \underline{n}_1^{'t} \\ \vdots \\ \underline{n}_\rho^{'t} \end{pmatrix} \tag{4}$$

then

$$S_m - R_m = H_m \Sigma H_m^{'} \qquad \text{for} \quad m = 1, \dots, p-1 \tag{5}$$

H, $H^{'t}$ can be thought as $((p-1) \times 1)$ block vectors with entries $(k \times \rho)$ matrices. H_m $(H_m^{'})$ results from H $(H^{'})$ after deletion of its last (p-m-1) block rows (columns). Σ is a $(\rho \times \rho)$ signature matrix with ± 1 on its main diagonal and zeros elsewhere.

In order to derive the stairwise Levinson type algorithm we define [4]:

a) R_{mk+i}^j: $(mk+i) \times (mk+i)$ submatrix of the block matrix R, whose top leftmost element is the (j,j) element of R.

$$R_{mk+i}^j = \begin{pmatrix} r(j,j) & \dots & r(j,mk+j+i-1) \\ \vdots & \ddots & \vdots \\ r(mk+j+i-1,j) & \dots & r(mk+j+i-1,mk+j+i-1) \end{pmatrix} \tag{6}$$

b) $\underline{r}_{mk+i}^{'(x,y)t}$: $1 \times (mk+i)$ vector. It is part of the x-th row of R with leftmost element $r(x,y)$.

$$\underline{r}_{mk+i}^{'(x,y)t} = (\, r(x,y) \quad \dots \quad r(x,mk+y+i-1) \,) \tag{7}$$

c) $\underline{r}_{mk+i}^{(x,y)}$: $(mk+i) \times 1$ vector. It is part of the y-th column of R with topmost element $r(x,y)$.

$$\underline{r}_{mk+i}^{(x,y)} = (\, r(x,y) \quad \dots \quad r(mk+x+i-1,y) \,)^t \tag{8}$$

d) S_{mk+i}: $(mk+i) \times (mk+i)$ principal submatrix of the block matrix S_{p-1}. At this point we should note that since S_{p-1} results from R after deleting its first block row and column – on a scalar level, after deleting its first k rows and columns – then:

$$S_{mk+i} \equiv R_{mk+i}^{k+1} \tag{9}$$

It can easily be seen from (2-5) that for $m = 0, \dots, p-2$ and $i = 1, \dots, k$:

$$S_{mk+i} \equiv R_{mk+i}^{k+1} = R_{mk+i}^1 + H_{mk+i} \Sigma H_{mk+i}^{'} \tag{10}$$

where H_{mk+i} $(H_{mk+i}^{'})$ results from H $(H^{'})$ after deleting its last $(p-m-1)k-i$ rows (columns).

e) $\underline{d}_{mk+i}$: $(mk+i) \times 1$ vector that results from $\underline{d}$ after deleting its last $(p-m)k-i$ elements.

One block order update is achieved in k single steps by the recursive solution of

$$R_{mk+i}^1 \underline{c}_{mk+i} = -\underline{d}_{mk+i} \qquad i = 1, \dots, k \tag{11}$$

The three basic internal variables of the stairwise Levinson type algorithm of [4] are given in Table 1. It is well known that Levinson type algorithms are not well suited for parallel implementation. Their drawback stems from the way the reflection coefficients of order $mk+i+1$ are computed [4]. They require the computation of a $(mk+i)$-dimensional inner product or matrix-vector product which in a parallel environment can be performed in at least $O(\log_2(mk+i))$ time units.

On the other hand, Schur type algorithms bypass

the above computational bottleneck by avoiding the direct computation of inner products. These, are in turn provided by a set of lattice type operations which can be performed with a high degree of concurrency, in time independent of order $mk+i$.

This can be achieved by defining a set of auxiliary error variables, whose values for certain combinations of their indices coincide with the required inner products. Let us concentrate for example in β^c_{mk+i} defined as [4]:

$$\beta_{mk+i} = \underline{r}'^{(mk+i,1)t}_{mk+i-1}\,\underline{c}_{mk+i-1} + d_{mk+i} \qquad (12)$$

Define for $mk+i \leq n \leq pk$ the error variable:

$$e^c_{mk+i}(n) = \underline{r}'^{(n,1)t}_{mk+i-1}\underline{c}_{mk+i-1} + d_n \qquad (13)$$

It is readily observed that $\beta^c_{mk+i} = e^c_{mk+i}(mk+i)$.

It turns out that for each time instant n, variables $e^c_{mk+i}(n)$ are propagated in order via lattice type operations. Furthermore, these operations can be performed concurrently for all the time instants n involved and this is the point where enhanced parallelism is achieved compared to Levinson type algorithms.

In Table 2 we give the definitions of some of the internal variables of the stairwise Schur type algorithm (due to space limitation), while their order updates are given in Table 3. (Similar results state for the rest internal variables).

The algorithm can be performed on a linear array processor composed by $O(pk)$ locally interconnected computing elements of the multiply-accumulate type in $O((\rho+k)pk)$ time units. Interprocessor communications involve only scalar interchanges (ρ-dimensional internal variables are passed element by element) making the resulting structure simple.

Table 1

Internal variables of the Levinson type Stairwise Algorithm

$1 \leq j \leq k$ and $1 \leq \mu \leq k$

$$\underline{a}^{\mu}_{mk+i} = \begin{cases} -(R^{\mu}_{mk+i})^{-1}\underline{r}^{(\mu,\mu-1)}_{mk+i} & \mu \neq 1 \\ -(R^{1}_{mk+i})^{-1}\underline{r}^{(k+1,k)}_{mk+i} & \mu = 1 \end{cases} \qquad (1.1)$$

$$\underline{b}^{j}_{mk+i} =$$

$$\begin{cases} -(R^{j-i}_{mk+i})^{-1}\underline{r}^{(j-i,mk+j)}_{mk+i} & 0 \leq i < j \\ -(R^{k+j-i}_{mk+i})^{-1}\underline{r}^{(k+j-i,(m+1)k+j)}_{mk+i} & j \leq i < k \end{cases} \qquad (1.2)$$

$$F_{mk+i} = -(R^{1}_{mk+i})^{-1}H_{mk+i} \qquad (1.3)$$

Table 2

Internal variables of the Schur type Stairwise Algorithm

$$e^c_{mk+i}(n) = \underline{r}'^{(n,1)t}_{mk+i-1}\underline{c}_{mk+i-1} + d_n \qquad (2.1)$$

$$e^F_{mk+i}(n) = \underline{r}'^{(n,1)t}_{mk+i-1}F_{mk+i-1} + \underline{h}^t_n \qquad (2.2)$$

$$e^{a(\mu)}_{mk+i}(n) =$$

$$= \begin{cases} \underline{r}'^{(n,\mu)t}_{mk+i-1}\underline{a}^{\mu}_{mk+i-1} + r(n,\mu-1) & \text{if } \mu \neq 1 \\ \underline{r}'^{(n,1)t}_{mk+i-1}\underline{a}^{1}_{mk+i-1} + r(n+k,k) & \text{if } \mu = 1 \end{cases} \qquad (2.3)$$

$$\hat{e}^{b(j)}_{mk+i}(n) =$$

$$\begin{cases} \underline{r}'^{(n,j-i+1)t}_{mk+i-1}\underline{b}^{j}_{mk+i-1} + r(n,mk+j) & j \geq i \\ \underline{r}'^{(n,k+j-i+1)t}_{mk+i-1}\underline{b}^{j}_{mk+i-1} + r(n,(m+1)k+j) & j < i \end{cases} \qquad (2.4)$$

$$\beta^c_{mk+i} = e^c_{mk+i}(mk+i) \qquad (2.5)$$

$$\beta^F_{mk+i} = e^F_{mk+i}(mk+i) \qquad (2.6)$$

$$\beta^{a(\mu)}_{mk+i} = e^{a(\mu)}_{mk+i}(mk+\mu+i-1) \qquad (2.7)$$

Table 3

Order updates of the variables of Table 2

for $n = mk + i + 1$ *to* pk *in parallel do*

$$e^{c}_{mk+i+1}(n) = e^{c}_{mk+i}(n) + \hat{e}^{b(i)}_{mk+i}(n)k^{c}_{mk+i} \quad (3.1)$$

$$e^{F}_{mk+i+1}(n) = e^{F}_{mk+i}(n) + \hat{e}^{b(i)}_{mk+i}(n)k^{F}_{mk+i} \quad (3.2)$$

$$e^{a(1)}_{mk+i+1}(n) = e^{a(1)}_{mk+i}(n) + \hat{e}^{b(i)}_{mk+i}(n)k^{a(1)}_{mk+i} \quad (3.3)$$

$$\hat{e}^{b(i)}_{mk+i+1}(n + k) = e^{b(i)}_{mk+i}(n) + $$

$$+ \left(e^{F}_{mk+i}(n) \quad e^{a(1)}_{mk+i}(n) \right) K_{mk+i} \quad (3.4)$$

end for n

for $j = 1$ *to* k $(j \neq i)$

$\mu = j \circ i$

for $n = mk + \mu + i$ *to* pk *in parallel do*

$$e^{a(\mu)}_{mk+i+1}(n) = e^{a(\mu)}_{mk+i}(n) + $$

$$+ \hat{e}^{b(j)}_{mk+i}(n)k^{a(\mu)}_{mk+i} \quad (3.5)$$

$$\hat{e}^{b(j)}_{mk+i+1}(n) = \hat{e}^{b(j)}_{mk+i}(n) + $$

$$+ e^{a(\mu)}_{mk+i}(n)k^{b(j)}_{mk+i} \quad (3.6)$$

end for n

end for j

REFERENCES:

[1] B. Friedlander, T. Kailath, M. Morf, L. Ljung: 'New Inversion Formulas For Matrices Classified in Terms of Their Distance From Toeplitz Matrices', J. of Linear Algebra and Appls., Vol. 27, pp 31-60, 1979.

[2] N. Kalouptsidis,D. Manolakis, G. Carayannis: 'Efficient Recursive Triangularization, Inversion and System Solution of Near– To Toeplitz matrices and Applications in Signal processing', Signal Processing 6, pp 235-259, 1984.

[3] G. Glentis, N. Kalouptsidis: 'Multichannel LS filtering and system identification and Block adaptive structures', to appear in IEEE Trans. on Signal Processing.

[4] A. Liavas, S. Theodoridis: 'Efficient Levinson type algorithm for block ρ-Toeplitz systems of equations', in Proc. 1991 IEEE ICASSP, Toronto, May 1991.

SIGNAL PROCESSING VI: Theories and Applications
J. Vandewalle, R. Boite, M. Moonen, A. Oosterlinck (eds.)

Using Hidden Variable Fractal Interpolation to Model One-Dimensional Signals*

Greg Vines and Monson H. Hayes, III

Georgia Tech Lorraine
2-3 Rue Marconi, 57070 Metz, France

Abstract

Hidden variable fractal interpolation has been proposed as a flexible way to model one-dimensional signals through the use of iterated function systems. While methods have been proposed for signal reconstruction and for determining map parameters for the hidden variable model, no algorithm exists for determining the necessary hidden variables. In this paper we propose and evaluate several methods for determining these hidden variables, including an optimal approach based on a least-squares solution to the problem. The various methods are tested on several data samples and results are provided showing an improvement in the SNR over modeling with the fractal interpolation method alone.

1 Introduction

Iterated Function Systems (IFSs) have recently received attention in the literature as a new technique for signal modeling [1, 2]. This interest stems from the fact that an IFS is simple in form and yet capable of producing complicated functions, many of which are fractal in nature [3]. IFSs have, in fact, been used to create images and other waveforms that closely resemble those found naturally [4]. In this paper we investigate the use of the Hidden Variable Fractal Interpolation (HVFI) technique to model one-dimensional signals.

Because IFSs are typically implemented with affine maps, the resulting data is self-affine. This property may not always be desirable. The use of the HVFI method allows a self-affine model to be created in a higher dimensional space, while the projection of this data is not constrained to be self-affine [3, 5]. The problem in using the HVFI method is how to determine the best set of data in the additional dimension.

*This work was supported in part by a grant from Rockwell International.

2 Background

The fractal interpolation method of modeling data is based on selecting points from the data samples and creating IFS maps to interpolate between these points. The maps may then be used to recreate the original data. The task is therefore two-fold: first to determine the appropriate interpolation points, and second to determine the best map parameters. The determination of the map parameters has typically been accomplished through a minimization of some error criterion. The selection of interpolation points has turned out to be a tougher problem. Present methods rely on iterative searches with no guarantee that an optimal set of maps will be chosen.

HVFI is a method where a data sequence is placed into a higher dimensional space and modeled there. We will work exclusively with one-dimensional sequences in this paper, which will be interpreted as a two-dimensional graph, residing in $x \times y$-space, and generate z data for these sequences to place them into three dimensions where they are modeled with an IFS. The original data can then be recovered by taking a projection of the reconstruction of the higher dimensional data. Through the selection of the third dimension's data, the HVFI model offers a great deal of flexibility.

2.1 Hidden Variable Fractal Interpolation

The HVFI method is based on using an IFS to interpolate between selected sample points in R^3. We are given the original data sequence, $x[n]$, which consists of N data samples for $N_1 \leq n \leq N_2$, and have to determine the additional dimension's data, $z[n]$, prior to modeling. A variety of methods will be examined to determine this $z[n]$ data. With a given $x[n]$ and $z[n]$, the jth map out of P maps is set up in the following manner

$$w_j \begin{bmatrix} n \\ x[n] \\ z[n] \end{bmatrix} = \begin{bmatrix} a_j & 0 & 0 \\ c_j & d_j & h_j \\ k_j & l_j & m_j \end{bmatrix} \begin{bmatrix} n \\ x[n] \\ z[n] \end{bmatrix} + \begin{bmatrix} e_j \\ f_j \\ g_j \end{bmatrix}, \quad (1)$$

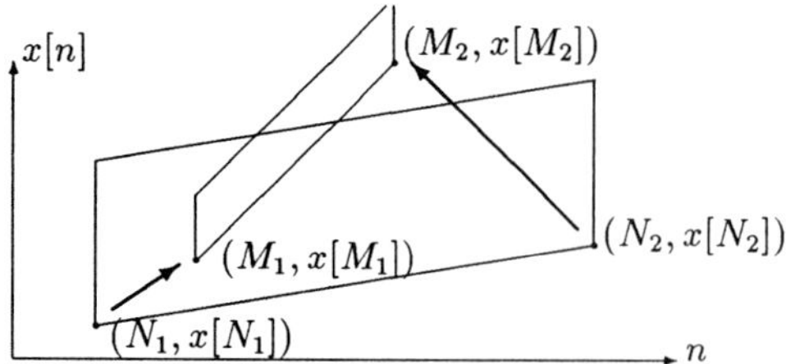

Figure 1: IFS Mapping

where the 0 entries are necessary to insure the resulting function is single valued. The jth map covers M_j of the data samples for $M_{j1} \leq n \leq M_{j2}$. The set of P maps are constructed to cover the entire data sequence from N_1 to N_2. Figure 1 illustrates how a typical map might be set up to generate points from M_1 to M_2. For notational simplicity the j subscript will be omitted in references to the individual maps from now on.

3 Map Parameters

In previous implementations of the fractal interpolation method, the interpolation points were used to eliminate some of the variables by forcing the endpoints of the maps to be exactly equal to the original data [6]. Experimental results have shown that a more accurate model is obtained if the interpolation points are not forced to be exactly correct [7] and we will take advantage of this increased accuracy. The map parameters are determined through a least squares minimization of the error between the data, $x[n]$ and $z[n]$, and the mapped data, $w_x[n]$ and $w_z[n]$,

$$\xi = \sum_{n=N_1}^{N_2} (w_x[n] - x[n])^2 + (w_z[n] - z[n])^2, \qquad (2)$$

where $w_x[n]$ and $w_z[n]$ are the map functions for the $x[n]$ and $z[n]$ data respectively. Using least squares to minimize this error gives the following sets of equations:

$$\sum_{i=0}^{M-1} \left[S \cdot S^T\right] \begin{bmatrix} c \\ d \\ h \\ f \end{bmatrix} = \sum_{i=0}^{M-1} S \cdot x[M_1 + i], \qquad (3)$$

and

$$\sum_{i=0}^{M-1} \left[S \cdot S^T\right] \begin{bmatrix} k \\ l \\ m \\ g \end{bmatrix} = \sum_{i=0}^{M-1} S \cdot z[M_1 + i] \qquad (4)$$

where

$$S = \begin{bmatrix} n & x[n] & z[n] & 1 \end{bmatrix}^T,$$

$$n = N_1 + \left(\frac{N-1}{M-1}\right) i, \qquad (5)$$

and ^{T} indicates the transpose.

Once the map parameters have been calculated, the map must be checked to see if it is a contraction mapping. This can be done by examination of the submatrix

$$A = \begin{bmatrix} d & h \\ l & m \end{bmatrix}. \qquad (6)$$

Essentially this matrix is performing a rotation and scaling of the $x[n]$ and $z[n]$ data during the mapping. If the determinant of A has a magnitude less than one, then it is a contraction mapping.

The remaining tasks in using the HVFI method are the determination of the $z[n]$ data and selection of the interpolation points. Several methods were developed to determine the $z[n]$ data based on a least squares minimization of the error for each map.

4 Hidden Data Determination

4.1 An Optimal Approach

The optimal $z[n]$ data is obtained in a least-squares sense through the error equation (2) for all of the maps. Because the $z[n]$ data is used by all of the maps together, it is not possible to work with each map individually. For each of the P maps there is an error equation of the form (2). Examination of (2) with regard to z shows that there are three distinct ranges depending on the value of i or n, where n is defined as in (5).

1. $M_1 \leq n \leq M_2$ and $i \neq \frac{(M_1-N_1)(M-1)}{M-N}$
2. $M_1 \leq n \leq M_2$ and $i = \frac{(M_1-N_1)(M-1)}{M-N}$
3. n outside of $\{M_1 \cdots M_2\}$

The range $M_1 \leq n \leq M_2$ is in terms of i given by

$$\frac{(M_1 - N_1)(M-1)}{N-1} \leq i \leq \frac{(M_2 - N_1)(M-1)}{N-1}. \qquad (7)$$

Taking the partial derivative of (2) with respect to z for each of these three cases yields three equations:

$$m \cdot z[n] - z[M_1 + i] = -k \cdot n - l \cdot x[n] - g, \qquad (8)$$

$$\begin{aligned} (h^2 + m^2 - m) \cdot z[n] + (1 - m) \cdot z[M_1 + i] = \\ (1-m)(k \cdot n + l \cdot x[n] + g) \\ +h(x[M_1 + i] - c \cdot n - d \cdot x[n] - f), \end{aligned} \qquad (9)$$

$$\begin{aligned} -(h^2 + m^2) \cdot z[n] + m \cdot z[M_1 + i] = \\ h(c \cdot n + d \cdot x[n] + f - x[M_1 + i]) \\ +m(k \cdot n + l \cdot x[n] + g). \end{aligned} \qquad (10)$$

For continuous functions, the IFS maps are constructed to be 'just touching', which means that adjacent maps share end points. For the discrete case this is not necessary and would result in an overdetermined system of equations. Therefore the maps are set up with adjacent, instead of overlapping, end points. With adjacent end points, the P maps provide N equations for the N unknowns. The nice feature of this system of equations is the sparseness of the matrix; there are only two non-zero coefficients in each row - one at n and the other at $M_1 + i$, except in the second case above, where $n = M_1 + i$. The system of equations was solved using an iterative method, both the Gauss-Seidel and the Conjugate-Gradient methods were implemented with similar results.

The algorithm to determine the map parameters and the $z[n]$ data proceeds as follows.

1. Initialize $z[n]$
2. Calculate the map parameters using the method from section 3
3. Calculate $z[n]$ data
4. Calculate error and compare to previous; if the change is small then stop, otherwise go to step 2

Obviously the last step offers some flexibility in that the iterations may be allowed to continue if the error is decreasing, or stopped after a certain number of iterations, and in addition, a check should be made for divergence. The initial value chosen for $z[n]$ will have some affect on the convergence of the algorithm. There is no guarantee of convergence for a given initial value. The only value which should not be used is $x[n]$ itself; which will lead to a singular set of equations when the map parameters are calculated.

4.2 Two Quick Iteration Methods

Rather than solve the system of equations for all cases, if instead $z[n]$ and $z[M_1+i]$ are treated as different variables then a simpler solution is derived. When the partial of (2) with respect to $z[n]$ and $z[M_1 + i]$ is set to zero, the following two equations are obtained:

$$\begin{aligned}(h^2 + m^2)z[n] + (-m)z[M_1 + i] = \\ h(c \cdot n + d \cdot x[n] + f - x[M_1 + i]) \\ +m(k \cdot n + l \cdot x[n] + g)\end{aligned} \tag{11}$$

and

$$(m)z[n] + (-1)z[M_1 + i] = k \cdot n + l \cdot x[n] + g. \tag{12}$$

These equations can be solved for either $z[n]$ or $z[M_1+i]$ resulting in a single equation to calculate $z[n]$ directly from $x[n]$. The two equations obtained are:

$$\begin{aligned}z[M_1 + i] = \frac{m}{h}(x[M_1 + i] - c \cdot n - d \cdot x[n] - f) \\ +k \cdot n + l \cdot x[n] + g,\end{aligned} \tag{13}$$

and

$$z[n] = \frac{x[M_1 + i] - c \cdot n - d \cdot x[n] - f}{h}. \tag{14}$$

While these two equations are not optimal, they can generate the necessary $z[n]$ much more quickly than solving the above system of equations.

The above iterative algorithm is again used to determine the best set of maps and corresponding $z[n]$ data.

4.3 Independent Hidden Variables

With the aforementioned methods one problem which occurs is the interdependence between $x[n]$ and $z[n]$ during the reconstruction process. Errors tend to propagate and in an interest of stopping this, the link from $x[n]$ to $z[n]$ was broken. This was implemented by setting the l variable in all of the maps equal to zero. With this modification equation (4) becomes

$$\sum_{i=0}^{M-1} \left[\hat{S} \cdot \hat{S}^T\right] \begin{bmatrix} k \\ m \\ g \end{bmatrix} = \sum_{i=0}^{M-1} \hat{S} \cdot z[M_1 + i], \tag{15}$$

where

$$\hat{S} = \begin{bmatrix} n & z[n] & 1 \end{bmatrix}^T.$$

This method was implemented with each of the methods for determining $z[n]$ and the results are given in the following section.

Another variation on this idea is to generate the $z[n]$ data as suggested previously, create an IFS to model the $z[n]$ data, use this IFS to recover $\hat{z}[n]$, and finally use $\hat{z}[n]$ when the map parameters are determined for $x[n]$. This method offers the advantage that $\hat{z}[n]$, which is used in determining the $x[n]$ map parameters, can be reconstructed without error. This method did not show any improvements over the others and will not be discussed further.

4.4 Non-Iterative Method

All of the aforementioned methods rely on iteratively calculating $z[n]$ and the map parameters in order to find a solution to the nonlinear system of equations. The computational requirements can be greatly reduced by avoiding these iterations. The system was tested after one iteration of the Independent Hidden Variable method as described in section 4.3 and the results are provided in the next section. The single iteration was set up by first solving for the map parameters, then determining the $z[n]$ data and finally with this new $z[n]$ data, recalculating the map parameters.

5 Results

Because the task of determining the interpolation points is still an unsolved problem, we used an exhaustive search for a fixed number of maps to allow comparison between the various methods proposed for determining $z[n]$.

Table 1 shows results for the different methods discussed including results from a regular IFS, without the $z[n]$ data, for comparison. The tests were conducted with a variety of test files and numbers of maps. Each of the test data sets is a line from an image file and was normalized prior to modeling to facilitate comparisons. Because the exhaustive search rapidly becomes computationally infeasible for more than a few maps, calculations were not made for more than 4 maps.

File Method	2 Maps SNR	3 Maps SNR	4 Maps SNR
1			
Regular IFS	20.0	23.2	24.5
Optimal	20.2	23.7	25.2
Quick: $z[n]$	20.3	23.3	25.1
Quick: $z[M_1+i]$	20.3	22.9	25.1
Independent $z[n]$	20.3	23.7	25.2
Non Iterative	20.3	23.7	25.2
2			
Regular IFS	22.6	25.6	26.8
Optimal	22.7	26.4	28.1
Quick: $z[n]$	22.8	25.2	27.5
Quick: $z[M_1+i]$	18.0	24.9	27.4
Independent $z[n]$	22.9	26.3	28.2
Non Iterative	22.4	26.3	28.2
3			
Regular IFS	25.4	31.0	33.3
Optimal	25.4	31.3	33.4
Quick: $z[n]$	25.3	30.9	33.1
Quick: $z[M_1+i]$	25.2	31.1	33.6
Independent $z[n]$	25.4	31.3	33.6
Non Iterative	25.4	31.2	33.6
4			
Regular IFS	26.4	30.4	33.0
Optimal	26.4	30.4	33.0
Quick: $z[n]$	26.2	30.4	33.1
Quick: $z[M_1+i]$	24.8	30.1	32.8
Independent $z[n]$	26.2	30.4	33.3
Non Iterative	26.3	30.4	33.3
5			
Regular IFS	11.9	15.6	16.9
Optimal	14.0	17.0	19.8
Quick: $z[n]$	14.0	17.3	19.6
Quick: $z[M_1+i]$	13.6	16.8	18.9
Independent $z[n]$	14.3	17.3	19.9
Non Iterative	14.2	17.0	19.9

Table 1: HVIFS Model Performance

6 Conclusion

Each of the methods provided similar results, which suggests using the simplest method to reduce the computational requirements. It turned out that the selection for the initial $z[n]$ made a difference and in general the best results were obtained with a constant value of 0.01.

The simple method for calculating $z[n]$ with independent hidden variables and initial $z[n] = 0.01$ gave the best results of all the methods presented. Of all the iterative approaches, this method also requires the least computations for solving a particular map. The non-iterative version of the independent hidden variables method also provided very good results with significantly fewer computations required. The difference in the results of these two methods is so small that the non-iterative method is probably the best approach overall.

References

[1] D. S. Mazel and M. H. Hayes. "Iterated function systems applied to discrete sequences". *IEEE Trans. ASSP*. To appear.

[2] Tor A. Ramstadt, Geir E. Øien, and Skjalg Lepsøy. "An inner product space approach to image coding by contractive transformations". In *Proc. ICASSP*, volume 4, pages 2773–2776, 1991.

[3] Michael Barnsley. *Fractals Everywhere*. Academic Press, Inc., New York, N. Y., 1988.

[4] B. B. Mandelbrot. *The Fractal Geometry of Nature*. W. H. Freeman and Co., New York, N. Y., 1982.

[5] M. F. Barnsley, J. Elton, D. Hardin, and P. Massopust. "Hidden-variable fractal interpolation functions". *SIAM Journal of Mathematical Analysis*, 20(5):1218–1242, September 1989.

[6] D. S. Mazel and M. H. Hayes. "Hidden-variable fractal interpolation of discrete sequences". In *Proc. ICASSP*, volume 5, pages 3393–3396, 1991.

[7] M. H. Hayes, G. Vines, and D. S. Mazel. "Using fractals to model one-dimensional signals". In *Proceedings of the Thirteenth GRETSI Symposium*, volume 1, pages 197–200, 1991.

SIGNAL PROCESSING VI: Theories and Applications
J. Vandewalle, R. Boite, M. Moonen, A. Oosterlinck (eds.)
1992 Elsevier Science Publishers B.V.

All-poles modelling using a linear combination of cumulants

Josep Vidal and José A. R. Fonollosa

Universitat Politècnica de Catalunya
E.T.S.E.Telecomunicació, Apdo. 30002, 08080 Barcelona, Spain
E-mail: pepe@tsc.upc.es Fax:34-3-4016447 Tel:34-3-4016457

We address the problem of estimating the parameters of a pure AR causal model excited with non-Gaussian i.i.d. samples of an unobservable process. Output samples may be corrupted with colored Gaussian noise. Other types of noise are allowed provided that they have a set of m-th order statistics whose value is zero, if the same set of statistics of the signal be different of zero. It is shown that each sample of the impulse response of the AR system that generates the process may be expressed as a linear combination of cumulant slices of any order. The resulting algorithm is shown to be well-behaved and to provide consistent estimates while reducing the complexity significantly with respect to other approaches.

I. Introduction

Last years have witnessed a great development of higher-order based techniques in system identification problems. Most of the efforts made in ARMA modeling in the past were based on the properties of the autocorrelation function as an optimum way to deal with Gaussian processes and linearities. However, in many applications were non-Gaussian processes or nonlinearities are present, second order approaches fail to find proper estimations. This is the case also if the system to identify is not minimum phase. Cumulants and their Fourier transform (polyspectra) do contain information about deviations of normality, nonlinearity and phase [4], and as such their use becomes very interesting. Another important property of higher-than-two order statistics (HOS) is that they give unbiased results even when the observed signal is corrupted with additive colored Gaussian noise. Therefore, the interest in their use has risen in other domains, as blind equalization, speech recognition, array processing, non-linear filtering, time-frequency spectral analysis, etc.

In this paper we will deal with the all-poles spectrum estimation of a non-Gaussian process corrupted with noise using HOS. For a complete review on the topic see [3]. The approach followed here makes use of the principles found in [7] and [8], that is combination of different cumulant slices in a linear algebra problem to render a consistent algorithm that guaranties uniqueness. Cumulants of different orders can be combined, and other cumulant slices added to improve the performance. The final algorithm requires much less computation than the HOS Yule-Walker equations while still giving low variance estimates.

II. AR impulse response determination

Consider a zero-mean and ergodic process $\{y(n)\}$, its m-th order cumulant function is defined as the coefficients of the Taylor series expansion of its characteristic function [4]. More precisely the third and fourth order cumulants of the process are defined as:

$$C_{3,y}(i,j) = E\{y(n)y(n+i)y(n+j)\}$$

$$\begin{aligned} C_{4,y}(i,j,k) = {} & E\{y(n)y(n+i)y(n+j)y(n+k)\} - \\ & E\{y(n)y(n+i)\}E\{y(n)y(n+j)\} - \\ & E\{y(n)y(n+i)\}E\{y(n)y(n+k)\} - \\ & E\{y(n)y(n+j)\}E\{y(n)y(n+k)\} \end{aligned}$$

By fixing m-2 lags of this function, a 1-dimensional sequence is obtained, usually called a 1-D cumulant slice. Assume that $\{y(n)\}$ is an AR(p) causal process:

$$y(n) = \sum_{i=1}^{p} a(i)y(n-i) + e(n) \qquad (1a)$$

corrupted by another process $\{z(n)\}$:

$$x(n) = y(n) + z(n) \qquad (1b)$$

and fitting the following hypotheses:

H1: The innovation process e(n) is stationary, i.i.d., non-Gaussian, with finite m-th order cumulants $\gamma_{m,e}$, and $\gamma_{1,e}=0$.

H2: H(z) is exponentially stable, that is, all poles are inside the unit circle.

H3: The process z(n) is independent of e(n), zero-mean, Gaussian and of unknown power spectrum. z(n) is allowed to be non-Gaussian if $\gamma_{m,z}=C_{m,x}(0,...,0)=0$ for some $m\geq3$ provided that $\gamma_{m,e}\neq0$ for the same m.

The property that makes HOS useful is that cumulants of any order are zero for a Gaussian process, so the cumulants of x(n) equal the cumulants of y(n) provided that z(n) be Gaussian.

The impulse response h(n) of the linear time-invariant system that generates y(n) satisfies the following recursion:

$$\sum_{i=0}^{p} a(i)h(n-i) = \delta(n) \qquad a_0 = 1 \quad (2)$$

Using this recursion and the Brillinger-Rosenblatt summation formula [5] that relates the m-th order cumulants of x(n) with the impulse response (for $m \geq 2$):

$$C_{m,x}(i_1, i_2, ..., i_{m-1}) = \gamma_m \sum_{n=-\infty}^{\infty} \prod_{k=0}^{m-1} h(n+i_k) \qquad i_0 = 0 \quad (3)$$

Linearly combining equation (3) with the AR parameters and using (2) we get:

$$\sum_{l=0}^{p} a(l)\, C_{m,x}(-l, i_2, ..., i_{m-1}) = \gamma_m \prod_{k=2}^{m-1} h(i_k) \qquad (4)$$

Repeating the operation (m-1) times for the (m-1) lag coefficients, γ_m can be obtained [9] :

$$\sum_{l_1=0}^{p} \cdots \sum_{l_{m-1}=0}^{p} a(l_1)...a(l_{m-1})C_{m,x}(-l_1,...,-l_{m-1}) = \gamma_m \qquad (5)$$

In [1] it was shown that the impulse response h(n) can be obtained from (4) taking the lags $i_3=i_4=...=i_{m-1}=0$:

$$\sum_{l=0}^{p} a(l)\, C_{m,x}(-l, i, 0..., 0) = \gamma_m h(i) \qquad h(0) = 1 \quad (6)$$

However this is not the only linear combination of cumulants samples that yields h(n). In fact, it is easy to see from (4) that any linear combination of the type:

$$W(i) = \qquad (7)$$

$$\sum_{l_1=0}^{p} \sum_{l_2} \cdots \sum_{l_{m-1}} a(l_1)w(l_2)...w(l_{m-1})C_{m,x}(-l_1,-l_2,...,-l_{m-2},i)$$

renders h(n), that is $W(i) = \mu h(i)$, provided that:

$$\mu = \gamma_m \prod_{i=2}^{m-2} \sum_{l_i} w(l_i)h(l_i) \neq 0 \qquad (8)$$

Relations (7) and (8) are also satisfied by any linear combination of slices of order equal or greater than two. We will try to go further showing that the weighted sum of cumulant slices (w-slice):

$$C_w(i) = w_2C_{2,x}(i) + \sum_{j=-M}^{N} w_3(j)C_{3,x}(i,j) + \sum_{j=-M}^{N} \sum_{l=-M}^{N} w_4(j,l)\, C_{4,x}(i,j,l) + ... \qquad (9)$$

may yield the impulse response h(n) without previous knowledge of a(i), if the weights are appropriately chosen. This is proven in the following proposition, letting constants M and N unspecified for the moment:

Proposition 1: If $C_w(i)$ is causal, then $C_w(i) = C_w(0)h(i)$.
Proof. Let us linearly combine the w-slice of (9) with the AR parameters:

$$\sum_{k=0}^{p} a(k)C_w(n-k) = w_2 \sum_{k=0}^{p} a(k)C_{2,x}(n-k) + \sum_{j=-M}^{N} w_3(j) \sum_{k=0}^{p} a(k)C_{3,x}(n-k,j) + \sum_{j,l=-M}^{N} w_4(j,l) \sum_{k=0}^{p} a(k)C_{4,x}(n-k,j,l) + ... \qquad (10)$$

combining (2), (4) and (10) the following equality is obtained:

$$\sum_{k=0}^{p} a(k)C_w(n-k) = [\gamma_2 w_2 + \gamma_3 \sum_{j>0} w_3(j)\, h(j) + \gamma_4 \sum_{j,l>0} w_4(j,l)\, h(j)h(l) + ...]\, h(0)\delta(n) \qquad (11)$$

Note that taking $n \geq 0$ in (11) the Yule-Walker set of equations for the w-slice $C_w(i)$ are obtained. If the weights w_2, $w_3(j)$, $w_4(j,l)$,... are chosen in such a way that:

$$C_w(i) = 0 \qquad i < 0$$
$$C_w(0) \neq 0 \qquad (12)$$

by similarity between (11) and (2) we get that $C_w(i)=C_w(0)h(i)$ ♦

This is an important theoretical result but does not help much in finding a proper set of weights because the conditions stated in (12) suppose an infinite set of linear equations to be solved. Fortunately, we can reduce the number of equations considering the following proposition:

Proposition 2: If $C_w(-i) = 0$ for the set $A:\{i = 1...P\}$, being P an upper bound of p, then $C_w(i) = C_w(0)h(i)$ for the same set A.
Proof: The proof is very simple and it follows from (11) taking $0 \leq n \leq P$ and by similarity with (2) ♦

It is worth noting that second order statistics can also be used at constructing the w-slice if the signal is known to be non corrupted by noise.

III. AR w-slice algorithm

We have just found a hint on how to estimate the impulse response through a limited size linear algebra problem. More specifically, proposition 2 gives the set of equations to be considered. In practice, estimation of the AR parameters needs only p+1 samples of h(n), so the resulting set can be condensed in the matrix equation:

$$\mathbf{S_a w} = \mathbf{1} \qquad (13)$$

where $\mathbf{S_a}$ is the anticausal w-slice matrix:

$$\mathbf{S_a} = \begin{pmatrix} C_{2,x}(-P) & C_{3,x}(-P,j) & \dots & C_{4,x}(-P,j,k) & \dots \\ \vdots & \vdots & & \vdots & \\ C_{2,x}(0) & C_{3,x}(0,j) & \dots & C_{4,x}(0,j,k) & \dots \end{pmatrix}$$

w is the weights vector:

$$\mathbf{w} = \begin{pmatrix} w_2 & w_3(j) & \dots & w_4(j,k) & \dots & w_5(j,k,l) & \dots \end{pmatrix}^t$$

and **1** the anticausal impulse response:

$$\mathbf{1} = (0, \dots 0, \quad 1)^t$$

Although in general the rows-rank of $\mathbf{S_a}$ is unknown and may depend on the actual parameters of the process, there is always at least one exact solution to (13). Considering (7) and (8) it may be seen that if the weights $w(l_i)$ equal the actual AR parameters, h(n) is obtained. This is a proof of existence of solution of the vector **w** but not of uniqueness. In fact we are not concerned with uniqueness since proposition 1 shows that any solution yielding (12) is a valid one. The use of the Moore-Penrose inverse matrix $\mathbf{S_a}^{\#}$ in this case ensures that, among the subspace of exact solutions, the minimum-norm one is obtained if the upscript and subscript N and M in (9) and (10) are specified in such a way that the solution $w(l_i)= a(l)$ be included. It means that if an upper bound P of p is known, the range of values are to be N>0 and M<-P. If SVD is used in the computation of $\mathbf{S_a}^{\#}$, the solution will always be well conditioned.

The final AR w-slice algorithm can be resumed in three steps:

S1: Computation of the minimum norm weights that yields a causal w-slice with $C_w(0)=1$:

$$\mathbf{w_m} = \mathbf{S_a}^{\#}\,\mathbf{1} \tag{14}$$

S2: Estimate the causal part of the impulse response using:

$$\hat{\mathbf{h}} = \mathbf{S_c}\,\mathbf{w_m} \tag{15}$$

where $\mathbf{S_c}$ denotes the causal counterpart of $\mathbf{S_a}$:

$$\mathbf{S_c} = \begin{pmatrix} C_{2,x}(P) & C_{3,x}(P,j) & \cdots & C_{4,x}(P,j,k) & \cdots \\ \cdot & \cdot & & \cdot & \\ \cdot & \cdot & & \cdot & \\ \cdot & \cdot & & \cdot & \\ C_{2,x}(0) & C_{3,x}(0,j) & \cdots & C_{4,x}(0,j,k) & \cdots \end{pmatrix}$$

and $\hat{\mathbf{h}}$ is the estimated impulse response:

$$\hat{\mathbf{h}} = (\hat{h}(P), \ldots, \hat{h}(1), 1)^t$$

S3: Solve (2) by simple backsubstitution, that is:

$$\begin{pmatrix} \hat{h}(0) & 0 & \cdots & 0 \\ \hat{h}(1) & \hat{h}(0) & \cdots & 0 \\ \cdot & \cdot & & \cdot \\ \cdot & \cdot & & \cdot \\ \hat{h}(P-1) & \hat{h}(P-2) & \cdots & \hat{h}(0) \end{pmatrix} \begin{pmatrix} \hat{a}_1 \\ \hat{a}_2 \\ \cdot \\ \cdot \\ \hat{a}_P \end{pmatrix} = \begin{pmatrix} -\hat{h}(1) \\ -\hat{h}(2) \\ \cdot \\ \cdot \\ -\hat{h}(P) \end{pmatrix} \tag{16}$$

Once the AR parameters have been estimated, the higher-order moments γ_m of e(n) can be computed from (5). It is worth pointing that the values of M and N are parameters of design, and as they are chosen to consider more cumulant lags, a progressive reduction in variance might be expected. However this reduction may not be so because it depends also on the variance of the estimated cumulants being included. Of course, the length of the data record as well as the theoretical minimum variance limit the improvement. The backsubstitution approach in S3 suffers from the effect of the accumulation of the variance of the estimated AR parameters as the order is increased. This can be solved choosing deliberately a big value of P and overdetermining the system in (16). In solving the overdetermination Total Least Squares or Least Squares can be used. Simulations will show the improvement in this case. This issue has little importance in those applications using the impulse response rather than the AR parameters. In these cases step S3 is skipped. Note that although order determination cannot be stated from the rank of $\mathbf{S_a}$ this decision can be done on the final vector **a**.

IV. Computational complexity

Once the methods have been described, some comments on their computational cost are in order.The simulations in section V will exhibit the behavior of the method AR w-slice presented in section III with respect to the High Order Yule-Walker equations. The main computational cost for each algorithm consists of two parts: 1) the sample mean estimation of the cumulants, which depends only on the number of data points and 2) the computation of the vector of parameters, which depends mainly on the order of the model. The computation of the first part is specially significant if the data record is large, and might exceed the second one. The computational burden of the second part is mainly due to the matrices size.

Solving a 3rd order Yule-Walker set of equations requires SVD on a pxp(p+1) matrix, if p+1 slices are used to ensure identifiability. The AR w-slice algorithm proposed in III, with the use of 3rd order statistics, solves a matrix of size (p+1)x(2p+1) if M=N=p in step S1. Computations of step S2 imply p(2p+1) floating point operations. Step S3 needs p(p+1)/2 flops if the system is not overdetermined. Although being conceptually different, the minimum norm solution to the AR w-slice algorithm and the LS solution - used in the Yule-Walker equations solution- are computed by means of the Moore-Penrose inverse matrix via SVD. In all cases the number of flops required for an mxn matrix (being m>n) is $mn^2 + \frac{17}{3}n^3$ [2]. Table 1 summarizes the total number of flops required in each case, once the cumulants have been computed. Clearly the computational demands of the AR w-slice proposed algorithm are significantly smaller than those of the other approach.

Table I. Computational cost of the pseudoinverse solution.

Algorithm	3rd order Yule-Walker	AR w-slice
Matrices size	p x p(p+1)	(p+1)x(2p+1)
Floating point operations	$p^4+\frac{20}{3}p^3$	$\frac{23}{3}p^3+22p^2+O(p)$

V. Simulations

The goal of these simulations is to illustrate the performance of the Yule-Walker HOS equations versus the AR w-slice algorithm proposed in section III . Both methods use in this case third-order statistics and, for meaningful comparisons, all are adapted to deal with the same non-redundant set of cumulants. All examples involve pure causal AR processes. The driving noise is formed with zero-mean, i.i.d. samples of an exponential process with $\gamma_2 = 1$. In each example 100 independent Monte Carlo runs are performed, and the sample mean of each realization is subtracted in every run.

Example 1. As previously mentioned, the use of backsubstitution on (16) accumulates the variance of the estimates as the order p increases. This problem may be solved by overestimating the order P when solving h(n), and then by constructing (16) as an overdetermined system. The objective of this example is to compare the use of the Least Squares and the Total Least Squares in the solution of the overdetermination. In a general case, LS solution is preferred if the noisy observations are present in the r.h.s. of the equation to be solved. TLS is better if the errors are present both in the system matrix and in the r.h.s. vector. In this case nothing can be said a priori since r.h.s. vector contains noisy observations and the system matrix consist of actual data (zeros in the upper part) and noisy observations (h(n) samples). First, non-overdetermined AR w-slice algorithm is run with M=N=3p and P=p, solving (16) by backsubstitution. Then we overdetermine the algorithm setting M=N=2p and P=2p, in such a way that the same set of non-redundant lags are used. Results in table II for an AR (4) process, show that overdetermination improves the results significantly and that insignificant differences are found between TLS and LS solutions. In the sequel, simulations involving the AR w-slice algorithm will be done taking M=N=2p and P=2p and LS solving of equation (16).

Table II. Ex. 1, 3000 samples, 100 MonteCarlo runs

Pars.	True values	AR w-slice + backsubstitution	Overestimated order AR w-slice + TLS	Overestimated order AR w-slice + LS
a(1)	-1.352	-1.327±0.102	-1.356±0.073	-1.349±0.073
a(2)	1.338	1.307±0.161	1.351±0.094	1.338±0.093
a(3)	-0.662	-0.642±0.195	-0.682±0.088	-0.670±0.087
a(4)	0.240	0.239±0.179	0.250±0.055	0.247±0.0540

Example 2. In general the performance of all algorithms depends on the bandwidth of the process being tested. This is so because all algorithms use a finite number of slices, whereas narrow band AR processes cumulants spread over a large set. The performance is also influenced by the central frequence of the process spectrum. So, rather than testing specific well-chosen values for the AR parameters we have preferred to carry out a more general simulation. Bandpass processes of the kind $H(z)=(1+a(1)z^{-1}+a(2)z^{-2})^{-1} = (1 - 2r\cos\theta z^{-1} + r^2 z^{-2})^{-1}$ are used. The damping factor r takes the value r=0.85. Twenty different values of θ, between 0 and 2π have been used and for each θ, 100 MonteCarlo realizations have been run with record lengths of 1000 samples. No noise is present in this case. Variances are shown in figures 1 and 2 along with the deviations of the mean values from the expected ones. As can be seen, the AR w-slice and Yule-Walker equations present a similar behavior in variance. In mean, the AR w-slice present a better performance. We do not present further simulations for the lack of space, but more can be found in [10].

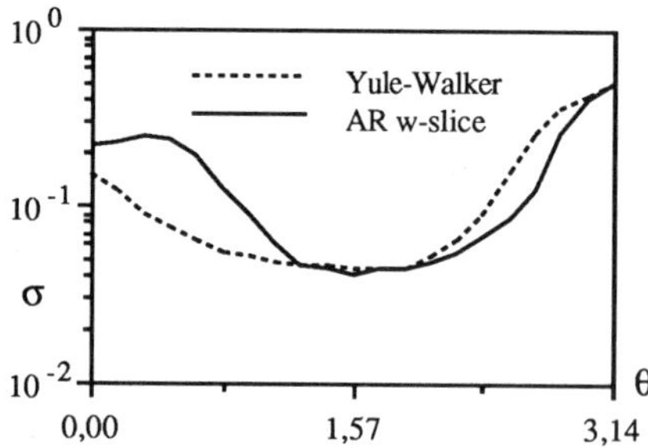

Figure 1. Standard deviation of a(1) versus θ (central frequency of the band) for both methods.

VI. Conclusions

We have studied how to combine cumulant slices in the estimation of an AR system from its output samples. It has been shown that, the proposed AR w-slice algorithm exhibits a behavior comparable to the Yule-Walker equations, but a much smaller computational burden. It represents a good trade-off between cost and statistical behavior, with the additional feature of being capable to deal with statistics of different orders within the same framework. Further work points to the analytical computation of the statistical behavior of the method in comparison with other AR approaches as well as its optimization using the approach in [6].

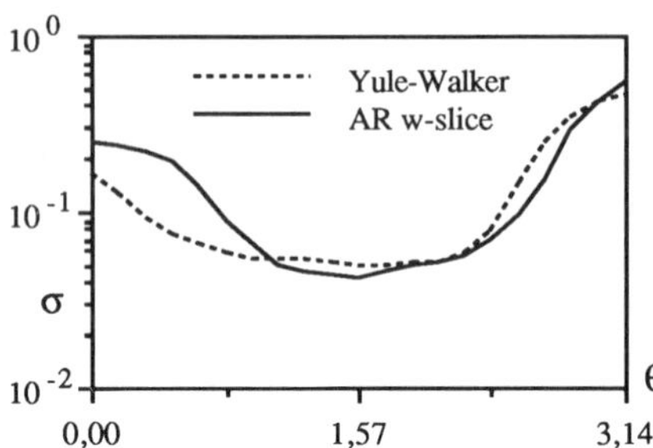

Figure 2. Standard deviation of a(2) versus θ (central frequency of the band) for both methods.

References

[1] G. B. Giannakis, "On the Identifiability of Non-Gaussian ARMA Models Using Cumulants", *IEEE Transactions on Automatic Control*, vol. 35, no. 1, January 1990.

[2] G. H. Golub and C. F. Van Loan, *Matrix Computations*, The John Hopkins University Press, Baltimore, MA, 1983, chap. 6, pp. 169-177

[3] J. M. Mendel, "Tutorial on Higher Order Statistics (Spectra) in Signal Processing and System Theory: theoretical results and some applications", *Proc. IEEE*, vol. 79, no. 3, pp. 278-305, March 1991.

[4] C.L. Nikias and M. R. Raghuveer, "Bispectrum Estimation: A Digital Signal Processing Framework", *Proc. IEEE*, vol. 75, pp. 869-891, July 1987.

[5] -, "Bispectrum Estimation: A Parametric Approach", *IEEE Trans. on Acoustics, Speech and Signal Processing, vol. ASSP-33, no. 4, October 1985.*

[6] B. Porat and B. Friedlander, "Performance Analysis of Parameter Estimation Algorithms based on Higher-Order Moments", *Int. J. Adaptive Contr. Signal Processing*, vol. 3, pp. 191-229, 1989.

[7] J. A. Rodríguez-Fonollosa, J. Vidal and E. Masgrau, "Adaptive System Identification Based on Higher-Order Statistics", *Proc. of the ICASSP'91*, Toronto, May 1991.

[8] J. A. Rodríguez-Fonollosa and J. Vidal, "Adaptive ARMA Identification Using Cumulants", *Proc. of the International Signal Processing Workshop on HOS*, Chamrousse, July 1991.

[9] J. K. Tugnait, "On Identifiability of ARMA Models of Non-Gaussian Processes Via Cumulant Matching", *Proc. of the International Signal Processing Workshop on HOS,* Chamrousse, July 1991.

[10] J. Vidal and J. A. R. Fonollosa, "Causal AR modeling using a linear combination of cumulnat slices", *submitted to Signal Processing, Special Issue on Higher Order Statistics (EURASIP).*

SIGNAL PROCESSING VI: Theories and Applications
J. Vandewalle, R. Boite, M. Moonen, A. Oosterlinck (eds.)
1992 Elsevier Science Publishers B.V.

TRANSFER FUNCTION OF LINEAR CONTINUOUS TIME-VARYING SYSTEMS VIA NON-COMMUTATIVE ALGEBRA

M. GUGLIELMI
Laboratoire d'Automatique de NANTES U.R.A. C.N.R.S. 823
Ecole Centrale de NANTES
1 rue de La Noë 44072, NANTES cedex 03, FRANCE
Tel: (33) 40 37 16 48
Fax: (33) 40 74 74 06

Keyword: linear continuous system, non-stationary, transfer function, non-commutative polynomials

Abstract:

In this paper, we pay attention to continuous systems described by linear differential equations with time-varying coefficients. Some properties of such systems can be obtained from the study of non-commutative polynomials of the formal variable $\lambda=\frac{d}{dt}$. In our approach, it is showed that it is possible to introduce the transfer function by the means of "skew" rational fraction. But, in this case, each rational fraction has two factorizations: the left one and the right one whose factors are in relation. From that, we can enlarge the transfer function concept for non-stationary systems. Then, we precise the association rules for serial or parallel systems from the computation rules over rational "skew" fractions.

1- INTRODUCTION:

During the past decade, considerable research has been done about linear stationary systems via the transfer function approach. By means of this representation, it may be solved some control problems like pole placement, reference model... The basic tool used is an algebraic set of operations over commutative polynomials of the formal LAPLACE variable in the continuous case, or of the z-variable for discrete systems.

Generalizations along this line to linear time-varying systems is difficult. One way to enlarge the transfer function concept is done through formal power series. For discrete systems, Kamen and al [1] have proposed a solution via an extension of formal power series with coefficients in a ring of time fonctions. In the continuous case, Emre and al [2] have showed that transfer marices can also be defined as formal power series and they proposed a general solution to time-vaying problems like observers... These approachs lead to important results like, for example, those obtained in realization theory. Unfortunately, this description is not always very easy to use (for example: interconnection of systems, relation with the original differential equation..). Here, we propose another approach for continuous single input-single output (SISO) systems described by A(p) y = B(p) u, in which A(p) and B(p) are non-commutative polynomials with coefficients in a subset of C^{∞}. Then, a necessary and sufficient condition to define the rational non-commutative fraction field is that the "skew" polynomial ring must verify the two ORE's conditions [3] (here "skew" means only that the variable does not commute with the coefficients). The problem is that each rational fraction has two different factorizations, the left one described by XY^{-1} and the right one by $Y^{-1}X$. In this way, we develop a theory which resembles the time-invariant case. We define the transfer function of a SISO time-varying system by the means of the isomorphism with the rational non-commutative field. The major consequence is that the tranfer function has two expressions: the right one and the left one which are in relations.

This paper is organised as follows: after the preliminaries in the second section, we present, in the third one, the elementary operations as the computation of the Greatest Common Left Divisor (G.C.L.D.) or the Least Common Right Multiple (L.C.R.M.) over non-commutative polynomials. These two polynomials are the basic elements since that, in general, they are sufficient to introduce all the other operations like Greatest Common Right Divisor and Least Common Left Multiple. In the final part of this section, we consider the field of the "skew" rational fractions. Then, in the fourth section, we enlarge the definition of the transfer function of time-varying systems. From that, we give, in the next section, some results about classical operations over systems (like serial and parallel combinations) for which the results are obtained from arithmetic operations over rational skew fractions . As in the stationary case, all these computations need the G.C.L.D. and the L.C.R.M..

2- DEFINITIONS:

In the whole paper, we define (Σ) as the S.I.S.O system described by an ordinary differential equation **(I)** with time-varying coefficients:

$$\boxed{\begin{array}{l} y^{(n)}(t)+a_1(t)y^{(n-1)}(t)+\dots\dots\dots +a_n(t)y^{(0)}(t)= \\ b_0(t)u^{(n)}(t)+b_1(t)u^{n-1})(t)+ \dots\dots\dots + b_n(t)u^{(0)}(t) \end{array}}$$

where: $y^{(i)}(t)=\dfrac{d^iy(t)}{dt^i}$, $u^{(i)}(t) =\dfrac{d^iu(t)}{dt^i}$

The functions $a_i(t)$ and $b_i(t)$ belong to an ordinary differential field **K.**

Recall that **K** is an ordinary differential field if : [4]

1°) it is a field

2°)there exists a unique derivation λ denoted $\lambda[a] = \dot{a}$ which verifies the usual properties:

if $\forall$ a and b $\in$ K then

$$\lambda[a + b] = \lambda[a] + \lambda[b] = \dot{a} + \dot{b}$$

and $\lambda[ab] = \dot{a}\, b + a\, \dot{b}$

We suppose, here, that all the coefficients a_i and b_i are in a subset of C^∞ functions of time without zero divisors.

We define λ as the operator $\dfrac{d}{dt}$. Then, all the properties of (I) may be obtained from those of the polynomial equation:

$$[\lambda^n+ a_1(t)\lambda^{n-1}+\dots + a_n(t)]\ y(t) =$$
$$[b_0(t)\lambda^n+ b_1(t)\lambda^{n-1}+ \dots + b_n(t)]\ u(t) \quad \textbf{(II)}$$

Or in the formal form:

$$[\mathcal{Q}(\lambda)]\{\mathbf{y(t)}\} = [\mathbf{P}(\lambda)]\{\mathbf{u(t)}\}$$

3- NON-COMMUTATIVE POLYNOMIALS :

Let us define $\Pi(\lambda)$ as the general polynomial set :

$$\Pi(\lambda) = \{\mathbf{P}(\lambda) = a_0(t)\lambda^n+ a_1(t)\lambda^{n-1}+ \dots + a_n(t) ,$$
$$\forall\ a_i(t)\in K \text{ for } i\in [0,n]\}$$

where K is a differential field.

When $a_0(t) = 1$, $\mathbf{P}(\lambda)$ is monic.

For the multiplication, we define the non-commutative operation:

$$\boxed{\lambda\, a = a\, \lambda + \dot{a} \quad \forall\, a\in K} \qquad \text{with } \dot{a} =\frac{da}{dt}$$

It is easy to show that $\mathbf{P(\lambda)}$ with the two internal laws (addition and multiplication) is a **non-commutative ring** [5]

3-1: Division and Greatest Common Right Divisor (G.C.R.D.):

It is possible to define the **right division** :

Definition:

For every pair $\{\mathbf{P_1}(\lambda) , \mathbf{P_2}(\lambda)\} \in \Pi(\lambda) \oplus \Pi(\lambda)$ with $d°P_1(\lambda) > d°P_2(\lambda)$, it exists a unique polynomial pair $(\mathcal{Q}(\lambda) , \mathbf{R}(\lambda))$ which satisfies:

$$\boxed{\mathbf{P_1}(\lambda) = \mathcal{Q}(\lambda)\ \mathbf{P_2}(\lambda) + \mathbf{R}(\lambda)}$$

with $\mathbf{d°R(\lambda) < d°P_2(\lambda)}$

and $\mathbf{d°\mathcal{Q}(\lambda) = d°P_1(\lambda) - d°P_2(\lambda)}$

Thus we have the EUCLID's algorithm:

$$\begin{array}{lcl} \mathbf{P_1}(\lambda) & = & \mathcal{Q}_1(\lambda)\ \mathbf{P_2}(\lambda) + \mathbf{P_3}(\lambda) \\ \mathbf{P_2}(\lambda) & = & \mathcal{Q}_2(\lambda)\ \mathbf{P_3}(\lambda) + \mathbf{P_4}(\lambda) \\ \dots & = & \dots \qquad\qquad \dots \\ \mathbf{P_{n-1}}(\lambda) & = & \mathcal{Q}_{n-1}(\lambda)\ \mathbf{P_n}(\lambda) \end{array}$$

The G.C.R.D. of $(\mathbf{P_1}(\lambda) , \mathbf{P_2}(\lambda))$ is the well defined monic polynomial: $\boxed{\mathbf{P_n}(\lambda)}$

3-2: Least Common Right Multiple (L.C.R.M.):

We can define the L.C.R.M. of $[\mathbf{P_1}(\lambda) , \mathbf{P_2}(\lambda)]$ as the lowest degree monic polynomial $\mathbf{M}(\lambda)$ which admits both $\mathbf{P_1}(\lambda)$ and $\mathbf{P_2}(\lambda)$ as right divisors :

$$\boxed{\mathbf{M(\lambda) = \mathcal{Q}_1(\lambda)\ P_1(\lambda) = \mathcal{Q}_2(\lambda)P_2(\lambda)}}$$

In the general case, ORE [5] showed that the existence of EUCLID's algorithm implies the existence of **L.C.R.M.**

III-3: Explicit formula of the L.C.R.M.:

From all the quotients obtained in the EUCLID's algorithm, the L.C.R.M. is calculated by the following formula:

$$\mathbf{M(\lambda) = [P_1(\lambda)\ ,\ P_2(\lambda)] =}$$

$$\mathbf{= a\ P_{n-1}(\lambda)P_n(\lambda)^{-1}P_{n-2}(\lambda)P_{n-1}(\lambda)^{-1} \dots\ P_2(\lambda)P_3(\lambda)^{-1}P_1(}$$

The constant (i.e: not dependent of λ) **a** is chosen so that $M(\lambda)$ is monic.

Remarks:

1 - The product of two polynomials $\mathbf{P_1(\lambda)\ P_2(\lambda)}$ is not usually right divisible by $\mathbf{P_1(\lambda)}$.

2 - In the general case, left-hand division differs from right-hand division in that it cannot always be performed. Here, the definition of the differential operator allows to assert the existence of the right-hand division of two polynomials as the Least Common Left Polynomial.[5].

3-4: Rational fraction:

With commutative polynomials, it always be considered the rational fraction field. This is not possible when the polynomials are "skew" (that means only that, for these polynomials, the coefficients do not commute with the variable). Nethertheless, a ring has a rational left field and a rational right one if the two ORE's conditions are satisfied [3]:

$$\forall\ P_1(\lambda)\ ,\ P_2(\lambda) \in \Pi(\lambda) \oplus \Pi(\lambda),$$
$$\exists\ \hat{P}_1(\lambda) \in \Pi(\lambda) \text{ and } \hat{P}_2(\lambda) \in \Pi(\lambda)$$
$$\text{such as} \quad \hat{P}_2(\lambda)\ P_1(\lambda) = \hat{P}_1(\lambda)P_2(\lambda)$$

(left condition)

$$\forall\ P_1(\lambda)\ ,\ P_2(\lambda) \in \Pi(\lambda) \oplus \Pi(\lambda),$$
$$\exists\ \tilde{P}_1(\lambda) \in \Pi(\lambda) \text{ and } \tilde{P}_2(\lambda) \in \Pi(\lambda)$$
$$\text{such as} \quad P_1(\lambda)\ \tilde{P}_1(\lambda) = P_2(\lambda)\ \tilde{P}_2(\lambda)$$

In this case, we may consider the field:

$$\mathbf{Frac} = \{\ \mathbf{P(\lambda)\ \mathcal{Q}(\lambda)^{-1},\ (\ P(\lambda), \mathcal{Q}(\lambda))} \in \Pi(\lambda) \oplus \Pi^*(\lambda)\ \}$$

$$= \{\ \mathbf{\mathcal{Q}_1(\lambda)^{-1}\ P_1(\lambda), (\ P_1(\lambda)\ ,\ \ \mathcal{Q}_1(\lambda))} \in \mathbf{P(\lambda)} \oplus \Pi^*(\lambda)$$

$$\text{with} \quad \Pi^*(\lambda) = \Pi(\lambda) - \{0\}$$

It is obvious to show that, with the two internal laws (addition and multiplication), it is a "skew" field. For details see Dixmier [6].

Notice that each rational fraction has two representations: the left one and the right one. Notice also that the two ORE's conditions are not necessary: see Krob [3].

Remark: for a polynomial $\mathbf{P(\lambda)} \in \mathbf{Frac}$, we may define the unique inverse $\mathbf{P(\lambda)^{-1}}$ by:

$$\mathbf{P(\lambda)^{-1}P(\lambda) = P(\lambda)\ P(\lambda)^{-1} = 1}$$

It is obvious that:

$$\boxed{\mathbf{[P(\lambda)P_1(\lambda)]^{-1} = P_1(\lambda)^{-1}\ P(\lambda)^{-1}}}$$

4- TRANSFER FUNCTION:

In this section, we present the definition of the transfer function based on the rational fraction field. One way to enlarge the transfer function concept for time-varying systems is to generalize the formal power series. In our approach, the transfer function can be defined from the rational fraction field. One problem is that each transfer funtion has two representations which are in relation.

4-1: Definition:

If (Σ) is a system defined by a differential equation (I), by using the operator $\frac{d}{dt}(= \lambda)$, we can describe (Σ) by the formal relation:

$$\boxed{\mathbf{[\mathcal{Q}(\lambda)]\{y(t)\} = [P(\lambda)]\{u(t)\}}}$$

As in the stationary case, the **left transfer function (L.T.F.)** of the system (Σ) may be defined by the rational fraction:

$$\boxed{\mathbf{H(\lambda) = \mathcal{Q}(\lambda)^{-1}P(\lambda)}}$$

But as $H(\lambda)$ has another representation in the rational field **Frac,** we say that the **right transfer function (R.T.F)** of (Σ) is the other representation of $H(\lambda)$.

$$\boxed{\mathbf{H(\lambda) = P_1(\lambda)Q_1^{-1}(\lambda)}}$$

Remark: The left representation is directly issued from the differential equations.

4-2: Relations between L.T.F. and R.T.F.:

$$\mathbf{P(\lambda)\ Q_1(\lambda) = Q(\lambda)P_1(\lambda)}$$

That means that, from the L.T.F. (R.T.F.) the other one is easily computed with the remainding factors of the L.C.L.M. (L.C.R.M.).

Example:

Let $\dot{y} + \frac{1}{t} y = \dot{u}$

$$Q(\lambda) = 1 + \frac{1}{t} \text{ and } P(\lambda) = 1$$

The L.C.L.M of $Q(\lambda)$ and $P(\lambda)$ is $\lambda^2 = (\lambda + \frac{1}{t})(\lambda - \frac{1}{t})$

so $\mathbf{Q_1(\lambda) = 1}$ and $\mathbf{P_1(\lambda) = (\lambda - \frac{1}{t})}$

4-3: Minimal transfer function :

As in the stationary case, we can define the **minimal transfer function.** First, we introduce the equivalence relation of two transfer function.

Theorem 1: Two transfer functions $H_1(\lambda)$ and $H_2(\lambda)$ are equivalent if and only if in their L.T.F. $Q_1^{-1}(\lambda)\ P_1(\lambda)$ and $Q_2^{-1}(\lambda)P_2(\lambda)$, it exists a polynomial $D(\lambda)$ such as:

$$P_1(\lambda) = D(\lambda)\ P_2(\lambda) \quad \text{and } Q_1(\lambda) = D(\lambda)\ Q_2(\lambda)$$

$\{$if $d°(P_1(\lambda)) > d°(P_2(\lambda)$ and $d°(Q_1(\lambda)) > d°(Q_2(\lambda))\}$

For demonstration see [7]

Example:

Let $\ddot{y} = \dot{u} + \frac{1}{t} u$

The L.T.F is $H(\lambda) = Q^{-1}(\lambda)\ P(\lambda) = \lambda^{-2}(\lambda + \frac{1}{t})$

but $\lambda^2 = (\lambda + \frac{1}{t})(\lambda - \frac{1}{t})$ So $H(\lambda) = (\lambda - \frac{1}{t})^{-1}$

That corresponds to the differential equation:

$$\dot{y} + \frac{1}{t} y = u$$

(the first differential equation is the derivative of the last one)

Theorem 2: Two transfer functions $H_1(\lambda)$ and $H_2(\lambda)$ are equivalent if and only if in their R.T.F. $P_1(\lambda)Q_1^{-1}(\lambda)$ and $P_2(\lambda)Q_2^{-1}(\lambda)$, it exists a polynomial $D(\lambda)$ such as:

$$P_1(\lambda) = P_2(\lambda)\ D(\lambda) \quad \text{and } Q_1(\lambda) = Q_2(\lambda)\ D(\lambda)$$

$\{$if $d°(P_1(\lambda)) > d°(P_2(\lambda)$ and $d°(Q_1(\lambda)) > d°(Q_2(\lambda))\}$

From that, we can define the minimal transfer function:

Definition: The transfer function of a linear time-varying system is minimal if, for the L.T.F. (R.T.F.) representation, the numerator and the denominator are left-coprime (right co-prime).

5- ALGEBRAIC RULES OF COMBINATION OF NON-STATIONARY SYSTEMS:

From the preliminary results, we consider, now, the elementary operations on SISO non-stationary systems. For the two associations (serial and parallel), it is possible to compute the transfer function of the compound system [7].

5-1: Transfer function for series combination:

For two systems Σ_1 and Σ_2 described by their L.T.F $\mathbf{Q_1(\lambda)^{-1}P_1(\lambda)}$ and $\mathbf{Q_2(\lambda)^{-1}P_2(\lambda)}$, it is possible to find the left transfer function of Σ of the series-combination of Σ_1 and Σ_2 :

We have: $\frac{y_1}{u} = Q_1(\lambda)^{-1}P_1(\lambda)$ and $\frac{y_2}{y_1} = Q_2(\lambda)^{-1}P_2(\lambda)$

Thus $\frac{y_2}{u} = Q_2(\lambda)^{-1}P_2(\lambda)\ Q_1(\lambda)^{-1}P_1(\lambda)$

If $M(\lambda) = \hat{P}_2(\lambda)\ P_2(\lambda) = \hat{Q}_1(\lambda)\ Q_1(\lambda)$

and
$$\frac{y_2}{y_1} = Q_2(\lambda)^{-1}P_2(\lambda)\ Q_1(\lambda)^{-1}P_1(\lambda) = \{\hat{P}_2(\lambda)\ Q_2(\lambda)\}^{-1}\ \hat{Q}_1(\lambda)\ P_1(\lambda)$$

Thus:

$$\boxed{\frac{y_2}{u}=\{\hat{P}_2(\lambda)Q_2(\lambda)\}^{-1}\hat{Q}_1(\lambda)P_1(\lambda)}$$

Example:

if $(\Sigma_1) \Rightarrow (\lambda + \frac{1}{t})^{-1}\lambda$ and $(\Sigma_2) \Rightarrow (\lambda - \frac{1}{t})^{-1}$

we have: $Q_1(\lambda) = \lambda + \frac{1}{t}$, $P_1(\lambda) = \lambda$

$Q_2(\lambda) = \lambda - \frac{1}{t}$; $P_2(\lambda) = \lambda$

So the L.C.R.M. of $(1 , \lambda + \frac{1}{t}) = \lambda + \frac{1}{t}$

$\hat{Q}_1(\lambda) = 1$; $\hat{P}_2(\lambda) = \lambda + \frac{1}{t}$

$$(\Sigma) \Rightarrow [(\lambda + \frac{1}{t}) (\lambda - \frac{1}{t})]^{-1}\lambda = \lambda^{-2} \lambda$$

and the L.T.F. of the (Σ) is equal to $\boxed{\lambda^{-1}}$

5-2: Transfer function for parallel combination:

The problem is easier to solve. For two systems Σ_1 and Σ_2 described by their L.T.F $Q_1(\lambda)^{-1}P_1(\lambda)$ and $Q_2(\lambda)^{-1}P_2(\lambda)$, it is possible to find the left transfer function of Σ of the parallel-combination of Σ_1 and Σ_2 :

$$\frac{y_1}{u} = Q_1^{-1}(\lambda)P_1(\lambda) \text{ and } \frac{y_2}{u} = Q_2^{-1}(\lambda)P_2(\lambda)$$

If $M(\lambda) = \hat{Q}_1(\lambda)\, Q_1(\lambda) = \hat{Q}_2(\lambda)\, Q_2(\lambda)$

Thus:

$$\boxed{\frac{y}{u} = M^{-1}(\lambda)[\hat{Q}_1(\lambda)\, P_1(\lambda) + \hat{Q}_2(\lambda)\, P_2(\lambda)]}$$

Example:

if $(\Sigma_1) = (\lambda + \frac{1}{t})^{-1}$ and $(\Sigma_2) = (\lambda - \frac{1}{t})^{-1}$

$$M(\lambda) = \text{L.C.R.M. } [Q_1(\lambda), Q_2(\lambda)] = \lambda^2 + \frac{1}{t}\lambda - \frac{1}{t^2}$$

$$= \lambda (\lambda + \frac{1}{t}) = (\lambda + \frac{2}{t})(\lambda - \frac{1}{t})$$

Thus:

and (Σ) has the L.T.F:

$$\boxed{\frac{y}{u} = [\lambda^2 + \frac{1}{t}\lambda - \frac{1}{t^2}]^{-1}[2 (\lambda + \frac{1}{t})]}$$

6- CONCLUSION:

By using theory of non commutative polynomials, we have been able to define the transfer function of a linear system with time varying coefficients in the continuous case. The major difference with the stationary case is that it appears that the transfer function has two factorizations. The relation between the factors of each one is given. The left representation is closed to the differential equation. For the minimal transfer function, we achieve a necessary and sufficient condition . Besides, we have presented the algebraic rules for compound systems.

We think that these results can be obtained without difficulty to discrete non-stationary systems and can be enlarged to mutivariable systems.

ACKNOWLEDGEMENTS:

I should like to thank Jan JEZEC for his critical and incisive comments which have done much to help improve the quality of this paper.

BIBLIOGRAPHY:

[1] KAMEN E.W., KHARGONEKAR P.P., POOLLA K.R.: A transfer-function approach to linear time-varying discrete-time systems. SIAM J. Control and Optimization Vol 23 N°4 July 1985 pp 550-565

[2] EMRE E., TAI H.M, SEO J.H.: Transfer matrices, Realization, and Control of Continuous-Time Linear Time- Varying Systems via Polynomial Fractional Representations. Linear Algebra and its Application 1990 pp 79-141

[3] KROB D: Quelques exemples de séries formelles utilisées en algèbre non commutative. Note interne Institut BLAISE PASCAL n° 90-2 Janvier 1990

[4] KOLCHIN E.R. Differential Algebra and Algebraic Groups, Academic Press, New-York, 1978

[5] ORE O.: Theory of non-commutative polynomials Annals of mathematics vol 34 (1933) pp. 480-508.

[6] DIXMIER J.: Algebres enveloppantes Gauthiers-Villars 1974

[7] GUGLIELMI M.: Some algebraic properties of the non-stationary systems. 4th Latino-American congress on automatics PUEBLA 26-30 Novembre 1990 AMCA editor pp601-608

[8] KAMEN E.W.: The poles and the zeros of a linear time-varying system. Linear Alg. and its application, Vol 98 (1988) pp. 263-289.

SIGNAL PROCESSING VI: Theories and Applications
J. Vandewalle, R. Boite, M. Moonen, A. Oosterlinck (eds.)

A TIME-VARYING PRONY MODEL

F.MOLINARO, F.CASTANIE (Member EURASIP)

GAPSE/ENSEEIHT, 2 rue Camichel, 31071 Toulouse, France

Abstract. The communication proposes a new Prony model with time-varying coefficients for modelisation of nonstationary signals. This method leads to an extension of several techniques of stationary spectral estimation to the nonstationary case. The determination of the time-varying parameters requires five steps : estimation of the time-varying AR parameters, estimation of the right poles of the linear time-varying system, modelisation of these poles, computation of new poles, and least-square estimation of the amplitudes.

1. INTRODUCTION

Parametric spectral estimation methods are well known in signal processing. They are, however limited, due to the fact that they necessarily assume a stationary signal. Many actual signals are nonstationary processes : speech, seismic or engine vibration... For these kinds of signals, in the last 20 years, some authors [1], [2], [3] have developed time-dependent models. Such models are an extension of the time-invariant ARMA model to the nonstationary case. ARMA time-varying coefficient are approximated by a weighted combination of basis functions [4].

This paper proposes an extension of the stationary Prony model [5],[6], to the nonstationary case. This time-varying Prony model is adapted for time-varying multicomponent signals.

In what follows the time-varying Prony model is introduced and defined. In Part 3 the different steps for parameter estimation are detailed. The model is applied to a simulation case in Part 4.

2. THE MODEL

The order p time-varying Prony model under consideration is defined by :

$$x(n) = \sum_{k=1}^{p} B_k Z_k^n(n) + e(n) \qquad n = 0, 1, .., N-1$$

with the particular following for $Z_k(n)$:

$$Z_k(n) = e^{\alpha_k(n) + j\omega_k(n)}$$

with $\alpha_k(n) = \sum_{j=0}^{N_\alpha} \alpha_{kj} n^j$ and $\omega_k(n) = \sum_{j=0}^{N_\omega} \omega_{kj} n^j$

$B_k = A_k e^{j\theta_k}$ is constant

A_k : constant amplitude.

θ_k : constant phase.

$\alpha_k(n)$: polynomial time-varying damping factor.

$f_k(n)$: polynomial time-varying frequency.

It should be noted that complex amplitudes are not time function in our model, the nonstationary part being limited to the poles.

2.1 Definition of time dependent right poles

Kamen [7] has defined the left shift operator z by :

$$z^i . z^j = z^{i+j}$$
$$z^i . x(n) = x(n+i)$$
$$a(n) . z^i . x(n) = a(n) . x(n+i)$$

$P_k(n)$ are time-dependent poles if they have satisfied the following equation :

$$\left[1 + \sum_{k=1}^{p} a_k(n) z^{-k}\right] = \left[\tilde{\prod_{k=1}^{p}}(1 - z^{-1} P_k(n))\right] \qquad (1)$$

where $\tilde{\Pi}$ denotes skew multiplication [7]. Let $P_p(n)$ be a right pole of the system, there exists $\beta_{p-1,k}(n)$ such that :

$$\tilde{\prod_{k=1}^{p}}(1 - z^{-1} P_k(n)) = \left(1 + \sum_{k=1}^{p-1} \beta_{p-1,k}(n) z^{-k}\right)(1 - z^{-1} P_p(n)) \qquad (2)$$

(1) and (2) gives :

$$a_1(n) = \beta_{p-1,1}(n) - P_p(n)$$
$$a_k(n) = \beta_{p-1,k}(n) - \beta_{p-1,k-1} P_p(n-k+1) \quad k = 2, .., p-1$$
$$a_p(n) = -\beta_{p-1,p-1}(n) P_p(n-p+1)$$

Expressing this system in the matrix form :

$$\begin{bmatrix} -1 & 1 & 0 & \cdots & 0 \\ 0 & -P_p(n-1) & 1 & \cdots & 0 \\ 0 & 0 & -P_p(n-2) & \cdots & 0 \\ . & . & . & \cdots & . \\ . & . & . & \cdots & . \\ 0 & 0 & 0 & \cdots & -P_p(n-p+1) \end{bmatrix} \begin{bmatrix} P_p(n) \\ \beta_{p-1,1}(n) \\ \beta_{p-1,2}(n) \\ . \\ . \\ \beta_{p-1,p-1}(n) \end{bmatrix} = \begin{bmatrix} a_1(n) \\ a_2(n) \\ a_3(n) \\ . \\ . \\ a_p(n) \end{bmatrix} \qquad (3)$$

$P_p(n)$ is recursively computed given its initial p values and the time varying coefficients $a_k(n)$.

With this formulation, p right poles $P_1(n), P_2(n), \ldots, P_p(n)$ can be determined. An advantage of this time-dependent pole factorization is that the value of each pole at any given instant is obtained independently of the other poles.

2.2 Justification of the model

The two following theorems justify this model.

Theorem 1: [7] If the Vandermonde determinant $V(0)$ is nonzero, then for any initial values $x(0), x(1), \ldots, x(p-1)$ there exist constants $B_1, B_2, \ldots, B_p$ such that the solution of the time-varying equation :

$$x(n) = -\sum_{k=1}^{p} a_k(n)x(n-k)$$

can be written in the form :

$$x(n) = \sum_{k=1}^{p} B_k \Phi_k(n)$$

$\Phi_k(n) = P_k(n-1)P_k(n-2)\ldots P_k(0)$
$P_1(n), P_2(n), \ldots, P_p(n)$ are p right poles
$V(0)$ is defined by :

$$V(0) = \begin{bmatrix} 1 & 1 & \cdots & 1 \\ P_1(0) & P_2(0) & \cdots & P_p(0) \\ P_1(1)P_1(0) & P_2(1)P_2(0) & \cdots & P_p(1)P_p(0) \\ \vdots & \vdots & \vdots & \vdots \\ P_1(p-2)\ldots P_1(0) & P_2(p-2)\ldots P_2(0) & \cdots & P_p(p-2)\ldots P_p(0) \end{bmatrix}$$

Theorem 2 : If the right poles are modeled in the particular form :

$$P_k(n) = e^{\alpha'_k(n) + j\omega'_k(n)}$$

with $\alpha'_k(n) = \sum_{j=0}^{N_\alpha} \alpha'_{kj} n^j$ and $\omega'_k(n) = \sum_{j=0}^{N_\omega} \omega'_{kj} n^j$

Then it comes :

$$x(n) = \sum_{k=1}^{p} B_k Z_k^n(n)$$

Proof : Namely by denoting $S_j = \sum_{k=0}^{n-1} k^j$, we obtain :

$$S_j = n\left(\sum_{k=0}^{j} s_{jk} n^k\right)$$

$$S_j = \frac{1}{j+1}[n^{j+1} - n - C_{j+1}^{j-1} S_{j-1} - C_{j+1}^{j-2} S_{j-2} - \ldots - C_{j+1}^{1} S_1]$$

with $s_{00} = 1$, every s_{ij} can be computed.

Then :

$$\Phi_k(n) = e^{\sum_{j=0}^{N_\alpha} \alpha'_{kj} S_j} . e^{j \sum_{j=0}^{N_\omega} \omega'_{kj} S_j}$$

$$= e^{n \sum_{j=0}^{N_\alpha} \alpha_{kj} n^j} . e^{jn \sum_{j=0}^{N_\omega} \omega_{kj} n^j}$$

$$= Z_k^n(n)$$

With :

$$\alpha_{kj} = \sum_{i=j}^{N_\alpha} \alpha'_{ki} s_{ij}$$

$$\omega_{kj} = \sum_{i=j}^{N_\omega} \omega'_{ki} s_{ij}$$

α_{kj} is a weighted combination of α'_{ki} $(i = 0, \ldots, N_\alpha)$
ω_{kj} is a weighted combination of ω'_{ki} $(i = 0, \ldots, N_\omega)$.

3. COMPUTATION OF THE PARAMETERS

According to the above properties, we can derive the following algorithm in five steps :

- Estimation of the time-varying AR parameters $a_k(n)$
- Estimation of the right poles $P_k(n)$
- Modelisation of the right poles
- Calculation of $Z_k(n)$
- Estimation of B_k

3.1 Estimation of the time-varying AR parameters

The time-dependent AR modeling [4] of $x(n)$ is represented by the following equation :

$$x(n) = -\sum_{k=1}^{p} a_k(n)x(n-k) + e(n)$$

Parameters $a_k(n)$ are a function of n. They are approximated by a weighted combination of a number q of functions $f_i(n)$.

$$a_k(n) = \sum_{j=0}^{q} a_{kj} f_j(n)$$

One may use different sets of functions : Legendre, prolate, Fourier [4]...

3.2 Estimation of the time varying right poles

Resolution of the matrix system of Eq.(3) gives :

$$P_p(n) = -a_1(n) - \sum_{j=0}^{p-2} \frac{a_{p-j}(n)}{\prod_{i=1}^{p-j-1} P_p(n-i)}$$

Computation of the right poles is initialized with a stationary Prony modeling on a fixed window at the beginning of the signal [6].

3.3 Modelisation of the poles

The pole model is given by :

$$P_k(n) = e^{\alpha'_k(n) + j\omega'_k(n)}$$

with

$$\alpha'_k(n) = \sum_{j=0}^{N_\alpha} \alpha'_{kj} n^j$$

$$\omega'_k(n) = \sum_{j=0}^{N_\omega} \omega'_{kj} n^j$$

For each pole we estimate the module with a least-square polynomial estimation, and the phase with the specific estimation described in [8].

3.4 Computation of $Z_k(n)$

With (see proof of theorem 2) :

$$\alpha_{kj} = \sum_{i=j}^{N_\alpha} \alpha'_{ki} S_{ij}$$

$$\omega_{kj} = \sum_{i=j}^{N_\omega} \omega'_{ki} S_{ij}$$

The $Z_k(n)$ can be determined easily.

$$Z_k^n(n) = e^{n \sum_{j=0}^{N_\alpha} \alpha_{kj} n^j} . e^{jn \sum_{j=0}^{N_\omega} \omega_{kj} n^j}$$

3.5 Estimation of B_k amplitudes

B_k amplitudes are estimated with the help of a conventional Least-Square procedure

$$B = [B_1 \ldots B_p]^T = (V^H V)^{-1} V^H X$$

V is a Van Der Monde matrix, whose term is given by :

$$V = [v_{nm}] = [Z_m^n(n)] \qquad n = 0, \ldots, N-1 \qquad m = 1, \ldots, p$$

V^H denotes its conjugated transposed matrix.

$X = [x(0) \ldots x(N-1)]^T$ is the signal vector.

4. APPLICATION OF A SIMULATED SIGNAL

To validate this model, a simulation signal is chosen. It is composed of two chirp signals and can be expressed by :

$$x(n) = s_1(n) + s_2(n) + w(n)$$

$$= \cos(\omega_{11} n + \omega_{12} n^2) + \cos(\omega_{21} n + \omega_{22} n^2) + w(n)$$

$w(n)$ being a white noise. Different SNR levels are also chosen. In order to illustrate the properties of our method, various cases of time-frequency trajectories are given : three signals (see *fig. 1,2,3*) with a change of linear modulation of $s_2(n)$, i.e. parameter ω_{22}.

Fig. 1 shows that *signal 1* is composed of two chirps with separate frequencies.

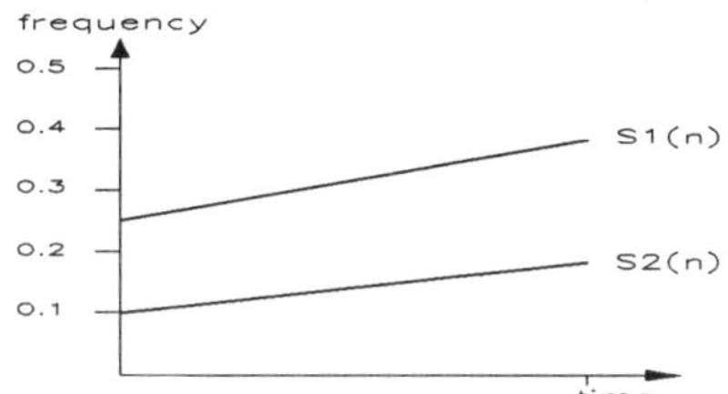

Fig. 1 - Signal 1 frequency modulations
$\omega_{11} = 0.25, \omega_{12} = 0.0005, \omega_{21} = 0.1, \omega_{22} = 0.0003$

In *signal 2* frequency linear modulation of $s_2(n)$ is such that the trajectories become very close to one another (see *Fig. 2*).

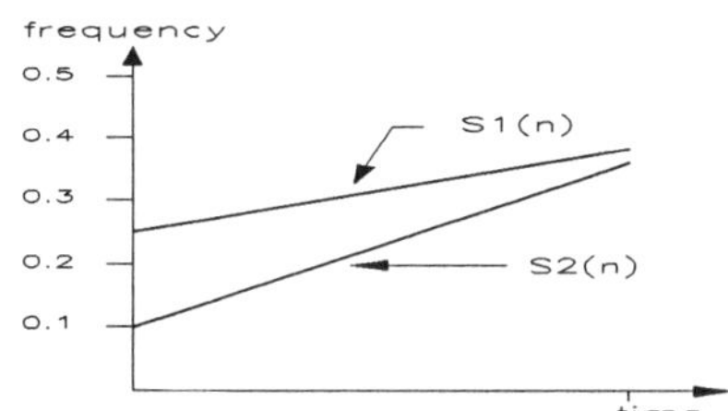

Fig. 2 - Signal 2 frequency modulations
$\omega_{11} = 0.25, \omega_{12} = 0.0005, \omega_{21} = 0.1, \omega_{22} = 0.001$

In case 3, the two trajectories intersect (see *Fig. 3*).

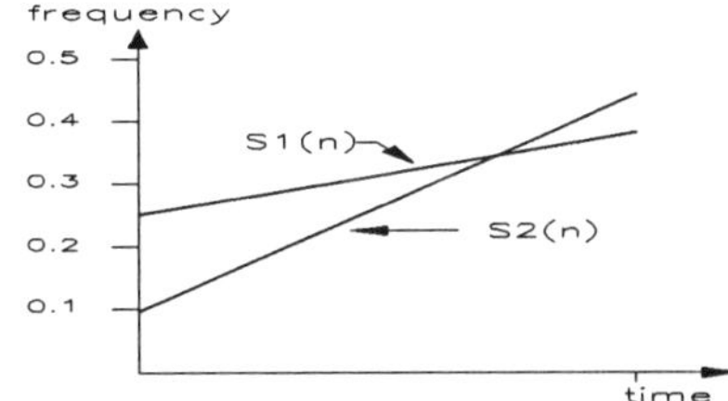

Fig. 3 - Signal 3 frequency modulations
$\omega_{11} = 0.25, \omega_{12} = 0.0005, \omega_{21} = 0.1, \omega_{22} = 0.0013$

Parameter estimation error for each signal is given in *Tab.1,2,3* respectively. Error is defined by :

$$E = \frac{\hat{\omega}_{ij} - \omega_{ij}}{\omega_{ij}}$$

SNR in db	Error in %			
	ω_{11}	ω_{12}	ω_{21}	ω_{22}
100	6.4 10^{-1}	3.1 10^{-1}	4.7 10^{-1}	1.6
20	8.4 10^{-1}	3.5	6.9 10^{-1}	2.8
10	7.6	16	49	63

Tab. 1 - Error signal 1 $= \frac{\hat{\omega}_{ij} - \omega_{ij}}{\omega_{ij}}$

SNR in db	Error in %			
	ω_{11}	ω_{12}	ω_{21}	ω_{22}
100	7.7 10^{-1}	8.6 10^{-1}	3.4	4.1 10^{-2}
20	3.5	23	9.4	8.1
10	23	47	14.7	6

Tab. 2 - Error signal 2 $= \frac{\hat{\omega}_{ij} - \omega_{ij}}{\omega_{ij}}$

SNR in db	Error in %			
	ω_{11}	ω_{12}	ω_{21}	ω_{22}
100	8.7	43	15.9	4.9
20	8.4 10^{-2}	4.7	6.4	10.6
10	2	7.1	29.1	12.9

Tab. 3 - Error signal 3 $= \frac{\hat{\omega}_{ij} - \omega_{ij}}{\omega_{ij}}$

Estimation of frequency parameters is better for separate frequencies (see *Tab.1,2,3*) . The intersection of the two chirps increases estimation error (see *Tab.3*). For a high SNR, i.e more than 10 db, this model gives good frequency estimation.

5. CONCLUSION

The present paper has introduced a Prony model with time-varying poles. This model is well suited to frequency modulated nonstationary multicomponent signals.

We have shown that it is possible to introduce a new model, based upon a linear combination of time-varying exponentials with constant amplitudes. With the restriction of the time-varying poles to a well specified class (namely damping factors and instantaneous frequencies being time polynomials), we have proposed a method and the corresponding algorithm to estimate the complete new parameter vector.

Special attention has been paid to the retrieval of frequency modulation laws in simulated signals.

The method has proved capable of giving accurate measurements of time-frequency trajectories. The properties of the corresponding amplitude estimators will be dealt with in a forthcoming submission.

REFERENCES

[1] T.S.RAO, "The fitting of nonstationary time-series models with time-dependent parameters", J. Royal Statist. Soc. Series B, vol. 32, n°2, pp 312-322, 1970.

[2] J.M.MENDEL, "Discrete Techniques of Parameters Estimation : The equation Error Formulation", vol. 1 New York : Marcel Dekker, ch.6, 1973.

[3] L.A.LIPORACE, "Linear estimation of nonstationary signals", J. Acoust. Soc. Amer., vol. 58, n°6, pp 1288-1295, 1975.

[4] Y.GRENIER, "Time-Dependent ARMA Modeling of Nonstationary Signals", IEEE Trans. on ASSP, vol. 31, n°4, pp 899-911, August 1983.

[5] R.PRONY, "Essai expérimental et analytique sur les lois de la dilatabilité des fluides élatiques", J. de l'école Polytechnique, n°1, Floréal et Prairial, An III, 1795.

[6] F.CASTANIE, E.DAYMIER, "Prony spectral analysis of stationary processes", Signal processing III : Theories and Applications, pp 283-286, 1986.

[7] E.W.KAMEN, "The Poles and Zeros of a linear Time-Varying System", Linear Algebra and its applications, pp 263-289, 1988.

[8] S.PELEG, B.PORAT, "Estimation and Classification of Polynomial-Phase Signal", IEEE Trans. on Inf. Theory, vol.37, n°2, pp 422-430, march 1991.

SIGNAL PROCESSING VI: Theories and Applications
J. Vandewalle, R. Boite, M. Moonen, A. Oosterlinck (eds.)

AAK MODEL REDUCTION FOR TIME-VARYING SYSTEMS

A.-J. van der Veen and P. Dewilde

Department of Electrical Engineering, Delft University of Technology
2628 CD Delft, The Netherlands

The paper summarizes a procedure to obtain the optimal low degree "generalized Hankel norm approximant" of a bounded upper operator which is described by a high-order (time-varying) system. In the classical Hardy space theory this is known as the "model reduction problem" and it has resulted in a solution which goes back to the work of Adamyan, Arov and Krein on the Schur-Takagi problem. Our approach extends the frontiers of that theory to cover general, not only Toeplitz, upper operators as well. The procedure is computational and utilizes time-varying state space representations of the operators.

1. INTRODUCTION

The singular value decomposition (SVD) has been used by Adamjan, Arov and Krein (AAK)[1] to obtain approximations in the context of complex function theory, in relation with the approximation of infinite-size Hankel matrices. The AAK paper is written in an analytic setting of a model reduction problem but the same problem can be put into operator theory, in particular of upper triangular Toeplitz operators (or Toeplitz matrices). In the present paper, the aim is to extend the model reduction theory to the time-varying context, by considering (bounded) upper operators

$$T = \begin{bmatrix} \ddots & \vdots & & & ⋰ \\ & \boxed{T_{00}} & T_{01} & T_{02} & \\ & & T_{11} & T_{12} & \\ & \mathbf{0} & & T_{22} & \cdots \\ & & & & \ddots \end{bmatrix}$$

which are no longer taken to be Toeplitz (the central $(0,0)$ element is distinguished by a box). T can be regarded as the transfer operator of a causal time-varying system which acts on (row) sequences, in which case the i-th row of T is the impulse response of T to an impulse applied at time i. The notion of a (physical) system is however not essential; the approximation theory applies to a general class of (bounded) upper operators which have a finite-dimensional *computational model*. The derivation of such models is the subject of a recently developed realization theory for such operators [2]. The approximation theory in this paper draws heavily onto [2], but in this short paper we omit most of those results and notation, in particular on the spaces to which each of the operators belong. A complete account is presented in the full paper [3], the outline of which runs more or less in parallel with [4]. For a given T, a time-varying state space realization (alias a computational model for T) has the form

$$\begin{aligned} x_{k+1} &= x_k A_k + u_k B_k \\ y_k &= x_k C_k + u_k D_k \end{aligned} \qquad (1)$$

where T is called a locally finite operator if the individual $\{A_k, B_k, C_k, D_k\}$ all have finite (but possibly time-varying) dimensions. We can assemble the matrices $\{A_k\}$, $\{B_k\}$ etc. as operators on spaces of sequences of appropriate dimensions, by defining $A = \text{diag}(A_k)$, $B = \text{diag}(B_k)$, $C = \text{diag}(C_k)$ and $D = \text{diag}(D_k)$. Together these operators define a realization $\mathbf{T}$ of T:

$$\begin{aligned} xZ^{-1} &= xA + uB \\ y &= xC + uD \end{aligned} \qquad \mathbf{T} = \begin{bmatrix} A & C \\ B & D \end{bmatrix}. \qquad (2)$$

The diagonal operators act on sequences $u = [\cdots, \boxed{u_0}, u_1, u_2, \cdots]$, $x = [\cdots, \boxed{x_0}, x_1, x_2, \cdots]$, and the shift operator Z on these sequences is defined by $xZ^{-1} = [\cdots, \boxed{x_1}, x_2, x_3, \cdots]$. The realization in (2) is equivalent to (1), but more convenient to handle in equations because the time-index has been suppressed. Shifted diagonal operators are $A^{(1)} = Z^{-1}AZ = \text{diag}(A_{k-1})$ and $A^{(-1)} = ZAZ^{-1} = \text{diag}(A_{k+1})$. An important aspect of these sequences is that the dimensions of their components can vary in time. Suppose that $x_k \in \mathcal{B}_k$, with $\mathcal{B}_k = \mathbb{C}^{N_k}$ an Euclidean space of dimension N_k, then we define $\mathcal{B}$ to be the space of sequences x with entries in $\mathcal{B}_k$, and hence $x \in \mathcal{B}$ and $A : \mathcal{B} \to \mathcal{B}^{(-1)}$. Even input- and output sequences can have varying dimensions. We will typically use $\mathcal{M}$ for input sequences and $\mathcal{N}$ for output sequences, and hence $T : \mathcal{M} \to \mathcal{N}$. If $(I - AZ)$ has a bounded inverse that is again upper (this requires the spectral radius of AZ

Research was supported in part by the commission of the EC under the ESPRIT BRA program 6632 (NANA2).

to be less than one), then the realization $\mathbf{T}$ is such that

$$T = D + BZ(I - AZ)^{-1}C.$$

We will assume throughout the paper that this is the case.

Hankel operators are obtained as follows. Denote a certain time instant (k say) as 'current time', apply all possible inputs in the 'past' with respect to this instant, and measure the corresponding outputs in 'the future', from the current time instant on. The operator mapping this input sequence to the restricted output sequence corresponds to an upper-right part of the matrix representation of T. Denote this part (or rather its mirrored one-sided infinite equivalent acting on the reversed input sequence) by H_k, then for example

$$H_2 = \begin{bmatrix} T_{12} & T_{13} & T_{14} & \cdots \\ T_{02} & T_{03} & & \\ T_{-1,2} & & \ddots & \\ \vdots & & & \end{bmatrix}.$$

In analogy with the time-invariant case, we call H_k the (time-varying) Hankel operator at time k, and H_T = diag(H_k) the time-varying Hankel operator. For the time-invariant case, all H_k are the same and the entries of each of them are constant along anti-diagonals. Neither of this is true in the time-varying setting, but nonetheless the main features carry over (this regards the rank of H_k and a certain shift-invariance property). The Hankel norm of an operator T is $\| T \|_H = \| H_T \| = \sup \| H_k \|$, the supremum over the norms of the individual Hankel operators. In terms of the Hankel norm, the main approximation result reads as follows.

Theorem 1. ([3]) *Let T be a bounded, strictly upper, locally finite operator and let Γ be a Hermitian diagonal operator. Let H_k be the Hankel matrix of $\Gamma^{-1}T$ at time instant k. Suppose that the singular values of each H_k decompose into two sets $\sigma_{-,k}$ and $\sigma_{+,k}$, with $\sigma_{-,k}$ containing all (N_k say) singular values larger than one and $\sigma_{+,k}$ containing the singular values smaller than one, and suppose furthermore that the infimum over k of $\sigma_{-,k}$ is itself larger than one, and the supremum over k of $\sigma_{+,k}$ is itself smaller than one. Let N be the sequence $\{N_k\}$.*

Then there exists a strictly upper locally finite operator T_a of system order at most N such that

$$\| \Gamma^{-1}(T - T_a) \|_H \leq 1.$$

A proof of this theorem appears in the full paper [3]. In this summarizing paper we will focus on the computational aspects involved in obtaining one such Hankel-norm approximant, and present recursive formulas by which a state space realization of T_a can be obtained.

2. SUMMARY OF THE PROCEDURE

The problem that we shall solve in this section is the model reduction problem for a strictly upper operator described by a strictly stable "higher order model". Let Γ be a diagonal and hermitian operator. We shall use Γ as a measure for the local accuracy of the reduced order model. It will also parametrize the solutions. We will look for a contractive operator E such that $E = (T^* - T'^*)\Gamma^{-1}$ where T' is an operator which is not necessarily upper triangular, but whose strictly causal part will assumed to be bounded and have state space dimensions of low order — as low as possible for a given Γ. Once we have such a contractive E, it is immediately verified that it satisfies

$$\| \Gamma^{-1}(T - T') \| = \| E \| \leq 1.$$

Let T_a be the strictly causal part of T'. Then

$$\| \Gamma^{-1}(T - T_a) \|_H \leq \| \Gamma^{-1}(T - T') \| \leq 1,$$

and T_a is a Hankel-norm approximant when T' is an operator-norm approximant. To find T' we start by determining a (minimal) factorization of T in the form $T = \Delta^* U$ where Δ and U are upper operators which have finite state space dimensions of the same size as that of T, and U is inner: $UU^* = U^*U = I$. A J-unitary operator Θ is an operator with block decomposition

$$\Theta = \begin{bmatrix} \Theta_{11} & \Theta_{12} \\ \Theta_{21} & \Theta_{22} \end{bmatrix} \tag{3}$$

and signature matrices

$$J_1 = \begin{bmatrix} I_{\mathcal{M}_1} & \\ & -I_{\mathcal{N}_1} \end{bmatrix}, \quad J_2 = \begin{bmatrix} I_{\mathcal{M}_2} & \\ & -I_{\mathcal{N}_2} \end{bmatrix}.$$

such that $\Theta^* J_1 \Theta = J_2$, $\Theta^* J_2 \Theta = J_1$. We look for a locally finite J-unitary Θ-operator with upper block operators chosen such that

$$\begin{bmatrix} U^* & -T^*\Gamma^{-1} \end{bmatrix} \Theta = \begin{bmatrix} A' & -B' \end{bmatrix} \tag{4}$$

consists of two upper operators. A solution to this problem exists if certain conditions on a Lyapunov equation associated to $\Gamma^{-1}T$ are satisfied (see below). Θ will again be locally finite. Then, because Θ is J-unitary, we have that $\Theta_{22}^*\Theta_{22} = I + \Theta_{12}^*\Theta_{12}$, and because Θ is locally finite, Θ_{22}^{-1} will exist (but not necessarily upper) and $\Sigma_{12} = -\Theta_{12}\Theta_{22}^{-1}$ will be contractive. From (4) we have

$$B' = -U^*\Theta_{12} + T^*\Gamma^{-1}\Theta_{22}. \tag{5}$$

Define the approximating operator T' as

$$T'^* = B'\Theta_{22}^{-1}\Gamma, \tag{6}$$

then $E := (T^* - T'^*)\Gamma^{-1} = -U^*\Sigma_{12}$. Because Σ_{12} is contractive and U unitary, we infer that $\|E\| \le 1$, so that $T'^* = B'\Theta_{22}^{-1}\Gamma$ is indeed an approximant with an admissible modeling error. In view of the target theorem 1, we have to show that the strictly causal part of T' has the stated number of states and to verify the relation with the Hankel singular values of $\Gamma^{-1}T$. This will done in the remaining part of this section.

Factorization of T

For a factorization $T = \Delta^* U$, we start from a realization of T in output canonical form, *i.e.*, such that $AA^* + CC^* = I$. For each time instant k, augment the state transition matrices $[A_k \;\; C_k]$ of T with as many extra rows as needed to yield a unitary (hence square) matrix $\mathbf{U}_k$:

$$\mathbf{U}_k = \begin{bmatrix} A_k & C_k \\ B_k^u & D_k^u \end{bmatrix}. \tag{7}$$

Assemble the individual matrices $\{A_k, B_k^u, C_k D_k^u\}$ in diagonal operators $\{A, B_u, C, D_u\}$, then the corresponding operator $\mathbf{U}$ is a state space realization for U; $U = D_u + B_u Z(I - AZ)^{-1}C$. U is well-defined and upper, and it is unitary because it has a unitary realization. It is straightforward to verify that $\Delta = UT^*$ is indeed upper. Note that the number of rows added to $[A_k \;\; C_k]$ is time-varying, so that U (and hence also Δ) has a time-varying number of inputs.

Construction of Θ

The next step is to construct a locally finite and block-upper J-unitary Θ that satisfies equation (5). Let $\mathcal{B}$ be the space in which the state sequences of the realization $\mathbf{\Theta}$ of Θ live. Θ will be J-unitary in the sense of (3) if $\mathbf{\Theta}$ satisfies

$$\mathbf{\Theta}^* \begin{bmatrix} J_{\mathcal{B}} & \\ & J_1 \end{bmatrix} \mathbf{\Theta} = \begin{bmatrix} J_{\mathcal{B}}^{(-1)} & \\ & J_2 \end{bmatrix} \tag{8}$$

where $\mathcal{B} = \mathcal{B}_+ \oplus \mathcal{B}_-$ is a certain decomposition of $\mathcal{B}$, and

$$J_{\mathcal{B}} = \begin{bmatrix} I_{\mathcal{B}_+} & \\ & -I_{\mathcal{B}_-} \end{bmatrix} \tag{9}$$

is a corresponding signature matrix. (In this context, we call $J_{\mathcal{B}}$ the state signature sequence.) Let $\{A, B, C, 0\}$ be the realization for T used in the previous section (it is in output canonical form), and let $\{A, B_u, C, D_u\}$ be the realization for the inner factor U of T. We submit that Θ satisfying (4) has a realization $\mathbf{\Theta}$ of the form

$$\mathbf{\Theta} = \left[\begin{array}{c|c} X & \\ \hline & I \end{array}\right] \left[\begin{array}{c|cc} A & C_1 & C_2 \\ \hline B_u & D_{11} & D_{12} \\ \Gamma^{-1}B & D_{21} & D_{22} \end{array}\right] \left[\begin{array}{c|c} (X^{(-1)})^{-1} & \\ \hline & I \end{array}\right] \tag{10}$$

which is a square matrix at each time instant k, and where X and C_i, D_{ij} are yet to be determined. Note that the state sequence space $\mathcal{B}$ is the same for $\mathbf{\Theta}$ and T. X is a boundedly invertible diagonal state transformation operator which is such that $\mathbf{\Theta}$ is J-unitary as in (8). Writing $\Lambda = X^* J_{\mathcal{B}} X$, the signature of Λ will determine $J_{\mathcal{B}}$, and it is straightforward to derive that $\Lambda = I - M$, where M is given by

$$M^{(-1)} = A^* MA + B^* \Gamma^{-1}\Gamma^{-1} B. \tag{11}$$

M is the solution of one of the Lyapunov equations associated to $\Gamma^{-1}T$, and can be determined recursively from the given realization of T via $M_{k+1} = A_k^* M_k A_k + B_k^* \Gamma_k^{-1}\Gamma_k^{-1} B_k$. It is closely related to the Hankel operator of $\Gamma^{-1}T$. In particular, the non-zero singular values of the Hankel operator of $(\Gamma^{-1}T)$ at time k are equal to the eigenvalues of the matrix M_k. Using $\Lambda = I - M$, it follows that a singular value of the k-th Hankel operator of $\Gamma^{-1}T$ that is larger than 1 corresponds to a negative eigenvalue of Λ_k. Hence $J_{\mathcal{B}}$, the signature of Λ, is determined by the Hankel singular values of $\Gamma^{-1}T$. With X known, the remaining unknown columns in (10) can be determined straightforwardly and independently for each time instant k as the J-unitary complement of the columns that are already known, such that the resulting $\mathbf{\Theta}$ is square and J-unitary at each point in time.

At this point we have covered the first part of theorem 1: we have constructed a J-unitary Θ and from it an operator T_a which is a Hankel-norm approximant of T. It remains to verify the complexity assertion, which stated that the dimension of the state space of T_a is at most equal to N: the number of Hankel singular values of $\Gamma^{-1}T$ that are larger than one, or the number of negative entries in the state signature $J_{\mathcal{B}}$ of Θ. Assume that the strictly causal part T_a of T' exists as a bounded operator. In view of its definition in (6), it can be shown that (under the conditions of theorem 1) the input state space of T_a is contained in the input state space of Θ_{22}^{-*}, and that the dimension of this space is at most equal to $N = \#_-(J_{\mathcal{B}})$, the number of negative entries in the state signature sequence of $\mathbf{\Theta}$. This proves the second part of theorem 1.

3. STATE REALIZATION OF T_a

The preceding section has given a general expression for a Hankel-norm approximant T_a. It is possible to give a state realization for T_a in terms of the state realizations of T, U and Θ, which have been derived in the preceding sections. Necessary for the construction is to obtain a realization for B' and for the upper part of Θ_{22}^{-*}. The former is straightforward from its definition (4); the latter is more difficult because Θ_{22}^{-1} is not causal, and requires the construction of a (unitary) realization $\mathbf{\Sigma}$ that is related to $\mathbf{\Theta}$ in the following way. Partition the state x of $\mathbf{\Theta}$ according to the signature $J_{\mathcal{B}}$ into $x = [x_+ \; x_-]$, and partition $\mathbf{\Theta}$ likewise:

$$\mathbf{\Theta} = \begin{array}{c} \\ x_+ \\ x_- \\ a_1 \\ b_1 \end{array} \begin{array}{c} \begin{array}{cccc} x_+^{(-1)} & x_-^{(-1)} & a_2 & b_2 \end{array} \\ \left[\begin{array}{cc|cc} \alpha_{11} & \alpha_{12} & \gamma_{11} & \gamma_{12} \\ \alpha_{21} & \alpha_{22} & \gamma_{21} & \gamma_{22} \\ \hline \beta_{11} & \beta_{12} & \delta_{11} & \delta_{12} \\ \beta_{21} & \beta_{22} & \delta_{21} & \delta_{22} \end{array}\right] \end{array}, \qquad (12)$$

then the corresponding $\mathbf{\Sigma}$ is defined by the relation

$$\begin{array}{rrcl} & [x_+ \; x_- \; a_1 \; b_1]\,\mathbf{\Theta} & = & [x_+^{(-1)} \; x_-^{(-1)} \; a_2 \; b_2] \\ \Leftrightarrow & [x_+ \; x_-^{(-1)} \; a_1 \; b_2]\,\mathbf{\Sigma} & = & [x_+^{(-1)} \; x_- \; a_2 \; b_1] \end{array}$$

(inputs of $\mathbf{\Sigma}$ have positive signature) and has partitioning

$$\mathbf{\Sigma} = \begin{array}{c} \\ x_+ \\ x_-^{(-1)} \\ a_1 \\ b_2 \end{array} \begin{array}{c} \begin{array}{cccc} x_+^{(-1)} & x_- & a_2 & b_1 \end{array} \\ \left[\begin{array}{cc|cc} F_{11} & F_{12} & H_{11} & H_{12} \\ F_{21} & F_{22} & H_{21} & H_{22} \\ \hline G_{11} & G_{12} & K_{11} & K_{12} \\ G_{21} & G_{22} & K_{21} & K_{22} \end{array}\right] \end{array}. \qquad (13)$$

From this definition, the entries in $\mathbf{\Sigma}$ can be determined as

$$\begin{aligned} \begin{bmatrix} F_{11} & H_{11} \\ G_{11} & K_{11} \end{bmatrix} &= \begin{bmatrix} \alpha_{11} & \gamma_{11} \\ \beta_{11} & \delta_{11} \end{bmatrix} - \begin{bmatrix} \alpha_{12} & \gamma_{12} \\ \beta_{12} & \delta_{12} \end{bmatrix} \begin{bmatrix} \alpha_{22} & \gamma_{22} \\ \beta_{22} & \delta_{22} \end{bmatrix}^{-1} \begin{bmatrix} \alpha_{21} & \gamma_{21} \\ \beta_{21} & \delta_{21} \end{bmatrix} \\ \begin{bmatrix} F_{12} & H_{12} \\ G_{12} & K_{12} \end{bmatrix} &= - \begin{bmatrix} \alpha_{12} & \gamma_{12} \\ \beta_{12} & \delta_{12} \end{bmatrix} \begin{bmatrix} \alpha_{22} & \gamma_{22} \\ \beta_{22} & \delta_{22} \end{bmatrix}^{-1} \\ \begin{bmatrix} F_{21} & H_{21} \\ G_{21} & K_{21} \end{bmatrix} &= \begin{bmatrix} \alpha_{22} & \gamma_{22} \\ \beta_{22} & \delta_{22} \end{bmatrix}^{-1} \begin{bmatrix} \alpha_{21} & \gamma_{21} \\ \beta_{21} & \delta_{21} \end{bmatrix} \\ \begin{bmatrix} F_{22} & H_{22} \\ G_{22} & K_{22} \end{bmatrix} &= \begin{bmatrix} \alpha_{22} & \gamma_{22} \\ \beta_{22} & \delta_{22} \end{bmatrix}^{-1} \end{aligned}$$

Note that the matrix $\mathbf{\Sigma}_k$ only depends on the entries of $\mathbf{\Theta}_k$ so that it can be computed independently for each time instant. $\mathbf{\Sigma}$ is the state realization of the operator Σ which is the scattering operator that can be associated to Θ; from equation (13) it is seen that the realization is not causal: the state x_- is *computed* instead of used.

Let $\{A, B, C, 0\}$ be a state realization for T as in theorem 1 with M defined by the recursion in (11), let $\{A, B_u, C, D_u\}$ be a realization for the inner factor U of T, the interpolating $\mathbf{\Theta}$ as in (10), and $\mathbf{\Sigma}$ corresponding to $\mathbf{\Theta}$ as above. The following additional recursions are needed to generate a state realization of T_a:

$$\begin{aligned} S^{(-1)} &= F_{21} + F_{22}(I - SF_{12})^{-1} SF_{11} \\ R &= F_{12} + F_{11}R^{(-1)}(I - F_{21}R^{(-1)})^{-1}F_{22} \\ C_r^* &= \left[G_{22} + G_{21}R^{(-1)}(I - F_{21}R^{(-1)})^{-1}F_{22}\right](I - SR)^{-1}. \end{aligned}$$

In terms of these quantities, a state realization $\{A_a, B_a, C_a, 0\}$ of T_a is given by

$$\begin{aligned} A_a^* &= F_{22}(I - SF_{12})^{-1} \\ B_a^* &= H_{22} + F_{22}(I - SF_{12})^{-1}SH_{12} \\ C_a &= C_r\left[-D_{12}^* D_u + C_2^*(I - M)C\right] + A_a Y^{(-1)} A^* MC \\ Y &= A_a Y^{(-1)} A^* + C_r C_2^*, \end{aligned}$$

in which a fourth recursively determined quantity (Y) is introduced. A check on the dimensions of A_a reveals that the state realization for T_a has indeed a state space dimension given by $N = \#_-(J_{\mathcal{B}})$: the number of Hankel singular values of T that are larger than 1. The realization is given in terms of four recursions: two for M and S that run forward in time, the other two for R and Y that run backward in time and that depend on S. One implication of this is that an optimal approximant of T can only be derived if the model of T is known for all time, which is of course not very surprising. It is possible to obtain results in finite time for special cases, such as (1) finite $(n \times n)$ matrices, in which case M, S, R, Y have zero dimensions outside an interval of size n, and (2) systems which are time-invariant outside a finite interval of size n, in which case the recursions become (eigenvalue) equations outside that interval, and can be solved analytically. This brings another remarkable feature of this algorithm into light: in the time-invariant case, an eigenvalue decomposition is needed to split Θ_{22}^{-*} into upper and lower parts, *i.e.*, to retrieve the stable and unstable poles of the approximant. In the current setting, the notion of a 'time-varying pole' is ill-defined, because *e.g.*, the state dimension can vary in time. However, the above recursions avoids this notion altogether, while at the same time Θ_{22}^{-*} is never explicitely formed.

REFERENCES

[1] V.M. Adamjan, D.Z. Arov, and M.G. Krein, "Analytic Properties of Schmidt Pairs for a Hankel Operator and the Generalized Schur-Takagi Problem," *Math. USSR Sbornik*, **15**(1):31–73, 1971 (transl. of *Iz. Akad. Nauk Armjan. SSR Ser. Mat. 6 (1971))*.

[2] A.J. van der Veen and P.M. Dewilde, "Time-Varying System Theory for Computational Networks," In P. Quinton and Y. Robert, editors, *Algorithms and Parallel VLSI Architectures, II*. Elsevier, 1991.

[3] P.M. Dewilde and A.J. van der Veen, "On the Hankel-Norm Approximation of Upper-Triangular Operators and Matrices," *submitted to Integral Equations and Operator Theory*, 1992.

[4] D.J.N. Limebeer and M. Green, "Parametric Interpolation, H_∞-Control and Model Reduction," *Int. J. Control*, **52**(2):293–318, 1990.

SIGNAL PROCESSING VI: Theories and Applications
J. Vandewalle, R. Boite, M. Moonen, A. Oosterlinck (eds.)

OPTIMUM INTERPOLATION FOR TIME-LIMITED AND BAND-LIMITED SIGNALS

Victor NEAGOE, Member EURASIP

Department of Electronics and Telecommunications, Polytechnic Institute of Bucharest, Bucharest 16, Romania 77206*

This paper proposes an optimum method of uniform sampling reconstruction for signals assumed to be both band-limited and also time-limited. The method is based on the estimation of the Legendre nonuniformly taken samples of the Fourier transform in order to be consistent with the given sequence of uniform time samples. An analytical formula of the maximum reconstruction error and the corresponding condition of convergence for interpolation are deduced. The computer simulations prove the important advantage of the proposed method over the classical Shannon interpolation.

1. INTRODUCTION

The Shannon sampling theorem and its variants [1],[5] are well known as performing the reconstruction of a band-limited signal, from the knowledge of its corresponding infinite series of uniformly taken samples. However, in practice we have only time-limited signals, so that the signal reconstruction must be performed from a finite sequence of samples, which leads to the truncation of Shannon series and implicitly to the occurence of corresponding truncation errors [1], [4].

Recently, Neagoe [2] has presented an optimum nonuniform sampling theorem for time-limited signals, by minimizing the maximum instantaneous interpolation error. The sampling moments are taken according to the Chebyshev polynomial roots and then the Discrete Cosine Transform (DCT) is applied on the vector of nonuniformly taken samples. However, although it is an optimum model for time-limited signals, firstly, it does not consider the constraint of band-limitation as a previous assumption for optimization and secondly, it seems a cumbersome task to build a practical device for nonuniform (asynchronous) signal sampling in the time domain.

Neagoe [3] also proposes a nonuniform sampling theorem for time-limited signals, to the aim of high accuracy preserving the integral signal characteristics (such as energy and Fourier Transform). It consists of the cascade of nonuniform sampling in the time domain according to Legendre polynomial roots followed by the Discrete Legendre Transform (DLT) applied on the vector of nonuniformly taken samples. We retain from the above method the idea of Legendre nonuniform sampling but in the frequency domain instead of the time domain and we further apply this idea in the now proposed model.

This paper presents a new method for signal interpolation giving several improvements over the previous two models of the author, namely it takes into consideration both the time-limiting and also the band-limiting conditions as a priori constraints for optimization. Moreover, it considers the usual uniform sampling in the time domain, which is very simple and practical, but the method estimates the nonuniformly taken samples of the Fourier Transform in the frequency domain, giving us a high accuracy interpolation formula, according to the Gauss quadrature rule.

2. THE NEW INTERPOLATION METHOD

Interpolation Theorem: Assume we have a real-valued signal, $g(t)$, considered to be both time-limited and also band-limited $\{g:[0,n] \rightarrow \mathbb{R},\ n \in \mathbb{N},\ g \in \mathcal{L}^2(0,n),\ G(f)=0$ for $|f| > W$, where $G(f)$ is the Fourier transform of $g(t)$, so that $g(t)=\int_{-W}^{W} G(f)e^{2\pi ift}\,dt$, $i^2=-1\}$ and also assume that the vector of uniformly taken signal samples $\underset{\sim}{g} = (g_j = g(j))_{j=0}^{n} = (g_0 g_1 \dots g_n)^T$ is given.
Then:
(a) The optimum interpolation formula minimizing the maximum instantaneous error is:

$$\hat{g}_n(t) = \sum_{j=0}^{n} \Psi_j(t) g_j \,, \tag{1}$$

where $\Psi_j(t)$ is a real-valued function, $\Psi_j: [0,n] \rightarrow \mathbb{R}$, given by

$$\Psi_j(t) = \sum_{j=0}^{n} \mu_{kj}\,(w_k)^t \,, \quad (j=0,1,\dots,n) \tag{2}$$

where $w_k=e^{i\pi r x_k}$, $(i^2=-1)$, $r=2W=1/\lceil 1/(2W) \rceil$ is the ratio between the present sampling period and the Shannon sampling period, $\{x_k\}_{k=0}^{n}$ is the set of ordered roots of the $(n+1)$th degree Legendre polynomial $(-1 < x_0 < x_1 < \dots < x_n < 1)$, and $\mu_{kj} \in \mathbb{C}$ is the general element of the matrix $\underset{\sim}{U}=(\mu_{kj})_{0 \leq k,j \leq n}$, k being the row index, and j the column index, so that $\underset{\sim}{U}$

* Author's mailing address: V. Neagoe, P.O. Box 16-37, Bucharest 16, Romania 77500, Phone: 40-0-714769.

is the inverse of the matrix $\underset{\sim}{A}$, i.e., $\underset{\sim}{U}=\underset{\sim}{A}^{-1}$, where $\underset{\sim}{A}$ is the "(n+1) x (n+1)" Van der Monde complex matrix

$$\underset{\sim}{A} = \begin{bmatrix} 1 & 1 & \cdots & 1 \\ w_0 & w_1 & \cdots & w_n \\ w_0^2 & w_1^2 & \cdots & w_n^2 \\ \vdots & \vdots & & \vdots \\ w_0^n & w_1^n & \cdots & w_n^n \end{bmatrix} . \tag{3}$$

(b) The maximum instantaneous error bound (for n even) is

$$\max_{t \in \left(\frac{n}{2}-\frac{1}{2}, \frac{n}{2}+\frac{1}{2}\right)} |g(t)-\hat{g}_n(t)| \leqslant \mathcal{E}(n) , \tag{4}$$

where

$$\mathcal{E}(n)=q(n)\left[A(n)+B(n)\cdot\phi(n)\right] , \tag{5}$$

and where

$$q(n)=\left(\frac{r\pi e}{4}\right)^{2n+2} \cdot \frac{\sqrt{n(n+1)}}{\left(1+\frac{1}{n}\right)^{2n+2}\cdot(2n+3)\sqrt{4n+5}} \cdot \sqrt{E} \tag{6}$$

$$E = \int_0^n g^2(t)dt \quad , \text{(signal energy)} \tag{7}$$

$$A(n) = \sqrt{\left(\frac{n-1}{2n}\right)^{4n+5} + \left(\frac{n+1}{2n}\right)^{4n+5}} \quad , \tag{8}$$

$$B(n) = \sqrt{\sum_{j=0}^{n} \left[\left(\frac{j}{n}\right)^{4n+5} + \left(1-\frac{j}{n}\right)^{4n+5}\right]} \tag{9}$$

$$\phi(n)= \max_{t \in \left(\frac{n}{2}-\frac{1}{2}, \frac{n}{2}+\frac{1}{2}\right)} \sqrt{\sum_{j=0}^{n} \Psi_j^2(n)} \tag{10}$$

(c) In order that interpolation converges ($\lim_{n\to\infty} \mathcal{E}(n)=0$), it is sufficient to have $r < 4/(\pi e) = 0.468$. It means to choose an average sampling rate 2.135 times faster than the Shannon sampling rate.

Hints for proof: We evaluate the signal g(t) from its inverse Fourier transform according to the Gauss quadrature rule [3]; it leads to

$$\hat{g}_n(t) = \sum_{k=0}^{n} (w_k)^t \; G_k \, , \tag{11}$$

where $G_k = W \cdot \lambda_k \cdot G(Wx_k)$, λ_k being the Cristophell coefficients, and the Legendre nonuniformly taken frequency samples of the Fourier transform $G(Wx_k)$ being estimated from a linear system of N=n+1 equations, in order to be consistent with the given set of uniformly taken time samples g_j, namely

$$g(j)=\sum_{k=0}^{n} G_k \, w_k^j \quad , \; (j=0,1,..n; w_k^j=(w_k)^j). \tag{12}$$

According to the previous assumptions, we can easily deduce that the relations (11) and (12) lead to (1) and (2); thus, the section (a) of the Interpolation Theorem can be proved.

The proof of the section (b) of the Theorem regarding the bound of the instantaneous error is not given here, requiring a large space; it will be given in the extended paper version.

The section (c) can be deduced immediately imposing the convergence condition on relations (4), (5) and (6)

$\left(\lim_{n\to\infty} q(n)=0 \;\text{ from }\; \lim_{n\to\infty} \mathcal{E}(n)=0 \right)$.

Remarks:

(i) We have deduced an analytical formula for inversion of the complex Van der Monde matrix; its proof will be given in the extended paper version. The formula of the inverse matrix $\underset{\sim}{U} = \underset{\sim}{A}^{-1}$ is

$$\underset{\sim}{U}=(\mu_{kj})_{\substack{0\leqslant k\leqslant n\\0\leqslant j\leqslant n}} = \left(\frac{(-1)^{k+j}\,\sigma_{n+1-k}(w_0,\ldots,w_{j-1},w_{j+1},\ldots w_n)}{\prod_{h=0}^{j-1}(w_j-w_h)\prod_{h=j+1}^{n}(w_k-w_h)}\right)_{\substack{0\leqslant k\leqslant n\\0\leqslant j\leqslant n}} \tag{13}$$

where $\sigma_p(w_0,w_1,\ldots,w_n)$ are symmetric polynomials of degree p, (p=0,1,...,n+1)

$$\begin{cases} \sigma_0(w_0,w_1,\ldots,w_n) = 1 \\ \sigma_1(w_0,w_1,\ldots,w_n) = w_0+w_1+\ldots+w_n \\ \sigma_2(w_0,w_1,\ldots,w_n) = w_0w_1+w_0w_3+\ldots+ \\ \quad +w_0w_n+w_1w_2+\ldots+w_1w_n+\ldots+w_{n-1}w_n \\ \vdots \\ \sigma_{n+1}(w_0,w_1,\ldots,w_n) = w_0w_1w_2\ldots w_n \end{cases} \tag{14}$$

(ii) For n even (N=n+1= the odd number of taken samples), the function $\Psi_j(t)$ (relation (2)) may be equivalently expressed as

$$\Psi_j(t)= \mu_{\frac{n}{2}} +2 \sum_{k=0}^{\frac{n}{2}-1}\left[\text{Real}(\mu_{kj})\cos(\pi r x_k t) - I_m(\mu_{kj})\sin(\pi r x_k t)\right] .$$

$$(j=0,1,\ldots,n;\; \mu_{\frac{n}{2}j} \in \mathbb{R}) \tag{15}$$

(iii) The real functions $\Psi_j(t)$, (j=0,1,...,n) satisfy the conditions

$$\Psi_j(h) = \delta_{jh} = \begin{cases} 1, & \text{for } j=h \\ 0, & \text{for } j\neq h \end{cases} \tag{16}$$

and from relations (1) and (16), we obtain $\hat{g}_n(j)=g_j$.

(iv) We may show that, under some constraints, $\phi(n)=1$.

(v) For n large (ideally, $n\to\infty$), we can obtain useful approximations of the expressions (8) and (9), namely $A(n)\cong 0$ and $B(n)\cong\sqrt{10}/2$.

3. INTERPOLATION PERFORMANCES AND CONCLUDING REMARKS

Table I shows the numerical results of the error bound of the proposed method (relations (4)-(10) by comparison with the corresponding error of Shannon interpolation (Brown bound, Piper bound, and Yao-Thomas bound [1], [4]) and also by comparison with the error bound of Neagoe nonuniform time sampling reconstruction model [2]. We have chosen r=0.45, satisfying the convergence condition (c) of the interpolation theorem near its limit ($r < 0.468$). We can appreciate the very good error bound performances of the proposed method, that are far much better than the ones obtained by Shannon formula (from 21 times better for N=n+1=9, to 515 times better for N=n+1=65, by considering as reference the Yao-Thomas bound). The performances of the proposed method are also better as compared to those of the previous method of the author based on nonuniform interpolation (from 1.6 times better for N=n+1=9 to 5.2 times better for N=n+1=65); moreover, it is much easier to build a uniform time sampling device than a nonuniform time sampling one according to Chebyshev polynomial roots. We can use the following error bound approximations for n large, to point out the strong convergence towards zero of the error bound for the new method by comparison with the other methods considered as reference:

$$\frac{\varepsilon^{(\text{Yao-Thomas})}_{\text{Shannon}}}{\sqrt{E_1}} \cong c_1(r)\cdot\frac{1}{n} \qquad (17)$$

$$\frac{\overset{(\text{Chebyshev})}{\text{Neagoe}_1}}{\sqrt{E_1}} \cong c_2(r)\cdot\frac{1}{n+1}\cdot\left(\frac{r\pi e}{4}\right)^{n+1} \qquad (18)$$

$$\frac{\varepsilon^{(\text{new})}_{\text{Neagoe}_2}}{\sqrt{E}} \cong c_3(r)\cdot\frac{1}{\sqrt{n+1}}\cdot\left(\frac{r\pi e}{4}\right)^{2(n+1)} \qquad (19)$$

We can remark the strong exponential convergence towards zero of the error bound for the presented method (the exponent being (2n+2)) by comparison with the exponential convergence for the Neagoe previous nonuniform sampling method (with exponent (n+1)), and with the weak n^{-1} convergence of the error bound for Shannon interpolation.

We have also evaluated by computer simulation the experimental interpolation performances of the new method. We have considered the signal

$$g(t) = A_0 + \sum_{i=1}^{5} A_i \cos(2\pi f_i t), \qquad (20)$$

whose parameters are given in Table II. The signal is sampled with a sampling period equal to 1 (r=2W=0.45), using N=n+1=9 samples. The experimental interpolation results are given in Table II. For the considered example, we deduce that the interpolation error becomes at least 110 times smaller by applying our proposed method, as compared to Shannon formula!

REFERENCES

[1] Jerri, A.J., Proceedings IEEE (1977), vol.65, pp. 1565-1691.

[2] Neagoe, V., IEEE Trans.Acoust., Speech, Signal Processing (1990), vol.ASSP-38, pp. 1812-1815.

[3] Neagoe, V., Optimum Discrete Representation of Signals by Preserving Their Integral Characteristics, in: Lacoume, J.L., Chehikian, A., Martin, N., and Malbos, J., (eds.), Signal Processing IV: Theories and Applications, Proceedings of EUSIPCO-88 (North Holland, Amsterdam, 1988), pp.1101-1104.

[4] Piper, H., IEEE Trans.Inform.Theory (1975), vol.IT-21, pp. 482-485.

[5] Shannon, C.E., Proceedings IRE (1949), vol.37, pp. 10-21.

TABLE II

COMPUTER SIMULATION PERFORMANCES OBTAINED FOR THE NEW INTERPOLATION METHOD BY COMPARISON WITH SHANNON FORMULA

n=8 (n+1=9 uniformly taken samples)

$$\left\{g: [0,n] \rightarrow \mathbb{R},\ G(f)=0 \text{ for } |f|>W, \text{ where } g(t)=\int_{-W}^{W} G(f)e^{2\pi i f t}\,df\right\},\ r=0.45 \quad (W=0.225)$$

$$g(t)=A_0 + \sum_{i=0}^{5} A_i\cos(2\pi f_i t)$$

$f_0=0$	$f_1=0.05$	$f_2=0.115$	$f_3=0.14$	$f_4=0.175$	$f_5=0.225$
$A_0=0.5$	$A_1=1$	$A_2=1/(2.3)$	$A_3=1/(2.8)$	$A_4=1/(3.5)$	$A_5=1/(4.5)$

$$\varepsilon_j = |\hat{g}_8(j+0.5) - g(j+0.5)|$$

j →		0	1	2	3	4	5	6	7
$10^4\cdot\varepsilon_j$ normalized interpolation error	The New Method	0.37	0.34	0.17	0.28	0.22	0.17	0.95	4.16
	Shannon Interpolation	3541	2003	1378	1040	827	677	560	463

TABLE I

NORMALIZED INSTANTANEOUS ERROR BOUND OF INTERPOLATION FOR THE PROPOSED METHOD BY COMPARISON WITH THE ERROR OF CLASSICAL SHANNON INTERPOLATION [1], [4], AS WELL AS WITH THE ERROR OF NEAGOE NONUNIFORM SAMPLING INTERPOLATION MODEL [2] (RESULTS OF ANALYTICAL EVALUATIONS)

$(r=2W=0.45)$

<table>
<tr><th colspan="2" rowspan="2">TYPE OF INTERPOLATION ERROR BOUND</th><th colspan="5">NUMBER OF TAKEN SAMPLES N=n+1</th></tr>
<tr><th>N=5</th><th>N=9</th><th>N=17</th><th>N=33</th><th>N=65</th></tr>
<tr><td rowspan="3">INTERPOLATION ERROR BOUND FOR SHANNON UNIFORM SAMPLING RECONSTRUCTION
$\varepsilon_{SH} = \max_{t \in (\frac{n}{2}-\frac{1}{2};\frac{n}{2}+\frac{1}{2})} \frac{|g(t)-\hat{g}_n(t)|}{\sqrt{E_1}}$
(for a signal g(t) uniformly sampled in N=n+1 points: j=0,1,..,n; n=even; sampling interval=1; r=1/(1/2w); W=signal bandwidth; G(f)=Fourier (g(t)); where G(f)=0, for $|f|\geqslant W$;
$E_1 = \int_{-W}^{W}|G(f)|^2\,df$)</td><td>BROWN BOUND</td><td>0.625904</td><td>0.250361</td><td>0.119219</td><td>0.058908</td><td>0.029367</td></tr>
<tr><td>PIPER BOUND</td><td>0.534243</td><td>0.213697</td><td>0.101760</td><td>0.050281</td><td>0.025067</td></tr>
<tr><td>YAO-THOMAS BOUND</td><td>0.329543</td><td>0.131817</td><td>0.062770</td><td>0.031015</td><td>0.015462</td></tr>
<tr><td colspan="2">INTERPOLATION ERROR BOUND FOR NEAGOE NONUNIFORM TIME SAMPLING RECONSTRUCTION [2]
$\varepsilon_{N_1} = \max_{t \in (0,n)} \frac{|g(t)-\hat{g}_n(t)|}{\sqrt{E_1}}$
(for a signal g(t) uniformly sampled in N=n+1 points over the interval [0,n], according to Chebyshev polynomial roots; r,W,G(f), and E_1 have the significances given above).</td><td>0.019037</td><td>0.009795</td><td>0.003942</td><td>0.001096</td><td>0.000156</td></tr>
<tr><td colspan="2">INTERPOLATION ERROR BOUND FOR THE PRESENTED OPTIMUM UNIFORM SAMPLING RECONSTRUCTION METHOD
$\varepsilon_{N_2} = \max_{t \in (\frac{n}{2}-\frac{1}{2},\frac{n}{2}+\frac{1}{2})} \frac{|g(t)-\hat{g}_n(t)|}{\sqrt{E}}$
(for a signal g(t), uniformly sampled in N=n+1 points: j=0,1,...,n; n=even; sampling interval=1; r,W,G(f) have the significances given above, while
$E = \int_0^n g^2(t)dt$)</td><td>0.009080</td><td>0.006079</td><td>0.002630</td><td>0.000561</td><td>0.000030</td></tr>
</table>

SIGNAL PROCESSING VI: Theories and Applications
J. Vandewalle, R. Boite, M. Moonen, A. Oosterlinck (eds.)

NONUNIFORM DISCRETE FOURIER TRANSFORM AND ITS APPLICATIONS IN SIGNAL PROCESSING

SANJIT K. MITRA, SONALI CHAKRABARTI, and EDUARDO ABREU

Department of Electrical and Computer Engineering
University of California, Santa Barbara, CA 93106

The *nonuniform discrete Fourier transform* (NDFT) of a sequence of length N is defined as samples of its z-transform at N arbitrarily located distinct points on the z-plane. The NDFT reduces to the conventional DFT when the sampling points are located on the unit circle at equally spaced angles. The properties of the NDFT and its applications in signal processing such as spectral analysis and FIR filter design are outlined. Extension of the concept of the NDFT to two dimensions is discussed.

1. INTRODUCTION

In many applications, the problem of computing samples of the z-transform X(z) of a sequence {x(n)} of length N is of interest. One popularly used approach is to compute the discrete Fourier transform (DFT) of the sequence,which corresponds to evaluating its z-transform at N equally spaced points on the unit circle. However, in certain applications such as spectral analysis, it is preferable to sample the z-transform X(z) at N points that are nonuniformly spaced. There have been various efforts to achieve such nonuniform sampling [1],[2].

A better approach, considered in this paper, is to generalize the definition of the DFT. The *nonuniform discrete Fourier transform* (NDFT) of a sequence {x(n)} of length N is defined as a sequence of length N that corresponds to samples of the z-transform X(z) at N distinct points located arbitrarily on the z-plane. The NDFT reduces to the DFT when the N points are chosen at equally spaced angles on the unit circle. The freedom offered by the NDFT in choosing the points on the z-plane can be exploited in many signal processing applications.

2. DEFINITION

The *nonuniform discrete Fourier transform* (NDFT) of a sequence {x(n)} of length N is defined as

$$X(z_k) = \sum_{n=0}^{N-1} x(n) z_k^{-n}, \quad k = 0,1,\ldots,N-1, \tag{1}$$

where $z_0, z_1, \ldots, z_{N-1}$ are distinct points located arbitrarily on the z-plane. This can be expressed in a matrix form as

$$\mathbf{X} = \mathbf{D}\,\mathbf{x} \tag{2}$$

where

$$\mathbf{X} = \begin{bmatrix} X(z_0) \\ X(z_1) \\ \vdots \\ X(z_{N-1}) \end{bmatrix}, \quad \mathbf{x} = \begin{bmatrix} x(0) \\ x(1) \\ \vdots \\ x(N-1) \end{bmatrix}, \tag{3a}$$

and

$$\mathbf{D} = \begin{bmatrix} 1 & z_0^{-1} & z_0^{-2} & \cdots & z_0^{-N+1} \\ 1 & z_1^{-1} & z_1^{-2} & \cdots & z_1^{-N+1} \\ 1 & z_2^{-1} & z_2^{-2} & \cdots & z_2^{-N+1} \\ \vdots & \vdots & \vdots & \ddots & \vdots \\ 1 & z_{N-1}^{-1} & z_{N-1}^{-2} & \cdots & z_{N-1}^{-N+1} \end{bmatrix}. \tag{3b}$$

Note that the NDFT matrix **D** is a Vandermonde matrix and therefore is non-singular provided the N sampling points are distinct. Thus the inverse NDFT is simply

$$\mathbf{x} = \mathbf{D}^{-1}\mathbf{X}. \tag{4}$$

The DFT has been widely used in analyzing the frequency content of signals. Typically, a continuous-time signal is sampled, a finite-duration window is applied to it to obtain a sequence of length N, and then an N-point DFT is used to compute samples of the spectrum of the windowed sequence. Often, misleading results are obtained using this procedure. For example, if the signal contains two closely spaced sinusoids, the resulting DFT may fail to resolve the corresponding peaks in the spectrum. For fixed length transforms, and some knowledge of the signal, the resolution can be improved by using the NDFT to compute the spectral samples. A simple example to illustrate this idea is shown in Figure 1.

The discrete-time Fourier transform of the windowed sequence is shown in Figure 1(a). Figure 1(b) shows the spectrum obtained by using a 64-point DFT. The unsatisfactory resolution obtained was improved by using a 64-point NDFT, where the spectrum was sampled more densely in the critical region.

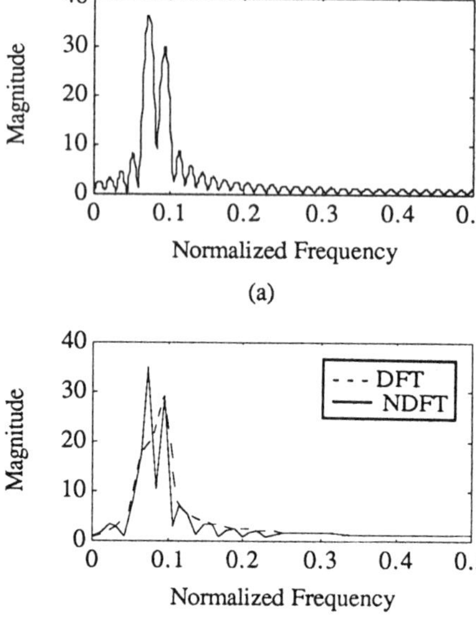

Figure 1. (a) Original spectrum, (b) 64-point DFT and NDFT

3. PROPERTIES OF THE NDFT

Some relevant properties of the NDFT are outlined in this section.

3.1. Basic Properties

Let $X(z_k)$ and $Y(z_k)$ denote the NDFT of the sequences x(n) and y(n) respectively. For notational convenience we denote an NDFT pair as

$$x(n) \xleftrightarrow{\text{NDFT}} X(z_k)$$

<u>Property 1</u>. Linearity:

$$\alpha x(n) + \beta y(n) \xleftrightarrow{\text{NDFT}} \alpha X(z_k) + \beta Y(z_k)$$

Properties 2-5 hold if the sampling points $z_0, z_1, \ldots, z_{N-1}$ are chosen in complex conjugate pairs.

<u>Property 2</u>.

$$x^*(n) \xleftrightarrow{\text{NDFT}} X^*(z_k^*)$$

<u>Property 3</u>.

$$\text{Re}\{x(n)\} \xleftrightarrow{\text{NDFT}} X_e(z_k) = \frac{1}{2}\left\{X(z_k) + X^*(z_k^*)\right\}$$

<u>Property 4</u>.

$$j\,\text{Im}\{x(n)\} \xleftrightarrow{\text{NDFT}} X_o(z_k) = \frac{1}{2}\left\{X(z_k) - X^*(z_k^*)\right\}$$

Property 5 holds if x(n) is real.

<u>Property 5</u>. Symmetry Properties.

$$X(z_k) = X^*(z_k^*)$$

$$\text{Re}\left\{X(z_k)\right\} = \text{Re}\left\{X(z_k^*)\right\}$$

$$\text{Im}\left\{X(z_k)\right\} = -\text{Im}\left\{X(z_k^*)\right\}$$

$$\left|X(z_k)\right| = \left|X(z_k^*)\right|$$

$$\arg\left\{X(z_k)\right\} = -\arg\left\{X(z_k^*)\right\}$$

3.2. Linear Convolution Using the NDFT

Let $\{y(n)\}$ be the linear convolution of the sequences $\{x(n)\}$ and $\{h(n)\}$. Using vector notation this convolution can be expressed as

$$\mathbf{y} = \mathbf{x} * \mathbf{h} \tag{5}$$

where the vectors **x** and **h** are of length N and **y** is of length 2N–1. Let **X** and **H** denote the N-point NDFT of **x** and **h** respectively, i.e.

$$\mathbf{X} = \mathbf{D}\,\mathbf{x}, \quad \mathbf{H} = \mathbf{D}\,\mathbf{h}, \tag{6}$$

and let

$$\mathbf{x}_c = \mathbf{D}^{-1}[\mathbf{X} \cdot \mathbf{H}] \tag{7}$$

where the dot "·" denotes the componentwise product of the vectors $\mathbf{X}_1$ and $\mathbf{X}_2$. Evaluating the z-transforms of both sides of Eq. (5) at $z = z_0, z_1, \ldots, z_{N-1}$ we get

$$\mathbf{X} \cdot \mathbf{H} = [\mathbf{D} \;\; \mathbf{C}]\,\mathbf{y} = \mathbf{D}\,[\mathbf{I} \;\; \mathbf{D}^{-1}\mathbf{C}]\,\mathbf{y}. \tag{8}$$

where **I** is the N×N identity matrix and

$$\mathbf{C} = \begin{bmatrix} z_0^{-N} & z_0^{-N-1} & \cdots & z_0^{-2N+2} \\ z_1^{-N} & z_1^{-N-1} & \cdots & z_1^{-2N+2} \\ \vdots & & & \vdots \\ z_{N-1}^{-N} & z_{N-1}^{-N-1} & \cdots & z_{N-1}^{-2N+2} \end{bmatrix}. \tag{9}$$

Substituting Eq. (8) in Eq. (7), we arrive at

$$\mathbf{x}_c = [\mathbf{I} \;\; \mathbf{D}^{-1}\mathbf{C}]\,\mathbf{y}. \tag{10}$$

From Eq. (10) we conclude that if $\mathbf{x}_c$ is made to have length 2N–1 by augmenting the sequences **x** and **h** with N–1 zeroes at the end, and then using an NDFT matrix of size (2N−1)×(2N−1), x_c will correspond to the linear convolution of **x** and **h**. Thus linear convolution can be obtained using the NDFT and the zero padding technique, similar to the DFT.

4. FIR FILTER DESIGN USING THE NDFT

In our proposed FIR filter design method, the desired frequency response is first evaluated at N nonuniformly spaced points around the unit circle. The filter coefficients are then obtained by an N-point inverse NDFT of these frequency samples. This procedure is mathematically equivalent to computing the coefficients of an interpolating polynomial of order N that has the desired values at the specified frequency locations [3]. This approach is thus a generalization of the classical frequency sampling approach but yields improved results.

As an example we considered the design of linear phase lowpass filters with nearly equiripple frequency responses. The desired response was obtained by the method described in [4]. In this approach, separate functions derived from Chebyshev polynomials are used to approximate the passband and stopband of the desired equiripple frequency response. In our proposed design method, since the frequency samples are nonuniformly spaced, they were chosen to be in the passband and stopband only. It was found that the best locations for the frequency samples were at the extrema of the desired equiripple response. Analytic expressions were found, specifying these locations precisely from the given filter specifications. In the NDFT formulation in Eq. (2), the Vandermonde matrix **D** was constructed using the frequency sample locations and **X** contained the sample values. The impulse response **x** was then obtained by solving this linear system of equations. In many cases, the resulting interpolated frequency responses were found to be very close to optimal responses obtained by using the Parks-McClellan algorithm for the same filter specifications. Moreover, the CPU design time needed was considerably less when compared to the Parks-McClellan algorithm.

4.1. A Design Example

A lowpass filter was designed with the following specifications: passband edge ω_p at 0.3π, stopband edge ω_s at 0.36π, passband ripple $\delta_p = 0.01$, and stopband ripple $\delta_s = 0.005$. Table 1 shows a comparison of the two filters, each of length 71, designed using the proposed approach and the Parks-McClellan algorithm respectively.

Table 1

Design method	A_p (dB)	A_s (dB)	CPU time (sec)
NDFT	0.097	44.589	2.08
Parks-McClellan	0.106	45.029	12.58

5. THE 2-D NDFT

The nonuniform discrete Fourier transform of a 2-D sequence can be defined in a manner similar to that of a 1-D sequence. Thus, the 2-D NDFT of a sequence {x(m,n)} of size MxN is defined as

$$\hat{X}(z_{1k},z_{2k}) = \sum_{m=0}^{M-1}\sum_{n=0}^{N-1} x(m,n)\, z_{1k}^{-m} z_{2k}^{-n},$$

$$k = 0,1,\ldots,MN-1, \tag{11}$$

where (z_{1k}, z_{2k}) represent distinct frequency locations, which can be chosen arbitrarily but in such a way that the inverse transform exists.

For the sake of clarity, this is illustrated by means of a simple example. Consider the case when M = N = 2. Then Eq. (10) can be expressed in a matrix form as

$$\hat{\mathbf{X}} = \mathbf{DX} \tag{12}$$

where

$$\hat{\mathbf{X}} = \begin{bmatrix} \hat{X}(z_{10},z_{20}) \\ \hat{X}(z_{11},z_{21}) \\ \hat{X}(z_{12},z_{22}) \\ \hat{X}(z_{13},z_{23}) \end{bmatrix}, \qquad \mathbf{X} = \begin{bmatrix} x(0,0) \\ x(0,1) \\ x(1,0) \\ x(1,1) \end{bmatrix},$$

$$\mathbf{D} = \begin{bmatrix} 1 & z_{20}^{-1} & z_{10}^{-1} & z_{10}^{-1}z_{20}^{-1} \\ 1 & z_{21}^{-1} & z_{11}^{-1} & z_{11}^{-1}z_{21}^{-1} \\ 1 & z_{22}^{-1} & z_{12}^{-1} & z_{12}^{-1}z_{22}^{-1} \\ 1 & z_{23}^{-1} & z_{13}^{-1} & z_{13}^{-1}z_{23}^{-1} \end{bmatrix}.$$

In general, **D** is a MN×MN 2-D Vandermonde matrix. In the special case when the frequencies are chosen on a nonuniform rectangular grid, i.e.,

$$z_{1,m} = z_{1,m+pM}, \quad m = 0,1,\ldots,M-1, p = 1,2,\ldots,N-1,$$

$$z_{2,n} = z_{2,n+qN}, \quad n = 0,1,\ldots,N-1, q = 1,2,\ldots,M-1,$$

Eq. (11) can alternatively be expressed in terms of two 1-D Vandermonde matrices, $\mathbf{D}_1$ and $\mathbf{D}_2$, of sizes M×M and N×N respectively, as

$$\hat{\mathbf{X}} = \mathbf{D}_1\mathbf{X}\mathbf{D}_2^{\,t}, \tag{13}$$

where

$$\hat{\mathbf{X}} = \begin{bmatrix} \hat{X}(z_{10},z_{20}) & \hat{X}(z_{10},z_{21}) \\ \hat{X}(z_{11},z_{20}) & \hat{X}(z_{11},z_{21}) \end{bmatrix},$$

$$\mathbf{X} = \begin{bmatrix} x(0,0) & x(0,1) \\ x(1,0) & x(1,1) \end{bmatrix},$$

$$\mathbf{D}_1 = \begin{bmatrix} 1 & z_{10}^{-1} \\ 1 & z_{11}^{-1} \end{bmatrix}, \mathbf{D}_2 = \begin{bmatrix} 1 & z_{20}^{-1} \\ 1 & z_{21}^{-1} \end{bmatrix}.$$

As can be seen, the 2-D DFT is a special case of the 2-D NDFT when the points are located on a uniform rectangular grid. In general the 2-D NDFT has MN degrees of freedom, which is far greater than the M+N degrees of freedom available when using a rectangular grid. This additional freedom can be harnessed in applications such as 2-D FIR filter design.

The 2-D NDFT can be used for the design of 2-D FIR filters by using a generalization of frequency sampling design technique. This method holds great potential for designing filters with unusual passband shapes and for achieving non-rectangular sampling geometries. Although several efforts for the design of 2-D FIR filters using nonuniform frequency sampling have been reported [5],[6], a general design procedure for 2-D nonseparable filters has not yet been found.

As an example we considered the design of 2-D half-band filters using this approach, e.g. diamond filters and 90° fan filters. Figure 2 shows the magnitude response and the contour plot of a 9×9 diamond filter designed using the proposed approach. Due to the octagonal symmetry and the half-band nature of the frequency response, only 6 of the 81 impulse response coefficients are independent. These were found by taking 6 frequency samples of unit value, located nonuniformly as shown in Fig. 2(b). The maximum ripple obtained is only 0.003. A 9×9 diamond filter with the same transition bandwidth, designed using the frequency transformation method, has a maximum ripple of 0.029.

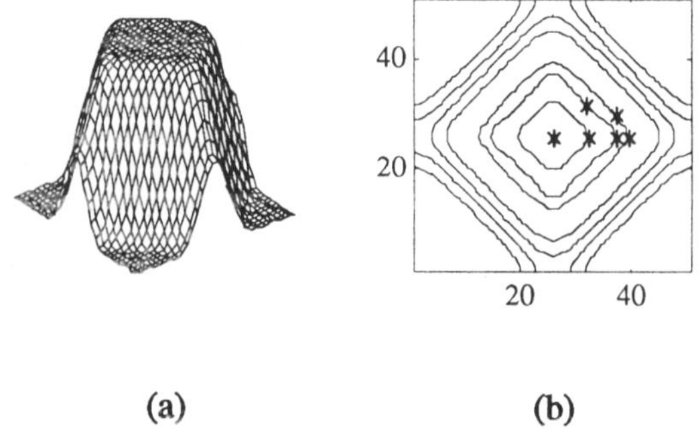

(a) (b)

Figure 2. (a) Magnitude response, and (b) Contour plot of a 9×9 diamond filter.

6. CONCLUDING REMARKS

The concept of the nonuniform discrete Fourier transform, which corresponds to sampling the z-transform at arbitrarily chosen points on the z-plane, was presented in this paper. The NDFT was defined and its properties were studied. The advantage offered by the NDFT in choosing the points on the z-plane was illustrated by an example in spectral analysis. In particular, this idea was applied to nonuniform frequency sampling design of FIR filters. In many cases the resulting filters were practically equal to optimal equiripple filters, and the CPU design time was considerably reduced. A filter design example was presented to illustrate this technique. The NDFT was also extended to 2-D. An example of diamond filter design was presented to demonstrate the potential of the 2-D NDFT in 2-D nonseparable FIR filter design. Finally, the concept of the NDFT can easily be extended to higher dimensions.

Acknowledgements

This work was supported in part by a University of California MICRO grant with matching support from Granger-Telettra, Inc.

References

[1] L. R. Rabiner, R. W. Schafer, C. M. Rader, "The chirp z-transform algorithm", *IEEE Trans. Audio and Electroacoustics*, vol AU-17, no. 2, pp. 86-92, June 1969.

[2] A. Oppenheim, D. Johnson, "Computation of spectra with unequal resolution using the fast Fourier transform", *Proc. IEEE*, vol 59, pp. 299-301, February 1971.

[3] L. R. Rabiner, B. Gold, C. A. McGonegal, "An approach to the approximation problem for nonrecursive digital filters", *IEEE Trans. Audio and Electroacoustics*, vol AU-18, pp. 83-106, June 1970.

[4] P. Jarske, Y. Neuvo, S. K. Mitra, "Improved frequency sampling FIR filter design", *Proc. EUSIPCO '88*, Grenoble, France, September 1988, pp. 691-694.

[5] W. J. Rozwod, C. W. Therrien, J. S. Lim, "Design of 2-D FIR Filters by Nonuniform Frequency Sampling", *IEEE Trans. Acoust., Speech, Signal Processing*, vol. ASSP-39, no. 11, pp. 2508-2514, November 1991.

[6] A. Zakhor, G. Alvstad, "Two-dimensional polynomial interpolation from nonuniform samples", *IEEE Trans. Acoust., Speech, Signal Processing*, vol. ASSP-40, no. 1, pp. 169-180, January 1992.

SIGNAL PROCESSING VI: Theories and Applications
J. Vandewalle, R. Boite, M. Moonen, A. Oosterlinck (eds.)

UNIQUENESS AND STABILITY OF DISCRETE ZERO-CROSSINGS AND MAXIMA WAVELET REPRESENTATIONS

Zeev Berman and John S. Baras

Electrical Engineering Department and Systems Research Center
University of Maryland, College Park, MD 20742 , USA
e-mail: berman@src.umd.edu

A general signal description, called an inherently bounded Adaptive Quasi Linear Representation (AQLR), motivated by two important examples, namely the wavelet maxima representation, and the wavelet zero-crossings representation, is introduced. It is shown, that the dyadic wavelet maxima (zero-crossings) representation is, in general, nonunique. Using the idea of the inherently bounded AQLR, a global BIBO stability is shown. For a special case, where perturbations are limited to the continuous part of the representation, a Lipschitz condition is satisfied.

1 Introduction[1]

S. Mallat in [4] and, together with Zhong, in [5], introduced zero-crossings and extrema of the wavelet transform as a multiscale edge representation. Two important advantages of these methods are low algorithmic complexity and flexibility in choosing the basic filter. Moreover, [4] and [5] propose reconstruction procedures and show accurate numerical reconstruction results from zero-crossings and maxima representations. In [4, 5], as in many other works in this area, the basic algorithms were developed using continuous variables. The continuous approach gives an excellent background to motivate and justify the use of either local extrema or zero-crossings as important signal features. Unfortunately, in the continuous framework, analytic tools to investigate the information content of the representation are not yet available. The knowledge about properties of the representations is mainly based on empirical reconstruction results. From the theoretical point of view, there are still important open problems, e.g. stability, uniqueness, and structure of a reconstruction set (a family of signals having the same representation).

Our objective is to analyze these theoretical questions using a model of an actual implementation. The main assumption is that the data is discrete and finite. The discrete multiscale maxima and zero-crossings representations are defined in a general set-up of a linear filter bank, however, the main goal is to consider a particular case where the filter bank describes the wavelet transform. Since reconstruction sets of both maxima and zero-crossings representations have a similar structure, a general form called the Adaptive Quasi Linear Representation (AQLR) is introduced. Moreover, many generalizations of the basic maxima and zero-crossings representations fit into the framework of the AQLR.

We first present conditions for uniqueness, then apply these conditions to the wavelet transform-based representation, and then obtain a conclusive result. It turns out, that neither the wavelet maxima representation nor the wavelet zero-crossings representation is, in general, unique.

The next subject is stability of the representation. This issue is of great importance because there are many known examples of unstable zero-crossings representations. In order to improve the stability properties, Mallat has included additional sums in the standard zero-crossings representation and, together with Zhong, they have introduced the wavelet maxima representation, as a stable alternative to the zero-crossings representation. Using the idea of the inherently bounded AQLR, global BIBO (bounded input, bounded output) stability is shown. For a special case, where perturbations are limited to the continuous part of the representation, a Lipschitz condition is satisfied.

[1]Research supported in part by NSF grant NSF CDR-85-00108 under the Engineering Research Centers Program.

2 Multiscale Maxima Representation

Consider $\mathcal{L}$, a linear space of real, finite sequences:

$$\mathcal{L} \triangleq \left\{ f : f = \{f(n)\}_{n=0}^{N-1} \; f(n) \in \Re \right\}.$$

Let X and Y denote operators on $\mathcal{L}$ which provide the sets of local maxima and minima, respectively, of a sequence $f \in \mathcal{L}$. The formal definitions are :

$$Xf = \{k : f(k+1) \leq f(k) \text{ and } f(k-1) \leq f(k)\}$$

$$Yf = \{k : f(k+1) \geq f(k) \text{ and } f(k-1) \geq f(k)\}$$

In this work, in order to avoid boundary problems, an N-periodic extension of finite sequences is assumed.

Let $W_1, W_2, \ldots, W_J, S_J$ be linear operators on $\mathcal{L}$. The multiscale local extrema representation, $R_m f$ is defined as:

$$R_m f \triangleq \left\{ \left\{ XW_j f, YW_j f, \{W_j f(k)\}_{k \in EW_j f} \right\}_{j=1}^{J}, S_J f \right\}.$$

where $EW_j f = XW_j f \cup YW_j f$.

Following [5], $R_m f$, will be called the multiscale maxima representation as well. In the particular case, when $W_1, W_2, \ldots, W_J, S_J$ correspond to a wavelet transform, $R_m f$ will be called the wavelet maxima representation.

The determination of the extrema point sets is highly nonlinear. However, for the given extrema sets, $XW_j f$ and $YW_j f$, the remaining data, called the sampling information, are obtained by a linear operation of sampling an image of a linear operator at fixed points.

Let T_{mf} denote the linear operator associated with the sampling information. Then, $R_m f$ is written in an alternative way as:

$$R_m f = \left\{ \{XW_j f, YW_j f\}_{j=1}^{J}, T_{mf} f \right\}. \qquad (1)$$

For a given representation Rf, a reconstruction set $\Gamma(Rf)$ is defined as a set of all sequences satisfying this representation, i.e.

$$\Gamma(Rf) \triangleq \{\gamma \in \mathcal{L} : R\gamma = Rf\}. \qquad (2)$$

It is clear that in order to satisfy a given maxima representation, a sequence $h \in \mathcal{L}$, in addition to obeying the sampling information $T_{mf} h = T_{mf} f$, needs to meet the requirement that $W_j h$ has local extrema at the points of $XW_j f$ and $YW_j f$. Loosely speaking, we have to assure that $W_j h$ is increasing after a minimum and before a maximum and it is decreasing otherwise. Straightforward analysis yields:

Theorem 1 *$R_m f$ is a given multiscale maxima representation. $h \in \Gamma(R_m f)$ if and only if*

$$T_{mf} h = T_{mf} f \qquad (3)$$

$$t(k) \cdot (W_j h(k+1) - W_j h(k)) > 0 \qquad (4)$$

The last inequality should be satisfied for $j = 1, 2, \ldots, J$ and for almost all k (if two consecutive k's belong to $EW_j f$ then the first is omitted here). $t(k)$ is called the type of k and can be either $+1$ or -1.

The maxima representations can be cast into the form $Rf = \{Vf, Tf\}$, where Vf is a set of points and T is a linear operator which may depend on Vf. However, the key feature of the maxima representation is the fact that the set Vf yields additional constraints, in the form of linear inequalities, which do not appear directly in Rf. Stimulated by this observation, we define the following general family of signal representations.

Definition 1 *$Rf = \{Vf, Tf\}$ is called an Adaptive Quasi Linear Representation (AQLR) if there exists a linear operator A and a sequence a such that:*

$$x \in \Gamma(Rf) \Leftrightarrow Tx = Tf \text{ and } Ax > a. \qquad (5)$$

A, a may depend on Vf, but they must be independent of Tf.

The reasoning behind the name "Adaptive Quasi Linear Representation" (AQLR) is as follows. This representation is adaptive since T, A, a depend on the sequence f (via the set Vf). It is quasi linear because it is based on a set of linear equalities and inequalities.

Clearly, the following is true.

Proposition 1 *Any multiscale maxima representation is an AQLR.*

The next definition is a generalization of an essential boundness property of the wavelet maxima representation.

Definition 2 *An AQLR is called inherently bounded if there exists a real $K > 0$ such that*

$$x \in \Gamma(Rf) \Rightarrow \|x\| \leq K\|Tf\|.$$

The coefficient K can depend on the parameters of the representation e.g. $N, J, W_1, \ldots, W_J, S_J$ but it must be independent of Vf and Tf.

Proposition 2 *The wavelet maxima representation is an inherently bounded AQLR.*

It turns out that the wavelet maxima representation, as defined here, provides bounds for $\|W_j h\|$, $\|S_J h\|$. These bounds, due to Parseval's equality, yield a bound for the original sequence h. For details, the reader is referred to [3].

3 The Multiscale Zero-Crossings

In defining the multiscale zero-crossings representation, we essentially follow [4], but minor changes are necessary due to our basic assumption that only a discrete signal version is available. Let Z be an operator which provides the set of zero-crossings of a given sequence $f \in \mathcal{L}$, i.e.

$$Zf \triangleq \{k : f(k-1) \cdot f(k) \leq 0\}. \tag{6}$$

Mallat in [4] has stabilized the zero-crossings representation by including the values of the wavelet transform integral calculated between consecutive zero-crossing points. Therefore, the multiscale zero-crossings representation, $R_z f$, is defined as:

$$R_z f \triangleq \left\{\{ZW_j f, U_j^{zf} f\}_{j=1}^{J}, S_J f\right\}. \tag{7}$$

where

$$U_j^{zf} f(k) = \sum_{l=k}^{n(k)-1} W_j f(l).$$

k and $n(k)$ are two consecutive zero-crossings of $W_j f$.

As in the maxima representation case, for fixed sets $ZW_j f$, the remaining data $U_j^{zf} f$ and $S_J f$ are obtained by a linear operator, denoted by T_{zf}. The zero-crossings representation can also be written as:

$$R_z f = \left\{\{ZW_j f\}_{j=1}^{J}, T_{zf} f\right\}. \tag{8}$$

We have the following characterization of the reconstruction set.

Theorem 2 *Let $R_z f$ be a given multiscale zero-crossings representation. $h \in \Gamma(R_z f)$ if and only if*

$$T_{zf} h = T_{zf} f \tag{9}$$

$$t(k) \cdot W_j h(k) > 0. \tag{10}$$

$t(k)$ is the type of k and can be either $+1$ or -1. The last inequality should be satisfied for almost all k (if $W_j f(k) = 0$ then k is omitted).

As an immediate consequence of Theorem 2 we have:

Proposition 3 *Any multiscale zero-crossings representation is an AQLR.*

Moreover:

Theorem 3 *The wavelet zero-crossings representation is an inherently bounded AQLR.*

4 Nonuniqueness

A representation $Rf = \{Vf, Tf\}$ is said to be unique, if the reconstruction set $\Gamma(Rf)$ consists of exactly one element. We have the following uniqueness characterization for AQLR's.

Lemma 1 *Let $Rf = \{Vf, Tf\}$ be an AQLR. Then Rf is unique if and only if the kernel of the operator T is trivial, i.e. $\mathcal{N}T = \{0\}$.*

The proof is clear from topological considerations. Nevertheless, an elementary but constructive proof is given in [3].

This claim has some significant consequences. Using the above lemma, an algorithm which tests for uniqueness can be developed. Perhaps the most important consequence of Lemma 1 is the fact that uniqueness of the representation Rf is equivalent to uniqueness of the underlying irregular sampling Tf. In other words, in the unique case, all the information about the signal is already contained in Tf. Additional constraints $Af > a$ are redundant. On the other hand, from the signal compression, understanding and interpretation point of view, it seems to be desirable that a little information would be specified explicitly by Tf and as much as possible information about a signal should be described implicitly by $Af > a$. Therefore, in our opinion, the most important and interesting features of AQLR's appear in the nonunique case.

Using the previous lemma, we are able to show that:

Theorem 4 *A discrete dyadic wavelet maxima (zero-crossings) representation based on a discrete low pass filter $H(\omega)$ is given. If $H(\pi) = 0$, $J \geq 3$, and N is a multiple of 2^J then there exists a sequence f which has a nonunique maxima (zero-crossings) representation.*

As a universal example of nonuniqueness the following sequence is proposed.

$$f(n) = \cos(2\pi \frac{n}{2^J}) \quad n = 0, 1, \ldots, N-1. \tag{11}$$

Observe that the same sequence is proposed for all dyadic wavelet transforms and for both the maxima representation and the zero-crossings representation.

The example of the nonunique maxima representation is described in [1, 2]. Now, let us consider the zero-crossings representation based on the wavelet transform defined in [4]. Let $J = 5$ and $N = 256$. Consider two sequences:

$$f(k) = \cos\left(\frac{2\pi k}{32}\right)$$

$$f_a(k) = \cos\left(\frac{2\pi k}{32}\right) + 0.1 \cdot \sin\left(\frac{2\pi k}{16}\right).$$

It can be shown that they have the same zero-crossings representation. Figure 1 describes these sequences, while Figure 2 gives their first level wavelet transforms (other levels wavelet transforms are very similar).

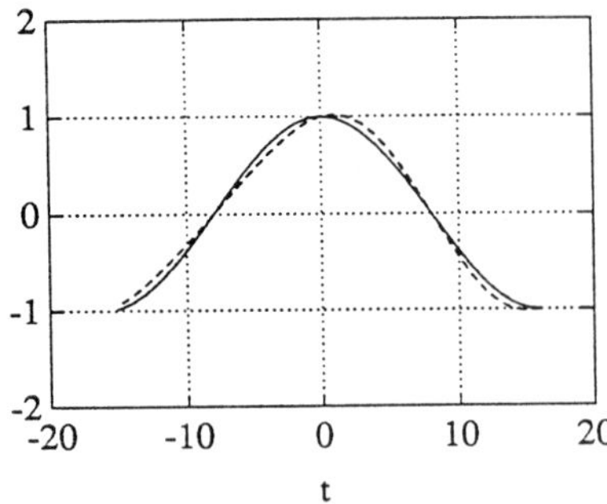

Figure 1: The sequences f and f_a

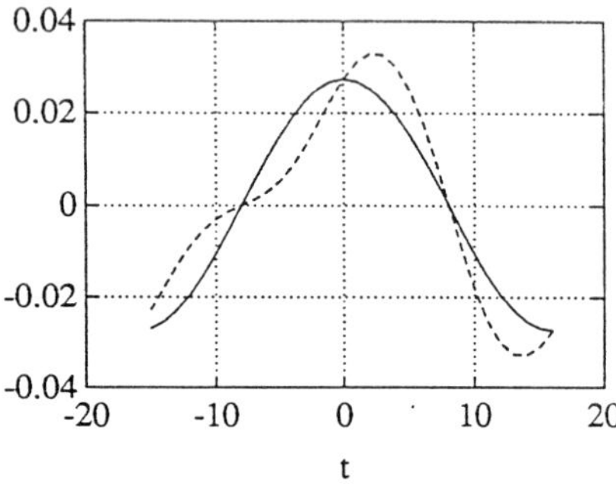

Figure 2: The sequences $W_1 f$ and $W_1 f_a$

5 Stability

To address the stability issue, the standard approach is to introduce the notion of perturbation of the representation, and of the reconstruction set. In addition, distance measures between distinct representations and between different reconstruction sets should be defined. In general, this is not an easy task. Observe that Vf, Tf may have different sizes for different representations. Fortunately, for inherently bounded representations, the following characterization of BIBO (bounded input, bounded output) stability is easily verified.

Proposition 4 *Let $Rf_i = \{Vf_i, T_if_i\}$ $(i = 1,2)$ be inherently bounded AQLR's. Then for all $K_I > 0$ there exists K_O such that (i=1,2):*

$$\|T_if_i\| \leq K_I \Rightarrow \|x_1 - x_2\| \leq K_O \quad \forall x_i \in \Gamma(Rf_i)$$

In many applications, the reasons for perturbations in a representation are arithmetic or quantization errors in a reconstruction algorithm. This kind of perturbations may change the continuous values of Tf but it preserves the discrete values of Vf. Therefore the perturbed representation, $(Rf)_p$, can be written as:

$$(Rf)_p = \{Vf, Tf + \Delta(Tf)\}. \tag{12}$$

Let Γ_p be the corresponding reconstruction set. In general, the distance, d, between two reconstruction sets, Γ and Γ_p, is defined as:

$$d(\Gamma, \Gamma_p) \triangleq \sup\{\|\gamma - \gamma_p\| : \gamma \in \Gamma, \gamma_p \in \Gamma_p\}.$$

Observe, that for an inherently bounded AQLR, $d(\Gamma, \Gamma_p)$ is always finite. The measure of the perturbation in the reconstruction set is the difference between $d(\Gamma, \Gamma_p)$ and the size of Γ which is defined as follows:

$$s(\Gamma) \triangleq d(\Gamma, \Gamma) = \sup\{\|\gamma_1 - \gamma_2\| : \gamma_1, \gamma_2 \in \Gamma\}. \tag{13}$$

$s(\Gamma)$ and $d(\Gamma, \Gamma_p)$ describe the largest possible Euclidian norm of a reconstruction error, from the original representation and from a perturbed one, respectively.

One remark is in order. In general, for an arbitrary $\Delta(Tf)$, the associated reconstruction set may be empty and then $d(\Gamma, \Gamma_p)$ would not be defined. In the sequel, it is assumed that this problem is treated by a reconstruction algorithm and hence $\Delta(Tf)$ yields a nonempty Γ_p. In this case, the following Lipschitz condition is satisfied.

Theorem 5 *For all inherently bounded AQLR, there exists $K > 0$ such that:*

$$d(\Gamma, \Gamma_p) \leq K \cdot \|\Delta(Tf)\| + s(\Gamma). \tag{14}$$

Conclusions

The described theoretical results about uniqueness and stability are new. In our opinion, the most significant contribution of this work is to create a framework to define and analyze a wide family of representations. Important examples are generalizations of a basic maxima representation obtained by using only a subset of local extreme points. Their properties are the subject of the undergoing research.

References

[1] Z. Berman. The uniqueness question of discrete wavelet maxima representation. Technical Report TR 91-48r1, University of Maryland, System Research Center, April 1991.

[2] Z. Berman. A reconstruction set of a discrete wavelet maxima representation. In *Proc. IEEE International Conference on Acoust., Speech, Signal Proc.*, San Francisco, 1992.

[3] Z. Berman and J. S. Baras. A theory of adaptive quasi linear representations. Technical Report TR 92-24, University of Maryland, System Research Center, January 1992.

[4] S. Mallat. Zero-crossing of a wavelet transform. *IEEE Trans. on Information Theory*, 37(4):1019–1033, July 1991.

[5] S. Mallat and S. Zhong. Complete signal representation with multiscale edges. Courant Institute of Mathematical Sciences, Technical Report 483, December 1989, to appear in IEEE Trans. on PAMI.

SIGNAL PROCESSING VI: Theories and Applications
J. Vandewalle, R. Boite, M. Moonen, A. Oosterlinck (eds.)

A Simple and Reliable Phase Unwrapping Algorithm

Eduard KRAJNÍK
Faculty of Electrical Engineering, Czech Technical University,
166 27 Prague 6, Czechoslovakia

Abstract

A great majority of phase unwrapping algorithms are based on the observation that the values of the unwrapped phase Φ differ at each frequency ω_k from the principal value of the phase $ARG(X(e^{j\omega_k}))$ by an integer multiple of 2π: $\Phi(\omega_k) = ARG(X(e^{j\omega_k})) + 2\pi m_k$ and determine the correct value of m_k. We propose a novel method in which the unwrapped phase valeus are computed recursively: $\Phi(\omega_{k+1}) = \Phi(\omega_k) + \Delta\Phi_k$. We first ensure that the frequencies ω_k and ω_{k+1} are such that $|\Delta\Phi_k| < \pi$ (a sufficient condition for this is given) and the phase increment $\Delta\Phi_k$ can be then unequivocally expressed by the principal value of the phase of the quotient of $\Phi(\omega_{k+1})$ and $\Phi(\omega_k)$. Thus, neither the detection of discontinuities nor the addition of multiples of 2π are needed. If the fast Fourier transform is used to evaluate $\Phi(\omega_k)$ a sufficient value of its size N is shown to be any integer not less than $2 \max_{\omega\in[0,2\pi]} |\Phi'(\omega)|$.

Introduction

Signal processing in the cepstral domain has proven to be useful in many deconvolution problems (seismology, speech, radar, sonar and others). The processing requires the computation of the cepstrum which is usually performed using the discrete Fourier transform and the complex logarithm function. In this process we encounter the problem of correctly choosing the phase values so that they will represent samples of the continuous phase function - the problem which is commonly called phase unwrapping. Though "exact" methods for phase unwrapping have been published recently [2, 3], the most popular method still seems to be Tribolet's adaptive integration algorithm [1], which gives an estimation to the unwrapped phase. The accuracy of the method depends critically on the FFT size N and the author claims that "it is not possible to estimate *a priori* how large the FFT size N should be in order to accurately unwrap the phase".

We propose a modification in the algorithm which not only enables *a priori* estimate of the FFT size N but also increases the accuracy up to the precision of FFT.

The algorithm

Let x_n be a sequence of a finite length L and $X(e^{j\omega})$ its Fourier transform; let $X(e^{j\omega}) \neq 0$ for all $\omega \in [0, 2\pi]$. Then there exists a continuous function (unwrapped phase) $\Phi(\omega)$ such that

$$X(e^{j\omega}) = |X(e^{j\omega})| \exp(j\Phi(\omega))$$

Let $\omega_k = 2k\pi/N$, $k = 0, 1, \ldots, N-1$, be the frequencies where the values $\Phi_k = \Phi(\omega_k)$ are required. If we are able to choose N such that the phase Φ never changes by more than π between any two adjacent frequencies ω_k, ω_{k+1}, we can unequivocally compute the phase increment $\Delta\Phi_k$ between ω_k and ω_{k+1}:

$$\Delta\Phi_k = ARG\left(\frac{X(e^{j\omega_{k+1}})}{X(e^{j\omega_k})}\right)$$

where ARG denotes the principal value of the phase. The unwrapped phase samples are then computed recursively:

$$\Phi_{k+1} = \Phi_k + \Delta\Phi_k$$

Unlike Tribolet's algorithm, the phase values are not approximated and their accuracy depends only on the accuracy of the values $X(e^{j\omega_k})$; the cummulative rounding error in the sequence Φ_k can easily be eliminated by the comparison of Φ_k and the nearest value of $ARG(X(e^{j\omega_k})) + 2\pi m_k$ for a few values of k. Also, neither the detection of discontinuities nor the addition of mutliples of 2π are needed. Of course, these advantages have to be payed for; but the price is reasonable: we have to ensure that the phase Φ never changes by more than π between any two adjacent frequencies ω_k, ω_{k+1}. This, in principle, can be achieved by a suitable chioce of N. A sufficient value can be derived by using the Mean Value Theorem:

$$\Phi(\omega_{k+1}) - \Phi(\omega_k) = \Phi'(\tau)(\omega_{k+1} - \omega_k)$$

for some $\tau \in (\omega_k, \omega_{k+1})$. Hence

$$|\Phi(\omega_{k+1}) - \Phi(\omega_k)| \leq M\,|\omega_{k+1} - \omega_k| = 2\pi M/N$$

where

$$M \geq \max_{\omega\in[0,2\pi]} |\Phi'(\omega)|$$

The condition $|\Phi(\omega_{k+1}) - \Phi(\omega_k)| < \pi$ thus will be satisfied for any $N > 2M$ and the problem is reduced to finding an upper bound M of the phase derivative Φ', preferably $\max_{\omega\in[0,2\pi]}|\Phi'(\omega)|$. The numerical computation of a good approximation to the global maximum of a function is difficult in general; it is impossible to find the maximum with a prescribed tolerance in a finite number of function evaluations, unless we have some *a priori* information about the function. Efficient algorithms do exist, however, for unimodal functions or C^2 functions provided an upper bound of the second derivative is known [4].

Let us now investigate the maximum of the phase derivative Φ'. We express the phase derivative in a similar way as in [1]. Setting $y_n = nx_n$ and denoting $Y(e^{j\omega})$ the Fourier transform of y_n, we have

$$\Phi'(\omega) = \frac{X_R(e^{j\omega})Y_R(e^{j\omega}) + X_I(e^{j\omega})Y_I(e^{j\omega})}{X_R^2(e^{j\omega}) + X_I^2(e^{j\omega})}$$

where

$$\begin{aligned} X(e^{j\omega}) &= X_R(e^{j\omega}) + jX_I(e^{j\omega}) \quad \text{and} \\ Y(e^{j\omega}) &= Y_R(e^{j\omega}) + jY_I(e^{j\omega}) \end{aligned}$$

After a rearrangement we get

$$\begin{aligned} \Phi'(\omega) &= \frac{\sum_{k=0}^{L}\sum_{l=0}^{L} k\,x_k\,x_l\,\cos(k-l)\omega}{\sum_{k=0}^{L}\sum_{l=0}^{L} x_k\,x_l\,\cos(k-l)\omega} \\ &= \frac{\sum_{k=0}^{L} a_k\,\cos k\omega}{\sum_{k=0}^{L} b_k\,\cos k\omega} \stackrel{\text{def}}{=} \frac{A(\omega)}{B(\omega)} \end{aligned}$$

where

$$\begin{aligned} a_0 &= \sum_{n=1}^{L} n\,x_n^2, \qquad b_0 = \sum_{n=1}^{L} x_n^2 \\ a_k &= \sum_{n=0}^{L-k}(2n+k)\,x_n\,x_{n+k} \quad \text{for } k = 1,\ldots,L \\ b_k &= 2\sum_{n=0}^{L-k} x_n\,x_{n+k} \qquad \text{for } k = 1,\ldots,L \end{aligned}$$

Since Φ' is the qoutient of two trigonometric polynomials A, B, it is not a unimodal function and though it is a C^2 function, a reasonable upper bound of its second derivative cannot be obtained. We therefore find

$$M_B = \min_{\omega\in[0,2\pi]} |B(\omega)| \quad \text{and} \quad M_A = \max_{\omega\in[0,2\pi]} |A(\omega)|$$

using the algorithm *globmin* given in the Appendix. We note that *globmin* can be used to find both a minimum and a maximum of a function since $\max f(t) = -\min(-f(t))$. The upper bounds of the second derivatives of A and B required in *globmin* are taken $\sum k^2\,|a_k|$ and $\sum k^2\,|b_k|$ respectively. Then we select $N > 2M_A/M_B$.

If X has a zero extremally close to the unit circle resulting N may be unacceptably large. In such a case, we face a problem whose numerical solution will always be dubious and certain compromises are necessary. A detailed discussion of this case will be given in [5].

The presented method gives a sufficient condition for simplification of phase unwrapping. The accuracy of the result is limited only by the accuracy of FFT.

Appendix

For the reader's convenience we present a minimization algorithm *globmin* with guaranteed convergence [4]. *Globmin* returns the global minimum of the external function func(x:real): real defined on $[a,b]$ and the point x where the function attains its minimum; c is an initial guess at x (any value in $[a,b]$ will do), m is an upper bound of the second derivative of func on $[a,b]$, the constants *macheps*, e, t are the relative machine precision and tolerances respectively; it is assumed that func(x) is computed with an absolute error less than e and the minimum is returned with an absolute error less than $t + 2e$.

```
FUNCTION globmin(a,b,c,m: real;
                 var x  : real) : real;

LABEL 1;

CONST macheps = 1e-15;
      e       = 1e-12;
      t       = 1e-7;

VAR y0, y1, y2, y3, yb, y, h     : real;
    a0, a2, a3, d0, d1, d2,r, s  : real;
    z0, z1, z2, m2, qs, p, q     : real;

BEGIN
  a0 := b;          x  := a0;
  y0 := func(b);  y2 := func(a);
  yb := y0;         y  := y2;
  a2 := a;
  IF y0 < y THEN y := y0 ELSE  x := a;
  IF (m > 0) AND (a < b) THEN BEGIN
    m2 := 0.5*(1+16*macheps)*m;
    IF (c <= a) or (c >= b) THEN
        c := 0.5*(a+b);
    y1 := func(c);   d0 := a2-c;    h := 9/11;
    IF y1 < y THEN BEGIN  x := c;
                          y := y1
    END;
    REPEAT
      d1 := a2-a0;   d2 := c-a0;    z2 := b-a2;
      z0 := y2-y1;   z1 := y2-y0;
      r  := d1*d1*z0-d0*d0*z1;
      qs := 2*(d0*z1-d1*z0);
      p  := r;         q  := qs;
      IF y >= y2  THEN  BEGIN
        IF q*(r*(yb-y2)+z2*q*((y2-y)+t)) <
           z2*m2*r*(z2*q-r) THEN
        BEGIN
          a3 := a2+r/q;       y3 := func(a3);
          IF y3 < y THEN BEGIN  x :=a3;
                                y := y3
          END
        END
      END;
      r := m2*d0*d1*d2;
      s := sqrt(((y2-y)+t)/m2);
      h := 0.5*(1+h);
      p := h*(p+2*r*s);       q := r+0.5*qs;
      r := -0.5*(d0+(z0+2.01*e)/(d0*m2));
      IF (r < s) or (d0 < 0) THEN r := a2+s
      ELSE r := a2+r;
      IF p*q > 0 THEN a3 := a2+p/q
      ELSE a3 := r;
1:    IF a3 < r THEN a3 := r;
      IF a3 >= b THEN  BEGIN  a3 := b;
                              y3 := yb
      END
      ELSE y3 := func(a3);
      IF y3 < y THEN   BEGIN   x := a3;
                               y := y3
      END;
      d0 := a3-a2;
      IF a3 > r THEN BEGIN
        p := 2*(y2-y3)/(m*d0);
        IF (abs(p) < (1+9*macheps)*d0) AND
           (0.5*m2*(d0*d0+p*p) >
           (y2-y)+(y3-y)+2*t) THEN BEGIN
           a3 := 0.5*(a2+a3);   h := 0.9*h;
           GOTO 1
        END
      END;
      a0 := c;     c  := a2;    a2 := a3;
      y0 := y1;    y1 := y2;    y2 := y3
    UNTIL a3 >= b
  END;
  globmin := y;
END;
```

References

[1] J. M. Tribolet, "A new phase unwrapping algorithm," *IEEE Trans. Acoust., Speech, Signal Processing,* vol. ASSP-25, pp. 170–177, April 1977.

[2] R. McGowan and R. Kuc "A direct relation between a signal time series and its unwrapped phase," *IEEE Trans. Acoust., Speech, Signal Processing,* vol. ASSP-30, pp. 717–726, Oct. 1982.

[3] D. G. Long, "Exact computation of the unwrapped phase of a finite-length time series," *IEEE Trans. Acoust., Speech, Signal Processing,* vol. ASSP-36, pp. 1787–1790, Nov. 1988.

[4] R. P. Brent, *Algorithms for Minimization without Derivatives,* 1973, Prentice-Hall, Englewood Cliffs, NJ.

[5] E. Krajník, "A simple phase unwrapping algorithm with guaranteed reliability," to be submitted to *Signal Processing,* 1992.

SIGNAL PROCESSING VI: Theories and Applications
J. Vandewalle, R. Boite, M. Moonen, A. Oosterlinck (eds.)

2-D FFT ALGORITHMS WITH PARALLEL BIT REVERSION ON HYPERCUBE COMPUTERS

G. Angelopoulos I. Pitas

Department of Electrical Engineering, University of Thessaloniki,
Thessaloniki 540 06, GREECE

This paper discusses the parallel computation of the 2-D Discrete Fourier Transform by using three different Fast Fourier Transform (FFT) algorithms. The examined algorithms are: Row-Column FFT, Vector Radix FFT and Polynomial Transform FFT. Hypercube computers with non-shared memory are considered. The proposed implementations allow both numeric computations and the bit-reversion operation to be performed in parallel.

1. INTRODUCTION

High performance parallel computers is an interesting alternative to the serial computation of 2-D FFTs. The RC FFT is usually proposed for parallel 2-D DFT calculation [4,5,6,7]. In [5,6] transposition of the data after the row FFTs is proposed. A RC FFT on a Transputer-based Perfect Shuffle machine is presented in [4]. Another parallel multidimensional FFT on *d-dimensional Perfect Shuffle* permutation network is presented in [8]. A parallel version of the VR FFT in a *Cube* network of PEs is presented in [6]. Parallel implementations of PTFFTs have not been reported, at least to the authors' knowledge. In this paper we propose suitable implementations of the above mentioned 2-D FFTs for parallel systems with non-shared memory and hypercube topology. The proposed implementations allow both numeric computations and the bit-reversion operation, which is required in these FFTs, to be performed in parallel.

The 2-D Discrete Fourier Transform of an $N_1 \times N_2$ - point digital signal $x(n_1, n_2)$, $0 \le n_1 < N_1$, $0 \le n_2 < N_2$, is defined by the equation:

$$X(k_1, k_2) = \sum_{n_1=0}^{N_1-1} \sum_{n_2=0}^{N_2-1} x(n_1, n_2)\, W_{N_1}^{n_1 k_1}\, W_{N_2}^{n_2 k_2} \qquad (1)$$

$W_{N_i} = exp(-j2\pi/N_i)$, $i = 1,2$. In most practical cases $N_1 = N_2 = N = 2^n$.

The 2-D DFT of a $N \times N$ - point sequence defined in equation (1) can be computed in two successive steps which consist the Row-Column (RC) FFT [1]:

$$X'(n_1, k_2) = \sum_{n_2=0}^{N-1} x(n_1, n_2)\, W_N^{n_2 k_2} \qquad (2)$$

$$X(k_1, k_2) = \sum_{n_1=0}^{N-1} X'(n_1, k_2)\, W_N^{n_1 k_1} \qquad (3)$$

Alternatively, the 2-D DFT can be calculated by working both along columns and rows simultaneously. The decimation in time version of the Vector Radix 2×2 (VR) FFT is derived by expressing the $N \times N$ - point DFT in terms of four $N/2 \times N/2$ - point DFTs [1]. This procedure can be repeated until we reach transforms of size 2×2. The resulting VR FFT algorithm requires 25% less complex multiplications than the RC FFT.

The DFT of a 2-D $N \times N$ - point sequence defined in equation (1) can also be calculated by using the following steps:

$$X_{n_1}(z) = \sum_{n_2=0}^{N-1} x(n_1, n_2)\, W^{-n_2}\, z^{n_2} \qquad (4)$$

$$\bar{X}_{(2k_2+1)k_1}(z) = \sum_{n_1=0}^{N-1} X_{n_1}(z)\, z^{2n_1 k_1}\; mod\,(z^N + 1) \qquad (5)$$

$$X[(2k_2+1)k_1, k_2] = \bar{X}_{(2k_2+1)k_1}(z)\; mod\,(z - W^{2k_2+1}) \qquad (6)$$

where $W = \exp(-j\pi/N)$. These relations represent the Polynomial Transform FFT (PTFFT) [2]. Relation (4) describes a multiplication of $x(n_1, n_2)$ by W^{-n_2}, followed by a polynomial transform of length N with root z^2. Equation (6) is equivalent to a multiplication by W^{n_2} and an 1-D DFT. A final unscrambling step is required, because the indices $(2k_2+1)k_1$ in (6) are scrambled. After this brief introduction to the 2-D FFT algorithms we proceed to their parallel implementation on a hypercube computer.

2. 2-D FFTS ON HYPERCUBE COMPUTERS

A hypercube computer of dimension p consists of $P = 2^p$ PEs located at the corners of the p-dimensional cube [3]. The edges of the hypercube represent bidirectional communication links. Each PE has p links. Any two PEs are at most p links apart. Each PE is labeled by a number from 0 to $P-1$ so that two PEs are connected if and only if the p-bit binary representations of their labels differ only in one bit. Furthermore, each of the p dimensions is characterized by a unique bit position in the binary representation of the PE labels. Figure 1 shows a hypercube computer with $P = 16$ PEs. Usually, a parallel machine is connected to a host computer via PE 0 for input/output operations. In the case of the 2-D FFTs, the distribution of the data to the PEs can be performed by embedding a pseudo binary tree on the hypercube [9]. The same procedure holds for the gathering of the results into the PE 0.

RC FFT. Let us consider a hypercube with P PEs and the binary representation of their labels $(l_{p-1}l_{p-2} \ldots l_1l_0)_2$. A 2-D signal of size $N \times N$, where $N \geq 2P$, can be divided into $2P$ stripes of size $N/(2P)$ rows. These stripes are labeled with $(p+1)$ - bit labels and are distributed to the local memories of the PEs. Each PE i stores on its local memory the two stripes labeled by i and $P+i$. According to the distribution of the signal described above, the computations of the FFTs of the rows can be performed in parallel without any interprocessor communication. Column DFT computation requires communications among the PEs, because each column is distributed on P PEs. The proposed algorithm has two phases. In the first phase, the first $p = \log_2 P$ stages of the 1-D FFTs are computed. After finishing each of these stages, data exchange is required that is done in pairs of PEs. Two PEs whose labels differ only in their l_{p-1} bit of their binary label representation $(l_{p-1}l_{p-2} \ldots l_1l_0)_2$ exchange data after the first stage. These two PEs are one link apart. The PE which has its l_{p-1} bit equal to zero exchanges its second stripe with the first stripe of the other member of the pair. After these communications, which are done in parallel in all PEs, the calculation of the butterflies of the second stage is performed in parallel. This process is continued for the rest of the stages of the first phase. In the second phase, the last $\log_2(N/P)$ stages of the radix-2 1-D FFT algorithm are performed without communications. This version of the parallel RC FFT algorithm requires the transfers of $(N^2/2)\log_2 P$ data items that are performed in $p = \log_2 P$ steps, one in each dimension of the hypercube.

DFT output unscrambling along the horizontal direction can be performed in parallel at each PE. The unscrambling along columns can be performed in parallel in two steps. In the first step, each PE performs the bit-reversed ordering of each column segment residing on this PE. In the second step, each PE performs a bit-reversed ordering of its label $(l_{p-1}l_{p-2} \ldots l_1l_0)_2$. Thus, the logical labels of the PEs change and each PE with new label $\mathtt{i} = (l_0l_1 \ldots l_{p-2}l_{p-1})_2$ holds the output data rows k_1 for which $\mathtt{i} = k_1 \; mod \; P$.

VR FFT. Let us consider a hypercube with P PEs, where P is a power of 4, and a 2-D signal of size $N \times N$ where $N \geq 2P^{1/2}$. Such a signal can be divided into $4P$ square blocks of size $(N/2P^{1/2}) \times (N/2P^{1/2})$ that cover the entire 2-D signal in a row-wise manner, as can be seen in Figure 2. These blocks are labeled with $(p+2)$ - bit labels. The p-bit labels $(l_{p-1}l_{p-2} \ldots l_1l_0)_2$ of the PEs can be split into labels of the form $(r_{p/2-1}r_{p/2-2} \ldots r_1r_0c_{p/2-1}c_{p/2-2} \ldots c_1c_0)_2$. For simplicity, these labels can be written as $(r_{q-1}r_{q-2} \ldots r_1r_0c_{q-1}c_{q-2} \ldots c_1c_0)_2$, where $q = p/2$. The distribution of the data into the PEs is done by storing the four square blocks $(xr_{q-1} \ldots r_0 \; xc_{q-1} \ldots c_0)_2$ on the PE $(r_{q-1} \ldots r_0c_{q-1} \ldots c_0)_2$, where x denotes the "don't care" bit value. Figure 2 shows the partitioning of an $N \times N$ signal into $4P = 16$ square blocks and the associated labeling.

The proposed algorithm has two phases. In the first phase, the first $q = p/2 = \log_4 P$ stages of the VR 2×2 FFT algorithm are performed. After finishing the computations in each of these stages, an exchange of data among PEs is required. The exchange of data is done among groups of PEs. Each group consists of four PEs. In the first stage of this phase, four PEs are clustered in the same group if their labels differ only in their r_{q-1} and c_{q-1} bits. There are $P/4$ such groups in a hypercube having P PEs. The four PEs of a group are interconnected in a 2-D sub-hypercube (square), as shown in Figure 3. They are distinguished by their r_{q-1} and c_{q-1} bits which form a binary number $r_{q-1}c_{q-1}$ for each PE of a group. The range of label $(r_{q-1}c_{q-1})_2$ is from 0 to 3. By labeling the four outputs of the butterflies with the labels a, b, c and d, the state of the data after the first phase of the algorithm is shown in Figure 3a. The goal of the communication is to gather the butterfly outputs having the same label to a single PE. Each of the four PEs in a group retains one quarter of its data and sends the three quarters of its data to the three other members of its group. This is achieved in two steps. In each step, exchange of data is performed through the links of a specific dimension of the 2-D hypercube that the four PEs construct. In the first step, exchange of data is performed between processors which have the same r_{q-1} bit label and different c_{q-1} bit label. These processors are characterized by a label of the form $r_{q-1}x$. In the second step, exchange of data is performed between processors which have the same c_{q-1} bit label and different r_{q-1} bit label, that is a label of the form xc_{q-1}. These two steps are shown in Figures 3b and 3c respectively. In the entire hypercube, the clustering of the PEs into groups of four PEs in this stage is represented by the labels $(xr_{q-2}\dots r_1r_0xc_{q-2}\dots c_1c_0)_2$. The PEs that exchange data in these two steps are characterized by the labels $(r_{q-1}r_{q-2}\dots r_1r_0xc_{q-2}\dots c_1c_0)_2$ and $(xr_{q-2}\dots r_1r_0c_{q-1}c_{q-2}\dots c_1c_0)_2$ respectively. In this way, each of the four PEs in a group receives $(3/4)N^2/P$ data from the three other members of its group. However, $(1/4)N^2/P$ data pass through a PE in order to achieve data transfers between non adjacent (diagonally lying) PEs in a group. Thus, the total number of transferred data in the entire hypercube is N^2. After these communications, which are done in parallel in all PEs, the calculation of the butterflies of the next stage is performed in parallel. This process is continued for the rest of the stages of the first phase of the VR FFT algorithm with the rest of the bits of the PE labels. The clustering of the PEs into groups of four requires that P is a power of 4. In the second phase, the last $\log_2(N/P^{1/2})$ stages of the VR FFT algorithm are computed on the PEs without communications. The total amount of data transfers is $N^2\log_4 P = (N^2/2)\log_2 P$ data items and they are performed in $p = \log_2 P$ steps, one in each dimension of the hypercube.

In the VR 2×2 decimation in time 2-D FFT algorithm, the output data must be unscrambled in bit-reversed order in both dimensions k_1 and k_2 of the 2-D signal, after the main (numerical) processing. The bit-reversion operation can be performed in parallel on a hypercube computer. Each PE performs the bit-reversed ordering of the data that are locally stored. The four square blocks that each PE holds construct a 2-D array of size $(N/P^{1/2})\times(N/P^{1/2})$. All PEs working in parallel and independently perform the bit-reversion operation on both dimensions of their arrays. Subsequently, each PE labeled $(r_{q-1}r_{q-2}\dots r_1r_0c_{q-1}c_{q-2}\dots c_1c_0)_2$ performs a bit-reversed ordering on each split part of its label and the new label becomes $(r_0r_1\dots r_{q-2}r_{q-1}c_0c_1\dots c_{q-2}c_{q-1})_2$. Thus, each PE, whose new row sub-label is $\mathtt{i} = (r_0r_1\dots r_{q-2}r_{q-1})_2$, holds the elements of the output data rows k_1, for which $\mathtt{i} = k_1 \, mod \, P^{1/2}$. Similarly, each PE, whose new column sub-label is $\mathtt{j} = (c_0c_1\dots c_{q-2}c_{q-1})_2$, holds the elements of the output data columns k_2, for which $\mathtt{j} = k_2 mod P^{1/2}$. Finally, the gathering of the results into the PE 0 is performed by embedding a pseudo binary tree on the hypercube [9].

PT FFT. A stripe partitioning of a 2-D signal into $2P$ stripes and the distribution of them into the PEs is assumed, as in the RC FFT. All multiplications by W^{-n_2} and W^{n_2} involved in the PTFFT can be performed in parallel without any communication. For the computation of the polynomial transform, communication among the PEs must be performed. The proposed algorithm has two phases. In the first phase, the first $\log_2 P$ stages of the polynomial transform are computed. The communications required in these stages are the same to the ones needed in the RC FFT algorithm. Similarly, the total amount of the transferred data is $(N^2/2)\log_2 P$ data items and the data tranfers are performed in $p = \log_2 P$ steps, one in each dimension of the hypercube. In the second phase, the last $\log_2(N/P)$ stages of the polynomial transform are performed without communications. The FFTs of rows can be performed in parallel without any communication. The gathering of the results into the PE 0 can be performed by embedding a pseudo binary tree on the hypercube [9].

3. DISCUSSION

All three algorithms described in the previous section have the same communication load. Thus the deciding factor for the choice among the three algorithms is their computational complexity. The VR FFT algorithm has been implemented on a parallel machine, because of its efficiency over the RC FFT and its simplicity over the PTFFT algorithm. It was implemented on the *Telmat T-Node 16-32* machine in hypercube configuration by using HELIOS C programming language. This machine consists up to 32 PEs (transputers T800) and has reconfigurable architecture. The total time (processing plus communications) for $P = 1$, $P = 16$ PEs and the achieved speedups for various values of N are shown in Table 1. The PE communications have been performed by using the *read - write* functions, that provide relatively slow interprocessor communication with large startup overheads. The total time includes the time required for the computation of the complex exponentials. As expected, the speedup increases with the signal size $N \times N$, because in this case the computational load is much larger than the communication load. Furthermore, in the case of large signal sizes, the effect of startup time in the communication time is reduced, because large data blocks are transferred in each communication operation.

Acknowledgements This work has been partially supported by ESPRIT PC project No4012 and by a project of the Greek Ministry of Technology (PENED 89).

References

[1] D. E. Dudgeon and R. M. Mersereau, *Multidimensional Digital Signal Processing*, Prentice-Hall, 1983.

[2] H. J. Nussbaumer, *Fast Fourier Transform and Convolution Algorithms*, Springer Verlag, 1981.

[3] John P. Hayes and Trevor Mudge, "Hypercube Supercomputers", *Special Issue on Supercomputer Technology, Proceedings of the IEEE*, Vol. 77, No. 12, Dec. 1989, pp. 1829-1841.

[4] H.Schomberg, "Image Processing on a Transputer-Based Perfect Shuffle Machine", *Microprocessing and Microprogramming 25*, North-Holland, 1989, pp. 277-280.

[5] Susan M. Miller and Harvey F. Silverman, "Optimized Implementation of the 2-D DFT on Loosely-Coupled Parallel Systems", *Proc. 1989 IEEE Internat. Conf. on Acoustics, Speech and Signal Processing*, Glasgow, 1989, pp. 1536-1539.

[6] Leah H. Jamieson, Philip T. Mueller, Jr., and Howard Jay Siegel, "FFT Algorithms for SIMD Parallel Processing Systems", *Journal of Parallel and Distributed Computing 3*, 1986, pp. 48-71.

[7] E. L. Zapata, F. F. Rivera, I. Benavides, J. M. Carazo and R. Peskin, "Multidimensional Fast Fourier Transform into SIMD Hypercubes", *IEE Proceedings*, Vol. 137, Pt. E, No. 4, July 1990, pp. 253-260.

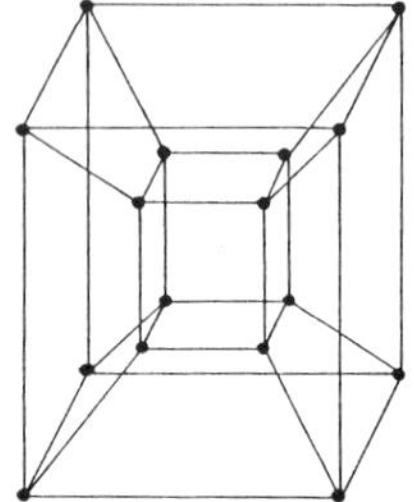

Figure 1: Hypercube with 16 processors.

r_1r_0 \ c_1c_0	00	01	10	11
00	0	1	2	3
01	4	5	6	7
10	8	9	10	11
11	12	13	14	15

Figure 2: Partitioning of a 2-D signal in 16 square blocks. The numbers in the boxes denote square block labels. $r_1r_0c_1c_0$ is their binary representation.

[8] M. Shamash and I. Gertner, "A New Parallel Algorithm for the Multidimensional Fourier Transform Processing", *Proc. 1990 IEEE Internat. Conf. on Acoustics, Speech and Signal Processing*, Albuquerque, 1990, pp. 1993-1996.

[9] S. Y. Lee and J. K. Aggarwal, "Exploitation of Image Parallelism via the Hypercube", *Hypercube Microprocessors*, SIAM, 1987, pp. 427-437.

TABLE 1

Total time (in *msec*) and speedups for $N \times N$ VR FFT on the *Telmat T-Node 16-32* machine in hypercube configuration.

N	$P = 1$	$P = 16$	Speed Up
32	100	70	1.43
64	470	110	4.27
128	2210	240	9.21
256	9810	860	11.41
512	43970	3510	12.53

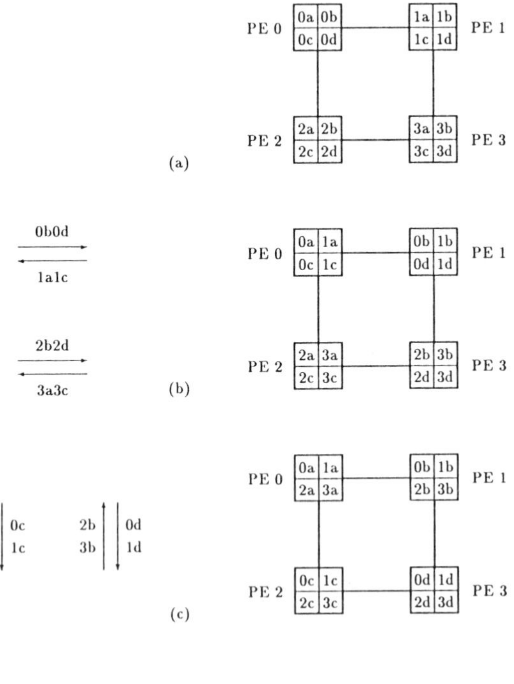

Figure 3: Data exchange in VR FFT among four PEs; (a) Initial state; (b) First exchange; (c) Second exchange.

SIGNAL PROCESSING VI: Theories and Applications
J. Vandewalle, R. Boite, M. Moonen, A. Oosterlinck (eds.)

Time-warped Polynomials for Signal Representation and Coding

Wilfried Philips

Laboratory for Electronics, St.-Pietersnieuwstraat 41, B9000 Gent, Belgium
E-Mail: philips@lem.rug.ac.be Tel: 32-91-64.3385 Fax: 32-91-64.35.94

This paper introduces a new class of discrete, orthogonal transforms, whose base functions are time-warped polynomials. The general properties of the base functions are described. The class generalizes at least three well-known transforms. A data compression application on Electrocardiograms based on one of the new transforms is presented. At a given signal quality, the new transform allows a compression more than twice that of the DCT.

1 Introduction

Orthogonal transforms are widely used in signal processing. Although infinitely many of them exist, only a few of them are used in practice [1]. These include the Discrete Cosine Transform (DCT) [1], the Walsh-Hadamard Transform, etc. The transform matrices of the Karhunen-Loève Transform (KLT) and the Singular Value Decomposition (SVD) are not fixed, but they depend on the signal's statistics and the signal's data respectively [1]. These transforms can only be described by enumerating all of their base functions, or by storing the data from which they were derived. The large overhead involved in describing the transform limits the practical useability of these adaptive transforms.

This paper describes a class of transforms that are completely determined by specifying a single function $f(x)$. The base functions of these transforms are *time-warped polynomials*, i.e., functions of the form $P_n(f(x))$, where $P_n(f)$ is a polynomial in f. By parametrizing the *warping function* $f(x)$, transforms depending on only a few numbers can easily be devised. By changing the parameters that determine $f(x)$, the transform can be adapted to a given signal. Since the overhead is relatively small, it is even possible to vary the transform parameters in time.

The new class of base functions is introduced in section 2. It includes the DCT and two different polynomial transforms. Section 3 describes the main properties of the time-warped base functions and gives some guidelines for choosing the right transform for a given application.

In section 4, a data compression method for Electrocardiograms (ECGs) based on a specific time-warped transform is compared to a DCT compression scheme. It is shown that for the same Root Mean Squared Error (RMSE), the compression factor of the time-warped transform is more than twice that of the DCT.

2 The New Transform

Let $S = \{x_i = -1 + 2i/L : i = 0 \ldots L\}$ be a set of sample points on the real line. The *inner product* of two functions $g(x)$ and $h(x)$ on S is defined as $g \cdot h = \sum_{x \in S} g(x)h(x)$ and $\|g(x)\| = (g \cdot g)^{1/2}$ is called the *norm* of $g(x)$. The set $\mathcal{B} = \{\phi_n : n = 0, \ldots, L\}$ is a *complete orthogonal base* on S if the $L+1$ functions ϕ_n are mutually orthogonal, i.e., if they satisfy $\phi_m \cdot \phi_n = \delta_{m,n}$ for all m, n for which $0 \leq m, n \leq L$.

Let $s(x)$ be a function on S and $s_N(x) = \sum_{n=0}^{N} A_n \phi_n(x)$, where $A_n = \phi_n \cdot s$. Due to the completeness of $\mathcal{B}$, $s_L(x) = s(x)$, for all $x \in S$. On the other hand, for $N < L$, $s_N(x)$ is that combination of the base functions $\phi_n(x), n = 0 \ldots N$ that minimizes the RMSE $e_N = \sqrt{\|s_n - s\|^2/(L+1)}$ between $s(x)$ and $s_n(x)$ on S.

If the set of base functions $\phi_n(x)$, $n = 0, \ldots, L$ is suitably chosen, the approximation $s_N(x)$ will be close to $s(x)$ on S, even if N is much smaller than L. In that case, the coefficients $A_n, n = 0 \ldots N$ form an efficient, though approximate, representation of the signal $s(x)$. This is exploited in data compression applications.

In adaptive transform coding, the orthogonal base used for representing the signal $s(x)$ depends on $s(x)$ itself. If the base is matched to a given signal, fewer coefficients are needed to reduce the RMSE for this signal to a given level. On the other hand, the overhead data needed to describe which transform was used by the coder, can significantly reduce the net compression factor. This overhead is small for transforms that depend on a few parameters only. Such transforms could be devised in several ways. As far as the author knows, the subject has not yet been investigated in detail in the signal processing literature. The time-warped polynomial bases that are introduced below, can be represented by a few parameters only. They can be computed at high speed and generalize some well-known transforms. This makes them very promising for adaptive coding.

Let the function $f(x)$ be defined on S. It is easily shown that the powers $f^n(x)$, $n = 0 \ldots L$ of $f(x)$, are linearly independent on S, if $f(x_i) \neq f(x_j)$ when $i \neq j$. In the following, $f(x)$ is a strictly increasing, continuous function with $f(-1) = 1$ and $f(1) = 1$. For such an $f(x)$, the independence property holds trivially. In practice, $f(x)$ is also slowly varying.

In general, the non-orthogonal functions $f^n(x)$ are not very suitable for signal representation. As described above, a set $\mathcal{B}$ of orthogonal base functions $\phi_n(x)$ should be used instead. These functions $\phi_n(x) = P_n\big(f(x)\big)$ are obtained by Gramm-Schmidt Orthogonalization (GSO) of the functions $f^n(x)$ in the order $n = 0, 1, \ldots L$.

Since $P_n(f)$ is a polynomial of precise degree n in f, the orthogonal base functions are called time-warped polynomials. It can be shown that the orthogonal functions $P_n\big(f(x)\big)$ satisfy a three-term recursion, which is similar to the recursion for orthogonal polynomials [3, p. 42].

Since the recursion is much faster than GSO, it should be used to generate the orthogonal base. A fast transform generation is essential in adaptive transform coding, where the orthogonal base must be generated at both the coder (transmitter) and the decoder (receiver). In such applications, the transform might even vary in time.

The orthogonal base $\mathcal{B}$ is completely determined by the warping function $f(x)$. It can be shown that $\mathcal{B}$ is also obtained for any warping function $f'(x) = af(x) + b$, where $a \neq 0$ and b are constants. In practice, a function that can be completely specified by a few parameters only will be chosen for $f(x)$. The orthogonal base $\mathcal{B}$ is then completely determined by the value of these parameters.

For specific choices of $f(x)$, two kinds of Discrete Legendre Transforms (DLT1 and DLT2) and the Discrete Cosine (DCT) are obtained:

- DLT1 [4] $f(x) = x$.
- DLT2 [5] $f(x_i) = z_i$, where $z_i, i = 0, \ldots L$ are the zeros of the (ordinary) Legendre polynomial $p_{L+1}(x)$ of degree L, where it is assumed that $z_i > z_j$ if $i > j$.
- DCT [1] $f(x_i) = \cos\Big((x+1/2)\pi/N\Big)$, where $N = L+1$.

The last two transforms are usually defined in different ways, but the definitions given here are in agreement with the usual ones.

3 General Properties

The theory of orthogonal polynomials [3, p. 44] predicts that $P_n\big(f(x)\big)$ has exactly n zeros in the real interval $[-1, 1]$, if the conditions of the previous sections are fulfilled. Therefore, the base functions of all degrees oscillate in $[-1, 1]$.

In general, the following approximation is valid [6]:

$$P_n\big(f(x)\big) \approx \sqrt{2/(\pi L)}\sqrt{\tfrac{df}{dx}/|\sin\theta|}\cos\big(n\theta + \Gamma(\theta)\big), \quad (1)$$

where $\theta = \arccos\big(f(x)\big)$ and $\Gamma(\theta)$ depends only upon f and θ.

The previous equation is valid if n is small compared to the number of sample points $L + 1$, because then the discrete orthogonality relation approximates a continuous one. On the other hand n should not be too small, because (1) is really an asymptotic relation for large n. The precise range of n-values for which equation (1) holds, depends on $f(x)$ and is difficult to determine in general. Experiments using discrete Legendre Polynomials [7] suggest that eq. (1) accurately predicts the behavior of $P_n\big(f(x)\big)$ for n-values as low as 5 and as high as $L/2$. Also, eq. (1) holds exactly for the DCT.

Equation (1) shows that the base functions are approximately *amplitude and frequency modulated sine waves.* The amplitude $A(x)$ and the angular frequency $\omega(x)$ are given by

$$\begin{aligned} A(x) &= \sqrt{2/(\pi L)}\sqrt{\tfrac{df}{dx}/|\sin\theta|} = \sqrt{2/(\pi L)}\sqrt{g(x)} & (2)\\ \omega(x) &= \left|\tfrac{d\theta}{dx}(n + \tfrac{d\Gamma}{d\theta})\right| & (3)\\ &= \tfrac{df}{dx}/|\sin\theta|\left|n + \tfrac{d\Gamma}{d\theta}\right| = g(x)\left|n + \tfrac{d\Gamma}{d\theta}\right| & (4) \end{aligned}$$

Both $A^2(x)$ and $\omega(x)$ are proportional to $g(x) = \frac{df}{dx}/|\sin\theta|$ if $n \gg |\frac{d\Gamma}{d\theta}|$.

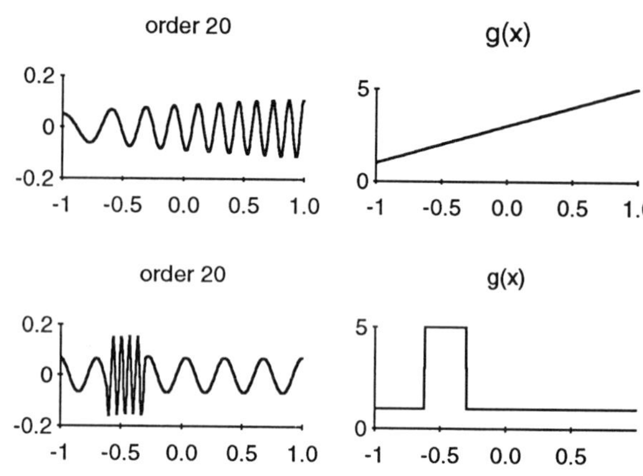

Figure 1: Some base functions and their corresponding $g(x)$-functions.

Figure 1 shows the base functions of degree 20, for different choices of $g(x)$. These examples suggest that $g(x)$ accurately predicts the local base function frequency, even for non-classical transforms.

The previous results can be used to qualitatively determine the best choice of $f(x)$, for any particular application (see next section).

4 ECG Data Compression

The top curves of figures 2 and 3 show original ECG signals. These signals were sampled at 500 Hz and at a resolution of 10 bits per sample. By convention, the ECG waves are labelled P,Q,R,S and T (see fig. 2). The ECG signal is highly non-stationary and quasi-periodic.

In the transform coding methods described in this article, the ECG signal is first segmented into intervals. In the following, S represents one such interval. The signal $s(x)$ on S is approximated by the weighted sum $s_N(x)$, as described in section 2. For best results, each interval should correspond to one ECG period [7]. Such an interval choice allows the coefficients A_n to be coded efficiently, since they will only change slowly from one interval to another. The coefficients A_n are quantized and coded using a variable word-length coding scheme. The same uniform quantizer is used for all coefficients. More detailed descriptions of the coefficient quantizer and encoder can be found in [7].

Different transforms, i.e., different sets of base functions $\phi_n(x)$, can be compared by computing the RMSE e_N or the Peak Error (PE) for specific values of N. Obviously, transforms producing a lower RMSE or a lower PE are to be preferred for data compression purposes. In this section, the DCT is compared to the DLT1 and to a third transform, that is called the Linear Spline-Warped Transform (LSWT). Its warping function $f(x)$ is a linear spline and is shown in fig. 4. Since the spline has unity slope in $[a, b]$, $f(x)$ is completely determined by the breakpoints a and b and the slopes α and β in the intervals $[-1, a]$ and $[b, 1]$. In the following, α and β are chosen equal for simplicity. The breakpoints of $f(x)$ coincide with the end and the beginning of succeeding QRS complexes. Therefore, by choosing $\alpha > 1$, the amplitude and frequency of the base functions can be increased in the vicinity of the high-frequency QRS complexes.

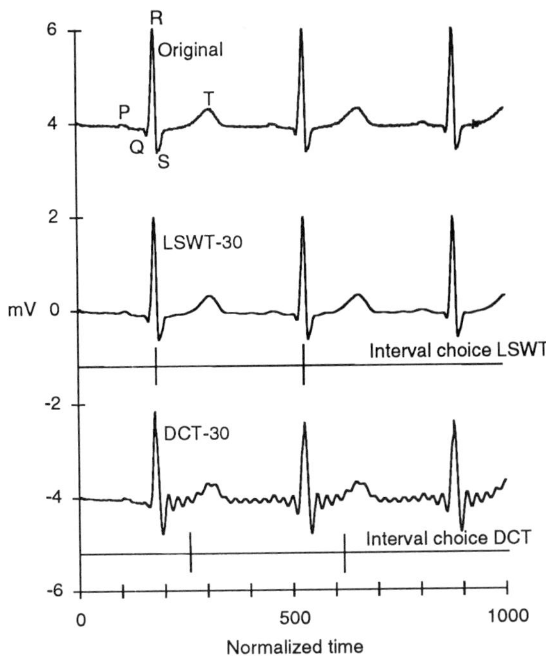

Figure 2: Original ECG and its approximations using 30 encoded coefficients. Numerical values for the compression ratio and the reconstruction errors can be found in table 1.

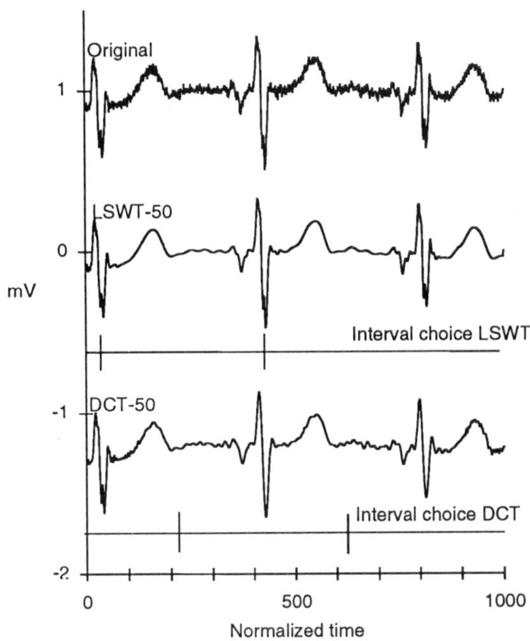

Figure 3: Original ECG and its approximations using 50 encoded coefficients. Numerical values for the compression ratio and the reconstruction errors can be found in table 1.

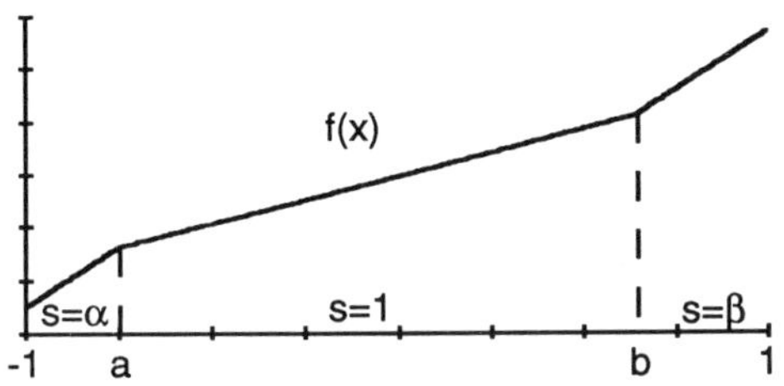

Figure 4: The LSWT warping function used to compress the ECGs of figures 2 and 3.

The DCT is widely used in practice because its RMSE closely approximates the minimal one, for stationary signal statistics [1]. However, the ECG is a highly non-stationary signal and therefore it is to be expected that the DCT is not close to optimal, for the compression of ECG data. In the remainder of this article, it will be shown that the DLT1 is better suited for representing the non-stationary ECG-signal.

For the DLT1, eq. (1) reduces to Laplace's approximation [7, 8], the amplitude becomes $A(x) = \sqrt{2/(\pi L \sin\theta)}$ and the angular frequency $\omega(x) = (n + 1/2)/\sin\theta$. Both $A(x)$ and $\omega(x)$ increase near the boundary of S, i.e., for $|x| = |\cos\theta|$ approaching 1. The base functions of the DCT, on the other hand, have a constant amplitude and frequency on the whole interval S.

The non-stationary behaviour of the discrete Legendre polynomials qualitatively matches that of the ECG-signal in a single R-R interval. This is because the high-frequency, high-amplitude QRS-complexes are located on the boundaries of the R-R intervals, while the other waves, which generally have lower frequency and amplitude, are located away from these boundaries (see figures 2 and 3). Therefore, for the DLT1, S is (approximately) an R-R interval. Numerical results [7] show that for the DLT1, both the RMSE and the PE increase sharply if S deviates substantially from an R-R-interval. For the DCT, the PE and RMSE are virtually independent of the interval alignment, because of the stationary characteristics of its base functions.

Figures 2 and 3 show some DCT and LSWT approximations. The (optimal) interval boundaries used, are indicated on the figures, while table 1 lists numerical values of the PE and the RMSE for these and other approximations. For the LSWT, a and b were chosen as described above. For a given ECG, the value of $\alpha = \beta$ that minimized e_N was used. This optimal value of α is computed by a gradient-descent type algorithm and depends on both the order N and on the ECG being coded. In the table, CR is the compression ratio, i.e., the ratio of the number of bits in the original signal and the number of coefficients needed to store the quantized coefficients and some overhead data, such as interval lengths.

The LSWT clearly results in the highest compression and in the lowest RMSE. For instance, at a CR of 22.1, the RMSE for the LSWT is $16\mu V$ for the ECG of fig. 2. At a comparable RMSE ($17\mu V$) the CR for the DCT is only 8.1. The values of the PE are less consistent, since it is generally more randomly distributed. However, the PEs for the DLT1 and the LSWT are significantly lower than for the DCT.

Results for the ECG of figure 2 ($\Delta = 1/16\ \mu$V, $a = -0.83$, $b = 0.83$, $\alpha = \beta = 2.7$)

	CR			RMSE			PE		
N	DCT	DLT1	LSWT	DCT	DLT1	LSWT	DCT	DLT1	LSWT
30	16.0	21.1	22.1	112	24	16	701	215	101
50	12.2	15.4	17.4	34	14	12	250	90	94
75	8.1	12.5	15.3	17	11	11	146	79	88
100	8.0	11.7	14.4	14	11	11	93	79	88

Results for the ECG of figure 3 ($\Delta = 1/64\ \mu$V, $a = -0.75$, $b = 0.65$, $\alpha = \beta = 3.3$)

	CR			RMSE			PE		
N	DCT	DLT1	LSWT	DCT	DLT1	LSWT	DCT	DLT1	LSWT
30	17.0	19.8	19.9	32	23	19	149	118	82
50	11.6	13.8	13.8	24	18	17	149	84	60
75	8.5	10.2	10.7	21	16	16	117	64	60
100	6.7	8.3	8.7	16	15	15	73	55	50

Table 1: Compression (CR), peak error (PE) and RMS error (RMSE) for the DLT1, the DCT, and the LSWT in function of the number of coefficients N+1. The quantization step Δ is the same for all transforms. The parameters of the LSWT warping function are listed.

Also, at very high compressions (e.g. for $N = 30$ in table 1) the PE of the LSWT is clearly better than that of the DLT1. For the ECG of fig. 2, the LSWT performs slightly worse as far as PE is concerned. However, a higher compression largely compensates the small difference (10%) in PE.

Numerical results for other ECG signals confirm that the LSWT leads to both a higher compression and a lower RMSE, for a given quantization step Δ. Therefore, it is the method of choice for ECG data compression. The LSWT can be further improved by allowing different values for α and β and by optimizing the positions a and b of the spline break points. This has not yet been examined. A slight disadvantage is that some computations are needed to determine the optimal spline parameters. These computations are only needed at the coder however. Furthermore, the spline parameters can be optimized for a given class of ECG signals (instead of for a single signal). This will tend to lower the mean compression for a given RMSE, but the results will never be worse than for the DLT1, which is still far superior than the DCT.

5 Conclusion

The paper has introduced a new set of transforms, that includes well-known transforms such as the DCT. The transform base functions can be generated quickly and are fully determined by a single (parametrized) function. The main properties of the base functions have been described. The usefulness of the base functions has been demonstrated in an ECG data compression application.

Acknowledgement

The author is grateful to the Belgian National Fund for Scientific Research, for supporting this work financially.

References

[1] R.J. Clarke, *Transform Coding of Images.* Academic Press, New York, 1985.

[2] R. C. Thompson and A. Yaqub, *Introduction to Linear and Abstract Algebra.* Glenview, Illinois: Scott, Foresman and Co, 1971, pp. 349-351.

[3] G. Szegö, *Orthogonal Polynomials.* American Mathematical Society Colloquium Publications, vol. 23, Fourth Edition, 1975.

[4] N. Morrison, *Introduction to Sequential Smoothing and Prediction.* New York: McGraw-Hill, 1969, ch. 3, pp. 55-71.

[5] V.E. Neagoe, "Optimal Discrete Representation of Signals by Preserving their Integral Characteristics," in *Proceedings of EUSIPCO-88*, Amsterdam: North Holland, 1988, pp. 1101-1104.

[6] D.S. Lubinsky, "The Approximate Approach to Orthogonal Polynomials for Weights on $(-\infty,\infty)$," in *Orthogonal Polynomials: Theory and Practice*, Edited by P. Nevai. Nato ASI series C, vol 294. London: Kluwer, 1990, pp. 293-310.

[7] W. Philips and G. De Jonghe, "Data Compression of ECGs by High-Degree Polynomial Approximation," to be published in *IEEE Trans. on Biomedical Engineering.*

[8] P. Frank and R. Von Mises, *Differentialgleichungen der Physik I.* Dover: Vieweg, 1930, pp. 434-440.

SIGNAL PROCESSING VI: Theories and Applications
J. Vandewalle, R. Boite, M. Moonen, A. Oosterlinck (eds.)

Design of Signal-Matched Wavelets - Theory and Results

T. Greiner, E. Eberle, M. Pandit

University of Kaiserslautern, PO-Box 3049, W-6750 Kaiserlautern, Germany

Abstract: We propose a new method to design wavelets. These wavelets contain signal characteristics and are designed by determining the impulse responses of the wavelet digital filters by a linearized optimization process. The achieved wavelet filters fulfill the wavelet conditions and yield better results in texture analysis. The 'Eigenfilters' of Ade and a Karhunen-Loeve related maximum variance criterion are the basis of the matching. The new filters are tested with the Brodatz textures.

1. Introduction

The wavelet transform is a relatively new tool in signal processing, which finds application in the analysis of non-stationary processes and the multiresolution decomposition of signals. In the latter application, which is the current context, orthogonal wavelets are employed for viewing images at various resolutions, thereby achieving a set of independent information levels.

2. Characteristics of the wavelet transform

We use the nomenclature of Mallat [8] and consider the wavelets $\Psi_{jk}(x)$ as the set of functions which are obtained by translation 'k' and dilation 'j' of the basic wavelet function $\Psi(x)$. For dyadic wavelets with a translation and dilation grid of a power of the basis two we get

$$\Psi_{jk}(x) = \left(\sqrt{2^{-j}}\cdot\Psi_{2^j}(x-2^{-j}\cdot k)\right)_{j,k\in z} \quad \text{with } \Psi_{2^j}(x)=2^j\cdot\Psi(2^j\cdot x) \tag{1}$$

The function $\Psi(x)$ is so chosen, that the wavelet functions $\Psi_{jk}(x)$ are orthogonal. Well known types of orthogonal wavelet functions are: the classical Haar functions and the wavelets (based on the work of Lemarie [7]) suggested by Mallat [8] and those proposed by Daubechies [3]. Furthermore a scaling function $\Phi(x)$ is introduced, which is related to $\Psi(x)$ in the frequency domain by a filter g(n)

$$\tilde{\Psi}(\omega)=G\left(\frac{\omega}{2}\right)\cdot\tilde{\Phi}\left(\frac{\omega}{2}\right) \tag{2}$$

whereby the tilde represents the Fourier transform.

The functions $\Psi_{jk}(x)$ and $\Phi_{jk}(x)$ form a pair of orthonormal bases and are used to resolve a function f(x) into a 'bandpass' and a 'lowpass' component, respectively. In the context of digital signal processing the functions $\Psi_{jk}(x)$ and $\Phi_{jk}(x)$ are used with the filter g(n) and a filter h(n). These filters form a Quadrature-Mirror-Filter (QMF) filter pair [11], whereby h(n) is a lowpass filter and g(n) is a highpass filter. The impulse responses themselves are determined on the wavelet and scaling functions using the following equations

$$\begin{aligned} h(n) &= \langle \Phi_{2^{-1}}(u),\Phi(u-n)\rangle \\ g(n) &= \langle \Psi_{2^{-1}}(u),\Phi(u-n)\rangle \end{aligned} \tag{3}$$

$\langle\ \rangle$ denotes the scalar product.

Multiresolution decomposition consists in convolving the signal with the scaling filter h(n) and the wavelet filter g(n), respectively. This is done iteratively by downsampling both filtered signal components and filtering again the lowpass signal with h(n) and g(n). The signal components achieved by filtering with g(n) are called 'Detail'-signals. Hence, a procedure is given, which leads sucessively to a multiresolution representation of the analyzed signal.

The filters h(n) and g(n) must fulfill several conditions. These are the 'Finite Energy' condition, the 'Mean' condition, the 'Orthogonality' condition and the 'Regularity' condition. Daubechies [3] formulated these conditions for discrete filters with finite support.

The 'Orthogonality' condition is mathematically formulated by

$$\sum_n g(n-2\cdot k)\cdot g(n-2\cdot l) = \delta_{kl} \tag{4a}$$

and the 'Regularity' condition demands

$$\sum_n n^k\cdot g(n)=0 \qquad k=0,1,\ldots,N-1 \tag{4b}$$

resulting, that H(z) has N zeroes at z = (-1). This constraint implies the convergence of the iterated filter

H(z) to a continous limit function $H_\infty(z)$ [9]

$$H_\infty(z) = \prod_{p=0}^{\infty} H(z^{2^p}) \qquad (5),$$

i. e. the corresponding scaling function. Since this condition is only necessary, but not sufficient, the convergence can not be guaranteed. The strategy is to put as many zeroes as possible to (-1). It depends on the application, whether a continous limit function is not achievable, the filters designed can be useful for multiresolution decomposition.

2.1 Two-dimensional wavelets

In this paper we investigate the wavelet decomposition for the pupose of texture analysis. Hence 2-d filters $H(\omega_1,\omega_2)$ are necessary. This is obtained by a separable 2-d filter $H(\omega_1,\omega_2)$, i. e. the filter can be expressed by two 1-d filters $H(\omega_1)$ and $H(\omega_2)$. Addionally a orientation selectivity is achieved. The result are a lowpass-lowpass filter $H(\omega_1)H(\omega_2)$, two mixed filters $H(\omega_1)G(\omega_2)$, $G(\omega_1)H(\omega_2)$ and a highpass-highpass filter $G(\omega_1)G(\omega_2)$ building a pyramid structure with 4 components on each level.

3. Signal-matched analysing wavelet filters

All known orthogonal wavelet functions are designed on the basis of mathematical criteria, i.e. solving a dilation equation. Mallat solves this equation for the continuous case and uses a discrete approximation, while Daubechies proposes a solution for the digital filter case directly. Whereas the procedures of Mallat and Debauchies do not take the signal to be analysed into consideration, the procedure given by Tewfik in [10] is signal oriented. Here, the wavelet is designed by minimizing the approximation error up to a given scale in the multiresolution decomposition and reconstruction process. Because of the complicated optimization process Tewfik restricts himself by using the upper bound of the error as the optimization criterion.

Each resolution level decomposes the image into a highpass and a lowpass part by filtering. Thus image characteristics should influence the filter pairs $h(n_1)$, $g(n_1)$ (horizontal filters) and $h(n_2)$, $g(n_2)$ (vertical filters). Hence, first we design the filters $h(n_i)$ and $g(n_i)$ instead of the wavelet function directly. Subsequently the wavelet function can be calculated from the corresponding filter pair as long as the filters converge.

In this paper we propose two approaches for designing wavelet filters for texture analysis which are matched to the textures to be analysed. The filters are designed by minimizing the mean square error between their impulse response and the impulse response of a given signal matched filter at the same time considering the wavelet conditions or by following a Karhunen-Loeve-Transform related maximum variance criterion.

4. Signal matched filters for texture analysis

In texture analysis, one basic filter design method is the 'Eigenfilter' method proposed by Ade [1].

This approach leads to a decomposition of the image into a series of uncorrelated images, each of which contains progressively less information. The impulse responses $f_{eigen}(n_1,n_2)$ are the normalized eigenvectors $\underline{v}_i$ of the estimated covariance matrix $\underline{C}_{ss}$ of the image-signal $s(n_1,n_2)$, ordered according to declining eigenvalues arranged as 2-d filters.

The design of 'Eigenfilters' can be traced back to the discrete Karhunen-Loeve-Transform, which is a signal matched transform reducing the number of given components by a maximum variance extraction in the first few transformed components $s'_i(n_1,n_2)$.

This approach is derived by multiplying the covariance matrix $\underline{C}_{ss}$ by a column vector $\underline{v}$, which can be used as a digital filter

$$\underline{v}^T \cdot \underline{C}_{ss} \cdot \underline{v} = \sigma^2_{s's'} \rightarrow \max. \qquad (6a)$$

with the normalization constraint $\underline{v}^T \cdot \underline{v} = 1$ (6b)

Again the vectors $\underline{v}$ are the normalized eigenvectors of $\underline{C}_{ss}$.

Aiming at separable solutions, the covariance matrix is estimated for the lines and rows of the image.

5. Optimization of wavelet filters under nonlinear constraints

The optimization process is formulated in the spatial domain. Because of the high nonlinearity of the optimizing process we use an iterative Lagrange-Newton approach [5] with a linearized equation system. A similiar method has been used for designing Linear-Phase QMF-filters achieving the 'Perfect-Reconstruction' property [6].

Two approaches are possible. We start with an approximation of the impulse response of one 'Eigenfilter' $f_{eigen}(n)$ by an optimized scaling filter $h_I(n)$ or a wavelet filter $g_I(n)$. Here two design steps are required; first the 'Eigenfilters' have to be designed and secondly, their impulse responses have to be approximated.

In contrast to this two step solution a scaling filter $h_{II}(n)$ or a wavelet filter $g_{II}(n)$ can be designed directly by maximizing the variance of the filtered image and fulfilling the wavelet conditions.

For both approaches the used wavelet constraints lead to highly nonlinear equations. We have to minimize

$$L(\underline{f}_{wavelet},\underline{\lambda}) = Q(\underline{f}_{wavelet}) - \sum_i \lambda_i * c_i(\underline{f}_{wavelet}) \tag{7a}$$

with

$$Q(\underline{f}_{wavelet}) = \sum_n (f_{wavelet}(n) - f_{eigen}(n))^2 \tag{7b}$$

= quality criterion (approach I)

or $Q(\underline{f}_{wavelet}) = \underline{f}^T_{wavelet} \cdot \underline{C}_{ss} \cdot \underline{f}_{wavelet}$ (7c)

= quality criterion (approach II)

and L()- modified error criterion
λ_i - Lagrange Multipliers
c_i - wavelet filter conditions
$\underline{f}_{wavelet}$ - filter to be determined
$\underline{C}_{ss}$ - Covariance matrix

The optimization problem can be solved iteratively, using a Taylor approximation of the nonlinear equations on each iteration step and truncating after the linear term.

For the optimization it is necessary to set

$$\nabla L(\underline{f},\underline{\lambda}) = 0 \tag{8}$$

The Taylor approximation results in

$$\nabla L(\underline{f}^{(k)} + \Delta\underline{f}^{(k)}, \underline{\lambda}^{(k)} + \Delta\underline{\lambda}^{(k)}) = \nabla L(\underline{f}^{(k)}, \underline{\lambda}^{(k)}) + \nabla^2 L(\underline{f}^{(k)}, \underline{\lambda}^{(k)}) \begin{pmatrix} \Delta\underline{f} \\ \Delta\underline{\lambda} \end{pmatrix} \tag{9}$$

The iteration is formulated with

$$\begin{bmatrix} \Delta\underline{f}^{(k)} \\ \underline{\lambda}^{(k+1)} \end{bmatrix} = \begin{bmatrix} \underline{W}^{(k)} & -\underline{A}^{(k)} \\ -\underline{A}^{(k)\,T} & 0 \end{bmatrix}^{-1} \cdot \begin{bmatrix} -\nabla Q^{(k)} \\ \underline{c}^{(k)} \end{bmatrix} \tag{10a}$$

$$\underline{f}^{(k+1)} = \underline{f}^{(k)} + \Delta\underline{f}^{(k)} \tag{10b}$$

$\underline{W}$ - Hesse-Matrix of the Lagrange equation
$\underline{A}$ - Jacoby-Matrix of the wavelet conditions

as long as $\Delta\underline{f}$ and $\Delta\underline{\lambda}$ exceed a specified level. The starting value are the impulse response of the selected 'Eigenfilter' for the first optimization approach or the corresponding normalized eigenvector for the second approach.

6. Results

The design proposed was carried out, filters which contain signal characteristics and additionally fulfill the orthogonal wavelet conditions were obtained. The filters were matched to the Brodatz textures No. 5 ("Expanded Mica") and No. 9 ("Grass lawn"). Both images were normalized getting the same mean and variance values. The designed filters are 6-tap filters with H(z) having 2 zeroes at z=(-1).

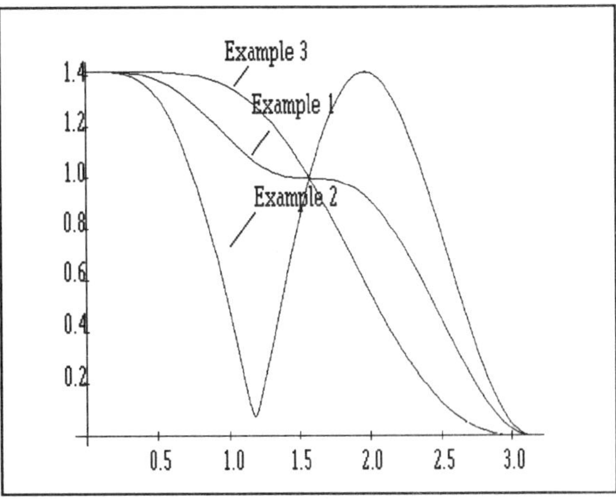

Fig. 1. Magnitude Responses

Three examples for the application of the filter design are now given. The first and second examples are based on the approximation of the impulse responses of the 'Eigenfilters' (first and fourth 'Eigenfilter' out of 6 possible 'Eigenfilters'). The third example is based on the maximum-variance criterion. In all cases h(n) was first determined. Fig. 1, 2 and 3 show the magnitude of $H(\omega)$ and the corresponding iterated scaling functions of the 1-d filter matched to the horizontal signal. In the first example the magnitude of $H(\omega)$ exhibits a drop at $(\omega=\pi/2)$. $H(\omega)$ of the second example shows an interesting lowpass-bandpass run. The scaling functions converge for examples 1 and 3, but not for 2.

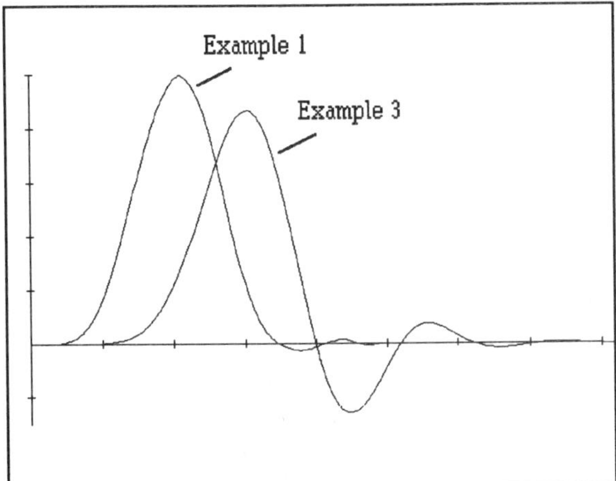

Fig. 2. Iterated scaling functions

The Tables 1 and 2 give a comparison of the results obtained with the signal matched filters and those obtained with the filters of Mallat and Daubechies. The percentage of the energy in the various levels obtained by filtering with $H(\omega_1)H(\omega_2)$, $H(\omega_1)G(\omega_2)$, $G(\omega_1)H(\omega_2)$ and $G(\omega_1)G(\omega_2)$ is shown. Mallat and Daubechies procedures yield similiar results for the first and second

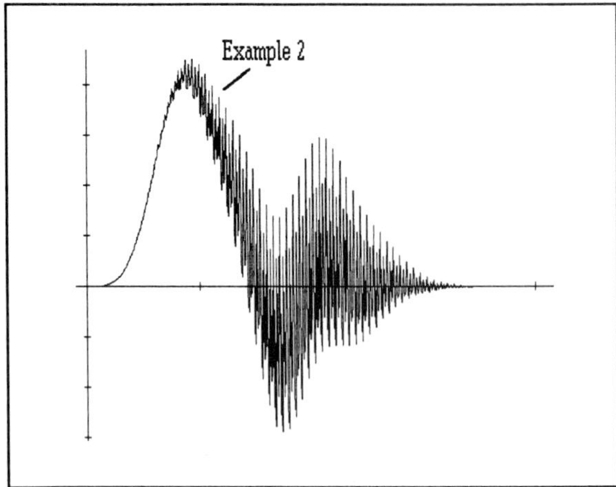

Fig. 3. Iterated scaling function (Example 2)

resolution steps. The result of the first filter proposed show distinct differences in the two textures. Because of the separable solutions this is true both for filtering with $H(\omega_1)G(\omega_2)$ and $G(\omega_1)H(\omega_2)$. The second filter example reveals the greatest differences. Even though one does not obtain a continous scaling function, the filter can be employed for texture analysis. The result of the third example exhibits the strongest resemblance with the classical case. By virtue of the maximum-variance criterion, more than 99% of the energy of the first texture is contained in the lowpass component. The analysis of other textures yields similiar results.

Pyramid level	Filter	Mallat	Daubechies	Example 1	Example 2	Example 3
1 st level	HH	96.02	97.88	97.11	92.76	99.32
	HG	1.91	1.83	1.83	4.40	0.44
	GH	2.01	0.24	0.94	2.22	0.20
	GG	0.06	0.05	0.12	0.62	0.04
2 nd level	HH	93.21	92.81	90.71	84.28	95.00
	HG	3.70	5.04	5.10	8.36	3.18
	GH	2.84	1.63	3.25	4.88	1.35
	GG	0.25	0.52	0.94	2.48	0.47
3 rd level	HH	87.11	80.30	78.20	70.07	85.91
	HG	6.80	12.54	11.30	12.59	8.14
	GH	4.78	5.42	6.83	11.48	3.50
	GG	1.31	1.74	3.67	5.86	2.45

Table 1. Energy contents of the different approaches for Brodatz Texture No. 5 achieved with the signal matched filters in comparison with the classical wavelet filters

Pyramid level	Filter	Mallat	Daubechies	Example 1	Example 2	Example 3
1 st level	HH	95.80	94.74	92.27	81.38	97.52
	HG	2.11	4.00	3.88	8.37	1.31
	GH	2.02	1.08	3.29	7.99	1.05
	GG	0.07	0.18	0.56	2.26	0.12
2 nd level	HH	87.90	84.27	81.10	66.60	85.76
	HG	5.49	8.30	8.88	13.50	6.49
	GH	6.05	5.64	7.93	13.66	6.33
	GG	0.56	1.79	2.09	6.24	1.42
3 rd level	HH	76.18	66.32	65.40	44.97	69.60
	HG	9.98	13.20	13.85	24.91	11.38
	GH	10.41	13.67	14.41	19.21	12.22
	GG	3.43	6.81	6.34	10.91	6.80

Table 2. Energy contents of the different approaches for Brodatz Texture No. 9 achieved with filters matched to Brodatz Texture No. 5

7. References

[1] F. Ade: "Characterization of Textures by 'Eigenfilters'", Signal Processing 5, pp. 451-457, 1983

[2] P. Brodatz: "Textures: A Photographic Album for Artists and Designers", Dover, New York, 1968

[3] I. Daubechies: "Orthonormal Bases for Compactly Supported Wavelets", Communications on Pure and Applied Mathematics, Vol. XLI, pp. 909-969, 1988

[4] D. E. Dudgeon, R. M. Mersereau: "Multidimensional Digital Signal Processing", Prentice-Hall, Englewood Cliffs, New Jersey, 1984

[5] R. Fletcher: "Practical methods of optimization", Vol. 2, pp. 138-145, John Wiley and Sons, 1991

[6] B. Horng, A. N. Willson: "Lagrange Multiplier Approaches to the Design of Two-Channel Perfect-Reconstruction Linear-Phase FIR Filter Banks", IEEE ASSP, Vol. 40., No. 2, pp. 364-374, 1992

[7] P. G. Lemarie: "Ondelletes a localisation exponentielles", J. Math. Pures Appl., 1988

[8] S. Mallat: " A Theory for Multiresolution Signal Decomposition: The Wavelet Representation", IEEE Trans. PAMI, Vol. 11, No. 7, pp. 675-693, 1989

[9] O. Rioul: "Dyadic Up-Scaling Schemes: Simple Criteria for Regularity", submitted to SIAM J. Math. Anal.

[10] A. H. Tewfik, P. E. Jorgensen: "On the Optimal Choice of a Wavelet for Signal Representation", submitted to the IEEE Trans. on Information Theory

[11] P. P.Vaidyanathan: "Multirate Digital Filters, Polyphase Networks and Applications: A Tutorial", Proceedings IEEE, Vol. 78, S. 56-92, January 1990

SIGNAL PROCESSING VI: Theories and Applications
J. Vandewalle, R. Boite, M. Moonen, A. Oosterlinck (eds.)

A NEW FAST ALGEBRAIC CONVOLUTION ALGORITHM

Torsten MINKWITZ & Reiner CREUTZBURG

University of Karlsruhe, Faculty of Informatics,Institute of Algorithms and Cognitive Systems, P. O. Box 6980, D-7500 Karlsruhe 1, Germany

phone: (+49)-721-6084325; fax: (+49)-721-696893; e-mail:minkwitz@ira.uka.de; creutzbu@ira.uka.de

This paper introduces a new fast algorithm to do convolutions, which is a reoccuring problem in signal processing. The concepts used are fully algebraic. Therefore the algorithm does not produce roundoff errors. For modestly large signal lengths it is faster than any other previously described method. The calculations in our new method are done in the field $\mathbb{Q}$ of rational numbers and in finite field extensions of $\mathbb{Q}$. The algorithm performs cyclic convolutions of lengths $N = 2^n$. This can be interpreted as a multiplication of polynomials modulo the polynomial $X^N - 1 \in \mathbb{Q}[X]$. Since $X^N - 1 = (X^{N/2} - 1)(X^{N/2} + 1)$ for N even, the solution to be sought is a fast algorithm to do convolutions modulo the irreducible polynomial $X^{N/2} + 1 \in \mathbb{Q}[X]$. Precisely that has been achieved by our work. The algorithm introduced here calculates the cyclic convolution of two signals at a cost of $O(N \log(N) \log\log(N))$ operations in $\mathbb{Q}$. The algorithm was successfully implemented and the measured execution times are given.

1 Introduction

The Chinese remainder theorem and existing FFT algorithms provide for mechanisms to develop divide and conquer strategies for algorithms in algebraic computing. Solving a problem of size $N = p \cdot q$ is reduced to solving the same problem p times for size q. If this reduction and its inverse can be done rapidly, then a faster algorithm can be achieved. Convolutions are particularly well suited for such an approach, because discrete Fourier transforms have the convolution property [1-8,10-15].

Theorem 1 *If A, B are vectors of length N over a field containing all N-th roots of unity, then if $*$ denotes the cyclic convolution and $\odot$ denotes pointwise multiplication:*

$$A * B = DFT^{-1}(DFT(A) \odot DFT(B)).$$

A convolution of size $N/2$ can always be computed by a cyclic convolution of size N. The higher order coefficients are set to zero and no actual overflow occurs. The FFT is a the fast method of computing the DFT. It is based on the fact that the roots of unity form a cyclic group C_N. Instead of taking the single group elements (roots of unity) of this large group, first the normal subgroup C_q $(N = p \cdot q)$ is taken and the DFT is computed for the smaller group C_N/C_q. The cosets are now used instead of the single roots of unity. Thereafter the same algorithm is applied recursively for the cosets. Substituting the DFTs with FFTs in (Theorem 1.) yields an efficient method to do convolutions. In real applications the coefficients will often be elements of the ring $\mathbb{Z}$ of integers. However, the maximum size DFT that can be done in $\mathbb{Z}$ is 2, since higher roots of unity are missing. Therefore the integers are embedded in the field $\mathbb{C}$ of complex numbers, which includes all roots of unity, or in a suitable number ring, which contains just the necessary roots of unity. The next section discusses some drawbacks of these approaches next to describing some alternative methods that avoid just those difficulties, but don't achieve a comparable efficiency for large problem sizes.

If the vectors are interpreted as polynomials, their convolution is a polynomial multiplication. This interpretation will be used in the later sections to introduce a new algorithm, which does not suffer from any of the disadvantages that the others have. It uses the concept of finite field extensions to use a FFT type divide and conquer strategy.

2 Previous Algorithms

A number of different methods have been designed to do convolutions [1-8,10-15]. If the coefficients are integers, two have been suggested in the introduction, which will be covered in this section. Besides the simple method which is tought in school for polynomial multiplications, there is also more efficient algorithms that don't require the concept of roots of unity. As an example the algorithm of Karatsuba is described here. Another algorithm that uses the convolution property of FFTs is the method described by Schönhage, which does convolutions over the ring of integers by mapping the vectors into large numbers and then doing fast multiplications of those large numbers.

2.1 Convolution in the field of complex numbers C

As was mentioned in the introduction, the convolution property of the FFT yields a fast algorithm since the necessary roots of unity exist in C. The calculations are done with floating point arithmetic. The problem comes from the fact that there is no exact mapping of all roots of unity into floating point numbers on a machine [8]. Also the use of floating point arithmetic results in roundoff errors. The convolution can therefore be only of limited precision. Another disadvantage of this method is that complex arithmetic with floating point numbers is usually slower than simple integer arithmetic. Therefore the speedup achieved by this algorithm is somewhat diminished.

2.2 Convolutions using NTTs

Number theoretic transforms (NTT) were introduced as a generalization of the DFT over residue class rings of integers.

The FNT (Fermat number transform) uses the transform length $N = 2^{d+1}$, $\alpha = 2$ as a primitive N-th root of unity modulo m and the Fermat number modulus

$$m = F_d = 2^{2^d} + 1 = 2^b + 1, \qquad (d \geq 0,\ b = 2^d).$$

Since the transform length is always a power of two, the well-known radix-2 FFT-algorithm can be applied to the FNT. Note that since $\alpha = 2$, the multiplications with roots of unity are only shifts on a binary machine.

2.3 Karatsuba's Algorithm

This method [10] uses a simple recursion scheme. The vectors are interpreted as polynomials. The formula for a polynomial multiplication of length N is:

$$A * B = \sum_{i=0}^{2N-2} \Big(\sum_{j+k=i} a_j \cdot b_k \Big) X^i \quad .$$

If N is even, then the vectors A and B consist of upper halfs A_1 and B_1, (coefficients a_i, b_i, $N/2 \leq \mathrm{i} < N$) and lower halfs A_0 and B_0, (coefficients a_i, b_i, $0 \leq \mathrm{i} < N/2$). Now the multiplication formula is

$$A*B = (A_1*B_1)X^N + (A_1*B_0 + A_0*B_1)X^{N/2} + (A_0*B_0) \quad ,$$

which is equivalent to

$$\begin{aligned} A * B \ = \ & (A_1 * B_1)(X^N + X^{N/2}) \\ & + ((A_1 - A_0) * (B_0 - B_1))X^{N/2} \\ & + (A_0 * B_0)(X^{N/2} + 1). \end{aligned}$$

The last formula shows how one multiplication of length N is reduced to only 3 multiplications of length $N/2$ instead of 4 in the previous formula. If $N/2$ is also even, the scheme can be applied recursively. Therefore for N a power of 2, instead of $O(N^2)$, this method computes length N polynomial multiplications with $O(N^{\log_2 3}) \approx O(N^{1.58..})$ elementary multiplications. For small N, this method is faster than the others described here.

2.4 Schönhage's Algorithm

Instead of explicitly doing a convolution, here the coefficients are mapped into a large integer and the convolution is done by multiplying those large integers. If the maximum coefficient is l bits long and the convolution is of length N, then the numbers to be multiplied are:

$$A' = \sum_{i=0}^{N-1} a_i \cdot 2^{i(2l+\log(N)+1)} \quad ,$$

$$B' = \sum_{i=0}^{N-1} b_i \cdot 2^{i(2l+\log(N)+1)} \quad .$$

These numbers are multiplied by the efficient Schönhage-Strassen algorithm and the result of the convolution can be read from the result of the multiplication of numbers. The (i)th coefficient is in the $2l + \log(N) + 1$ bits beginning at the $(i(2l + \log(N) + 1))$th bit. For a fixed maximum size of resulting coefficients, this method has a cost of $O(N \cdot \log(N) \cdot \log\log(N))$ operations. The disadvantage it has in comparison with the algorithm introduced in the following section is the cumbersome arithmetic used in the Schönhage-Strassen algorithm [14,15].

3 Convolution with finite field extensions

Instead of looking at direct convolutions, the rest of this paper will deal with cyclic convolutions of lengths $N = 2^n$, n an integer, which are polynomial multiplications modulo the polynomial $X^N - 1$, [3,4,12]. These can be used to do direct convolutions of length $N/2$. The splitting of $X^N - 1$ yields (see also fig. 1)

$$X^N - 1 = (x - 1) \cdot \prod_{i=0}^{n-1} (X^{2^i} + 1) \quad .$$

If a fast method is known to do convolutions modulo $X^M + 1$ for M a power of two, the Chinese remainder theorem suggests a fast algorithm to do cyclic convolutions.

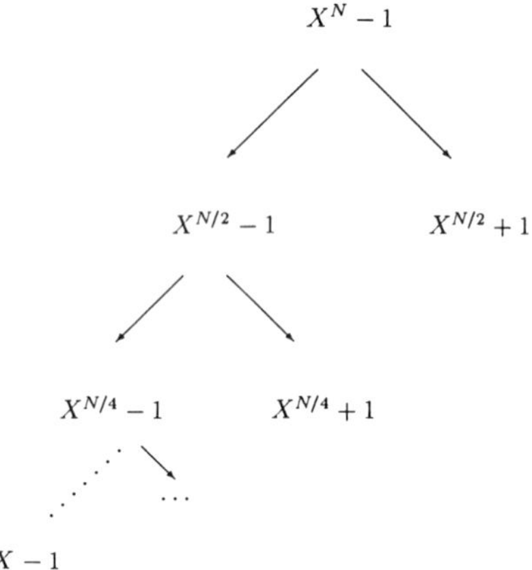

Fig. 1. Structure for Chinese remaindering

3.1 Extending the field of rational numbers $\mathbb{Q}$

The field $\mathbb{Q}$ of rational numbers does not include the necessary roots of unity to do cyclic convolutions of lengths greater than 2, [12]. Therefore $X^M + 1$ is irreducible over $\mathbb{Q}$ for M a power of two and $M > 2$.

Definition 1 a) *$n \in \mathbb{Z}$, then $\lfloor n \rfloor^*$ denotes the mapping to the next smaller power of two of n.*

b) *$n \in \mathbb{Z}$, then $\lceil n \rceil^*$ denotes the mapping to the next greater power of two of n.*

In order to improve the situation, calculations will be done in $\mathbb{Q}(\omega^{\lfloor\sqrt{M}\rfloor^*})$, where ω is a $(2M)$th root of unity. $X^M + 1$ is reducible over this field. It now splits into

(*)

$$X^M + 1 = \prod_{i=1}^{\lceil\sqrt{M}\rceil^*} (X^{\lfloor\sqrt{M}\rfloor^*} - \omega^{\lfloor\sqrt{M}\rfloor^* + \frac{2M}{\lfloor\sqrt{M}\rfloor^*}(i-1)}) .$$

The chinese remainder theorem shows that doing multiplications modulo $X^M + 1$ can be performed by calculating the remainders modulo the factors of $X^M + 1$, doing separate multiplications modulo these and, by interpolation, calculating the product. Since the original coefficients were rationals, the remainders modulo the factors in (*) are all conjugate. Therefore only one remainder needs to be considered (v. fig. 2). This remainder has $\lfloor\sqrt{M}\rfloor^*$ coefficients and can be aquired without any real calculations. To multiply in this modulo, a double length cyclic convolution is performed to do a full multiplication. The remainder is calculated thereafter. The cyclic convolution is a polynomial multiplication modulo $X^{2\lfloor\sqrt{M}\rfloor^*} - 1$. Since the $2\lceil\sqrt{M}\rceil^*$-th roots of unity exist in the selected field extension, the DFT exists and with the convolution property and an FFT-type algorithm, the cyclic convolution can be calculated efficiently.

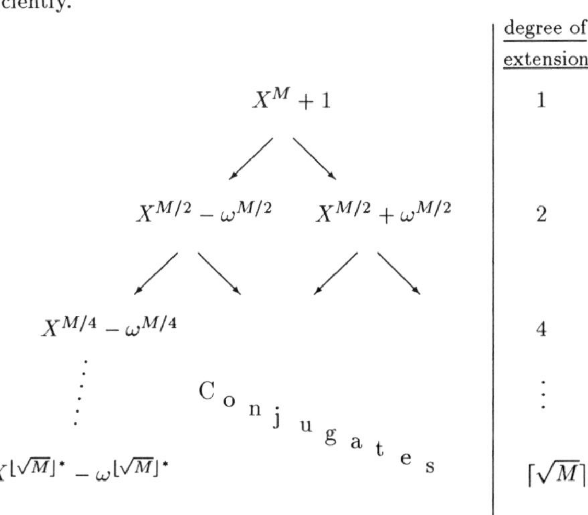

Fig. 2. Scheme for splitting of $X^M + 1$ through field extension.

With the help of the FFT, the transformed coefficients can be multiplied separatly. The coefficients have dimension $\lceil\sqrt{M}\rceil^*$ over $\mathbb{Q}$. So they are polynomials in $\psi = \omega^{\lfloor\sqrt{M}\rfloor^*}$ of degree $\lceil\sqrt{M}\rceil^* - 1$. Since $\psi^{\lceil\sqrt{M}\rceil^*} \equiv -1$, the same algorithm as before applies for the multiplication of the coefficients, as it did for polynomials modulo $X^M + 1$, however, this time of square root length.

3.2 Investigation of time expense

The cyclic convolution is a multiplication of polynomials modulo $X^N - 1$. That problem is reduced to multiplying once modulo $X^{N/2} - 1$ and once modulo $X^{N/2} + 1$ by the Chinese remainder theorem. Just scanning the input takes $O(N)$ operations to do, so the algorithm will at least have that cost. Therefore reducing the problemsize to half results in a runtime which is at most half of that of the original length. Since the above reduction and its inverse can be performed in $O(N)$ steps, the investigations can be limited to the expense of multiplying modulo $X^M + 1$, $M = N/2$. Twice the operations for that problem plus the $O(N)$ from the reduction are an upper bound for the total cost.

The reduction modulo $X^{\lfloor\sqrt{M}\rfloor^*} - \omega^{\lfloor\sqrt{M}\rfloor^*}$ takes $O(M)$ steps, since it is just a copying of the input (v. example 1.). The FFT is known to have a cost of $O(n \cdot \log(n))$ for input of length n. Therefore the double length FFTs need $O(\sqrt{M} \cdot \log(M))$ operations in the field extension. Since the coefficients are of length $\sqrt{M}$, every one of those operations requires $O(\sqrt{M})$ steps (only additions, subtractions and shifts occur). So the total expense so far is

$$O(\sqrt{M} \cdot \sqrt{M} \cdot \log(M)) = O(M \cdot \log(M)).$$

What is left is the complexity of multiplying $2 \cdot \sqrt{M}$ coefficients of length $\sqrt{M}$. Since the same algorithm applies, the complexity is again

$$O(2 \cdot \sqrt{M} \cdot \sqrt{M} \cdot \log(\sqrt{M})) = O(M \cdot \log(M)).$$

So every recursion brings about this complexity. It stops after about $\log\log(M)$ recursion steps. Therefore, the total complexity is:

$$O(M \cdot \log(M) \cdot \log\log(M)).$$

4 Results

All convolution algorithms were tested on a Sun-3 computer. The programs are implemented in machine code. Every input coefficient is coded in a 32 bit integer. The resulting coefficients are 64 bit integers. The times are measured in milliseconds.

N	School Method	Karatsuba	New Algorithm
4		0.089	0.323
8		0.288	1.024
16	1.083	0.983	2.499
32	4.133	3.116	6.899
64	16.332	9.832	16.833
128	64.997	30.332	41.998
256	259.989	93.329	94.996
512	1036.625	284.988	248.323
1024	4149.834	853.299	573.310
2048	16582.670	2599.896	1316.614
4096	66264.016	7799.688	2816.554
8192	265022.732	23365.732	6449.742
16384	1060074.262	70197.192	14016.106
32768		210624.908	30098.796

These times were measured for programs written in the C programming language. C lacks 64 bit integers as basic data types. Therefore, the values stem from running the code for 32 bit input and output. The code for the FFTs uses the floating point coprocessor M68881 of Motorola. A floating point number is also coded in 32 bits. The FFT method in the last column assumes the input to be real. The times are measured in milliseconds.

N	Karatsuba	New Alg.	compl.FFT	real FFT
4	0.099	0.366	3.833	1.983
8	0.349	0.833	9.333	5.016
16	1.117	2.183	21.999	11.666
32	3.583	6.199	52.331	26.832
64	11.666	17.165	119.995	62.831
128	34.832	45.331	269.989	133.328
256	104.996	108.329	601.642	296.654
512	323.320	269.989	1349.946	664.973
1024	966.628	614.975	2949.882	1383.278
2048	2949.882	1516.606	6399.782	3033.212
4096	8816.314	3349.866	13782.782	6516.406
8192	26448.942	7899.684	29532.152	14032.772
16384	79696.812	17182.646	63447.462	30098.796
32768	239707.078	37098.516	135227.924	64514.086

5 Summary

The results of the implementation show that the new algorithm is suitable for a good range of signal lengths. The advantage of using integers instead of floating point numbers makes it far better than the FFT for all practical problem sizes. However it must be mentioned that 32 bits are not enough to represent the resulting signals for very large lengths. One either has to use machine code to add 64 bit arithmetic to the C programming language or one has to accept roundoff errors by using fixed point arithmetic. The 64 bit machine code implementation, which produced the values of the first table is really a C program with only the main loops being coded in machine language. The structure of the algorithm suggests an easy way to implement it on a parallel hardware and possible direct hardware implementations in VLSI. The method does not require a fast floating-point unit and is therefore efficiently usable on a wide range of different machines.

6 References

[1] Agarwal, R. C. & J. W. Cooley: *New algorithms for digital convolution.* IEEE ASSP-**25**, 392-410 (1977).

[2] Aho, A. V.; J. E. Hopcroft & J. D. Ullman: *The Design and Analysis of Computer Algorithms.* Addison-Wesley: London 1974.

[3] Beth, Th.: *Verfahren der schnellen Fouriertransformation.* Teubner: Stuttgart 1984.

[4] Beth, Th.; W. Fumy & R. Mühlfeld: *Zur Algebraischen Diskreten Fourier-Transformation.* Arch. Math. **40**, 238-244 (1983)

[5] Bhattacharya, M. & R.C. Agarwal: *Number theoretic techniques for computation of digital convolution,* IEEE ASSP-**32**, No. 3, 507-511 (1984).

[6] Burrus, C. S. & T. W. Parks: *DFT/FFT and Convolution Algorithms.* John Wiley & Sons: New York 1985.

[7] Creutzburg, R. & M. Tasche: *Parameter determination for complex number-theoretic transforms using cyclotomic polynomials.* Math. Comp. **52**, 189-200 (1989)

[8] Elliott, D. F. & K. R. Rao: *Fast Transforms - Algorithms, Analyses, Applications.* Academic Press: New York 1982.

[9] Jacobson, N.: *Basic Algebra I, II.* Freeman: New York 1985.

[10] Karatsuba, A. & Y. Ofman: *Multiplication of multidigit numbers on automata,* Sov. Physiks Dokl. **7**, 714-716 (1963).

[11] Knuth, D.: *The Art of Computer Programming, Vol. 2.* Addison-Wesley: London 1981.

[12] Maisel, W.: *Die ADFT und ein schneller Algorithmus zur Multiplikation ganzer komplexer Zahlen.* Dissertation, Karlsruhe 1989

[13] Nussbaumer, H. J.: *Fast Fourier Transform and Convolution Algorithms.* Springer: Berlin 1982.

[14] Schönhage, A.: *Asymptotically fast algorithms for the multiplication and division of polynomials with complex coefficients,* in *Computer Algebra,* EUROCAM 1982, Springer Lect. Notes Comput. Sci. **144**, 3 - 15.

[15] Schönhage, A. & V. Strassen: *Schnelle Multiplikation großer Zahlen.* Computing **7**, 281-292 (1971).

SIGNAL PROCESSING VI: Theories and Applications
J. Vandewalle, R. Boite, M. Moonen, A. Oosterlinck (eds.)

SIMPLE, OPTIMAL REGULARITY ESTIMATES FOR WAVELETS

Olivier Rioul
CNET Centre Paris B, CRPE
38–40 rue du Gal Leclerc, 92131 Issy-Les-Moulineaux
France.

The new criterion of regularity is of increasing interest in applications involving wavelet decomposition schemes. In this paper, regularity is fully characterized on filter taps, resulting in easily implementable, optimal regularity estimates which can be used for any filter.

1. INTRODUCTION

Perhaps the biggest potential of wavelet theory has been claimed for signal compression schemes [1,5] in which the signal is decomposed into several resolution levels using a "discrete wavelet transform (DWT)" [3,8]. In fact, the DWT was soon recognized to be equivalent to an octave-band tree filter bank which was proposed for some time in subband coding of images [9]. In this particular context, the novelty of wavelet theory comes down to the choice of the filters present in a two-band filter bank: "Wavelet" filters are *regular*.

In order to provide an intuitive feel for what regularity represents, consider the following iterated interpolations with low-pass filter impulse response $G(z)$ which are obviously present in DWT's [8].

$$Y(z) = X(z^2)G(z). \tag{1}$$

Iterating (1) j times yields

$$Y^j(z) = X(z^{2^j})G^j(z) \tag{2}$$

where

$$G^j(z) = G(z)G(z^2)G(z^4)\dots G(z^{2^{j-1}}). \tag{3}$$

The sequence g_n^j corresponding to (3) is the equivalent impulse response at jth stage of the reconstruction. Now, for special choices of $G(z)$, the temporal shape of the g_n^j's, plotted against $n2^{-j}$ (i.e., with the same temporal extent), rapidly converges to a "regular" limit function $\varphi(t)$ as $j \to \infty$ (see Fig. 1). However, for "bad" choices of g_n, $\varphi(t)$ may be highly irregular; the iterated scheme may even diverge, even though $G(z)$ is a "good" half-band low-pass filter [8]. Note that filters are assumed FIR here, $\varphi(t)$ is compactly supported. A first definition of the regularity order of $\varphi(t)$ is the number of times it is continuously differentiable; this is clearly a smoothness requirement on the temporal waveforms of the g_n^j's.

The band-pass impulse responses present in a DWT are obtained with the same iterated interpolation procedure

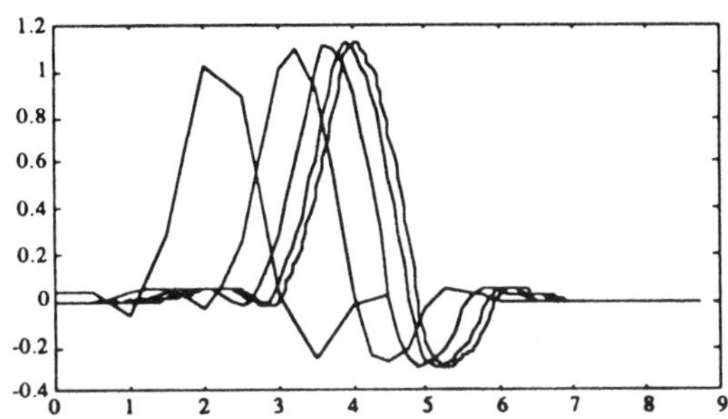

Figure 1. An example of rapidly converging iteration scheme: The g_n^j's are plotted against $n2^{-j}$ for j = 1, 2, 3, 4 and 5 iterations, for one of Daubechies "orthonormal" filter of length 10 [3].

as (2), which is initialized with the high-pass filter impulse response h_n [8]. The resulting limit function $\psi(t)$ is the continuous-time wavelet prototype [3,8]. Here we restrict ourselves to the convergence of the g_n^j's toward $\varphi(t)$, because $\varphi(t)$ and $\psi(t)$ share the same regularity properties [3,6].

Several intuitive arguments have been raised which hint that this property should be useful in image coding applications [1,8]. First, requiring that the signal is analyzed by smooth "basis functions" g_n^j and h_n^j ensures that no artificial discontinuity—not due to the signal itself—appears in the transform coefficients, which are inner products of the signal with these basis functions. That is, regularity would lead to a "better" representation of the signal by the transform coefficients. Second, any quantization error made in a coefficient at some resolution level results, at reconstruction, in an error signal that is proportional to the basis function corresponding to this resolution level. It is therefore natural to require that this perturbation be smooth, rather than discountinuous: A discountinuous perturbation is likely to strike the eye more than a smooth one for the same m.s.e. distortion level.

However, understanding the role of regularity in a DWT-based compression scheme requires precise evaluation of it. One difficulty is that it is a mathematical notion which is expressed on $\varphi(t)$ rather than on filters taps g_n. Therefore, the characterization of regularity on any set of coefficents g_n is a difficult problem, which was first addressed in the wavelet context by Daubechies [3]. A number of regularity order estimates, most of them based on the spectrum $|G(e^{i\omega})|^2$, have been investigated [2,3,4]. Unfortunately, these estimates turn out to be suboptimal in general and sometimes computationally expensive.

This paper presents a complete characterization of regularity on the filter taps g_n in simple terms, restricting to the one-dimensional case. This method is original in that all regularity properties of $\varphi(t)$ are translated into equivalent[1] properties of the discrete-time sequences g_n^j.

2. CONTINUITY

It can be shown [6] that as long as the resulting limit function is regular, the type of convergence of the g_n^j is "uniform," which is a strong type of convergence. Uniform convergence of the g_n^j's is in fact equivalent to the existence and *continuity* of $\varphi(t)$ [6]. Continuity (or uniform convergence) [6] is equivalent to the following intuitive conditions.

$$G(1) = 2, \quad (4)$$
$$G(-1) = 0, \quad (5)$$
$$\lim_{j \to \infty} \max_n |g_{n+1}^j - g_n^j| = 0. \quad (6)$$

Condition (4) is simply a normalization requirement on $G(z)$, while (5) is crucial for convergence and regularity, as explained in section 3.1. The basic requirement (6) is that the difference between two successive values of g_n^j tend to zero *uniformly* in n. Hence, no jumps or discontinuities should appear anywhere in the iterated sequences g_n^j as j increases, and the limit function is continuous.

However, even when $\varphi(t)$ is required to be continuous, it may not appear to be smooth at all, as shown in Fig. 2. It is therefore natural to require more, namely that $\varphi(t)$ possess $N > 0$ continuous derivatives. This is done next.

3. DERIVATIVES

The limit function $\varphi(t)$ has regularity order N if its Nth derivative, $d^N\varphi(t)/dt^N$, is continuous. To characterize this on g_n^j, consider the first-order finite difference sequence δg_n^j, defined as the sequence of the *slopes* of the "discrete curve" g_n^j plotted against $n2^{-j}$.

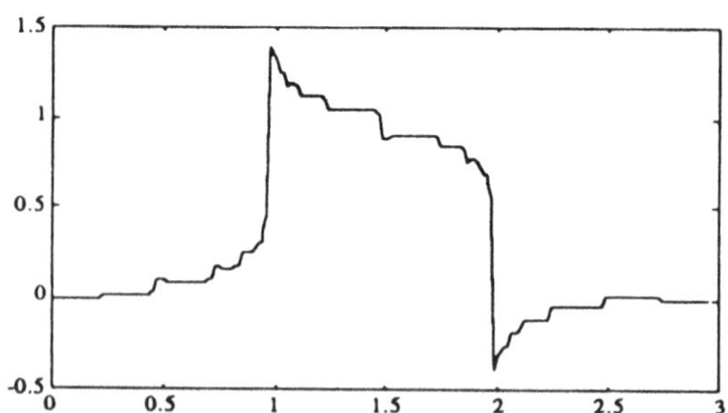

Figure 2. An example of limit function generated by an orthogonal "wavelet" filter. The optimal Sobolev regularity is negative but the Hölder regularity order is in fact 0.0146..., which implies that this limit function is continuous (see section 4).

$$\delta g_n^j = \frac{g_n^j - g_{n-1}^j}{2^{-j}}. \quad (7)$$

The correponding z-transform is $\Delta G^j(z) = 2^j(1 - z^{-1})G^j(z)$. Applying N times the operator δ yields the finite difference of g_n^j of order N, $\delta^N g_n^j$, given by $\Delta^N G^j(z) = 2^{jN}(1 - z^{-1})^N G^j(z)$.

Since the role of the derivative of $\varphi(t)$ of order N is played in the discrete-time domain by $\delta^N g_n^j$, it can be shown [6] that *regularity order N is simply characterized by uniform convergence of $\delta^N g_n^j$.*

3.1. The role of zeroes at $z = -1$ in $G(z)$.

In fact, the Nth derivative of $\varphi(t)$ can obtained from the same iterated interpolation procedure as (2), where N zeroes in $G(z)$ have been removed [6]. As a side result of this and (5), $G(z)$ should have at least $N+1$ zeroes at $z = -1$ to achieve regularity order N. Note that adding one zero at $z = -1$ in $G(z)$ will increase its regularity order by one since removing one amounts to "differentiate." Therefore, zeroes at $z = -1$ have a favorable effect for regularity. This was used by Daubechies in [3] to design regular, orthonormal "wavelet" filters, by imposing as many zeroes at $z = -1$ as possible in $G(z)$ for a given filter length. Note that imposing such zeroes in $G(z)$ amounts to requiring that the frequency response $G(e^{j\omega})$ is "flat" about half the sampling frequency $\omega = \pi$.

However, the effect of zeroes at $z = -1$ may be killed by the *other* zeroes present in $G(z)$. The rest of this paper aims at quantifying the "destructive effect" of zeroes in $G(z)$ that are not located at $z = -1$ in order to quantify regularity accurately.

[1] except for very few pathological cases which are never encountered in practical systems [7]; we here state general results and refer the interested reader to [6] for further details.

4. HÖLDER AND SOBOLEV REGULARITY

We first extend regularity orders to arbitrary, real-valued numbers. A popular extension [2,3] uses a spectral approach to regularity which regards it as a spectral localization. This definition is typically based on Sobolev spaces [2,6]. However, it masks the effect of regularity on the temporal waveform of $\varphi(t)$ and does not use *phase* information of $G(e^{j\omega})$. This may be inappropriate: Fig. 2 shows an example of $\varphi(t)$ for which the best Sobolev exponent r is negative, although it can shown that $\varphi(t)$ is in fact continuous.

These limitations are overcome in the following definition of Hölder regularity. The function $\varphi(t)$ is regular of order α, $0 < \alpha \leq 1$, if

$$|\varphi(t+h) - \varphi(t)| < c|h|^{\alpha}. \tag{8}$$

This controls the way infinitesimal *slopes* of $\varphi(t)$, $|\varphi(t+h) - \varphi(t)|/|h|$, grow as h becomes indefinitely small. For higher regularity orders $r = N + \alpha$, $0 < \alpha \leq 1$, the same definition is used on the Nth derivative of $\varphi(t)$. This definition is more compatible with continuity and differentiability because it can be shown that [6] if $\varphi(t)$ is a limit function of g_n^j, then $\varphi(t)$ possess N continuous derivatives ***if and only if*** it has some Hölder regularity order r greater than N. We have seen that this property is not shared by Sobolev regularity. In the following we therefore concentrate on Hölder regularity.

There is a slight irritation in that $\varphi(t)$ possess N continuous derivatives only when its Hölder regularity is $r = N + \varepsilon$, where $\varepsilon > 0$ is arbitrarily small. To simplify our presentation, we drop the ε in the sequel and regard regularity orders within an arbitrarily small constant.

4.1. Hölder regularity order $0 < \alpha \leq 1$

To characterize Hölder regularity α, $0 < \alpha \leq 1$, on g_n^j, we can do an analogy with (8), replacing $\varphi(t)$ by g_n^j with $t = n2^{-j}$ and $h = 2^{-j}$. This gives

$$|g_{n+1}^j - g_n^j| < c2^{-j\alpha}. \tag{9}$$

This property, along with (4), (5), is indeed equivalent to (8) [6]. This gives an intuitive interpretation of Hölder regularity: The *slopes* of g_n^j plotted against $n2^{-j}$, $|g_{n+1}^j - g_n^j|/2^{-j}$, grow less than $2^{j(1-\alpha)}$ as $j \to \infty$. For example, bounded slopes means that regularity order is 1, i.e., $\varphi(t)$ is almost continuously differentiable. And less regularity allows slopes to increase indefinitely: This explains why $\varphi(t)$, although continuous, may sometimes be quite "nasty" as in Fig. 2.

4.2. Arbitrary regularity orders

Since derivatives of $\varphi(t)$ correspond to finite differences of g_n^j, a natural discrete-time characterization [6] of regularity order $r = N + \alpha$, $0 < \alpha \leq 1$, is (4), (5), and (9) written for $\delta^N g_n^j$, i.e.,

$$|\delta^N g_{n+1}^j - \delta^N g_n^j| < c2^{-j\alpha}. \tag{10}$$

A remarkable fact is that (10) can be extended to negative values of α [6]. That is, even if (10) "fails," i.e., gives a ***negative*** regularity order for $\delta^N g_n^j$, it can be used to prove that g_n^j has some (positive) regularity if $N > \alpha$. It is therefore worthwhile to consider negative regularity orders. In particular, assume that $G(z)$ has exactly N zeroes at $z = -1$. The maximum number $\alpha \leq 0$ for which (10) holds is then the exact amount of regularity lost due to the destructive effect—discussed in section 3.1—of the zeroes in $G(z)$ that are not located at $z = -1$ [7]

4.3. Regularity and rate of convergence

In practical systems involving a discrete implementation of the DWT, the number of iterations j is limited. It is therefore questionable to study the limit function as $j \to \infty$. However, the rate of convergence of g_n^j to $\varphi(t)$ is faster as regularity is high (the difference tends to 0 as $2^{-j\alpha}$ [6]). The convergence is even faster for higer regularity orders (see Fig. 1).

5. OPTIMAL REGULARITY ESTIMATES

A regularity estimate r is here said to be ***optimal*** if $\varphi(t)$ is at least regular of order $r - \varepsilon$ and is ***not*** regular of order $r + \varepsilon$, where $\varepsilon > 0$ is arbitrarily small.

A simple algorithm [6], which was independently derived using the "Littlewood-Paley theory" by Cohen and Daubechies [2], gives the optimal Sobolev regularity order. However this is not optimal for Hölder regularity in general: Hölder regularity is always greater than Sobolev regularity by at most 1/2 [6]. This gives suboptimal Sobolev lower and upper bounds for Hölder regularity. Sobolev regularity depends on the modulus of the spectrum while two filters that differ only by their phase have Hölder regularity orders that differ by at most 1/2. In the following we provide sharp lower and upper bound estimates based on characterization (10).

5.1. Lower bound

Since (10) must be satisfied for infinitely many j's and with an unknown constant c, this is impossible to check in practice. Fortunately, this task can be reduced to a finite-time computer search [6,7]:

Algorithm 1 (Lower bound on Hölder regularity). Let $N > 0$ be the exact number of zeroes at $z = -1$ in lowpass interpolation FIR filter $G(z)$, normalized such that

$G(1) = 2$. If $G(z)$ only has zeroes at $z = -1$, stop. The Hölder regularity order is N. To estimate the amount of regularity lost due to the other zeroes, compute $F(z)$, defined as

$$G(z) = 2^{-N}(1 + z^{-1})^N F(z). \tag{11}$$

Let j be any positive integer. Compute the (positive) number

$$\beta_j = \frac{1}{j} \log_2 \max_{0 \le n < 2^j} \sum_k |f^j_{n-2^j k}|, \tag{12}$$

where f^j_n is given by

$$F^j(z) = F(z)F(z^2)\dots F(z^{2^{j-1}}). \tag{13}$$

The Hölder regularity order of $G(z)$ is at least $N - \beta_j$.

A matrix formulation can be shown [6] to be equivalent to a Hölder regularity estimate which was derived by Daubechies and Lagarias [2,4] using a very different approach. While the method they describe in [4] is only managable for very short filters $G(z)$, Algorithm 1 gives nearly optimal results (as j increases) for any filter: In fact, $N - \beta_j$ tends (at most as $1/j$) to the *optimal* Hölder regularity order as $j \to \infty$ [6]. In practice, the exact (optimal) regularity order r is generally obtained to two decimal places after $j = 20$ iterations. This algorithm can be easily implemented by recursive calls to the same small subroutine [7].

5.2. Upper bound

One possible drawback of Algorithm 1 is its exponentially increasing numerical complexity [7]. Now, assume that one retains only the values $n = 0$ and $2^j - 1$ in the computation of the maximum in (12): This results of course in a much faster algorithm. The obtained estimate clearly gives an *upper bound* of Hölder regularity as $j \to \infty$ since β_j is under-estimated. We give here the matrix formulation of this algorithm, which simplifies to the computation of a spectral radius of one matrix [6]:

Algorithm 2 (Sharp Hölder regularity upper bound). Let $G(z)$, $F(z)$, $N > 0$ be as in Algorithm 1 and let $K > 1$ be the length of $F(z)$. Form the matrix $\mathbf{F} = (\mathbf{F}_{i,j})$, $0 \le i, j \le K - 2$, defined by

$$\mathbf{F}_{i,j} = f_{2i-j+1} \tag{14}$$

and compute its spectral radius ρ. The Hölder regularity order of $G(z)$ is bounded by $N - \max(|f_0|, |f_{K-1}|, \rho)$.

The resulting estimates are very close to the optimal Hölder regularity order, as seen in Fig. 3.

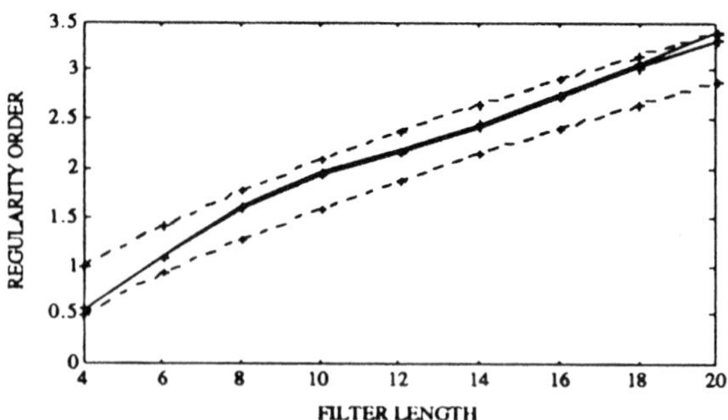

Figure 3. Comparison of regularity estimates: Sobolev lower and upper bound (dashed). Hölder upper and lower bounds (solid) for Daubechies filters given in [3].

6. Conclusion

The method presented here, which characterizes regularity on discrete-time sequences, was found to be powerful: We have provided regularity estimates that are, in contrast with earlier ones [2,3,4], easily implementable, optimal, and of general applicability. Local regularity [4] can also be studied as alternatives to global regularity using this method. [7].

REFERENCES

[1] M. Antonini, M. Barlaud, P. Mathieu, and I. Daubechies, "Image coding using vector quantization in the wavelet transform domain," in *Proc. ICASSP'90*, (Albuquerque, NM), pp. 2297–2300.

[2] A. Cohen and I. Daubechies, "Non-separable bidimensional wavelet bases," *Revista Matematica Iberoamericana*, 1992. To appear.

[3] I. Daubechies, "Orthonormal bases of compactly supported wavelets," *Comm. Pure Applied Math.*, vol. XLI, no. 7, pp. 909–996, 1988.

[4] I. Daubechies and J. Lagarias, "Two-scale difference equations II. Local regularity, infinite products of matrices and fractals," *SIAM J. Math. Anal.*, 1990. To appear.

[5] S. G. Mallat and S. Zhong, "Compact image coding from edges with wavelets," in *Proc. ICASSP'91*, (Toronto, Ontario), pp. 2745–2748.

[6] O. Rioul, "Simple regularity criteria for subdivision schemes," *SIAM J. Math. Anal.*, 1991. To appear.

[7] O. Rioul, "Regular wavelets: A discrete-time approach," in preparation.

[8] O. Rioul and M. Vetterli, "Wavelets and signal processing," *IEEE Signal Processing Magazine*, vol. 8, no. 4, pp. 14–38, Oct. 1991.

[9] J. W. Woods, Ed., *Subband Image Coding*, Kluwer Academic, 1991.

SIGNAL PROCESSING VI: Theories and Applications
J. Vandewalle, R. Boite, M. Moonen, A. Oosterlinck (eds.)

CERTAIN ASPECTS OF FIXED POINT FFT AND WFTA ROUNDING ERRORS COMPARISON

Ewa ŁUKASIK

Institute of Computing Sciences, Technical University of Poznań,
PL-60-965 Poznań, ul. Piotrowo 3a, Poland

Ryszard STASIŃSKI

Institute of Electronics and Telecommunications, Technical University of Poznań,
PL-60-965 Poznań, ul. Piotrowo 3a, Poland

After the revision of to date results of fixed point FFT and WFTA precision comparison some new conclusions are presented based on the experimental investigation. Several aspects of the problem are sketched out. The stress is laid on statistical parameters of the rounding errors. The useful measure is introduced: difference of accurate bits of FFT and WFTA.

1. INTRODUCTION

Numerical errors introduced by limited word length became a factor of special interest for algorithms used in signal processing. Their value coincides with the structure and computational speed of algorithms, as well as with the hardware design. Fixed point realizations still find wide acceptance in practice, mainly when there is a need of fast realizations. Algorithms discussed in this paper concern computing the Discrete Fourier Transform given by the formula:

$$X(k)=\sum_{n=0}^{N-1} x(n)\cdot w_N^{nk} \qquad (1)$$

$$w_N= \exp(-j2\pi/N)$$

$$n,k \in \{0,..,N-1\}$$

There is a large literature treating of the Fast Fourier Transform (FFT) rounding errors. The analysis of the errors of Winograd Fourier Transform Algorithm (WFTA), which is the fastest one, has been the subject of the works reported in [1] and then in [4,5,6]. In this paper we present a comparison of errors introduced by thoroughly prepared (optimized) fixed point WFTA and FFT algorithms for real data in two's complement arithmetics.

The errors of special interest are:

- random errors resulting from word length reduction after multiplication and/or addition/ subtraction and shift, assuming, that input signals are not deterministic,
- errors when input signals are deterministic, three various methods of quantization.

2. THE CHOICE OF ALGORITHMS

Complex radix-2 FFTs, in both versions - decimation in time (DIT) and decimation in frequency (DIF) are the most widely used algorithms for computing Discrete Fourier Transform. On the other hand split-radix FFT [3] is the one with minimal computational complexity among all FFTs and minor rounding errors, esp. in its real valued DIF version. The structure of both algorithms is modular and easy to implement. The main problem concerning their fixed point realizations accuracy is only multiplication by the coefficient equal to 1, which can be easily skipped over, and the place of scaling. These algorithms were chosen as the representatives for the comparison with WFTA from the point of view of rounding errors magnitude.

The Winograd Fourier Transform Algorithms, being in general more complicated than FFT, have a very important advantage of the least number of multiplications among all the DFT algorithms. This feature is still very important because generally it makes them faster then other DFTs especially taking into account the fact, that in all kinds of implementations the cost of executing multiplication is bigger than that of addition. Although the structure of WFTAs causes that rounding errors are unavoidably bigger [7] but there are more possibilities to alter their structure in order to diminish them [5]. The multiplications in WFTA are placed in the middle of the algorithm. In 'classical' WFTAs [2] all the multiplications are the scalar ones. In the algorithms examined in this paper some of the multiplications are scalar and some of them are polynomial products (PP) modulo z^2+1. This was one of the measures undertaken in order to construct WFTA-like algorithms with diminished values of rounding errors in fixed point realizations. Basic modules for these algorithms were as follows:

- $N=2^t$ - point module, the simplest, requiring the use of PP mod z^2+1 starting from t=4, causing the smallest rounding errors;

- N=15- point module, constructed with the 3-point and 5-point small DFTs; 5-point one is based on the 4-point circular convolution, in which PP mod z^2+1 has to be computed; in 3-point DFT all the multiplications are scalar;
- N=17- point module, being the most complicated and the most erroneous one; it is built of a 16-point circular convolution, where PP mod z^8+1 has to be developed in five stages to get PP mod z^2+1 inside; a rather big expansion of the algorithm is observed (effect of algorithm lengthening and growing its dimension in internal stages), resulting in greater rounding errors.

The above modules may form two subclasses of algorithms, having the following sizes:

1. $N=2^t \cdot 15$ - point DFTs, eg N=30, 60, 120, 240,
2. $N=2^t \cdot 255$ - point DFT, eg.N= 510, 1020, 2040, 4080 ...

These sizes are compatible to radix-2 and split radix FFT algorithms sizes i.e. $N=2^s$ and they are consistent with those specified in [7] as the least erroneous. Smaller computational complexity is their advantage over FFT.

In fixed point realizations all three mentioned above algorithms of computing DFT have some characteristic features:

- all data and coefficients are the fractional numbers,
- scalings by the factor 2 is implemented at each stage of addings to ensure keeping the intermediate results as fractions (that results in $1/N \cdot X(k)$ instead of original definition (1)),
- overscaling (dividing a sum of r components by the number z>r) never occurs.

3. FFT AND WFTA ROUNDING ERRORS COMPARISON

3.1. Random rounding errors

The important factor of fixed point rounding error comparison is the quantization method. For the analysis presented in this paper three of them were used:

- rounding (R),
- truncation (T),
- random rounding (RR), where the value 0.5 is randomly rounded or truncated.

The most suitable aspect of FFT and WFTA rounding errors comparison is mean square error. It may be computed analytically using the method presented in [6] or by simulations for a large number of experiments. The second method was chosen to perform the comparison. For three DFT algorithms presented above the same experiments were carried out. For the Gaussian noise as the input signal the given algorithm was used twice: first in its floating point version, the results X(k) being treated as the exact ones and then in simulated fixed - point version, that operated on 16 - bit data words, giving the erroneous result $[X(k)]$. The squared difference:

$$[\Delta X(k,i)]^2 = \{1/N \cdot X(k,i) - 1/N \cdot [X(k,i)]_Q\}^2 \quad (2)$$

k- number of DFT sample, k = 0,1,...,N-1
i- number of experiment, i = 0,1,...,M-1

was the error measure. For big number of experiments (M=1024) we get the information about the statistical features of each $[X(k)]_Q$ error by computing the mean value of all M values of $[\Delta X(k,i)]^2$. The results are presented in rising order on graphs for the representatives of mentioned above sizes of FFT and WFTA, i.e. N=128 and 120, N=1024 and 1020 points (Fig.1). Additionally we introduce two criteria for algorithms comparison:

- maximal mean-square error of all the results (P_{mx}),
- average mean-square error of all the results (P).

The most implementation-oriented measure of the accuracy is the difference in number of accurate bits of FFTs and WFTAs for both criteria and three quantization methods, presented in table 1.

Table 1. Comparison the WFTAs and FFTs precision expressed as difference of number of accurate bits

N	error	Arithmetics/ Algorithm	R	RR	T
120/128	P	WFTA/FFT	-0.27	0.08	0.09
		WFTA/FFTs-r	-0.26	0.08	0.36
	P_{mx}	WFTA/FFT	-0.40	0.19	0.14
		WFTA/FFTs-r	-0.51	0.18	0.24
255/256	P	WFTA/FFT	0.72	1.11	1.35
		WFTA/FFTs-r	0.88	1.09	1.64
	P_{mx}	WFTA/FFT	0.07	1.20	1.37
		WFTA/FFTs-r	-0.03	1.16	1.41
1020/1024	P	WFTA/FFT	0.98	1.39	1.62
		WFTA/FFTs-r	1.16	1.37	1.91
	P_{mx}	WFTA/FFT	0.24	1.64	1.53
		WFTA/FFTs-r	0.08	1.63	1.62

R-rounding, RR-random rounding, T-truncation, s-r - split-radix

The main conclusions of this comparison are:

- optimization of WFTA algorithms leads to the best results for rounding; - in general the differences do not exceed 2 bits,
- experimental results for WFTA are concordant with the model presented in [7], whereas the model for FFTs is too pessimistic. It does not include the effect of noise attenuation caused by multiplication by fractional coefficients in every step of the algorithm. In WFTA such a noise reduction performs only once.

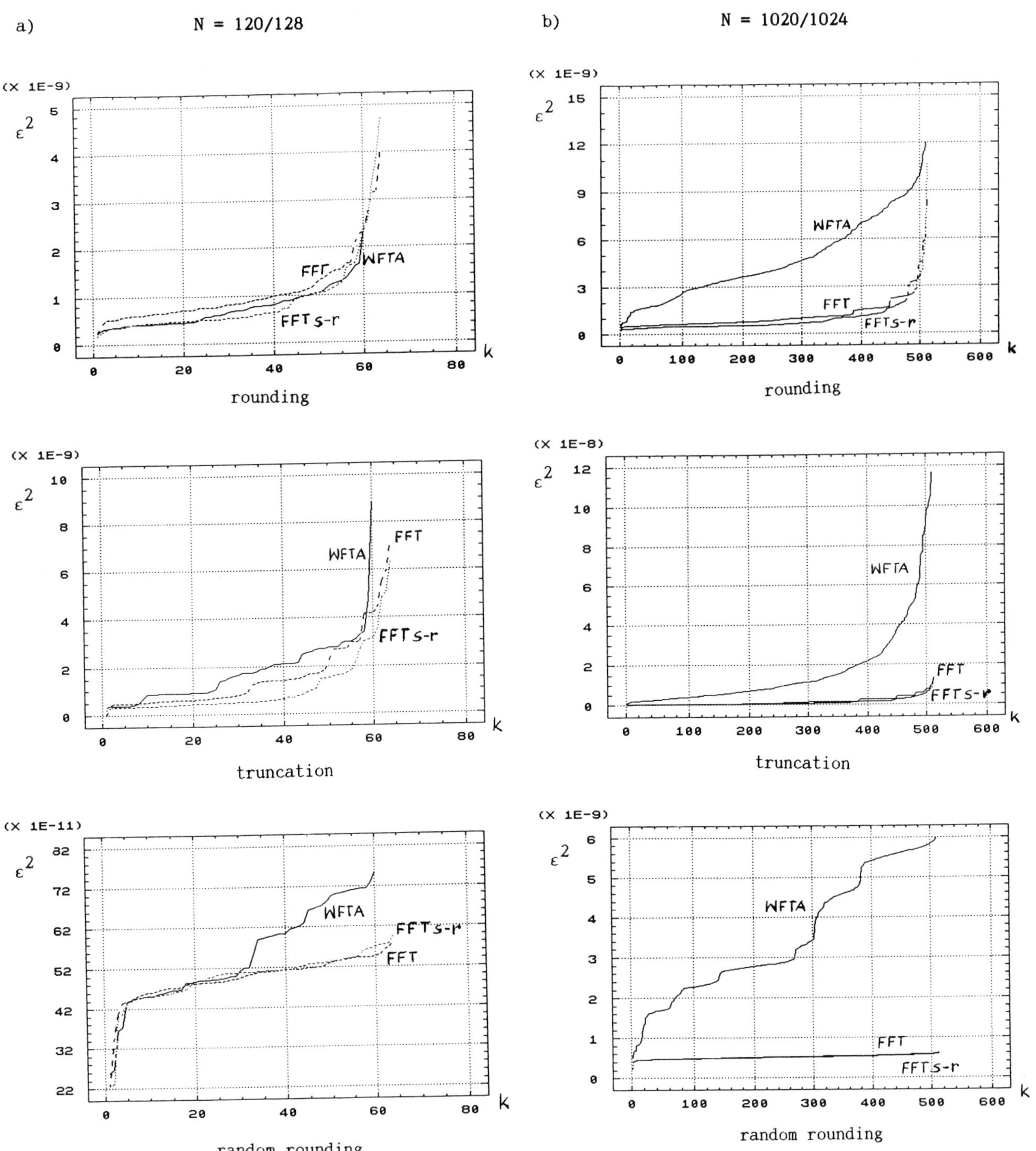

Fig.1. Mean-square errors (ε^2) of N output samples of three various DFT algorithms and three quantization methods in rising order: a) N=120/128, b) N=1020/1024

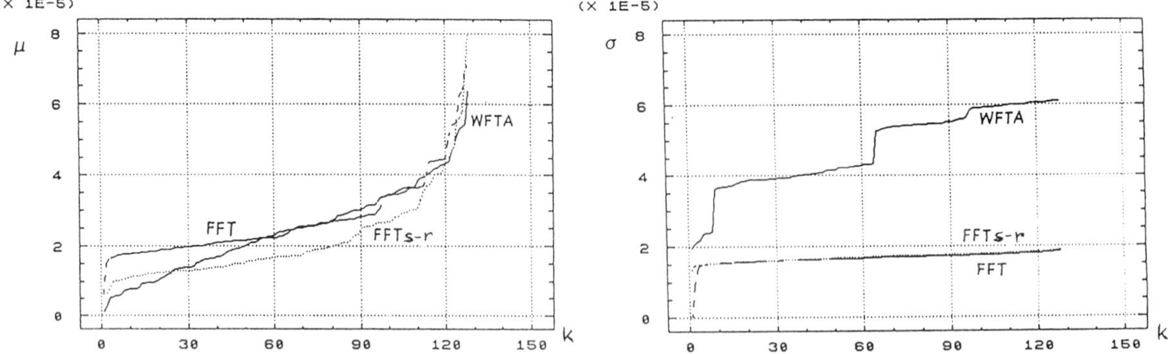

Fig.2. Error mean value (μ) and standard deviation (σ) for three DFT algorithms, N=255/256, rounding.

Fig.2. presents the example of mean value (offset) and standard deviation of FFTs and WFTA (N=255, rounding). Since the offset is the significant error component (for truncation even more) it may be compensated at the output of the algorithm. However standard deviation, being rather stable for all output samples, remains higher for WFTA than for both FFTs for the same reasons as mentioned above.

3.2. Other aspects of FFT and WFTA accuracy comparison

A question may be raised, concerning the worst-case input function causing the biggest rounding error at the DFT output. Some useful functions for algorithms rounding error comparison are sine, cosine or impulse functions. However they do not involve in computation all the algorithm graph paths and nodes, being the rounding error sources, i.e. some of them are omitted and final error does not take maximal value. Only for random input function this may not occur. Indeed the highest rounding errors we have got for realizations of Gaussian noise as input functions. A sequence of 510 random N-sample realizations was examined for all N to catch these giving maximal error values. In Table 2 the results are listed. Evidently errors of WFTA are bigger than of any of FFTs.
These results confirm previous conclusions.

4. CONCLUSIONS

All efforts leading to the minimization of rounding errors in WFTA did not give results compensating the structural features determining higher values of rounding errors of WFTA than for FFT.
These structural features are:
- lengthening and growing internal dimensions of the algorithm,
- little influence of coefficients on the noise reduction.

They are involved in all kinds of errors. The results are slightly better than general ones given in [1].

Table 2. Rounding errors for N-point realiza - tions of Gaussian noise as the input

N	error	WFTA	FFT	FFTs-r
120/128	ρ	1.04	1.14	1.06
	ρ_{mx}	2.48	2.80	3.04
255/256	ρ	2.09	1.18	1.11
	ρ_{mx}	6.67	3.28	3.22
1020/1024	ρ	2.52	1.21	1.14
	ρ_{mx}	9.46	4.10	4.43

$\rho = \sqrt{P}/Q, \quad \rho_{max} = \sqrt{P_{max}}/Q, \quad Q = 2^{-(b-1)}$

REFERENCES

[1] Patterson R.W., McClellan J.H., 'Fixed point error analysis of Winograd -Fourier transform algorithm', IEEE Trans. ASSP, vol. ASSP-26, 1978, pp.447-455.

[2] Nussbaumer H.J. 'Fast Fourier Transform and Convolution Algorithms', Springer - Verlag, 1981

[3] Duhamel P., 'Implementation of "Split-Radix" FFT Algorithms for Complex, Real, and Real-Symmetric Data', IEEE Trans. ASSP, vol. ASSP-34,no 2, 1986, pp. 285-295.

[4] Stasiński R., Łukasik E., 'Minimization of rounding errors in WFTA programs', Proc ICASSP, 1988, New York, pp. 1423-1426.

[5] Łukasik E., 'Experimental results in minimizing rounding errors in fixed-point WFTA programs'- Proc. EUSIPCO-90, Barcelona, pp. 1543-1546.

[6] Łukasik E., 'Analytical computation of rounding errors in fixed point Winograd Fourier Transform Algorithm', Proceedings of the Int. Conf. DSP-91, Florence 1991, pp. 55-58.

[7] Stasiński R., 'On the reduction of nesting algorithms rounding errors', submitted to EUSIPCO-92.

SIGNAL PROCESSING VI: Theories and Applications
J. Vandewalle, R. Boite, M. Moonen, A. Oosterlinck (eds.)
1992 Elsevier Science Publishers B.V.

A NOVEL FAST FIR FILTERING ALGORITHM AND ITS IMPLEMENTATION*

Sau-Gee Chen & Ren-Jer Tsai

Department of Electronics Engineering, National Chiao Tung University
Hsinchu, Taiwan, ROC

A class of novel convolution algorithms are proposed which reduce the number of multiplications by more than 50%, and in some cases decrease the number of additions by more than 50% per output point, compared with the direct convolution. As such, their roundoff noises are greatly reduced. Under the same direct-form L_1-norm scaling scheme, the problem of increased dynamic range is irrelevant. The algorithm is well applied to IIR, and 2-D convolutions with more computation reductions both in multiplication and addition counts. When implemented, the algorithms maintains architectural regularity.

1. INTRODUCTION

There are numerous fast convolution algorithms in the literature [1,2,3,4], which can be categorized as two different classes. The first one is the well-known transform-domain FFT approach. Though asymptotically the FFT-based algorithms are optimal, they require dedicated hardware implementation for bit-reversed data routing, in addition that every single output sample may not be available at the instant.

The second one is done directly in the temporal domain. Some of the popular algorithms include the iterated FIR filtering [2,5] that builds long convolution iteratively by using short Winograd convolution algorithms. Asymptotically, the complexity for the filtering operations with N by N filter kernel costs $N^{1.585}$ multiplications per output point which is inferior to FFT approach for long convolution length. Another time-domain fast convolution is done by multirate processing [3,4] of decimating output signals. the original N multiplications required for a single output is reduced to $3N/4$ multiplications due to the decomposition of the convolution as three shorter convolutions each with size of $N/2$ that produce two output points at the same time. The algorithm can also be iteratively decomposed to its fullest, but accordingly there are disadvantages of using multi-clocking system.

Here a novel temporal-domain filtering algorithm is proposed which reduces number of multiplications from N to $N/2$ at the expense of increased additions from $N-1$ to $3N/2+1$. The saving of $N/2$ multiplication operations dwarfs the penalty of $N/2$ extra addition operations. Moreover, the new algorithm preserves the regular FIR filtering structures. Compared with the multirate convolution algorithm, the algorithm reduces $N/4$ multiplications, while costs $3N/4$ more additions. More reductions in multiplication and even addition counts are obtained when the 1-D algorithm is extended to 2-D convolutions. For instance, in the new 2-D algorithm for a 2-D quadrisymmetric convolution, the numbers of multiplications and additions required are only respectively 1/8 and 3/8 that of by direct convolution.

In what follows, we shall first elaborate on the new 1-D convolution algorithm. Then, extension of the algorithm to various convolutions are detailed. Finally, example realizations for the algorithms and their implementation considerations are discussed.

2. THE ALGORITHM

Given the input sequence $x(n)$ and filter coefficients $h(n)$, the convolution output sequence $y(n)$ can be reformulated as three parts:

$$\begin{aligned} y(n) &= \sum_{k=0}^{N-1} x(n-k)h(k) \\ &= \sum_{k=0}^{N/2-1} [x(n-2k)+h(2k+1)][x(n-2k-1)+h(2k)] \\ &\quad - \sum_{k=0}^{N/2-1} h(2k)h(2k+1) - \sum_{k=0}^{N/2-1} x(n-2k)x(n-2k-1) \end{aligned}$$

The first part has complexity of $N/2$ multiplication and $3N/2-1$ addition operations per output point. The second one can be precalculated and fixed during the rest of entire convolution operation, thus contributes only one addition operation.

* This work was supported in part by National Science Council under Contract NSC-81-0404-E009-130.

The term

$$\sum_{k=0}^{N/2-1} x(n-2k)x(n-2k-1)$$

can be computed by defining a recursive sum-of-product equation as

$$p(n) = p(n-2) + x(n)x(n-1) - x(n-N)x(n-N-1)$$
$$\text{where } x(k) = 0 \text{ for } k < 0,\ p(k) = 0 \text{ for } k < 0$$

thus contributes only one multiplication and two additions. The total computational complexity is thus $N/2+1$ multiplications and $3N/2+3$ additions.

3. 2-D ALGORITHMS

3.1 Nonseparable 2-D Convolution Algorithm

A nonseparable 2-D convolution operation can be considered as a stacking of 1-D convolutions, and the operation can be decomposed as

$$y(i,j) = \sum_{k=0}^{N-1}\sum_{l=0}^{M-1} x(i-k, j-l)h_k(l)$$
$$= [\sum_{k=0}^{N-1}\sum_{l=0}^{M/2-1} z_1(i,j,k,l)z_2(i,j,k,l)] - p(i,j) - C_h$$

where

$$z_1(i,j,k,l) = x(i-k, j-2l) + h_k(2l+1),$$
$$z_2(i,j,k,l) = x(i-k, j-2l-1) + h_k(2l)$$
$$p(i,j) = \sum_{k=0}^{N-1}\sum_{l=0}^{M/2-1} x(i-k, j-2l)x(i-k, j-2l-1)$$
$$= p(i-1,j) + p_{new}(i,j) - p_{old}(i,j)$$
$$p_{new}(i,j) = \sum_{l=0}^{M/2-1} x(i, j-2l)x(i, j-2l-1)$$
$$= p_{new}(i, j-2) + x(i,j)x(i,j-1) - x(i, j-M)x(i, j-M-1)$$
$$p_{old}(i,j) = \sum_{l=0}^{M/2-1} x(i-N, j-2l)x(i-N, j-2l-1)$$
$$= p_{new}(i-N, j)$$
$$C_h = \sum_{k=0}^{N-1}\sum_{l=0}^{M/2-1} h_k(2l)h_k(2l+1) = \text{constant}$$

If signals are processed rowwise, then for an $M*N$ 2-D filter kernel, the number of multiplications and additions required for a single output point will asymptotically be $MN/2+1$ and $3MN/2+5$ respectively.

3.2 2-D Separable Convolution Algorithm

A 2-D separable convolution can be executed as a concatenation of two 1-D algorithms, i.e. by the nested algorithm as follows,

$$y(n,m) = \sum_{j=0}^{N-1}\sum_{k=0}^{M-1} x(n-j, m-k)h_1(k)h_2(j)$$
$$= \sum_{j=0}^{N-1} z(n-j,m)h_2(j)$$
$$= \sum_{k=0}^{N/2-1} [z(n-2j,m) + h_2(2j+1)][z(n-2j-1,m) + h_2(2j)]$$
$$- \sum_{k=0}^{N/2-1} h_2(2k)h_2(2k+1) - \sum_{k=0}^{N/2-1} z(n-2j,m)z(n-2j-1,m)$$
$$\text{where } z(n-j,m) \equiv \sum_{k=0}^{M-1} h_1(k)x(n-j, m-k)$$

which amounts to $(N/2+1)(M/2+1)+1$ multiplications and $(3MN+5N)/2+2$ additions. Compared with the direct convolution, there is a close to 4-fold reduction in multiplications, at the expense of 1.5-fold increase in additions.

3.3 2-D Quadrisymmetry Convolution Algorithm

Frequently, we require that a 2-D filter be symmetric so as to have linear-phase response. The algorithm gives a much reduced computation for quadrisymmetry is as follows,

given $h(k,l)=h(M-k-1,l)=h(k,N-l-1)=h(M-k-1,N-l-1)$

$$y(i,j) = \sum_{k=0}^{M-1}\sum_{l=0}^{N-1} h(k,l)x(i-k, j-l)$$
$$= \sum_{k=0}^{M/2-1} y_k(i,j) + \sum_{k=M/2}^{M-1} y_k(i,j)$$
$$y_k(i,j) \equiv \sum_{l=0}^{N/2-1} h(k,l)[x(i-k, j-l) + x(i-k, j-N+l+1)]$$

By taking advantage of the symmetry, the equation can be further simplified as

$$y(i,j) = \sum_{k=0}^{M/2-1} y_k(i,j) + \sum_{k'=0}^{M/2-1} y_{k'}(i+2k'-M+1, j)$$

The second term of RHS had already been computed for the earlier indices i. The first term of RHS can be similarly decomposed as before,

$$\sum_{k=0}^{M/2-1} y_k(i,j) = [\sum_{k=0}^{M/2-1}\sum_{l=0}^{N/4-1} x_1(i,j,k,l)x_2(i,j,k,l)] - C_h - p(i,j)$$
$$x_1(i,j,k,l) = x_3(i,j,k,l) + h(k, 2l+1)$$
$$x_2(i,j,k,l) = x_4(i,j,k,l) + h(k, 2l)$$
$$x_3(i,j,k,l) = x(i-k, j-2l) + x(i-k, j-N+2l+1)$$
$$x_4(i,j,k,l) = x(i-k, j-2l-1) + x(i-k, j-N+2l+2)$$
$$p(i,j) = \sum_{k=0}^{M/2-1}\sum_{l=0}^{N/4-1} x_3(i,j,k,l)x_4(i,j,k,l)$$
$$= p(i-1,j) + p_{new}(i,j) - p_{old}(i,j)$$

$$p_{new}(i,j) = \sum_{l=0}^{N/4-1} x_3(i,j,0,l)x_4(i,j,0,l)$$

$$p_{old}(i,j) = \sum_{l=0}^{N/4-1} x_3(i,j,M/2,l)x_4(i,j,M/2,l) = p_{new}(i-M/2,j)$$

The algorithm costs $MN/8+N/4$ multiplications and $3MN/8+3N/4+M/2+2$ additions. Close to eightfold reduction in multiplications and 2.7-fold reduction in additions are obtained, compared with the direct convolution.

4. IMPLEMENTATION

In hardware, the algorithms can be realized by most of the conventional FIR filter structures. One of many possible realizations for the 1-D convolution algorithm is shown in Figure 1. Figures 2 and 3 depict possible structures for an 2-D 4×4 quadrisymmetry filter. All the structures maintain high degree of regularity.

4.1 Word Length

For the 1-D realizations, assumed (B+1)-bit fixed-point representation and stationary white uncorrelated rounding noises [6] for the multiplications, then the variance of roundoff noise of this structure is equal to

$$\sigma_f^2 = (N/2+1) \times 2^{-2B}/12$$

which is half that of direct-form realization. To remedy the more severe overflow problem than the direct-form realization, the same L1-norm scaling scheme as that of direct form is sufficient to eliminate possible overflows, i.e. choose a scaling constant that confines $x(n)$ as follows,

$$\max\{|x(n)|\} < \frac{1}{\sum_i |h(i)|}$$

then it is enough that the adders are implemented with (B+1) bits as the direct form. The overflow bits in the addition operations are don't-care. In the multiplier implementations, (B+2)-bit operand size have to be allowed for the possible carry bit from the addition before multiplication and multiplication results are rounded to (B+1) bits as usual. Similar argument is also applied to 2-D convolver design.

5. CONCLUSION

Several efficient convolution algorithms are proposed in this work, which achieve more than 50% reductions in multiplication counts, and in some cases also the reduction of addition counts. The approach is well applied to 1-D or 2-D IIR filtering operations. The algorithms can be easily adapted to systolic-array-like realizations. But care must be taken in deciding the scanning schemes for 2-D input signals, which deserved further study. The algorithms are under further investigation when they are extended to other cases of filter symmetries [7]. Likewise, the algorithms can be combined with multirate approaches for possible more efficient filter realizations.

REFERENCES

1. H. J. Nussbaumer, Fast Fourier Transform and Convolution Algorithms, Springer-Verlag, Berlin/New York, 1981.
2. R. E. Blahut, Fast Algorithms for Digital Signal Processings, Addison-Wesley, Reading, MA, 1985.
3. Z. J. Mou and P. Duhamel, "Fast FIR Filtering: Algorithms and Implementations," Signal Processing, pp. 377-384, Dec. 1987.
4. M. Vetterli, "Running FIR and IIR Filtering Using Multirate Filter Banks," IEEE Trans. on ASSP, Vol. ASSP-36, pp. 730-738. May 1988.
5. K. C. Chew, R. C. North and W. H. Ku, "An Iterated FIR Filtering Using Radix-4 Serial Arithmetic For Gallium Arsenide Technology," Proc. of ISCAS, 1991, Singapore, pp. 500-503.
6. A. V. Oppenheim and R. W. Schafer, Discrete-Time Signal Processing, Prentice-Hall International, 1988.
7. Z. J. Mou, "New 2-D Symmetric FIR Filter Structures," Proc. of IEEE VSPC'91, 1991, pp. 98-101.

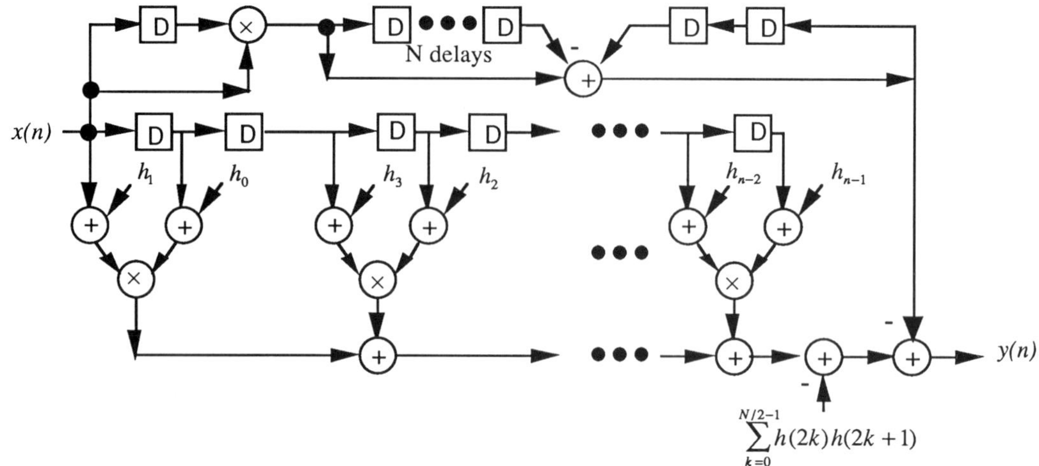

Figure 1. 1-D filter structure based on the new convolution algorithm

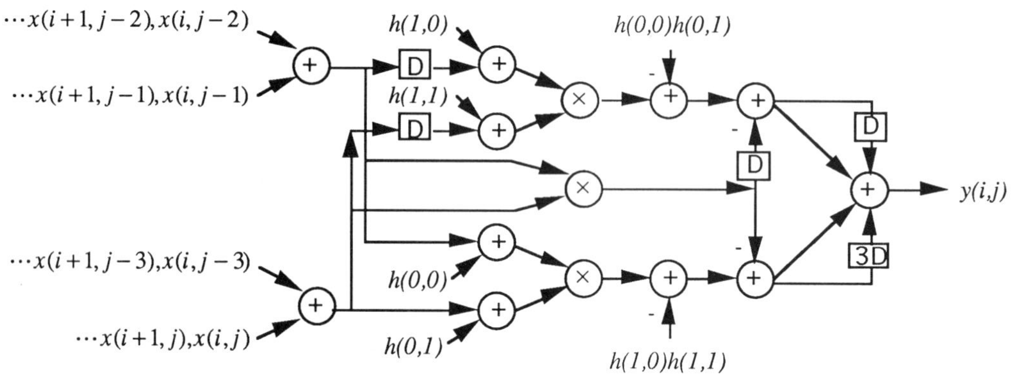

Figure 2. Filter structure for a 2-D 4*4 quadrisymmetric convolution based on the new 2-D algorithm

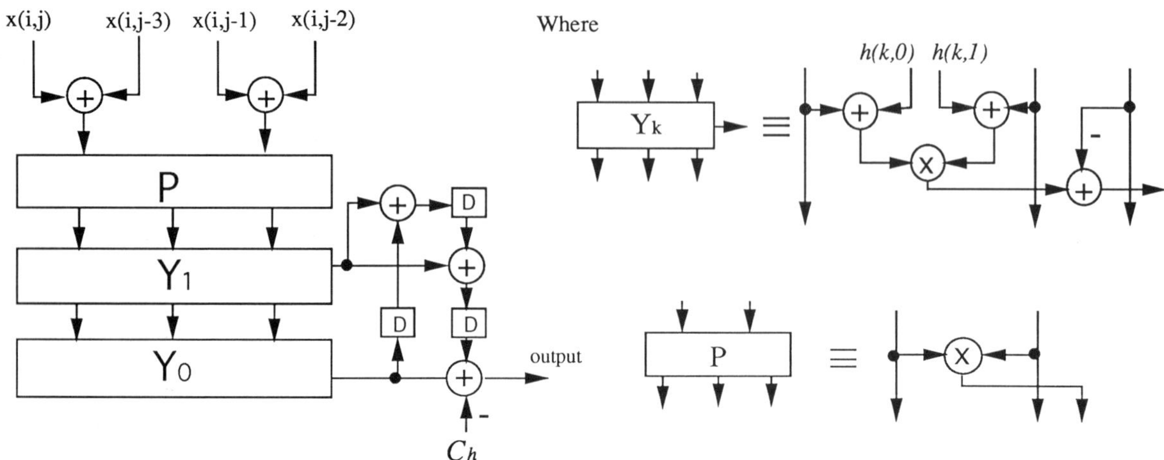

Figure 3. Another realization for 4*4 quadrisymmetric filter

SIGNAL PROCESSING VI: Theories and Applications
J. Vandewalle, R. Boite, M. Moonen, A. Oosterlinck (eds.)

A study of the filter roundoff noise in subband coding systems

RENARD Jean-Pierre (*), LEICH Henri (**)

(*) *Lernout and Hauspie Speechproducts*
Rozendaalstraat 14
8900 Ieper (Belgium)

(**) *Faculté Polytechnique de Mons*
Boulevard Dolez 31
7000 Mons (Belgium)

Tel : 32 65 374132 Fax : 32 65 374300

ABSTRACT

The influence of filter roundoff noise in subband coding systems is examined in this paper. First of all, we present a methodology which allows to determine the energy propagation of signals in tree-structured systems. Next, the output level of filter roundoff noise is determined as a function of the number of subbands, for the fixed-point arithmetic (FXP) on one hand, and for the floating-point arithmetic (FLP) on the other hand. Results of simulation program implemented on a PC are presented in order to verify the predeterminations. Finally, practical case of listening tests where roundoff noise influence is obvious, is presented, with some possible solutions to eliminate it.

1. INTRODUCTION

Subband coding systems are nowadays widely used for sound [1,2,3] and speech processing [4]. They indeed allow a significant bit-rate reduction without altering the quality of the processed signal. These systems are based on digital signal processors (DSP) which work with a fixed-point (FXP) or floating-point (FLP) arithmetic.

Our aim is to determine the noise performances of the subband coding systems according to the arithmetic of the DSP used, and the complexity of the whole system. This problem is important as experiments show that an higher complexity yields better bit-rate reduction. Moreover, DSP producers are continuously improving the processing capacity of their chips, in terms of accuracy and speed. This allows then the realization of more complex algorithms.

Section 2 presents the architecture of the subband coding systems studied and gives the methodology which allows the determination of the signal energy propagation in these multicadence systems. In the next section, this methodology is used to compute the roundoff noise as a function of the different parameters of the subband system, as the processor arithmetic, the number of subbands,... Then, results of simulations made on a PC in order to verify theoretical predeterminations are presented and commented. Section 4 describes a practical example where roundoff noise is put in evidence by listening tests.

Conclusions are given in the last section.

2. ARCHITECTURE OF SUBBAND CODING SYSTEMS AND DETERMINATION OF ENERGY PROPAGATION

2.1 Architecture of subband coding systems

A subband coding system can schematically be divided into three main phases: the analysis, the subband processing and the synthesis.

The analysis consists in the splitting of the input signal in a certain number of signals each of which represents a determined part of the original signal spectrum, and which are called "subbands". This operation is done, in principle, by a band-pass filters bank which isolates the different spectrum parts, followed by a decimation process which preserves the original bit-rate.

At this step, the subband processing allows to reduce the bit-rate needed to store or transmit the signal. This phase is not taken into account in this paper.

Finally, the synthesis reconstructs the signal from the subbands. It is the dual operation of the analysis and consists in an interpolation process followed by a band-pass filters bank.

The analysis and the synthesis are obviously elaborated so as to preserve the original signal if there is no subband processing. This implies that "aliasing" and "imaging" effects generated respectively by the decimation and the interpolation processes, must be eliminated. This is supposed to be achieved by employing "quadrature mirror filters" (QMF) [5].

In this paper, we only consider systems which architecture is based on the "dichotomous tree structure". In these systems, the input signal spectrum is split, at the first stage, in two half-band signals. At next stage, each of these last signals are "spectrally" divided in two again, and so on. If N stages are implemented, 2^N subbands are created. This structure is the most often used in subband coders [2,3].

It also allows a very symmetric implementation of the analysis and synthesis which is, practically, obtained by the combination of elementary "direct" cells of figures 1 and 2.

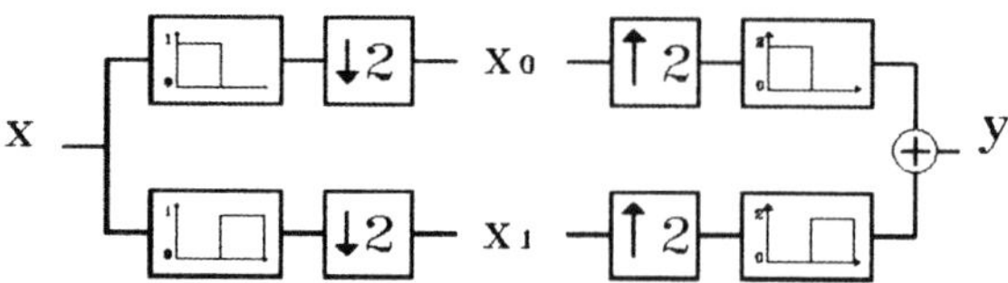

Direct analysis cell

Figure 1

Direct synthesis cell

Figure 2

The QMF filters used to implement the subband system are always of finite impulse response (FIR) as they easily yield a linear phase transmittance. The QMF property may be exploited to obtain an optimum implementation in terms of arithmetics operations number. Special filters (p_{0a}, p_{1a}, p_{0s}, p_{1s}) called "polyphase" are then used. It leads to the "polyphase" cells shown in figures 3 and 4:

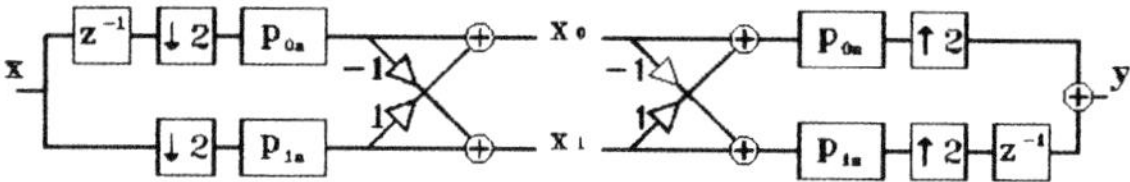

Polyphase analysis cell

Figure 3

Polyphase synthesis cell

Figure 4

2.2 Determination of the energy propagation

2.2.1 Energy estimator

In order to prepare the study of roundoff noise in subband systems, it is important to determine the energy of the different signals of the dichotomous tree structure. The particularity of this study lies in the multicadence nature of the system (many different sampling frequencies coexist).

Generally, energy E_x of a digital signal $\{x_i\}$ can be calculated with the Parseval relation [6,7]:

$\Phi = 2\pi f / f_E$ (f: analog frequency; f_E: sampling frequency)

$$E_x = \sum_{i=-\infty}^{+\infty} x_i^2$$

$$= \frac{1}{\pi} \int_{\Phi=0}^{\pi} |X(e^{j\Phi})|^2 \, d\Phi \qquad (1)$$

$X(e^{j\Phi})$ is the transmittance of $\{x_i\}$.

This relation (1) does not allow to compare signals at different sampling frequencies. However, if a correcting factor is taken into account, comparison is possible. For example, if E_x is the "digital" energy at the sampling frequency f_E and E_{x1} at f_{E1}, the correcting factor K is given as follows:

$$K = f_E / f_{E1}$$

$$E_x = K \, E_{x1}$$

In practise, as the processed signals are of finite duration, we do not compute the global energy. We prefer a mean energy estimator (on n samples) called "variance" σ_x^2 which is given by the following formula :

$$\sigma_x^2 = \frac{1}{n} \sum_{i=1}^{n} x_i^2$$

It can be shown that the variance can be freely used to compare (mean) energy of signals.

2.2.2 Propagation of the energy in the cells

The energy computation of the signals issued from the filters cells is straightforward if these filters are considered as ideal. However, in the cases of the decimation and the interpolation, the determination of the input-output energy relation is not obvious. Relations between transmittances of the input and output signals (see [10]) lead to the following results for the interpolation:

d: input signal at $f_{E/2}$ $\quad (\Phi' = 2\pi f / (f_E/2))$

y: output signal at f_E $\quad (\Phi = 2\pi f / f_E)$

$K = 1/2$

$E_y = E_d/2$

For the decimation, the relations are the following:

x: input signal at f_E $\quad (\Phi = 2\pi f / f_E)$

d: output signal at $f_{E/2}$ $\quad (\Phi' = 2\pi f / (f_E/2))$

$K = 2$

$E_d = E_x + 2\,COV_x$

where:

COV_x: correlation between the low frequency half-band spectrum and the high frequency half-band spectrum.

Practical experiments made on different types of music signals have shown that COV_x is always negligible. It is then easy to compute the energy relations between all the signals of each direct and polyphase cells.

3. ROUNDOFF NOISE IN SUBBAND SYSTEMS

In this section, we are going to determine the characteristics of the roundoff noise in the subband systems and its influence on the signal to noise ratio.

3.1 Generality on the roundoff noise

The implementation of the subband analysis and synthesis implies the utilization of DSP. These processors use the fixed-point arithmetic (FXP) or the floating-point arithmetic (FLP). Whatever the arithmetic chosen, the finite length of represented word leads to errors, called "roundoff errors". For that reason, each multiplication operator constitutes an error source in both FXP and FLP. However, addition operator only creates errors in FLP.

Real operator can be modelized [8] as ideal operator followed by a roundoff error source of variance σ_{FXP}^2 in the FXP case, $\sigma_{FLP}^2 \sigma_s^2$ in the FLP case (σ^2_s is the variance of the signal being rounded). σ_{FXP}^2 and σ_{FLP}^2 are only depending on the length of representation (word for FXP, mantissa for FLP).

The influence of the roundoff noise on the signal to noise ratio at the output of the system is going to be investigated. This signal to noise ratio (*SNR*) is defined as follows:

$$SNR = 10 \log \frac{E_y}{E_e}$$

E_y = energy of the output signal

E_e = energy of the error signal at the system output.

3.2 Roundoff noise of systems implemented with the direct cells

The roundoff errors sources of the two direct cells are located at the output of each filter. Note that each FIR-QMF (direct or polyphase) filter introduces only one source because we suppose that the filter is implemented following the direct canonical structure [7] and that the DSP used is able to perform successive "multiply and add" with high accuracy (those operations are made with longer word/mantissa length of representation).

By hypothesis [8], all the roundoff sources are not correlated with the signal.

It can thus be shown that the *SNR* for FXP and FLP arithmetics in the case of *N* stages subband system, is given by the following relations :

$$SNR_{FXP} = 10 \log \frac{\sigma_x^2}{\sigma_{FXP}^2\, 4(2^N - 1)} \quad \text{(dB)} \qquad (2)$$

$$SNR_{FLP} = -10 \log (2N\, \sigma_{FLP}^2) \quad \text{(dB)} \qquad (3)$$

3.3 Roundoff noise of systems implemented with the polyphase cells

For the polyphase cells, noise sources are located at the filters output. More sources exist in the FLP case and are situated at the output of each addition operator.

Following the same consideration as in section 3.2, it is possible to demonstrate:

$$SNR_{FXP} = 10 \log \frac{\sigma_x^2}{\sigma_{FXP}^2\, 5(2^N - 1)} \quad \text{(dB)} \qquad (4)$$

$$SNR_{FLP} = -10 \log (4N\, \sigma_{FLP}^2) \quad \text{(dB)} \qquad (5)$$

By comparing relations (2) and (4), on one hand, and relations (3) and (5), on the other hand, we conclude that the roundoff noise performances of the subband system depends of the implementation chosen.

3.4 Simulations

In order to verify the *SNR* presented in section 3.2 and 3.3, we have developed a simulation program on a PC. This program is able to compute noise variances by comparing results obtained on one hand, by an "ideal system" where arithmetic is realized with FLP "double" accuracy, and on the other hand, by a "real system" where arithmetic is realized with FLP "float" accuracy (σ_{FLP}^2 = -151.93 dB) or FXP "int" accuracy (σ_{FXP}^2 = -10.79 dB).

In the tables 1 and 2, we compare the results obtained using the relations (2) to (5) and by simulations (σ_x^2 = 70.57 dB)

N	DIRECT		POLYPHASE	
	Theory	Experiments	Theory	Experiments
1	75.35	75.36	74.37	74.37
2	70.57	70.53	69.60	69.49
3	66.89	66.90	65.92	65.75
4	63.58	63.56	62.61	62.46
5	60.43	60.46	59.46	59.32

Table 1 (FXP)

N	DIRECT		POLYPHASE	
	Theory	Experiments	Theory	Experiments
1	148.94	148.90	145.93	145.01
2	145.93	145.44	142.92	142.21
3	144.17	144.21	141.16	140.66
4	142.92	142.94	139.91	139.52
5	141.95	141.91	138.94	138.52

Table 2 (FLP)

These last tables show that experimental results are in perfect concordance with the theoretical predeterminations, excepted for the polyphase structure in FLP. A more detailed study shows that this difference comes from the analysis cells and is an effect of the correlation between signals processed by the addition operators.

4. LISTENING TESTS AND FIXED-POINT ARITHMETIC

The methodology presented in section 3 shows how to compute the output *SNR* when no subband processing occurs. This *SNR* can thus be considered as an upper limit of the *SNR* obtainable in subband coders.

As the *SNR* decreases as the number *N* of stages of the dichotomous tree structure increases, it may exist a limit above which noise in audio subband coders becomes audible.

We have put this problem in evidence for FXP arithmetic and polyphase implementation with a particular piece of music (N° 65 of the SQAM-EBU: R. STRAUSS).

The experiments have been led with a restitution station constituted by digital amplifier and headphones. The amplifier is calibrated in order that when we impose a certain gain "*G*", the acoustic level *L* of restitution in the headphone is related to the numeric variance σ_x^2 of the piece of music by the following relation :

$$L(dB) = G(dB) + \sigma_x^2 \ (dB)$$

Listening tests have shown that, when *G* is equal to 31 dB, a shrill noise becomes audible when 5 stages are implemented. This noise is not audible when 4 stages or less are used.

Obviously, the noise origin comes from roundoff errors. Moreover, the shrill nature of the noise is explained with the help of the perceptual property of the ear known as the auditory threshold [9]. Indeed, only a small part of the noise spectrum of a 5 stages system (from 2500 Hz to 5000 Hz) is above the auditory threshold (cf. figure 5).

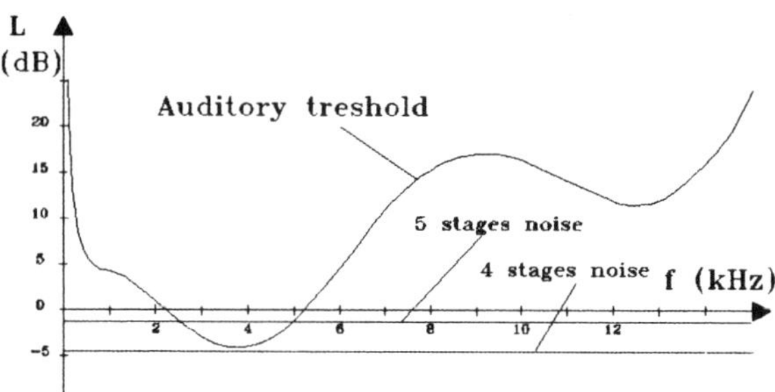

Filter roundoff noise spectrum

Figure 5

To avoid these problems, many solutions can be found. We may use bigger length of representation, or use the FLP arithmetic. It is even possible to create subbands with different "spectral width" in order that the spectrum noise matches the auditory threshold.

5. CONCLUSIONS

A methodology is presented for the computation of the output filter roundoff noise of subband coding system based on "dichotomous tree structure". The computation of the output roundoff noise is made as a function of the number of subbands for the FXP and FLP representation, for the direct and polyphase implementation of the system. This allows to compare the influence of each feature. Moreover, in the case of fixed-point representation, listening tests show that the roundoff noise can be considered as an upper limit of coding noise performances of the subband systems.

REFERENCES

[1] G. THEILLE, G STOLL and M. LINK : " Low Bit-rate Coding of High-quality Audio Signal: An introduction to the MASCAM System", Advanced digital techniques for UHF satellite sound broadcasting,EBU,August 1988.

[2] Y-F DEHERY : "MUSICAM Source Coding", Proceedings of the 10th International AES Conference, London, 7th-9th September 1991.

[3] S.M.F. SMYTH and J.V. MC CANNY : " High Quality Music Coding at 4 bits", IEEE Electronic Letters, January 1988.

[4] CCITT G722 Norm

[5] C. GALAND : "Codage en sous-bandes: Théorie et application à la compression numérique du signal de parole", Thèse de l'Université de Nice, 1983.

[6] R.E. CROCHIERE and L.R. RABINER : "Multirate Digital Signal Processing", Prentice Hall, Inc., Englewood Cliffs, New Jersey, 1983.

[7] BOITE, H. LEICH : "Les Filtres Numériques", Editions Masson.

[8] R. BOITE, XIAN-LIANG HE, J-P. RENARD : "On the Floating-point Realization of Digital Filters", Eusipco 88, Grenoble

[9] E. ZWICKER, R. FELDKELLER : "Psychoacoustique : L'oreille récepteur d'information", 1981, Editions Masson.

SIGNAL PROCESSING VI: Theories and Applications
J. Vandewalle, R. Boite, M. Moonen, A. Oosterlinck (eds.)

DESIGN OF FINITE WORDLENGTH 2-D IIR FILTERS USING SIMULATED ANNEALING

Jan Radecki, Janusz Konrad and Eric Dubois

Institut National de la Recherche Scientifique
INRS-Télécommunications
16 Place du Commerce, Verdun
Québec, Canada, H3E 1H6

Abstract

This paper proposes a new approach to the design of two-dimensional (2-D) infinite impulse response (IIR) filters with finite precision coefficients. An objective function is proposed which combines magnitude, phase, step response and stability errors. This function being multidimensional and, in general, non-convex is minimized using simulated annealing. Development of this method constitutes the first step in a feasibility study of the application of 2-D IIR filters to the processing of video signals. Initial results on the design of low-pass filters are very encouraging and compare favourably with similar finite impulse response (FIR) designs.

1. INTRODUCTION

In this paper we propose a new approach to the design of finite wordlength 2-D IIR filters. This approach rests on a multiple-term objective function minimized using *simulated annealing* and is influenced by a potential application of the method to the design of filters for video processing.

As has been demonstrated in [1], 2-D non-separable digital filtering is a very attractive alternative to 1-D filtering (analog or digital) in certain video processing tasks. For example, in an NTSC encoder/decoder a 2-D digital filter is capable of precise band-limitation of luminance and chrominance components, thus substantially reducing component crosstalk in comparison with 1-D filters [1]. To date only FIR filters have been used for such applications. In this contribution we investigate a possibility of designing 2-D IIR filters instead. Such filters have a potential for substantial savings in filter complexity and thus in implementation costs.

The approach we propose is an extension of our earlier work on the design of finite precision FIR [2] and IIR [3,4] filters. It is based on simultaneous minimization of four error terms. Similarly to various other methods [5,6,7], the first three terms deal with magnitude response, group delay (or phase) and stability. The principle term describing selective properties of the filter is specified through constraints on the magnitude response in the frequency domain. Since IIR filters are in general characterized by a nonlinear phase response and since the human visual system (HVS) is sensitive to phase errors, the second error term ensures that this nonlinearity is as small as possible. The third error term enforces filter stability with a given minimum stability margin, which is particularly important in fixed-point hardware implementations. Unlike in other methods [5,6,7], however, our objective function contains the fourth error term which controls spatial properties of the filter by measuring step response overshoots. Such overshoots appear as ringing at sharp intensity transitions in an image, and are very objectionable to the HVS. The objective function resulting from the above constraints is multidimensional and, in general, non-convex. Moreover, due to the nature of calculations involved in evaluating step response overshoots, the fourth error term cannot be expressed analytically. Thus, to avoid local minima and to deal with step response overshoots we use *simulated annealing* to carry out minimization. This is in contrast with optimization methods used in other design methods of 2-D IIR filters.

2. STATEMENT OF THE PROBLEM

Recently Wan and Fahmy [8] have given necessary and sufficient conditions for quadrantal symmetry of a transfer function. Their result states that the denominator of a transfer function possessing quadrantal symmetry must be separable. Since it is usually required that magnitude response of a filter used in image (video) processing be of quadrantal symmetry (no preference in frequency orientations), the above result suggests that optimal separable-denominator filters can be designed.

Thus, we consider a 2-D IIR filter with separable-denominator transfer function defined as follows:

$$\boldsymbol{H}(z_1, z_2) = \frac{\sum_{m=0}^{M}\sum_{n=0}^{N} a(m,n)\cdot z_1^{-m}\cdot z_2^{-n}}{\prod_{k=1}^{K}[\sum_{m=0}^{2} b(m,k)\cdot z_1^{-m}]\cdot\prod_{l=1}^{L}[\sum_{n=0}^{2} c(n,l)\cdot z_2^{-n}]}. \tag{1}$$

The cascaded form above is characterized by a relatively low coefficient sensitivity to errors, and together with denominator separability it permits simple calculation of the stability margin. Our design method, however, is general enough to handle arbitrary form of the transfer function $\boldsymbol{H}(z_1, z_2)$, also with a non-separable denominator. The major concern is the cost of computing the stability error.

Assuming that $b(0,k)=c(0,l)=1$ for $k=1,...,K, l=1,...,L$ we define the vector of unknowns $\mathbf{a}$ containing all $(M+1)(N+1)+2(K+L)$ coefficients. Requiring that coefficients be represented by b bits and that the maximum absolute value of a coefficient be A, it follows that the quantization step q and the coefficient domain $\mathcal{S}$ should be defined as

$$q = \frac{A}{2^{b-1}}, \qquad \mathcal{S} = \{a : a = -A + q \cdot i, i = 1, ..., 2^b\}.$$

Thus, the vector $\mathbf{a}$ belongs to the finite set $\mathcal{S}^{(M+1)(N+1)+2(K+L)}$.

Let $\boldsymbol{\omega} = (\omega_1, \omega_2) \in R^2$ be a 2-D frequency, $\boldsymbol{H}(e^{j\omega_1}, e^{j\omega_2}) = H(\boldsymbol{\omega}) \cdot e^{j\phi(\boldsymbol{\omega})}$ be the frequency response of filter (1) and

$$\tau_1(\boldsymbol{\omega}) = -d\phi(\boldsymbol{\omega})/d\omega_1, \quad \tau_2(\boldsymbol{\omega}) = -d\phi(\boldsymbol{\omega})/d\omega_2$$

be its partial group delays with respect to ω_1 and ω_2, respectively. Based on the discussion above, our design method for 2-D IIR filters rests on the following minimization:

$$\min_{\{\mathbf{a}\}} \mathcal{E}(\mathbf{a}), \quad \mathcal{E}(\mathbf{a}) = \mathcal{E}_m + \lambda_\tau \cdot \mathcal{E}_\tau + \lambda_s \cdot \mathcal{E}_s + \lambda_o \cdot \mathcal{E}_o, \quad (2)$$

where λ's are weights and $\mathcal{E}$'s are error terms defined as follows:

1. magnitude response error $\mathcal{E}_m$:
$$\mathcal{E}_m = \max_{\boldsymbol{\omega} \in \boldsymbol{\Omega}} [\gamma_m(\boldsymbol{\omega}) \cdot |H(\boldsymbol{\omega}) - D(\boldsymbol{\omega})|],$$
where $D(\boldsymbol{\omega})$ is the desired magnitude response, $\gamma_m(\boldsymbol{\omega})$ is a frequency-dependent weighting function, and $\boldsymbol{\Omega}$ is a discrete set of frequencies $\boldsymbol{\omega}$ over which magnitude and group delay errors are evaluated,
2. group delay error $\mathcal{E}_\tau$ (instead of minimizing phase nonlinearity the maximum spread of the group delay is minimized):
$$\mathcal{E}_\tau = \max\Big\{ \max_{\boldsymbol{\omega} \in \boldsymbol{\Omega}} [\gamma_\tau(\boldsymbol{\omega}) \cdot \tau_1(\boldsymbol{\omega})] - \min_{\boldsymbol{\omega} \in \boldsymbol{\Omega}} [\gamma_\tau(\boldsymbol{\omega}) \cdot \tau_1(\boldsymbol{\omega})], \max_{\boldsymbol{\omega} \in \boldsymbol{\Omega}} [\gamma_\tau(\boldsymbol{\omega}) \cdot \tau_2(\boldsymbol{\omega})] - \min_{\boldsymbol{\omega} \in \boldsymbol{\Omega}} [\gamma_\tau(\boldsymbol{\omega}) \cdot \tau_2(\boldsymbol{\omega})]\Big\},$$
where $\gamma_\tau(\boldsymbol{\omega})$ is a frequency-dependent weighting function used to deemphasize the importance of transition and stop bands,
3. stability error $\mathcal{E}_s$:
$$\mathcal{E}_s = \begin{cases} q/((1-\delta) - |z_k|_{max}) & \text{for } |z_k|_{max} < 1-\delta, \\ \zeta & \text{for } |z_k|_{max} \geq 1-\delta, \end{cases}$$
where $|z_k|_{max}$ is the magnitude of the pole closest to the unit circle, δ is a required stability margin, ζ is a large positive value penalizing introduction of poles outside the circle with radius $1-\delta$,
4. overshoot error $\mathcal{E}_o$ (maximum overshoot of the unit step response):
$$\mathcal{E}_o = \max_{\boldsymbol{n}_1, \boldsymbol{n}_2 \in \boldsymbol{N}} |\eta(\boldsymbol{n}_1) - \eta(\boldsymbol{n}_2)|,$$
where η is filter response to the 2-D corner unit step (first quadrant and positive axes are 1, otherwise 0) and $\boldsymbol{N}$ is a discrete set of 2-D spatial locations $\boldsymbol{n} \in Z^2$ (Z – the set of integers) over which the error is evaluated.

Since the error $\mathcal{E}_o$ is related to the transition bandwidth of the magnitude response of the filter, the overshoot constraint is in a partial conflict with the magnitude constraint, and as such must be used with caution. Step response overshoots are also related to group delay variations (different delays for different frequencies). Our intent is to partially trade-off overshoots in the step response for magnitude ripples. Such a trade-off is expected to be beneficial to image quality since a magnitude response ripple becomes dispersed over the whole image and thus is less objectionable than a well localized spatial overshoot. To achieve this goal the weight λ_o must be carefully selected.

3. SOLUTION

The cost function $\mathcal{E}(\mathbf{a})$ under minimization is non-quadratic (no linear system can be solved), non-differentiable (no gradient-based technique can be used) and multi-variable. In fact, it is likely to have multiple minima, thus requiring global optimization. A method which we propose to use here is *simulated annealing* [9], which under certain conditions is able to localize the global minimum of an arbitrary cost function by executing a random walk in the space of solutions.

In the implementation of simulated annealing we have used the *Metropolis algorithm* [10] as follows. If $\mathbf{a}_t$ is the coefficient vector at time t and $\mathbf{a}_t^*$ is the vector after a change of one coefficient, then $\Delta\mathcal{E}_t = \mathcal{E}(\mathbf{a}_t^*) - \mathcal{E}(\mathbf{a}_t)$ is a cost function change due to a change of one coefficient. At every time instant t temperature T is chosen ($T \to 0$ with $t \to \infty$) and the following decision is made:

1. $\Delta\mathcal{E}_t \leq 0 \Rightarrow$ accept $\mathbf{a}_t^*$ unconditionally: $\mathbf{a}_{t+1} = \mathbf{a}_t^*$,
2. $\Delta\mathcal{E}_t > 0 \Rightarrow$ accept $\mathbf{a}_t^*$ with probability $p = e^{-\frac{\Delta\mathcal{E}_t}{T}}$.

The process is repeated until T is sufficiently low, or as long as a certain criterion imposed on $\mathcal{E}(\mathbf{a})$ or on the acceptance ratio has not been met.

4. RESULTS

We have tested the above method by designing 2-D lowpass IIR filters of various orders and coefficient precisions, as well as different pass-band shapes such as diamond, hexagonal, circularly symmetric. Below we present some results obtained for chrominance filters used in modulation and demodulation of the NTSC composite signal. Those filters are used to band-limit the chrominance I and Q signals in order to reduce component cross-talk [1], and thus are called I- and Q-filters, respectively. We have designed those filters with 8-bit coefficient precision, and we have compared them with corresponding 8-bit FIR filters proposed in [1].

To deemphasize the importance of exact magnitude value in the transition band and of the group delay constancy in the transition and stop bands, the weights γ_m and γ_T have been appropriately shaped. The overshoot error $\mathcal{E}_o$ has been implemented thus far by finding the tallest peak (deepest valley) closest to the step response origin and then by searching its neighbourhood, but staying on the same side of the step response rise, for minimum (maximum) accessible by a monotonic descent (ascent).

Table 1 shows principle parameters of the designed filters. All filters have been designed with coefficient precision of 8 bits. The IIR filters have been designed using the method proposed here, while the FIR filters have been obtained by the method proposed in [2]. The IIR filters have 33 coefficients altogether ($M = N = 4$, $K = L = 2$), while the FIR filters have 165 coefficients (15 by 11) out of which 48 are independent. In our experience, higher order 8-bit lowpass I and Q FIR filters do not provide significant improvement due to additional coefficients being quantized to zero, which is not the case with IIR filters. Fig.1 shows the desired magnitude response $D(\omega)$ of the I-filter, while Fig.2 shows magnitude responses of filters designed to approximate $D(\omega)$. Note somewhat larger departure of the IIR filter's magnitude response from the desired reponse, which is confirmed by larger pass-band ripples and smaller stop-band attenuation (Table 1). Fig.3 shows the partial group delay τ_1 of the IIR filter. Note the almost constant value of τ_1 in the pass-band (center of the plot). In fact, the maximum group delay spread in the pass-band is well below 1 sample (Table 1). Fig.4 shows responses of both filters to the 2-D corner unit step (Section 2). Qualitatively, it seems that the IIR filter has smaller overshoots. The value of $\mathcal{E}_o$ from Table 1 confirms this observation.

	I-filter		Q-filter	
	IIR	FIR	IIR	FIR
total # of coeff.	33	165	33	165
ripple (pass) [dB]	0.73	0.48	1.40	0.46
atten. (stop) [dB]	19.5	26.2	16.1	25.9
$\tau_1^{max} - \tau_1^{min}$ (pass)	0.30	—	0.61	—
$\tau_2^{max} - \tau_2^{min}$ (pass)	0.52	—	0.65	—
$1 - \lvert z_k \rvert_{max}$	0.60	—	0.79	—
$\mathcal{E}_o$	0.104	0.118	0.140	0.156

Table 1 Comparison of parameters of 8-bit FIR and IIR filters for I and Q specifications

We have tested the designed filters on real video images, especially scrutinizing areas of sharp luminance transitions. The images produced by IIR and FIR filters were almost indistinguishable. A tiny difference in image "softness" was due to slightly different magnitude responses of the filters, which can be compensated for. The IIR filters, however, have performed better in terms of step response overshoots. Spurious "ghosts" were slightly less visible, which indeed is a desirable property of a video filter.

In conclusion, our study indicates that 2D IIR filters may become a viable alternative to FIR filters in video processing, thus reducing computational costs involved.

REFERENCES

[1] E. Dubois and W. Schreiber, "Improvements to NTSC by multidimensional filtering," *SMPTE J.*, vol. 97, pp. 446–463, June 1988.

[2] J. Radecki, J. Konrad, and E. Dubois, "A comparison of simulated annealing and N-step Newton methods for designing 1-D and 2-D finite wordlength FIR filters," in *Proc. Canadian Conf. Electr. Comp. Eng.*, pp. 53.3.1–53.3.4, Sept. 1990.

[3] J. Radecki, J. Konrad, and E. Dubois, "Design of finite wordlength IIR filters with prescribed magnitude, group delay and stability properties using simulated annealing," in *Proc. IEEE Int. Conf. on Acoust. Speech Signal Process.*, pp. 1637–1640, May 1991.

[4] J. Konrad, J. Radecki, and E. Dubois, "On the design of finite wordlength IIR filters for video applications," in *Proc. IEEE Int. Conf. on Acoust. Speech Signal Process.*, Mar. 1992.

[5] J. Woods, J.-H. Lee, and I. Paul, "Two-dimensional IIR filter design with magnitude and phase error criteria," *IEEE Trans. Acoust. Speech Signal Process.*, vol. ASSP-31, pp. 886–893, Aug. 1983.

[6] M. Ahmadi, M. Boraie, V. Ramachandran, and C. Gargour, "Design of 2-D recursive digital filters with constant group delay characteristics using separable denominator transfer function and a new stability test," *IEEE Trans. Acoust. Speech Signal Process.*, vol. ASSP-33, pp. 1316–1318, Oct. 1985.

[7] J.-H. Lee and Y.-M. Chen, "A new method for the design of two-dimensional recursive digital filters," *IEEE Trans. Acoust. Speech Signal Process.*, vol. ASSP-36, pp. 589–598, Apr. 1988.

[8] Y. Wan and M. Fahmy, "N-dimensional symmetries and their applications in digital filters," *Signal Processing*, vol. 19, pp. 103–117, 1990.

[9] S. Kirkpatrick, C. Gelatt Jr., and M. Vecchi, "Optimization by simulated annealing," *Science*, vol. 220, pp. 671–680, May 1983.

[10] N. Metropolis, A. Rosenbluth, M. Rosenbluth, H. Teller, and E. Teller, "Equation of state calculations by fast computing machines," *J. Chem. Phys.*, vol. 21, pp. 1087–1092, June 1953.

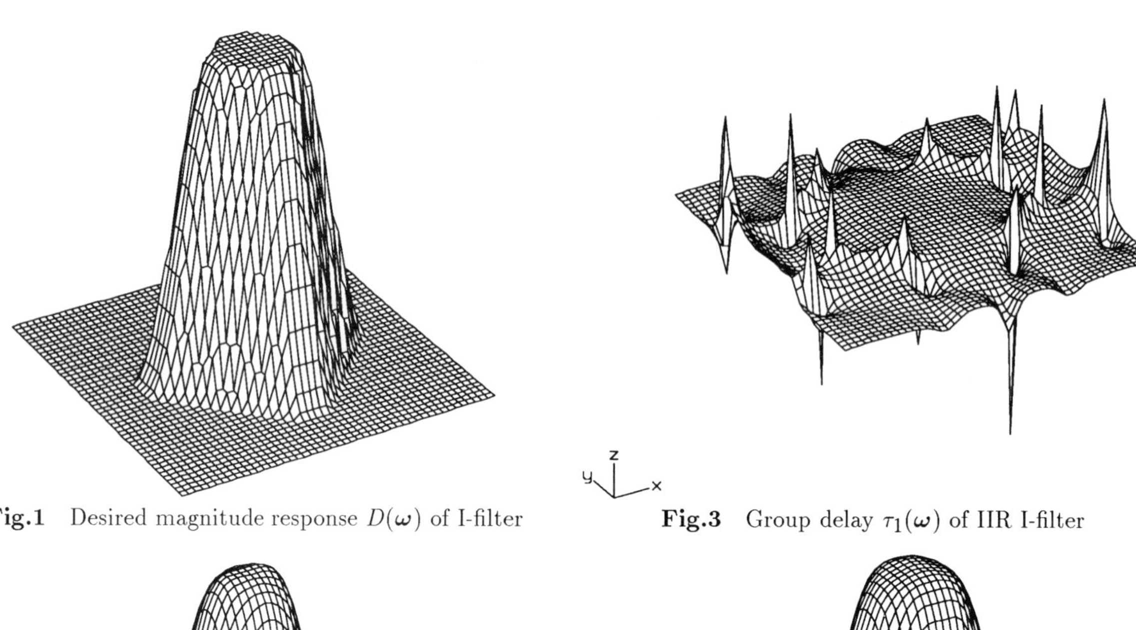

Fig.1 Desired magnitude response $D(\omega)$ of I-filter

Fig.3 Group delay $\tau_1(\omega)$ of IIR I-filter

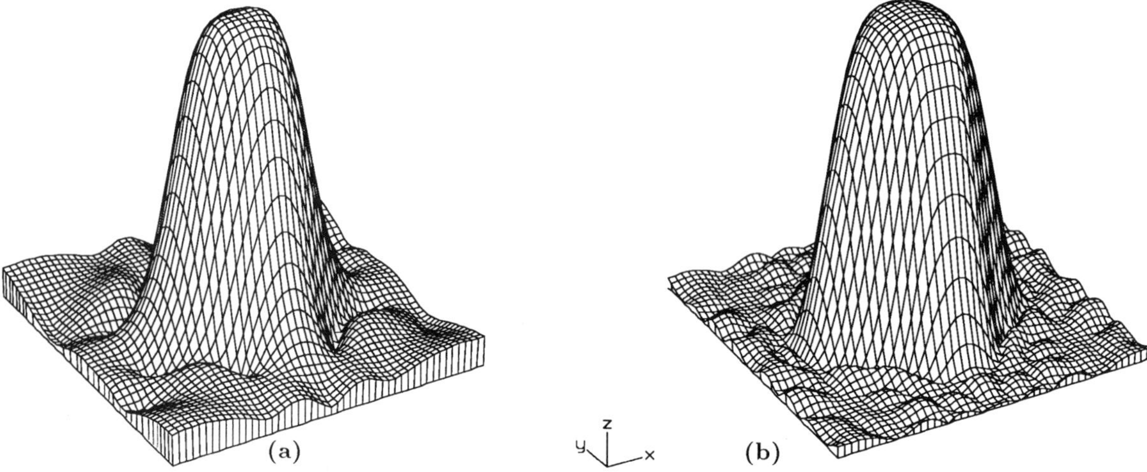

Fig. 2 Magnitude response of IIR (a) and FIR (b) I-filter

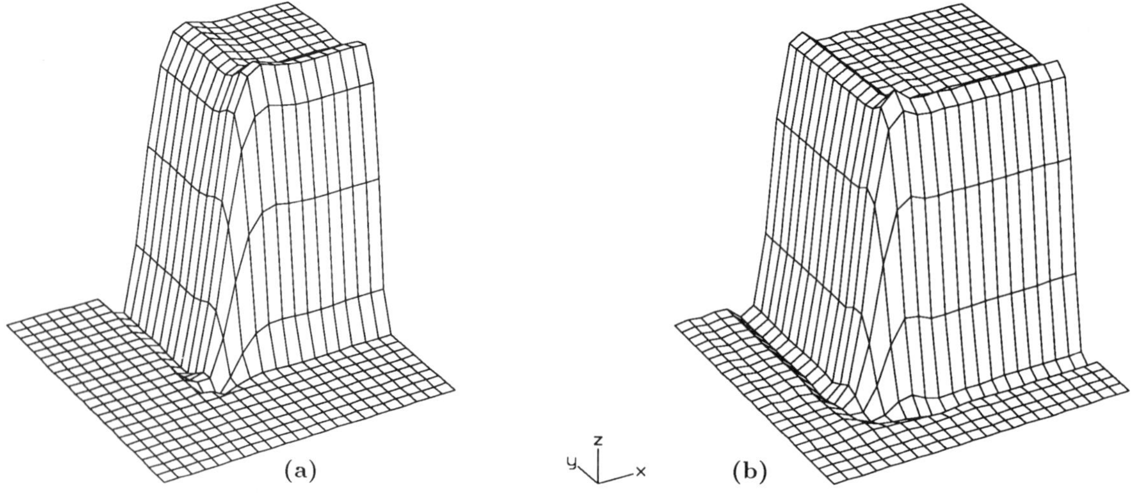

Fig. 4 Step response of IIR (a) and FIR (b) I-filter

SIGNAL PROCESSING VI: Theories and Applications
J. Vandewalle, R. Boite, M. Moonen, A. Oosterlinck (eds.)

A simple design method for FIR filters with integer coefficients using a personal computer

Shogo NAKAMURA, Yasumasa MORISHITA, Yoshihiko HORIO
Akiyoshi KAWAHASHI, Yukio KADOWAKI *

Department of Electronic Engineering, Tokyo Denki University, 2-2 Kanda Nishiki-cho, Chiyoda-ku, Tokyo, 101, Japan

*LSI R &D Center Electronics Devices Division Ricoh Co., LTD, 13-1 Himemuro-cho, Ikeda, Osaka, 563, Japan

FIR filters with linear phases are useful in various fields. However, when we desire the sharper filter characteristics, its hardware and computational complexities inevitably increase so that an available frequency band is limited. Therefore, obtaining a fast filtering architecture of the FIR filters becomes a valuable task.
This paper mainly describes a simple design method of a lower order FIR filter with integer coefficients by using a conventional personal computer.

1. Introduction

There were several efforts to get FIR filters with integer coefficients.[1],[2] Lim proposed a design method using mixed integer linear programming in [1] and provided his nice program. However, since the program is large and takes a lot of computational times, it is oriented to a big computer such as a super-computer. Sometimes, a simple design approach is desirable in many applications of the real fields. Therefore, as another approach, the method to achieve the linear phase FIR filter with good properties by realizing a hierachical building block structure by using a lot of lower order FIR subfilters has been proposed.[3],[4] This paper mainly describes a simple design method of the above lower order FIR subfilter with integer coefficients by using a conventional simple personal computer.

Althogh the proposed approach is not optimal, it is considered that the results obtained are sufficiently good for practical applications. It is also shown by two different examples that this approach is useful.

2. Outline of the proposed FIR filter

Let us explain the properties of the proposed filter of the hierachical building block structure. This basically has the strucute of tapped-cascaded subfilters with linear phases. The linear phase FIR filter of a characteristic which is better than that of the subfilter, can be obtained by changing tapping coefficients appropriately. When the above mentioned filter is regarded again as a subfilter, the filter design can be achieved by using the building block structure such as shown in Fig.1.

Since the processings of a lot of identical subfilters and the tapping coefficients can be executed in parallel, this kind of filter is suitable for a high speed processing and an LSI implementation. For the high speed processing, the arithmetic operations of the subfilter and the tapping parts must be simple. In addition, the use of the multiplier is not desirable from the point of view of LSI implementation because it needs large chip areas in the LSI. So, the adder-based architecture was adopted.

3. A simple design approach

The proposed design is based on a linear equation set which comes from frequency specifications of a desired filter. First, the frequency sampling points which give the desired value of 1 in the pass-band and the value of 0 in the stop-band, are estimated by using the transition band width of a given filter specifications. Thus the frequency sampling points are selected with the same interval of the transition band width from the pass band and the stop band edges such as shown in Fig.2.

$$\omega_{dif}=\omega_{stop}-\omega_{pass} \qquad (1)$$

In Fig.2, $N_p(1)$ means the nearest frequency sampling point of the pass-band edge and $N_s(1)$ also means the nearest frequency sampling point of the stop-band edge. $N_p(M_p)$ and $N_s(M_s)$ represent the closest sampling points at ω=0 and ω=0.5, respectively.

If $n(=M_p+M_s)$ sampling points are chosen, they are substituted in the following frequency transfer function so that the n linear equations are obtained.

$$H(e^{j\omega k})=h_0+2\sum_{i=1}^{n} h_i\cos(i\omega k) \qquad (2)$$

$$k=1,2,3, \quad ....,n$$

To solve this simultaneous equation, one more equation is needed. Another frequency sampling point ω_t is taken at the center of the transition band of the filter. Next, the conditional number of this simultaneous equation is calculated. The conditional number Cond(*) is defined by the following relation;

$$\mathrm{Cond}(A)=\|A\|\,\|A^{-1}\| \qquad (3)$$

where A^{-1} is the inverse matrix of A, A is the coefficients matrix of the simultaneous equation such as:

$$A=\begin{pmatrix} 1 & 2\cos(\omega_1) & 2\cos(2\omega_1) ..... 2\cos(n\omega_1) \\ 1 & 2\cos(\omega_2) & 2\cos(2\omega_2) ..... 2\cos(n\omega_2) \\ \cdots & \cdots & \cdots \\ 1 & 2\cos(\omega_n) & 2\cos(2\omega_n) ..... 2\cos(n\omega_n) \\ 1 & 2\cos(\omega_t) & 2\cos(2\omega_t) ..... 2\cos(n\omega_t) \end{pmatrix}$$

(4)

and $\|A\|$ is a norm of A which is defined such as:

$$\|A\| = \mathrm{MAX}_i\{\sum_{j=1}^{n}|a_{ij}|\} \qquad (5)$$

where a_{ij} is an element of matrix A.

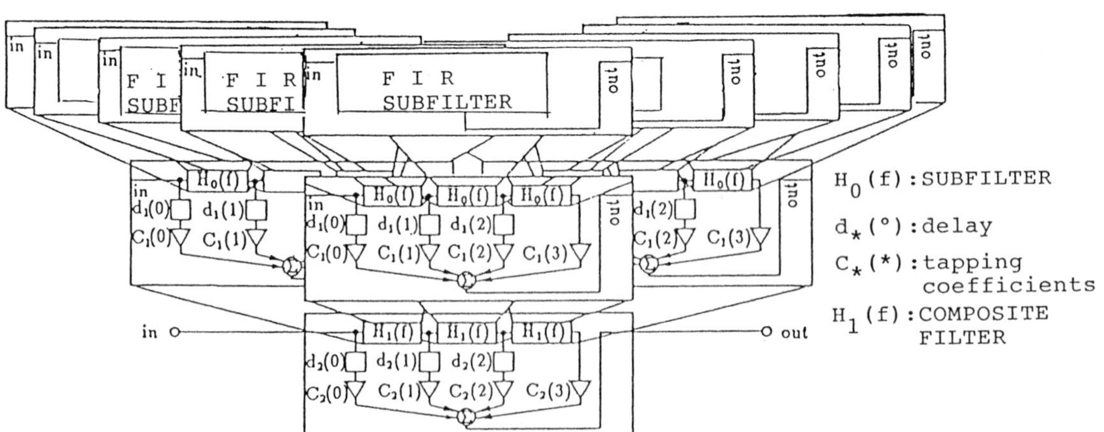

Fig.1 A structure of a hierachical structure transversal filter

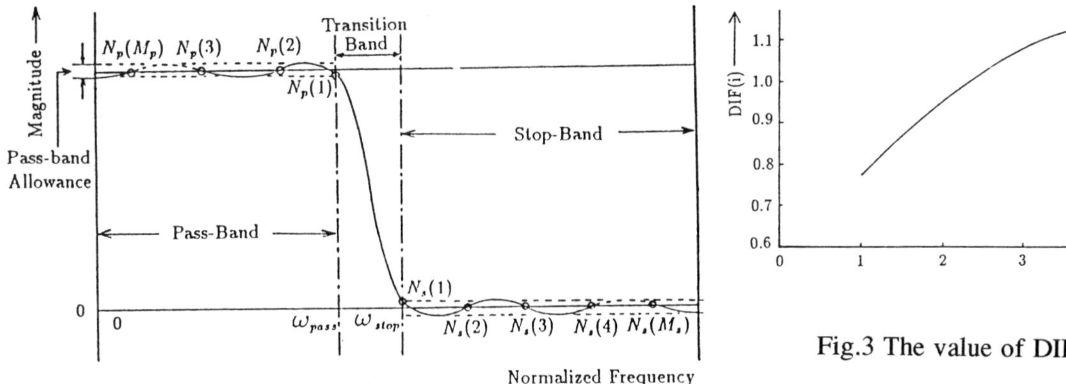

Fig.2 An example of initial frequency sampling points.

Fig.3 The value of DIF(i)

Now, the original frequency sampling points except for ω_t are modified by changing the interval ω_{dif} between the frequency sampling points by the eq.(6).

$$N_p(i+1)=N_p(i)+DIF(i)*FACT*\omega_{dif}$$
$$N_s(j+1)=N_s(j)+DIF(j)*FACT*\omega_{dif}$$
$$i=1,2,...,N_p(M_p-1)\,,$$
$$j=1,2,...,N_s(M_s-1) \qquad (6)$$

where the DIF(i) or DIF(j) was obtained experimentally such as shown in Fig.3 from the result of the equi-ripple FIR filter design due to the Remez Exchange Algorithm [5] and FACT is a constant number to vary from 0.9 to 1. The conditional number of eq.(2) is repeatedly calculated by using a new frequency sampling points depending on the different values of FACT. When the minimum conditional number is obtained, the corresponding frequency sampling points are fixed in order to solve the coefficients $\{h_i\}$ of eq.(2).

Solving eq.(2) by using the above coefficients, the coefficients $\{h_i\}$ of the subfilter can be obtained.

Finally, the coefficients obtained are rounded to several bits. Since this procedure is very simple, the use of a conventional simple personal computer is allowed for the design of the FIR filter.

4. A design example

Figures 4 and 5 show examples designed with different band edges. The relation between the conditional number and the deviations from the desired values can be found from two figures.

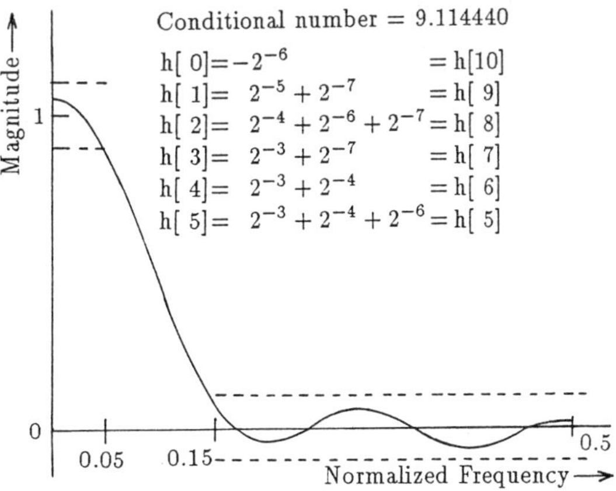

Fig.4 (a) The result of a small conditional number.

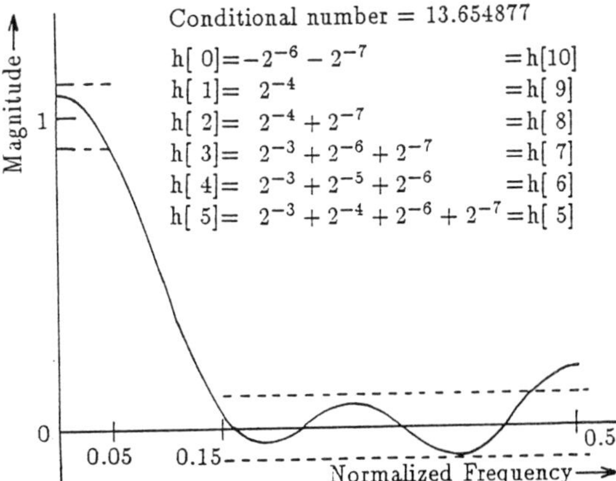

Fig.4(b) The result of a large conditional number.

Thus, it is observed that the small conditional number offers less quantization effect. The specifications of the filters in Fig.4 is such as:

Filter length : 11, max. deviation : 0.1
Pass-band edge : 0.05
Stop-band edge : 0.15

The other filter in Fig.5 is described below such as :

Filter length : 11, max. deviation : 0.1
Pass-band edge : 0.15
Stop-band edge : 0.25

Each frequency-magnitude property is plotted by using the filter coefficients quantized to 7 bits which are also represented in the figures.

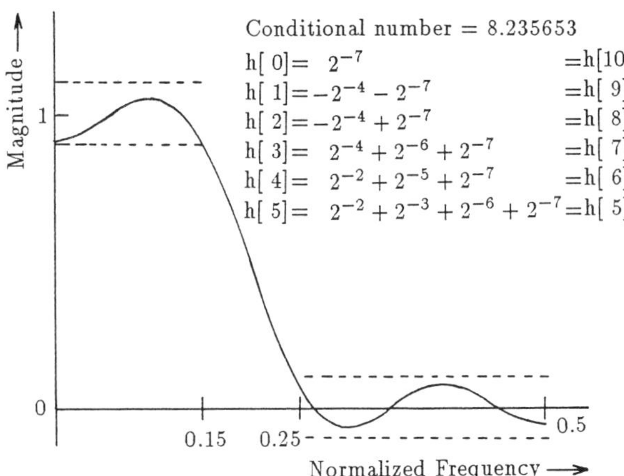

Fig.5(a) The result of a small conditional number.

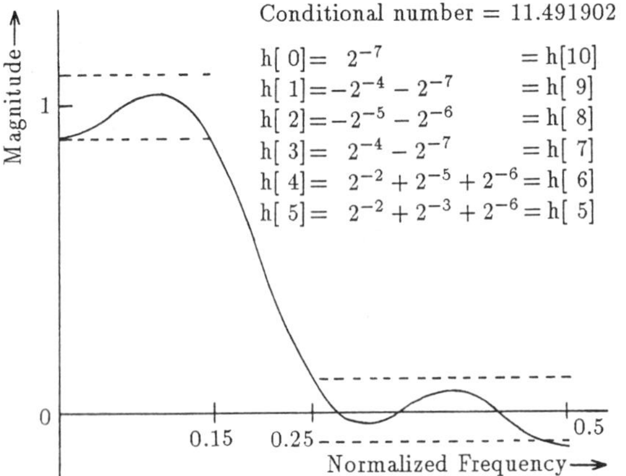

Fig.5(b) The result of a large conditional number.

6. Conclusion

The authors proposed an approach to simplify the design of the FIR filter. The advantage of the proposed method is that since the basic architecture is a tapped cascaded subfilter structure, the direct design problem of the FIR filter is replaced by design of the simple subfilter, and the design of the subfilter with integer coefficients can be simply performed by using a conventional personal computer.

7. Acknowledgment

The authors wish to express their gratitude to the Research Institute for Technology, Tokyo Denki University and Ricoh Co. LTD. for their financial support.

8. References

[1]Y.C.Lim : "Design of Discrete-Coefficient-Value Linear Phase FIR Filters with Optimum Normalized Peak Ripple Magnitude"IEEE Trans. on CAS, vol.37, no.12, pp1480-1486, Dec. 1990

[2]Y.C.Lim and B. Liu : "Design of Cascade Form FIR Filters with Discrete Valued Coefficients"IEEE Trans. on ASSP, vol.36, no.11,pp.1735-1739, Nov. 1988

[3] S.Nakamura,Y. Horio, and S. Yasuda : "A realization of programmable FIR filter without a multiplier" Proceedings of EUSIPCO-88 fourth, pp.707-710, France, Sept. 5-8, 1988

[4] S.Nakamura, Y.Kadowaki, and S. Matsuoka : "An LSI implementation of generalized transversal filter" IEEE Trans. on SP, pp.227-230, vol. 39, no.1, Jan., 1991

[5]L.R.Rabiner,J.H.McClellan, and T.W.Parks : "FIR digital filter design techniques using weighted Chebyshev approximation" IEEE Trans. Audio & Electroacoustics, AU-21, no.6, pp.81-117, Dec., 1973

SIGNAL PROCESSING VI: Theories and Applications
J. Vandewalle, R. Boite, M. Moonen, A. Oosterlinck (eds.)

SYNTHESIS METHODS FOR WAVE DIGITAL FILTERS EXHIBITING ARBITRARY AMPLITUDE CHARACTERISTICS

M. Yaseen* and **T. Henk**

Dept. of Telecommunications and Telematics,
Technical Univ. of Budapest, HUNGARY

Design conceptions are delivered for several wave digital filter (WDF) structures satisfying arbitrary amplitude specifications given at all frequencies , i.e., the transition band can also be specified beside the passband and stopband. The approximation problem is solved using the interpolation methods combined with the Remez-exchange algorithm. The different conceptions are compared from the wave realization cost, sensitivity, dynamic range and nonlinearity points of view.

1. INTRODUCTION

WDF's exhibit excellent properties, including coefficient sensitivity, roundoff noise overflow level, stability and nonlinear effects under finite-arithmetic conditions [1,2]. The robustness of WDF's is insured by their passive [3,4] reference structures and specially the losslessness of the filter two-port [1]. The losslessness within passivity is insured if the approximated transmission is unity and zero in the passband and stopband respectively. Otherwise there is no direct relationship between losslessness and low coefficient sensitivity. This is why these properties are usually not investigated in the transition band and also in the passband and stopband if the transmission is not unity and zero respectively, but are in the focus of this paper.

In many applications , e.g., data transmission, sampling, etc., the amplitude characteristic $A(\omega)$ is specified at all frequencies , i.e., also in the transition band while the transmission is also a given function in the passband. To solve this filter design problem optimally, design conceptions are delivered for several WDF structures satisfying arbitrary amplitude specifications given at all frequencies in form of lower and upper amplitude limit functions :

$$A_l(\omega) \le A(\omega) \le A_u(\omega) \quad . \qquad (1)$$

Approximation methods are developed for the considered WDF structures by applying the interpolation method combined with the Remez-exchange algorithm. The design conceptions are compared from the wave digital realization cost, sensitivity, dynamic range and nonlinearity point of view to find an optimal solution.

2. DESIGN CONCEPTIONS

The first design conception is relying on a structure composed of only one reactant filter. This one filter can be restricted to possess odd degree lattice structure if the relationship,

$$A\ (\omega) \le A\ (0) \qquad (2)$$

holds for the approximated amplitude characteristic. In such a case, bireciprocal filter structure can be employed if further

$$A^2(\omega)+A^2(\pi.F-\omega)=A^2(0) \qquad (3)$$

holds, where F is the sampling rate. In case of

$$A(\omega) > A(0) \text{ in some band,} \qquad (4)$$

the constructed filter must be of ladder structure.

The second design conception is relying on a structure composed of cascaded filter and amplitude equalizer [5]. The third design conception is relying on a structure composed of two cascaded filters. The first filter in the second or third conception is designed to satisfy the biggest part of the selectivity as a prototype filter and can be restricted to possess odd degree lattice structure or maybe bireciprocal one while an amplitude equalizer or a second filter of lower order insures the final shape of the characteristics. In case of two cascaded filters, the second filter can be lattice only if

$$A(\omega) \le A(0)\ . \qquad (5)$$

* On leave from Dept. of Electrical Eng. , Univ. of Assiut, Assiut, EGYPT.

3. APPROXIMATION AND REALIZATION

The approximation problem is solved by applying interpolation methods combined with the Remez-exchange algorithm similarly to Ref. 6. The lattice and bireciprocal solutions may become of nonminimum phase [7]. If the specification is given in the digital p or $j\omega$ frequency domain , it is transformed to the reference ψ or $j\phi$ frequency domain by the bilinear transformation [1] :

$$\psi = (z-1)/(z+1),\; z= e^{pT},\; \phi = \tan(\omega T/2),\quad T = \text{sampling period} . \tag{6}$$

The approximation procedure is summarized for each conception as follows.

3.1. One Reactant Filter

According to this conception the given specification is approximated by only one filter. The approximation is carried out by constructing the characteristic function $\Psi(\psi)$:

$$\Psi(\psi) = h(\psi) \,/\, f(\psi) = S_{11}(\psi)/S_{21}(\psi), \tag{7}$$

where $S_{11}(\psi)$ is the reflectance and $S_{21}(\psi)$ is the transmittance.

In case of lowpass lattice filter, $h(\psi)$ is an odd polynomial with degree n, which is the degree of the filter and it is an odd number:

$$h(\psi) = \psi \sum_{i=0}^{\{n-1\}/2} c(i)\, \psi^{2i} , \tag{8}$$

with its roots are not restricted to lie on the $j\phi$-axis. On the other hand, $f(\psi)$ is an even polynomial of degree n-1 :

$$f(\psi) = \sum_{i=0}^{\{n-1\}/2} d(i)\, \psi^{2i} = \prod_{i=1}^{\{n-1\}/2} \{\psi^2+B^2(i)\} , \tag{9}$$

so, it is restricted to possess $j\phi$-axis roots only with transmission zeros B(i) .

The specification for $\Psi(j\phi)$ is obtained from that of the transmittance by :

$$|\Psi(j\phi)|^2 = 1/|S_{21}(j\phi)|^2 -1 \tag{10}$$

The design procedure is summarized as follows :

1) Calculate the specification of the $\Psi(j\phi)$ from that of the transmittance and select a degree for the filter.
2) Select proper values for the transmission zeros B(i).
3) At a selected set of frequencies determine c(i) by interpolating $|\Psi(j\phi)|$ in the passband and transition band if specified.
4) By applying the Remez-exchange algorithm, move the interpolation frequencies and recalculate c(i) for optimal passband and transition band if specified.
5) With the obtained c(i) , apply the Remez-exchange algorithm for new optimal transmission zeros B(i).
6) Go back to step 3 to re-satisfy the passband and transition band if specified. Continue till the specification is satisfied in the three bands. If the selected degree is not sufficient select new one and go to step 2.

In case of bireciprocal structure, the function $\Psi(\psi)$ is decomposed :

$$\Psi(\psi) = \Psi_1(\psi)\, \Psi_2(\psi), \tag{11}$$

$$\Psi_1(\psi)=\frac{\psi \prod_{i=1}^{k/2} \{B^2(i)\psi^2+1\}}{\prod_{i=1}^{k/2} \{\psi^2+B^2(i)\}},\quad \Psi_2(\psi)= \frac{\sum_{i=0}^{m/2} c(i)\psi^{2i}}{\psi^m \sum_{i=0}^{m/2} c(i)\psi^{-2i}}, \tag{12}$$

such that the first of them is restricted to possess roots on the $j\phi$-axis only while the second is not restricted on the $j\phi$-axis. In (12), k and m are related to the filter degree n by

$$n = k+m+1 , \tag{13}$$

furthermore B(i) and c(i) are redefined. The design procedure is similar to that of the lattice structure.

In case of

$$A(\omega) > A(0) \text{ in some band,} \tag{14}$$

the one filter structure must be of ladder type. In such a case it is approximated by constructing its squared magnitude transmittance :

$$S_{21}(\psi)S_{21}(-\psi) = \frac{\prod_{i=1}^{m} \{\psi^2+B^2(i)\}^2}{\sum_{i=0}^{n} c(i)\, \psi^{2i}} = \frac{N(\psi)}{D(\psi)},\quad \psi = j\phi \tag{15}$$

where B(i) are the corresponding transmission zeros, furthermore B(i) and c(i) are redefined again. The design procedure for B(i) and c(i) is similar again to that of the lattice structure . The transmittance of the filter is obtained by the factorization of $N(\psi)$ and the Hurwitz factorization of $D(\psi)$.

To realize a lattice filter from its $\Psi(\psi)$, the following expression is considered :

$$h(\psi) + f(\psi) = g_1(\psi)\, g_2(-\psi) \ , \qquad (16)$$

where the denominator $g(\psi)$ of the transmittance $S_{21}(\psi)$ is related to $g_1(\psi)$ and $g_2(\psi)$ by

$$g(\psi)=g_1(\psi)\, g_2(\psi) \ . \qquad (17)$$

So, by considering the Hurwitz and anti-Hurwitz factors of $h(\psi)+f(\psi)$, the polynomials $g_1(\psi)$ and $g_2(\psi)$ respectively can be determined.

The two all-pass functions of the lattice branches are calculated as :

$$S_1(\psi)= -g_1(-\psi)/g_1(\psi),\ S_2(\psi)=g_2(-\psi)/g_2(\psi) \qquad (18)$$

and they are realized by factorizing them into cascaded sections of first and second orders [2]. In the bireciprocal case however fourth order sections are also needed [8].

In case of ladder structure , the use of Fujisawa's theorems [9] results in a lossless ladder two-port terminated with a resistance.

3.2. Prototype Filter Cascaded with Equalizer

According to this conception, explicit formulas [2] can be used to design the prototype filter . The loss characteristic of this filter is subtracted from the original specification to get new specification for the required amplitude equalizer. The equalizer is approximated through the construction of its squared magnitude transmittance :

$$S_{21}(\psi)S_{21}(-\psi) = \frac{\sum_{i=0}^{m} n(i)\, \psi^{2i}}{\sum_{i=0}^{m} d(i)\, \psi^{2i}} = \frac{N(\psi)}{D(\psi)}$$

$$= f(\psi)f(-\psi)/g(\psi)g(-\psi) \ , \ \psi = j\phi \qquad (19)$$

where $g(\psi)$ is a strictly Hurwitzian polynomial ,while there is no restriction for $f(\psi)$. The approximation procedure is summarized as follows:

1) Select the degrees for the filter and the equalizer.
2) Design an odd degree lattice filter to satisfy the biggest part of selectivity.
3) Get the specification of the equalizer.
4) At a selected set of frequencies interpolate the squared magnitude of the equalizer for having n(i) and d(i).
5) Move the interpolation frequencies for optimal amplitude characteristic for new n(i) and d(i).
6) Investigate a new lattice filter with changed passband and stopband requirements but with the same degree to find the optimal lattice filter and go to step 3. Continue till the specification is satisfied in the passband, stopband and transition band if specified. If the degrees are not sufficient , select new degrees and go to 2.
7) The transmittance of the equalizer is obtained by any factorization for $N(\psi)$ and the Hurwitz factorization of $D(\psi)$.

To realize the equalizer its transmittance is equal to the reflectance of a lossy one port [4]. The realization is carried out by cascaded sections of first and second orders.

3.3. Two Cascaded Filters

According to this conception , the given specification is approximated such that the first filter with higher order and prototype specification is constructed to satisfy the biggest part of selectivity. This filter can be restricted to possess odd-degree lattice structure or maybe bireciprocal one. Explicit formulas [2] can be used to design this filter. The loss characteristic of this filter is subtracted from the original specification to obtain a new specification for a low order second filter. The second filter can be restricted to be lattice if

$$A(\omega) \le A(0) \ , \qquad (20)$$

otherwise it must be of ladder type. The approximation procedure is similar to that of the cascaded filter and equalizer conception.

4. SIMULATION

Considering a model consisting of A/D, digital filter and D/A, the fixed point arithmetic is used with 12 digits for the truncated coefficient wordlength and for A/D and the register wordlength. The magnitude truncation or 2's complement truncation is applied at the adaptor outputs as appropriate [2]. The developed structures are investigated by means of typical examples from scaling, dynamic range, sensitivity and nonlinearity point of view.

5. DESIGN EXAMPLE

A monotonous specification of a data transmission demodulator filter is approximated by 9-th order bireciprocal filter , F=64 kHz. The loss response with the specification can be seen in Fig. 1.a . The WD realization is shown in Fig 1.b with the following values for the adaptor coefficients :

$\gamma_1= -0.1108907$ $\gamma_2=0.4170397$ $\gamma_3= -0.2912512$
$\gamma_4= 0.66324614$

The dynamic range of this structure is 63.35 dB and its signal to noise ratio in the passband is 62.0 dB.

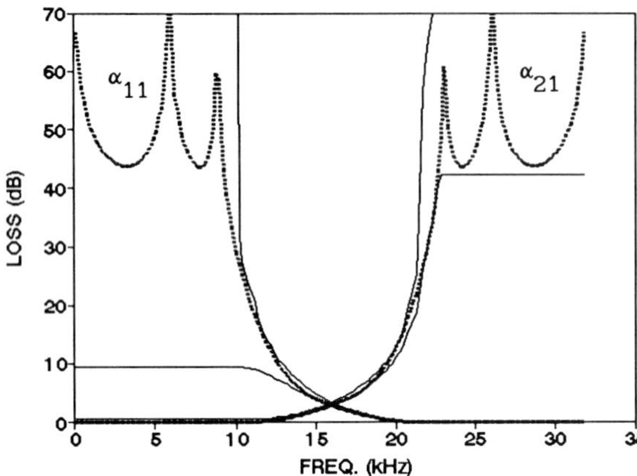

Fig. 1.a. Loss characteristics with the specification of a 9-th order bireciprocal filter, $\alpha_{ij} = -20 \log|S_{ij}|$.

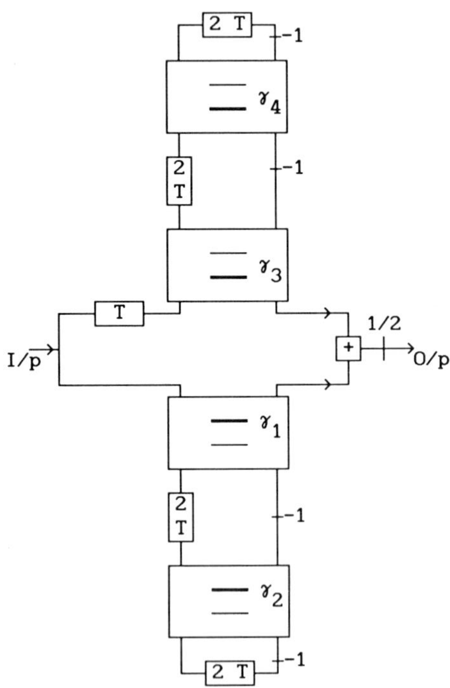

Fig. 1.b. The corresponding wave digital filter realization.

6. CONCLUSIONS

Several design conceptions are delivered for WDF's exhibiting arbitrarily specified loss characteristics and they are compared by means of typical examples. Comparing them from the number of adaptors point of view, the conception of one bireciprocal or two lattices and the conception of bireciprocal + ladder or lattice+ladder are the optimal solutions for $A(\omega) \leq A(0)$ and for $A(\omega) > A(0)$ in some band respectively. Comparing them from dynamic range point of view, the conception of two filters gives the optimal value for both $A(\omega) \leq A(0)$ and for $A(\omega) > A(0)$ in some band respectively. Comparing them from the nonlinearity point of view , the conception of two filters gives the optimal behavior . The use of equalizers results in more attenuation in the stopband but at the cost of higher number of adaptors. The two filter solution enables us in case of $A(\omega) \leq A(0)$ to design a cascaded two lattice structure with total even degree which almost is lower than the degree of one lattice filter structure. The use of two filter conception enables us to restrict the biggest part of the filter structure to be lattice in case of $A(\omega) > A(0)$ in some band.

REFERENCES

[1] Fettweis, A., Wave Digital Filters :Theory and Practice, Proc. of the IEEE , Vol. 74, Feb. 1986, pp. 270-327.

[2] Gazsi, L., Explicit Formulas for Lattice Wave Digital Filters , IEEE Trans. Circuits and Systems, vol. CAS-32, Jan. 1985, pp. 68-88.

[3] Fettweis, A., Some General Properties of Signal-Flow Networks , in : Network and Signal Theory edited by Skwirzynki, J. K. and Scanlan, J. O., (Peter Peregrinus Ltd, London 1973), pp. 48-59.

[4] Belevitch, V., Classical Network Theory, (San Francisco, CA , Holden Day, 1968).

[5] Sauvagard, U., A Ten-Channel Equalizer for Digital Audio-Applications , IEEE Trans. Circuits and Systems, Vol. CAS-36, Feb. 1989, pp. 276-280.

[6] Földvári-Orosz, J., Henk, T. and Simonyi , E., Simultaneous Amplitude and Phase Approximation for Lumped and Sampled Filters , I. J. of Circuit Theory and Applications , Vol. 19, pp. 77-100, Jan.-Feb. 1991.

[7] Leeb, F., Henk, T., Simultaneous Approximation for Bireciprocal Lattice Wave Digital Filters, Proc. of the ECCTD'89 ,(Brighton, U. K.), 5-9 Sept. 1980, pp. 472-476.

[8] Wegener, W., Wave Digital Filters with Reduced Number of Multipliers and Adders , AEÜ, Band 33, 1979, Heft 6, pp. 239-243.

[9] Baher, H., Synthesis of Electrical Networks , (John Willey and Sons, 1984).

SIGNAL PROCESSING VI: Theories and Applications
J. Vandewalle, R. Boite, M. Moonen, A. Oosterlinck (eds.)

EFFICIENT TEMPORAL BAND-LIMITATION BASED ON MOTION ADAPTIVE SPATIAL FILTERING

JONG-HUN KIM, SEONG-DAE KIM

DEPT. OF ELECTIRCAL ENGINEERIG, VISUAL COMMUNICATIONS LAB., KAIST
373-1, KUSEONG-DONG, YUSEONG-KU, DAEJEON 305-701, KOREA

Most of coding applications have been developed without concerning about temporal aliasing which causes visual artefacts such as motion judder, stroke-width modulation, etc.. This paper describes a band-limiting temporal filter which is not affected by the aliasing. The proposed filter, *Motion Adaptive Spatial Filter*, is employed on spatial domain in direction of movement. By performming temporal band-limitation efficiently, it will be applicalbe to image sequence coding as a prefilter.

1. INTRODUCTION

Unlike a noise removal recursive temporal filter[3], we have different point of view how we can perform a temporal band-limitation efficiently. Recent progress on HDTV which combines spatial and temporal subsampling uses an adaptation of their resolutions according to the movement. If a lowpass temporal filter is applied, more improvement of their quality will be achieved. Moreover, most of coding applications have been studied without the concern of visual artefacts resulting from temporal aliasing. Therefore, it is necessary to investigate a band-limiting temporal filter not affected by the aliasing.

In words,. this paper focuses on the elimination of the temporal aliasing components. To devise the filter, *de-aliasing temporal filter*, we assume that moving objects in image sequence have only constant-velocity rigid translational motion[1]. And, we restrict to non-interlace scanning, since interlace scanning can be easily converted to non-interlace one by de-interlacing techniques[4]. In the following, for simplicity, lowpass temporal filter will be mainly considered, since other filters can be anticipated by the use of lowpass characteristics.

2. EFFECTS ON TEMPORAL DOMAIN FILTERING

Video signal can be regarded as a three-dimensional one which has horizontal, vertical and temporal components. Let $f_3(x,y,t)$ be a three-dimensional continuous video signal. Assuming that its moving components have only constant-velocity rigid translational motion $\mathbf{v} = (v_x, v_y)$, Fourier transform of $f_3(\cdot)$ is represented by,

$$F_3(f_x,f_y,f_t) = F_2(f_x,f_y) \cdot \delta(f_x v_x + f_y v_y + f_t), \quad (1)$$

where $F_2(f_x,f_y)$ is a two-dimensional image function, and $\delta(f_x v_x + f_y v_y + f_t)$ is a tilted plane in 3D frequency space described by the equation $f_x v_x + f_y v_y + f_t = 0$, i.e., baseband exists only on the plane. (1) indicates that the greater spatial details or velocity are, the more resultant temporal components will be. Thus, sharp pictures may produce a greater proportion of high temporal frequencies than soft pictures.

(1) is description of a continuous case. Since the used pixels for filtering are 3D sampled data, sampling of $f_3(\cdot)$ is expressed by multiplying $f_3(x,y,t)$ by 3D array of delta functions. Spectrum of the sampled signal is then given by the convolution of Fourier transform $f_3(\cdot)$ and delta function. Thus, it is known that spectrum aliasing between the repeated spectras may occur according to increase of the velocity. If the repeated spectras include aliasing components, visual artefacts appear. Especially, moving areas composed of spatial high frequency components may distort psychovisual effects.

To derive procedure of the de-aliasing temporal filtering, we made two assumptions : first, baseband spectrum has no spatial aliasing components, and secondly there exists only purely horizontal motion. Filtered results based on temporal domain samples are shown in Fig. 1(a). They contain spatial high frequency components of adjacent spectras which is temporal aliasing. We previously stated that the presence of spatial high frequencies results in high temporal frequencies. However, Fig. 1(a) presents that the spatial high frequency components produce temporal low fre-

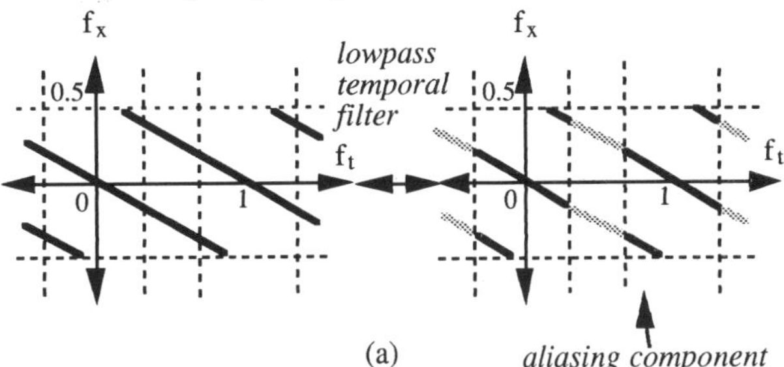

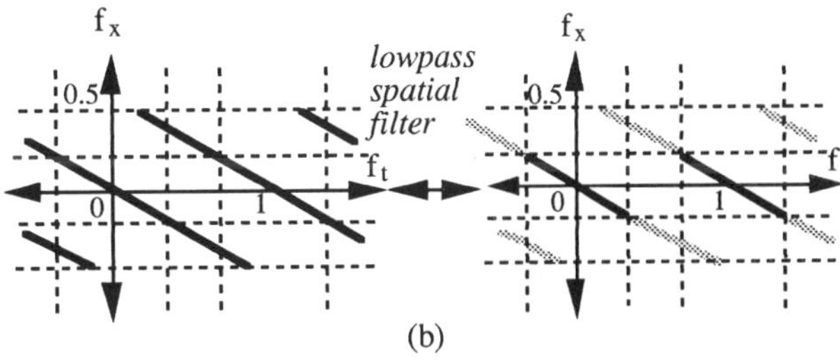

Fig. 1 Comparisons of temporal bandlimitation.
(a) Temporal domain filtering,
(b) Spatial domain filtering (MASF).

quency ones. In other words, the display causes a disturbance between the spatial high frequency components and the low frequency ones. Therefore, reducing of the effects introduced by temporal aliasing (e.g. motion judder, stroke-width modulation) is desirable.

The attenuation afforded to spatial high frequencies has a great effect in reducing temporal aliasing. Although it is allowable, it may cause severe loss of spatial informations. But on the other hand, higher temporal sampling rate is able to achieve an attenuation of temporal aliasing. If given sampling rate is however low, it requires temporal sampling rate conversion which is known to be ill-posed problem. As a result, it is necessary to develop a sophisticated and general approach.

3. DE-ALIASING TEMPORAL FILTER

Using the relstionship between the location of baseband spectrum and the velocity, temporal band-limitation can be described by a spatial filter, i.e., lowpass spatial filtering with its cutoff frequency corresponding to the temporal one provides same effects. From the location of baseband spectrum in (1), a relation between spatial and temporal frequencies is given by,

$$f_{cutoff}^{\ s} = \frac{1}{|\mathbf{v}|} \cdot f_{cutoff}^{\ t}. \tag{2}$$

Compared with a temporal band-limitation using temporal domain pixels of Fig. 1(a), the MASF has de-aliasing characteristic. Temporally band-limited results which do not include temporal aliasing are shown in Fig. 1(b).

This result, de-aliasing characteristic, can be applicable to any types of band-limitations. For examples, though filter types are given by highpass or bandpass, the MASF will preserve the de-aliasing characteristic. They are illustrated in Fig. 2.

Previously mentioned investigation is one-dimensional case which has only horizontal velocity. Extending the velocity into two-dimension, spatial filtering in terms of the velocity is equal to the filtering along movement trajectory, which results in *Motion Adaptive Spatial Filter (MASF)*. Two-dimensional case is shown in Fig. 3.

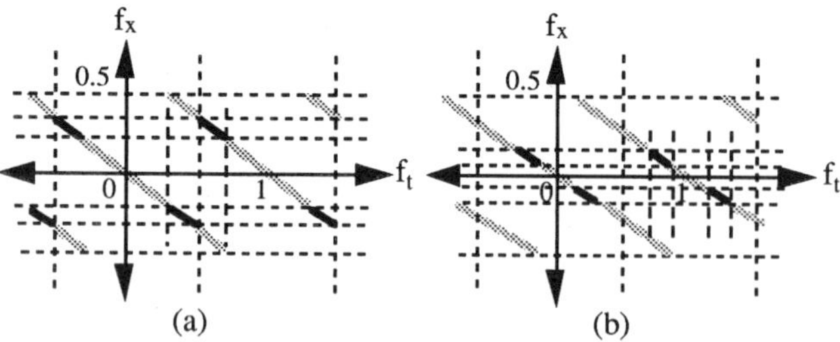

Fig. 2 Examples of temporal bandlimitation based on MASF. (a) Highpass filtering, (b) Bandpass filtering.

With the motivation of the spatial filtering using a velocity, an implementation of a MASF is undertaken. Let $h(\cdot)$ be lowpass impulse response, then the temporally band-limited form of $f(\cdot)$ is given by

$$g(x,t) = \int_{-\infty}^{\infty} h(\tau)\cdot f(x,t-\tau)\,d\tau, \tag{3}$$

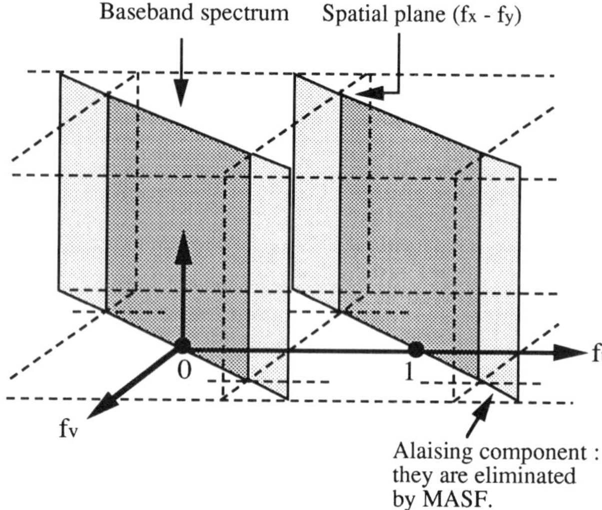

Fig. 3 Two-dimensional illustration of MASF. : Lowpass temporal filtering along movement trajectory of fv.

where a linear phase filter is used concerning the group-delay of response. From an assumption of motion,

$$f(x,t-\tau) = f(x-v_x\tau,t). \tag{4}$$

Thus, (3) is represented by

$$g(x,t) = \int_{-\infty}^{\infty} h(\tau)\cdot f(x-v_x\tau,t)\,d\tau. \tag{5}$$

And its Fourier transform is expressed by

$$G(f_x f_t) = H(f_x v_x)\cdot F(f_x,f_t). \tag{6}$$

$H(\cdot)$ had initially temporal band-limiting charcateristics. Filtering characteristics however vary by a conversion of filtering domain. Since the filtering operation on spatial domain is changed according to the velocity, $H(\cdot)$ performs a lowpass spatial filtering whose cutoff frequency is adapted by the velocity. Therefore, direct realization of (5) results in de-aliasing temporal filter (MASF).

(5) is a continuous description of the MASF. Similar results hold in the discrete case : the integral is replaced by summation and $d\tau$ is represented by combination of $\Delta\tau$ and k. (5) is then given by

$$g(\mathbf{x},n) = \sum_{k=-N}^{N} h(k\Delta\tau)\cdot f(\mathbf{x}-\mathbf{v}(\mathbf{x},n)\Delta\tau\cdot k,n), \tag{7}$$

where velocity and filtering positions are replaced by vector form. And filter length is $2N+1$, $\Delta\tau$ is selected to satisfy $|\mathbf{v}(\cdot)\cdot\Delta\tau| \le |\Delta\mathbf{x}|$. If $\Delta\tau$ is different from the required value, it may cause spatial aliasing.

Let ΔT be a frame to frame interval, $\mathbf{v}(\cdot)\Delta T$ is then equal to $\mathbf{D}(\cdot)$ which is a displacement between current and previous frames. If $\alpha(\mathbf{x},n)$ is defined as a ratio of $\Delta\tau\cdot\alpha(\mathbf{x},n) = \Delta T$,

$$\frac{\mathbf{D}(\mathbf{x},n)}{\alpha(\mathbf{x},n)} = \mathbf{d}(\mathbf{x},n), \quad |\mathbf{d}(\mathbf{x},n)| = |\Delta\mathbf{x}| = 1,$$

(7) is modified as a following description.

$$g(\mathbf{x},n) = \sum_{k=-N}^{N} h(k)\cdot f(\mathbf{x}-\mathbf{d}(\mathbf{x},n)\cdot k, n), \qquad (8)$$

where $\Delta\tau$ is normalized to 1, and $h(\cdot)$ is determined by (2). (8) is a direct realization of (5). However, due to $\alpha(\cdot)$ and $\mathbf{d}(\cdot)$, actual data for filtering may be non-grid pixels. Concerning the computation and efficiency, bi-linear filter which is weighted average with nearest neighbor pixel is used.

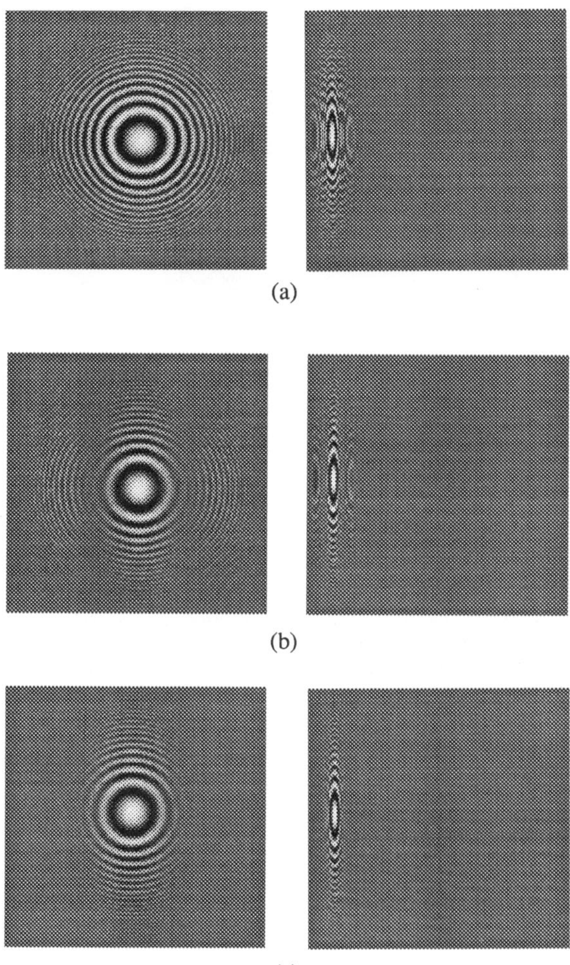

Fig. 4 De-aliasing characteristic of MASF.
Each plane indicates Horizontal-Vertical and Vertical-Temporal planes, respectively. The velocity is assumed to be 5 pixel / frame interval, and normalized temporal cutoff frequency is 0.25.
(a) Original planes,
(b) Filtered planes based on temporal domain filtering,
(c) Filtered planes based on spatial domain filtering.

4. EXPERIMENTAL RESULTS

A zone-plate can be used to study the frequency response of a filter. It has the appearance of concentric rings whose spacing decreases away from the center. And the center of rings is periodically repeated in horizontal or vertical directions. To evaluate a de-aliasing characteristic of the proposed filter, a moving zone-plate without periodicity is generated.

Assuming that velocity function of the moving zone-plate is constant, equation (8) as a temporal filter is evaluated. To verify the de-aliasing characteristic of the MASF, lowpass filtering is campared with direct filtering of Fig. 1(a). Fig. 4 shows that the MASF posseses efficient temporal band-limitation. The MASF results in blurring toward moving direction. The blurring is necessary consequence of temporal band-limitation without severe loss of spatial informations and are not perceived at typical displays[2] (30Hz ···).

For an application of real image sequence, "Miss America" is used. Fig. 5 presents filtered results as a function of temporal cutoff frequency, where motion vector based on BMA is used[1]. Fig. 5(d) gives the result with temporal cutoff frequency-15Hz which is half of the temporal sampling frequency. It indicates maximum bound to eliminate the aliasing. Thus, there are no spectral components of adjacent spectra in any Nyquist region. This filtered sequence is applicable to image coding.

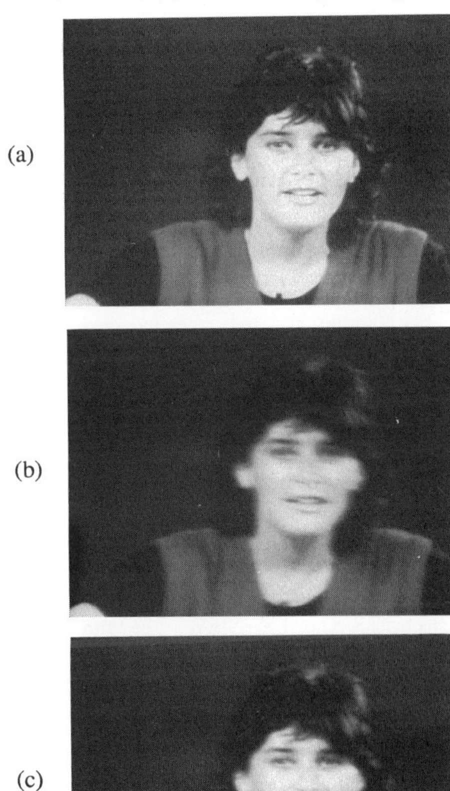

(d) 

Fig. 5 Temporally band-limited results.
:Original temporal sampling frequency is 30Hz, (a) Original frame, (b) (c) (d) indicate filtered frames when normalized temporal cutoff frequency is 0.125, 0.25 and 0.5, respectively.

Clearly, lowpass filteirng results in the increase of redundancy for image coding. Thus, data compression efficiency may be impreved. Results applied to a motion compensated predictive coder is shown in Fig. 6. As can be seen, the PSNR (Peak Signal to Noise Ratio) for the filtered sequence (Fig. 5(d)) is reduced wheras the entropy is high. Consequently, the MASF can be used as prefilter for image sequence coding.

5. CONCLUSIONS

In this paper, effects of temporal aliasing and dealiasing temporal filter are investigated. Unlike a conventional temporal filter, the proposed filter, MASF, achieves efficient temporal band-limitation. Additionally, temporal noises on moving areas are eliminated since they can be regarded to high temporal frequency components. To construct a temporal filter which is not affected by aliasing components, we performed temporal filtering on spatial domain along movement trajectory. The MASF creates blurring on moving areas (e.g. motion blur) resulting from temporal band-limitation and thereby reduces visual artefacts at sequential display.

The MASF achieves temporal band-limitation with arbitrary cutoff frequency, even half of temporal sampling frequency. The filtered sequence is applicable to an input for image sequence coding. i.e., The MASF will be feasibel as a prefilter for coding applications.

ACKNOWLEDGEMENTS

This work was supported by Video Research Center of Daewoo Electronics Co. Ltd. B.W. Lee and G.H. Jang made helpful comments on the manuscript.

REFERENCES

[1] H.G. Musmann, P. Pirsch, H.J. Grallert, "Advances in picture coding", Proceedings of IEEE, Vol.73, 1985, pp. 523-548.

[2] T. Fujio, "High-definition television systems", Proceedings of IEEE, Vol.73, 1985.

[3] W. Chen, D. Hein, "Recursive temporal filtering and frame rate reduction for image coding", IEEE J.of SAC, Vol.SAC-5, No.7, 1987, pp. 1155-1165.

[4] P. Haavisto, J. Juhola, and Y. Neuvo, "Fractional frame rate up-conversion using weighted median filters", IEEE Trans. CE, Vol.35, No.3, 1989, pp. 272-278.

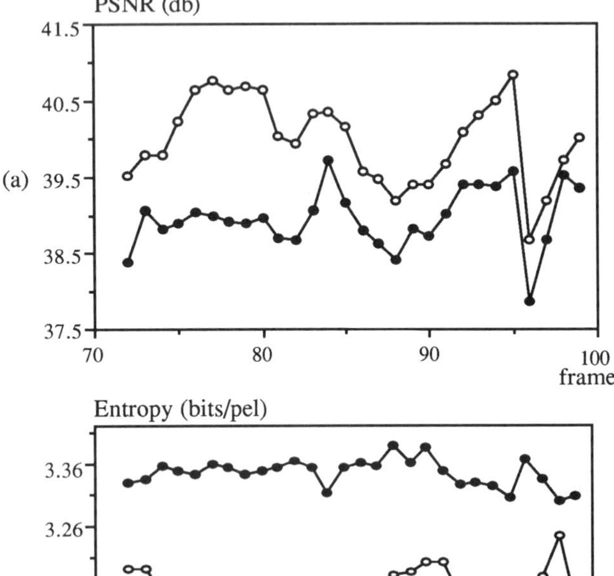

Fig. 6 PSNR and entropy curve of motion compensated predictive coder. (a) PSNR, (b) Entropy.
—●— Original sequence
—○— Filtered sequence

SIGNAL PROCESSING VI: Theories and Applications
J. Vandewalle, R. Boite, M. Moonen, A. Oosterlinck (eds.)

Generalized Theory of Multidimensional M-Band Filter Bank Design

I.A. Shah A.A.C. Kalker

Philips Research Laboratories,
P.O. Box 80.000, 5600 JA Eindhoven, The Netherlands
Tel: (+31) 40 744174
Fax: (+31) 40 744675
Net: kalker@prl.philips.nl, shah@prl.philips.nl

Abstract

The paper presents a generalized theory of designing M-band N_1-dimensional (N_1-D) perfect reconstruction filter banks, from N_2-dimensional (N_2-D) M-band filter banks using transformations, where N_1, N_2 and M are arbitrary integers. We will state theorems on the preservation of the perfect reconstruction property when transforming from N_1-D to N_2-D. The theory is applied in the design of bi-orthogonal filters suitable for quincunx subsampling.

1 Introduction

Multidimensional subband coding systems are becoming popular for image representation and coding. This makes designing multidimensional filters for use in filter banks important. Such designs are well-known for 1-D filters, but are more difficult to design for higher dimensions. Various authors have designed 2-D 2-band filter banks starting from a 1-D 2-band filter bank [2] [3] [4]. Recently, the theory for multidimensional QMF designs was presented in [5].

In this paper we present a generalized theory for transforming an M-band N_1-D filter bank into an M-band N_2-D filter bank for arbitrary integers N_1, N_2 and M. The theorems governing the transformations will be stated. They will be illustrated by the transformation of 1-D bi-orthogonal filters to 2 dimensions.

Section 2 presents theory of the transformations by stating the appropriate theorems. Section 3 will work out the conditions imposed by the theorems for the case of 1-D to 2-D transforms, where the 2-D subsampling matrix is quincunx. Also an example of a 2-D bi-orthogonal filter-bank design is given. In Section 4 extensions are made to the case of M-band designs for $M \geq 2$.

2 General theory of filter transformation

In a previous paper [5] we have shown how a 1-D QMF bank can be transformed into a 2-D QMF bank using a method introduced by McClellan [1]. In this section we recall the McClellan transform and show its generalization.

2.1 The McClellan transform

Given is a zero-phase filter $H(z)$ and a 2-D kernel filter $K(\vec{z})$ where $\vec{z} = (z_1, z_2)$. The zero-phase condition on $H(z)$ implies that there exists a filter $H_S(x)$ such that

$$H(z) = H_S(\frac{1}{2}(z + z^{-1})). \tag{1}$$

The McClellan transform $H^K(\vec{z})$ of $H(z)$ by $K(\vec{z})$ is now defined as

$$H^K(\vec{z}) = H_S(K(\vec{z})), \tag{2}$$

i.e. the 1-D *kernel* $S(z) = \frac{1}{2}(z + z^{-1})$ is replaced by the 2-D kernel $K(\vec{z})$. In order to control the shape of the resulting filter it is required that the values of $K(\vec{z})$ on the unit bi-circle $B = \{(z_1, z_2) : |z_1| = 1, |z_2| = 1\}$ form a subset of the range of $S(z)$ where z ranges over the unit circle. One easily checks that $K(\vec{z})$ has to be zero-phase (in order to be real valued) and to be bounded between -1 and +1 on B.

2.2 Notations and basic facts

In the previous section we have shown that 1-D filters can be transformed into 2-D filters by extraction and insertion of kernels. In our previous paper [5] we have also shown that if we apply this transformation to each filter of a 1-D QMF filter bank the resulting 2-D filter bank is also alias free and perfect reconstructing. Certain restrictions on the 2-D kernel however were needed. More specifically we found that the kernel must live on the non-zero coset of the sampling raster.

Analyzing the proof needed for the QMF filter bank we found that the property which makes things work is the *preservation of the polyphase structure* of the analysis/synthesis bank. In order to formalize the notion of preservation of polyphase structure we have to introduce some notations and definitions.

Let $\mathcal{L}$ be an N dimensional sampling raster represented by a matrix L. The set of cosets $C(\mathcal{L})$ of $\mathcal{L}$ has $\lambda = |det(L)|$ elements and is endowed with a canonical additive group structure. Dual to $C(\mathcal{L})$ the set $W(\mathcal{L})$ is defined by

$$W(\mathcal{L}) = \{w : w^L = \mathbf{1}\}, \tag{3}$$

where $\mathbf{1}$ is the M dimensional vector $(1, 1, \cdots, 1)$ and where exponentation by a matrix is defined as usual [6]. The set $W(\mathcal{L})$ will be used to characterize the polyphase components of an M-D filter $F(z)$.

Definition 1 *Let $F(z)$ be a filter. Let l be a coset of $\mathcal{L}$. Then the polyphase component $F_l(z)$ of $F(z)$ is defined as that part of $F(z)$ which lives on l (i.e. has support on l).*

It is clear that $F(z)$ is equal to the sum of its polyphase components. Moreover we have a characterization of polyphase components in terms of the set $W(\mathcal{L})$.

Theorem 1 *Let $F(z)$, $P_l(z)$, $\cdots$, $P_\lambda(z)$ be filters such that*

$$F(z) = \sum_l P_l(z) \tag{4}$$

and such that

$$P_l(w \otimes z) = w^{-l} P_l(z), \tag{5}$$

where $\otimes$ denotes componentwise multiplication and where exponentation by a coset is defined as usual [6]. Then we have

$$P_l(z) = F_l(z), \tag{6}$$

i.e. equation (4) defines the polyphase decomposition of $F(z)$.

Let (F, G) be an M band analysis/synthesis filter bank with channels $i = 1, \cdots, M$. By considering the *polyphase matrices* of the filter bank one can check for alias cancellation and also determine the transfer function $T(z)$ of the filter bank . For the analysis bank $\{F^i(z)\}_{i=1,\cdots,M}$ the polyphase matrix $(\mathbf{F}_{i,l}(z))$ is defined by

$$\mathbf{F}_{i,l}(z) = (F^i)_l(z), \tag{7}$$

and a similar formula holds for the synthesis bank $\{G^i(z)\}_{i=1,\cdots,M}$.

$$\mathbf{G}_{i,l}(z) = (G^i)_l(z). \tag{8}$$

Let $\{\mathbf{G}^\dagger_{l,i}(z)\}$ be defined by

$$\{\mathbf{G}^\dagger_{l,i}(z)\} = \mathbf{G}_{i,-l}(z), \tag{9}$$

then we have the following theorem.

Theorem 2 *The filter bank (F, G) is alias free and has transfer function $T(z)$ if and only if*

$$(\mathbf{G}^\dagger_{l,i} \circ \mathbf{F}_{i,l})(z) = T(z)I \tag{10}$$

where $\circ$ denotes matrix multiplication and where I is the identity matrix.

With this formulation at hand a natural way of transforming a filter bank is given by extracting a kernel from or inserting a kernel into the components of the polyphase matrices. If such an insertion or extraction is valid (see below), alias cancellation is guaranteed. In that case the transfer function changes by extraction or insertion of the same kernel. The problem is the retainment of the property of being a polyphase matrix. The next section discusses conditions on the kernel to solve this problem.

2.3 Valid transformations

Before stating a theorem about valid transformation we need some preliminaries. First of all we need a formal definition of a valid transformation. Let $\mathcal{L}_i$, $i = 1, 2$, be two sampling rasters of dimension N_i. Let τ be an isomorphism from $C(\mathcal{L}_1)$ to $C(\mathcal{L}_2)$. Finally let $K(z)$ be a N_2-D filter (kernel) with N_1-D values. With these notations we are able to give the definition of a valid transformation.

Definition 2 *Let $F(z)$ be a N_1-D filter with 1-D values. We say that the transformation*

$$F(z) \rightarrow F(K(z)), \tag{11}$$

where

$$F_K(z) = F(K(z)) \tag{12}$$

is a valid transformation with respect to τ if for all $l \in C(\mathcal{L}_1)$ the following identity holds.

$$F_l(K(z)) = (F_K)_{\tau(l)}(z). \tag{13}$$

The following theorem can now easily be proved.

Theorem 3 *Let (F, G) be an alias free (with respect to the sampling raster $\mathcal{L}_1$) M-band analysis/synthesis bank with transfer function $T(z)$. Let kernel substitution with $K(z)$ be valid with respect to τ for every filter in the total filter bank. Then the filter bank (F_K, G_K) is also alias free (with respect to the sampling raster $\mathcal{L}_2$) and its transfer function is given by $T(K(z))$.*

For the above theorem to be of any use two problems have to be solved: given a filter bank which kernels provide for valid transformations, and secondly, which of these kernels generate filter banks useful for applications? We postpone the latter question to the following section and treat the other question first. The following theorem holds.

Theorem 4 *Given two sampling rasters $\mathcal{L}_i$, $i = 1, 2$, an isomorphism τ and a kernel filter $K(z)$ as above, then the following assertions holds.*

1. *There exist an isomorphism σ from $W(\mathcal{L}_2)$ to $W(\mathcal{L}_1)$ such that for all $w \in W(\mathcal{L}_2)$ and all $l \in C(\mathcal{L}_1)$ the equality*
$$\sigma(w)^l = w^{\tau(l)} \tag{14}$$
holds. The morphism σ is uniquely determined by τ. Moreover for every isomorphism σ a corresponding isomorphism τ can be found.

2. *Substitution by $K(z)$ is valid with respect to τ for every $F(z)$ (of the correct dimension) if and only if $K(z)$ satisfies*
$$K(w \otimes z) = \sigma(w) \otimes K(z) \tag{15}$$
for all $w \in W(\mathcal{L}_1)$.

For a specific case this theorem can be reformulated in a less abstract way.

Theorem 5 *Let $K(z)$ be an kernel with 1-D values, i.e. $N_1 = 1$. Then substitution by $K(z)$ is a valid transformation with respect to some τ if and only if*

1. *$C(\mathcal{L}_2)$ is cyclic of order $\lambda_1 = \lambda_2 \stackrel{\text{def}}{=} \lambda$.*

2. *$K(z)$ is equal to one of its polyphase components K_l such that l generates the group $C(\mathcal{L}_2)$*

3 Examples and Design Applications

3.1 The Quincunx Case

In this example we will apply the theorems to transform a 1-D 2-band filter bank consisting of zero-phase filters to a 2-D 2-band filter bank with a quincunx subsampling raster. The transform consists of two phases: first extraction of the kernel $S(z) = \frac{1}{2}(z + z^{-1})$, then the insertion of a 2-D kernel filter $K(z)$.

In order to check that the extraction of $S(z)$ is a valid transformation we can apply Theorem 5 where we use subsampling by 2 on both sides of the transformation. The group structure of the coset space is in this case isomorphic to the cyclic space of order 2. Consequently there exists only one generating element, viz. the non-zero coset. Extraction of $S(z)$ is a valid transformation as $S(z)$ lives on this non-zero coset . Another way of checking is given by Theorem 4, i.e. we have to verify

$$S(w_{1d} \otimes z) = \sigma(w_{1d})S(z), \tag{16}$$

where $\sigma(w_{1d}) \stackrel{\text{def}}{=} w_{1d}$ and where w_{1d} ranges over $\{1, -1\}$. This equality is easily checked.

A possible matrix for quincunx subsampling is given by

$$L_{2d} = \begin{bmatrix} 1 & 1 \\ -1 & 1 \end{bmatrix} \tag{17}$$

and the corresponding $w's$ are given by the set:

$$W(\mathcal{L}_{2d}) = \left\{ \begin{pmatrix} 1 \\ 1 \end{pmatrix}, \begin{pmatrix} -1 \\ -1 \end{pmatrix} \right\}. \tag{18}$$

The quincunx coset space is a cyclic group of order two. Again by Theorem 5 substitution by $K(\vec{z})$ is valid if $K(\vec{z})$ lives on the non-zero coset of the quincunx sampling raster. As above one can also check validity by applying Theorem 4.

$$K(w_{2d} \otimes \vec{z}) = \sigma(w_{2d})K(\vec{z}). \tag{19}$$

The isomorphism σ is unique given that we are dealing with cyclic groups of order two. Its explicit representation is given by

$$\begin{pmatrix} 1 \\ 1 \end{pmatrix} \to 1, \begin{pmatrix} -1 \\ -1 \end{pmatrix} \to -1. \tag{20}$$

As both phases of the transform are valid transformations the alias cancellation property of the filter bank will be retained. If, moreover, the original 1-D filter bank is perfect reconstructing then the resulting 2-D filter bank will also be perfect reconstructing. Starting from a 1-D bi-orthogonal zero-phase filter bank we can now do an actual 2-D filter-bank design.

3.2 Design Application

The transformation from 1-D to 2 and 3-D QMF filters was discussed in [5]. Here we will use the above technique to transform a set of 1-D bi-orthogonal filters into 2-D "Fan" shaped filters, suitable for use in a filter bank with quincunx subsampling.

We start with zero-phase 1-D filters. We perform a valid transformation by extracting the 1-D kernel $S(z)$. We then insert a valid 2-D kernel $K(\vec{z})$ given by

$$K = \begin{pmatrix} 0 & -1/4 & 0 \\ 1/4 & 0 & 1/4 \\ 0 & -1/4 & 0 \end{pmatrix}. \tag{21}$$

It can be shown that $K(\vec{z})$ is valid by checking that $K(w_{2d} \otimes \vec{z}) = \sigma(w_{2d})K(\vec{z})$. Substituting the values for the 2-band quincunx case we get $K(-\vec{z}) = -K(\vec{z})$. Figures 1 and 2 shows the frequency response of the upper and lower branch of the resulting fan-shape filters.

4 Extensions to M-band Designs

All the theory has been presented in a general way, and is valid for arbitrary dimensions and number of

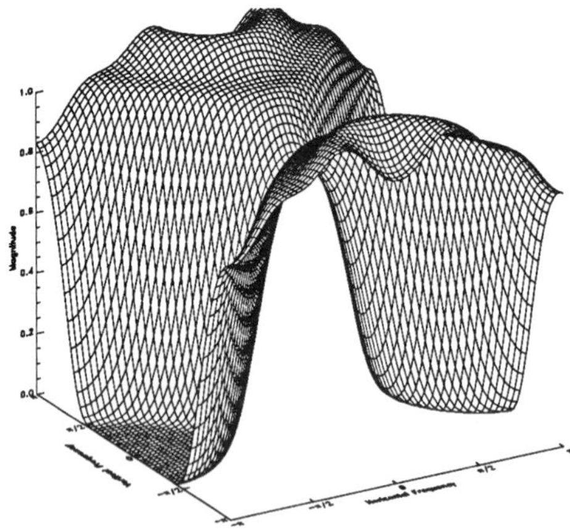

Figure 1: Lower branch of fan-shaped filter

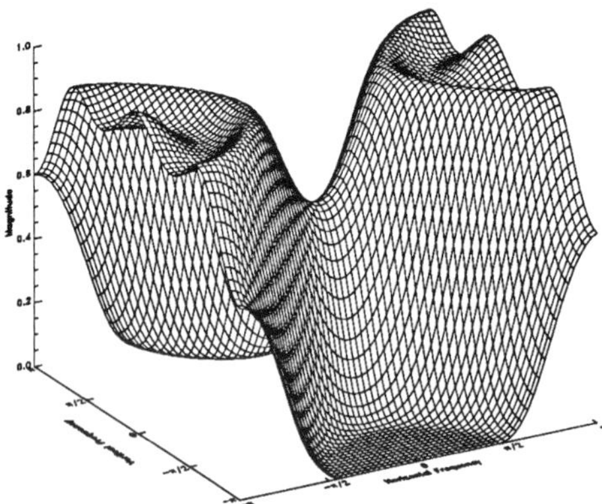

Figure 2: Upper branch of fan-shaped filter

bands. In this section we concentrate on the application of the theory to the design of 2-D filters with M bands for M ≥ 2.

The procedure is the same as in the two band case. First extract from each M_1-D filter $H(z)$ a valid kernel $S(z)$, that lives on an allowed coset and then insert a valid N_2-D kernel. The resulting N_2-D filter bank will preserve the alias cancellation and perfect reconstruction properties of the N_1-D filter bank. To give a specific example, we consider the transformation of a 1-D 3-band filter bank into a 2-D 3-band filter bank. The extraction of the valid 1-D kernel in this case would mean finding a filter $H_S(z)$ such that

$$H(z) = H_S(\frac{1}{2}(z + z^{-2})) \tag{22}$$

and then replacing $\frac{1}{2}(z + z^{-2})$ by a valid 2-D kernel $K(\vec{z})$. The practical problem with such a procedure is that finding $H_S(z)$ is more difficult than the 2-band case. It should be noted that the extraction of the 1-D kernel controls the shape of the 2-D filter, which is of importance in practical filter design. An alternative procedure could be to perform a direct 2-D kernel insertion into $H(z)$ and not extract a 1-D kernel at all. If the kernel insertion is valid, this procedure will result in a 2-D filter bank with alias cancellation and perfect reconstruction. However there will be little control over the frequency response of the resulting individual 2-D filters.

5 Summary

In this paper we have presented generalized theory of transforming M-band N_1-D filter banks into N_2-D filter banks, while preserving the N_1-D properties. The theory has been used to design 2-band 2-D biorthogonal filters. The design of M-band (M ≥ 2) filter banks is considered.

References

[1] J.H. McClellan, *The Design of Two-Dimensional Digital Filters by Transformations*, 7th Annual Princeton Conf. Inf. Sci. and Syst., Proc., pp. 247–251,1973.

[2] J. De Lameillieure, G. Schamel, U.Goelz, M. Leptin, *Two-Dimensional diagonal passband filters for subband coding with a quincunx sampling pattern*, Proc. of the IEEE/ProRISC Symposium on Circuits, Systems and Signal Processing, pp. 75–78, 1991.

[3] J. Kovačević, M. Vetterli, *Nonseperable Multidimensional Perfect Reconstructing Filter Banks and Wavelet Bases for $\mathcal{R}^n$*, IEEE Trans. on information theory, 1992 (to be published).

[4] C. Guillemot, A. Enis Cetin, R. Ansari, *M-channel Nonrectangular Wavelet Representations for 2-D Signals: Basis for Quincunx Sampled Signals*, Proc. of ICASSP 1991, pp. 2813-2816, 1991.

[5] I. Shah, A. Kalker, *Theory and Design of Multidimensional QMF Sub-Band Filters From 1-D Filters Using Transforms*, Proc. of the 4th Int. Conf. on Image Processing and applications, 1992 (to be published).

[6] E. Viscito, J.P. Allebach, *The Analysis and Design of Multidimensional FIR Perfect Reconstruction Filter Banks for Arbitrary Sampling Lattices*, IEEE Trans. on Circuits and Systems, Vol. 38, No.1, pp. 29-41, January 1991.

SIGNAL PROCESSING VI: Theories and Applications
J. Vandewalle, R. Boite, M. Moonen, A. Oosterlinck (eds.)

A NEW WLS CHEBYSHEV APPROXIMATION METHOD FOR THE DESIGN OF FIR DIGITAL FILTERS WITH ARBITRARY COMPLEX FREQUENCY RESPONSE

Chong-Yung Chi and Shyu-Li Chiou

Department of Electrical Engineering, National Tsing Hua University, Hsinchu, Taiwan, ROC.

Abstract: In this paper, motivated by Chi–Kou's weighted least–squares (WLS) Chebyshev approximation method [1] for the design of FIR digital filters with linear phase, we propose a new self–initiated iterative WLS Chebyshev approximation method for the design of FIR digital filters with arbitrary complex frequency response. The proposed method not only inherits all the advantages of Chi–Kou's method but also is computationally more efficient than Chi–Kou's method for the case of linear phase FIR filter design. Two design examples are provided to demonstrate the good performance of the proposed method.

1. Introduction

Recently, Chi and Kou [1] proposed an efficient weighted least–squares (WLS) Chebyshev approximation method for the design of linear phase equiripple FIR digital filters which performs as well as the well–known Parks–McClellan's algorithm [2] with some extra advantages. For instance, the former does not need the initial guess for a suitable set of extremal frequencies or filter coefficients; it is applicable in the design of any type linear phase FIR filters with no need of nontrivial modifications required by the latter. However, they are not applicable in the design of FIR digital filters with arbitrary complex frequency response.

Mason and Chit [3] proposed a time–domain least mean square (LMS) approximation method for the design of FIR filters with arbitrary complex frequency response. In this paper, we propose a new WLS Chebyshev approximation method based on Chi–Kou's approximation method. The proposed approximation method is not only applicable in the case of FIR filter with arbitrary complex frequency response but also computationally more efficient than Chi–Kou's approximation method for the linear phase case. It can be viewed as a frequency–domain counterpart to Mason–Chit's LMS method based on similar philosophies. Furthermore, the former has better performance than the latter and is much simpler than some well–known complex approximation methods such as [4,5].

In Section 2, we present the new WLS Chebyshev approximation method. Then we show two design examples to support the proposed approximation method in Section 3. Finally, we draw some conclusions.

2. The WLS Chebyshev Approximation Method

Assume that the desired complex frequency response $H_d(f)$ is conjugate–symmetric, i.e., $H_d(f) = H_d^*(-f)$, and that the domain $B_{NT} \subset [0,0.5]$ over which $H_d(f)$ is well defined includes p disjoint nontransition bands as follows:

$$B_{NT} = B_1 \cup B_2 \cup \cdots \cup B_p$$

where

$$B_m = \{ f | f_{m1} \le f \le f_{m2} \},$$

f_{m1} and f_{m2} denote the specified cutoff frequencies in the mth frequency band B_m. Then the union of the transition bands, denoted B_{TS}, is given by

$$B_{TS} = \{f \mid 0 \le f \le 1/2,\ f \notin B_{NT}\}.$$

Assume that the filter to be designed is an (M–1)th–order FIR filter with real filter coefficients h(n). The frequency response is then

$$H(f) = \sum_{n=0}^{M-1} h(n)\, e^{-j2\pi fn}. \qquad (1)$$

We define the complex approximation error E(f) as

$$E(f) = H_d(f) - H(f) = E_r(f) + j\, E_i(f) \qquad (2)$$

where $E_r(f)$ and $E_i(f)$ are real part and imaginary part of E(f), respectively. Let $W_e(f)$, $f \in B_{NT}$, be a piecewise–constant function associated with the desired relative approximation error ratio among p nontransition bands, defined as

$$W_e(f) = \rho_m, \text{ if } f \in B_m \qquad (3)$$

where $\rho_1>0$, $\rho_2>0$, $\cdots$, $\rho_p>0$, $\max\{\rho_1, \rho_2, \ldots, \rho_p\}=1$, and the ratio $(1/\rho_1):(1/\rho_2): \cdots:(1/\rho_p)$ denotes the desired relative approximation error ratio among B_1, B_2, ..., B_p. Our object is to find a set of filter coefficients h(n) by the well–known WLS method such that H(f) is equiripple with $\delta_1:\delta_2 \cdots:\delta_p = (1/\rho_1):(1/\rho_2): \cdots:(1/\rho_p)$ where δ_m is the maximum error in B_m. Next, let us present the WLS estimator on which the new approximation design method is based.

For notational simplicity, let $W_e(k)$, E(k), $E_r(k)$ and $E_i(k)$ also denote $W_e(f=k/2N)$, E(f=k/2N), $_r(f=k/2N)$ and $E_i(k/2N)$, respectively, where N is the total number of uniform samples in the interval [0, 0.5]. Thus, by (1) and (2), we can express $E_r(k)$ and $E_i(k)$ for k = 0,1, ..., N–1, in the following

This work was supported by the National Science Council under Grant NSC81–0404–E–007–010.

linear vector form:

$$\begin{bmatrix} \underline{E}_r \\ \underline{E}_i \end{bmatrix} = \begin{bmatrix} Re\{\underline{H}_d\} \\ Im\{\underline{H}_d\} \end{bmatrix} - \begin{bmatrix} D_1 \\ D_2 \end{bmatrix} \underline{h} \quad (4)$$

where

$\underline{h} = [h(0), h(1), ..., h(M-1)]'$,

$\underline{E}_r = [E_r(0), E_r(1), ..., E_r(N-1)]'$,

$\underline{E}_i = [E_i(0), E_i(1), ..., E_i(N-1)]'$,

$\underline{H}_d = [H_d(0), H_d(1), ..., H_d(N-1)]'$,

$Re\{\underline{H}_d\}$ and $Im\{\underline{H}_d\}$ denote real part and imaginary part of $\underline{H}_d$, respectively, and D_1 as well as D_2 are N×M matrices with the (i,j)th element $[D_1]_{ij} = \cos[(i-1)(j-1)\pi/N]$ and $[D_2]_{ij} = -\sin[(i-1)(j-1)\pi/N]$, respectively. The sum of weighted error squares is defined as

$$J(\underline{h}) = \sum_{k=0}^{N-1} w(k)|E(k)|^2 = \underline{E}_r' W \underline{E}_r + \underline{E}_i' W \underline{E}_i \quad (5)$$

where $W = diag\,[w(0), w(1), ..., w(N-1)]$ with $w(k) \geq 0$ for all $0 \leq k \leq N-1$. It is well–known that the WLS estimate $\hat{\underline{h}}$, which minimizes J, is given by [6]

$$\hat{\underline{h}} = [D_1' W D_1 + D_2' W D_2]^{-1} \cdot [D_1' W Re\{\underline{H}_d\} + D_2' W Im\{\underline{H}_d\}]. \quad (6)$$

A well–known property of WLS estimators is as follows:

(P1) *The larger the weight w(k), the smaller is the associated error |E(k)|.*

The new approximation method is an iterative algorithm based on (P1) for finding the optimum w(k) such that |E(k)| is equiripple with the desired approximation error ratio among the nontransition bands, or, equivalently, $|W_e(k)E(k)|$ for $(k/2N)\epsilon B_{NT}$ is equiripple. Before presenting the new design method, we define some notations for ease of latter use:

- Error ripple $E_m^i(k)$:

$$E_m^i(k) = |E(k)|,\ k/2N \in B_m^i,\ i = 1, 2, ..., q, \quad (7)$$

where q is the total number of error ripples in B_m and

$$B_m^i = \{ f \mid f_m^{i-1} \leq f \leq f_m^i \} \subset B_m$$

where $f_m^0 = f_{m1}$, $f_m^q = f_{m2}$, and $f_m^i \in B_m$, i = 1, 2, ..., q–1, are the frequencies at each of which |E(f)| is a local minimum.

- Amplitude e_m^i of error ripple E_m^i:

$$e_m^i = \max\,\{E_m^i(k),\ k/2N \in B_m^i\}. \quad (8)$$

- Piecewise–constant function R(k), $k/2N\epsilon B_{NT}$:

$$R(k) = W_e(k)\cdot e_m^i = \rho_m \cdot e_m^i,\ \text{ if } k/2N \in B_m^i. \quad (9)$$

The new design method, which is shown in Fig. 1, begins with the initial weighting function

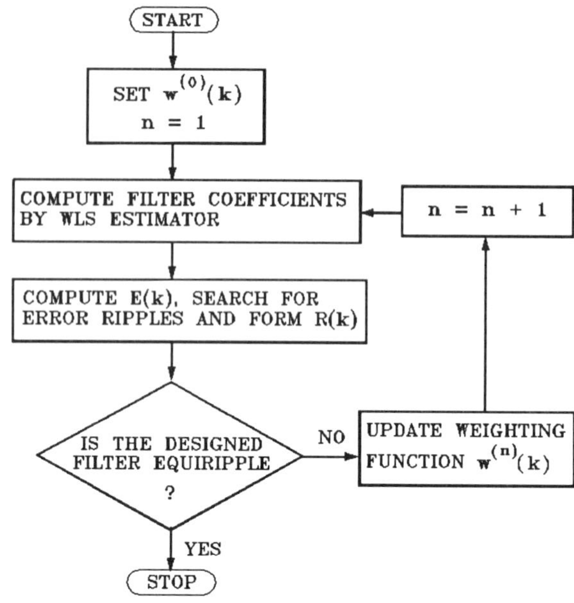

Fig. 1. The optimum WLS approximation method.

$$w^{(0)}(k) = \begin{cases} W_e(k), & k/2N \in B_{NT} \\ 0, & k/2N \in B_{TS}. \end{cases} \quad (10)$$

Assume that we ended up with the weighting function $w(k) = w^{(n-1)}(k)$ at the (n–1)th iteration. For the nth iteration, the WLS estimate, $\hat{\underline{h}}$, is computed by (6), in which $w(k) = w^{(n-1)}(k)$. Then the associated E(k) is computed by (4). Next, we search for $E_m^i(k)$ and e_m^i, for all i and m, and form R(k). Then we check whether $|W_e(k)\cdot E(k)|$ for $k/2N\epsilon B_{NT}$ is equiripple by

$$(R_{max} - R_{min})/R_{max} \leq \sigma \quad (11)$$

where $R_{max} = \max\{R(k),\ k/2N\epsilon B_{NT}\}$, $R_{min} = \min\{R(k),\ k/2N\epsilon B_{NT}\}$ and σ is a preassigned small positive constant. If $|W_e(k)\cdot E(k)|$ for $k/2N\epsilon B_{NT}$ is not equiripple yet, we update the weighting function by

$$w^{(n)}(k) = \begin{cases} w^{(n-1)}(k)R(k)/w_{max}, & k/2N\epsilon B_{NT} \\ 0, & k/2N\epsilon B_{TS} \end{cases} \quad (12)$$

where

$$w_{max} = \max\,\{w^{(n-1)}(k)\cdot R(k),\ k/2N\epsilon B_{NT}\} \quad (13)$$

normalizes $w^{(n)}(k)$ such that $0<w^{(n)}(k)\leq 1$ for all $k/2N\epsilon B_{NT}$.

Although the proposed approximation method was illuminated by the case of real filter coefficients due to $H_d(f)=H_d^*(-f)$, it is applicable in the case of complex filter coefficients by replacing the linear model (4) and the WLS estimate (6) with

$$\underline{E} = \underline{H}_d - D\,\underline{h} \qquad (14)$$

and

$$\hat{\underline{h}} = [D^H W D]^{-1} D^H W \underline{H}_d, \qquad (15)$$

respectively, where D^H is the complex conjugate transpose of N×M matrix D whose elements can be easily determined from the definition of H(f).

The FIR filter to be designed may have to satisfy certain implicit constraints. For instance, type I and type II linear phase FIR filters require h(n) = h(M−1−n) which can be easily taken into account in the proposed approximation method as follows. Assume that M is even (type II). Then, H(f) and $H_d(f)$ can be expressed as

$$H(f)=\exp\{-j2\pi fL\} \sum_{n=0}^{M/2-1} 2\,h(n)\cos(2\pi f(L-n)) \qquad (16)$$

and

$$H_d(f) = d(f)\exp\{-j2\pi fL\} \qquad (17)$$

where L=(M−1)/2 and d(f) is real. The approximation error E(f) can be redefined as

$$E(f) = d(f) - \sum_{n=0}^{M/2-1} 2\,h(n)\cos(2\pi f(L-n)) \qquad (18)$$

which leads to the following real linear vector model:

$$\underline{E} = \underline{d} - D\,\underline{h} \qquad (19)$$

where $\underline{d}$ = [d(0), d(1), ..., d(N−1)]', $\underline{h}$ = [h(0), h(1), ..., h(M/2−1)]' containing only nonredundant filter coefficients and D is an N×(M/2) matrix whose elements can be easily determined from (18). The WLS estimate is then

$$\hat{\underline{h}} = [D' W D]^{-1} D' W \underline{d}. \qquad (20)$$

3. Design Examples

Two design examples, which were obtained with N = 1000 and convergence parameter σ = 0.01, are presented below to support the proposed WLS Chebyshev approximation method.

Example 1. Lowpass filter with constant group delay τ_d (taken from [4,5]):

The desired frequency response $H_d(f)$ is as follows:

$$H_d(f) = \begin{cases} e^{-j2\pi\tau_d f}, & f \in B_1=[0,0.06] \text{ (passband)} \\ 0, & f \in B_2=[0.12,0.5] \text{ (stopband)} \end{cases}$$

where τ_d=12. The filter to be designed is a 2–band filter (p=2) of order M−1=30 and

$$W_e(f) = \begin{cases} \rho_1=0.1, & f\in B_1 \\ \rho_2=1, & f\in B_2. \end{cases}$$

The total iterations spent by the proposed approximation method is 11. The magnitude response and group delay response of the designed filter are shown in Fig. 2(a) and Fig. 2(b), respectively. One can see, from Fig. 2(b), that in the passband, the group delay is nearly constant although it oscillates around 12 with maximum deviation of 1.068. Figure 2(c) shows the absolute approximation error |E(f)|

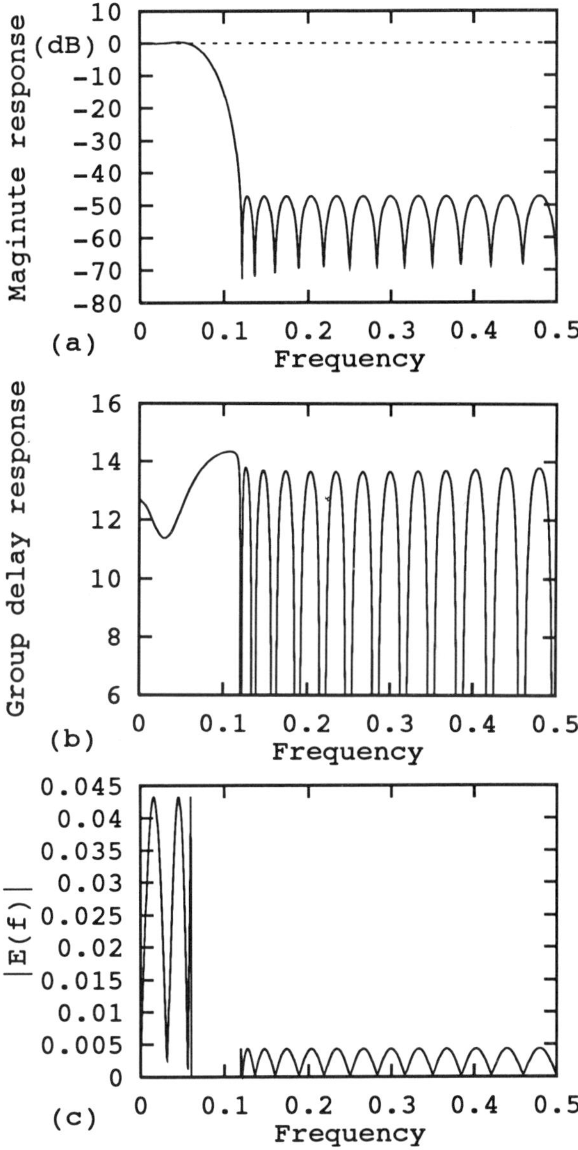

Fig. 2. The designed lowpass filter of order M−1=30 for B_1=[0,0.06] (passband), B_2=[0.12,0.5] (stopband) and τ_d=12 (group delay). (a) Magnitude response; (b) group delay response; (c) absolute approximation error.

which is equiripple with maximum errors δ_1=0.0441 and δ_2=0.00443 , in the passband and the stopband, respectively. Note that $\delta_1/\delta_2 \approx (1/\rho_1)/(1/\rho_2) = 10$. These results are also comparable with the corresponding results reported in [4] and [5].

Example 2. Type II linear phase shaping filter (taken from [3]):

The desired amplitude response d(f) is as follows:

$$d(f) = \begin{cases} e^{-10(0.25-f)\ln(10)}, & f \in B_1=[0,0.25] \\ 0, & f \in B_2=[0.28,0.5]. \end{cases}$$

The filter to be designed is a 2–band filter of order M–1=49 and $W_e(f) = 1, \quad f \in B_{NT}=B_1 \cup B_2$. The total iterations spent by the proposed approximation method is 10. Figures 3(a) and 3(b) show magnitude response and $|E(f)|$, respectively. From Fig. 3(b), one can see that $|E(f)|$ is equiripple with maximum error $\delta_1 = \delta_2 = \delta = 0.0353$. For the same example, it was reported in [3] that Mason–Chit's method spends 412 (>>10) iterations and the corresponding equiripple $|E(f)|$ has maximum error $0.064 \cong 1.8 \cdot \delta$. Hence, this design example supports that the performance of the proposed method is much superior to that of Mason–Chit's method from both viewpoints of iteration number and maximum error.

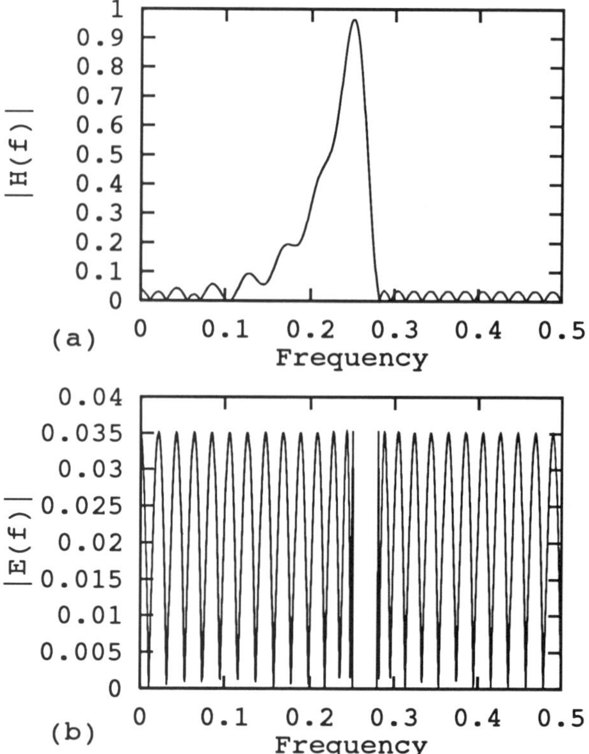

Fig. 3. The designed type II linear phase 2–band shaping filter of order M–1=49 for $B_1=[0,0.25]$ and $B_2=[0.28,0.5]$. (a) Magnitude response and (b) absolute approximation error.

4. Conclusions

Let us summarize the characteristics of the proposed WLS Chebyshev approximation method for the design of FIR filters with arbitrary complex frequency response in the following:

(1) Contrast to Mason–Chit's time–domain LMS method [3] which is also for the design of FIR filters with arbitrary frequency response, the proposed method is a frequency–domain WLS method based on similar philosophies. The latter has better performance than the former although the designed optimum filters by both methods are equiripple.

(2) Contrast to Chi–Kou's method which is for the design of FIR filters with linear phase, the proposed method adjusts the filter coefficients by simultaneously considering approximation errors in each frequency band and desired error ratio among all frequency bands. Therefore, the proposed method is computationally more efficient than Chi–Kou's approximation method from the viewpoint of total iterations needed.

(3) The proposed method possesses the same advantages as Chi–Kou's approximation method as follows. It allows the exact specification of the cutoff frequencies; it does not need any initial guess for a suitable set of extremal frequencies or filter coefficients; it can be directly applied to the design of any type linear phase FIR filters without need of extra modifications.

(4) The FIR filter to be designed may have to satisfy certain implicit constraints (such as particular symmetries of linear phase FIR filters) which can be easily taken into account in the proposed approximation method by properly reformulating the linear model (4) such that $\underline{h}$ contains only nonredundant filter coefficients. However, as long as N is large enough, the same designed filter will be obtained no matter whether implicit constraints on filter coefficients are considered or not.

References

[1] C.–Y. Chi and Y.–T. Kou, "A new self–initiated optimum WLS approximation method for the design of linear phase FIR digital filters." Proc. IEEE 1991 International Symposium on Circuits and Systems, vol. 1, pp. 168–171, Singapore, June 1991.

[2] T. W. Parks and J. H. McClellan, "Chebyshev approximation for nonrecursive digital filters with linear phase," IEEE Trans. Circuit Theory, vol. CT–19, pp. 189–194, March 1972.

[3] J. S. Mason and N. N. Chit, "New approach to the design of FIR digital filters," IEE Proceedings, Pt. G, vol. 134, no. 4, pp. 167–180, Aug. 1987.

[4] X. Chen and T. W. Parks, "Design of FIR filters in the complex domain," IEEE Trans. Acoustics, Speech, and Signal Processing, vol. ASSP–35, pp. 144–153, Feb. 1987.

[5] K. Preuss, "On the design of FIR filters by complex Chebyshev approximation," IEEE Trans. Acoustics, Speech, and Signal Processing, vol. ASSP–37, pp. 702–712, May 1989.

[6] J. M. Mendel, Lessons in digital estimation theory, Prentice–Hall, Englewood Cliffs, New Jersey, 1987.

SIGNAL PROCESSING VI: Theories and Applications
J. Vandewalle, R. Boite, M. Moonen, A. Oosterlinck (eds.)

MODULATED FILTER BANKS IN MULTIDIMENSIONS WITH ARBITRARY SAMPLING LATTICES

David B. H. Tay and Nick G. Kingsbury

Signal Processing and Communications Laboratory, Department of Engineering,
University of Cambridge, Trumpington Street, Cambridge, CB2 1PZ, England.

An analysis of DFT modulated filter banks in multidimensions with arbitrary sampling lattices is presented. The filters are obtained by modulating a prototype filter. An efficient polyphase network implementation will be derived and the issue of aliasing cancellation and perfect reconstruction is addressed. The shape aspects of the frequency subbands are discussed.

1. INTRODUCTION, NOTATIONS AND DEFINITIONS

Over the last decade there has been a great deal of interest in the area of multirate digital filter banks. These systems perform analysis/synthesis of signals to/from subband components and find wide applications in areas such as bandwidth compression, signal scramblers and spectrum analysis. Most of the work concerning multirate digital filter banks has dealt with one dimensional signals and systems but recently there has been an increasing interest in multidimensions [1],[2]. One important class of filter banks is the DFT (Discrete Fourier Transform) modulated filter bank and it has a very efficient polyphase implementation [3]. This class of filter bank in one dimension has been treated by various authors [4],[5]. A special case in two dimensions where the sampling lattice used was rectangular was presented in [6], but the issue of aliasing cancellation and perfect reconstruction was however not addressed. In this paper we extend and generalize the DFT modulated filter bank to multidimensions with an arbitrary sampling lattice.

We briefly review some definitions and a notation that are useful for compact description of multidimensional multirate systems (see [1] for a more complete treatment).
An N-dimensional sequence is denoted by $x(\mathbf{n})$ where $\mathbf{n} = [\ n_1 \ \ldots \ n_N\ ]^T$ is a vector of integer variables. The z-transform variable $\mathbf{z} = [\ z_1 \ \ldots \ z_N\ ]^T$ is a vector of complex variables. We define $\mathbf{z}^{\mathbf{n}} = \prod_{i=1}^{N} z_i^{n_i}$ and the z-transform by $X(\mathbf{z}) = \sum_{\mathbf{n}\in\Lambda} x(\mathbf{n})\,\mathbf{z}^{-\mathbf{n}}$ (Λ is the set of all integer vectors $\mathbf{n}$). The Fourier transform $X^F(\boldsymbol{\omega})$ ($\boldsymbol{\omega} = [\ \omega_1 \ \ldots \ \omega_N\ ]^T$) is obtained by evaluating $X(\mathbf{z})$ at $\mathbf{z} = [\ e^{j\omega_1} \ \ldots \ e^{j\omega_N}\ ]^T$.

A sublattice $\Lambda_{\mathbf{D}}$ of Λ is defined to be the set of points $\mathbf{m} = \mathbf{Dn}$ ($\mathbf{n} \in \Lambda$). The matrix $\mathbf{D}$ of integer elements is known as the sampling (or periodicity) matrix. A coset of $\Lambda_{\mathbf{D}}$ is the set of points obtained by shifting $\Lambda_{\mathbf{D}}$ by an integer vector $\mathbf{m}$ (coset vector). There are exactly $D = |\det(\mathbf{D})|$ distinct cosets that are represented by the $\mathbf{m}_0, \ldots, \mathbf{m}_{D-1}$ coset vectors. The subsampling operation is defined as $y(\mathbf{n}) = x(\mathbf{Dn})$ (only points in $\Lambda_{\mathbf{D}}$ are retained) with effective subsampling rate D. The upsampling is defined as $y(\mathbf{n}) = x(\mathbf{D}^{-1}\mathbf{n})$ for $\mathbf{n} \in \Lambda_{\mathbf{D}}$ and $y(\mathbf{n}) = 0$ elsewhere. If $\mathbf{d}_l$ are the columns of $\mathbf{D}$ we define $\mathbf{z}^{\mathbf{D}} = [\ \mathbf{z}^{\mathbf{d}_1} \ \mathbf{z}^{\mathbf{d}_2} \ \ldots \ \mathbf{z}^{\mathbf{d}_N}\ ]^T$. The z-transform input/output relationships of the subsampling and upsampling operations are respectively :-

$$Y(\mathbf{z}) = \frac{1}{D}\sum_{l=0}^{D-1} X(\mathbf{W}(\mathbf{k}_l) \circ \mathbf{z}^{\mathbf{D}^{-1}}) \qquad (1)$$

$$Y(\mathbf{z}) = X(\mathbf{z}^{\mathbf{D}}) \qquad (2)$$

where $\mathbf{k}_l$ are the coset vectors of $\mathbf{D}^T$ (sublattice $\Lambda_{\mathbf{D}^T}$) and $\mathbf{W}(\mathbf{k}_l) = [\ \exp(-j2\pi\mathbf{k}_l^T\mathbf{d}'_1) \ \ldots \ \exp(-j2\pi\mathbf{k}_l^T\mathbf{d}'_N)\ ]^T$ ($\mathbf{d}'_l$ are the columns of $\mathbf{D}^{-1}$). The symbol '$\circ$' denotes the element-wise product i.e. $\mathbf{x} \circ \mathbf{y} = [\ x_1y_1 \ \ldots \ x_Ny_N\ ]^T$.

The DFT pair with periodicity matrix $\mathbf{D}$ is defined as :

$$X(\mathbf{k}) = \sum_{l=0}^{D-1} x(\mathbf{m}_l)\exp[j2\pi\mathbf{k}^T\mathbf{D}^{-1}\mathbf{m}_l] \qquad (3)$$

$$x(\mathbf{n}) = \frac{1}{D}\sum_{l=0}^{D-1} X(\mathbf{k}_l)\exp[-j2\pi\mathbf{k}_l^T\mathbf{D}^{-1}\mathbf{n}] \qquad (4)$$

$X(\mathbf{k})$ and $x(\mathbf{n})$ are both periodic with periodicity matrix $\mathbf{D}^T$ and $\mathbf{D}$ respectively (e.g. $x(\mathbf{n}+\mathbf{Dp}) = x(\mathbf{n})$ for $\mathbf{p} \in \Lambda$).

2. MODULATED FILTER BANKS

The DFT modulated filter bank is closely related to the Short-Space Fourier Transform (SSFT) [4] which is fundamental to spectral analysis. There are two representations of the SSFT, namely (i) low-pass and (ii) band-pass. We shall employ the band-pass representation which is given by :

$$X(\mathbf{m},\boldsymbol{\omega}) = \sum_{\mathbf{n}\in\Lambda} h(\mathbf{n})\exp[j\boldsymbol{\omega}^T\mathbf{n}]\,x(\mathbf{m}-\mathbf{n})$$

The discrete SSFT is obtained by sampling $X(\mathbf{m},\boldsymbol{\omega})$ at frequencies $\boldsymbol{\omega} = 2\pi\mathbf{D}^{-T}\mathbf{k}_l$ and is critically decimated when $\mathbf{m}$ is subsampled by $\mathbf{D}$. It is obtained directly from a modulated filter bank with sampling matrix $\mathbf{D}$ (there are exactly D channels). The filter in each channel is obtained by *frequency translating* (modulating) a prototype filter $H(\mathbf{z})$ (impulse response $h(\mathbf{n})$) :

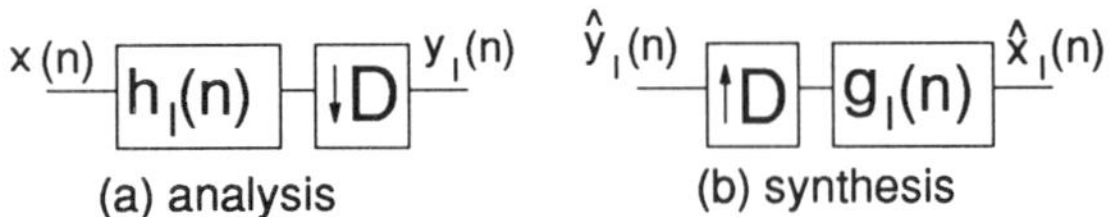

Figure 1: A Typical Channel of Filter Bank - (a) Analysis Section (b) Synthesis Section

$$H_l(\mathbf{z}) = H(\mathbf{W}(\mathbf{k}_l) \circ \mathbf{z}) \quad (5)$$

$$h_l(\mathbf{n}) = \exp[j2\pi \mathbf{k}_l^T \mathbf{D}^{-1}\mathbf{n}]\, h(\mathbf{n}) \quad (6)$$

where $l = 0, \ldots, D-1$ is the channel number label. The ideal passband of $H(\mathbf{z})$ is related to $\mathbf{D}$ (to be discussed later). A typical channel of the filter bank which performs analysis and synthesis of signals is shown in Fig. 1(a) and Fig. 1(b) respectively. Note that in the synthesis section the $\hat{x}_l(\mathbf{n})$ $(l = 0, \ldots, D-1)$ signals are added up in the reconstruction process.

2.1. FREQUENCY DOMAIN ANALYSIS

With the analysis and synthesis sections connected back-to-back the input/output relationship can be obtained using (1) and (2) and it is :

$$\hat{X}(\mathbf{z}) = 1/D\, \mathbf{g}^T(\mathbf{z})\, \mathbf{H}_{AC}\, \mathbf{x}_M(\mathbf{z}) \quad (7)$$

where $\mathbf{g}(\mathbf{z}) = [\ G_0(\mathbf{z}) \ \ldots \ G_{D-1}(\mathbf{z})\ ]^T$, $\mathbf{x}_M(\mathbf{z}) = [\ X(\mathbf{W}(\mathbf{k}_0) \circ \mathbf{z}) \ \ldots \ X(\mathbf{W}(\mathbf{k}_{D-1}) \circ \mathbf{z})\ ]^T$ and $\mathbf{H}_{AC}$ is the AC (*aliasing-component*) matrix with its (l,p) element - $[\mathbf{H}_{AC}]_{l,p} = H_l(\mathbf{W}(\mathbf{k}_p) \circ \mathbf{z})$. When the analysis filters are modulated versions of a prototype as in (5), we have :

$$[\mathbf{H}_{AC}]_{l,p} = H(\mathbf{W}(\mathbf{k}_l + \mathbf{k}_p) \circ \mathbf{z}) = [\mathbf{H}_{AC}]_{p,l}$$

i.e. $\mathbf{H}_{AC} = \mathbf{H}_{AC}^T$ - the AC matrix is *symmetric*. Without loss of generality we can set $\mathbf{k}_0 = [\ 0 \ \ldots \ 0\ ]^T$. Then $\mathbf{W}(\mathbf{k}_0) = [\ 1 \ \ldots \ 1\ ]^T$. The first element of $\mathbf{x}_M(\mathbf{z})$ is then the signal component $X(\mathbf{z})$ while other elements are aliased versions of $X(\mathbf{z})$. We can easily verify from (7) that the condition for aliasing cancellation is :

$$\mathbf{g}(\mathbf{z}) = D\, T(\mathbf{z})\, \mathbf{H}_{AC}^{-T}\, [\ 1 \ \ 0 \ \ldots \ 0\ ]^T \quad (8)$$

where $T(\mathbf{z})$ is an arbitrary transfer function. Equation (8) gives the synthesis filters that will achieve aliasing cancellation given the analysis filters.

The polyphase decomposition of $H(\mathbf{z})$ with respect to $\mathbf{D}$ is $H(\mathbf{z}) = \sum_{p=0}^{D-1} \mathbf{z}^{-\mathbf{m}_p} E_p(\mathbf{z}^{\mathbf{D}})$ where $E_0(\mathbf{z}), \ldots, E_{D-1}(\mathbf{z})$ are the polyphase components of $H(\mathbf{z})$. It can be verified that $\mathbf{W}^{\mathbf{D}}(\mathbf{k}_l) = [\ 1 \ \ldots \ 1\ ]^T$. Then we have : $H_l(\mathbf{z}) = \sum_{p=0}^{D-1} \mathbf{W}^{-\mathbf{m}_p}(\mathbf{k}_l)\ \mathbf{z}^{-\mathbf{m}_p} E_p(\mathbf{z}^{\mathbf{D}})$. We can then express $\mathbf{h}(\mathbf{z}) = [\ H_0(\mathbf{z}) \ \ldots \ H_{D-1}(\mathbf{z})\ ]^T$ as :

$$\mathbf{h}(\mathbf{z}) = \mathbf{\Omega}\, \mathrm{diag}\{\mathbf{z}^{-\mathbf{m}_0}, \ldots, \mathbf{z}^{-\mathbf{m}_{D-1}}\}\, \mathbf{e}(\mathbf{z}^{\mathbf{D}}) \quad (9)$$

where $\mathbf{e}(\mathbf{z}) = [\ E_0(\mathbf{z}) \ \ldots \ E_{D-1}(\mathbf{z})\ ]^T$ and $\mathbf{\Omega}$ is referred to as the *Multidimensional Fourier Matrix* with elements $[\mathbf{\Omega}]_{l,p} = \mathbf{W}^{-\mathbf{m}_p}(\mathbf{k}_l) = \exp[j2\pi \mathbf{k}_l^T \mathbf{D}^{-1} \mathbf{m}_p]$.

The AC matrix can be expressed in terms of $\mathbf{h}(\mathbf{z})$ as :

$$\mathbf{H}_{AC} = [\ h(\mathbf{z}) \mid h(\mathbf{W}(\mathbf{k}_1) \circ \mathbf{z}) \mid \ldots \mid h(\mathbf{W}(\mathbf{k}_{D-1}) \circ \mathbf{z})\ ]$$

With (9) we arrive at the *diagonalization* of the AC matrix :

$$\mathbf{H}_{AC} = \mathbf{\Omega}\, \mathrm{diag}\{\mathbf{z}^{-\mathbf{m}_0} E_0(\mathbf{z}^{\mathbf{D}}), \ldots, \mathbf{z}^{-\mathbf{m}_{D-1}} E_{D-1}(\mathbf{z}^{\mathbf{D}})\}\, \mathbf{\Omega}^T$$

It can be verified that $\mathbf{\Omega}^T \mathbf{\Omega}^* = \mathbf{\Omega}^* \mathbf{\Omega}^T = D\,\mathbf{I}$ so that $\mathbf{\Omega}^{-1} = 1/D\,(\mathbf{\Omega}^*)^T$. By formal inversion of the AC matrix with the aid of its diagonal form, we obtain the solution to (8). After some matrix algebra manipulations we obtain :

$$G_l(\mathbf{z}) = T(\mathbf{z}) \sum_{p=0}^{D-1} \mathbf{W}^{\mathbf{m}_p}(\mathbf{k}_l)\, \mathbf{z}^{\mathbf{m}_p}\, E_p^{-1}(\mathbf{z}^{\mathbf{D}}) \quad (10)$$

By defining (i) $R_l(\mathbf{z}) \equiv \prod_{p \neq l} E_p(\mathbf{z})$;
(ii) $G(\mathbf{z}) = \sum_{p=0}^{D-1} \mathbf{z}^{\mathbf{m}_p} R_p(\mathbf{z}^{\mathbf{D}})$; (iii) $T'(\mathbf{z})$ by the equation :

$$T(\mathbf{z}) \equiv T'(\mathbf{z}) \prod_{p=0}^{D-1} E_p(\mathbf{z}^{\mathbf{D}}) = T'(\mathbf{z})\, R_l(\mathbf{z}^{\mathbf{D}})\, E_l(\mathbf{z}^{\mathbf{D}}) \quad (11)$$

we can express (10) as :

$$G_l(\mathbf{z}) = T'(\mathbf{z})\, G(\mathbf{W}(\mathbf{k}_l) \circ \mathbf{z}) \quad (12)$$

i.e. the synthesis filters can be expressed as modulated versions of a prototype filter $G(\mathbf{z})$ with a common factor $T'(\mathbf{z})$ which can be intepreted as a *post-filter* after the reconstruction process. For arbitrary FIR (finite impulse response) analysis filters (with FIR $E_p(\mathbf{z})$), the synthesis filters required for aliasing cancellation are in general IIR (infinite impulse response) due to the $E_p^{-1}(\mathbf{z}^{\mathbf{D}})$ factors in (10). If, however, $T'(\mathbf{z})$ in (11) is FIR, then the synthesis filters will also be FIR as seen from (12). If only aliasing cancellation is required $T(\mathbf{z})$ can be arbitrary but for perfect reconstruction we require $T(\mathbf{z}) = c\,\mathbf{z}^{\mathbf{q}}$ (c - constant factor, $\mathbf{q}$ - integer vector). From (11), we see that in general this would require $T'(\mathbf{z})$ to be IIR. The synthesis filters are then IIR unless $E_p(\mathbf{z}) = c_p\, \mathbf{z}^{\mathbf{q}_p}$ (arbitrary $c_p, \mathbf{q}_p$). However, this gives trivial FIR analysis and synthesis filters and the filter bank would basically perform only a normal DFT (see Section 2.2 and Fig. 2). From here onwards we shall deal only with the case where both the analysis and synthesis filters are obtained from modulated versions of prototype filters as in (5) (or (6)) and (12) (with $T'(\mathbf{z}) = 1$).

2.2. TIME DOMAIN ANALYSIS

The output of a typical channel of the analysis filter bank (Fig. 1(a)) is :

$$y_l(\mathbf{n}) = \sum_{\mathbf{n}' \in \Lambda} h(\mathbf{n}') \exp[j2\pi \mathbf{k}_l^T \mathbf{D}^{-1} \mathbf{n}']\, x(\mathbf{D}\mathbf{n} - \mathbf{n}') \quad (13)$$

The polyphase components of $h(\mathbf{n})$ and $x(\mathbf{n})$ are respectively

$$e_p(\mathbf{n}) = h(\mathbf{D}\mathbf{n} + \mathbf{m}_p) \quad \text{and} \quad x_p(\mathbf{n}) = x(\mathbf{D}\mathbf{n} - \mathbf{m}_p) \quad (14)$$

for $p = 0, \ldots, D-1$ (note the difference between the definitions of polyphase components of $h(\mathbf{n})$ and $x(\mathbf{n})$). The

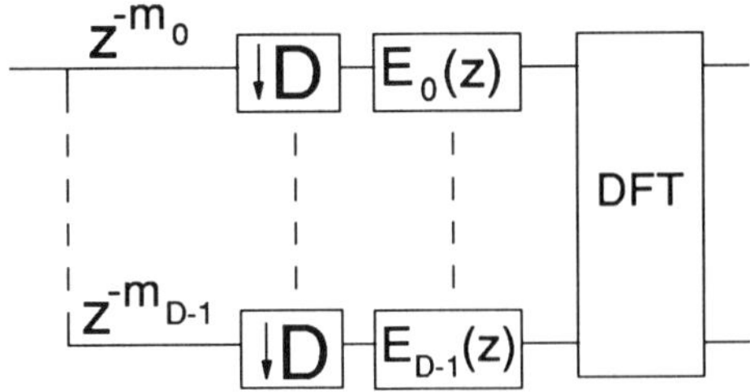

Figure 2: Polyphase Network Implementation of Analysis Filter Bank

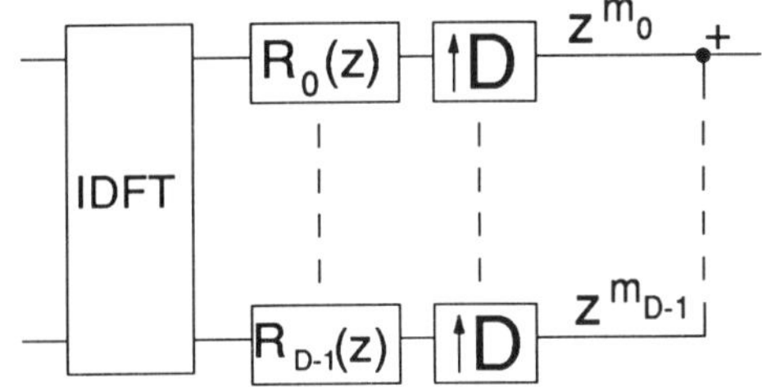

Figure 3: Polyphase Network Implementation of Synthesis Filter Bank

polyphase components are obtained by shifting the signal by $\mathbf{m}_p$ (or $-\mathbf{m}_p$) and subsampling with $\mathbf{D}$. By defining the change of variables : $\mathbf{n}' = \mathbf{D}\mathbf{r} + \mathbf{m}_p$, we have the following change in summation :

$$\sum_{\mathbf{n}' \in \Lambda} \Longrightarrow \sum_{\mathbf{r} \in \Lambda} \sum_{p=0}^{D-1} \tag{15}$$

Equation (13) then becomes

$$\begin{aligned} y_l(\mathbf{n}) &= \sum_{\mathbf{r} \in \Lambda} \sum_{p=0}^{D-1} h(\mathbf{D}\mathbf{r} + \mathbf{m}_p)\, x(\mathbf{D}(\mathbf{n} - \mathbf{r}) - \mathbf{m}_p) \\ &\qquad . \exp[\, j2\pi \mathbf{k}_l^T \mathbf{D}^{-1} (\mathbf{D}\mathbf{r} + \mathbf{m}_p)\,] \end{aligned}$$

Using the fact that $\exp[\, j2\pi \mathbf{k}_l^T \mathbf{r}\,] = 1$ with (14), we have :

$$y_l(\mathbf{n}) = \sum_{p=0}^{D-1} \exp[\, j2\pi \mathbf{k}_l^T \mathbf{D}^{-1} \mathbf{m}_p\,] \, (\sum_{\mathbf{r} \in \Lambda} e_p(\mathbf{r})\, x_p(\mathbf{n} - \mathbf{r}) \,)$$

The summation in $\mathbf{r}$ represents a convolution (filtering) of $x_p(\mathbf{n})$ with $e_p(\mathbf{n})$. The summation in p represents a DFT of the filtered polyphase components (cf. eqn. (3)). Hence we obtain the polyphase network implementation as shown in Fig. 2. The processing of the signals is carried out at a reduced sampling rate.

The reconstructed signal from the synthesis filter bank (Fig. 1(b)) is :

$$\hat{x}(\mathbf{n}) = \sum_{l=0}^{D-1} \sum_{\mathbf{n}' \in \Lambda} \hat{y}_l(\mathbf{n}')\, g_l(\mathbf{n} - \mathbf{D}\mathbf{n}') \tag{16}$$

where $g_l(\mathbf{n}) = \exp[\, j2\pi \mathbf{k}_l^T \mathbf{D}^{-1} \mathbf{n}\,]\, g(\mathbf{n})$; and $g(\mathbf{n})$ is the prototype synthesis filter. The polyphase components of $g(\mathbf{n})$ and $\hat{x}(\mathbf{n})$ are respectively :

$$r_p(\mathbf{n}) = g(\mathbf{D}\mathbf{n} - \mathbf{m}_p) \tag{17}$$

$$\hat{x}_p(\mathbf{n}) = \hat{x}(\mathbf{D}\mathbf{n} - \mathbf{m}_p) \tag{18}$$

Using (17) and (16) in (18), we have :

$$\begin{aligned} \hat{x}_p(\mathbf{n}) &= \sum_{l=0}^{D-1} \sum_{\mathbf{n}' \in \Lambda} \hat{y}_l(\mathbf{n}')\, g_l(\, \mathbf{D}(\mathbf{n} - \mathbf{n}') - \mathbf{m}_p\,) \\ &= \sum_{\mathbf{n}' \in \Lambda} r_p(\mathbf{n} - \mathbf{n}') (\sum_{l=0}^{D-1} \hat{y}_l(\mathbf{n}') \exp[-j2\pi \mathbf{k}_l^T \mathbf{D}^{-1} \mathbf{m}_p\,]) \end{aligned}$$

The summation in l represents an inverse DFT (IDFT) of the subband signals $\hat{y}_l(\mathbf{n}')$ (cf. eqn. (4)). The summation in $\mathbf{n}'$ represents a filtering of the output of the DFT operation with $r_p(\mathbf{n})$. We obtain an efficient polyphase network implementation as shown in Fig. 3. Note that upsampling by $\mathbf{D}$, shifting by $\mathbf{m}_p$ and summing of the signals corresponds to interleaving the polyphase components $\hat{x}_p(\mathbf{n})$ to obtain $\hat{x}(\mathbf{n})$.

Computation of DFT and IDFT : A fast algorithm that is similar to the Cooley-Tukey algorithm (in 1D) can be formulated for the multidimensional DFT with arbitrary periodicity lattice. Instead of factorizing the integer N (= length of 1D FFT), the factorization of the periodicity matrix $\mathbf{D}$ is used to derive the algorithm (see [7] for more details). A different algorithm which is more suitable for highly parallel processing is formulated by using the *Smith Normal Form* of the periodicity matrix : $\mathbf{D} = \mathbf{U}\,\boldsymbol{\Delta}\mathbf{V}$. All the matrices have integer elements and $\boldsymbol{\Delta}$ is diagonal while $|\det \mathbf{U}| = |\det \mathbf{V}| = 1$ (unimodular matrices). The effect of the matrices $\mathbf{U}$ and $\mathbf{V}$ is to rearrange (reindex) the input and output data of the DFT respectively. The DFT then becomes separable (due to the diagonal nature of $\boldsymbol{\Delta}$) and can be computed using 1D DFT along the rows and columns (see [8] for more details).

3. FREQUENCY SUBBANDS - SHAPE ASPECTS

From its definition the Fourier transform $X^F(\omega)$ of a discrete signal $\mathbf{x}(\mathbf{n})$ is periodic in each of the ω_l variable with period 2π. Hence $X^F(\omega)$ needs to be defined only in the *unit frequency cell* - $U = \{\omega : \omega \in [-\pi, \pi]^N\}$. If a signal $X^F(\omega)$ is subsampled by sampling matrix $\mathbf{D}$, then it can be verified from (1) that aliasing is avoided if the signal's passband is confined to $S \equiv \{\omega : \omega = \mathbf{D}^{-T}\omega' \,;\, \omega' \in U\}$.

The ideal passband of the prototype filter $H^F(\omega)$ used in the modulated filter bank should be S. The ideal passband of the modulated filter $H_l^F(\omega)$ (ie. $H_l^F(\omega) = H^F(\omega - 2\pi\mathbf{D}^{-T}\mathbf{k}_l)$ from (5)), is obtained by translating S by the *modulation vector* $\mathbf{k}_l' = 2\pi\mathbf{D}^{-T}\mathbf{k}_l$:

$$S + \mathbf{k}_l' \equiv \{\omega : \omega = \omega' + \mathbf{k}_l' \,;\, \omega' \in S\} \tag{19}$$

Note that regions that are translated out of U are 'wrapped back' into U due to the $(2\pi)^N$ periodicity of the Fourier

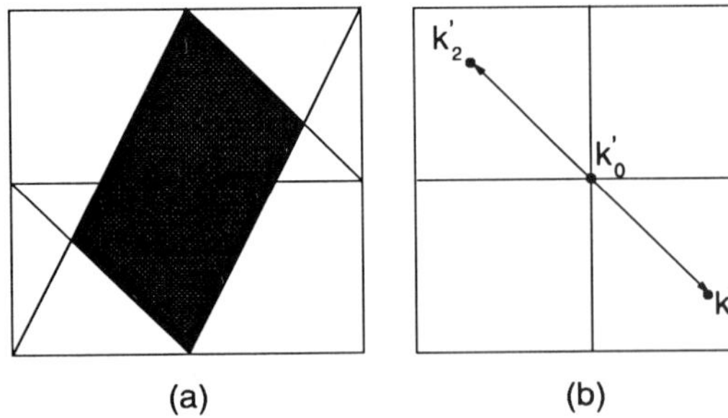

Figure 4: Frequency Domain Diagram of Example 1 - (a) Pass-Band of Prototype Filter (b) Modulation Vectors

transform. The complete set of coset vectors $\mathbf{k}_l$ for $\mathbf{D}$ can be obtained from $\{\mathbf{k} : \mathbf{k} = \mathbf{D}^T\mathbf{x} = \text{integer vector} \; ; \; \mathbf{x} \in [0,1)^N\}$. The passbands of the complete set of modulated filters $H_l^F(\omega)$ $(l = 0, ..., D-1)$ cover the whole frequency band ie. $\bigcup_{l=0}^{D-1}(S + \mathbf{k}'_l) = U$. The filter bank thus analyzes (decomposes) a full band (U) signal into subbands defined by (19). Subsampling by $\mathbf{D}$ maps the ideal subbands defined by (19) into U. In practice there will be a degree of aliasing as the filters cannot have the ideal passbands described. The aliasing can however be completely eliminated in the synthesis process as was shown in Sec. 2.1. One important aspect of using an arbitrary sampling lattice is that the subband shapes need not be rectangular. This can also introduce a directional aspect which is not present in one-dimensional systems. We will illustrate these aspects in two-dimensions with two examples.

The first example is described by the sampling matrix and coset vectors :

$$\mathbf{D} = \begin{bmatrix} 2 & 1 \\ -1 & 1 \end{bmatrix} \quad \frac{1}{\pi}\mathbf{k}_{0,1,2} = \begin{bmatrix} 0 \\ 0 \end{bmatrix}, \begin{bmatrix} 1 \\ 0 \end{bmatrix}, \begin{bmatrix} 2 \\ 0 \end{bmatrix}$$

with $|\det(\mathbf{D})| = 3$. The ideal passband S is shown in Fig. 4(a) as the shaded region inside the square which is U. The modulation vectors $\mathbf{k}'_l$ are shown in Fig. 4(b).

The second example is known as *hexagonal* sampling and is described by the sampling matrix and coset vectors :

$$\mathbf{D} = \begin{bmatrix} 1 & 1 \\ 2 & -2 \end{bmatrix} \quad \frac{1}{\pi}\mathbf{k}_{0,1,2,3} = \begin{bmatrix} 0 \\ 0 \end{bmatrix}, \begin{bmatrix} 1 \\ 0 \end{bmatrix}, \begin{bmatrix} 1 \\ -1 \end{bmatrix}, \begin{bmatrix} 2 \\ -1 \end{bmatrix}$$

with $|\det(\mathbf{D})| = 4$. The ideal passband S and modulation vectors $\mathbf{k}'_l$ are shown in Fig. 5(a) and Fig. 5(b) respectively. We see that the subband S has more frequency components along the horizontal direction than the vertical direction.

Design of the prototype filter : Several general design techniques for multidimensional filters described in [6],[9] (e.g. frequency sampling technique) could be used to design the prototype filter. However, a recently reported technique described in [10] is very suitable for designing filters with a

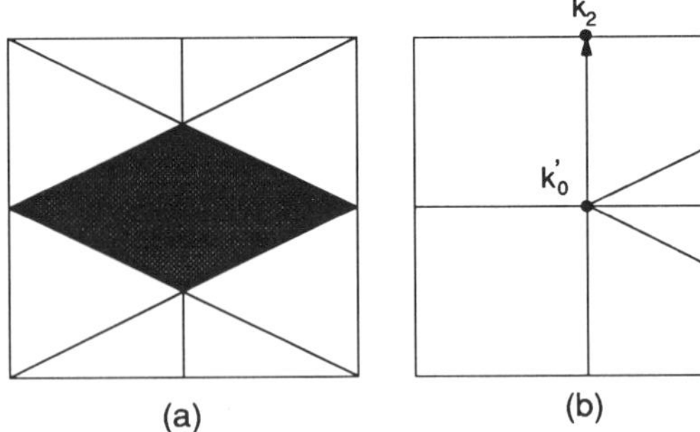

Figure 5: Frequency Domain Diagram of Example 2 - (a) Pass-Band of Prototype Filter (b) Modulation Vectors

passband described by S. A 1D filter $h_1(n)$ with a nominal passband $[-\pi/|\det(\mathbf{D})|, \pi/|\det(\mathbf{D})|]$ is first designed using any of the 1D design techniques. An N-dimensional separable filter is then formed : $h_S(\mathbf{n}) = \prod_{i=1}^{N} h_1(n_i)$. The resultant filter is obtained by sampling $h_S(\mathbf{n})$ with the sampling matrix $\hat{\mathbf{D}} = \text{adj}\,(\mathbf{D}) = |\det(\mathbf{D})|\; \mathbf{D}$ ie. $h(\mathbf{n}) = h_S(\hat{\mathbf{D}}\mathbf{n})$.

4. CONCLUSIONS

The DFT modulated filter bank (in multidimensions with arbitrary sampling lattice) can achieve aliasing cancellation with FIR analysis and synthesis filters but in general IIR synthesis filters are required for perfect reconstruction. It has an efficient polyphase network implementation that employ the DFT. The shape of the frequency subbands need not be rectangular and the shape depends on the sampling lattice used.

REFERENCES

[1] Viscito E. and Allebach J., IEEE Trans. Cct and Sys. (1991) CAS38.

[2] Karlsson G., and Vetterli M., IEEE Trans. ASSP (1990) ASSP38.

[3] Bellanger M. G., Bonnerot G. and B. M. Coudreuse, IEEE Trans. ASSP (1976) ASSP24.

[4] Crochiere R. E. and Rabiner L. R., Multirate Digital Signal Processing (Prentice-Hall, N. J., 1983).

[5] Swaminathan K. and Vaidyanathan P. P., IEEE Trans. Cct. and Sys. (1986) CAS13.

[6] Wackersreuther G., IEEE Trans. ASSP (1986) ASSP34.

[7] Dudgeon D. E. and Mersereau R. M., Multidimensional Digital Signal Processing (Prentice-Hall, N.J., 1984).

[8] Mersereau R. M., Brown III E. W. and Guessoum A., Row-Column Algorithms for the Evaluation of Multidimensional DFT's on Arbitrary Periodic Sampling Lattices, Int. Conf. Acoust., Speech, Signal Processing, (IEEE 1983), pp. 26B.5

[9] Lim J. S., Two-Dimensional Signal and Image Processing (Prentice-Hall, N. J., 1990).

[10] Chen T. and Vaidyanathan P. P., Elec. Lett. (1991) 27

SIGNAL PROCESSING VI: Theories and Applications
J. Vandewalle, R. Boite, M. Moonen, A. Oosterlinck (eds.)

OVERSAMPLED DIGITAL LEAPFROG FILTERS

F. Corthay, F. Pellandini

Institute of Microtechnology, University Neuchâtel, Tivoli 28,
CH-2003 Neuchâtel-Serrières, Switzerland.

A new digital filter architecture is presented which is based on the leapfrog simulation of passive analog LC filters. This filter type exploits a high oversampling in order to minimize the hardware of the devices. Indeed, due to the oversampling, the filter state variables are coded on a single bit and can thus be interpreted as pulse rates.

Two leapfrog synthesis methods are proposed: they correspond to the ladder and the lattice structures. An analysis method allows to evaluate the effects of the quantization of both coefficients and variables. The integrated circuit implementation of a filter can be realized by the abutment of a small set of basic bit slices.

The proposed filters provide an efficient alternative to classical systems for the processing of pulse frequency modulated signals. The interfaces between this kind of signals and the analog or the digital world are performed by simple readily available building blocks. Due to the high oversampling, the oversampled digital leapfrog filters are limited to applications with band edge frequencies of some kHz.

1. INTRODUCTION

Oversampled Digital LeapFrog Filters (ODLFs) rely on the simulation of an analog passive LC prototype in order to inherit its interesting properties. The analog active LeapFrog (LF) simulation of a reactance in its ladder expansion has been shown to be performed by a chain of integrators [1]. Within ODLFs, the digital implementation of the integrators in the LF structure is carried out by Up-Down Counters (UDCs) and Rate Multipliers (RMs).

The use of these basic operators reveal the inheritance from Quasi-Continuous Digital Filters (QCDFs) [2], [3], [4]. The QCDFs also implement integrator based structures using UDCs and RMs, but their internal signals are defined as pulse rates which are not synchronized to a system clock. This difference leads to a more difficult characterization of the QCDF signals and, above all, serious realization problems due to the asynchronous configuration of the pulses. A novel realization scheme of the RM allowed to refine the signal processing model of this operator and so to extend the insight of the ODLF structure.

With this, ODLFs can be described as a strictly digital filter class. Together with the proposed structure, a complete design method is presented, starting with the LF synthesis, scoping the analysis of transfer function deviation, and concluding with an integrated circuit realization [5].

The study of the ODLFs relies on the model of the workings of its operators. To these operators corresponds a specific coding of the signals. A clear outline of both operators and signals allows a straightforward approach of the complete ODLF investigation.

The proposed filtering method is derived from the fundamental choice to constrict the filter hardware at the expense of speed. It offers an insight to the processing of pulse rate coded signals related to various domains such as quartz sensors or sigma-delta ($\Sigma\Delta$) modulators.

The investigation of ODLFs starts with the presentation of the operators. Then, the synthesis shows the method used to derive an ODLF from its analog prototype. The analysis allows to determine the requirements for the appliance of the filter. An Application Specific Integrated Circuit (ASIC) design method is sketched. Finally, the whole topic is illustrated with the example of a benchmark filter design.

2. ODLF OPERATORS

The digital LF synthesis leads to different possible structures. In the Direct-transform Digital Integration (DDI) synthesis method [6], the integrators of the active LF are realized by accumulators. The integrator gains are set by a digital multiplier. ODLFs are DDI structures whose sampling rate is high enough to allow the coding of the filter variables on a single bit.

Coding the signals on a single bit trims the accumulators down to counters so that a LF accumulator with a positive and a negative input is implemented by an Up-Down Counter (UDC). The UDC implements the LF integrator function:

$$H(s) \cong \frac{f_s}{2^{n_{bits}}} \cdot \frac{1}{s} \quad (1)$$

The multipliers setting the integrator gains also have to code their output to a single bit. This is efficiently implemented by Rate Multipliers (RMs). RMs are basically made out of a quantizer, which codes the bit-parallel input to a single bit, and a one bit multiplier which multiplies the quantizer output with a pulse rate coefficient input. The quantization algorithm of the RM makes use of dithering: a dither signal originating from a free running counter is added to the input coming from an UDC and the most significant bit of the sum is kept as the quantizer output. The RM coefficient is restricted to the domain [0, 1].

In review, the cascade of an UDC and a RM is the ODLF equivalent of the integrator of the active LF. Inputs and outputs of this digital replication are coded in the form of pulse rates. Likewise, the filter coefficients too are provided to the RMs as pulse rates. Using unsigned arithmetic, these pulse rates are normalized in the domain [0, 1], where the pulse rate coding the amplitude of 1 is equal to the sampling frequency of the filter.

ODLF synthesis is achieved in the same manner as analog integrator based filter synthesis, only the integrators of the active device are implemented by UDC-RM couples.

3. LF SYNTHESIS

For all-pole filters, the classical LF structure has been introduced by Girling and Good [1]. With this structure, the active filter simulation of a doubly terminated LC ladder two-port is performed by a chain of integrators. Figure 1 displays the schematic appearance of the corresponding fifth order ladder ODLF.

The integrator based simulation of an analog passive filter owning finite zeroes is not as straightforward as the ladder LF synthesis. An interesting approach was proposed by Johns & Al. [7]. In this solution, an integrator chain corresponding to a singly loaded ladder determines the filter poles and the output is formed by the weighted sum of the state variables, so setting the zeroes of the filter. However, this structure leads to an ODLF implementation which is not canonic in terms of the number of operators.

For the most common amplitude approximation functions, the simulation of an analog lattice prototype shows to be preferable. The two lattice reactances are simulated by two ladder LF chains and the filter output is derived from the difference of the two chains, as depicted in figure 2. This procedure is similar to the synthesis of Lattice Wave Digital Filters (LWDFs) and a complementary output is also found for ODLFs as it is the case for LWDFs.

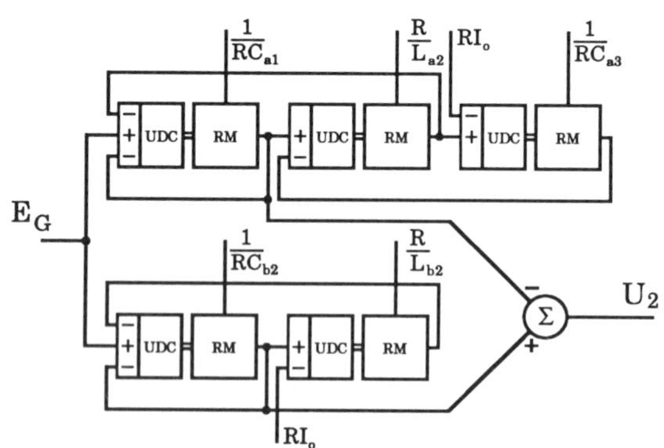

Figure 2 Lattice ODLF.

Additionally to the UDCs and RMs, the lattice structure requires a bit rate subtracter. In order to keep the output in the form of a bit rate, this operator can be realized as proposed by O'Leary & Al. [8]. Furthermore, a constant negative input RI_0 is fed to the last integrator of the LF chains in order to keep all state variables positive.

4. QUANTIZATION ERROR ANALYSIS

The analysis of the transfer function deviations is realized with the help of the state-space description of the filter. Two deviation sources are taken into account: the quantization of the coefficients and the quantization of the filter variables. As the oversampling is high and the device is based on integrator structures, the state-space description is made in the continuous time s-domain.

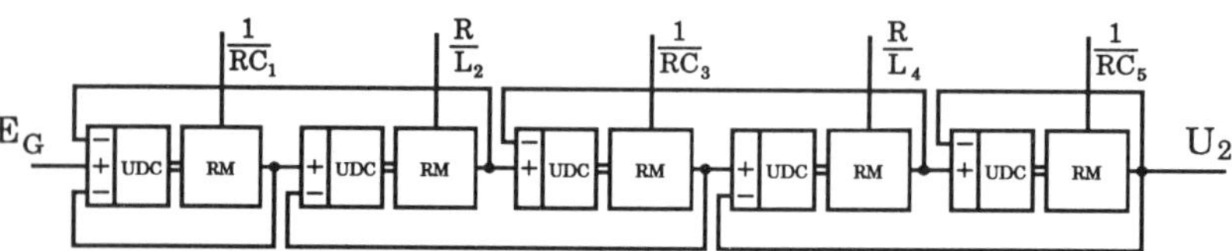

Figure 1 Ladder ODLF.

The state-space matrices of the ladder and the lattice filters can be obtained by the inspection of their schematic representation. As the LF links only interconnect neighboring integrators in a chain, the matrix A, whose eigenvalues are the poles of the system, is tridiagonal.

The quantization of the filter coefficients leads to a deterministic distortion of the transfer function. The quantized coefficients are inserted in the state-space matrices in order to estimate the transfer function of the actual filter and to check wether it still meets the specifications. Algorithms for coefficient quantization repeatedly execute this function in order to find an optimal coefficient set. Furthermore, the state-space matrices can be used to examine the sensitivity of the transfer function to coefficient values [9].

As opposed to coefficient quantization, the quantization of the signals inside a filter does not produce a deterministic distortion of the transfer function. Large scale distortions of the signals inside the filter originate from overflow of the UDCs. This effect is avoided by a proper scaling of the UDCs, which is achieved by adding least significant bits to the counters. The range which has to be provided to the counter outputs is estimated from the transfer functions between the input and the state variables. Small scale distortions of the signals emanate from the quantization inside the RMs. These noise sources can be inserted in the state-space description of the filter and their cumulative effect is estimated as a the maximal deviation of the transfer function.

The number of bits allocated to the UDCs determines the ODLF sampling rate. In effect, adding one bit to the UDCs requires to double the sampling rate in order to leave the transfer function described by equation (1) unchanged.

Analysis is based on reasonable assumptions, but filter simulation still remains a necessary tool for the verification of the proper functioning of the device.

5. INTEGRATED CIRCUIT REALIZATION

As well the UDC as the RM are basically made out of a counter. ASIC realization of these operators is successfully performed by the abutment of bit slices. Without too strenuous constraints, the operators can be designed for being abutted together so a complete LF chain is realized by the abutment of a limited set of basic slices.

The realization of the basic slices using synchronous counters has been developed [10]. More recently, the layout of the slices for asynchronous counters has proved to require even less silicon surface. Using a 1.6 μm technology, the core of the fifth order lattice filter represented in figure 2 requires an approximate area of 0.87 mm^2 (ca 1.45 mm x 0.6 mm)

6. BENCHMARK FILTER DESIGN EXAMPLE

The CCITT specifications G712 PCM have been extensively used as a filter benchmark [11]. They correspond to a lowpass filter. The passband ripple is of ± 0.125 dB from 0 up to 3 kHz. The stopband attenuation is -14 dB from 4 kHz to 4.6 kHz and -32 dB higher.

As most implementations of this benchmark, a fifth order elliptic function is chosen for the filter approximation. This function is mapped into the lattice structure of figure 2.

Coefficient quantization has provided a suitable coefficient set: [9/16, 1/4, 1, 1/2, 9/16]. In this context, it is interesting to point out that the coefficient quantization does not affect the size of the RMs; the precision of the coefficients actually determines the size of the logic providing the pulse rate coefficients. In this point of view, it is also favorable to have identical coefficients.

Analysis of the transfer function deviation due to the quantization of the variables requires a 10 bit dynamic range for the counter outputs to keep the transfer function and its expected deviation within the specifications. Due to different DC working points of the state variables, this necessitates 11 bit operators. This choice of 11 bit operators leads to a sampling rate of 78 MHz for a cutoff frequency of 3 kHz.

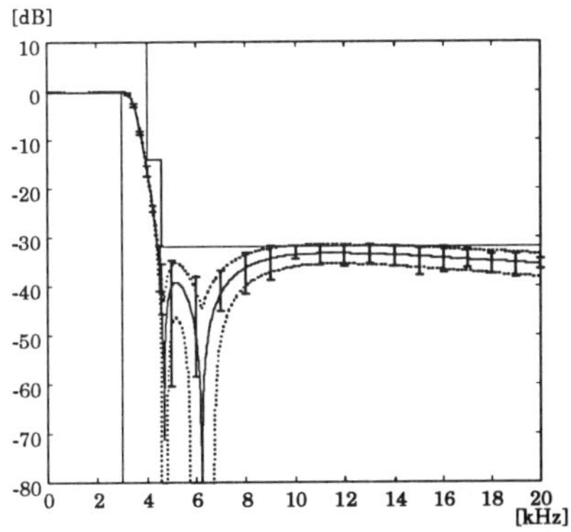

Figure 3 Transfer function deviation.

Simulations with sine waves of different frequencies applied to the ODLF input are used to estimate the transfer function deviation of the device and to compare it with the expected values provided by the analysis. Figure 3 allows to compare the frequency responses issued from analysis and simulation.

The transfer function deviations of figure 3 have been estimated for a 10 bit device clocked at 39 MHz and there is a slight overlap of the frequency response on the specifications. The choice of the simulation of a simpler device arose from the large calculations and memory requirements which are bound to the simulation of any oversampled device.

Nevertheless, the comparison shows a satisfactory correspondence between the theoretical transfer function deviation provided by the analysis and the actual deviation obtained from the simulations.

7. CONCLUSIONS

Based on methods used for analog integrator based active filters, the design procedure of a new oversampled digital filter structure has been investigated.

In summary, the design methodology covers the topics of synthesis, analysis and realization of the filter class. It provides a valuable insight to the matter of oversampled digital filters.

The present work has focused on the topic of filters with variables coded on a single bit. The leading goal was the hardware simplicity of the device. As it is the case for ΣΔ modulators, an optimum can be found between the number of bits of the signals, which determines the oversampling, and the cost of the realization.

The derivation of a LF structure results in a low coefficient sensitivity and the canonic realization of the filter with respect to the number of operators. The description of a signal processing model of the RM allows to provide a valuable estimation of the effects of quantization inside the ODLFs.

The high oversampling is the major limitation to the applicability of the ODLFs. As a result, they are appropriate for the processing of low frequency signals. Important to note is that the oversampling of the ODLFs depends exponentially from the number of bits of the operators, this is from the required signal to noise ratio. For this reason, the use of higher order filters is mostly of interest for specifications bearing narrow transition bands rather than severe attenuations. In addition, the ODLF delivers in its passband an output frequency corresponding to the input rate. As such, it can be regarded as a digital Phase Locked Loop (PLL) and applied for tasks such as digitally controlled frequency synthesis or the demodulation of frequency modulated signals.

ACKNOWLEDGEMENTS

This work has been supported by the Swiss National Foundation for Scientific Research under project numbers FN 20-5627.88 and FN 20-28536.90.

REFERENCES

[1] F. E. J. Girling, F. E. Good, “Active filters 12 - The leapfrog or active-ladder synthesis”, Wireless World, vol. 76, pp 341-345, July 1970.

[2] P.-A. Farine, F. Pellandini, “On the behavior of quasi-continuous digital filters”, International conference on digital signal processing, pp 27-31, Florence, Italy, September 1984.

[3] P.-A. Farine, F. Pellandini, “Technique de filtrage numérique quasi-continu. Avantages et limitations”, Dixième colloque sur le traitement du signal et ses applications, pp 449-454, GRETSI, Nice, France, May 1985.

[4] P.-A. Farine, “Filtrage numérique caractérisé par un traitement quasi-continu des signaux”, Ph.d thesis, University Neuchâtel, Switzerland, 1986.

[5] F. Corthay, “Oversampled digital leapfrog filters”, Ph.d thesis, University Neuchâtel, Switzerland, 1992.

[6] L. T. Bruton, “Low-sensitivity digital ladder filters”, IEEE Trans Circuits and Syst., vol. CAS-22, no 3, pp 168-176, March 1975.

[7] D. A. Johns, W. M. Snelgrove, A. S. Sedra, “Orthonormal ladder filters”, IEEE Trans. Circuits and Syst., Vol. CAS-36, No 3, pp 337-343, March 1989.

[8] P. O’Leary, F. Maloberti, “Bit stream adder for oversampling coded data”, Electronics letters, Vol. 26, No 20, pp 1708-1709, September 1990.

[9] W. M. Snelgrove, A. S. Sedra, “Synthesis and analysis of state-space active filters using intermediate transfer functions”, IEEE Trans. Circuits and Syst., Vol. CAS-33, No 3, pp 287-301, March 1986.

[10] F. Corthay, F. Pellandini, “Realization of quasi-continuous digital filters”, Douzième colloque GRETSI, pp 83-86, Juan-les-Pins, France, June 1989.

[11] L. Claesen, F. Catthoor, D. Lanneer, G. Goossens, S. Note, J. Van Meerbergen, H. De Man, “Automatic synthesis of signal processing benchmark using the CATHEDRAL silicon compilers”, IEEE 1988 custom integrated circuits conference, 1988.

SIGNAL PROCESSING VI: Theories and Applications
J. Vandewalle, R. Boite, M. Moonen, A. Oosterlinck (eds.)

USING THE ROOT DATASIEVE TO RECOGNISE AND LOCATE PATTERNS IN A DATA SEQUENCE

J. Andrew Bangham

Schools of Biology and Information Systems, University of East Anglia, Norwich, NR4 7TJ, UK.

Abstract

The root datasieve is an arrangement of median filters that is suitable for decomposing a digital signal to a set of rectangular pulses or rects. These characterise a one dimensional pattern in terms of spatial, as opposed to amplitude, features. It performs a multiscale decomposition. This is illustrated by analysing a simple spatial pattern, a set of fixed width bands, and using the result to design a non-linear matched filter, or sieve. The matched sieve is remarkably tolerant of variations in band amplitude but can, never-the-less, select out the pattern from a data signal that includes non-stationary and random noise.

Introduction

Median [1] and related stack [2,3,4] filters are attractive because of their edge preserving and statistically robust properties. Recent analysis [5] has shown their relationship with morphological filters [7, 8] and positive boolean functions applied to the thresholded binary data streams [6]. The datasieve is based on median filters.

It has recently been argued [10,11] that the root datasieve is an effective method for removing random noise, E, from a signal D *whilst also* preserving an underlying pulsetile signal, U, where D=U+E. The critical feature of the root datasieve is the fidelity with which it preserves edges and discriminates between pulses of different lengths. Let the length be m. In this simple one dimensional case, m is a measure of a spatial characteristic.

The datasieve is a multistage device in which successive stages, each is a median filter, filter out rectangular pulses (rects) of increasing m. However, it is not immediately clear from a theoretical viewpoint that the resulting rects would be sufficiently independent of each other to make them a reliable 'signature' for a spatial pattern. So a practical test has been devised.

This paper shows that by matching the passband properties of a root datasieve to the spatial properties of a test signal, a matched filter can be designed that selectively passes only the specified patterns. To distinguish it from linear matched filters this matched non-linear filter will be called a 'matched sieve'.

Methods

Imagine a pattern, U, of thick and thin bands on a cloth fabric, a tartan perhaps as illustrated in Fig.1G. Within each repeating unit the bands are of fixed but differing width and are uniformly distant from one another. If a one dimensional signal is obtained by making a single scan of the pattern, the bands form a signal such as that shown in Fig. 1F, the band amplitudes vary but the width and spacing does not. Fig. 1A. shows, by way of a synthetic example, the intensities that may occur as result of scanning an image with non-uniform lighting, sheen and dirt. The associated amplitude signal is shown by Fig. 1B. It is the spatial characteristics and edges of this pattern unit that make it distinctive, not the intensity. Selecting this pattern is, therefore, analogous to recognising features in an image.

Spatial analysis is performed by a root datasieve. It is cascade of root median filters. A standard median filter runs the data through a window of 2m+1 samples and finds the median at each position and a root is achieved by repeatedly filtering through a 2m+1 median filter until there is no further change. A root datasieve is an arrangement of root median filters in which the signal is median filtered with window of 3, then the output is filtered through a median filter with a window of 5 and so forth to a window of 2m+1. Despite being superficially complex, this strategy has an order complexity than depends on $m^{1.8}$ [10,11] because each stage implicitly subsamples the previous stage.

By subtracting the output of each stage (m) of the root datasieve from its input, a profile of what is 'removed' by each stage is obtained (Fig.2). These are rectangular pulses called, for the moment, rects(). Rects are characterised by the relative position at which they occur (x), an amplitude (y) and width (m). In other words, each stage of the sieve removes rects of a different scale, where scale is represented by m. This is analogous to multiscale analysis by wavelets [12].

Results

Fig.2 A shows the amplitude profile of the underlying signal U. Fig.2 B shows the corresponding spatial decomposition using the root datasieve. Each unit of the pattern is represented by six rects. If the rect at m=21, rect(x,y,21), is used as a positional reference the others occur offset and with spatial characteristics, m, as follows: rect(x+4, y, 12), rect(x+21, y, 9), rec(x-7, y, 7), rect(x-12, y, 5) and rect(x+30, y, 4). Fig. 2 B shows the result of decomposing the raw test data (shown in Fig. 1B) into rects. A matched sieve has to select from these the appropriate subset

{(rect(x1,y1,m1), (rect(x2,y2,m2), (rect(x3,y3,m3), (rect(x4,y4,m4), (rect(x5,y5,m5)); x2=x1+4, x3=x1+21, x4=x1-7, x5=x1-12, x6=x1+30, m1=21, m2=12, m3=9, m4=7, m5=5, m6=4}

Notice that the pulse amplitude is not constrained, but it could have been. The result of filtering the underlying signal Fig. 1F through this device is shown in Fig.1 E. As expected they are identical. The question is how well does the device distinguish this pattern from others?

A preliminary result from an exact match sieve is shown in Fig.1 D. The result is similar to E but two units of the pattern are missing. However, it is possible to increase the tolerance of the matched filter. The simplest method is to accommodate a match if at any position at least, say, four rects are present. It is also possible toa accept a slight mismatch in x and m. The result of accepting a match within $x\pm2$, $m\pm2$ for only four out of six rects is shown in Fig. 1C. To ease comparison with the original underlying patterns the output from the tolerant matched sieve is plotted as a dotted line over the original in Fig. 2 A. They are very similar.

Conclusions

It has been shown that the properties of the non-linear root datasieve, and by implication the single pass datasieve [10], can be used as a pattern recognizer. In this example the filter design was manual however adaptive sieves are also possible [cf. 12]. These results also lead to the possibility of designing two dimensional matched sieves that may take the place of linear spatial filters in models of primal vision processes [14].

References

1 Tukey, J.W. (1977) Exploratory Data Analysis. Addison-Wesley, London.

2 Gallagher, L.C. and Wise, G.N. (1981) A theoretical analysis of the properties of median filters. IEEE ASSP-21, 1136-1141

3 Fitch, J.P., Coyle, E.J. and Gallagher, N.C.Jr., (1984) Median filtering by threshold decomposition. IEEE Trans. Acoust. Speech, Signal Processing ASSP 32:1183-1188

4 Coyle, E.J., Lin, JH. and Gabbouj, M. (1989) Optimal Stack Filtering and the Estimation and Structural Approaches to Image Processing. IEEE ASSP-37:2037-2066

5 Maragos, P. and Schafer, R.W. (1987) morphological filters. II Their relation to median, order-statistic, and stack filters. IEEE ASSP 35:1170-1184.

6 Yli-Harja, Astola, J. and Neuvo, Y. (1991) Analysis of the properties of median and weighted median filters using threshold logic and stack filter representation. IEEE SP 39:395-410

7 Serra, J. (1982) Image analysis and mathematical morphology. Academic press, N.Y.Preston, K.Jr. and Duff, J.B. (1984) Modern

cellular Automata, theory and Applications. Plenum Press, London.

8 Salembier, P. (1991) Multiresolution decomposition and adaptive filtering with rank order based filter - application to surface detection. ICASSP Toronto.

9 Preston, K.Jr. and Duff, J.B. (1984) Modern cellular Automata, theory and Applications. Plenum Press, London.

10 Bangham, J.A. (1990) Properties of an improved variant on the median filter. in 'Communication, Control and Signal Processing in 'Comm, Control and Sig Proc' Bilcon'90, ed. E.Arikan, Elsevier press, p1591-1597

11 Bangham, J.A. (1992) Properties of a series of nested median filters, namely the datasieve. IEEE SP *in press*.

12 Rioul, O. and Vetterli, M. (1991) Wavelets and Signal Processing. IEEE SP Magazine October:14-38

13 Yin, L, Astola, J. and Neuvo, Y. (1991) Adaptive Boolean Filters for Noise Reduction, in Artificial Neural Networks. Kohonen, T, Makisara, K. Simula, O. and Kangas (Eds) Elsevier Science, North Holland.

14 Marr, D (1982) Vision. pub. W.H.Freeman, San Francisco.

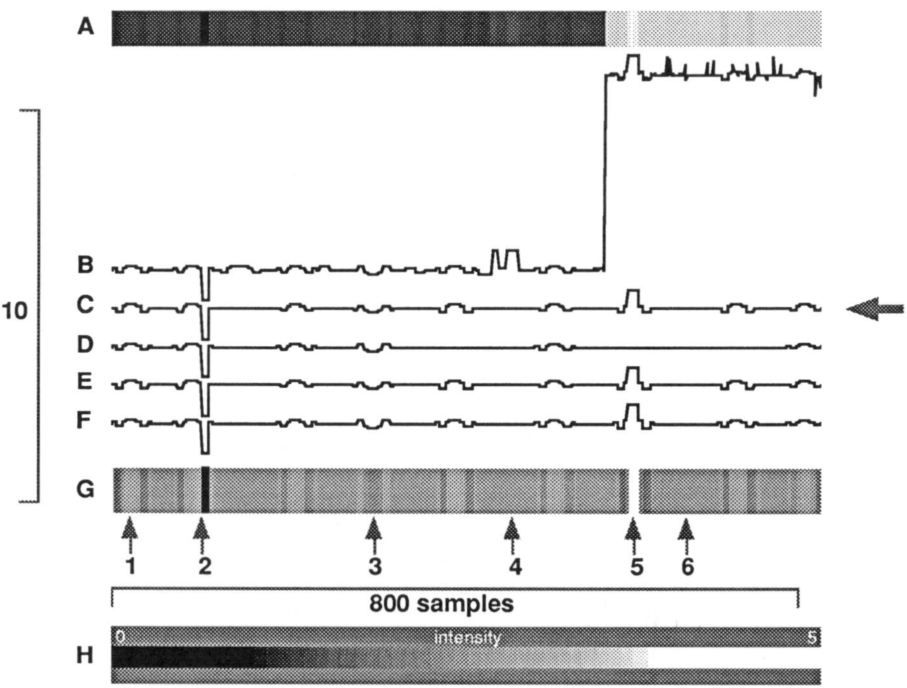

Figure 1. Amplitude traces B to F are all to the same scale, likewise the two intensity profiles. A linear intensity bar is shown in H. **A)** Shows the intensity profile of the test signal. It could be a section through an image. **B)** Shows the amplitude of the test signal. Notice that the underlying patterns are interspersed both with noise patterns that are similar to the test patterns and with random and impulsive noise and a changing baseline. **C)** Shows the result of filtering the test signal (B) through the matched sieve accepting a loose match of m±2, x±2 and n-1 matching features. **D)** The same as (C) but using a matched sieve that only passes features that match exactly. **E)** Shows the result of filtering the underlying patterns through the matched sieve (exact match). **F)** Shows the underlying pattern. **G)** Similar to A but showing the underlying intensity pattern. The arrows point to patterns with bands of different amplitudes (1,2,3,5) and to noise (4,6).

Figure 2 on next page.

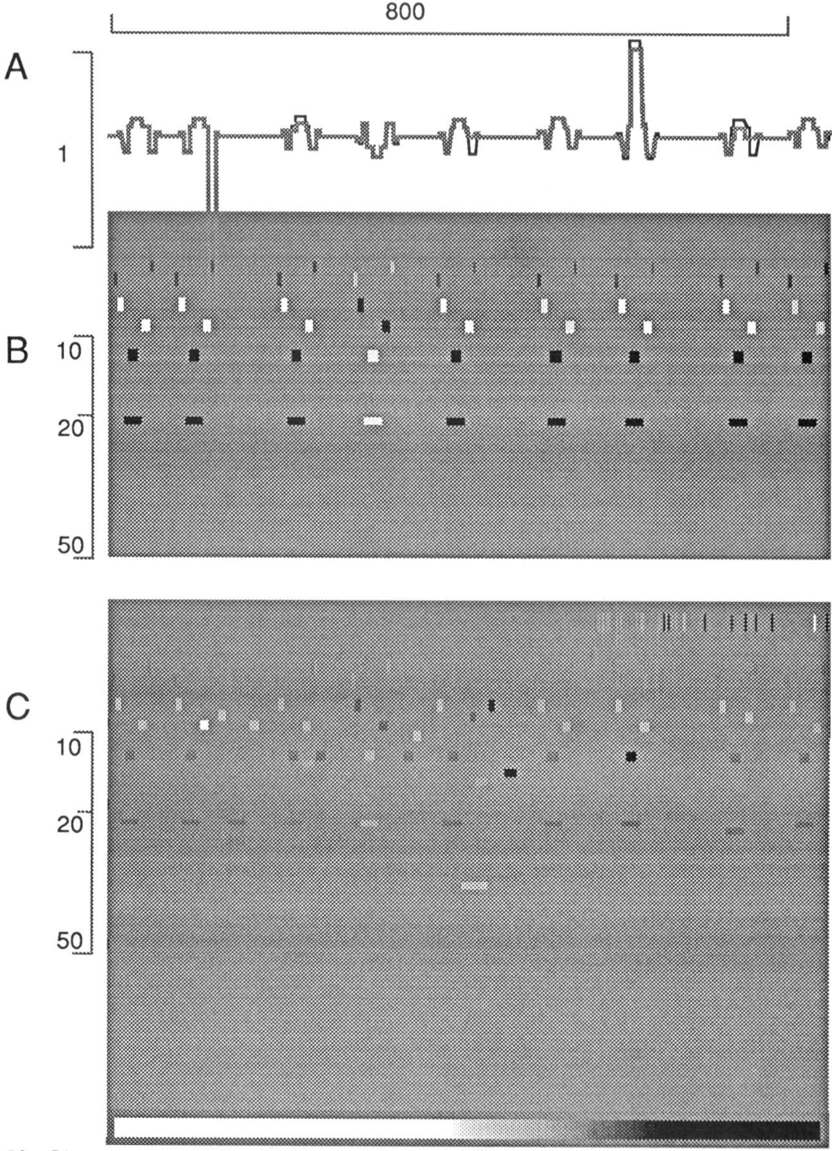

Figure 2 **A)** Shows a one dimensional pattern. The solid line shows the original underlying pattern U and the hatched line shows the result of filtering the test signal D through a matching sieve. **B)** Shows a multiscale root datasieve decomposition of the original set of patterns, U. The abscissa is the same as in A. Scale, m, is plotted logarithmically down the ordinate and intensity represents rect amplitude. Grey is zero. **C)** Similar to B, but shows the result of decomposing the test signal. Notice that the underlying patterns can be discerned from the noise. The intensity of

SIGNAL PROCESSING VI: Theories and Applications
J. Vandewalle, R. Boite, M. Moonen, A. Oosterlinck (eds.)

A BEZOUT RESULTANT BASED STABILITY TEST FOR 2-D DIGITAL RECURSIVE FILTERS

M. BENIDIR* and M. BARRET**

* *Université de Paris-Sud, L2S-ESE, Plateau de Moulon, 91192 Gif-sur-Yvette, France*
** *ESE, Établissement de Metz, 4 Rue É. Belin, Technopôle 2000, 57078 Metz, France*

A simple algorithm based on the Bezout resultant of two polynomials is proposed for testing the stability of 2-D digital recursive filters. The method consists in the calculation of the determinant of a matrix whose elements are polynomials. This determinant is 1-D polynomial and we have to test if it has zeros on the unit circle or not. The proposed method is compared to those given in recent works.

1. INTRODUCTION

The bidimensional (2-D) recursive digital filters have a wide range of applications in different areas such as geophysics, processing of radar and sonar data, medical imagery, etc... The stability constraint of these filters is a major problem and there is a growing need for the development of procedures for testing this stability. The fundamental stability theorems (necessary and sufficient conditions) have been established by Huang, Justice, Shanks, Anderson, Jury, DeCarlo, Fettweis [1]–[10]. The procedures for testing these conditions introduce many computations and numerical algorithms that reduce the volume of numerical computations have recently been introduced. In this paper, we propose a method based on the computation of the Bezout resultant of two bidimensional polynomials for testing the stability of 2-D digital recursive filters. The complexity of the proposed method is studied and several examples are discussed.

2. THE BEZOUT RESULTANT

We consider a first-quadrant 2-D linear shift-invariant digital filter, described by its transfer function $G(z_1, z_2) = A(z_1, z_2) / B(z_1, z_2)$ where $A(z_1, z_2)$ and $B(z_1, z_2)$ are polynomials in the independant complex variables z_1 and z_2. In the following, we assume that the transfer function $G(z_1, z_2)$ is irreducible and $G(z_1, z_2)$ has no *non essential singularities of the second kind* (NSSK). In this case, the filter is stable *if, and only if,* [1] [2] its denominator satisfies the following condition :

$$B(z_1, z_2) \neq 0, \quad (z_1, z_2) \in \bar{U}^2 \tag{1}$$

where $\bar{U} = \{z \,/\, |z| \leq 1\}$ denotes the closed unit disk. However, condition (1) is complicated to test and it is shown [1] [2] [8] that this condition can equivalently be replaced by the conjonction of the three following conditions (Strintzis theorem [5])

$$B(z_1, 0) \neq 0, \quad z_1 \in \bar{U} \tag{2}$$

$$B(1, z_2) \neq 0, \quad z_2 \in \bar{U} \tag{3}$$

$$B(z_1, z_2) \neq 0, \quad (z_1, z_2) \in T^2 \tag{4}$$

where $T = \{z \,/\, |z| = 1\}$ denotes the unit circle.

Conditions (2) and (3) are very simple to test by using classical procedures, well-known for the 1-D case. Condition (4), however, is still complicated to test. For practical applications, we have to reduce the volume of computations introduced by (4). For simplicity we suppose that $B(z_1, z_2)$ has real coefficients, the proposed method can easily be extended to any complex polynomial. A classical method to determine if two 1-D polynomials have any common factors consists in computing the resultant of these polynomials. In order to extend this method the 2-D case, we have to define the Bezout resultant. For this, let us associate to

$$B(z_1, z_2) = \sum_{k=0}^{n} \sum_{l=0}^{m} a_{k,l} z_1^k z_2^l = \sum_{l=0}^{m} a_l(z_1) z_2^l \tag{5}$$

its reciprocal, defined by

$$B^*(z_1, z_2) = z_1^n z_2^m B(z_1^{-1}, z_2^{-1}) = \sum_{l=0}^{m} \overset{(-)}{a}_{m-l}(\dot{z}_1) z_2^l \tag{6}$$

where for $0 \le l \le m$

$$a_l(z_1) = \sum_{k=0}^{n} a_{k,l} z_1^k \quad \text{and} \quad a_l^{(-)}(z_1) = z_1^n a_l(z_1^{-1}). \qquad (7)$$

We can suppose without loss of generality that $m \le n$. The zeros of $B(z_1, z_2)$ on the unit bicircle (T^2) are common zeros of $B(z_1, z_2)$ and $B^*(z_1, z_2)$. Then it is natural to introduce the Bezout resultant $\mathcal{R}(z_1)$ of

$$B(z_1, z_2) = 0 \quad \text{and} \quad B^*(z_1, z_2) = 0. \qquad (8)$$

This method has already been exposed by Liénard-Chipart [5] for monodimensional polynomials. In this paper, we extend their results to bidimensional polynomials. Consider the $2m$ polynomials constructed from $B(z_1, z_2)$ and $B^*(z_1, z_2)$ as follows : for $1 \le k \le m$

$$B_k(z_1, z) = a_m(z_1)z^{m-k} + \ldots + a_k(z_1) \qquad (9.a)$$

$$B_k^*(z_1, z) = a_0^{(-)}(z_1)z^{m-k} + \ldots + a_{m-k}^{(-)}(z_1). \qquad (9.b)$$

Comment Relations (9.a-b), extended to $k = 0$, give $B_0(z_1, z) = B(z_1, z)$ and $B_0^*(z_1, z) = B^*(z_1, z)$. When z_1 is a constant, (9.a-b) are the Bezout polynomials of (8).

Now consider the polynomial expended in power of z_2

$$F(z_1, z_2, z) = \frac{B(z_1, z_2)B^*(z_1, z) - B^*(z_1, z_2)B(z_1, z)}{z_2 - z}$$

$$= F_1(z_1, z) + F_2(z_1, z)z_2 + \ldots + F_m(z_1, z)z_2^{m-1} \qquad (10)$$

Lemma 1. (Proof in the Appendix) The degrees with respect to z of polynomials $F_k(z_1, z)$, $1 \le k \le m$, in (10) are not greater than $m - 1$ and the $F_k(z_1, z)$ are given by

$$F_k(z_1, z) = B_k(z_1, z)B^*(z_1, z_2) - B_k^*(z_1, z)B(z_1, z_2). \qquad (11)$$

The $c_{k,1}(z_1)$ appearing in the following form of $F_k(z_1, z)$:

$$F_k(z_1, z) = c_{k,1}(z_1) + c_{k,2}(z_1)z + \ldots + c_{k,m}(z_1)z^{m-1} \qquad (12)$$

satisfy

$$c_{k,l}(z_1) = c_{l,k}(z_1) \quad \text{for } 1 \le l, k \le m. \qquad (13)$$

Definition The Bezout resultant $\mathcal{R}(z_1)$ of (8) is defined by

$$\mathcal{R}(z_1) = \det\{\mathcal{M}(z_1)\} \qquad (14)$$

where

$$\mathcal{M}(z_1) = \begin{pmatrix} c_{1,1}(z_1) & \ldots & c_{1,m}(z_1) \\ \vdots & & \vdots \\ c_{m,1}(z_1) & \ldots & c_{m,m}(z_1) \end{pmatrix}. \qquad (15)$$

The determinant $\mathcal{R}(z_1)$ is a polynomial and the location of its zeros is related to this of $B(z_1, z_2)$ as follows.

Lemma 2 (Proof in the Appendix) With the previous notations, let λ be a complex number such that either $a_m(\lambda) \ne 0$ or $a_0^{(-)}(\lambda) \ne 0$. Then, $\mathcal{R}(\lambda) = 0$ if, and only if, $B(\lambda, z)$ and $B^*(\lambda, z)$ have a common zero. The hypothesis $a_m(\lambda) \ne 0$ or $a_0^{(-)}(\lambda) \ne 0$ can be omitted in the only if part.

Theorem 1 The polynomial $B(z_1, z_2)$ doesn't vanish on the closed unit bidisk $\bar{U}^2$ if, and only if, the three following conditions hold.

$$B(z_1, 0) \ne 0 \quad \text{for all } z_1 \in U \qquad (16)$$
$$B(1, z_2) \ne 0 \quad \text{for all } z_2 \in U \qquad (17)$$
$$\mathcal{R}(z_1) \ne 0 \quad \text{for all } z_1 \in T. \qquad (18)$$

Proof Assuming that (1) holds, (16) and (17) are obvious. Let us suppose there exists $\lambda \in T$ such that $\mathcal{R}(\lambda) = 0$. If $a_m(\lambda) = a_0^{(-)}(\lambda) = 0$, then $a_0(\lambda) = \lambda^n \overline{a_0^{(-)}(\lambda)} = 0$. Thus $B(\lambda, 0) = 0$ and this contradicts the hypothesis. If either $a_m(\lambda) \ne 0$ or $a_0^{(-)}(\lambda) \ne 0$, then (Lemma 2) there exists ζ such that

$$B(\lambda, \zeta) = B^*(\lambda, \zeta) = 0. \qquad (19)$$

Thus

$$\overline{B^*(\lambda, \zeta)} = B(\lambda, \bar{\zeta}^{-1}) = 0. \qquad (20)$$

As either (λ, ζ) or $(\lambda, \bar{\zeta}^{-1}) \in \bar{U}^2$ there is a contradiction. Conversely, according to Lemma 1, if $B(z_1, z_2) = 0$ on T^2, then $\mathcal{R}(z_1) = 0$ on T. The end of the proof is a direct consequence of abovementioned Strintzis theorem .

The first two conditions of theorem 2 are easy to test using classical stability tests for 1-D digital filters. When $\mathcal{R}(z_1)$ is expanded in powers of z_1, the third condition can easily be tested by the extended Schur-Cohn procedure proposed in [13]. In another way, we will see in appendix that $\mathcal{R}(z_1)$ has a degree not greater than $2nm$ and satisfy :

$$\mathcal{R}(z_1) = z_1^{2nm}\mathcal{R}(z_1^{-1}) \tag{21}$$

then it is possible without many computations to find the polynomial $Q(u)$ of degree nm wich satisfies :

$$z_1^{-nm}\mathcal{R}(z_1) = Q(u) \quad \text{where} \quad u = z_1 + z_1^{-1}. \tag{22}$$

It is well-known that $\mathcal{R}(z_1)$ doesn't vanish on the UC if, and only if, $Q(u)$ doesn't vanish on the real segment [–2 ; 2]. This last condition can be tested by Sturm theorem [11]. The main difficulty is to compute $\mathcal{R}(z_1)$. We can easily extend the results of [10] to show that the z_1-matrix $\mathcal{M}(z_1)$ is equal to the product of the z_1-matrix : $\mathcal{F}(z_1)$ of dimension $m \times 2m$ and $\mathcal{G}(z_1)$ of dimension $2m \times m$, given by

$$\mathcal{F}(z_1) = \begin{pmatrix} a_1(z_1) & -a_{m-1}^{(-)}(z_1) & a_2(z_1) & -a_{m-2}^{(-)}(z_1) & \dots & -a_0^{(-)}(z_1) \\ a_2(z_1) & -a_{m-2}^{(-)}(z_1) & a_3(z_1) & -a_{m-3}^{(-)}(z_1) & & 0 \\ \vdots & \vdots & \vdots & \vdots & & \vdots \\ a_m(z_1) & -a_0^{(-)}(z_1) & 0 & 0 & \dots & 0 \end{pmatrix} \tag{23}$$

and

$$\mathcal{G}(z_1) = \begin{pmatrix} a_m^{(-)}(z_1) & a_{m-1}^{(-)}(z_1) & \dots & a_1^{(-)}(z_1) \\ a_0(z_1) & a_1(z_1) & \dots & a_{m-1}(z_1) \\ 0 & a_m^{(-)}(z_1) & \dots & a_2^{(-)}(z_1) \\ 0 & a_0(z_1) & \dots & a_{m-2}(z_1) \\ \vdots & \vdots & & \vdots \\ 0 & 0 & \dots & a_0(z_1) \end{pmatrix} \tag{24}$$

The column k, $k > 2$, of $\mathcal{F}(z_1)$ is obtained from its column $k-2$ by shifting one row to the top and adding a zero on the last row. The row k, $k > 2$, of $\mathcal{G}(z_1)$ is obtained from its row $k-2$ by shifting one column to the right and adding a zero at the left of the row k.

The symmetries of $\mathcal{M}(z_1)$, with respect to its two diagonals (see the appendix), can reduce the volume of its computation. Below, we will propose a method to determine $\mathcal{R}(z_1)$. First, let us consider some examples.

3 APPLICATIONS

Example 1 : Let $B(z_1, z_2) = \sum_{k=0}^{n} a_{k,0} z_1^k + z_2 \sum_{k=0}^{n} a_{k,1} z_1^k$. Then

$$\mathcal{R}(z_1) = \sum_{l=0}^{n-1}\left(\sum_{j=0}^{l} a_{l-j,1}\, a_{n-j,1} - \sum_{j=0}^{l} a_{l-j,0} a_{n-j,0}\right)(z_1^l + z_1^{2n-l}) + \left(\sum_{j=0}^{n} a_{n-j,1}^2 - \sum_{j=0}^{n} a_{n-j,0}^2\right) z_1^n \tag{25}$$

Example 2 [8] : $B(z_1, z_2) = 8 + 2z_1 + (6 + 5z_1)z_2 + (1 + 2z_1)z_2^2$

$Q(u) = -294(6 - u)$, then the 2-D filter is stable.

Example 3 : $B(z_1, z_2) = 36 + 60z_1 + 37z_1^2 + 13z_1^3 + z_1^4 + (1 + 3z_1 + 2z_1^2 + 5z_1^3 + 4z_1^4)z_2$

$Q(u) = -2196 - 3302u - 1996u^2 - 511u^3 - 32u^4$.

$Q(-2) = 0$ and $B(-1, 1) = 0$: the 2-D filter is unstable.

Example 4 [6] : $B(z_1, z_2) = z_1^3 z_2^3 + (z_1^2 + 2z_1^3)z_2^2 + (z_1 + 2z_1^2 + 4z_1^3)z_2 + 1 + 2z_1 + 4z_1^2 + 8z_1^3$

$Q(u) = -8(7614 + 8898u + 14522u^2 + 13397u^3 + 10332u^4 + 7052u^5 + 3392u^6 + 1392u^7 + 448u^8 + 64u^9)$ and the 2-D filter is stable.

4 CALCULATION OF $\mathcal{R}(z_1)$

B. T. O'Connor and T. S. Huang [7] state, a congruence method is the most efficient to compute $\mathcal{R}(z_1)$. The total number of calculations needed is approximatively $O[2mn(2nm^2 + m^3)]$. If n is close to m this gives $O(6m^5)$ operations. This is really too much for large values of m. But, for $m = 2$, 3 or 4, the proposed method is competitive. For $m = 2$, the number of operations for computing $\mathcal{M}(z_1)$, using the symmetries of this matrix is $O(3n^2)$ multiplications and $O(3n^2)$ additions. To compute the autoreciprocal polynomial $\mathcal{R}(z_1)$ it takes $O(3n^2)$ multipli-cations and $O(3n^2)$ additions. To compute $\mathcal{R}(z_1)$ from $B(z_1, z_2)$ it takes $O(12n^2)$ operations. For $m = 3$, the calculation of this matrix needs $O(12n^2)$ operations, to compute $\mathcal{R}(z_1)$ from $B(z_1, z_2)$ it needs $O(58n^2)$ operations. For $m = 4$, the calculation of the matrix needs $O(20n^2)$ operations. To compute $\mathcal{R}(z_1)$ from $B(z_1, z_2)$ it needs $O(238n^2)$ operations.

Appendix

Proof of Lemma 1 : From (10) we deduce

$$F(z_1, z_2, z) = B^*(z_1, z)\frac{B(z_1, z_2) - B(z_1, z)}{z_2 - z} - B(z_1, z)\frac{B^*(z_1, z_2) - B^*(z_1, z)}{z_2 - z} \quad (26)$$

$$\frac{B(z_1, z_2) - B(z_1, z)}{z_2 - z} = B_m(z_1, z)z_2^{m-1} + \ldots + B_1(z_1, z). \quad (27)$$

We obtain a similar relation replacing B by B^*, then (11) results from (10) and (27). We see in (10) that $F(z_1, z_2, z) = F(z_1, z, z_2)$, then we deduce (12) and (13) from (27).

Proof of Theorem 1 : Replacing the first column of (15) by the linear combination, given in (12), implies that $\mathcal{R}(z_1)$ is equal, for all complex z, to the determinant of the matrix (15) when the first column is replaced by $[F_1(z_1, z), \ldots, F_m(z_1, z)]^T$. If $B^*(z_1, z)$ and $B(z_1, z)$ have a common zero (λ, ζ) then $F_k(\lambda, \zeta) = 0$ for $1 \leq k \leq m$. Thus $\mathcal{R}(\lambda) = 0$. Conversely, if $a_m(\lambda) = \overset{(-)}{a_0}(\lambda) = 0$, then the first column of $\mathcal{G}(\lambda)$ is null and $\mathcal{R}(\lambda) = 0$. Assume either $a_m(\lambda) \neq 0$ or $\overset{(-)}{a_0}(\lambda) \neq 0$. If $\mathcal{R}(\lambda) = 0$, we deduce of (12) and (15) that there are m complex numbers $\mu_1, \ldots, \mu_m$ which depend on λ and are not all equal to zero such that for all complex z $\mu_1 F_1(\lambda, z) + \ldots + \mu_m F_m(\lambda, z) = 0$. From (11) we can deduce that there exist two polynomials in z, $P_\lambda(z)$ and $Q_\lambda(z)$, with degrees not greater than $m - 1$, such that $P_\lambda(z)B(\lambda, z) + Q_\lambda(z)B^*(\lambda, z) = 0$, then (Gauss theorem) $B(\lambda, z)$ and $B^*(\lambda, z)$ have a common prime factor. As the coefficient beeing at the intersection of the kth row and the $(l + 1)$th column of $\mathcal{M}(z_1)$ is equal to :

$$c_{k,l+1}(z_1) = \sum_{i=0}^{\mathrm{Min}(m-k,l)} a_{k+i}(z_1)\overset{(-)}{a}_{m+i-l}(z_1) - \sum_{i=\mathrm{Max}(0,l-m+k)}^{l} a_i(z_1)\overset{(-)}{a}_{m-k+i-l}(z_1) \quad (28)$$

Then $c_{k,l+1}(z_1)$ is of degree $\leq 2n$, so $\mathcal{R}(z_1)$ is of degree $\leq 2nm$. We now show that $\mathcal{R}(z_1)$ is autoreciprocal. If $a_m(z_1)$ and $\overset{(-)}{a_0}(z_1)$ have no common zero, and if $\mathcal{R}(0) \neq 0$, then $\mathcal{R}(\zeta_1) = 0$ if, and only if, there exists $\zeta_2 \neq 0$ such that $B(\zeta_1, \zeta_2) = B^*(\zeta_1, \zeta_2) = 0$. It is clear that $B(\zeta_1^{-1}, \zeta_2^{-1}) = B^*(\zeta_1^{-1}, \zeta_2^{-1}) = 0$ and then $\mathcal{R}(\zeta_1^{-1}) = 0$. As the determinant is continuous, this result can be extend to any polynomial $B(z_1, z_2)$. We can also verify from (28) that $c_{m-l,m+1-k}(z_1) = z_1^{2n} c_{k,l+1}(z_1^{-1})$.

References

[01] D. E. Dudgeon, R. Mersereau, *Multidimensional Digital Signal Processing*, Prentice Hall, Inc., Englewood Cliffs, 1984.

[02] N.K.Bose, *Digital Filters Theory and Applica-tions*, North Holland, 1985.

[03] D. H. Kanellakis, N. J. Theodorou, "Simple sufficient test for stability of 2-dimensional recursive digital filters", IEEE Proc., vol 134, Pt G, No 5, Oct. 1987, pp. 246-247.

[04] M. Benidir, "Extension of some sufficient conditions for the stability of twodimensional recursive digital filters", EUSIPCO-90, Barcelona.

[05] M.G. Strintzis, "Test of stability of multidimensional filters", IEEE Trans. CAS-24, No 8, pp. 432-437, 1977.

[06] X. Hu, "2-D filter stability tests using polynomial array for $F(z_1, z_2)$ on $|z_1| = 1$", IEEE Trans. Circ. and Syst., vol. 38, No 9, pp. 1092-1095, Sept. 1991.

[07] B. T. O'Connor, T.S. Huang, *Two-Dimensional Digital Signal Processing I*, Topics in Applied Physics, vol 42, Springer-Verlag, New-York, 1981.

[08] S. Basu, A. Fettweis, "Simple Proofs of some Discrete Domain Stability Related Properties of Multidimensional Polynomials", Int. J. of Cir. Th. and App., vol 15, pp. 357-370, 1987.

[09] G. Gu, E. B. Lee, "A Numerical Algorithm for Stability Testing of 2-D Recursive Digital Filters", IEEE Trans. on Circ. and Sys., vol 37, No 1, Jan. 1990.

[10] E. I. Jury, "Modified Stability Table for 2-D Digital Filters", IEEE Trans. on Circ. and Sys., vol. 35, No 1, Jan. 1988.

[11] F. R. Gantmacher, *The Theory of Matrices*, Chelsea Publishing Company, New-York, Vol. 2, chap. 15, 1960.

[12] Liénart, Chipart, "Sur le signe de la partie réelle des racines d'une équation algébrique", J. Math. Pures et Appliquées (6ème série), t. X, PP. 291-346, 1914.

[13] M. Benidir, B. Picinbono, "Extensions of the stability criterion for ARMA filters", IEEE Trans. ASSP, vol. ASSP-35, pp. 425-431, April 1987.

SIGNAL PROCESSING VI: Theories and Applications
J. Vandewalle, R. Boite, M. Moonen, A. Oosterlinck (eds.)

FIR DIGITAL FILTERS USING A LEAST-SQUARES DERIVED RECURSIVE SECTION

D. H. Horrocks and D. R. Bull

School of Electrical, Electronic and Systems Engineering, University of Wales,P.O. Box 904, Cardiff CF1 3YH, UK.

This paper presents a new approach to the realisation of finite impulse response digital filters. The technique utilises a transversal section in cascade with a small recursive section designed such that the preceeding non recursive part has minimum computational complexity in the least squares sense. This provides a flexibility in filter response characteristics not offered by other similar methods. The paper outlines the design methodology and presents an example filter designed using this method.

1. Introduction

FIR digital filters have the advantages of linear phase and guaranteed stability. The former is achieved with symmetry in the filter impulse response while the latter is due to the fact that in a non-recursive realisation, all poles lie at the origin in the z-plane. In spite of these advantages, the large number of coefficient multiplications often required in such filters result in conventional direct-form implementations which are computationally complex.

Over the past few years numerous structures and design methods have been proposed which reduce or eliminate the multiplication burden in FIR filters. Such approaches are appropriate for application specific systems or filters which form a component in a larger integrated circuit with a requirement for high throughput coupled with low implementation complexity. Approaches based on the conventional transversal architecture, shown in figure 1, include the use of signed digit coding for coefficient representation [1]. Other methods have relied on benefits obtained from restricting coefficient values, either by limiting wordlength or, more usefully, by allowing an increased dynamic range but restricting the number of non-zero digits available [2]. These approaches do however often result in a deterioration in the filter frequency response characteristics. A powerful technique which results in the complete elimination of multipliers with no loss of coefficient accuracy has been presented by Bull and Horrocks [3] and involves the use of a primitive operator graph.

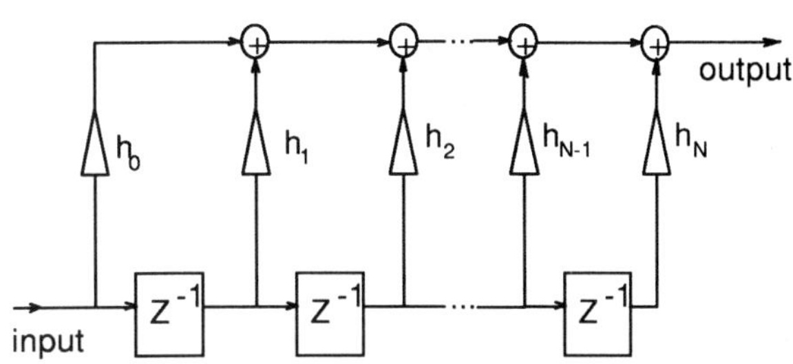

Figure 1 Conventional FIR structure

Competing approaches to FIR filter implementation have attempted to reduce implementation complexity through the use of a recursive realisation. These include the familiar frequency sampling structure where a comb filter is cascaded with a bank of resonators whose poles selectively cancel the zeros of the preceeding section. If appropriately positioned, the poles can result in simple resonator coefficients. An alternative method employs a single low order recursive section, $B(z)^{-1}$, as shown in Figure 2. This is chosen such that the modified FIR section

$H'(z)$ is realised with coefficients which are reduced to small integers while the cascade provides the same overall transfer function H(z), where

$$H'(z) = H(z) \cdot B(z) \tag{1}$$

Benvenuto et al [4] have described such an arrangement with simple recursive accumulator sections of the form $B(z) = (1-z^{-1})^k$. Coefficients, h'_j, are limited to {-1,0,+1} and optimised using dynamic programming techniques. In earlier work by Van Gerwen et al [5], B(z) is selected from a family of five resonators with poles placed around the unit circle and with coefficient values for both non-recursive and recursive sections limited to small powers of 2. Such structures are economical in hardware but suffer from reduced performance if the centre frequency is not close to that of the cascaded resonator. This family was later extended to eleven by Lynn [6], who fills the gaps present in centre frequency coverage while still retaining the advantages of small integer coefficients.

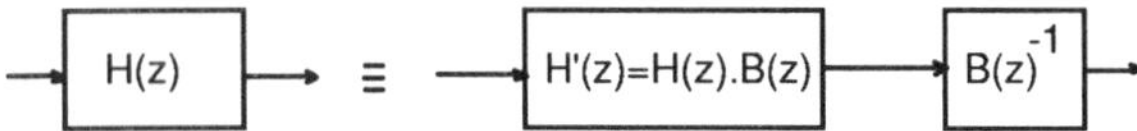

Figure 2 Modified FIR Structure

The aim of this contribution is to present a new approach for the choice of B(z) based on a least-mean-square (LMS) approach. In this sense the recursive section is optimal for the given overall filter response and does not depend on the identification and selection from a limited set of pre-defined B(z) functions.

2. The Design Method

The required overall transfer function for an Nth order FIR filter is given by:

$$H(z) = h_0 + h_1 z^{-1} + \cdots + h_N z^{-N} \tag{2}$$

and the Mth order recursive section has the transfer function:

$$B(z)^{-1} = 1/(b_0 + b_1 z^{-1} + \cdots + b_M z^{-M}) \tag{3}$$

where $b_0=1$ is assumed without loss of generality.

Thus, to achieve the required overall transfer function, (2), the recursive section, (3), must be placed in cascade with a modified FIR section, $H'(z)$, that satisfies the following relationship:

$$H'(z) = H(z).B(z) = h'_0 + h'_1 z^{-1} + \cdots + h'_{N+M} z^{-N-M} \tag{4}$$

The aim is to choose the coefficients $\{b_j\}$ of the recursive section so that the coefficients, $\{h'_j\}$ in the modified FIR section have minimum amplitudes. A convenient way to measure these amplitudes is by the sum of squares,

$$P = \sum_j (h'_j)^2 \tag{5}$$

Differentiating this expression with respect to each of the recursive-section coefficients $\{b_j\}$ produces a set of linear equations involving the autocorrelation coefficients, r_k, of the required coefficient sequence $\{h_j\}$. The equations are similar in form to the normal equations arising from linear prediction theory and other optimal signal processing problems [7]. To illustrate, for the case N=4, they are:

$$\begin{bmatrix} r_0 & r_1 & r_2 & r_3 & r_4 \\ r_1 & r_0 & r_1 & r_2 & r_3 \\ r_2 & r_1 & r_0 & r_1 & r_2 \\ r_3 & r_2 & r_1 & r_0 & r_1 \\ r_4 & r_3 & r_2 & r_1 & r_0 \end{bmatrix} \cdot \begin{bmatrix} b_0 \\ b_1 \\ b_2 \\ b_3 \\ b_4 \end{bmatrix} = \begin{bmatrix} 0 \\ 0 \\ 0 \\ 0 \\ 0 \end{bmatrix} \tag{6}$$

where

$$r_k = \sum_{i=0}^{N-k} h_{i+k} \cdot h_i \tag{7}$$

and $b_0 = 1$.

These equations could be solved for the optimal $\{b_j\}$ and then the associated $\{h'_j\}$ obtained from equation (1). However most FIR filters are symmetric or antisymmetric and permit a folded direct-form that saves on multipliers. In order to preserve this desirable property B(z) must correspondingly be symmetric or antisymmetric. The effect of this constraint on the set of equations for M=4 is:

$$\begin{bmatrix} 2r_0 \pm 2r_4 & 2r_1 \pm 2r_3 & r_2 \pm r_2 \\ 2r_1 \pm 2r_3 & 2r_0 \pm 2r_2 & r_1 \pm r_1 \\ r_2 \pm r_2 & r_1 \pm r_1 & r_0 \end{bmatrix} \cdot \begin{bmatrix} b_0 \\ b_1 \\ b_2 \end{bmatrix} = \begin{bmatrix} 0 \\ 0 \\ 0 \end{bmatrix} \qquad (8)$$

where the signs +/- are chosen according to whether the filter is symmetric or antisymmetric respectively. The form of these equations for other values of M is obvious.

Because of symmetry/antisymmetry, b_M=+1, or -1 and because b_M equals the product of the roots of B(z) it follows that at least some of the poles of the recursive section will lie on or outside the unit circle. However, because poles from the recursive section are exactly cancelled by corresponding zeros in the modified FIR section, H'(z), instability due to excitation of unstable poles by the filter input signal can be avoided. This can be ensured simply by placing the H'(z) section before the recursive section, $B(z)^{-1}$, in the cascade. To prevent instability in the recursive part due to noise induced by signal quantisation, a further condition is that coefficients $\{b_j\}$ be integer. As a consequence, coefficients $\{h'_j\}$ are also integer. In this contribution simple rounding is performed to convert $\{b_j\}$ to integers.

Given the desired FIR response, H(z), and resonator order M, the design method proceeds as follows:

i) calculate the autocorrelation coefficients r_k from equation (7),

ii) solve the set of equations (8) for $\{b_j\}$ and round to integer values.

iii) Calculate the modified FIR coefficients, $\{h'_j\}$, using equation (1).

3. A Design Example

The impulse-response coefficients for a typical bandpass filter with N = 120 and maximum coefficient magnitude of 64 are illustrated in figure 3. Resonator sections were generated over the range M=1 to 15. The recursive section resulting in H'(z) with the smallest coefficient dynamic range is given in equation (9),

$$B(z)^{-1} = 1/(1 - 2z^{-1} + 2z^{-3} - 2z^{-5} + z^{-6}) \qquad (9)$$

This section increases the order of the system slightly to 126, but produces coefficients in H'(z) which have a maximum value of only 3. A comparison between the coefficient dynamic ranges of H'(z) and H(z) can be observed in figure 3.

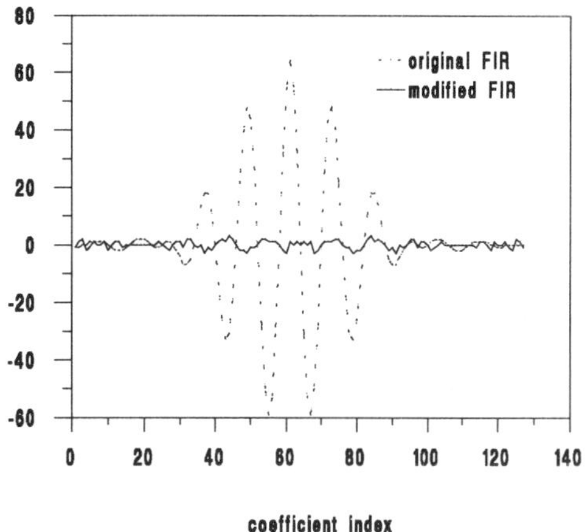

Figure 3 FIR Coefficients

4. Conclusions

A method has been presented for the design of resonator structures for the difference coefficient approach to reducing the coefficient amplitudes in

FIR digital filter structures. Since FIR structures are typically of high order any reduction in computational complexity obtained is significant. The method presented here produces a recursive section that is optimal for the given overall transfer function, H(z), and does not depend on selection from predefined functions as do other methods.

References

[1] Peled, A., "On the Hardware Implementation of Digital Signal Processors", IEEE Trans. ASSP-24, No. 1, Feb 1976, pp 76-86.

[2] Lim Yong Ching and Constantinides, A. G., "New Integer Programming Scheme for Non-recursive Digital Filters", Electronics Letters, Vol. 15, No. 25, Dec 1979, pp 812-3.

[3] Bull, D. R. and Horrocks, D. H., "Primitive Operator Digital Filters", IEE Proc. , Part G, Vol.138 , No.3 ,June 1991, pp 401-412.

[4] Benvenuto, N. et al, "Realisation of Finite Impulse Response Filters Using Coefficients +1, 0 and -1", IEEE Trans COM-33, No. 10, October 1985, pp 1117-1125.

[5] Van Gerwen, P. J., et al, "A New Type of Digital Filter for Data Transmission", IEEE Trans Com-23, 1975, pp 222-234.

[6] Lynn, P. A., "FIR Digital Filters Based on Difference Coefficients: Design Improvements and Software Implementation", IEE Proc. Vol. 127, Pt E, no. 6, November 1980, pp 253-258.

[7] Orfanides, S. J., "Optimal Signal Processing", Macmillan Publishing Co.,New York,1988.

SIGNAL PROCESSING VI: Theories and Applications
J. Vandewalle, R. Boite, M. Moonen, A. Oosterlinck (eds.)

EXPLICIT DESCRIPTION OF LATTICE STATE-SPACE WAVE DIGITAL FILTERS

M. Ansorge and F. Pellandini

Institute of Microtechnology, University of Neuchâtel, Rue Tivoli 28,
CH-2003 NEUCHATEL-Serrières, Switzerland.

This paper provides explicit formulas for Lattice State-Space Wave Digital Filters derived from reference filters based on cascaded all-pass sections and on cascaded unit elements. The achieved results show that the latter filters are the most interesting from an implementational point of view. Their complexity increases only linearly with the filter order compared to other state-space Wave Digital Filters, and their structure is particularly well suited for efficient realizations on vector processors and array processors.

1. INTRODUCTION

Wave Digital Filters (WDFs) are obtained from analog reference filters and are characterized by excellent numerical properties with respect to coefficient accuracy requirements, dynamic range, and all aspects of stability under finite-arithmetic conditions [1].

WDFs can be designed using many different filter structures [1], among which the *lattice* configurations show to be especially interesting for a broad range of applications, since they provide canonic and regular filter structures which can be designed in a straightforward way [2]. In addition, Lattice WDFs are well suited for the realization of half-band (or bireciprocal) [2, 3], branching, and quasi linear phase filters.

The WDFs can advantageously be transformed into an equivalent state-space form leading to efficient hardware and software implementations (high regularity, maximum parallelism, high throughput, etc.) with a reduced round-off noise. By properly deriving the state-space form, so-called *Essentially Equivalent State-Space* WDFs can be obtained which preserve all the properties of the initial filter [4].

The present paper provides the explicit (symbolic) formulas of Lattice State-Space Wave Digital Filters (LSSWDFs) derived from reference filters where both lattice-branches are realized either by cascaded all-pass sections or by cascaded unit elements [1].

The usefulness of explicit filter descriptions is twofold. First, they give an essential insight into the properties of the filters to determine their overall performance and complexity characteristics. Second, they can efficiently be used for the finite wordlength optimization of filter coefficients, leading to an improvement of the computation accuracy and a reduction of the processing time.

The explicit formulas of a given filter can usually be obtained using either general purpose algebraic computation tools like Mathematica [5], or specific signal flow graph processing tools as SFD [6]. However, in case of regular filter structures such as lattice WDFs, it is even more effective to determine the formulas in a *generic* form by a proper analysis of the wave flow graph.

In the following sections, the basic properties of lattice WDFs are first reviewed. Then the explicit formulas of LSSWDFs are derived from the state-space realization of the lattice-branches. Finally, the achieved symbolic filter descriptions are discussed and compared with respect to their complexity.

2. LATTICE WAVE DIGITAL FILTERS

Lattice Wave Digital Filters (LWDFs) are basically derived from lossless symmetrical lattice reference filters [1]. A classical lattice two-port inserted between terminations with equal resistance R is represented in Fig. 1, Z_1 and Z_2 being the canonic lattice impedances. For reasons of simplicity, only the case with a single voltage source E applied to the two-port will be considered.

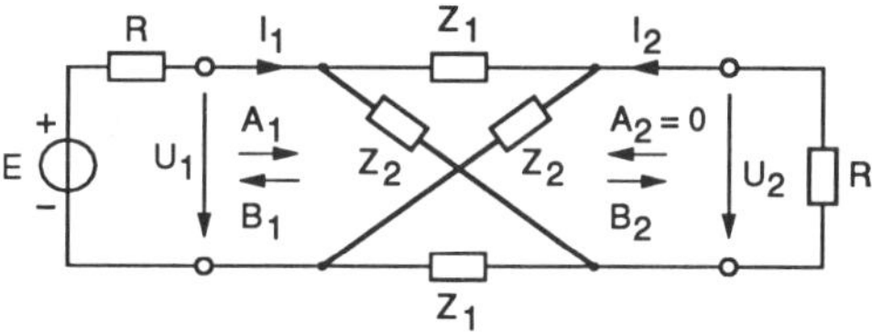

Figure 1 : Symmetrical lattice reference filter.

The transmittance S_{21} of the two-port is given by :

$$S_{21} = B_2 / A_1 = (S_2 - S_1) / 2 \qquad (1)$$

A_1 being the incident wave at port 1, and B_2 the reflected wave at port 2. The terms S_1 and S_2 are the (canonic) reflectances of Z_1 and Z_2, respectively. S_1 and S_2 correspond to all-pass functions, since Z_1 and Z_2 are pure reactances. They can be expressed as :

$$S_i = (Z_i - R) / (Z_i + R) \; ; \qquad i = 1, 2. \qquad (2)$$

The realization of the lattice filters depends essentially on the realization of the canonic lattice impedances for which many structures have been proposed in the literature, eg [1]. Two configurations based on cascaded all-pass sections and on cascaded unit elements have shown to be particularly interesting and will be discussed in this paper.

In the next sections, the realization of a single lattice reflectance will be considered and all terms (ie variables, coefficients, indices, reactance order N) are implicitly related to it.

3. LATTICE STATE-SPACE WDFS

LSSWDFs are derived from reference WDFs by compacting the initial wave flow graph in function of the signal dependencies and by keeping only the signals corresponding to the filter delays [7, 4, 6]. These signals are then referred to as the state variables.

3.1 State-space description

The state-space description is defined in (3, 4), where the vector **W** contains the state variables, while X and Y specify the input and output variables. **A**, **B**, $\mathbf{C}^T$, and D are the usual state-space terms which can be merged into a single state-space matrix **M** [1, 4].

$$\mathbf{W}(k+1) = \mathbf{A} \cdot \mathbf{W}(k) + \mathbf{B} \cdot X(k) \qquad (3)$$

$$Y(k) = \mathbf{C}^T \cdot \mathbf{W}(k) + D \cdot X(k) \qquad (4)$$

$$\begin{pmatrix} W_1(k+1) \\ W_2(k+1) \\ W_3(k+1) \\ Y(k) \end{pmatrix} = \begin{pmatrix} \alpha_{1,1} & \alpha_{1,2} & \alpha_{1,3} & \alpha_{1,4} \\ \alpha_{2,1} & \alpha_{2,2} & \alpha_{2,3} & \alpha_{2,4} \\ \alpha_{3,1} & \alpha_{3,2} & \alpha_{3,3} & \alpha_{3,4} \\ \alpha_{4,1} & \alpha_{4,2} & \alpha_{4,3} & \alpha_{4,4} \end{pmatrix} \cdot \begin{pmatrix} W_1(k) \\ W_2(k) \\ W_3(k) \\ X(k) \end{pmatrix} \qquad (5)$$

(Partitions: **A** = upper-left 3×3 block, **B** = upper-right column, $\mathbf{C}^T$ = lower-left row, D = lower-right element.)

3.2 Realization with cascaded all-pass sections

For LWDFs with cascaded first and second-order all-pass sections, the basic reflectances S_1 and S_2 can directly be designed using the approach proposed by Gazsi [2], leading to the structure represented in Fig. 2. The first-order section only exists for an odd reactance order N, while the number of second-order sections is specified by m_{max}:

$$m_{max} = \begin{cases} N/2 & \text{for N even} \\ (N-1)/2 & \text{for N odd} \end{cases} \qquad (6)$$

Compared to the approach given in [2], the same two-port adaptor is systematically used in the whole structure, since the scaling of state-space filters can be performed very easily in an ultimate step [4]. The considered two-port adaptor is illustrated in Fig. 3 and defined in (7), γ_m being the adaptor coefficient.

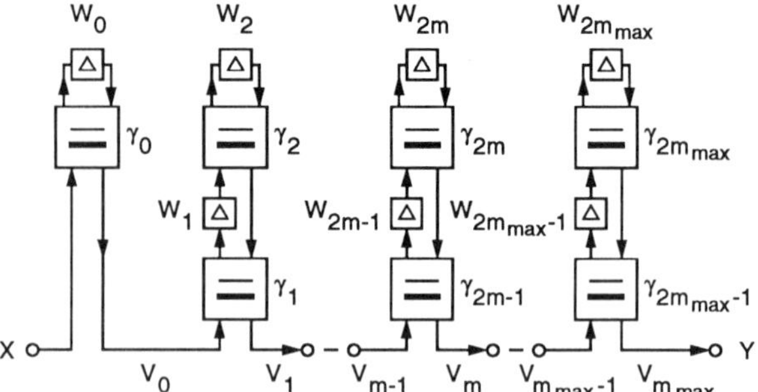

Figure 2 : Lattice react. with cascaded all-pass sect.

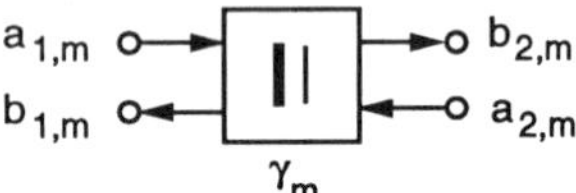

Figure 3 : Two-port parallel adaptor

$$\begin{pmatrix} b_{1,m} \\ b_{2,m} \end{pmatrix} = \begin{pmatrix} -\gamma_m & (1+\gamma_m) \\ (1-\gamma_m) & \gamma_m \end{pmatrix} \cdot \begin{pmatrix} a_{1,m} \\ a_{2,m} \end{pmatrix} \qquad (7a)$$

$$-1 < \gamma_m < 1 \qquad (7b)$$

There exists a *global* signal dependency which is due to the direct link between the auxiliary variables V_m (Fig. 2). This signal dependency increases with N and has an important effect on the complexity of the state-space representation. Solving the linear recurrence defined by the variables V_m, the explicit formulas of the lattice reactance can be determined for the state-space form. The final results are given in Listing 1, and the corresponding equations are provided in the appendix. As expected, the complexity of the obtained expressions grows with N, see (A6). At the end, the state-space coefficients α_{ij} can be determined by matching the achieved expressions with (5).

```
IF ODD(N) THEN
  Apply_Formulas (A1,A2)
ELSE    { EVEN(N) }
  Apply_Formula (A3);

{ 1st second-order section }
IF (m_max ≥ 1) THEN
  Apply_Formulas (A4,A7 @ m=1);

{ 2nd second-order section.}
IF (m_max ≥ 2) THEN
  Apply_Formulas (A5,A7 @ m=2);

{ Remaining second-order sections }
m := 3;
WHILE (m ≤ m_max) DO
  BEGIN
    Apply_Formulas (A6,A7);
    m := m + 1;
  END;    { m ≤ m_max }
Apply_Formula (A8);
```

Listing 1 : Design algorithm for cascaded all-pass sect.

3.3 Realization with cascaded unit elements

It has been shown by Richards that any reactance of degree N can be realized by means of either an open-circuited or a short-circuited cascade of Unit Elements (UEs) [8, 9]. By properly retiming the obtained structure with an alternating delay distribution, an efficient configuration with only *local* signal dependencies can be derived. The resulting lattice reactance is illustrated in Fig. 4, where k_0 is defined as (k_0 = 1) for open-ended, and (k_0 = -1) for short-circuited reactances.

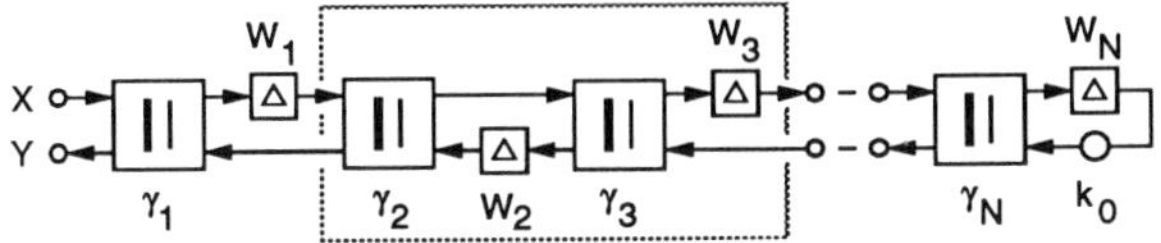

Figure 4 : Lattice reactance with cascaded UEs

Again, the wave flow graph analysis, which is much simpler in this case, provides the explicit formulas which can be merged to get the formal design algorithm given in Listing 2. The state-space coefficients α_{ij} are then derived from the obtained expressions which are here independent of N (B1-B9).

```
IF (N = 1) THEN
  Apply_Formulas (B1,B2)
ELSE
  BEGIN  { N > 1}
    { 1st UE handled separately }
    Apply_Formulas (B3,B4);

    { Intermediate sections with 2 UEs }
    m := 1;
    WHILE (2*m+1 < N) DO
      BEGIN
        Apply_Formulas (B5,B6);
        m := m+1;
      END;   { WHILE (2*m+1 < N) }

    { Final UE sections }
    IF ODD(N) THEN
      Apply_Formulas (B7,B8)
    ELSE   { EVEN(N) }
      Apply_Formula (B9);
  END;   { N > 1}
```

Listing 2 : Design algorithm for cascaded UEs

4. PERFORMANCE COMPARISON

Comparing the explicit formulas which have been determined, the following observations can be made.

First, the number of significant state-space coefficients α_{ij} increases with a factor of $(N^2+12N)/4$ for lattice reactances based on cascaded all-pass sections, in contrast to a factor of 4N for lattice reactances with cascaded UEs. The situation is illustrated in Fig. 5 for *complete* (canonic and non-bireciprocal) state-space WDFs. For comparison purposes, the situation is also represented for ladder state-space WDFs, which have a growing factor of $(N_F+1)^2$, N_F being the filter order.

Second, the complexity of the expressions defining the coefficients α_{ij} increases linearly with N for lattice reactances based on cascaded all-pass sections, while it remains constant for reactances with cascaded UEs. This aspect is important with regard to the finite wordlength optimization of the filter coefficients, as explained in [4, 6].

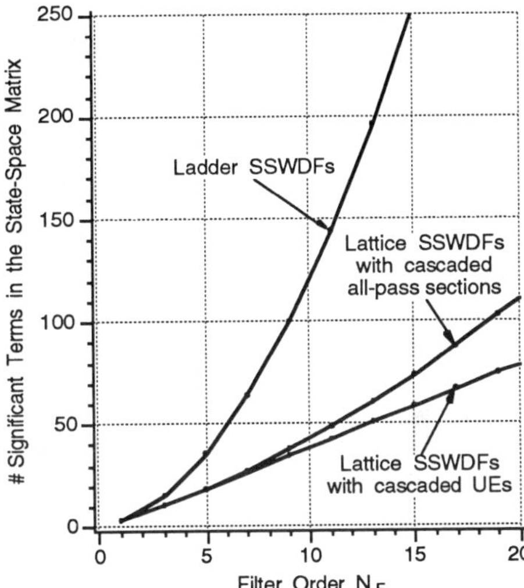

Figure 5 : WDF state-space matrix density

Finally, Fig. 6 shows that the distribution of the significant coefficients α_{ij} within the state-space matrix is more structured and regular for lattice reactances with cascaded UEs. This topic is significant with respect to the implementation efficiency.

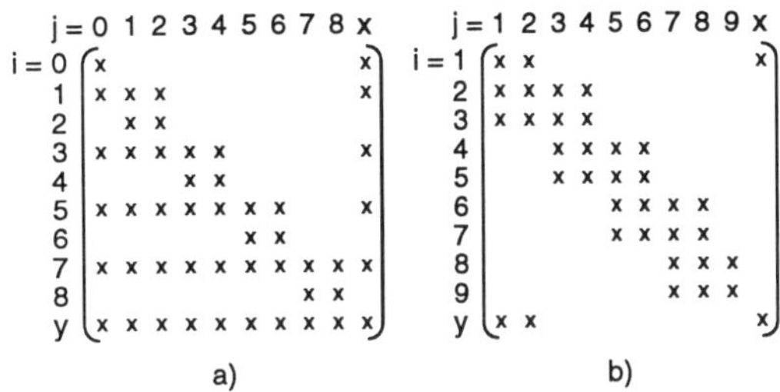

Figure 6 : State-space lattice reactances :
a) with cascaded all-pass sect.
b) with cascaded UEs

From the former comparison criteria, the last two are the most important for practical applications, since the filter order N_F usually remains in a limited range. Clearly, LSSWDFs derived from reference WDFs with cascaded UEs are more efficient. From an implementational point of view, these filters are particularly well suited for effective realizations on vector processors [4], array processors [10], and systolic arrays for band matrix processing.

5. CONCLUSION

In this paper, the explicit formulas of Lattice State-Space Wave Digital Filters derived from reference filters based on cascaded all-pass sections and cascaded unit elements have been determined. The achieved results, which will be further refined in the future with a set of application examples, show that the second alternative is the most interesting from an implementational point of view.

Finally, it should be noticed that the presented approach can easily be extended to the design of bireciprocal LSSWDFs.

ACKNOWLEDGEMENTS

This project was supported by the Commission for the Promotion of Applied Scientific Research (Grant CERS C-2014.1).

REFERENCES

[1] A. Fettweis, "Wave Digital Filters : Theory and Practice", *Proc. IEEE*, Vol. 74, No. 2, pp. 270-327, Feb. 1986.

[2] L. Gazsi, "Explicit Formulas for Lattice Wave Digital Filters", *IEEE Trans. on Circuits and Syst.*, Vol. CAS-32, No. 1, pp. 68-88, Jan. 1985.

[3] W. Wegener, "Wave Digital Filters with Reduced Number of Multipliers and Adders", *Arch. für Elek. Übertragung*, Vol. 33, No. 6, June 1979, pp. 239-243.

[4] U. Sjöström, *On the Design and VLSI Implementation of Digital Signal Processing Algorithms: an Approach Using Wave Digital State Space Filters and Distributed Arithmetic*, Ph. D. Thesis, University of Neuchâtel, CH, 1991.

[5] S. Wolfram, *Mathematica*, 2nd Ed., Addison-Wesley Publ. Co., Redwood City, CA, USA, 1991.

[6] M. Ansorge, U. Sjöström, I. Defilippis, and F. Pellandini, "Symbolic Design of Ladder Wave Digital Filters", *Proc. GRETSI-91*, Juan-Les-Pins, France, Sept. 1991, pp. 949-952.

[7] R. E. Crochiere and A. V. Oppenheim, "Analysis of Linear Digital Networks", *Proc. IEEE*, Vol. 63, No. 4, April 1975, pp. 581-595.

[8] H. Baher, *Synthesis of Electrical Networks*, John Wiley & Sons, Chichester, UK, 1984.

[9] S. N. Güllüoglu, *Diskrete Optimierung von Wellendigitalfiltern, insbesondere von Brücken-Wellendigitalfiltern mit Zweitoradaptoren*, Ph. D. Thesis, Ruhr-Universität Bochum, FRG, 1986.

[10] S. S. Lawson and S. Summerfield, "The Design of Wave Digital Filters Using Fully Pipelined Bit-Level Systolic Arrays", *J. of VLSI Processing*, Vol. 2, No. 1, Sept. 1990, pp. 51-64.

APPENDIX

Lattice reactances with cascaded all-pass sections

$$W_0(k+1) = (1-\gamma_0)\cdot X(k) + \gamma_0\cdot W_0(k) \quad \text{(A1)}$$

$$V_0 = (1+\gamma_0)\cdot W_0(k) - \gamma_0\cdot X(k) \quad \text{(A2)}$$

$$V_0 = X(k) \quad \text{(A3)}$$

$$W_1(k+1) = (1-\gamma_1)\cdot V_0 + \gamma_1\cdot\left[(1+\gamma_2)\cdot W_2(k) - \gamma_2\cdot W_1(k)\right] \quad \text{(A4)}$$

$$W_3(k+1) = (1-\gamma_3)\cdot\left[\xi_{1,1}\cdot V_0 + \xi_{2,1}\cdot W_1(k) + \xi_{3,1}\cdot W_2(k)\right] + \gamma_3\cdot\left[(1+\gamma_4)\cdot W_4(k) - \gamma_4\cdot W_3(k)\right] \quad \text{(A5)}$$

$$W_{2m}(k+1) = (1-\gamma_{2m})\cdot W_{2m-1}(k) + \gamma_{2m}\cdot W_{2m}(k) \quad \text{(A7)}$$

$$Y(k) = V_{m_{max}} \quad \text{(A8)}$$

$$\xi_{1,m} = -\gamma_{2m-1} \quad \text{(A9a)}$$

$$\xi_{2,m} = -(1+\gamma_{2m-1})\cdot\gamma_{2m} \quad \text{(A9b)}$$

$$\xi_{3,m} = (1+\gamma_{2m-1})\cdot(1+\gamma_{2m}) \quad \text{(A9c)}$$

Lattice reactances with cascaded unit elements

$$W_1(k+1) = (1-\gamma_1)\cdot X(k) + k_0\cdot\gamma_1\cdot W_1(k) \quad \text{(B1)}$$

$$Y(k) = -\gamma_1\cdot X(k) + k_0\cdot(1+\gamma_1)\cdot W_1(k) \quad \text{(B2)}$$

$$W_1(k+1) = (1-\gamma_1)\cdot X(k) - \gamma_1\cdot\gamma_2\cdot W_1(k) + \gamma_1\cdot(1+\gamma_2)\cdot W_2(k) \quad \text{(B3)}$$

$$Y(k) = -\gamma_1\cdot X(k) - \gamma_2\cdot(1+\gamma_1)\cdot W_1(k) + (1+\gamma_1)\cdot(1+\gamma_2)\cdot W_2(k) \quad \text{(B4)}$$

$$\begin{aligned} W_{2m}(k+1) = &-\gamma_{2m+1}\cdot(1-\gamma_{2m})\cdot W_{2m-1}(k) + \\ &-\gamma_{2m}\cdot\gamma_{2m+1}\cdot W_{2m}(k) + \\ &-\gamma_{2m+2}\cdot(1+\gamma_{2m+1})\cdot W_{2m+1}(k) + \\ &+(1+\gamma_{2m+1})\cdot(1+\gamma_{2m+2})\cdot W_{2m+2}(k) \end{aligned} \quad \text{(B5)}$$

$$\begin{aligned} W_{2m+1}(k+1) = &(1-\gamma_{2m+1})\cdot(1-\gamma_{2m})\cdot W_{2m-1}(k) + \\ &+\gamma_{2m}\cdot(1-\gamma_{2m+1})\cdot W_{2m}(k) + \\ &-\gamma_{2m+1}\cdot\gamma_{2m+2}\cdot W_{2m+1}(k) + \\ &+\gamma_{2m+1}\cdot(1+\gamma_{2m+2})\cdot W_{2m+2}(k) \end{aligned} \quad \text{(B6)}$$

$$\begin{aligned} W_{N-1}(k+1) = &-\gamma_N\cdot(1-\gamma_{N-1})\cdot W_{N-2}(k) + \\ &-\gamma_{N-1}\cdot\gamma_N\cdot W_{N-1}(k) + k_0\cdot(1+\gamma_N)\cdot W_N(k) \end{aligned} \quad \text{(B7)}$$

$$\begin{aligned} W_N(k+1) = &(1-\gamma_N)\cdot(1-\gamma_{N-1})\cdot W_{N-2}(k) + \\ &+\gamma_{N-1}\cdot(1-\gamma_N)\cdot W_{N-1}(k) + k_0\cdot\gamma_N\cdot W_N(k) \end{aligned} \quad \text{(B8)}$$

$$W_N(k+1) = k_0\cdot\left[(1-\gamma_N)\cdot W_{N-1}(k) + \gamma_N\cdot W_N(k)\right] \quad \text{(B9)}$$

$$\begin{aligned} W_{2m-1}(k+1) = (1-\gamma_{2m-1})\cdot &\left\{ \prod_{\nu=1}^{m-1}\xi_{1,\nu}\cdot V_0 + \sum_{\mu=1}^{m-2}\left(\prod_{\nu=\mu+1}^{m-1}\xi_{1,\nu}\right)\cdot\left[\xi_{2,\mu}\cdot W_{2\mu-1}(k) + \xi_{3,\mu}\cdot W_{2\mu}(k)\right] + \right. \\ &\left. + \xi_{2,m-1}\cdot W_{2m-3}(k) + \xi_{3,m-1}\cdot W_{2m-2}(k) \right\} + \gamma_{2m-1}\cdot\left[(1+\gamma_{2m})\cdot W_{2m}(k) - \gamma_{2m}\cdot W_{2m-1}(k)\right] \end{aligned} \quad \text{(A6)}$$

SIGNAL PROCESSING VI: Theories and Applications
J. Vandewalle, R. Boite, M. Moonen, A. Oosterlinck (eds.)

DESIGN OF TWO-DIMENSIONAL VIDEO FILTERS WITH SPATIAL CONSTRAINTS

Valéry OUVRARD, Pierre SIOHAN

CCETT, 4 rue du Clos Courtel, BP 59, 35512 Cesson-Sévigné, FRANCE
Tel: (33) 99.12.43.05 - Télex: 740 284 F - Fax: (33) 99.12.40.98

A method based on linear programming is used to design optimal minimax 2-D FIR filters which have to satisfy frequency and space-time constraints. The ringing is controlled for edges having different shapes and orientations.

1. INTRODUCTION

Owing to the current fast technological evolution the use of multidimensional digital filters is becoming, nowadays, more and more frequent, even for high speed processing such as those involved in video applications. In most cases the video filters have to be designed in order to achieve an acceptable trade-off between different types of degradations. For instance when designing low-pass decimation filters one tries to keep a maximum resolution while producing a minimum aliasing [1]. Another disturbing distortion is related to the oscillations called "ringing" that appear at the filter output when the input signal contains sharp transitions [2]. These ringing effects are directly related to the filter step response. Thus, for Finite Impulse Response (FIR) filters, the interest of which is to allow a perfect linearity of the phase-frequency characteristic, the control of the step response results in a simple set of linear constraints. Therefore linear programming formulations can be derived for minimax FIR design problems with constraints on the step response. Originally proposed for 1-D FIR filter design [3, pp.180-183], this technique has been also applied to 2-D FIR filters [4]-[5]. However if the mathematical extension to the 2-D case of the linear programming formulation is straightforward an important difference between the problems in one or two dimensions comes from the fact that the 2-D unit step does not represent the whole set of all the possible sharp transitions. In this paper we consider step functions which not only correspond to a quadrantal edge, but also to horizontal, vertical and diagonal edges. An analysis and design results corresponding to these different step functions are given.

2. PROBLEM STATEMENT

2.1. Edge orientation and 2-D unit sequences

Sharp transitions in a video signal may cause ringing [2], to avoid or, at least, limit this visually disturbing effect , filter design techniques including a control of the step response can be used. A straightforward solution to this problem has already been proposed for 1-D FIR digital filters [3]. The extension to 2-D FIR non-separable filters proposed in [4]-[5] has globally the same features, and both problems can be formulated as a linear program. However an important difference between the two problems is related to the definition of the step function. If in 1-D, the unit step is the unique mathematical model to describe a sharp transition this is no longer true in the 2-D case for the usual definition of the 2-D unit sequence [6, pp. 8], which only corresponds to a specific shape of transition.

2-D Step functions	2-D Step responses
Horizontal step $u_1(n_1, n_2) = 1$ if $n_2 \geq 0$ $u_1(n_1, n_2) = 0$ if $n_2 < 0$	$v_1(I_1) = \sum_{n_2=0}^{I_1-1} \sum_{n_1=0}^{N_1-1} h(n_1, n_2)$ $\Lambda_1 = \{I_1, 1 \leq I_1 \leq N_2\}$
Vertical step $u_2(n_1, n_2) = 1$ if $n_1 \geq 0$ $u_2(n_1, n_2) = 0$ if $n_1 < 0$	$v_2(I_1) = \sum_{n_1=0}^{I_1-1} \sum_{n_2=0}^{N_2-1} h(n_1, n_2)$ $\Lambda_2 = \{I_1, 1 \leq I_1 \leq N_1\}$
Quadrantal step $u_3(n_1, n_2) = 1$ if $n_1 \geq 0$ and $n_2 \geq 0$ $u_3(n_1, n_2) = 0$ otherwise	$v_3(I_1, I_2) = \sum_{n_1=0}^{I_1-1} \sum_{n_2=0}^{I_2-1} h(n_1, n_2)$ $\Lambda_3 = \{ (I_1, I_2), 1 \leq I_1 \leq N_1, 1 \leq I_2 \leq N_2\}$
Diagonal step $u_4(n_1, n_2) = 1$ if $n_1 + n_2 \geq 0$ $u_4(n_1, n_2) = 0$ otherwise	$v_4(I_1) = \sum_{n_1=0}^{I_1-1} \sum_{n_2=0}^{I_1-n_1-1} h(n_1, n_2)$ $\Lambda_4 = \{I_1, 1 \leq I_1 \leq N_1+N_2-1\}$

Table I. Examples of 2-D step functions, and their corresponding step responses.

In Table 1, where it assumed that the 2-D FIR filter has support over the region (n_1, n_2), $0 \leq n_1 \leq N_1-1$, $0 \leq n_2 \leq N_2-1$, the 2-D unit step is denoted by $u_3(n_1, n_2)$ and we call it the quadrantal step because it is non-zero over one quadrant in the (n_1, n_2)-plane. Besides the 2-D unit step we have added three other step functions, the common characteristics of which are, firstly, to be non-zero over one half-plane in the (n_1, n_2)-plane, and, secondly, to correspond, for the horizontal and vertical steps, to directions to which the human visual system is particularly sensitive. Let card Λ_k, with $1 \le k \le 4$, denote the number of samples for the four different step functions. From Table I, it can be seen that the step response control will, in general, yield a greater number of constraints for the quadrantal step, for which card $\Lambda_3 = N_1 N_2$, than for the three other step functions, which yield respectively: card $\Lambda_1 = N_1$, card $\Lambda_2 = N_2$, card $\Lambda_4 = N_1+N_2-1$. In Section III it will be shown that, if we choose to limit the ripples for the half-plane step functions, which yield a smaller number of constraints, it is easier to reach an acceptable trade-off between ringing and frequency selectivity.

2.2. Linear programming formulation

Let $h(n_1, n_2)$, $0 \le n_1 \le N_1-1$, $0 \le n_2 \le N_2-1$, denote the impulse response of the 2-D FIR digital filter. If $h(n_1, n_2)$ is symmetrical the filter has a linear phase characteristic with respect to the frequency and its frequency response can be written as:

$$H_0(e^{j\omega_1}, e^{j\omega_2}) = H(\omega_1, \omega_2)\, e^{-j(\frac{N_1-1}{2}\omega_1 + \frac{N_2-1}{2}\omega_2)} \quad (1)$$

In the following we will assume that the filter has an eightfold symmetry, and that, consequently, $N_1 = N_2 = N$, $H(\omega_1, \omega_2)$ being a real trigonometric polynomial such as:

$$H(\omega_1, \omega_2) = \sum_{i=1}^{R} b(i)\, g_i(\omega_1, \omega_2) \quad (2)$$

The R independent coefficients b(i) depend linearly from the $h(n_1, n_2)$'s, these relations and the expression of the basis functions $g_i(\omega_1, \omega_2)$ can be found in [6, pp. 128]. The use of symmetry allows us to reduce the number of variables which thus decreases from N^2 to a number R given by:

$$R = \begin{cases} \dfrac{(N+1)(N+3)}{8} & \text{if N odd} \\ \dfrac{N(N+2)}{8} & \text{if N even} \end{cases} \quad (3)$$

In the frequency domain we, similarly, assume that the desired frequency characteristic $D(\omega_1, \omega_2)$ is octagonally symmetric, i. e. $D(\omega_2, \omega_1) = D(\omega_1, \omega_2)$. We, now, want to obtain a set of coefficients {b(i)} which minimizes the weighted Chebyshev norm between the filter and the desired frequency response. Let $W(\omega_1, \omega_2)$ be a positive weighting function the problem is to find the coefficients b(i) which minimize the following expression:

$$\|W(H-D)\| = \max_{(\omega_1, \omega_2) \in F} W(\omega_1, \omega_2)\,|D(\omega_1, \omega_2) - H(\omega_1, \omega_2)| \quad (4)$$

where F is a compact subset of the 2-D frequency plane.

After a discretization of the frequency domain F using a finite number of K_p discretization points the continuous minimax problem (4) may be approximated by a discrete problem, for which, a linear programming formulation is given by:

$$\begin{cases} \text{Find the R coefficients b(i) minimizing } \delta\ (\delta > 0) \\ \text{Under the } 2K_p \text{ following constraints } (1 \le k \le K_p) \\ \displaystyle\sum_{i=1}^{R} b(i) g_i(\omega_{1k}, \omega_{2k}) - \frac{\delta}{W(\omega_{1k}, \omega_{2k})} \le D(\omega_{1k}, \omega_{2k}) \\ \displaystyle -\sum_{i=1}^{R} b(i) g_i(\omega_{1k}, \omega_{2k}) - \frac{\delta}{W(\omega_{1k}, \omega_{2k})} \le -D(\omega_{1k}, \omega_{2k}) \end{cases} \quad (5)$$

If we want to obtain an accurate result for this later problem, the number of discretization points, K_p, has to be much greater than the number of variables R. Therefore, as in [1], [4] and [5], we choose to solve the dual problem of (4), the result of which can be obtained more quickly.

In Figure 1 the desired function for a diamond-shaped low-pass filter is represented. This typical shape of filter may be required in video applications such as scan conversion or HDTV encoding [1].

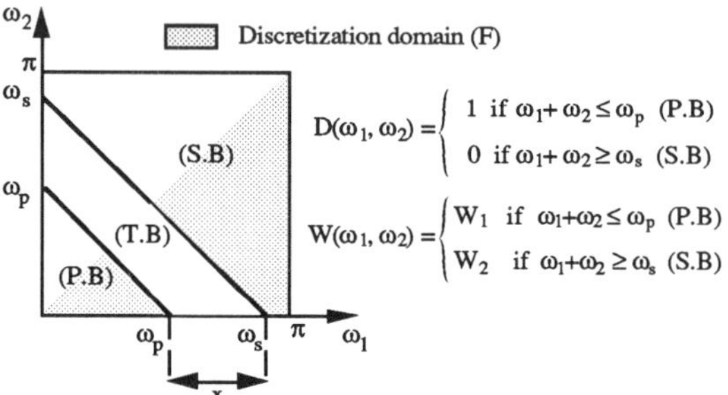

Fig. 1. Desired function $D(\omega_1, \omega_2)$ for the diamond-shaped filter.

2.3. Linear program including step response constraints

Let $L_k(\underline{l})$ and $U_k(\underline{l})$ denote the lower and upper bounds, respectively, for the functions $v_k(\underline{l})$, $\underline{l}$ being, as shown in Table I, an integer index l_1 or l_2, or an integer vector (l_1, l_2). For a given step function $v_k(\underline{l})$ the step response constraints are such as:

$$L_k(\underline{l}) \le v_k(\underline{l}) \le U_k(\underline{l}),\ \underline{l} \in I_{Sk} \subseteq \Lambda_k \quad (6)$$

I_{Sk} is the subset of Λ_k where the inequalities (6) have to be satisfied. It can be seen from the different expressions of $v_k(\underline{l})$ that they are linear with respect to the filter coefficients $h(n_1, n_2)$. Therefore as the variables b(i) are such as $b(i) = c.h(n_1, n_2)$

with c = 1/2, 2 or 4, the linear constraints we add to the linear program (5) are such as:

$$- \sum_{i=1}^{R} \alpha(i, k, \underline{I}) b(i) \leq - L_k(\underline{I})$$
$$\sum_{i=1}^{R} \alpha(i, k, \underline{I}) b(i) \leq U_k(\underline{I}) \quad (7)$$

where $\alpha(i, k, \underline{I})$ are real numbers and k is an integer value less or equal 4.

A graphical illustration of the step response constraints is given in Figure 2 for k = 1 or 2. In this example it can be noted that owing to the impulse response symmetry it is sufficient to limit the constraints to the (N-1)/2 first samples. Furthermore if we do not specify the raise time period we can take $I_{Sk} = \{1, 2, ..., J_1\}$. In the examples presented in the next section we allow identical ripple values for the overshoot and undershoot.

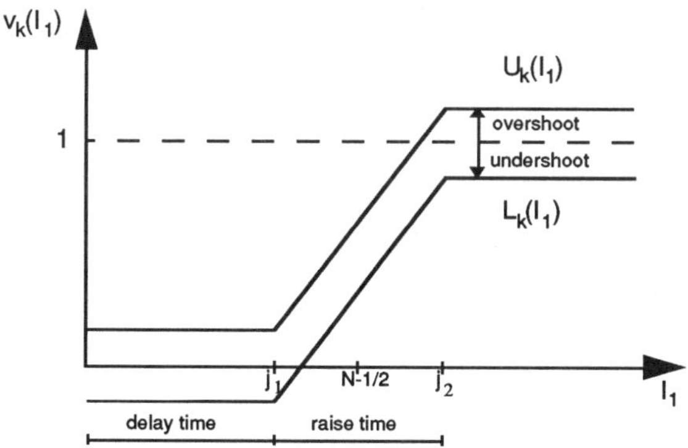

Fig. 2. One-dimensional step response constraints.

3. EXAMPLES

The method used to control ringing outlined from the description occurring in Section 2, in other words one justify carrying out the filter design without spatial constraints, once this result is obtained we can determine the ripple for each step response given in Table 1. Therefore we can advocate several solutions in order to carry out a test, for example to determine which response produces the maximum ringing, the next stage is to reduce it, for example by a factor of two, however at the same time trying to avoid an increase in the maximum value for the other step response ripples. We can also imagine the case for lowering the maximum oscillations of each response. We should also emphasize that for each filter obtained, the optimal frequency error should not go beyond a fixed limit. The used filters will be diamond-shaped filters (cf. Fig. 1), the value x is the width of the transition band along the horizontal direction. Table II shows, for a filter size 13x13 and for filter specifications corresponding to $\omega_p = 0.6\pi$, $x = 0.4\pi$, the maximum ripple for each step response (horizontal and vertical responses are the same) and the filter optimal approximation error. The weights W_1, W_2 in the passband and stopband regions, respectively, are constant and chosen such as $W_1 = 1.$, $W_2 = 0.1$, thus the optimal error, denoted δ_{opt}, is given by $\delta_{opt} = \max(\delta_1, 0.1\delta_2)$ where the δ_i $(1 \leq i \leq 2)$ represent the approximation error in each of the two specified regions. The first line of Table II contains the results without spatial constraints, while for the three following examples constraints are imposed, either to keep a maximum ripple which is less or equal than the initial value, or to reduce it to a given inferior value which is indicated by an asterix (*).

δ_{opt}	θ_v	θ_d	θ_{qb}	θ_{qh}
4.23E-3	8.04E-2	9.17E-2	6.11E-2	1.65E-1
1.78E-2	8.E-2	4.5E-2 *	6.E-2	1.6E-1
2.94E-2	8.E-2	8.E-2 *	5.E-2 *	8.E-2 *
1.34E-2	6.E-2 *	6.E-2 *	6.E-2	1.65E-1

Table II. Results for a 13x13 diamond-shaped filter with and without spatial constraints.

θ_v: maximum ripple for the vertical step response
θ_d : maximum ripple for the diagonal step response
θ_{qb}: maximum ripple for the quarter step response (lower part)
θ_{qh}: maximum ripple for the quarter step response (higher part)

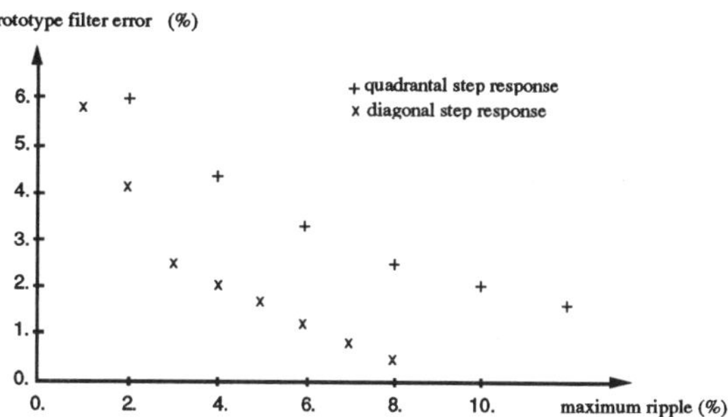

Fig. 3. Frequency approximation error obtained by varying the ripple for the filter of Table II.

The results represented in Fig. 3 are obtained with the same filter and an identical simulation procedure, i. e. the maximum step response ripples we authorize are those given in the first line of Table II, then we compare the evolution of the approximation error with respect to the quadrantal and diagonal step response ripples. It can be seen that, owing to the smaller number of constraints involved, the approximation error is always less higher with the diagonal step than with the quadrantal step. It can be added that the results corresponding to the horizontal (and vertical) step, not reported here, are quite near to those obtained with the diagonal step. The frequency responses and the diagonal step responses corresponding to the first two lines of Table II are reported in Fig. 4-7.

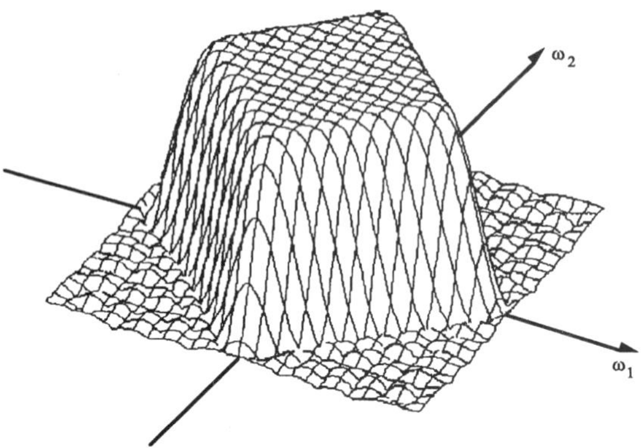

Fig. 4. Frequency response for the filter of Table II without spatial constraints (line 1 of Table II).

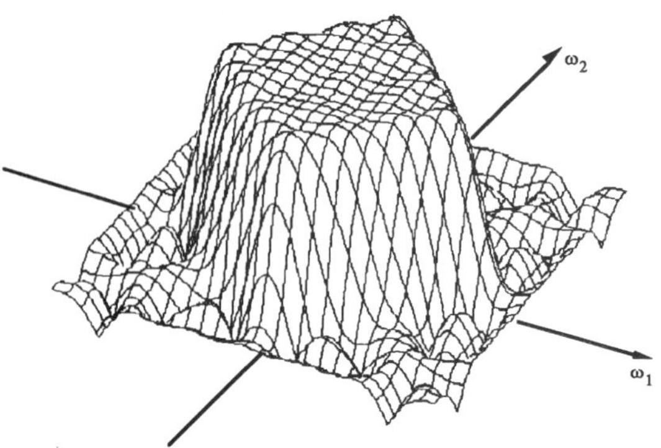

Fig. 5. Frequency response for the filter of Table II (line 2 of Table II).

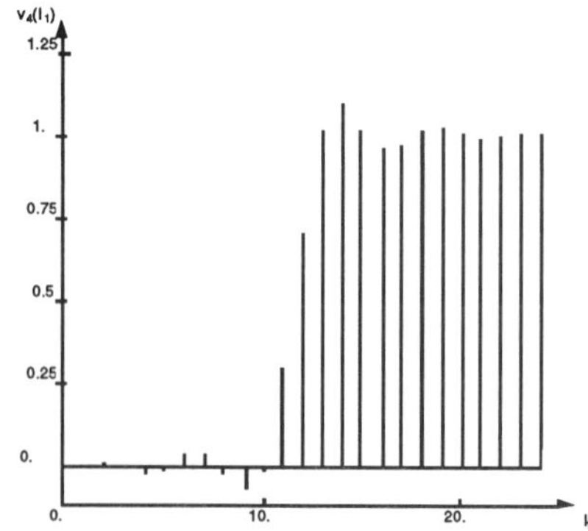

Fig. 6. Diagonal step response for the filter of Table II without spatial constraints (line 1 of Table II).

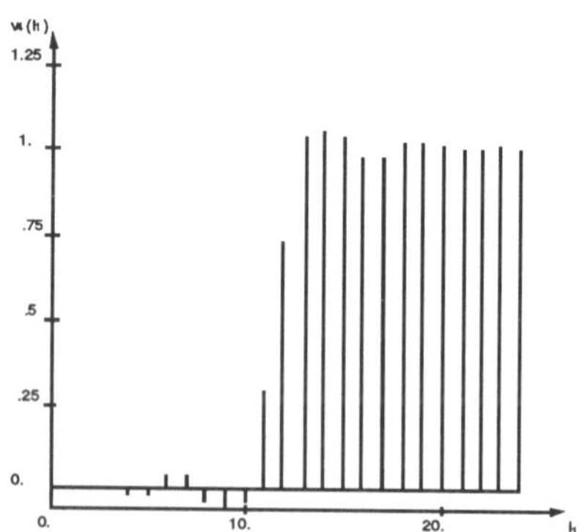

Fig. 7. Diagonal step response for the filter of Table II (line 2 of Table II).

4. CONCLUSION

A 2-D FIR filter design method is proposed which takes into account frequency and space-time constraints. A control of the ringing is ensured for edges having different shapes and orientations.

REFERENCES

[1] P. Siohan, P. Bernard, Optimal filter design for HDTV systems, in Proc. of the third Int. Workshop on HDTV, Torino, Aug.1989.

[2] G. A. Reitmeir, The effects of analog filtering on the picture quality of component digital television systems, SMPTE J., Oct. 1981, pp. 949-955.

[3] B. Gold, L. R. Rabiner, Theory and Application of Digital Signal Processing (Prentice-Hall Signal Processing Series, 1978).

[4] A. Biasolo et al., Design of Multidimensional Video FIR Filters based on Linear Programming, 33rd Ann. Inter. Symp. on Opti. and Optoelec. Applied Science and Eng., San Diego, Aug. 1989, pp. 1153-1160.

[5] A. Biasolo et al., Computer Aided Design of Multidimensional FIR filters for Video Applications, IEEE Trans. Cons. Elec., Aug. 1989, pp. 290-295.

[6] D. E. Dudgeon, R. M. Mersereau, Multidimensional Digital Signal Processing (Prentice-Hall Signal Processing Series, 1984).

SIGNAL PROCESSING VI: Theories and Applications
J. Vandewalle, R. Boite, M. Moonen, A. Oosterlinck (eds.)

Improved Frequency-Sampling Two-Dimensional FIR Filter Design

MICHAEL LIGHTSTONE and SANJIT K. MITRA

Department of Electrical and Computer Engineering
University of California at Santa Barbara
Santa Barbara, CA 93106, USA

A new method is presented for the fast and efficient design of two-dimensional (2-D) linear phase FIR filters. In this approach, the concept of the structural subband decomposition is applied to the frequency sampling design of 2-D FIR filters. The filter is implemented as a parallel connection of KL sparse subfilters cascaded with interpolators. An algebraic relationship is then developed to determine the frequency samples of the KL subband filters from the original specified $M \times N$ frequency samples. An efficient overall filter design results by individually designing the shorter length subfilters.

1 Introduction

Recently it has been shown that any FIR filter can be decomposed into a parallel connection of cascades of an interpolator and a sparse subfilter with each cascade contributing to the overall frequency response only within a given band of frequencies [1]. In the 2-D case, an FIR filter of size $M \times N$ is decomposed into a parallel implementation (bank) of KL sparse 2-D FIR subfilters cascaded with interpolators. An elegant algebraic method is presented to determine the frequency samples of the subfilters in terms of the original specified frequency samples. The impulse responses of the subfilters can then be determined with much shorter length inverse DFT's using the conventional frequency sampling approach [2]. In addition, for very narrow-band or very wide-band filters, many of the subbands contribute little to the overall frequency response. Thus, a large number of these subbands may be dropped from the overall structure with the remaining branches still matching the desired frequency specifications reasonably well, resulting in a significant reduction of the hardware requirements. If the original response is not narrow-band or wide-band, the number of discardable branches is smaller. However, the contributions of the remaining branches are not equal, and significant hardware reductions can result from selectively quantizing the wordsize of the coefficients for the subfilters.

2 2-D Structural Subband Decomposition [2]

Consider a 2-D sequence $h(m,n)$ of size $M \times N$ with z-transform $H(z_1,z_2)$. If $M = KP$ and $N = LQ$, then the $K \times L$ structural subband decomposition of $H(z_1,z_2)$ is given by

$$H(z_1,z_2) = \begin{pmatrix} 1 & z_1^{-1} & \ldots & z_1^{-(K-1)} \end{pmatrix} \mathbf{R} \cdot \mathbf{G}(z_1^K, z_2^L) \cdot \mathbf{S} \begin{pmatrix} 1 \\ z_2^{-1} \\ \vdots \\ z_2^{-(L-1)} \end{pmatrix} \quad (1)$$

where $\mathbf{R} = [\mathrm{r_{ij}}]$ is a $K \times K$ nonsingular matrix, $\mathbf{S} = [\mathrm{s_{ij}}]$ is an $L \times L$ nonsingular matrix, and $\mathbf{G}(\mathrm{z_1^M, z_2^N})$ is a $K \times L$ matrix whose (i,j) element is given $G_{i,j}(z_1^K, z_2^L)$. Note that $H(z_1,z_2)$ can also be rewritten as

$$H(z_1,z_2) = \sum_{i=0}^{K-1} \sum_{j=0}^{L-1} I_i(z_1) G_{i,j}(z_1^K, z_2^L) F_j(z_2) \quad (2)$$

where $I_i(z_1)$ and $F_j(z_2)$ are given by

$$I_i(z_1) = \sum_{k=0}^{K-1} r_{ik} z_1^{-k}, \qquad i = 0,1,\ldots,K-1, \quad (3)$$

$$F_j(z_2) = \sum_{k=0}^{L-1} s_{kj} z_2^{-k}, \qquad j = 0,1,\ldots,L-1. \quad (4)$$

Realizations of $H(z_1, z_2)$ based on the structural subband decomposition given by Eqs. (1) and (2) are shown, respectively, in Figures 1 and 2. Note that the delays in the implementation of the subfilters in Figure 1 can be shared, leading to a canonic realization of the overall structure. As can be seen from Figure 2, the structural subband decomposition can be viewed as as a generalization and 2-D extension of the interpolated FIR structure [3], where $I_i(z_1)$ and $F_j(z_2)$ are the interpolators, and $G_{i,j}(z_1, z_2)$ is a 2-D shaping filter of size $P \times Q$. Thus, the transfer function of an arbitrary 2-D FIR filter of size $M \times N$ (where M and N are divisible by K and L, respectively) can be decomposed into a bank of KL subbands and structurally implemented as a parallel combination of the KL subfilters with PQ nonzero coefficients and their respective interpolators.

3 Determination of the Subband Frequency Samples

A very attractive method for 2-D frequency-sampling design is introduced by transforming the 1-D analytic response given by Jarske et al [4] to a 2-D analytic response using the McClellan frequency transformation [5]

$$\cos(\omega) = -\frac{1}{2} + \frac{1}{2}\cos(\omega_1) + \frac{1}{2}\cos(\omega_2) + \frac{1}{2}\cos(\omega_1)\cos(\omega_2). \quad (5)$$

This function, when sampled, provides approximately circularly symmetric 2-D FIR filters with passband and stopband characteristics very near that of the 1-D design with minimal computation. Though similar, this procedure is not equivalent to the frequency transformation method proposed by McClellan which is applied to an already designed 1-D impulse response and guarantees the amplitude characteristics of the 1-D filter will be preserved in the 2-D filter [5]. However, this approach does guarantee that the 1-D amplitude characteristic will be preserved along the horizontal and vertical frequencies, and in practice actually improves the amplitude characteristic along the diagonal frequencies.

In order to develop the frequency-sampling design of 2-D FIR filters based on the structural subband decomposition, the $P \times Q$ frequency samples for each of the KL shaping filters $G_{i,j}(z_1^K, z_2^L)$ must be determined in terms of original $M \times N$ frequency samples of $H(z_1, z_2)$. The corresponding impulse responses of the shaping filters may then be computed separately. To determine the frequency samples of each shaping filter, the z-transform of each shaping filter must be evaluated at $P \times Q$ evenly spaced points in the ω_1, ω_2 plane. If we define

$$\hat{H}(m,n) = H(W_M^m, W_N^n), \quad m = 0, 1, \ldots, M-1, \; n = 0, 1, \ldots, N-1 \quad (6)$$

$$\hat{G}_{i,j}(k,l) = H(W_P^k, W_Q^l), \quad k = 0, 1, \ldots, P-1, \; l = 0, 1, \ldots, Q-1 \quad (7)$$

where $W_N = e^{-j2\pi/N}$, we can obtain from Eq. (1) for $k = 0, \ldots, P-1$, and $l = 0, \ldots, Q-1$:

$$\hat{\mathbf{H}} = \mathbf{W}\mathbf{R}\hat{\mathbf{G}}\mathbf{S}\mathbf{V}^T \quad (8)$$

where $\hat{\mathbf{H}}$ and $\hat{\mathbf{G}}$ are $K \times L$ matrices whose (i,j) elements are given by $\hat{H}(k+(i-1)P, l+(j-1)Q)$ and $\hat{G}_{i,j}(k,l)$, respectively, and $\mathbf{V}$ and $\mathbf{W}$ are $L \times L$ and $K \times K$ matrices whose (i,j) entries are given by $W_N^{(j-1)(l+(i-1)Q)}$ and $W_N^{(j-1)(k+(i-1)P)}$, respectively. Note that $\mathbf{V}$ can be written as $\mathbf{V} = \mathbf{AB}$ where $\mathbf{A}$ is an $L \times L$ DFT matrix and $\mathbf{B}$ is the diagonal matrix given by

$$\mathbf{B} = diag\begin{pmatrix} 1 & W_N^l & \ldots & W_N^{(L-1)l} \end{pmatrix}. \quad (9)$$

Similarly, $\mathbf{W}$ can be written as $\mathbf{W} = \mathbf{CD}$ where $\mathbf{C}$ is a $K \times K$ DFT matrix, and $\mathbf{D}$ is given by

$$\mathbf{D} = diag\begin{pmatrix} 1 & W_M^l & \ldots & W_M^{(K-1)l} \end{pmatrix}. \quad (10)$$

The frequency samples of the subfilters can now be solved in terms of the frequency samples of the original $M \times N$ filter by solving the matrix equation of Eq. (7) for $k = 0, \ldots, P-1$, and $l = 0, \ldots, Q-1$, resulting in:

$$\hat{\mathbf{G}} = \mathbf{R}^{-1}\mathbf{B}^{-1}\mathbf{A}^{-1}\hat{\mathbf{H}}\mathbf{C}^{-1}\mathbf{D}^{-1}\mathbf{S}^{-1}. \quad (11)$$

Note that because $\mathbf{A}$ and $\mathbf{C}$ are DFT matrices, and because $\mathbf{B}$ and $\mathbf{D}$ are diagonal, their inverses are easy to determine. In addition, $\mathbf{A}$, $\mathbf{C}$, $\mathbf{S}$, and $\mathbf{R}$ are all independent of k and l, so they need only be computed once. Furthermore, instead of actually computing $\mathbf{A}^{-1}$ and $\mathbf{C}^{-1}$ and performing matrix multiplications, a standard inverse FFT algorithm can be utilized. Thus, Eq. (11) provides a simple algebraic method to find the frequency samples of each shaping filter for the 2-D structural subband decomposition.

It is evident from the design procedure that several possibilities exist for the choices of the subband decomposition transform matrices, **S**, and **T** with the only restriction being that the matrices be nonsingular. Two possibilities with integer valued coefficients are the Hadamard transform and binomial transform [6]. Their corresponding advantages and disadvantages are examined in [7,8]. A third choice for the transform matrix examined in this paper is the DCT. Its primary advantage is that the energy compaction property that is useful in image coding can be utilized in filter design as a method for reducing the average wordsize allocated to the filter coefficients of specific subfilters with little effect on the overall filter performance. Thus, for many filters (not necessarily narrow-band or wide-band), the DCT-based subband decomposition provides an implementation with an overall savings in the bit allocation of the filter coefficients as compared with a direct-form realization (as exemplified in Section 4). While the DCT does not have integer-valued elements, in practice, the wordsize allocated to each matrix element can be significantly reduced to just a few bits with little effect on the energy compaction property of the transform.

4 A Design Example

A design example is now presented to illustrate the advantage of DCT-based subband structures and variable bit allocation in the subfilters for 2-D frequency-sampling filter design. The original 1-D filter specifications are for a lowpass filter with $\omega_p = 0.4\pi$, $\omega_s = 0.52\pi$, $\delta_p = 0.067$, and $\delta_s = 0.018$. From these specifications the length of the 1-D filter is determined, and simple analytic functions are obtained for the passband, the transition band, and the stopband. Using the standard frequency-sampling approach, the 1-D impulse response of the designed filter is found from an inverse DFT of 24 samples evenly spaced from 0 to 2π. Next, the 1-D analytic function is mapped to a new 2-D analytic function using the frequency transformation of Eq. (5) and sampled at 24×24 evenly spaced points in the ω_1, ω_2 plane. The 2-D impulse reponse is then calculated from a 24×24 2-D inverse DFT. The resulting 2-D circularly symmetric filter specifications are $\omega_p = 0.42\pi, \omega_s = 0.56\pi, \delta_p = 0.067$, and $\delta_s = 0.018$. Next, using the DCT-based subband decomposition, 6×6 frequency samples are calculated for each of the 16 subfilters for a 4×4 band decomposition. In order to take advantage of the DCT's energy compaction property, the fixed bit allocation scheme for the subband coefficients in Figure 3 was used. Figure 4 shows a 2-D mesh plot of the resulting frequency response which was obtained with an average of 6.3 bits per filter coefficient and the addition of two 4×4 DCT matrices quantized to 4 bits per matrix element. An equivalent response from a direct implementation of the 24×24 impulse response requires a 13-bit wordsize for each filter coefficient. A 1-D plot of the worst case magnitude response at each radial frequency is shown in Figure 5 for both a direct implementation and the DCT-based subband approach.

References

[1] S.K. Mitra and A. Mahalanobis, "Efficient FIR filter design and implementation using a structural subband decomposition," *Proc. 1990 Bilkent International Conference*, Ankara, Turkey, July 1990, pp. 1005-1022.

[2] L.R. Rabiner, B. Gold, and G. A. McGonegal, "An approach to the approximation problem for nonrecursive digital filters," *IEEE Trans.*, vol. AU-18, June 1970, pp. 83-106.

[3] Y. Neuvo, C-Y Dong, and S.K. Mitra, "Interpolated finite impulse response filters," *IEEE Trans.*, vol. ASSP-32, June 1984, pp. 563-570.

[4] P. Jarske, Y. Neuvo, and S.K. Mitra, "Improved frequency sampling design of FIR filters," *Proc. European Signal Processing Conference - EUSIPCO 88.*, Grenoble, France, September 1988, pp. 691-694.

[5] J.H. McClellan, "The design of two-dimensional digital filters by transformation," *Proc. 7th Ann. Princeton Conference on Inform. Sci. and Syst.*, Princeton, N.J., March 1973, pp. 247-251.

[6] P.P. Vaidyanathan, "Multirate digital filters, filter banks, polyphase networks, and applications: a tutorial," *Proc. IEEE*, vol. 78, Jan. 1990, pp. 56-93.

[7] S.K. Mitra, M. Lightstone, and I. Lin, "Efficient frequency sampling FIR filter design using structural subband decomposition," *Proc. Int. Conf. on Digital Signal Processing*, Florence, Italy, September 1991, pp. 3-7.

[8] M. Lightstone, "Efficient frequency-sampling design of one-and two-dimensional FIR filters using structural subband decomposition," M.S. Thesis, University of California at Santa Barbara, March 1992, pp. 17-18.

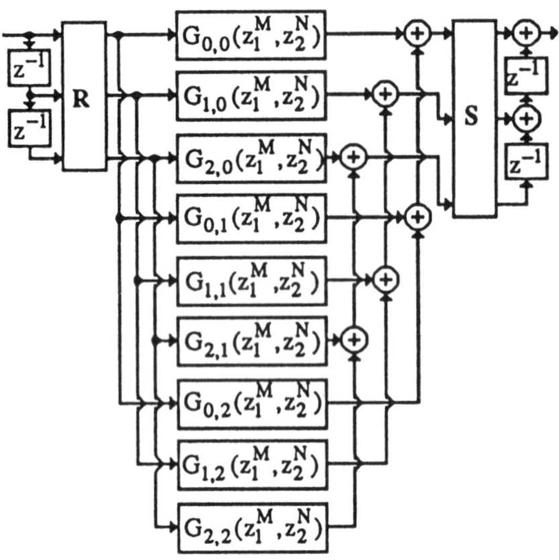

Figure 1: Realization of 2-D structural subband decomposition based on the decomposition of Eq.(1) for K=L=3.

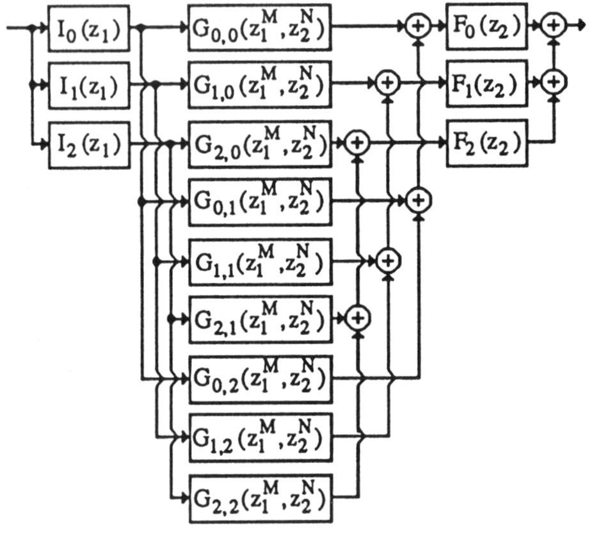

Figure 2: Realization of 2-D structural subband decomposition based on the decomposition of Eq.(2) for K=L=3.

10	11	8	6
11	10	8	5
8	8	5	0
6	5	0	0

Figure 3: Bit allocation scheme for subband filters.

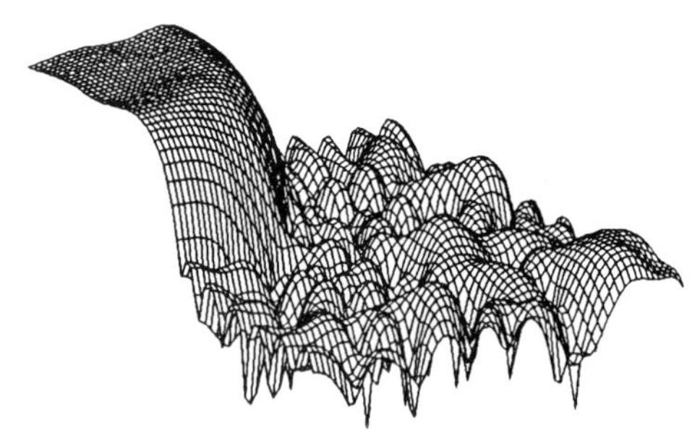

Figure 4: Frequency response of 2D filter obtained using DCT-based decomposition and implemented with an average of 6.3 bits per filter coefficient.

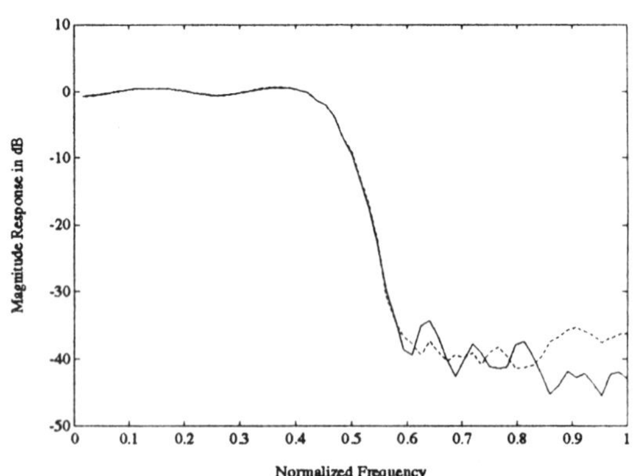

Figure 5: Frequency response comparing the worst case magnitude response at each radial frequency for direct form (solid) implementation with 13 bits per coefficient and DCT-based decomposition (dashed) with 6.3 bits per coefficient.

SIGNAL PROCESSING VI: Theories and Applications
J. Vandewalle, R. Boite, M. Moonen, A. Oosterlinck (eds.)

OPTIMAL IMPLEMENTATION OF MULTI-CHANNEL FILTER BANKS

Erik I. Verriest[1] and Gang Li[2]

School of Electrical Engineering, Georgia Institute of Technology,
Atlanta, GA 30332, USA

A new parametrization, called *implementation*, of a matrix fraction description in terms of certain polynomial operators is proposed, Its parameters appear in a straightforward way in the implementation of the given input-output behavior. The effective computational sensitivity with respect to these implementation parameters is easily analyzed, and in fact optimized.

1 Introduction

This paper extends the results of the polynomial operator representation, introduced in [1] to the matrix fraction descriptions [2]. The latter give a useful description of multi-channel processors, which lead to state space implementations. The filter-banks are of wide interest in multi-scale signal processing, among other applications. It is well known that state space implementations for a given system are not unique. However, in a finite precision computational environment, all theoretically equivalent state space realizations may have very different performance. We have adressed these problems from a differential geometric point of view, and the optimal realizations were characterized in terms of geometric quantities [3]. For a uniform metric we were led to the so called class of generalized balanced realizations as optimal. This is consistent with the work of Mullis and Roberts [4], who used different probabilistic arguments. This differential geometric framework, although very general and at the same time very concise has the serious disadvantage that it is based on infinitessimal analysis. This language is not quite right for discrete valued problems, which are intrinsically of a combinatorial nature. In particular, there is no reason to assume that a coefficient equal to 0 or 1 can only be imprecisely implemented digitally. In fact a large class of rational numbers can be implemented exactly, and this is not taken into account with an infinitessimal sensitivity analysis. For this reason there is a need to separate precisely implementable coefficients, and imprecisely implementable ones.

An alternative theory, in the frequency domain, was introduced by us [1] which can take such a dichotomy between exactly implementable and imprecisely implementable numbers into account. It extends in turn an idea by Middleton and Goodwin [5] of implementing systems by using a representation in the δ-operator rather than the shift operator. They showed superior numerical properties. The proposed generalization expresses both the numerator and denominator polynomial for the siso-case, not in terms of one new indeterminate (the δ), but in terms of n optimally chosen polynomials.

This paper will extend these ideas further to the multi-variate or multi-channel processors, by deriving the generalized implementations from the matrix fraction description of the system (or filter) to be implemented. For clarity, we describe below the methodology for the single input filter bank, i.e. a single-input multi-output system, but in a framework that makes its multi-variate extension transparent.

2 Polynomial Operator Representations

For discrete time systems, the transfer matrix or matrix fraction description (MFD) is usually specified in terms of the delay operator z^{-1}, from which implementations directly follow in the so-called canonical form (sometimes referred to as the "direct" forms). We shall have a closer look at such realization schemes, and generalize these concepts.

Consider the strictly proper $p\times m$ right matrix fraction description [2]

$$\begin{aligned} H(z) &= N(z)D(z)^{-1} \\ &= [N_0\Psi(z)][D_{hc}S_0(z) + D_{lc}\Psi(z)]^{-1} \end{aligned} \quad (1)$$

where $S_0(z) = \mathrm{Diag}(s^{n_1},\ldots,s^{n_m})$, and $\Psi(z) =$ $=$ Blockdiag $([z^{n_i-1},\ldots,z,1]';i=1,\ldots,m)$, and D_{hc}, D_{lc} and N_{lc} are coefficient matrices of size respectively $m\times m$, $m\times n$, and $p\times n$. The n_i are the column degrees

[1] Presently at Georgia Tech Lorraine, F-57070 Metz, France

[2] Université Catholique de Louvain, B-1348, Louvain-la-Neuve, Belgium

of the MFD, and $\sum_{i=1}^{m} n_i = n$. Moreover, we shall assume that the denominator matrix $D(z)$ is in column-reduced form. A preliminary transformation to column reduced form can always be performed. The higher order coefficient matrix D_{hc} is then invertible. Without loss of generality, we shall assume that this coefficient matrix is the identity matrix (which can be achieved by an invertible transformation of the inputs). In this representation, the polynomial matrix $\Psi(z)$ is associated with the partial states. The states are represented by the operators z^k acting on these partial states. We shall now consider a polynomial representation of the transfer MFD, by selecting a different set of *representation polynomials*. So let $\mathcal{T}$ be the matrix transforming the polynomial matrix $[S_0(z), \Psi(z)']'$ to $[Z_0'(z), \hat{Z}(z)']'$, where we require for certain reasons of 'causality' that the column degrees of $Z_0(z)$ are the same as those of $S_0(z)$, and that the higher order coefficient matrix of $Z_0(z)$ is also the identity. Let also the column degrees of $\hat{Z}(z)$ be strictly less than the corresponding column degrees of $Z_0(z)$. With these constraints, $\mathcal{T}$ is of the form

$$\mathcal{T} = \begin{bmatrix} I_m & \tau \\ & T \end{bmatrix} \tag{2}$$

so that the higher order coefficient matrices of the numerator and the denominator polynomial matrices are respectively $0_{p\times m}$ and I_m, and hence are redundant. The transformation matrix is fully parametrized by the $m \times n$-vector τ and the $n \times n$ matrix T. It is not necessary that the submatrix T, and hence $\mathcal{T}$ be invertible, as was assumed in [6].

Example: The scalar transfer function (a trivial MFD)

$$\begin{aligned} H(z) &= [z+1][z^2+2z+3]^{-1} \\ &= [(0,1,1)(z^2,z,1)'][(1,2,3)(z^2,z,1)']^{-1} \end{aligned}$$

is also represented by $[(0,1,0)(z^2+2z+3, z+1, 0)'] \cdot$ $\cdot[(1,0,0)(z^2+2z+3, z+1, 0)']^{-1}$ corresponding to a singular transformation

$$\mathcal{T} = \begin{bmatrix} 1 & 2 & 3 \\ 0 & 1 & 1 \\ 0 & 0 & 0 \end{bmatrix}.$$

At this point we should remark that with each transfer matrix, several right MFD's may correspond, all satisfying the assumptions. Their numerator and denominator matrices must then be related by a unimodular transformation on the right. Should one want this, a unique MFD can be obtained for each transfer matrix for instance in Popov- or polynomial echelon form [2]. The generalized right MFD is thus specified in terms of *coefficients* (acting as coordinates) and *polynomials* (acting as frames). Let $\mathbf{H}_0$ be the class of these representations. A particular element will be denoted by the quadruple $\mathbf{H}_0[(\mathcal{D}, \mathcal{N}), (Z_0(z), \hat{Z}(z))]$ with $(\mathcal{D}, \mathcal{N}) \in \mathbf{R}^{m\times n} \times \mathbf{R}^{p\times n}$ and $(Z_0(z), \hat{Z}(z))$ as specified.

A multi-channel filter bank is a system with one input and $p > 1$ outputs. Let n be its McMillan degree. Its right fraction description involves only a scalar denominator, i.e $N(z)d(z)^{-1}$, where the degree of $d(z)$ is n. Denoting the n dimensional coefficient-vector of the denominator and the $p \times n$ numerator coefficient matrix in generalized polynomial operator form respectively by α and β, we get for a particular filter bank representation:

$$\mathbf{H}_0[(\alpha,\beta),(z_0(z),Z(z))] \in$$

$$\mathbf{R}^n \times \mathbf{R}^{n\times p} \times M\mathbf{R}_n[z] \times (\mathbf{R}_{n-1}[z])^n \tag{3}$$

where $M\mathbf{R}_n[z]$ is the set of monic polynomials of order n in z. For the above scalar example, the two generalized polynomial representations are in the standard form:
$\mathbf{H}_0[((2,3)',(1,1)'),(z^2;z,1)]$, and in the generalized right fraction description:
$\mathbf{H}_0[((0,0)',(1,0)'),(z^2+2z+3;z+1,0)]$.

Clearly, the transformation to a rational vector function in z is in terms of the coordinates and the polynomial frame simply

$$[\beta' Z(z)][z_0(z) + \alpha' Z(z)]^{-1}$$

In the terminology of [3], this map H

$$H : \mathbf{H}_0 = \mathbf{R}^{n+np} \times M\mathbf{R}_n[z] \times (\mathbf{R}_{n-1}[z])^n \longrightarrow (\mathrm{Rat}_n[z])^p$$

where $\mathrm{Rat}_n[z]$ is the set of the strictly proper rational vector functions of z of degree n, defines an *observable*. Alternatively, rather than specifying the polynomials $z_o(z)$ and $Z(z)$, one could also specify the transformation $\mathcal{T}$, i.e. the matrix T, together with the vector τ, which generate $[z_0(z), Z(z)']$ from the "standard" $[z^n, \Psi(z)']$. This class of parametrizations will be denoted by $\mathbf{H}_1[(\alpha,\beta),(\tau,T)]$.

An equivalence, for which the observable $H(z)$ (the filter bank right MFD) is invariant, can be defined in the class $\mathbf{H}_0$ of polynomial operator representations,

Definition: Equivalence in $\mathbf{H}_0$:
$\mathbf{H}_0[(\alpha_1,\beta_1),(z_{01},Z_1)] \sim \mathbf{H}_0[(\alpha_2,\beta_2),(z_{02},Z_2)]$ iff there exists vectors τ_i and matrices T_i , for $i = 1,2$ such that

$$\begin{aligned} z_{oi}(z) &= z^n + \tau_i' \Psi(z) && (4) \\ Z_i(z) &= T_i \Psi(z) && (5) \\ T_1'\alpha_1 + \tau_1 &= T_2'\alpha_2 + \tau_2 && (6) \\ T_1'\beta_1 &= T_2'\beta_2 && (7) \end{aligned}$$

It is clear that the common vector $d = T_i'\alpha_i + \tau_i$ and matrix $N = T_i'\beta_1$ uniquely specify the right MFD $H(z) = N(z)d(z)^{-1}$. In the next section, the key role of α, β, z_o, and Z will be demonstrated in the implementation of the i-o functions represented by $\mathbf{H}_0$.

3 Generalized Reachable Implementations

Consider the *polynomial operator parametrization* of the filter bank $\mathbf{H}_0[(\alpha,\beta),(z_o,Z)]$ as a right fraction description [2]. The relation between input and outputs is then

$$y_t = [\beta' Z(z)][z_0(z) + \alpha' Z(z)]^{-1} u(t) \tag{8}$$

It will be shown that this leads to a *natural* implementation, that is akin to the construction of the canonical realization via Kelvin's principle. First separate Eq. (8) into

$$y_t = \beta' Z(z) \varsigma_t \tag{9}$$

$$[z_0(z) + \alpha' Z(z)]^{-1} u_t = \varsigma_t \tag{10}$$

In turn (10) leads to

$$z_0(z)\varsigma_t = u_t - \alpha_1 z_1(z)\varsigma_t - \ldots - \alpha_n z_n(z)\varsigma_t \tag{11}$$

The polynomial $z_o(s)$ plays a privileged role, and is referred to as the *core-polynomial*. Its has degree n, which is higher than the degree of the polynomials $z_1(z),\ldots,z_n(z)$ by assumption. Consequently, with this core-polynomial, a *core-realization* of

$$z_0(z)\varsigma_t = p_t \tag{12}$$

is derived in a feed-out companion form. Specifically, this is obtained by writing (12) as

$$z^n \varsigma_t = p_t - (\tau_1 z^{n-1}\varsigma_t + \ldots + \tau_n \varsigma_t) \tag{13}$$

This "core of the core" is implemented as the usual chain of n delay-elements. The chain output, ς_t, is the partial state, and the quantity $\chi_t = [z^{n-1},\ldots,z,1]'\varsigma_t$ is the *core state*. It follows from (13) that the core-realization of (11) consists of a weighted feedback of the core-states χ_t superposed on p_t, and fed to the input of the core-core (the chain of delay-elements).

The state-space representation of this feed-out core implementation with core-polynomial $z_0(z)$ is then given in reachable canonical form

$$\chi_{t+1} = \begin{bmatrix} -\tau_1 & \ldots & \ldots & -\tau_n \\ 1 & & & 0 \\ \vdots & \ddots & & \vdots \\ 0 & \ldots & 1 & 0 \end{bmatrix} \chi_t + \begin{bmatrix} 1 \\ 0 \\ \vdots \\ 0 \end{bmatrix} p_t \tag{14}$$

$$\varsigma_t = [0,\ldots,0,1]\chi_t \tag{15}$$

Next, p_t is assembled by identifying the right hand sides of (11) and (12)

$$p_t = u_t - \alpha_1 z_1(z)\varsigma_t - \ldots - \alpha_n z_n(z)\varsigma_t \tag{16}$$

Defining the *implementation-state* x_t by

$$\begin{bmatrix} z_1(z) \\ \vdots \\ z_{n-1}(z) \\ z_n(z) \end{bmatrix} \varsigma_t = T \begin{bmatrix} z^{n-1} \\ \vdots \\ z \\ 1 \end{bmatrix} = T\chi_t = x_t \tag{17}$$

yields the core-input p_t as a linear combination of the input and the implementation states

$$p_t = u_t - \alpha' x_t. \tag{18}$$

Finally, the filter bank outputs follow from (9) and (17)

$$y_t = \beta' Z(z)\varsigma_t = \beta' x_t. \tag{19}$$

The full implementation of the right fraction description $\mathbf{H}_0[(\alpha,\beta),(z_o,Z)]$ is then given by (14), (16), (17), and (19), which are summarized below, using the equivalent parametrization $\mathbf{H}_1[(\alpha,\beta),(\tau,T)]$

Core Update	$\chi_{k+1} = A_c(\tau)\chi_k + b_c p_k$
State Transformation	$x_k = T\chi_k$
Input Assembly	$p_k = u_k - \alpha' x_k$
Read Out	$y_k = \beta' x_k$

where χ_k is the core state, x_k the realization state, p_k the core input, $A_c(\tau)$ a top companion form matrix with characteristic polynomial $z^n + \tau'\Psi(z)$, and $b_c = [1,0,...,0]'$.

All elements of the parametrization α, β, τ and T are found as coefficients in the *implementation*. We further exercise care in the use of the term implementation, in order to distinguish it from the usual notion of a *realization*, which is specified by a triple, rather than a quadruple.

Discussion: i) The implementation requires $p+2$ inner products: $<\alpha, x>$, $<\beta, x>$ and $<\tau, x>$, together with the matrix vector product $T\chi$. The latter should have as many "fixed" (i.e. zeros or ones) elements as possible, since these do not contribute any perturbation terms. ii) The implementation and its states can be interpreted as follows: The core consists of a bank of n filters, with transfer functions $z_1 z_0^{-1},\ldots,z_n z_0^{-1}$. All these are driven by p. Their outputs are the implementation-states. Feedback of the implementation-states is performed around the core with gains α, while the read-out of the implementation-states is accomplished with the multi-channel linear combiner β. The actual number of independent filters in the core filter bank equals the rank of T.

4 Implementation Sensitivity

The reason for distinguishing between realization and

implementation is that the latter allows a more direct analysis of the sensitivity, as it involves directly the parameter values that are *implemented*, and not a function thereoff. Conversely, changes in these parameters show up directly in the associated observable (the transfer function vector H).

Indeed, consider the set $\mathbf{H}_1$, and the associated observable H, the multi-channel MFD (or transfer function vector). Since $\mathbf{H}_1$ is parametrized by α, β, τ, and T, a frequency independent sensitivity measure is obtained by the sum of the H_2-norms of the gradients of the transfer matrix. For a parameter ρ this H_2^2 norm gives the measure [6]

$$M_\rho = \frac{1}{2\pi j}\mathrm{Tr}\oint \frac{\partial H(z)}{\partial \rho}\frac{\partial H(z)}{\partial \rho}^H \frac{dz}{z} \tag{20}$$

The contour integral is over the unit circle $|z| = 1$, and the superscript ^{H} denotes the Hermitian conjugate. In fact, this M_ρ is the sum of the sensitivities in each channel, with scalar transfer function $H_i(z) = N_i(z)/d(z)$. Noting that on the unit circle, $z^* = z^{-1}$, and that $\Psi(z)'\Psi(z^{-1}) = n$, one finds for the sensitivity in each channel (the detailed derivations are in [1])

$$M^{(i)} = \mathrm{Tr}TW^{(i)}T' + \mathrm{Tr}(c_1^{(i)}\beta_i\beta_i' - 2c_2^{(i)}\beta_i\alpha' + c_3^{(i)}\alpha\alpha') + \mathrm{Tr}W_\alpha^{(i)} \tag{21}$$

where $W^{(i)}$ and $W_\alpha^{(i)}$ are certain gramian matrices computable from the transfer function $H_i(z)$. The constants $c_k^{(i)}$ are also invariant to the particular implementation. The last sensitivity term $M_\tau^{(i)}$ is implementation independent, and therefore just adds a constant term to the overall sensitivity.

The optimal implementation is obtained by constraint minimization of the quantity $M = \sum_{i=1}^p M^{(i)}$ which can be rewritten in the form $M = Tr\{TWT' + [\beta, \alpha]\, C \begin{bmatrix} \beta' \\ \alpha' \end{bmatrix}\} + k$ with the constraints

$$d_{lc}' = T'\alpha + \tau \tag{22}$$
$$N_{lc}' = T'\beta \tag{23}$$

where W and C are positive definite matrices, and k a positive constant, all depending only on the i-o characteristics of the filter bank. This formulation avoids the necessity of the existence of T^{-1}. The solution (e.g. based on SVD) to this standard constraint matrix minimization problem gives the optimal implementation parameters, which are for the *siso*-case (so that N is now an n-vector). The c_i are the elements of C.

$$\beta^* = [\frac{N'WN}{\Gamma}]^{\frac{1}{4}} e_1 \tag{24}$$
$$\alpha^* = \frac{c_2}{c_3}\beta^* \tag{25}$$
$$\tau^* = d - \frac{c_2}{c_3}N \tag{26}$$
$$t^* = [\frac{\Gamma}{N'WN}]^{\frac{1}{4}} N' \tag{27}$$
$$T^* = e_1 t^* \tag{28}$$
$$\Gamma = \frac{c_1 c_3 - c_2^2}{c_3} \tag{29}$$

It follows from this that the extremal implementation corresponds to a controller canonical realization of the scalar transfer function $\frac{t^*(z)}{\tau^*(z)}$, with its output fed back over a gain α_1^* to the input. Moreover this output is then scaled by β_1^* to form the system output. It is easily checked that the optimal filter bank implementation does not coincide with the optimal implementation of each individual channel transfer function. In fact, it is well known that the latter is not even a minimal realization in general.

References

[1] Verriest, E.I., Li, G. and Gevers, M., "Computationally Optimal Implementations via Polynomial Operators", in Proc. 1991 MTNS, Kobe Japan, June 1991.

[2] Kailath, T. *Linear Systems* , Prentice Hall, 1980.

[3] Verriest, E.I., and Gray, W.S., "A Geometric Approach to the Minimum Sensitivity Design Problem", submitted to SIAM Journal of Control and Optimization, September 1991.

[4] Mullis, C.T., and Roberts, R.A., "Synthesis of Minimum Roundoff Noise Fixed-Point Digital Filters," IEEE Trans. Circuits Syst., Vol. CAS-23, pp. 551-562, 1976.

[5] Middleton, R.H. and Goodwin, G.C. *Digital Control and Estimation*, Prentice Hall, Englewood Cliffs, 1990.

[6] Li, Gang, "Finite Precision Aspects in the Parametrization of Control, Estimation, and Filtering Problems", Ph.D. Dissertation, Université Catholique de Louvain, Belgium, 1990.

Computing the cardinality of root signal sets of morphological filters

Qiaofei Wang, Moncef Gabbouj, and Yrjö Neuvo
Signal Processing Laboratory
Tampere University of Technology
Tampere, Finland

Abstract: Morphological opening, closing, open-closing, and clos-opening are idempotent operations, i.e., they reduce an input signal to a root in a single pass. In this paper, we apply the theory of state description for the root signal set of these morphological filters. Some properties of root signals are discussed. A complete system of equations has been established which allows us to calculate the number of root signals for morphological filters by structuring element of width k and signals of length L with m quantization levels. For binary signals, simplified recursive equations have been obtained.

1. INTRODUCTION

The basic morphological operations are dilation, erosion, opening, and closing [1,2]. Morphological opening and closing are important algorithms in signal processing. They eliminate specific signal structures smaller than some structuring element without distorting the global features of the signal. Of particular significance is the fact that morphological opening and closing possess idempotency property, i.e., their reapplication does not cause further changes to the previously filtered result. In other words, one pass of opening or closing yields a root signal which is invariant to further passes of the same filter. Moreover, the operations employing iteratively applied openings and closings are also idempotent. They are open-closing (opening followed by closing by the same structuring element) and clos-opening (closing followed by opening by the same structuring element). They in general behave similarly to the median filter. The root signals of median filters have been extensively studied [3-5]. It has been proved [2] that the root of the median is bounded below by the corresponding open-closing and above by the corresponding clos-opening; and the roots of open-closing and clos-opening are roots of the median filter.

A state model describing the binary root signal set of median filters was proposed by Arce and Gallagher [5]. The states in a median root propagate following a tree structure as the signal length increases. Summing over all states at certain stage gives the number of roots associated with the filter for the specified window width and the signal length. By developing a state model in the space of multi-level signals, the authors of [3] have established a complete system of equations for finding the number of root signals associated with any number of quantization levels and any window width median filter.

For morphological opening and closing, however, the set of the states is not complete, which means that the set of the states can not generate itself in the following stage when the signal length increases. The problem has been solved for binary signals in [6] by taking the root number as a special state and recursive equations have been obtained for finding the number of binary roots of opening, closing, open-closing, and clos-opening. For multilevel signals, we will sove this problem in this paper by introducing the concept of virtual states.

2. SOME PROPERTIES OF 1D ROOT SIGNALS

2.1. Definitions

In this paper, morphological operations are considered as function-processing filters. In the function-processing system, it is convenient to define the signal length which is a key parameter in the analysis of root signals. Definitions are given as follows [1],

$$\begin{cases} \text{DILATION}: & (f \oplus \check{K})(x) = \sup\{f(y) : y \in K_x\} \\ \text{EROSION}: & (f \ominus \check{K})(x) = \inf\{f(y) : y \in K_x\} \\ \text{OPENING}: & f \circ K = (f \ominus \check{K}) \oplus K \\ \text{CLOSING}: & f \bullet K = (f \oplus \check{K}) \ominus K \end{cases} \tag{1}$$

where $\check{K} = \{-z : z \in K\}$ denotes the symmetric set of K with respect to the origin, and $K_x = \{z + x : z \in K\}$ denotes the translation of K by x. For sampled signals, the structuring element K is a discrete and finite set which is viewed as a moving window. Hence, erosion (dilation) of a sampled function by a finite set K is equal to the moving local minimum (maximum).

2.2. Threshold Decomposition Property and PBF Expressions

Let $\underline{X} = [X_1, X_2, \ldots, X_L]^T$ (where ^{T} denotes transpose) be a multilevel, nonnegative signal vector. Without loss of generality, we require that $X_i \in \{0, 1, \ldots, m-$

1}. This signal vector can be decomposed into $m-1$ binary vectors $\underline{x}^t, t = 1, 2, \ldots, m-1$, by thresholding. This thresholding operation is T^t, so that

$$x_i^t = T^t(X_i) = \begin{cases} 1 & \text{if } X_i \geq t; \\ 0 & \text{otherwise.} \end{cases} \quad (2)$$

Note that

$$\underline{X} = \sum_{t=1}^{m-1} T^t(\underline{X}) = \sum_{t=1}^{m-1} \underline{x}^t. \quad (3)$$

Definition 1: An ordered set of sequences $\underline{X}_1, \underline{X}_2, \ldots, \underline{X}_k$, is said to obey the stacking property if

$$\underline{X}_1 \geq \underline{X}_2 \geq \cdots \geq \underline{X}_k. \quad (4)$$

Definition 2: A binary function g is said to possess the stacking property [7] if and only if

$$g(\underline{x}) \geq g(\underline{y}) \quad \text{whenever } \underline{x} \geq \underline{y}. \quad (5)$$

It has been shown [7] that a necessary and sufficient condition for a binary function to possess the stacking property is that it is a positive Boolean function (PBF).

Definition 3: The stack filter S_g is defined by a binary filter $g(\underline{x})$ as follows,

$$S_g(\underline{X}) = \sum_{t=1}^{m-1} S_g(\underline{x}^t) = \sum_{t=1}^{m-1} g(\underline{x}^t). \quad (6)$$

We say that S_g obeys the threshold decomposition [4] or the weak superposition property. On the other hand, a stack filter S_g is determined and, hence, represented by the PBF g.

Since morphological filters by flat structuring elements are stack filters, we can find their PBF expressions. Let g_o, g_c, g_{oc}, and g_{co} denote the PBF of opening, closing, open-closing, and clos-opening by structuring element K, respectively. Then we have [8],

$$\begin{cases} g_o(\underline{x}) = \bigvee_{a\in\check{K}} (\bigwedge_{b\in K_a} x_b) \\ g_c(\underline{x}) = \bigwedge_{a\in\check{K}} (\bigvee_{b\in K_a} x_b) \\ g_{oc}(\underline{x}) = \bigwedge_{a\in\check{K}} (\bigvee_{b\in(K\oplus\check{K})_a} (\bigwedge_{c\in K_b} x_c)) \\ g_{co}(\underline{x}) = \bigvee_{a\in\check{K}} (\bigwedge_{b\in(K\oplus\check{K})_a} (\bigvee_{c\in K_b} x_c)) \end{cases} \quad (7)$$

It is important to note that g_o is a dual function of g_c and g_{oc} is a dual function of g_{co}.

Example: Consider morphological opening by structuring element

$$K = \{-1, 0, 1\}. \quad (8)$$

Then the corresponding Boolean function is,

$$g_o(\underline{x}) = (x_{-2} \wedge x_{-1} \wedge x_0) \vee (x_{-1} \wedge x_0 \wedge x_1) \vee (x_0 \wedge x_1 \wedge x_2). \quad (9)$$

2.3. 1D Root Structures

In 1D case, assume the structuring element K is a convex set of length k. Some important 1D structures are defined as follows.

Constant Neighborhood: It consists of at least k consecutive identically valued points.

Edge: A monotonic region which contains no constant neighborhood. If the region is monotonically non-decreasing, it is called a *positive edge*. If the region is monotonically non-increasing, it is called a *negative edge*.

Based on the results found in [2,6,9], the following properties can be proved using the threshold decomposition property [10].

Property 1: A root of opening (closing) is composed of constant neighborhoods and edges. A positive (negative) edge and a following negative (positive) edge must be separated by a constant neighborhood.

Property 2: If $\underline{R}$ is a root signal of opening, then $\underline{P} = (m-1)\underline{\mathbf{1}} - \underline{R}$ is a root of closing by the same structuring element. Hence, the numbers of the roots of opening and closing by the same structuring element are equal to each other for a given signal length.

Property 3: A root of open-closing is a root of clos-opening by the same structuring element and *vice versa*. Hence, the number of the roots of open-closing is equal to the number of the root of clos-opening for a given signal length.

Property 4: A root of open-closing and clos-opening is composed of constant neighborhoods and edges. Two edges must be separated by a constant neighborhood.

Property 5: A root of opening, closing, open-closing, and clos-opening by a particular structuring element is a root of opening, closing, open-closing, and clos-opening by a smaller structuring element, respectively.

3. OPENING AND CLOSING

Suppose the last k digits of the signal are $a_1, a_2, \cdots, a_k$ and the k digits preceding the last $k-j$ $(1 \leq j \leq k-1)$ digits are $b_1, b_2, \cdots, b_k$. The states are defined as follows.

Static state $S_L(i)$: When the values of the last k digits of the signal are identical and equal to i, i.e., $a_1 = a_2 = \cdots = a_k = i$, for $0 \leq i \leq m-1$, we say that it is in a static state.

Downward transition state $D_L(j,i)$: If and only if $a_{k-j} > a_{k+1-j} = a_{k+2-j} = \cdots = a_k = i$, for $1 \leq j \leq k-1$ and $0 \leq i \leq m-1$, we say that it is in a downward state.

Upward transition state $\tilde{U}_Q(j,i)$: If and only if $b_{k-j} < b_{k+1-j} = b_{k+2-j} = \cdots = b_k = m-1-i$, for $1 \leq j \leq k-1$, $0 \leq i \leq m-1$, and $1 \leq Q \leq L-k+j$, we say that it is in an upward transition state.

Since the upward transition state $\tilde{U}_Q(j,i)$ does not appear at the end of a root signal, we call it a virtual state. The recursive relationship among the states can be obtained by inspection.

$$\left\{\begin{array}{l} S_{L+1}(i) = S_L(i) + D_L(k-1,i) + \\ \sum_{r=m-i}^{m-1} (S_{L+1-k}(m-1-r) + \\ \sum_{l=1}^{k-1} (\tilde{U}_{L+1-k}(l,r) + D_{L+1-k}(l,m-1-r))) \\ D_{L+1}(1,i) = \sum_{r=i+1}^{m-1} (S_L(r) + \sum_{l=1}^{k-1} D_L(l,r)) \\ D_{L+1}(j,i) = D_L(j-1,i), \text{ for } 1 < j < k \\ \tilde{U}_{L+1-k}(1,i) = \sum_{r=i+1}^{m-1} (S_{L-k}(m-1-r) + \\ \sum_{l=1}^{k-1} (\tilde{U}_{L-k}(l,r) + D_{L-k}(l,m-1-r))) \\ \tilde{U}_{L+1-k}(j,i) = \tilde{U}_{L-k}(j-1,i), \text{ for } 1 < j < k \end{array}\right. \tag{10}$$

The total number of static states and downward states is equal to the number of roots.

$$R^m(L+1) = \sum_{i=0}^{m-1} (S_{L+1}(i) + \sum_{l=1}^{k-1} D_{L+1}(l,i)) \tag{11}$$

Some numerical results are given in Fig. 1.

For binary signals ($m = 2$), we can simplify (10) and (11) to get a difference equation for $R^2(L+1)$. In binary case, we note that $D_L(j,1) = 0$, for $1 \le j \le k-1$. By (10) and (11),

$$\begin{aligned} \tilde{U}_L(k-1,0) &= S_{L+1-k}(0) + \sum_{l=1}^{k-1} D_{L+1-k}(l,0) \\ &= S_{L-k}(0) + D_{L-k}(k-1,0) + \\ &\quad \sum_{l=2}^{k-1} D_{L-k}(l-1,0) + S_{L-k}(1) \\ &= R(L-k) \end{aligned} \tag{12}$$

where we note that $D_{L-k}(k-1,0) + \sum_{l=2}^{k-1} D_{L-k}(l-1,0) = \sum_{l=1}^{k-1} D_{L-k}(l,0)$. By (11),

$$\begin{aligned} R^2(L+1) &= S_{L+1}(0) + S_{L+1}(1) + \sum_{l=1}^{k-1} D_{L+1}(l,0) \\ &= S_L(0) + D_L(k-1,0) + S_L(1) + \\ &\quad \tilde{U}_L(k-1,0) + \sum_{l=2}^{k-1} D_L(l-1,0) + S_L(1) \\ &= R^2(L) + R^2(L-k) + S_L(1) \end{aligned} \tag{13}$$

Thus we can calculate the difference,

$$\begin{aligned} & R^2(L+1) - R^2(L) \\ & = R^2(L) + R^2(L-k) + S_L(1) - \\ & (R^2(L-1) + R^2(L-1-k) + S_{L-1}(1)) \\ & = R^2(L) + R^2(L-k) - R^2(L-1) \end{aligned} \tag{14}$$

In the above equation, we have used the relation $S_L(1) - S_{L-1}(1) = \tilde{U}_{L-1}(1,0) = R^2(L-1-k)$. Adding $R^2(L)$ to the both sides of (14), we have,

$$R^2(L+1) = 2R^2(L) + R^2(L-k) - R^2(L-1) \tag{15}$$

This equation checks with the result found in [6].

4. OPEN-CLOSING AND CLOS-OPENING

It has been proved that a 1D root of open-closing or clos-opening is also the root of the corresponding median filter [2]. However, a root of median filter is not necessarily a root of open-closing or clos-opening, if we use first and last sample carrying on appending strategy for median filters. Therefore, the positive or negative impulse structures at ends will be removed by morphological filters but preserved by median filters because of the appending strategy. For example, "10011001" is a root of window width three median filter, but it is not a root of open-closing by structuring element of length two.

Suppose the last k digits of the signal are $a_1, a_2, \cdots, a_k$ and the k digits preceding the last $k-j$ $(1 \le j \le k-1)$ digits are $b_1, b_2, \cdots, b_k$. The states are defined as follows.

Static state $S_L(i)$: When the values of the last k digits of the signal are identical and equal to i, i.e., $a_1 = a_2 = \cdots = a_k = i$, for $0 \le i \le m-1$, we say that it is in a static state.

Downward transition state $\tilde{D}_Q(j,i)$: If and only if $b_{k-j} > b_{k+1-j} = b_{k+2-j} = \cdots = b_k = i$, for $1 \le j \le k-1$, $0 \le i \le m-1$, and $1 \le Q \le L-k+j$, we say that it is in a downward state.

Upward transition state $\tilde{U}_L(j,i)$: If and only if $b_{k-j} < b_{k+1-j} = b_{k+2-j} = \cdots = b_k = m-1-i$, for $1 \le j \le k-1$, $0 \le i \le m-1$, and $1 \le Q \le L-k+j$, we say that it is in a upward state.

The upward and downward transition states are not allowed to appear at the end of the root signal. Both of them are virtual states.

Since the root structures of open-closing and clos-opening are same as those of median filters (except the end points), the mechanism of deriving the recursive relationships among the states are similar to that found in the state model of median filters [3].

$$\left\{\begin{array}{l} S_{L+1}(i) = S_L(i) + \sum_{r=i+1}^{m-1} (S_{L+1-k}(r) + \\ \sum_{l=1}^{k-1} \tilde{D}_{L+1-k}(l,r)) + \sum_{r=m-i}^{m-1} (S_{L+1-k}(m-1-r) + \\ \sum_{l=1}^{k-1} \tilde{U}_{L+1-k}(l,r)) \\ \tilde{D}_{L+1-k}(1,i) = \sum_{r=i+1}^{m-1} (S_{L-k}(r) + \sum_{l=1}^{k-1} \tilde{D}_{L-k}(l,r)) \\ \tilde{D}_{L+1-k}(j,i) = \tilde{D}_{L-k}(j-1,i), \text{ for } 1 < j < k \\ \tilde{U}_{L+1-k}(1,i) = \sum_{r=i+1}^{m-1} (S_{L-k}(m-1-r) + \\ \sum_{l=1}^{k-1} \tilde{U}_{L-k}(l,r)) \\ \tilde{U}_{L+1-k}(j,i) = \tilde{U}_{L-k}(j-1,i), \text{ for } 1 < j < k \end{array}\right. \tag{16}$$

Since only static states are allowed to be at the end of

the root signal, the number of roots is equal to the sum of them at the given stage.

$$R^m(L+1) = \sum_{i=0}^{m-1} S_{L+1}(i) \tag{17}$$

Some numerical results are given in Fig. 2.

For binary signals ($m = 2$), we can simplify (16) and (17) to a single difference equation for $R^2(L+1)$. In binary case, we note that $\tilde{D}_L(j,1) = \tilde{U}_L(j,1) = 0$, for $1 \le j \le k$. By (16) and (17),

$$\begin{aligned} R^2(L+1) &= S_{L+1}(0) + S_{L+1}(1) \\ &= S_L(0) + S_{L+1-k}(1) + S_L(1) + S_{L+1-k}(0) \\ &= R^2(L) + R^2(L+1-k) \end{aligned} \tag{18}$$

This equation checks with the result found in [6].

System symmetry:

We note that $S_L(i) = S_L(m-1-i)$ for $i = 0, \cdots, q$ where q is equal to $(m-2)/2$ for m even and $(m-1)/2$ for m odd. Therefore, the number of downward states $\tilde{D}_L(j,i)$ and upward states $\tilde{U}_L(j,i)$ are equal for any specified j and i. The symmetries can be exploited to reduce the amount of computation.

5. CONCLUSIONS

In this paper, we have constructed the 1D root structures of morphological filters and obtained some root properties. The state models for morphological filters have been developed and a complete system of equations has been established by which we can calculate the number of roots of opening, closing, open-closing, and clos-opening for a given signal length.

References

[1] P.A. Maragos and R.W. Schafer, "Morphological filters—Part I: their set-theoretic analysis and relations to linear shift-invariant filters," *IEEE Trans. on Acoustics, Speech, and Signal Processing*, vol. ASSP-35, no. 8, pp. 1153-1169, August 1987.

[2] P.A. Maragos and R.W. Schafer, "Morphological filters—Part II: their relations to median, order-statistic, and stack filters," *IEEE Trans. on Acoustics, Speech, and Signal Processing*, vol. ASSP-35, no. 8, pp. 1170-1184, August 1987.

[3] J.P. Fitch, E.J. Coyle and N.C. Gallagher, Jr., "Root properties and convergence rates of median filters," *IEEE Trans. on Acoustics, Speech, and Signal Processing*, vol. ASSP-33, no. 1, pp. 230-240, Feb. 1985.

[4] J.P. Fitch, E.J. Coyle and N.C. Gallagher, Jr., "Median filtering by threshold decomposition," *IEEE Trans. on Acoustics, Speech, and Signal Processing*, vol. ASSP-32, no. 6, pp. 1183-1188, Dec. 1984.

[5] G.R. Arce and N.C. Gallagher, Jr., "State description for the root-signal set of median filters," *IEEE Trans. on Acoustics, Speech, and Signal Processing*, vol. ASSP-30, no. 6, pp. 894-902, Dec. 1982.

[6] Qiaofei Wang, Moncef Gabbouj, and Yrjö Neuvo, "State description for the root signal sets of morphological filters," accepted to *1992 Int. Symp. on Circuits and Systems*, San Diego, California, USA, May 10-13, 1992.

[7] P.D. Wendt, E.J. Coyle and N.C. Gallagher, Jr., "Stack filters," *IEEE Trans on Acoustics, Speech, and Signal Processing*, vol. ASSP-34, no. 4, pp. 898-911, Aug. 1986.

[8] L. Koskinen, J. Astola, and Y. Neuvo, "Morphological filtering of noisy images," *Proc. of Visual Communications and Image Processing '90*, Lausanne, Switzerland, Oct. 1-4,1990, pp. 155-165.

[9] R.L. Stevenson and G.R. Arce, "Morphological filters: statistics and further syntactic properties," *IEEE trans. on Circuit and systems*, vol. CAS-34, no. 11, pp. 1292-1305, November 1987.

[10] Qiaofei Wang, Moncef Gabbouj, and Yrjö Neuvo, "Root properties of morphological filters," submitted to *Signal Processing*, Dec. 1991.

number of levels m=3

structuring element length k	L=2	3	4	5	6	7	8	9	10
2	3	9	22	51	121	292	704	1691	4059
3	1	3	9	22	46	91	183	383	819
4	1	1	3	9	22	46	86	153	274
5	1	1	1	3	9	22	46	86	148

Fig. 1 Number of roots for opening and closing.

number of levels m=3

structuring element length k	L=2	3	4	5	6	7	8	9	10
2	3	3	9	17	37	77	163	343	723
3	0	3	3	3	9	17	27	49	91
4	0	0	3	3	3	3	9	17	27
5	0	0	0	3	3	3	3	3	9

Fig. 2 Number of roots for open-closing and clos-opening.

SIGNAL PROCESSING VI: Theories and Applications
J. Vandewalle, R. Boite, M. Moonen, A. Oosterlinck (eds.)

A NEW MOMENT-BASED METHOD FOR LOW NOISE ORDERING OF CASCADE FORM FIR DIGITAL FILTER

1. Anil KHARE
2. T. YOSHIKAWA

Nagaoka University of Technology
Dept. of Electrical Engineering
Kamitomioka 1603 - 1,
Nagaoka-shi, Niigata-ken 940-21, JAPAN

The effect of the finite word length on the moment of the impulse response of a digital filter is investigated. It is shown that the quantization error in the moment values is almost linear with a slope of 3dB/bit and is well related to the output roundoff noise. An algorithm is also proposed to order the cascade sections of an FIR filter with low output roundoff noise.

1. INTRODUCTION

Realization of digital filters using finite word length leads to deviation in the filter characteristics from the designed values[1-5]. In signal processing, moment is known as one of the effective methods to express the various properties of the continuous and discrete time signals in time domain[6,7,9]. In this paper, we investigate the effect of the finite word length on the moment of the impulse response of a digital filter. It is shown that the quantization error in the moment values is almost linear with a slope of 3dB/bit and is well related to the output roundoff noise. An algorithm is also proposed to order the cascade sections of an FIR filter with low output roundoff noise using the deviation in moment criterion.

2. THE MOMENTS OF THE IMPULSE RESPONSE

2.1 Basic Definitions

For a general discrete signal and finite duration signal of duration N, x(n), the respective moments of order r are defined in time domain as

$$m_r = \sum_{n=-\infty}^{\infty} (n)^r x(n) \qquad (1)$$

$$m_r = \sum_{n=0}^{N-1} (n)^r x(n) \qquad (2)$$

If $X(e^{j\omega})$ represents the Fourier transform of x(n), r-th order moment is given by the r-th order derivative of $X(e^{j\omega})$ with respect to ω, evaluated at $\omega=0$.

$$m_r = j^r \left. \frac{\partial^r X(e^{j\omega})}{\partial \omega^r} \right|_{\omega=0} \qquad (3)$$

2.2 Moments of direct form and cascade form FIR filter

Let h(n) be the impulse response of an FIR digital filter of order N-1 and let H(z) denote its system function.
If this filter is assumed to be realized in direct form, the moments of the impulse response are given by the following expression.

$$m_r = \sum_{n=0}^{N-1} (n)^r h(n) \qquad (4)$$

For cascade form of realization, the system function H(z) can be expressed as the product of second order sections as follows:

$$H(z) = \prod_{i=1}^{N_s} Hi(z) \qquad (5)$$

Here N_s is the number of second order sections[8].
If we express $H_i(z)$ as

$$H_i(z) = K\,(1-a_i z^{-1})(1-a_i^* z^{-1}) \tag{6}$$

the impulse response of the i-th section $h_i(n)$ is given by

$$h_i(0) = K \tag{7}$$

$$h_i(1) = K(a_i + a_i^*) \tag{8}$$

$$h_i(2) = K a_i a_i^* \tag{9}$$

Obviously the length of the impulse response for each second order section is 3. Using the above values for the impulse response of the individual sections, we can easily obtain the moments of the second order section as

$$m_r(i) = \sum_{n=0}^{2} (n)^r h_i(n) \tag{10}$$

Here $m_r(i)$ is the r-th order moment of the i-th section of the filter realized in cascade form. Even if the i-th order section is not a second-order section, it is possible to consider moments up to order two. We have observed that the results obtained even by considering moments only up to order two do not differ greatly when all the moments up to the required order are considered[9].

2.3 Coefficient quantization and deviation in moment values

If we assume the finite precision arithmetic for the filter coefficients, obviously the moment values given by eq.(4) and eq.(10) will deviate from their ideal values. We define the deviation in moment values Δm_r as follows.

$$\Delta m_r = \mathrm{ABS}\left[\frac{Q[m_r] - m_r}{m_r}\right] \tag{11}$$

Here $Q[m_r]$ is the moment of order r when the coefficients are assumed to be quantized. ABS indicates the absolute value. Fig.1 shows the deviation in moment for different values of the coefficient word length. We note from this figure that the deviation in the moment values is almost linear with a slope of 3dB/bit. Fig.2 shows (figures on right) the deviation in moment vs. output roundoff noise variance for the direct-form realization. From the above relationships, the deviation in moment is seen to be a good indication of the deterioration in the filter characteristics in time-domain for finite word length of the coefficients.

For the i-th second order section in cascade form of realization, we define $\Delta m(i)$ as the overall indication of the change of characteristics in the following way.

$$\Delta m(i) = \sum_{r=0}^{2} \Delta m_r(i) \tag{12}$$

From the frequency domain definition of the moment, we see that $\Delta m(i)$ provides a measure for the change in the sum of zeroth, first and second order derivatives of the Fourier transform of each section, evaluated at zero frequency. Since $\Delta m(i)$ has a simple relationship to the impulse response coefficients in time domain and $\Delta m(i)$ is almost linearly related to the roundoff noise variance for direct-form realization, we conclude that $\Delta m(i)$ is also a good time-domain measure of the roundoff noise for the i-th section.

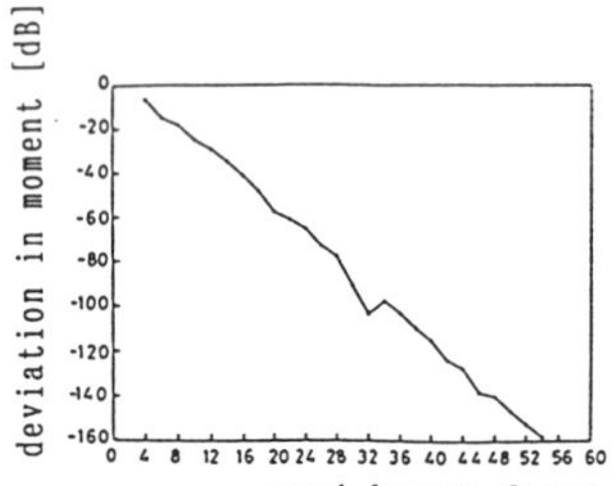

Fig.1 Deviation in moment vs. word length

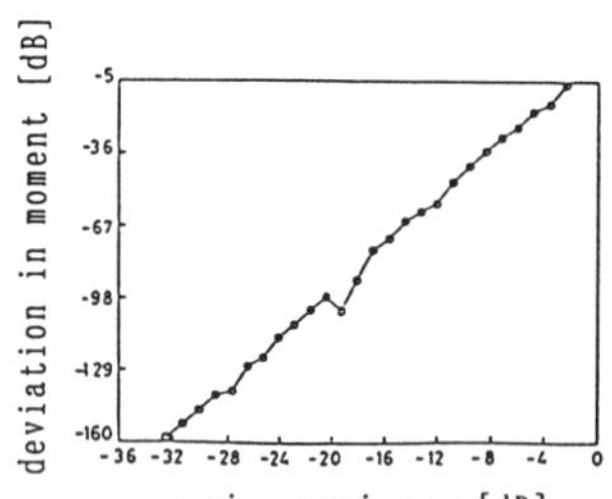

Fig.2 Deviation in moment vs. noise variance

3. NOISE FIGURE CRITERION

In communication theory, we have the concept of noise figure which relates to the internally-generated noise of a two-port analog receiver[10]. If the spectral densities are uniform over the frequency range of interest, the average noise figure F is defined as

$$F = \frac{(S/N)_i}{(S/N)_o} \qquad (13)$$

where $(S/N)_i$ is the ratio of total input available signal and noise power and $(S/N)_o$ is the ratio of total output available signal and noise power.

For a cascade of several stages of gain g_a's, the overall noise figure of the cascade structure is given by

$$F = F_1 + \frac{F_2 - 1}{g_{a1}} + \frac{F_3 - 1}{g_{a1} g_{a2}} + \ldots \qquad (14)$$

For ideal systems with no internally-generated noise, F becomes equal to unity or 0 dB. The above criterion in terms of the equivalent noise temperature leads to the ordering which has the lowest equivalent temperature section in the front-end. While this concept of noise figure cannot be completely adapted to the cascade of digital filters, there are certain noteworthy similarities between the two systems. We have seen that each second order section of the filter realized in cascade form with fixed-point finite precision arithmetic, there exists a roundoff noise due to rounding after multiplication by the filter's coefficients. This noise is totally internal to the cascade section and there are several such sections to realize the filter. Therefore, if we can suitably define an equivalent measure of the noise figure for the sections of the cascade form digital filter, we can use the same concept in ordering these sections according to their "noise figures". We define such a measure as deviation in moment $\Delta m(i)$ for the i-th section of the cascade form digital filter.

4. AN ALGORITHM FOR ORDERING THE CASCADE SECTIONS OF FIR FILTER

If we now consider the quantization of the impulse response coefficients, a qualitative measure of the roundoff noise is claimed to be the deviation in moment, $\Delta m(i)$, defined by Eq.(12). The total output roundoff noise depends upon the method of scaling and the ordering and is very complicated to analyze theoretically. However as is well-known, most orderings have low output roundoff noise and it is possible to find out a good ordering which may not be the "best" yet may have roundoff noise comparable to the lowest possible roundoff noise in the best ordering. The fact that our algorithm does not involve any explicit scaling, it leads to the result that given any appropriate scaling method which ensures the proper operation of the filter, our method will give a good ordering in terms of the output roundoff noise.

The proposed algorithm is as follows.

1. Given a system function H(z), decompose it into second order sections $H_i(z)$.
2. From the knowledge of zeros of the system, calculate the impulse response for each section, using the highest precision available in a computer.
3. Using the impulse response coefficients of step 2 and the definition of moments given by eq.(10), calculate the moments of each second order section.
4. Repeat step 2 for the desired word length and calculate the quantized impulse response coefficients.
5. From the knowledge of quantized coefficients, calculate the quantized values of moments for all the sections.
6. Calculate Δm defined by Eq.(12) for all the sections.
7. Perform the ordering using the rule that if $\Delta m(i) < \Delta m(j)$, section i precedes section j. In the case of equality, choice is arbitrary.

5. EXAMPLE CALCULATION

We consider the same example filter as discussed in reference[5]. Fig.3 shows the zero locations for this filter. Each section is given a number for identification for the purposes of ordering.

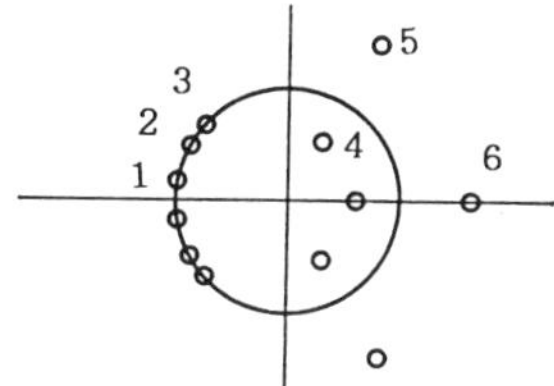

Fig.3 Zero locations of example filter

The specifications of this extraripple filter are as follows:

duration = 13

$f_p = 0.279$ $\quad f_s = 0.376$

$\delta_p = 0.1$ $\quad \delta_s = 0.01$

Here f_p and f_s are normalized pass and stop band frequencies. δ_p and δ_s are the deviations of the magnitude of its frequency response from the desired values.

While many different word lengths have been considered to obtain the low-noise ordering according to our method, we discuss the representative results only. Deviation in moment values for each section for 8 bit, 16 bit and 18 bits against 64 bit word length which for all practical cases is equivalent to infinite word length, have been calculated. Using the algorithm described above, we obtain the ordering as shown in table 1.

TABLE 1 LOW NOISE ORDERING BASED ON DEVIATION IN MOMENT

word length/ordering	roundoff noise
8 bit {2, 5, 6, 1, 3, 4}	$1.5778Q^2$
16 bit {2, 5, 6, 1, 3, 4}	$1.5778Q^2$
18 bit {6, 2, 5, 1, 3, 4}	$3.0945Q^2$

The best ordering as obtained by considering all possible combinations is {2 6 3 4 5 1} which has roundoff noise $1.0983Q^2$. The highest roundoff noise value for this filter is $192.8610Q^2$ which corresponds to the ordering {6, 4, 5, 3, 1, 2}. It is clear from these results that the roundoff noise for the ordering found by the proposed method is very close to the best ordering and is almost one-tenth of the highest roundoff noise. The fact that the proposed method may give different orderings for different word lengths is due to the reason that the deviation in moment is not exactly linear to the changes in word length. So different word lengths may produce different changes in the moment values resulting in slightly different orderings but the values of the output roundoff noise for these orderings are maintained at a low level. As mentioned before, these results are also verified for several different filters and in each case, the resulting ordering from our new algorithm gives very low roundoff noise, within ten percent of the worst-case value of the roundoff noise.

6. CONCLUSIONS

Due to quantization moments of all the individual impulse responses deviate from their ideal values. Though using a rigorous mathematical analysis, this deviation in moments may be shown as related to the roundoff noise, this deviation has been utilized as a qualitative measure for the output roundoff noise and cascade sections have been ordered according to their deviation in moments. We have used the concept of noise figure in analog signal processing and proposed an equivalent argument to find out a low noise ordering based on the deviation in the moment values of each individual section. The resulting ordering then has been compared with the actual analysis and it has been found that the ordering according the new algorithm indeed results in very low roundoff noise. The method has also been verified for several different FIR filters and results for one such filter have been discussed mainly.

REFERENCES

[1] Liu B.,"Effect of Finite Word Length on the Accuracy of Digital Filters - A Review", IEEE Trans. Circuit Theory vol.CT-18,Nov.1971,pp.670-677

[2] Fettweis A.,"On sensitivity and round-off noise in wave digital filters", IEEE Trans. ASSP, vol.ASSP-22,Oct.1974,no.5,pp.383-384

[3] Antoniou A. and Rezk M.G., "A Comparison of Cascade and Wave Fixed-Point Digital-Filter Structures" IEEE Trans. Circuits and Systems, vol.CAS-27,no.12, Dec.1980, pp.1184-1194

[4] Chan D.S.K. and Rabiner L.R.,"Theory of Roundoff Noise in Cascade Realizations of Finite Impulse Response Digital Filters", Bell System Tech.Journal, vol.52,no.3, Mar.1973, pp.329-345

[5] Chan D.S.K.and Rabiner L.R.,"An Algorithm for Minimizing Roundoff Noise in Cascade Realizations of Finite Impulse Response Digital Filters," Bell System Tech.Journal, vol.52, no.3, Mar.1973,pp.347-385

[6] Kendall M.G. and Stuart A., The Advanced Theory of Statistics. London,England,Charles Griffin & Co. Ltd.,vol.1(1977),ch. 1,2,3,4

[7] Khare A. and Yoshikawa T.,"Relation between impulse response moments and locations of poles and zeros of digital filter", Electronics Letters, vol.27, no.12, 6th June 1991,pp.1084-1086

[8] Oppenheim A.V. and Schafer R.W., Digital Signal Processing. Englewood Cliffs,NJ:Prentice Hall,1975, ch.10

[9] Khare A. and T.Yoshikawa,"A Novel Representation based on Moments for Finite Duration Discrete Signals", Proc.ICCS'90,Intl Conf on Commn Syst,Singapore,vol.2,pp

[10]Taub H.and Schilling D.L., Principles of Communication Systems. McGraw Hill, chap.14

SIGNAL PROCESSING VI: Theories and Applications
J. Vandewalle, R. Boite, M. Moonen, A. Oosterlinck (eds.)

MULTIRATE DISCRETE FOURIER TRANSFORM

Yoshiaki Tadokoro, Shin-ichi Yamanaka and Atsuhiko Ishihara

Dept. of Inf.& Comp. Sciences, Toyohashi University of Technology, Toyohashi, 441 Japan.

This paper proposes a new computational algorithm which enables us to compute the discrete Fourier transform (DFT) faster than the fast Fourier transform (FFT). In this algorithm which is referred to as the multirate discrete Fourier transform (MR-DFT), the extreme values of each signal frequency component are sampled and accumulated by alternate additions and subtractions. Since the main operations are additions and subtractions, we can implement this algorithm simultaneously with the sampling operation.

I. INTRODUCTION

The DFT is one of the most important mathematical aids to signal processing [1]. Various algorithms for the DFT have been developed [2], [3], [4]. In many applications it is required that the DFT is computed as fast as possible. The FFT approach is an effective algorithm and widely used [2]. But generally the FFT can be computed only after the end of the sampling operation as shown in Fig.1(a). If we want to compute the DFT faster than the FFT approach, the DFT must be computed simultaneously with the sampling operation as shown in Fig.1(b). If the N-point DFT is implemented with parallel computation, N multipliers and N accumulators are necessary.

We propose a new approach for the computation of the DFT, that is, MR-DFT, without any multiplications of twiddle factors. In this algorithm, the extreme values of each signal frequency component are sampled and accumulated by alternate additions and subtractions. This algoritnm can be implemented with N accumulators and one multiplier using parallel processing.

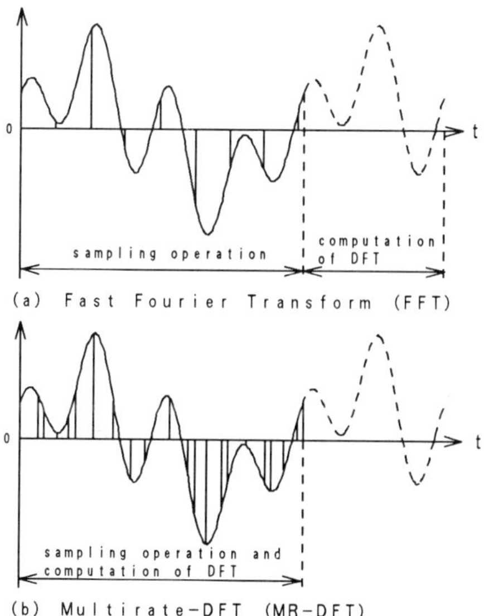

Fig.1. Computation methods of the DFT.

The basic-type MR-DFT algorithm mentioned above has compensation terms. This algorithm has the same features as the arithmetic Fourier transform (AFT) [5], but the MR-DFT is developed from other viewpoint, that is, the multirate signal processing [6].

To exclude the compensation operations of the basic-type MR-DFT, the frequency-shift-type MR-DFT algorithm is developed. In the MR-DFT, the generation of sampling pulses is important. Two methods for the generation of sampling pulses are proposed and the effects of the sampling time errors are considered.

II. ALGORITHM OF THE MR-DFT

A. Principle of the MR-DFT

First, we consider an input signal $x(t)$.

$$x(t) = a_0 + \sum_{p=1}^{4}(a_p \cos 2\pi f_p t + b_p \sin 2\pi f_p t) \qquad (1)$$

where $f_p = pf_1$ and f_1 shows a fundamental frequency.

The extreme values of cosine and sine waveforms of each frequency component in the input signal $x(t)$ are sampled. Then these sampled values are added and subtracted alternately, and accumulated. The accumulated values are shown in the right edge of Fig.2. From these accumulated values, Fourier coefficients can be computed. But the direct component a_0 can be computed by only additions as the same manner in the DFT.

$$a_0 = \frac{1}{8}\sum_{n=0}^{7} x\left(\frac{n}{2f_4}\right)$$

$$a_1 = \frac{1}{2}\sum_{n=0}^{1}(-1)^n x\left(\frac{n}{2f_1}\right) - a_3$$

$$b_1 = \frac{1}{2}\sum_{n=0}^{1}(-1)^n x\left(\frac{n}{2f_1} + \frac{1}{4f_1}\right) + b_3 \qquad (2)$$

$$a_k = \frac{1}{2k}\sum_{n=0}^{2k-1}(-1)^n x\left(\frac{n}{2f_k}\right) \qquad (k = 2,\ 3,\ 4)$$

$$b_k = \frac{1}{2k}\sum_{n=0}^{2k-1}(-1)^n x\left(\frac{n}{2f_k} + \frac{1}{4f_k}\right)$$

B. Basic-type MR-DFT

We consider a band-limited periodic signal $x(t)$.

$$x(t) = a_0 + \sum_{p=1}^{r}(a_p \cos 2\pi f_p t + b_p \sin 2\pi f_p t), \qquad (3)$$

To compute Fourier coefficients a_k and b_k, first of all, $x(t)$ is sampled at the sampling time $t_n = n/(2f_k)$ and $t_n = n/(2f_k) + 1/(4f_k)$ $(n = 0,\ 1,\ \ldots,\ 2k-1)$, respectively.

$$x\left(\frac{n}{2f_k}\right) = a_0 + \sum_{p=1}^{r}\left(a_p \cos\left(\pi \frac{p}{k} n\right) + b_p \sin\left(\pi \frac{p}{k} n\right)\right) \qquad (4)$$

$$x\left(\frac{n}{2f_k} + \frac{1}{4f_k}\right) = a_0 + \sum_{p=1}^{r}\left(a_p \cos\left(\pi n + \frac{\pi}{2}\right)\frac{p}{k} + b_p \sin\left(\pi n + \frac{\pi}{2}\right)\frac{p}{k}\right) \qquad (5)$$

Next, these sampled data are added and subtracted alternately and these accumulated values are divided by the sample number 2k.

$$\frac{1}{2k}\sum_{n=0}^{2k-1}(-1)^n x\left(\frac{n}{2f_k}\right) = \frac{1}{2k}\left(2ka_k + 2k\sum_{q=1}^{u} a_{(2q+1)k}\right) \qquad (6)$$

$$\frac{1}{2k}\sum_{n=0}^{2k-1}(-1)^n x\left(\frac{n}{2f_k} + \frac{1}{4f_k}\right) = \frac{1}{2k}\left(2kb_k + 2k\sum_{q=1}^{u}(-1)^q b_{(2q+1)k}\right) \qquad (7)$$

Where $u = \left[\frac{1}{2}\left(\frac{r}{k} - 1\right)\right]$, (integer).

From Eqs.(6) and (7), the algorithm of the basic-type MR-DFT can be derived.

$$a_0 = \frac{1}{2r}\sum_{n=0}^{2r-1} x\left(\frac{n}{2f_r}\right)$$

$$a_k = \frac{1}{2k}\sum_{n=0}^{2k-1}(-1)^n x\left(\frac{n}{2f_k}\right) - \sum_{q=1}^{u} a_{(2q+1)k}$$

$$(k = 1,\ 2,\ \ldots,\ r) \qquad (8)$$

$$b_k = \frac{1}{2k}\sum_{n=0}^{2k-1}(-1)^n x\left(\frac{n}{2f_k} + \frac{1}{4f_k}\right) - \sum_{q=1}^{u}(-1)^q b_{(2q+1)k},$$

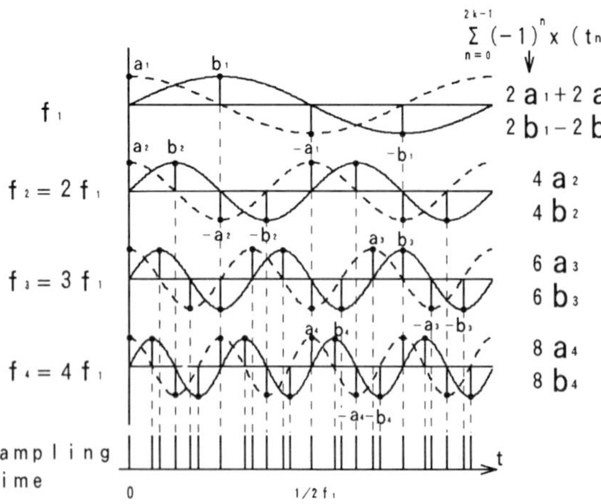

Fig.2. Principle of the multirate-DFT (MR-DFT).

This algorithm shows that if $k \leqq [r/3]$, Fourier coefficients a_k and b_k must be compensated by odd multiple components of these Fourier coefficients.

C. Frequency-shift-type MR-DFT

To eliminate the problem that the basic-type MR-DFT algorithm has compensation terms, $x(t)$ is modulated by the frequency f_m that satisfies the following condition, as shown in Fig.3.

$$f_m > f_r/2 - 3f_1/2. \qquad (9)$$

The modulated signal $x'(t) = x(t)\cos 2\pi f_m t$ has no longer three times component.

The signal $x'(t)$ is sampled at $t_n = n/(2f_{k+m})$ and $t_n = n/(2f_{k+m}) + 1/(4f_{k+m})$, and processed in the same manner as the basic-type MR-DFT. But, the accumulated values are divided by the half number of sampled data.

As a result, the algorithm of the frequency-shift-type MR-DFT can be represented by the following equations:

$$a_0 = \frac{1}{2r}\sum_{n=0}^{2r-1} x\left(\frac{n}{2f_r}\right)$$

$$a_k = \frac{1}{j}\sum_{n=0}^{2j-1}(-1)^n x'\left(\frac{n}{2f_j}\right)$$

$$b_k = \frac{1}{j}\sum_{n=0}^{2j-1}(-1)^n x\left(\frac{n}{2f_j} + \frac{1}{4f_j}\right) \qquad (10)$$

where

$$j = k + m \qquad (k = 1,\ 2,\ \ldots,\ r), \qquad (11)$$

$$x'(t) = x(t)\cos(2\pi f_m t). \qquad (12)$$

III. IMPLEMENTATION OF THE MR-DFT

If $f_m = f_{r/2}$, as shown in Fig.4, the lower half part of signal frequency components can be computed by the frequency-shift-type MR-DFT and another higher half part can be computed by the basic-type one. The frequency modulation can be carried out by an analog-digital multiplication using a digital-to-analog converter as shown in Fig.5. Alternate operations of additions and subtractions can be realized in the following computation :

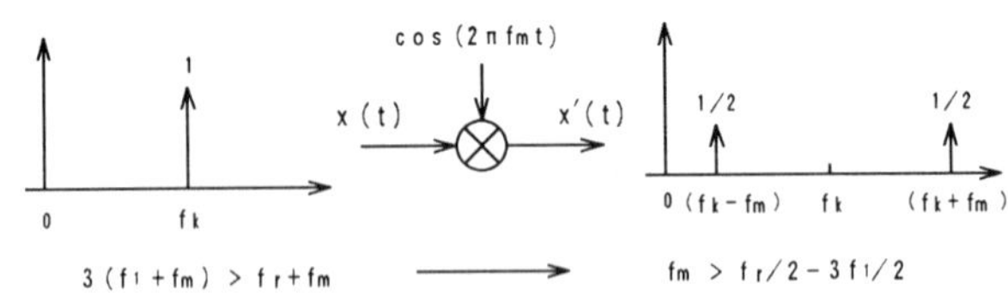

Fig.3. Principle of frequency shift.

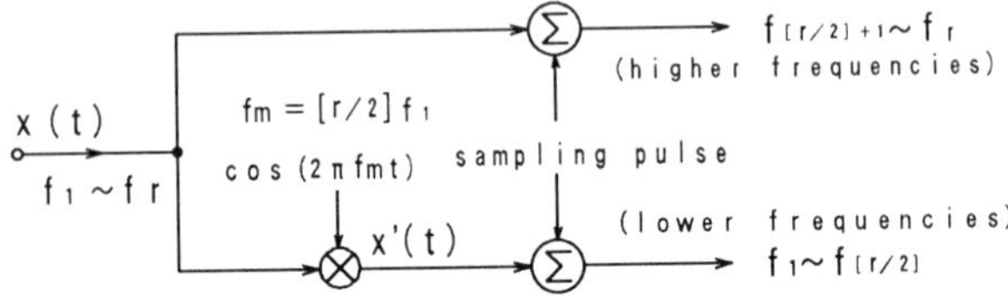

Fig.4. Implementation of the MR-DFT without compensation.

$$Acc(n) = x(n) - Acc(n-1), \tag{13}$$

where Acc(n) and x(n) show the values of accumulator and sampled data, respectively. After the last accumulation, the value of accumulator can be written by

$$Acc(N-1) = \sum_{n=0}^{N-1} (-1)^n x(n). \tag{14}$$

The multiplier for mean operation can be used in time sharing, because the last operations of each accumulator end at almost different times.

Two methods for the generation of sampling pulses are shown in Fig.6. One of them is based on using read only memories (ROMs) and its resolution time is determined by the clock frequency of an address counter. Another is realized by the consumed time of the program [7]. The minimum sampling period is 20 states and the sampling periods greater than 20 states can be made with the resolution of 4 states when the CPU of Z80 is used.

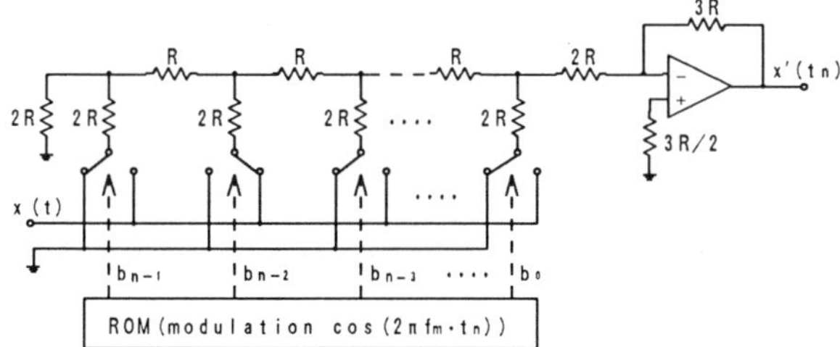

Fig.5. Modulator using a digital-to-analog converter

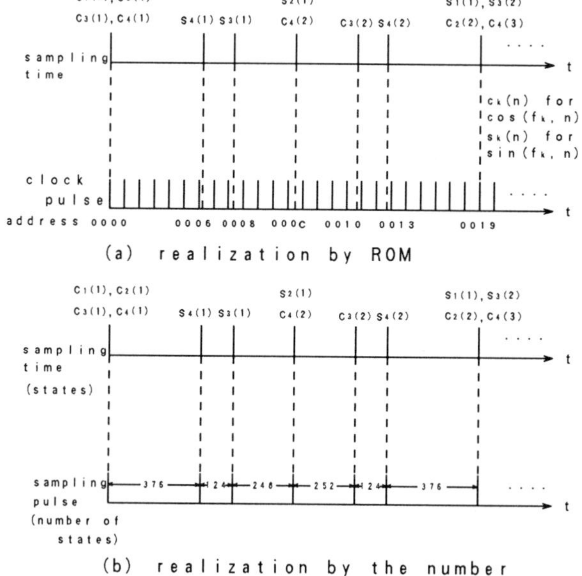

Fig.6. Generation of the sampling pulses (in the case of Fig.2).

IV. PERFORMANCE OF THE MR-DFT

A. Frequency responses

The operations of the basic-type MR-DFT can be represented by the following an FIR digital filter.

$$H(z) = (1 - z^{-1} + z^{-2} - \cdot \cdot \cdot \cdot - z^{-(N-1)})/N \tag{15}$$

Fig.7 represents the amplitude characteristics of the frequency responses of the MR-DFT and the conventional DFT in the case r = 4. The amplitude characteristics for computing the frequency component f_u in the MR-DFT are equivalent to those of the conventional DFT with the sampling frequency $f_{su} = 2f_u$.

B. Effects of sampling time errors

The performance of the MR-DFT is determined by the accuracy of sampling time when the amplitude quantization errors can be ignored.

For the case of f_1 = 100 Hz and f_r = 6.4 kHz, we evaluate the amplitude errors ($\triangle a'_k/a_k$) and the orthogonality errors (a'_p/a_k, $p \neq k$) using sampling pulses in Fig.6(a) and four sampling-time patterns shown in Fig.8, where a_k is the true amplitude, a'_k, a'_p are calculated ones and $\triangle a'_k = a'_k - a_k$.

The relationship between the maximum amplitude errors ($\triangle a'_k/a_k$) and the clock frequencies (f_c) of the ROM address counter is shown in Fig.9. The computer simulations are

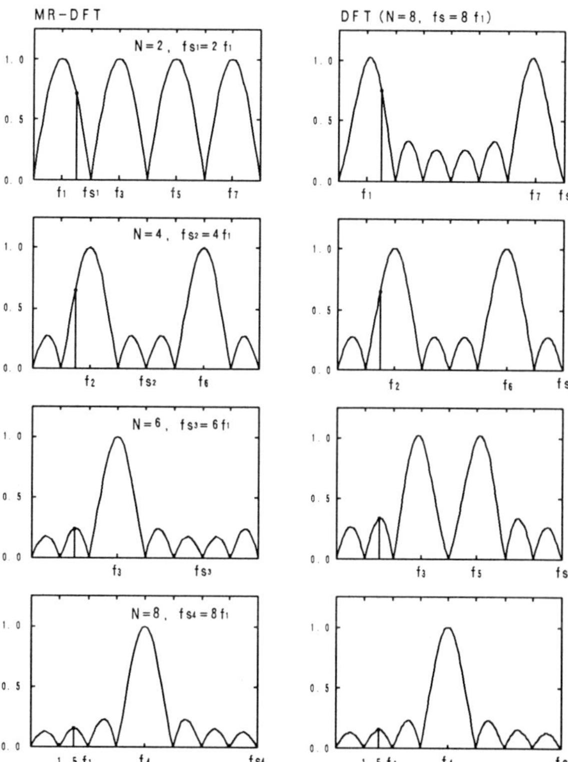

Fig.7. Comparison of the MR-DFT and the conventional DFT spectra.

made with the arithmetic operation of 32 bit floating point. On the other hand, Fig.10 shows the relationship between the maximum orthogonality errors (a'_p/a_k, $p \neq k$) and the clock frequencies (f_c). In these figures, the sampling pattern (4) shows the best results.

C. The number of operations

Table 1 shows the number of operations when an input signal composed of r frequency components is analyzed and all Fourier coefficients are computed. Those of the FFT are also written, but generally the FFT cannot be computed if the sampling operation does not end. Comparing to the conventional DFT, the MR-DFT has smaller additions by the magnitude of 3/4 and smaller multiplications by that of 1/2r. But the number of sample data in the MR-DFT is larger, that is, about $3r^2/2$, comparing to 2r in the DFT.

V. CONCLUSIONS

We propose a new computational algorithm of the DFT that enables us to compute Fourier coefficients simultaneously with the sampling operation. The MR-DFT is implemented by both the basic-type MR-DFT and the frequency-shift-type one. The performance of the MR-DFT is shown in the view points of amplitude characteristics of the frequency responses, the effects of sampling time errors, and the number of operations.

The MR-DFT algorithm is attractive, because it can compute Fourier coefficients effectively faster than the FFT algorithm.

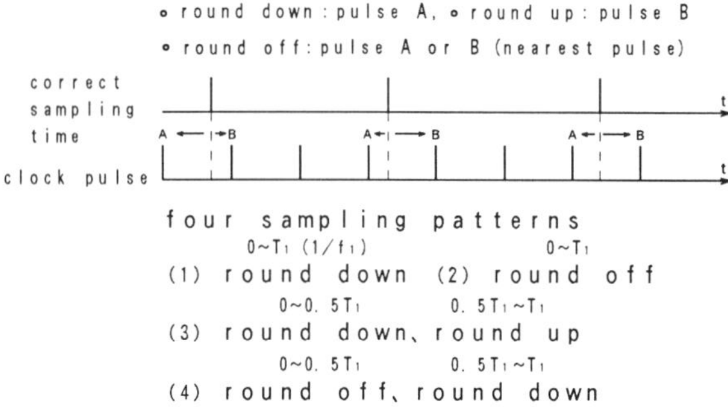

Fig.8. Four sampling-time patterns.

Table 1 Comparison of the number of operations (r : number of frequencies).

	MR-DFT	DFT	(FFT)
addition	$3r^2+4r$	$4r^2-2r$	$6r\log_2 2r$
multiplication	$2r$	$4r^2$	$4r\log_2 2r$

REFERENCES

[1]A.V.Oppenheim and R.W.Shafer, Digital Signal Processing. Prentice-Hall, 1975.

[2]D.F.Elliot and K.R.Rao, Fast Transform Algorithm, Analysis,Applications. Academic, 1982.

[3]Y.Tadokoro and T.Higuchi,"Discrete Fourier transform via the walsh transform," IEEE Trans.ASSP-26, pp.236-240, 1978.

[4]Y.Tadokoro and K.Abe,"Notch Fourier transform," IEEE Trans. ASSP-35, pp.1282-1288, 1987.

[5]D.W.Tufts and G.Sadasive,"The arithmetic Fourier transform," IEEE ASSP Magazine, pp.13-17,1988.

[6]R.E.Crochiere and L.R.Rabiner, Multirate Digital Signal Processing, Prentice-Hall,1983.

[7]Y.Tadokoro and Y.Haneda,"A dual-tone multifrequency receiver using synchronous additions and subtractions," IEEE Trans. COM-35, pp.414-418, 1987.

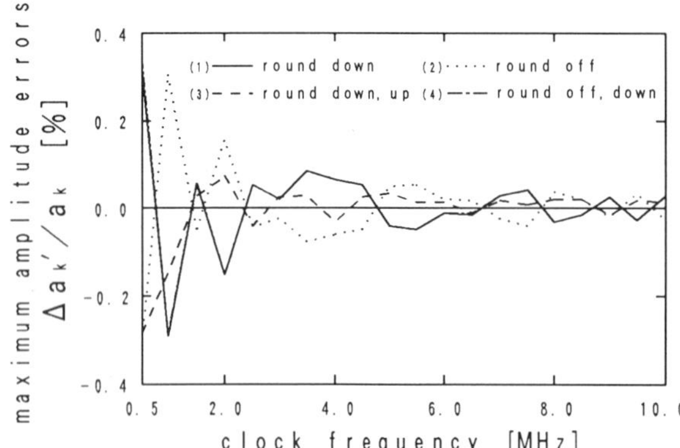

Fig.9. Relationship between the maximum amplitude errors ($\Delta a'_k/a_k$) and clock frequencies (f_c) of the ROM address counter.

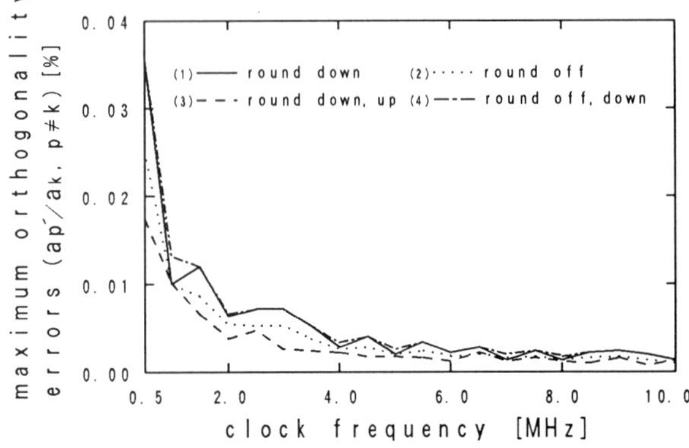

Fig.10. Relationship between the maximum orthogonality errors (a'_p/a_k, $p\neq k$) and clock frequencies (f_c) of the ROM address counter.

SIGNAL PROCESSING VI: Theories and Applications
J. Vandewalle, R. Boite, M. Moonen, A. Oosterlinck (eds.)

ANALYSIS OF QUANTIZATION NOISE IN FREQUENCY SAMPLING FIR FILTERS AND ITS REDUCTION VIA ERROR FEEDBACK

Daniel Gold and Hanoch Ur

Faculty of Engineering, Department of Electrical Engineering - Systems, Tel-Aviv University, Tel-Aviv 69978, Israel

This paper considers first the problem of analysis of quantization noise in frequency sampling complex FIR filters. The error feedback method, in which the discarded part of the word is appropriately fedback to the filter, so as to reduce the total noise quantization, is then applied to these filters. Reduction is indeed achievable under certain conditions and the amount of improvement is a function of the filter characteristics and the specific implementation. This is in contrast to the conventional, non recursive FIR filters.

1. INTRODUCTION

In this paper we first examine the quantization noise in frequency sampling FIR filters and then show that it can be reduced via error feedback (EFB).

The quantization operation, in multiplication in digital filters, usually done by rounding the product to the filter wordlength, causes a roundoff error. This roundoff error is usually described as a white noise sequence, which satisfies appropriate assumptions and its probability density function is assumed to be uniform over the quantization interval: $\Delta=2^{-b}$, in a b+1 bits system [1,2,3,4]. The nonlinear quantization effect is then replaced by a statistical linear model. The variance of the noise signal e(n) is $\sigma_e^2 = 2^{-2b}/12$ for roundoff. In general the signal and the multipliers may be complex (see later), thus four noise sources are created. The noise which was created at the multipliers is filtered by the filter and may cause a strong noise at the filter output.

The EFB technique suggested to reduce this effect is as follows: the error (the least significant bits), created after the multiplication, by quantizing the double length intermediate result to the original wordlength of the digital filter, is saved in order to correct the product at the following sampling instants. Here insight can be obtained by considering the technique in the frequency domain and regarding it as error spectrum shaping (ESS). This shaping is done by introducing zeroes into the noise transfer function from its (equivalent) sources to the output. These zeroes are located at the pass-band of the filter, and decrease the poles' influence. In this way, the quantization noise (which is no longer white) is decreased in the pass band, possibly increased in the stop band, where, however, it is attenuated.
The technique, EFB, or ESS, has been analyzed in various digital systems: IIR filters, ROM filters [2,5,6,7,8,9], and state space digital filters, [9,10].

The corresponding technique for reducing errors in structures containing zeroes is feed-forward [13]. It is not done, however, in conventional FIR structures because there it is tantamount to double precision arithmetic. Nevertheless, FIR filters of the frequency sampling type, considered in this work have a recursive structure and are thus amenable to EFB.
The frequency sampling FIR filter [1], Fig. 1, has a transfer function:

$$H(Z) = (1 - Z^{-N}) \cdot \frac{1}{N} \cdot \sum_{k=0}^{N-1} \frac{H(W_N^{-k})}{1-W_N^{-k} \cdot Z^{-1}} , \qquad (1)$$

$W_N^{-k} = e^{j(2\pi k/N)}$ are equidistant frequencies (frequency samples) at which the transfer function of the filter H is specified. They, as seen, are also poles of H(Z), cancelled by the zeroes of $1-Z^{-N}$. In order to avoid stability problems the pole locations and the zeroes are moved from W_N^{-k}, on the unit circle, to rW_N^{-k}, r is usually taken as $1-\epsilon$, $0<\epsilon<<1$, (often $\epsilon=2^{-m}$), so H(z) becomes:

$$H(z) = (1 - r^N Z^{-N}) \frac{1}{N} \sum_{k=0}^{N-1} \frac{H(rW_N^{-k})}{1 - rW_N^{-k} \cdot z^{-1}} . \qquad (2)$$

Hereafter $H(rW_N^{-k})$ will be denoted, for convenience, by H(k).

This paper deals with the general case of complex (signals and coefficients) frequency sampling FIR filters. This complex representation is very useful in applications involving signals and their quadrature, such as in communication systems, radar systems, electronic warfare systems, etc. [14]. Investigation of the special, but common, case of linear phase real coefficients structures indicates that EFB doesn't significantly improve the quantization error behavior. The analysis for this special case will be published later.

2. The quantization noise in Frequency Sampling realization of complex FIR filters

After each multiplication, in the filtering process, a roundoff operation is performed, this operation will be represented by a noise source [1]. (Fig. 1) where,

- $n_A(n)$ is the noise source created by the $-r^N$ multiplication and roundoff.
- $n_B(k,n)$ is the noise source created by the rW_N^{-k} multiplication and roundoff in the k-th parallel branch.
- $n_c(n)$ is the noise source created after rounding the sum of the multiplications by H(k)/N. Note that by rounding *after summation in double precision* of the intermediate multiplication results, only one noise source is introduced. This is instead of N noise sources, which is the case when rounding in each parallel branch before summation. This scheme is not always implementable, depending on the system.

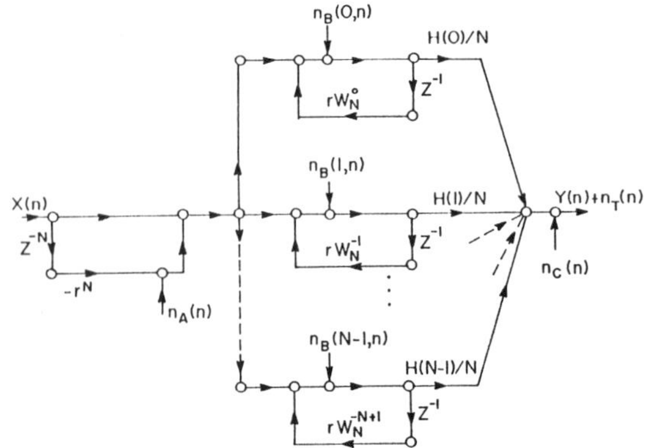

Fig. 1: A frequency sampling FIR filter and its quantization noise sources.

We now calculate the contribution of the various noise sources to the overall output noise. We will assume that the input sequence is complex. Let us note that usually the input signal is scaled in order to avoid the risk of overflow. Absence of overflow is assumed in the following:

The contribution of $n_A(n)$ to σ_T^2, the output noise variance is:

$$\sigma_A^2 \simeq \frac{2\sigma_e^2}{N(1-r^{2N})}\left[\sum_{k=0}^{N-1}|H(k)|^2\right]. \tag{3}$$

It is seen that the relative contribution of σ_A^2 depends on the filter characteristics and thus may be significant. One way to eliminate the effect of $n_A(n)$ is to feed $(-)n_A(n)$ directly to the adders in the junctions of the feedback loops, using the principle of error feedforword [15]. It should be noted that this operation is equivalent to double precision processing until the recursive elements, there is no need, however to implement the noise feed forword technique in branches, where the coefficients H(k), are equal to zero, or small enough, thus saving hardware/software. Thus the effect of $n_A(n)$ is eliminated and it will not be taken further into consideration.

The total output noise variance is:

$$\sigma_T^2 = \sigma_c^2 + \sum_{k=0}^{N-1}\sigma_B^2(k) = 2\sigma_e^2\left[1+\frac{1}{N^2(1-r^2)}\sum_{k=0}^{N-1}|H(k)|^2\right]. \tag{4}$$

3. Error Feedback Implementation

In this section we will apply error feedback to the frequency sampling filter. As our structure contains first order loops (one delay) only, first order error feedback is used [15, 11], to compensate each parallel path pole. The quantization noise is fedback to the multiplier input, while delayed and multiplied by a coefficient, which may be complex. The feedback is implemented only for the quantization noises in the loops, and not for the $n_c(n)$ source which is not part of a recursive circuit.

The error feedback coefficients can be complex, and expressed as $r_k e^{j\phi_k}$, k=0,1....N-1, Fig. 2. For the error feedback realization let us denote the quantization noise after the feedback has been added by $n_B'(k,n)$, Fig. 2, presents the error feedback realization. The total output noise variance is:

$$\sigma^2 = 2\sigma_e^2\{1+\frac{1}{N^2(1-r^2)}\sum_{k=0}^{N-1}|H(k)|^2[(1-r^2) + [r^2+r_k^2-2rr_k\cos(2\pi k/N-\phi_k)]]\}. \tag{5}$$

The next step is to insure the quantization noise reduction by properly choosing the error feedback coefficients: r_k and ϕ_k, k=0,1....N-1. By comparing the output noise variance in error feedback configuration (5) to the noise variance in the original configuration (4), it can be shown that

$$r_k < 2\cos(2\pi k/N-\phi_k). \tag{6}$$

(6) is the first condition for ensuring noise reduction.

In order to avoid additional quantization which can be created

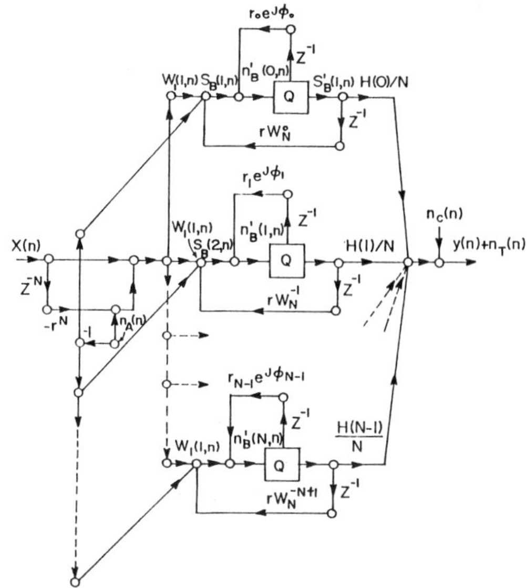

Fig. 2: EFB realization.

in the error feedback circuitry by error feedback coefficient multiplication, we form another condition:

$$\mathrm{Re}\{r_k e^{j\phi_k}\} = r_k \cos\phi_k = \pm 2^m \text{ or } 0 ,$$

$$\mathrm{Im}\{r_k e^{j\phi_k}\} = r_k \sin\phi_k = \pm 2^m \text{ or } 0 ,\ 0 \leq m = \text{integer} . \quad (7)$$

We note that r_k and ϕ_k appear in (5) as part as the expression p_k which is defined as:

$$p_k \triangleq r_k^2 + r^2 - 2rr_k \cos(2\pi k/N - \phi_k) \cong r_k^2 + 1 - 2r_k \cos(2\pi k/N - \phi_k).$$

The output noise variance is decreased as p_k becomes small (for each k).

The necessary procedure for choosing the error feedback coefficient (r_k, ϕ_k) is to find (r_k, ϕ_k) which satisfied conditions, (6) and (7). If several pairs of (r_k, ϕ_k) are found, then choose the pair which minimizes p_k. It should be noted that adding error feedback is worthwhile only in the pass band branches. There is no practical need to add error feedback circuity in the stop band branches, where the noise is well attenuated and its contribution is small due to the small value of the corresponding terms of H(k).

4. Simulation Results

The theoretical amount of improvement due to ESS implementation is defined as the ratio between the output noise variance without the error feedback (4) and with the error feedback (5). Let this improvement factor be denoted as P_T. In order to confirm the theoretical results, a simulation program was written in Pascal language. During the simulation, the actual improvement factors, P_S, were calculated.

Figure 3, a÷b present an example of graphic outputs of the error output without and with error feedback. Since the real and imaginary parts look alike, only the real part is presented. The filter was a BPF with N=128, r=0.999, passband - 25% and 12 bit wordlength.

Extensive simulation results confirmed the theoretical ones.

Note: proper scaling of the input signal is required to avoid overflow. "Overscaling", however, results in a reduced signal, while the quantization noise remains the same, thus the signal to noise ratio, with or without error feedback, is reduced. A large dynamic range, i.e. no "overscaling", is implied [3] in the assumptions mentioned earlier for error feedback analysis. During simulation a proper scaling was generated for each filter.

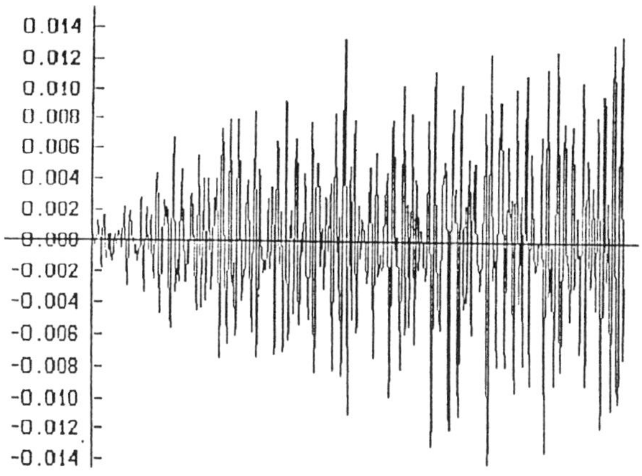

a: the error output without EFB.

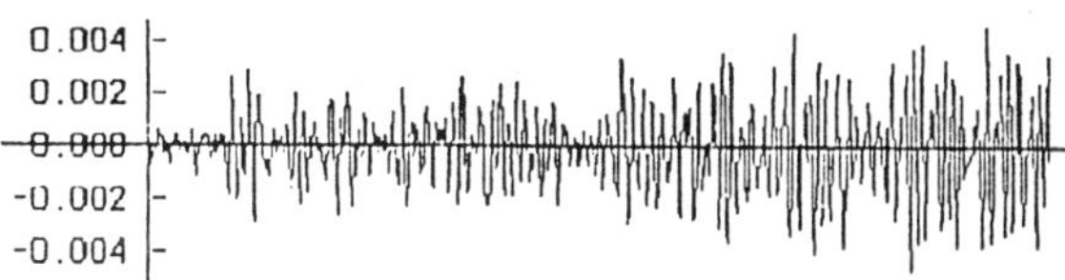

b: the error output with EFB.

Fig. 3: Simulation results for a BPF N=128, r=0.999, BP=25%, 12 bits, real part only.

REFERENCES

[1] A.V. Oppenheim and R.W. Schafer, "Digital Signal Processing", Prentice-Hall, Englewood Cliffs, New Jersey, 1975.

[2] L.R. Rabiner and B. Gold, "Theory and application of digital signal processing", Prentice-Hall, 1975.

[3] C.W. Barnes, B.N. Tran and S.H. Leung, "On the statistics of fixed-point roundoff noise", IEEE Trans. Acoust. Speech, Signal processing, Vol. ASSP-33, pp. 595-606, June 1985.

[4] B. Widrow, "A study of rough amplitude quantization by means of Nyquist Sampling theory", IRE Trans. Circuits Theory, pp.266-276, Dec. 1956.

[5] Tran-Thong and B. Liu, "Error spectrum shaping in narrow-band recursive filters", IEEE Trans. Acoust. Speech, Signal processing, Vol. ASSP-25, pp.200-203, Apr., 1977.

[6] D.C. Munson, Jr. and B. Liu, "Narrow-band recursive filters with error spectrum shaping", IEEE Trans. Circuits Syst., Vol. CAS-28, pp.160-163, Feb. 1981.

[7] D. Williamson and S. Sridharan, "Residue feedback in digital filters using fractional feedback coefficients", IEEE Trans., Acoust., Speech, Signal processing, Vol. ASSP-33, pp. 477-483, Apr. 1985.

[8] J. Szazupak and S.K. Mitra, "Recursive digital filters with low roundoff noise", Circuits Theory and Appl., Vol.5, pp.275-286, 1977.

[9] T. Thong, "Finite wordlength effects in ROM digital filters", IEEE Trans. Acoust. Speech, Signal Processing, pp. 436-437, Oct. 1976.

[10] S.Y. Hwang, "Roundoff noise in state-space digital filtering: A general analysis", IEEE Trans. Acoust. Speech, Signal Processing, Vol. ASSP-24, pp.256-262, June 1976.

[11] T. Classen and L. Kristiasson, "Improvement of overflow behavior of 2nd order digital filters by means of error feedback", Electron. Lett., Vol.10, pp.240-241, June 1974.

[12] H.A. Spang and P.M. Schultheiss, "Reduction of quantizing noise by use of feedback", IRE Trans. Commun. Syst., Vol. CS-10, pp. 373-380, Dec. 1962.

[13] T.L. Chang and S.A. White, "An error cancellation digital filter structure and its distributed arithmetic implementation", IEEE Trans. Circuits Syst., Vol. CAS-28, pp.339-342, Apr. 1981.

[14] A.W. Rihazcek, "Principles of High-Resolution Radar", Mark Resources, Inc., Marina del Rey, CA, 1977.

[15] P.P. Vaidyanathan, "On error spectrum shaping in state space digital filters", IEEE Trans. Circuits Syst., Vol. CAS-32, pp.88-92, Jan 1985.

SIGNAL PROCESSING VI: Theories and Applications
J. Vandewalle, R. Boite, M. Moonen, A. Oosterlinck (eds.)

SENSITIVITY AND ROUNDOFF NOISE IN WAVE DIGITAL FILTERS VERSUS OPTIMAL STATE-SPACE REALIZATIONS

José L. Sanz-González, Enrique Rodrigo-Aparicio and Francisco Lema
Universidad Politécnica de Madrid
ETSI Telecomunicación-UPM
Ciudad Universitaria, 28040 Madrid, Spain.

This paper compares under quantization conditions the performances of elliptic, Chebyshev and Butterworth filters realized by wave digital filters and by parallel structures composed of second order state-space sections (optimal and direct sections). The classical LC ladder filters were the reference analog structure for the wave digital filters.

1. INTRODUCTION.

As is well known, a great number of structures exists for realizing linear discrete-time systems by digital implementation. However, most structures can be included in two families: the state-space structures and the wave digital filters.

The state-space structures [1] are derived from the state equations

$$\mathbf{x}(k+1) = \mathbf{A}\cdot\mathbf{x}(k) + \mathbf{b}\cdot u(k)$$
$$y(k) = \mathbf{c}\cdot\mathbf{x}(k) + d\cdot u(k)$$

where $u(k)$ is the scalar input, $y(k)$ is the scalar output and $\mathbf{x}(k)$ the nth-order state vector; matrices $\mathbf{A}$, $\mathbf{b}$, $\mathbf{c}$ and d are $n\times n$, $n\times 1$, $1\times n$ and 1×1 real constant matrices, respectively.

The transfer function $H(z)$ is expressed by $H(z) = \mathbf{c}\cdot(z\mathbf{I}-\mathbf{A})^{-1}\cdot\mathbf{b} + d$ and the structure coefficients are the elements of $\mathbf{A}$, $\mathbf{b}$, $\mathbf{c}$ and d. The application of similarity transformations modifies structure coefficients, but the transfer function $H(z)$ remains invariant.

The wave digital filters [2] are derived from the analog filters, specially from the classical LC filters. Several of the good properties of classical filters are transferred to the corresponding wave digital filters.

The transfer function $H_a(s)$ of the analog filter is related to the transfer function $H(z)$ of the corresponding wave digital filter by means of the bilinear transformation: $s=(z-1)/(z+1)$. The signal variables considered in the analog filter are voltage (or current) wave quantities; the corresponding discrete-time signal flow graph relates incident and reflected wave signals. The discrete-time structure is obtained directly from the analog filter structure with some specific and precise rules of transformation [2].

2. QUANTIZATION.

The finite register length of digital realizations modifies the characteristics of the filters designed for infinite precision arithmetic. The transfer function degradation caused by quantization of filter coefficients, and the roundoff noise due to the operation result quantizations, are usually studied independently. However, some researchers concluded (by theoretic and experimental proofs) that low-sensitivity structures with respect to the coefficients have also low roundoff noise power, and vice versa.

For state-space structures and fixed-point numerical representation, there is a tradeoff between minimum roundoff noise power and minimum number of multiplications in the implementation of a specific transfer function. Roughly speaking, for an nth-order filter, $(n+1)(n+1)$ is the number of multiplications per output sample for the minimum roundoff noise structure, and $2(n+1)$ for the direct structure (minimum number of operations); however direct structures have high roundoff noise power when poles are close to $z=1$ (or -1). In [3] and [4] results of roundoff noise power were given for several kinds of filters realized in parallel and cascade connections of second order sections (direct sections against optimal state-space sections); in [5] results are given for floating-point arithmetic.

For wave digital filters, it is well known the low sensitivity of the transfer function with respect to the coefficients. Also, the number of coefficients can be minimum (n coefficients for an nth-order filter), and the dynamic range acceptable. Consequently, it is expected that the roundoff noise will be low.

3. COMPUTER RESULTS.

We have done a comparative analysis of the quantization effects on wave digital filters and state-space structures in the case of fixed-point numerical representation. The filters under consideration were Butterworth, Chebyshev and elliptic low-pass filters, and the cutoff frequency was varied from 0.05π to 0.95π. The reference analog structures for wave digital filters were the classical ladder LC-filters (LC chains). The results were obtained by formulas evaluation and by computer simulation.

3.1 *Sensitivity.*

Considering the sensitivity of the transfer function amplitude with respect to the filter coefficients, the number of bits for wave digital filter coefficients ranges from 6 to 8 with filter orders from 5 to 13. On the other hand, for the parallel structures composed of optimal state-space sections, the number of bits ranges from 10 to 16, under the same conditions (see Fig.1 and Table 1). The number of bits required for the coefficients increases as the order of the filter increases and, also, as the cutoff frequency approaches to 0 or to π, but this number is strongly insensitive to the other filter characteristics. The number of bits required for the wave digital filter coefficients is less sensitive to the filter parameters than the number of bits for state-space structures, as it can be seen from Table 1.

3.2 *Roundoff Noise.*

Now, we are referred to Fig.2. The roundoff noise power of optimal state-space filters is independent of the bandwidth (or cutoff frequency value). On the other hand, for wave digital Butterworth and Chebyshev filters, the roundoff noise power depends on the bandwidth and increases as the cutoff frequency approaches to 0 (narrow bandwidth) or to π (broad bandwidth), being the minimum in the interval $(0.4\pi, 0.7\pi)$; for elliptic filters, the noise power are flat and minimum from 0 and 0.4π. Considering the same overflow probability (about 10^{-6}), these minimum roundoff noise power values for wave digital filters are similar to those of the optimal state-space filters (e.g. about -82 dB for 5th-order filters and register wordlength of 16 bits). However, for wave digital filters, the roundoff noise power increases monotonically as the cutoff frequency decreases from 0.4π to 0 (except for elliptic filters), or increases from 0.7π to π; on the other hand, for optimal state-space filters, the roundoff noise power is maintained constant over the whole interval $(0, \pi)$ of cutoff frequencies.

The dynamic range for wave digital filters (unscaled filters) decreases as the cutoff frequency (ω_c) approaches to 0 or to π, but for optimal state-space filters (l_2-scaled filters) the dynamic range is independent of ω_c.

4. CONCLUSIONS.

The most important conclusions of this paper are summarized here.

For similar transfer function degradations, the wave digital filters require less bits for the coefficients than the optimal state-space filters.

For narrow and broad bandwidth low-pass filters (except for elliptic filters), wave digital filters (obtained from LC ladder structures) have greater roundoff noise power than optimal state-space filters (it is considered the same overflow probability and white Gaussian signal at the input).

REFERENCES.

[1] R.A. Roberts and C.T. Mullis, *Digital Signal Processing*. Reading, MA: Addison-Wesley, 1987.

[2] A. Fettweis, "Wave Digital Filters: Theory and Practice", Proc. IEEE, Vol. 74, pp. 270-327, Feb. 1986.

[3] J.L. Sanz-González and J.C. Herranz Torés, "Some results on low roundoff noise digital filter realizations", Proc. 1990 Bilkent International Conference on New Trends in Communication, Control and Signal Processing, pp. 1059-1065, Ankara (Turkey), (Editor: E. Arikan, Elsevier), July 1990.

[4] J.L. Sanz-González, J.C. Herranz Torés and E. Calero-Pérez. "Optimal structures for high speed and low roundoff noise digital filters". Proc. IEEE 1991 International Conference on Acoustics, Speech and Signal Processing (ICASSP'91), Vol. 3 pp. 1897-1900. Toronto, Canada, May 1991.

[5] J. L. Sanz-González, F. López-Ferreras and D. Andina. "Floating-point realizations for minimum roundoff noise digital filters". Proc. EUSIPCO'92 (This issue).

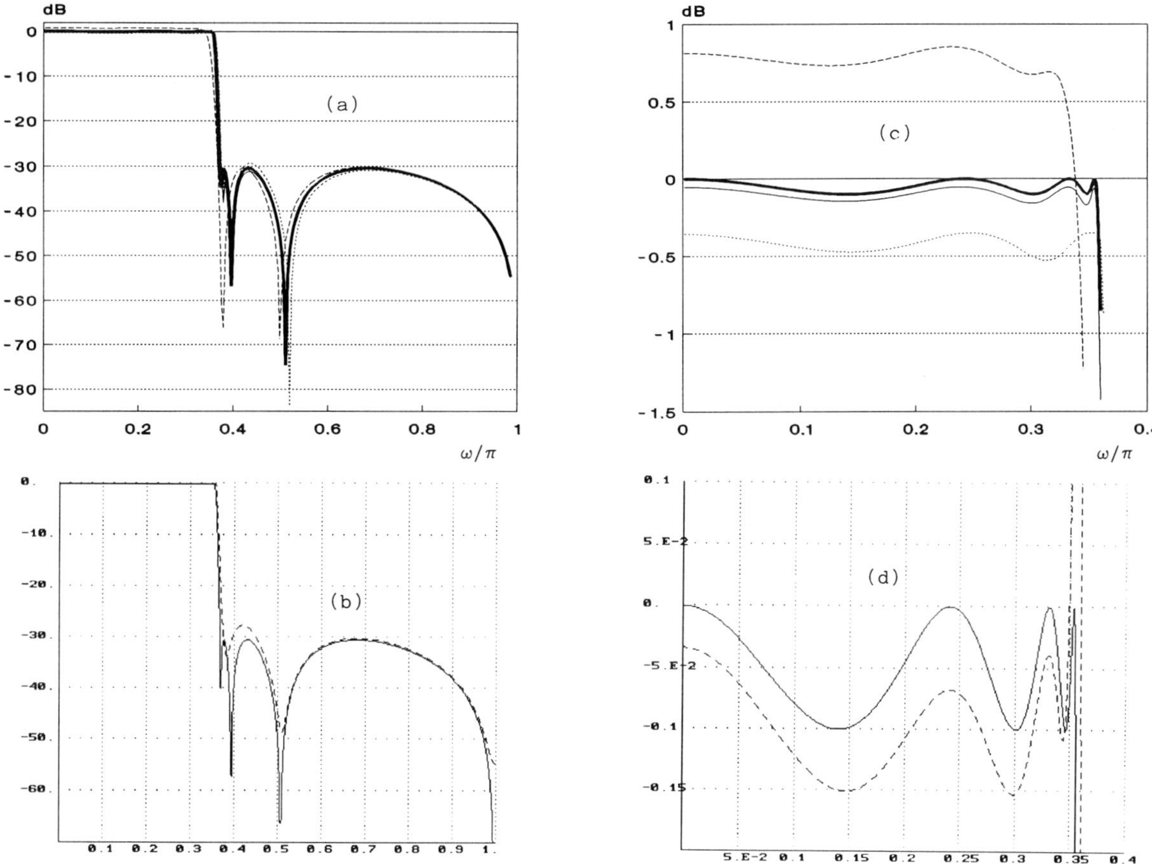

Fig.1. Frequency response (Magnitude) of a 5th-order elliptic low-pass filter. In (a) and (c) wave digital filter responses with ▬ ideal, —— 6-bits ···· 5-bits and - - - 4-bits coefficients. In (b) and (d) the parallel structure responses (considering optimal state-space sections) with —— ideal and - - - 8-bits coefficients. In (a) and (b) the overall frequency responses and in (c) and (d) the passband responses.

	NUMBER OF BITS IN THE FILTER COEFFICIENTS					
Filter Order	Elliptic		Chebyshev		Butterworth	
	Wave	Parallel	Wave	Parallel	Wave	Parallel
5	6	10	6	10	6	10
7	7	12	6	11	6	11
9	8	13	7	12	6	13
13	9	15	8	13	6	16

Table 1. The minimum number of bits required by de coefficients of the low-pass filters with similar frequency responses in passband and stop-band than Fig.1. The filter order depends on the transition band. The degradation of the passband frequency response delimits the lower number of bits.

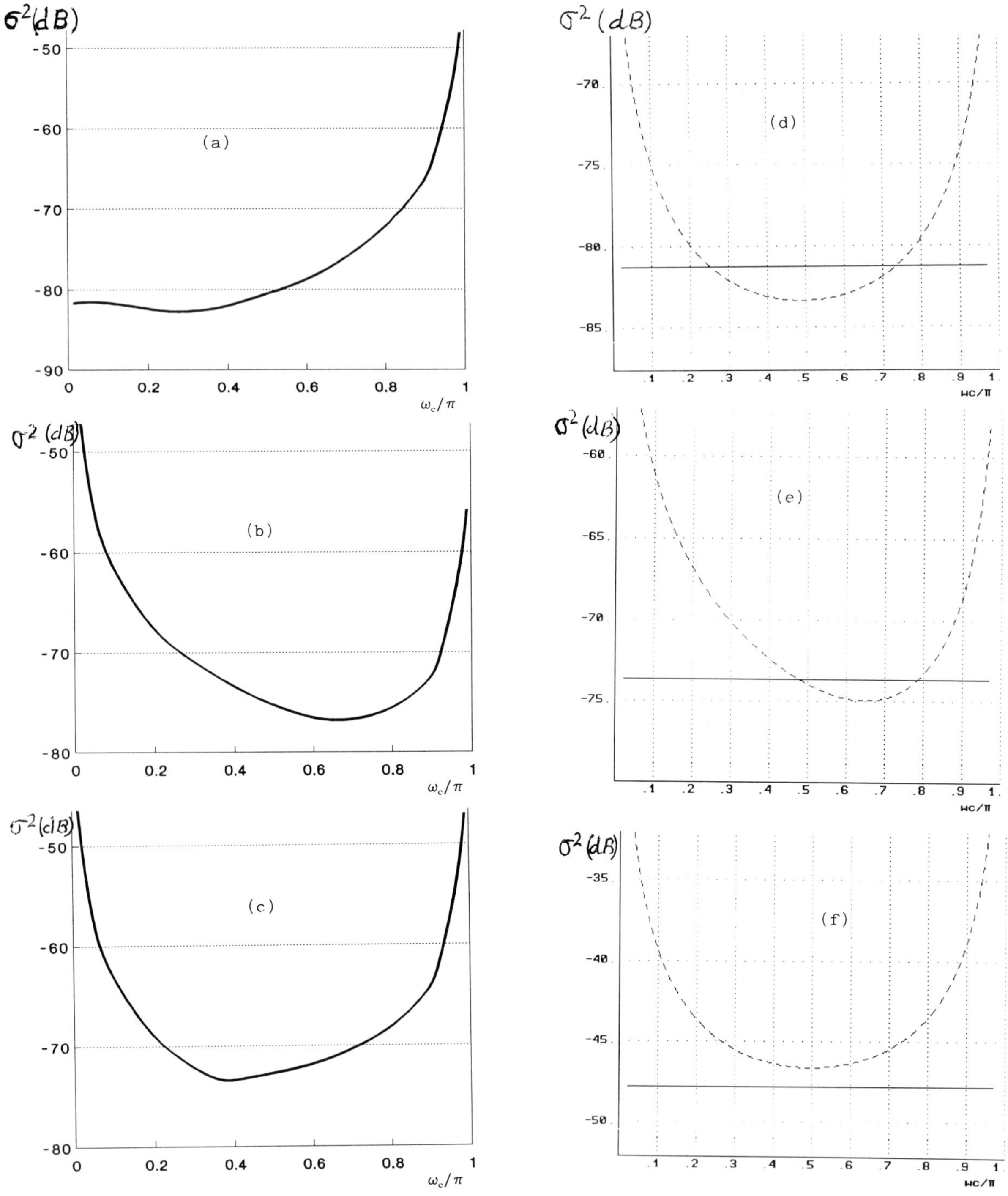

Fig.2. Output roundoff noise power (σ^2) versus cutoff frequency (ω_c) for low-pass filters and 16-bit fixed-point arithmetic. In (a), (b) and (c) the noise power curves of wave digital filters, and in (d), (e) and (f) the noise curves of parallel structures with second-order sections (—— optimal sections and - - - direct sections). Also, (a) and (d) curves correspond to 5th-order elliptic filters, (b) and (e): 13th-order Chebyshev filters, and (c) and (f): 13th-order Butterworth filters.

SIGNAL PROCESSING VI: Theories and Applications
J. Vandewalle, R. Boite, M. Moonen, A. Oosterlinck (eds.)

LUMPED, SAMPLED-DATA AND DIGITAL FILTERS WITH MAXIMUM NUMBER OF CONSTRAINTS ON AMPLITUDE AND PHASE CHARACTERISTICS

TAMÁS HENK and MOHAMMED ABO-ZAHHAD*

Department of Telecommunications and Telematics, Technical University of Budapest, H-1111 Budapest, Stoczek U. 2, HUNGARY
Tel. +361 1 810 300 ext 2441 and Fax: +361 1 1 666 808

For optimal filters exhibiting specified amplitude and phase (or group delay) characteristics, it is required that all the free parameters of the transfer function be used for the approximation. To achieve this requirement, the number of constraints on the amplitude and phase characteristics is defined first on general interpolation bases. Constant or arbitrarily prescribed low-pass or band-pass group delay and amplitude characteristics are approximated in the maximally-flat, ripple or mixed sense for lumped filters, whereas high-pass and band-rejection characteristics are also considered for distributed or sampled data filters. For optimal filters the relationship between the number of the free parameters and the number of the amplitude and phase constraints for a given degree is derived for the non-reciprocal and for the reciprocal lossy as well as for the reciprocal reactant cases. Some published separate and simultaneous approximation methods are referred, evaluated and compared on the above basis. It is pointed out that some of these methods don't satisfy the above requirements, although the optimal solution would exhibit higher performances.

1. INTRODUCTION

The simultaneous amplitude and phase approximation is an old filter design problem [1]-[4]. Considering the development of these methods Rhodes introduced new aspects into the construction of transfer functions exhibiting simultaneous amplitude and phase requirements [5]. On these bases a number of contributions have been published for further results [6]-[11]. Some of them however don't investigate how much constraints on the amplitude and phase characteristics can be prescribed altogether for a given degree and the number of the prescribed constraints is less than could be [7]-[10].

In this paper the relationships between the number of free parameters and the number of amplitude and phase conditions for reciprocal and non-reciprocal lumped, sampled-data and digital filters are given. On this basis a simple method is given to check the optimal property of these filters.

2. THE DESIGN CONCEPTION

The complex frequency variable λ is introduced for the frequency variable of lumped filters or for the reference frequency variable of sampled data filters [12] or for the Richard's variable of distributed filters. The transmission of filters $S_{12}(\lambda)$ is considered for which the boundary condition $|S_{12}(\lambda)| \le C$ holds, where $C \le 1$ for structurally passive filters and C is an arbitrary real number otherwise. As examples, reciprocal lossless (reactant) or lossy, non-reciprocal reflectional terminated lossless and also non-reciprocal lossless filters are structurally passive. Although the structurally passive condition ensures good realization properties [12], $C > 1$ is not excluded from this paper since in some referred methods $C > 1$ [7]-[9]. The scattering matrix of the lossless cases will be characterised by the canonical polynomials $f(\lambda)$, $g(\lambda)$, and $h(\lambda)$ [13], whereas the notation

$$S_{12}(\lambda) = \frac{f(\lambda)}{g(\lambda)} \qquad (1)$$

will be used also for other cases. The polynomial $g(\lambda)$ is strictly Hurwitzian with degree d and the degree of $f(\lambda)$ and $h(\lambda)$ is not greater than d, $f(\lambda)$ is even or odd for reciprocal lossless filters, whereas $h(\lambda)$ is odd or even for symmetrical lossless filters.

* On leave from Department of Electrical and Electronic Engineering, Faculty of Engineering, University of Assiut, Assiut, Egypt.

Simultaneous approximation of linear or arbitrary phase, constant or non-constant passband amplitude and zero stopband amplitude characteristics will be considered in maximally -flat, combined flat and ripple, ripple, or equiripple sense for low-pass, band-pass, high-pass and band-rejection specifications. For lumped filters however, the linear phase specification can't be approximated in an infinite frequency band, therefore band-pass or combined low-pass and band-pass filters have to be applied instead of high-pass or band-rejection filters respectively.

Considering linear phase filters the specified delay for low-pass and band-pass filters and also the mid-band phase-shift in the band-pass case are often free parameters for filter design [14]. So the appropriate filter design conception is to choose a sufficiently high filter degree d, to fix the passband(s) for the amplitude, the band(s) for the phase-approximation, the stopband(s), the amplitude ripple in the passband(s) and in the stopband(s), whereas to have the phase-error ripple, the delay (and also the mid-band phase shift for band-pass filters) as a result of the design [11, 15-17] (first design conception). It is more common however to fix the delay, and to have also the stopband and/or the passband amplitude ripples as a result instead [1]-[10] (second design conception).

3. DEFINITION OF THE NUMBER OF CONSTRAINTS

The above approximation problems can be solved by linear interpolation methods. So the number of the phase and amplitude constraints are introduced on an interpolation basis accommodating to both the first and the second design conceptions.

Let us define the sets of the interpolating frequencies ω_{hi}, ω_{pi} and ω_{si} where values and derivatives of the phase, the passband and stopband amplitude characteristics is specified respectively. The secondary set of frequencies ν_{hl}, ν_{pl} and ν_{sl} are further introduced comprising each frequency ω_{hi}, ω_{pi} and ω_{si} respectively as many times as the total number of the specified value and derivatives is at this frequency.

To define the number of constraints for each characteristic, the number of the specifications, i.e. the number of the elements in the sets ν_{hl}, ν_{pl} and ν_{sl} are considered. More exactly, the specified zeros of the interpolation error is written in the form:

$$\lambda \prod_{l=1}^{h} (\lambda^2 + \nu_{hl}^2) \quad , \qquad (2\text{-a})$$

$$\prod_{l=1}^{p} (\lambda^2 + \nu_{pl}^2) \quad , \qquad (2\text{-b})$$

and

$$\lambda^{s_o} \prod_{l=1}^{s_f} (\lambda^2 + \nu_{sl}^2) \qquad (2\text{-c})$$

for the phase, for the squared passband amplitude and for the zero valued stopband amplitude characteristics respectively [11]. In equations (2), p and h denote the number of constraints for the passband amplitude passband phase respectively. Also, denoting the degree difference of $g(\lambda)$ and $f(\lambda)$ by s_i, the number of stopband amplitude constraints is defined by $s = s_o + s_i + 2\, s_f$.

4. THE MAXIMUM NUMBER OF CONSTRAINTS

Fixing the filter degree d, boundary conditions can be set up for p , s and h for different filters determining the range in which p , s and h can be chosen to satisfy a given specification. On the other hand, p , s and h have to satisfy also an optimality condition ensuring that all the free parameters of the transfer function is used for the approximation. Some previous results are available in References [1]-[5] for special cases. The general conditions can be stated as: **the boundary and optimality conditions on the passband and stopband amplitude and on the phase constraints for reciprocal reactant (lossless) and for reciprocal lossy or non-reciprocal filters are summarised in Table 1.**

In degenerate cases not only the specified but also extra zeros of the interpolation error occur [7],[11], these zeros however are not counted in s , p and h .

Observe that $f(\lambda)$ becomes even or odd also for non-reciprocal filters if s=d , so the conditions become identical to the conditions of the reciprocal reactant filter. If on the other hand p=d such that the specified passband amplitude is constant, then $h(\lambda)$ becomes odd or even in the reciprocal reactant cases and so symmetrical lossless filter is involved for odd d. If for example a d-th degree Cauer filter is considered, then s=p=d and h=1 due to the delay or the stopband ripple as a free parameter according to the first or the second design conception respectively, both optimality conditions are satisfied and $h(\lambda)$ is odd.

Proof: Consider first the case of $d\underline{th}$ degree non-reciprocal transfer function to be:

$$H(\lambda) = \frac{N_m(\lambda)}{D_d(\lambda)} = \frac{\sum_{i=0}^{m} a_i \lambda^i}{\sum_{i=0}^{d} b_i \lambda^i} \qquad (3)$$

The relation between the degree of the transfer function and the number of the amplitude and phase constraints, for non-reciprocal case, is given by:

$$s + p + h = 2d + 1 \qquad (4)$$

Equation (4) can be proved as follows.

a) The formulation given by (3) insures that there are (d-m) stopband amplitude conditions satisfied at $\lambda=\infty$. i.e., initially s=d-m.

b) There are (d+m+2) coefficients in equation (3), but only (d+m+1) are independent, because any of the (d+m+2) coefficients may be assigned an arbitrary value without any loss in generality. The remaining (d+m+1) coefficients are satisfied such that the same number of amplitude and phase conditions are satisfied.

c) From (a) and (b), the total number of the passband amplitude conditions, the stopband amplitude conditions and the phase conditions are (2d+1).

Consider second the case of reciprocal reactant transfer function. It can be deduced from equation (3) with:

$$\left.\begin{array}{llll} a_i=0, & i=1,\ 3,\ ..,\ m-1 & \text{for } m & \text{even} \\ a_i=0, & i=0,\ 2,\ ..,\ m-1 & \text{for } m & \text{odd} \end{array}\right\} \qquad (5)$$

In this case the relationship between the degree and the amplitude and phase constraints is given by:

$$s + 2(p + h) = 3d + 2 \qquad (6)$$

Equation (6) can be proved as follows.

1) As in the non-reciprocal case, the number of stopband conditions results from the formulation given by equations (3) and (5) , is (d-m) where all are at $\lambda=\infty$.

2) The number of free parameters in the denominator is (d+1) and the numerator has m zeros. If x from these zeros is forced to be on the $j\omega$-axis to satisfy some stopband amplitude specifications, then the total number of stopband conditions is (d-m+x). Consequently, the number of passband amplitude and phase conditions are (d+1+a/2), where a=m-x is the number of numerator zeros not on the $j\omega$-axis.

3) From (1) and (2) one can argue that s+2(p+h) is equal to 3d+2.

5. EVALUATIONS OF SOME METHODS

In previous section it has been shown that the number of free parameters for non-reciprocal transfer functions is 2d+1. This number can be distributed between the passband amplitude, stopband amplitude and the phase requirements under the conditions given in Table (1). But generally further considerations are needed for stable and bounded solutions. Evaluation of some published solutions shows that those of [7]-[9] are non-optimal. This is because not all the degrees of freedom are used to improve the required characteristics. The most severe case is that given in [9]. It is non-optimal even if no conditions are imposed on the stopband amplitude characteristic; i.e., if m=0.

Similarly it has been founded that there are some non-optimum reciprocal reactant filters. The solutions given in [7]-[10] are the most severe cases. However the solution given in [10], is optimum if r=0 and this solution corresponds to the low-pass case with two transmission zeros at infinity.

6. CONCLUSION

The relationships between the number of amplitude and phase constraints and the number of free parameters for reciprocal reactant and non-reciprocal optimal filters are given. Evaluating by means of Table 1 some published simultaneous approximation methods, it turns out that some of them are not optimal in the sense that they don't use all the free parameters available in a transfer function with fixed degree, without any extra realization benefit. Examples confirm however that their features can be enhanced by using all the parameters.

Table 1. Conditions on passband-stopband amplitude and phase constraints

<table>
<tr><th rowspan="2">Type of filter</th><th rowspan="2">Condition on $f(\lambda)$</th><th rowspan="2">Specified $\lvert S_{12}(\lambda)\rvert$</th><th colspan="3">Boundary Conditions</th><th rowspan="2">Optimality conditions</th></tr>
<tr><th>p</th><th>s</th><th>h</th></tr>
<tr><td rowspan="2">Reciprocal reactant</td><td rowspan="2">even or odd</td><td>constant</td><td>$0 \le p \le d$</td><td rowspan="4">$0 \le s \le d$</td><td rowspan="2">$1 \le h \le d$</td><td rowspan="2">$s+2(p+h)=3d+2$</td></tr>
<tr><td>non-constant</td><td>$0 \le p \le \begin{cases} \frac{3d-1}{2} \\ \frac{3d}{2} \end{cases}$</td></tr>
<tr><td rowspan="2">Reciprocal lossy or Non-reciprocal</td><td rowspan="2">—</td><td>constant</td><td>$0 \le p \le d$</td><td rowspan="2">$1 \le h \le 2d$</td><td rowspan="2">$s+p+h = 2d+1$</td></tr>
<tr><td>non-constant</td><td>$0 \le p \le 2d$</td></tr>
</table>

REFERENCES

[1] Takebe, T., Kato, K., and Nishikawa, K., Simultaneous Flat Approximations for Amplitude and Group Delay in Transfer Functions, Electronics and Communications in Japan, vol. 54-A, no.1, pp.47-53, 1971.

[2] Gazsi, L., Simultaneously Maximally Flat Approximation of Amplitude and Delay Characteristics in the Low-pass Case, Non-Published communication, Budapest, 1972.

[3] Yoshida, N., Transfer Function of Maximally Flat Group Delay Low-Pass Filters with Equal-Ripple Attenuation in the Stopband and Flat Attenuation in the Passband, IEEE Trans., Circuits and Systems, CAS-23, pp.81-84, Feb. 1976.

[4] Wellekens, G.J., and Godard, A.N., Simultaneous Flat Approximations of the Ideal Low-Pass Attenuation and Delay for Recursive Digital, Distributed, and Lumped Filters, IEEE Trans., Circuits and Systems, CAS-24, pp. 221-230, May 1977.

[5] Rhodes, J.D., Theory of Electrical Filters (Jhon Wiley, London, 1976).

[6] Rhodes, J.D., and Zabalawi, I.H., Selective Linear Phase Filters Possessing A Pair of j-axis Transmission Zeros, Int. J. of Circuit Theory and Appl., Vol.-10, pp.251-263, 1982.

[7] Fahmy, M.F. and Rhodes, J.D., Finite Band Approximation for Distributed and Digital Selective Linear Phase Transfer Functions, Int. J. of Circuit Theory and Appl., pp.57-70, March 1975

[8] Jarry, B. and Verdier, A., Low-pass Approximations of a Constant Amplitude with Arbitrary Phase and Delay, Int. J. of Circuit Theory and Appl., vol.13, pp.61-71, 1985.

[9] Baher, H. and O'Malley, M., Generalized Approximation Techniques for Selective Linear Phase Digital and Non-Reciprocal Lumped Filters, IEEE Trans. Circuits and Systems, CAS-33, pp.1159-1169., Dec. 1986.

[10] Fahmy, M.F., Abo-Zahhad, M. and Sobhy, M.I., Design of Selective Linear Phase Bandpass Switched-Capacitor Filters with Equiripple Passband Amplitude Response, IEEE Trans. Circuits and Systems, CAS-35, pp. 1220-1229, Oct. 1988.

[11] Földvári-Orosz, J., Henk, T. and Simonyi, E., Simultaneous Amplitude and Phase Approximation for Lumped and Sampled Filters, Int. J. of Circuit Theory and Appl., 19, pp. 77-100, Jan.-Feb. 1991.

[12] Fettweis, A., Wave Digital Filters, Theory and Practice, Proc. of the IEEE, vol. 74, February 1986, pp. 270-327.

[13] Belevitch, V., Classical Network Theory, (Holden-Day, 1968).

[14] Abo-Zahhad, M., Odd-Degree Selective Bandpass Digital Filters Interpolating Linear Phase and Constant Group Delay, Int. J. of Circuit Theory and Appl., 19, pp. 375-387, 1991.

[15] Abo-Zahhad, M. and Henk, T., Simultaneous Amplitude and Phase Approximations for Sampled-data Reciprocal Reactant Transfer Functions : Applications on Low-pass Filter Design, Proc. Al-Azhar 2nd Int. Conf., Cairo, pp. 109-120, Dec. 1991.

[16] Abo-Zahhad, M. and Henk, T. and Fahmy, M.F., Design of Digital, Sampled-data and Non-reciprocal Lumped Low-pass Filters with Constant Amplitude and Linear Phase Characteristics, Proc. Al-Azhar 2nd Int. Conf., Cairo, pp. 121-132, Dec. 1991.

[17] Leeb, F. and Henk, T., Simultaneous Approximation for Bireciprocal Lattice Wave Digital Filters Proc. of the ECCTD'89, Brighton, U.K., 5-8 Sept. 1989, pp.472-476.

SIGNAL PROCESSING VI: Theories and Applications
J. Vandewalle, R. Boite, M. Moonen, A. Oosterlinck (eds.)

DIGITAL FILTERS BASED ON COMPLEX TRANSMISSION LINE NETWORKS

Akira TAKAHASHI, Nobuo NAGAI and Nobuhiro MIKI

Research Institute for Electronic Science, Hokkaido University, Sapporo 060, JAPAN

This paper deals with complex wave digital filters based on complex transmission line networks with imaginary resistances. Such networks can realize a single band pass response with asymmetrical transitions on the both sides of the pass band. Therefore an order reduction can be achieved if symmetrical response is not required. Even not one real, if there exist a reflection free port, a series adaptor based on voltage dimensional waves or a parallel adaptor based on current dimensional waves holds losslessness after quantization. Furthermore the structures of these two adaptors are identical. Therefore a new ladder construction is proposed for modularity as an alternative combination of voltage dimensional series sections and current dimensional parallel sections.

1. INTRODUCTION

This paper introduces some properties of complex wave digital filters based on complex transmission line networks with imaginary resistances. Complex wave digital filters can realize highly flexible responses [1], and useful to perform digital simulations of complex systems such as quantum electronics devices [2]. The incorporation of imaginary resistances [3] into transmission line networks [4] bring the following properties. 1) With a reference to such a network, a single band-pass transfer function is easily derived from a real low-pass filter through a new type of reactance spectral transformation. The transformation enables a flexibility to control the upper transition and the lower transition be arbitrary asymmetrical. If the design tolerance may accept such a response, the necessary filter order can be reduced in comparison with the frequency shifting transformation shown by Crystal and Ehrman [5]. 2) Even in complex cases, wave digital filters [6] possess structural correspondences to their reference network with wide aspects of numerical robustness. Even not one real [7], if there exist a reflection free port, a series adaptor based on voltage dimensional waves or a parallel adaptor based on current dimensional waves holds losslessness after quantization. Furthermore the structures of these two adaptors are identical. A new ladder construction is proposed for modularity as an alternative combination of voltage dimensional series sections and current dimensional parallel sections.

2. REFERENCE NETWORK

Lossless commensurate transmission line (LCTL) networks are basically constructed with elements shown in Fig.1 [4]. A 'unit element' is a lossless uniform element with two ports, which can be regarded as a delay unit in the sense of reflectance. In a 'commensurate' network, all elements have a unified 'round propagation delay time' T.

The correspondences of three complex frequency variables, λ for LCTL networks, z^{-1} for digital filters and s for Laplace transformations are expressed as

$$\lambda = \tanh\frac{sT}{2} = \frac{1-z^{-1}}{1+z^{-1}}. \qquad (1)$$

Because of the above relationship, a realizable function of λ and a realizable function of z^{-1} are compatible with each other, if T is identified with the sampling period.

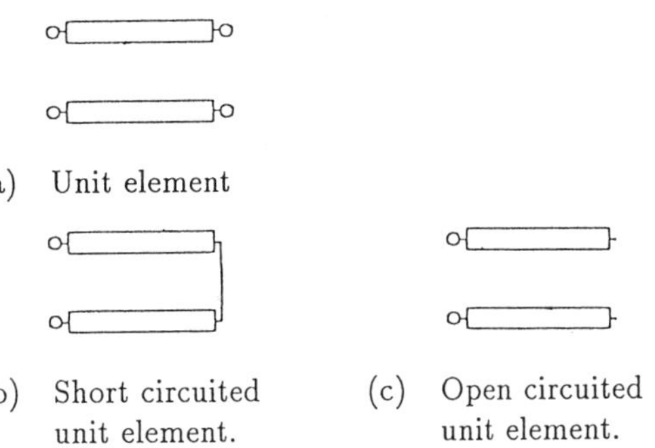

Figure 1: Basic elements of LCTL networks.

A short circuited unit element with characteristic resistance R has a voltage-current relationship in the frequency domain, similar to a lumped inductance but in the sense of λ,

$$V = \lambda RI \qquad (2)$$

on its input terminal pair. Besides an open circuited unit element with characteristic conductance G follows a relationship, similar to a capacitance,

$$I = \lambda GV. \tag{3}$$

Let R and X be real part and imaginary part of impedance Z. Similarly, G and S denote real part and imaginary part of admittance Y. In a complex circuit, all signals are complex-valued even in the time domain. Let $v(t)$ and $i(t)$ correspond to time signals of V and I, respectively. Then an instantaneous effective power $p(t)$ is defined by (4). The superscript star denotes complex conjugate.

$$p(t) = Re[v(t)i^*(t)] \tag{4}$$

A complex time-invariant lumped element, characterized as (5) is named an 'imaginary resistance' [3].

$$v(t) = jXi(t) \qquad V = jXI \tag{5}$$

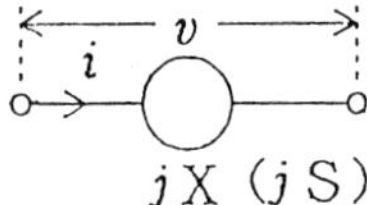

Figure 2: Imaginary resistance.

If necessary, negative value of X is available. This element is lossless in the sense of (4). If a value of an imaginary resistance is expressed by an inverse relationship of (5) such as (6), it is named an 'imaginary conductance'.

$$i(t) = jSv(t) \qquad I = jSV \tag{6}$$

3. ASYMMETRICAL TRANSFORMATION

If imaginary resistances are properly incorporated to low-pass LCTL networks, complex asymmetrical band-pass responses are obtained. Fig.3 shows one of the simplest case, where

$$jX = -jR\tan\frac{\Omega_\lambda T}{2}. \tag{7}$$

The original response and the modified response are illustrated in Fig.4, as a continues line and a break line, respectively. The attenuation level for $\omega = 0$ is mapped to $\omega = \Omega_\lambda$, while the positions of $\omega = \pm\pi/T$ are unchanged. Therefore a real low-pass response is transformed to an asymmetrical complex response.

If an imaginary conductance (8) is inserted parallel to an open-circuited unit element, the identical spectral transformation is achieved.

$$jS = -jG\tan\frac{\Omega_\lambda T}{2} \tag{8}$$

Where G is the characteristic conductance of the open-circuited unit element.

From (1),(2),(3),(7) and (8), such a transformation can be regarded as replacement of the complex frequency variable

$$\lambda \longmapsto \lambda - j\tan\frac{\Omega_\lambda T}{2}. \tag{9}$$

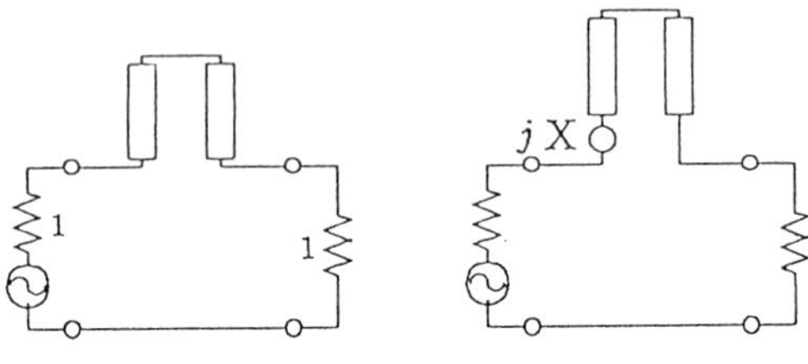

$(T = 1, \quad \Omega_\lambda = 2.)$

(a) Original filter. (b) Transformed filter.

Figure 3: Incorporation of imaginary resistance.

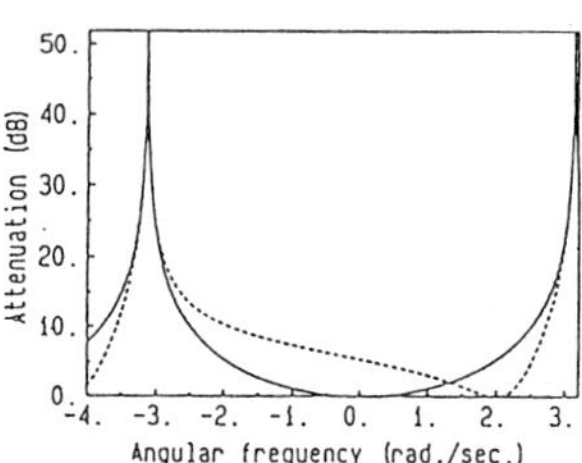

—Original response. ---Transformed response.

Figure 4: Asymmetrical spectral transformation.

A higher order example is shown in Fig.5. Because of an imaginary resistance is lossless, the proposed transformation can be regarded as one of the special case of reactance spectral transformations. Therefore the magnitude ripples are inherited straightforward. If the initial specification may accept such an asymmetrical response, the necessary filter order is reduced in comparison with a popular symmetrical design.

The asymmetrical transformation described here, does not have the freedom itself to be shifted along the frequency axis. But the conventional shifting transformation (10) [5] can be applied to the network by replacing all the unit elements into complex ones including distributed imaginary resistances [8]. Therefore an arbitrary allocation of the pass band can be achieved.

$$z \longmapsto ze^{-j\Omega_z} \quad \text{or}$$

$$\lambda \longmapsto \frac{\lambda - j\alpha}{1 - j\alpha\lambda}, \qquad \alpha = \tan\frac{\Omega_z}{2}. \tag{10}$$

At last, through the relationship (1), the identical frequency response can be reproduced as a digital transfer function. By tracing back the above chain of the spectral transformations, an asymmetrical complex design problem is reduced to a classical low-pass design problem.

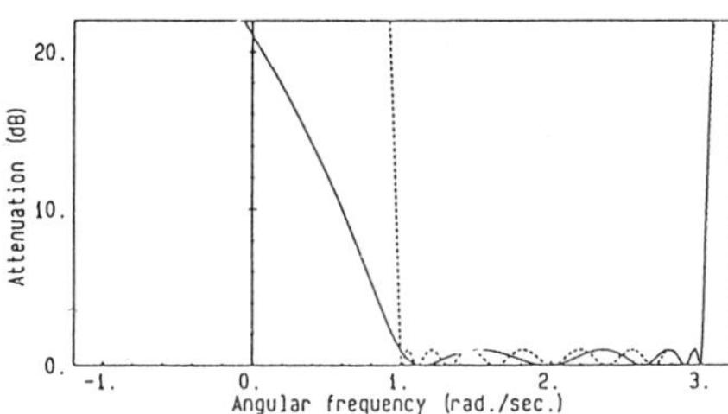

— 5th order asymmetrical filter.

--- 9th order symmetrical filter.

Figure 5: Design example.

As a result, starting from a normalized low pass response with pass band $\lambda = -j \sim j$, the whole transformation can be expressed as,

$$\lambda \longmapsto \frac{1}{m}\left(\frac{\lambda - j\alpha}{1 - j\alpha\lambda} - jX\right). \tag{11}$$

The parameter m describes the pass band expansion of normalized low pass response.

If there is any need to digitally simulate the structures of the prototype networks, such as a multiport filter, complex wave digital filters become useful.

4. Wave quantities and adaptors

Wave digital filters are digital filters which simulate LCTL networks in the sense of incident and reflected wave quantities [6]. An individual reflectance is expressed as a unimodular multiplier. An element interconnection is simulated by a digitally realized scattering matrix, called an 'adaptor'. A wave digital filter is sufficiently stable, if all the building blocks are lossless or passive.

A impedance value for a port, which is evaluated under a condition

$$\lambda = 1 \qquad (z^{-1} = 0) \tag{12}$$

is called 'port impedance'.

Complex power wave quantities for a complex port are defined as

$$A = \frac{V + ZI}{2\sqrt{R}} \qquad B = \frac{V - Z^*I}{2\sqrt{R}} \tag{13}$$

in the frequency domain, where Z and R are a port impedance and its resistance part, respectively. These wave quantities directly lead to the incident power P_A and the reflected power P_B,

$$P_A = AA^*, \qquad P_B = BB^*. \tag{14}$$

Therefore coefficient multipliers of a building block in a power wave digital filter can be easily quantized to guarantee passivity, $P_A - P_B \geq 0$.

The definition (13) originates from voltage dimensional waves (16). Other power waves (15) based on current dimensional waves (17) are also available.

$$A = \frac{I + YV}{2\sqrt{G}} \qquad B = \frac{I - Y^*V}{2\sqrt{G}} \tag{15}$$

Y and G are a complex port admittance and its conductance part.

$$A = \frac{V + ZI}{2} \qquad B = \frac{V - Z^*I}{2} \tag{16}$$

$$A = \frac{I + YV}{2} \qquad B = \frac{I - Y^*V}{2} \tag{17}$$

If the number of the independently quantized multipliers is equal to the number of the freedom in a reference network, the losslessness is exactly reproduced, even after the multipliers are quantized. If all the entries of a scattering matrix are purely real-valued, the corresponding adaptor is exactly lossless even after quantization in the same reason as a real wave adaptor. A restricted condition for such losslessness is known as 'one realness' [7], which excludes complex impedance under the evaluation (12). Although the networks introduced in **3.** do not possess this property, at least passivity after quantization can be guaranteed easily by using power wave quantities. But in general, a power wave filter needs a greater number of multipliers in comparison with a voltage dimensional wave filter or a current dimensional wave filter because of the nonlinear relationship of the entries of each scattering matrix [6].

Even not one real, if there exist a reflection free port, a series adaptor based on voltage dimensional waves (16) or a parallel adaptor based on current dimensional waves (17) holds losslessness after quantization. In ad-

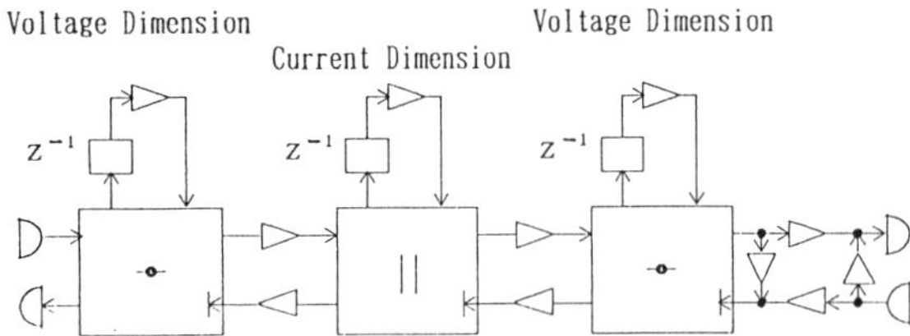

Figure 6: Proposed ladder realization.

dition, the signal flows of these adaptors are identical [10][11][12].
Therefore we propose a hybrid structure for a ladder configuration, as an alternative combination of voltage dimensional series sections and current dimensional parallel sections (Fig.6). The wave conversion between (16) and (17) can be performed by a complex multiplier for each direction because both of the definitions are linear combinations of the voltage and the current of a port. Including the conversion multipliers, such wave digital filter needs a smaller number of multipliers than a power wave filter, and its stability is easily guaranteed by simple truncations of conversion multipliers while all the resistant part of port impedance is positive. All the adaptors in Fig.6 have the same structure. Instead of an adaptor with no reflection free port, a combination of a matched adaptor and a quadruple pair of multipliers are used.

5. CONCLUSIONS

Properties of complex digital filters based on complex transmission line networks are shown. The described networks and the equivalent wave digital filters can accept a new frequency transformation and suitable to perform simulations of complex systems. For modularity, a new ladder structure is introduced.

REFERENCES

[1] T.Sekiguchi and S.Takahashi(1988). Complex Digital Signal Processing. *Proc. IEICE Workshop on Circuits and Systems (Karuizawa)*, 1-8(*Japanese*).

[2] M.Suzuki, N.Ohtani, N.Nagai and N.Miki(1990). A Complex Valued Wave Digital Filter Which Simulates Time Evolution of One Dimensional Electron. *Proc. Asia-Pacific Microwave Conference(Tokyo)*, 359-362.

[3] V.Belevitch(1968). *Classical network theory.* Holden Day, San Francisco.

[4] A.Matsumoto(1970). *Microwave filters and circuits.* Academic Press, N.Y.

[5] T.H.Crystal and L.Ehrman(1968). The design and applications of digital filters with complex coefficients. *IEEE Trans. Audio and Electroacoustics*, **AU-16**, 315-320.

[6] A.Fettweis(1986). Wave digital filters: theory and practice. *IEEE Proc.*, **74**, 270-320.

[7] A.Fettweis(1981). Principles of complex wave digital filters. *International Journal of Circuit Theory and Applications*, 119-134.

[8] N.Nagai(1990). *Linear Circuits, Systems and Signal Processing.* Marcel Dekker, N.Y.

[9] M.Suzuki, I.Sakagami, N.Miki and N.Nagai(1980). A Consideration of Time Domain Analysis of Distributed Constant Networks. *IECE Trans.*, **Vol.63-B No.11**, 1063-1070(*Japanese*).

[10] A.Takahashi, N.Nagai, M.Suzuki and N.Miki(1990). Ladder Type Wave Digital Filters with Complex Port References. *IEICE Tec. Rep.*,**CAS-89-161**, 41-48(*Japanese*).

[11] A.Takahashi, N.Nagai, M.Suzuki and N.Miki(1991). Complex Digital Filters with Asymmetrical Bandpass Characteristics based on Complex Commensurate Transmission Line Networks. *Proc. Mathematical Theory of Networks and Systems(Kobe)*, 433-434.

[12] G.B.Scarth and G.O.Martens(1991). Complex Wave Digital Ladder Filters Using Complex Port References. *Proc. IEEE ISCAS(Singapore)*, 2447-2450.

SIGNAL PROCESSING VI: Theories and Applications
J. Vandewalle, R. Boite, M. Moonen, A. Oosterlinck (eds.)

HINTS ON THE RELATIONSHIPS BETWEEN SUBBAND CODING, WAVELET AND VAGUELETTE DECOMPOSITIONS AND COSINE SINE ALTERNED TRANSFORMATIONS

Stéphane MAES (‡)

Universite Catholique de Louvain, Unité FYMA, Unité TELE, chemin du Cyclotron,2, B-1348 Louvain-la-Neuve, Belgium

The equivalence between subband coding and wavelet-vaguelette decompositions is described. For this purpose, CSAT are introduced. The approach is suitable to all perfect reconstruction subband decompositions and to hybrid subband coding schemes. Hybrid DCT-subband coding is also presented for optimal decompositions in time and frequency.

1. INTRODUCTION

The wavelets have been introduced by Morlet and Grossmann [1,2] for efficient analyses of analog signals. Subband coding schemes [3] were already widely used in signal processing. Different filter bank design have been developed for perfect (or almost perfect) reconstruction [4]. Quadrature Mirror Filters (QMF) are, by definition, filter banks adequate for perfect reconstruction subband (PR) coding schemes. Mallat [5] has shown that discrete wavelet transforms with a dilation factor equal to 2 are hierarchical subband coding decomposition (HSD) with CQMF (conjugate QMF); i.e.the decomposition filters are the conjugates of the reconstruction filters. I.Daubechies [6] has studied all the FIR-CQMF filters in one dimension.

Non hierarchical decomposition (uniform, best basis) are also well known in signal processing. A mathematical formulation (the wavelet packets) has been presented by Wickerhauser [7]. It is a straightforward extension of Mallat's theory.

The purpose of this paper is to show that any PR-HSD is a wavelet decomposition. Our approach requires the introduction of the CSAT (Cosine Sine Alterned Transformation) and it may be extended to analyses with a rational dilation factor. We call those decompositions:"vaguelette analyses". It is a generic term which embodies all the basis (or tight frame) functions associated as coherent states [1] with PR subband coding schemes.

2. WAVELET EXTRACTION

In this section, we are concerned by analyses with an integer dilation factor. We present an original construction different from Cohens's approach [8]. The conclusions are similar for the Smith-Barnwell class of filters (see [4] for classification):HSD are orthogonal decompositions into biorthogonal wavelets. However the approach is also suitable for all the lattice filter banks [4] and hybrid subband coding schemes [9].

2.1. The Smith-Barnwell Class

Let us consider the Smith-Barnwell class [4] which includes CQMF. The dilation factor is choosen equal to 2. The decomposition filters are $H_1^*(\omega)$ and $G_1^*(\omega)$. The reconstruction filters are $H_2(\omega)$ and $G_2(\omega)$. They are choosen to cancel at reconstruction the aliasing errors which have been introduced by downsampling. In order to have a PR-HSD, we impose:

$$\sum_{k=-\infty}^{+\infty} \left|H_1^*\left(\frac{\omega}{2}\right)\chi(\omega+k2\pi)\right|^2 \chi(\omega) = \chi(\omega) \qquad (1)$$

where $\chi(\omega)$ is the characteristic function of $[-\pi,\pi]$;

$$|H_1^*(0)| = 1 \qquad (2)$$

(1) ensures the validity of a HSD for any signal with a frequency band lower than π. It is equivalent to Cohens's (P)-condition [8].
(2) ensures a minimal discrimination between the low and high frequency subbands obtained after decomposition.

For the Smith-Barnwell Class, it is also mandatory to introduce the power complementarity:

(‡) Research Assistant of the FNRS (National Fund for Scientific Research, Belgium).

$$|H_1^*(\omega)|^2 + |G_1^*(\omega)|^2 = 1 \qquad (3)$$

Without (3) it would not be possible to relate easily PR-HSD and wavelet orthogonal transform: We introduce the CSAT as defined by Y.Meyer [10].

(3) implies that if we consider the filter bank $BF' = \left\{ |H_1^*(\omega)|, e^{j\omega}|G_1^*(\omega)| \right\}$ we may associate four different spaces:

$$V'_{-1} = \underset{\perp}{\hat{span}} \left\{ \frac{1}{\sqrt{\pi}} |H_1^*(\omega)|, \frac{\sqrt{2}}{\sqrt{\pi}} |H_1^*(\omega)| \cos k(\omega + \frac{\pi}{2}) \right\}_{k=1,2,...}$$

$\underset{\perp}{\hat{span}}\{.\}$ stands for "generated by $\{.\}$ in the frequency domain".

$$W'_{-1} = \underset{\perp}{\hat{span}} \left\{ \frac{2}{\sqrt{\pi}} |G_1^*(\omega)| 1_+(-\omega) \sin 2k(\omega+\pi), \frac{2}{\sqrt{\pi}} |G_1^*(\omega)| 1_+(\omega) \sin 2k(\omega-\frac{\pi}{2}) \right\}_{k=1,2,...}$$

V''_{-1} and W''_{-1} are similarly obtained: sine are associated with $|H_1^*(\omega)|$ and cosine (as well as the constant function) are associated with $|G_1^*(\omega)|$.

If we define V_0 as the space generated by the low frequency coherent states of the HSD [11], then it results from [10] that we may write:

$$V_0 = V'_{-1} \underset{\perp}{\oplus} W'_{-1} = V''_{-1} \underset{\perp}{\oplus} W''_{-1}$$

By definition of V_0, $K = \prod_{k=2}^{\infty} |H_1^*(\omega/2^k)|$ is an isometry on V_0. Therefore: $V_0 = KV'_{-1} \underset{\perp}{\oplus} KW'_{-1} = KV''_{-1} \underset{\perp}{\oplus} KW''_{-1}$. We may also write:

$$V_0 = \left[KV'_{-1} \cup KV''_{-1}\right] + \left[KW'_{-1} \cup KW''_{-1}\right] \text{ i.e. :}$$

$$V_0 = \tilde{V}_{-1} + \tilde{W}_{-1}$$

For example, if we define $\phi(\omega) = \prod_{k=1}^{\infty} |H_1^*(\omega/2^k)|$, we have

$$\tilde{V}_{-1} = \text{span} \left\{ \frac{1}{2\sqrt{\pi}} \phi(\frac{x}{2}), \frac{1}{\|T_k \phi(\frac{x}{2})\|} T_k \phi(\frac{x}{2}) \right\}_{k=\pm1,\pm2,...} \qquad (4)$$

with $T_k f(x) = f(x-k)$. span $\{.\}$ means the same as $\hat{span}\{.\}$ but in the time domain. We have a similar expression for $\tilde{W}_{-1}$ with

$$\hat{\psi}(\omega) = e^{j\frac{\omega}{2}} |G_1^*(\frac{\omega}{2})| \prod_{k=2}^{\infty} |H_1^*(\omega/2^k)|.$$ V_0 appears no more as decomposed into two supplementary spaces. However, we can extract V_{-1} and W_{-1} such that $V_0 = V_{-1} \underset{\perp}{\oplus} W_{-1}$ and

$$V_{-1} = \underset{\perp}{\text{span}} \left\{ \frac{1}{\|\phi(\frac{x}{2})\|} \phi(\frac{x}{2}), \frac{1}{\|T_k \phi(\frac{x}{2})\|} T_k \phi(\frac{x}{2}) \right\}_{k=\pm2,\pm4,..}$$

W_{-1} is similar with ψ instead of ϕ.

ϕ and ψ are the coherent states of the HDS with BF'. The unitary transforms $Ph_H(\omega)$ and $Ph_G(\omega)$ are applied respectively on V_{-1} and W_{-1} in order to implement the HDS $\left\{ H_1^*(\omega), G_1^*(\omega) \right\}$:

$$Ph_H(\omega) = \prod_{k=1}^{\infty} e^{j \arg\left(H_1^*(\omega/2^k)\right)}$$

$$Ph_G(\omega) = e^{j\left(\arg\left(G_1^*(\omega/2)\right) - \arg\left(H_1^*(\omega/2)\right) - \frac{\omega}{2}\right)} Ph_H(\omega)$$

A block diagram interpretation is presented in figure 1.The associated coherent states (vaguelettes) $Ph_H\phi$ and $Ph_G\psi$ are no more orthogonal basis functions. They are the biorthogonal wavelets found by A.Cohen [8].

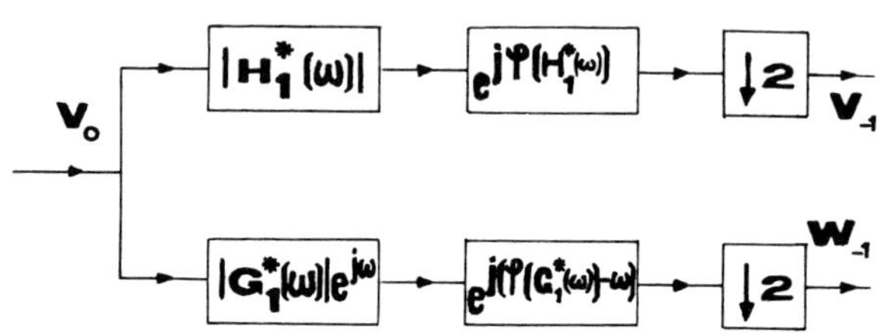

figure 1

The whole approach is easy to repeat for any integer dilation factor and for any type of decomposition (uniform, best basis). For rational dilation factors it is not possible to go farther than (4): the HSD decomposes V_0 into V_{-1} and W_{-1}: $V_0 = (V_{-1}^{(i)} \oplus W_{-1}^{(k)})$

$$V_{-1}^{(i)} \subset \text{span} \left\{ T_k \left\{ Ph_{H_i} \phi_i \right\} \right\} \qquad (5) \quad \text{and}$$

$$W_{-1}^{(k)} \subset \text{span} \left\{ T_k \left\{ Ph_{G_k} \psi_k \right\} \right\} \qquad (6)$$ but it is generally impossible to extract vaguelette bases out of (5) and (6): orthogonal bases are not "coherent" bases. This is related to the non commensurability between $\mathbb{Z}$ networks and $\alpha\mathbb{Z}$ networks with $\alpha \in \mathbb{Q}\setminus\mathbb{Z}$. It is in agreement with Auscher's results [12].

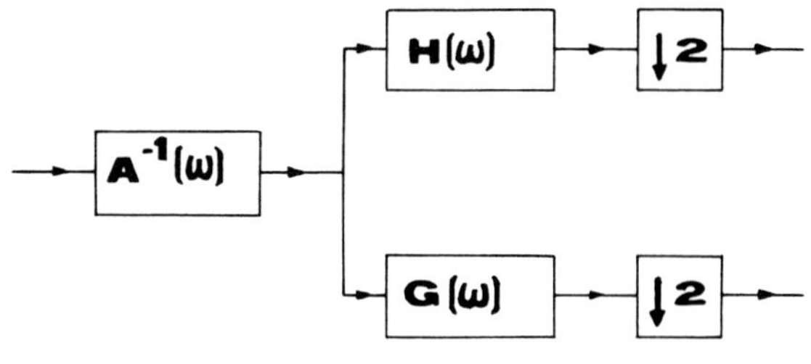

figure 2

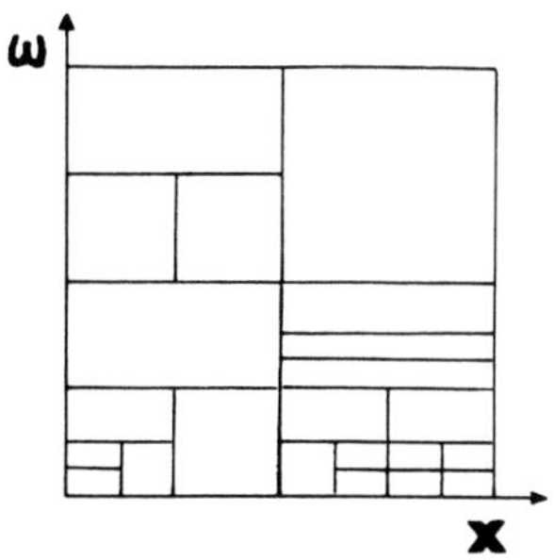

figure 3

2.2 Lattice filters and hybrid subband decompositions

In [4] other filter banks are introduced: the lattice filters. Most of them belongs to the Smith-Barnwell class and have already been treated. Others are for example non power complementary filters. To have a PR, we impose

$$|H_1^*(\omega)|^2 + |G_1^*(\omega)|^2 = (A^{-1}(\omega))^2 > 0$$

$$AFB = \left\{A(\omega)H_1^*(\omega),\ A(\omega)G_1^*(\omega)\right\}$$ belongs to the Smith-Barnwell class. Therefore, we can reproduce the analysis. V_0, V_{-1}, W_{-1} are the spaces associated to AFB. If P_V denotes the orthogonal projection on V, we have: $f \in V_0$,

$f = P_{V_{-1}}(A^{-1}f) + P_{W_{-1}}(A^{-1}f)$. ϕ and ψ of AFB are also the vaguelettes of this HSD. (See figure 2)

If $A_\infty^{-1} = \prod_{k=1}^{\infty} A^{-1}(\omega/2^k)$ is reversible, we may also consider the non orthogonal vaguelettes $A_\infty^{-1}\phi$ and $A_\infty^{-1}\psi$.

It introduces hybrid subband decompositions where the filter banks are classical and a reversible transformation (A^{-1}) is applied on the resulting subband at each step of decomposition. For example, DCT-subband hybrid best-basis decomposition (i.e A^{-1}=DCT) leads to optimal decomposition in time and frequency:decomposition of the signal or of its DCT is decided, provided that it reduces the number of associated bits. This is illustrated in figure 3. DCT is not only an excellent decorrelator, it is also a practical way to switch from the tie domain to the frequency domain. The construction presented hereabove also emphasize the duality that exists between subband coding and DCT.

2.3 Other classes

In [4], "almost perfect reconstruction" filter banks obtained by different design methods are also presented. Those analyses have to be understood as approximation of the vaguelette decomposition associated to the closest (in L^2) PR filter bank.

3. CONCLUSIONS

We have shown that any PR (hybrid) subband coding is in fact a vaguelette decomposition. Studies of the associated vaguelettes, allow to better understand the subband coding performance (regularity, stability,symetry,locality...) Second generation subband coding schemes (with edge charaterization, subband prediction, non uniform sampling,...) especially require such a background theory.

ACKNOWLEDGEMENTS

We want to thank all the members of TELE and FYMA especially Professor J.P.Antoine and Professor P.Delogne, for their support.

REFERENCES

[1] Grossmann, A. et al., J. Math. Phys. 26, 10 (1985), 2473.

[2] Grossmann, A. et al., Ann. Inst. 45, 3 (1986), 293.

[3] Crochiere, R.E. et al., Multirate Digital Signal Processing (Prentice Hall, 1983).

[4] Vaidyanathan, P.P., Proc. IEEE, 78, 1 (1990), 56.

[5] Mallat, S.G., IEEE Trans. PAMI, 11, 7 (1989), 674.

[6] Daubechies, I., Com. on Pure and Appl. Math., XLI (1988), 909.

[7] Wickerhauser, M.V., Picture Compression using Best Basis Subband Coding, Dpt.Math. Yale Univ. (1990).

[8] Cohen A., Ondelettes, Analyses Multirésolutions et Traitement Numérique du Signal, Thesis (Univ. Paris IX Dauphine, 1990).

[9] Comes S., Maes, S. Van Droogenbroeck, M., Recent and Prospective Decorrelation Techniques for Image Processing, to appear in Proc. ISSSE'92, (URSI, Paris 1992).

[10] Meyer, Y. et al., C.R. Acad. Sc.Paris, t.312, 1 (1991), 259.

[11] Perelomov,A., Generalized Coherent States and Their Applications (Springer-Verlag, 1986).

[12] Auscher, P., Ondelettes Fractales et Applications, Thesis (Univ. Paris IX Dauphine, 1989).

Prime Factor Real-Valued Fourier, Cosine and Hartley Transforms

Shing-chow Chan* and Ka-leung Ho**

*Department of Electronic Engineering, City Polytechnic of Hong Kong, Hong Kong.
**Department of Electrical & Electronic Engineering, University of Hong Kong, Hong Kong.

Abstract: Prime Factor Algorithm is an efficient algorithm for computing DFTs of lengths which are co-primes. However, for real-valued input (RPFA), the indexing is very complicated. In this paper, we examine and solve these problems for the vertical-split FFT algorithms including the RPFA. Two FORTRAN programs, inplementing the RPFA-I and RPFA-II, together with their timing results on the VAX 8820 computer are presented. The same approach is then applied to the Discrete Cosine and Hartley transforms to obtain new PFA-DCT and PFA-DHT algorithms. The PFA-DCT is particular useful in the realization of multirate modulated filter banks and transmultiplexers.

I. Introduction

Prime Factor Algorithm [1,2] is an efficient algorithm for computing DFTs of lengths which are co-primes. However, for real-valued input (RPFA), the indexing is very complicated [3]. In this paper, we examine and solve these problems for the vertical-split FFT algorithms including the RPFA. Two FORTRAN programs implementing the RPFA-I and RPFA-II, together with their timing results on the VAX 8820 computer are presented. The RPFA-II which is usually believed to be faster is found to be inferior to the RPFA-I if indexing problems are taken into account.

Yang and Narasimha [10] have demonstrated that it is possible to obtain a PFA-DCT from the PFA FFT through the mapping of Narasimha and Peterson, and Makhoul. Recently, Lee [11] has presented a direct derivation of the algorithm and suggested to implement the input and output mapping in form of tables. The indexing is very complicated and a general method for obtaining the short DCT modules is lacking. Using the general approach for the RPFA, a unified treatment of the prime factor decomposition of DCT (PFA DCT) is presented in this paper which avoids all these shortcomings. By similar development, a prime factor discrete Hartley transform (PFA DHT) has also been derived.

II. Prime Factor Real-Valued FFT

The PFA decomposes a length-N ($=N_1N_2...N_m$ & are coprimes) DFT into a m-dimensional DFT (MDFT):

$$X(k_1,...,k_m) = X(<\tilde{r}_1k_1M_1+...+\tilde{r}_mk_mM_m>_N)$$

$$= \sum_{n_1=0}^{N_1-1} ... \sum_{n_m=0}^{N_m-1} x(n_1,...,n_m)\, W_{N_1}^{n_1k_1}..W_{N_m}^{n_mk_m} \qquad (1)$$

$$x(n_1,...,n_m) = x(<r_1n_1M_1+...+r_mn_mM_m>_N)$$

using the Prime Factor Mapping:

$$n=<r_1n_1M_1+..+r_mn_mM_m>_N \qquad k=<\tilde{r}_1k_1M_1+..+\tilde{r}_mk_mM_m>_N \qquad (2)$$

where $M_i= N/N_i$ and $<r_i\tilde{r}_iM_i>_{N_i} = 1$. (<.> means mod N)

The MDFT is computed in a *row-column manner* using the Winograd short DFT modules. At the ith stage:

$$X^{(i)}(k_1,..,k_i,n_{i+1},..,n_m)$$

$$= \sum_{n_i=0}^{N_i-1} X^{(i-1)}(k_1,..,k_{i-1},n_i,..,n_m)\, W_{N_i}^{n_ik_i} \qquad (3)$$

$$X^{(m)}(k_1,..,k_m) = X(k_1,..,k_m); \quad X^{(0)}(n_1,..,n_m) = x(n_1,..,n_m)$$

If the multidimensional indices are arranged in lexicographically, the corresponding DFT matrix W_N can be written as:

$$\mathbf{W}_N = (\mathbf{B}_{N_m}\mathbb{W}_{N_m}) \otimes ... \otimes (\mathbf{B}_{N_1}\mathbb{W}_{N_1}) \qquad (4)$$

where $\mathbb{W}_{N_i}$ is the real DFT modules, $\mathbf{B}_{N_i}$ is the combination matrix, and ⊗ is the Kronecker matrix product. The vertical split FFT can be obtained by appropriate permutating the factors in (4). The two RPFAs are given by:

RPFA-I: $\mathbf{W}_N = (\mathbf{B}_{N_m} \otimes ... \otimes \mathbf{B}_{N_1})\left[(\mathbb{W}_{N_m}) \otimes ... \otimes (\mathbb{W}_{N_1})\right]$ (5)

RPFA-II: $\mathbf{W}_N = (\mathbf{B}_{N_m}\mathbb{W}_{N_m}) \otimes ... \otimes (\mathbf{B}_{N_1}\mathbb{W}_{N_1})$ (6)

If the multiplications are nested together, we obtain two similar real-valued Winograd Fourier transform algorithms (RWFTA).

It can be shown [13] that the following conjugate symmetry exists at each stage of the algorithm

$$X^{(i)}(k_1,..,k_i,n_{i+1},..,n_m)$$

$$= \left[X^{(i)}(N_1-k_1,..,N_i-k_i,n_{i+1},..,n_m)\right]^* \qquad (7)$$

$$0 \le k_j \le N_j-1 \quad j = 1,...,i \qquad \text{for } 1 \le i \le m$$
$$0 \le n_j \le M_j-1 \quad j = i+1,...,m$$

The intermediate symmetry can be used to reduce the arithmetic operations and storage. Since the computation of RPFA is similar to RWFTA, only the former will be discussed here.

III. Indexing and Unscrambling for RPFA FFT

RPFA-II:

In RPFA-II. both real and complex small N transforms are required. This real-valued in-place PFA was first introduced in [4] and a more detail formulation was given by Pitas and Strintzis [5].

The indexing and storage scheme in [5] have two drawbacks:

1) An unscramber is needed at the end because the real and imaginary parts are stored according to the PFM.
2) The real and imaginary parts are not simply related and considerable modulo operations are required in indexing the data.

Improved indexing and storage scheme can be obtained by storing the real and imaginary parts in locations which are conjugate to each other as shown in Fig. 1.

The application of the efficient indexing scheme in [6,7] to the RPFA-II has two problems:

1) It is necessary to determine whether the real parts or the imaginary parts are being referenced during the updating.
2) It is necessary to determine whether the data to be transformed is real or complex.

For example, the real and imaginary parts of group i-j at the ith stage are stored as follows ($\mathcal{X}$ is the array of storage):

$$Re[X^{(i)}(k_1^{(s)},..,k_j^{(s)},k_{j+1},..,k_i,n_{i+1},..,n_m)] \longrightarrow \mathcal{X}(k_1^{(s)},..,k_j^{(s)},k_{j+1},..,n_m) \tag{8}$$

$$Im[X^{(i)}(k_1^{(s)},..,k_j^{(s)},k_{j+1},..,k_i,n_{i+1},..,n_m)] \longrightarrow \mathcal{X}(k_1^{(s)},..,k_j^{(s)},N_{j+1}-k_{j+1},..,N_m-n_m) \tag{9}$$

It can be seen that

1) The transform will be real if all the first i indices are equal to $k_\ell^{(s)}$.
2) If the first j indices are equal to $k_\ell^{(s)}$, and k_{j+1} ($\neq$ 0) lies within $[1,p_{j+1}]$, then the real part of the transform is being referenced. Otherwise, the imaginary part is being referenced.

It is found that the entire implementation can be greatly simplified if we restrict N_1 to be the even factor whenever the transform is even.

In order to determine whether the transform is real or complex and determine whether the real or imaginary parts are being referenced, we need to maintain the values of k_ℓ. To avoid excessive number of modulo operations, $k_{\ell+1}$ is only updated whenever k_ℓ is equal $k_\ell^{(s)}$.

For group 1, all the data to be transformed are real and we have to perform two real transforms at the same time to ensure in-place operation. In order to reduce the length of the program, the odd-length DFT modules are modified so that they can transform complex, a pair of real transforms and a single real transform. Since the even-length modules, if invoke, must deal with real transform only, their implementation is considerably simplified. The exchange operation between the real and imaginary part are merged with the last stage of the algorithm.

RPFA-I:

The indexing of the first stage of RPFA-I is exactly the same as the complex PFA except that only real transforms are performed. Though the combination can be carried out by successively applying the combination operators in turn to the intermediate sequence and make use of the complex conjugate symmetry to avoid redundant operations, it is inconvenient and inefficient because the index map has to be set up for each factor. Due to the simple structure of the combination operators, it is found that the 2^m (less than 2^m for combination of lower order) output points $X(\tilde{n}_1,...,\tilde{n}_m)$, can be obtained from the 2^m input points $x(\tilde{n}_1,...,\tilde{n}_m)$, where $\tilde{n}_i$ = n_i or N_i - n_i, and only one pass is required. Thus only one indexing through the data is required.

Without loss of generality, let's consider the combination of the group $X(n_1^{(s)},n_2,n_3,0)$ (dimension 2) as shown in Fig. 2. The procedure starts with the four initial addresses denoted by I and are incremented along the solid line followed by the dotted line. Combination subprograms of different dimensions are written and properly organized to accomplish the entire combination. This is illustrated in Fig. 3 for the RPFA FFT up to four factors. Unscrambling the PFA FFT outputs can be done by permutating the output or input pointers of the modules or by modifying their constants. For the real-valued PFA, the latter two approaches are applicable depending on whether the applications require a fix-length transform or not. The output permutation approach will introduce a lot of inconvenience. Comparisons of the PFA-FFT, RPFA-I and RPFA-II are shown in Table 1. The RPFA-I is faster than the RPFA-II except for N = 1008, where the timings are very close to each other. The RPFA-I is also faster than the PFA FFT, except for the length N = 30. The speedup is seen to increase with the transform length because of the overheads required to support the storage scheme. For more than one transforms, these overheads can be shared using the pipeline extension technique introduced in [14].

IV. Prime Factor Hartley Transform

Since the DHT kernel is not separable, the PFM decomposed DHT cannot be computed by the simple row-column method, as in the PFA FFT. Therefore, it is more appropriate to obtain the PFA DHT via its relation to the DFT. Using the relation:

$$(1+j)X(k) = \mathbb{X}(k) = H(k) + j\ H(N\text{-}k)$$

and the prime factor decomposition, we obtain:

$$\mathbb{X}(k_1,...,k_m)=(1+j)\sum_{n_1=0}^{N_1-1} .. \sum_{n_m=0}^{N_m-1} x(n_1,..,n_m)W_{N_1}^{n_1k_1}..W_{N_m}^{n_mk_m}$$

$$= \sum_{n_m=0}^{N_m-1}\Big[..\Big[\sum_{n_1=0}^{N_1-1} x(n_1,..,n_m)\Big(cas(2\pi n_1k_1/N_1) +jcas(-2\pi n_1k_1/N_1)\Big)\Big]..\Big]W_{N_m}^{n_mk_m} \tag{10}$$

The equation implies that the Hartley transform can be computed via a hybrid Fourier-Hartley transform as in the power of two cases. The transform is computed in m different stages. At the first stage, we have:

$$\mathcal{X}^{(1)}(k_1,..,n_m)=\sum_{n_1=0}^{N_1-1} x(n_1,..,n_m)\left[cas(2\pi n_1k_1/N_1)+jcas(-2\pi n_1k_1/N_1)\right]$$

and at the ith stage the following:

$$\mathcal{X}^{(i)}(k_1,..,k_i,..,n_m)=\sum_{n_i=0}^{N_i-1} \mathcal{X}^{(i-1)}(k_1,..,n_i,..n_m)\, W_{N_i}^{n_ik_i}$$

and $\mathcal{X}(k_1,..,k_m) = \mathcal{X}^{(m)}(k_1,..,k_m)$ (11)

Careful examination reveals that the following symmetry exists at each stage of the algorithm [13]:

$$\mathcal{X}^{(i)}(k_1,..,k_i,n_{i+1},..,n_m)=j\left[\mathcal{X}^{(i)}(N_1-k_1,..,N_i-k_i,n_{i+1},..,n_m)^*\right] \quad (12)$$

Since the symmetry has the same form as the real-valued PFA FFT, the same storage scheme is also applicable here. It should be noted that similar argument holds true for the vertical split discrete Hartley transform which is obtained by permutating the factors of the small-N modules.

V. Prime Factor Discrete Cosine Transform

The classical method of Narasimha and Peterson [5] and Makhoul for computing the DCT from the real-valued FFT can be written as:

$$X_c^{II}(k) = \text{Re}\left[W_{4N}^{k} \sum_{n=0}^{N-1} \tilde{x}(n)\, W_N^{nk} \right] = \text{Re}[X_{\mathcal{P}}(k)] \quad (13)$$

where

$$\tilde{x}(n) = x(2n), \qquad n = 0,1,...,\lfloor\tfrac{N-1}{2}\rfloor$$
$$\tilde{x}(N-n-1) = x(2n+1), \qquad n = 0,1,...,\lfloor\tfrac{N}{2}\rfloor-1 \quad (14)$$

The sequence $X_{\mathcal{P}}(k)$ is called the phase modulated DFT (PM-DFT) of $\tilde{x}(n)$. The corresponding matrix factorization is then given by:

$$V_N = Q_N C_N^{II} \quad (15)$$

where

V_N is the matrix associated with the transformation from $\tilde{x}(n)$ to $X_{\mathcal{P}}(n)$,

Q_N is the combination matrix and C_N^{II} is the DCT-II matrix

If n is product of prime numbers, applications of the PFM to the PFM-DFT gives:

$$X_{\mathcal{P}}(k) = X_{\mathcal{P}}(k_1,...,k_m) = e^{j\pi h(k)/2}\, W_{4N_m}^{k_m} \sum_{n_m=0}^{N_m-1}\Bigg[\cdots \left[W_{4N_1}^{k_1} \sum_{n_1=0}^{N_1-1} x(n_1,...,n_m)\, W_{N_1}^{n_1k_1}\right]\cdots\Bigg] W_{N_m}^{n_mk_m} \quad (16)$$

The factor $e^{j\pi h(k)/2}$ arises from the fact that the term $\tilde{r}_1k_1M_1+...+\tilde{r}_mk_mM_m$ can take values outside 4N. If the multidimensional indices are arranged in lexicographical order, then the Kronecker matrix product representation of V_N is:

$$V_N = D_N\left[V_{N_m}\otimes...\otimes V_{N_1}\right] \quad (17)$$

where D_N is a diagonal matrix corresponding to the post-multiplication with $e^{j\pi h(k)/2}$. Two PFA-DCT algorithms are given by

$$\text{PFA-DCT}^{I}\text{: } V_N = D_N(Q_{N_m}\otimes\bullet\bullet\bullet\otimes Q_{N_1})\left[(C_{N_m}^{II})\otimes...\otimes(C_{N_1}^{II})\right]$$

$$\text{PFA-DCT}^{II}\text{: } V_N = D_N\left[(Q_{N_m}C_{N_m}^{II})\otimes...\otimes(Q_{N_1}C_{N_1}^{II})\right] \quad (18)$$

For the nesting approach, the Winograd discrete Cosine Transform Algorithm, similar its FFT counterpart, can be obtained. The indexing mapping in [12,15] provides short DCT modules with minimal number of multiplications.

Similar to the DFT, symmetry exists at each stage of the algorithm [13]:

$$X^{(i)}(k_1,..,k_i,n_{i+1},..,n_m) = (-j)^{\tau(k_1)+..+\tau(k_i)}\, X^{(i)}(N_1-k_1,..,N_i-k_i,n_{i+1},..,n_m) \quad (19)$$

$0 \le k_l \le N_l-1,\ l=1,...,i;\quad 0 \le n_\ell \le M_\ell-1,\ \ell = i+1,..,m$

This is identical to (7) except for the term $(-j)^{\tau(k_1)+..+\tau(k_i)}$. Fortunately, it does not affect the general form of the storage scheme and the computation algorithm. Thus the results obtained in Section II is applicable here, except that the unscrambling is slightly different.

We shall make use of the concept of "rotated transform". If both the input and output maps are given by the *Ruritanian map*, then $X_{\mathcal{P}}(k)$ reduces to:

$$X_{\mathcal{P}}(k_1,..., k_m) = e^{j\pi h(k)/2}\, W_{4N_m}^{k_m} \sum_{\tilde{n}_m=0}^{N_m-1}\left[..\left[W_{4N_1}^{k_1} \sum_{\tilde{n}_1=0}^{N_1-1} x(M_1^{-1}\tilde{n}_1,..,M_m^{-1}\tilde{n}_m)\, W_{N_1}^{\tilde{n}_1k_1}\right]..\right] W_{N_m}^{\tilde{n}_mk_m} \quad (20)$$

where $\tilde{n}_i = M_in_i$, i = 1,...,m.

On the other hand, if the CRT map is used, we have

$$X_{\mathcal{P}}(k_1,..., k_m) = e^{j\pi h(k)/2}\, W_{4N_m}^{\tilde{r}_mk_m} \sum_{n_m=0}^{N_m-1}\left[..\left[W_{4N_1}^{\tilde{r}_1k_1} \sum_{n_1=0}^{N_1-1} x(n_1,...,n_m)\, W_{N_1}^{\tilde{r}_1n_1k_1}\right]..\right] W_{N_m}^{\tilde{r}_mn_mk_m} \quad (21)$$

Since M_i is prime to N_i, $\tilde{n}_i= M_in_i$ is simply a permutation of n_i. Thus the unscrambling in (20) can be obtained by permutating the input index according to $\tilde{n}_iM_i^{-1}$ before taking the DFT. On the other hand, (21) requires the permutation of the output indices which will be very inconvenient. The fastest way is to use CRT and to modify the constants of the short DCT modules. Since the indexing of PFA-DCT^{I} is considerable simpler than PFA-DCT^{II}, the former is the better choice.

VI. CONCLUSION

In this paper, a unified treatment of the

prime factor DFT, DHT, and DCT algorithms has been performed. All these algorithms share the same common properties and can be computed with the same general structure. Results on RPFA show that the RPFA-II, which is usually believed to be faster, is inferior to RPFA-I if indexing consideration is taken into account. This is demonstrated by a timing comparison on the VAX 8820 computer.

REFERENCES

[1] D. P. Kolba and T. W. Parks, "A prime factor FFT algorithm using high speed convolution," IEEE Trans. ASSP-25, pp. 281-294, Aug. 1977.

[2] C. S. Burrus and P. W. Eschenbacher, "An in-place, in-order prime factor FFT algorithm," IEEE Trans. ASSP-29, pp. 806-817, Aug. 1981.

[3] C. S. Burrus, "Efficient Fourier transform and convolution algorithm," in Advanced Topics in Signal Processing, J. S. Lim and A. V. Oppenheim Eds, Prentice-Hall, Englewood Cliffs, 1988, pp. 199-215.

[4] M. T. Heideman, C. S. Burrus and H. W. Johnston: "Prime factor FFT algorithms for real-valued series," Proc IEEE ICASSP 84, Mar 19-21, 1984.

[5] I. Pitas and M. G. Strintzis, "General in-place calculation of discrete Fourier transforms of multidimensional sequences," IEEE Trans. ASSP-34, pp. 565-572, June 1986.

[6] C. Temperton, "Implementation of a self-sorting in-place prime factor FFT algorithm," Journal of computational physics vol. 5, No.3, pp. 283-299, May 1985.

[7] S. C. Chan and K. L. Ho, "On indexing the Prime Factor Fast Fourier Transform Algorithm," IEEE Trans. CAS-38, pp. 951-953, Aug. 1991.

[8] N. J. Narasimha and A. M. Peterson, "On the computation of the discrete cosine transform," IEEE Commun. vol. 26, pp. 934-946, 1978.

[9] J. Makhoul, "A fast cosine transform in one and two dimensions," IEEE Trans. ASSP-28, pp. 27-34, 1980.

[10] P. P. N. Yang and M. T. Narasimha, "Prime factor decomposition of the discrete cosine transform and its hardware implementation," Proc. IEEE ICASSP, March 1985, pp. 772-775.

[11] B. G. Lee, "Input and output index mappings for a prime factor decomposition computation of discrete cosine transform," IEEE Trans. ASSP-37, pp. 237-224, Feb. 1989.

[12] S. C. Chan and K. L. Ho, "Efficient index mapping for computing the discrete cosine transform," Electron. Lett., vol. 25, no. 22, pp. 1499-1500, Oct. 1989.

[13] S. C. Chan: Fast Transform Algorithms and Their Applications. PhD. Dissertation, University of Hong Kong, submitted 1991.

[14] S. C. Chan, Y. K. Wong and K. L. Ho, "Software Implementation of 2D FFT on DSP 96002 Digital Signal Processor," to appear in Proc. EUSIPCO-'92, Brussels, Aug., 1992.

[15] S. C. Chan and K. L. Ho, "Fast algorithms for computing the discrete cosine transform," to appear in IEEE Trans. Circuits and Systems, Mar. 1992.

$$\mathrm{Re}[X^{(i)}(k_1,\dots,k_i,\dots,n_m)] \longrightarrow \mathfrak{X}(k_1,\dots,k_i,\dots,n_m)$$
$0 \le k_1 \le p_1;\quad 0 \le k_l \le N_l-1,\quad l = 2,\dots,i;\qquad 0 \le n_\ell \le M_\ell-1,\quad \ell = i+1,\dots,m.$

Group i

$$\mathrm{Im}[X^{(i)}(k_1,\dots,k_i,\dots,n_m)] \longrightarrow \mathfrak{X}(N_1-k_1,\dots,N_i-k_i,\dots,N_m-n_m)$$
$1 \le k_1 \le t_1;\quad 0 \le k_l \le N_l-1,\quad l = 2,\dots,i;\qquad 0 \le n_\ell \le M_\ell-1,\quad \ell = i+1,\dots,m.$

Group i-1

$$\mathrm{Im}[X^{(i)}(k_1^{(s)},k_2,\dots,k_i,\dots,n_m)] \longrightarrow \mathfrak{X}(k_1^{(s)},N_2-k_2,\dots,N_i-k_i,\dots,N_m-n_m)$$
$1 \le k_1 \le t_2;\quad 0 \le k_l \le N_l-1,\quad l = 3,\dots,i;\qquad 0 \le n_\ell \le M_\ell-1,\quad \ell = i+1,\dots,m.$

Group 1

$$\mathrm{Im}[X^{(i)}(k_1^{(s)},\dots,k_{i-1}^{(s)},k_i,\dots,n_m)] \longrightarrow \mathfrak{X}(k_1^{(s)},\dots,k_{i-1}^{(s)},N_i-k_i,\dots,N_m-n_m)$$
$1 \le k_i \le t_i;\quad 0 \le n_\ell \le M_\ell-1,\quad \ell = i+1,\dots,m.$

FIG. 1 PROPOSED STORAGE SCHEME

N.B. $n_i^{(s)} = \begin{cases} 0 & N_i = 2p_i + 1 \\ 0, p_i & N_i = 2p_i \end{cases}$; $t_i = \begin{cases} p_i - 1 & N_i \text{ odd} \\ p_i & N_i \text{ even} \end{cases}$

Fig. 2 Generation of locations to be recombined for group $X(n_1^{(s)},n_2,n_3,0)$ $n_1^{(s)} = 0$ (or $N_1/2$ if N_1 is even)

```
      GOTO (101,112,113,114), M
      GOTO 108
112   B = MOD(Q(1) + Q(2) + 1,N)
      GOTO 102
113   B = MOD(Q(1) + Q(3) + 1,N)
      GOTO 103
114   B = Q(1) + Q(4) + 1
      IF (B.GT.N)  B = B - N
      CALL CB2(X,N,P(1),P(4),Q(1),Q(4),.FALSE.)
      B = B + Q(2)
      IF (B.GT.N)  B = B - N
      CALL CB3(X,N,P(1),P(2),P(4),Q(1),Q(2),Q(4),.FALSE.)
      B = B - Q(1)
      IF (B.LT.1)  B = B + N
      CALL CB2(X,N,P(2),P(4),Q(2),Q(4),N_EVEN)
      B = B + Q(3)
      IF (B.GT.N)  B = B - N
      CALL CB3(X,N,P(2),P(3),P(4),Q(2),Q(3),Q(4),N_EVEN)
      B = B + Q(1)
      IF (B.GT.N)  B = B - N
      CALL  CB4(X,N,P(1),P(2),P(3),P(4),Q(1),Q(2),Q(3),Q(4),.FALSE.)
      B = B - Q(2)
      IF (B.LT.1)  B = B + N
      CALL CB3(X,N,P(1),P(3),P(4),Q(1),Q(3),Q(4),.FALSE.)
      B = B - Q(1)
      IF (B.LT.1)  B = B + N
      CALL CB2(X,N,P(3),P(4),Q(3),Q(4),N_EVEN)
      B = MOD(B - Q(4) + Q(1),N)
      IF (B.LT.1)  B = B + N
103   CALL CB2(X,N,P(1),P(3),Q(1),Q(3),.FALSE.)
      B = B + Q(2)
      IF (B.GT.N)  B = B - N
      CALL CB3(X,N,P(1),P(2),P(3),Q(1),Q(2),Q(3),.FALSE.)
      B = B - Q(1)
      IF (B.LT.1)  B = B + N
      CALL CB2(X,N,P(2),P(3),Q(2),Q(3),N_EVEN)
      B = MOD(B - Q(3) + Q(1),N)
      IF (B.LT.1)  B = B + N
102   CALL CB2(X,N,P(1),P(2),Q(1),Q(2),.FALSE.)
```

FIG. 3 COMBINATION OPERATION FOR FOUR FACTOR TRANSFORM
N.B. CBi - combination routine of dimension i.

	Typical Execution Time (ms)		
Length	**PFA-FFT**	**RPFA-I**	**RPFA-II**
8	0.096	0.071	0.082
12	0.160	0.155	0.194
15	0.218	0.190	0.260
16	0.181	0.126	0.160
30	0.424	0.428	0.512
35	0.598	0.408	0.521
60	0.893	0.738	0.898
72	1.117	0.768	0.830
112	1.924	1.231	1.302
120	2.092	1.432	1.775
140	2.591	1.760	2.076
252	4.964	3.330	3.380
280	5.662	3.602	4.266
504	10.972	7.080	7.252
560	11.884	7.644	8.804
1008	23.240	15.380	15.300

Table 1 Timing results for the algorithms on VAX 8820

SIGNAL PROCESSING VI: Theories and Applications
J. Vandewalle, R. Boite, M. Moonen, A. Oosterlinck (eds.)

On the Design of Chebyshev Filters with Complex Specifications

J.S. Mason, A. Terchi and N.N. Chit
University College of Swansea
SWANSEA, SA2 8PP, UK

Two different approaches have been proposed recently offering solutions to *complex* Chebyshev filter design. The first is based on a Remez exchange algorithm, and the second is based on a double adaptive system identification algorithm. This paper lends practical evidence to the optimality of solutions from the DAS approach via three classic linear phase examples, designed using a completely general complex function specification and approximation.

1 Introduction

The design of FIR digital filters approximating a complex-valued frequency response in both magnitude and phase arise in circumstances, such as in the design of equalisation networks and frequency selective filters with approximately linear phase of arbitrary slope. There have been a number of proposals for designing such filters, usually based on separate approximations to the two functions, magnitude and phase, and optimising each separately. Considering the two as a complex-valued function, leads naturally to an approximation based on complex polynomials.

Such an approximation, resulting in real-valued filter coeficients, has for a long time lacked a practical solution. Rivlin and Shapiro [1] mention that in general the Chebyshev approximation is a least-squares approximation with appropriate costs. The information required is the error extrema locations and their suitable costs. Lawson [2] proposes a procedure for deriving these values for the complex case but the procedure proved to be impracticable, even after further refinement by a number of researchers.

Very recently however, two very different approaches have been proposed, almost concurrently, both offering solutions to this problem. The first of these is presented by Preuss, [3] [4], who describes an extension of the popular real-valued Remez exchange algorithm. Schulist [5] claims improved convergence of this algorithm, and in the vein of Remez exchange extensions, Alkhairy [6] reports a new algorithm appropriate for the complex case.

The second solution is attributed to the Authors, Chit and Mason, [7] [8] [9]; their double adaptive system (DAS) approach comes from the field of systems identification. The merit here is in its generality: the DAS package is succesful in tackling linear-phase, minimum-phase, as well as that of complex specifications.

Both the extended Remez exchange and the DAS approaches are based on iterative procedures approximating the complex function directly, and the resultant approximations are claimed optimum in a complex Chebyshev sense. Here we focus on the DAS approach, which is an LMS procedure within an iterative loop which finds the error extrema and their suitable costs.

Optimality

All Chebyshev optimum solutions have a number of extrema that is has lower and upper bounds, [1]. More is known about the characteristics for the real case, and it is far easier to check whether or not a given solution approaches Chebyshev optimality or not. Less is known, however, about the complex case, and thus it is more difficult to be certain that a designed solution has a sufficient number of extrema to be a true Chebeyshev approximation. It is not enough to merely derive an equiripple characteristic. An equiripple solution which includes overlapping zeros is an obvious example which is unquestionably sub-optimum in that alternative solution, with the same peak value and an equiripple characteristic but reduced extrema count, always exists.

To give an additional level of confidence that the DAS approach is capable of giving optimum complex Chebyshev approximations we design examples for which the solutions are well documented, namely that with linear phase specifications. We consider the three classic cases of optimal filters, extraripple, scaled ripple and equiripple. These are difficult to achieve in their detail, and as such provide a good test for any design procedure. Obviously it is acknowledged that such examples reduce to a real-valued approximation, and hence can be designed by the standard Remez exchange algorithm. It remains of interest however, to see how close the complex approximation procedure meets the detail of these special cases.

2 The DAS Approach

The DAS approach is shown pictorially in Fig. 1, and full details are given by Chit in [7]. The input signal u_n to both the realized and the ideal filter, and the desired output signal y_n are given by

$$u_n = K \sum_{i=1}^{N_f} C(w_i) \sin(w_i n) \qquad (3.1)$$

$$y_n = K \sum_{i=1}^{N_f} C(w_i) G(w_i) \sin(w_i n + \Theta(w_i)) \qquad (3.2)$$

where K is a scalar controlling the absolute level of the input and output signals and N_f is the number of frequency domain samples needed to adequately represent the spectrum. The iterative update of C is designed to give meet both G and Θ in a minimax complex sense.

3 Design Examples

We design three linear-phase filters exhibiting the classic extraripple, scaled extraripple and equiripple characteristics. These filters present a stern task to any design procedure, if for example the subtle differences at the band edges are to be revealed. These three examples are taken from Reference [11] and represent three similar lowpass filters with slightly different transition band edges, giving the three subtly different ripple characteristics.

Example (1):
extraripple, $f_p = 0.2076$ and $f_s = 0.2513$

Example (2):
scaled extraripple, $f_p = 0.2095$ and $f_s = 0.25377$

Example (3):
equiripple, $f_p = 0.2280$ and $f_s = 0.27551$

and for all three the remaining design data are the same:

$$N_z = 24, \quad \tau = 12, \quad \delta_p = \delta_s = 0.05.$$

4 Discussion and Conclusion

The imposition of exact symmetry on the filter coefficients leads to exact linear phase (ie zero phase error), and with the DAS algorithm such constraints are very easily imposed, since the algorithm actually models the filter action directly ie in the time domain. However, here the purpose is to not to resort to such constraints, but to demonstrate convergence towards optimum solutions using a general algorithm, and a general complex specification.

Details of the responses of the designed filters are presented below. It is found that the phase errors tend to zero, the coefficients tend to be symmetrical, and the complex errors tend to the gain errors. Responses of the designed filters are presented below. In all three cases. the maximum deviation is 0.05, as reported originally by [11]. The general unconstrained algorithm, given a specification in a complex form, is therfore capable of producing solutions which are known to be optimum.

References

[1] T.J. Rivlin and H.S. Shapiro, 'A Unified Approach to Certain Problems of Approximation and Minimization', J. Soc. Indust. Appl. Math., vol. 9, No. 4, pp. 670-699, December 1961.

[2] C.L. Lawson, 'Contributions to the Theory of Linear Least Maximum Approximations', Ph.D. Thesis, UCLA, 1961.

[3] K. Preuss, 'A Novel Approach for Complex Chebyshev Approximations with FIR Filters using the Remez Exchange Algorithm' ICASSP-87, vol. 2, pp. 872-875, April 1987.

[4] K. Preuss, IEEE ASSP, vol. 37, No.5, pp. 702-712, May 1989.

[5] M.Schulist, 'Improvememts of a Complex FIR Filter Design Algorithm' Signal Processing, vol. 20, no. 1, pp. 81-90, May 1990.

[6] A.Alkhairy, K.Christian and J.Lim, 'Design of FIR Filters by Complex Chebyshev Approximation' ICASSP-91, vol.3 , pp. 1985-1988, March 1991.

[7] N.N. Chit, 'Weighted Chebyshev Complex-valued Approximation for FIR Digital Filters', Ph.D. Thesis, University of Wales, Swansea, UK., July 1987.

[8] J.S. Mason and N.N. Chit, 'New Approach to the Design of FIR Digital Filters', Proceedings of the IEE, vol. 134, Pt. G, No.4, pp. 167-180, August 1987.

[9] N.N. Chit and J.S. Mason, 'Complex Chebyshev Approximation for FIR Digital Filters', IEEE ASSP, January 1991.

[10] J.H. McClellan, T.W. Parks, and L.R. Rabiner, IEEE Trans. Audio and Electroacoustics, vol. AU-21, No.6, pp. 506-526, December 1973.

[11] L.R. Rabiner, J.H. McClellan and T.W. Parks, Proceedings of the IEEE, vol. 63, No.4, pp. 595-610, April 1975.

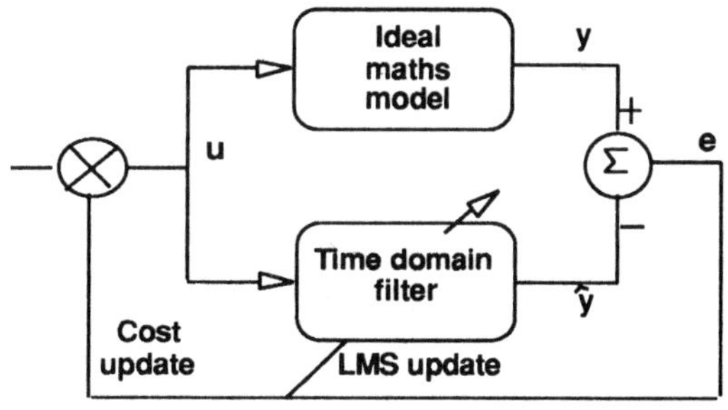

Fig.1 Structure of the Double Adaptive Systems Approach. The difference between the ideal and the realised filter is used in both loop adaptations.

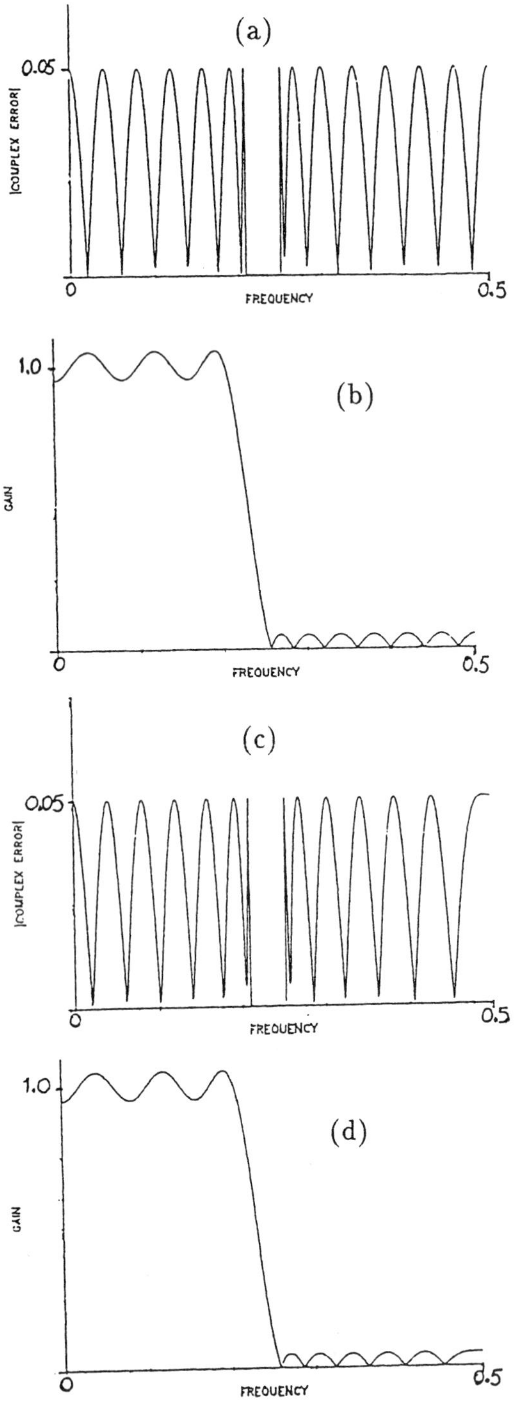

Fig.2 |complex error| and gain response of 24th order LPH filters:
(a & b) extraripple,
(c & d) scaled extraripple.
In both cases |complex error| and gain error tend to 0.05

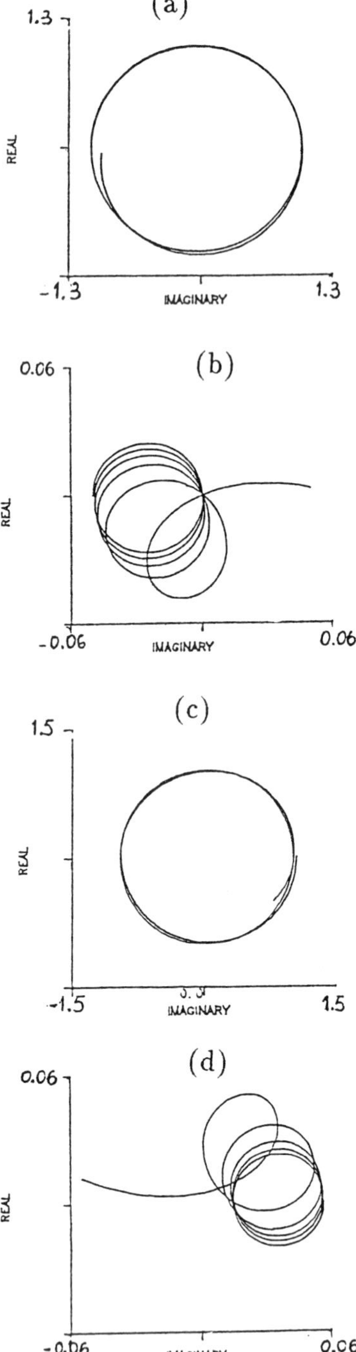

Fig.3 Complex-domain representation within passband for extraripple and scaled extraripple:
(a & c) trace of $H(w)$,
(b & d) trace of $E(w)$.

SIGNAL PROCESSING VI: Theories and Applications
J. Vandewalle, R. Boite, M. Moonen, A. Oosterlinck (eds.)

Active Noise Cancellation using a Modified Form of the Filtered-X LMS Algorithm

Elias Bjarnason

Institut für Netzwerk- und Signaltheorie
Technische Hochschule Darmstadt
Merckstr. 25, D-6100 Darmstadt, Germany

Abstract

The Filtered-X LMS algorithm [1] is an alternate form of the LMS algorithm for use when there is a transfer function in the auxiliary path following the adaptive filter and/or in the error path. To ensure convergence of the algorithm, the input to the error correlator has to be filtered by a copy of the auxiliary-error path transfer function[3]. A major drawback is that bounds for the stepsize are very difficult to calculate and that the stepsize has to be chosen small to guarantee convergence. A modification is introduced to improve the situation. Simulation results comparing the two algorithms and the results of real time measurements in a duct using a floating point signal processor system are presented.

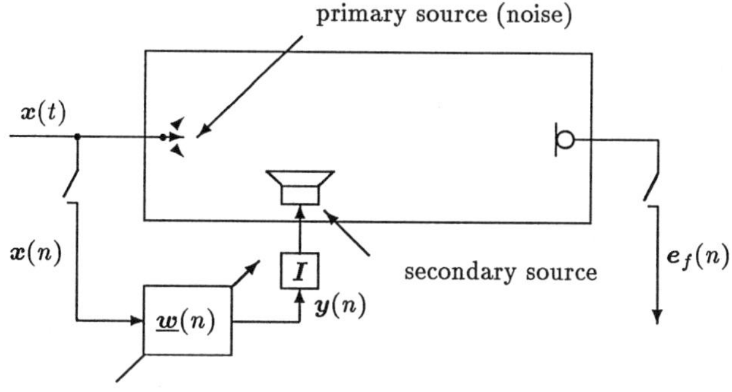

Figure 1: Active noise cancellation system without feedback

1 Introduction

The active control of sound and vibration involves the introduction of a number of controlled "secondary" sources driven such that the field generated by these sources interferes destructively with the field caused by the original "primary" source. The extent to which such destructive interference is possible depends on the geometric arrangement of the primary and secondary sources and their environment, and on the spectrum of the field produced by the primary source [2]. Due to the time-variance of the acoustic path the system has to be adaptive. Fig. 1 shows a simple active noise cancellation system without feedback using a loudspeaker as a secondary source. Here, $\underline{w}$ describes the adaptive controller and $\boldsymbol{I}$ the interpolation filter used to reconstruct the estimated signal $\boldsymbol{y}(t)$.

Fig. 2 shows a model of the active cancellation system. Here, h_x describes the system in the auxiliary path, modeling anti-aliasing filter, loudspeaker and acoustic path from loudspeaker to microphone. The system in the error path h_e models the path common to desired and estimated signal, including error microphone and

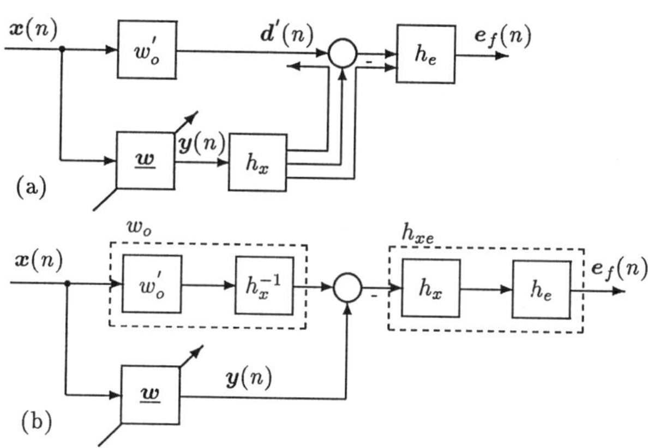

Figure 2: (a) h_x can be moved through the summing junction to form an equivalent system. (b) h_x and h_e can be combined to form a new error path system function h_{xe}. In the same way $w_o^{'}$ and the inverse system of the auxiliary path h_x^{-1} can be combined to w_o which is the system to be identified

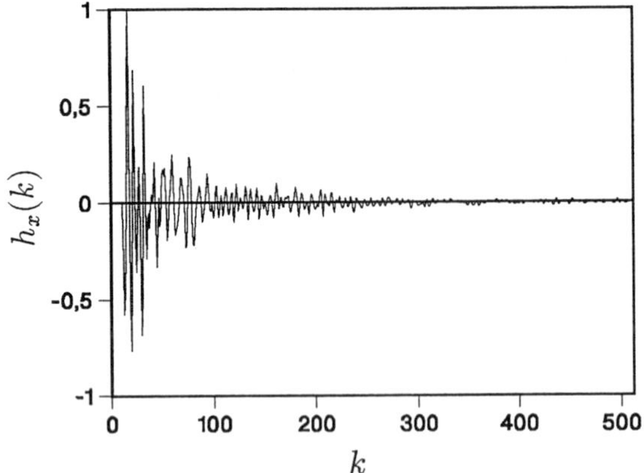

Figure 3: Measured impulse response h_x of auxiliary path

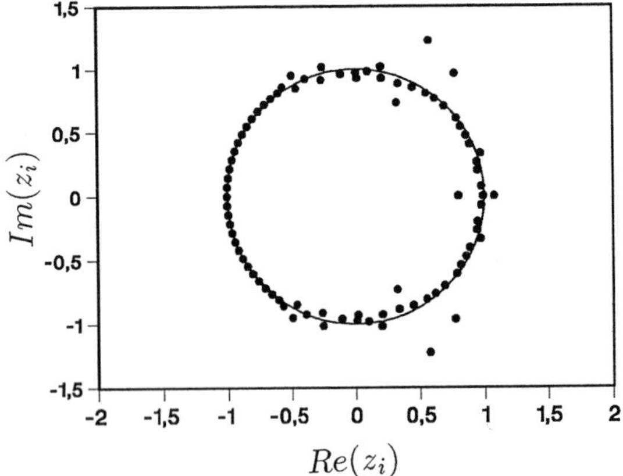

Figure 4: Root diagramm of auxiliary path ($H_x(z_i) = 0$)

anti-aliasing filter. Because of the system in the auxiliary path, w_o has to include the inverse system of h_x ($W_o'(z)/H_x(z)$ is the system to be identified). Measurements show (see Fig. 3 and 4) that h_x can be a nonminimum phase system. This results in making w_o noncausal and therefore degrading the cancellation effect. The drawback of the non causal parts is decreased due to inherent delay in acoustic system. The filter h_e does not affect the required model w_o but it does affect the behavior of the adaptive algorithm as does h_x.

2 The Filtered-X LMS Algorithm

2.1 The General Form

The existence of a filter in the auxiliary and/or the error path has been shown to generally degrade the performance of the LMS algorithm. Thus the convergence rate is lowered, the residal power is increased and the algorithm can even become unstable. In order to stabilize the algorithm an identical filter to the filter h_{xe} in the auxiliary-error path (the combined auxiliary and error path) is used to filter the input $\boldsymbol{x}(n)$ (see Fig. 5). The recursive relation for updating the tap-weight vector in this case is:

$$\underline{\boldsymbol{w}}(n+1) = \underline{\boldsymbol{w}}(n) + \mu \underline{\boldsymbol{x}}_f(n) \boldsymbol{e}_f(n), \tag{1}$$

where $\underline{\boldsymbol{x}}_f$ und $\underline{\boldsymbol{w}}$ are vectors of dimension N and:

$$\begin{aligned}
\boldsymbol{x}_f(n) &= \boldsymbol{x}(n) * h_x(n) * h_e(n) &&= \boldsymbol{x}(n) * h_{xe}(n),\\
\boldsymbol{y}(n) &= \underline{\boldsymbol{w}}^T(n)\underline{\boldsymbol{x}}(n),\\
\boldsymbol{y}_f(n) &= \boldsymbol{y}(n) * h_x(n) * h_e(n) &&= \boldsymbol{y}(n) * h_{xe}(n),\\
\boldsymbol{d}_f(n) &= \boldsymbol{d}'(n) * h_e(n) &&= \boldsymbol{d}(n) * h_{xe}(n),\\
\boldsymbol{e}_f(n) &= \boldsymbol{d}_f(n) - \boldsymbol{y}_f(n).
\end{aligned}$$

It is assumed that $\boldsymbol{x}(n)$ and $\boldsymbol{d}(n)$ are jointly stationary and that their first moments are zero.

Defining the weight-error vector at time n as

$$\underline{\varepsilon}(n) = \underline{w}_o - \underline{\boldsymbol{w}}(n), \tag{2}$$

taking the expectation, using the independance assumption, and modeling the auxiliary-error path with a transversal filter h_{xe} of order M results in:

$$\begin{aligned}
E[\underline{\varepsilon}(n+1)] = E[\underline{\varepsilon}(n)] - \mu\Big(\sum_{i=0}^{M-1}\sum_{j=0}^{M-1} h_{xe}(i)h_{xe}(j)\\
E[\underline{\boldsymbol{x}}(n-j)\underline{\boldsymbol{x}}^T(n-i)]E[\underline{\varepsilon}(n-i)]\Big).
\end{aligned} \tag{3}$$

Here, bounds for the stepsize have been calculated for the simple case of a delay in the auxiliary-error path [4,5].

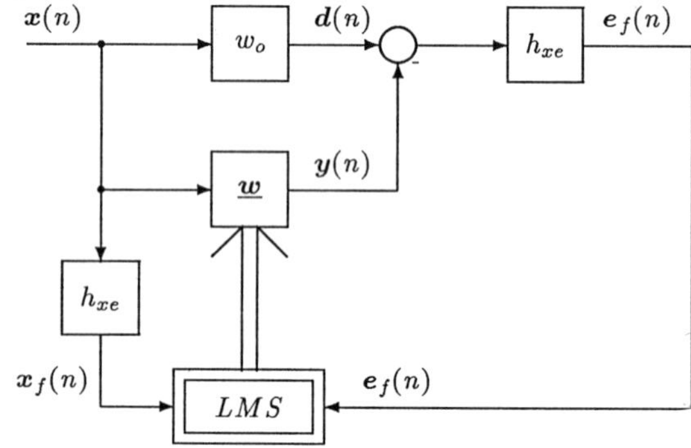

Figure 5: System with general algorithm

For very small μ the expectation value of the weight-error can be approximated:

$$E[\underline{\varepsilon}(n+1)] \approx (\underline{I} - \mu \sum_{i=0}^{M-1} \sum_{j=0}^{M-1} h_{xe}(i) h_{xe}(j) \\ E[\underline{x}(n-j)\underline{x}^T(n-i)])E[\underline{\varepsilon}(n)] = \\ (\underline{I} - \mu \underline{R}_{x_f x_f}) E[\underline{\varepsilon}(n)]. \quad (4)$$

This is the same equation as for the LMS algorithm with filtered input $\boldsymbol{x}_f(n)$, thus guaranteeing the convergence of the Filtered X-LMS algorithm for very small μ.

2.2 The Modified Form

Assuming that the whole system is time invariant, calculating the mean square of the filtered error results in:

$$E[e_f^2(n)] = \sigma_{d_f}^2 - \underline{p}_{d_f x_f}^T \underline{w} - \underline{w}^T \underline{p}_{x_f d_f} + \underline{w}^T \underline{R}_{x_f x_f} \underline{w}, \quad (5)$$

$$\underline{p}_{d_f x_f} = E[\boldsymbol{d}_f(n) \underline{x}_f(n)].$$

Then differentiating the mean square of the filtered error with respect to the weight vector $\underline{w}$:

$$\underline{\nabla} = \frac{dE[e_f^2(n)]}{d\underline{w}} = 2\underline{R}_{x_f x_f} \underline{w} - 2\underline{p}_{d_f x_f}, \quad (6)$$

and using the method of steepest descend, the weight update equation is given by:

$$\begin{aligned} \underline{w}(n+1) &= \underline{w}(n) - \tfrac{1}{2}\mu \underline{\nabla}(n) \\ &= \underline{w}(n) + \mu[\underline{p}_{d_f x_f} - \underline{R}_{x_f x_f} \underline{w}(n)]. \end{aligned} \quad (7)$$

Replacing the unknowns with instantaneous estimates of their values results in the modified form of the Filtered-X LMS algorithm:

$$\underline{w}(n+1) = \underline{w}(n) + \mu \underline{x}_f(n)[\boldsymbol{d}_f(n) - \underline{x}_f^T(n) \underline{w}(n)]. \quad (8)$$

Observing that $\boldsymbol{d}_f(n)$ is not available in the case of active noise cancellation but knowing the filtered error $\boldsymbol{e}_f(n)$ and the auxiliary-error path function h_{xe} the modified form (see Fig. 6) of the Filtered-X LMS al-

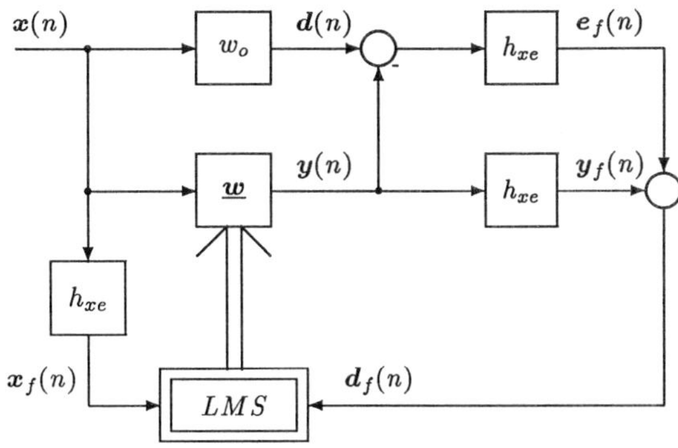

Figure 6: System with modified algorithm

gorithm can be rewritten to:

$$\underline{w}(n+1) = \underline{w}(n) + \mu \underline{x}_f(n)[\boldsymbol{e}_f(n) + \boldsymbol{y}_f(n) - \underline{x}_f^T(n) \underline{w}(n)]. (9)$$

In [5] the same kind of modification is introduced for the Delayed LMS algorithm.

For the modified form the expectation of the weight-error vector is:

$$E[\underline{\varepsilon}(n+1)] = (\underline{I} - \mu \sum_{i=0}^{M-1} \sum_{j=0}^{M-1} h_{xe}(i) h_{xe}(j) \\ E[\underline{x}(n-j)\underline{x}^T(n-i)])E[\underline{\varepsilon}(n)] \\ = (\underline{I} - \mu \underline{R}_{x_f x_f}) E[\underline{\varepsilon}(n)]. \quad (10)$$

From Equation (10) it can be seen, that the modified form of the Filtered-X LMS algorithm has the same behavior as the LMS algorithm with filtered input $\boldsymbol{x}_f(n)$, except in the case of time variant systems.

2.3 Comparison between the General and the Modified Form

Comparing Eq. (3) and (10) it can be seen that in Eq. (3) the weight-error vector update is a function of weighted past values of the weight-error vector, but in Eq. (10) only the actual value of the weight-error vector and thus all information learned from past updates is used. Therefore, the modified form has a higher convergence rate and the same predictable behavior as the LMS with a filtered input. On the other hand the modified form requires a larger amount of computation

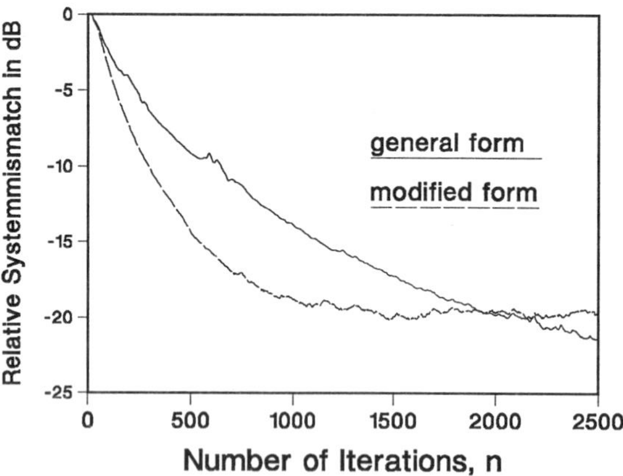

Figure 7: Comparison of the highest convergence rate in each case between the general and the modified form for a white Gaussian input ($N = 50, M = 20, S/N = 20dB$)

($3N + 2M$ instead of $2N + M$) and storage capacity. In Fig. 7 a comparison is shown between the normalized form of the two algorithms. For both algorithms the stepsize was chosen to optimize the dynamic behavior of the algorithm.

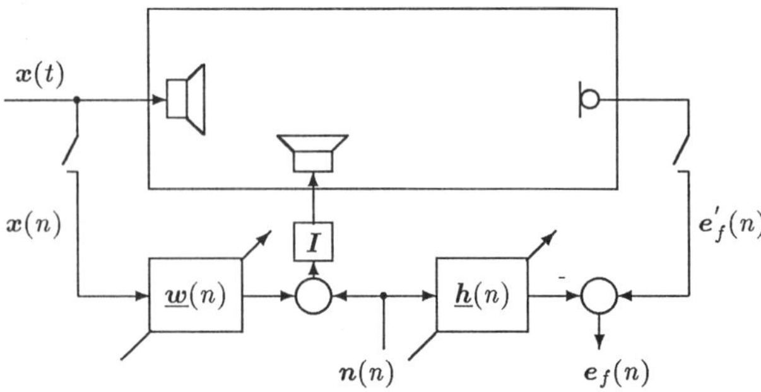

Figure 8: Realtime implementation

2.4 Implementation

The approach shown in Fig. 8 was implemented using the DSP32C signal processor from AT&T. For the acoustical system a duct with 40cm diameter and 2m length was used. Both algorithms were implemented in their normalized form, including adaptive modeling of the auxiliary-error path in series or parallel using the NLMS algorithm with independent low-level random source $n(n)$ (see [3]). Typical results for the noise reduction after the adaptation using a broadband noise as input are shown in Fig. 9. The curve was obtained by subtracting the canceled from the uncanceled log spectrum. The minima at about 400 Hz and 800 Hz are due to acoustical resonances of the duct. The cancellation system showed to be effective for broadband noise as well as for narrowband noise.

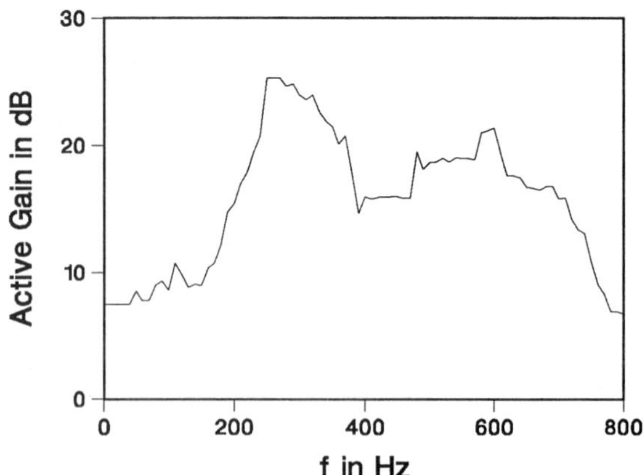

Figure 9: Noise reduction with an active attenuation system using the normalized modified form of the Filtered-X LMS algorithm

3 Conclusion

A modified form of Filtered-X LMS algorithm has been introduced. This algorithm has the same properties as the LMS algorithm with a filtered input. The advantages of the modified form over the general form are the known bounds for the stepsize and the higher convergence rate. This has to be paid for with increased complexity. A complete active attenuation system has been described and implemented. The system can be operated fully adaptive employing an adaptive filter using the NLMS algorithm with independent low-level random noise source to determine the auxiliary-error path transfer function [3], and using the general or the modified form of the Filtered-X NLMS algorithm for the active noise cancellation.

References

[1] B. Widrow, D. Shur and S. Shaffer, "On Adaptive Inverse Control," in Proc. 15th Asilomar Conf. Circuits, Sys., Comput., 1981, pp. 185-189, 1981.

[2] S. J. Elliott, I. M. Stothers and P. A. Nelson, "A Multiple Error LMS Algorithm and Its Application to the Active Control of Sound and Vibration," IEEE Trans. Acoust., Speech, Signal Processing, vol. ASSP-35, pp. 1423-1434, 1987.

[3] L. J. Eriksson and M. C. Allie, "Use of Random Noise for On-Line Transducer Modeling in an Adaptive Active Attenuation System," J. Acoust. Soc. Amer., vol. 89, pp. 797-802, 1989.

[4] G. Long, F. Ling and J. G. Proakis, "The LMS Algorithm with Delayed Coefficient Adaption," IEEE Trans. Acoust., Speech, Signal Processing, vol. ASSP-37, no. 9, pp. 1397-1405, Sep. 1989.

[5] M. Rupp, R. Frenzel, "Analysis of LMS and NLMS Algorithms with Delayed Coefficient Update Under the Presence of Spherically Invariant Processes," IEEE Trans. Acoust., Speech, Signal Processing, to be published.

SIGNAL PROCESSING VI: Theories and Applications
J. Vandewalle, R. Boite, M. Moonen, A. Oosterlinck (eds.)

A PIPELINED ADAPTIVE LATTICE FILTER ARCHITECTURE: THEORY AND APPLICATIONS*

Naresh R. Shanbhag   Keshab K. Parhi

Dept. of Electrical Engineering
University of Minnesota
Minneapolis, MN 55455

Abstract

The stochastic gradient adaptive lattice filter is pipelined by the application of the *relaxed look-ahead*. This form of look-ahead maintains the functional behavour instead of the input-output mapping. The sum and product relaxations are employed to pipeline the lattice filter. The hardware complexity of the proposed pipelined filter is of the same order as the sequential filter. Thus, the new architecture is attractive from an implementation point of view. Convergence analysis results are presented to illustrate the trade-off offered by relaxed look-ahead. Application of the lattice filter to pipelined video predictive coding is demonstrated.

1 Introduction

Algorithm transformation techniques [1], for increasing the concurrency in digital signal processing architectures, have been well studied. Pipelining [2] and parallel processing [3] are two major techniques for developing high-speed digital signal processing architectures. Pipelining of fixed-coefficient filters via the *look-ahead computation* [2] has already been developed and has also been applied to pipeline adaptive lattice filters [4, 5]. The conventional lookahead technique transforms a given serial algorithm into an equivalent (in the sense of input-output mapping) pipelined one. This, however, results in an enormous increase in the hardware complexity. In case of adaptive filters, this complexity is even higher due to the coefficient-update loop. For pipelining adaptive filters, it is possible to give up this exact input-output mapping if the average convergence behavior is not changed substantially. In this paper, we use a *relaxed form of look-ahead* as opposed to the conventional look-ahead for pipelining the stochastic gradient lattice adaptive filter [6]. The relaxed look-ahead is an approximation to the conventional look-ahead which does not alter the convergence behavior substantially.

The relaxed look-ahead has been succesfully employed to develop a pipelined LMS adaptive filter architecture [7] and the adaptive pulse-code modulation codec [8]. In this paper, we choose a particular stochastic gradient lattice adaptive algorithm [6] to pipeline via this technique. Note that relaxed look-ahead can be employed to pipeline any of the adaptive lattice algorithms in [6].

A brief discussion of the conventional look-ahead and its relaxed form is provided in section 2. In section 3, we derive the pipelined stochastic gradient lattice architecture (PIPSGLA). In section 4, we discuss its convergence properties. In section 5, a pipelining example and image compression using pipelined adaptive pulse code modulation (ADPCM) is demonstrated.

2 The Relaxed Form of Look-ahead

We first illustrate the conventional lookahead [2] with the help of an example. Consider the following first-order recursion,

$$x(n+1) = ax(n) + u(n), \qquad (2.1)$$

where $x(n)$ is the output, $u(n)$ is the input and a is a constant. Iterating (2.1) M_1 times, we get

$$x(n+M_1) = a^{M_1}x(n) + \sum_{i=0}^{M_1-1} a^i u(n+M_1-1-i). \qquad (2.2)$$

Equation (2.2) represents the application of look-ahead to (2.1). Due to the nature of the transformation both (2.1) and (2.2) represent the same system. However, (2.2) can process data M_1 times faster than (2.1). This is because the M_1 latches introduced into the system can be used to pipeline the multiply-add operation. The second term in (2.2) represents the extra hardware overhead necessary. This overhead can be quite expensive for higher order systems and for fine-grained pipelining (i.e. large values of M_1). However, if (2.1) were an adaptation algorithm, then we may apply the *sum relaxation* to write the second term in (2.2), as follows

$$x(n+M_1) = a^{M_1}x(n) + M_1 u(n+M_1-1). \qquad (2.3)$$

The system described by (2.3) would be a close approximation of the system in (2.2) if a is close to unity and

* This research was supported by the army research office by contract number DAAL03-90-G-0063.

$u(n)$ is approximately constant. The a^{M_1} term in (2.3) can be precomputed if a is a constant. If a in (2.3) is time-varying (i.e. we have $a(n)$ instead of a), but the magnitude of $a(n)$ is close to unity, then we replace $a(n)$ by $(1 - a'(n))$, where $a'(n)$ is close to zero. Then, we apply the *product relaxation* to write (2.3) as

$$x(n+M_1) = (1-M_1a'(n))x(n)+M_1u(n+M_1-1). \quad (2.4)$$

Thus, (2.3) and (2.4) represent possible application of relaxed look-ahead to (2.1). Depending on the application at hand, different types of approximations may be employed. Unlike (2.2), (2.3-2.4) are not equivalent to (2.1) with respect to its input-output mapping. Therefore, in case of adaptive algorithms, convergence analysis of (2.3-2.4) needs to be done. In general, the relaxed form of look-ahead is not an algorithm transformation technique in the usual sense [1]. This is because it modifies the algorithm to which it is applied. However, it can be considered an algorithm transformation technique in the stochastic sense if the average output profile is maintained.

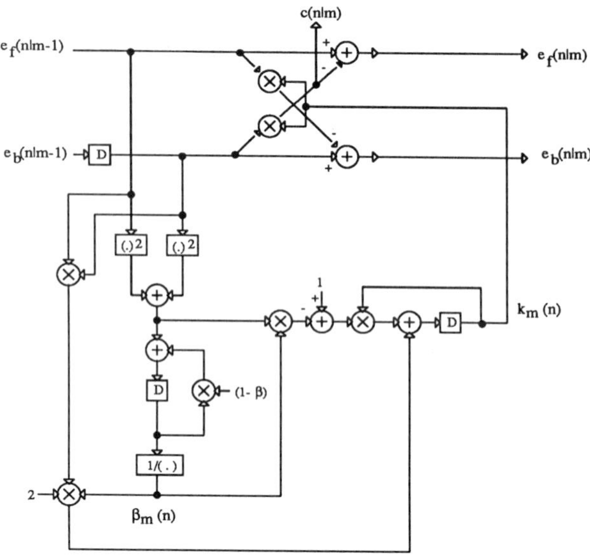

Figure 1: The SSGLA architecture.

3 The PIPSGLA Architecture

As usual, the indices n and m denote the time instance and the lattice stage number, respectively. The serial stochastic gradient lattice algorithm (SSGLA) chosen for pipelining is described as follows [6]

$$k_m(n+1) = [1 - \beta_m(n)P(n|m-1)]k_m(n) + 2\beta_m(n)Q(n|m-1) \quad (3.1)$$

$$S(n+1|m-1) = (1-\beta)S(n|m-1)+P(n|m-1) \quad (3.2)$$

$$\beta_m(n) = \frac{1}{S(n|m-1)} \quad (3.3)$$

$$e_f(n|m) = e_f(n|m-1) - k_m(n)e_b(n-1|m-1) \quad (3.4)$$

$$e_b(n|m) = e_b(n-1|m-1) - k_m(n)e_f(n|m-1) \quad (3.5)$$

where

$$P(n|m) = e_f^2(n|m) + e_b^2(n-1|m) \quad (3.6)$$

$$Q(n|m) = e_f(n|m)e_b(n-1|m). \quad (3.7)$$

The reflection coefficient $k_m(n)$ is updated according to (3.1), while the input power estimate $S(n|m)$ is updated using (3.2). Thus, (3.1) and (3.2) are the two recursive loops in the algorithm. The equations (3.3) and (3.4) are the order-updates. The SSGLA architecture in Fig.1 shows intermediate outputs $c(n|m)$, which would be employed to develop an ADPCM codec. Critical path analysis of Fig.1 shows a computation time (T_c) of

$$T_c = 4T_m + (N+2)T_a, \quad (3.8)$$

where T_m and T_a are two operand multiply and add times. The loop computation time T_L is

$$T_L = T_m + T_a. \quad (3.9)$$

The pipelining of SSGLA is done by first inserting interstage pipelining latches. This step is trivial as the stages are connected in a non-recursive fashion. If the desired speed-up is not achieved, then the two recursive loops need to be pipelined. Let M_S and M_L represent the number of interstage and loop pipelining latches, respectively. To do this we apply relaxed look-ahead to (3.1) and (3.2). Like (2.1), (3.2) is a first order recursion. Therefore, applying the sum relaxation (see (2.3)) we can simply write down the final equation by inspection

$$S(n'+M_L) = (1-\beta)^{M_L}S(n'|m-1)+P(n'|m-1). \quad (3.10)$$

where $n' = n - M_S$. Similarly, we can show that application of the product and sum relaxations (see (2.4)) to (3.1) results in the following

$$k_m(n'+M_L) = [1 - M_L\beta_m(n')P(n'|m-1)] k_m(n') + 2M_L\beta_m(n')Q(n'|m-1). \quad (3.11)$$

Note that the loop multiplier coefficients in (3.1) and (3.2) are close to unity. Therefore, the form of relaxation used to derive (3.10) and (3.11) is justified. The rest of the equations describing PIPSGLA are identical to SSGLA and hence are not provided.

The complete PIPSGLA is shown in Fig.2, where it can be seen that the hardware increase is two multipliers per stage and the pipelining latches (see Fig.2). This is remarkably lower than that obtained by conventional look-ahead [4, 5].

4 Convergence Analysis

The convergence analysis, for a single lattice stage, was carried out under the independence assumptions [6]. These assumptions become more true as the number of lattice stages increases. The inputs to the stage under

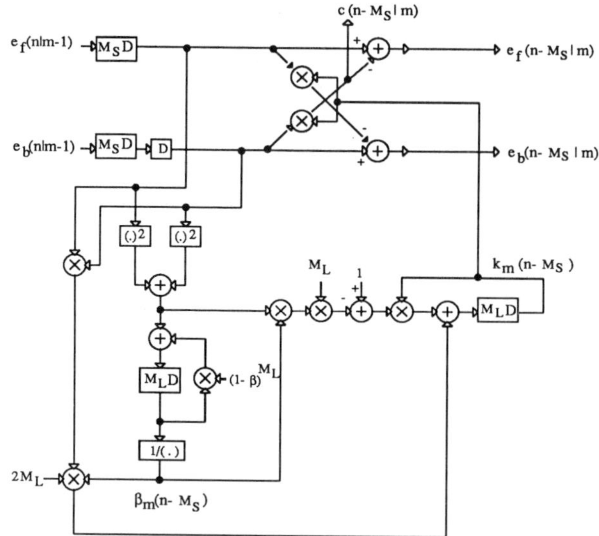

Figure 2: The PIPSGLA architecture.

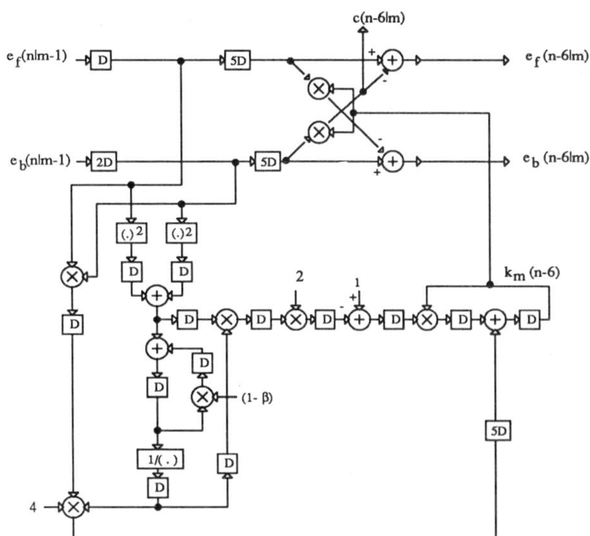

Figure 3: The pipelining example.

consideration is assumed to be stationary. The results of this analysis are presented below. The bounds on β for convergence of weights is

$$0 \leq \beta \leq \frac{2}{M_L^2}, \tag{4.1}$$

which is tighter than that of SSGLA (for which the upper bound is 2). However, for most applications the value of β is smaller that the upper bound.

The convergence time-constant for PIPSGLA (τ_{PIP}) equals

$$\tau_{PIP} = \frac{1}{2\beta M_L} \tag{4.2}$$

and this value is $1/M_L$ times that of SSGLA. This is to be expected as (3.11) has an effective adaptation constant which is M_L times that of SSGLA in (3.1).

The misadjustment for PIPSGLA ($\mathcal{M}_{PIP}$) is given by

$$\mathcal{M}_{PIP} = \frac{M_L\gamma(1-k_{m,opt}^2)}{(2-M_L\gamma)(2+k_{m,opt}^2)} \tag{4.3}$$

where $\gamma = 1-(1-\beta)^{M_L}$. Assuming β is much smaller than unity, (4.3) implies that $\mathcal{M}_{PIP}$ is M_L^2 times that of SSGLA.

5 Examples

In this section, we first present a pipelining example to demonstrate the increase in speed. Then, we employ image coding using ADPCM to compare the performance of SSGLA and PIPSGLA.

5.1 Pipelining Example

To illustrate the speed-up due to pipelining, we assume $T_a = 20$, $T_m = 40$ and the filter order $N = 2$. Therefore, from (3.8), it follows that $T_c = 240$. For a speed-up of $M = 6$, the clock-period of PIPSGLA should be 40. This can be achieved with $M_L = 2$ and $M_S = 6$. Retiming these latches (Fig.3) results in the desired clock-period.

5.2 Image Coding

A pipelined ADPCM codec is developed by employing PIPSGLA as a predictor [6]. The pipelined ADPCM coder architecture is shown in Fig.4, where $Ncols$ is the number of pixels in a row of the image. The predictor output is formed by summing the $c(n|m)$ outputs of each of the lattice stages L_0, L_1 and L_2. The input is in a row-by-row raster scan format. The $Ncols - M_S - 1$ latches at the input of the lattice predictor are for enabling prediction of the current pixel with three nearest pixels from the previous line.

The input (Fig.5) is a 256×256 frame, 8-bit per pixel (bpp) with 256 gray levels. The quantizer has 3 bits. The adaptation constant β was 0.009 for both the serial and pipelined architectures. The pipelined architecture (Fig.4) had $M_S = 35$ and $M_L = 3$, which corresponded to a speed-up of 12. The reconstructed images for the serial (Fig.6) and the pipelined architectures (Fig.7) show no perceptual differences. The signal-to-noise ratio (SNR) for the serial architecture was 23.39 dB, while that of the pipelined architecture was 23.33 dB.

Thus, we see that large speed-ups are achievable without any appreciable degradation in the convergence behavior.

6 Summary and Conclusions

The stochastic gradient lattice filter has been pipelined via the application of *relaxed look-ahead* [7]. The relaxed look-ahead results in an enormous savings in hardware. As it is not a one-to-one mapping like its deterministic

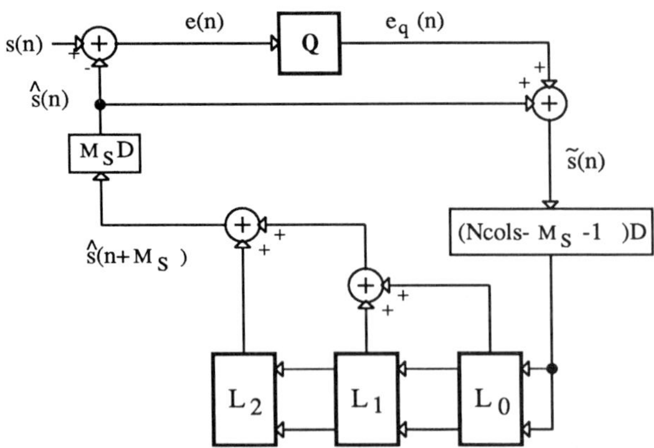

Figure 4: The pipelined ADPCM coder.

counterpart, the relaxed look-ahead offers a multitude of architectural possibilities. The PIPSGLA is one such choice but the user may select any other form of relaxation depending on the application at hand.

Figure 5: Original image.

Figure 6: Reconstructed image for serial codec.

Figure 7: Reconstructed image for pipelined codec.

References

[1] K.K. Parhi, "Algorithm transformation techniques for concurrent processors", *Proceedings of the IEEE*, vol. 77, pp. 1879-1895, Dec. 1989.

[2] K.K. Parhi and D.G. Messerschmitt, "Pipeline interleaving and parallelism in recursive digital filters - Part I : Pipelining using scattered look-ahead and decomposition", *IEEE Trans. on Acoustics, Speech and Signal Proc.*, vol. 37, pp. 1099-1117, July 1989.

[3] K.K. Parhi and D.G. Messerschmitt, "Pipeline interleaving and parallelism in recursive digital filters - Part II : Pipelined incremental block filtering", *IEEE Trans. on Acoustics, Speech and Signal Proc.*, vol. 37, pp. 1118-1134, July 1989.

[4] K.K. Parhi and D.G. Messerschmitt, "Concurrent cellular VLSI adaptive filter architectures", *IEEE Trans. on Circuits and Systems*, vol. 34, pp. 1141-1151, Oct. 1987.

[5] T. H.-Y. Meng and D.G. Messerschmitt, "Arbitrarily high sampling rate adaptive filters", *IEEE Trans. on Acoustics, Speech and Signal Proc.*, vol. 35, pp. 455-470, April 1987.

[6] M.L. Honig and D.G. Messerschmitt, *Adaptive Filters : Structures, Algorithms, and Applications.* Boston: Kluwer, 1984.

[7] N.R. Shanbhag and K.K. Parhi, "A pipelined LMS adaptive filter architecture", *Proc. Asilomar Conf. on Sig., Syst. and Comput.*, Nov. 1991, Pacific Grove (CA).

[8] N.R. Shanbhag and K.K. Parhi, "A high-speed architecture for ADPCM codec", *Proc. IEEE Intl. Symp. on Circuits and Systems*, San Diego, 1992.

SIGNAL PROCESSING VI: Theories and Applications
J. Vandewalle, R. Boite, M. Moonen, A. Oosterlinck (eds.)

BLIND DECONVOLUTION OF MULTIVARIATE SIGNALS USING ADAPTIVE FIR LOSSLESS FILTERS

Philippe LOUBATON*, Philip A. REGALIA**

* Ecole Nationale Supérieure des Télécommunications, Département Signal, 46 rue Barrault, 75364 PARIS CEDEX 13, FRANCE. ** Institut National des Télécommunications, Département Electronique et Communications, 9 rue Charles Fourier, 91011 EVRY CEDEX, FRANCE.

The problem of adaptive blind deconvolution of multivariate signals is considered. A deflation procedure is proposed. First, a classical contrast is maximized in order to extract the first component of the p-variate excitation. Next, an adaptive FIR lossless filter produces a (p–1)-variate signal on which the preceding algorithms can be applied to extract the other components of the excitation.

1. INTRODUCTION.

In a number of applications, a certain scalar stationary random process $y = (y_n)_{n\in\mathbb{Z}}$ (the observation) is the output of an unknown filter driven by an independent identically distributed (iid) sequence of random variables $w = (w_n)_{n\in\mathbb{Z}}$ representing the useful signal. The blind deconvolution of y consists of extracting the excitation w from y, or equivalently of identifying the inverse of the transfer function h(z) of the unknown filter. As it is well known, this problem does not make sense when the signal w is Gaussian, except in the case where h(z) is assumed to be minimum (or maximum) phase. By contrast, when w is assumed to be non Gaussian, it is possible to establish under very mild technical assumptions ([5]) that the representation of y as $y(n) = [h(z)].w(n)$ is unique up to time delays and scaling factors. Therefore, the adaptive blind deconvolution of non Gaussian scalar signals makes sense, and it has been investigated in numerous works (e.g. [1], [4], [7]). The concept of contrast (called in [4] objective function) plays a central role in the above mentionned question. A contrast is a function Ψ, defined on an appropriate set $\mathcal{X}$ of random variables, which satisfies the following requirements : if $x\in\mathcal{X}$, then $\Psi(x)$ depends only on the probability distribution of x ; $\Psi(\alpha x) = |\alpha|\Psi(\alpha x)$ for $\alpha\in\mathbb{R}$ and $x\in\mathcal{X}$; if $(x_i)_{i\in\mathbb{N}}$ are independent copies of $x\in\mathcal{X}$, then $\Psi(\sum_{i\geq 0} \alpha_i x_i) \leq \Psi(x)$ with equality if and only if (iff) $\alpha_i = \alpha\delta(i-i_0)$. It turns out that a possible approach to perform the adaptive deconvolution of y is to adapt a filter g(z) so as to maximize $\Psi([g(z)].y(n))$ for some contrast Ψ. One of the simplest implementation is obtained when Ψ is given by

$$\Psi(x) = \frac{|c_4(x)|}{(E(x^2))^2} \tag{1-1}$$

where $c_4(x)$ denotes the fourth-order cumulant of x (i.e. $c_4(x) = E(x^4) - 3(E(x^2))^2$ if $E(x) = 0$). In fact, by using a classical whitening of the observations, it is possible to assume without restrictions that $(y_n)_{n\in\mathbb{Z}}$ is a unit variance uncorrelated sequence. Then, if the excitation w is known to have a strictly negative (resp. strictly positive) fourth-order cumulant, the maximization of Ψ is equivalent to the minimization (resp. the maximization) of $E([g(z)].y(n))^4$ under the constraint $\|g\|^2 = \int_{[-\pi,\pi]} |g(e^{i\omega})|^2 d\omega = 1$. The key point is that this problem does not possess spurious local extrema [7], so that it is possble to use a stochastic gradient algorithm in order to adapt a unit norm filter so as to minimize or maximize $E([g(z)].y(n))^4$. However, in practice, the optimization has to be performed in the class of FIR filters of fixed degree N, and it is known that if N is not large enough, then the previously introduced extremum problem may exhibit spurious local extrema.

In this paper, we adress the above adaptive blind deconvolution problem, but in the case where the observation is a p-variate signal $Y = (Y(n))_{n\in\mathbb{Z}}$ which can be written as

$$Y(n) = \sum_{k=1}^{p} [H_k(z)].W_k(n) \tag{1-2}$$

where the $H_k(z)$ are unknown $p\times 1$ transfer functions and where the excitations $(W_k)_{k=1,p}$ are statistically independent non Gaussian centered iid sequences of variance 1. To our knowledge, this problem has not yet been considered in previous works. However, it should be noted that some recent works devoted to the separation of statistically independent narrow band sources by means of a sensor array (e.g.[2], [3]) are directly related to the multivariate deconvolution problem considered here, but in the very special case where the unknown transfer function $H(z) = (H_1(z), ..., H_p(z))$ is supposed to be a constant matrix. In the frame of this paper, we shall extensively use the contrast Ψ defined by (1-1), so that we shall assume that the fourth-order cumulants $(c_4(W_i))_{i=1,p}$ of the excitations are non zero, and that the numbers of positive and negative cumulants are known ; for ease of exposition, we shall suppose that the $(c_4(W_i))_{i=1,q}$ are strictly negative, and that the $(c_4(W_i))_{i=q+1,p}$ are strictly positive. It is however worth mentionning that this last hypothesis is not strictly necessary, since the deflation procedure to be presented below may be used to estimate the numbers of signals of positive and negative fourth-order cumulant.

The first step of our deconvolution scheme consists of a classical whitening of $(Y(n))_{n\in\mathbf{Z}}$ which can be realized by means of an adaptive multivariate linear prediction error filter. Therefore, it is possible to assume without restrictions that $(Y(n))_{n\in\mathbf{Z}}$ is a (second-order) white noise sequence (i.e. that $E(Y(n)Y^T(m)) = I_p\,\delta(n-m)$), and that the transfer function H(z) is lossless, in the sense that $H(z)H^T(z^{-1}) = I_p$. The signals $(W_i)_{i=1,q}$ whose fourth-order cumulants are negative are extracted by using the following two step procedure. A unit norm large enough degree 1xp FIR filter g(z) is adapted by means of a stochastic gradient algorithm so as to minimize $J(g) = E([g(z)].Y(n))^4$ on the unit ball $\mathcal{B}_p = \{g \,/\, \|g\|^2 = \int_{[-\pi,\pi]} \|g(e^{i\omega})\|^2\, d\omega = 1\}$. By studying the properties of J on $\mathcal{B}_p$, we shall show that, up to the truncation effects, the signal $r(n) = [g(z)].Y(n)$ converges to a signal W_j such that $c_4(W_j) < 0$. On the other hand, a 1xp FIR filter whose transfer function V(z) satisfies $V(z)V^T(z^{-1}) = 1$, is adapted in such a way that $[V(z)].Y(n) - r(n)$ converges to zero. The adaptation algorithm is based on the fact that these transfer functions admit simple parametrizations based on lossless lattice structure ([8]). Moreover, for such a transfer function V, the above mentionned lossless lattice structure makes possible to exhibit a FIR (p–1)xp transfer function U(z) for which $(U^T(z), V^T(z))^T$ is lossless. Then, the p–1 variate signal $Y'(n) = [U(z)].Y(n)$ converges to a p–1 variate second order white noise sequence, uncorrelated with W_j, which coincides with the output of a (p–1)x(p–1) lossless filter driven by $W'(n) = (W_1(n),..., W_{j-1}(n), W_{j+1}(n),..., W_p(n))^T$. It turns out that the preceding algorithms can be applied to Y'(n) in order to extract an another signal with negative fourth-order cumulant. Clearly, this deflation procedure allows the extraction of the signal W_i for i = 1, q. The positive fourth-order cumulant excitations are extracted similarly by exchanging minimization by maximization in what preceds.

This paper is organized as follows. The section 2 is devoted to the adaptive minimization of J(g). The deflation procedure based on the adaptive lossless filter is presented in section 3. Finally, a simple example illustrating these results is given in section 4.

2. ADAPTIVE MINIMIZATION OF J.

In this section, we justify that a stochastic gradient algorithm based on the minimization of $J(g) = E([g(z)].Y(n))^4$ on the unit ball $\mathcal{B}_p$ allows the deconvolution a negative fourth-order cumulant excitation signal. For this purpose, we need to study the extrema of J on $\mathcal{B}_p$. First, we remark that as $E([g(z)].Y(n))^2 = 1$ if $g\in\mathcal{B}_p$, then the study of J(g) is equivalent to that of the function $g \rightarrow c_4([g(z)].Y(n))$. As in [1] and [7], we put $f(z) = g(z)H(z)$; as H is lossless, the correspondence $g \rightarrow f$ is a unitary operator. Therefore, the study of the extrema of J on $\mathcal{B}_p$ is equivalent to that of the extrema of the function K defined on $\mathcal{B}_1$ by $K(f) = c_4([f(z)].W(n))$. The analytic expression of K is very simple. In fact, if $f(z) = \sum_k f_k z^{-k}$, and if we denote by $(f_{k,l})_{l=1,p}$ the components of the row vector f_k, then we get immediatly that

$$K(f) = \sum_{k\in\mathbf{Z}} \sum_{l=1,p} c_4(W_l)\, f_{k,l}^4 \qquad (2\text{-}1)$$

In order to study the extrema of K on $\mathcal{B}_p$, we follow [7] and we consider the Lagrangian $\mathcal{L}(f,\lambda)$ defined by $\mathcal{L}(f,\lambda) = K(f) - 2\lambda(\|f\|^2 - 1)$.The extrema belong to the set $\mathfrak{T}$ of all points f of $\mathcal{B}_p$ for which there exists λ such that

$$\frac{\partial\mathcal{L}(f,\lambda)}{\partial f_{k,l}} = 0 \quad \forall k\in\mathbf{Z}, \forall l = 1, p. \qquad (2\text{-}2)$$

The following nice result can be seen as a generalization of what is presented in [7]. Due to the lack of space, the proof is omitted.

Proposition 2-1.

Let f be an element of $\mathfrak{T}$. If at least two coefficients $f_{k,l}$ are non zero, then, f is a saddle point of K.

If f is the sequence defined by $f_{k,l} = \delta(k-k_0, l-l_0)$ for $l_0 \le q$, then f is a local minimum of K ; if $l_0 \ge q+1$, then f is a local maximum of K. □

From this, we get immediatly that the local minima of J on $\mathcal{B}_p$ are the filters $z^{-n} H_i^T(z^{-1})$ for $n\in\mathbf{Z}$ and i = 1, q, and that the local maxima of J are the filters $z^{-n} H_i^T(z^{-1})$ for $n\in\mathbf{Z}$ and i = q+1, p. In practice, the deconvolution filters g(z) over which the minimization is performed are finite impulse response filters with a fixed number of coefficients N ; in this case, there is no restriction to assume that they are causal. Therefore, the real problem is that of the adaptive minimization of J over the class of unit norm degree N FIR causal filters, for which no positive result similar to Proposition 2-1 holds. However, Proposition 2-1 suggests that if N is chosen large enough, then the local extrema of the restricted minimization problem should represent reasonnable approximations of the optimal filters $z^{-n}H_i^T(z^{-1})$. Consequently, the stochastic gradient algorithm corresponding to the minimization of J on $\mathcal{B}_p$ should converge to a reasonnably good FIR causal approximation of one of the filters $z^{-n}H_i^T(z^{-1})$ for i = 1, q. Finally, if we denote by $\mathcal{Y}(n)$ the px(N+1) dimensional vector defined by $\mathcal{Y}(n) = (Y^T(n), ..., Y^T(n-p))^T$, let us recall that the most classical stochastic gradient algorithm can be written as

$$g(n+1) = \frac{g(n) - \mu_n\, r(n)\, \mathcal{Y}(n)}{\|g(n) - \mu_n\, r(n)\, \mathcal{Y}(n)\|} \tag{2-3}$$

where $g(n) = (g_0(n),......, g_p(n))^T$ represents the vector of the coefficients of the filter $g(n,z) = \sum_{k=0,N} g_k(n)\, z^{-k}$ at time n and where r(n) is defined as $r(n) = \sum_{k=0,N} g_k(n) Y(n-k)$.

3. THE DEFLATION ALGORITHM.

Before presenting the adaptive deflation algorithm, it is appropriate to recall ([8]) that every 1xp transfer function $V(z) = \sum_{k=0,N} V_k\, z^{-k}$ satisfying $V(z)V^T(z^{-1}) = 1$ admits the lossless lattice realization presented fig. 1, where the $(\theta_k)_{k=1,(N+1)(p-1)}$ are uniquely defined angles belonging to $[-\pi/2, \pi/2]$. Therefore, such a transfer function is parametrized by its corresponding vector $\theta = (\theta_1,..., \theta_M)^T$, where we have put $M = (N+1)(p-1)$. It is also important to note that the (p–1)xp transfer function U(z) defined fig. 1 by $Z_1(n) = [U(z)].Y(n)$ is such that $[U^T(z), V^T(z)]^T$ is lossless.

Suppose that the adaptation scheme presented in section 2 converges to a FIR filter $g_*(z)$ representing a good approximation of a certain filter $z^{-m}H_j^T(z^{-1})$ for $j \leq q$; in this case, g_* will approximatively satisfy $g_*(z)g_*^T(z^{-1}) = 1$. Let us now adapt V(z) so as to minimize $E(z_2(n) - r(n))^2$. Then, after convergence, V(z) will approximate the filter $z^{-m}H_j^T(z^{-1})$, and $z_2(n)$ will "coincides" with $W_j(n-m)$. Therefore, the p–1 dimensional signal $Z_1(n)$ associated to the evolution of V(z) through fig. 1 will converge to a signal uncorrelated with W_j, which will represent the output of a (p–1)x(p–1) lossless filter driven by the (p–1) variate excitation $(W_1(n),...,W_{j-1}(n),W_{j+1}(n),...,W_p(n))^T$. Consequently, it will be possible to use the previously introduced algorithms on $Z_1(n)$. This procedure allows clearly the extraction of the signals W_i for $i \leq q$.

At this point, it is necessary to derive a stochastic gradient algorithm associated to the minimization of $E(z_2(n) - r(n))^2$, or equivalently to the maximization of $E(z_2(n)r(n))$ (recall that $z_2(n)$ and r(n) have a variance equal to 1). For this purpose, let us begin by giving the state-space description of the input-output relation between Y(n) and $(Z_1^T(n), z_2(n))^T$. If X(n) is the N(p–1) dimensional vector defined fig. 1, then it is easily seen that

$$(Z_1^T(n), X^T(n+1), z_2(n))^T = Q(\theta)\, (X^T(n), Y^T(n))^T, \tag{3-1}$$

where $Q(\theta)$ represents the orthogonal matrix given by $Q(\theta) = Q(\theta_1)....Q(\theta_M)$ and where $Q(\theta_k)$ is the Givens rotation

$$Q(\theta_k) = \begin{pmatrix} I_{k-1} & & & & \\ & \cos\theta_k & & -\sin\theta_k & \\ & & I_{M-k} & & \\ & \sin\theta_k & & \cos\theta_k & \end{pmatrix}. \tag{3-2}$$

In particular, if $q(\theta)$ represents the last row of $Q(\theta)$, then $z_2(n) = q(\theta)\, (X^T(n), Y^T(n))^T$. Consequently, the stochastic gradient algorithm can be written as $\theta(n+1) = \theta(n) + \mu_n\, r(n)\, (dz_2(n)/d\theta)_{\theta=\theta(n)}$. But, this algorithm poses two problems. First, the calculation of $(dz_2(n)/d\theta)_{\theta=\theta(n)}$ requires the calculation of the derivative of the state-space signal with respect to each parameter, leading to an enormous computational complexity. Second, the function $E(z_2(n)r(n))$ may admit spurious local maxima, so that the stochastic gradient algorithm cannot be guaranteed to converge to the global maximum of $E(z_2(n)r(n))$. These two remarks lead us to consider an algorithm in which $(dz_2(n)/d\theta)_{\theta=\theta(n)}$ is replaced by $(dq(\theta)/d\theta)_{\theta=\theta(n)}(X^T(n), Y^T(n))$. By using the fact that $(dq(\theta)/d\theta)$ coincides up to a diagonal matrix with positive elements with the M first rows of $Q(\theta)$ [6], and taking into account (3-1), we get that the modified adaptation algorithm can be written as

$$\theta(n+1) = \theta(n) + \mu_n\, r(n) \begin{bmatrix} Z_1(n) \\ X(n+1) \end{bmatrix}. \tag{3-3}$$

Now, we have to study the possible convergence points of the algorithm (3-3). For this purpose, we shall suppose that the limit filter $g_*(z)$ of the adaptation scheme presented in section 2 coincides exactly with the optimal filter $z^{-m}H_j^T(z^{-1})$; in this case, g_* satisfies $g_*(z)g_*^T(z^{-1}) = 1$. Obviously, this will not be exactly true in practice, but we think that the derivation of a positive result concerning the algorithm (3-3) under this hypothesis is sufficient to motivate its utilisation. The possible convergence points of (3-3) are those filters V whose parameters θ satisfy $E_\theta[r(n)((Z_1^T(n), X^T(n+1))] = 0$. Let $(U(z), V(z)) = (D_1 + C_1(zI-A)^{-1}B,\ D_2 + C_2(zI-A)^{-1}B)$ be the minimal state-space representation of the corresponding lossless structure of fig. 1. Then, it is easily seen that $E_\theta[r(n)((Z_1^T(n), X^T(n+1))] = 0$ iff the p(N+1) dimensional row vector $g_* = (g_{*,0}, ..., g_{*,N})$ is orthogonal to the rows of the commandability matrix $(B, AB, ..., A^{N-1}B, 0)$, and to the rows of the matrix $(D_1, C_1B,.....,C_1A^{N-1}B)$. But, by elementary properties of lossless filters, this holds iff g_* belongs to the space generated by the vectors $(V_0, ...,V_N)$, $(0,V_0, ...,V_{N-1})$, ..., $(0, 0,.....,V_0)$. By using the fact that $g_*(z)$ satisfies $g_*(z)g_*^T(z^{-1}) = 1$, it is possible to establish the following result.

Proposition 3-1.

Suppose $g_{*,0} = g_*(\infty)$ is non zero. Then, the stationary points of the algorithm (3-3) correspond to the filters $V(z) = g_*(z)$ and $V(z) = -g_*(z)$. □

This result implies that if the first coefficient of the adaptive filter g(n,z) defined in section 2 does not converge to zero, then the deflation procedure based on the adaptive generation of $Z_1(n)$ will be successfull. On the other hand, if $g_{*,0} = 0$ in Proposition (3-1), then (3-3) may possess another stationary points. This means that the algorithm (3-3) should converge to an indesirable point if $||g(n,\infty)||$ converges to zero ; however, based on simulations, we have observed that it is not the case, so that the spurious stationary points corresponding to the case where $g_{*,0} = 0$ are probably unstable. Nevertheless, the convergence speed of (3-3) is degraded when $||g(n,\infty)||$ converges to zero. A simple solution to overcome this drawback should be to test the value of $||g(n,\infty)||$, and to replace r(n) by r(n+1) in (3-3) if $||g(n,\infty)||$ is sufficiently small.

4. A SIMPLE EXAMPLE.

In order to illustrate what preceds, we consider the following simple example. p is equal to 2, $W_1(n)$ takes the values 1 and –1 with probability 1/2 and $W_2(n)$ is uniformly distributed on the interval $[-\sqrt{3}, \sqrt{3}]$. The observation is the output of the 2x2 FIR filter whose transfer function F(z) is given by $F(z) = (I + F_1 z^{-1})(I + F_2 z^{-1}) Q(\pi/4)$,where $Q(\pi/4)$ is the 2x2 Givens rotation of angle $\pi/4$, and where F_1 and F_2 are the matrices

$$F_1 = \begin{pmatrix} 2.25 & 0.75 \\ 0.75 & 2.25 \end{pmatrix}, \quad F_2 = \begin{pmatrix} 1 & 1 \\ 2 & 1 \end{pmatrix}.$$

The zeros of F are –2, –4, $\sqrt{2}-1$ and $-(\sqrt{2}+1)$; therefore, F has 3 zeros outside the unit disk, from which we deduce that F is non minimum phase. In order to simulate the whitening filter, we have generated the signal [A(z)F(z)].W(n), where A(z) represents the order 10 normalized prediction error filter associated to the 11 first exact correlation coefficients of the observations. The degrees of the filters g(z) and V(z) are both equal to 7. The figure 2 represents the evolution of the 4 first angles corresponding the adaptation of V(z), and the figure 3 represents the signal $z_2(n)$ after convergence of the filter V(z).

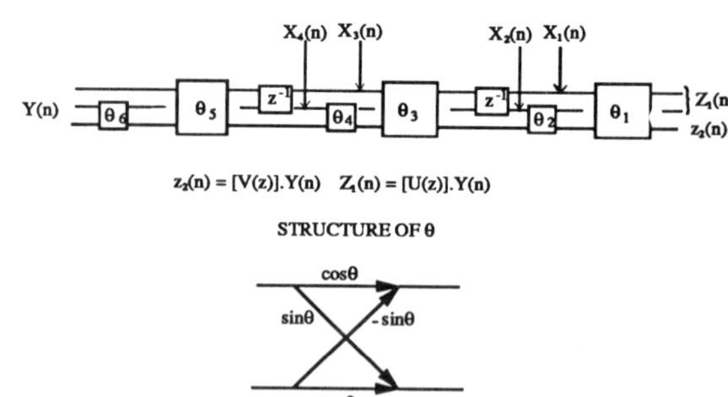

Fig. 1. Lossless Structure for p=3 and N=2

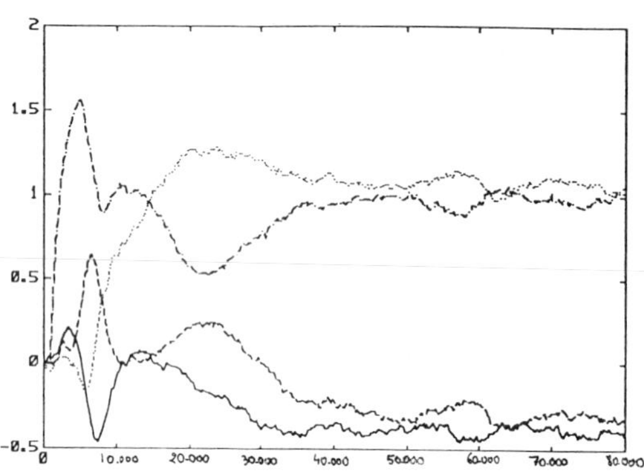

Figure 2

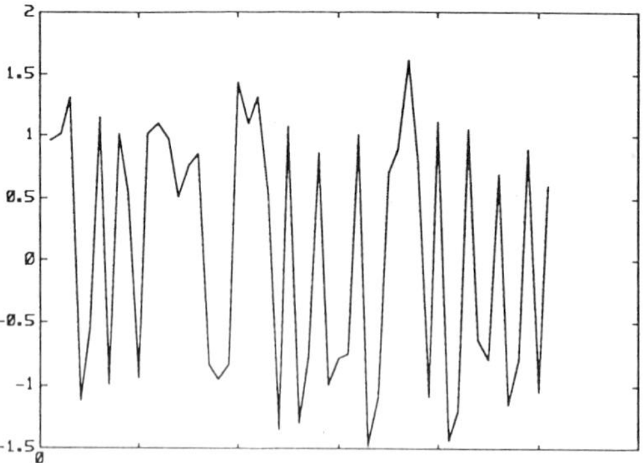

Figure 3

REFERENCES.

[1] A. Benveniste, M. Goursat, G. Ruget, "Robust identification of nonminimum phase system ", IEEE Trans. Automatic Control, vol.25, n°3, 1980, pp. 385-399.

[2] J.F. Cardoso, "Source separation using higher-order moments", Proc. ICASSP-89, pp. 2109-2112.

[3] P. Comon, " Independent component analysis", Proc. International Workshop on higher order statistics, Chamrousse, July 10-12 1991, pp. 111-120.

[4] D. Donoho, " On minimum entropy deconvolution", in Applied Time Series Analysis II, Academic Press, 1981, pp. 565-608.

[5] D. Findley, "The uniqueness of moving average representations with iid random variables for non Gaussian stationary time-series", Biometrika, 73, 1986, pp. 520-521.

[6] P.A. Regalia, "Stable and efficient lattice algorithms for adaptive IIR filtering", IEEE Transactions on Signal Processing, vol. 40, n°2, February 1992, pp. 375-388.

[7] O. Shalvi, E. Weinstein, " New criteria for blind deconvolution of nonminimum phase systems (channels)", IEEE Transactions on Information Theory, vol. 36, n°2, March 90, pp. 312-321.

[8] P.P. Vaidyanathan, "Passive cascaded-lattice structures for low-sensitivity FIR filter design, with applications to filter banks", IEEE Transactions on Circuit and Systems, vol. 33, November 1986, pp. 1045-1064.

Relation between reduced dimension time and frequency domain adaptive algorithm

G.P.M. Egelmeers P.C.W. Sommen
Eindhoven University of Technology
P.O. Box 513, 5600 MB Eindhoven

Abstract

Convergence characteristics of update algorithms for adaptive filters are influenced by statistical properties of the input signal. This dependency can by removed by decorrelation of the input signal. Usually this decorrelation is performed, in time- or frequency-domain, by using N degrees of freedom, with N the number of adaptive weights. In literature also techniques are available that can perform this decorrelation with M ($\leq N$) degrees of freedom, where the number M can be chosen more properly in resemblance to the statistical properties of the input signal. These techniques are the Partitioned Block Frequency Domain Adaptive Filter (PBFDAF) in frequency domain and the Block Orthogonal Projection (BOP) algorithm in time domain. The aim of this paper is to investigate the relationship between these two algorithms. One of the main novel results is that decorrelation is performed in PBFDAF with low complexity but not in accordance with the BOP approach.

1 Introduction

In this paper the signal estimation scheme, as depicted in fig. 1, is used as a vehicle. This is a form of system identification problem, from which the system to be identified is the optimal Wiener solution for the adaptive filter. It is assumed that all signals are discrete in time, having an inter sample distance of T seconds, and furthermore that the unknown system H has an impulse response $\underline{h}$, that can be modelled by a transversal structure with N coefficients h_i for $i = 0, \ldots, N-1$. The input signal $x[k]$ produces via the unknown system H a signal $e[k]$, that has to be estimated by the adaptive filter. A signal $s[k]$, uncorrelated with $x[k]$, is added to $e[k]$ resulting in the measurable signal $\tilde{e}[k]$. The adaptive filter makes a replica $\hat{e}[k]$ of the signal $e[k]$. Theoretically the residual signal $r[k] = \tilde{e}[k] - \hat{e}[k]$ in steady state will become equal, in average, to the signal $s[k]$.

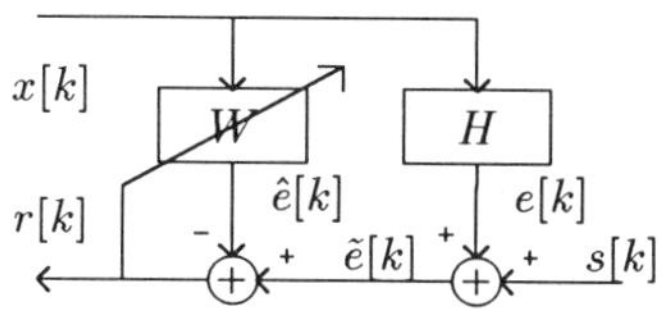

Figure 1: *General signal estimation scheme*

From literature [1] it is known that convergence behavior of adaptive filters is strongly influenced by statistical properties of the input signal. This dependency can be removed by decorrelating the input signal.

In frequency domain this decorrelation is performed by normalizing the power of separate frequency components resulting in a Block Frequency Domain Adaptive Filter (BFDAF). One of the problems with this technique is the large processing delay for large N. In order to reduce this delay by a factor $Q = N/M$ the impulse responce of the adaptive filter can be partitioned. This leads to the PBFDAF algorithm [2], where decorrelation is performed by normalizing over less frequency components.

On the other hand a well known method to decorrelate the input signal in time domain is the Recursive Least Squares (RLS) algorithm. This method uses an estimate of the inverse of the MxM autocorrelation matrix of the input signal, with the dimension M equal to the number of adaptive weights N. The complexity of the RLS method can be reduced by choosing the dimension M in more resemblance to the statistical properties of the input signal, resulting in the BOP algorithm [3].

A strong relationship between the PBFDAF and BOP algorithm is expected. In order to investigate this relationship this paper is organized as follows: Section 2 describes the PBFDAF algorithm. In Section 3 the BOP algorithm is first rewritten and the result is transformed to frequency domain. From this result it follows that decorrelation of BOP and PBFDAF is sligthly different. In Section 4 the decorrelation methods of BOP and PBFDAF are compared. The results are verified by simulations in Section 5.

2 PBFDAF algorithm

In order to reduce the processing delay of a BFDAF by a factor $Q = N/M$, the adaptive weigth vector is partitioned into Q vectors of dimension M. Each of these smaller vectors is updated by separate BFDAFs. Combining Discrete Fourier Transforms (DFTs) and performing the delay between different parts of the impulse response and their summation in frequency domain, results in a PBFDAF structure [2] as depicted in fig. 2.

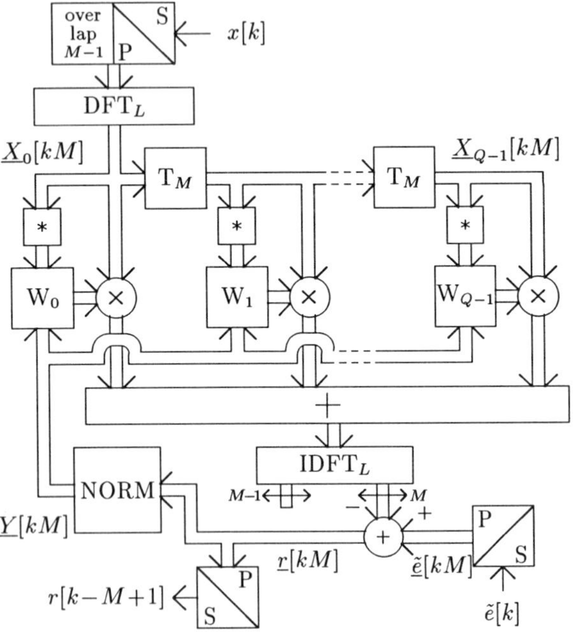

Figure 2: *Partitioned structure:* $L = 2M-1$, $Q = N/M$

In order to perform the overlap-save method with block processing techniques, a Serial/Parallel (S/P) convertor fetches the input signal samples with an overlap of $M-1$ samples. Once every M samples a block of $L = 2M-1$ input signal samples is transformed to frequency domain resulting in a vector $\underline{X}_0[kM]$:

$$\underline{X}_0[kM] = \mathcal{F} \cdot (x[kM-L+1], \ldots, x[kM])^T \tag{1}$$

with $(\cdot)^T$ the transpose of a vector and $\mathcal{F}$ the $L \times L$ DFT matrix, from which the $(p,q)^{\text{th}}$ element is defined as $(\mathcal{F})_{p,q} = e^{-j2\pi pq/L}$. The vector $\underline{X}_0[kM]$ is input to a length $Q-1$ delay line, having delays of $T_M = M \cdot T$ seconds. The convolution of each separate part of the weigth vector with the corresponding part of the input signal vector is performed in frequency domain by multiplying $\underline{X}_i[kM]$ with $\underline{W}_i[kM]$ for $i = 0, \ldots, Q-1$. Furthermore the convolution of the complete weigth vector with the input signal vector is given by summing up all these multiplications and transforming this result back to time domain with an inverse DFT.

Substracting the last M values of this result from the vector $\underline{\tilde{e}}[kM]$ yields the residual signal vector $\underline{r}[kM] = (r[kM-M+1], \ldots, r[kM])^T$. This vector is converted by a Parallel/Serial (P/S) convertor to the residual signal samples $r[k-M+1]$, so the processing delay is $M-1$ samples.

To perform decorrelation, as depicted in fig. 3, the residual signal vector $\underline{r}[kM]$ is extended with $M-1$ zeros and transformed to frequency domain as:

$$\underline{R}[kM] = \mathcal{F} \cdot \begin{pmatrix} \mathbf{0} \\ \mathbf{I}_M \end{pmatrix} \cdot \underline{r}[kM] \tag{2}$$

with $\mathbf{I}_M$ the $M \times M$ identity matrix and $\mathbf{0}$ a matrix containing all zeros having appropriate dimension. The frequency domain vector $\underline{R}[kM]$ is multiplied by the adaptation constant $2\alpha/M$, and each frequency component l is multiplied by the lth component of the input signal power vector $\underline{P}^{-1}[kM]$. This l^{th} element equals $L/\mathcal{E}\{|(\underline{X}_0[kM])_l|^2\}$, with $\mathcal{E}$ the mathematical expectation. The normalization yields the L dimensional vector

$$\underline{Y}_P[kM] = \frac{2\alpha}{M}\underline{P}^{-1}[kM] \odot \underline{R}[kM] \tag{3}$$

as depicted in figure 3, where the operator $\odot$ is defined as an element by element multiplication of its two operand vectors.

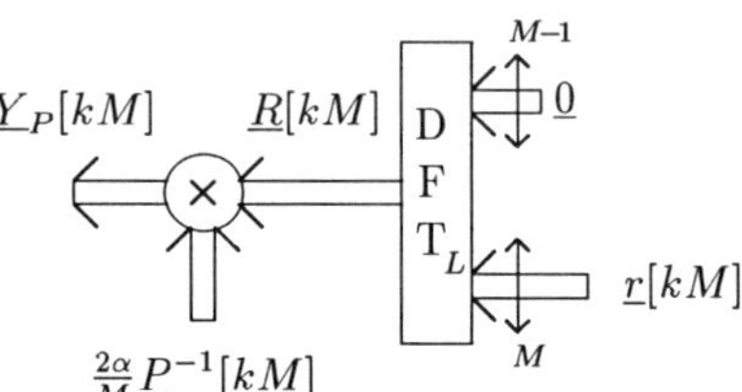

Figure 3: *Normalization of PBFDAF*

This vector $\underline{Y}_P[kM]$ is used for the update of each weight vector $\underline{W}_i[kM]$ as depicted in fig. 4. Note that in fig. 2 and fig. 4 $\underline{Y}[kM] = \underline{Y}_P[kM]$. These figures will be used later on for the BOP approach with $\underline{Y}[kM] = \underline{Y}_B[kM]$. The estimate of the gradient for each part i of the filter is calculated by multiplying the vector $\underline{Y}_P[kM]$ with the conjugate $((\cdot)^*)$ of the (delayed) input signal vector $\underline{X}_i[kM]$. In order to obtain a correct overlap-save result a window $\mathbf{V}$ is applied to this product, defined as:

$$\mathbf{V} = \mathcal{F} \cdot \begin{pmatrix} \mathbf{I}_M & \mathbf{0} \\ \mathbf{0} & \mathbf{0} \end{pmatrix} \mathcal{F}^{-1}. \tag{4}$$

Finally the update of each adaptive weigth vector $\underline{W}_i[kM]$ for $i = 0, \ldots, Q-1$ is given by:

$$\underline{W}_i[(k+1)M] = \underline{W}_i[kM] + \mathbf{V} \cdot (\underline{X}_i^*[kM] \odot \underline{Y}_P[kM]). \tag{5}$$

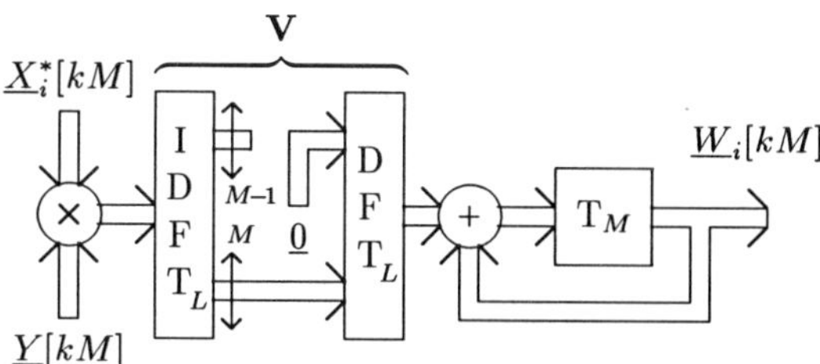

Figure 4: *Update part* W_i

3 BOP algorithm

In [3] it is shown that the BOP algorithm uses, in contrast to the RLS method, an $M \times M$ matrix $\breve{\mathbf{R}}[kM]$ to decorrelate the input signal. Furthermore, as a result of the block processing approach, the adaptive weight vector $\underline{w}[kM] = (w_{N-1}[kM], \ldots, w_0[kM])^T$ is updated only once every M samples according to:

$$\underline{w}[(k+1)M] = \underline{w}[kM] + \frac{2\alpha}{M}\chi[kM] \cdot \breve{\mathbf{R}}^{-1}[kM] \cdot \underline{r}[kM]. \quad (6)$$

In this update equation the input signal samples are collected in the $N \times M$ matrix

$$\chi[kM] = (\underline{x}[kM-M+1], \ldots, \underline{x}[kM]) \quad (7)$$

containing M vectors of dimension N defined for $i = 0, \ldots, M-1$ as:

$$\underline{x}[kM-i] = (x[kM-i-N+1], \ldots, x[kM-i])^T. \quad (8)$$

Furthermore the estimate of the $M \times M$ autocorrelation matrix is defined as:

$$\breve{\mathbf{R}}[kM] = \frac{1}{N}\chi^t[kM] \cdot \chi[kM]. \quad (9)$$

The aim of this paper is to investigate the relationship between the BOP and PBFDAF algorithm. For this the N dimensional adaptive weight vector $\underline{w}[kM]$ is first partitioned in $Q = N/M$ separate weight vectors $\underline{w}_i[kM] = (w_{(i+1)M-1}[kM], \ldots, w_{iM}[kM])^T$, of dimension M. The update equation (6) can now be rewritten for $i = 0, \ldots, Q-1$ as:

$$\underline{w}_i[(k+1)M] = \underline{w}_i[kM] + \frac{2\alpha}{M}\chi_i[kM]\breve{\mathbf{R}}^{-1}[kM]\underline{r}[kM] \quad (10)$$

with the "partitioned" $M \times M$ matrix $\chi_i[kM]$ containing the appropriate part of the $N \times M$ input signal matrix $\chi[kM]$. In order to transform this update equation to frequency domain the following two steps are used:

1. With $L = 2M - 1$ the $L \times L$ matrix $\chi_i^c[kM]$ is defined as a mirrored circulant expansion of the $M \times M$ matrix $\chi_i[kM]$ and these matrices are related as:

$$\chi_i[kM] = \mathbf{J}_M \cdot \begin{pmatrix} \mathbf{I}_M & \mathbf{0} \end{pmatrix} \cdot \chi_i^c[kM] \cdot \begin{pmatrix} \mathbf{0} \\ \mathbf{I}_M \end{pmatrix} \quad (11)$$

   with $\mathbf{J}_M$ the $M \times M$ mirror matrix containing only ones on the mirrored main diagonal, thus $(\mathbf{J}_M)_{p,q} = 1$ for $|p-q| = M-1$.

2. The $L \times L$ circulant matrix $\chi_i^c[kM]$ can be diagonalized with the DFT matrix $\mathcal{F}$ as follows:

$$\mathcal{F}^{-1} \cdot diag\{\underline{X}_i^*[kM]\} \cdot \mathcal{F} = \chi_i^c[kM] \quad (12)$$

   with $diag\{\cdot\}$ representing the diagonal matrix with as diagnonal the vector between the curly brackets.

Using these steps it is possible to transform the update equation (10) to the frequency domain. The transformation is constructed as to obtain an equal residual signal vector $\underline{r}[kM]$ for both algorithms. This results in the following transformation $\mathbf{T}$:

$$\mathbf{T} = \mathcal{F} \cdot \begin{pmatrix} \mathbf{I}_M \\ \mathbf{0} \end{pmatrix} \cdot \mathbf{J}_M. \quad (13)$$

Applying $\mathbf{T}$ to both sides of (10) results in the following update equation:

$$\underline{W}_i[(k+1)M] = \underline{W}_i[kM] + \mathbf{V} \cdot (\underline{X}_i^*[kM] \odot \underline{Y}_B[kM])$$
$$\underline{Y}_B[kM] = \frac{2\alpha}{M}\mathcal{F} \cdot \begin{pmatrix} \mathbf{0} \\ \mathbf{I}_M \end{pmatrix} \breve{\mathbf{R}}^{-1}[kM]\underline{r}[kM]. \quad (14)$$

From these equations it follows that the BOP algorithm can be implemented with the structure as depicted in fig. 2 using the update as given in fig. 4. The difference between the BOP and PBFDAF algorithm is the way of normalization to obtain vector $\underline{Y}_B[kM]$.

4 Normalization methods

In this section the normalization methods of the BOP and the PBFDAF algorithm are compared to obtain insight into their differences. The PBFDAF normalized vector $\underline{Y}_P[kM]$ is obtained by a component-wise multiplication in frequency domain according to eq. 3. Such a component-wise multiplication is the frequency-domain equivalent of the product of a circulant matrix and a vector, so:

$$\underline{Y}_P[kM] = \frac{2\alpha}{M}\mathcal{F} \cdot \mathbf{R}_P^{-1}[kM] \cdot \begin{pmatrix} \mathbf{0} \\ \mathbf{I}_M \end{pmatrix} \cdot \underline{r}[kM] \quad (15)$$

with $\mathcal{F} \cdot \mathbf{R}_P^{-1}[kM] \cdot \mathcal{F}^{-1} = diag\{\underline{P}^{-1}[kM]\}$.

The $L \times L$ circulant pseudo-autocorrelation matrix $\mathbf{R}_P[kM]$ contains the (pseudo-)autocorrelation vector $\underline{\rho}_P[kM] = (\rho_P[0], \ldots, \rho_P[L-1])^T$. The elements of $\underline{\rho}_P[kM]$ are given by $\rho_P[\tau] = \frac{L-\tau}{L}\rho[\tau] + \frac{\tau}{L}\rho[L-\tau]$ with $\rho[\tau] = \mathcal{E}\{x[k]x[k-\tau]\}$.

Equation 14 can now be written as:

$$\underline{Y}_B[kM] = \frac{2\alpha}{M}\mathcal{F} \cdot \mathbf{R}_P^{-1}[kM] \cdot \begin{pmatrix} \tilde{\mathbf{0}} \\ \tilde{\mathbf{I}}_M \end{pmatrix} \cdot \underline{r}[kM] \quad (16)$$

$$\begin{pmatrix} \tilde{\mathbf{0}} \\ \tilde{\mathbf{I}}_M \end{pmatrix} = \mathbf{R}_P[kM] \cdot \begin{pmatrix} \mathbf{0} \\ \mathbf{I}_M \end{pmatrix} \cdot \breve{\mathbf{R}}^{-1}[kM]. \quad (17)$$

The difference between the normalization of BOP and PBFDAF is concentrated in equation 17. BOP normalization uses this equation, while PBFDAF approximates the matrix $\tilde{\mathbf{I}}_M$ by the $M \times M$ identity matrix $\mathbf{I}_M$ and $\tilde{\mathbf{0}}$ by the zero matrix $\mathbf{0}$.

5 Simulation results

In order to gain insight into the magnitude of the approximation made in the PBFDAF case of the previous section, both algorithms were simulated. Some typical results are presented in figure 5. As input an Auto-Regressive signal of order 1 (AR1) signal is used: $x[k] = 0.3122 \cdot n[k] + 0.95 \cdot x[k-1]$ with $n[k]$ a zero mean (white) noise signal with variance equal to 1. The additive noise signal $s[k]$ is also (zero mean), but with variance 0.01. The adaptive filter length N is 32 and the adaptian constant $\alpha = 0.0001$. Choosing the parameter $M = 1$ results in the BNLMS algorithm (no decorrelation), while choosing $M = N$ and using the BOP algorithm results in the IDEAL case (full decorrelation). Values of M between 1 and N yield no relevant differences between the BOP and PBFDAF algorithm. The case of $M = 4$ is depicted in figure 5 for both algorithms.

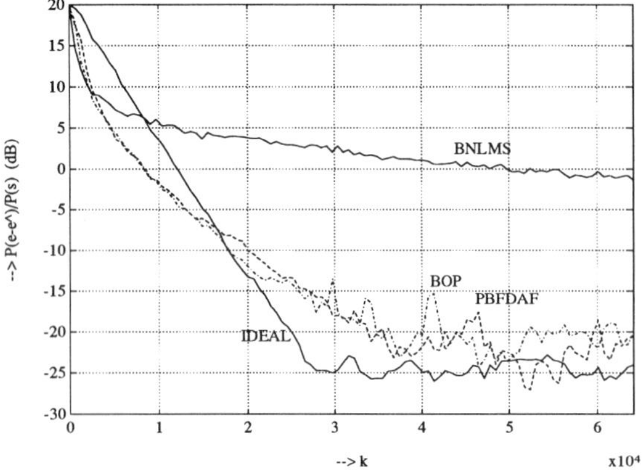

Figure 5: *Results BOP and PBFDAF*

6 Conclusions

The approximation of PBFDAF with respect to BOP (section 4) does not seem to introduce relevant deviations in the the convergence curves of the PBFDAF algorithm compared to the BOP algorithm. As the complexity of the PBFDAF algorithm is for most pratical cases smaller than the complexity of the BOP algorithm, the first one seems to be preferable. However, as the fraction M has to be a divisor of the total adaptive filter length N in the PBFDAF case, it is not always possible to replace the BOP algorithm by a PBFDAF algorithm with equal parameters. More investigations of the exact nature of the difference between both algorithms have to be carried out in future.

References

[1] Haykin S., "Adaptive Filter Theory", USA, Englewood Cliffs (New Jersey): Prentice Hall, 1986, ISBN 0-13-004052-5 025.

[2] Soo J.S. and Pang K.K., "A new structure for Block FIR Adaptive Digital Filters", Proc. IEEE Conf., Sydney, sept. 1987, pp 364- 367.

[3] Sommen P.C.W., "On the Orthogonal Projection method for Adaptive Filters", Proc. Sym. MTNS-89, vol.: 3, Basel: Birkhäuser, 1990, Progress in systems and control theory 5, pp 283-290.

SIGNAL PROCESSING VI: Theories and Applications
J. Vandewalle, R. Boite, M. Moonen, A. Oosterlinck (eds.)

AN ALTERNATIVE SCHEME FOR ADAPTIVE NONLINEAR PREDICTION USING RADIAL BASIS FUNCTIONS

JM Vesin
Laboratoire de Traitement des Signaux
Ecole Polytechnique Fédérale de Lausanne, 1015 Lausanne Switzerland

Abstract. Radial basis functions (RBF) constitute a valuable alternative to polynomials and multi-layer perceptrons in the context of nonlinear signal prediction. They are characterized by the fact that they are linear in some of their parameters (the external parameters), and nonlinear in the other ones (the internal parameters). This paper describes an adaptive RBF prediction scheme, based on LMS, but incorporating a recursive form of the internal parameter selection technique used in batch mode prediction.

1. Introduction

There has been a recent interest in the nonlinear predictive analysis of time series, for instance in domains such as speech processing and the study of chaotic signals. Although polynomials possess the universal approximation capability, their use in nonlinear prediction presents some problems such as lack of stability and difficulties in the choice of the relevant terms [1]. Some authors [2, 3] have proposed the use of neural networks, mainly in the context of multilayer perceptrons (MLP) equipped with the well-known back-propagation algorithm, due to the fact that they too can approximate any continuous multidimensional function. However, one of the major drawbacks of this scheme is its very slow speed of convergence. More recently Radial Basis Functions (RBF), which have been used for some time in the context of multidimensional interpolation [4], and possess too the universal approximation capability [5], have attracted attention for nonlinear prediction purposes, for the first time to our knowledge in [6] on chaotic time series and speech data, and in [7] on speech data. In this last paper, a significant improvement upon linear prediction was reported in an experiment. Known procedures exist in the literature to select the internal parameters of the RBF so that their amplitude parameters can be estimated in a least squares sense in batch mode prediction. However, poor speed of convergence is observed in the case of adaptive prediction if a nonlinear LMS scheme [8] is used, due to the fact that the internal and amplitude parameters get mixed in the update equations. We first summarize briefly in this paper some known results about RBF, then we present an adaptive scheme which includes the advantages of RBF internal parameter selection. It is shown on real speech data to provide a faster convergence with a smaller computation load when compared to a classical nonlinear LMS algorithm.

2. RBF parameter selection

$\{x_n\}$, $n = 1, ..., N$, being the time series of interest, and defining $X_{n-1} = [x_{n-1}, x_{n-2}, ..., x_{n-p}]^T$, the prediction estimate $\hat{x}_n$ of x_n of order p with a RBF network will be expressed as

$$\hat{x}_n = f(X_{n-1}) = \sum_{i=1}^{m} \lambda_i f_i(\|X_{n-1} - C_i\|) \qquad (1)$$

that is, as the weighted sum of nonlinear functions of the Euclidian distance of the data vector X_{n-1} to a set of predetermined vectors (centers) $\{C_i\}$, $i = 1, ..., m$. Several choices are possible for the functions $f_i(.)$ [9]. The most common one is

$$f_i(r) = \exp\left(-\frac{r^2}{\beta_i^2}\right) \qquad (2)$$

If prediction is done in block mode, one generally tries to minimize the squared error E over N samples, that is

$$E = \sum_{j=1}^{N} (x_j - \hat{x}_j)^2 \qquad (3)$$

In [7] the estimation of the parameters $\{\lambda_i\}$, $\{\beta_i\}$, and $\{C_i\}$, $i = 1, ..., m$, was performed by a gradient descent over E. But due to the nonlinear dependence of E on the $\{\beta_i\}$ and $\{C_i\}$, convergence to a global minimum is not assured. A better strategy [9] seems to select first the centers $\{C_i\}$ using a data-dependent clustering algorithm such as Kohonen self-organizing procedure [11] which in its simplest form can be expressed as

$$C_k(n) = C_k(n-1) + \alpha(n)[X_{n-1} - C_k(n-1)] \qquad (4)$$

where the center C_k at step $(n-1)$ is such that

$$||X_{n-1} - C_k|| = \min_i \{||X_{n-1} - C_i||\} \quad (5)$$

and $\alpha(n)$ is a small coefficient decreasing with n. The operation described by eq. (4) and (5) can be re-iterated on the set of data until convergence is obtained. Once the centers are selected, the $\{\beta_i\}$ parameters can be chosen in one of the two following ways: 1) if $\beta_i = \beta$ for all $i = 1, ..., m$, then a convenient rule-of -the-thumb value for β is [12]

$$\beta = \max_{i,j}\{||C_i - C_j||\} \quad (6)$$

2) otherwise, the $\{\beta_i\}$ can be obtained by minimizing with respect to them the following quantity [10]

$$D = \sum_{i=1}^{m} \left[\sum_{j=1}^{m} \left(\frac{||C_i - C_j||}{\beta_i} \right)^2 f_i(||C_i - C_j||) - P \right]^2 \quad (7)$$

with P an overlap parameter. The minimization of E in eq.(3) with respect to the $\{\lambda_i\}$ can then be performed in an optimal way using an efficient method such as the singular value decomposition. We have found experimentally by Monte-Carlo tests on real and simulated data (results are not shown here due to the lack of space) that another good choice for the $\{\beta_i\}$ is

$$\beta_i = \frac{1}{N} \sum_{j=1}^{N} ||X_j - C_i|| \quad (8)$$

Intuitively, this formula is justified by the fact that the clustering algorithm described by eqs. (5) and (6) will concentrate the centers in the vicinity of the modes of the probability density of $\{X_n\}$. The range of action, parametrized by the β_i, of the functions $f_i(.)$ corresponding to these clustered centers should be smaller in order not to "blur" their contributions in that part of the state space. Eq. (8) expresses this dependence of the $\{\beta_i\}$ upon the relative location of the $\{C_i\}$ with respect to the data vectors $\{X_n\}$.

3. Proposed adaptive algorithm

For the time being, let us denote the global parameter vector at time n by $\theta(n) = [\lambda_1, ..., \lambda_m, \beta_1, ..., \beta_m, c_{11}, ..., c_{1p}, ..., c_{mp}]^T$, where c_{jk} (dependence on n not shown) represents the jth coordinate of the center C_k. In a classical LMS frame, the objective is to minimize $\mathbf{E}\{(\varepsilon_n)^2\}$, where $\varepsilon_n = x_n - f(X_{n-1})$. The gradient of this statistical quantity is replaced by its instantaneous estimate, namely

$$\nabla(n) = \frac{\partial(\varepsilon_n)^2}{\partial\theta(n)} = 2\,\varepsilon_n \frac{\partial\varepsilon_n}{\partial\theta(n)} = -2\,\varepsilon_n \frac{\partial f(X_{n-1})}{\partial\theta(n)} \quad (9)$$

and the parameter vector update equation is

$$\theta(n+1) = \theta(n) - \mu\nabla(n) = \theta(n) + 2\mu\varepsilon_n \frac{\partial f(X_{n-1})}{\partial\theta(n)} \quad (10)$$

with μ a coefficient governing the rate of convergence.

Let us split now $\theta(n)$ into two vectors $\lambda(n) = [\lambda_1, ..., \lambda_m]^T$ and $\gamma(n) = [\beta_1, ..., \beta_m, c_{11}, ..., c_{1p}, ..., c_{mp}]^T$. The idea is to update $\gamma(n)$, that is the internal parameters of the RBF, using recursive schemes corresponding to the "good" choices presented in the section **2**. Thus only the center C_k closest to X_{n-1} will be updated with eq.(4), which is already in a recursive form (now a constant update coefficient, α_C, instead of $\alpha(n)$, will be employed). If a unique parameter β is considered, it is then updated using eq.(6). Note that only the distances involving the updated center C_k have to be re-computed and compared to the old value of β. If different β_i are considered, eq.(8) is easily put in the following recursive form

$$\beta_i(n+1) = (1 - \alpha_\beta)\,\beta_i(n) + \alpha_\beta\,||X_{n-1} - C_i|| \quad (11)$$

with $0 < \alpha_\beta << 1$. The objective is still to minimize $\mathbf{E}\{(\varepsilon_n)^2\}$, but now $\varepsilon_n = x_n - [f(X_{n-1}) + \Delta f(X_{n-1})]$, where the term $\Delta f(X_{n-1})$ stands for the modification of $f(.)$ due to the modification $\Delta\gamma(n) = \gamma(n+1) - \gamma(n)$. Using

$$\Delta f(X_{n-1}) \approx (\Delta\gamma(n))^T \frac{\partial f(X_{n-1})}{\partial\gamma(n)} \quad (12)$$

the instantaneous gradient estimate becomes:

$$\nabla(n) = \frac{\partial(\varepsilon_n)^2}{\partial\lambda(n)} = 2\,\varepsilon_n \frac{\partial\varepsilon_n}{\partial\lambda(n)}$$
$$= -2\,\varepsilon_n \left[\frac{\partial f(X_{n-1})}{\partial\lambda(n)} + (\Delta\gamma(n))^T \frac{\partial^2 f(X_{n-1})}{\partial\gamma(n)\partial\lambda(n)} \right] \quad (13)$$

and the update equation for $\lambda(n)$ remains:

$$\lambda(n+1) = \lambda(n) - \mu\,\nabla(n) \quad (14)$$

This scheme can be interpreted in the following way: If the centers $\{C_i\}$ and the parameters $\{\beta_i\}$ were held constant ($\Delta f(X_{n-1}) = 0$), then eqs. (13) and (14) would correspond to a classical LMS scheme on $\lambda(n)$, that is, to an approximate gradient descent on a paraboloid. Since $\{C_i\}$ and $\{\beta_i\}$ vary, the paraboloid moves slowly and the additional term in the right-hand side of eq.(13) constitutes a correction to the update value taking account of this displacement. Note that, due to the fact that $\Delta\gamma(n)$ is a sparse vector and that the coordinates of one center only have to be updated, eq. (14) necessitates a lot less computations than eq. (10) does.

4. Experimental illustration

Nonlinear prediction with a RBF network has been performed on two speech signals sampled at 20 KHz corresponding to the sounds /*f*/ and /*a*/. These signals are shown respectively in figure 1 and 3.
For both of them, a prediction order p=3 was chosen by inspection of the singular values of their estimated covariance matrix. In all experiments a RBF network of m=10 functions was used. The same value μ=0.2 was used for the gradient coefficient in the standard LMS scheme (eq. (10)) and in the proposed scheme (eq. (14)).
In all experiments the center update coefficient was taken to be α_C=0.2. In the case of multiple $\{\beta_i\}$ the update coefficient value was α_β =0.02.
In figures 2, 4, and 5 the prediction error signal obtained with the classical LMS scheme is represented by a dashed line and the prediction error signal obtained with our method by a solid line.

Figure 2 show the prediction error signals corresponding to the signal of figure 1 with a single parameter value β.
Figure 4 shows the prediction error signals corresponding to the signal of figure 3 with a single parameter value β.
Figure 5 shows the prediction error signals corresponding to the signal of figure 3 with multiple parameter values $\{\beta_i\}$
Faster convergence and smaller residual error for our algorithm are clearly visible in all examples.

5. Conclusion

We have presented here an adaptive algorithm for time series prediction using RBF that incorporates an interesting feature of these functions in block mode prediction, namely the possibility to separate the computation of their external and internal parameters. It was shown on examples that this algorithm leads to faster convergence and lower computational load when compared to classical LMS. An interesting improvement would be a strategy to switch off the internal parameter update when the prediction power is below a fixed threshold.

References

[1] V.John Mathews, "Adaptive Polynomial Filters," *IEEE Signal Proc. Mag.*, Vol. 8, No. 3, July 1991, pp. 10-26.
[2] K.S. Narendra and K. Parthasarathy, "Identification and Control of Dynamical Systems Using Neural Networks," *IEEE Trans. on Neural Networks*, Vol. 1, No 1, 1990, pp. 4-27.
[3] H. Reininger and D. Wolf, "Nonlinear Prediction of Stochastic Processes Using Neural Networks," *Proc. EUSIPCO-90*, Barcelona, Spain, Sept. 1990, pp. 1623-1626.
[4] M.I.D. Powell, "Radial Basis Functions for Multivariable Interpolation: a Review," *IMA Conference on Algorithms for the Approximation of Functions and Data*, RMCS, Shrivenham, 1985.
[5] J. Park and I.W. Sandberg, "Universal Approximation Using Radial-Basis-Function Networks," *Neural Computation*, Vol. 3, No. 2, 1991, pp. 246-257.
[6] D. Lowe and A. Webb, "Adaptive Networks, Dynamical Systems, and the Predictive Analysis of Time Series," *Proc. 1st IEE Int. Conf. on Artificial Neural Networks*, London, UK, Oct. 1989, pp. 95-99.
[7] M. Niranjan and V. Kadirkamanathan, "A Nonlinear Model for Time Series Prediction and Signal Interpolation," *Proc. ICASSP-91*, Toronto, Canada, May 1991, pp. 1713-1716.
[8] B. Widrow and M.A. Lehr, "30 Years of Adaptive Neural Networks: Perceptron, Madaline, and Backpropagation," *Proc. IEEE*, Vol. 78, No. 9, Sept. 1990, pp. 1415-1442.
[9] T. Poggio and F. Girosi, "Networks for Approximation and Learning," *Proc. IEEE*, Vol. 78, No. 9, Sept. 1990, pp. 1481-1497.
[10] P. Downes, "Application of Chaotic time Series Analysis to signal Characterization," *Proc. ICASSP-91*, Toronto, Canada, May 1991, pp. 3129-3132.
[11] T. Kohonen, "The Self-Organizing Map," *Proc. IEEE*, Vol. 78, No. 9, Sept. 1990, pp.1464-1480.
[12] S. Chen, G.J. Gibson, C.F.N. Cowan and P.M. Grant, "Reconstruction of binary Signals Using an Adaptive Radial-Basis-Function Equalizer," *Signal Processing*, Vol. 22, 1991, pp. 77-93.

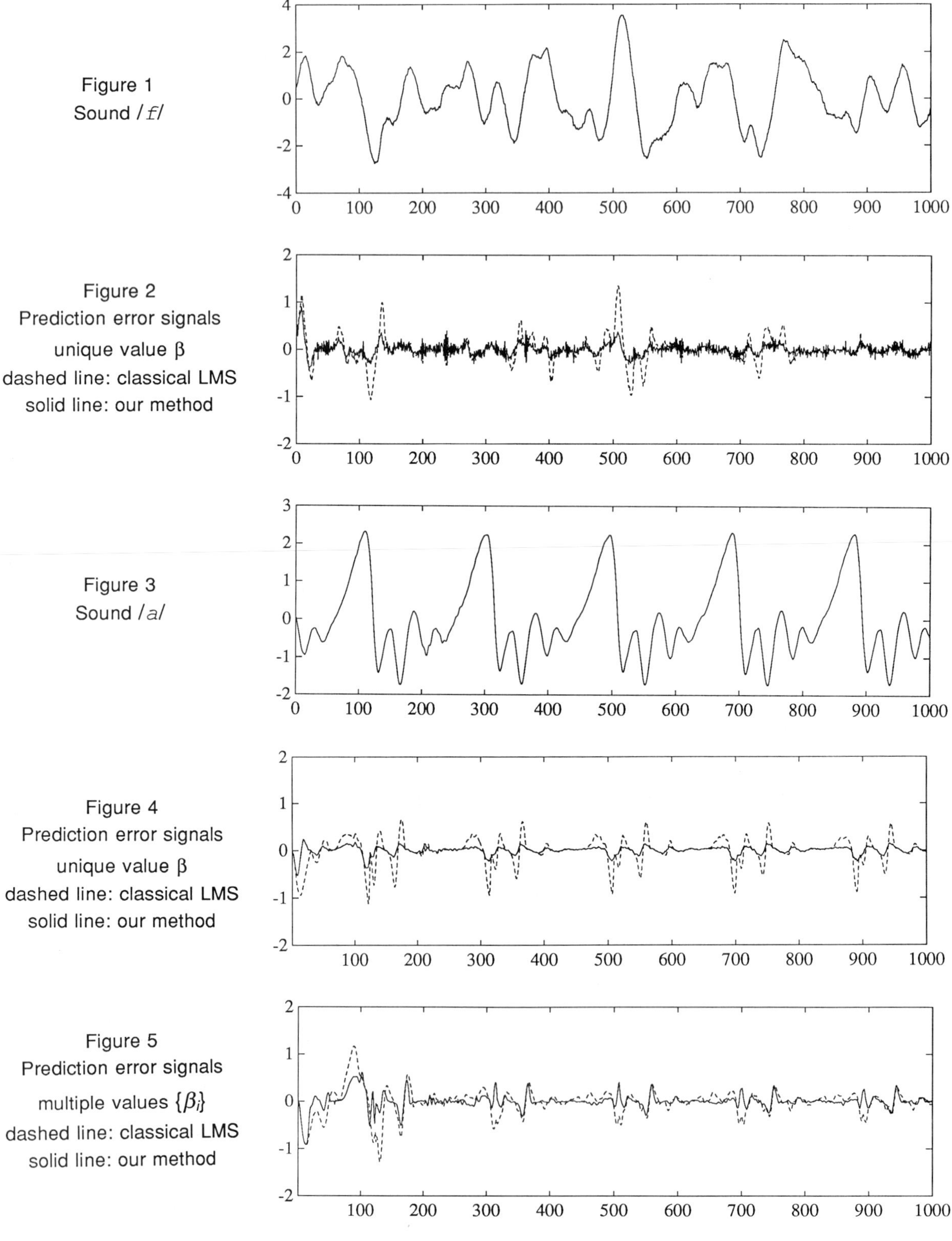

Figure 1
Sound /*f*/

Figure 2
Prediction error signals
unique value β
dashed line: classical LMS
solid line: our method

Figure 3
Sound /*a*/

Figure 4
Prediction error signals
unique value β
dashed line: classical LMS
solid line: our method

Figure 5
Prediction error signals
multiple values $\{\beta_i\}$
dashed line: classical LMS
solid line: our method

SIGNAL PROCESSING VI: Theories and Applications
J. Vandewalle, R. Boite, M. Moonen, A. Oosterlinck (eds.)

Convergence analysis of stochastic gradient adaptive algorithms for arbitrary error surfaces

José M. Páez Borrallo and Francisco J. Casajús Quirós
ETSI Telecomunicación, Universidad Politécnica de Madrid
Ciudad Universitaria s/n, 28040, Madrid, SPAIN

Abstract. The paper presents a general analysis of the second-order convergence of gradient adaptive algorithms for arbitrary error surface without considering the reason for its choice. It shows how the absence of any design criteria of the adaption step may cause undesired misconvergences, or even divergences, for the cases of non-quadratic error surfaces. The convergence analysis is then particularized for some members of the L_k family (LMS and LMF) and a special case of decision-directed algorithm for Binary statistic. Some computer simulations are also given.

I. Introduction

As it is widely known, the performance of any stochastic gradient adaptive algorithm is strongly linked to the shape of its error cost function. This is usually a memory-less, even-shaped and convex function of the measured error signal. Therefore, its minimum is supposed to be reachable following a steepest descent line (its negative gradient) conveniently scaled by a constant adaption step. This step is generally the only user-dependent parameter involved in the algorithm and plays an important role in the design and satisfaction of the convergence conditions. In fact, it controls simultaneously the speed and degree of convergence of the algorithm (unfortunately two opposite objectives). In summary, the family of stochastic gradient adaptive algorithms is generally expressed by:

$$\mathbf{w}(n+1) = \mathbf{w}(n) - \mu \, \mathbf{grad}_{\mathbf{w}} \{ C[e(n)] \} \qquad (1)$$

where $\mathbf{w}$ is the vector of parameters to update in each iteration n, μ, the adaption step or convergence parameter, e the current measured error signal and $C(\bullet)$ the associated cost function. In our work, we will make use of a transversal adaptive filter working in a plant identification environment (see figure 1). With this structure, expression (1) simplifies to

$$\mathbf{w}(n+1) = \mathbf{w}(n) + \mu f[e(n)]\mathbf{x}(n) \qquad (2)$$

where $f[\bullet]$ is an odd-shape function of $(\bullet)$ and $\mathbf{x}$ is the vector of input data. Here, it is important to remark the separability of the instantaneous gradient into two factors $f[e(n)]$ and $\mathbf{x}(n)$. This separation will be helpful in the study of final stages of the convergence where both processes are generally assumed to be nearly uncorrelated. This assumption allows us to focus the second-order convergence analysis (disregarding the input power) on the arbitrary shape of the cost function.

Perhaps, the most relevant member of such family of algorithms is the extensively studied LMS algorithm due to Widrow and Hoff, which searches for the unique minimum of an instantaneous quadratic error surface (the best possible choice for Gaussian statistics) leading to a linear gradient. However there exist other possibilities in the choice of the error criterion (more complex than the quadratic one) depending on several aspects such as noise statistics, presence of inhomogeneities in the data, or the use of some problem-oriented techniques to exploit features and symmetries of the signals in order to improve the original error reference. Examples of these arise in communication applications, for instance, the CMA family of adaptive algorithms that exploits signals with constant modulus, the L_k-norm family (k>2), better suited to signals with short-tail probability density functions, the decision-directed algorithms, which try to improve the error reference removing part of the interfering far-end signal, and some others. These non-standard adaptive algorithms have in common non-linear gradients of the error reference that leads, in general, to more complex convergence conditions than that of the LMS algorithm (just a constant upper bound for the adaption step).

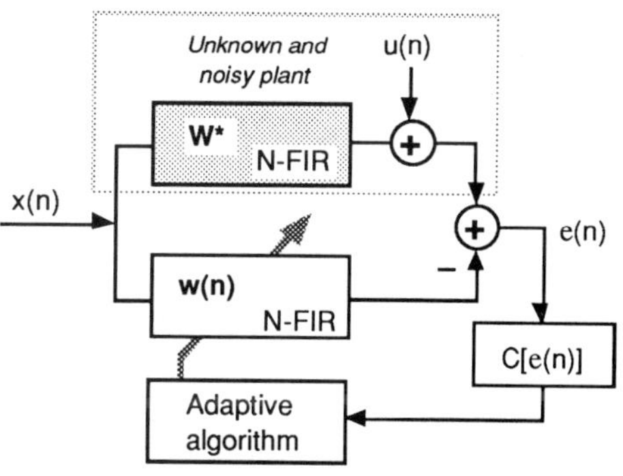

Figure 1

To give some insight into the behavior of algorithms with non–linear gradients, consider, as an example, the unidimensional curve $C(z)=|z|^3$ as an error surface of z (figure 2). The maximum allowed correction, before divergence, for an updating algorithm at any generic point z is $Cr_{max}=2|z|$ (the maximum distance to the opposite side) and the absolute gradient at z is $gradC(z)=3|z|^2$ (greater than Cr_{max} for certain values of z). It implies a convergence dependence on the starting position $z(0)$ when using an algorithm based on the gradient of this deterministic cost function to update any set of parameters. That is, the gradient has to be sufficiently scaled to satisfy the following convergence condition:

$$Cr_{max}(z) = 2|z| > \mu|gradC(z)| \ \forall z \ (\mu > 0) \qquad (3)$$

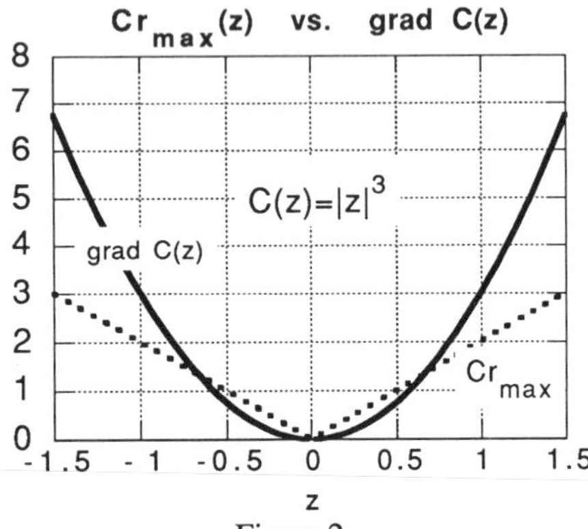

Figure 2

It shows that μ has, in general, to be selected according to the $2|z(0)|/gradC[z(0)]$ starting ratio. For the quadratic surface, both functions, $Cr_{max}(\bullet)$ and $gradC(\bullet)$ are linear, and therefore only a slope adjustment is needed (μ is therefore independent of the starting condition).

Thus, having in mind the possibility of a undesired convergence behavior shown in the latter example, the aim of this paper is to present, regardless of the error surface shape or special error measure, a general formulation for the second order statistics of the convergence that allows to establish recommendations in the choice of a constant adaption step that guarantees convergence to the desired steady-state value. We will make use of usual approximations and assumptions (zero mean i.i.d. inputs, slow convergence, uncorrelation between tap vector and input, etc) in our analysis.

2. Second order analysis

Denoting $\varepsilon(n)$ as the identification error, that is, $e(n)$ is splitted as $e(n)=\varepsilon(n)+u(n)$, we can write:

$$\varepsilon(n) = [\mathbf{w}^* - \mathbf{w}(n)]^t \mathbf{x}(n) = \mathbf{v}^t(n)\mathbf{x}(n) \qquad (4)$$

and using (1) and (4) we arrive to

$$\mathbf{v}^t(n+1)\mathbf{v}(n+1) = \mathbf{v}^t(n)\mathbf{v}(n) - \\ -2\mu f(n)\varepsilon(n) + \mu^2 f^2(n)\mathbf{x}^t(n)\mathbf{x}(n) \qquad (5)$$

Considering now the usual assumptions above mentioned, we can derive:

$$\sigma_\varepsilon^2(n) = E\left\{\left[\mathbf{v}^t(n)\mathbf{x}(n)\right]^2\right\} = \sigma_x^2 E\left\{\mathbf{v}^t(n)\mathbf{v}(n)\right\} \qquad (6)$$

Therefore, taking expectations in (5) and multiplying both sides by σ_x^2 we obtain the following recursive equation for the identification error variance

$$\sigma_\varepsilon^2(n+1) = \sigma_\varepsilon^2(n)P(\mu,n) \qquad (7)$$

where $P(\mu,n)$ is a time-dependent quadratic polynomial in μ (required to be positive defined in μ) given by:

$$P(\mu,n) = 1 - \mu S_1[\sigma_\varepsilon(n)] + \mu^2 S_2[\sigma_\varepsilon(n)] \qquad (8)$$

Polynomial $P(\mu,n)$ depends on the statistics of the involved signals and shape of the measured gradient. Here S_1 and S_2 are the following mean values

$$S_1(n) = 2\sigma_x^2 \frac{E\{\varepsilon(n)f(n)\}}{\sigma_\varepsilon^2(n)} \qquad (9a)$$

$$S_2(n) = N\sigma_x^4 \frac{E\{f^2(n)\}}{\sigma_\varepsilon^2(n)} \qquad (9b)$$

The main subject under study is to guarantee the global second order convergence, that is, $\sigma_\varepsilon^2(n+1)<\sigma_\varepsilon^2(n)$. The analysis, here presented, is therefore focused on a systematic analysis of the positiveness and local boundeness of $P(\bullet,\bullet)$, which, assuming the surface shape (or its derivatives) are known in a stationary environment, allows us to establish general conclusions about local and global convergence conditions.

Thus, since $S_2(n)$ is a quotient of quadratic means, and therefore, it is always positive, (8) turns out to be a positive parabolic function of μ. Nevertheless, to show its validity in any positive range of σ_ε^2, we have also to prove that its discriminant, defined as $D = S_1^2(n) - 4S_2(n) \le 0$; $\forall \sigma_\varepsilon^2 > 0$ or equivalently $S_1^2(n)/4S_2(n) \le 1$. Hence, considering the ratio

$$\beta = \frac{S_1^2(n)}{4S_2(n)} = \frac{E^2\{\varepsilon(n)f(n)\}}{N\sigma_\varepsilon^2(n)E\{f^2(n)\}} \qquad (10)$$

and according to the Cauchy-Schwarz inequality, $E^2\{\mathbf{XY}\} \le E\{\mathbf{X}^2\}E\{\mathbf{Y}^2\}$, we obtain $\beta \le 1/N$, where $\forall N \ge 1 \Rightarrow \beta \le 1$ and therefore, it proves that (8) is not negative.

3.- Convergence analysis

$P(\mu,n)$ allows an easy analysis of two important aspects of any adaptive algorithm:

a) the *convergence condition*, by means of solving for μ the inequality:

$$P(\mu,n)<1 \;;\; \forall n \in [0,\infty) \qquad (11)$$

b) the *final misadjustment* knowledge ($\sigma_\varepsilon^2(\infty)$) by considering the convergence stopped when:

$$1-\mu S_1(\infty)+\mu^2 S_2(\infty)=1 \Rightarrow \sigma_\varepsilon^2(\infty) \qquad (12)$$

From (11) the convergence condition follows

$$(0)<\mu<\min\left[\frac{S_1(n)}{S_2(n)}\right] \quad \forall n \in [0,\infty) \qquad (13)$$

that is, μ has to be upper bounded by a function of $\sigma_\varepsilon^2(n)$ in the whole convergence range of the algorithm.

Regarding (12), the constant adaption step μ is obtained as

$$\mu=\frac{S_1[\sigma_\varepsilon^2(\infty)]}{S_2[\sigma_\varepsilon^2(\infty)]} \qquad (14)$$

and therefore from (13) and (14) we arrive to a general dynamic mean condition which guarantee convergences to any selected final value $\sigma_\varepsilon^2(\infty)$:

$$\frac{S_1(\infty)}{S_2(\infty)}<\frac{S_1(n)}{S_2(n)} \quad \text{for } \sigma_\varepsilon^2(\infty)<\sigma_\varepsilon^2(n) \;\forall n \qquad (15)$$

That is, with *a priori* knowledge of the μ–bounding function, $B(n)=S_1(n)/S_2(n)$, in the desired convergence range, $\sigma_\varepsilon^2(0)\to\sigma_\varepsilon^2(\infty)$, it is possible to predict, in a mean sense, any possible misconvergence of the adaptive algorithm.

To illustrate the last assertion, we make use of a graphic example. Let figure 3 be a generic bounding function $B[\sigma_\varepsilon^2(n)]$, where $\sigma_\varepsilon^2(0)$ and $\sigma_\varepsilon^2(\infty)$ are the initial and the steady state values of the considered convergence range.

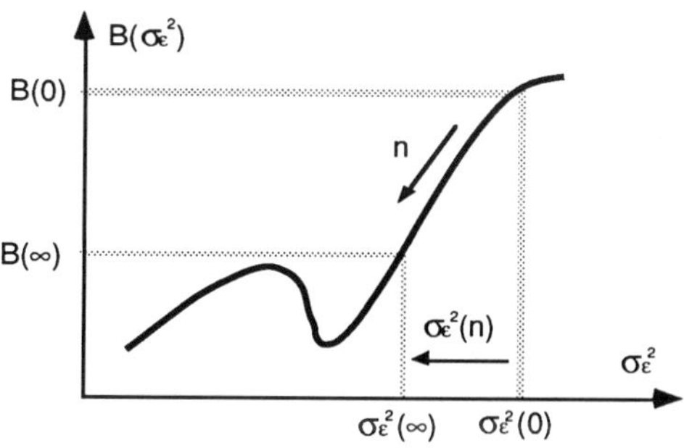

Figure 3

For this particular situation the practical convergence analysis should be as follows: a) $\sigma_\varepsilon^2(0)$ is known (or at least an accurate estimate or guess of it). It determines the convergence starting point. b) The desired $\sigma_\varepsilon^2(\infty)$ is selected. It fixes $\mu=B(\infty)$. c) The inequality $B(\infty)<B(n)$ $\forall n$ is satisfied.

Thus, we can conclude here that mean convergence from $\sigma_\varepsilon^2(0)$ towards $\sigma_\varepsilon^2(\infty)$ is guaranteed. However, if we tried any other choice of $\sigma_\varepsilon^2(\infty)$ located between the locals minimum and maximum of figure 3, in a non–monotonic part of the curve, the algorithm would lead to a possible local misconvergence because the condition $B(\infty)<B(n)$ is violated in some parts of the curve.

4. Some particular error surfaces

Let us consider now the three following error criteria:

a) $C(e)=e^2$ → LMS algorithm
b) $C(e)=|e|^4$ → LMF algorithm
c) $C(e)=[e\text{-sgn}(e)]^2$ → Decision-directed algorithm

Their corresponding μ–bounding functions are shown in figure 4. We have assumed Binary statistic in the analysis for both processes, $x(n)$ and $u(n)$, because only in this case does the use of decision-directed algorithms make sense. Here we can observe the increasing monotonic character of $B(\sigma_\varepsilon^2)$ for the LMS algorithm (solid line). It implies that the convergence condition will be always satisfied for any selected steady state value $\sigma_\varepsilon^2(\infty)<\sigma_\varepsilon^2(0)$. Although not shown in the figure, the corresponding bounding function for the LMA algorithm ($C(e)=|e|$) also has monotonic character with σ_ε^2. In the case of the LMF algorithm (dashed line), if the initial error variance $\sigma_\varepsilon^2(0)$ is located to the left of the local maximum, the effective bounding function behaves as a monotonic curve again and the convergence condition will be always satisfied. Further, if $\sigma_\varepsilon^2(0)$ is on the right of the maximum, the steady state value $\sigma_\varepsilon^2(\infty)$, had to be selected in such a way that $\mu=B[\sigma_\varepsilon^2(\infty)]<B[\sigma_\varepsilon^2(0)]$. It means that the convergence of this kind of algorithms (termed generally as L_k) for $k>2$, is conditioned by the initial settings of the adaptive filter. Finally, the bounding function corresponding to the decision-directed algorithm shows an inflection that could lead to undesired misconvergences before reaching the selected steady state.

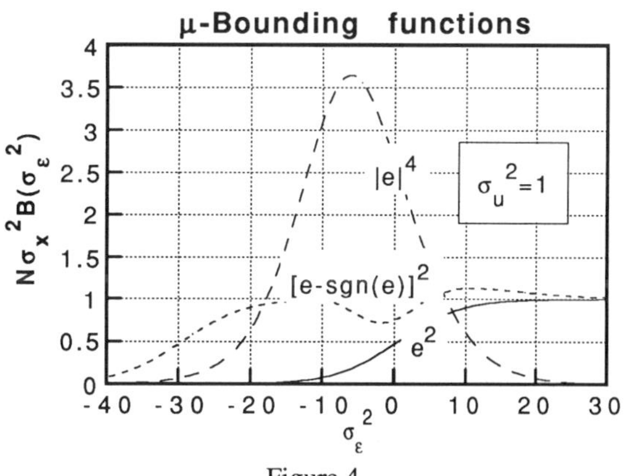

Figure 4

To show the real (strictly speaking, simulated) behavior of such misconvergences in the case of the above mentioned decision-directed algorithm, consider the situation of figure 5. Here, the designer would like a steady state value of $\sigma_\varepsilon^2(\infty)_S$=–30 dB (this choice for $\sigma_\varepsilon^2(\infty)_S$ fixes μ). Since this steady point is located on the left of the local minimum of $B(\bullet)$, there exists a risk of stopping the convergence around $\sigma_\varepsilon^2(\infty)_R \approx 3$ dB whenever the initial settings of the adaptive filter, **w**(0), forces $\sigma_\varepsilon^2(0)$ to be on the right side of the local minimum in the bounding curve.

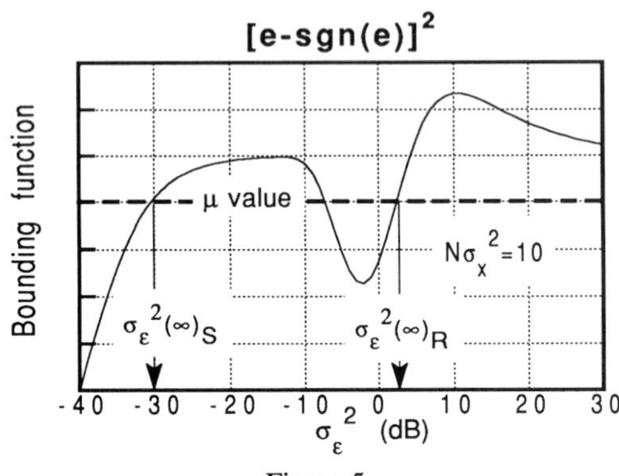

Figure 5

Figure 6 shows three simulated realizations (#1, 2 and 3 in dashed lines against the theoretical result, solid line) of the above situation. The initial error power, $\varepsilon^2(0)$, is around 12 dB for all of them. After the initialization of the adaptive process, the convergence first stops in the vicinity of $\sigma_\varepsilon^2(\infty)_R \approx 3$ dB after an undetermined period of time. Suddenly, without any kind of indication, and because of the noisy character of the stochastic gradient, it leaves that pseudo-stationary point to converge to the selected steady state $\sigma_\varepsilon^2(\infty)_S$.

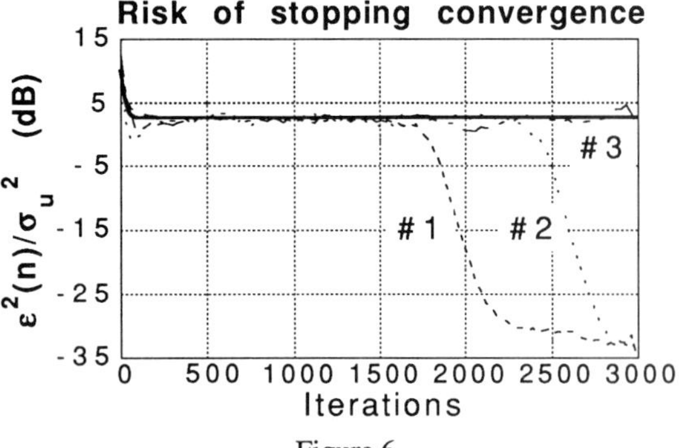

Figure 6

These results, for the decision-directed algorithm, are quite representative of the behavior of adaptive algorithms with non-monotonic μ-bounding functions in the desired convergence range $\sigma_\varepsilon^2(0) \rightarrow \sigma_\varepsilon^2(\infty)$.

In the latter example, the designer should know both, the position of the local minimum and an accurate guess of $\sigma_\varepsilon^2(0)$, before proceeding with the choice of μ. In this way it is possible to mantain the inequality $B[\sigma_\varepsilon^2(\infty)]<B[\sigma_\varepsilon^2(0)]$ $\forall n$ and to guarantee the desired convergence.

5. Conclusions

The work here summarized presents how particular shapes of non-quadratic error criteria can lead to second-order misconvergences of their corresponding stochastic gradient adaptive algorithms. In some ill-conditioned cases these misconvergences will even result in proper divergences. The work shows that this undesired behavior for any adaptive algorithm can be predicted and avoided, in a mean sense, by making prior use of a running bounding function for the adaption step μ. This function depends on the shape of the selected error criterion an the involved data statistic. It is presented *versus* the identification error variance, $B[\sigma_\varepsilon^2]$, and allows one to design the right adaption step in order to guarantee the convergence from the initial guessed situation $\sigma_\varepsilon^2(0)$ to the selected steady state value $\sigma_\varepsilon^2(\infty)$. In other words, it would detect possible misconvergence when a μ is selected without considering any design criteria related to the steady state value.

6. References

[1] Widrow, B., Stearns, S.D.: *Adaptive Signal Processing*, New jersey, Prentice Hall, 1985..

[2] Ljung, L., Söderström, T.: *Theory and Practice of Recursive Identification*, Cambrigde, MA, MIT Press, 1983

[3] Walach, E., Widrow, B.: *"The Least Mean Fourth (LMF) Adaptive Algorithm and its Family"*; IEEE Trans. on IT, vol. IT-30, no. 2, March 1984.

[4] Figueiras, A.R., Páez, J.M., García, R., Hernández, L.A.: *" Lk-norm adaptive transversal filters"*, Traitement du Signal, vol.4, no.5, pp. 447-458, 1987

[5] Gibson, J.D., *et al.*: *"MVSE adaptive filtering subject to a constraint on MSE"*, IEEE Trans. on CAS, May 1988.

[6] Mazo, J.E.: *"Analysis of decision-directed equalizer convergence"*, Bell Syst. Tech. Journal, pp. 1857-1876, Dec. 1980.

[7] Macchi, O., Eweda, E.: *" Convergence analysis of self-adaptive equalizers"*, IEEE Trans. on Information Theory, vol. IT-30, pp.161-176, Mar. 1984

[8] Lorenzo, F., *"Cancelación adaptativa de ecos en transmision de datos basada en el conocimiento estadístico de la señal observada"*, Doctoral Thesis, Univ. Politécnica de Madrid, March 1990

[9] Páez, J.M.; Lorenzo, F.: *"Convergence of L1 and L2 Decision-directed Adaptive Echo Cancellers for Baseband Data transmission"* , submitted to Signal Processing. Eurasip

SIGNAL PROCESSING VI: Theories and Applications
J. Vandewalle, R. Boite, M. Moonen, A. Oosterlinck (eds.)

NONLINEAR EFFECTS IN ADAPTIVE FILTERS OF THE LMS TYPE

H. J. BUTTERWECK

EE-Department, Eindhoven University of Technology, Eindhoven
P.O. Box 513, NL-5600 MB Eindhoven, The Netherlands

An adaptive filter can be viewed as a nonlinear, time-invariant system with a deterministic input-output relation. For the LMS algorithm various exact and approximative forms of this relation are discussed. Typical nonlinear effects such as the generation of combination frequencies under time-harmonic excitation are predicted.

1. INTRODUCTION

In a widely adopted view, adaptive filters are treated as linear systems with time-varying parameters. The parameter variations are strong during an adaptation phase and approach a weak, but nonzero amplitude under the stationary conditions of statistical equilibrium.

Viewed as a system whose input-output relations are governed by deterministic laws, an adaptive filter should more correctly be described by the attributes *time-invariant* and *nonlinear*. Indeed, applying an arbitrary common delay to the input signals amounts to an equal delay in the output signal without any change of its waveform. Further, the input-output relations contain cross-products between pairs of input signals and can exhibit a nonlinear behaviour in terms of the individual input signals.

In this paper we examine the LMS algorithm for an M-length adaptive filter with input signal x_k, reference signal d_k, and output signal y_k. It is governed by the relations

$$\begin{aligned} y_k &= \underline{w}_k^t \underline{x}_k, & (1)\\ \underline{w}_{k+1} &= \text{function of } \underline{w}_k, \underline{x}_k, d_k, & (2) \end{aligned}$$

where $\underline{x}_k = (x_k, x_{k-1}, \cdots, x_{k-M+1})^t$ and $\underline{w}_k = (w_{k0}, w_{k1}, \cdots, w_{k,M-1})^t$ stand for the *input vector* (representing a finite past of the scalar input signal x_k) and the *weight vector*. The adaptation mechanism (2) will be discussed in Section 2.

Like many other adaptive structures the LMS-filter can be viewed as a black box with two inputs (x_k, d_k) and one output (y_k), whose signal transformation can be formally described in the form

$$y_k = F[x_k, d_k; x_{k-1}, d_{k-1}; \cdots; x_{k-i}, d_{k-i}; \cdots]. \quad (3)$$

The function F of the present and past values of x and d has a *nonlinear* character. Furthermore it does not explicitly depend upon the time k of observation so that, if x and d are delayed with a certain time, y undergoes the same delay under preservation of its waveform. These properties find expression in the attributes *nonlinear* and *time-invariant* for the systems under discussion. It should be mentioned that many adaptive systems do not admit a closed-form solution for F. The LMS algorithm belongs to the class for which F can be formulated, cf. (7)-(10) and (1)-(2).

2. DETERMINISTIC ANALYSIS

For the LMS algorithm the *updating* of the weight vector is governed by the relation

$$\begin{aligned} \underline{w}_{k+1} &= \underline{w}_k + e_k \underline{x}_k, & (4)\\ e_k &= d_k - y_k, & (5) \end{aligned}$$

where e_k is called the error signal. With (1) we then have

$$\underline{w}_{k+1} = \underline{w}_k - \underline{x}_k \underline{x}_k^t \underline{w}_k + d_k \underline{x}_k. \quad (6)$$

Observe that (4) differs from the usual notation in current literature, where $e_k \underline{x}_k$ is replaced by $2\mu e_k \underline{x}_k$. In our notation the adaptation parameter μ and the factor 2 are absorbed (after splitting into equal factors) into $\underline{x}_k$ and e_k. This implies that the rate of adaptation is directly determined by the amplitude of $\underline{x}_k$.

The whole dynamics of an LMS filter is governed by (6). In accordance with common terminology in the theory of dynamic systems, (6) is called a *state equation*, while its solution $\underline{w}_k$ is referred to as a *state vector*. In this terminology the variable y_k is an output signal which is determined with the auxiliary relation (1).

Although the operator $(d_k, x_k) \rightarrow \underline{w}_k$ is basically nonlinear, this is not the case for the difference equation (6), that for a given $\underline{x}_k$ is *linear with time-varying coefficients* and can thus be solved in closed form. We study the evolution of the system between $k = 0$ and some $k = n$ ($n > 0$), starting with an *initial state* $\underline{w}_0$ and ending up with a *final state* $\underline{w}_n$. Repeated application of (6) for $k = 0, 1, 2, \cdots$, yields the superposition sum

$$\underline{w}_n = C_{n,-1}\underline{w}_0 + \sum_{i=0}^{n-1} C_{ni} d_i \underline{x}_i, \qquad (7)$$

where

$$\begin{aligned} C_{ni} &= B_{n-1}B_{n-2}\cdots B_{i+1} \quad (i \leq n-2), && (8) \\ C_{n,n-1} &= I \text{ (unit matrix)}, && (9) \\ B_k &= I - \underline{x}_k \underline{x}_k^t. && (10) \end{aligned}$$

Observe that $\underline{x}_k \underline{x}_k^t$ is a dyadic product and that, in general, the order of the B-matrices in (8) is not interchangeable. Due to the time dependence of the coefficients $\underline{x}_k \underline{x}_k^t$ in (6), the solution (7) is not of the convolution type.

3. STOCHASTIC ANALYSIS

Let d_k and x_k now represent stationary *stochastic signals* and let the adaptation process be finished such that $\underline{w}_k$ is stationary, too. Let further the pair d_k, x_k be partially correlated according to

$$d_k = \underline{w}^{ot} \underline{x}_k + n_k, \qquad (11)$$

where n_k is a stationary signal uncorrelated with x_k and $\underline{w}^o$ is an M-length fixed vector. Thus d_k can be decomposed into a part uncorrelated with x_k and another one fully correlated through a causal convolutional operator, whose memory has a maximum length equal to M. The more general case of an infinite memory and also that of a noncausal operator are not treated here.

The above decomposition implies a relation between the autocorrelation $r_l^{xx} = E[x_k x_{k-l}]$ of x and the cross-correlation $r_l^{dx} = E[d_k x_{k-l}]$ of d and x. This reads

$$r_l^{dx} = \sum_{j=0}^{M-1} w_j^o r_{l-j}^{xx}. \qquad (12)$$

Often the inverse problem has to be solved, in which $\underline{w}^o$ has to be determined from given correlations. For this purpose only the first M values of r_l^{xx} and r_l^{dx} ($l = 0, 1, \cdots, M-1$) are required which are conveniently represented by the $M \times 1$ correlation vector $\underline{p} = (r_0^{dx}, r_1^{dx}, \cdots)^t$ and the $M \times M$ correlation matrix R (with $R_{mn} = r_{|m-n|}^{xx}$). Then we arrive at the *Wiener-Hopf relation*

$$\underline{p} = R\underline{w}^o \qquad (13)$$

with the *Wiener solution* $\underline{w}^o = R^{-1}\underline{p}$.

In our context it is important that under the assumptions stated at the outset, the weight vector $\underline{w}_k$ oscillates *around the Wiener solution* $\underline{w}^o$. This assertion can be verified by considering the state equation

$$\underline{v}_{k+1} = \underline{v}_k - \underline{x}_k \underline{x}_k^t \underline{v}_k + n_k \underline{x}_k \qquad (14)$$

for the difference vector $\underline{v}_k = \underline{w}_k - \underline{w}^o$ which can be derived from (6) using (11). The mathematical formulation of the assertion then reads† $\overline{\underline{w}_k} = \underline{w}^o$ or $\overline{\underline{v}_k} = \underline{0}$. Its proof is facilitated by adding the (rather mild) condition that n_k is symmetrically distributed (implying $\overline{n_k} = 0$) and statistically independent of x_k (which is more than *uncorrelated with* x_k). Then from (14) one readily concludes that also $\underline{v}_k$ is symmetrically distributed (i.e. a certain $\underline{v}_k$ and its sign-reversed counterpart are equiprobable) so that $\overline{\underline{v}_k} = \underline{0}$.

4. APPROXIMATE SOLUTION FOR THE WEIGHT FLUCTUATIONS

In this section the *weight fluctuations* around the Wiener solution are analyzed. Since the exact solution of (14) in a form similar to (7) is difficult to handle, several small-signal approximations are discussed. With increasing amplitude of x_k these have to be more and more refined (0th, 1st, 2nd-order solutions). With the exception of the zero-order solution they appear in a mixed deterministic-stochastic form and can, in a strict sense, only be used for stochastic excitations. If time-harmonic signals can be approximately viewed as narrow-band stochastic signals, the results are meaningful also in that domain.

The *zero-order solution* is found by neglecting $\underline{x}_k \underline{x}_k^t \underline{v}_k$ in (14) which is permitted for small x-values. The solution

$$\underline{v}_k = \sum_{j=-\infty}^{k-1} n_j \underline{x}_j \qquad (15)$$

of the resulting difference equation represents a system that simply adds up past values of the product $n\underline{x}$. In this model we encounter the most elementary nonlinearity of the LMS algorithm, viz., the product operation applied to n and $\underline{x}$. For two time-harmonic signals with frequencies Ω_n and Ω_x the combination frequencies $|\Omega_n \pm \Omega_x|$ are generated, of which, due to the basic peculiarity of any summation of the type (15), only the difference frequency $|\Omega_n - \Omega_x|$ has considerable amplitude and thus deserves further consideration. With this frequency in $\underline{v}_k$ the frequencies $\Omega_x \pm |\Omega_n - \Omega_x|$ occur in the output signal y_k as determined by (1). While the linear theory with $\underline{w}_k = \underline{w}^o$ predicts the single output frequency Ω_x, the simple nonlinear model under discussion yields two concomitant side frequencies (which become sidebands in the case of a more general signal n_k) with a total bandwidth $2|\Omega_n - \Omega_x|$.

A drawback of this most rudimentary approach to the nonlinear behaviour of the LMS algorithm is its inherent instability. This becomes evident if $n\underline{x}$ contains a d.c.-term or, in the case of a stochastic signal, a nonvanishing spectral contribution at zero frequency. The latter, however, is the normal situation, which does not occur if and only if the power spectra of n_k and x_k do not overlap. (In that case, however, there is no need for an adaptive filter.) In a frequency-domain description of the summation (15) we are faced with a degenerate first-order system with system function $H(z) = (z-1)^{-1}$, which has a pole at $z = 1$ and has to be rated amongst the (marginally) unstable systems.

This problem is avoided in the *first-order approximation*, where the pole at $z = 1$ is split up into a number of poles *inside* the unit circle. This is accomplished by replacing the dyadic product $\underline{x}_k\underline{x}_k^t$ in (14) with the correlation matrix R being the average value of that product. Then (14) passes into a (vector) difference equation with constant coefficients which can be solved with elementary methods. The term $R\underline{v}_k$ in the difference equation obviously represents a damping mechanism.

With d_k and $\underline{w}_k$ in (7) replaced with n_k and $\underline{v}_k$ and neglecting the influence of an infinitely remote initial value of $\underline{v}$, we find

$$\underline{v}_k = \sum_{i=-\infty}^{k-1} (B_{k-1} \cdots B_{i+1}) n_i \underline{x}_i. \tag{16}$$

In the first-order approximation we have

$$B_k = I - \underline{x}_k\underline{x}_k^t \approx I - R = \overline{B} \tag{17}$$

so that (16) passes into

$$\underline{v}_k \approx \sum_{i=-\infty}^{k-1} \overline{B}^{k-i-1} n_i \underline{x}_i. \tag{18}$$

Thus we have an input-output relation of the convolution type, reflecting the time-invariant character of the first-order approximation. Compared with (15) past values of $n\underline{x}$ are no longer uniformly weighted, but in accordance with an exponentially decaying memory. So, very remote input signals contribute negligibly to the output signal. This implies that now a *constant* signal $n\underline{x}$ leads to a finite result and that the unstable character of the previous solution is removed.

Simulations reveal that the present approximation leads to surprisingly accurate results. This can be explained by the fact that in (16) successive values of $B_k = I - \underline{x}_k\underline{x}_k^t$ have to be multiplied which, for sufficiently small x values, amounts to an *addition or averaging* of successive dyadic products $\underline{x}_k\underline{x}_k^t$. The assumption in (17) obviously anticipates such averaging and leads to less final errors than might be expected from the rather rough approximation (17).

The vectorial nature of (18) asks for a *modal* treatment. Let λ_m denote the eigenvalues of R and $\underline{q}_m$ the orthonormal set of eigenvectors ($m = 1, 2, \cdots, M$). Due to special properties of R, the λ_m are real and nonnegative, while the freedom exists to choose $\underline{q}_m$ in real form [1]. The two vectors $\underline{v}_k$ and $\underline{x}_k$ are now advantageously written as linear combinations of eigenvectors, reading as

$$\underline{v}_k = \sum_{m=1}^{M} v_{km}\underline{q}_m, \qquad v_{km} = \underline{v}_k^t\underline{q}_m, \tag{19}$$

$$\underline{x}_k = \sum_{m=1}^{M} x_{km}\underline{q}_m, \qquad x_{km} = \underline{x}_k^t\underline{q}_m. \tag{20}$$

The coefficients v_{km} and x_{km} can be interpreted as normal-mode components of the pertinent vectors. From (18) and (17) it can be concluded that

$$v_{km} = \sum_{i=-\infty}^{k-1} \xi_m^{k-i-1} n_i x_{im}, \tag{21}$$

$$\xi_m = 1 - \lambda_m. \tag{22}$$

Obviously, in this model no mode coupling occurs, i.e. the mth component of $\underline{v}_k$ is determined only by the mth component of $\underline{x}_k$. In a more convenient notation this relationship reads as a convolution

$$v_{km} = h_{km} * (n_k x_{km}), \tag{23}$$

where the impulse response h_{km} of the mth mode is given by

$$h_{km} = \begin{cases} 0 & \text{for } k < 1, \\ \xi_m^{k-1} & \text{for } k \geq 1. \end{cases} \tag{24}$$

A specific mode m ($m = 1, ..., M$) is thus associated with a first-order system having an exponentially decaying impulse response and a system function

$$H_m(z) = \sum_{k=1}^{\infty} h_{km} z^{-k} = (z - \xi_m)^{-1} \tag{25}$$

with a single pole at $z = \xi_m < 1$.

Observe that with decreasing amplitude of $\underline{x}_k$ also λ_m decreases while ξ_m approaches unity. Then the bandwidth of the pertinent transmission decreases, v_{km} contains less high frequencies and thus exhibits slower variations, while the impulse response and herewith the system's memory grow in duration.

For time-harmonic excitations with frequencies Ω_n, Ω_x the basic results of the zero-order solution are not affected. Only for $\Omega_n \approx \Omega_x$ the difference frequency $|\Omega_n - \Omega_x|$ becomes so low that the pole shift from the unit circle (at $z = 1$) to its interior becomes tangible. Not only is the amplitude of the low-frequency sinusoid (frequency $|\Omega_n - \Omega_x|$) reduced due to the damping mechanism, but there occurs also an additional phase shift. Clearly, there remains a low-pass filtering, also in the more general case of a non-sinusoidal excitation. The result is a slowly varying $\underline{v}_k$ on which rather weak oscillations due to the high-frequency components of $(n\underline{x})$ are superimposed.

5. THE SECOND-ORDER SOLUTION FOR THE WEIGHT FLUCTUATIONS

In the rare cases where, due to a relatively high input level, the first-order solution is not sufficiently accurate a further iteration can be applied to improve the result. The essence of such a *second-order solution* is a refinement of the damping term in (14)

$$\begin{aligned} \underline{x}_k\underline{x}_k^t\underline{v}_k &= R\underline{v}_k + (\underline{x}_k\underline{x}_k^t - R)\underline{v}_k \\ &\approx R\underline{v}_k + (\underline{x}_k\underline{x}_k^t - R)\underline{v}_k^{(1)}. \end{aligned} \qquad (26)$$

Here $\underline{v}_k^{(1)}$ is the result of the first-order approximation. The error of the approximation (26) is small of second order because the term $(\underline{x}_k\underline{x}_k^t - R)\underline{v}_k$ itself is small and because the difference between $\underline{v}_k$ and $\underline{v}_k^{(1)}$ is small. Further one has to keep in mind that the term under discussion occurs successively and tends to be averaged out.

By virtue of the approximation (26) the feedback in the difference equation (14) is cut open so that the term under consideration is transformed into an additional source term. With $\underline{v}_k$ replaced by $\underline{v}_k^{(2)}$ we then obtain

$$\underline{v}_{k+1}^{(2)} = \underline{v}_k^{(2)} - R\underline{v}_k^{(2)} + [(R - \underline{x}_k\underline{x}_k^t)\underline{v}_k^{(1)} + n_k\underline{x}_k]. \quad (27)$$

A detailed discussion of the solution of this equation is beyond the scope of this paper. Observe that also here modes are preserved: for $\underline{x}_k = x_{km}\underline{q}_m$ ($\underline{q}_m$ eigenvector of R) we have $\underline{v}_k^{(1)} = v_{km}^{(1)}\underline{q}_m$ (see previous section) so that the complete source term in (27) points in the direction of $\underline{q}_m$ and hence also $\underline{v}_k^{(2)}$. Note, however, that due to the strong internal cohesion of $\underline{x}_k = (x_k, x_{k-1}, \cdots)^t$ this is not more than a formal statement, because a pure mode cannot be excited. Indeed, even for a deterministic signal x_k a constant direction of the vector $\underline{x}_k$ can hardly be maintained.

More interesting is the response to two time-harmonic signals with frequencies Ω_x, Ω_n. With $n_k\underline{x}_k$ and $R\underline{v}_k^{(1)}$ essentially oscillating with frequency $|\Omega_n - \Omega_x|$, one part of $\underline{v}_k^{(2)}$ oscillates with the same frequency. The term $\underline{x}_k\underline{x}_k^t\underline{v}_k^{(1)}$ consists of a slowly varying part with frequency $|\Omega_n - \Omega_x|$ and a rapid variation with frequency $\Omega_n + \Omega_x$ which is weakly "transmitted" to $\underline{v}_{k+1}^{(2)}$. Thus all contributions to $\underline{v}_k^{(2)}$ essentially oscillate with the difference frequency $|\Omega_n - \Omega_x|$, as was already found for the former, more primitive solutions.

6. CONCLUSIONS

The weight vector and the output signal of an LMS adaptive filter are shown to be nonlinearly dependent on the input and the reference signal. For small signal levels several approximations of the nonlinearities are discussed whose complexity increases with the required accuracy. Under time-harmonic excitation combination frequencies are generated in an adaptive filter which have to be reckoned with in applications (e.g. in the audio field) where a high degree of linearity is required.

ACKNOWLEDGEMENTS

The author wishes to thank A.C. den Brinker and A.W.J. Oomen for the extensive discussions which have substantially contributed to the notions developed in this paper.

FOOTNOTES

† The upper bar denotes time or (due to ergodicity) ensemble averaging.

REFERENCES

[1] Haykin, S., Adaptive filter theory. (Prentice Hall, Englewood Cliffs, N.J., 1986).

SIGNAL PROCESSING VI: Theories and Applications
J. Vandewalle, R. Boite, M. Moonen, A. Oosterlinck (eds.)

FEEDFORWARD AND FEEDBACK IN A SYMMETRIC ADAPTIVE NOISE CANCELER: "STABILITY ANALYSIS IN A SIMPLIFIED CASE"

Stefaan Van Gerven*and Dirk Van Compernolle†

K.U. Leuven - ESAT
Kardinaal Mercierlaan 94
B-3001 Heverlee (Belgium)
TEL: (+32) 16 / 22.09.31
FAX: (+32) 16 / 22.18.55

In this paper we present a stability analysis for the problem of blind signal separation using a symmetric adaptive decorrelation (SAD) algorithm. In the first part the signal separation problem is briefly stated and the feedforward implementation of the algorithm is derived from the interpretation of the classical Widrow LMS noise canceler as a decorrelator. In the new algorithm decorrelation is done between the signal estimate and a "signal free" noise estimate. In a second part we present a feedback implementation of the algorithm. In a third part we consider a simplified case and subject the derived algorithm to a more rigorous analysis. The asymptotic stability is investigated by a linearisation of the algorithm around the desired solution. Apart from limitations on the adaptation gains, convergence and stability are only feasible for a subclass of signal mixtures. These limitations seem to be inherent to the signal separation problem as they arise in both implementations.

PART I: DERIVATION

The Signal Separation Problem. The goal of signal separation techniques is to extract the unknown statistically independent signals $s_i(k)$ out of a set of measurements $y_i(k)$, being unknown convolutive mixtures of the original signals. We will restrict ourselves to a two channel formulation in which the direct couplings equal 1 (cfr. Fig.1) and hence that can be written as:

$$\begin{aligned} y_1(k) &= s_1(k) + h_1(k) * s_2(k) \\ y_2(k) &= s_2(k) + h_2(k) * s_1(k) \end{aligned}$$

Following notations are used in this paper ($i = 1, 2$):

$$\begin{aligned} \xi_i(k) &= E[u_i^2(k)] \\ C_{uy}(m) &= E[u(k)y(k-m)] \end{aligned}$$

Some theoretical derivations are only valid for white noise signals, i.e.:

$$\begin{aligned} E[s_i(k)s_i(k-m)] &= \delta(m)\sigma_i^2 \\ E[s_1(k)s_2(k-m)] &= 0 \ \forall m \end{aligned}$$

The LMS Noise Canceler as a Decorrelator. The solid lines in Fig.1 show a classical adaptive noise canceler, in which a signal estimate of $s_1(k)$ is obtained in $u_1(k)$. The well known LMS performance criterion "minimizing the energy in the signal estimate" corresponds to "decorrelating signal estimate and noise reference", since it can easily be shown that:

$$\frac{\partial \xi_1(k)}{\partial w_1^k(m)} = -2C_{u_1 y_2}(m)$$

The adaptation step can even be reinterpreted from "gradient descent on the quadratic error surface" to "Newton search of zeros in the crosscorrelation". For white noise signals with no signal leakage (i.e. $H_2(z) = 0$), this is illustrated by:

$$\begin{aligned} C_{u_1 y_2}(m) &= (h_1(m) - w_1(m))\sigma_2^2 \\ \nabla_m &= \frac{\partial C_{u_1 y_2}(m)}{\partial w_1(m)} = -\sigma_2^2 \\ w_1^{(k+1)}(m) &= w_1^{(k)}(m) - \gamma \nabla_m^{-1}(u_1(k)y_2(k-m)) \\ &= w_1^{(k)}(m) + \mu(u_1(k)y_2(k-m)) \end{aligned}$$

The exact knowledge of the gradient sign and approximate knowledge of its magnitude allow for a robust Newton search of the zeros in the crosscorrelations, yielding an "Adaptive Decorrelation (AD)" algorithm which is identical to the LMS algorithm !! The stability requirement $\gamma < 1$ allows for an appropriate choice of $\mu < 1/\sigma_2^2$. For non-white noise signals the above formulas are no longer exact but the interpretation of the algorithm is still the same.

*Research Assistant of the I.W.O.N.L.

†Research Associate of the National Fund for Scientific Research of Belgium (N.F.W.O.)

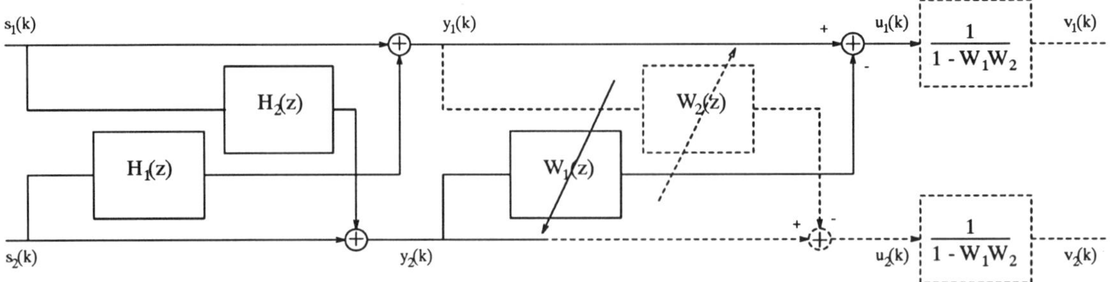

Figure 1: Signal Separation and Noise Cancelling in Feedforward Implementation

From Noise Cancelling to Signal Separation. The effectiveness of an LMS adaptive noise canceler [1] is proportional to the noise-to-signal ratio in the noise reference, limiting its applicability to "small" $H_2(z)$. Decorrelation of $y_2(k)$ with $u_1(k)$ is still obtained but no longer yields the desired solution. Decorrelating the signal estimate with a "signal free" noise estimate could extend the usage of the LMS algorithm. Looking at Fig. 1 the desirable solution is obvious. The original signals $s_1(k)$ and $s_2(k)$ are obtained in $v_1(k)$ and $v_2(k)$ if $W_1(z) = H_1(z)$ and $W_2(z) = H_2(z)$. If $s_1(k)$ and $s_2(k)$ are independent, then this will also be the case for $u_1(k)$, $u_2(k)$ and $v_1(k)$,$v_2(k)$. The symmetric adaptive decorrelation (SAD) algorithm is directly infered from the LMS/AD algorithm by replacing $y_2(k)$ by $u_2(k)$ and by adding symmetric formulas for $w_2(k)$. The full SAD algorithm has been presented in [2] and looks as follows:

$$\begin{aligned} w_1^{(k+1)}(m) &= w_1^{(k)}(m) + \mu_1(u_1(k)u_2(k-m)) \\ w_2^{(k+1)}(n) &= w_2^{(k)}(n) + \mu_2(u_2(k)u_1(k-n)) \end{aligned}$$

Important constraints of this algorithm were discussed in [3] and some implementation requirements are listed below:

- all blocks must be realizable (causal)
- no unstabilities may occur in the postprocessing step, i.e. $(1 - W_1(z)W_2(z))$ must have all its zeros within the unit circle, this condition equally applies to the generating filters $H_1(z)$ and $H_2(z)$

In order to satisfy the first criterion we will restrict ourselves to causal finite length FIR filters, and request that the product of the z^0 coefficients differs from 1. The stability constraint cannot trivially be satisfied, since the coefficients must remain in the stable region with every iteration step. It could therefore be argued to omit the post-processing step but this will lead to signal distortion. It therefore looks more like a fundamental constraint in the signal separation problem. However let's consider the typical situation of an LMS noise canceler with a considerable signal leakage. The leakage function $|H_2(z)|$ will still < 1 and the condition will most probably be fulfilled (if $H_1(z)$ is not an amplifier).

Robustness of Gradient Estimates. In [2] it is shown that an important difference exists between the one-sided (AD) and symmetric (SAD) algorithms with respect to the quality of the gradient estimate. In the AD-algorithm each crosscorrelation coefficient is function of only one filter tap and the gradient is constant for stationary interference. In the SAD-algorithm the gradient depends also on the goodness of the estimate of the coefficients in the other filter. As a consequence the coupled equations give rise to an inherently unstable region, independent of the adaptation step. This unstable behaviour will NOT occur if all coefficients were set to zero, which helps us in choosing initial values.

PART II: FEEDBACK IMPLEMENTATION

Implementation of the Algorithm. Fig.2 shows the feedback implementation which directly gives the solutions $v_1(k)$ and $v_2(k)$. The coefficient adaptation formula is identical to the formulas used in the feedforward case, except that v appears instead of u:

$$\begin{aligned} w_1^{(k+1)}(m) &= w_1^{(k)}(m) + \mu_1(v_1(k)v_2(k-m)) \\ w_2^{(k+1)}(n) &= w_2^{(k)}(n) + \mu_2(v_2(k)v_1(k-n)) \end{aligned}$$

Stability. Looking at the feedback loop in the algorithm it is clear that $|W_1(z)W_2(z)|$ must be < 1 $\forall z$ on the unit circle. This stability condition is identical to the one derived for the feedforward implementation. Furthermore we assume one of either $w_1(0)$ or $w_2(0)$ to be 0 for implementation of the algorithm. This approach is very similar to the "blind signal separation" described in [4], but their work has been limited to scalar mixing instead of the filter (convolution) mixing treated here. Their (scalar) adaptive connections are placed in a feedback way, resulting in an algorithm with identical unstabilities and spurious solutions, as the ones described in our work. For convolutive mixtures the problem

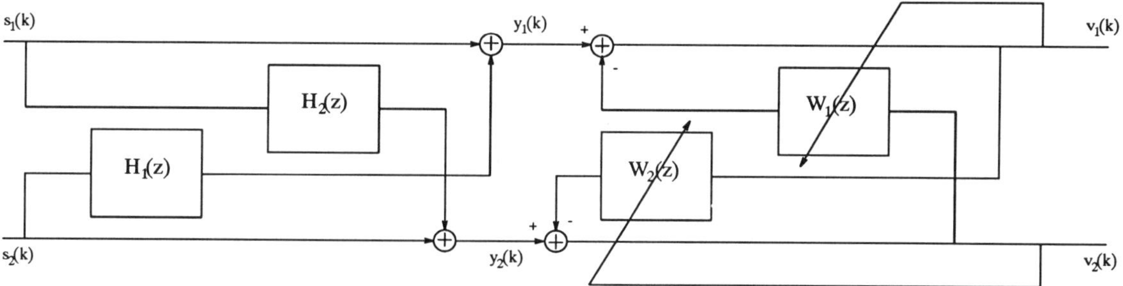

Figure 2: Feedback Implementation of a Symmetric Adaptive Noise Canceler

has also been adressed by [5] but without stability analysis. The choice of the adaptation constants μ_1 and μ_2 is very similar to the way this is done in the classical LMS noise canceler. For equal μ-values the feedforward and feedback algorithm show the same convergence speed. Choosing μ inversely proportional to the energy of the incoming signals $y(k)$ is a somewhat conservative but a "save" choice.

PART III: ANALYSIS OF A SIMPLIFIED CASE

Simplification. From now on we will restrict ourselves to a simplified case using only single tap filters [6], i.e.:

$$\begin{aligned} H_1(z) &= h_1(m)z^{-m} \\ H_2(z) &= h_2(n)z^{-n} \end{aligned}$$

Stability. Unstabilities due to positive feedback in the error propagation will be analysed in this paragraph. In the neigbourhood of the desired solution one can linearise the algorithm. Defining

$$\begin{aligned} \triangle_1^k(m) &= w_1^k(m) - h_1^k(m) \\ \triangle_2^k(n) &= w_2^k(n) - h_2^k(n) \\ \theta &= h_1(m)h_2(n) \end{aligned}$$

and replacing the signal products by their expected values, the matrix formulation looks as follows:

$$\begin{bmatrix} \triangle_1^{k+1}(m) \\ \triangle_2^{k+1}(n) \end{bmatrix} \approx \begin{bmatrix} 1-\mu_1\sigma_2^2 & \mu_1\theta\sigma_1^2 \\ \mu_2\theta\sigma_2^2 & 1-\mu_2\sigma_1^2 \end{bmatrix} \begin{bmatrix} \triangle_1^k(m) \\ \triangle_2^k(n) \end{bmatrix}$$

This linearisation is valid for both implementations of the algorithm. The absolute value of the eigenvalues of the linearised algorithm must be < 1. This leads to the following conditions:

$$\begin{aligned} \theta^2 &< 1 \ \mathit{or}\ |h_1(m)h_2(n)| < 1 \\ \mu_1 &> 0 \ \mathit{and}\ \mu_2 > 0 \\ 2(\mu_1\sigma_2^2 + \mu_2\sigma_1^2) &< 4 + \mu_1\sigma_2^2\mu_2\sigma_1^2(1-\theta^2) \end{aligned}$$

The third condition depends on θ but can be fulfilled if $\mu_1\sigma_2^2 + \mu_2\sigma_1^2 < 2$, independent of θ. Conditions 2 and 3 then define a tiangular region in the μ_1 μ_2 plane. Condition 1 is identical to the postprocessing and feedback constraint derived earlier and is a property of the mixtures. It is interesting to note that if one of the filters equals zero and therefore must not be estimated, the algorithm collapses to the classical LMS noise canceler and so do the stability conditions. This analysis is only true, however, in the immediate neighbourhood of the "desired solution", where $H_1(z)H_2(z) \approx W_1(z)W_2(z) \approx H_1(z)W_2(z) \approx H_2(z)W_1(z)$.

Alternate Equilibrium Points. The desired solution is an obvious equilibrium point of the algorithm. But in general "decorrelation" has multiple solutions and therefore the adaptive algorithm has multiple equilibrium points. An analytical expression for the cross power spectrum can easily be derived and the alternate solutions are shown to be:

$$\begin{aligned} w_1'(m) &= \frac{1}{h_1(m)}\frac{\sigma_1^2 + h_1^2(m)\sigma_2^2}{\sigma_2^2 + h_2^2(n)\sigma_1^2} \\ w_2'(m) &= \frac{1}{h_2(n)}\frac{\sigma_2^2 + h_2^2(n)\sigma_1^2}{\sigma_1^2 + h_1^2(m)\sigma_2^2} \\ w_1'(m)w_2'(n) &= \frac{1}{h_1(m)h_2(n)} \end{aligned}$$

The last equation reveals that $|w_1'(m)w_2'(n)| > 1$ whenever $|h_1(m)h_2(n)| < 1$ and vice versa. Stability analysis around this second equilibrium shows that it is UNSTABLE whenever the desired solution is stable, and that this solution is a stable equilibrium if the desired one is unstable.

Simulations. Simulations were carried out with both stable and unstable conditions for the filtercoefficients $h_1(m)$ and $h_2(n)$. Different signal energy ratio's were used with both $s_1(k)$ and $s_2(k)$ being white noise signals. Alternate solutions were theoretically calculated and convergence from zero initial conditions was controlled. The results of these simulations are summarized in Tab.1 and confirm:

- the alternate solution is unstable whenever the desired one is stable and vice-versa
- starting from zero initial conditions, the adaptive algorithm will converge to the desired solution if stable and to the alternate solution if the desired one is unstable
- for an intrinsically stable situation (i.e. $|h_1(m)h_2(n)| < 1$) the decorrelation criterion convergences to the correct solution

experiment		1	2	3	4
energy	σ_1^2	40	40	30	30
	σ_2^2	30	30	40	40
coeff.	h_1	1	2	1	2
	h_2	0.5	1	0.5	1
altern.	w_1'	1.75	1.143	1.473	1.357
solution	w_2'	1.143	0.437	1.357	0.368
corr. solution		stable	unstab.	stable	unstab.
convergence		corr.	alter.	corr.	alter.

Table 1: Simulation results for SAD stability analysis

Simulations with both $s_1(k)$ and $s_2(k)$ being speech signals gave satisfying results. Almost 80% of the energy of the unwanted signals was cancelled, resulting in a improved intelligibility, although distortions of the desired signals were observed.

Identical simulations were carried out with the feedback structure and it shows the same convergence speed, but a slightly better steady state behaviour. As for the speech signals less distortions of the desired signals were introduced.

CONCLUSION AND FUTURE WORK

In this paper we have reviewed two symmetric adaptive decorrelation (SAD) algorithms for signal separation. Important convergence and stability constraints were found, but many typical real-life applications fulfill these conditions. Therefore the algorithm is a very useful extension of the classical LMS noise canceler. For small adaptation constants, inherently unstable mixtures converge to a "phantom" solution. This non-uniqueness of the solution can probably be solved by using higher order statistics, although it is suggested in [3] that this might lead to unsurmountable unstabilities. Preliminary simulations still show the existence of a "phantom" solution, but here further work is needed.

REFERENCES

[1] **B.Widrow and S.D.Stearns.** *Adaptive Signal Processing.* Prentice-Hall, 1985.

[2] **D. Van Compernolle** and **S. Van Gerven.** Signal Separation in a Symmetric Adaptive Noise Canceler by Output Decorrelation. In *Proc. Int. Conf. Acoust., Speech & Signal Processing,* San Francisco, March 1992. To Appear.

[3] **D. Van Compernolle and S. Van Gerven.** Mathematical and Physical Constraints in Blind Signal Separation. In *IEEE Benelux & ProRISC Proceedings of the Workshop on Circuits, Systems and Signal Processing,* pages 187–192, Houthalen, April 8-9 1992.

[4] **C. Jutten and J. Herault .** Blind separation of sources, part i: An adaptive algorithm based on neuromimetic architecture. *Signal Processing,* 24.1:1–10, 1991.

[5] **M.J. Al-Kindi and J. Dunlop .** Improved adaptive noise cancellation in the presence of signal leakage on the noise reference channel. *Signal Processing,* 17.3:241–250, 1989.

[6] **S. Van Gerven.** Verslag van de activiteiten van het eerste mandaat van het doctoraatswerk: Verbetering van de opnamekwaliteit van spraaksignalen m. b. v. multimicrofoonopnamen. IWONL-verslag 1, K. U. Leuven, E.S.A.T., Augustus 1991.

SIGNAL PROCESSING VI: Theories and Applications
J. Vandewalle, R. Boite, M. Moonen, A. Oosterlinck (eds.)

HIGH ORDER LEARNING IN TEMPORAL REFERENCE ARRAY BEAMFORMING

Miguel A. Lagunas, A. Pagès-Zamora, A.Pèrez-Neira

ETSI de Telecomunicacion, Departamento de TSC
Apdo. 30002, 08080 BARCELONA, SPAIN

A filter bank able to acquire and track complex sinusoids in the presence of noise and doppler rate is presented. Firstly, the so-called extended Kalman filter (EKF) which estimates magnitude, phase and frequency of one sinusoid is derived. Afterwards, the final proposed scheme is examined. It consists in EKFs whose outputs are driven to the inputs crossconnecting them as inhibitory cells. Some simulation resuts are enclosed proving the excellent performance of the procedure making it a valid alternative to classical spectral estimation and time-frequency algorithms.

1. INTRODUCTION

The motivation of this work is to increase the efficiency of the communication link between satellite and users. The goal was faced to be reached from the subject of adaptive array beamforming with temporal reference [1].

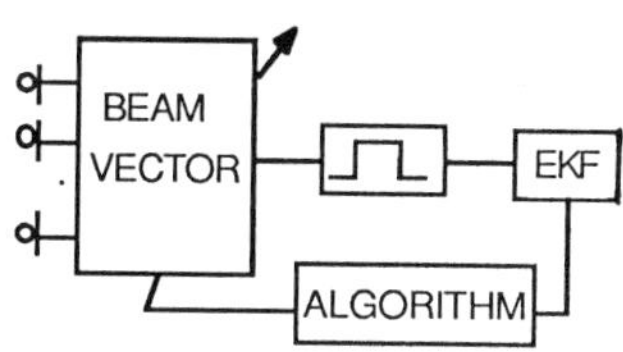

Fig.1. Basic structure of the receiver.

In these methods, the receiver system forms a dedicated beam to the user which is asking for it. For that purpose, the user sends a pure tone in a pre-assigned frequency band. From filtering the output of the beamformer, a noisy sinusoid is obtained in the receiver. Once this signal has been regenerated by an EKF is driven to modify the weigths of the array (beamvector) by means of different algorithms. The result is a beam which steers to the desired source and nulls the interferences as much as possible [2].

The work deals with the problem of collision produced when two users ask simultaneously for a dedicated beam (notice that in solving this problem, the performance of the communications link will be increased). To overcome this problem two beams are nedeed ,time or frequency multiplexed, each of them steering to one different user. For this reason, the receiver has to be able to regenerate both reference signals.

Some other works can be found in the litterature dealing with similar problems [3], [4]. In these works, one of the signals is assumed to be the desired one and the other an interference. Therefore, in such a system one signal must be stronger than the other; furthermore, normally either magnitude or frequency are assumed to be a priori known with no doppler rate in anyone of them. In fact, this was not our situation since the reference signals are ,actually, more or less at the same signal to noise ratio (SNR) and they show up with unknown doppler and doppler rate.

In order to formulate the problem, when a collision appears, the signal received in the satellite after front end amplification and filtering will be :

$$x(n) = e_1(n)\exp[j\phi_1(n)] + e_2(n)\exp[j\phi_2(n)] + \omega(n)$$

being e1 and e2 the magnitude of each pure tone, ϕ 1 and ϕ 2 the instantantaneous phases and ω is a zero mean, white gaussian noise (wgn). The frequency of both signals ,f1 and f2, is in a range of 1KHz up to 3KHz. The video bandwidth is 4KHz ,normalized 0.5 hereafter, with a central frequency of 2KHz, being the sampling rate of 8000 samples per second. The doppler effect is in the range of +/- 1'5KHz with a maximum doppler rate of 4KHz per second.

The SNR of each reference was defined as :

$$SNR_i = \text{average}\left[e_i^2(n)\right]/s^2 \quad ;s^2 = 1$$

It has been taken the average value because the magnitudes may fluctuate in time +/-3dB during tracking [5].

*This work was supported by the National Research Plan of Spain, PRONTIC 302/89. Underwater Communications.

2. EXTENDED KALMAN FILTER

In this section, the EKF is presented as a previous step to achieve the reported scheme.

In this application, the EKF is required to estimate magnitude, phase and frequency of a single source signal from a noisy measure of it.The measured signal consists in a pure tone with severe doppler plus wgn.

Readers can wonder why an analog phase-locked loop (PLL) hasn't been chosen to acquire and track a pure tone located in a given frequency band. First of all, it is required a digital processing since the control of the beamformer is carried out by a ASIC processor together with DSPs. On the other hand, it is necessary to obtain a signal that estimates not only phase and frequency but also the magnitude.

The received signal ,in a single source scenario, is assumed to be :

$$z(n) = e(n)\exp[\phi(n)] + \omega(n)$$

The model of the system ,from which the received signal comes, is described by the next state equations:

$$\underline{x}_{n+1} = \underline{\underline{F}}\,\underline{x}_n + \underline{v}_n \qquad ;\underline{x}_n^T = [e(n), \phi(n), f(n)]$$

Being $\underline{x}_n$ the state vector which contains the discret values of magnitude, phase and frequency of z(n), $\underline{\underline{F}}$ the so-called transition matrix which governs the evolution of the state vector, and $\underline{v}_n$ the innovation vector, a random vector which introduces uncertainty in the evolution of the state parameters.

$$\underline{\underline{F}} = \begin{bmatrix} 1 & 0 & 0 \\ 0 & 1 & 2\pi \\ 0 & 0 & 1 \end{bmatrix} \qquad E\left[\underline{v}_n \underline{v}_n^T\right] = \underline{\underline{Q}}(\text{diagonal})$$

The EKF is described by the following expressions:

$$\hat{\underline{x}}_{n+1} = \underline{\underline{F}}\,\hat{\underline{x}}_n + \underline{\underline{K}}_n \underline{\epsilon}_n$$

$$\hat{\underline{z}}_n^T = [\hat{e}(n)\cos\hat{\phi}(n)\ ,\ -\hat{e}(n)\sin\hat{\phi}(n)]$$

being $\hat{\underline{x}}_n$ the state estimation of the system, $\underline{\underline{K}}_n$ the so-called gain matrix, $\underline{\epsilon}_n$ the signal error between the measured signal ,z_n, and the estimated one,$\hat{z}_n$.(Notice that z_n, $\hat{z}_n$ and ϵ_n are 2-dimensional vectors composed by quadrature and phase components of the original signals).

Up to here, it has been exposed the general Kalman filter [6]. The version named extended Kalman filter comes from the first-order appproximation done to set the signal error ,ϵ_n, in a linear dependence of the state error, $\tilde{x}_n$(the difference between the actual state ,x_n, and the estimated one ,$\hat{x}_n$).

$$\underline{\epsilon}_n = \underline{\underline{H}}\,\tilde{\underline{x}}_n + \underline{\omega}_n \qquad ;\tilde{\underline{x}}_n = \underline{x}_n - \hat{\underline{x}}_n = \left[\tilde{e}(n), \tilde{\phi}(n), \tilde{f}(n)\right]$$

Being $\underline{\underline{H}}$ equal to :

$$\underline{\underline{H}} = \begin{bmatrix} \cos\hat{\phi}(n) & -\hat{e}(n)\sin\hat{\phi}(n) & 0 \\ -\sin\hat{\phi}(n) & -\hat{e}(n)\cos\hat{\phi}(n) & 0 \end{bmatrix}$$

This expression results from assuming small enough the parameters of state error.

From the orthogonality [6] between the state error prediction ,$\tilde{\underline{X}}_{n+1}$, and the signal error ,$\epsilon_n$, the gain matrix ,$\underline{\underline{K}}_n$, is obtained and ,therefore, the EKF is completely defined.

In summary, the algorithm which implements the EKF is given hereafter ,assuming initial values $\underline{x}o$ and $\underline{\Sigma}o$:

$$\underline{\epsilon}_n = \underline{z}_n - \hat{\underline{z}}_n \ ;\hat{\underline{z}}_n^T = [\hat{e}(n)\cos\hat{\phi}(n)\ ,\ -\hat{e}(n)\sin\hat{\phi}(n)]$$

$$\underline{\underline{E}} = E\left[\underline{\epsilon}_n \underline{\epsilon}_n^T\right] = \underline{\underline{H}}_n \underline{\underline{\Sigma}}_n \underline{\underline{H}}_n^T + \underline{\underline{L}} \ ;\underline{\underline{L}} = E\left[\underline{\omega}_n \underline{\omega}_n^T\right] \equiv \underline{\underline{I}}$$

$$\underline{\underline{K}}_n = \underline{\underline{F}}\,\underline{\underline{\Sigma}}_n \underline{\underline{E}}^{-1} \qquad \underline{\underline{\Sigma}}_{n+1} = \underline{\underline{F}}\,\underline{\underline{\Sigma}}_n \underline{\underline{F}}_n^T - \underline{\underline{K}}_n \underline{\underline{E}}\,\underline{\underline{K}}_n^T - \underline{\underline{Q}}$$

$$\hat{\underline{x}}_{n+1} = \underline{\underline{F}}\,\hat{\underline{x}}_n + \underline{\underline{K}}_n \underline{\epsilon}_n$$

where $\underline{\underline{\Sigma}}_n$ is the state error covariance matrix and $\hat{\underline{z}}_n$ the signal that estimates z_n in magnitude, phase and frequency.

3. MULTITONE TRACKER.

From the EKF of the previous section working over a single line, a novel system which tracks and acquires completely two or more pure tones is derived.

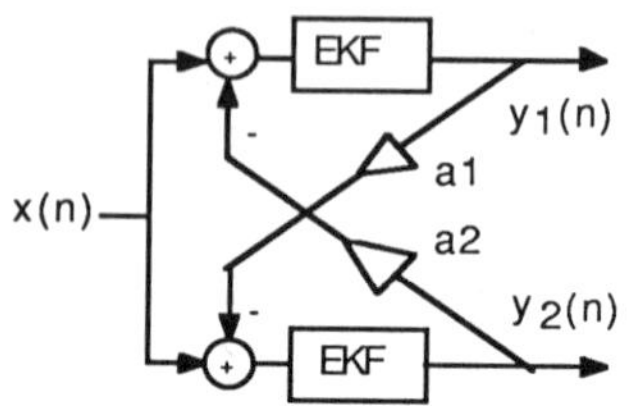

Fig.2. Multitone tracker for two signals.

It is composed by two EKF in parallel, with feedback from the output of each EKF to the input of the other one. These feedback signals are multiplied by two adaptive weigths, a1(n) and a2(n). Notice that the filter has a symmetry which could force both outputs ,y1(n) and y2(n), to be equal. This symmetry is broken thanks to the different values of the weights.

The evolution of a1(n) and a2(n) is controlled in order to make the outputs as orthogonal as possible. Intuitively, this rule seems to be correct since the desired signals have different frequency and ,therefore, they are orthogonal between each other.

In order to achieve the desired orthogonality between y1(n) and y2(n), it is necessary to minimize the scalar product between both outputs in a given interval of time T.

$$\varphi = \left(\int_T y_1(t)\ y_2^*(t) \right)^2 \qquad \text{Eq(1)}$$

It is desired a1 and a2 drive φ to a minimum. That is why it is obtained the derivate of function φ with respect to each weigth. We will focuss the case of a1 since the reader may conclude in the same manner the corresponding learning rule for a2.

$$\frac{d\varphi}{da_1} = \int\int_T y_1(t)\, y_2(\tau)\, y_2^*(t) \frac{dy_1^*(\tau)}{da_1} dt\, d\tau$$

Being the output of the upper EKF denoted by y1(t), we may say that y1 obeys to:

$$y_1 = g(x - a_1 y_2) \equiv g(z)$$

being g(.) a non-linear function. It is possible to express :

$$\frac{dy_1^*}{da_1} = \frac{dy_1^*}{dz^*} \frac{dz^*}{da_1} = (-y_2^*) \frac{dy_1^*}{dz^*}$$

In order to obtain a closed form of Eq(1) some approximations have to be carried over. Firstly, it is considered (dy_1^*/dz^*) as a constant and ,secondly, it is taken the instantaneous value of the integral assuming a small interval of time T. From these approximations, a linear rule ,based on gradient methods, has been chosen to update the weigths :

$$\frac{d\varphi}{da_1} = -\left|y_2(n)\right|^2 y_2^*(n)\ y_1(n)$$

$$a_1(n+1) = a_1(n) + m_1 \left|y_2(n)\right|^2 y_2^*(n)\ y_1(n) \qquad \text{Eq(2)}$$

being m1 a parameter which controls the trade-off between the speed of the weigths in getting a value near the optimum and the final accuracy. From Eq(2) is possible to infer the expression from which a2 is updated :

$$a_2(n+1) = a_2(n) + m_2 \left|y_1(n)\right|^2 y_1^*(n)\ y_2(n)$$

The structure formed by the EKFs in parallel with inhibitory cells was inspired in the pioneer work of C.Jutten [7] and further works of the same author [8]. In these works the author proposed an unsymmetrical rule for learning of the inhibitory cells. Afterwards ,in the field of high order signal processing applied to blind equalization [9], the four order objective for orthogonalizing two signals was reported. Herein, the authors prove that both approaches are connected and that unsymmetrical third order learning rules are strongly related with minimization of the cross-correlation between two signals.

One of the most important aspects of the method is the simplicity in the extension of the filter in order to acquire and track more than two signals. For instance, let us assume a signal which contains four different tones:

$$x(n) = \sum_{i=1}^{4} e_i(n) \exp\left[j\phi_i(n)\right] + \omega(n)$$

The filter can be dessigned as the neuron structure formed by four EKF of Fig.3. :

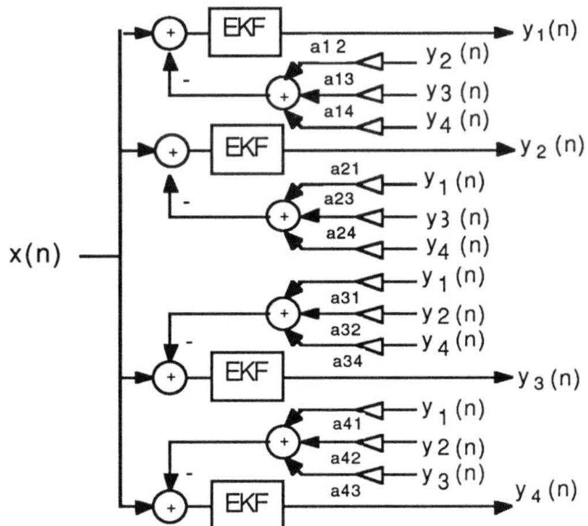

Fig.3. Multitone tracker of four signals

Each output is driven to the input of the other three EKFs multiplied by three different weights. The weight aij represents the weight which multiplies the output j and the result is driven to input i. Going on the same idea developed above, each weight aij is adaptive and its learning is done in order to make the output j and output i as orthogonal as possible. Therefore, as Eq(2) :

$$a_{ij}(n+1) = a_{ij}(n) + m_{ij} \left|y_j(n)\right|^2 y_j^*(n)\ y_i(n)$$

4. SIMULATIONS

The first group of simulations corresponds to the filter of Fig.4. The plots ,in this figure, represent magnitude and frequency ,versus discret time (samples), of the outcoming signals.The input signal contains two tones of (-3 dB) and (0 dB) of SNR. The doppler rate was sinusoidal with a maximum of +/- 1.5KHz and rate ,on average, of 3.2KHz/sec.

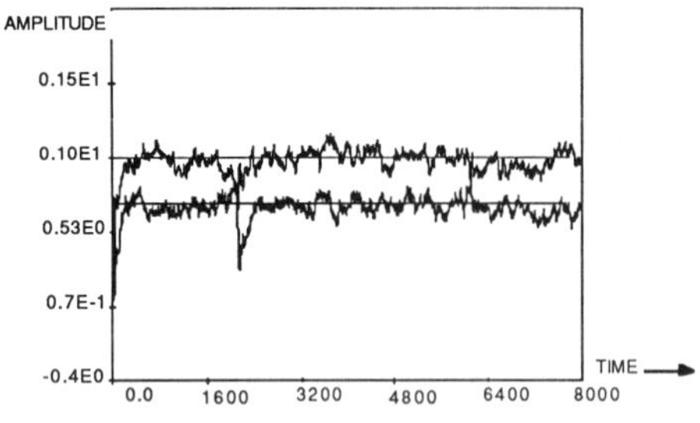

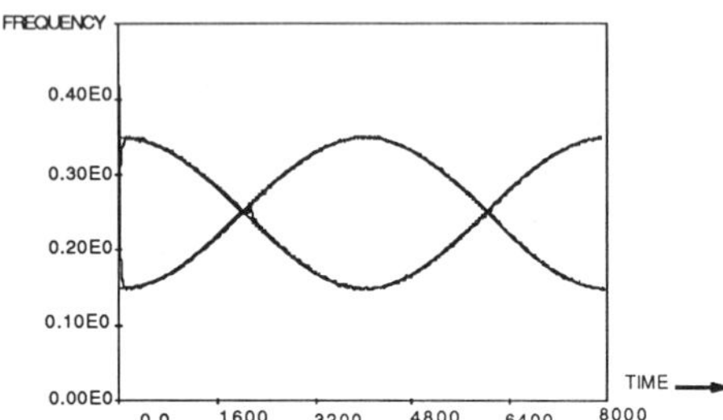

Fig.4, Magnitude and frequency versus time.

The figures have been obtained with m1=m2=1E-3 and the diagonal values of matrix Q has been set to 1E-4, 1E-5 and 1E-5. It can be observed that ,in the same signal, frequency is acquired before magnitude and the estimation is also better. The reason for this behaviour is the fact that the received signal has more noise in magnitude than in phase (that's why ,generally, phase modulations are more suitable than magnitude ones).

The following plots are magnitude and frequency from the filter of Fig.3. The diagonal values of matrix Q are set to 1E-6, 1E-6 and 1E-7. The values of mij are in the range from 1.5E-2 to 2E-2.

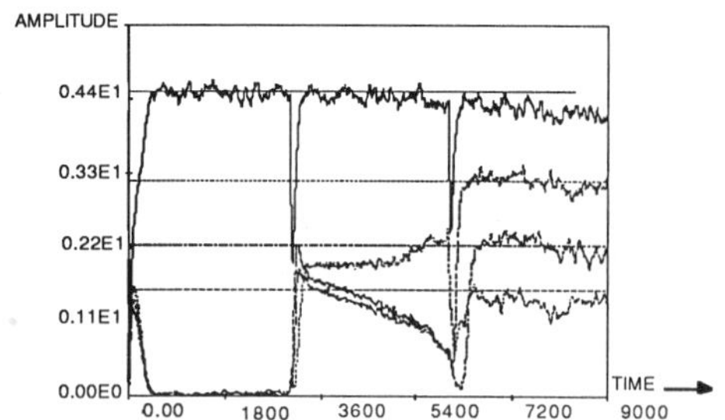

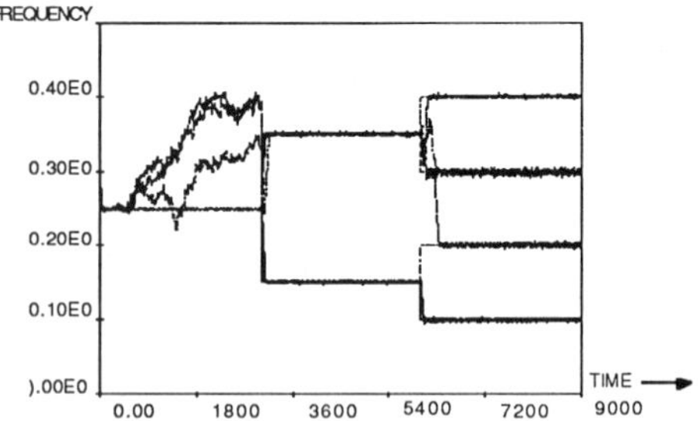

Fig.5, Magnitude and frequency versus time.

It is distinguished three different parts :with one, two and four incoming signals. When there are four inputs the SNR of them is of 10, 7, 4 and 1. It is important to point out that at the output are obtained as many signals as input signals are. For instance, firstly there is only one input signal which is acquired and tracked by only one output since the rest of the outputs have a null magnitude.

REFERENCES

[1] J.Fernandez, M.Lagunas, G.Vazquez, "Breadbording of a digital beamforming network" ERA-ESA Technical report, European Spatial Agency, Doc. #1897/01. Contract 8714/90/ NL/US(SC), The Netherlands, Dec 1990.

[2] M.A. Lagunas, M. Nàjar, A. Pèrez-Neira, "Comparative Analysis of High Resolution DOA Estimation and Time Reference Beamforming".ESA-ESTEC,PO 112052.

[3] F.A. Cassara, H. Schakter and G.H. Simowitz, "Acquisition behaviour of the Cross-Coupled PLL FM demodulator". IEEE Trans. on Communications, COM 28, no.6, pp 897-904, June 1980.

[4] R.L.Kirlin and Y.Su. "Maximum-likekihood multiple source tracking utilitsing multiple coupled PLLs". Proc. ICASSP 90, pp2851-2854, 1990.

[5] P.J.Parker B.D.O. Anderson. "Frequency Tracking of nonsinusoidal periodic signals in noise". Signal Processing, Vol.20, No.2 June 1990, pp.127-152.

[6] B.D.O. Anderson, J.B. Moore, "Optimal Filtering". Englewood Cliffs, NJ : Prentice Hall, 1979.

[7] C.Jutten, "Calcul Neuromimétique et traitement du signal". Doctoral thesis USM of Grenoble, June 1987.

[8] P.Comon, C.Jutten and J.Herault, "Blind Separation of sources, Part II: Problems statement". Signal Processing, Vol.24, No.1 July 1991, pp.11-20.

[9] Y. Chen, C.L. Nikias, J.G. Proakis, "Crimno: Criterion with memory nonlinearity for blind equalization", HOS Proc. Chamrouse Jule 1991, Ed. J.L. Lacoume, pp 87-90.

SIGNAL PROCESSING VI: Theories and Applications
J. Vandewalle, R. Boite, M. Moonen, A. Oosterlinck (eds.)

An Adaptive Morphological Filtering Method

Pauli Kuosmanen, Lasse Koskinen, and Jaakko Astola
Tampere University
Department of Mathematical Sciences
Tampere, Finland

Abstract

This paper introduces a new type of adaptive filters. It is based on mathematical morphology, especially on soft morphological filters. In this paper, we also present a generalization of soft morphological filters. In the same way as standard and soft morphological filters this generalization can be expressed and analyzed in the framework of stack filters. Examples demonstrating the usefulness of the new adaptive morphological filtering method are provided. Finally, a biomedical application where promising results are obtained with the new method is presented.

1. Introduction

The field of adaptive systems design has been the subject of considerable effort of research in the area of control and signal processing for more than 30 years, cf. *e.g.* Honig *et al.*[4]. During this time adaptive filters have been applied to many applications including medical ones, Linkers[10]. In this paper we present a new adaptive morphological filtering method and a biomedical application.

As to morphological methods, they have proven useful in many image processing applications Serra[14], Haralick *et al.*[3], Maragos *et al.*[11,12], and Dougherty *et al.*[2]. The main reason for their success is that they can well take into account the geometrical shape of the analyzed objects. On the other hand they have high sensitivity to noise or defects in the image, Koskinen *et al.*[6,7], and Stevenson *et al.*[13]. Recently Koskinen *et al.*[8] introduced soft morphological operations, which have a better performance than standard morphological operations in noisy conditions.

The idea of soft morphological operations is to relax the standard definitions a little, in such a way that a degree of robustness is achieved while most of the desirable properties of standard morphological operations are maintained. Standard discrete morphological filters, which process signals by sets, are based on local maximum and minimum operations. In soft morphological filters maximum and minimum are replaced by more general weighted order statistics.

2. Generalized Soft Morphological Operations

We now define generalized soft morphological operations. They are most naturally defined in the framework of weighted order statistics.

Let A and B, $A \subset B$, be finite subsets of $\mathbb{Z}^2$ and let k and i be natural numbers such that $1 \leq k < |B|$, $1 \leq i < k$. For $x \in \mathbb{Z}^2$ and $T \subset \mathbb{Z}^2$ we denote the "translated set" by T_x, that is,

$$T_x = \{x + s : s \in T\}.$$

The symmetric set of T is the set

$$T^s = \{-t : t \in T\}.$$

We denote repetition operation of any object, in our case real numbers, by $\diamondsuit$, that is $k\diamondsuit x = \overbrace{x,\ldots,x}^{k\ times}$. A multiset is a collection of objects, where repetition of objects is allowed. For example $\{1,1,1,2,3,3\} = \{3\diamondsuit 1, 2, 2\diamondsuit 3\}$ is a multiset.

The basic generalized soft morphological operations are generalized soft erosion and generalized soft dilation:

Generalized soft erosion of f by $[B, A, k, i]$ is denoted by $f \ominus [B, A, k, i]$ and is defined by

$f \ominus [B, A, k, i](x) = k^{th}$ smallest value of the multiset

$$\{k\diamondsuit f(a) : a \in A_x\} \cup \{i\diamondsuit f(b) : b \in (B - A)_x\}.$$

Generalized soft dilation of f by $[B, A, k, i]$ is denoted by $f \oplus [B, A, k, i]$ and is defined by

$$f \oplus [B, A, k, i](x) = k^{th} \text{ largest value of the multiset } \{k \Diamond f(a) : a \in A_x\} \cup \{i \Diamond f(b) : b \in (B - A)_x\}.$$

The set B is called the structuring set, A its centre, $B - A$ its boundary, k the order index of its centre and i the order index of its boundary.

Thus, the definition of generalized soft morphological operations is similar to the definition of soft morphological filters. The differences are that the condition $1 \leq k \leq \min\{|B|/2, |B - A|\}$ is relaxed and that the soft boundary has also an order index.

Example 1. Let the structuring set be $B = \{(-1, 0), (0, 1), (0, 0), (0, -1), (1, 0)\}, A = \{(0, 0)\}$ and the order indexes $k = 4, i = 1$. Then the generalized soft erosion by $[B, A, k, i]$ is defined by $f \ominus [B, A, k, i](x) = 4^{th}$ smallest value of the multiset $\{f(x_1 - 1, x_2),\ f(x_1, x_2 + 1),\ f(x_1, x_2),\ f(x_1, x_2),\ f(x_1, x_2),\ f(x_1, x_2),\ f(x_1, x_2 - 1),\ f(x_1 + 1, x_2)\}$. The output of this filter at point $x = (x_1, x_2)$ is $f(x)$ unless all the values of set $\{f(b) : b \in (B - A)_x\}$ are smaller than $f(x)$, when the output is the largest value of set $\{f(b) : b \in (B - A)_x\}$.

Generalized soft opening and closing are defined in the usual way:

Generalized soft opening of f by $[B, A, k, i]$ is denoted by $f_{[B,A,k,i]}$ and is defined by

$$f_{[B,A,k,i]}(x) = (f \ominus [B, A, k, i]) \oplus [B^s, A^s, k, i](x).$$

Generalized soft closing of f by $[B, A, k, i]$ is denoted by $f^{[B,A,k,i]}$ and is defined by

$$f^{[B,A,k,i]}(x) = (f \oplus [B, A, k, i]) \ominus [B^s, A^s, k, i](x).$$

Generalized clos-opening and open-closing are defined to be closing followed by opening and opening followed by closing respectively.

3. Some Properties of Generalized Soft Morphological Filters

Generalized soft morphological filters have some attractive properties. Kuosmanen *et al.*[9] constructed a class of idempotent generalized soft morphological filters: generalized soft morphological operations by $[B, A, |B| - 1, 1]$, where B is the structuring element and $A = \{(0, 0)\}$, (two-dimensional) or $A = \{0\}$ (one-dimensional). Kuosmanen *et al.*[9] also showed that generalized soft morphological filters preserve shapes and details of the original image well and filter out noise at the same time.

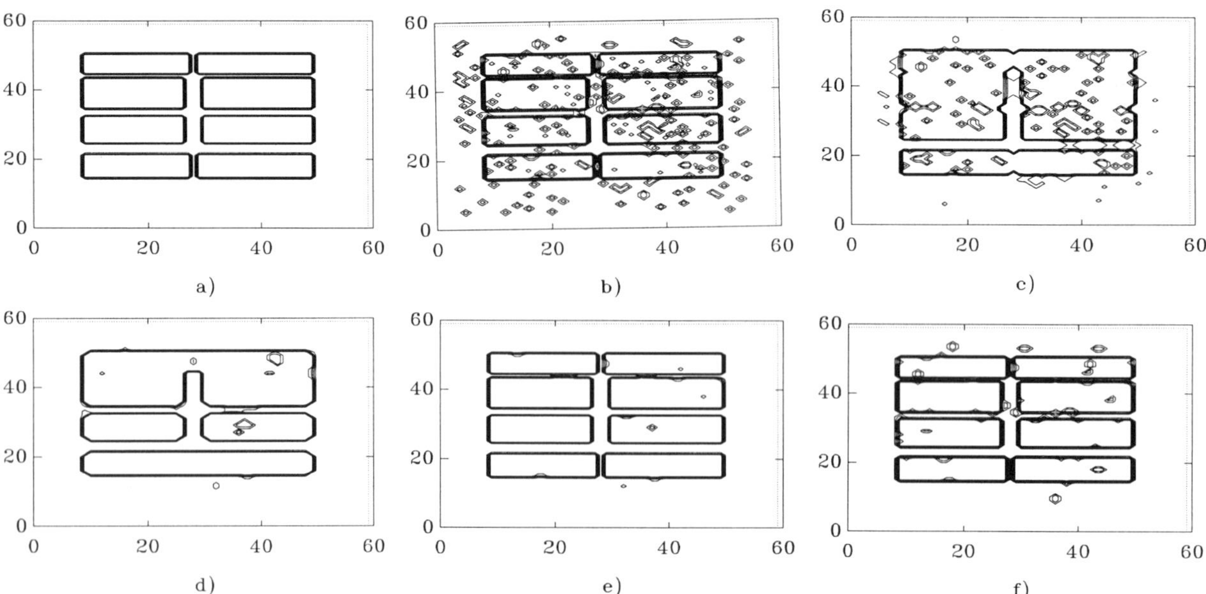

Figure 1.

Figure 1 illustrates differences in 2-dimensional generalized soft closing when B and A are the same as in our example 1. (a) is the contour plot of the original image, (b) is the contour plot of a corrupted image. In (c) – (f) are images after closing by B, where the order index k varies from 1 to 4 respectively and $i = 1$ in all cases. Thus (c) is the image after standard closing, which is known to be idempotent. Generalized soft closing in (f) belongs in the class of idempotent generalized soft morphological filters Kuosmanen *et al.*[9] constructed.

We can see from figure 1, that the same structuring set with the same centre acts differently when the order index of the centre varies. Koskinen *et al.*[8] pointed out that soft closing is less sensitive to impulsive noise than standard closing, which can be seen by comparing (c) and (d). Both standard closing (c) and soft closing with index $k = 2$ (d) had difficulties with valleys of width 1, which they filled. Soft closing (d) also lost the corners of the original image.

4. A New Adaptive Morphological Filter

Once the structuring element and its hard centre are fixed the parametres which determine the generalized soft morphological filter are the order indexes k and i. The structuring element and its hard centre are determined using a priori knowledge about the filtering problem. Also the allowed variation on order indexes can be determined before if one aim of the filtering is to preserve some details of the input. As an analytical optimization of the indexes k and i would require unreasonably careful modelling of the problem (data permitting such modelling is seldom available) an adaptive scheme for choosing the indexes is proposed.

The scheme is based on minimization of a cost function over the filters having the same structuring set. The choise of the cost function depends on the problem. Our experiments have shown that mean absolute error or Hamming distance between the filtered and desired signal perform well while mean squared error tends to destroy shapes. Also cost fuctions which reveal the difference of the shape of the desired and filtered data/image can be used.

One aim of the filtering process is often to remove outliers. Barnett *et al.*[1] divided outliers into two types. The first type is when an isolated measurement error is superimposed on an otherwise reasonable realization of the process. Its effect can be obvious and it does not reflect in the values of adjacent observations. The other type is when a more inherent discordancy arises and is reflected by the structure of the process in neighbouring (usually later) observations. The detection of outliers becomes more problematic then.

Figure 2 illustrates also the difficulties and importance of the study of outliers. In figure 2 a part of a process is presented. Althought it may look out stationary and have some outliers is does not have any. In fact the process is

$$X_t = 0.5 * X_{t-1} + 0.9 * X_{t-1} * Y_{t-1} + Y_t,$$

where Y_t are independent $N(0, 1)$.

If the expected number of outliers in the measured data is known, it is possible to design an adaptive generalized soft morphological filtering system, which optimizes the percentage of removed possible outliers. The optimization process is performed over the class of idempotent generalized soft morphological filters.

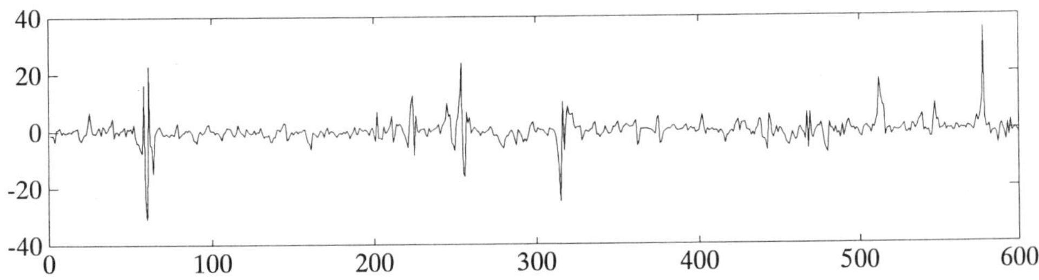

Figure 2.

5. Application

The new filtering method has been in use in a biomedical experiment at Tampere University. The purpose of the experiment was to examine the levels and variability in blood pressure of overweight persons before and after a reducing diet. The pulse rates were also examined. Kalli *et al.*[4] stated that blood pressure signal processing needs non–linear smoothing, which is capable of preserving sharp changes in the data and still able to filter out noise and other fast components superimposed on the data. The measurement of blood pressure has been done by using a new continous, non-invasive tehcnique in which the amount of samples varies from 50 to more than 100 in a minute depending on the pulse rate.

In this experiment there exist many criteria for the goodness of the output after filtering process. For example, there is a need for such a filter, which removes the most possible outliers (of type one), when the percentage of them is known. Our experiments showed that an adaptive method is necessary for achieving this and the method presented before worked well.

The proposed adaptive morphological methods turned out to be promising biological signal processing methods, because the filters are able to preserve and identify trends and patterns of different length from longterm records. The freedom for the choice of the cost function makes it suitable for different situations where the criteria for the goodness of the output varies.

References

[1] V. Barnett and T. Lewis, "Outliers in Statistical Data", John Wiley & Sons Ltd., Norwich, 1984.

[2] E. Dougherty and C. Giardiana, "Morphological Methods in Image and Signal Processing", Prentice Hall Englewood Cliffs, 1988.

[3] R. Haralick, S. Stenberg, and X. Zhuang, "Greyscale morphology", in *Proceedings IEEE Conf. Comput. Vis Pattern Recogn.*, Miami, USA, pp. 543–550, June 1986.

[4] M. L. Honig and D. G. Messerschmitt, *Adaptive Filters: Structures, Algorithms, and applications,* Kluwer Academic Publishers, Norwell, MA, 1984.

[5] S. Kalli, P. Heinonen, V. Turjanmaa, and N. Saranummi, "Identifying Patterns and Profiles in 24–hour Blood Pressure Recordings", in *Proceedings of the Fifth International Symposium on Ambulatory Monitoring,* pp 113–118, Padua 1985, Cleup Editore.

[6] L. Koskinen, J. Astola, and Y. Neuvo, "Statistical Properties of Discrete Morphological Filters", in *Proceedings IEEE Int. Symp. Circ., Syst.*, ISCAS-1990, pp. 1219–1222, May 1990.

[7] L. Koskinen, J. Astola, and Y. Neuvo, "Analysis of Noise Attenuation in Morphological Image Processing", SPIE/SPSE Symp. on Nonlinear Image Processing II, San Jose, USA, s. 102–113, February 1991.

[8] L. Koskinen, J. Astola, and Y. Neuvo, "Soft Morphological Filters" , SPIE Symp. on Image Algebra and Morphological Image Processing II, San Diego, USA, s. 262–270, July 1991.

[9] P. Kuosmanen, L. Koskinen, and J. Astola, "Detail Preserving Morphological Filtering", in *Proceedings of the 11th International Conference on Pattern Recognition,* September 1992 (Submitted).

[10] D. A. Linkers, "Short–time–series Spectral Analysis of Biomedical Data", *IEE Proceedings Analysis part A,* Vol. 129, no. 9, pp 663–672, 1982.

[11] P. Maragos, and R. Schafer, "Morphological Filters - Part I, Their Set Theoretic Analysis and Relations to Linear Shift-invariant Filters", *IEEE Trans. Acoust., Speech, and Signal Processing,* Vol. ASSP-35, pp. 1153–1169, August 1987.

[12] P. Maragos, and R. Schafer, "Morphological Filters - Part II, Their Relations to Median, Order-Statistics, and Stack Filters", *IEEE Trans. Acoust., Speech, and Signal Processing,* Vol. ASSP-35, pp. 1170–1184, August 1987.

[13] R. Stevenson, and G. Arce, "Morphological Filters: Statistics and Further Syntactic Properties", *IEEE Trans. on Circuits and Systems,* Vol. 34, pp. 898–911, August 1986.

[14] S. Serra, *Image Analysis And Mathematical Morphology,* Academic Press, London, 1988.

SIGNAL PROCESSING VI: Theories and Applications
J. Vandewalle, R. Boite, M. Moonen, A. Oosterlinck (eds.)

A REALIZATION METHOD OF 2-D SEPARABLE-DENOMINATOR DIGITAL FILTER BASED ON REDUCED-DIMENSIONAL DECOMPOSITION

Syu-ji Kawasaki, Nozomu Hamada

Department of Electrical Engineering, Faculty of Science and Technology, KEIO University, 223, Hiyoshi, Yokohama, Japan

abstract The class of 2-D separable-denominator(SD) digital filter is attractive for designing and implementing the 2-D signal processing. This paper presents a new technique for synthesizing 2-D SD digital filter based on the reduced-dimensinal decomposition(RDD). The proposed filter structure is the cascade of a SIMO(single-input/ multi-output) output tap 1-D filter and a MISO input tap 1-D filter. Thus the multiple outputs of the first filter are injected to the inputs of the second filter. The systematic synthesis procedure is derived by giving the explicit input/output relationships for two dual 1-D filters. The result carried out a specific canonical structure of a RDD idea proposed by Lin, et.al[1].

1. INTRODUCTION

In digital processing of multidimensional data, such as video or image, seismic, radar, sonar, etc., multidimensional digital filter is an important tool, and many efficient filter structure have been proposed. Especially, the filter that is represented by separable-denominator(SD) transfer function is attractive in the sense ;
1. its stability is easily checked by 1-D stability criteria
2. it can be implemented by simple way

In the last decade, a general treatment of 2-D digital filter is developed via the reduced-dimensional decomposition(RDD) by T.Lin, et al[1], or several other reserchers. In those works, Roesser's state-space representation model is used and their balanced realization, low sensitivity or minimal roundoff noise structure, and controllability/observability aspects are studied.

On the other hand, Taguchi and Hamada[2] proposed lattice form structure for realizing 2-D SD digital filter. They have done the work with equivalent transformation of direct form state-space model into the lattice form realization. The obtained structure can be also derived as a specific realization model of RDD point of views.

And the conventional realization methods based on RDD stand on such concepts as, singular value decompo sition[5], LU decomposition[6], etc.

This paper gives more systematic and simpler method of realizing 2-D SD digital filter than the conventional RDD realizations using simple algebraic calculations, and reveals that the practical RDD realization procedure is brought by dividing whole system into three blocks, single-input/multi-output(SIMO) 1-D filter, linear matrix operation block, and MISO 1-D filter.

2. THE REALIZATION METHOD

Any 2-D SD transfer function can be written as

$$H(z_1,z_2) = \frac{N(z_1,z_2)}{D(z_1,z_2)}$$

$$= \frac{[1\ z_2^{-1}\ \cdots\ z_2^{-L}] \begin{bmatrix} n_{00} & \cdots & n_{0K} \\ \vdots & \ddots & \vdots \\ n_{L0} & \cdots & n_{LK} \end{bmatrix} \begin{bmatrix} 1 \\ z_1^{-1} \\ \vdots \\ z_1^{-K} \end{bmatrix}}{\left(\sum_{l=0}^{L} d_{2l}\, z_2^{-l}\right)\left(\sum_{k=0}^{K} d_{1k}\, z_1^{-k}\right)}$$

$$\triangleq \frac{Z_2^T\, N\, Z_1}{D_2(z_2)\, D_1(z_1)} \quad (1)$$

(T: transpose operation)

If one would like to realize this 2-D transfer function via a method of matrix decomposition, such as the singular value decomposition[5], the Jordan decomposition[6], LU decomposition[7], or maximum rank decomposition[1], though each algebraic meaning of these decompositions are different, all these methods will reduce (1) into the inner product form

$$\begin{bmatrix} \eta_0(z_2) \\ \eta_1(z_2) \\ \vdots \\ \eta_r(z_2) \end{bmatrix}^T \bullet \begin{bmatrix} \zeta_0(z_1) \\ \zeta_1(z_1) \\ \vdots \\ \zeta_r(z_1) \end{bmatrix} \quad (2)$$

where r is the rank of matrix N. These methods are called RDD, and are used to realize 2-D SD digital filter with r 1-D filter pairs $\zeta(z_1), \eta(z_2)$(Fig.1). In these methods, denominators $D_1(z_1), D_2(z_2)$ are realized within each block respectively. And (2), the inner product of two 1-D transfer function vector can be understood as the extension of simple product of two scalar 1-D transfer functions

$$H(z_1,z_2) = H_2(z_2)H_1(z_1) \quad (3)$$

But in these methods, the number of multipliers are relatively large.

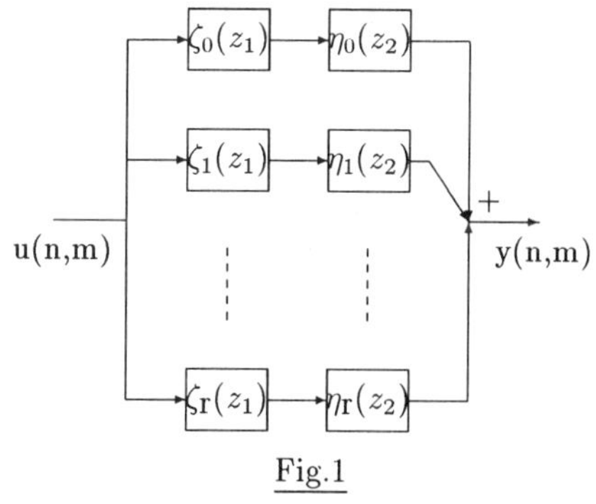

Fig.1

The proposed realization method is briefly summerized as follow ; that is, we realize the desired 2-D SD digital filter with two 1-D digital filters and intermediate matrix operation. Denominators $D_1(z_1), D_2(z_2)$ are realized in each 1-D filter block respectively, and this is the same idea as the method of inner product of two 1-D transfer function vector, as expressed in (2).

Numerator polynomial $N(z_1, z_2)$ is, however, realized more flexible fashion ; it is done such that the whole 2-D SD digital filter will be realized by cascade connection of three blocks ;
1. the first block : 1-D digital filter with delay element only z_1, transfer function vector $\boldsymbol{P}\boldsymbol{Z}_1/D_1(z_1)$ where $\boldsymbol{P}$ is a (K+1)×(K+1) matrix
2. the intermediate block : memoryless coeffiecients multiplier with (L+1)×(K+1) matrix $\boldsymbol{M}$
3. the last block : 1-D digital filter with delay element only z_2, transfer function vector $\boldsymbol{Z}_2^T\boldsymbol{Q}/D_2(z_2)$ where $\boldsymbol{Q}$ is a (L+1)×(L+1) matrix

And it is illustrated in Fig.2. In Fig.2, x_k (k=0,1,...,K) are i-th internal state-variable of the first block, and $\widehat{x}_l$ (l=0,1,...,L) are l-th input to the last block, i.e.,

$$\widehat{\boldsymbol{x}}(n,m) = \boldsymbol{M}\boldsymbol{x}(n,m) \tag{4}$$

where $\boldsymbol{x} = [x_0, ..., x_K], \widehat{\boldsymbol{x}} = [\widehat{x}_0, ..., \widehat{x}_K]$.

Now, let us assume we could realize 1-D SIMO first block by using some specific structure. Then, its transfer function vector from input $U(z_1, z_2) \triangleq Z_{1,2}[u(n,m)]$ to $X_k(z_1) \triangleq Z_1[x_k(n,m)]$ can be written in the form

$$\begin{bmatrix} \zeta_0(z_1) \\ \zeta_1(z_1) \\ \vdots \\ \zeta_K(z_1) \end{bmatrix} = \frac{1}{D_1(z_1)} \begin{bmatrix} P_0(z_1) \\ P_1(z_1) \\ \vdots \\ P_K(z_1) \end{bmatrix}$$

$$= \frac{1}{D_1(z_1)} \begin{bmatrix} p_{00} & \cdots & p_{0K} \\ \vdots & \ddots & \vdots \\ p_{K0} & \cdots & p_{KK} \end{bmatrix} \begin{bmatrix} 1 \\ z_1^{-1} \\ \vdots \\ z_1^{-K} \end{bmatrix} = \frac{\boldsymbol{P}\,\boldsymbol{Z}_1}{D_1\,(z_1)} \tag{5}$$

Similarly, if we could realize 1-D MISO last block in some specific structure, its transfer function from $\widehat{X}_l(z_1) \triangleq Z_1[\widehat{x}_l(n,m)]$ to $Y(z_1, z_2) \triangleq Z_{1,2}[y(n,m)]$ can be written as

$$[\eta_0(z_2)\ \ \eta_1(z_2) \cdots \eta_L(z_2)]$$
$$= \frac{1}{D_2(z_2)}\ [Q_0(z_2)\ \ Q_1(z_2) \cdots Q_L(z_2)]$$

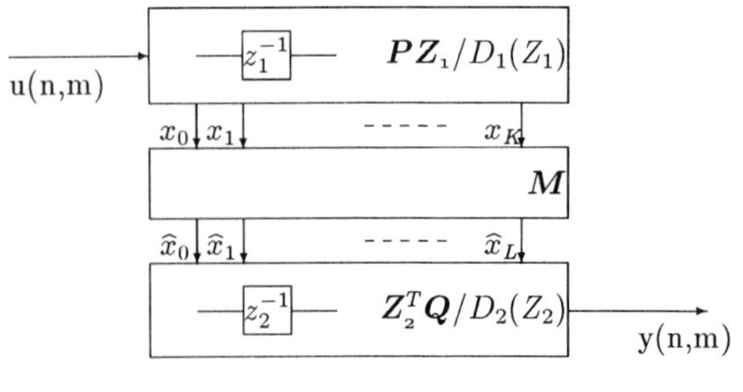

Fig.2

$$= \frac{1}{D_2(z_2)} \begin{bmatrix} 1 & z_2^{-1} \cdots z_2^{-L} \end{bmatrix} \begin{bmatrix} q_{00} & \cdots & q_{L0} \\ \vdots & \ddots & \vdots \\ q_{0L} & \cdots & q_{LL} \end{bmatrix}$$

$$= \frac{\boldsymbol{Z}_2^T\boldsymbol{Q}}{D_2(z_2)} \tag{6}$$

Therefore we get the realization of $H(z_1, z_2)$ of whole 2-D system as the cascade of three blocks ;

$$\left(\frac{1}{D_2(z_2)} \begin{bmatrix} Q_0(z_2) \\ Q_1(z_2) \\ \vdots \\ Q_L(z_2) \end{bmatrix} \right)^T \boldsymbol{M} \left(\frac{1}{D_1(z_1)} \begin{bmatrix} P_0(z_1) \\ P_1(z_1) \\ \vdots \\ P_K(z_1) \end{bmatrix} \right)$$

$$= \frac{\boldsymbol{Z}_2^T\boldsymbol{Q}\boldsymbol{M}\boldsymbol{P}\boldsymbol{Z}_1}{D_2(z_2)D_1(z_1)} \tag{7}$$

Now the intermediate coefficient is determined by

$$\boldsymbol{M} = \boldsymbol{Q}^{-1}\boldsymbol{N}\boldsymbol{P}^{-1} \tag{8}$$

here we assumed both $\boldsymbol{Q}$ and $\boldsymbol{P}$ are regular matrices.

In the conventional inner product methods described above require r×(K+L) multipliers in general. But in our method, K×L multipliers are needed. If $\boldsymbol{N}$ is full rank, r=K, and if K=L, then multipliers required in our method become as half as the required in conventional methods.

3. REALIZATIONS BY DIRECT AND LATTICE FORM

In this section, we realize 2-D SD digital filter applying the method described section 1. Direct form and Gray-Markel lattice form are adopted as realization structure.

Direct form

Here we consider the mutually dual pair, controller canonical form and observer canonical form[12] shown in Fig.3 to realize 2-D SD digital filter. And let us adopt the controller canonical form as indicated in Fig.3(a) with dashed line for 1-D SIMO first block. Its transfer function vector become as

$$\frac{1}{D_1(z_1)} \begin{bmatrix} P_0(z_1) \\ P_1(z_1) \\ \vdots \\ P_K(z_1) \end{bmatrix} = \frac{1}{D_1(z_1)} \begin{bmatrix} 1 \\ z_1^{-1} \\ \vdots \\ z_1^{-K} \end{bmatrix} = \frac{\boldsymbol{I}_K\boldsymbol{Z}_1}{D_1(z_1)} \tag{9}$$

And, on the other hand, if we adopt observer canonical form as indicated in Fig.3(b) with dashed line for 1-D MISO last block, its transfer function vector become as

$$\frac{1}{D_2(z_2)} \begin{bmatrix} Q_0(z_2) \\ Q_1(z_2) \\ \vdots \\ Q_L(z_2) \end{bmatrix} = \frac{1}{D_2(z_2)} \begin{bmatrix} 1 \\ z_2^{-1} \\ \vdots \\ z_2^{-L} \end{bmatrix} = \frac{\boldsymbol{Z}_2^T\boldsymbol{I}_L}{D_2(z_2)} \tag{10}$$

where $\boldsymbol{I}_K$, $\boldsymbol{I}_L$ are (K+1)-th, (L+1)-th order identity matrices respectively. Then in this occasion, intermediate coefficient matrix $\boldsymbol{M}$ become

$$\boldsymbol{M} = \boldsymbol{N} \tag{11}$$

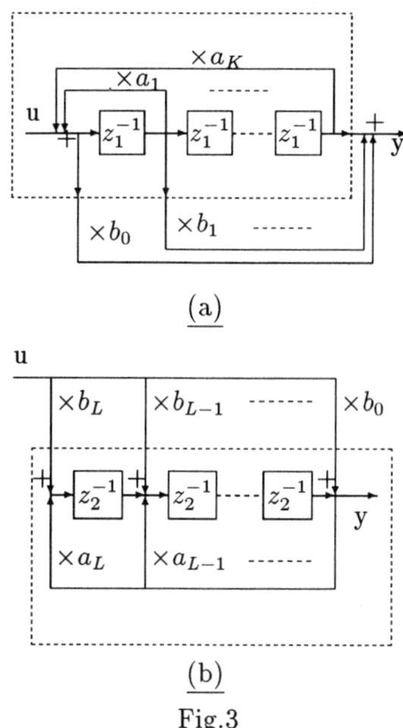

(a)

(b)

Fig.3

This realization form may be refered as 2-D direct form (SD), because all multiplier coeffiecients in this filter is directly correspond to the coeffiecients of desired 2-D transfer function. And, similar realization patterns may be obtained by adopting another dual pair, controllerbility canonical form and observability canonical form[12].

Lattice form

Here, we adopt ordinary ARMA type lattice filter(output-tap type)[8][9][10] for 1-D SIMO first block, and input-tap type lattice filter for 1-D MISO last block (Fig.4).

Accoding to well-known Levinson algorithm in the linear prediction theory, the transfer function vector of (1-D) ordinary output-tap type lattice filter can be written as

$$\frac{1}{D_1(z_1)}\begin{bmatrix} P_0(z_1) \\ P_1(z_1) \\ \vdots \\ P_K(z_1) \end{bmatrix} = \frac{1}{A_K^{(1)}(z_1)}\begin{bmatrix} z_1B_0^{(1)}(z_1) \\ z_1B_1^{(1)}(z_1) \\ \vdots \\ z_1B_K^{(1)}(z_1) \end{bmatrix}$$

$$= \frac{1}{A_K^{(1)}(z_1)}\begin{bmatrix} b_{00}^{(1)} & & & \mathbf{o} \\ b_{10}^{(1)} & b_{11}^{(1)} & & \\ \vdots & & \ddots & \\ b_{K0}^{(1)} & \cdots\cdots & & b_{KK}^{(1)} \end{bmatrix}\begin{bmatrix} 1 \\ z_1^{-1} \\ \vdots \\ z_1^{-K} \end{bmatrix}$$

$$= \frac{\boldsymbol{B}^{(1)}\boldsymbol{Z}_1}{A_K^{(1)}(z_1)} \qquad (12)$$

where $P_k(z_1)$ (k=0,1,...,K) are the numerator polynomials of transfer function between input and each k-th output tap, and are k-th order in z_1^{-1}. Here, $b_{ij}^{(1)}$ is the coeffiecient of z_1^{-j} of $zB_i^{(1)}$ determined by Levinson algorithm

$$\begin{bmatrix} A_{k+1}^{(1)}(z_1) \\ z_1B_{k+1}^{(1)}(z_1) \end{bmatrix} = \begin{bmatrix} 1 & \Delta_k \\ \Delta_k & 1 \end{bmatrix}\begin{bmatrix} A_k^{(1)}(z_1) \\ z_1B_k^{(1)}(z_1) \end{bmatrix} \qquad (13)$$

and especially $b_{k0}^{(1)}$ (k=0,1,...,K) are equal to unity, and $z_1B_0^{(1)}(z_1) = 1$.

On the other hand, transfer function vector of input-tap type lattice filter can be written as (see Appendix for derivation)

$$\frac{1}{D_2(z_2)}\begin{bmatrix} A_0^{(2)}(z_2) \\ A_1^{(2)}(z_2) \\ \vdots \\ A_L^{(2)}(z_2) \end{bmatrix} = \frac{1}{A_L^{(2)}(z_2)}\begin{bmatrix} c_0A_0^{(2)}(z_2) \\ c_1z_2^{-1}A_1^{(2)}(z_2) \\ \vdots \\ c_Lz_2^{-L}A_L^{(2)}(z_2) \end{bmatrix}$$

$$= \frac{1}{A_L^{(2)}(z_2)}\begin{bmatrix} a_{00}^{(2)} & a_{01}^{(2)} & \cdots & a_{0L}^{(2)} \\ & a_{11}^{(2)} & & \vdots \\ & & \ddots & \vdots \\ \mathbf{o} & & & a_{LL}^{(2)} \end{bmatrix}\begin{bmatrix} 1 \\ z_2^{-1} \\ \vdots \\ z_2^{-L} \end{bmatrix}$$

$$= \frac{\boldsymbol{Z}_2^T\boldsymbol{A}^{(2)}}{A_L^{(2)}(z_2)} \qquad (14)$$

where $A_l^{(2)}(z_2)$ (l=0,1,...,L) are the numerator polynomials of transfer function between l-th input tap and output, and are all L-th order. Here, $c_0A_0^{(2)}(z_2) = A_L^{(2)}(z_2)$, and $a_{ll}^{(2)}$ (l=0,1,...,L) are equal to unity.

In this lattice form relization, we can easily calculate coefficient matrix $\boldsymbol{M}$ as $[\boldsymbol{A}^{(2)}]^{-1}\boldsymbol{N}[\boldsymbol{B}^{(1)}]^{-1}$, because both $\boldsymbol{B}^{(1)}$, $\boldsymbol{A}^{(2)}$ are triangular matrices.

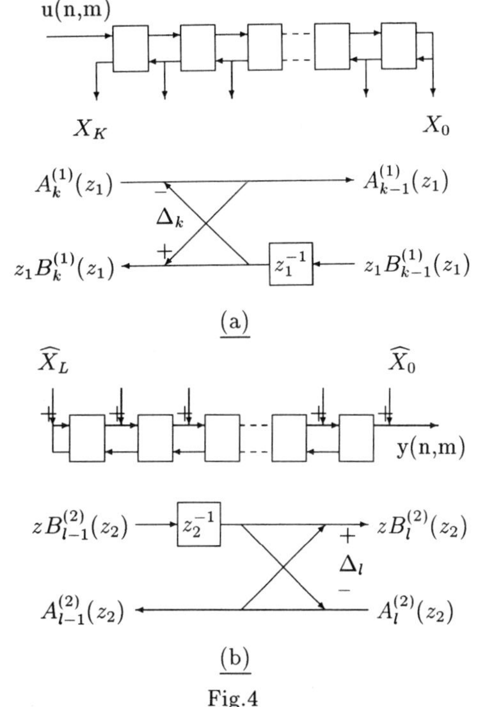

(a)

(b)

Fig.4

4. Concluding remarks

In this paper, we presented an approach for realzing 2-D SD denominator digital filter. The concept of the proposed method is based on the RDD by which three blocks, i.e. z_1-directional operation block, intermediate block, and z_2-directional operation block were assembled into 2-D system.

This method can be said of the approach that most utilized the advantages of SD transfer function, because the first and the last block are constructed independently so as to realize each denominator polynomials $D_1(z_1)$, $D_2(z_2)$.

APPENDIX

Derivation of (14):

Basic idea of finding transfer function between u_i and y is as follow. That is, we divide whole input-tap lattice filter into three parts $F_i(z)$, $G_i(z)$, and $zB_{L-i}(z)/A_{L-i}(z)$ as shown in Fig.5.

If we take state $V_i(z)$ at the position shown in Fig.5, then

$$\frac{V_i}{U_i} = \frac{1}{1 - F_i \dfrac{zB_{L-i}}{A_{L-i}}} \tag{15}$$

and

$$\frac{Q_i}{D_2(z_2)} = \frac{Y}{U_i} = G_i \frac{V_i}{U_i} = \frac{G_i}{1 - F_i \dfrac{zB_{L-i}}{A_{L-i}}} \tag{16}$$

where zB_{L-i}/A_{L-i} are transfer function of (L-i)-th order lattice filter of right side of input u_i, and zB_{L-i} are reciprocal polynomials of A_{L-i}. We can find zB_{L-i}/A_{L-i} from $A_L(z)$ by using Levinson's stepdown recursion. Then, using Levinson's recursion, transfer function F_i and G_i become

$$F_i = \frac{zB_{L-i}V_i}{A_{L-i}V_i} = \frac{zB_{L-i}}{A_{L-i}}, G_i = \frac{A_L V_i}{A_{L-i}V_i} = \frac{A_L}{A_{L-i}} \tag{17}$$

F_i and G_i has recursive relation such as

$$F_i = \frac{z^{-1}(F_{i-1} - \Delta_{L-i+1})}{1 - \Delta_{L-i+1}F_{i-1}} \tag{18}$$

$$\begin{aligned} G_i &= \frac{A_{L(i)}}{A_{L-i}} = \frac{A_L}{A_{L-i+1}} \frac{A_{L-i+1}}{A_{L-i}} \\ &= G_{i-1}(z^{-1} + \Delta_{M-(i-1)}F_i) \end{aligned} \tag{19}$$

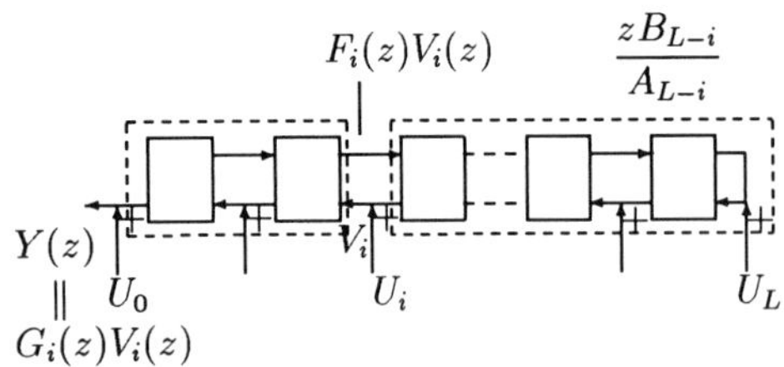

Fig.5

Using (18), (19), we can exactly simplify the calculation of F_i, G_i. From $F_i, G_i, zB_{L-i}/A_{L-i}$ as mentioned above, we can determine transfer function $Q_i/D_2(z_2)$ of $(u_i \mapsto y)$ by (7). For example, $Q_0/D_2, Q_1/D_2, Q_2/D_2, Q_3/D_2$ become as

$$\begin{aligned} \frac{Q_0}{D_2} &= \frac{A_L}{A_L} = 1 \\ \frac{Q_1}{D_2} &= \frac{z^{-1} + \Delta_L(-\Delta_L z^{-1})}{1 - (-\Delta_L z^{-1})\dfrac{z(-\Delta_L A_L + zB_L)/(1-\Delta_L^2)}{(A_L - \Delta_L zB_L)/(1-\Delta_L^2)}} \\ &= \frac{z^{-1}(A_L - \Delta_L zB_L)}{A_L} = \frac{(1-\Delta_L^2)z^{-1}A_{L-1}}{A_L} \\ \frac{Q_2}{D_2} &= \frac{(1-\Delta_L^2)(1-\Delta_{L-1}^2)z^{-2}A_{L-2}}{A_L} \\ \frac{Q_3}{D_2} &= \frac{(1-\Delta_L^2)(1-\Delta_{L-1}^2)(1-\Delta_{L-2}^2)z^{-3}A_{L-3}}{A_L} \end{aligned}$$

It is easy to show

$$\frac{Q_i}{D_2} = \frac{\left(\prod_{i=0}^{i-1}(1-\Delta_{L-i}^2)\right) z^{-i} A_{L-i}}{A_L} \tag{20}$$

References

[1] T.Lin, M.Kawamata, T.Higuchi. "Decomposition of 2-D Separable-Denominator Systems: Existence, Uniqueness, and Applications.", IEEE Trans. Circuit Syst., vol.CAS-34, No3, pp.292-296, March 1987

[2] A.Taguchi, N.Hamada. "Realization of Separable-Denominator Two-Dimensional Digital Filters with Lattice Structures" IECE, Japan, trans. Vol.J68-A No.12, '85/November

[3] M.Y.Dabbagh, W.E.Alexander. "Multiprocessor Implementation of 2-D Denominator-Separable Digital filter for Real-Time Processing." IEEE, Trans. Acoust. Speach, Signal Processing, vol. ASSP-37, pp.872-881, June 1989.

[4] A.N.Venetsanopoulos, B.G.Mertzios. "Decomposition of Multidimensional Filters." IEEE, Trans. Circuits Syst., vol.CAS-30, pp.915-917, Dec. 1983.

[5] S.Treitel, J.L.Shanks. "The Design of Multistage Separable Planar Filters." IEEE, Trans. Geosci.Electron, vol.GE-9, pp.10-27, Jan. 1971.

[6] A.N.Venetsanopoulos, B.G.Mertzios. "A General implementation Technique for two-dimensional digital filters." in Proc. 6th Summer Symp. Circuit Theory, Prague, Czecho, July. 1982, pp.176-180.

[7] C.L.Nikias, A.P.Chrysafis, A.N.Venetsanopoulos. "The LU decomposition Theorem and its implication to the realization of two-dimensional digital filters." IEEE Trans. Acoustic, Speech, Signal Processing, vol.ASSP-33, pp.694-711, June. 1985.

[8] A.H.Gray, J.D.Markel. "Digital Lattice and Ladder Filter Synthesis", IEEE Trans.on Audio Electroacoust., vol.AU-21, No.6, pp.491-500, Dec.1973.

[9] A.H.Gray. "Passive Cascaded Lattice Digital Filters", IEEE Trans.on Circuit Syst., vol.CAS-27, No.5, pp.337-344, May.1980.

[10] A.H.Gray, J.D.Markel. "Linear Prediction of Speech", Berlin, Germany: Springer-Verlag,1976.

[11] S.Y.Kung, B.C.Levy, M.Morf, and T.Kailath. "New results in 2-D systems theory, Part II." Proc. IEEE, vol.65, pp.945-961, June. 1977.

[12] T.Kailath. "Linear Systems", PLENTICE HALL (1980).

SIGNAL PROCESSING VI: Theories and Applications
J. Vandewalle, R. Boite, M. Moonen, A. Oosterlinck (eds.)

ADAPTIVE PREDICTION AND INTERPOLATION FILTERS FOR RADAR DATA PROCESSING

Michał Tuszyński and Andrzej Wojtkiewicz

Institute of Electronic Fundamentals, Warsaw University of Technology
Nowowiejska 15/19, 00-665 Warsaw, POLAND

This paper discusses application of adaptive prediction and interpolation filters for Moving Target Indicator (MTI) filtering in coherent Doppler radars. The first part of the paper considers use of adaptive interpolation filters, instead of adaptive prediction filters as MTI filters in Doppler radars. It is shown that these filters can achieve higher improvement factors (IF) than standard prediction filters. The second part of the paper presents an overview of adaptive filtering of nonuniformly sampled signals and considers use of such filters in radar systems.

1. INTRODUCTION

Suppression of clutter and man-made interference in coherent Doppler radars widely used in air traffic control systems is still an important research subject [1]. Moving Target Indicator (MTI) filters used in Doppler radars for suppression of clutter should reject echoes from stationary and slowly moving objects (ground clutter) and precipitation echoes (weather clutter) while passing through useful signals from airplanes (target signals).

Doppler frequencies of airplane signals may be much higher than the average sampling frequency. If the sampling was uniform (constant pulse repetition frequency, PRF), the MTI filter would have so-called "blind speeds" - deep nulls in the frequency response; this means that some target (aircraft) signals would be suppressed by the MTI filter and pass undetected. Periodically nonuniform sampling (staggered PRF), which "fills" these nulls and flattens the frequency response in the passband, is often used to avoid associated loss in detection probability.

As mentioned above, an MTI filter should not only reject low-frequency signals (ground clutter) but also reject weather clutter, which may have a quite considerable mean Doppler frequency. If the clutter spectrum was known exactly, an optimal filter could be designed. In practice, however, a clutter spectrum is not known and, furthermore, this spectrum changes in range, azimuth and time. Obviously, some form of adaptation to a clutter is necessary to achieve near-optimal clutter rejection. The first part of the paper introduces improvement factor and optimal MTI filters and discusses links between optimal MTI filters and Pisarenko harmonic decomposition. The next part of the paper considers use of adaptive interpolation filters, instead of adaptive prediction filters for MTI filtering. It is shown that interpolation filters can achieve higher improvement factors (IF) than prediction filters. This part also presents recursive formulas for computing interpolation filter coefficients and suggestions for realizations of adaptive interpolation filters in a given data case. The last part of the paper presents an overview of adaptive filtering of nonuniformly sampled signals and considers use of such filters in radar systems. This part presents basic properties of adaptive filters working with such signals and an algorithm for fast inversion of a time-varying correlation matrix.

2. ADAPTIVE MTI FILTERS

MTI filters are usually analyzed under assumption that the target signal may be adequately modeled as a stochastic complex harmonic signal. The stochastic clutter signal is usually modeled as an AR process of a low order [1].

The purpose of an MTI filter is to reject clutter and pass through useful target signal in such a way as to maximize the improvement factor (IF), defined as: "the signal-to-clutter ratio at the output of the clutter filter divided by the signal-to-clutter ratio at the input of the clutter filter, averaged uniformly over all target radial velocities of interest" [2]:

$$IF = \frac{SCR_{out}}{SCR_{in}} \qquad (1)$$

where SCR (Signal to Clutter Ratio) is a ratio of powers of a target and a clutter signal at the MTI filter input and output correspondingly.

The optimal MTI filter is a filter which maximizes improvement factor defined by (1). It was proved [3] that the coefficient vector of the optimal MTI filter is equal to the eigenvector corresponding to the smallest eigenvalue $\lambda_{min}(N)$ of the $N \times N$ clutter correlation matrix $\boldsymbol{R}_N$. It is interesting to note that the optimal MTI filter is the same as the *eigenfilter* in Pisarenko harmonic decomposition (PHD) [4]. This gives better insight into behavior of an optimal filter, but it seems has not been noticed before.

The roots of the transfer function of the optimal MTI filter are located on the unit circle. The frequencies of these roots coincide with the frequencies of the sinusoids in the solution of the PHD model for the correlation matrix $\boldsymbol{R}_N$. Let us remind that the PHD technique assumes that the random process may be modelled as a sum of stochastic sinusoids in white noise.

The PHD model may also be used for extending the autocorrelation sequence, in a way similar to, for instance, maximum entropy extension. It turns out [5], that this extension by PHD is equivalent to the minimization of the maximum IF, that is to setting $\lambda_{min}(N+1) = \lambda_{min}(N)$. It is easy to see that if indeed the clutter signal follows PHD model, then the optimal filter (eigenfilter) simply eliminates individual sinusoids; actually, the eigenfilter is a prediction polynomial associated with the positive semidefinite autocorrelation matrix $\boldsymbol{R}_N - \lambda_{min}(N)\boldsymbol{I}$.

The conclusion of the above discussion is that the maximization of the IF implies, in a way, the PHD model of the autocorrelation function. However, this may not be very compatible with a physical nature of a clutter signal; it seems that the AR (or ARMA) model is nearer to the physical reality. Moreover, the implementation of an adaptive optimal MTI filter is difficult, because it requires solving an eigenvalue problem in real time.

Many researchers [1] proposed using adaptive linear prediction filters. It was found that in many cases the IF of the prediction filter is not much less than the IF of the optimal MTI filter. Nevertheless it might be useful, both from practical and theoretical points of view, to find an adaptive filter which is relatively easy to implement in real time, has computational complexity comparable to a prediction filter and to have the IF greater than a prediction filter. In the next part of the paper we show that the interpolation filter meets all of these objectives.

3. ADAPTIVE INTERPOLATION FILTERS

Prediction filters use only past samples of the signal to estimate the current sample while interpolation filters [6,7] use both past and future samples for linear estimation of the current sample. Interpolation filters, although not causal, can nevertheless be used as MTI filters, because a radar system can usually tolerate delay in the MTI filter output signal.

The linear interpolation is defined as optimal (in the mean square sense) estimation of a sample of a stochastic signal $y(n)$ by K preceding and L following samples. The interpolation error is the output signal of a non-causal FIR interpolation filter $G_{K,L}(z)$:

$$\tilde{e}_{K,L}(T) = y(T) - \tilde{y}_{K,L}(T) = \sum_{i=-L}^{K} g_{K,L}(i)\, y(T-i), \quad g_{K,L}(0)=1 \tag{2}$$

In a given correlation matrix case, the coefficient vector of the interpolation filter $G_{K,L}(z)$ is easily found as a normalized $K+1$-th row of the inverse of the correlation matrix [6]. It may be proved [6] that the normalization factor is equal to the interpolation error variance $\sigma^2_{K,L}$. In the infinite interpolation case $(K,L \to \infty)$ the frequency response of the interpolation filter $G(\omega)$ is equal to the inverse of the input signal power spectrum $S(\omega)$:

$$G(\omega) = \frac{1}{S(\omega)} \tag{3}$$

It is easy to see that the coefficients $g(i)$ of the infinite order interpolation filter may be found as a convolution of the infinite order prediction filter coefficients $a(i)$:

$$g(i) = \frac{\hat{\sigma}^2}{\tilde{\sigma}^2}\,[a(i) * a^*(-i)] \tag{4}$$

where $\tilde{\sigma}^2$ is the interpolation error variance and $\hat{\sigma}^2$ is the prediction error variance. The ratio of the variances is [6]:

$$\frac{\tilde{\sigma}^2}{\hat{\sigma}^2} = \frac{1}{\dfrac{1}{2\pi}\displaystyle\int_{-\pi}^{\pi} \frac{d\omega}{S(\omega)}} = 1 + \sum_{i=1}^{\infty} |a(i)|^2 \tag{5}$$

Let us consider now interpolation of an AR process of the order p. In this case all of the higher order coefficients of the prediction filter are equal to 0, and it follows (4) that the coefficients $g(i)$ for $|i| > p$ are also equal to 0.

The IF of the MTI filter $H(z)$ may be written as:

$$IF = \left[\frac{1}{2\pi}\int_{-\pi}^{\pi} |H(\omega)|^2 d\omega\right] \frac{\sigma_x^2}{\sigma_y^2} \tag{6}$$

where σ_x^2 is the variance of the input signal and σ_y^2 is the variance of the MTI filter output signal. We may now use (5) and (6) to compute a ratio of improvement factors for interpolation filter IF_{int} and prediction filter IF_{pred}:

$$\frac{IF_{int}}{IF_{pred}} = \left[\frac{\displaystyle\int_{-\pi}^{\pi} |G(\omega)|^2 d\omega}{\displaystyle\int_{-\pi}^{\pi} |A(\omega)|^2 d\omega}\right] \frac{\hat{\sigma}^2}{\tilde{\sigma}^2} = \sum_{i=-p}^{p} |g(i)|^2 \tag{7}$$

Obviously, this ratio is not less than one ($g(0)=1$).

In practice, this ratio of interpolation and prediction IF will depend on the nature of the input signal and orders of both filters. Simulations performed so far indicate that the IF of the interpolation filter of sufficient order is usually greater than IF of the prediction filter of the same order. It seems that in practice interpolation filters should perform better than prediction filters. For example, in one simulation we compared the IF of the optimal, interpolation and prediction filters of the fourth order for a gaussian shaped sixth order AR process. The improvement factors of the optimal, interpolation and prediction filters were 39.66 dB, 39.65 dB and 38.11 dB.

Let us also remark that the maximization of the IF is not equivalent to the maximization of the detection probability. The interpolation filters, as well as prediction filters, may also be used in different, optimal detection schemes, for instance IBDA [1].

There remain two important problems: first, how to implement adaptive interpolation filter and second, how to filter nonuniformly sampled signals; the latter problem is considered in the next section.

Coefficients of the interpolation filter in the given correlation matrix case are easily found as a row of the inverse of the correlation matrix [6]. We will show that by application of a Gohberg-Semencul decomposition [8,9] it is possible to find simple order recursions for lattice implementation of the interpolation filter.

The important Gohberg-Semencul theorem [9,10] states that an inverse of a Toeplitz matrix may be written as a difference of two products of triangular matrices:

$$\boldsymbol{R}^{-1} = \boldsymbol{A}\boldsymbol{A}^T - \boldsymbol{B}\boldsymbol{B}^T \tag{8}$$

where elements of $\boldsymbol{A}$ and $\boldsymbol{B}$ are variance-normalized coefficients of a stationary, N-th order prediction filter.

When the correlation matrix is not known and must be estimated from data, two correlation matrix estimates are often used: so-called prewindowed and covariance matrices. As these matrices no longer have Toeplitz structure, the Gohberg-Semencul theorem cannot be applied directly. In a recent paper [10] the Gohberg-Semencul formula was generalized to include these estimates. Let $\hat{\boldsymbol{R}}_t$ denote the prewindowed covariance matrix estimated at the time instant t. The inverse matrix $\hat{\boldsymbol{R}}_t^{-1}$ may then be written as:

$$\hat{\boldsymbol{R}}_t^{-1} = \boldsymbol{A}_t\boldsymbol{A}_t^T - \boldsymbol{B}_t\boldsymbol{B}_t^T \tag{9}$$

where the variance-normalized matrix $\boldsymbol{A}_t$ is:

$$\boldsymbol{A}_t = \begin{bmatrix} a_N^t(0) & 0 & \cdots & 0 & 0 \\ a_N^t(1) & a_N^{t-1}(0) & \cdots & 0 & 0 \\ & & \cdots & & \\ a_N^t(N-2) & a_N^{t-1}(N-3) & \cdots & a_N^{t-N+1}(0) & 0 \\ a_N^t(N-1) & a_N^{t-1}(N-2) & \cdots & a_N^{t-N+1}(1) & a_N^{t-N}(0) \end{bmatrix} \tag{10}$$

and the matrix $\boldsymbol{B}_t$ is:

$$\boldsymbol{B}_t = \begin{bmatrix} 0 & 0 & \cdots & 0 & 0 \\ b_N^{t-1}(N-1) & 0 & \cdots & 0 & 0 \\ & & \cdots & & \\ b_N^{t-1}(2) & b_N^{t-2}(3) & \cdots & 0 & 0 \\ b_N^{t-1}(1) & b_N^{t-2}(2) & \cdots & b_N^{t-N}(N-1) & 0 \end{bmatrix} \tag{11}$$

The $a_N^t(k)$ and $b_N^t(k)$ are variance-normalized coefficients of time-varying N-th order forward and backward prediction filters. In case of Toeplitz matrices time indexes may be ignored. The generalized Gohberg-Semencul decomposition (9) makes possible expressing coefficients of interpolation filters by coefficients of fixed order time-varying prediction filters. The transfer function of the interpolation filter may be found recursively from [11]:

$$C_{K+1,L-1}^t(z) = C_{K,L}^{t-1}(z) + a_N^t(K+1)z^{(K+1)}A_N^t(z) - b_N^{t-1}(L)z^{-L}B_N^{t-1}(z) \tag{12}$$

where $C_{K,L}^t(z) = c_{K,L}^t(0)\,G_{K,L}^t(z)$ denotes variance-normalized time-varying interpolation filter transfer functions, $c_{K,L}^t(0)$ is a free coefficient of the $C_{K,L}(z)$, and where $A_N^t(z)$ and $B_N^t(z)$ are transfer functions of the variance-normalized time-varying forward and backward predictors. Let us note that the transfer function of an interpolation filter may also be recursively computed from transfer function of a lower order interpolation filter [8,11]:

$$C_{K,L+1}^t(z) = C_{K,L}^t(z) + a_{N+1}^t(L+1)z^{-(L+1)}A_{N+1}^t(z) \tag{13}$$

3. ADAPTIVE PROCESSING OF NONUNIFORMLY SAMPLED SIGNALS

As mentioned in the introduction, Doppler radars often work with staggered PRF, that is with periodically nonuniform sampling. Adaptive processing of such signals is not, to our knowledge, described in the literature.

Let us assume that an analog signal $x(t)$ is sampled at time instants $t_{np}=nT_0+t_p$, $n=0,\pm1,\pm2,\ldots$, $p=0,1,\ldots,M-1$, $t_{p+1}>t_p>0$, $t_0=0$, $t_{rM}=rT_0$, $r=0,\pm1,\pm2,\ldots$. Sampling instants belonging to one T_0 period may also be written as $t_p=p\tau+\delta_p$, where δ_p is a displacement from the average sampling period $\tau=T_0/M$ ($\delta_0=\delta_{rM}=0$). The clutter signal vector is:

$$\boldsymbol{x}_m = [x(t_{m+N}), x(t_{m+N-1}), \ldots, x(t_m)]^T \tag{14}$$

The time-varying $N\times N$ correlation matrix $\boldsymbol{R}_N^{(m)}$ is:

$$\boldsymbol{R}_N^{(m)} = \mathscr{E}\{\boldsymbol{x}_m\boldsymbol{x}_m^{*T}\} \tag{15}$$

It may be proved that in this case the coefficients of the optimal, prediction and interpolation filters are expressed as in a case of uniform sampling; the only difference is that the time-varying matrix $\boldsymbol{R}_N^{(m)}$ is substituted for constant matrix $\boldsymbol{R}_N$.

The periodically time-varying correlation matrix (15) is no longer Toeplitz, but the structure of this correlation matrix is very special. A displacement rank $\nabla \boldsymbol{R}$:

$$\nabla \boldsymbol{R} = \text{rank}\,(\boldsymbol{R} - \boldsymbol{Z}\boldsymbol{R}\boldsymbol{Z}^T) \tag{16}$$

where $\boldsymbol{Z}$ is a low shift matrix (a matrix which has ones on a first lower subdiagonal), is often used for measuring "closeness" of a matrix $\boldsymbol{R}$ to a Toeplitz matrix. A displacement rank of a Toeplitz matrix is, generally, equal to 2. It is well known that there exist efficient algorithms for inverting matrices of low displacement rank.

Unfortunately, the displacement rank $\nabla \boldsymbol{R}_N^{(m)}$ of the time-varying correlation matrix may be very large. We found, however, that because of the special structure of the time-varying correlation matrix, it has a very low time-varying displacement rank. The *time-varying displacement rank* $\nabla_t \boldsymbol{R}^{(t)}$ of a time-varying matrix $\boldsymbol{R}^{(t)}$ has been introduced in a recent paper [10], and is defined as:

$$\nabla_t \boldsymbol{R}^{(t)} = \text{rank}\,(\boldsymbol{R}^{(t)} - \boldsymbol{Z}\boldsymbol{R}^{(t-1)}\boldsymbol{Z}^T) \tag{17}$$

It is easy to see that the time-varying displacement rank of a matrix $\nabla_t \boldsymbol{R}_N^{(m)}$ (15) is, generally, equal to 2:

$$\nabla_t \boldsymbol{R}_N^{(t)} = \text{rank}\,(\boldsymbol{R}_N^{(m)} - \boldsymbol{Z}\boldsymbol{R}_N^{(m-1)}\boldsymbol{Z}^T) = 2 \tag{18}$$

Using techniques presented in [10] it is relatively easy to write a generalized Gohberg-Semencul formula (9) for a case of nonuniform sampling, where $a_N^t(k)$ and $b_N^t(k)$ are now variance-normalized coefficients of time-varying N-th order forward and backward prediction filters for the time-varying correlation matrix $\boldsymbol{R}_N^{(m)}$. Extension of this technique may also allow modification of given data algorithms for nonuniformly sampled signals.

Let us conclude this part of the paper with presentation of a lattice prediction filter for nonuniformly sampled signals. Let us assume that we know the $N-1$-th order time-varying forward and backward predictors $A_{N-1}^{(m)}(z)$, $B_{N-1}^{(t-1)}$. The N-th order forward and backward predictors can be expressed as a combination of known predictors:

$$\begin{aligned} A_N^{(m)}(z) &= A_{N-1}^{(m)}(z) - k_N^{(m)} z^{-1} B_{N-1}^{(m-1)}(z) \\ B_N^{(m-1)}(z) &= z^{-1} B_{N-1}^{(m-1)}(z) - k_N^{(m)*} A_{N-1}^{(m)} \end{aligned} \tag{19}$$

These order update recursions form a basis of a lattice implementation of a prediction filter for nonuniformly sampled signals. Let us point out that, generally, it is not possible to find any time-update recursions, because there may not be any correspondence between correlation matrices for various time instants. Fortunately, in case of a periodically nonuniform sampling we have $\boldsymbol{R}_N^{(m)} = \boldsymbol{R}_N^{(m+nM)}$, where M is a length of a sampling pattern, so the coefficient vectors of the predictors are also periodical $A_N^{(m)}(z) = A_N^{(m+nM)}(z)$.

4. CONCLUSIONS

This paper discussed application of adaptive prediction and interpolation filters for MTI filtering in coherent Doppler radars. We proposed use of adaptive interpolation filters, instead of adaptive prediction filters as MTI filters in Doppler radars. It was shown that these filters can achieve higher IF than prediction filters.

The second part of the paper presented basic properties of adaptive prediction and interpolation filters working with nonuniformly sampled signals. We think that this research direction is very perspective, for instance it may allow identification of wideband (wider than Nyquist range) random signals.

REFERENCES

[1] Haykin S.: "Radar signal processing", IEEE ASSP Mag., April 1985, pp. 2-18.

[2] "IEEE Standard Dictionary of Electrical and Electronics Terms", 3rd ed., IEEE Inc., New York, 1984.

[3] Capon J.: "Optimum weighing functions for the detection of sampled signals in noise", IEEE Trans. on IT, vol. IT-10, pp. 152-159, April 1964.

[4] Hayes M., Clements M.: "An Efficient Algorithm for Computing Pisarenko's Harmonic Decomposition Using Levinson's Recursion", IEEE Trans. on ASSP, vol. 34, No. 3, pp. 485-491, 1986.

[5] Clements M., Isabelle S.: "Reconstruction of a Positive Definite Toeplitz Matrix from Its Sequence of Minimum Eigenvalues", IEEE Trans. on ASSP, vol. 36, No 11, pp. 1784-1786, 1988.

[6] Kay S.: "Some results in linear interpolation theory", IEEE Trans. on ASSP, vol. 31, No. 3, pp. 746-749, 1983

[7] Picinbono B., Kerilis A.: "Some properties of prediction and interpolation errors", IEEE Trans. on ASSP, vol. 36, No. 4, pp. 525-531, 1988

[8] Tuszyński M., Wojtkiewicz A.: "Adaptive interpolation filters", Proc. XIII Nat. Conf. on Circuit Theory and Electronic Circuits, pp. 509-514, Bielsko-Biała 1990 (in Polish).

[9] Iohvidov I.S.: "Hankel and Toeplitz matrices and forms: algebraic theory", Boston, Basel, Stuttgart: Birkhauser, 1982

[10] Desbouvries F., Gueguen C.: "A Gohberg-Semencul formula for linear time-varying systems", Proc. EUSIPCO-90, pp. 445-448, Barcelona 1990

[11] Tuszyński M., Wojtkiewicz A.: "Interpolation filters in radar detection", Proc. XIV Nat. Conf. on Circuit Theory and Electronic Circuits, pp. 502-507, Waplewo 1991.

EFFICIENT ALGORITHMS FOR SECOND ORDER POLYNOMIAL SYSTEM IDENTIFICATION

G.O. GLENTIS and N. KALOUPTSIDIS

UNIVERSITY OF ATHENS, DEPT. OF INFORMATICS

PANEPISTIMIOPOLIS,TYPA BUILDINGS, ATHENS 157 71, GREECE.

ABSTRACT. In this paper an efficient methodology for general second order system identification is presented. The proposed technique utilizes a suitable transformation to convert the problem into a multichannel FIR filtering format.

1. INTRODUCTION

Recent work has led to the development of algorithms for estimating the parameters of an unknown plant using finite Volterra polynomial input- output representations, [1]- [2].

The estimation task is to determine at each time instant a suitable estimate of a finite dimensional parameter which, once specified, completely characterizes the plant. The selection of the estimate is based on the comparison between the actual output sample and a judiciously choosen predicted value on the basis of data up to that instant. We shall use the minimum variance predictor specified by the conditional expectation of the unknown system output at each time, given all past information. In LS identification the unkonwn parameter is chosen to minimize the total squared error between the systems output and predictor value over a finite horizon. Batch or off-line identification techniques provide the optimal solution based on a finite number of observations. Sequential or adaptive techniques adapt the optimal solution at each step, taking into accout the newcoming information. Therefore they are capable to track fast time varying dynamics.

In this paper an efficient method for general second order system identification is presented. The proposed technique utilizes a suitable transformation toconvert the problem into a multichannel FIR filtering format. It then applies recently proposed fast schemes, in batch as well as in adaptive format, [3]- [8]. The performance of the proposed technique is illustated by simulation.

2. PROBLEM FORMULATION

A second order nonlinear polynomial system has the form

$$y(n) = -\sum_i a_y(i)y(n-i) - \sum_i a_x(i)x(n-i) -$$

$$-\sum_i\sum_j a_{xy}(i,j)x(n-i)y(n-j) -$$

$$-\sum_i\sum_j a_{yy}(i,j)y(n-i)y(n-j) -$$

$$-\sum_i\sum_j a_{xx}(i,j)x(n-i)x(n-j) \quad (1)$$

where $x(n)$ is the input, $y(n)$ is the output and the summations are understood over a finite number of terms. The above input output expression admits a finite parametrization with parameters $a_x(i)$, $a_y(i)$, $a_{xy}(i,j)$, $a_{yy}(i,j)$ and $a_{xx}(i,j)$.

We group together all coefficients $a(i,j)$ of (1) associated to products $y(n-i)x(n-j)$, $x(n-i)x(n-j)$ and $y(n-i)y(n-j)$ that satisfy the condition

$$y(n-i)x(n-j) : i-j = d^{yx},$$

$$x(n-i)x(n-j) : i-j = d^{xx},$$

$$y(n-i)y(n-j) : i-j = d^{yy}.$$

where d^{yx}, d^{xx} and d^{yy} are constant integers. d^{yx} takes values on the interval $[d_1^{yx}, d_p^{yx}]$, where d_1^{yx} may be negative, while, due to parsimony, d^{xx} and d^{yy} take values on $[d_1^{xx}, d_q^{xx}]$ and $[d_1^{yy}, d_q^{yy}]$, where $0 \leq d_1^{xx}$ and $0 \leq d_1^{yy}$, respectively. To simplify notation we suppose that summation in (1) contains all consecutive products, $y(n-d_\ell^{yx}-i)x(n-i)$, $i = 1,2\dots m_\ell^{yx}$, $x(n-d_\ell^{xx}-i)x(n-i)$, $i = 1,2\dots m_\ell^{xx}$ and $y(n-d_\ell^{yy}-i)y(n-i)$, $i = 1,2\dots m_\ell^{yy}$. We reorginize coefficients through the format $a(i,j) \to a^\ell(k)$, where $a^\ell(k)$ denotes the entry of the array $a(i,j)$ located on the ℓ diagonal at the k position.

Then, model (1) can be viewed as a multichannel ARX system with inputs the signals

$$\left\{\begin{array}{ll} y(n), \quad x(n) & \\ u_\ell^{yx}(n) = y(n-d_\ell^{yx})x(n), & 1 \leq \ell \leq q \\ u_\ell^{xx}(n) = x(n-d_\ell^{xx})x(n), & 1 \leq \ell \leq p \\ u_\ell^{yy}(n) = y(n-d_\ell^{yy})y(n), & 1 \leq \ell \leq r \end{array}\right\} \tag{2}$$

as

$$y(n) = -\sum_{i=1}^{m^y} a_y(i)y(n-1) - \sum_{i=1}^{m^x} a_x(i)x(n-i) -$$

$$-\sum_{\ell=1}^{p}\sum_{k=1}^{m_\ell^{yx}} a_{yx}^\ell(k)y(n-d_\ell^{yx}-k)x(n-k) -$$

$$-\sum_{\ell=1}^{r}\sum_{k=1}^{m_\ell^{yy}} a_{yy}^\ell(k)y(n-d_\ell^{yy}-k)y(n-k) -$$

$$-\sum_{\ell=1}^{q}\sum_{k=1}^{m_\ell^{xx}} a_{xx}^\ell(k)x(n-d_\ell^{xx}-k)x(n-k)$$

or

$$y(n) = -\sum_{i=1}^{m^y} a_y(i)y(n-1) - \sum_{i=1}^{m^x} a_x(i)x(n-i) -$$

$$-\sum_{\ell=1}^{p}\sum_{k=1}^{m_\ell^{yx}} a_{yx}^\ell(k)u_l^{yx}(n-k) - \sum_{\ell=1}^{r}\sum_{k=1}^{m_\ell^{yy}} a_{yy}^\ell(k)u_l^{yy}(n-k) -$$

$$-\sum_{\ell=1}^{q}\sum_{k=1}^{m_\ell^{xx}} a_{xx}^\ell(k)u_l^{xx}(n-k) \tag{3}$$

We rewrite (3) in compact form as

$$y(n) = -\boldsymbol{\theta}^t\boldsymbol{\phi}(n) \tag{4}$$

where

$$\boldsymbol{\theta} = [\mathbf{a}_y^t(m_y)\mathbf{a}_x^t(m_x)\mathbf{a}_{yx}^{1t}(m_1^{yx})\dots\mathbf{a}_{yx}^{tp}(m_p^{yx})$$

$$\mathbf{a}_{yy}^{t1}(m_1^{yy})\dots\mathbf{a}_{yy}^{tr}(m_r^{yy})\mathbf{a}_{xx}^{t1}(m_1^{xx})\dots\mathbf{a}_{xx}^{tq}(m_q^{xx})]^t \tag{5}$$

and

$$\boldsymbol{\phi}(n) = [\mathbf{y}_{m^y}^t(n-1)\mathbf{x}_{m^x}^t(n)\mathbf{u}_{m_1^{yx}}^{yxt}(n)\dots\mathbf{u}_{m_p^{yx}}^{yxt}(n)$$

$$\mathbf{u}_{m_1^{yy}}^{yyt}(n)\dots\mathbf{u}_{m_r^{yy}}^{yyt}(n)\mathbf{u}_{m_1^{xx}}^{xxt}(n)\dots\mathbf{u}_{m_q^{xx}}^{xxt}(n)]^t \tag{6}$$

where

$$\mathbf{u}_{m_\ell^{yx}}^{yx}(n) = [u_l^{yx}(n)u_l^{yx}(n-1)\dots u_l^{yx}(n-m_l^{yx}+1)]^t$$

$$\ell = 1,2\dots q$$

$$\mathbf{u}_{m_\ell^{yx}}^{yy}(n) = [u_l^{yx}(n)u_l^{yx}(n-1)\dots u_l^{yx}(n-m_l^{yx}+1)]^t$$

$$\ell = 1,2\dots r$$

and

$$\mathbf{u}_{m_\ell^{xx}}^{xx}(n) = [u_l^{xx}(n)u_l^{xx}(n-1)\dots u_l^{xx}(n-m_l^{xx}+1)]^t$$

$$\ell = 1,2\dots q.$$

3. ESTIMATION

The above transformation converts the Volterra predictor estimation to a multichannel FIR filtering problem with different number of delay elements for each input channel. The resultingl LS multichannel filtering has been studied in [3]- [8]. Let $\mathrm{z}(n) = [\mathrm{x}^t(n)\mathrm{y}^t(n)]^t$ be a record of input output data over the interval $0 \leq n \leq N$, $\mathrm{y}(n) \in \Re^s$, $\mathrm{x}(n) \in \Re^k$. We seek to determine the optimal predictor $\boldsymbol{\theta}$ which minimizes the total squared error between the estimated output $\hat{y}(n)$ and the system's output $y(n)$

$$\min_{\boldsymbol{\theta}} V_N(\boldsymbol{\theta}) = \frac{1}{N}\sum_{n=0}^{N} \lambda^{N-n}(\mathrm{e}^t(n|\boldsymbol{\theta})\mathrm{e}(n|\boldsymbol{\theta})) \quad (8)$$

where $\mathrm{e}(n|\boldsymbol{\theta}) = \mathrm{y}(n) - \hat{\mathrm{y}}(n|\boldsymbol{\theta})$ is the instantaneous error, and λ is the exponential forgetting factor, $0 < \lambda \leq 1$. $\hat{\mathrm{y}}(n|\boldsymbol{\theta})$ is given by (4), ie., $\hat{\mathrm{y}}(n|\boldsymbol{\theta}) = -\boldsymbol{\theta}^t\boldsymbol{\phi}(n)$.

Minimization of (8) leads to the linear system of equations

$$\mathcal{R}(N)\boldsymbol{\theta}(N) = -\boldsymbol{\phi}(N) \quad (9)$$

where $\mathcal{R}(N)$ is the sampled autocorrelation matrix

$$\mathcal{R}(N) = \sum_{n=0}^{N} \lambda^{N-n}\boldsymbol{\phi}(n)\boldsymbol{\phi}^t(n) = \lambda\mathcal{R}(N-1) + \boldsymbol{\phi}(N)\boldsymbol{\phi}^t(N) \quad (10)$$

and $\mathcal{D}(N)$ is the sampled cross correlation vector

$$\mathcal{D}(N) = \sum_{n=0}^{N} \lambda^{N-n}\boldsymbol{\phi}(n)y(n) = \lambda\mathcal{D}(N-1) + \boldsymbol{\phi}(N)y(N) \quad (11)$$

The least squares second order identification problem can therefore solved through conversion to the dimensions varying multichannel format.

Several algorithms can be utilized to deal with the latter problem. These can be divided into two basic classes, order recursive **batch** algorithms and **adaptive** algorithms. Efficient order recursive algorithms have benn derived in [3]- [5]. They solve (9) utilizing suitable permutations that are utilized to unshuffle the shift invariant properties at the data level. Matrix (10) is a block matrix with entries near to Toeplitz submatrices and in fact this property serves for the development of two term Levinson recursions. The Levinson algorithm can be transformed to a Schur type counterpart which offers significant advantange if parallel processing environment is available, since it reduces processing time by an order of magnitude. Implementation of the above algorithms on VLSI array processors is discussed in [10]. In order to cope with time varying dynamics, fast adaptive transversal and lattice algorithms can be used. The derivation is based on the shift invriant properties of the pertinent data. Detailed description of the algorithmic structures is givem in [6]-[9].

4. SIMULATION

Let us consider a second order polynomial filter of the form

$$y(n) = -\sum_{i=1}^{m^x} a_x(i)x(n-i) -$$

$$\sum_{\ell=1}^{q}\sum_{i=1}^{m_l^{xx}} a_{xx}^l(i)x(n-d_l^{xx}-i)x(n-i) \quad (12)$$

where, $m^x = 5$, $q = 5$,

$$[d_1^{xx} d_2^{xx} d_3^{xx} d_4^{xx} d_5^{xx}] = [0\ 1\ 2\ 3\ 4]$$

$$[m_1^{xx} m_2^{xx} m_3^{xx} m_4^{xx} m_5^{xx}] = [54321].$$

We define the auxiliary signals as dictaded by (2)

$$u_1^{xx}(n) = x(n)x(n),\ u_2^{xx}(n) = x(n-1)x(n),$$

$$u_3^{xx}(n) = x(n-2)x(n), u_4^{xx}(n) = x(n-3)x(n),$$

$$u_5^{xx}(n) = x(n-4)x(n),$$

Then, eq. (12) can be transformed to a multichannel filter of 6 input channels and varying filter orders

$$y(n) = -\boldsymbol{\theta}^t\boldsymbol{\phi}(n)$$

where $\boldsymbol{\theta}$ and $(\phi)(n)$ are defined by eqs. (5) and (6).

The objective of this experiment is to identify the system parameter vector $\boldsymbol{\theta}$, given the input $x(n)$ and the output $v(n) = y(n) + e(n)$, where $e(n)$ is white noise. The optimal filter is obtained as the solution of linear system (10). A multichannel fast Kalman adaptive algorithm of the FAEST family, developed in [5]-[8] was utilized. Figure 1 illustates the performance of the proposed method, measured by the relative estimation error (learning curve).

5. CONCLUSIONS

An efficient methodology for general second order system identification has been presented. Extensions to higher order terms is readily derived.

REFERENCES

[1] M. Schetzen, The Volterra and Wiener Theories of nonlinear systems, Wiley, 1980.

[2] V.J. Mathews, Adaptive polynomial filters, IEEE ASSP Magazine, July 1991.

[3] G. Glentis and N. Kalouptsidis, Efficient order recursive algorithms for multichannel LS filtering, precented in IEEE IC-ASSP, New Mexico, 1990.

[4] G. Glentis and N. Kalouptsidis, Efficient Order Recursive Algorithms for Multichannel Least Squares Filtering, to appear, IEEE ASSP.

[5] G. Glentis and N. Kalouptsidis, Efficient Algorithms for the Solution of Block Linear Systems with Toeplitz Entries, to appear, Linear Algebra and its Applications.

[6] G. Glentis and N. Kalouptsidis, Fast adaptive algorithms for multichannel LS filtering, precented in IEEE IC-CAS, New Orleans, 1990.

[7] G. Glentis and N. Kalouptsidis, Fast Adaptive Algorithms for Multichannel Filtering and System Identification, to appear, IEEE ASSP.

[8] G. Glentis and N. Kalouptsidis, Efficient Adaptive Algorithms for Multichannel Least Squares Filtering Using a Channel Decomposition Technique, to appear, Computers and Electrical Engineering, Special Issue on Adaptive Signal Processing.

[9] I. Bouras, G. Glentis and N. Kalouptsidis, Efficient implementation of block toeplitz solvers in VLSI array processors, IMAS/PDCOM Corfu, Greece 1991.

Figure 1 learning curves

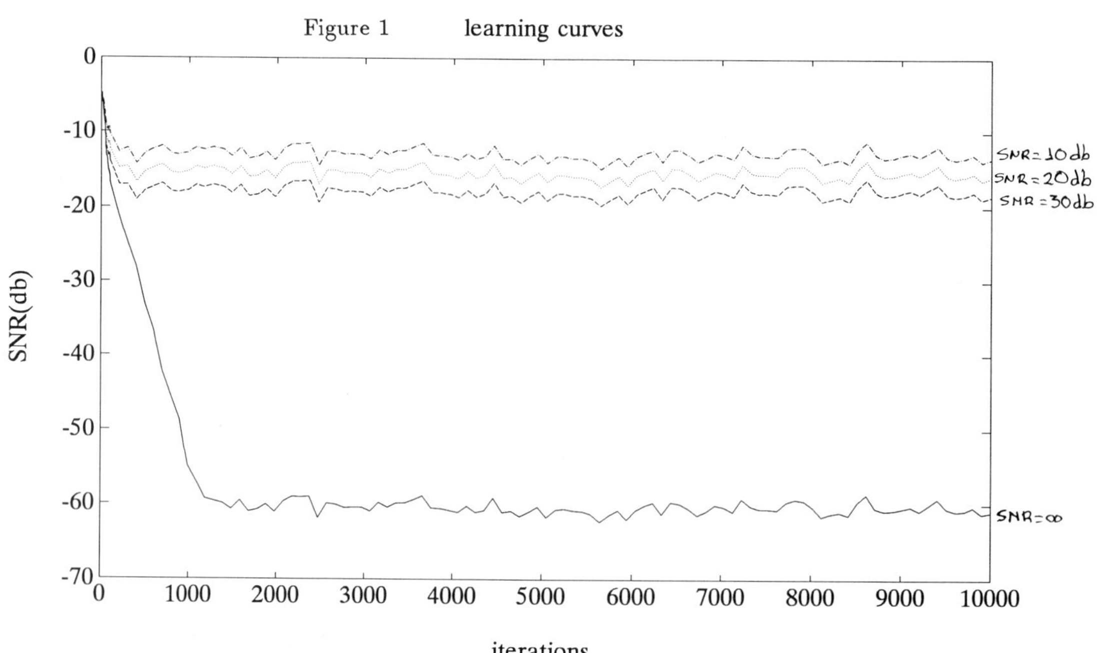

SIGNAL PROCESSING VI: Theories and Applications
J. Vandewalle, R. Boite, M. Moonen, A. Oosterlinck (eds.)

Fast Linear Phase Adaptive Filtering using Constrained Least Squares Algorithm

L. S. Resende[(1)], J. M. T. Romano[(1)] and M. G. Bellanger[(2)]

(1) DECOM/FEE - Universidade de Campinas - C.P. 6101
13081 Campinas, SP - BRAZIL
(2) Conservatoire National des Arts et Métiers - 292, Rue Saint-Martin
75141 PARIS CEDEX 03 - FRANCE

In this work, an original and fast algorithm is proposed for linear phase adaptive filtering. The approach consists in imposing the symmetry of the impulse response of the filter by means of a set of linear constraints, and then, in treating the problem as a particular case of constrained adaptive filtering. Hence, the solution is attained by using the fast least-squares algorithm for adaptive filters with linear constraints, recently proposed by the authors in a previous work. The algorithm presents a fast and accurate convergence, as it can be observed in the simulations results obtained in simple examples of adaptive equalization and prediction.

I - INTRODUCTION

In several applications of digital signal processing techniques, it is suitable to preserve a linear phase characteristic of FIR digital filters. This characteristic prevents phase distortions in the passband and implies that the $\mathcal{Z}$-transfer function of the filter is a symmetrical or anti-symmetrical polynomial.

Channel equalization, systems identification, frequency estimation and line enhancement are some of the typical applications where the linear phase property may be of great interest. In many cases, the parameters of the filter must be obtained by an adaptive procedure, in order to provide useful methods for real-time operations and non-stationary environments.

Different approaches have been presented in the literature, leading to least-squares algorithms for linear phase adaptive filters [1-5]. This paper proposes a novel algorithm, departing from an original approach. First, the symmetry of the impulse response of the filter is imposed by means of a set of linear constraints over its coefficients. Hence, the problem is reduced to a particular case of adaptive filtering with linear constraints, for which a fast algorithm has been derived in [6].

In order to well establish the proposed approach, the derivation of the constrained adaptive algorithm is briefly presented first. Afterwards, the linear phase constraint is introduced and the final result is analyzed. Finally, the performance and potential of the method is discussed and illustrated by some simulations results.

II - CONSTRAINED LEAST-SQUARES ALGORITHM

The problem may be posed by considering the scheme of the adaptive transversal filter in figure 1, where the coefficients are subject to a set of linear constraints. Then, if the least squares criterion over the error signal is assumed, the following formulation holds:

$$\textit{Minimize} \quad \mathcal{J}[\mathbf{h}(n)] = \sum_{p=1}^{n} [d(p) - \mathbf{x}^t(p)\mathbf{h}(n)]^2 \tag{1}$$

$$\textit{such that} \quad \mathbf{C}^t\mathbf{h}(n) = \mathbf{f}, \tag{2}$$

where

$$\mathbf{x}(p) = [x(p)\ x(p-1)\ \ldots\ x(p-N+1)]^t,$$

$$\mathbf{h}(n) = [h_0(n)\ h_1(n)\ \ldots\ h_{N-1}(n)]^t$$

and the NxK matrix **C** and the K-element vector **f** establish the set of constraints.

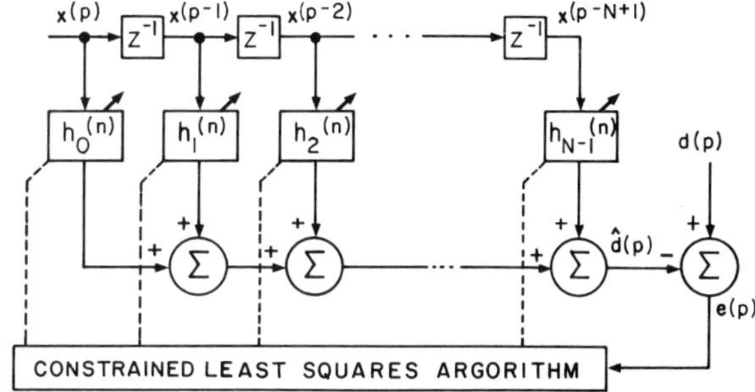

Figure 1: Adaptive Transversal Filter.

The optimal coefficients $\mathbf{h}(n)$, supposing n input samples, are obtained by means of the Lagrange multipliers method [7]. After some calculations, we obtain [6]:

$$\mathbf{h}(n) = \mathbf{R}_{xx}^{-1}(n)\mathbf{p}_{xd}(n) + \\ + \mathbf{R}_{xx}^{-1}(n)\mathbf{C}\left[\mathbf{C}^t\mathbf{R}_{xx}^{-1}(n)\mathbf{C}\right]^{-1}\left[\mathbf{f} - \mathbf{C}^t\mathbf{R}_{xx}^{-1}(n)\mathbf{p}_{xd}(n)\right] \tag{3}$$

where $\mathbf{R}_{xx}(n)$ is the autocorrelation matrix and $\mathbf{p}_{xd}(n)$ the cross-correlation vector, estimated by:

$$\mathbf{R}_{xx}(n) = \sum_{i=1}^{n} W^{n-i}\, \mathbf{x}(i)\mathbf{x}^t(i) \tag{4}$$

$$\mathbf{p}_{xd}(n) = \sum_{i=1}^{n} W^{n-i}\, \mathbf{x}(i)d(i) \tag{5}$$

Using this result, our objective is to obtain a recursive solution for $\mathbf{h}(n+1)$, when the new data $x(n+1)$ and $d(n+1)$ are available. According to equation (3), the optimal coefficients at instant n+1 are given by:

$$\mathbf{h}(n+1) = \mathbf{R}_{xx}^{-1}(n+1)\mathbf{p}_{xd}(n+1) + $$

$$+ \Gamma(n+1)\left[\mathbf{C}^t\Gamma(n+1)\right]^{-1}\left[\mathbf{f} - \mathbf{C}^t\mathbf{R}_{xx}^{-1}(n+1)\mathbf{p}_{xd}(n+1)\right] \tag{6}$$

where we have defined:

$$\Gamma(n+1) = \mathbf{R}_{xx}^{-1}(n+1)\mathbf{C} \tag{7}$$

Since the inverse autocorrelation matrix can be updated by [8]:

$$\mathbf{R}_{xx}^{-1}(n+1) = \frac{1}{W}\left[\mathbf{R}_{xx}^{-1}(n) - \mathbf{g}(n+1)\mathbf{x}^t(n+1)\mathbf{R}_{xx}^{-1}(n)\right] \tag{8}$$

where $\mathbf{g}(n+1)$ is the adaptation gain in recursive least-squares (RLS) procedure, given by [8]

$$\mathbf{g}(n+1) = \frac{\mathbf{R}_{xx}^{-1}(n)\mathbf{x}(n+1)}{W + \mathbf{x}^t(n+1)\mathbf{R}_{xx}^{-1}(n)\mathbf{x}(n+1)} \tag{9}$$

From (7) and (8), a recurrence can be easily obtained for $\Gamma(n+1)$:

$$\Gamma(n+1) = \frac{1}{W}\left[\Gamma(n) - \mathbf{g}(n+1)\mathbf{x}^t(n+1)\Gamma(n)\right] \tag{10}$$

In this expression, only the matrix $\Gamma(n)$ depends on the constraints. In order to reduce the computational complexity, the vector $\mathbf{g}(n+1)$ may be calculated by means of the fast least-squares (FLS) algorithm [8].

Thus, an updating is attained for the coefficients, using expressions (6) and (10). In this procedure, it is necessary to calculate the inverse of the matrix $\mathbf{C}^t\Gamma(n+1)$ at each iteration. So, the efficiency can be substantially increased if a recurrence is also derived for this inverse matrix.

Premultiplying equation (10) by $\mathbf{C}^t$, we obtain:

$$\mathbf{C}^t\Gamma(n+1) = \frac{1}{W}\left[\mathbf{C}^t\Gamma(n) - \mathbf{C}^t\mathbf{g}(n+1)\mathbf{x}^t(n+1)\Gamma(n)\right] \tag{11}$$

By an appropriate use of the matrix inversion lemma in the above expression, the following recursion holds for the inverse matrix [6]:

$$\left[\mathbf{C}^t\Gamma(n+1)\right]^{-1} = W\left\{\left[\mathbf{C}^t\Gamma(n)\right]^{-1} + \right.$$

$$\left. + \mathbf{l}(n+1)\mathbf{x}^t(n+1)\Gamma(n)\left[\mathbf{C}^t\Gamma(n)\right]^{-1}\right\} \tag{12}$$

where the vector $\mathbf{l}(n+1)$ is defined by

$$\mathbf{l}(n+1) = \frac{\left[\mathbf{C}^t\Gamma(n)\right]^{-1}\mathbf{C}^t\mathbf{g}(n+1)}{1 - \mathbf{x}^t(n+1)\Gamma(n)\left[\mathbf{C}^t\Gamma(n)\right]^{-1}\mathbf{C}^t\mathbf{g}(n+1)} \tag{13}$$

Finally, by replacing equations (5), (8), (10) and (12) in (6), we obtain, after some algebraic manipulations, the following recursive expression for the coefficients of the filter:

$$\mathbf{h}(n+1) = \mathbf{h}(n) + \mathbf{g}(n+1)e(n+1) + $$

$$- W\,\Gamma(n+1)\mathbf{l}(n+1)e(n+1) \tag{14}$$

It is interesting to observe that the above result corresponds to the least-squares adaptive procedure modified by an additional correction term depending on the constraints. The whole proposed algorithm is summarized in Table I:

- New input data : $x(n+1)$, $d(n+1)$
- Output error : $e(n+1) = d(n+1) - \mathbf{h}^t(n)\mathbf{x}(n+1)$
- Adaptation gain : $\mathbf{g}(n+1)$

Introduction of the constraints:

$$\mathbf{l}(n+1) = \frac{\left[\mathbf{C}^t\Gamma(n)\right]^{-1}\mathbf{C}^t\mathbf{g}(n+1)}{1 - \mathbf{x}^t(n+1)\Gamma(n)\left[\mathbf{C}^t\Gamma(n)\right]^{-1}\mathbf{C}^t\mathbf{g}(n+1)}$$

$$\left[\mathbf{C}^t\Gamma(n+1)\right]^{-1} = W\left\{\left[\mathbf{C}^t\Gamma(n)\right]^{-1} + \right.$$

$$\left. + \mathbf{l}(n+1)\mathbf{x}^t(n+1)\Gamma(n)\left[\mathbf{C}^t\Gamma(n)\right]^{-1}\right\}$$

$$\Gamma(n+1) = \frac{1}{W}\left[\Gamma(n) - \mathbf{g}(n+1)\mathbf{x}^t(n+1)\Gamma(n)\right]$$

$$\mathbf{h}(n+1) = \mathbf{h}(n) + \mathbf{g}(n+1)e(n+1) - W\,\Gamma(n+1)\mathbf{l}(n+1)e(n+1)$$

Table I: Least-Squares Algorithm with Multiple Constraint

As mentioned in [6], the computational complexity of the above procedure is proportional to (K+1)(N+K).

The initialization of the algorithm must be done according to the least-squares criterion [8,chap.6]. For the FLS algorithm, used to obtain $\mathbf{g}(n+1)$, it is reasonable to assume all variables to be null for n=0, setting the prediction error energy to a positive value E_0. This is equivalent to admit the autocorrelation matrix initialized by

$$\mathbf{R}_{xx}(0) = E_0\ \mathrm{diag}[1, W^{-1}, W^{-2}, \ldots, W^{-(N-1)}] \tag{15}$$

Then, this must be considered for the variables depending on the constraints:

$$\Gamma(0) = \mathbf{R}_{xx}^{-1}(0)\mathbf{C} \tag{16}$$

Finally, the filter coefficients have the following initial conditions:

$$\mathbf{h}(0) = \Gamma(0)\left[\mathbf{C}^t\Gamma(0)\right]^{-1}\mathbf{f} \tag{17}$$

The algorithm presented in this section is based on the minimization of the error signal, obtained by $d(n) - \mathbf{h}^t(n)\mathbf{x}(n+1)$. The same procedure can be developed for the case of a constrained prediction error filter, where the error signal is given directly by $\mathbf{h}^t(n)\mathbf{x}(n+1)$ [6].

III - APPLICATION TO LINEAR PHASE ADAPTIVE FILTERING

The algorithm presented in Table I can be used to update a linear phase adaptive filter if the constraints, **C** and **f**, are chosen in such a way to impose the symmetry on the filter coefficients. This can be easily achieved by posing:

$$\mathbf{C} = \begin{bmatrix} 1 & 0 & \cdots & 0 \\ 0 & 1 & \cdots & 0 \\ \vdots & \vdots & \ddots & \vdots \\ 0 & 0 & \cdots & 1 \\ \hline 0 & 0 & \cdots & 0 \\ \hline 0 & 0 & \cdots & -1 \\ \vdots & \vdots & \cdot^{\cdot^{\cdot}} & \vdots \\ 0 & -1 & \cdots & 0 \\ -1 & 0 & \cdots & 0 \end{bmatrix}_{N\times K} = \begin{bmatrix} \mathbf{I}_K \\ \hline \mathbf{0}^t \\ \hline \pm\mathbf{J}_K \end{bmatrix} \qquad (18)$$

for N odd or

$$\mathbf{C} = \begin{bmatrix} \mathbf{I}_K \\ \hline \pm\mathbf{J}_K \end{bmatrix} \qquad (19)$$

para N even and **f**=**0** in both cases.

Thus, it is clear that the filter is linear phase if the set of constraints $\mathbf{C}^t\mathbf{h}=\mathbf{f}$ is posed. The sign of matrix **J** in expressions (18) and (19) can be negative or positive, providing symmetrical or anti-symmetrical coefficients, respectively.

In the particular case of linear phase prediction, it is also necessary to impose that one of the filter coefficients be non-null, in order to avoid the trivial solution **h**=**0**. This can be achieved, for example in forward prediction, by increasing the matrix C with a $(K+1)^{\underline{th}}$ column given by

$$\mathbf{c}^t_{K+1} = [\ 1\ 0\ \ldots\ 0\ ] \qquad (20)$$

and by posing:

$$\mathbf{f}^t = [\ 0\ \ldots\ 0\ 1\ ]$$

This case is specially interesting in signal analysis, since the proposed algorithm provides an efficient way to obtain the composite sinusoidal modeling, or harmonic decomposition, of the signal under analysis.

IV - SIMULATIONS RESULTS

In order to evaluate the performance of the proposed algorithm, we have first supposed a simple case of adaptive equalization, described by the scheme of the Figure 2. The input signal a(n) is a binary sequence (a(n)=±1) and the channel response is assumed to be raised cosine with linear phase, that is, it has a finite and symmetrical impulse response [7,chap.5]. Additive white noise v(n), with variance σ_v^2=0.001, is also considered.

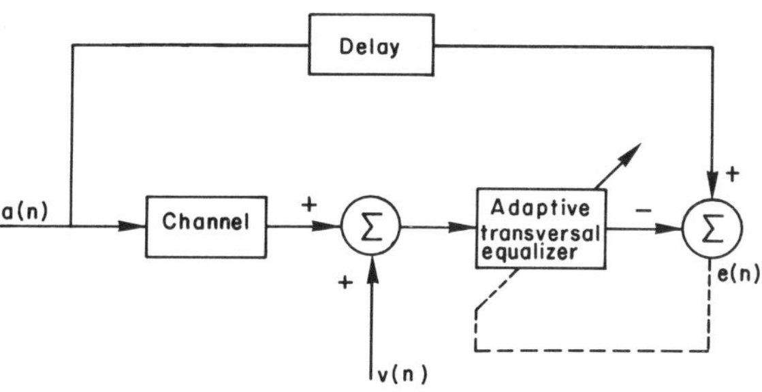

Figure 2: Adaptive Channel Equalization

We have applied the algorithm of table I to an equalizer of order 6 (N=7) where the matrix **C** is given by (18) and **f**=**0**. Since the channel is also linear phase, the equalizer must only compensate the amplitude distortion, keeping the linear phase property.

The Figure 3a illustrates the fast convergence of the equalizer coefficients, after 30 iterations. The symmetry property is also observed during the updating. The evolution of the error signal is presented in Figure 3b.

In the second example, we have considered the problem of linear phase predictors. The input signal is composed of three sinusoids in white noise, where the frequencies are given by $\omega_1=\pi/4$, $\omega_2=7\pi/16$ and $\omega_3=15\pi/16$ and the signal-to-noise ratio (SNR) is 3dB. A predictor of order 6 has been used and the matrix **C** is given by (18), where K=3, increased by the $(K+1)^{\underline{th}}$ column to impose the first coefficient to be one.

The Figure 4a shows the evolution of the coefficients for 150 iterations and the symmetry property can be noted during all the updating. The frequency response of the filter after convergence is presented in Figure 4b. Three notches are observed near the frequencies ω_1, ω_2 and ω_3, since the linear phase condition constrains the zeros of the filter to be on the unit circle. A bias regarding the input frequencies occurs due to the presence of noise.

In all the above simulations, we have used W=0.99 for the forgetting factor and E_0=0.01 for the initial error energy.

V - CONCLUSION

The fundamental aim of this work is the proposition of the constrained least-squares algorithm of Table I and its application to the specific case of linear phase adaptive filters. Departing from the idea of linear constraints, we attain a simple and elegant way to preserve the linear phase property by constraining the filter coefficients to be symmetrical or anti-symmetrical during the updating process. The use of exact least-squares recursion provides an efficient dynamic behavior to the proposed method.

The performance was illustrated by simple cases of

adaptive equalization and line detection. Clearly, the algorithm may be used in a general context of system identification, where the linear phase property is desired, or in adaptive prediction, providing the harmonic decomposition of the input signal.

Finally, it is interesting to improve the performance by reducing the computational complexity of the proposed algorithm. This may be achieve by using some interesting properties of symmetry observed in the matrix involved in the algorithm. The studies concerning these aspects are now in course.

REFERENCES

[1] B. Friedlander and M. Morf; "Least Squares Algorithms for Adaptive Linear-Phase Filtering". IEEE Trans. on Acoust., Speech, and Signal Processing, vol. ASSP-30, pp. 381-390, Jun. 1982.

[2] S. L. Marple Jr.; "Fast Algorithms for Linear Prediction and System Identification Filters with Linear Phase". IEEE Trans. on Acoust., Speech, and Signal Processing, vol. ASSP-30, pp. 942-953, Dec. 1982.

[3] N. Kalouptsidis; "Fast Adaptive Least Squares Algorithms for Power Spectral Estimation" IEEE Trans. on Acoust., Speech, and Signal Processing, vol. ASSP-35, pp. 661-670, May 1987.

[4] N. Kalouptsidis and S. Theodoridis; "Fast Sequential Algorithms for Least Squares FIR Filters with Linear Phase". IEEE Trans. on Circuits and Systems, vol. CAS-29, pp. 425-432, Apr. 1988.

[5] J. S. Sunwoo and C. K. Un; "Recursive Modified Gram-Schmidt Algorithm for Linear Phase Filtering". Signal Processing, vol. 22, pp. 43-51, Jan. 1991.

[6] L. S. Resende, J. M. T. Romano and M. B. Bellanger; "A Fast Least Squares Algorithm for Constrained Adaptive Filtering", IEEE International Conference on ASSP, San Francisco, USA, Mar. 1992.

[7] S. Haykin; "Adaptive Filter Theory". Prentice-Hall; New Jersey; 1986.

[8] M. G. Bellanger; "Adaptive Digital Filters and Signal Analysis". Marcel Dekker, Inc.; New York and Basel; 1987.

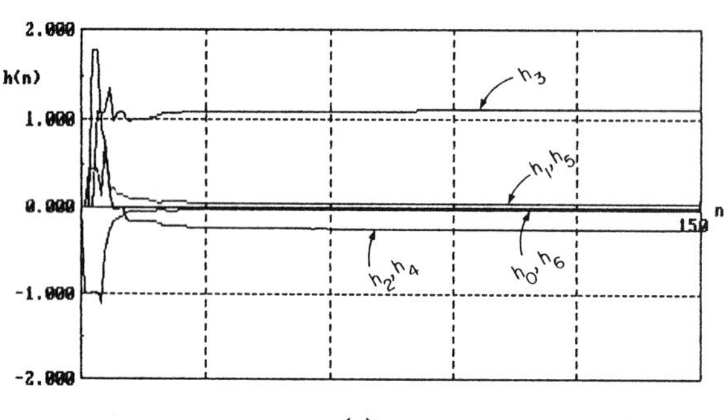

(a)

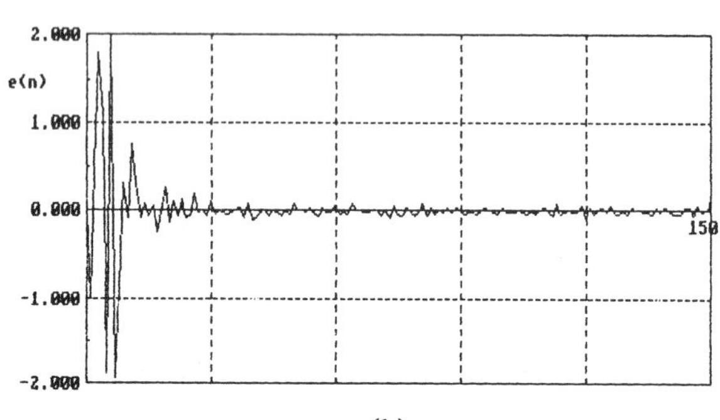

(b)

Figure 3: Case of Adaptive Equalization:
(a) Evolution of the coefficients;
(b) Evolution of the error signal.

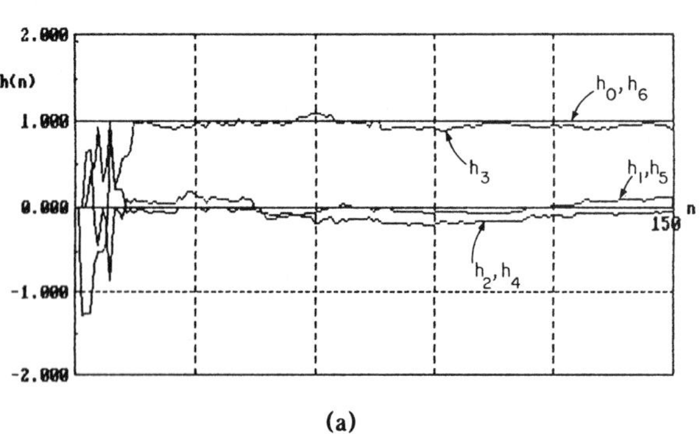

(a)

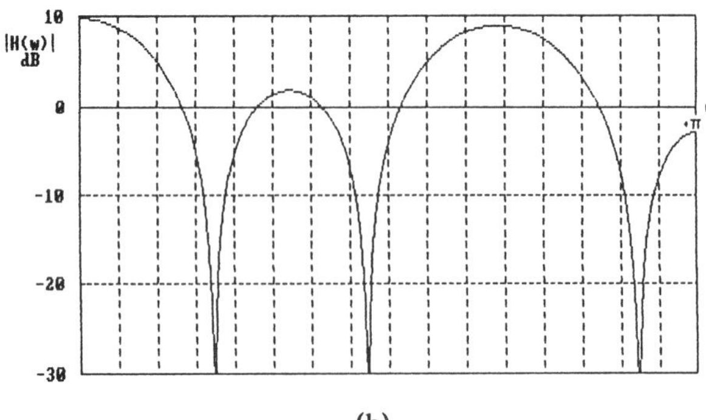

(b)

Figure 4: Case of Adaptive Prediction:
(a) Evolution of the coefficients;
(b) Frequency response after convergence.

SIGNAL PROCESSING VI: Theories and Applications
J. Vandewalle, R. Boite, M. Moonen, A. Oosterlinck (eds.)

A Blind Adaptive Equalizer Based on a Lattice/All-pass Configuration

B. Jelonnek, K. D. Kammeyer
Telecommunications Group, Technical University of Hamburg-Harburg
Eissendorfer Str. 40, D-2100 Hamburg 90, FRG

Networks for the blind channel equalization can be designed as a combination of a prewhitening filter (adjusted by means of 2nd-order statistics) and an inverse all-pass system (adjusted by 4th-order statistics). In this paper an appropriate FIR approximation of inverse all-pass systems is presented. On the other hand the inverse all-pass modelling can be avoided if the minimum phase prewhitening filter is replaced by a lattice configuration where the maximum phase (backward prediction error) output is used for decorrelation. In that case conventional recursive all-pass networks can be applied for phase correction. It will be shown that the convergence speed can be considerably increased due to the reduced number of parameters to be identified. Under additive noise conditions a solution near the optimum MSE-solution is obtained.

1 Introduction

Blind Equalization, i.e. adaptive channel correction without use of a trainings sequence or decided data is a new field of research since the recent few years. It is well-known that the traditional 2nd order statistics are "phase blind" in the sense that the transmission path is always interpreted as a pure minimum-phase system (or pure maximum phase, respectively) if only the received signal is used for channel estimation. Consider as an example the classical application of a prediction-error filter as a prewhitening system. The design of this filter is based on the autocorrelation samples of the received signal – the result is a minimum-phase inverse system model.

The fundamental idea to perform a blind adaptive equalization including the phase characteristics is the application of *higher order statistics*. There exist two fundamental problems: 1st channel identification (see [1],[2],[3]) and 2nd inverse channel estimation (see e.g. [4],[5]). The latter problem will be regarded in the present paper.

2 FIR-modelling of inverse systems

The basic idea for the design af a blind equalizer is to make use of all the information that can be derived from 2nd order statistics. Consider, for example, a complex-valued Moving-Average (MA-) channel model with a z-plane zero configuration shown in fig. 1a (mixed phase). If a FIR prediction-error filter is applied pole approximations (denoted as ⊗) are introduced as shown in fig. 1b. The resulting pole-zero configuration of the system composed of the channel and the prediction-error filter is demonstrated in fig. 1c. Obviously the resulting

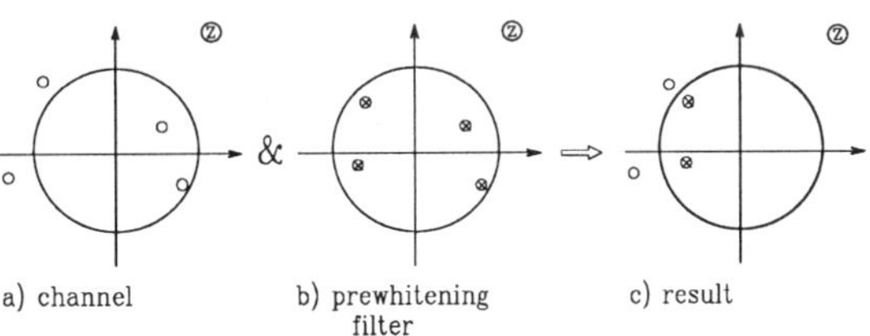

Figure 1: Pole-zero configuration under prewhitening conditions
⊗ approximated pole; ○ zero

system contains a number of all-pass pole/zero pairs which are related to the non-minimum phase channel zeros: A proper equalization requires the additional application of inverse all-pass systems after the prewhitening filter which has to be designed on the basis of higher order statistics. Before the discussion of an appropriate blind adaptive algorithm let us at first discuss the problem of the FIR approximation of poles which is necessary for the prewhitening filter design as well as for inverse all-pass modelling. Consider the 1st order transfer function

$$H_{IIR}(z) = \frac{z}{z - z_\infty}; \qquad |z_\infty| > 1 \tag{1}$$

under the specific condition of the pole outside the unit circle. Eq. (1) can be described by the following series expansion

$$H_{IIR}(z) = -\sum_{k=1}^{\infty} z_\infty^{-k} z^k \tag{2}$$

which converges, since $|z_\infty|^{-1} < 1$. Thus the system can be interpretated as *causal and non-stable* or *non-causal and stable*, equivalently. In the 2nd case we get the impulse response

$$h_{IIR}(k) = z_{\infty}^{k}, \qquad k = -1, -2, \dots \tag{3}$$

which can be approximated by a causal FIR system by time truncation and time shift. This causal impulse response (length ℓ) can be written as

$$h_{FIR}(k) = h_{IIR}(k - \ell) - z_{\infty}^{-\ell} h_{IIR}(k). \tag{4}$$

After the z-transform we obtain the transfer function of the approximated pole

$$H_{FIR}(z) = H_{IIR}(z) \left[z^{-\ell} - z_{\infty}^{-\ell} \right], \tag{5}$$

the second term describes ℓ zeros in the z-plane, equidistantly spaced on a circle with the radius $|z_{\infty}|$. One of these zeros (located at the position of the approximated pole) is canceled by the term $H_{IIR}(z)$. The result of the pole approximation is shown in fig. 2a. This form of approximation is based on a rectangular windowing of the true impulse response. For comparison fig. 2b shows the result of a Least-Squares- (LS-) solution where the zeros are removed from their positions on a circle. Finally fig. 2c shows the FIR approximation of an inverse all-pass system based on a LS design.

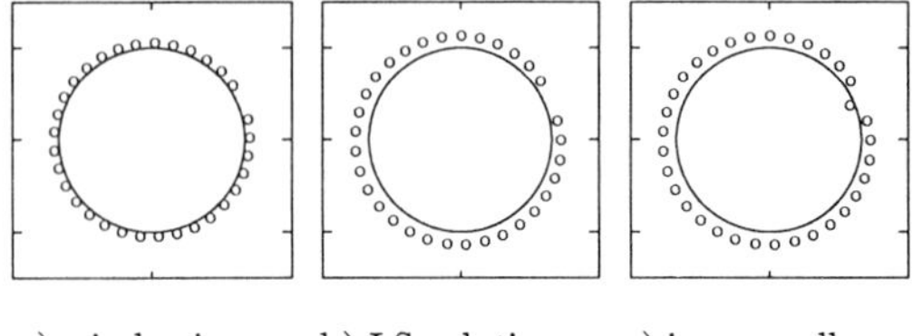

a) windowing   b) LS-solution   c) inverse all-pass

Figure 2: Approximation of a pole outside the unit circle $\ell = 32$

3 Phase Correction Networks

As discussed above a proper equalization is obtained by the combination of a prediction error filter and a phase correction FIR-filter. Fig. 3 shows this configuration. In [5] the phase correction network was realized by a conventional FIR transversal filter. In the present paper we intend to make use of the specific a-priori knowledge, that the phase correction networks is composed of at most n complex-valued 1st order (inverse) all-pass systems, where n denotes the order of the (MA-) channel. For low-order channels this leads to a great advantage with regard to the adaptive adjustment, since the convergence speed strongly depends on the number of parameter to be adapted: In many applications the

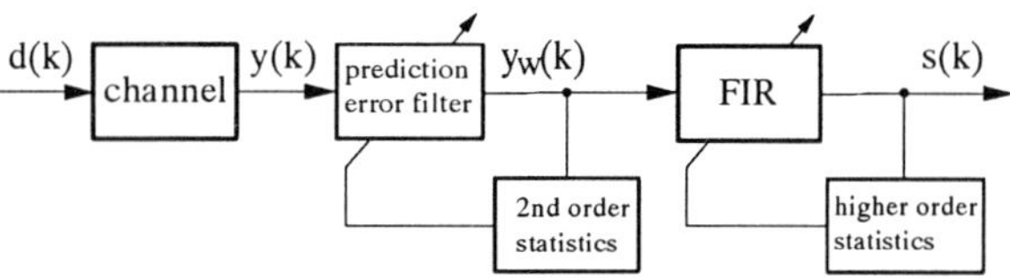

Figure 3: Equalizer with prewhitening filter and phase correction network

approximation degree ℓ of the phase correction network is much higher than the channel order n.

As far as regarded the phase correction network requires the realization of inverse all-pass systems which is possible in the form of the FIR approximation discussed in the last section. On the other hand the inverse all-pass phase correction is a consequence of the minimum-phase prediction-error filter. In contrast we can apply the well-known *lattice prewhitening filter* and use the *backward predictor output* which shows *maximum phase characteristics*. The consequences are illustrated by fig. 4:

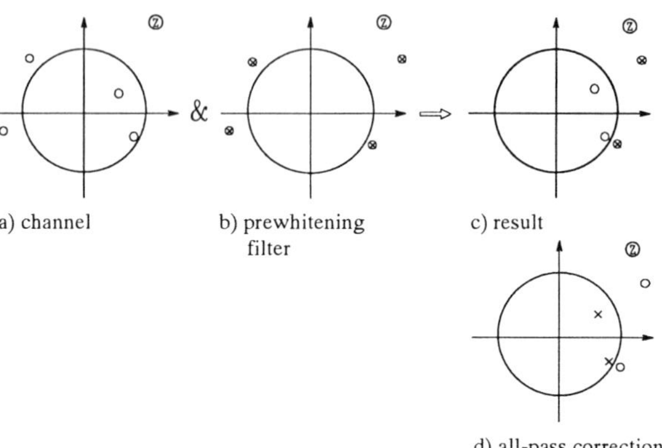

Figure 4: All-pass channel correction using a maximum phase prewhitening filter

The phase correction filter can be designed as a cascade of 1st order all-pass systems (instead of inverse) that can be realized by a recursive network with considerably reduced hardware complexity. Fig. 5 shows the resulting equalizer configuration.

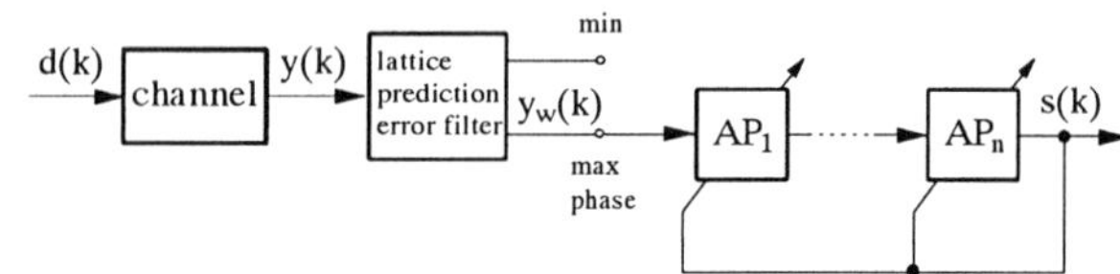

Figure 5: Lattice-all-pass equalizer

4 Adaptive blind equalization

The adaptive design of the prewhitening filter is solved by the power minimization of the forward (and backward) prediction error; of course this is a blind algorithm. A proper phase correction can only be performed under the application of higher order statistics. In [5] a very simple cost function is derived which is based on the specific definition of 4th order cumulants c_4^s:

$$c_4^s = E\{|s(k)|^4\} - 2\left[E\left\{|s(k)|^2\right\}\right]^2 - |E\{s^2(k)\}|^2 \stackrel{!}{=} max$$
$$\text{if} \quad E\left\{|s(k)|^2\right\} = const; \qquad (6)$$

where $s(k)$ denotes the complex envelope of the equalizer output signal and $E\{\cdot\}$ means the expectancy value. The criterion (6) is closely related to the self recovering equalizer introduced by Godard [4].

The cumulant criterion is used in [5] for the adaptive adjustment of the ℓ coefficients of the FIR phase correction filter; in the present paper it is applied for the calculation of the n poles z_{∞_ν} (and the related zeros) of the all-pass network. For the iterative solution the well-known stochastic gradient algorithm is applied. After some steps of calculation we obtain the following update equation:

$$z_{\infty_\nu}(i+1) = z_{\infty_\nu}(i) - \delta \cdot sgn(c_4^d)|s(k)|^2 \left[s^*(k)\frac{\partial s(k)}{\partial z_{\infty_\nu}} + s(k)\frac{\partial s^*(k)}{\partial z_{\infty_\nu}}\right] \qquad (7)$$

where δ is a small positive constant value and c_4^d is the 4th-order cumulant of the input data. The symbol $*$ denotes the complex value. In (7) the derivative $\partial s(k)/\partial z_{\infty_\nu}$ is still unknown. The calculation can be carried out in the z-plane (provided that transient effects are suffciently small). Finally we obtain the gradient network shown in fig. 6. The main disadvantage of the

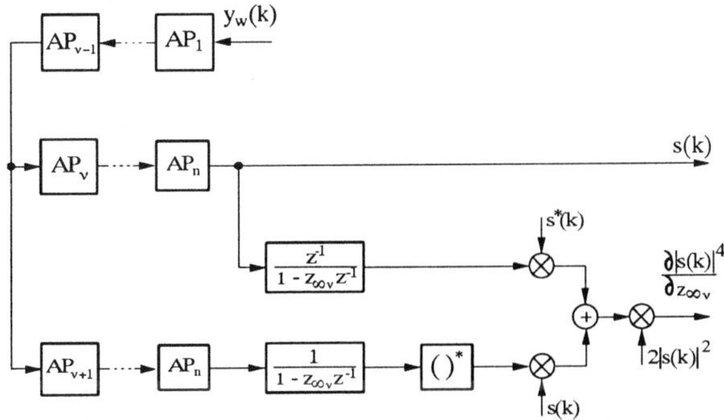

Figure 6: Gradient network for the calculation of $\partial|s(k)|^4/\partial z_{\infty_\nu}$

adaptive allpass structure explained so far is the high nonlinearity of the cost function (due to the cascade realization of the phase correction filter). This results in secondary minima and thus in convergence problems under unfavourable start conditions. As a more intuitive approach we use the minimum phase solution (which can easily be derived from the lattice filter) as an initial position. Under these conditions a proper adaptive equalization is achieved for sufficiently low channel order. Fig. 7 shows an example for the adaptive equalization of a 4th order mixed phase channel (a). Fig. 7b demonstrates the trajectories of the all-pass poles during the adaption process: Obviously the poles related to the minimum -phase channel remain at their initial positions while the poles related to the maximum-phase channel are removed on the unit circle or into the origin, respectively, where their influence vanishes. For conclusion the convergence

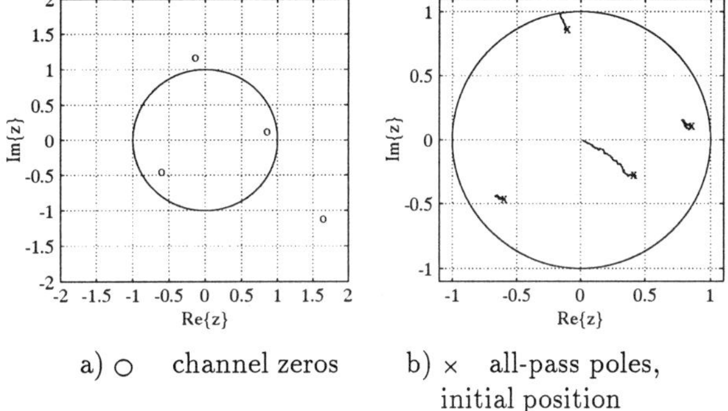

a) ○ channel zeros b) × all-pass poles, initial position

Figure 7: Demonstration of the adaptive process

behaviour of the lattice/all-pass equalizer is compared to the FIR-solution [5]. As an example we pick the channel characteristics according to fig. 7a. Fig. 8 shows the power of the intersymbol interference (ISI) during the adaption process for different values of δ. It can be seen that the convergence speed of the all-pass structure is superior to the FIR solution due to the small number of parameters, ($n = 4$), in comparision to the $\ell = 32$ FIR filter coefficients.

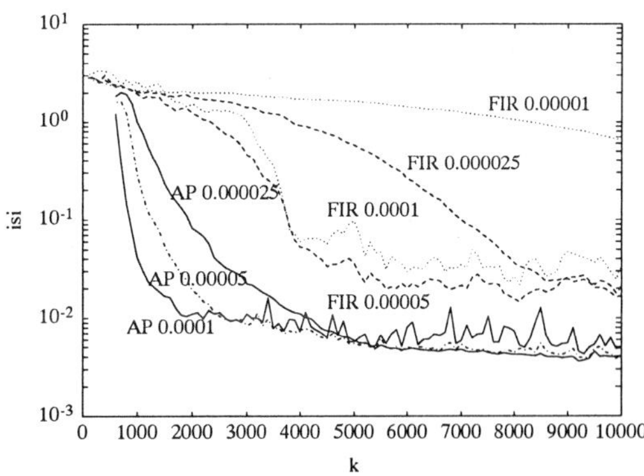

Figure 8: Performance of different blind equalizers (Parameter $\equiv \delta$)

5 Equalizer performance under additive gaussian noise

In this section the influence of additive noise is considered. It is well-known that the 2nd-order autocorrelation samples are corrupted by additive noise whereas the 4th-order cumulants are insensitive as far as the noise is gaussian. The 1st effect results in a misadjustment of the prewhitening filter; this is illustrated in fig. 9a. Fig. 9b indicates that a proper equalization is not achieved in presence of additive noise. On the other hand an exact modelling of the inverse system (which is related to the so-called "zero-forcing" solution) is not desirable because both – the residual ISI and the power of the additive noise at the equalizer output – have to taken into account. An optimum solution in this sense is the well-known "mean-square error" (MSE) solution. The following examples illustrate that the blind adaptive design of the lattice/all-pass equalizer reveals a solution near the MSE. Fig. 10a shows the resuting power of the error at the equalizer output for zero-forcing, lattice/all-pass and MSE equalization. Fig. 10b gives the related bit error rate as a function of the signal-to-noise ratio. As an example a 1st order channel was used with a zero of $z_0 = 0.9$.

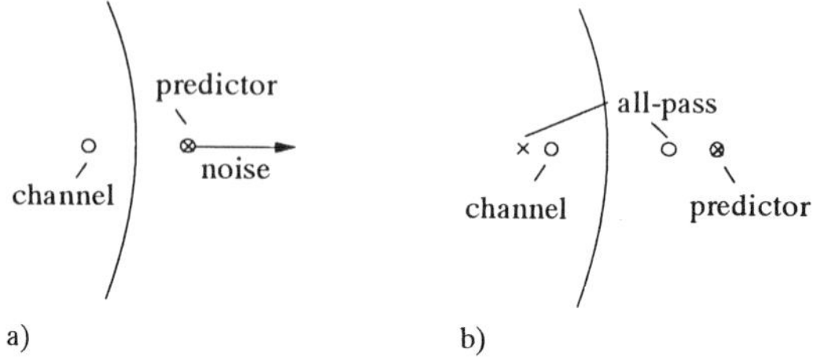

Figure 9: Pole-zero conditions under noise conditions

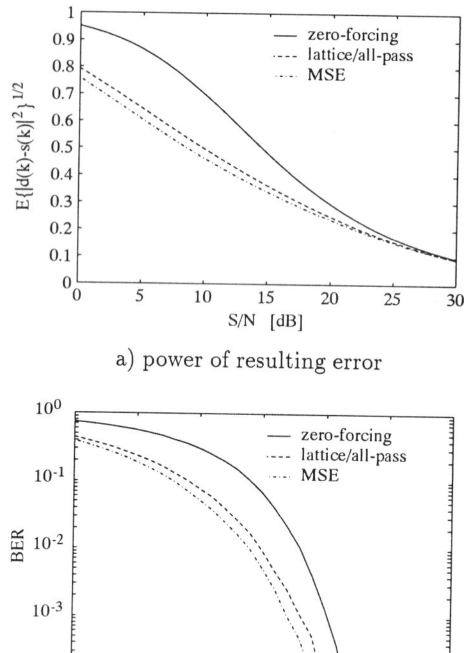

Figure 10: Equalizer performance under gaussian noise

Conclusions

In this paper it was shown that the blind channel equalization problem can be solved efficiently by a certain lattice/all-pass configuration. If the maximum-phase lattice output is used for decorrelation the phase correction network can be designed as a conventional recursive all-pass system. For the blind adaptation a 4th-order scalar cumulant criterion is applied. It was shown that the convergence speed is considerably increased due to the small number of parameters to be identified (all-pass poles). Some simulation results demonstrate that under additive noise conditions a solution near the optimum MSE-solution is obtained.

Mr. M. Benthin is gratefully acknowledged for his support in carrying out the simulations and for many helpful discussions.

References

[1] G. B. Giannakis and J. M. Mendel. Identification of nonminimum phase systems using higher order cumulants. *IEEE Trans. Acoust., Speech, Signal Processing*, 37(3):360–377, 1989.

[2] K. D. Kammeyer and B. Jelonnek. A cumulant zero-matching method for the blind system identification. *International Signal Processing Workshop on Higher Order Statistics*, 1991. Chamrousse, July 10-12.

[3] B. Jelonnek and K. D. Kammeyer. Improved methods for the blind system identification using higher order statistics. *accepted IEEE Trans. on Signal Processing to be published*, Dec. 1992.

[4] D. Godard. Self-recovering equalization and carrier tracking in two-dimensional data communication systems. *IEEE Trans. Acoust., Speech, Signal Processing*, Com-28(11):1867–1875, 1980.

[5] O. Shalvi and G. Weinstein. New criteria for blind deconvolution of nonminimum phase systems (channels). *IEEE Trans. Inform. Theory*, 36(2):312–321, 1990.

SIGNAL PROCESSING VI: Theories and Applications
J. Vandewalle, R. Boite, M. Moonen, A. Oosterlinck (eds.)

ON THE REDUCTION OF NESTING ALGORITHMS ROUNDING ERRORS

Ryszard Stasiński

Department of Electronics & Communications
Technical University of Poznań
Piotrowo 3A, PL-60-965 Poznań, Poland

In the paper a simple rounding error model is used for evaluating worst-case rounding errors of nesting algorithms. The list of potentially the best from this point of view nesting DFT algorithms (WFTAs) is provided, showing that the best results are obtained for some of the following data sizes: $N = 2^r \cdot 3^s \cdot 5^t \cdot 17^u, r = 1,2,3,4,5,6; s = 1,2; t,u = 0,1$. The results are compared with those for the radix-4 and split-radix FFT, and it is shown that for practical N values the error level of WFTAs need not be more than 1 bit higher than those for the FFT. The paper contains also remarks concerning the construction of low-error nesting convolution algorithms.

1 Introduction

The nesting technique is used for reducing multiplication count in separable multidimensional computations. The most important computational problems of this sort are DFT, circular and linear convolution computation for multidimensional data, or for data sizes being highly composite numbers [1], [2], [3]. Till now it has existed only one work in which a comparison of nested and not-nested (i.e. row-column) algorithms rounding errors was done [4]. The paper concerned the comparison of WFTA with FFT, and showed that WFTA requires 1–2 bits longer accumulator registers, if the same accuracy of WFTA and FFT results were needed. The results are partly confirmed in [5], a paper based on premises given in [6].

Recently it has been shown how to explain the result, and that the effect is characteristic for nested algorithms in general [7] (see also section 2 of this paper). Namely, in the case of FFT an output error is a combination of those from $0(N)$ error sources, while the whole FFT structure contains $0(N log N)$ error sources. The situation changes dramatically when the FFT is used for computing a convolution, when all $0(N log N)$ FFT sources contribute to an output error of the algorithm. Unfortunately, nesting algorithms have structures similar to those of the FFT-based convolution ones, while the row-column ones that of FFT. Nevertheless, a detailed analysis of circular convolution and DFT algorithms shows that for some small-size algorithm structures a smaller part of error sources is "visible" from its worst-case output than for other ones.

In the paper the technique from [7] is used for finding potentially the best from the point of view of rounding errors nesting DFT and convolution algorithms for practical data sizes. Section 2 contains general description of the algorithms of interest, definition of the proposed error model, and general comparison of row-column and nested algorithms errors. In section 3 the results of section 2 and of the paper [7] are used for founding the best nesting DFT algorithms, so, the section provides verification of conclusions from [7]. The problem of generating the best nested convolution algorithms is outlined in section 4. The evaluations are necessary for restricting the range of algorithms for which computer simulations of their rounding-error performance are made [5]. It is shown that the choice of two algorithms classes in [5]: for $N = 15 \cdot 2^s$, and for $N = 255 \cdot 2^s, s = 0,1,\ldots$, has been correct.

2 Theoretical background

Nesting algorithms are constructed from the so-called small-N modules having a particular structure [1], [2], [3]. In matrix form the structure can be described as follows:

$$\mathbf{T} = \mathbf{EHD} \tag{1}$$

where the $\mathbf{D}, \mathbf{E}$ matrices contain only integer elements, and the $\mathbf{H}$ matrix is either block-diagonal, or diagonal one. Namely, all multiplications of the algorithm are concentrated in the central stage denoted by multiplication by the $\mathbf{H}$ matrix.

The nesting technique can be applied if a computational problem can be transformed into a separable multidimensional one. Separability means that computations can be made in each dimension independently. So, if the transform **T** is used in computations in a dimension of a r-dimensional algorithm, the calculations are the same as for the matrix **T** by a vector product, except for the fact that each element of a vector is not a scalar but a r-1-dimensional matrix of data. For accomplish the computations the scheme is repeated for each of r dimensions. This is the so-called *row-column* algorithm. When *nesting* the computations in other dimensions are not performed after computations for the first chosen dimension, but are merged with multiplications by elements of the **H** matrix (1). If applied repeatidly to the following dimensions the nesting results in a sort of "double row-column" algorithm: it starts with the row-column calculations for all **D** matrices, then we have a stage or stages resulting from merging of all **H** matrices, and it finishes with the row-column structure for **E** matrices in all dimensions.

For evaluating rounding errors of nesting algorithms let us consider the following simple error model [7]: all quantizations after additions, subtractions, and multiplications produce the same error equal to one, once generated an error propagates unchanged across the algorithm flowgraph (hence, e.g. error damping by multipliers is neglected), finally, correlation between errors, if any, is ignored. Quantizations are here such operations as roundings, truncations, etc. The error value in an algorithm flowgraph node is then equal to the number of error sources linked with the node by a path, hence, error counting is as simple as paths founding.

Let us take into consideration the same distribution of errors as assumed in [7] — additions and subtractions linked with multiplications by matrices **E** and **H** (1) are errorless (no quantization). The overflow is avoided by quantizing results of all additions and subtractions due to multiplication by the matrix **D**, and by proper scaling of **H** matrix multiplier values. Of course, this assumption is quite natural for a fixed-point arithmetic, but not for a floating-point one.

Let us analyze now the error of a row-column algorithm. Assume that the size of the whole problem is $N = N_0 \cdot N_1 \cdot \ldots \cdot N_{r-1}$, and that the problem is transformed into an r-dimensional one of size $N_0 \times N_1 \times \ldots \times N_{r-1}$. As it was stated above, the algorithm can be described as a series of generalized multiplications by $\mathbf{T_i}$ matrices (1), $i = 0, 1, \ldots, r-1$. In this way an output error of the algorithm consists of an internal error due to multiplication by the $\mathbf{T_{r-1}}$ matrix being proportional to N_{r-1}, and to sum of all N_{r-1} input node errors of the flowgraph depicting multiplication by the $\mathbf{T_{r-1}}$ matrix. Notice that the formula is true for any algorithm stage, hence, we have the following formula on the error cumulation:

$$e_r = \epsilon_{r-1} + N_{r-1} \cdot e_{r-1} =$$

$$= \epsilon_{r-1} + N_{r-1} \cdot \epsilon_{r-2} + N_{r-1} \cdot N_{r-2} \cdot e_{r-2} = \ldots$$

$$= \epsilon_{r-1} + N_{r-1} \cdot \epsilon_{r-2} + \ldots + \frac{N}{N_0} \cdot \epsilon_0 + N \cdot e_0 \qquad (2)$$

where e_i is the error in a node at the input to stage i (i.e. e_r is an error at an output node of the algorithm), and ϵ_i is the error linked with multiplication by the $\mathbf{T_i}$ matrix. Notice that the output error is

$$e_r = 0(N_{r-1} + N_{r-1} \cdot N_{r-2} + \ldots + N) = 0(N) \qquad (3)$$

while the whole number of error source in the algorithm is proportional to

$$\sum_{i=0}^{r-1} \frac{N}{N_i} \cdot \epsilon_i = 0(rN) \qquad (4)$$

as each generalized matrix multiplications involves $\frac{N}{N_i}$ multiplications by the matrix $\mathbf{T_i}$ on "scalar" data vectors, compare [7]. Notice also that if they are any input errors e_0 to the algorithm, all of them contribute to any output error e_r.

The last observation implies that nesting algorithms could not have smaller errors than the row-column ones. Namely, it seems that an output error of the nesting algorithm consists of all errors due to multiplications by $\mathbf{D_i}$, and combination of $\mathbf{H_i}$ matrices (1), compare [7]. Fortunately it appears that for some small modules the $\mathbf{E_i}$ matrices are sparse ones, hence, they "transmit" a part of input errors, only. The best from this point of view are output operations of circular convolution algorithms for $N = p^s$ when s is as high as possible, while p as small as possible, p is a prime number [7]. Operations due to **E** matrices of small DFT modules consist mainly of output operations of circular convolutions, hence, this is an important hint for constructing of these modules, too.

3 Errors of Winograd DFT algorithms

With no doubt the most complicated nesting algorithms are those for the DFT, named Winograd-Fourier transform algorithms (WFTAs) [1], [2], [3]. In [7] the problem of constructing the best from the rounding error point of view DFT modules for these algorithms was stated. It was found that the best DFT modules are those for $N = 2^s$, then those for Fermat prime numbers (3,5,17,...) and $N = 3^s$, then those for $N = 5^s$ and prime numbers of the form $N = 2^s \cdot 3^t + 1$, while the worse results are obtained for prime numbers for which $N - 1$ is a product of first powers of other prime numbers. Somewhat modified results from [7] for the most interesting DFT modules are summarized in Table I. The modifications concentrate on the following two aspects:

- The modules are modified for making **E** (1) matrices as sparse as possible. In particular, computations linked with x(0) and X(0) data samples are done in accordance with Fig.1d from [7], and not with Fig.1c as it has been assumed in the previous paper. As it was found empirically the modification indeed results in error reduction of whole nesting algorithms.
- Polynomial products in **H** (1) matrices are computed using reasonable algorithms, and not directly from the definition formulae as in [7]. The only difference with normally applied algorithms consists in fact that polynomial products modulo $Z^2 + 1$, and $Z^2 \pm Z + 1$ are computed using definition formula, hence, 4 multiplications and 2 additions, and not 3 multiplications and additions as in usual case.

The last assumption is linked with the fact that all described here algorithms are compared with the radix-4 (hence, split-radix) FFT in which complex multiplications are computed using the definition formula, too. It was found in [7] that the error for this FFT is smaller than

$$E_{FFT}(N) = \frac{5}{3} \cdot N \qquad (5)$$

Table I

Worst-case error values for small-N DFT modules

N	Error	N	Error	N	Error
2	1	8	10	19	137
3	5	9	21	25	101
4	3	13	61	27	113
5	11	16	22	32	50
7	27	17	57	64	114

The above mentioned DFT modules (and some worse ones) have been combined in the Winograd's algorithms, and their errors calculated. Polynomial products in **H** (1) matrices have been merged using polynomial transforms, of course, if applicable [3]. Similarly as in the case of **E** matrices, it has been assumed that inverse (output) polynomial transforms following the residual polynomial products are errorless. When residual polynomial products are multidimensional ones, they are computed using the row-column method, which has resulted e.g. in reduced maximum error for the $N = 25$-point module, compare Table I in [7]. In accordance with (2) the best method of constructing row-column structures consists in taking modules with the smallest errors as the first ones. When completing computations due to **E** (1) matrices it is also important to use "the most sparse" matrices as the first ones for separating paths in the algorithm flowgraph as early as possible.

In Table II the best received Winograd's algorithms are shown. The best means here that they have the smallest error to N ratio among all nested algorithms, see the last column. The $E_{FFT}(N)$ value is calculated on the basis of (5). As can be seen, the best are families of algorithms for data sizes of the form $N = q \cdot 2^s, s = (1), 2, 3, \ldots$, where $q = 3, 15, 255,$ and 765. There are also sample algorithms for $q = 9, 45,$ and 51. In this context the choice of algorithm to be analyzed in [5] seems to be very fortunate, although based on somewhat different premises [6].

In Figure 1 the data for the best Winograd's algorithm families for $q = 15, 45, 51, 255,$ and 765 are plotted together with data for two frequently used algorithm families — for $q = 63,$ and 315 [1], [2]. As can be seen, the error level for the latter families is approximately $\frac{1}{4}$ to $\frac{1}{2}$ bit above that for the best algorithms. It is interesting that this gap between error levels is filled with data for several other algorithm families — for $q = 75, 153, 17, 25,$ and even for 27, see also Table I. When compared with FFT the error level for WFTAs grows faster, and for $N > 10000$ and the best WFTAs it becames more than 1 bit greater than for the FFT. This N value is approximately 4 times greater than for frequently used family of algorithms for $q = 315$.

4 Note on nested convolution algorithms

The rounding-error properties of convolution algorithms are much easier to predict than those of DFT ones. With no doubt the best modules are those for $N = 2^s$, for $s \leq 4$, and for $N = 3, 9$, then those for $N = 5$ and so on. Nevertheless, when considering convolution algorithms in general one observation must be strongly underlined here. Namely, for obtaining the best results convolution algorithms should be constructed of convolution modules rather than of DFT ones, although in many cases the use of latter modules results in more efficient algorithms. Namely, it is unlikely that a cascade of forward and inverse small-N DFT modules produces smaller rounding errors than one convolution algorithm of the same size.

5 Conclusion

In the paper the results of exhaustive search for the best from the rounding-error point of view nesting DFT and convolution algorithms are reported. It is shown that the best DFT algorithms of this type exist for N: $2^s \cdot 3, 2^s \cdot 15, 2^s \cdot 255,$ and $2^s \cdot 765$, $0 < s \leq 6$, and to some other data sizes composed of DFT modules for $N =$

Fig. 1: WFTA errors

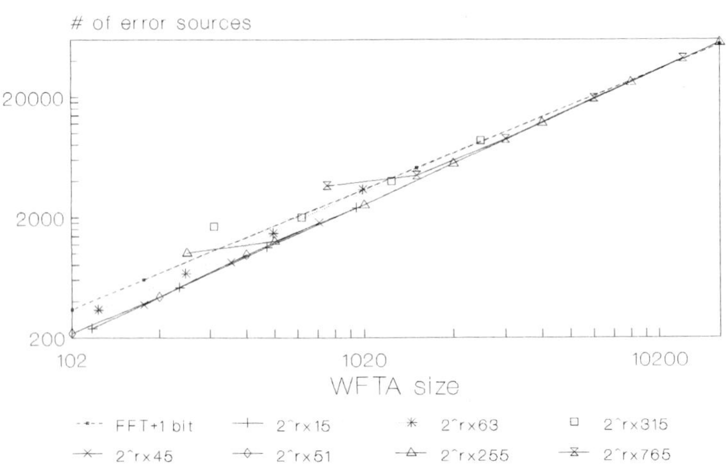

$2^s, 3, 9, 5,$ and 17, too. It is shown that for practical data sizes the error level of resultant DFT algorithms need not be more than 1 bit above that for the radix-4 and split-radix FFTs. It is somewhat less than it has been obtained in simulations reported in [5]. It seems to be linked with the fact that the error model discards error damping introduced by stages of multiplications in FFTs. On the other hand, the given by the model error growth prediction for the latter algorithms seems to be correct [5]. Finally, the paper provides some hints concerning the construction of error-efficient nesting convolution algorithms — compare [7].

References

[1] R.E. Blahut, "Fast algorithms for digital signal processing", Addison-Wesley, Reading, Mass., 1985.

[2] H.J. Nussbaumer, "Fast Fourier transform and convolution algorithms", Springer-Verlag, 1981.

[3] R. Stasiński, "Prime factor DFT algorithms for new small-N DFT modules", IEE Proc., Pt. G, vol. 134, pp. 117-126, 1987.

[4] R.W. Patterson, J.H. McClellan, "Fixed-point error analysis of Winograd-Fourier transform algorithms", IEEE Trans. Acoust., Speech, Signal Proces., Vol. ASSP-28, pp. 447-455, 1978.

[5] E. Lukasik, R. Stasiński: "Certain aspects of fixed-point FFT and WFTA rounding-error comparison", this volume (Proc. EUSIPCO-92), 1992.

[6] R. Stasiński, E. Lukasik, "Minimization of rounding errors in WFTA programs", Proc. ICASSP-88, pp. 1423-1426, 1988.

[7] R. Stasiński, "Rounding-error optimized small-N DFT modules", ECCTD'91, pp. 689-698, 1991.

Table II

The lowest worst-case errors of Winograd's algorithms

N	N divisors	Error	$E_{FFT}(N)$	Error/N
12	4×3	16	20	1.25
24	8×3	38	40	1.58[1]
36	4×9	60	60	1.67
48	16×3	82	80	1.71
60	$4 \times 3 \times 5$	104	100	1.73
96	32×3	178	160	1.85
120	$8 \times 3 \times 5$	238	200	1.98
192	64×3	386	320	2.01
204	$4 \times 3 \times 17$	432	340	2.12
240	$16 \times 3 \times 5$	522	400	2.175
360	$8 \times 9 \times 5$	834	600	2.32
480	$32 \times 3 \times 5$	1122	800	2.34
1020[2]	$4 \times 3 \times 5 \times 17$	2520	1700	2.47
2040	$8 \times 3 \times 5 \times 17$	5550	3400	2.72
3060	$4 \times 9 \times 5 \times 17$	8844	5100	2.89
4080	$16 \times 3 \times 5 \times 17$	12090	6800	2.96
6120	$8 \times 9 \times 5 \times 17$	19218	10200	3.14
8160	$32 \times 3 \times 5 \times 17$	26374	13600	3.23
12240	$16 \times 9 \times 5 \times 17$	41438	20400	3.38
16320	$64 \times 3 \times 5 \times 17$	56374	27200	3.46

[1] More than 1.56 for the 32-point module, see Table I.
[2] and the 510-point module, for which Error= 1260.

SIGNAL PROCESSING VI: Theories and Applications
J. Vandewalle, R. Boite, M. Moonen, A. Oosterlinck (eds.)

WAVELET TRANSFORM BASED ADAPTIVE FILTERING

Nurgün Erdöl and Filiz Başbuğ

Department of Electrical Engineering, Florida Atlantic University, Boca Raton, Florida 33431

Wavelet transforms in the discrete time domain are incorporated into an adaptive filter that uses the Widrow-Hoff least-mean square (LMS) algorithm to update the weights of a linear combiner. It is shown that the use of the wavelet transform in place of the discrete Fourier transform in the transform domain adaptive structure [1,7] significantly reduces the condition number ($\lambda_{max}/\lambda_{min}$) of the input signal. Experiments with the D4 wavelet [12] in a predictor configuration are encouraging.

1. INTRODUCTION

The most common implementation of adaptive filters utilizes a time domain tapped-delay-line (TDL) form and the Widrow-Hoff least-mean square (LMS) algorithm [5] for the adaptation of filter weights. In case of a large spread in the eigenvalues, however, some form of orthogonalization of the signals that are inputs to the adaptive weights can result in faster adaptation than is possible with LMS alone [6]. The first orthogonalization scheme [7] was based on the discrete Fourier transform (DFT). Others use adaptive lattice filters [8] and Gram-Schmidt orthogonalization methods [9].

Here we investigate the use of wavelet transforms (WT) in adaptive filtering. The method proposed is very much analogous to the DFT based adaptive filtering of Narayan and Peterson [1,7]. However, it has all the advantages of wavelet analysis over Fourier analysis when a weighted sum of sinusoids does not adequately represent the time varying modes of the signals involved. In discrete wavelet analysis [10-13] signals are represented by a weighted sum of the translates and dilates of a mother wavelet. Multiresolution analysis [12,13] provides a means of constructing orthonormal bases of wavelets spanning the space $L^2(\mathbb{R})$ of square integrable functions. These wavelets can be grouped by their dilation (scaling) constant into disjoint subsets spanning proper and orthogonal sub spaces of $L^2(\mathbb{R})$. These subspaces that correspond to different scales are said to represent signals at different resolution levels.

In the adaptive filtering scheme using wavelet transforms the projections of the input signal on to the above mentioned orthogonal subspaces are used as inputs to a linear combiner. The weights of the linear combiner can hence be updated by the orthogonalized LMS algorithm [6] while normalizing the power at each resolution to achieve faster and uniform convergence of all weights to be optimal. Although very much like the DFT based adaptive system [1,7], the WT based system is superior in its ability to track rapid changes in the input signal. Use of real valued wavelet requires no complex arithmetic. Due to the recursive properties of wavelet analysis at different resolution levels, the amount of computation is much less than that using the FFT.

2. BACKGROUND INFORMATION

The LMS algorithm [5] is a computationally simple algorithm that is used to obtain iteratively the weight vector $\mathbf{W}_j$ that minimizes the difference between a desired signal S_j and its estimate $y_j = \mathbf{W}^T{}_j\mathbf{X}_j$ in the mean-square sense. The weight vector is updated according to

$$\mathbf{W}_{j+1} = \mathbf{W}_j + 2\mu\varepsilon_j\mathbf{X}_j \tag{1}$$

where the subscript j denotes time and ε_j is the error signal equal to the difference $S_j - y_j$. If the adaptation constant $\mu > 0$ is less than $1/tr(\mathbf{R}_x)$, where $\mathbf{R}_x$ is the input autocorrelation matrix, then W_j converges to the Wiener solution $\mathbf{W}^*$ in the mean [2]. The slowest time constant for the weights is given by

$$\tau \geq tr(\mathbf{R}_x)/2\lambda_{min} \tag{2}$$

where λ_{min} is the smallest eigenvalue of the autocorrelation matrix $\mathbf{R}_x$. This property slows down the convergence rate when there is a large spread in the eigenvalues. This deficiency can be overcome by preprocessing the input signal $\mathbf{X}_j$ so that it is orthogonalized [6]. Specifically, if the new autocorrelation matrix is the identity matrix, then all the eigenvalues being unity assures the only and fastest time constant

$$\tau = \frac{1}{2\mu} \tag{3}$$

The orthonormal wavelets are used to process the input signal. A function $x(t) \in L^2(\mathbb{R})$ can be expressed as a linear combination of the orthonormal wavelets as

$$\Psi_{jk}(t) = 2^{j/2}\Psi(2^{j/2}t - k) \quad \text{for all integers } j,k \tag{4}$$

$$x(t) = \sum_{j,k \in Z} 2^{j/2} d_k^j\, \Psi(2^j t - k) \tag{5}$$

The restriction of $x(t)$ to a given resolution j is

$$x_j(t) = 2^{j/2} \sum_{k \in Z} d_k^j\, \Psi(2^j t - k) \tag{6}$$

where the wavelet coefficients $d_k{}^j$ are given by the inner product of $x(t)$ and the wavelet, i.e.,

$$d_k^j = < x(t),\ 2^{j/2}\,\Psi(2^j t - k) > \tag{7}$$

In practice the above series will usually be truncated and an approximation to x(t) will be expressed as

$$x^{(J)}(t) = \sum_{j=0}^{-J} \sum_{k} 2^{j/2}\, d_k^j\, \Psi(2^j t - k) \tag{8}$$

where J is a positive integer. Equation (8) shows that $x^{(J)}(t)$ is the projection of x(t) onto the collection of orthogonal subspaces $W_0 \cup W_{-1} \cup ... \cup W_{-J}$ with W_0 and W_{-J} representing respectively the highest and lowest resolution subspaces. The error incurred by such approximations is addressed in [14] and warrants further analysis. For the purposes of this work, however, we assume that $x^{(J)}(t)$ is acceptably close to x(t).

3. "SCALE" BASED ADAPTIVE SYSTEMS

Here we outline a discrete-time adaptive system that uses the orthogonalized version of an input sequence as input to a linear combiner as shown in Figure 1. For this construction we refer to the filter bank interpretation of wavelet transforms in the discrete time domain [15,16]. Let $\Psi(2^{-J}t)$ be represented by its at least 2^J times oversampled values on the "t" scale as g(n). Specifically, let

$$g(n) = \begin{cases} \Psi(2^{-J} n) & n = 0,1,...,N-1 \\ 0 & otherwise \end{cases} \tag{9}$$

where $N = 2^K$ form some integer $K >> J$. Then

$$g_j(n) = g(2^j n) = \Psi(2^{j-J} n)$$

$$n = 0,1,...2^{-j}N - 1,\ j = 0,1,...,J \tag{10}$$

represents g(n) subsampled by 2^j. Then let $\{g(n);\ n = 0,1,...,N-1\}$ be the discrete time representation of a wavelet oversampled by a factor of 2^J and $N = 2^K$, $K >> J > 0$. Define

$$\alpha_k^j = \sum_{n=n_o}^{2^{-j}N-1+n_o} x_n\, g_{jk}(n) \tag{11}$$

$$= 2^{j/2} \sum_{n=n_o}^{2^{-j}N-1+n_o} x_n\, g(2^j n - 2^J k)$$

where $n_o = 2^{J-j}k$, and let the synthesized signal be given by

$$x_n^{(J)} = C \sum_{j=0}^{J} \sum_{k} \alpha_k^j\, g_{jk}(n) \tag{12}$$

where C is a positive constant of normalization. Substituting (11) into (12) yields

$$x_n^{(J)} = \sum_{j=0}^{J} \sum_{m} x_m\, r_j(n-m) \tag{13}$$

where

$$r_j(n-m) = C \sum_{k} g_{jk}(n)\, g_{jk}(m) \tag{14}$$

and

$$r_j(n) = r_o(2^J n) = r_j(-n) \tag{15}$$

From Figure 1 it is seen that $r_j(n)$ denotes the real symmetric finite impulse response of the filter banks used to process input vector $X_n = \{x_n\ x_{n-1},...x_{n-N+1}\}^T$. Filter outputs $v_j(n)$,

$$v_j(n) = \sum_{m} r_j(m)\, x_{n-m} \tag{16}$$

are multiplied by the weights $w_j(n)$. The output signal y(n)

$$y(n) = \sum_{j} w_j(n)\, v_j(n) \tag{17}$$

is subtracted from the desired signal x(n) to yield the error signal ε(n). Weights $w_j(n)$ are updated by the Widrow-Hoff LMS algorithm as

$$w_j(n+1) = w_j(n) + 2\mu\, \varepsilon(n)\, x_n \tag{18}$$

The filter outputs can be expressed in matrix notation as

$$\mathbf{V}_n = \mathbf{R}\,\mathbf{X}_n \tag{19}$$

where

$$\mathbf{V}_n = \left[v_o(n),\ v_1(n),...,v_J(n) \right]^T,\ \mathbf{X}_n = \left[x_n x_{n-1},...,x_{n-N+1} \right]^T$$

and

$$\left[R \right]_{jm} = r_j(m) \quad for\, j = 0,1,...,J \ \ and \ \ m = 0,1,...,N-1$$

We note that the covariance of the new input is

$$\mathbf{R}_v = E\left[\mathbf{V}_n \mathbf{V}_n^T \right] = \mathbf{R}\,E\left[\mathbf{X}_n \mathbf{X}_n^T \right] R^T = \mathbf{R}\,\mathbf{R}_x\,\mathbf{R}^T \tag{20}$$

If we impose that $\mathbf{R}_v = \mathbf{I}$, i.e., all eigenvalues of $\mathbf{R}_v$ be identical and equal to 1 then we get

$$\mathbf{R}\,\mathbf{R}^T = \mathbf{R}_x^{-1} \tag{21}$$

Indeed, equation (21) does not give us a practical way to find **R**. However, if the discrete time wavelets are orthonormal in the scaling and the translation parameters, then $v_j(n)$ are orthogonal in j.

This can be seen by observing from equations (12-14) that $x_n^{(J)}$ is the j-sum of $v_j(n)$. Hence, an alternative to equation (16) is

$$v_j(n) = C \sum_k a^j_k g_{jk}(n) \tag{22}$$

Substituting eqn. (22) in (20) and using

$$c \sum_n g_{jk}(n) g_{\ell i}(n) = \delta(j-\ell)\delta(k-i)) \tag{23}$$

proves that R_v = diagonal ($C\,\sigma_j^2$) where

$$\sigma_j^2 = \sum_k (a^j_k)^2 \tag{24}$$

The new eigenvalues are hence given by σ_j^2 which represent the signal energy at the jth resolution level or the energy of the output of the jth filter in Figure 1. These filters denoted by their impulse responses $r_j(n)$ are related to each other by scaling specifically

$$r_{j+1}(e^{j\omega}) = r_j(e^{j\omega/2}) + r_j(-e^{j\omega/2}) \tag{25}$$

where $rj(ej\omega)$ is the Fourier transform of the sequence $r_j(n)$ and $r_j(e^{jo}) = 0$.
From eqn. (25) (see also [15,16]) it can be seen that the frequency domain support of the filters gets larger and shifts to the high frequency ranges as the resolution level increases. For a class of signals with 1/f like spectra the constant Q characteristics of the above filters implies that the energy of the filter outputs will tend toward a uniform distribution. Our initial experimental results supporting this argument are given in Table 1.

4. EXPERIMENTAL RESULTS

The condition number $\lambda max/\lambda min$, is computed as a measure of the eigenvalue spread for the input signal x(n) and the transformed signal $v_j(n)$. The condition number is indicated by ξ_{xs} and ξ_{vs} for the input and the transformed input respectively and s denotes the size of the correlations matrix. The results are given below

Table 1
Condition Numbers as a Function of the Order of the Adaptive Filter

S	2	3	4	5
ξ_{xs}	9.8755	31.6764	60.1742	91.9855
ξ_{vs}	8.0154	8.0375	8.0564	8.0564

It is clear that the transform decreases the eigenvalue spread for the experimental signal.

5. CONCLUSIONS

A wavelet transform based adaptive filter is proposed and analyzed for its property to orthogonalize an input data sequence. The orthogonal sequence produced by a bank of real symmetric FIR filters are applied to a linear combiner whose weights are updated by the LMS algorithm for Widrow and Hoff. Such orthogonalization has the effect of fast and uniform convergence to the optimal (Wiener) solution.

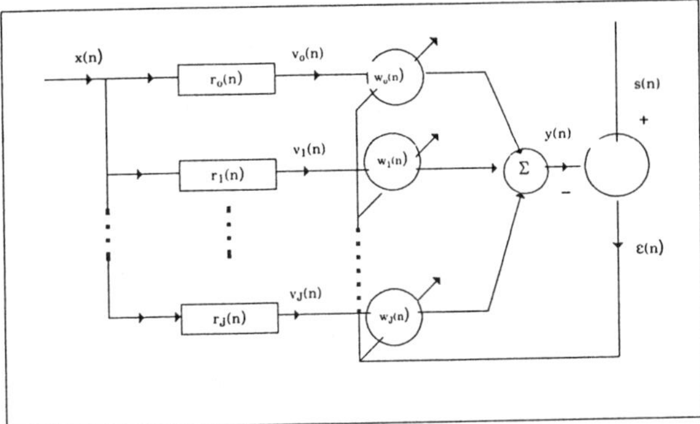

Figure 1. Wavelet Transform Based Adaptive System.

REFERENCES

[1] S. S. Narayan, A. M. Peterson and M. J. Narasimha, "Transform domain LMS algorithm," IEEE Trans. Acous., Speech Signal Processing, Vol. ASSP-31, No. 3, June 1983.

[2] B. Widrow, et al., "Adaptive noise canceling: Principles and its applications," Proc. IEEE, vol. 63, pp. 1692-1716, December 1975.

[3] R. W. Lucky, "Techniques for adaptive equalization of digital communication systems," Bell Syst. Tech. J., vol. 45, pp. 255-286, February 1966.

[4] L. Griffiths, "Rapid measurement of digital instantaneous frequency," IEEE Trans. Acoust., Speech, Signal Processing, vol. ASSP-23, pp. 209-222, April 1975.

[5] B. Widrow, "Adaptive Filters," in Aspects of Network and System Theory, R. Kalman and N. DeClaris, Eds. New York: Holt, Rinehart and Winston, pp. 563-587, 197__.

[6] B. Widrow and E. Walach, "On the statistical efficiency of the LMS algorithm with nonstationary inputs," IEEE Trans. Information Theory-Special Issue on Adaptive Filtering, vol. 30, no. 2, part 1, pp. 211-221, March 1984.

[7] S. S. Narayan and A. M. Peterson, "Frequency domain least-mean square algorithm," Proc. IEEE, vol. 69, pp. 124-126, January 1981.

[8] L. S. Griffiths, "An adaptive lattice structure for noise-cancelling applications," Proc. ICASSP-78, p. 87.

[9] R. A. Monzingo and T. W. Miller, Introduction to Adaptive Arrays, New York: John Wiley, sec. 8.3., 1980.

[10] C. E. Heil and D. F. Walnut, "Continuous and discrete wavelet transforms," SIAM Review, vol. 32, pp. 628-666, December 1989.

[11] G. Strang, "Wavelets and dilation equations: A brief introduction," SIAM Review, vol. 31, pp. 614-627, December 1989.

[12] I. Daubechies, "Orthonormal bases of compactly supported wavelets," Comm. Pure Appl. Math, vol. XLI, pp. 909-996, 1988.

[13] S. G. Mallat, "Multiresolution approximations and wavelet orthonormal bases of $L^2(R)$," Trans. Am. Math. Soc., vol. 315, pp. 69-87, September 1989.

[14] A. W. Tewfik and P. E. Jorgensen, "On the optimal choice of a wavelet for signal representation" Submitted to the IEEE Trans. Information Theory.

[15] M. J. T. Smith and T. P. Barnwell, "Exact reconstruction for tree-structured subband coders," IEEE Trans. Acoust., Speech and Signal Proc., vol. ASSP-34, pp. 434-441, June 1988.

[16] D. Rioul and M. Vetterli, "Wavelets and signal processing", IEEE Signal Processing Magazine, pp. 14-38, October 1991.

SIGNAL PROCESSING VI: Theories and Applications
J. Vandewalle, R. Boite, M. Moonen, A. Oosterlinck (eds.)

An Adjustable Constraint Approach for Robust Adaptive Beamforming

Luis Castedo†, *Ching-Yih Tseng*‡ *and Lloyd J. Griffiths*‡

†DSSR, ETSI Telecomunicación-UPM, Ciudad Universitaria s/n, 28040 Madrid.

‡University of Southern California, Los Angeles, CA 90089-2564.

This paper presents a new beamforming scheme which uses a reference signal in linearly-constrained minimum variance (LCMV) adaptive beamformers. The proposed approach is able to track variations in the signal of interest (SOI) steering vector and incorporates additional constraints that can be used to control the sidelobe level. Since the constraints are adjusted to match the SOI steering vector, the signal cancellation behavior due to perturbation errors in the conventional LCMV beamformers is alleviated. An efficient implementation structure is also presented to update the beamformer weights in an adaptive fashion. The method is intended for communications applications where a reference signal can be extracted from the array output.

1 Introduction

In Linearly Constrained Minimum Variance (LCMV) beamformers a single point constraint which fixes a unity gain at the desired signal direction is typically used. This constraint makes use of the steering vector at the signal direction. However, such a beamforming system is known to be very sensitive to perturbation errors because the steering vector used in the constraint design does not match the true steering vector perfectly. In practice, it is almost impossible to precisely obtain the true steering vector because the element locations are only known within a certain degree of accuracy. Other uncertainty factors may also include random gain and phase errors occurring at the antenna elements and the variation of the desired signal direction. Even with only a small amount of perturbation errors, the output signal-to-interference plus noise ratio (SINR) can be severely degraded due to signal cancellation. Feldman and Griffiths [1] pointed out that the problem of signal cancellation can also occur if there is not sufficient amount of snapshots to estimate the covariance matrix, even though the true steering vector is perfectly known. It was shown that this finite sample data effect in fact represents a kind of perturbation error similar to other earlier mentioned errors. When the input SNR is high or the number of array elements is large, an excessively number of snapshots is generally required so that the sample data effect can be ignored.

This paper describes a new method for LCMV beamforming under perturbed conditions in which an adjustable constraint is used. It is assumed that the beamformer contains a fixed constraint that determine the quiescent response (the optimal response under white noise input). A typical requirement is to impose a low sidelobe level so that interferences can be nulled out in a short period of time. The adjustable constraint, on the other hand, is continuously updated according to the changes in the estimation of the signal of interest (SOI) steering vector. The estimate is computed using the input data and a reference signal which is recovered from the array output. The advantages of the method are that a less number of snapshots is required to calculate the optimum weights, pointing and calibration errors are compensated and the beamformer is able to track the desired signal source.

2 Background

The beamformer output $y(n)$ can be expressed as the inner product of a data vector $\mathbf{x}(n)$ and a weight vector $\mathbf{w}(n)$,

$$y(n) = \mathbf{w}^H(n)\mathbf{x}(n) \tag{1}$$

where the superscript ^{H} denotes Hermitian transpose. For a narrowband array with N elements, $\mathbf{x}(n)$ and $\mathbf{w}(n)$ are N-dimensional complex vectors. The observed data vector which contains SOI, noise and interferences can be expressed as

$$\mathbf{x}(n) = d(n)\mathbf{a}(\theta) + \mathbf{u}(n) \tag{2}$$

where $d(n)$ is the SOI, $\mathbf{a}(\theta)$ is the corresponding steering vector and $\mathbf{u}(n)$ represents the undesired signal portion that may include both noise and interferences.

In LCMV beamformers, the optimization problem is formulated as

$$\min_{\mathbf{w}} \mathbf{w}^H\mathbf{R}_{xx}\mathbf{w} \text{ subject to } \mathbf{C}^H\mathbf{w} = \mathbf{f} \tag{3}$$

where $\mathbf{R}_{xx}$ denotes the covariance matrix of $\mathbf{x}(n)$, the constraint matrix $\mathbf{C}$ contains M column vectors while the response vector $\mathbf{f}$ specifies the corresponding constraint value for each vector. The well known optimal solution to this problem is given by

$$\mathbf{w}_{opt} = \mathbf{R}_{xx}^{-1}\mathbf{C}(\mathbf{C}^H\mathbf{R}_{xx}^{-1}\mathbf{C})^{-1}\mathbf{f} \tag{4}$$

Under white noise only conditions, the optimal response is known as the quiescent response which is specified by the weights

$$\mathbf{w}_q = \mathbf{C}(\mathbf{C}^H\mathbf{C})^{-1}\mathbf{f} \tag{5}$$

An alternate solution to (3) which is the basis of the Generalized Sidelobe Canceller (GSC) structure [2] is also known as

$$\mathbf{w}_{opt} = \mathbf{w}_q - \mathbf{B}(\mathbf{B}^H\mathbf{R}_{xx}\mathbf{B})^{-1}\mathbf{B}^H\mathbf{R}_{xx}\mathbf{w}_q \tag{6}$$

where $\mathbf{B}$ is the blocking matrix of size $N \times (N - M)$ whose columns span the orthogonal subspace of the column space of $\mathbf{C}$.

In this paper, we consider a two constraint system in which the first constraint is used to impose a desired quiescent response and the second to prevent SOI cancellation at a given direction of arrival. These two constraints are outlined below while the details of the constraint design process can be found in [3]. We first assume that a desired quiescent response with low sidelobes has been designed using the algorithm in [4] and the corresponding weights are specified by $\mathbf{w}_d$. Also, let $\mathbf{a}(\theta)$ be the steering vector in the SOI direction. The two constraints are given by

$$\mathbf{w}_d^H \mathbf{w} = \mathbf{w}_d^H \mathbf{w}_d \tag{7}$$

$$\mathbf{a}^H(\theta)[\mathbf{I} - \mathbf{w}_d(\mathbf{w}_d^H \mathbf{w}_d)^{-1} \mathbf{w}_d^H]\mathbf{w} = 0 \tag{8}$$

Since both $\mathbf{w}_d$ and $\mathbf{a}(\theta)$ are obtained based on an assumed array, they are inaccurate when there is a mismatch between the true and assumed arrays. The error induced by $\mathbf{w}_d$ is generally negligible since it only alters the desired quiescent response slightly, provided that the amount of perturbation is not enormous. The error caused by $\mathbf{a}(\theta)$ is more significant, however, since even a small perburbation can result in a complete cancellation of the SOI, especially for the cases of high SNR. As shown in [1], signal cancellation can also occur, even when the array is ideal, if the optimal weights are computed based on insufficient number of snapshots. The purpose of this paper is to present an approach which prevents signal cancellation by iteratively estimating $\mathbf{a}(\theta)$ from the data and continuously updating the second constraint.

3 Adjustable Constraint

In certain communications applications, reference signals may be transmitted for different purposes such as acquisition, synchronization or user identification. These signals are commonly multiplexed with the transmitted information either in time, frequency or code [5]. At the beamformer output, the reference and information signals can be separated using any conventional demultiplexing technique. The reference signal is then regenerated which can be used to estimate the steering vector. The $\mathbf{p}$-vector (that is, the cross-correlation vector between the reference signal $r(n)$ and the array input $\mathbf{x}(n)$) is

$$\mathbf{p} = E[r^*(n)\mathbf{x}(n)] = E[r^*(n)d(n)]\mathbf{a}(\theta) + E[r^*(n)\mathbf{u}(n)] \tag{9}$$

If $r(n)$ is correlated with $d(n)$ and uncorrelated with the noise and the interferences, the second term on the right hand side of (9) equals zero. Hence, (9) reduces to be

$$\mathbf{p} = \gamma_{rd}\mathbf{a}(\theta) \tag{10}$$

where γ_{rd} is the correlation between $r(n)$ and $d(n)$. It should be noted that $r(n)$ does not need to be a perfect replica of $r(n)$ in order to get a $\mathbf{p}$-vector proportional to the steering vector. A $r(n)$ which is correlated with $d(n)$ and uncorrelated with $\mathbf{u}(n)$ is sufficient.

For the purpose of this paper, we consider a simple case where an unmodulated carrier is transmitted in addition to the information signal [5]. Thus, the desired signal is

$$d(n) = A_d e^{j(\omega n + \phi_d)} \tag{11}$$

where A_d, ω and ϕ_d are respectively the amplitude, frequency and phase of the carrier. In this expression, the frequency is assumed known and fixed while A_d and ϕ_d are random. A reference signal can be obtained if we can generate a sinusoid whose phase is correlated with the phase of the unmodulated carrier. One way of doing this is by means of a band-pass filter and a Phase-Locked-Loop (PLL). The PLL basically consists of a multiplier, a loop filter and a voltage-controlled oscillator (VCO). The output of the VCO constitutes the reference signal

$$r(n) = A_r e^{j(\omega n + \phi_r)} \tag{12}$$

where A_r and ϕ_r are the amplitude and the phase of the VCO. Due to the feedback loop structure of the PLL, ϕ_r continuously tracks the input carrier phase ϕ_d.

To incorporate the reference signal into the LCMV beamformer, $\mathbf{a}(\theta)$ is replaced by the $\mathbf{p}$-vector in equation (8)

$$\mathbf{p}^H[\mathbf{I} - \mathbf{w}_d(\mathbf{w}_d^H \mathbf{w}_d)^{-1} \mathbf{w}_d^H]\mathbf{w} = 0 \tag{13}$$

Making use of the constraint in (7), (13) reduces to be

$$\mathbf{p}^H \mathbf{w} = \mathbf{p}^H \mathbf{w}_d \tag{14}$$

The scale factor γ_{rd} is irrelevant because the corresponding constraint value equals zero. Formulating the second constraint in terms of the P-vector as in (13) has a significant advantage that it can be updated easily by use of the maximum likelihood estimate of $\mathbf{p}$ from the data

$$\hat{\mathbf{p}} = \frac{1}{N} \sum_{n=1}^{N} r^*(n)\mathbf{x}(n) \tag{15}$$

The problem of signal cancellation due to perturbation errors in LCMV beamformers becomes worse as the SNR increases. In that case, the utilization of the $\mathbf{p}$-vector instead of the steering vector improves the beamformer performance. This is because the $\mathbf{p}$-vector involves an implicit estimation of the perturbation errors and this estimate becomes more accurate as the SNR increases. Consequently, an accurate knowledge of the array characteristics and SOI direction is not necessary if the adjustable constraint approach is used. In addition, the use of the $\mathbf{p}$-vector also reduces the finite data sample effect since the errors in the estimate of the covariance matrix are correlated with the errors in the $\mathbf{p}$-vector estimate, [6]. The number of snapshots which is required to reach within 3 dB of the optimal SINR is subsequently reduced.

4 Adaptive Implementation

A direct way to implement the adjustable constraint is to compute the optimal weights using the closed-form solution in (4) where $\mathbf{R}_{xx}$ and $\mathbf{p}$ are estimated from the data. For adaptive implementation, it is more desirable to update the weights in a simple fashion which does not require matrix inversion in each iteration. In this section, such an adaptive implementation structure is derived.

The adaptive implementation is based on solving two optimization subproblems in which the first is defined as

$$\min_{\mathbf{w}} \mathbf{w}^H \mathbf{R}_{xx}\mathbf{w} \text{ s. t. } \mathbf{w}_d^H \mathbf{w} = \mathbf{w}_d^H \mathbf{w}_d \qquad (16)$$

and the second is

$$\min_{\mathbf{w}} \mathbf{w}^H \mathbf{R}_{xx}\mathbf{w} \text{ s. t. } \mathbf{w}_d^H \mathbf{w} = 0 \text{ and } \mathbf{p}^H \mathbf{w} = 1 \qquad (17)$$

It was shown in [7] that the optimal solution to the minimization of $\mathbf{w}^H \mathbf{R}_{xx}\mathbf{w}$ subject to (7) and (14) can be expressed as

$$\mathbf{w}_{opt} = \alpha_1 \mathbf{w}_{opt1} + \alpha_2 \mathbf{w}_{opt2} \qquad (18)$$

where $\mathbf{w}_{opt1}$ and $\mathbf{w}_{opt2}$ are the optimal solutions to (16) and (17), respectively. The combination coefficients α_1 and α_2 can be determined by noting that $\mathbf{w}_{opt}$ has to satisfy both (7) and (14)

$$\begin{aligned} \alpha_1 &= 1 \\ \alpha_2 &= \mathbf{p}^H(\mathbf{w}_q - \mathbf{w}_{opt1}) \end{aligned} \qquad (19)$$

With this decomposition, $\mathbf{w}_{opt}$ can be implemented via computing $\mathbf{w}_{opt1}$ and $\mathbf{w}_{opt2}$.

Adaptive implementation of $\mathbf{w}_{opt1}$ can be readily obtained through the use of the GSC structure in which

$$\mathbf{w}_{opt1} = \mathbf{w}_q - \mathbf{B}\mathbf{w}_{aopt1} \qquad (20)$$

where $\mathbf{B}$ is the blocking matrix of size $N \times (N-1)$ and $\mathbf{w}_q$ equals $\mathbf{w}_d$ in this case. The $\mathbf{w}_{aopt1}$ vector is the only data dependent part which is given by

$$\mathbf{w}_{aopt1} = \mathbf{R}_{aa}^{-1}\mathbf{p}_q \qquad (21)$$

where

$$\begin{aligned} \mathbf{R}_{aa} &= E[\mathbf{x}_a(n)\mathbf{x}_a^H(n)], \quad \mathbf{x}_a(n) = \mathbf{B}^H\mathbf{x}(n), \\ \mathbf{p}_q &= E[y_q^*(n)\mathbf{x}_a(n)], \quad y_q(n) = \mathbf{w}_q^H\mathbf{x}(n) \end{aligned} \qquad (22)$$

Since (21) has the form of the Wiener solution, any unconstrained adaptive algorithm such as the LMS or RLS is directly applicable to implement $\mathbf{w}_{aopt1}$ adaptively.

Adaptive implementation of $\mathbf{w}_{aopt2}$ is somewhat more complicated since the second constraint is also data dependent. However, because the first constraint simply requires that $\mathbf{w}$ be orthogonal to $\mathbf{w}_d$, this constraint can be eliminated by letting

$$\mathbf{w} = \mathbf{B}\mathbf{w}_a \qquad (23)$$

and reformulating the minimization in terms of $\mathbf{w}_a$

$$\min_{\mathbf{w}_a} \mathbf{w}_a^H \mathbf{R}_{aa}\mathbf{w}_a \text{ s. t. } \mathbf{p}^H\mathbf{B}\mathbf{w}_a = 1 \qquad (24)$$

This is a single constraint system and its solution is given by

$$\mathbf{w}_{aopt} = \frac{\mathbf{R}_{aa}^{-1}\mathbf{B}^H\mathbf{p}}{\mathbf{p}^H\mathbf{B}\mathbf{R}_{aa}^{-1}\mathbf{B}^H\mathbf{p}} \qquad (25)$$

As a result, $\mathbf{w}_{opt2}$ can be expressed as

$$\mathbf{w}_{opt2} = \frac{\mathbf{B}\mathbf{w}_{aopt2}}{\mathbf{p}_r^H\mathbf{w}_{aopt2}} \qquad (26)$$

where

$$\mathbf{w}_{aopt2} = \mathbf{R}_{aa}^{-1}\mathbf{p}_r, \ \mathbf{p}_r = E[r^*(n)\mathbf{x}_a(n)] \qquad (27)$$

From this equation, it is clear that $\mathbf{w}_{aopt2}$ can be implemented in an adaptive fashion. Finally, substituting (19), (20) and (26), into (18) we can express $\mathbf{w}_{opt}$ in the decomposed form

$$\mathbf{w}_{opt} = \mathbf{w}_q - \mathbf{B}(\mathbf{w}_{aopt1} - k\mathbf{w}_{aopt2}) \qquad (28)$$

where

$$k = \frac{\mathbf{p}_r^H\mathbf{w}_{aopt1}}{\mathbf{p}_r^H\mathbf{w}_{aopt2}} \qquad (29)$$

Figure 1 shows the implementation scheme which is derived from this decomposition. Note that according to equations (21) and (27), $\mathbf{w}_{aopt1}$ and $\mathbf{w}_{aopt2}$ are the optimal solutions to two unconstrained standard adaptive filter problems. The input data vector is $\mathbf{x}_a(n)$ for both of them, while the desired response is $y_q(n)$ for $\mathbf{w}_{aopt1}$ and $r(n)$ for $\mathbf{w}_{aopt2}$.

The scheme of figure 1 can be seen as a LCMV beamformer with a TR beamforming connected in a parallel form. The role of the variable parameter k is to ensure that the output of both beamformers are summed appropriately regardless of the reference signal amplitude and the norm of $\mathbf{w}_q$. This implementation structure also suggests that the underlining problem can be considered either as a LCMV beamformer where a point constraint is designed using a reference signal, or a TR beamformer where one constraint is used for the quiescent pattern control.

5 Simulation Results

Computer simulations were carried out to show the performance of the proposed approach. We considered a 25-element linear equispaced array where each element has a random phase perturbation uniformly distributed from −2.5° to 2.5°. The signal environment consisted of a 10 dB desired signal in the broadside direction, a strong interference of 30 dB from 20° and a 0 dB white noise background.

Figure 2 shows the radiation pattern for a LCMV beamformer with two constraints as mentioned earlier. The desired quiescent response in this case was selected to be a beam pattern with uniform 30 dB sidelobe. Notice that the antenna response has a null in the interference direction. However, it exhibits an unacceptable sidelobe level due to the signal cancellation phenomenon caused by the perturbation errors.

Better performance was achieved using the adjustable constraint approach. The $\mathbf{p}$-vector was estimated using the reference signal extracted from the PLL. Since this estimate took perturbation errors into account , signal cancellation was greatly reduced. The corresponding optimal beamformer response is shown in figure 3. Notice that the beamformer can now steer a main lobe towards the SOI direction and also place a null in the jammer direction. Simulations show a larger sidelobe level than the specified 30 dB level due to errors in the covariance matrix and noise in the PLL output.

The convergence curves of the SINR using recursive implementation for the proposed approach and the conventional LCMV beamformer are shown in figure 4. In this implementation, a recursive least-squares (RLS) adaptive algorithm with a forgetting factor of 0.995 is used for both adaptive loops. Notice how phase perturbation errors as small as 2.5° can seriously degrade the LCMV beamformer adaptive response while the adjustable constraint approach is insensitive to these errors.

References

[1] D. D. Feldman and L. J. Griffiths, "A Constraint Projection Approach for Robust Adaptive Beamforming", *Proc. ICASSP91,* pp. 1381-1384, Toronto, Canada, May 1991.

[2] L. J. Griffiths and C. W. Jim, "An Alternative Approach to Linearly Constrained Adaptive Beamforming", *IEEE Trans. on Antennas and Propagation,* AP-30, pp. 27-34, Jan. 1982.

[3] C. Y. Tseng and L. J. Griffiths, "A Novel Approach for Designing Linear Constraints in Adaptive Arrays", *Proc. of the Asilomar Conference on Signals, Systems and Computers,* pp. 124-128, Pacific Grove, CA, Nov. 1990.

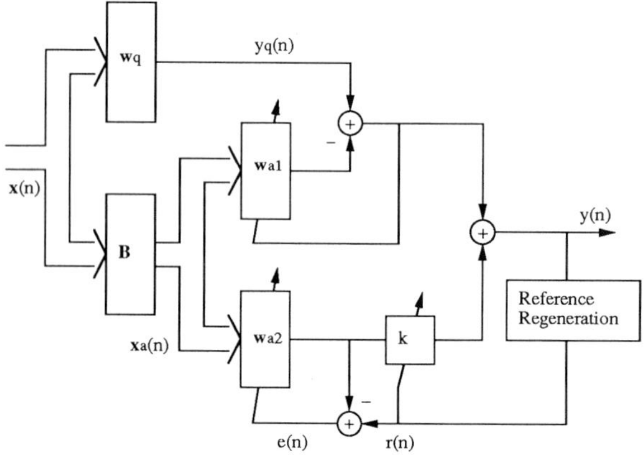

Figure 1: Scheme for recursive adaptation

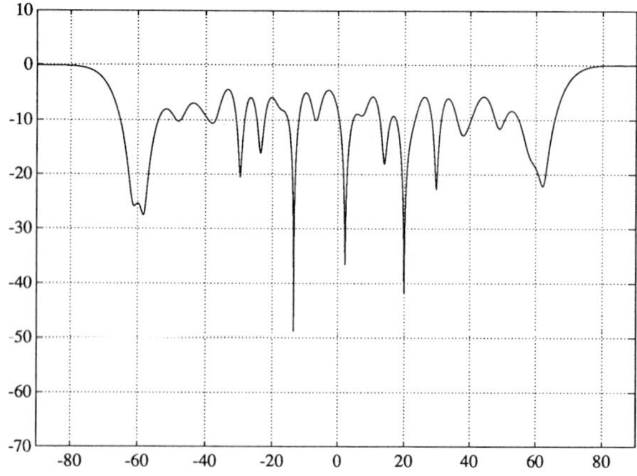

Figure 2: Antenna response with an LCMV beamformer

[4] C. Y. Tseng and L. J. Griffiths, "A Simple Algorithm to Achieve Desired Patterns for Arbitrary Arrays", *to appear in IEEE Trans. on Acoust. Speech and Signal Processing,* Nov. 1992.

[5] M. A. Lagunas, L. Corbera and F. Corominas, "Time Reference Systems in Adaptive Array Beamforming", *Proc. ESA Workshop on Advanced Beamforming Networks for Space Applications,* ESA-ESTEC, Noordwijk, The Netherlands, Nov. 1991.

[6] T. W. Miller, "The Transient Response of Adaptive Arrays in TDMA Systems", Ph.D. dissertation, Ohio State University, Ohio, Illinois.

[7] C. Y. Tseng and L. J. Griffiths, "On the implementation of adaptive filters with adjustable linear constraints", *Proc. ICASSP90,* pp. 1449-1452, Albuquerque, New Mexico, April 1990.

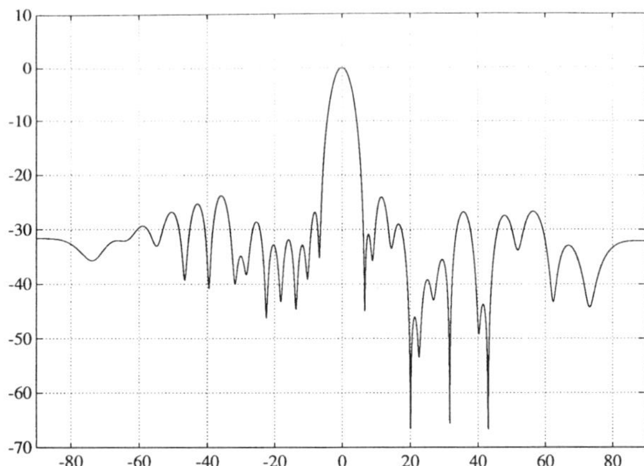

Figure 3: Antenna response with an adjustable constraint

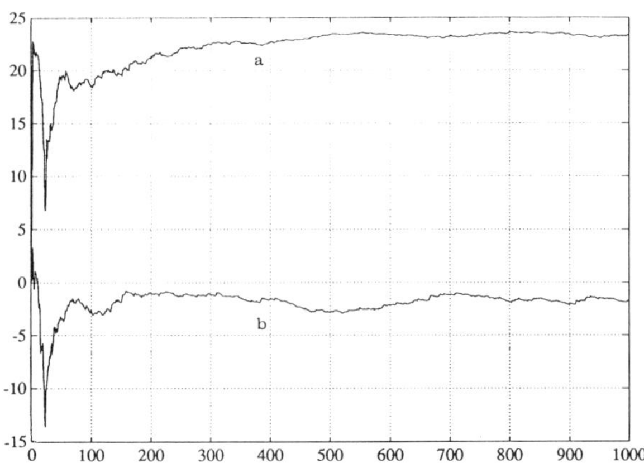

Figure 4: SINR behavior: a) Adjustable constraint b) LCMV beamformer

SIGNAL PROCESSING VI: Theories and Applications
J. Vandewalle, R. Boite, M. Moonen, A. Oosterlinck (eds.)

TRANSIENT AND STEADY-STATE BEHAVIOR OF THE 2D FAST RLS LATTICE ALGORITHM

X. LIU , L. MANDRIDAKE , M. NAJIM

Equipe Signal et Image and GRECO_TDSI
ENSERB, 351, Cours de la Libération, F33405 TALENCE Cedex, France

This paper mainly deals with the analysis of the performances of a new algorithm: 2D Fast, Recursive-Least-Squares Lattice(2D FLRLS) algorithm. The transient performances, initialization and steady-state finite-precision effects are investigated. The preliminary results compared with 2D fast RLS transversal algorithm are included in this paper.

1. INTRODUCTION

The application of adaptive algorithms for the two-dimensional problems have been more and more noticeable. Because of complexity of the 2D problems, the 2D adaptive algorithms, which require quite limited prior knowledge, show a wide application foreground. Many successful extensions of 1D linear prediction, autoregressive and lattice modeling to the 2D case have been reported [8][10][11]. The latest developments in VLSI technology which had a significant impact on the realization of real time applications of 1D adaptive and lattice algorithms have also made it possible to consider architectures and implementation for two or higher dimensional signals and suggest the feasibility of massive parallelism in the implementation schemes.

Recently, we presented a lattice version of the 2D Recursive Least Squares Algorithm(2D FLRLS)[13], which provides an exact growing-order least squares solution to the deterministic normal equations for the AR and MA models. It can be also extended to ARMA model.We have reported an application for the restoration of noisy images in the GRETSI91[12]. It showed improved performances compared to the TDJPL-NLMS [8][9].

This paper is organized as follows: the summary of 2D FLRLS algorithm is included in section 2. In section 3, we examine the performances of the 2D FLRLS and 2D FRLS[11] algorithms. The conclusions are included in section 4.

2. 2D PREWINDOWED RLS LATTICE ALGORITHM

2.1. 2D Prewindowed RLS Estimation

We consider a general 2D adaptive digital filter which is acting on the input 2D L×K data x(n1,n2) and provide an optimum estimation of the desired response d(n1,n2). The d(n1,n2) is assumed to be the sum of a Gaussian noise N(n1,n2) and output of the unknown filter. The output of the adaptive filter is as follows:

$$\hat{y}(n_1,n_2)= \sum_{(i,j)\in R} \hat{a}_{i,j}x(n_1-i,n_2-j)$$

R is the filter support region. Here, we choose a quarter plan (as shown in fig. 1). It is defined as:

$$R^{(N,\,M)}=\{i,j:(0\le i \le N, 0\le j\le M)\cap((i,j)\neq(0,0)\}$$

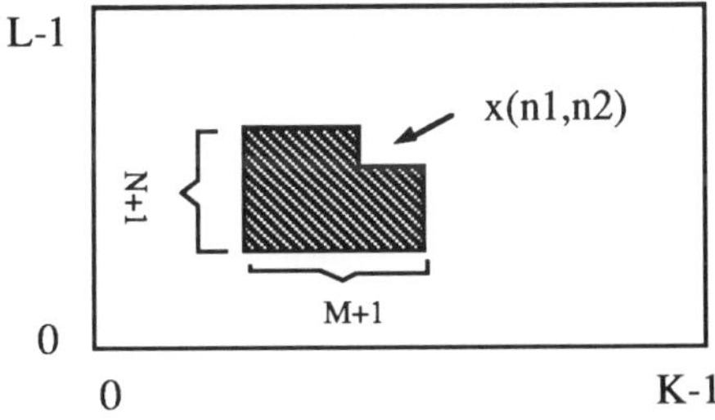

Fig. 1 First quadrant filter support region

The partial order of computation is defined as a linear scanning index along column where the data is considered to update in a helicoidal fashion. So, we introduce a linear scanning index n such that:

$$n=n_2\times L+n_1$$

We consider the 2D pre-windowed least-squares prediction problem, which estimates the coefficients of the filter by minimizing the accumulated squared error with exponential weighting:

$$\xi(n)=\sum_{k_1=0}^{n}[\lambda^{n-k_1}e(k_1)]^2$$

where $e(k_1)=d(k_1)-\hat{y}(k_1)$

The minimum mean-square error(MMSE) becomes $E(N(n1,n2)^2)$.

We exploit the local analogy of multichannel within the QP support region. Then, the mathematical concepts of vector space, orthogonal projection, and subspace decomposition can be used to derive this new algorithm. The detailed derivation can be found in [13].

2.2 Summary of 2D FLRLS Algorithm

We start recursions by setting all initial conditions to zero except the error covariance matrix as $\frac{1}{\delta}I_{(M+1)\times(M+1)}$ (δ is a small positive constant) and angle parameters as 1.

For n=(n1,n2):

$$\underline{x}^f(n)=[\ x_2(n+1)\cdots x_{M+1}(n+1)\ \ x_1(n)]$$

$$\underline{x}_0^b(n)=[\ x_1(n)\ \ x_2(n)\cdots x_{M+1}(n)\ ]$$

$$\overline{\underline{x}}^f(n)=[\ x_2(n+1)\cdots x_{M+1}(n+1)\ ]$$

$$\overline{\underline{x}}_0^b(n)=[\ x_2(n)\cdots x_{M+1}(n)\ ]$$

For m=0,

The gain of the for/backward reflection coefficients are:

$$\overline{g}_0^f=\frac{(\lambda\overline{\varepsilon}_0^{B,n-1})^{-1}(\overline{\underline{x}}_0^b(n))^T}{1+\overline{\underline{x}}_0^b(n)(\lambda\overline{\varepsilon}_0^{B,n-1})^{-1}(\overline{\underline{x}}_0^b(n))^T}$$

$$\overline{g}_0^b=\frac{(\lambda\overline{\varepsilon}_0^{F,n-1})^{-1}(\overline{\underline{x}}^f(n))^T}{1+\overline{\underline{x}}^f(n(\lambda\overline{\varepsilon}_0^{F,n-1})^{-1}(\overline{\underline{x}}^f(n))^T}$$

The inverse of the for/backward error covariance matrix are:

$$(\overline{\varepsilon}_0^{B,n})^{-1}=(\lambda\overline{\varepsilon}_0^{B,n-1})^{-1}-\overline{g}_0^f\,\overline{\underline{x}}_0^b(n)(\lambda\overline{\varepsilon}_0^{B,n-1})^{-1}$$

$$(\overline{\varepsilon}_0^{F,n})^{-1}=(\lambda\overline{\varepsilon}_0^{F,n-1})^{-1}-\overline{g}_0^b\,\overline{\underline{x}}^f(n)(\lambda\overline{\varepsilon}_0^{F,n-1})^{-1}$$

The for/backward reflection coefficients matrix are:

$$\overline{K}_0^b(n)=\overline{K}_0^b(n-1)+\ \overline{g}_0^b(\underline{x}_0^b(n)-\overline{\underline{x}}^f(n)\overline{K}_0^b(n-1))$$

$$\overline{K}_0^f(n)=K_0^f(n-1)+\ \overline{g}_0^f(\underline{x}^f(n)-\overline{\underline{x}}_0^b(n)(n)\overline{K}_0^f(n-1))$$

The angle parameter is:

$$\gamma_0(n-1)=1-\overline{\underline{x}}_0^b(n)(\overline{\varepsilon}_0^{B,n})^{-1}(\overline{\underline{x}}_0^b(n))^T$$

The for/backward error are:

$$\underline{e}_0^b(n)=\underline{x}_0^b(n)-\overline{\underline{x}}^f(n)\overline{K}_0^b(n)$$

$$\underline{e}_0^f(n)=\underline{x}^f(n)-\overline{\underline{x}}_0^b(n)\overline{K}_0^f(n)$$

The joint process updates as follows:

$$\overline{H}_0(n)=\overline{H}_0(n-1)+\ \overline{g}_0^f(d(n)-\overline{\underline{x}}_0^b(n)\overline{H}_0(n-1))$$

$$e_0(n)=d(n)-\overline{\underline{x}}_0^b(n)\overline{H}_0(n)$$

For m=1, ... , N

The gain of the for/backward reflection coefficients are:

$$g_{m-1}^f=\frac{(\lambda\varepsilon_{m-1}^{B,n-2})^{-1}(\underline{e}_{m-1}^b(n-1))^T}{\gamma_{m-1}(n-1)+(\underline{e}_{m-1}^b(n-1))(\lambda\varepsilon_{m-1}^{B,n-2})^{-1}(\underline{e}_{m-1}^b(n-1))}$$

$$g_{m-1}^b=\frac{(\lambda\varepsilon_{m-1}^{F,n-1})^{-1}(\underline{e}_{m-1}^f(n))^T}{\gamma_{m-1}(n-1)+(\underline{e}_{m-1}^f(n))(\lambda\varepsilon_{m-1}^{F,n-1})^{-1}(\underline{e}_{m-1}^f(n))^T}$$

The inverse of the for/backward error covariance matrix are:

$$(\varepsilon_{m-1}^{B,n-1})^{-1}=(\lambda\varepsilon_{m-1}^{B,n-2})^{-1}-g_{m-1}^f\underline{e}_{m-1}^b(n-1)(\lambda\varepsilon_{m-1}^{B,n-2})^{-1}$$

$$(\varepsilon_{m-1}^{F,n})^{-1}=(\lambda\varepsilon_{m-1}^{F,n-1})^{-1}-g_{m-1}^b\underline{e}_{m-1}^f(n)(\lambda\varepsilon_{m-1}^{F,n-1})^{-1}$$

The a priori for/backward predicted error are:

$$\underline{e}_m^b(n/n-1)=\underline{e}_{m-1}^b(n-1)\ -\ \underline{e}_{m-1}^f(n)K_{m-1}^b(n-1)$$

$$\underline{e}_m^f(n/n-1)=\underline{e}_{m-1}^f(n)\ -\ \underline{e}_{m-1}^b(n-1)K_{m-1}^f(n-1)$$

The for/backward reflection coefficients matrix are:

$$K_{m-1}^b(n)=K_{m-1}^b(n-1)+\ g_{m-1}^b\underline{e}_m^b(n/n-1)$$

$$K_{m-1}^f(n)=K_{m-1}^f(n-1)+\ g_{m-1}^f\underline{e}_m^f(n/n-1)$$

The angle parameter is:

$$\gamma_m(n)=\gamma_{m-1}(n-1)-\ \underline{e}_{m-1}^f(n)(\varepsilon_{m-1}^{F,n})^{-1}(\underline{e}_{m-1}^f(n))^T$$

The for/backward error are:

$$\underline{e}_m^b(n)=\underline{e}_{m-1}^b(n-1)\ -\ \underline{e}_{m-1}^f(n)K_{m-1}^b(n)$$

$$\underline{e}_m^f(n)=\underline{e}_{m-1}^f(n)\ -\ \underline{e}_{m-1}^b(n-1)K_{m-1}^f(n)$$

The joint process updates as follows:

$$H_m(n)=H_m(n-1)+\ g_{m-1}^f(e_{m-1}(n)-\underline{e}_{m-1}^b(n)H_m(n-1))$$

$$e_m(n)=e_{m-1}(n)-\underline{e}_{m-1}^b(n)H_m(n-1)$$

2.3. Limited-Precision Effects of 2D FRLS Algorithms

The limited-precision effects in digital implementation of 1D adaptive algorithms have been researched[14][15]. Though 2D RLS algorithms have more complex forms compared with their 1D forms, they also include the matrix inverse implicit. It is possible for this matrix to become indefinite from accumulation of quantization errors as in 1D cases. For example, effects of initialization, when the parameter δ is too small make this matrix close to singularity at $(n_1,n_2)=(N,M)$. Another effects involves the exponential increase of the effects of the numerical errors when exponential weighting is used. 2D FRLS[11] diverge in both cases(as we will see later).

For the 2D FLRLS, the recursions at spatial position are also recursive in order, and quantities are constructed from lower order quantities until the desired order is achieved. It increases the magnitude of quantization errors as the recursions progress and are limited in a finite number. It can be propagated in a numerically stable fashion, except in cases where the order is very large.

Due to the limited space in this paper, we will not give a detailed analysis. We only provide in the next section some simulations.

3. SIMULATIONS AND RESULTS

We implement the 2D FRLS[11] and 2D FLRLS algorithms in a VAX Station 3100. We use a 32-bit floating arithmetic and 64×64 2D artificial data in all cases.

Test results are presented for transient performances, initialization and steady-state finite-precision effects.

Fig 2 shows the basic test configuration. The input x(n1,n2) is generated by passing white noise through a 2D low_pass digital filter. The unknown system chosen is a 5×5 digital filter. We select the relative error between the estimated coefficients and coefficients of "unknown" system as the performance indicator. It is defined as follows:

$$Q(n_1,n_2)=\frac{||H(n_1,n_2)-\widehat{H}(n_1,n_2)||}{||H(n_1,n_2)||}$$

Fig.3 shows the convergence of both algorithms after proper initialization ($\delta=10^{-3}$).

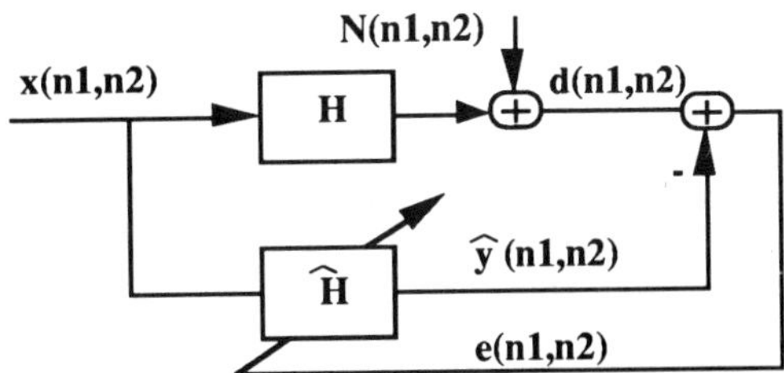

Fig. 2 Basic test configuration

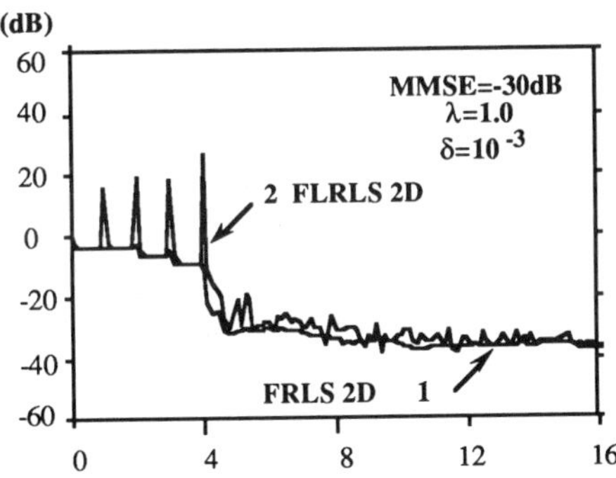

Fig.3 Convergence of the algorithms

A poor choice of the initialization constant δ leads the 2D FRLS[11] to instability on step $(n_1,n_2)=(M,N)$. Fig. 4 gives an example of such a phenomenon. The 2D FLRLS algorithm remains to converge in this case.

Fig.5 compares both algorithms when the exponential weighting is used. The 2D FRLS algorithm exhibits an effect of steady-state divergence. The 2D FLRLS remains stable.

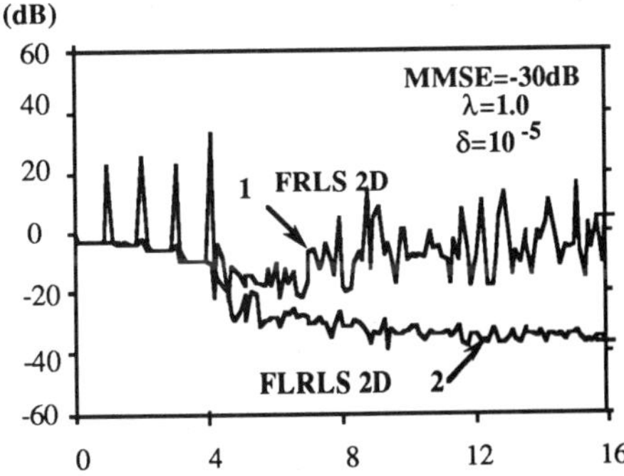

Fig. 4 Influence of the initialization parameter

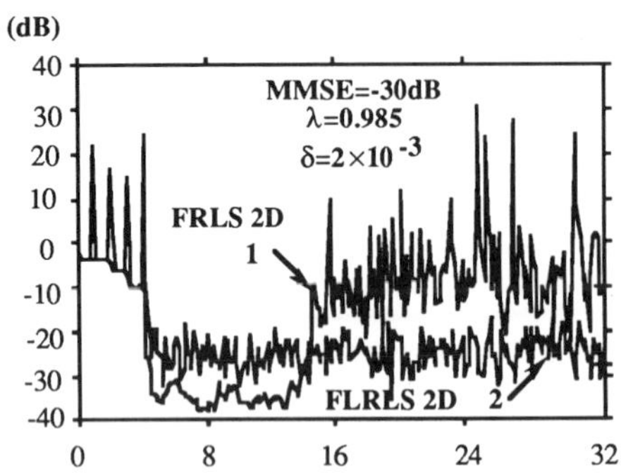

Fig. 5 Steady-state finite-precision effects

4. CONCLUSIONS

The new 2D FLRLS algorithm requires about $11(M+1)K_1$ operations per iteration, where $(M+1)$ is the number of channels, K_1 is the total number of data used in the 2D filter. A reduction in the computational cost is obtained when compared with the standard RLS algorithm($1.5K_1^2$).

Simulations show that the 2D FLRLS algorithm can update in numerical stable fashion, while the 2D FRLS[11] exhibits a numerical unstable behavior. Though the new algorithm has more complexity in computation than its transversal form($6(M+1)K_1$ operations)[11], we can expect better numerical properties, stage independence and pipelineable implementation. The research on the effects of limited-precision is under way and we will report further results.

REFERENCES

[1] M. Morf: “Ladder forms in estimation and system identification” Proc. 11th Asilomar Conf. Circuits Systems and Computers, Monterey, pp 424_429, CA1977

[2] B. Friedlander, D.T.L. Lee and M. Morf: " Recursive least squares ladder estimation algorithms" IEEE. Trans. on ASSP, vol. ASSP.29, pp.627-641, June 1981.

[3] J. M. Cioffi and T. Kailath: " Fast, recursive-least-squares transversal filters for adaptive filtering" IEEE Trans. on ASSP, vol. ASSP.32, pp. 304-337, April 1984.

[4] G. Carayannis, D. Manolakis, N. Kalouptsidis: " A fast sequential algorithm for least-squares filtering and prediction" IEEE Trans. on ASSP, vol. ASSP_31, pp. 1394-1402, 1983.

[5] J. M. Cioffi: " The fast adaptive rotor's RLS algorithm" IEEE Trans. on ASSP, vol. ASSP.38, pp. 631-653, April 1990.

[6] M.G. Bellanger: “The potential of QR adaptive filter variables for signal analysis” Proc. IEEE ICASSP, Glasgow(1989) pp. 2166-2169,

[7] Special issue of the journal " Traitement du Signal" vol. 6, No. 5, Decembre 1989.

[8] H. Youlal, M. Janati-I. and M. Najim: “ On 2-D adaptive gradient lattice algorithms” EUSIPCO-EURASIP(1986), Vol. II, pp. 705-708.

[9] H. Youlal, M. Janati-I. and M. Najim: " Two-dimensional joint process lattice LMS for adaptive restoration of images" To be published IEEE Trans. on Image Processing, July, 1992.

[10] Y.S. Boutalis and S.D. Kollias: "A fast multichannel approach to adaptive image estimation" IEEE Trans. on ASSP, vol.37, pp. 1090-1098, July 1989.

[11] A.M. Sequeira and C.W. Therrien: " A new 2-D fast RLS algorithm" Proc. IEEE ICASSP, Albuquerque.(1990).

[12] X. Liu, M. Najim, B. Ténèze and H. Youlal: “A New 2D Fast Lattice RLS Algorithm: Application for Restoration of Images” GRETSI_91 SEPT. 16-20, 1991

[13] X. Liu, P. Baylou, M. Najim: “A New 2D Fast Lattice RLS Algorithm” ICASSP-92, San Francisco, March 23-26, 1992.

[14] J. M. Cioffi: "Limited-Precision Effects in Adaptive Filtering" IEEE Trans. on CAS. vol.34, pp. 821-833, July 1987.

[15] R. Atay, P. Baylou and M. Najim "Implementation of the stabilized Fast Transversal Filter Algorithm on Fixed Point DSP" ICASSP92, San Francisco, March 23-26, 1992.

SIGNAL PROCESSING VI: Theories and Applications
J. Vandewalle, R. Boite, M. Moonen, A. Oosterlinck (eds.)

A COMPARATIVE FIXED-POINT ERROR ANALYSIS OF THE SCHUR AND THE SPLIT SCHUR ALGORITHMS

By Nicholas Glaros and George Carayannis*

A new analytical methodology is introduced here for fixed-point error analysis of two Toeplitz solving algorithms, the Schur algorithm and the lately introduced split Schur algorithm. The theoretical results obtained are consistent with experimentation. Besides the intrinsic symmetry of the error propagation recursive formulae, the technique presented here is capable of explaining many practical situations: For signals having a small eigenvalue spread the Schur algorithm behaves better than the split Schur in the fixed-point environment. The intermediate coefficients of the split Schur algorithm leading to the PARCORs cannot serve as alternatives to the reflection coefficients in error sensitive applications, etc. It is demonstrated that the error-weight vectors of the Schur propagation mechanism follow Levinson-like (2nd order) recursions, while the same vectors of the split Schur propagation mechanism follow split Levinson-like (3rd order) recursions.

I. INTRODUCTION

The Schur algorithm [1] is used for solving symmetrical Toeplitz systems, particularly those related to the autocorrelation method of linear prediction and the least squares FIR filtering techniques. A less redundant in complexity version of the Schur algorithm is the split Schur algorithm [2]. Both algorithms can be realized using fixed-point (FP) arithmetic as they process a priori bounded variables. In [3], [4] some first-order FP error analysis of the Schur and the split Schur algorithms can be found. However, the results given cannot be directly applied for comparing the two algorithms, because they are restricted to the first two reflection coefficients. In this paper, the numerical performance under FP conditions of the Schur and the split Schur algorithms for computing the reflection coefficients is examined by introducing an appropriate set of error-weight vectors and error related scalar quantities. These parameters provide a tool for an in-depth analytical study of the FP error propagation mechanism of each algorithm, and demonstrate a data-dependent superiority of the Schur algorithm over the split Schur algorithm.

II. FP ERROR ANALYSIS OF THE SCHUR AND SPLIT SCHUR ALGORITHMS

A Schur-type algorithm, which is a slightly modified version of the Le Roux-Gueguen algorithm [1], [2] has been used as a basis to analyze the FP effects of the Schur algorithm. This algorithm is outlined in (2.1) [5]

for $i = 1, 2, \ldots, p$

$$e(1,i) = r(i) \tag{2.1a}$$
$$f(1,i) = r(i-1) \tag{2.1b}$$

for $m = 1, 2, \ldots, i-1$

$$e(m+1,i) = e(m,i) + k(m)\cdot f(m,i) \tag{2.1c}$$
$$f(m+1,i) = f(m,i-1) + k(m)\cdot e(m,i-1) \tag{2.1d}$$

$$k(i) = -e(i,i)/f(i,i) \tag{2.1e}$$

$r(i)$ is the ith normalized autocorrelation coefficient, $k(i)$ stands for the ith reflection coefficient, and $f(i,i)$ indicates the ith stage mean square prediction error. The FP computation of the products and quotient involved in recursions (2.1c), (2.1d) and (2.1e) produces the local rounding errors $\gamma_e(m,i)$, $\gamma_f(m,i)$ and $\delta(i)$ respectively. The round-off error ε due to the quantization of an infinite precision real number χ into B-bits is assumed to be a zero-mean independent random variable with variance $\sigma_\varepsilon^2=q_\varepsilon^2/12$, $q_\varepsilon=2^{-F}$, q_ε being the quantization width and F, $(0\le F\le B-1)$, being the number of fractional bits in the B-bits FP representation of χ. This statistical model will be adopted for all local rounding errors in the algorithms and the error in the autocorrelation coefficient FP computation. Thus, if σ_e^2, σ_f^2, σ^2 and σ_r^2 denote the variance of γ_e, γ_f, $\delta(i)$ and the autocorrelation coefficient FP error, then

$$\sigma_r^2 = \sigma_e^2 = \sigma_f^2 = \sigma^2 = q^2/12, \quad q = 2^{-B+1} \tag{2.2}$$

B being the word length used in the FP implementation of the algorithms. The quantities defined in Table I have been found very useful in formulating general closed-form expressions for the reflection coefficient FP error variances. By means of these parameters

*The authors are with the Division of Computer Science, Department of Electrical Engineering, National Technical University of Athens, GR-157 73 Zographou, Athens, GREECE.

and by using simple algebra, it can be shown that the refl. coefficient first-order FP error variances in the Schur algorithm satisfy [5]

$$\underline{u}_i^T \cdot \underline{m}_i + Var(R_i) = 0 \quad , \quad 1 \le i \le p \qquad (2.3)$$

where the *ix1* vector $\underline{m}_i \equiv [\, m_{i,1}\; m_{i,2} \cdots m_{i,i}]^T$, displays the statistical correlation of $\Delta k(i)$, i.e., the FP error in the *i*th reflection coefficient, with all the other reflection coefficient FP errors up to the *i*th stage, that is, $m_{n,j} \equiv E[\Delta k(n) \cdot \Delta k(j)]$; $\underline{u}_i \equiv [u_{i,1}\; u_{i,2} \cdots u_{i,i}]^T$, $u_{i,i} \equiv -1$, is an *ix1* error-weight vector and R_i is a scalar function of all rounding errors local to the Schur algorithm as well as the autocorrelation coefficient FP errors. Complicated relationships exist, which when combined with the recursions shown in Table I, compute the components of $\underline{u}_i$ and the variance of R_i. These relationships are given in [5], and they will not be presented here for the sake of brevity.

The symmetric form of the split Schur algorithm can be described as follows [5]

$$y(0,0) = r(0)/2 \qquad (2.4a)$$
$$k(0) = 1 \qquad (2.4b)$$

for i = 1, 2, ..., p

$$y(0,i) = r(i) \qquad (2.4c)$$
$$y(1,i) = r(i) + r(i-1) \qquad (2.4d)$$

for m=1, 2, ..., i-1

$$y(m+1,i) = y(m,i)+y(m,i-1) - \lambda(m) \cdot y(m-1,i-1) \qquad (2.4e)$$

$$\lambda(i) = y(i,i) \,/\, y(i-1,i-1) \qquad (2.4f)$$

$$k(i) = 1 - \frac{\lambda(i)}{1 + k(i-1)} \qquad (2.4g)$$

The FP realization of the relations (2.4e), (2.4f) and (2.4g) results in the local quantization errors $\gamma_y(m,i)$, $\delta_\lambda(i)$ and $\delta(i)$ respectively. The variances σ_y^2 and σ_λ^2 of the γ_y and δ_λ are expressed as

$$\sigma_y^2 = q_y^2 \,/\, 12 \,, \qquad q_y = 2^{-B+2} \qquad (2.5a)$$

$$\sigma_\lambda^2 = q_\lambda^2 \,/\, 12 \,, \qquad q_\lambda = 2^{-B+3} \qquad (2.5b)$$

In Table II the FP error-related parameters for the split Schur algorithm are introduced. With Table II's definitions, analytical expressions for the evaluation of the first-order FP error variance in the computation of the *i*th λ-coefficient and the *i*th reflection coefficient by the split Schur algorithm can be derived [5]. The final results are

$$\underline{\tau}_i^T \cdot \underline{\varphi}_i + Var(\Psi_i) = 0 \quad , \quad 1 \le i \le p \qquad (2.6)$$

$$\underline{x}_i^T \cdot \underline{w}_i + Var(U_i) = 0 \quad , \quad 1 \le i \le p \qquad (2.7)$$

The various terms appearing in (2.6), (2.7) have the same significance with the ones involved in (2.3), namely: $\underline{\tau}_i$, $\underline{x}_i$ are *ix1* error-weight vectors, as the $\underline{u}_i$ one, with the convention $\tau_{i,i} \equiv x_{i,i} \equiv -1$; Ψ_i , U_i are error expressions involving the local round-off errors and the autocorrelation coefficient FP errors; finally the *ix1* column vectors $\underline{\varphi}_i$ and $\underline{w}_i$ correspond to the *i*th column of the covariance matrix of the error vector $\underline{\Delta\lambda}_i \equiv [\Delta\lambda(1)\; \Delta\lambda(2) \cdots \Delta\lambda(i)]^T$, which contains the total FP errors in the computation of the λ-coefficients, and to the *i*th column of the covariance matrix of the error vector $\underline{\Delta k}_i \equiv [\Delta k(1)\; \Delta k(2) \cdots \Delta k(i)]^T$ in the split Schur algorithm. Note that the terms $\varphi_{i,i}$ and $w_{i,i}$ give the *i*th λ-coefficient and the *i*th reflection coefficient FP error variance.

III. COMPARISON OF THE TWO ALGORITHMS

In (2.3), (2.6), (2.7), the error variances $m_{i,i}$, $\varphi_{i,i}$, $w_{i,i}$ exhibit moving average dependence on the *i-1* past covariances $m_{i,n}$, $\varphi_{i,n}$, $w_{i,n}$, $1 \le n \le i-1$, with coefficients $u_{i,n}$, $\tau_{i,n}$, $x_{i,n}$ and direct dependence on the variance of the error variables R_i , Ψ_i , U_i respectively. Let us compare the error variances in the two algorithms by considering specific values for the reflection coefficients and studying their effects on the $u_{i,n}$, $\tau_{i,n}$, $x_{i,n}$ coefficients as well as the variance of the R_i, Ψ_i, U_i error quantities. As shown in Table I the order *m+1* error-weight vectors in the FP Schur algorithm are defined on the basis of the order *m* similar vectors by using Levinson-like recursions. On the other hand, in Table II the order *m+1* error-weight vectors in the split Schur algorithm are defined on the basis of both the order *m* and the order *m-1* similar vectors by using a split Levinson-like recursions. Given that property of the error-weight vectors in the FP Schur and FP split Schur algorithm, the potential numerical behaviour of the vectors $\underline{u}_i$, $\underline{\tau}_i$, $\underline{x}_i$ and the variances of R_i , Ψ_i , U_i as all reflection coefficients tend to zero simultaneously ($|k(\nu)| \longrightarrow 0$, $1 \le \nu \le i$) is found to be [5]

$$u_{i,n},\; \tau_{i,n},\; x_{i,n} \longrightarrow 0, \quad 1 \le n \le i-1, \quad 2 \le i \le p \qquad (3.1)$$

$$Var(R_i) \longrightarrow \sigma_r^2 + (i-1) \cdot \sigma_e^2 + \sigma^2, \quad 1 \le i \le p \qquad (3.2)$$

$$Var(\Psi_i) \longrightarrow 4 \cdot \sigma_r^2 + (i-1)^2 \cdot \sigma_y^2 + \sigma_\lambda^2 \,, \quad 1 \le i \le p \qquad (3.3)$$

$$Var(U_i) \longrightarrow Var(\Psi_i) + \sigma^2 \,, \quad 1 \le i \le p \qquad (3.4)$$

The previous relations state that when the first *i* reflection coefficients are too small, then, for both algorithms, the error variance in the *i*th stage depends mostly on the variance of the error expressions R_i, U_i, Ψ_i, which involve only the local rounding errors and the FP errors in the autocorrelation coefficient computation. This is because the weight of each of the covariance vectors $\underline{m}_i$, $\underline{w}_i$, $\underline{\varphi}_i$ in this case is quite close to zero. Therefore, the reflection coefficient and the λ-coefficient FP errors generated in earlier stages of the algorithms propagate too little into later stages and the variances of R_i, U_i, Ψ_i become then very good approximations of the true reflection coefficient and λ-coefficient error variances. Comparing (3.2) to (3.3), it turns out that the variance of Ψ_i, which is related to the ith stage λ-coefficient computation, is much bigger than the variance of R_i corresponding to the ith stage reflection coefficient in the Schur algorithm. In particular, $Var(\Psi_i)$ is $O(i^2)$ dependent from the order of the algorithm's stage, whereas the respective dependence of $Var(R_i)$ is $O(i)$. As it is concluded from (3.4), the FP error variances in the λ-coefficient computation are approximately the same as the ones in the reflection coefficient computation for the split Schur algorithm, if the previous signal environment is taken into account. Consequently, the Schur algorithm produces smaller reflection coefficient FP error variances than the split Schur algorithm when the signal under process is noise or it has low correlation between successive samples. It is important to note that (3.4) appears to be valid only when the condition about the smallness of the reflection coefficients is strictly met, since the variance of the $\Delta k(i)$ error produced by the FP split Schur algorithm has been found to be substantially different from that of the $\Delta\lambda(i)$ error in most of the problems arising in practice.

Let us now consider the case of $|k(i-1)| \longrightarrow 1$, $i>2$, with all the reflection coefficients of the i-2 previous stages properly valued within the (-1,1) range. This situation arises in real applications where sinusoidal signals or signals with highly low-pass characteristics are processed. Then, during the ith stage of computations, the mean square prediction error $f(i,i)$ will arbitrarily approach to zero. Since all the left-hand quantities of (3.1)-(3.4) depend on the reciprocal of $f(i,i)$, it follows that the ith reflection coefficient FP error variance becomes extremely large for both algorithms when the reflection coefficient of the previous stage is absolutely close to 1. The previous condition on the reflection coefficients values also produces very high λ-coefficient FP error variances.

Some numerical examples, which validate the previous findings are given in fig. 1, 2. Two standard types of input signals were used: a zero mean, white Gaussian pseudorandom noise sequence with unit variance, and a pure sinusoidal signal of radian frequency 0.85π. The exact mean-squared error of each coefficient, averaged over ten sets of data, is depicted in Fig. 1a and 2a for each type of the signals used respectively. The corresponding theoretical values, obtained by means of (2.3), (2.6), (2.7), are plotted in Fig. 1b and 2b. It is observed that the theoretical and experimental results show an adequate match, with the first ones being pessimistic in all cases. The superiority of the Schur over the split Schur algorithm also appears to be decreasing as the autocorrelation matrices corresponding to the signal under process tends to become ill-conditioned.

IV. CONCLUSIONS

It was proved that the Schur algorithm is more accurate than the split Schur algorithm when the underlying signal is totally unpredictable, or, equivalently, the reflection coefficients are absolutely very close to zero. The validity of this result was experimentally extended to signals with low predictability. Both algorithms were found to perform poorly when the Toeplitz autocorrelation matrix is close to ill-conditioning. Furthermore, the FP arithmetic properties of the λ-coefficients were found to be inferior than those of the reflection coefficients computed by the Schur algorithm and no better than those of the reflection coefficients computed by the split Schur algorithm. This result implies that the λ-coefficients cannot serve as numerical substitutes of the reflection coefficients in real applications employing the split Schur algorithm.

REFERENCES

[1] J. Le Roux and C. Gueguen, "A fixed-point computation of partial correlation coefficients", *IEEE Trans. ASSP*, vol. ASSP-25, June 1977.

[2] P. Delsarte and Y. Genin, "On the splitting of classical algorithms in linear prediction theory", *IEEE Trans. ASSP*, vol. ASSP-35, pp. 645-653, May 1987.

[3] C. J. Zarowski and H. C. Card, "Finite Precision Arithmetic and the Schur algorithm", *IEEE Trans. ASSP*, vol. ASSP-38, Aug. 1990.

[4] C. J. Zarowski and H. C. Card, "Finite Precision Arithmetic and the Split Schur algorithms", *IEEE Trans. ASSP*, vol. ASSP-39, No 8, pp. 1805-1811, Aug. 1991.

[5] N. Glaros and G. Carayannis, "Exact and First-order Error Analysis of the Schur and split Schur algorithms: Theory and Practice", submitted for publication to *IEEE Trans. on Signal Processing.*

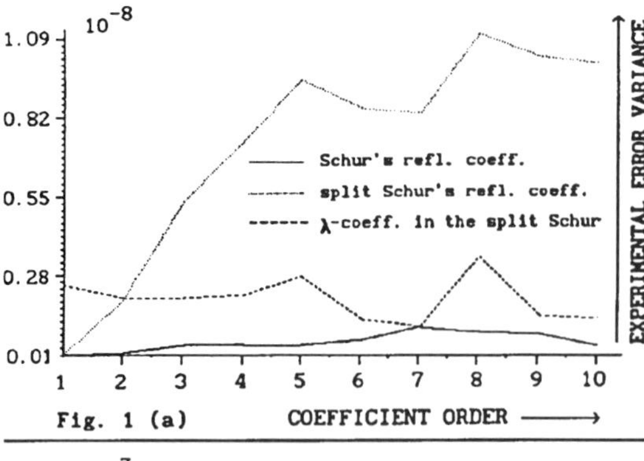

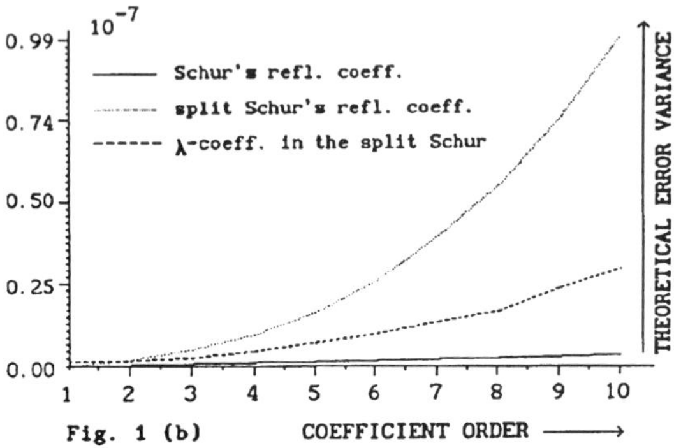

Fig. 1. Comparison of the FP error variances induced by **white noise data** in the Schur and split-Schur algorithms. *(Word Length B = 16).* **(a) Experimental , (b) Theoretical .**

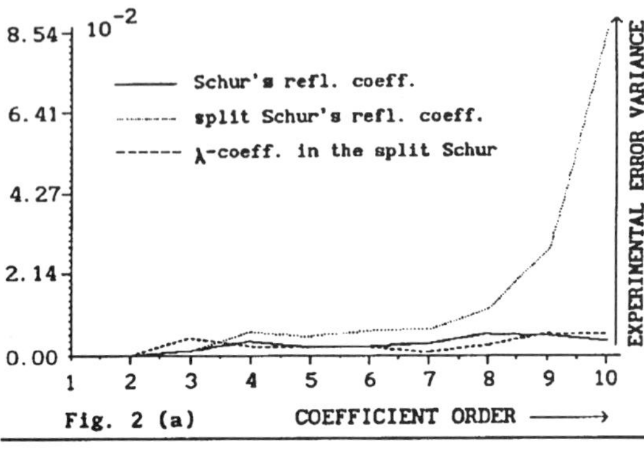

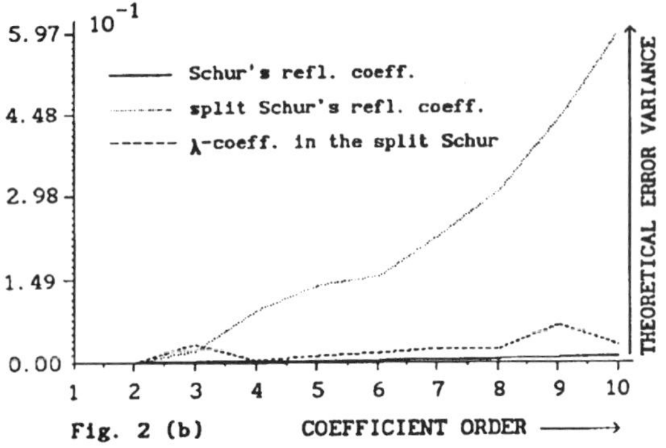

Fig. 2. Comparison of the FP error variances induced by **Sinusoidal** data in the Schur and split-Schur algorithms. *(Word Length B = 20).* **(a) Experimental , (b) Theoretical .**

TABLE I : *Definition of the parameters used for the analytical description of the Schur algorithm's FP error propagation mechanism.*

$$\underline{A}_m \equiv [A_{m,0} \quad A_{m,1} \quad \cdots \quad A_{m,m}]^T \quad , \quad m \geq 1$$

$$\underline{B}_m \equiv [B_{m,0} \quad B_{m,1} \quad \cdots \quad B_{m,m}]^T \quad , \quad m \geq 1$$

$$\underline{A}_1 \equiv [1 \quad 0]^T \quad , \quad \underline{B}_1 = [0 \quad 1]^T$$

$$\underline{A}_{m+1} \equiv \begin{bmatrix} \underline{A}_m \\ 0 \end{bmatrix} + k(m) \cdot \begin{bmatrix} \underline{B}_m \\ 0 \end{bmatrix} \quad , \quad 1 \leq m \leq p$$

$$\underline{B}_m \equiv J_m \cdot \underline{A}_m \quad , \quad 1 \leq m \leq p+1$$

$$\underline{E}_{m,n} \equiv [E_{m,0,n} \quad E_{m,1,n} \quad \cdots \quad E_{m,m,n}]^T \quad , \quad m \geq 0$$

$$\underline{F}_{m,n} \equiv [F_{m,0,n} \quad F_{m,1,n} \quad \cdots \quad F_{m,m,n}]^T \quad , \quad m \geq 0$$

$$\underline{G}_{m,n} \equiv [G_{m,0,n} \quad G_{m,1,n} \quad \cdots \quad G_{m,m,n}]^T \quad , \quad m \geq 0$$

$$\underline{H}_{m,n} \equiv [H_{m,0,n} \quad H_{m,1,n} \quad \cdots \quad H_{m,m,n}]^T \quad , \quad m \geq 0$$

$$\underline{E}_{0,n} \equiv [E_{0,0,n}] \equiv [1] \quad , \quad \underline{G}_{0,n} \equiv [G_{0,0,n}] \equiv [0]$$

$$\underline{F}_{0,n} = [F_{0,0,n}] = [0] \quad , \quad \underline{H}_{0,n} = [H_{0,0,n}] = [1]$$

$$\underline{E}_{m+1,n} \equiv \begin{bmatrix} \underline{E}_{m,n} \\ 0 \end{bmatrix} + k(m+n+1) \cdot \begin{bmatrix} \underline{F}_{m,n} \\ 0 \end{bmatrix} \quad , \quad m \geq 0 \ , \ 1 \leq m+n+1 \leq p$$

$$\underline{G}_{m+1,n} \equiv \begin{bmatrix} \underline{G}_{m,n} \\ 0 \end{bmatrix} + k(m+n+1) \cdot \begin{bmatrix} \underline{H}_{m,n} \\ 0 \end{bmatrix} \quad , \quad m \geq 0 \ , \ 1 \leq m+n+1 \leq p$$

$$\underline{F}_{m,n} \equiv J_m \cdot \underline{G}_{m,n} \quad , \quad \underline{H}_{m,n} \equiv J_m \cdot \underline{E}_{m,n} \quad , \quad m \geq 0 \ , \ 1 \leq m+n+1 \leq p$$

$$\xi_{i,m} \equiv f(m,i) \cdot \Delta k(m) + \gamma_e(m,i) \quad , \quad \xi_{i,0} \equiv \Delta r(i) \quad , \quad 1 \leq i \leq p$$

$$\zeta_{i,m} \equiv e(m,i-1) \cdot \Delta k(m) + \gamma_f(m,i) \quad , \quad \zeta_{i,0} \equiv \Delta r(i-1) \quad , \quad 1 \leq i \leq p$$

$$\underline{\xi}_{i,m,n} \equiv [\xi_{i,n} \quad \xi_{i-1,n} \cdots \quad \xi_{i-m,n}]^T \quad , \quad 0 \leq n \leq i-1 < p \ , \ m \geq 0$$

$$\underline{\zeta}_{i,m,n} \equiv [\zeta_{i,n} \quad \zeta_{i-1,n} \cdots \quad \zeta_{i-m,n}]^T \quad , \quad 0 \leq n \leq i-1 < p \ , \ m \geq 0$$

$$C_m^i \equiv \sum_{n=1}^{m-1} (\underline{E}^T_{m-n-1,n} \cdot \underline{\xi}_{i,m-n-1,n} + \underline{G}^T_{m-n-1,n} \cdot \underline{\zeta}_{i,m-n-1,n}), \quad 2 \leq m \leq i \leq p$$

$$D_m^i \equiv \sum_{n=1}^{m-1} (\underline{F}^T_{m-n-1,n} \cdot \underline{\xi}_{i,m-n-1,n} + \underline{H}^T_{m-n-1,n} \cdot \underline{\zeta}_{i,m-n-1,n}), \quad 2 \leq m \leq i \leq p$$

$$C_1^i \equiv D_1^i \equiv 0 \quad , \quad 1 \leq i \leq p$$

TABLE II : *Definition of the parameters used for the analytical description of the split Schur algorithm's FP error propagation mechanism.*

$$\alpha_{i,m} \equiv - y(m-1,i-1) \cdot \Delta\lambda(m) + \gamma_y(m,i) \quad , \quad 1 \leq m \leq i-1 < p$$

$$\alpha_{i,-1} \equiv \Delta r(i) \quad , \quad 1 \leq i \leq p$$

$$\underline{\alpha}_{i,m,n} \equiv [\alpha_{i,n} \quad \alpha_{i-1,n} \cdots \quad \alpha_{i-m,n}]^T \quad , \quad m \geq 0 \ , \ n \geq -1$$

$$\underline{s}_{m,n} \equiv [s_{m,0,n} \quad s_{m,1,n} \cdots \quad s_{m,m,n}]^T \quad , \quad n \geq -1 \ , \ m \geq 0$$

$$\underline{s}_{0,n} \equiv [1] \quad , \quad \underline{s}_{1,n} \equiv [1 \quad 1]^T \quad , \quad n \geq -1$$

$$\underline{s}_{m+1,n} \equiv \begin{bmatrix} \underline{s}_{m,n} \\ 0 \end{bmatrix} + \begin{bmatrix} 0 \\ \underline{s}_{m,n} \end{bmatrix} - \lambda(m+n+1) \cdot \begin{bmatrix} 0 \\ \underline{s}_{m-1,n} \\ 0 \end{bmatrix} \quad , \quad n \geq -1 \ , \ m \geq 1, \ 1 \leq m+n+1 \leq p$$

$$\underline{v}_m \equiv \underline{s}_{m,-1} \quad , \quad 0 \leq m \leq p+1$$

$$Z_m^i \equiv \sum_{n=1}^{m-1} \underline{s}^T_{m-n-1,n} \cdot \underline{\alpha}_{i,m-n-1,n} \quad , \quad 2 \leq m \leq i \leq p$$

$$Z_0^0 \equiv - \Delta r(0)/2 \quad , \quad Z_0^i \equiv Z_1^i \equiv 0 \quad , \quad 1 \leq i \leq p$$

SIGNAL PROCESSING VI: Theories and Applications
J. Vandewalle, R. Boite, M. Moonen, A. Oosterlinck (eds.)

Multidimensional Wave Digital Filters by Use of Transformation Principle

Wenzhe Li

Ruhr-Universität Bochum, Lehrstuhl für Nachrichtentechnik
Universitätsstr.150, P.O.Box 102148, D-4630 Bochum 1, FR Germany

An approach has been presented to the design of multidimensional (MD) wave digital filters (WDF) with various passband behaviors by use of transformation principle. Two kinds of filters, 2-D hexagonal filter and 3-D octaeder filter have been designed by using this method. The realizations of these filters are given to show how the frequency response and the condition for massive parallelism in the direction of recursion can be satisfied at the same time. Two numerical examples are calculated also.

1 Introduction

MD WDFs enjoy popularity as 1-D WDFs for their low sensitivity and good stability properties [1, 2]. For MD WDFs with a so-called open loss-behavior, fan filters for example, there exists already an analytical solution [3]; but for those with a closed loss-behavior, still no general solutions have been found [1, 2, 4]. The approach of using complex network theory seems to be a solution for the latter [9], unfortunately difficulty arises due to the undesired parasitic passbands. Hence the problem, how such a MD WDF i.e. a suitable MD passive lossless two-port can be derived or synthesised, remains still open.

The purpose of this paper is to present an approach to the design of MD WDFs with closed loss-behavior. This approach uses a transformation principle given in [5, 6] which has been originally used in physical systems that are simulated by MD WDF algorithm with massive parallelism. Such transformation principle then will be applied to those passive lossless two-ports given in [7], with which two kinds of WDFs are designed. It shows therewith that such a MD WDF on the one hand realizes the desired frequency response and on the other hand ensures an algorithm that offers massive parallelism in the direction of recursion. Some realization considerations are discussed, and two numerical examples, a 2-D hexagonal filter and a 3-D octaeder filter are given also to show the design procedure.

2 Principle of transformation

The transformation method given in [5] can be explained as the coordinate transformation bewteen a k'-D causal system and a k-D noncausal system with $k' \geq k$. In order to use such transformation for filter design, the following discussions are needed.

To design a k-D filter, one should begin with a k'-D causal filter. Then, the k-D filter can be obtained by using coordinate transformation in the k'-D causal filter. The k-D filter obtained in this way is still causal, but can be considered as a k-D noncausal filter with standard sampling, i.e. sampling with sampling matrix of $\mathbf{T} = diag(T_1, \cdots, T_k)$. Since the transformation, as will be seen in the following, causes a change in the sampling matrix, and thus a change in the sampling lattice, the transformed k-D filter is equivalent to a k-D causal filter that works with another new sampling matrix. To this sense, the k-D filter is embedded simply in the k'-D causal filter that will be designated as the auxiliary filter also, and the passivity of the k-D filter thus obtained can be ensured if the k'-D filter itself is (internally) passive.

Let $\mathbf{t}'$ and $\mathbf{t}$ denote the coordinate vectors of the k'-D filter system and the k-D filter system respectively, with

$$\begin{aligned} \mathbf{t}' &= (t'_1, \cdots, t'_{k'})^T = \mathbf{T}'\mathbf{n}', & (1a)\\ \mathbf{t} &= (t_1, \cdots, t_k)^T = \mathbf{T}\mathbf{n}, & (1b)\\ & \mathbf{n}' \in \mathbf{Z}^{k'}, \mathbf{n} \in \mathbf{Z}^k, k' \geq k, \end{aligned}$$

$\mathbf{p}' = (p'_1, \cdots, p'_{k'})^T$, $\mathbf{p} = (p_1, \cdots, p_k)^T$ the complex frequency vectors and $\underline{\psi}' = (\psi'_1, \cdots, \psi'_k)$, $\underline{\psi} = (\psi_1, \cdots, \psi_k)$ the equivalent complex frequency vectors; $\mathbf{T}'$ and $\mathbf{T}$ being the sampling matrice of the two systems for standard sampling. Using the known bilinear-transformation for variables $\mathbf{z}' = e^{\mathbf{p}'^T\mathbf{T}'}$ and $\mathbf{z} = e^{\mathbf{p}^T\mathbf{T}}$, it follows

$$\begin{aligned} \underline{\psi}' &= tanh(\frac{1}{2}\mathbf{T}'^T\mathbf{p}'), & (2a)\\ \underline{\psi} &= tanh(\frac{1}{2}\mathbf{T}^T\mathbf{p}) & (2b) \end{aligned}$$

with notation $tanh(\mathbf{x}) = (tanh(x_1), \cdots, tanh(x_n))^T$, $\mathbf{x} = (x_1, \cdots, x_n)^T$. The relation between the two filter systems then can be expressed as

$$\mathbf{n} = \mathbf{M}\mathbf{n}', \qquad (3)$$

where $\mathbf{M}$ is a $k \times k'$ integer matrix with $rank\mathbf{M} = k$, for which there exists at least a right inverse matrix $\mathbf{M}^{-R}$, such that $\mathbf{M}\mathbf{M}^{-R} = \mathbf{1}_k$ is valid, $\mathbf{1}_k$ being the unit matrix of k order. Considering (3) in (1), the coordinates of the two filter systems are related by

$$\begin{aligned} \mathbf{t} &= \mathbf{T}\mathbf{M}\mathbf{T}'^{-1}\mathbf{t}', & (4a)\\ \mathbf{t}' &= \mathbf{T}'\mathbf{M}^{-R}\mathbf{T}^{-1}\mathbf{t}. & (4b) \end{aligned}$$

Observing the product $\mathbf{p}'^T\mathbf{t}'$, one has according to (4)

$$\mathbf{p}'^T\mathbf{t}' = (\mathbf{T}^{-T}\mathbf{M}^{-TR}\mathbf{T}'^T\mathbf{p}')^T\mathbf{t} = \mathbf{p}^T\mathbf{t}, \tag{5}$$

where

$$\mathbf{p} = \mathbf{T}^{-T}\mathbf{M}^{-TR}\mathbf{T}'^T\mathbf{p}', \tag{6}$$

or written in another form

$$\mathbf{T}'^T\mathbf{p}' = \mathbf{M}^T\mathbf{T}^T\mathbf{p}. \tag{7}$$

Using this relation in (2) yields

$$\underline{\psi}' = tanh(\frac{1}{2}\mathbf{M}^T\mathbf{T}^T\mathbf{p}) = tanh(\frac{1}{2}\hat{\mathbf{T}}^T\mathbf{p}) \tag{8}$$

with

$$\hat{\mathbf{T}} = \mathbf{TM}. \tag{9}$$

Accordingly, as is implied in (7) and (8), the k-D filter is now embedded in the k'-D auxiliary fliter, and in fact with a new sampling matrix $\hat{\mathbf{T}}$. If the k'-D auxiliary filter is (internally) passive and the transformation matrix $\mathbf{M}$ is well defined according to some definite criteria stated in [6], one can expect to obtain, with respect to the auxiliary filter, a k-D filter that reflects itself in a causal behavior in the auxiliary filter. In this way, the passivity of the auxiliary filter would be preserved in the k-D filter. This property is advantageous especially if the k'-D auxiliary filter is a WDF, since then the internal passivity (or MD passivity [6]) of the k'-D auxiliary filter and those good stability properties thus derived can be retained in the transformed k-D filter, and the transformed k-D filter obtained remains to be a WDF.

It can be seen from the above discussions that the matrix M plays an important role for the whole transformation. It should be chosen in a proper way such that (internal) passivity property can be preserved and massive parallelism can be achieved in the k-D filter. In [6] there are altogether three essential criteria associated with the determination of M:

1. Any change $\Delta_k > 0$ in t_k alone must create changes $\Delta t'_\kappa > 0$ in the $t'_\kappa, \kappa = 1$ to k'. Moreover, there must exist at least one k-dimensional subspace of t'_1 to $t'_{k'}$ such that the above requirement holds. In other words, there must exist at least one right inverse $\mathbf{M}^{-R}$ such that any change $\Delta_k > 0$ in t_k alone creates changes $\Delta t'_\kappa > 0$ in the $t'_\kappa, \kappa = 1$ to k'.
2. Any change of one of the t'_κ for which $\Delta t'_\kappa > 0, \kappa \in \{1, 2, \cdots, k'\}$ must create a change $\Delta_k > 0$.
3. The structure of the transformation should be as simple as is reasonably feasible.

Note that the second requirement ensures the possibility to massive parallelism. From an immediate observation, it requires that all elements of the last row of $\mathbf{M}$ must be positive.

3 Filter design and realizations

3.1 2-D hexagonal filter

The transformation matrix $\mathbf{M}$ for this filter is chosen as

$$\mathbf{M} = \begin{pmatrix} -1 & 1 & 0 \\ 1 & 1 & 2 \end{pmatrix}, \tag{10}$$

accordingly one has $k = 2$ and $k' = 3$, i.e. the design of the 2-D filter should be started with a 3-D filter. It is easy to test that $\mathbf{M}$ satisfies all the requirements listed in the last section. And the corresponding new sampling matrix $\hat{\mathbf{T}}$ to the 2-D filter is calculated according to (9) as

$$\hat{\mathbf{T}} = \begin{pmatrix} -T_1 & T_1 & 0 \\ T_2 & T_2 & 2T_2 \end{pmatrix} = (\mathbf{T}_1 \vdots \mathbf{T}_2 \vdots \mathbf{T}_3), \tag{11}$$

where $\mathbf{T}_i, i = 1, 2, 3$ are shift vectors to $\hat{\mathbf{T}}$, with which the sampling lattice can be determined directly. Note that for the given $\hat{\mathbf{T}}$ the corresponding periode matrix Ω calculated from $\hat{\mathbf{T}}\Omega^T = 2\pi\mathbf{1}_2$ is not unique because of the nonuniqueness of $\hat{\mathbf{T}}^{-R}$ for $k' > k$, but all Ω describe the same reciprocal lattice.

Here, the case by setting $T_1 = \sqrt{3}T_2, T_2 = T$ in (11) is of interest because accordingly a hexagonal sampling can be derived; the sampling lattice and the corresponding reciprocal lattice are shown in Fig.1, where the Nyquist region may possess a hexagonal shape also.

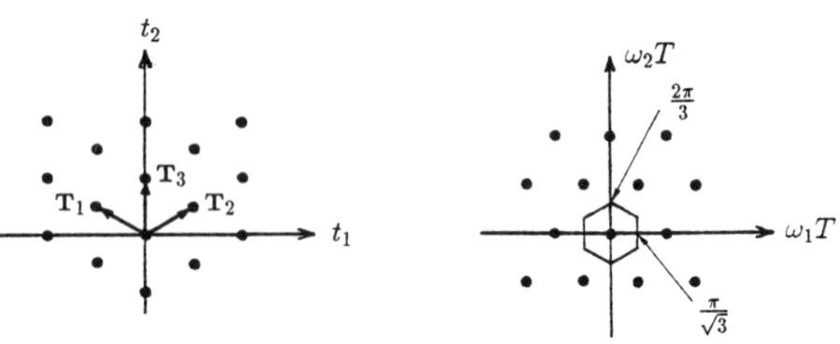

Fig.1: Sampling and reciprocal lattices according to the sampling matrix of (11).

Employing (11) in (8) and considering (2), one obtains for the transformation in frequency domain

$$\psi'_1 = \frac{\psi_2 - \psi_1}{1 - \psi_1\psi_2} = tanh(T\frac{p_2 - \sqrt{3}p_1}{2}), \tag{12a}$$

$$\psi'_2 = \frac{\psi_2 + \psi_1}{1 + \psi_1\psi_2} = tanh(T\frac{p_2 + \sqrt{3}p_1}{2}), \tag{12b}$$

$$\psi'_3 = \frac{2\psi_2}{1 + \psi_2^2} = tanh(p_2 T). \tag{12c}$$

With this transformation the design of a 2-D hexagonal lowpass filter is considered as follows. Since quadrant symmetry is required in the frequency response, the 3-D filter needed should possess the same symmetry also[5]. In Fig.2(a) a suitable 3-D two-port whose frequency response is lowpass with an approximately cubic passband[7] has been chosen as the reference circuit for the 3-D filter. It consists of three 1-D subnetworks $N'(\psi'_1)$ to $N'(\psi'_3)$, each of which is a lossless two-port also and in fact a symmetric lattice two-port; they are isolated by gyrator-inductor circuits $N''(\psi'_3)$ from each other. A good cubic passband response in the 3-D filter would be obtained if each subnetwork is designed as 1-D lowpass by use of approximations such as Butterworth, Chebyshev and so on[5].

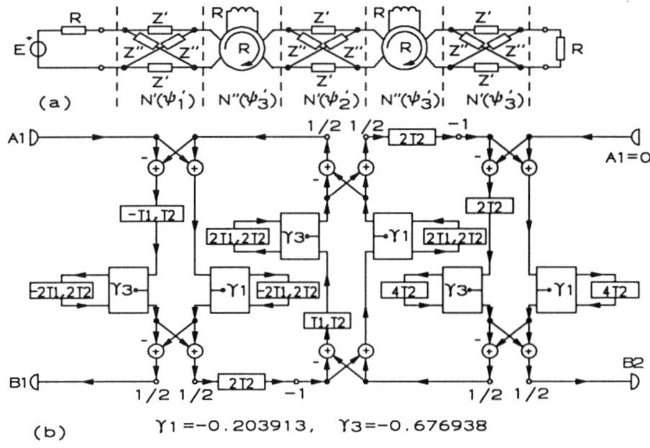

Fig.2: 2-D hexagonal WDF, (a) reference two-port before transformation; (b) filter realization after transformation.

Carrying out the above transformation in the 3-D lowpass filter, a 2-D hexagonal lowpass filter is then obtained, with the corresponding WDF realization shown in Fig.2(b). Although negative shift elements of T_1-type exist in the filter, they appear always together with T_2(e.g. $(-T_1, T_2)$). For the algorithm of the filter this means that a shift is made instead of in $-t_1$ and then t_2 direction in the $t_2 = -t_1$ direction, shifts of such type should be called vector-shifts, and shifts of T_1 and T_2 can be considered as a special case. Consequently, the 2-D filter with vector-shifts has been embedded in the 3-D causal WDF, massive parallelism can be achieved in the t_2-direction since illustratively there exist only shifts of types $T_2, (-T_1, T_2)$ and (T_1, T_2) in the transformed filter.

As a numerical example a 2-D hexagonal filter is designed and its frequency response is shown in Fig.3, in which each subfilter(subnetwork) is calculated according to Cauer approximation by use of the method in [10] for the bireciprocal case with ripple of $0.00033dB$ in passband $0 - 0.4\pi$ and attenuation of $41.24dB$ in stopband $0.63\pi - \pi$.

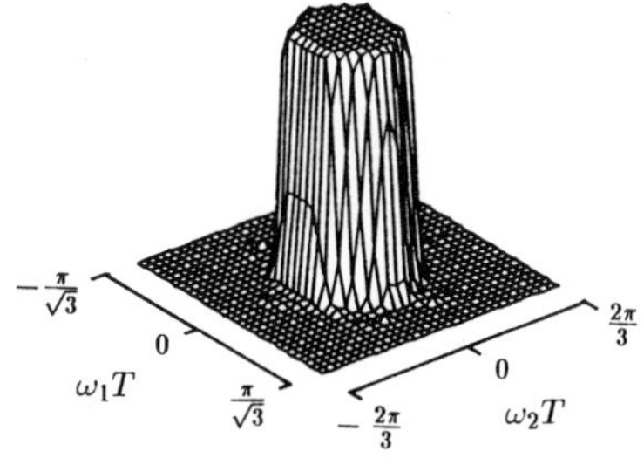

Fig.3: Frequency response of the 2-D hexagonal filter.

Note that the structure of this filter ist advantageous to hardware realization using signal processor[8] because only two-port adaptors are needed in filter. Besides, since the direction of recursion of the 2-D filter can be chosen as t_2, the existence of the shifts between the subfilters provides more pipeline possibilities, and thus more parallelism to the realization of the filter.

The design procedure presented above is relatively simple and no optimization has to be needed. But this design is limited to 2-D filter whose passband extends to $\frac{2\pi}{3\sqrt{3}}$(the maximum edge frequency of the hexagonal passband); if the edge frequency of the passband is greater than that, parasitic passbands may appear in the corner of the Nyquist region. In this case, the two-port of Fig.2(a) should be modified as a 4-D two-port with $N'(\psi'_3)$ replaced by $N'(\psi'_4)$ and $\psi'_4 = \psi'_1 + \psi'_2$. This means that the transformation should be first made in reference domain for the 4-D two-port, which then yields a new 3-D two-port. Again, the 2-D hexagonal filter can be obtained by using the transformation of (12) to this 3-D two-port.

3.2 3-D octaeder filter

The transformation matrix for this filter is

$$\mathbf{M} = \begin{pmatrix} -1 & -1 & 1 & 1 \\ -1 & 1 & 1 & -1 \\ 1 & 1 & 1 & 1 \end{pmatrix}, \tag{13}$$

which yields $k = 3$ und $k' = 4$, i.e. a 4-D auxiliary filter is needed for the design of the 3-D filter. If standard sampling has been done for the 4-D filter, the sampling matrix of the 3-D filter can be derived from (9) as

$$\hat{\mathbf{T}} = \begin{pmatrix} -T_1 & -T_1 & T_1 & T_1 \\ -T_2 & T_2 & T_2 & -T_2 \\ T_3 & T_3 & T_3 & T_3 \end{pmatrix} = (\mathbf{T}_1 \vdots \mathbf{T}_2 \vdots \mathbf{T}_3 \vdots \mathbf{T}_4), \tag{14}$$

$\mathbf{T}_i, i = 1, 2, 3, 4$ being again shift vectors. Accordingly, the corresponding sampling lattice is shown in Fig.4(a), for which $T_1 = T_2 = T_3 = T$ has been assumed. Obviously, the sampling thus obtained corresponds to a so-called *Body-centered orthorhombic* sampling [11], and the equivalent transformation in frequency domain is derived as

$$\psi'_1 = tanh(\frac{p_3T - p_2T - p_1T}{2}), \tag{15a}$$

$$\psi'_2 = tanh(\frac{p_3T + p_2T - p_1T}{2}), \tag{15b}$$

$$\psi'_3 = tanh(\frac{p_3T + p_2T + p_1T}{2}), \tag{15c}$$

$$\psi'_4 = tanh(\frac{p_3T - p_2T + p_1T}{2}). \tag{15d}$$

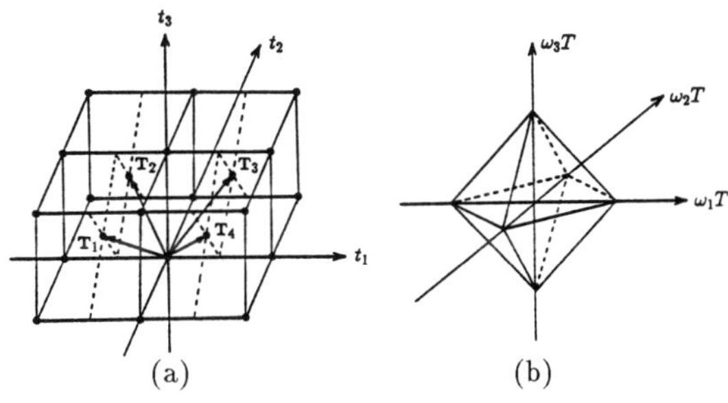

Fig.4: (a) sampling lattice according to (14); (b) passband region of the 3-D octaeder filter.

The desired passband shape(octaeder) of the 3-D filter is illustrated in Fig.4(b). In order to obtain such a response, the 4-D auxiliaray filter should be designed as a lowpass filter with hypercubic passband by using the method in [7], anti-metric ladder two-port is used for each subfilter. Applying the above transformation to the 4-D filter, a 3-D octaeder filter is obtained and its WDF realization is shown in Fig.5. As an example a filter has been calculated and its frequency response is shown in Fig.6, for which each subfilter is Chebyshev lowpass of 6 orders with ripple of $0.0109dB$ in passband and attenuation $\geq 26dB$ in stopband. Again for this filter, massive parallelism can be achieved in the t_3-direction.

4 Conclusions

k-D lowpass WDFs with various passband behaviors can be derived from a k'-D causal WDF(lossless two-port) with (hyper-)cubic passband by use of transformation principle. Two kinds of filters, 2-D hexagonal filter and 3-D octaeder filter have been designed by this method. It shows thereby that the designed filters on the one hand realize the desired frequency response and on the other hand ensure a algorithm that offers massive parallelism in the direction of recursion. Two numerical examples are calculated and the corresponding WDF realizations have been given also which show the effectiveness of this approach.

Acknowledgement

The author is deeply indebted to Prof. A. Fettweis for his helpful suggestion and wishes to thank DAAD (Deutscher Akademischer Austauschdienst) for the support of this work.

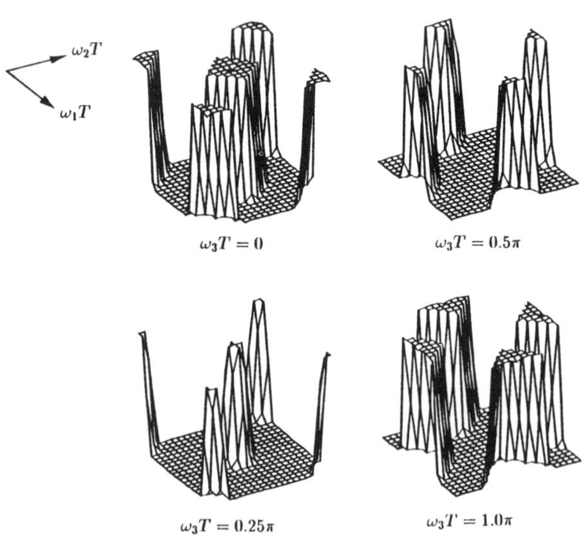

Fig.6: Frequency response of the 3-D octaeder filter.

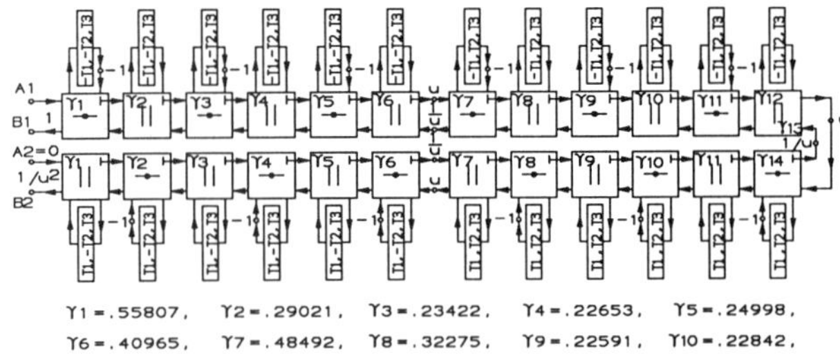

Fig.5: WDF realization of the 3-D octaeder filter.

References

[1] A.Fettweis,"Multidimensional Wave Digital Filters - Problems and Progress," in *IEEE Int. Symp. Circuits Syst.*, (San Jose,CA), pp.506-509,May 1986.

[2] A.Fettweis, "Wave Digital Filter: Theory and Practice," *Proc. IEEE,* vol.74,pp.270-327, Feb,1986.

[3] G.Linnenberg," Über die diskrete Verarbeitung mehrdimensionaler Signale unter Verwendung von Wellendigitalfiltern," Ph.D. dissertation, Ruhr University, Bochum, FR Germany,1984.

[4] P.A.Lennarz,"Realisierung von 2-D-Digitalfiltern mit kreissymmetrischem Tiefpassverhalten durch Kaskadierung von 2-D-Wellendigitalfiltern," Ph.D. dissertation, Ruhr University, Bochum, FR Germany,1983.

[5] A.Fettweis, "New Results in Wave Digital Filtering," in *Int. Symp. Signals, Syst. and Electr.*,(Erlangen, W.Germany), pp.17-23,18-20 Sept.1989.

[6] A.Fettweis and G.Nitsche, "Transformation approach to numerically integrating PDEs by means of WDF principles," *Multidimensional Syst. and Signal Processing,* No.2, pp127-159,1991.

[7] Wenzhe Li, "Mulltidimensional Lossless Two-ports: Approximation and WDF Realizations," (to be submitted for publication).

[8] A.Fettweis, "Modified Wave Digital Filters for Improved Implementation by commercial Digital Signal Processor," *Signal Processing,* No.16,(North-Holland),pp.193-207,1989.

[9] A.Fettweis, "Multidimensional Digital Filters with Closed Loss Behavior Designed by Complex Network Theory Approach," *IEEE Trans.Circirts Syst.*, Vol.34,No.4,pp.338-344,Apr.1987.

[10] L.Gazsi, "Explicit Formulae for Lattice Wave Digital Filters," *IEEE Trans. Circuits Syst.,* vol.32,pp68-88,Jan.1985.

[11] E.Dubois, "The Sampling and Reconstruction of Time-Varying Imagery with Application in Video Systems," *Proc. IEEE,* vol.73,No.4,pp.502-522, April,1985.

SIGNAL PROCESSING VI: Theories and Applications
J. Vandewalle, R. Boite, M. Moonen, A. Oosterlinck (eds.)

Adaptation of Grey Level Structuring Elements for Morphological filters with Application to Shape Detection

Philippe Salembier *

Harvard Robotics Laboratory
Harvard University
Cambridge, MA 02138, USA

Signal Processing Laboratory
Swiss Federal Institute of Technology
CH-1015, Lausanne, Switzerland

Abstract: This paper deals with algorithms for the adaptive optimisation of grey level structuring elements of morphological filters in discrete spaces. The adaptation process is similar to the classical least mean square algorithm used for linear filters and its goal is to minimise a statistical criterion such as the mean square (or mean absolute) error between the filter output and a desired signal. It is shown how a grey level structuring element can be optimised in the case of erosion and dilation. Then, adaptation formulas are derived for an arbitrary composition of erosions and dilations (optimisation of opening, closing, open_close or alternating sequential filters). Moreover, the results are further generalised to filters obtained by combination of various filters outputs using minimum and maximum operations (optimisation of maximum of openings or morphological centres). One of the major advantages of the proposed approach is that it leads to simple adaptation formulas and allow fast convergence rate. Finally, the approach is illustrated on grey level shape detection problems. Adaptation is used here to get the best structuring element for a particular detection problem.

1 Introduction

Morphological filters [1] have become very popular in signal and image processing because of their attractive features such as impulse noise suppression, edge preservation or shape-oriented approach to signal analysis. They analyze the geometrical structure of a signal by locally comparing it with a predefined elementary shape called a structuring element (SE). Most of the time, the choice of a specific filter is performed with the so-called structural approach. It consists in analyzing and selecting a relevant set of signal structures, and in choosing a filter which is able to properly interact with these structures. It leads very often to rely on the geometrical properties of the signal in order to choose a specific SE. Even if this approach has been successfully applied in many practical cases, a large number of researches are devoted to alternative solutions. Indeed, the structural approach often requires a fair amount of practical experience and intuition.

Recently, alternative approaches have been proposed [2,3]. They formulate the design as an estimation problem. The filter is optimised to minimise a criterion such as the mean square error (MSE) or the mean absolute error (MAE) between its output and a desired signal. In a similar way to that of linear filter optimisation, this approach yields two different types of solutions. The first one attempts to find an optimal fixed filter [2]. However, it has several drawbacks: it assumes stationarity of the various signals which is almost never the case in image processing. It also requires a certain knowledge about the noise characteristics which is often difficult to obtain. Another drawback is the computational complexity of the approach.

To avoid these difficulties of fixed optimum filters, a second approach involving adaptive optimisation problem has been proposed [3,4]. The algorithms reach the optimum point step by step and, at least partially, overcome the drawbacks mentioned above. This study is in fact an extension of an earlier work on adaptive rank order based filters [5] where algorithms have been proposed to adapt the filter mask or rank in order to minimise the MSE or the MAE. As pointed out in [3,5], this filter class includes as special case the erosion and the dilation with flat SE and all morphological filters obtained by composition of these two basic operators. The goal of this paper is to present adaptation algorithms in the context of morphological filters with grey level (not flat) SE. The solutions are valid for any composition of erosions and dilations, and also for combinations of filters with minimum and maximum operations. This allows adaptation of filters such as maximum of openings or morphological centres, etc.

The organisation of this paper is as follows: the next section defines more precisely the adaptation problem. The third section deals with formulations of the basic erosion and dilation operators with grey level SE. Then, the fourth section is devoted to the optimisation of erosion and dilation. The results are extended in the fifth section to filters obtained by composition of erosion and dilation and by combination with minimum and maximum operations. Finally, practical problems of shape detection are addressed in section six.

2 The optimisation problem

The adaptation problem can be stated as follows: let x_i denote an original signal, d_i a desired signal and F a morphological filter with a SE defined by the set of real values $\{m_j\}_{j\in M}$ and y_i the filter output. The optimisation tries to find the best set of parameters $\{m_j\}$ to minimise the MSE or MAE between the filter output and the desired process. The MSE optimisation can be statistically achieved in an iterative way using the LMS approach given by [6]:

$$m_k' = m_k - \mu \frac{\partial[(d_i-y_i)^2]}{\partial m_k} \approx m_k + 2\,\mu\,(d_i-y_i)\frac{\partial y_i}{\partial m_k} \quad (2.1)$$

where μ is the convergence parameter. As pointed in [5] the MAE optimisation can be approximated by replacing the (d_i-y_i) term in (2.1) by its sign.

With this approach, the only difficulty consists in computing the partial derivative of the filter output with respect to the parameter to optimise. The next section is

The author is now with the Department of Signal Theory and Communications, Universitat Politecnica de Catalunya, ETSIT, Apdo 30002, 08080-Barcelona, Spain.

devoted to the definitions of erosion and dilation and to the proposition of alternative formulations more suitable for the computation of the partial derivative mentioned above.

3 Morphological filters

Most morphological filters are obtained by composition of erosions and dilations and by combinations with minimum and maximum. For discrete space and grey level SE, the output y_i of an erosion (resp. dilation) is the minimum value of $(x_{i+j}-m_j)$ (resp. maximum value of $(x_{i-j}+m_j)$):

$$E(x_i) = \mathrm{Min}\ \{x_{i+j}-m_j,\ j\in M\} \quad (3.1)$$

$$D(x_i) = \mathrm{Max}\ \{x_{i-j}+m_j,\ j\in M\} \quad (3.2)$$

To solve the optimisation problem, an alternative formulation of these basic operators is proposed now. Note that this formulation is derived in a similar way to that used in [3,5] to deal with rank order based filters and flat SE. The basic idea of the approach is to express the erosion and the dilation with easily differentiable functions. Let S be the sum of signs of $(x_{i+j}-m_j-y_i)$ for all j. For an erosion, the filter output is the minimum of the set $\{x_{i+j}-m_j\}$, thus all terms $(x_{i+j}-m_j-y_i)$ are strictly positive except one of them which is equal to zero (we assume that only one such point exists. It can be shown that this approximation allows simplification of the algorithm without significant loss in performances). If the sign of 0 is defined as being 0, S is equal to the size of the SE (N) minus 1:

$$S = \sum_{j\in M} \mathrm{sgn}\ (x_{i+j}-m_j-y_i) = N-1 \quad (3.3)$$

This equation gives the following implicit formulation of the erosion: the output y_i of an erosion at location i, is the value of the set $\{x_{i+j}-m_j\}_{j\in M}$ such that:

$$f(x_{i+j},y_i,m_j) = \sum_{j\in M} [\mathrm{sgn}\ (x_{i+j}-m_j-y_i) - 1]+1 = 0 \quad (3.4)$$

The alternative definition of the dilation can be derived in a similar way. Let us see how these implicit formulations can be used to get the derivative of the filter output with respect to the parameter to optimise.

4 Adaptive erosions and dilations

The partial derivative of the filter output with respect to the m_k can be obtained by computing the total derivative of the implicit function f. It gives the following result:

$$\frac{\partial y_i}{\partial m_k} = -\frac{\partial f}{\partial m_k} / \frac{\partial f}{\partial y_i} \quad (4.1)$$

The computation of the derivative of y_i with respect to m_k is reduced to the computation of two derivatives of f with respect to m_k and y_i. In the case of an erosion, the first one can be extracted from (3.4):

$$\frac{\partial f_E}{\partial m_k} = \frac{\partial \mathrm{sgn}(x_{i+k}-m_k-y_i)}{\partial m_k} = -2\ \delta(x_{i+k}-m_k-y_i) \quad (4.2)$$

For the second one, (3.4) gives:

$$\frac{\partial f_E}{\partial y_i} = \sum_{j\in M} \frac{\partial \mathrm{sgn}(x_{i+j}-m_j-y_i)}{\partial y_i} = -2 \sum_{j\in M} \delta(x_{i+j}-m_j-y_i) \quad (4.3)$$

This last expression can be simplified. Indeed, in the case where all $(x_{i+j}-m_j-y_i)$ are different from zero except one of them (the output is assumed to come from a single location), the summation of (4.3) reduces to a single term:

$$\frac{\partial f_E}{\partial y_i} = -2 \quad (4.4)$$

Finally, combining these equations gives the partial derivative of y_i with respect to m_k:

$$\text{Erosion:} \quad \frac{\partial y_i}{\partial m_k} = -\ \delta(x_{i+k}-m_k-y_i) \quad (4.5)$$

This formula has a very simple interpretation. It means that the derivative of the filter output y_i with respect to a particular parameter m_k is equal to zero for all k except for the (single) one which corresponds to the location which has produced the output. In this case, the derivative estimate is simply -1.

The same approach can be used for the dilation with obvious modifications. It gives:

$$\text{Dilation:} \quad \frac{\partial y_i}{\partial m_k} = \delta(x_{i-k}+m_k-y_i) \quad (4.6)$$

Let us see now how these formulas can be extended to more complex and useful filters.

5 Extension to morphological filters

The approach of section 4 can be generalised to deal with arbitrary compositions of erosions and dilations: suppose that a filter F(.) is the composition of T elementary operators (erosions or dilations), and that the pth operator in the composition is denoted by $F_{p,\varepsilon_p}(.)$ where p is the position of the operator in the composition and ε_p its type defined by ε_p =1 for an erosion and ε_p =-1 for a dilation. The global filter is therefore:

$$y_i=F(x_i)=F_{T,\varepsilon_T}(F_{T-1,\varepsilon_{T-1}}(\ldots F_{p,\varepsilon_p}(\ldots F_{1,\varepsilon_1}(x_i)))) \quad (5.1)$$

Let x_i^p denote the signal at the output of the pth filter:

$$x_i^p = F_{p,\varepsilon_p}(\ldots F_{1,\varepsilon_1}(x_i)) \text{ with } x_i^0 = x_i,\ x_i^T = y_i \quad (5.2)$$

The adaptation rule involves the trajectory of the final output y_i in the various filtering levels. Indeed, the output of an erosion or a dilation with grey level SE is always one of the input values at some location plus or minus the value of the SE at that location. Thus, the final output at location i denoted by y_i comes from a particular location $i+\Delta_{T-1}$ at the level T-1. In turn, this value comes from a particular location $i+\Delta_{T-2}$ at the level T-2. Let us denote by $i+\Delta_p$ the location at level p (the output value is supposed to come from a single location) which corresponds to the output value y_i and $m_{\alpha(\Delta_p)}$ the value of the SE which was used at that location:

$$x_{i+\Delta_p}^p - \varepsilon_p m_{\alpha(\Delta_p)} = y_i \text{ for all p, with } \Delta_T = 0 \quad (5.3)$$

With these notations, it can be shown [4] that the gradient estimate of y_i is given by the following formulas:

$$\frac{\partial y_i}{\partial m_k} = -\sum_{p=1}^{T} \varepsilon_p\ \delta(x_{i+\Delta_p+\varepsilon_p k}^{p-1} - \varepsilon_p m_k - x_{i+\Delta_p}^p) \quad (5.4)$$

This rule can be used to easily derive an adaptive algorithm for an arbitrary composition of erosions and dilations. As an example, the adaptation formulas for an opening and a closing are respectively given by:

$$\text{Open:} \quad \frac{\partial y_i}{\partial m_k} = -\delta(x_{i+\Delta_1+k}-m_k-x_{i+\Delta_1}^1)+\delta(x_{i-k}^1+m_k-y_i) \quad (5.5)$$

$$\text{Close:} \quad \frac{\partial y_i}{\partial m_k} = +\delta(x_{i+\Delta_1-k}+m_k-x_{i+\Delta_1}^1)-\delta(x_{i+k}^1-m_k-y_i) \quad (5.6)$$

This approach already allows to deal with a large class of morphological filters. However, it cannot be directly used for more sophisticated filters such as maximum of openings or morphological centres. In the case of combinations of subfilters with min or max operations, it can be shown [4] that the derivative of the filter output is equal to the derivative of the subfilter which has produced the filter output. This very simple result is particularly useful to deal with filters such as maximum of openings, minimum of closings or morphological centres and more generally any toggle mapping.

6 Applications to shape detection

This section describes two examples of grey level shape detection where adaptation proves to be a very useful tool. The first example concerns the extraction of a given shape in a noisy environment. A typical example is presented in the upper left part of figure 1. This image is composed of black grey level shapes randomly located on a noisy background. Suppose that, the ideal shape detection result is known for one or a few original images. Then, a detection system can be optimised with these pairs of original and ideal images. Finally, the optimum detection system can be used to automatically process a very large number of original images. The goal of this section is to discuss two such examples of supervised optimisation.

In the case of the original image of figure 1, the dark grey level shapes can be extracted by a closing. However, the closing may also extract noise components. For this detection task, one of the most natural choice is to take as SE the shape to detect itself. The image obtained after a closing with the grey level shape to extract is shown in the lower left part of figure 1. One can see that the shapes are actually enhanced but the noise has still a very strong influence.

Let us assume that the ideal detection result is known. This desired image is shown in the upper right part of figure 1. A grey level SE is optimised in order to minimise the MSE between the image obtained after a closing on the original image and the desired image. From our experience, it seems that this optimisation problem involves several local minima. As the LMS algorithm is trapped by these local minima, one should start the optimisation procedure with a good initial guess. In our case, the shape to detect is selected as initial SE. From the practical point of view, a grey level SE of size 15*15 is optimised following the MSE criterion and the image is scanned several times (about 10 times with μ=1.e-6) in order to get convergence.

Figure 2 presents the original SE, that is the shape to detect, and the resulting SE after optimisation. The optimised SE is flatter than the original one to remove noise more efficiently. However, the global shape of the original SE is preserved. The optimisation makes a compromise between shape detection and noise cancellation. The image obtained after closing with the optimised SE is shown in the lower right part of figure 1. As one can see, the contrast is efficiently improved. Finally, the MSE between the various images of figure 1 and the desired one are computed. The resulting figures are as follows:

- original image / desired image: MSE=6438
- closing with shape to detect / desired image: MSE=3454
- closing with optimised SE / desired image: MSE=2364

Finally, a similar experiment is done with an original image composed of two different shapes. A typical example made of elongated dark spots and of dark rings is shown in the upper left part of figure 3. Here the goal is to eliminate as much as possible one shape while preserving the other one. The ideal result is shown in the upper right part of figure 3.

As previously, a detection can be performed by a closing with the shape to detect. The resulting image is shown in the lower left part of figure 3. The ring shapes are less dark than in the original image but are still quite visible. To improve this result, an optimisation procedure is done. The initial SE is the shape to detect and a grey level SE of size 13*13 is optimised to minimise the MSE between the output of a closing and the desired image. The initial and optimised shapes are presented in figure 4. As can be seen, the optimised SE is somewhat flatter than the original one, and there is a small depression in its centre to efficiently remove the ring shapes. The image obtained after a closing with the optimised SE is presented in the lower right part of figure 3. This result clearly demonstrates the benefit of using the optimised SE. Finally, the following table gives the MSE:

- original image / desired image MSE = 524
- closing with shape to detect / desired image MSE = 419
- closing with optimised SE / desired image MSE = 213

Conclusions

In this paper, several adaptive algorithms for the optimisation of grey level SE of morphological filters in discrete spaces have been proposed. The adaptation process is similar to the classical LMS algorithm which minimises the MSE (or MAE) between the filter output and a desired signal. It is shown how a grey level SE can be optimised in the case of erosion and dilation, arbitrary compositions of erosions and dilations (optimisation of opening, closing, open_close or alternating sequential filters) and also filters obtained by combination of various filters outputs using min/max operations (optimisation of maximum of openings or morphological centres). One of the major advantages of the proposed approach is that it leads to simple adaptation formulas and allow fast convergence rate. Finally, the approach is illustrated on grey level shape detection problems. Adaptation is used here to get the best structuring element for a particular detection problem.

References

[1] J.Serra, Image analysis and mathematical morphology, *Academic Press*, V.1(1982), V.2(1988)

[2] E.R.Dougherty, Optimal mean-square N-observation digital morphological filters, Part1: optimal binary filters, Part 2: optimal gray-scale filters, *To appear in Journal of CVGIP, also as Technical note of the Center for Imaging Lab., Rochester Inst. of Tech.*

[3] P.Salembier, Multiscale image modeling and analysis using rank order based filters - Application to defect detection -, *PhD Thesis*, E.E. Dept. EPFL, Lausanne, Switzerland, June 91.

[4] P.Salembier, Structuring element adaptation for morphological filters, *Journal of Visual Com. and Image Representation, June 92.*

[5] P.Salembier, Adaptive rank order based filters, *Signal Processing, Vol 27, No1*, April 92.

[6] B.Widrow, S.D.Stearns, Adaptive signal processing, *Englewood Cliffs, NJ: Prentice Hall.*

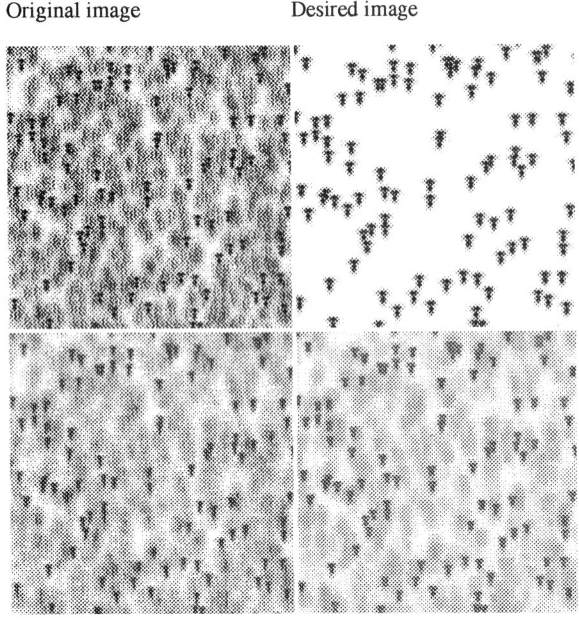

Figure 1: Noise cancellation with adaptive closing

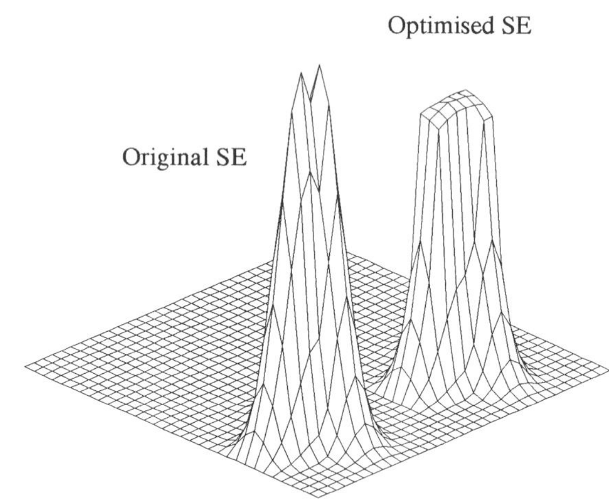

Figure 2: Struct. el. used in figure 1

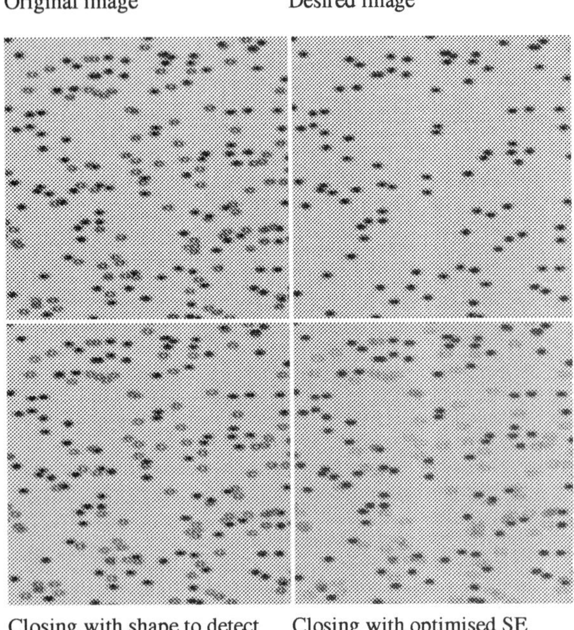

Figure 3: Shape detection with adaptive closing

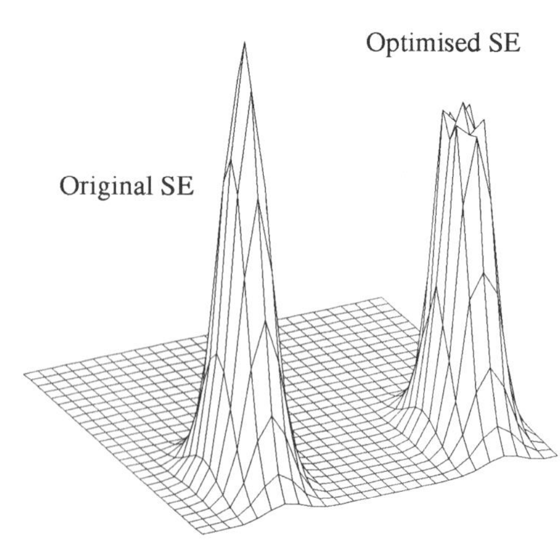

Figure 4: Struct. el. used in figure 3

SIGNAL PROCESSING VI: Theories and Applications
J. Vandewalle, R. Boite, M. Moonen, A. Oosterlinck (eds.)

Use of Bitonic Sorters for Rank Order Filtering

Lale Akarun, Richard A. Haddad and Cem Ersoy

Polytechnic University, Electrical Engineering Dept., 333 Jay Street, Brooklyn, NY 11201, USA

Bitonic sorters are widely used in sorting networks. We show that a new technique for rank order filtering, the decimated rank order filter (DROF), is closely related to bitonic sorters. The DROF can remove more impulses than a rank order filter of the same window size, with better edge performance and at lower computational cost.

1 Introduction

Rank order filters have been preferred to linear filters for impulsive noise removal from signals that contain sharp edges. The median filter is the most widely used rank order filter. It is noted for its ability to remove impulsive noise and to preserve sharp changes in the signal. However, there is a trade-off between noise removal capability and edge retention. When impulses occur in the neighborhood of edges, those edges are no longer preserved. Another trade-off is noise removal capability versus computational cost: complexity increases with the window size, and large window sizes are necessary for the removal of wide impulses.

The decimated rank order filter (DROF) [1] alleviates these problems: it improves noise removal while reducing edge smoothing and computational complexity. The DROF uses decimation to split long impulses into shorter ones which can then be removed by rank order filters of smaller window size. The small window sizes ensure good edge performance and low computational complexity. The decimators and the rank order filters are combined in a way that is similar to the filter-bank structure used in multirate signal processing [2].

The decimated sorter is obtained by replacing the rank order filters of the DROF by sorting elements. The decimated sorter, in general, does not sort any given sequence. However, under certain conditions, the operation performed by the decimated sorter is equivalent to sorting. We show that this condition is that the input signal be bitonic, i.e, a signal that consists of two sections, one, monotonically increasing, the other, monotonically decreasing. Therefore, the decimated sorter is a bitonic sorter. Bitonic sorters are well known in sorting networks [3], [4] and in switching [5], [6]. We show that the bitonic sorter of [5] is a special case of the decimated sorter.

Section 2 describes bitonic sorters, their structure and their areas of application. In Section 3, we introduce the decimated rank order filter and discuss its properties. In Section 4, we define the decimated sorter as an extension of the DROF. We show that the decimated sorter is equivalent to the bitonic sorter described in Section 2. Section 5 presents our conclusions.

2 Bitonic Sorters

A sequence of W numbers, $\langle x_1, \ldots, x_W \rangle$ is bitonic if it is first monotonically increasing up to some point x_m, $1 \leq m \leq W$, and monotonically decreasing after that point (or vice versa). Sorting a bitonic sequence is the same problem as merging two sorted lists: If the two sorted lists, one in increasing and the other in decreasing order, are juxtaposed, they form a bitonic sequence. Batcher devised a bitonic sorting network for merging two monotonic sequences into a single sorted sequence [5]. The structure of a bitonic sorter for $W = 8$ is shown in Fig. 1. In this structure, $W/2$ comparators (2-point sorters) are used to break the problem down. Batcher also proved that this structure is recursive, i.e, the $\frac{W}{2}$-item sorter in Fig. 1 is also a bitonic sorter. Large bitonic sorters are constructed from a number of smaller bitonic sorters: This modularity makes the architecture very suitable for VLSI applications. When the recursive structure is fully exploited and only comparators are used, $\log_2 W$ layers of $\frac{W}{2}$ comparators are needed. The depth of the network is $\log_2 W$. The comparators in each layer can operate in parallel, resulting in a delay of only $\log_2 W$ time units.

Cormen *et al.* [4] define a *comparison network* as a network consisting of wires and comparators. A *sorting network*, then, is defined as a comparison network whose outputs are sorted for every input sequence. The bitonic sorter is not a sorting network; it is a comparison network that sorts a limited class of inputs. However, since the bitonic sorter can merge two sorted lists, it is possible to construct a sorting network by recursively combining bitonic sorters [4]. The depth of a sorting network constructed in that way is $\log_2^2 n$.

A *permutation network* is a network that allows connection of its n inputs to its n outputs according to any

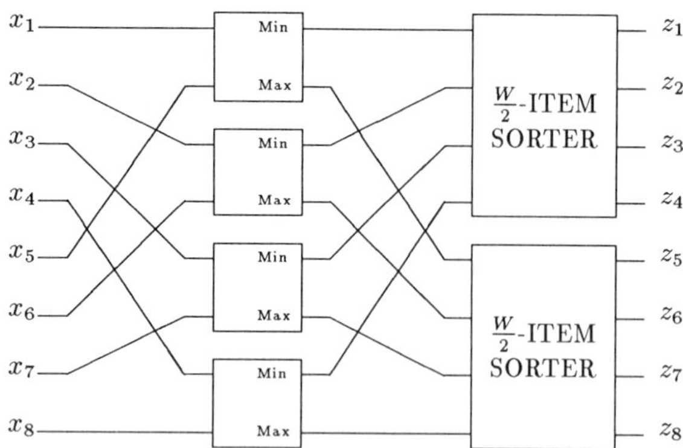

Figure 1: Construction of a bitonic sorter for W numbers ($W = 8$).

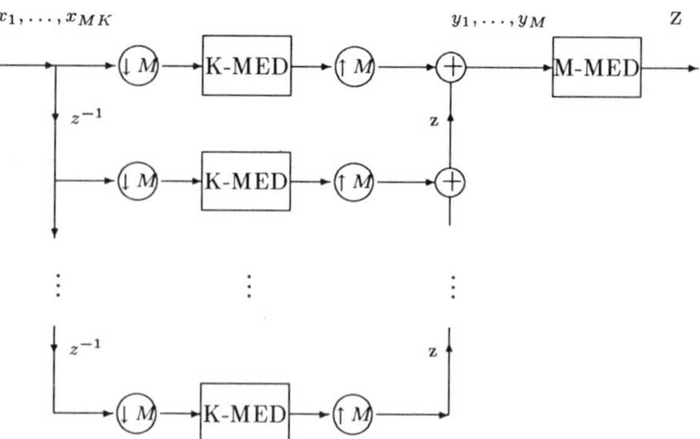

Figure 2: Decimated Median Filter for $W = Mk$ points

one of the $n!$ possible permutations. If each comparator in a sorting network is replaced with a 2-input, 2-output switch, a permutation network is obtained. Permutation networks are used for switching in ATM (asynchronous transfer mode) networks. Batcher [5] first proposed the use of bitonic sorters for switching. Hui [6] reviews the recent developments in this area.

3 Decimated Rank Order Filters

The output of a rank order filter $y_r = ROF\{x_1, \ldots, x_W\}$ is defined as the rth smallest among $\{x_1, \ldots, x_W\}$. For W odd, $r = (W-1)/2$ gives the median, whereas $r = 1$ and $r = W$ give the minimum and the maximum, respectively.

The median filter is the most commonly used rank order filter. It is preferred to linear filters when noise is of impulsive nature and the original signal contains sharp edges that are to be preserved. The noise reduction capability of a median filter increases with increasing window width. However, with non-constant background, wider window sizes also lead to smearing of details and edges in the signal. Another disadvantage of wide windows is the computational burden they bring. The decimated median filter removes a large number of impulses with better detail and edge retention than the standard median filter and with decreased computational cost.

3.1 Decimated Median Filter

We use decimation to deal with the problems induced by longer window lengths. The idea of linking decimation and rank order filtering arises from the following observation [7]: Large median filter windows are usually desired to remove impulses of longer length. However, long impulses can be split down into shorter ones by decimation. For example, if odd indexed samples and even indexed samples are separated to produce two individual subsequences, these sequences contain impulses of half the length of the impulse size in the original signal. The smaller impulses can then be removed by median filters of smaller window size, which do not cause the edge smearing problems the larger window sizes lead to. The overall computational complexity is also reduced. The structure of the Decimated Median Filter (DMF) is given in Fig. 2, where the output z is defined as follows:

$$\begin{aligned} z &= \text{Med}\{y_1, \ldots, y_M\} \\ y_j &= \text{Med}\{x_j, x_{j+M}, \ldots, x_{j+(K-1)M}\}, \qquad j = 1, \ldots, M \end{aligned} \tag{1}$$

where the sequence $\langle x_1, \ldots, x_W \rangle$ is obtained from the signal $x(n)$ by taking the samples inside a sliding window of $W = KM$ points centered at n. The decimators split up long impulses and the K-point median filters in the first layer remove them without detail and edge smoothing. The combining median filter in the second layer filters out any residual impulses. The K-point median filters in the first layer operate at a rate that is $(1/M)^{th}$ of the original sampling rate; therefore, each K-point filter operates once every M^{th} sampling instant. At any sampling instant, only one of the M filters is active, all others are idle. On the other hand, the M-point median filter in the second layer operates at the input sampling rate. This leads to a computational improvement factor, I of the DMF over the standard median:

$$I = \frac{M^2 K^2}{M^2 + K^2} \tag{2}$$

where the use of $O(n^2)$ algorithms is assumed. (This is a reasonable assumption since $n = W$ is small.)

The performance of the filter, on the other hand, is enhanced. The DMF removes wide impulses while leading to less edge shifting than the standard median filter. Moreover, the DMF has the potential for removing more impulses than the standard median filter if they occur in a certain pattern. The maximum number of impulses removed by the DMF, MIR, is:

$$\text{MIR} = \frac{(M-1)K}{2} + \frac{(M+1)(K-1)}{4} \qquad (3)$$

which is larger than $(KM-1)/2$, the maximum number of impulses removed by a standard median filter of the same window length, $W = KM$. The example in Fig. 3 illustrates these advantages. Fig. 3a shows a test signal that contains sharp edges and additive positive impulse noise. The results of filtering the noisy signal with a standard 9-point median filter (SM) and a DMF with $M = K = 3$ are given in Fig. 3b. It is seen that the DMF removes more impulses than the standard median filter. The DMF also leads to less edge shifting. Fig. 3c displays the errors produced by both filters. The mean absolute error (MAE) figures are 0.87 for the DMF and 1.18 for the SM. The computational improvement factor, as computed in Equation 2 is $I = 4.5$.

Since the filters considered are nonlinear, it is not possible to analyze their response to noise and predict their performance in removing noise from signals. A simplifying approach assumes that the signal will either be constant or a step edge in the given window. The output distribution is usually derived for these two cases under a white noise assumption [8]. The derivation of the output distribution of the DMF is simple for independent identically distributed (iid) inputs. The inputs of the K-point median filters in Fig. 2 are distinct; therefore, the outputs are also independent and identically distributed. The formulation of the output distributions for constant and step edge backgrounds have been detailed in [1]. Using those formulas, the output mse (mean square error) has been computed using numerical integration for iid noise with Laplacian distribution (Fig. 4). The filters compared are a DMF with $M = K = 3$, and a standard median filter with $W = 9$. The mse is plotted against edge height, h for a step edge at the middle of the window. When $h = 0$, the step edge is reduced to a constant and the filters perform similarly. For edge heights $h > 2$, the DMF shows a lower mse figure. This result agrees with our observation: The DMF performs similar to the SM in constant background regions but surpasses it near edges.

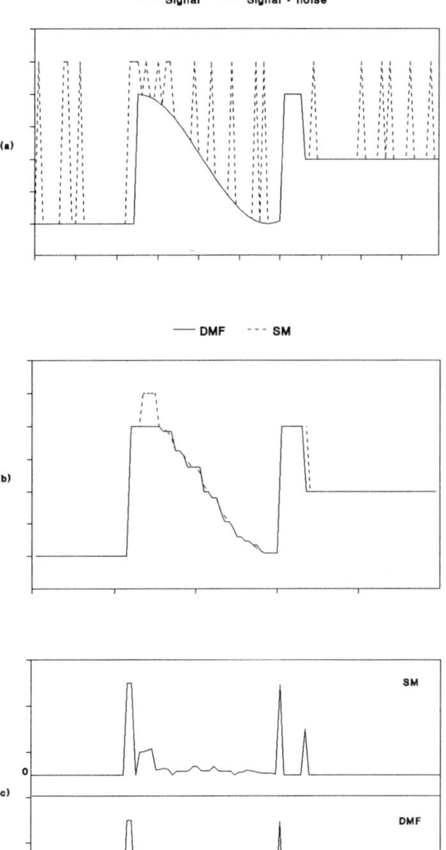

Figure 3: (a) An example signal with 15 % impulsive noise (b) The noisy signal as filtered by DMF ($M = K = 3$) and standard median (SM) filter ($W = 9$) (c) The error produced by each filter

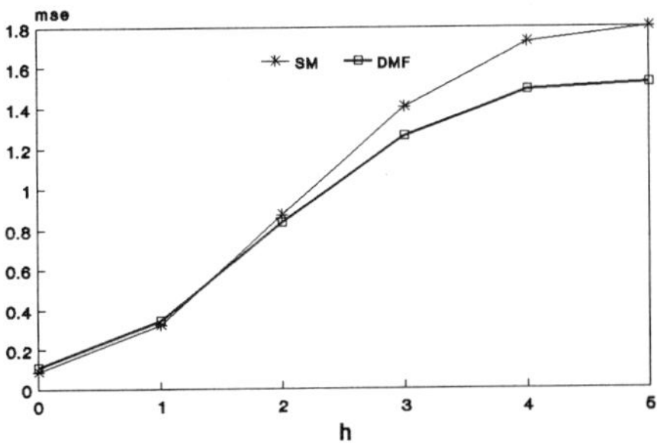

Figure 4: mse (mean square error) figures for SM and DMF for a step edge signal and Laplacian noise (0,1)

3.2 Decimated Rank Order Filter

To obtain decimated rank orders other than the median, we replace the median filters of Fig. 2 with other rank order filters. The output of the DROF, z is:

$$\begin{aligned} z &= \mu\{y_1, \ldots, y_M\} \\ y_j &= \kappa\{x_j, x_{j+M}, \ldots, x_{j+(K-1)M}\}, \qquad j = 1, \ldots, M \end{aligned} \qquad (4)$$

where (κ, μ) denote a pair of rank order filters, the former, K-input, and the latter, M-input. With this representation, (MED,med) represents the filter of Fig. 2.

If $M = K = 3$, it is possible to generate all decimated rank orders in a window of nine points with 3-point minimum, maximum and median operations. It is easy to see that (MIN,min) gives the minimum of the 9 points, or decimated rank order number one. Likewise, (MIN,-med), (MIN,max), (MED,min), (MED,med), (MED,-max), (MAX,min), (MAX,med), and (MAX,max) generate decimated rank orders 2 through 9.

4 Decimated Sorters

If the structure in Fig. 2 is generalized to obtain all decimated rank orders, the decimated sorter is obtained. The input to the decimated sorter is the sequence $\langle x_1, \ldots, x_W \rangle$. The input to the jth K-point sorter in the first layer is the sequence $\langle x_j, x_{j+M}, \ldots, x_{j+(K-1)M} \rangle$. The output $\langle y_r^j \rangle$ of the jth sorter is such that $y_1^j \le y_2^j \le \ldots \le y_K^j$. The output sequence $\langle z_1, \ldots, z_W \rangle$ consists of the juxtaposition of M subsequences that are obtained by sorting the M sequences $\langle y_r^1, \ldots, y_r^M \rangle$:

$$\begin{gathered} \{z_{(r-1)M+1}, \ldots, z_{rM}\} = \{y_r^1, \ldots, y_r^M\} \\ z_{(r-1)M+1} \le z_{(r-1)M+2} \le \ldots \le z_{rM}, \qquad r = 1, \ldots, K \end{gathered} \tag{5}$$

The sequences $\langle z_{(r-1)M+1}, \ldots, z_{rM} \rangle$ are obtained by taking the rth largest output from each of the K-point sorters and sorting them individually. For example, $\langle z_1, \ldots, z_M \rangle$ consists of the minima from each of the K-point sorters. Note that even though the subsequences $\langle z_1, \ldots, z_M \rangle, \ldots, \langle z_{(K-1)M+1}, \ldots, z_{KM} \rangle$ are sorted in themselves, $\langle z_1, \ldots, z_{KM} \rangle$ is not necessarily sorted. Thus, in general, the decimated sorter is not a sorting network.

Theorem 1: When the input sequence $\langle x_1, \ldots, x_W \rangle$ is bitonic, the output of the decimated sorter is sorted.

This theorem tells us that the decimated sorter is a bitonic sorter. Comparison of Fig. 2 and Fig. 1 indeed reveals that the two structures are identical when $K = 2$, $M = W/2$, and the medians are replaced by sorters.

Theorem 1 also enables us to see the relationship between the root signals of the DMF and those of the SM. The root signals are defined as signals that are unchanged by median filtering [9]. The root signals of a W-point median filter consist of constant neighborhoods of at least $(W+1)/2$ points separating monotonically increasing or decreasing edges. Inside any W-point window, such a signal can have at most three constant neighborhoods, and hence, only one minimum or maximum. Therefore, W samples of the root signals of a W-point median filter are bitonic. By theorem 1, the DMF is equivalent to the SM for a bitonic sequence. Therefore, the root signals of a median filter remain unchanged by decimated median filtering. We can state that the root signals of a median filter are also root signals of a DMF of the same window size.

Theorem 2: Given any bitonic input sequence, the output of the first level of the decimated sorter, $\langle y_r^l \rangle$ is also bitonic for any given r.

The proofs of both theorems are given in [1]. Theorem 2 implies that we can use a recursive structure for the decimated sorter; that is, we can replace the M-point sorter with another decimated sorter. If we pick K as small as possible ($K = 2$), we obtain the bitonic sorter of Fig. 1, in which case the depth of the whole network is $\log_2 W$. Alternatively, when larger values of K are used, the network is shallower but the computation at each level is more involved.

5 Conclusion

The DROF is very suitable for adaptive architectures due to its modular structure. It achieves better noise removing performance than the ROF at a lower computational cost. Parallel research in bitonic sorters gives valuable insights and suggests that the modularity can lead to sucessful VLSI implementations.

References

[1] L. Akarun and R. A. Haddad, "Decimated rank order filtering,". Submitted to the *IEEE Transactions on Signal Processing*.

[2] R. E. Crochiere and L. R. Rabiner, *Multirate Digital Signal Processing*. Englewood Cliffs, NJ: Prentice-Hall, 1983.

[3] D. E. Knuth, *Sorting and Searching*. Reading, MA: Addison-Wesley, 1973.

[4] T. H. Cormen, C. E. Leiserson, and R. L. Rivest, *Introduction to Algorithms*. Cambridge, Mass.: MIT Press, 1990.

[5] K. E. Batcher, "Sorting networks and their applications," in *AFIPS Proc. of Spring Joint Computer Conf.*, pp. 307–314, 1968.

[6] J. Hui, "Switching integrated broadband services by sort-banyan networks," *Proceedings of the IEEE*, vol. 79, pp. 146–154, February 1991.

[7] L. Akarun and R. A. Haddad, "Adaptive decimated median filtering," *Pattern Recognition Letters*, pp. 57–62, January 1992.

[8] T. A. Nodes and N. C. Gallagher, "The output distribution of median type filters," *IEEE Transactions on Communications*, vol. COM-32, pp. 532–541, May 1984.

[9] N. C. Gallagher and G. L. Wise, "A theoretical analysis of the properties of median filters," *IEEE Transactions on Acoustics, Speech and Signal Processing*, vol. ASSP-29, pp. 1136–1141, December 1981.

SIGNAL PROCESSING VI: Theories and Applications
J. Vandewalle, R. Boite, M. Moonen, A. Oosterlinck (eds.)

NOISE REDUCTION WITH CORRELATED SIGNALS USING HIGHER ORDER STATISTICS

Ch.SERVIERE - D.BAUDOIS

CEPHAG-ENSIEG
BP 46 38402 Saint-Martin d'Hères cedex
FRANCE

We study noise cancelling applications when observed references contain a part of signal. We use hypothesis of decorrelation and statistical independence between sources and cancel second-order and third-order moments of linear combinations of observations. These constrainsts lead to solve a second order equation where solutions represent complex gains of the linear filters to estimate. Applications are reduction of rotating machines noise and separation of random signals with non zero third-order moments.

1 Introduction

This paper deals with the problem of noise cancelling in presence of noise reference with non classical hypothesis. Noise cancelling using a noise reference has been first introduced by Widrow [1]. It consists in the estimation of an useful signal [of time sequence S(k)] from two observed signals. The first one is the noised observation [X(k)=S(k)+N(k)] where N(k) is an additive noise and the second one R(k), called "noise reference", is supposed to be correlated with N(k). S(k), N(k) and R(k) are assumed to be stationary random processes with zero-mean. Suppose the relation between N(k) and R(k) may be modelized by an unknown filter F. The solution is well-known [1] and consists first in the identification of F, by a FIR filter H. Values of coefficients of H are resulting from the minimization of a mean-square error. Their expressions only depend on second-order moments of observations X(k) and R(k). Many methods have been proposed to compute them for exemple with gradient methods [1] or lattice ones [2]. The estimation of N(k), is then obtained in filtering R(k) by H. [Figure 1].

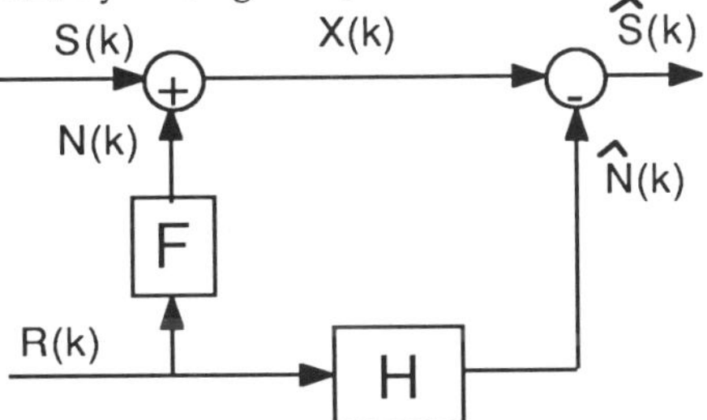

Figure 1 : Noise cancelling principle

H may be characterized in time-domain by its impulse response or in frequency-domain by its complex vector $\underline{h}$ composed of p complex weights.

(1) $\underline{h}^T = [\, h(1), ..., h(p)\,]$

S(k) and R(k) are supposed not to be correlated. Consequently the values of $\underline{h}$ are resulting from the minimization of a mean-square error (2) at data block m and for each frequency bin i :

(2) $$E\left\{\left|X_m(i) - h(i) . R_m(i)\right|^2\right\}$$

where $X_m(i)$ and $R_m(i)$ are obtained by p-point discrete Fourier transforms (DFT) of the m-th data block of the signals X(k) and R(k), for the i-th frequency bin. We assume that $X_m(i)$ and $R_m(i)$ are zeromean signals.

The solution, called "Wiener filter", is computed with the help of second-order moments. (3)

$$h(i) = \frac{E\left\{X_m(i).R^*_m(i)\right\}}{E\left\{\left|R_m(i)\right|^2\right\}} = \frac{\gamma_{XR}(i)}{\gamma_R(i)}$$

where $*$ represents the complex conjugate.

The identification of F by H is then perfect and the two complex vectors $\underline{f}$ and $\underline{h}$ are equal. In practice, h (i) is estimated at data block m, by using average periodogram methods.

2 Noise cancelling in presence of correlated inputs

In many applications, the noise reference R(k) is not accessible and a part of the useful signal S(k) is also recorded by the reference sensor. We observe then a signal Y(k) such that Y(k) is correlated with S(k). The previous method (3) replacing R(k) by Y(k) is no more usable because the two complex vectors $\underline{f}$ and $\underline{h}$ will not be equal. Performances of the estimation of S(k) will decrease and the degradation is studied in [3].

The new noise cancelling system leads to a problem of sources separation in which we generally modelized the contribution of the signal to Y(k) by an unknown linear filter G [Figure 2].[4] [5] [6] [7]

The noise cancelling system is then modelized by two inputs X(k) and Y(k) such that we have in frequency domain at data block m and for each frequency bin i :

(4) $X_m(i) = S_m(i) + f(i)\, R_m(i)$
$Y_m(i) = R_m(i) + g(i)\, S_m(i)$

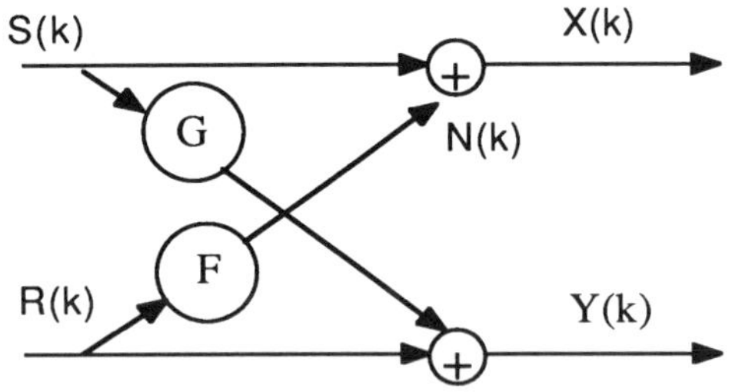

Figure 2 : Noise cancelling system

We now face with a problem of sources separation and have two linear filters F and G to identify, which means two complex values f(i) and g(i) to identify for each frequency bin i. The proposed approach is similar to [4] [5] [6] [7].

Define first two linear combinations (5) of $X_m(i)$ and $Y_m(i)$.

(5) $X'_m(i) = X_m(i) - \alpha(i)\, Y_m(i)$
$Y'_m(i) = Y_m(i) - \beta(i)\, X_m(i)$

or

$X'_m(i) = S_m(i)\, [1- \alpha(i).g(i)] + R_m(i)\, [f(i)- \alpha(i)\,]$

$Y'_m(i) = Y_m(i)\, [g(i) - \beta(i)\,] + R_m(i)\, [1-\beta(i)\, f(i)]$

The proposed method consists in the identification of two complex coefficients $\alpha(i)$ and $\beta(i)$ such that each of the linear combinations (5) only depends on the signal $S_m(i)$ or the reference $R_m(i)$. Suppose $\alpha o\ (i)$ and $\beta o\ (i)$ solutions, we have :

(6) $\alpha o\ (i) = f(i)$ and $\beta o\ (i) = g(i)$
or $\alpha o\ (i) = 1\ /\ g(i)$ and $\beta o\ (i) = 1\ /\ f(i)$

We remark that two couples of solutions for $\alpha o(i)$ and $\beta o(i)$ exist because the problem is totally symmetrical with regards to $S_m(i)$ and $R_m(i)$.

3 Case of signals with non zero third-order moments

We suppose that real and imaginary parts of the complex variables $X_m(i)$ or $Y_m(i)$ have non even probability densities. The third-order moments will generally not be zero. The proposed hypothesis is verified for exemple when one source contains sinusoids which is the case for reduction of noise of rotating machine or for particular random noises detailed in part 3.2.

3-1 Case of sinusoids in one source

We suppose that one source, S(k) for exemple, is the sum of a random noise Ns(k) and a sinusoid of determinist phase ϕs and determinist frequency ωs. Ns(k) and R(k) are supposed to be random noises, third-order stationary and zero-mean. No hypothesis is made on their probability densities.

(7) $S(k) = As \sin(2\pi\, \omega s\, k + \phi s) + Ns(k)$

R(k) is also supposed to be statistically independent of Ns(k).

The expression of the p-point discrete Fourier Transform of X(k) on data block m is following :

$$X_m(i) = \frac{As}{2j} \exp j\left[\phi s + 2\pi\omega s\, mp + \pi(p-1)(\omega s - \frac{i}{p})\right] \frac{\sin(\pi(\omega s - \frac{i}{p})p)}{\sin(\pi(\omega s - \frac{i}{p}))}$$

$$- \frac{As}{2j} \exp -j\left[\phi s + 2\pi\omega s\, mp + \pi(p-1)(\omega s + \frac{i}{p})\right] \frac{\sin(\pi(\omega s + \frac{i}{p})p)}{\sin(\pi(\omega s + \frac{i}{p}))}$$

$$+ \sum_{k=mp}^{k=(m+1)p-1} Ns(k) \exp -2\pi jk\frac{i}{p} + f(i)\, R_m(i)$$

We assume that p is large enough in order to approximate ωs by k/p with k integer and we propose us to separate the sources for frequency bin i such that : (9) $S_m(i) = A(i) + Ns_m(i)$ with non-zero A(i).

Consequently observations at data block m are :

(10) $X_m(i) = A(i) + Ns_m(i) + f(i)\, R_m(i)$
$Y_m(i) = R_m(i) + g(i)\, (A(i) + Ns_m(i))$

We assume that $Ns_m(i)$ and $R_m(i)$ are statistically independent and zero-mean.

We search $\alpha(i)$ and $\beta(i)$ such that the two linear combinations $[X_m(i) - \alpha(i)\, Y_m(i)]$ and $[Y_m(i) - \beta(i)\, X_m(i)]$ only depends on signal $S_m(i)$ or reference $R_m(i)$.

In these conditions, the two linear combinations are uncorrelated (11) :

$$E\left\{\left[X_m(i) - \alpha(i).Y_m(i)\right]\left[Y_m(i) - \beta(i).X_m(i)\right]^*\right\} = 0$$

We verify that $\alpha o(i)$ and $\beta o(i)$ are solutions of (11) with particular form of signal $S_m(i)$ (9).

Unfortunately this equation is not sufficient to find the researched solution $\alpha o(i)$ and $\beta o(i)$. It only allows to obtain a family of orthogonal signals $X'_m(i)$ and $Y'_m(i)$. We will find a complementary information at third-order. Suppose $\alpha o(i)=1/g(i)$ and $\beta o(i)=1/f(i)$. Then the linear combinations $X'_m(i)$ and $Y'_m(i)$ are :

(12) $X'_m(i) = (f(i) - 1/g(i))R_m(i)$
$Y'_m(i) = (g(i) - 1/f(i))\, (A(i) + Ns_m(i))$

Then the third-order crossed moment beween $X'_m(i)$ and $Y'_m(i)$ is assumed to be zero because $R_m(i)$ and $Ns_m(i)$ are supposed to be statistically independent and zero-mean (13).

$$E\left\{\left[X_m(i) - \alpha(i).Y_m(i)\right]\left|Y_m(i) - \beta(i).X_m(i)\right|^2\right\} = 0$$

We will prove that these two equations are

sufficient to compute coefficients $\alpha o(i)=1/g(i)$ and $\beta o(i)=1/f(i)$ which are solutions of system (11), (13). Develop equations (11) and (13) in function of $\alpha(i)$ and $\beta(i)$: (14)

$$\alpha(i)\beta^*(i)\gamma_{YX}(i) - \alpha(i)\gamma_Y(i) - \beta^*(i)\gamma_X(i) + \gamma_{XY}(i) = 0$$

(15)

$$- \alpha(i)|\beta(i)|^2 E\{Y_m(i)|X_m(i)|^2\} + \alpha(i)\beta^*(i)E\{X_m^*(i)[Y_m(i)]^2\}$$
$$+ \alpha(i)\beta(i)E\{X_m(i)|Y_m(i)|^2\} - \alpha(i)E\{Y_m(i)|Y_m(i)|^2\}$$
$$- \beta(i)E\{Y_m^*(i)[X_m(i)]^2\} - \beta^*(i)E\{Y_m(i)|X_m(i)|^2\}$$
$$+ |\beta(i)|^2 E\{X_m(i)|X_m(i)|^2\} + E\{X_m(i)|Y_m(i)|^2\} = 0$$

Replace in (15) $\alpha(i)$ by its expression in (14). We obtain the equivalent system (16) (17) where equation (17) only depends on variables $\beta(i)$ and $\beta^*(i)$: (16)

$$\alpha(i)\beta^*(i)\gamma_{YX}(i) - \alpha(i)\gamma_Y(i) - \beta^*(i)\gamma_X(i) + \gamma_{XY}(i) = 0$$

$$\text{(17)} \quad \begin{aligned} &A(i) + B(i)\beta(i) + C(i)\beta^*(i) + D(i)|\beta(i)|^2 + \\ &E(i)(\beta^*(i))^2 + F(i)\beta^*(i)|\beta(i)|^2 = 0 \end{aligned}$$

where complex coefficients A(i), ..., F(i) are linear combinations of second and third-order moments of signals $X_m(i)$ and $Y_m(i)$:

(18)

$$A(i) = \gamma_{XY}(i).E\{Y_m(i)|Y_m(i)|^2\} - \gamma_Y(i).E\{X_m(i)|Y_m(i)|^2\}$$
$$B(i) = \gamma_Y(i).E\{Y_m^*(i)[X_m(i)]^2\} - \gamma_{XY}(i).E\{X_m(i)|Y_m(i)|^2\}$$
$$C(i) = \gamma_{YX}(i).E\{X_m(i)|Y_m(i)|^2\} + \gamma_Y(i).E\{Y_m(i)|X_m(i)|^2\}$$
$$- \gamma_X(i).E\{Y_m(i)|Y_m(i)|^2\} - \gamma_{XY}(i).E\{X_m^*(i)[Y_m(i)]^2\}$$
$$D(i) = - \gamma_{YX}(i).E\{Y_m^*(i)[X_m(i)]^2\} + \gamma_{XY}(i).E\{Y_m(i)|X_m(i)|^2\}$$
$$+ \gamma_X(i).E\{X_m(i)|Y_m(i)|^2\} - \gamma_Y(i).E\{X_m(i)|X_m(i)|^2\}$$
$$E(i) = \gamma_X(i).E\{X_m^*(i)[Y_m(i)]^2\} - \gamma_{YX}(i).E\{Y_m(i)|X_m(i)|^2\}$$
$$F(i) = \gamma_{YX}(i).E\{X_m(i)|X_m(i)|^2\} - \gamma_X(i).E\{Y_m(i)|X_m(i)|^2\}$$

Equation (17) may be factorized in (19) :

$$\text{(19)} \left[\beta(i) - a_1(i)\right]\left[\beta^*(i) - a_2(i)\right]\left[\beta^*(i) - a_3(i)\right] = 0$$

where $a_1(i)$, $a_2(i)$ and $a_3(i)$ are the three complex roots of equation (17).

By identifying equations (17) and (19), we obtain the value of the first root $a_1(i)$.

$$\text{(20)} \quad a_1(i) = - \frac{E(i)}{F(i)}$$

Value of $a_1(i)$ may be computed by replacing $X_m(i)$ and $Y_m(i)$ in function of A(i), $Ns_m(i)$, $R_m(i)$, f(i) and g(i). We also use hypothesis of independence and zero-mean of signals $Ns_m(i)$ and $R_m(i)$.

We then remark that root $a_1(i)$ is of no interest because it depends on both values g(i) and f(i). On the contrary we will prove that one of the two other roots is exactly equal to $\beta o(i)=1/f(i)$.

By identifying (17) and (19), we also remark that roots $a_2(i)$ and $a_3(i)$ may be computed by solving an equation of second order in $\beta^*(i)$ (21) :

$$\text{(21)} \quad F(i)(\beta^*(i))^2 + D(i)\,\beta^*(i) + B(i) = 0$$

where coefficients F(i), D(i) and B(i) are those of equations (18) and only depend on observations $X_m(i)$ and $Y_m(i)$. By replacing $X_m(i)$ and $Y_m(i)$ by their expressions in function of A(i), $Ns_m(i)$, $R_m(i)$, f(i) and g(i), we may prove that the two roots are (22).

$$\begin{aligned} a2(i)=[&A(i)(g(i))^*\gamma_R(i)\gamma_{Ns}(i)+A(i)f(i)(\gamma_R(i))^2 \\ &+(g(i))^*\gamma_R(i)E\{Ns_m(i)|Ns_m(i)|^2\}- \\ &|A(i)|^2(g(i))^*f(i)E\{R_m(i)|R_m(i)|^2\}- \\ &(g(i))^*f(i)\gamma_{Ns}(i)E\{R_m(i)|R_m(i)|^2\}]/ \\ [&\gamma_R(i)\gamma_{Ns}(i)A(i)+A(i)|f(i)|^2(\gamma_R(i))^2+ \\ &\gamma_R(i)E\{Ns_m(i)|Ns_m(i)|^2\}- \\ &|A(i)|^2 f(i)E\{R_m(i)|R_m(i)|^2\}- \\ &f(i)\gamma_{Ns}(i)E\{R_m(i)|R_m(i)|^2\}] \end{aligned}$$

(23)

$$a_3(i) = \frac{1}{f^*(i)}$$

By replacing previous roots in (16), we obtain two couples of solutions for $\alpha(i)$ and $\beta(i)$ ($\alpha 2(i)$, $\beta 2(i) = a_2^*(i)$) and ($\alpha 3(i)$, $\beta 3(i) = a_3^*(i)$). The two couples of solutions verify system (11),(13) but only the second one is able to separate sources. We distinguish the two solutions by adding an other constrainst which is (24) : $E\{ Xm(i) - a(i)\, Ym(i) \} = 0$

The unique solution of the new system of equations (11), (13) and (24) is the researched solution : $\alpha o(i)=1/g(i)$ and $\beta o(i)=1/f(i)$

Consequently, values of the two complex vectors f(i) and g(i) may be easily computed by solving the equation of second order in $\beta^*(i)$ (21).

In practice the coefficients B(i), D(i) and F(i) of equation (21) may be computed by averages on data blocks.

We treated the case where signal S(k) contains one sinusoid. The problem is similar with regards to both sources. So if reference R(k) contains one sinusoid, equation (21) is similar but one of the roots will be equal to $f^*(i)$ instead of $1/f^*(i)$.

If one source contain several sinusoids of different frequencies, the problem will be equivalent at previous case and will be treated independently on each frequency bin.

3.2 Case of random signals with non zero third-order moments

We suppose now that S(k) and R(k) are random noises. In several cases one source, $S_m(i)$ or $R_m(i)$, may have non-zero third-order moments although p-point discrete Fourier transforms

$S_m(i)$ or $R_m(i)$ are generally supposed to lead to Gaussian complex variables if p is large enough (central limit theorem see [8]). Consider for exemple S(k) as the sum of sinusoid periods modulated by a random variable A with non even probability density and zero-mean :
at data block m, we have S(k) = A(m) exp(2pjki/p)
Consequently : $S_m(i) = p.A(m)$
If A(m) is a random variable with non even probability density, non-zero third-order moment and zero-mean, so is $S_m(i)$. This kind of signals is often used in communication problems. Remark that they cannot be separated by existing methods such that [4] [5] which attempt to cancel expressions like E{f(X'm(i))g(Y'm(i))} with f and g odd : these methods necessary need the assumption that signals have even probability densities. Others methods using informations at fourth-order (moments or cumulants) are able to separate them [7] [9] but are much more complicated than the following proposed method.
As in part 3.1, we search $\alpha o(i)$ and $\beta o(i)$ such that the two linear combinations $[X_m(i)-\alpha(i)Ym(i)]$ and $[Y_m(i)-\beta(i)Xm(i)]$ are uncorrelated and independent. Consequently $\alpha o(i)$ and $\beta o(i)$ are non unique solutions of (11) and (13).

$$E\left\{\left[X_m(i)-\alpha(i).Y_m(i)\right].\left[Y_m(i)-\beta(i).X_m(i)\right]^*\right\}=0$$

$$E\left\{\left[X_m(i)-\alpha(i).Y_m(i)\right].\left|Y_m(i)-\beta(i).X_m(i)\right|^2\right\}=0$$

They are also solutions of (16) and (17) :

$$\alpha(i)\beta^*(i)\gamma_{YX}(i)-\alpha(i)\gamma_Y(i)-\beta^*(i)\gamma_X(i)+\gamma_{XY}(i)=0$$

$$A(i)+B(i)\beta(i)+C(i)\beta^*(i)+D(i)|\beta(i)|^2+$$
$$E(i)(\beta^*(i))^2\ F(i)\ \beta^*(i)|\beta(i)|^2=0$$

where the complex coefficients A(i), ..., F(i) are developped in (18) (part 3.1). By replacing X and Y in function of $S_m(i)$ or $R_m(i)$ we remark that roots of (21) are equal to g*(i) and 1/f*(i).

(21) $F(i)\ (\beta^*(i))^2 + D(i)\ \beta^*(i) + B(i) = 0$

Coefficients F(i), D(i) and B(i) are detailed in (18).

4 Application

We validate proposed methods with following simulations. We show on figure 3 the evolution of the estimation of real and imaginary parts of complex value f(i) in function of data blocks m. Blocks own 64 samples and are 50 per cent overlapped. Signal contains a sinusoid in frequency bin i, Ns(k) and R(k) are white gaussian noises. We remark that filter converges to right values (0,35;-0,6) after 2000 samples. Similarly, we see on figure 4 the evolution of values f(i) and g(i). Signal and reference are white gauusian noises. We save positive values of real parts and center signals. We remark that filters converge to right values (1;0,7) and (0;2) after 3000 samples.

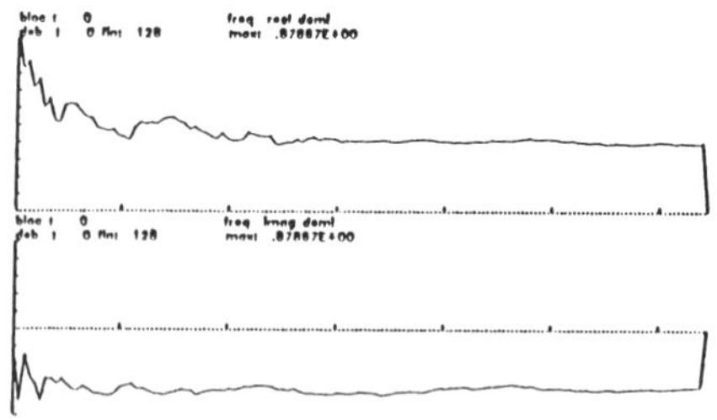

Figure 3 Estimation of coefficient f(i)

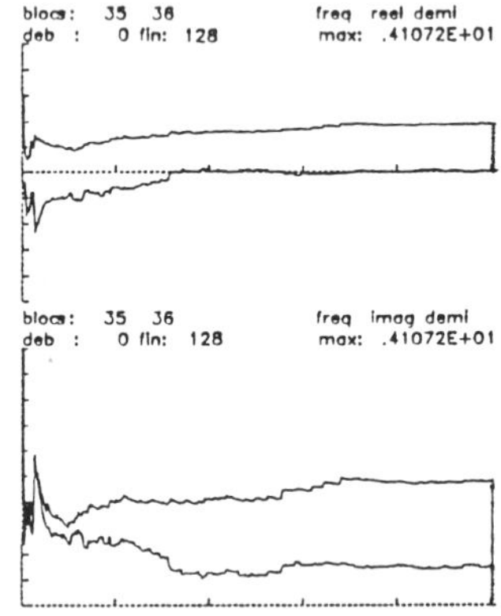

Figure 4 Estimation of coefficients f(i) and g(i)

5 Conclusion

We study noise cancelling applications when observed reference contains a part of signal. The proposed method needs non-zero third-order moments of sources, using hypothesis of statistical independence. It is easy to apply and finally consits in solving a second order equation in frequency-domain. Applications are reduction of rotating machines noise and separation of random signals with non zero third-order moments (communication problems).

References

[1] Widrow and al "Adaptive noise cancelling : principles and applications", Proc IEEE, vol 63, 1975

[2] Friedlander "Lattice filters for adaptive processing", Proc.IEEE, vol 70, pp829, August 1982

[3] Widrow, Stearns " Adaptive signal processing ", Prentice Hall, p311

[4] Jutten, Herault "Blind separation of sources, Part I : An adaptive algorithm based on neuromimetic architecture", vol.24, n°1, July 1991, Signal Processing

[5] Comon, Jutten, Herault "Blind separation of sources, Part II : Problems statement", vol 24, n°1, July 1991, Signal Processing

[6] Sorouchyari "Blind separation of sources, Part III, vol 24, n°1, July 1991, Signal Processing

[7] Comon "Higher-order separation : application to detection and localization ", EUSIPCO 90, Barcelona, Sept. 18-21, 1990, pp 277

[8] Brillinger "Time series data analysis and theory", 1981 Holden-day, p94

[9] Gaeta, Lacoume "Source separation vs hypothesis", International workshop on higher order statistics, July 1991, Chamrousse

SIGNAL PROCESSING VI: Theories and Applications
J. Vandewalle, R. Boite, M. Moonen, A. Oosterlinck (eds.)
1992 Elsevier Science Publishers B.V.

ON THE USE OF INSTANTANEOUS PHASE VARIANCE FOR CLUTTER SUPPRESSION

Antonio Albiol (*), Luis Vergara (**)

(*) Departamento de Comunicaciones, ETSI Telecomunicación, Universidad Politécnica de Valencia, Camino de Vera s/n, 46022 Valencia, SPAIN.

(**) ETSI Telecomunicación, Universidad Politécnica de Madrid, Ciudad Universitaria s/n, 28040 Madrid, SPAIN.

1.INTRODUCTION

This paper presents a method to suppress the clutter corrupting a desired signal. In radar this is mainly done by considering that the desired target has a relatively high speed and can thus be detected by Doppler processing. However in the case of stationary targets this method can no longer be used. For example in ultrasonics, the same problem arises when a flaw or defect enclosed by a strongly scattering medium such as grainy metal, has to be detected. In this case Doppler processing can not be used because the desired target (flaw) has no motion.

The signal to clutter ratio can be improved by compounding statistically independent echoes from the same region of the medium. These independent echoes can be obtained via frequency diversity [1],[2], or spatial diversity scanning the transducer. Usually only envelope information of the involved signals is considered.

In this paper we present a new processor to improve the detection of flaws using ultrasounds which makes use of both the envelope and the instantaneous frequency information of the data record.

The stationary clutter problem is very important in non-destructive evaluation of materials. To prove the validity of the method, it has been applied to real ultrasonic signals.

2.SIGNAL MODEL

2.1 Grain noise

In [4] the grain clutter has been modelled to be gaussian. This is based on the central limit theorem and the fact that in the resolution cell there is a very high number of scattering centres (grains). In [3], this model is revised for the cases where it is not possible to consider such a high number of scattering centres in the resolution cell, proposing a K-distribution for the envelope in those cases. In this paper gaussian statistics for the clutter are to be considered. This implies a Rayleigh distribution for the envelope and a uniform distribution for the instantaneous phase as well as independence between envelope and phase for the clutter statistics. The method would be applicable also to K-distributed clutter although a detailed mathematical justification would be much more complicated.

The transmitted waveform is normally of the form

$$s(t) = p(t) \cdot e^{j\omega_0 t} \tag{1}$$

where $p(t)$ is a short duration envelope, and ω_0 is the carrier frequency. The echo received from the grainy structure of the material is the convolution of the reflectivity function with the transmitted pulse. If we assume that the grains are much smaller than the pulse duration and the beamwidth then the reflectivity can be considered a white random process. The received power spectral density of the grain noise is then

$$\Phi_n(\omega) = K \cdot |P(\omega - \omega_0)|^2 \tag{2}$$

So, when only grain noise is present, the received signal can be considered a gaussian band-pass random process which has a power spectral density proportional to the square of the Fourier transform of the transmitted pulse.

2.2 Signal present

When a flaw or concentrated scatterer produces an echo, the received signal is

$$a \cdot p(t-t_0) \cdot e^{j\omega_0 (t-t_0)} + n(t) \tag{3}$$

where a is the amplitude associated with the flaw echo, t_0 is the delay related to its range, and $n(t)$ is the grain noise. We can see that locally, when a flaw is present, we have a sinusoid added to the grain noise. It is interesting to notice that a matched filter is of no use to improve the SNR in this case since the desired signal and noise share the same frequency band.

In the case of a target being present the pdf of the envelope is Rician, and the phase pdf is [5]:

$$f_{\psi}(\psi) = \frac{1}{2\pi}\exp(-u) + \tag{4}$$

$$+\sqrt{\frac{u}{\pi}}\cos\psi\,\exp(-u\sin^2\psi)\left[1-\frac{1}{2}\mathrm{erfc}(u\cos\psi)\right]$$

with

$$u = \frac{a^2}{2\sigma_N^2}$$

where a is the amplitude of the target echo and σ_N^2 is the variance of the in-phase and quadrature clutter components.

Just as an initial justification of the method, we show in the next figure the variation of the envelope mean as well as the inverse of the phase variance as functions of the SNR ($s = a/\sigma$). The variations are relative to the corresponding values when no desired signal is present.

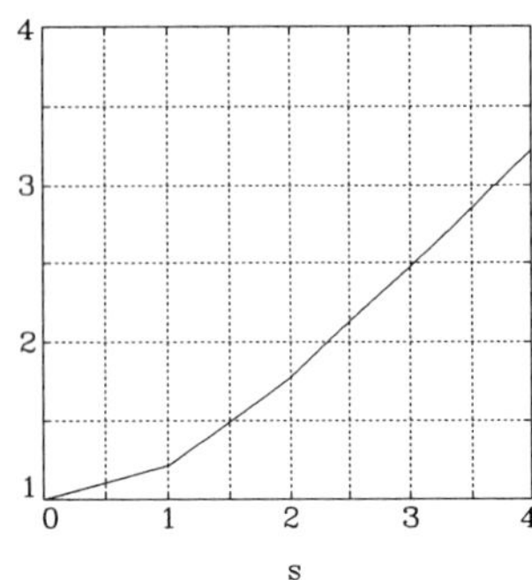

Fig. 1-a: Variation of envelope mean with s, relative to its value when no signal is present (s = 0).

As can be seen from the curves in fig. 1 the inverse of the phase variance increases with s faster than the mean of the envelope. This is specially interesting for very low signal to clutter ratios and is the basic idea to be exploited in the proposed processor. However, care must be taken with the above curves, since the variance of the instantaneous phase is a quadratic parameter, and the envelope mean is a linear one. If we take this into account, we can observe that the information contained in the phase is at least comparable to that present in the envelope. What we propose in

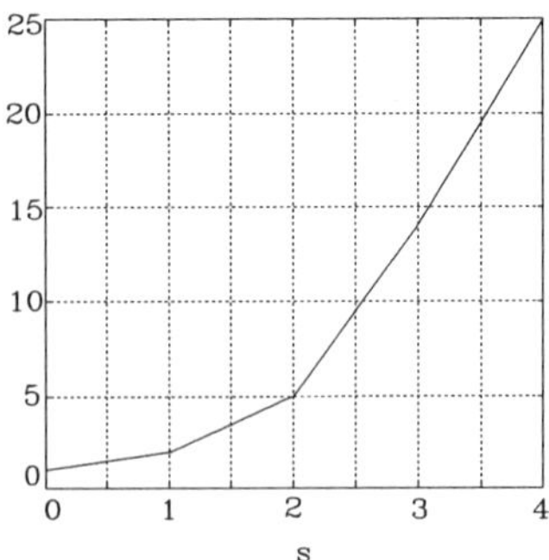

Fig. 1-b: Variation of phase variance inverse with s, relative to its value when no signal is present (s = 0).

this paper is to make use of the phase information as well as the envelope information to improve the SNR and the probability of detection.

The use of instantaneous phase as a detecting parameter has however practical problems. The instantaneous phase mentioned above is the one obtained from the analytic signal. A small error in the estimation of the central frequency produces an error that increases with time in the estimated instantaneous phase. We have used the instantaneous frequency (i.f.) as an alternative. Since all the signal processing has been discrete, the finite phase difference rather than the actual i.f. has been used. Figure 2 shows the variation of the inverse of i.f. variance relative to its value when no signal is present.

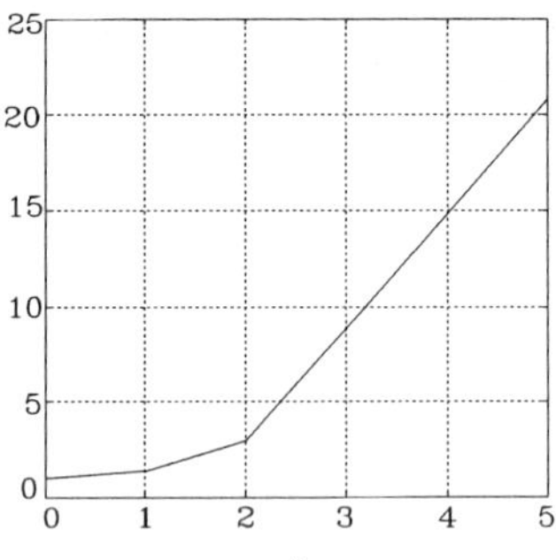

Fig. 2: Variation of inverse of phase difference variance with s, relative to its value when no signal is present (s = 0).

We can also see that the variance of the i.f. is a parameter which can be used to detect flaws in grain noise.

3.PROPOSED PROCESSOR

First of all the signal is band-pass filtered to reject unwanted out-of-band components, which may be due to thermal noise or very low frequency fluctuations. Processing of the signals begins with the computation of the analytic signal. This is done using a FIR filter of length 51. From the analytic signal (a complex signal), we obtain the envelope as the magnitude and the phase as the angle. The phase obtained from the analytic signal (in the interval $[-\pi,\pi]$), is then unwrapped using an algorithm which removes the 2π jumps in the phase. The i.f. is obtained by first-differencing the unwrapped phase and dividing by 2π. The mean of the i.f., f_0, is obtained based in all the register length. Finally the squared error with respect to the mean is obtained. The process is summarized in figure 3.

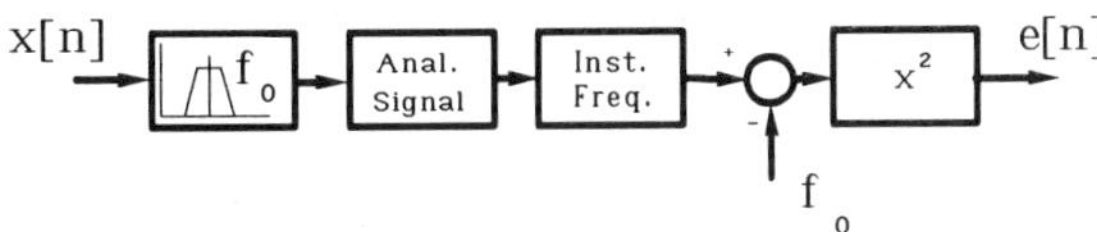

Fig. 3: Summary of processing of the received echo.

The underlying idea of the processor is to exploit the fact that when a flaw is present, locally, the envelope mean increases and also the i.f. variance decreases.

These can be obtained from within the same record, since the variance will be low during all the pulse duration. But since the pulse amplitude varies, so will the variance vary. Also the i.f. is not a white signal, so the samples involved in the estimation will not be independent. This all leads to the conclusion that the variance estimate obtained by means of averaging the squared i.f. error in a window, is a biased estimate of the actual variance. Nevertheless it can still be useful. The window length is the only parameter of this method and the optimum value depends on the signal bandwidth of the RF signal, the envelope shape and the sampling frequency; we have found that a number of samples corresponding to approximately half duration of the transmitted pulse (approximately gaussian-shaped envelopes were used for the transmitted pulse) is a good choice. The output signal is the inverse of the local variance estimate; if we want to take into account both, the envelope and i.f. information, we can form the quotient between envelope and local variance; in this way both the envelope and phase information carried by the received signal are exploited by the algorithm.

Another way for getting the different required samples to estimate the i.f. variance is from different signal records (we will call them scans). When ultrasonic non-destructive evaluation is being carried out the transducer is scanned along a certain line. Due to the finite beamwidth a certain point in the material is illuminated from different locations getting independent echoes in each scan. If we have different scans, we can on one hand average the envelope and on the other hand estimate the variance by averaging the squared i.f. error in different scans. However it has been seen that the use of an ordered statistics filter improves the probability of detection with respect to the use of the mean squared error. The best statistic has been found to be the maximum squared error.

4.RESULTS

This section shows some of the results obtained with the proposed method applied to real ultrasonic signals. The material under test was a block of austenitic steel. The defect to detect was a cut in the backside of the block under test. A 45 degrees incidence angle was used to avoid the backside specular echo. The central frequency transmitted was about 1 Mhz.; the 6 dB bandwidth was 330 Khz., And the propagation velocity was 5960 m/s. The sampling frequency was 6.25 Mhz.

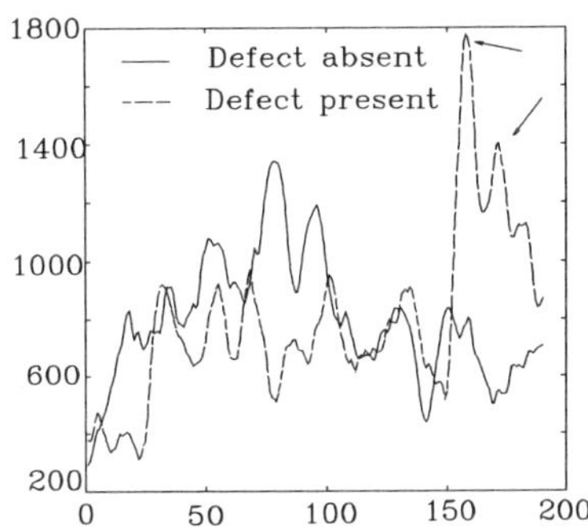

Fig. 4-a : Average of the envelope of nine consecutive scans, with and without defect.

The functional of the i.f. variance considered for figures 4 has been the mean squared frequency error. From figures 4, the detection possibilities present in the phase information become apparent and are at least as good as those of the envelope. Notice also, that if both informations, envelope end phase are somehow combined (dividing the envelope by the variance estimate for instance), the SNR can be further improved.

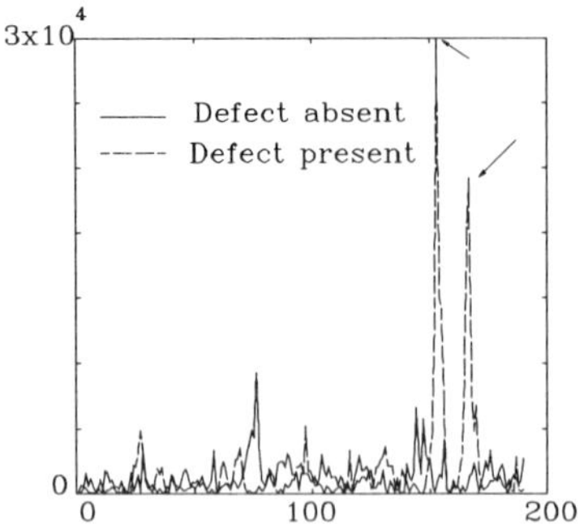

Fig. 4-b : Inverse of variance estimate using nine consecutive scans, with and without defect. The arrows indicate the position of the flaw. The two peaks are due to the geometry of the defect which had two orthogonal sides, behaving each as a scattering centre.

ROC Curves (figure 5) for the detector based in the i.f. variance have been obtained and are compared with those of an envelope detector. The functional of the i.f. error considered has been an ordered statistics filter. Figure 5-a shows the curve for the median (4:7) and the maximum (7:7) filter. We see the better behaviour of the maximum filter; this is due to the fact that it is in the region of large errors where the probability distributions with flaw present and absent are most different. In this figure we can see that the detector behaviour is similar to that of a conventional envelope detector. This shows that discarding phase information means discarding a lot of information. If information from both envelope and phase were combined the performance of the detector could be improved with respect to the case of using only envelope information.

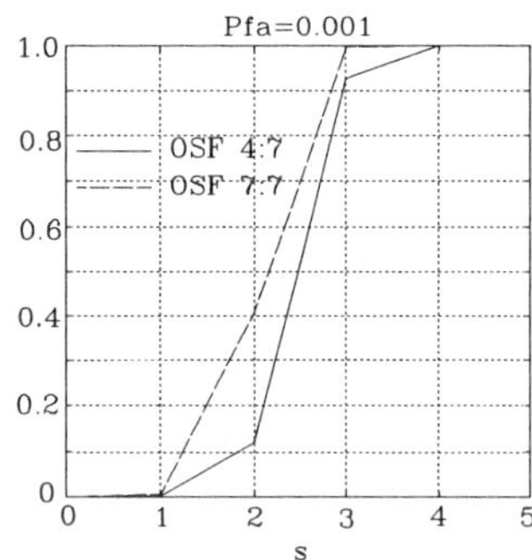

Fig. 5-a : Probability of detection using 7 scans and i.f. error for different SNR (s)and a probability of false alarm of 0.001.

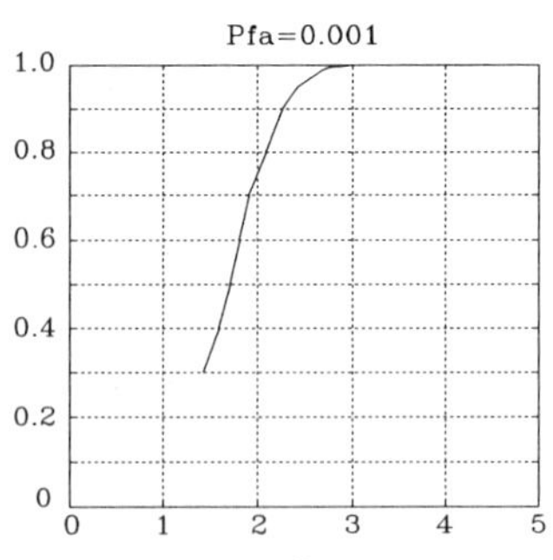

Fig. 5-b : Probability of detection using 7 scans and envelope only (Incoherent integration) for different SNR (s)and a probability of false alarm of 0.001.

5.CONCLUSIONS

A new method for the detection of signals in clutter, which makes use of the instantaneous phase information of the received signals has been developed. The method has been succesfully applied to ultrasonic B-scans signals for non-destructive testing. The method can be of interest in other related areas like medical echo-diagnostic, radar or sonar.

REFERENCES

[1]P. Mohana Shankar et Al. "Split-Spectrum Processing: Analysis of Polarity Thresholding Algorithm for Improvement of Signal-to-Noise Ratio and Detectability in Ultrasonic Signals". *IEEE Transactions on Ultrasonics, Ferroelectrics, and Frequency Control*. Vol 36, No 1, Jan. 1989.

[2]Israel Amir et Al. "Analysis and comparison of Some Frequency Compounding Algorithms for the Reduction of Ultrasonic Clutter". *IEEE Transactions on Ultrasonics, Ferroelectrics, and Frequency Control*. Vol 33, No 4, July. 1986.

[3]Luis Vergara-Domínguez & José Manuel Páez-Borrallo. "Backscattering grain noise modelling in ultrasonic non-destructive testing". *Waves in Random Media (Inst. of Physics)*, Vol. 1, Jan. 1991.

[4]Jafar Saniie & Tao Wang. "Statistical evaluation of backscattered ultrasonic grain signals". *Journal of Acoustic Society of America*,July 1988.

[5]B.R. Levin. "Fundamentos de radiotécnia estadística". Ed. Marcombo, Barcelona (Spain), 1984.

SIGNAL PROCESSING VI: Theories and Applications
J. Vandewalle, R. Boite, M. Moonen, A. Oosterlinck (eds.)

Study of Generalized Linear Interpolation Approximation for Multi-Dimensional Waveforms

Takuro KIDA Hiroshi MOCHIZUKI

Department of Information Processing, Graduate School at Nagatsuta
Tokyo Institute of Technology
Nagatsuta 4259, Midori-ku, Yokohama-shi, 227 Japan
Tel.+81-45-922-1111 ext.2522 Telefax.+81-45-921-1156

Abstract

In this paper, we consider the optimum approximation of band-limited multi-dimensional waveforms using samples of outputs of linear shift-invariant filters. We use a set of waveforms such that all the weighted integrals of the corresponding Fourier spectrums in the prescribed sub-bands are bounded. We establish a theorem showing that, among these waveforms, there exists a waveform with respect to which the sample values actually contributing to the interpolation are all zero and the absolute value of the corresponding approximation error reaches the upper limit.

As a direct consequence, we show that the presented linear approximation is superior to all other approximations using the same measure of error and sample values.

1 Introduction

Multi-dimensional signal processing is one of the important areas in digital signal processing. In the literature [1], we have proposed a certain optimum approximation for one-dimensional waveforms. When we apply the result to multi-dimensional waveforms, we may misunderstand that we can deal with the problem in the same way as the case of one-dimensional waveforms since a multi-dimensional waveform can be expressed simply by the product of one-dimensional waveforms. However, let us consider the case of two-dimensional waveforms. As the arrangements of the sampling points, we can choose orthogonal lattice, hexagonal lattice and octagonal lattice, etc. Even if we use the orthogonal lattice, the supports of the Fourier spectrums of the waveforms and/or those of the interpolation functions are not always separable with respect to the variables. Hence, if we can transform the hexagonal lattice or the octagonal lattice into the orthogonal lattice by using a skillful transformation of coordinates, in this case we cannot use the result of the one-dimensional case. This is one of the reasons why we present a unified treatment of multi-dimensional interpolation approximation in this paper.

2 Some definitions and notations

We begin with a summary of symbols and notations used in this paper.

(a) $X = (x_1, x_2, \cdots, x_n)$ indicates a n-dimensional real vector.

(b) $R^n = \{X|$ all the elements are real $\}$

(c) $K^n = \{X|$ all the elements are integer $\}$

(d) $X_K = X_{k_1,k_2,\cdots,k_n}$ means a vector with the integer subscripts k_m $(m = 1 \sim n)$. Also, $X_{KP} = X_{k_1,k_2,\cdots,k_n,p_1,p_2,\cdots,p_n}$ where $K = (k_1, k_2, \cdots, k_n)$ and $P = (p_1, p_2, \cdots, p_n)$.

(e) $\sum_{K\in J}\{\cdot\} = \sum_{k_1}\sum_{k_2}\cdots\sum_{k_n}\{\cdot\}$ indicates the summation taken over all the vectors in the subset J of K^n.

(f) $X * Y = (x_1y_1, x_2y_2, \cdots, x_ny_n)$ where $X = (x_1, x_2, \cdots, x_n)$ and $Y = (y_1, y_2, \cdots, y_n)$.

(g) The transpose of a vector or a matrix Y is expressed as Y^t. The complex conjugate of Y is expressed as $\overline{Y}$.

(h) $f(X) = f(x_1, x_2, \cdots, x_n)$

(i) $\int_J f(X)dX = \int\int\cdots\int f(x_1, x_2, \cdots, x_n)dx_1dx_2\cdots dx_n$

(j) $L^p(\Theta) = \{f(X)| \int_\Theta |f(X)|^p dX < +\infty\}$ where Θ is a subset of R^n.

3 Sample points and approximation formula

3.1 Set of sample points

We consider the prescribed vector $N = (N_1, N_2, \cdots, N_n)$ whose elements N_i are integers. Further, let $\Lambda(N)$ be the set of all the vectors in K^n whose elements k_i $(i = 1 \sim n)$ satisfy $1 \le k_i \le N_i$ $(i = 1 \sim n)$.

In the following, we assume that the symbol P always means a vector in K^n. We define also that K always indicates a vector in $\Lambda(N)$. Further, let

(a) H = a given $n \times n$ real regular matrix

(b) $T = (T_1, T_2, \cdots, T_n)$ be the prescribed vector with positive elements T_m $(m = 1 \sim n)$.

Now, we consider the following set of linear equations.

$$X_{KP} \cdot H - P * T - A_k = 0 \quad (X_{KP} \in R^n) \tag{1}$$

In eq.(1), X_{KP} is a solution vector. K and P are the subscript vectors. A_k is the prescribed vector having the form of $A_k = (A_{k_1}^{(1)}, A_{k_2}^{(2)}, \cdots, A_{k_n}^{(n)})$ and satisfies $0 \le A_p^{(m)} < A_q^{(m)} < T_m$ $(1 \le p < q < n;\ m = 1 \sim n)$.

Changing K and P one by one over all the elements of $\Lambda(N)$ and K^n, we obtain the infinite set of the n-D parallel polyhedra Γ_i $(i = 1, 2, \cdots)$ whose vertices correspond to the n-D points X_{KP}. We assume that Γ_i and Γ_j $(i \ne j)$ do not have common vectors except their boundaries. Further, we denote by Λ the prescribed subset of $\Lambda(N)$.

In this paper, we define the set of n-D sample points as the set of vertices X_{KP} ($K \in \Lambda$, $P \in K^n$). Selecting appropriate Λ, H and A_K, we can realize many types of arrangements of n-D sample points such as hexagonal or octagonal lattices. Letting $H^{-1} = G^t$ in eq.(1), we get the following equation.

$$X_{KP} = (P * T + A_k)G^t \tag{2}$$

3.2 Set of waveforms

Let Θ be a simply-connected closed domain of R^n expressed as $\Theta = \Theta_1 \cup \Theta_2$ ($\Theta_1 \cap \Theta_2 = 0$). We define positive bounded functions $W_1(U)$ and $W_2(U)$ as follows.

$$W_1(U) \in L^p(\Theta_1) \quad (U \in \Theta_1) \tag{3}$$

$$W_1(U) = 0 \quad (U \notin \Theta_1) \tag{4}$$

$$W_2(U) \in L^p(\Theta_2) \quad (U \in \Theta_2) \tag{5}$$

$$W_2(U) = 0 \quad (U \notin \Theta_2) \tag{6}$$

Moreover, we consider $F_{(1)}(U)$ and $F_{(2)}(U)$ satisfying the following conditions, where A_1 and A_2 are prescribed positive numbers.

$$\int_{\Theta_1} \frac{|F_{(1)}(U)|^p}{W_1(U)^p} dU \le A_1 \tag{7}$$

$$\int_{\Theta_2} \frac{|F_{(2)}(U)|^p}{W_2(U)^p} dU \le A_2 \tag{8}$$

Since $W_1(U)$ and $W_2(U)$ are bounded, $F_{(1)}(U)$ and $F_{(2)}(U)$ apparently belong to $L^p(\Theta_1)$ and $L^p(\Theta_2)$, respectively, where $F_{(1)}(U)$ and $F_{(2)}(U)$ are band-limited in $U \in \Theta_1$ and $U \in \Theta_2$, respectively.

We define $f_{(1)}(X)$ and $f_{(2)}(X)$ as the inverse Fourier transforms of $F_{(1)}(U)$ and $F_{(2)}(U)$, respectively. Further, we denote by $f(X) = f(x_1, x_2, \cdots, x_n)$ the sum of $f_{(1)}(X)$ and $f_{(2)}(X)$. Letting $A_1 \gg A_2 \simeq 0$ in eqs.(7)and(8), for example, $f(X)$ corresponds to ordinary low-pass type signals. Let

$$f(X) = \frac{1}{(2\pi)^n} \int_{\Theta} F(U) e^{jUX^t} dU \tag{9}$$

We abbreviate the relation of eq.(9) as $f(X) \longleftrightarrow F(U)$. The set of these $f(X)$ is written as $V(\Theta)$.

3.3 Approximation formula

We define the simply-connected bounded regions $B_{KP}(\in R^n)$ ($K \in \Lambda$; $P \in Z^n$) satisfying (a) and (b).

(a) B_{KP} contains X_{KP}.

(b) For arbitrary vectors P and Q ($P \neq Q$) in K^n, the region B_{KQ} is identical to the parallel translation of B_{KP} by $X_{KQ} - X_{KP}$.

For each B_{KP} ($K \in \Lambda$; $P \in Z^n$), we consider a bounded function $\psi_{KP}(X)$ ($K \in \Lambda$; $P \in Z^n$) satisfying $\psi_{KP}(X) = 0$ ($X \notin B_{KP}$). Further, we consider the approximation formula of $f(X)$ belonging to $V(\Theta)$ as follows.

$$g(X) = \sum_{K \in \Lambda} \sum_{P \in Z^n} f_K(X_{KP}) \psi_{KP}(X) \tag{10}$$

In eq.(10), we call $\psi_{KP}(X)$ the interpolation functions. Afterwards we prove that these interpolation functions can be realized as the impulse responses of the finite number of linear shift invariant filters.

4 Integral expression of the measure of approximation error

We define the approximation error $e(X)$ between $f(X)$ and $g(X)$ as follows.

$$e(X) = |f(X) - g(X)| \quad (X \in W_i) \tag{11}$$

The definition of W_i is given in Sec. 5.1. Make g(X) by eq.(10) and change $f(X)$ in $V(\Theta)$ under the condition that the interpolation functions $\psi_{KP}(X)$ are fixed. Moreover, consider an envelope $e_{max}(X)$ of the corresponding approximation errors $e(X)$ as follows.

$$e_{max}(X) = \sup_{f(X) \in V(\Theta)} \{e(X)\} \quad (X \in W_i) \tag{12}$$

From now on, we adopt $e_{max}(X)$ as the measure of approximation errors. The interpolation functions $\psi_{KP}(X)$ minimizing $e_{max}(X)$ are called the optimum interpolation functions. Moreover, at an arbitrarily specified point $X = Y$, if $f(X)$ exists in $V(\Theta)$ with respect to which $e(Y) = e_{max}(Y)$ is valid, we write one of these $f(X)$ as $f(X;Y)$ and call it the extremal waveform (for the fixed point Y).

By using the well known Hölder's inequality,

$$\begin{aligned} e(X) &= \frac{1}{(2\pi)^n} \left| \int_{\Theta_1} F_{(1)}(U) \Xi(U;X) dU \right. \\ &\quad + \left. \int_{\Theta_2} F_{(2)}(U) \Xi(U;X) dU \right| \\ &\le \frac{1}{(2\pi)^n} \left[\int_{\Theta_1} \frac{|F_{(1)}(U)|^p}{W_1(U)^p} dU \right]^{1/p} \\ &\quad \cdot \left[\int_{\Theta_1} W_1(U)^q |\Xi(U;X)|^q dU \right]^{1/q} \\ &\quad + \frac{1}{(2\pi)^n} \left[\int_{\Theta_2} \frac{|F_{(2)}(U)|^p}{W_2(U)^p} dU \right]^{1/p} \\ &\quad \cdot \left[\int_{\Theta_2} W_2(U)^q |\Xi(U;X)|^q dU \right]^{1/q} \\ &\le \frac{A_1^{1/p}}{(2\pi)^n} \left[\int_{\Theta_1} W_1(U)^q |\Xi(U;X)|^q dU \right]^{1/q} \\ &\quad + \frac{A_2^{1/p}}{(2\pi)^n} \left[\int_{\Theta_2} W_2(U)^q |\Xi(U;X)|^q dU \right]^{1/q} \\ &\quad (1/p + 1/q = 1;\ X \in W_i) \end{aligned} \tag{13}$$

holds, where

$$\Xi(U;X) = e^{jUX^t} - \sum_{K \in \Lambda} \sum_{P \in Z^n} \psi_{KP}(X) H_K(U) e^{jUX_{KP}^t} \tag{14}$$

The conditions that the inequalities in ineq.(13) hold at $X = Y$ are expressed as

$$\begin{aligned} \frac{F_{(\mu)}(U;Y)}{W_\mu(U)} &= \alpha_\mu(Y) W_\mu(U)^{q-1} \\ &\quad \cdot \overline{\Xi(U;Y)} \cdot |\Xi(U;Y)|^{q-2} \\ &\quad (U \in \Theta_\mu;\ Y \in W_i;\ \mu = 1,2) \end{aligned} \tag{15}$$

$$\int_{\Theta_\mu} \frac{|F_{(\mu)}(U;Y)|^p}{W_\mu(U)^p} dU = A_\mu \quad (Y \in W_i;\ \mu = 1,2) \tag{16}$$

where $\alpha_\mu(Y)$ is the scaling function which independent of U, and is chosen so that $F_{(\mu)}(U;Y)$ satisfies eq,(16), that is

$$\alpha_\mu(Y) = \frac{A_\mu^{1/p}}{\left\{\int_{\Theta_\mu} W_\mu(U)^q |\Xi(U;Y)|^q dU\right\}^{1/p}} \quad (\mu = 1,2) \qquad (17)$$

The Fourier spectrum of the optimal waveform $f(X;Y)$ is given by $F_{(1)}(U;Y) + F_{(2)}(U;Y)$ in eqs(15)and(16). Hence, by a similar consideration to the case of one-dimensional waveforms [1], we obtain the following theorem.

Theorem 1

$$\begin{aligned} e_{max}(X) &= \frac{A_1^{1/p}}{(2\pi)^n}\Big[\int_{\Theta_1} W_1(U)^q|\Xi(U;X)|^q dU\Big]^{1/q} \\ &+ \frac{A_2^{1/p}}{(2\pi)^n}\Big[\int_{\Theta_2} W_2(U)^q|\Xi(U;X)|^q dU\Big]^{1/q} \qquad (18) \end{aligned}$$

We call eq.(18) the integral expression of $e_{max}(X)$.

Although the detail discussion is omitted, $e_{max}(X)$ is convex for all the $\psi_{KP}(X)$. Hence, using an ordinary numerical optimization procedure, we easily obtain the "global" optimum $\psi_{KP}(X)$ at each X.

5 Optimality of the proposed approximation

5.1 Interpolation functions minimizing $e_{max}(X)$

Letting $Q \in Z^n$, we get the following equation of arbitrary K $(\in \Lambda)$ and P $(\in Z^n)$.

$$X_{K,P+Q} - X_{KP} = (Q * T)G^t \qquad (19)$$

As shown in eq.(19), the difference between $X_{K,P+Q}$ and X_{KP} is independent of P when T and G are fixed. Therefore the pattern of arrangement of $X_{K,P+Q}$ is the parallel translation of that of X_{KP}. By using the periodicity for the arrangement of the sample points, the following equation holds for arbitrary K $(\in \Lambda)$ and P and Q $(\in Z^n)$.

$$e^{jU[X_{KP}+(Q*T)G^t]^t} = e^{jUX^t_{K,P+Q}} \qquad (20)$$

Transform $e_{max}(X)$ of eq.(18) by using eq.(20) and perform a similar discussion as the one-dimensional case [1]. Then, we can obtain the following theorem.

Theorem 2 *Let the interpolation functions minimizing the measure of error $e_{max}(X)$ be $\psi_{KP}(X)$ $(K \in \Lambda;\ P \in Z^n)$, then*

$$\psi_{K,P+Q}(X) = \psi_{KP}[X-(Q*T)G^t] \ \ (K \in \Lambda;\ P,Q \in Z^n) \ \ (21)$$

holds. Thus, putting

$$\psi_{KQ}(X) = \psi_{K\theta}(X) = \psi_K(X - A_k G^t) \qquad (22)$$

we get the following equation.

$$\psi_{KP}(X) = \psi_K[X - (P*T)G^t - A_k G^t] \qquad (23)$$

where θ is a zero vector.

(End of the theorem)

Theorem 2 shows that the interpolation functions minimizing $e_{max}(X)$ are expressed as the finite number of parallel transformations of functions. Thus, the optimum interpolation functions can be realized as the impulse responses of the finite number of linear shift invariant filters.

Now we discuss the set of sample points actually contributing to the approximation. Firstly we consider the coordinate vectors corresponding vertices of a n-D unit cube.

$$\begin{aligned} 1_1 &= (0,0,\cdots,0) = \theta \\ 1_2 &= (1,0,\cdots,0) \\ 1_3 &= (0,1,\cdots,0) \\ &\cdots \\ 1_d &= (1,1,\cdots,1)\ (d=2^n) \end{aligned} \qquad (24)$$

Moreover, letting P be an arbitrary vector in R^n, we define $P_i = P + 1_i$ $(i = 1 \sim d)$. Further, by fixing $K = \theta$ and changing P with $P_1, P_2, \cdots, P_d$, in X_{KP}, we get the lattice points $X_1, X_2, \cdots, X_d$ $(d = 2^n)$. A parallel polyhedron having these lattice points as its vertices is written as Δ_P. Using the periodicity of the arrangement of the sample points, we can prove that the space R^n is filled with Δ_P *'s* without the overlaps except their boundaries.

Now, we notice the above described regions B_{KP} and write the boundaries of B_{KP} as C_{KP}. Moreover, changing K and P within Λ and Z^n, respectively, R^n is divided into disjoint regions by C_{KP} without overlaps except boundaries. We denote these regions by W_i $(i = 1,2,\cdots)$.

Eq.(19) shows that $X_{K,P+Q}$ is a parallel translation of X_{KP} by $X' = (Q*T)G^t$. Therefore, by using the argument of sec.3.3, $B_{K,P+Q}$ is a parallel translation of B_{KP} by X'. Thus, apparently the boundary $C_{K,P+Q}$ is a parallel translation of C_{KP} by X'.

This fact shows that a parallel translation of the region W_i by X' is also a congruent region (we call it W_j). Since Q in $X' = (Q*T)G^t$ is an arbitrary vector belonging to Z^n, the set of W_i is classified into the groups where the regions are the parallel translations one another. If W_i is the parallel translation of W_j, we write this fact, as $W_i \sim W_j$.

Let $\Omega_P = \{W_1, W_2, \cdots, W_\nu\}$ be the set of W_i having a common point with Δ_P except boundaries (by changing the subscripts appropriately). Moreover, when Ω_P contains the regions such that $W_i \sim W_j$ $(i \neq j)$, we leave in Ω_P only one of them. The new Ω_P obtained after this operation is written, by changing subscripts appropriately, $\Omega_P = \{W_1, W_2, \cdots, W_m\}$. Further, we can easily prove that Ω_{P+Q} is a parallel translation of Ω_P by $X' = (Q*T)G^t$ and the space R^n is filled with disjoint Ω_P *'s* except common boundaries.

Now, at a certain point X, and the fixed K, we consider the set of (K,P) expressed as

$$\Xi_{P_i} = \{(K,P) | X \in W_i,\ W_i \in \Omega_P,\ W_i \subset B_{KP}\} \qquad (25)$$

It should be noted, in this case, the sample values $f_K(X_{KP})$ $((K,P) \in \Xi_{P_i})$ actually contribute to the approximation.

5.2 Property of the optimum waveforms

As the important property of the optimum waveforms described in sec.4,1 the following theorem holds.

Theorem 3

$$f_K(X_{KP};Y)=0 \quad ((K,P)\in \Xi_{Pi},\ Y\in W_i) \tag{26}$$

(Proof) By the condition of the optimum interpolation functions, expanding eq.(18) with respect to $\psi_{KP}(X)$ and $\overline{\psi_{KP}(X)}$ and taking the partial derivative with respect to the real part and the imaginary part of $\psi_{KP}(X)$ and letting them be equal to zero and combining the results, then we can get

$$\begin{aligned}
&\frac{\partial e_{max}(X)}{\partial \psi_{KP}(X)} \\
&= B\cdot\Bigg[\frac{A_1^{1/p}}{(2\pi)^n}\Big\{\int_{\Theta_1} W_1(U)^q \\
&\quad\cdot\big[e^{-jU(X-X_{KP})^t}-\sum_{L\in\Lambda}\sum_{Q\in Z^n}\overline{\psi_{LQ}(X)H_L(U)e^{jU(X_{KP}-X_{LQ})^t}}\big] \\
&\quad\cdot H_K(U)dU\cdot\Big[\int_{\Theta_1} W_1(U)^q|\Xi(U;X)|^q dU\Big]^{-1/p}\Big\} \\
&+\frac{A_2^{1/p}}{(2\pi)^n}\Big\{\int_{\Theta_2} W_2(U)^q \\
&\quad\cdot\big[e^{-jU(X-X_{KP})^t}-\sum_{L\in\Lambda}\sum_{Q\in Z^n}\overline{\psi_{LQ}(X)H_L(U)e^{jU(X_{KP}-X_{LQ})^t}}\big] \\
&\quad\cdot H_K(U)dU\cdot\Big[\int_{\Theta_2} W_2(U)^q|\Xi(U;X)|^q dU\Big]^{-1/p}\Big\}\Bigg] \\
&= 0
\end{aligned} \tag{27}$$

where B is a nonzero real number. On the other hand, from eq.(9) we get

$$\begin{aligned}
&f_K(X_{KP};Y) \\
&= \frac{1}{(2\pi)^n}\int_{\Theta_1} F_{(1)}(U;Y)H_K(U)e^{jUX_{KP}^t}dU \\
&+ \frac{1}{(2\pi)^n}\int_{\Theta_2} F_{(2)}(U;Y)H_K(U)e^{jUX_{KP}^t}dU
\end{aligned} \tag{28}$$

Substituting eq.(15) into eq.(28) and by using eq.(17), we get

$$\begin{aligned}
&f_K(X_{KP};Y) \\
&= \frac{A_1^{1/p}}{(2\pi)^n}\Big\{\int_{\Theta_1} W_1(U)^q \\
&\quad\cdot\big[e^{-jU(X-X_{KP})^t}-\sum_{L\in\Lambda}\sum_{Q\in Z^n}\overline{\psi_{LQ}(X)H_L(U)e^{jU(X_{KP}-X_{LQ})^t}}\big] \\
&\quad\cdot H_K(U)dU\cdot\Big[\int_{\Theta_1} W_1(U)^q|\Xi(U;X)|^q dU\Big]^{-1/p}\Big\} \\
&+\frac{A_2^{1/p}}{(2\pi)^n}\Big\{\int_{\Theta_2} W_2(U)^q \\
&\quad\cdot\big[e^{-jU(X-X_{KP})^t}-\sum_{L\in\Lambda}\sum_{Q\in Z^n}\overline{\psi_{LQ}(X)H_L(U)e^{jU(X_{KP}-X_{LQ})^t}}\big] \\
&\quad\cdot H_K(U)dU\cdot\Big[\int_{\Theta_2} W_2(U)^q|\Xi(U;X)|^q dU\Big]^{-1/p}\Big\}
\end{aligned} \tag{29}$$

Comparing eq.(27) with eq.(29), we get the following equation.

$$f_K(X_{KP};Y)=0 \quad ((K,P)\in \Xi_{Pi},\ Y\in W_i) \tag{30}$$

(Q.E.D.)

Let $R(X)=R[\cdots,f_K(X_{KP}),\cdots;\ X]$ be another approximation formula and let

$$e^{(1)}_{max}(X)=\sup_{f(X)\in\Lambda}|f(X)-R(X)|$$

We assume $R[\cdots,0,\cdots;\ X]\equiv 0$. Further, we assume that $e^{(1)}_{max}(Y) < e_{max}(Y)$ holds at a fixed Y. Then, if $f(X)$ is equal to the optimal function $f(X;Y)$, we obtain $|f(Y)| = |f(Y)-R(Y)| \le e_{max}(Y) < e_{max}(Y) = e(Y) = |f(Y)-g(Y)| = |f(Y)|$ which makes a contradiction. Hence, at all Y, $e^{(1)}_{max}(Y)\ge e_{max}(Y)$ holds. Namely, when we adopt other arbitrary approximation formulas using the same sample values , the upper bounds of approximation error rather increase. Thus, as long as the measure of error and the set of waveforms are not changed, the proposed approximation formula is the optimum among all other approximation formulas. Extending the discussion, similar arguments are possible for the case that Fourier spectrums are divided into more than two bands. In this paper, we have discussed when the weighted L^p integrals are bounded. In particular, when $p=2$ we can still more extend the discussion mathematically [2]. However, when $p\ne 2$ we cannot extend the discussion any further, because an bilinear form can solely be defined by an integral expression in this case.

It should be noted, from theorem 2, the interpolation functions minimizing $e_{max}(X)$ are expressed as the parallel transformations of the finite number of functions . Thus, the interpolation functions can be realized as the impulse responses of the finite number of linear shift invariant filters.

6 Conclusion

Presented linear approximation for multi dimensional waveforms is superior to all other approximations using the same set of waveforms, sample values and measure of error.

This shows that the superiority of nonlinear signal processing using, for example, neural devices mainly depends on the choice of the set of waveforms.

Acknowledgement

We would like to thank Professor Kenneth Jenkins of Illinois University and Professor A. Fettweis of Ruhr Universität Bochum and Mr. Somsak Sa-nguankotchakorn of Tokyo Institute of Technology for their useful advice in writing this paper.

References

[1] Kida T. and Mochizuki H. *Study of the superiority of generalized linear interpolation approximation,* Trans. IEICE, J73-A, 11, pp.1788-1797 (1990-11)

[2] Kida T. and Mochizuki H. *Generalized Interpolatory Approximation of Multi-Dimensional Signals Having the Minimum Measure of Error,* the 7th IEEE workshop on multi-dimensional signal processing, Lake Placid, New York (1991-9)

SIGNAL PROCESSING VI: Theories and Applications
J. Vandewalle, R. Boite, M. Moonen, A. Oosterlinck (eds.)
1992 Elsevier Science Publishers B.V.

DETAIL-PRESERVING FILTERS WITH IMPROVED LOWPASS CHARACTERISTICS†

Irek DEFÉE, Riku SOININEN
and Yrjö NEUVO

Signal Processing Laboratory,
Department of Electrical Engineering,
Tampere University of Technology,
P. O. Box 553, SF–33101 Tampere, Finland.

A new scheme for high-quality, detail-preserving and aliasing-free filtering is proposed. The scheme uses combination of a nonlinear detail-preserving filter and linear lowpass filter. An algorithm is devised for switching between the nonlinear and linear filter in order to eliminate aliasing errors and preserve details whenever possible.

1. INTRODUCTION

Increasing resolution and faithfulness of images processed and displayed by electronic means leads to a growing demand for high-quality image filtering. Ideal filters should robustly attenuate noise and contaminations, without removing details or introducing blurring or artefacts. Classical linear filters are effective in attenuating Gaussian noise and have well-defined frequency response. On the other hand, they suffer from blurring of edges and details, which is unacceptable in high-quality applications. Also, linear filters can not be applied in the case of impulsive noise contamination.

To alleviate problems with linear filters, many nonlinear filter schemes based on different concepts were proposed [1]. Probably the most thoroughly investigated and used among them are filters based on the median operation. Median and median-type filters have simple implementation and can be designed to preserve image details [2]. They operate robustly in impulsive noise and are moderatly effective in Gaussian noise. All these desirable features made median filters strong contenders for high-quality image processing applications.

When considering the high-quality demands, however, one should realize that the impact of frequency response of nonlinear filters on the quality of processing has not been given sufficient attention. The most prominent illustration of this problem is that inadequate attenuation of higher signal frequencies may lead to the generation of severe aliasing artefacts. The aliasing is described by shifting of high frequency spectra to the baseband. One can talk about aliasing-blurring dichotomy when investigating properties of linear and nonlinear filters. Linear filters can be easily designed to eliminate aliasing components, but at the price of blurring. Nonlinear filters may not exhibit blurring but their frequency behavior can not be shaped so effectively and as a result aliasing artefacts may appear. It can be thus seen that a fresh approach is required to find algorithms resistant to all the quality impairments.

In this paper, a new scheme is proposed to solve the aliasing-blurring dichotomy. The scheme is based on a combination of linear and nonlinear filters and a decision structure. The decision structure is designed in such a way that it switches between the linear and nonlinear filters depending on the presence of a signal component which could give rise to serious aliasing artefacts. In this way, filtering with the elimination of both blurring and aliasing is achieved.

†Work supported in part by the RACE II Project No.2056 (AMICS).

2. COMBINED FILTERING SCHEME

2.1. Blurring and Aliasing Problem

The aliasing-blurring dichotomy is illustrated in Fig. 1. Linear filter causes blurring of sharp edges (Fig. 1b), but is able to eliminate effectively high frequencies (Fig. 1d). Median filtering, on the other hand, does not introduce blurring (Fig. 1b), but high frequencies are not sufficiently attenuated (Fig. 1d). The effect of insufficient attenuation of high frequencies by the median filtering can be seen in Fig. 2 where the original "Young Couple" image has been filtered by a multilevel median [2] filter operating in a 3×3 window and decimated by two. A very annoying aliasing artefacts can be seen in the shirt area (Fig. 2b). Such artefacts would not be created by properly designed linear lowpass filter, but at the cost of blurring.

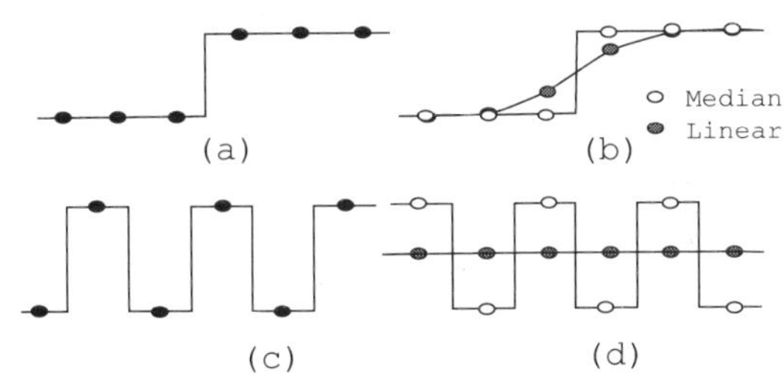

Fig. 1. Step edge (a) and a high-frequency signal (c), filtered by linear and 3-point median filters [(b), (d)].

From these observations, we can conclude that both linear and nonlinear (median in this case) filters have desirable features, but also some drawbacks which can not be tolerated in critical applications like HDTV, printing industry and multimedia. The goal of this paper is to introduce a new filtering structure which would be able to preserve only positives of both filter types, without bringing up their drawbacks. This consideration leads us to an idea of a combined structure with both linear and median filters operating in a switched mode, depending on a local signal shape. Such structure would satisfy the requirements of the high quality applications.

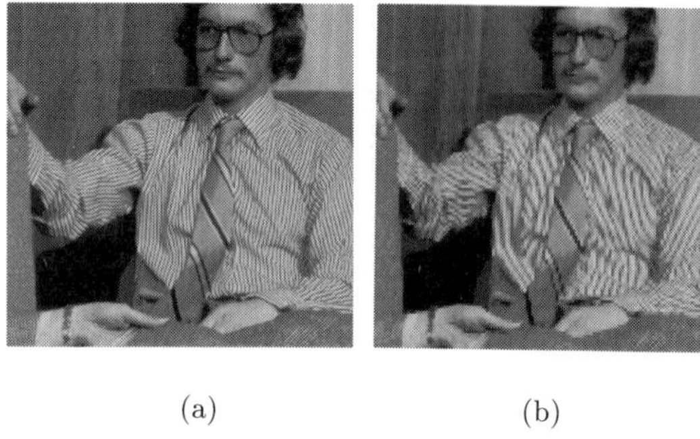

(a) (b)

Fig. 2. "Young Couple" (a), filtered with a 3×3 multilevel median filter, and decimated by two, (b).

2.2. New Filtering Structure

Blurring and aliasing could be eliminated if lowpass characteristics of median-type filters were improved. To achieve this, we developed a new structure which combines operation of linear lowpass and nonlinear median-type detail-preserving filters, [2]. This structure is shown in Fig. 3.

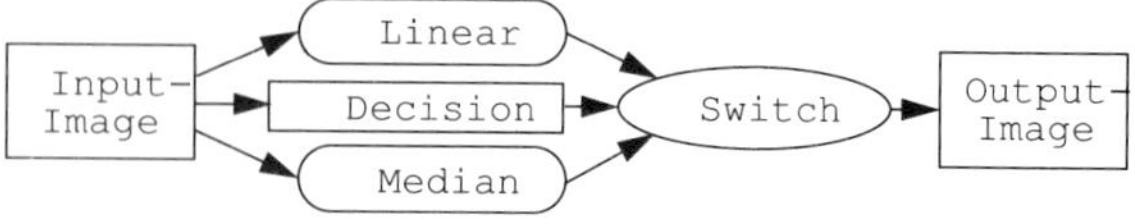

Fig. 3. New filtering structure.

The system shown in Fig. 3 should operate in such a way that it would switch to linear filter in case of the presence of local signal segment which could cause aliasing, and it would switch to a nonlinear filter in other cases, to avoid blurring. The main problem obviously is to find an algorithm for switching between the linear and nonlinear filter. In [3], a scheme has been proposed in which the switching is based on the output of a linear bandpass filter tuned to the frequency at which median filter has lowest attenuation. However, the use of the linear filter for the switching decision may lead to problems since such filter does not provide sufficient information for the discrimination between the highly-localized and high-frequency signal segments. It is felt that another algorithm is needed to detect only those signal parts which are responsible for aliasing.

3. SWITCHING ALGORITHM

3.1. Basic Algorithm

In developing of the switching algorithm we use the fact that monotonic (e.g. edges), or slowly varying signal segments do not contribute to the creation of aliasing errors. Such errors are caused only by the signal segments having sufficiently fast variation. The local signal variation can be described by the presence of local signal maxima and minima within certain specific distance. Assume for example that decimation by two is going to be performed. Then, aliasing can be generated from those signal segments where the distance between two local signal maxima or minima is three pixels or less, since they correspond to signals with frequencies higher than one fourth of the original sampling rate, which is maximum allowed in the decimated signal (Fig. 4a). However, this is not solving the problem completely since certain signal segments can be classified as "low frequency segments contaminated by high-frequency" (Fig. 4b). They should rather be preserved than eliminated, particularily if the high-frequency contamination is small.

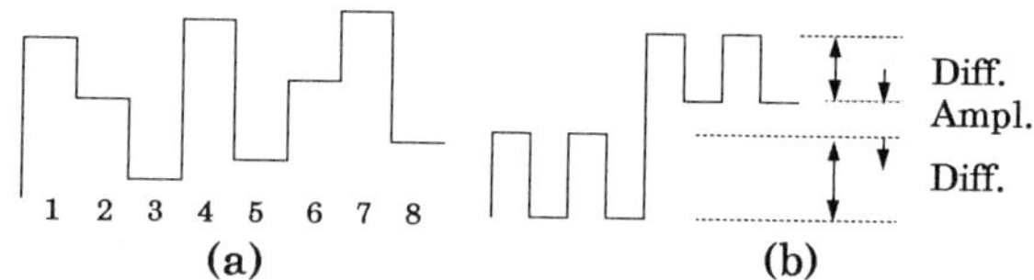

Fig. 4. High-frequency signal segments.

For the decimation factor of two, the condition for the detection of aliasing-generating signal segment means that, at any given point, we have to check whether there are two local maxima or minima within the distance of three or less points from it. An algorithm searching for this has been devised as follows.

We take a length-5 window and check if a subset of the two first samples is not mixing with the subset of the three next samples, i.e. the first two samples are greater than, or smaller, than the rest. Then we check also if the three first samples are not mixing with the subset of the two last samples. We move next to the previous point and perform the same test and after that we perform this test also in the point next to the center point.

If any of those tests shows us that there is no mixing, we check the variation amplitude of the signal on both sides of the point (Fig. 4b). This is because we may have monotonic signals "contaminated" by noise. If the amplitude of the contamination is lower than the difference between the varying parts (Fig. 4b), we classify the signal as "contaminated" monotonic signal. Also, if all those tests shows that there is no mixing, we classify the signal as "contaminated" monotonic signal, even without the amplitude check. In Fig. 5., we see the signal divided into two non-mixing subsets of length two and three.

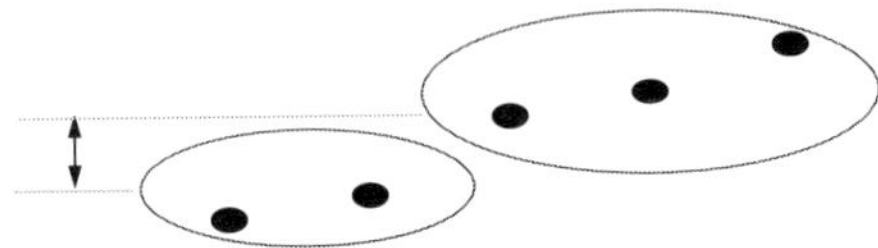

Fig. 5. Example grouping in length-5 window.

Now we can see that for an edge like in Fig. 4b, we always have at least one case of non-mixing subsets (the higher part of the edge and the lower part of the edge), and if the

contamination is low enough, i.e. the amplitude is lower than the height (difference) of the edge, we have a "contaminated" monotonic signal. Then, if we have a pure high frequency signal the following example shows that we always detect it.
If we take point 4 of the high frequency signal segment in Fig. 4a and divide the 5-point set into the two subsets (2,3) and (4,5,6) we see that there is clear mixing of samples between those subsets. Similarly, we also have mixing between subsets (2,3,4) and (5,6). Next we move to point 3 and see that there is mixing between (1,2) and (3,4,5) as well as between (1,2,3) and (4,5). The same check is still done in the point next to the center point, i.e. the point 5. The result is again mixing between (3,4) and (5,6,7) as well as between (3,4,5) and (6,7).

What if point 2 would have been lower that point 5? It would have meant that between subsets (2,3) and (3,4,5) there wouldn't have been mixing. Then we would have calculated the variation of the signal in (2,3) and (4,5,6) and found out that the variation would have been higher comparing to the difference in variation between the two subsets. The result would again have been classification as high frequency signal segment by the above procedure.

In practice, we only have to calculate the mixing for the last of the three points, because the status of the two first points, which have already been calculated previously, can be stored in the memory. So, we actually have to check the mixing only for one point at each direction (ver./hor./diag.) and check memory for the two others, which makes this scheme very simple.

Our basic algorithm is now as follows: If the high frequency signal segment is detected, then linear filtering is applied. In the opposite case, median filter is used.

3.2. Detection of Outliers

In our basic switching algorithm, signals classified as "low frequency", can still have impulsive noise contamination (Fig. 6), which may require removal by median filter.

Fig. 6. Impulsive contamination in low frequency segment.

In order to remove impulsive noise contamination from the low frequency signal segments, we need some kind of a signal-dependent way of detecting it. The idea used is that we take a 3x3 mask and check if the value of the center point of the mask is higher than the output of a linear lowpass filter in the same point. If it is so and if at least 7 of the other points covered by the mask have lower values than the center point value, we classify the point as an outlier. This means that if the size of an outlier is less than 3 pixels, it will be filtered out. Otherwise, it will be classified as a significant structure in the image and it will be preserved.

If we find an outlier and therefore use median filtering, we must also use median in all the surrounding 8 points. Otherwise, possible linear filtering at the next point could still cause some annoying effects, because the impulse may still have noticable effect if linear filter is used. Negative outliers are of course checked in the same way.

The final algorithm with outlier detection is illustrated in Fig. 7.

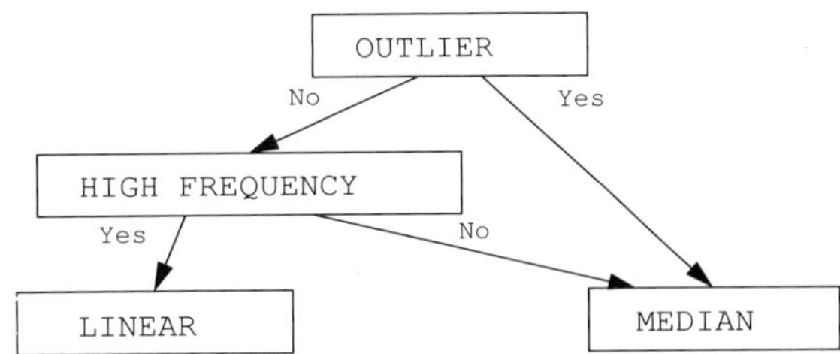

Fig. 7. The final algorithm.

4. FILTER IMPLEMENTATION

The filtering scheme from Fig. 3. has been implemented using carefully selected median and linear part. In the 2-D implementation, the ordering tests can be performed in two (vertical/horizontal) or four (vertical, horizontal and 2 diagonal) directions. In the implementation, a multilevel median filter, based on a 5-point cross- and diamond-shaped medians in a 3x3 mask has been used, because of its excellent detail preservation [2]. In this algorithm we can use of course any other non-linear filter, but for this specific implementation we found the multilevel median to be the best. As linear filters, three different filters were used: an optimal Burt filter in a 5x5 mask [4], an optimized 13x13 halfband filter [5], and a simple 3x3 averaging filter.

5. TESTS

5.1. Visual Tests

Visual tests were made to check if there is visible aliasing left in images after filtering. In Fig. 8a, we see the "Young Couple" filtered using our algorithm (with high frequency check in 4 directions) and decimated by two. In Fig. 8b, we show where the algorithm used median and linear filters. White areas indicate the use of median filter and black areas indicate the use of linear lowpass filter.

(a) (b)

Fig. 8. "Young couple" after filtering (a). Areas of operation of median and linear filter (b).

Clear difference can be seen, when compared to median filtered "Young couple". After our algorithm there is no aliasing left and sharp edges are still preserved, whereas after median filtering we have quite annoying visible aliasing left in the image (Fig. 2b).

5.2. Numerical Tests

5.2.1. Entropy

A good measure showing statistical information content in images is entropy. Entropy can be considered for example as the minimum amount of bits/pixel needed in compressing the image. The basic equation for the entropy in a 256 gray scale value image is:

$$E = \sum_{k=0}^{255} (-p_k \times log_2 p_k) \quad (1)$$

where p_k is probability of a gray scale value in the image. The difference image is calculated by subtracting the filtered image from the original image. The lower the entropy of the difference image is, the less information is lost in filtering. Entropies for the difference images after filtering of standard test images "Lena" and "Zone", are presented in Table 1.

Table 1.

Entropies.

Filter	Lena (256x256)	Zone (256x256)
Multilevel Median	2.261928	4.663237
3x3 aver. filter	4.605356	7.675303
Burt filter, 5x5	3.891643	7.477057
Opt. 13x13 halfband	4.490483	7.540577
Our alg. with 3x3	2.514802	6.224827
Our alg. with 5x5	2.461626	6.364391
Our alg. with 13x13	2.510714	6.313446

Table 2.

Gaussian noise supression.

Filter	Ratio of variances
Multilevel median	0.310
3x3 averaging filter	0.118
Burt filter, 5x5	0.247
Opt. 13x13 halfband	0.190
Our alg. with 3x3	0.251
Our alg. with 5x5	0.273
Our alg. with 13x13	0.262

5.2.2. Gaussian Noise Supression

To evaluate the Gaussian type noise supression capability of the algorithm, the ratio between the variance of the processed noise and the original noise is calculated. The Gaussian noise of mean 0 and standard deviation 10 is used in this test. The results for multilevel median, the three different linear filters and the basic algorithm using those three linear filters, are presented in Table 2.

Linear filters have the best Gaussian noise supression capabilities as expected and the median filter performs worst. Our algorithm gives results between them, which is predictable.

5.2.3. Impulsive Noise Supression

For impulsive noise supression, the mean absolute error (MAE) for noisy points is calculated. The impulsive noise used is "salt-and-pepper" -type, i.e. randomly located black and white impulses. The results can be seen in Table 3.

Table 3.

Impulsive noise suppression (MAE).

Amount of noisy pixels	Our Algorithm using 3x3 averaging	Pure Multilevel Median	Pure Linear 3x3 averaging
652	0.237730	0.237730	15.242332
1298	0.273498	0.273498	15.472265
3205	0.906396	0.906396	15.843369
6241	3.457299	3.503445	17.232655
11863	11.453174	11.481075	19.614685

When the noise level is high, our algorithm works slightly better than multilevel median filter because closely located impulses can be detected as high frequency segments, where pure multilevel median treats such impulse groups as parts of the image and preserves them. This is quite interesting result, because median is supposed normally to operate much better in impulsive noise than linear filter, but yet a combination of median and linear filter gives even better results.

6. CONCLUSION

Both linear and median filters have their drawbacks. Linear filters cause blurring on edges and details and median filters cause aliasing in high frequency areas. In this paper a new filtering scheme is proposed, which combines best features of the filters, leaving out the drawbacks of them. The new filter has proved to be efficient in the impulsive noise supression and even Gaussian noise supression capability is improved comparing to the median. Results of visual and numerical tests confirm for the expected behavior of the new filter.

REFERENCES

[1] Pitas, I. and Venetsanopoulos, A.N., "Nonlinear Digital Filters," Kluwer Academic Publishers, 1990.

[2] A. Nieminen, P. Heinonen and Y. Neuvo, "A New Class of Detail-Preserving Filters for Image Processing," *IEEE Trans. Pattern Anal. Mach. Intell.*, vol. PAMI-9, pp. 74–90, No.1, January 1987.

[3] H. J. Dreier, "Line Flicker Reduction by Adaptive Signal Processing," in: Signal Processing of HDTV, L. Chariglione, ed., Elsevier Science Publisher, B. V., 1990, pp. 695–701.

[4] P. Burt and E. Adelson, "The Laplacian Pyramid As a Compact Image Code," *IEEE Trans. Commun.*, vol. COM-33, pp. 532–540, No. 4, January 1983.

[5] P. Meer, E.S. Baugher and A. Rosenfeld, "Frequency Domain Analysis and Synthesis of Image Pyramid Generatin Kernels," *IEEE Trans. Pattern Anal. Mach. Intelligence*, vol. PAMI-9, pp. 512–522, No.4, July 1987.

SIGNAL PROCESSING VI: Theories and Applications
J. Vandewalle, R. Boite, M. Moonen, A. Oosterlinck (eds.)

SYNTHESIS OF INHERENTLY STABLE AND LOW-SENSITIVE 2-D IIR FILTERS USING THE ℓ_1-PASSIVITY CONCEPT

MAREK DOMAŃSKI

Politechnika Poznańska, Instytut Elektroniki i Telekomunikacji
ul. Piotrowo 3a, 60-965 Poznań, Poland*

The concept of ℓ_1-passivity has been obtained by extension of the well-known concept of passivity of digital systems. It has been proved that the stability properties implied by the ℓ_1-passivity are similar but even little bit stronger as those implied by the classical passivity related to the ℓ_2-norm. Both classical passivity and ℓ_1-passivity are powerful tools to obtain low-sensitive filters. The paper presents some techniques to design ℓ_1-passive recursive 2-D digital filters. Some low-sensitive filter structures are described.

1. ℓ_1-PASSIVE 2-D DIGITAL FILTERS

The very good properties of the wave digital filters [1] are well established also in the two-dimensional (2-D) case. The passivity related to boundedness of the weighted ℓ_2-norm of the state vector is the keystone of the theory of such filters. Recently it has been shown that the use of some more general functions of the state vector, of the input, and of the output also leads to reasonable definitions of the (generalized) passivity [2-4]. Among the great many possible measures that based on the ℓ_1-norm is of the particular interest.

A 2-D digital filter (cf. Fig.1) is said to be ℓ_1-passive iff

$$\sum_{i=1}^{N} |a_i(\underline{n})| - |b_i(\underline{n})| \geq 0 \qquad (1)$$

for all $\underline{n} = (n_1, n_2)$ where n_1, n_2 are integers, e.g., coordinates of a pixel in a digital image) and for all admissible $\underline{a}(\underline{n}) = [a_1(\underline{n}), \ldots, a_N(\underline{n})]^T$, where $\underline{a}(\underline{n})$ and $\underline{b}(\underline{n})$ are the vectors of the signals which are incident to the shift-free network and reflected from it, respectively. .

We could define also ℓ_1-lossless linear digital systems by setting the equality in Eq. (1). Nevertheless, it has been proved that a linear non-degenerate ℓ_1-lossless system with real signals does not exist [4,5]. The ℓ_1-lossless linear digital systems have been considered hitherto in the case of nonnegative signals only [3,4].

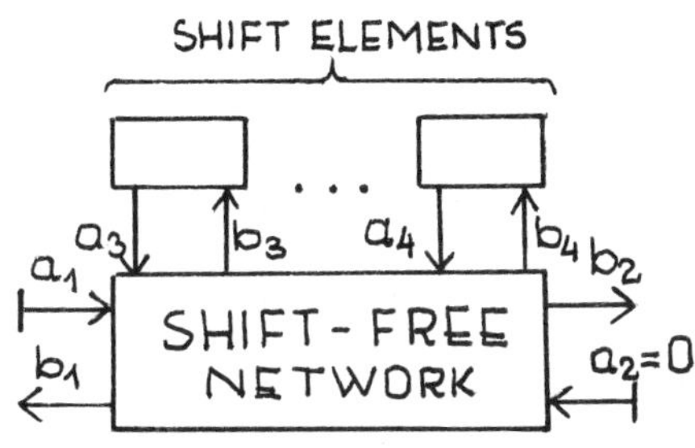

Fig. 1. A digital filter.

Now, we briefly report some properties of the ℓ_1-passive systems:

1) A linear system is ℓ_1-passive iff there exist $g_i > 0$ $(i=1,\ldots,N)$ such that

$$\sum_{i=1}^{N} g_i |s_{ij}| \leq g_j \qquad (2)$$

for each $j=1,\ldots,N$ and each $\underline{n}$, where $\mathbb{S} = [s_{ij}] \in \mathcal{R}^{N\times N}$ describes the shift-free network $\underline{b}(\underline{n}) = \mathbb{S}\cdot\underline{a}(\underline{n})$. (3)

Let assume $g_i=1$ $(i=1,\ldots,N)$. It can be

*The paper has been prepared in part on leave to Lehrstuhl für Nachrichtentechnik, Ruhr-Universität Bochum, Germany under support of the Alexander von Humboldt-Foundation.

shown that such an assumption does not influence the generality of any further considerations.

2) There is $$M(\omega_1,\omega_2) \le 1 \,, \tag{4}$$
where $M(\omega_1,\omega_2)$ is the magnitude response of a ℓ_1-passive system. This property gives an opportunity to obtain low-sensitive structures. Some filters considered in this paper exhibit zero sensitivity of their transfer functions at certain characteristic frequencies. The proper design techniques place such characteristic frequencies in the passbands and reduce the magnitude sensitivity in the passbands.

3) The ℓ1-passivity implies the BIBO-stability of 2-D linear shift invariant systems while the classical passivity (related to the weighted ℓ_2-norm) does not. This phenomenon is related to the possibility of existence of the non-essential singularities of the second kind on the boundary of the unit bidisc which often lead to severe testing problems in the systems which are not passive. On the other hand, the non-essential singularities of the second kind on the boundary of the unit bidisc are rather isolated points, and therefore, in real conditions of the finite precision arithmetic, they are not crucial.

4) A 2-D ℓ_1-passive digital systems is output ℓ_1-stable.

The definition of the output ℓ_1-stability [3] is a little bit complicated and will be omitted here.

The above property implies that an ℓ_1-passive 2-D system remains stable also under finite precision arithmetic conditions so far as Condition (2) holds for a shift-variant system being equivalent to the real system with constant coefficients and finite wordlength. That is, the zero-input parasitic oscillations are suppressed under such a condition.

The theory of ℓ_1-passivity has already become fruitful in the design of the low-sensitive 1-D and 2-D FIR filters [6,7]. The aim of this paper is to propose some new inherently stable 2-D recursive filter structures and corresponding new design techniques which result in low-sensitive filters.

2. THE DIRECT SYNTHESIS

A given transfer function $H(\underline{z})$, is to be implemented in the structure shown in Fig. 2. The transfer function

$$H(\underline{z}) = \frac{P(\underline{z})}{Q(\underline{z})} = \frac{\underline{Z}_1^T \mathbb{P} \underline{Z}_2}{\underline{Z}_1^T \mathbb{Q} \underline{Z}_2} \,, \tag{5}$$

where $\underline{Z}_i = [1,\ z_i^{-1},\ \dots,\ z_i^{-M_1}]^T$, $i=1,2$, $\underline{z}=(z_1,z_2)$, $L_1,L_2,M_1,M_2 \in \mathcal{Z}^+$, $\mathbb{P}=[p_{ij}] \in \mathcal{R}^{L_1 \times L_2}$, $\mathbb{Q}=[q_{ij}] \in \mathcal{R}^{M_1 \times M_2}$

has to be rearranged as

$$H(\underline{z}) = \frac{(1-\beta)F_1(\underline{z})F_3(\underline{z})}{1-\beta F_1(\underline{z})F_2(\underline{z})} \,, \tag{6}$$

where $0<\beta<1$, and $F_1(\underline{z})$, $F_2(\underline{z})$, $F_3(\underline{z})$ can to be implemented using the low-sensitive nonrecursive networks from Fig. 3.

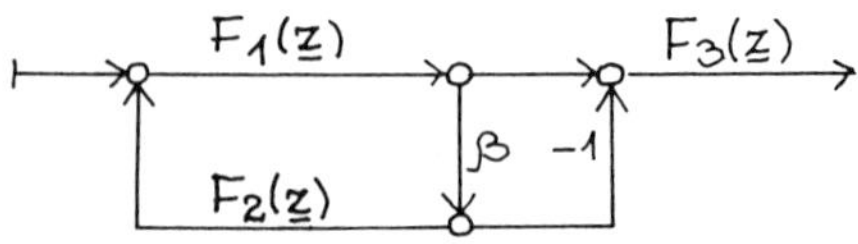

Fig.2. Simple structure for direct synthesis of ℓ_1-passive 2-D filters.

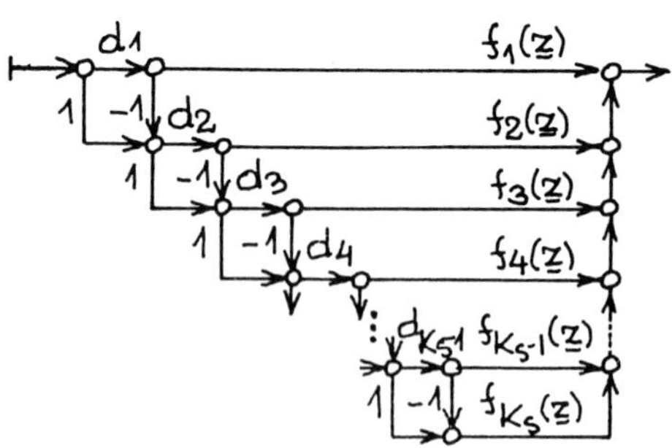

Fig.3. The low-sensitive nonrecursive building block.

There is $$F_j(\underline{z}) = \sum_{i=1}^{K_s} a_i f_{ij}(\underline{z}) \,, \tag{7}$$
$j=1,2,3$, $a_i>0$, $a_1+a_2+\dots+a_{K_s}=1$, $k_{ij}=\pm 1$, $f_{ij}(\underline{z}) = k_{ij} z_1^{-m_{1ij}} z_2^{-m_{2ij}}$, $m_{1ij}, m_{2ij} \in \mathcal{Z}^+$.
A transfer function $H(\underline{z})$ can be rea-

lized in the considered structute with the accuracy to a constant factor if

$$q_{00}=1,\ \sum_{i=1}^{M_1}\sum_{j=0}^{M_2}|q_{ij}|+\sum_{j=1}^{M_2}|q_{0j}|<1\ . \qquad (8)$$

The system remains ℓ_1-passive as longas the coefficients "d" (cf. Fig.3) are in the range $0<d_i<1$, i.e., the system exhibits the property of $\ell 1$-passivity also for rounded coefficients. This observation proves stability for any coefficients rounded according to the common sense. Low sensitivity of the amplitude characteristic can be mentioned in several practical examples already examined.

EXAMPLE 1. Given are

$$\mathbb{P}=\begin{bmatrix}0.0780 & 0.0386 & 0.0084\\ 0.0386 & 0.0260 & 0.0010\\ 0.0084 & 0.0010 & 0.0000\end{bmatrix},$$

$$\mathbb{Q}=\begin{bmatrix}1.0000 & 0.0000 & 0.0000\\ -0.3120 & -0.1544 & -0.0336\\ -0.1544 & -0.1040 & -0.0040\\ -0.0336 & -0.0040 & 0.0000\end{bmatrix}.$$

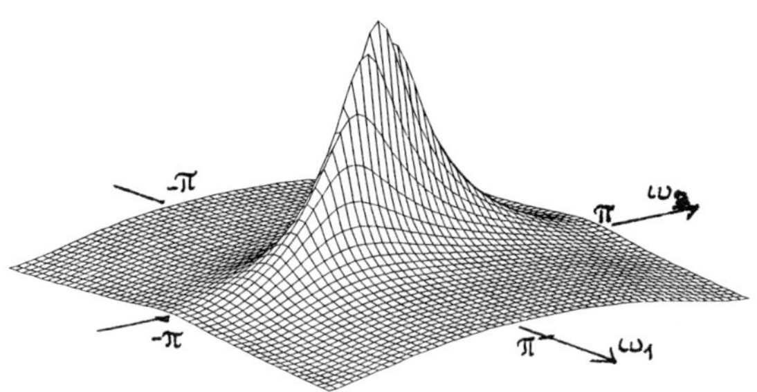

Fig.4. The amplitude response.

The corresponding low-sensitive network is given in Fig. 5, where $d_1=0.390$, $d_2=0.6328$, $d_3=0.5804$, $d_4=0.8936$, $\beta=0.8$. The network exhibits an attenuation zero in $\omega_1=\omega_2=0$ where the transfer function is independent from the above coefficients.

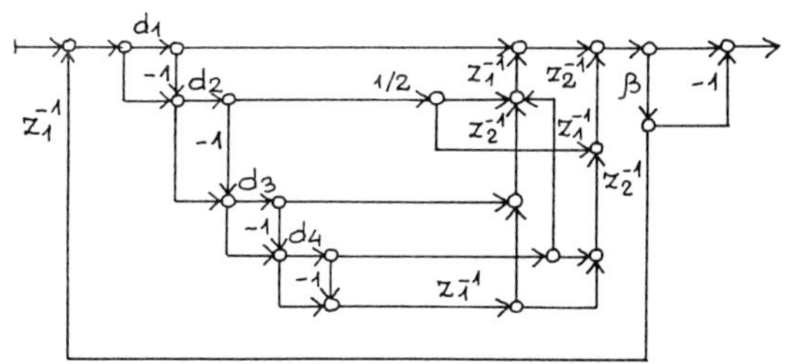

Fig.5. The filter from Example 1.

3. THE MULTILOOP TECHNIQUE

Another technique possible [8] consists in application of the multiloop structure. The filter obtained exhibits a characteristic frequency where the transfer function is independent from the values of the coefficients in the filter structure. Thus the amplitude sensitivity which is a continuous function of the frequency tends to be small in the neighborhood of the characteristic frequency which is possibly located in a passband.

Let consider synthesis of an all-pole filter $1/Q(\underline{z})$. We use the three groups of disjoint loops (cf. Fig. 6). There are M_{p1} loops $\beta_{11}z_1^{-1}$ (group 1), M_{p2} loops $\beta_{21}z_2^{-1}$ (group 2), and some loops (group 3) being disjoint with those above. There is $M_{p1}\le M_1$, $M_{p2}\le M_2$.

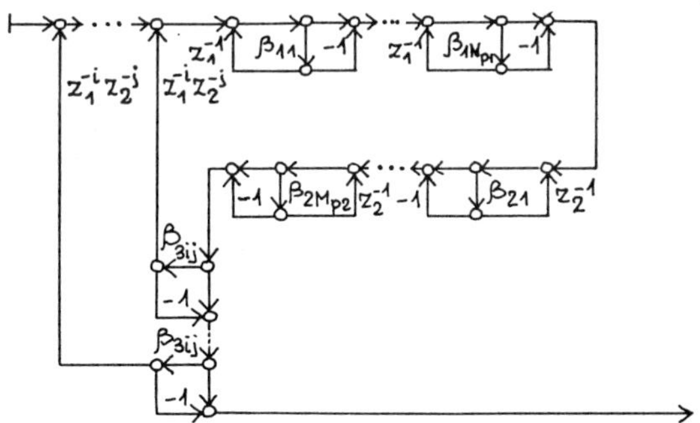

Fig.6. The multiloop structure.

If only the loops of the types 1 and 2 are used, we have $0<\beta_{kl}<1$, $q_{10}<0$, $q_{01}<0$

$\beta_{11}+\ldots+\beta_{1M_{p1}}=-q_{10}$, $\beta_{21}+\ldots+\beta_{2M_{p2}}=-q_{01}$,

and summing up over all the ordered pairs of respektive β's we get

$$\sum\tfrac{1}{2}\beta_{11}\cdot\beta_{11}"=q_{20},\ \sum\tfrac{1}{2}\beta_{21}\cdot\beta_{21}"=q_{02},$$
$$\sum\beta_{11}\cdot\beta_{21}"=q_{11}\ .$$

The expressions for the other coefficients q_{ij} we get from the Mason theorem. It is useful to get

$\beta_{11}=\ldots=\beta_{1M_{p1}}=\beta_1$, $\beta_{21}=\ldots=\beta_{2M_{p2}}=\beta_2$,

where $0<\beta_{kl}<1$ $(k=1,2,\ \ l=1,\ldots,M_{pk})$.

Then, using the loops of type 3 (with transfer functions $t_{3ij}z_1^{-i}z_2^{-j}$, $t_{3ij}\ge 0$) we get $-\beta_1M_{p1}-t_{310}=q_{10}$, $q_{10}<0$, $q_{01}<0$,

$-\beta_2M_{p2}-t_{301}=q_{01}$, $\frac{1}{2}M_{p1}(M_{p1}-1)\beta_1^2-t_{320}=q_{20}$,

$\frac{1}{2}M_{p2}(M_{p2}-1)\beta_2^2-t_{302}=q_{02}$,

$M_{p1}M_{p2}\beta_1\beta_2-t_{311}=q_{11}$,

$-\frac{1}{2}M_{p1}(M_{p1}-1)M_{p2}\beta_1^2\beta_2-t_{321}=q_{21}$,

$-\frac{1}{2}M_{p1}M_{p2}(M_{p2}-1)\beta_1\beta_2^2-t_{312}=q_{12}$, and so on. The design problem is to find out the numbers M_{p1}, M_{p2}, and β's. Then, some other groups of loops can be added. The technique can be extended in order to implement simultaneously the numerator $P(\underline{z})$.

EXAMPLE 2. Let

$$\mathbb{P} = \begin{bmatrix} 0.02500 & 0.01799 & 0.02500 \\ 0.01799 & 0.05837 & 0.01799 \\ 0.02500 & 0.01799 & 0.02500 \end{bmatrix},$$

$$\mathbb{Q} = \begin{bmatrix} 1.00000 & -0.58701 & 0.08614 \\ -0.58701 & 0.33142 & -0.05056 \\ 0.08614 & -0.05056 & 0.00177 \end{bmatrix}.$$

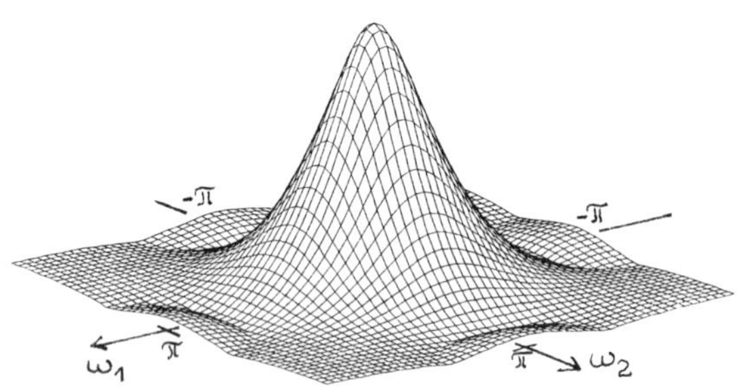

Fig.7. The amplitude response.

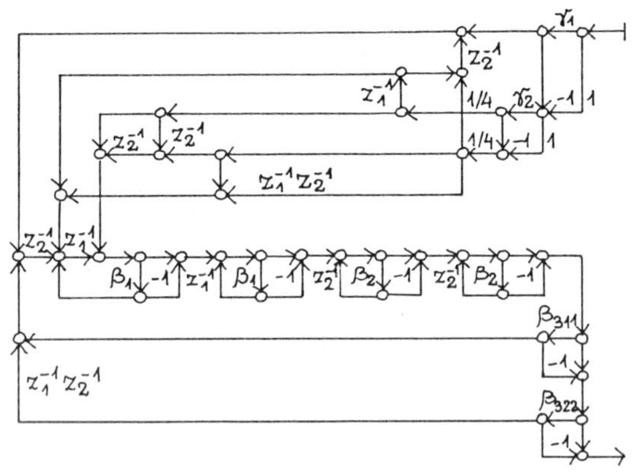

Fig.8. The filter obtained.

There is $\gamma_1=0.25340$, $\gamma_2=0.58158$, $\beta_1=\beta_2=0.29350$, $\beta_{311}=0.05282$, $\beta_{322}=0.02392$ in the obtained filter.

4. CASCADE STRUCTURE

We briefly report the synthesis in the cascade structure [9] which consists of the digital two-ports (cf. Fig.9) implementing the lines of infinite attenuation (cf. Fig. 10).

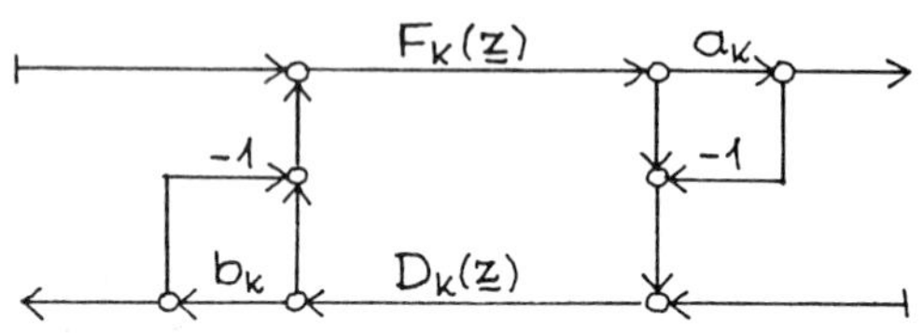

Fig. 9. A digital two-port.

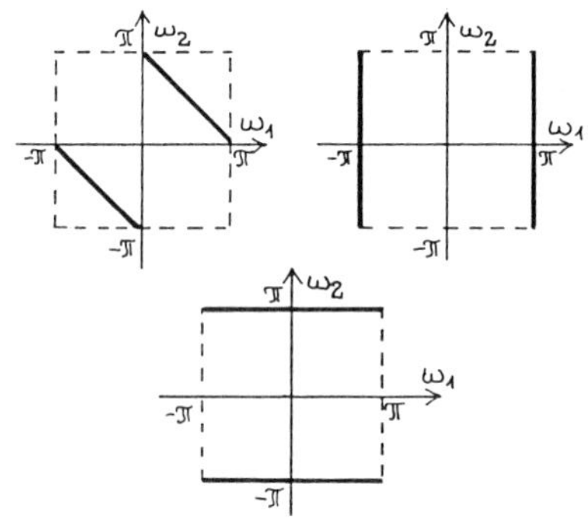

Fig.10. Lines of infinte attenuation.

REFERENCES

[1] Fettweis, A., Proc. IEEE 74(1986) 270.

[2] Domański, M., 2-D digital ℓ_1-pseudo-passive filters, in: Signal Processing III (North-Holland, Amsterdam, 1986) pp. 721-724.

[3] Domański, M., Fettweis, A., Int. Journal Circuit Theory and Appl. 17 (1989) 103.

[4] Domański, M., Digital ℓ_p-passive 2-D filters (Politechnika Poznańska, Poznań 1990, in Polish).

[5] Domański, M., 2-D ℓ_1-passive digital filters, to be published.

[6] Domański, M., Proc. IEE G 138 (1991) 229.

[7] Domański, M., Treizieme Coll. Traitement du Signal et des Images (Juan-les-Pins 1991), pp. 205-208.

[8] Domański, M., Proc. 6th Int. Symp. Networks, Systems, Signal Proc. (Zagreb 1989) pp. 216-219.

[9] Domański, M., Proc. Int. Symp. Signals, Syst., Electronics (URSI, Erlangen, 1989) pp. 807-810.

H_∞ OPTIMAL MULTICHANNEL LINEAR DECONVOLUTION FILTERS, PREDICTORS AND SMOOTHERS

M.J. GRIMBLE

INDUSTRIAL CONTROL UNIT, UNIVERSITY OF STRATHCLYDE, U.K.

Abstract

A solution is presented to the H_∞ optimal deconvolution filtering, smoothing and prediction problems for multivariable, discrete, linear signal processing problems. A weighted H_∞ cost-function is minimized where the dynamic weighting function can be chosen for robustness improvement. The signal and noise sources can be correlated and signal channel dynamics can be included in the system model. An example is presented to demonstrate the advantages of the approach.

1. Introduction

The H_∞ deconvolution filtering problem involves the estimation of the signal input to a communication channel where only noisy output measurements are available. The channel can be represented by a dynamical system which therefore distorts the gain and phase characteristics of the signal. The cost function involves the H_∞ norm of the weighted estimation error power spectrum. The measurement noise can be correlated with the signal source and can be coloured.

The H_∞ filtering problem was first considered by Grimble (1987[1]) and by Elsayed and Grimble (1989[2] ; 1990[3]). The importance of the H_∞ filter in signal processing problems was recognized early in its development. However, the role of the H_∞ in the robust control problem was not realized until more recently (Doyle et al 1989[4]). The H_∞ smoothing problem was described by Grimble (1991[5]).

2. Signal Processing Problem

The system, signal and noise models are assumed to be time-invariant and discrete-time and to be represented in transfer function or polynomial form. The dependence upon the indeterminate z^{-1} is often not shown explicitly to simplify notation. The z^{-1} term will represent the unit-delay operator in the time-domain expressions, or will represent the z-transform complex number in the completing the square optimization argument which follows.

The signal processing problem is shown in Fig. 1. The signal to be estimated $\{s(t)\}$ is assumed to enter a transmission channel $W(z^{-1})$ and to then be corrupted by white $\{v(t)\}$ and coloured noise $\{n(t)\}$ measurement noise, respectively. The signal source $W_s(z^{-1})$ and coloured noise $W_n(z^{-1})$ models are driven by the white noise sources $\{\xi(t)\}$ and $\{\omega(t)\}$, respectively. All noise sources are assumed to be zero mean.

The covariance matrix for the white measurement noise $\{v(t)\}$ is denoted by $R = R^T = E\{v(t)v^T(t)\} \geq 0$ and $\{v(t)\}$ is statistically independent of the other noise terms. The measurement noise $\{n(t)\}$ may, however, be correlated with the signal source, and the covariance matrix for the signal $[\xi(t)^T, \omega^T(t)]^T$ is defined as:

$$E\left\{\begin{bmatrix}\xi(t)\\ \omega(t)\end{bmatrix}[\xi^T(t),\ \omega^T(t)]\right\} = \begin{bmatrix}I & G\\ G^T & I\end{bmatrix} \tag{1}$$

There is no loss of generality in assuming the white noise sources $\{\xi(t)\}$ and $\{\omega(t)\}$, in equation (1), are normalized to the identity matrix.

2.1 System equations

By inspection of Fig. 1 the following system equations apply:

Signal model: $s = W_s\xi$

Channel output : $y = Ws$

Measurement noise: $n = W_n\omega$

Observations: $z = v + W_n\omega + WW_s\xi$ (2)

where $s(t) \varepsilon R^{n_s}$, $y(t) \varepsilon R^{n_y}$ and $z(t) \varepsilon R^{n_z}$ and n_y and n_z denotes the number of measured outputs.

After manipulation the system equations become:

Estimate:

$$\hat{s} = (H_f z^{-\ell})z = (H_f z^{-\ell})(v + W_n\omega + WW_s\xi) \tag{3}$$

Estimation error :

$$\begin{aligned} e \triangleq s - \hat{s} &= W_s\xi - (H_f z^{-\ell})(v + W_n\omega + WW_s\xi) \\ &= (I - H_f z^{-\ell}W)W_s\xi - (H_f z^{-\ell})(v + W_n\omega) \end{aligned} \tag{4}$$

Weighted error :

$$\begin{aligned} e_o \triangleq W_p(s - \hat{s}) &= W_pW_s\xi - W_p(H_f z^{-\ell})(v + W_n\omega + WW_s\xi) \\ &= W_p(I - H_f z^{-\ell}W)W_s\xi - W_p(H_f z^{-\ell})(v + W_n\omega) \end{aligned} \tag{5}$$

2.2 Summary of Assumptions

(2i) Noise sources are assumed to be zero mean. The measurement noise source $\{v(t)\}$ is statistically independent of the other noise sources and has a covariance matrix R. The covariance matrix for the white noise signal $[\xi(t)^T,\ \omega(t)^T]^T$ is denoted by $\begin{bmatrix}I & G\\ G^T & I\end{bmatrix}$.

(2ii) The signal channel W and the coloured measurement noise model W_n can be assumed to be asymptotically stable.

(2iii) There can be no unstable hidden modes in the individual plant subsystems or in the cascade system WW_s.

2.3 Polynomial Matrix Descriptions

The system models have the following polynomial matrix descriptions:

$$[W\ \ W_n] = A^{-1}[C\ \ C_n] \tag{6}$$

$$W_s = A_s^{-1}C_s \tag{7}$$

where A, C, C_n, A_s and C_s are polynomial matrices in z^{-1} of compatible dimensions. Introduce the left-coprime pair C_1 and A_1 which satisfy:

$$CA_s^{-1} = A_1^{-1}C_1 \tag{8}$$

and assume, without loss of generality, $A_1(0) = I$ and $A_s(0) = I$. Note for later use from Assumption (2iii) in §2.2 the model CA_s^{-1} can include no unstable hidden modes.

2.4 Spectral Factorization

The solution of the H_∞ deconvolution filtering problem (Shaked, 1976[6]) requires the introduction of a spectral-factor Y_f which satisfies the following equation:

$$Y_fY_f^* = WW_sW_s^*W^* + W_nW_n^* + R + WW_sGW_n^* + W_nG^TW_s^*W^* \tag{9}$$

Replacing the system models by their polynomial matrix counterparts, note that $Y_f \varepsilon R(z^{-1})^{n_y x n_y}$ may be written as :

$$Y_f = (A_1A)^{-1}D_f \tag{10}$$

where D_f is a polynomial matrix left spectral-factor defined using,

$$D_fD_f^* = C_1C_sC_s^*C_1^* + A_1C_nC_n^*A_1^* + A_1ARA^*A_1^* + C_1C_sGC_n^*A_1^* + A_1C_nG^TC_s^*C_1^* \tag{11}$$

The Hurwitz spectral-factor D_f exists, if and only if,

$$\text{rank } [[C_1C_s, \; A_1C_n] \begin{bmatrix} I & G \\ G^T & I \end{bmatrix}, A_1AR] = n_y$$

The spectral factor D_f is strictly Hurwitz, if and only if, $[C_1, C_s, \; A_1C_n] \begin{bmatrix} I & G \\ G^T & I \end{bmatrix}^{1/2}$ and $A_1AR^{1/2}$ have no left common factors with zeros on the unit-circle in the z-plane.

Assumption (2iv) : The noise and signal source models are assumed to be such that D_f exists and is strictly Hurwitz. •

3. The H_2 Cost Minimization Problem

In this section the problem of the minimization of the variance of the weighted estimation error is considered (Grimble and Johnson, 1988[7]). These results are needed before attention can turn to the H_∞ optimal estimation problem.

The H_2 optimal deconvolution problem involves the minimization of the estimation error:

$$e(t \mid t-\ell) = s(t) - \hat{s}(t \mid t-\ell) \tag{12}$$

where $\hat{s}(t \mid t-\ell)$ denotes the optimal linear estimate of the signal $\{s(t)\}$ at time t, given observations $\{z(t)\}$ up to time $(t-\ell)$. The scalar $\ell = 0$ (filtering), $\ell < 0$, (fixed lag smoothing) and $\ell > 0$, (prediction). The usual estimation error cost-function which is minimized has the form:

$$J = \text{trace}\{E\{e(t \mid t-\ell)e^T(t \mid t-\ell)\}\} \tag{13}$$

where $E\{.\}$ denotes the unconditional expectation generator.

Introduce the asymptotically stable n_s square weighting function matrix: $W_p = A_p^{-1}B_p$, where A_p and B_p denote square polynomial matrices. Then the H_2 cost-function to be minimized becomes:

$$J = \text{trace } \{E\{W_pe(t \mid t-\ell)(W_p\, e(t \mid t-\ell))^T\}\}$$

and this can be written in complex integral form:

$$J = \text{trace } \{ \frac{1}{2\pi j} \oint_D \{W_p\Phi_{ee}W_p^*\} \frac{dz}{z} \} \tag{14}$$

where D denotes the unit-circle contour in the z-plane and the adjoint $W_p(z^{-1})^* = W_p^T(z)$.

The following theorem summarizes the main results.

Theorem 3.1 : *H_2 Optimal Deconvolution Estimator*

Consider the signal processing system and assumptions described in §2 and illustrated in Fig. 1. The optimal deconvolution filter can be calculated.

Spectral factor :

$$D_fD_f^* = C_1C_sC_s^*C_1^* + A_1C_nC_n^*A_1^* + A_1ARA^*A_1^* + C_1C_sGC_n^*A_1^* + A_1C_nG^TC_s^*C_1^* \tag{15}$$

where $CA_s^{-1} = A_1^{-1}C_1$ and from the assumptions D_f is strictly Hurwitz.

Diophantine Equations:

The polynomial matrix solution (F_o, G_o, S_o), with F_o of smallest degree, is required:

$$A_sF_o + G_oD_f^*z^{-g} = C_s(C_s^*C_1^* + GC_n^*A_1^*)z^{\ell-g} \tag{16}$$

$$-CF_o + S_oD_f^*z^{-g+\ell_o} = ((C_nC_n^* + ARA^*)A_1^* + C_nG^TC_s^*C_1^*)z^{\ell-g}$$

where for filtering and prediction $\ell_o \triangleq \ell$ and for smoothing $\ell_o \triangleq 0$, and g denotes the smallest positive integer which ensures these equations involve polynomials in the indeterminate z^{-1}.

Robustness weighting diophantine equation:

If the cost-weighting W_p is dynamic the diophantine equation solution (F_1, N), with F_1 of smallest degree, is required using:

$$A_pF_1 + ND_f^*z^{-g} = B_pF_o \tag{18}$$

Optimal estimator:

$$H_f = (A_s^{-1}G_o + B_p^{-1}N)D_f^{-1}A_1A \tag{19}$$

•

Proof : Presented in Grimble (1982[8]). •

Lemma 3.1 : *Properties of the Deconvolution Filter, Predictor and Smoother*

The properties of the optimal deconvolution filter, defined in Theorem 3.1, can be detailed as follows.

Minimal cost :

$$J_{min} = \text{trace } \{ \frac{1}{2\pi j} \oint_D \{F_1D_f^{*-1}D_f^{-1}F_1^* + W_pT_oW_p^*\} \frac{dz}{z} \}$$

where (20)

$$T_o = W_s(I - (W_s^*W^* + GW_n^*)(Y_fY_f^*)^{-1}(WW_s + W_nG^T))W_s^* \tag{21}$$

Implied equation determining stability:

$$C_1G_oz^{-\ell} + A_1S_oz^{-\ell+\ell_o} = D_f \tag{22}$$

•

4. H_∞ Multivariable Estimator

The H_∞ multivariable estimation problem can be solved using the H_2 optimization results in the following lemma derived by Kwakernaak (1984[9]). Note that W_σ represents a weighting filter.

Lemma 4.1 : ***Auxiliary H_2 Minimization Problem***

Consider the auxiliary problem of minimizing the H_2 criterion:

$$J_\sigma = \frac{1}{2\pi j} \oint_{|z|=1} \text{trace}\{W_\sigma(z^{-1})X(z^{-1})W_\sigma^*(z^{-1})\} \frac{dz}{z} \quad (23)$$

Suppose that for some real-ratio matrix: $W_\sigma^*(z^{-1})W_\sigma(z^{-1}) \geq 0$, the cost function J_σ is minimized by a function $X(z^{-1}) = X^*(z^{-1})$, for which $X(z^{-1}) = \lambda^2 I_r$ (a real constant matrix on $|z| = 1$). Then the function $X(z^{-1})$ also minimizes:

$$J_\infty = \sup_{|z|=1} \|X(z^{-1})\|_2 = \sup_{|z|=1} \{\sigma_{max}\{X(z^{-1})\}\} \quad (24)$$

where $\|X(z^{-1})\|_2$ denotes the spectral norm.

•

Proof : The proof is similar to that in Kwakernaak, 1984[9].

•

A solution has been presented in §3 on an H_2 minimization problem which is similar to that referred to in the above lemma. To relate the two problems let the dynamic weighting function W_p be written as the product of two terms:

$$W_p = W_\sigma W_{po} \quad , \quad W_p \,\varepsilon\, R(z^{-1})^{n_s x n_s} \quad (25)$$

Then the function X above can be related to the weighted estimation error as:

$$X \triangleq W_{po}\Phi_{ee}W_{po}^* \quad , \quad X \,\varepsilon\, R(z^{-1})^{n_s x n_s} \quad (26)$$

The W_σ represents the weighting term, still to be determined, which appears in Lemma 4.1. The W_{po} will be assumed to represent the dynamic weighting selected by the designer to frequency shape the estimation error spectrum, so that the cost-function to be minimized is given as:

$$J_\infty = \sup_{|z|=1} \{\sigma_{max}\{X(z^{-1})\}\} = \sup_{|z|=1} \{\sigma_{max}\{W_{po}\Phi_{ee}W_{po}^*\}\}$$

The problem is now to determine the function W_σ which will ensure the conditions of the Lemma are satisfied.

4.1 *Derivation of the weighting filter W_σ*

Assume that the scalar λ in the above lemma is for the moment known. To ensure the conditions of the above lemma are satisfied, the W_σ must now be found which leads to an equalizing solution

$$X(z^{-1}) = \Lambda\Lambda^T = \lambda^2 I_r \text{ on } |z| = 1.$$

The matrix Λ introduced in the above expression may be defined as $\Lambda = \lambda I$ at the optimum. However, once the maximum singular value has been minimized and a solution for the controller obtained, it is sometimes possible to improve upon the solution by defining a more general Λ. The improvement can, for example, be in terms of reducing the maximum values of the remaining singular values.

Comparison of (20) and (23), evaluated at the H_2 optimum, therefore gives:

$$F_1 D_f^{*-1} D_f^{-1} F_1^* + W_\sigma W_{po} T_o W_{po}^* W_\sigma^* = W_\sigma \Lambda\Lambda^T W_\sigma^*$$

or

$$F_1 D_f^{*-1} D_f^{-1} F_1^* = W_\sigma(\Lambda\Lambda^T - W_{po}T_o W_{po}^*)W_\sigma^* \quad (27)$$

where

$$F_1 \,\varepsilon\, P(z^{-1})^{n_s x n_y}$$

Introduction the stable minimum-phase transfer-function $S \,\varepsilon\, R(z^{-1})^{n_s x n_s}$ which satisfies:

$$SS^* = \Lambda\Lambda^T - W_{po}T_o W_{po}^* \quad (28)$$

and assume that Λ is chosen so that the transfer S is invertible. From (27) and (28) now obtain:

$$F_1 D_f^{*-1} D_f^{-1} F_1^* = W_\sigma SS^* W_\sigma^*. \quad (29)$$

Define the spectral-factor D_{fo} which satisfies:

$$D_{fo}^* D_{fo} = D_f D_f^* \quad (30)$$

and let the left-coprime pair (F, D) satisfy:

$$D^{-1}F = F_1 D_{fo}^{-1}, \qquad F \,\varepsilon\, P(z^{-1})^{n_s x n_y} \quad (31)$$

Then, to ensure the equalizing solution is obtained, from (29) and (31):

$$D^{-1}FF^*D^{*-1} = W_\sigma SS^* W_\sigma^* \quad (32)$$

Let F_s denote a Hurwitz polynomial matrix which satisfies:

$$F_s F_s^* = FF^* \quad (33)$$

then from (32):

$$W_\sigma = D^{-1}F_s S^{-1}. \quad (34)$$

To ensure a suitable definition $F_s \,\varepsilon\, P(z^{-1})^{n_s x n_s}$ is obtained the following assumption is required:

Assumption (4i): The number of signal channels $n_s \leq$ number of observation channels $n_z = n_y$.

The linear equation which enables F_s to be computed can now be found. Introduce the left-coprime polynomial matrices $(\bar{S}, \bar{F}_s)$ as:

$$\bar{S}^{-1}\bar{F}_s = F_s S^{-1} W_{po} \quad (35)$$

then from (34) and (35):

$$W_\sigma W_{po} = A_p^{-1}B_p = D^{-1}\bar{S}^{-1}\bar{F}_s. \quad (36)$$

Hence, identify the desired weighting polynomials which ensure an equalizing solution is obtained, as:

$$A_p = \bar{S}D \quad \text{and } B_p = \bar{F}_s \quad (37)$$

and if W_σ is required this may be found from (36) as:

$$W_\sigma = A_p^{-1}B_p W_{po}^{-1} = (\bar{S}D)^{-1}\bar{F}_s W_{po}^{-1} \quad (38)$$

4.2 *Robust weighting diophantine equation*

To calculate $(N, \bar{F}_s)$ a more appropriate form of equation (18) is required. Substituting for the weightings in (37) the robustness diophantine equation

(18) becomes:

$$\bar{S}\, DF_1 + ND_f^* z^{-g} = \bar{F}_s F_o$$

but from (31) : $DF_1 = FD_{fo}$ and hence obtain:

$$\bar{S}\, FD_{fo} + ND_f^* z^{-g} = \bar{F}_s F_o \qquad (39)$$

The optimal estimator follows from (19) but note that B_p^{-1} enters the expression and hence from (33), (35) and (37) the matrix F must be of normal full rank. The reason for introducing Assumption (4.i) should now be apparent.

Theorem 4.1 : ***H_∞ Estimator for Multichannel Estimation Problems***

Consider the system shown in Fig. 1 and assume that the cost-function:

$$J_\infty = \sup_{|z|=1} \{\sigma_{max}\{W_{po}\Phi_{ee}W_{po}^*\}\}$$

is to be minimised where Φ_{ee} denotes the spectrum of the estimation error $e(t \mid t-\ell) = s(t) - s(t \mid t-\ell)$ where $\ell = 0$, $\ell < 0$ and $\ell > 0$ for filtering, smoothing and prediction problems, respectively. The H_∞ optimal estimator may be computed by first calculating D_f from (15) and then obtaining (G_o, S_o, F_o, g) from (16) and (17). Also define the transfer T_o from (21), then the stable minimum phase transfer S is defined using:

$$SS^* = \Lambda\Lambda^T - W_{po}T_oW_{po}^* = \lambda^2 I - W_{po}T_oW_{po}^* \qquad (40)$$

The left-coprime pair (F, D) can be found using:

$$D^{-1}F = F_1 D_{fo}^{-1} \text{ where } D_{fo} \text{ satisfies : } D_{fo}^* D_{fo} = D_f D_f^* \qquad (41)$$

Let F_s denote the Hurwitz spectral factor satisfying : $F_s F_s^* = FF^*$ and introduce the left-coprime polynomial matrices $(\bar{S}, \bar{F}_s)$ using:

$$\bar{S}^{-1}\bar{F}_s = F_s S^{-1} W_{po} \qquad (42)$$

The H_∞ optimal linear estimator can then be found by first calculating the solution (N, F), with F of smallest degree, of the linear equation:

$$\bar{S}\, FD_{fo} + ND_f^* z^{-g} = \bar{F}_s F_o \qquad (43)$$

***Optimal estimator*:**

$$H_f = (A_s^{-1}G_o + \bar{F}_s^{-1}N)D_f^{-1}A_1 A \qquad (44)$$

***Optimum function and minimum cost*:**

$$X_{min} = \lambda_o^2 I \quad \text{and} \quad J_\infty = \lambda_o^2 \qquad (45)$$

where λ_o^2 is the smallest scalar λ^2 such that equations (40) to (43) are satisfied. •

Proof : The theorem follows by collecting the previous results and by invoking the results of Theorem 3.1. •

5. Conclusions

A solution to the optimal H_∞ multichannel deconvolution estimation problem was presented which included coloured measurement noise which could be correlated with the signal to be estimated. The results apply to filtering, prediction and fixed-lag smoothing problems by appropriate specification of the integer ℓ.

The deconvolution problem has not previously been considered from a H_∞ norm minimization viewpoint but it should offer potential advantages for uncertain systems.

References

1. Grimble, M.J. 1988, *H_∞ design of optimal linear filters*, MTNS Conference, Pheonix, Arizona, published in C.I. Byrnes, C.F. Martin and R.E. Saeks, North Holland, Amsterdam, pp. 553-540.
2. ElSayed, A., and M.J. Grimble, 1989, *A new approach to the H_∞ design of optimal digital linear filters*, IMA Journal of Mathematical Contr., and Information, 6, pp. 233-251.
3. Grimble, M.J. and ElSayed, A., 1990, *Solution of the H_∞ optimal linear filtering problem for discrete-time systems*, IEEE Trans. on Acoustics Speech and Signal Processing, Vol. 38, No. 7, pp. 1092-1104.
4. Doyle, J.C., K. Glover, P.P. Khargonekar, and B.A., Francis, 1989, *State-space solutions to standard H_2 and H_∞ control problems*, IEEE Trans. on Auto. Contr., 34, 8, pp. 831-846.
5. Grimble, M.J., 1991, *H_∞ fixed-lag smoothing filter for scalar systems*, IEEE Trans. on Signal Processing, Vol. 39, No. 9, pp. 1955-1963.
6. Shaked, U., 1976, *A general transfer function approach to linear stationary filtering and steady-state optimal control problems*, Int. J. Control, 24, 6, pp. 741-770.
7. Grimble, M.J. and Johnson, M.A., 1988, *Optimal Control and Stochastic Estimation Theory and Applications, Volumes 1 and 2*, John Wiley.
8. Grimble, M.J., 1991, *Multichannel optimal linear deconvolution filters, predictors and smoothers*, Industrial Control Unit Research Report No. 345.
9. Kwakernaak, H., 1984, *Minimax frequency domain optimization of multivariable linear feedback systems*, Presented at the 1984 World Congrees, Budapest, Hungary.

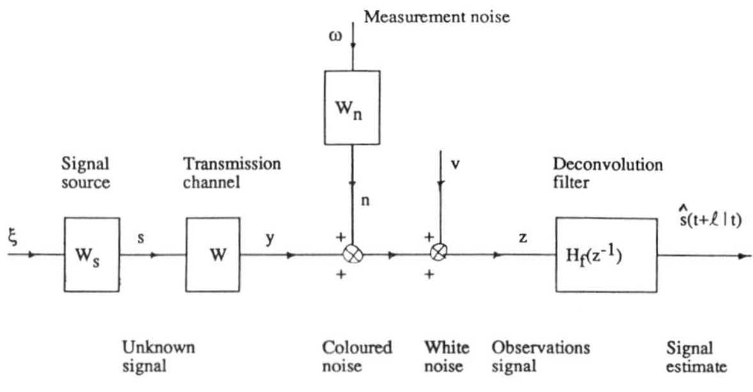

Fig. 1: Optimal H_2 and H_∞ Deconvolution Filtering Problem System Models

SIGNAL PROCESSING VI: Theories and Applications
J. Vandewalle, R. Boite, M. Moonen, A. Oosterlinck (eds.)

Underdetermined Growing and Sliding Window Covariance Fast Transversal Filter RLS Algorithms

DIRK T.M. SLOCK
Institut EURECOM
2229 route des Crêtes
Sophia Antipolis
F-06560 Valbonne, FRANCE

Abstract—**We consider the *covariance* formulation, or hence a rectangular window for the least-squares (LS) criterion. We are specifically interested in the case when the window length is shorter than the filter length, leading to an *underdetermined* LS problem. We develop *Fast Recursive LS* (RLS) algorithms in *Transversal Filter* form (FTF) for the two cases of an underdetermined *growing* or *sliding* window (UGWC/USWC). The UGWC case allows for a fast initialization of GWC algorithms, obviating the need for a soft-constraint initialization. The USWC case obviously represents an adaptive algorithm, while the growing window case can easily be converted into one by constantly sweeping the window length from one up to some finite value. These two classes of underdetermined LS algorithms can be situated in between the Normalized Least-Mean-Square (NLMS) algorithm and the conventional RLS algorithms. An inherent *prewhitening* mechanism renders their convergence less sensitive to the coloring of the input signal spectrum than is the case for the NLMS algorithm. Their underdetermined LS character additionally endows them with relatively fast tracking characteristics. Relations to existing algorithms, modifications and simplifications will also be discussed.**

1. Introduction

The tracking characteristics of the (N)LMS algorithm are inherent in the iterative nature of the algorithm; the rate of convergence depends on the eigenvalue spread of the input covariance matrix. For RLS algorithms, the tracking is independent of the eigenvalue spread and can be made arbitrarily fast by appropriate choice of the (effective) window length. Fast RLS algorithms and especially the FTF algorithms [1] take only $7N/8N$ (stabilized form) flops with prewindowing (and exponential weighting), or $14N/15N$ flops in the SWC formulation [7],[8]. This should be compared to the $2N$ flops for the LMS or Normalized LMS algorithms.

Here we propose a new class of algorithms which are intermediate between LMS and fast RLS in complexity, and in tracking performance (or sometimes the best). Before we get more specific, let us consider the following motivations.

1.1. Stochastic Considerations

A general stochastic gradient algorithm is characterized by the following update equations:

$$\begin{aligned} \epsilon_k^p &= d_k + W_{k-1} X_k \\ W_k &= W_{k-1} - \mu\, \epsilon_k^p X_k^H R^{-1} \end{aligned} \qquad (1)$$

where ideally $R = \mathrm{E}\, X_k X_k^H$ is used to compensate for the non-white spectrum of the input signal x_k. The NLMS algorithm omits this decorrelation ($R = \|X_k\|^2\, I$), while the RLS algorithm uses the sample covariance matrix R_k as an estimate for $\frac{1}{\mu} R$. One should notice that in the RLS algorithm, the size of the decorrelation matrix is automatically put equal to the length of the (FIR) filter that is being adapted. In many applications however, this coupling may not be well motivated. Consider for instance the problem of acoustic echo cancellation. Here, FIR filters of very great lengths are being used. On the other hand, the input signal is typically a speech signal for which in speech coding, prediction filters with an order of typically only 10 are used (to whiten the signal). So one could consider using a banded matrix for R^{-1} in (1) with a (one-sided) bandwidth of only 10. This has been proposed in [3], in which non-fast and fast recursive algorithms are then developed. The resulting algorithm could be considered to be of the symmetric Instrumental Variable type. Note that this approach corresponds very closely to first sending the input signal through an adaptive prediction filter of order 10, and then using the whitened input signal as input to the (N)LMS algorithm. This last approach has been proposed in [4]. One may remark that in this last approach, when considering tracking non-stationarities, the lag of the (N)LMS algorithm has to be added to the lag in the prewhitening process. In the Fast Newton Transversal Filters (FNTF) of [3] on the other hand, some peculiarities have been observed [5] when the input signal is nonstationary, due to the fact that the optimal low-order prediction filter may change considerably over the length of the adapted FIR filter.

1.2. Deterministic Considerations

Nevertheless, the algorithms described above yield a significant improvement in convergence speed (in particular a substantial reduction in sensitivity to eigenvalue disparity of R) when the input signal is strongly colored. However, it is useful to also take a look at some deterministic aspects. Indeed, based on only the above stochastic considerations, it is not possible to explain why the RLS algorithm converges significantly faster than the (N)LMS algorithm (e.g. by a factor of 5) even in the white input signal case! This advantage of the RLS algorithm is due to the fact that the RLS algorithm solves a (overdetermined) set of equations exactly at each time step (N equations are sufficient to get an unbiased estimate of the N FIR filter coefficients). Although asymptotically for large window lengths, the rectangular window and the exponential window give equivalent performance for corresponding window lengths, there is reason to believe that the rectangular window leads to better tracking performance for short windows in strongly non-stationary environments.

The SWC RLS algorithm provides the least-squares solution to $L \geq N$ equations, which are obtained by writing out the error signal at L consecutive time steps:

$$\min_{W_k} \left\{ \left\| d_{L,k} - X_{N,L,k} W_k^H \right\|^2 \right\} \qquad (2)$$

When $L < N$, the above problem is underdetermined. A unique solution can still be found though by taking the minimum-norm solution. In fact, the NLMS algorithm (with $\overline{\mu} = 1$) provides this minimum-norm solution for $L = 1$. We shall consider algorithms for the cost function (2) with $1 \leq L \leq N$, which hence cover a continuous range between the SWC RLS algorithm and the NLMS algorithm, and which have the deterministic projection interpretation of making L consecutive error samples exactly zero.

As far as stochastic considerations are concerned, it turns out that in the Underdetermined Growing and Sliding Window Covariance (UGWC/USWC) RLS algorithms considered here, there is also inherently a replacement of the input signal by its Lth order prediction residual. From this point of view, it is possible to view some recently proposed algorithms as approximations to the USWC RLS algorithms.

2. A Unified Least-Squares Criterion

We shall mostly adhere to the notation of [1],[7],[8]. The following criterion will prove to encompass quite a number of different adaptation techniques. Consider

$$\min_{W_{N,k}} \left\{ \left\| d_{L,k} - X_{N,L,k} W^H_{N,k} \right\|^2_{S^{-1}_k} + \left\| W^H_{N,k} - W^H_{N,k-M} \right\|^2_{T^{-1}_k} \right\} \tag{3}$$

where $\|v\|^2_S = v^H S v$ and S_k, T_k are Hermitian positive definite matrices. Before discussing various applications of this formulation, let us consider the minimization of this criterion (3). Setting the gradient equal to zero, and applying the Matrix Inversion Lemma (MIL) to solve the resulting linear equation in $W_{N,k}$, we get

$$\begin{aligned} W_{N,k} &= W_{N,k-M} + \left(d^H_{L,k} - W_{N,k-M} X^H_{N,L,k}\right) \\ &\quad \times \left(X_{N,L,k} T_k X^H_{N,L,k} + S_k\right)^{-1} X_{N,L,k} T_k \ . \end{aligned} \tag{4}$$

The Hessian of the quadratic criterion (3) is

$$H_k = T^{-1}_k + X^H_{N,L,k} S^{-1}_k X_{N,L,k} \tag{5}$$

to which we have applied the MIL to obtain (4). Usually, $M = 1$, unless we consider block processing. In all applications we consider here, the idea is to equalize the nature of both terms that constitute H_k. In a first class of algorithms, we take $L = 1$ and $S_k = I$, which renders the nature of the second term in (5) of a covariance type (be it of low rank). Hence, T^{-1}_k is chosen to be representative of the covariance matrix of $X_N(k)$. In the stochastic Newton algorithms, we choose (apart from a scalar multiple) $T^{-1}_k = R_N = \mathrm{E}\, X_N(k) X^H_N(k)$. In the absence of second order statistics, the RLS algorithms with exponential weighting factor λ uses $T^{-1}_k = \lambda R_{N,k-1}$ where $R_{N,k} = \sum_{i=0}^{k} \lambda^{k-i} X_N(i) X^H_N(i)$. In the Fast Newton algorithm [3], $R_{N,k}$ is replaced by the Maximum Entropy extension of a banded restriction of $R_{N,k}$ (with banded inverse).

In a second class of algorithms, one takes $T_k = I$, which has all eigenvalues equal to a constant. Hence, now S_k should be chosen such that the eigenvalues of the second term in (5) are either constant or zero (since the term is not of full rank). The way to achieve this exactly is to make the second term in (5) proportional to a projection matrix and hence to choose $S_k = \mu_k X_{N,L,k} X^H_{N,L,k}$. With $L < N$, we have an "underdetermined" problem (strictly speaking, we only get an underdetermined problem as $\mu_k \to 0$). The filter solution (4) can now be rewritten as

$$\begin{aligned} W_{N,k} &= W_{N,k-M} + \\ &\tfrac{1}{1+\mu_k} \left(d^H_{L,k} - W_{N,k-M} X^H_{N,L,k}\right) R^{-1}_{L,N,k} X_{N,L,k} \end{aligned} \tag{6}$$

where $R_{L,N,k}$ is the sample covariance matrix for a SWC problem with filter length L and window length N interchanged. One immediately recognizes the NLMS algorithm when $M = L = 1$. Using (6), one can establish the following relation between the *a posteriori* and the *a priori* error vectors

$$d^H_{L,k} - W_{N,k} X^H_{N,L,k} = \frac{\mu_k}{1+\mu_k} \left(d^H_{L,k} - W_{N,k-M} X^H_{N,L,k}\right) \ . \tag{7}$$

One may note that the convergence of (6) is governed by a product of matrices of the form $I - \frac{1}{1+\mu_k} P_{X^H_{N,L,k}}$, from which one can see a prewhitening effect of order $L-1$ transpire. In particular for $\mu_k = 0$, this matrix becomes the projection matrix $P^{\perp}_{X^H_{N,L,k}}$, leading to L a posteriori errors being zeroed.

3. Underdetermined Sliding Window Covariance FTF

Here we take $M = 1$ and fixed $L < N$. Using the relation (7) between a priori and a posteriori error vector, and the shift structure of the data vectors, one can establish the following recursion

$$\begin{bmatrix} d_{L,k} - X_{N,L,k} W^H_{N,k-1} \\ 0 \end{bmatrix} = \begin{bmatrix} \epsilon^H_N(k|k-1) \\ 0 \\ -\frac{\mu_{k-1}}{1+\mu_{k-1}} \epsilon^H_N(k-L|k-2) \end{bmatrix} + \frac{\mu_{k-1}}{1+\mu_{k-1}} \begin{bmatrix} 0 \\ d_{L,k-1} - X_{N,L,k-1} W^H_{N,k-2} \end{bmatrix} \tag{8}$$

where $\epsilon_N(k|i) = d(k) - W_{N,i} X_N(k)$. This is one recursion that is part of the joint-process section, and it takes N operations, for the computation of $\epsilon_N(k|k-1)$. On the other hand, we also have the well-known order update relations for $R^{-1}_{L,N,k}$. We get by multiplying with $X_{N,L+1,k}$ and using the shift structure of this matrix,

$$\begin{aligned} & R^{-1}_{L+1,N,k} X_{N,L+1,k} \\ &= \begin{bmatrix} R^{-1}_{L,N,k} X_{N,L,k} \\ 0 \end{bmatrix} + B^H_{L,N,k} \beta^{-1}_{L,N}(k) r^H_{L,N,k} \\ &= \begin{bmatrix} 0 \\ R^{-1}_{L,N,k-1} X_{N,L,k-1} \end{bmatrix} + A^H_{L,N,k} \alpha^{-1}_{L,N}(k) e^H_{L,N,k} \ . \end{aligned} \tag{9}$$

Let

$$g_k = \frac{1}{1+\mu_k} \left(d^H_{L,k} - W_{N,k-1} X^H_{N,L,k}\right) R^{-1}_{L,N,k} X_{N,L,k} \tag{10}$$

Then combining (8) and (9), one can work out a recursion for g_k and the remaining part of the joint-process section can be written as

$$g_k = \frac{\mu_{k-1}}{1+\mu_k} g_{k-1} + \xi_1 r^H_{L,N,k} + \xi_2 e^H_{L,N,k} \tag{11}$$

$$W_{N,k} = W_{N,k-1} + g_k \tag{12}$$

where ξ_1 and ξ_2 are quantities that take $O(L)$ operations to compute. Apart from the $O(L)$ operations, the joint-process part takes strictly speaking $4N$ multiplications and $4N$ additions, but in fact one should count these as $5N$ multiply-accumulates. A small simplification occurs when $\mu_k \equiv 0$. In that case we have

$$g_k = \epsilon_N(k|k-1) \alpha^{-1}_{L-1,N}(k) e^H_{L-1,N,k} \ . \tag{13}$$

4. Underdetermined Growing Window Covariance FTF

In this case we let L and k grow. We shall consider two possibilities, corresponding to whether we take $k - M$ to be some fixed time instant (the beginning of a sweep) or $M = 1$. For the first possibility, which we shall refer to as UGWC, let the sweep start at time instant $k_0 = k - M$ and we shall let μ be constant within one sweep (of $L = M$ increasing from 1 to some maximum value, and $k = k_0 + L$). In this case, it turns out to be convenient to work with the two components that make up the error vector separately. So we can rewrite (6) as

$$W_{N,k} = W_{N,k_0} + \tfrac{1}{1+\mu} d^H_{L,k} R^{-1}_{L,N,k} X_{N,L,k} - \tfrac{1}{1+\mu} W_{N,k_0} P_{X^H_{N,L,k}} . \quad (14)$$

To find a recursion for $W_{N,k}$ in this case, we shall exploit one more recursion, namely the one for the projection operators, viz.

$$P_{X^H_{N,L+1,k}} = P_{X^H_{N,L,k}} + P_{r_{L,N,k}} = P_{X^H_{N,L,k-1}} + P_{e_{L,N,k}} \quad (15)$$

of which we need the simultaneous order and time update. This, together with (9) and the shift structure of $d_{L,k}$, allows us to find the recursion

$$W_{N,k+1} = W_{N,k} + \tfrac{1}{1+\mu}\left(d^H_{L+1,k+1} - W_{N,k_0} X^H_{N,L+1,k}\right) \alpha^{-1}_{L,N}(k+1) e^H_{L,N,k+1} \quad (16)$$

So the joint-process part takes $2N$ operations.

In the second possibility, denoted as UGWC2, we let $M = 1$. The derivation parallels more that of the USWC algorithm, and in particular we can establish the following recursion, using the relation (7),

$$d^H_{L+1,k+1} - W_{N,k} X^H_{N,L+1,k+1} = \left[\epsilon_N(k+1|k) \;\; \tfrac{\mu_k}{1+\mu_k}\left(d^H_{L,k} - W_{N,k-1} X^H_{N,L,k}\right)\right] \quad (17)$$

Using this recursion, which is again part of the joint-process part, and the recursion (9) (simultaneous order and time update part), we get

$$g_{k+1} = \frac{\mu_k}{1+\mu_{k+1}} g_k + \frac{1}{1+\mu_{k+1}} \quad (18)$$

$$\times \left(d^H_{L+1,k+1} - W_{N,k} X^H_{N,L+1,k+1}\right) A^H_{L,N,k+1} \alpha^{-1}_{L,N}(k+1) e^H_{L,N,k+1}$$

$$W_{N,k+1} = W_{N,k} + g_{k+1} \quad (19)$$

where $g_{k_0} = 0$. Here, the joint-process part takes $3N$ multiplications and $3N$ additions, or $4N$ multiply-accumulates.

5. Prediction Parts

The joint-process parts of the above algorithms require vectors of N prediction error samples. To derive updates for these prediction error vectors, we rely on the biorthogonalization procedure explained in [9]. Basically, this procedure consists of a double application of the well-known update identity for projection operators. It leads to a formulation based on 2×2 rotations Φ (circular) and Ψ (hyperbolic). We shall not provide the details for these rotation matrices. However, the following general facts hold. These rotations can be organized so that they take $2N$ multiply-accumulates. And the entries of these rotations can be computed in $O(L)$ operations.

5.1. Underdetermined Sliding Window Covariance (USWC) FTF

The prediction part needs to provide $e_{L,N,k}$ and $r_{L,N,k}$ for all k (when $\mu_k \equiv 0$, change L by $L-1$ in the procedure below). This can be accomplished with the following four rotations

$$\begin{bmatrix} [e^H_{L,N-1,k-1} \;\; 0] \\ \delta^H_{L+1,N,k-1} \end{bmatrix} = \Psi^{(1)} \begin{bmatrix} e^H_{L,N,k-1} \\ \delta^H_{L,N,k-2} \end{bmatrix} \quad (20)$$

$$\begin{bmatrix} e^H_{L,N,k} \\ \gamma^H_{L+1,N,k} \end{bmatrix} = \Phi^{(1)} \begin{bmatrix} [0 \;\; e^H_{L,N-1,k-1}] \\ \gamma^H_{L,N,k-1} \end{bmatrix} \quad (21)$$

$$\begin{bmatrix} [r^H_{L,N-1,k-1} \;\; 0] \\ \delta^H_{L,N,k-1} \end{bmatrix} = \Phi^{(2)} \begin{bmatrix} r^H_{L,N,k-1} \\ \delta^H_{L+1,N,k-1} \end{bmatrix} \quad (22)$$

$$\begin{bmatrix} r^H_{L,N,k} \\ \gamma^H_{L,N,k} \end{bmatrix} = \Psi^{(2)} \begin{bmatrix} [0 \;\; r^H_{L,N-1,k-1}] \\ \gamma^H_{L+1,N,k} \end{bmatrix} \quad (23)$$

So the prediction part takes $8N$ multiply-accumulates.

5.2. Underdetermined Growing Window Covariance (UGWC) FTF

The prediction part needs to provide e_{l-1,N,k_0+l}, $l = 1, 2, \ldots, L$. This can be accomplished with the following three rotations

$$\begin{bmatrix} [e^H_{l-1,N-1,k_0+l} \;\; 0] \\ \delta^H_{l,N,k_0+l} \end{bmatrix} = \Psi^{(1)} \begin{bmatrix} e^H_{l-1,N,k_0+l} \\ \delta^H_{l-1,N,k_0+l-1} \end{bmatrix} \quad (24)$$

$$\begin{bmatrix} e^H_{l-1,N,k_0+l+1} \\ \gamma^H_{l,N,k_0+l+1} \end{bmatrix} = \Phi^{(1)} \begin{bmatrix} [0 \;\; e^H_{l-1,N-1,k_0+l}] \\ \gamma^H_{l-1,N,k_0+l} \end{bmatrix} \quad (25)$$

$$\begin{bmatrix} e^H_{l,N,k_0+l+1} \\ r^H_{l,N,k_0+l+1} \end{bmatrix} = \Psi^{(3)} \begin{bmatrix} e^H_{l-1,N,k_0+l+1} \\ r^H_{l-1,N,k_0+l} \end{bmatrix} \quad (26)$$

So the prediction part takes $6N$ multiply-accumulates.

6. Prewindowing Simplification

It is possible to choose S_k such that

$$X_{N,L,k} X^H_{N,L,k} + S_k = (1+\mu) R_{L,k} \quad (27)$$

where $R_{L,k}$ is the covariance matrix of size L for the prewindowing problem. In order to have forgetting, on would like to introduce an exponential weighting. Because the order and time dimensions are interchanged here w.r.t. the usual situation, one actually has to resort to data sequence weighting [10]. The projection matrix $P_{X^H_{N,L,k}}$ we had before is now replaced by

$$X^H_{N,L,k} R^{-1}_{L,k} X_{N,L,k} = [I_N 0]\, P_{X_{L,k}} \,[I_N 0]^H \quad (28)$$

which is a $N \times N$ submatrix of a projection matrix onto a L-dimensional subspace of prewindowed data vectors. If, with the data sequence weighting introduced, $\lambda^N \ll 1$, then this submatrix is almost a projection matrix.

With the choice of S_k as in (27), we don't have the simple relation between a priori and a posteriori error vector as we had in (7). The way to proceed now is to consider both parts of the error vector separately. This is immediately applicable to the UGWC FTF algorithm. In its prediction part, (24) can be dropped now. So the joint-process and prediction parts of the UGWC algorithm together take $6N$ operations with the prewindowing choice.

7. Further Discussions

7.1. Relation to Previous Work

It turns out that the USWC RLS algorithm has already been proposed as early as 1984 [11], where it was called the Affine Projection Algorithm (APA). This algorithm seems to have been largely confined to a community of Japanese researchers [12],[13]. The APA has a computationaly complexity of $O(NL) + O(L^2)$. Perhaps the limited popularity of the USWC RLS algorithm is due to this complexity, which becomes comparable to $O(N^2)$ (conventional RLS) when L becomes comparable to N. In [12], the Block Orthogonal Projection Algorithm (BOPA) was proposed, which minimizes the USWC criterion (with $\mu = 0$), but only every L samples. This can be considered equivalent as running the UGWC algorithm constantly for sweeps of L samples. Again, the algorithm proposed in [12] is a non-fast algorithm, and furthermore, our UGWC approach provides useful filter estimates also at the intermediate time instants during a sweep. In [13], the same USWC criterion is optimized, but at a rate intermediate between the APA and the BOPA rates.

In [14], a new adaptive filtering algorithm is proposed, based on conjugate gradient techniques. It turns out that the resulting algorithm is exactly equivalent to the USWC algorithm, but is again not fast. In [15], a stochastic gradient technique is applied to the USWC criterion. The algorithm is made fast by exploiting divide and conquer techniques.

In the USWC algorithm with $\mu = 0$, the gradient has the direction of $e^H_{L-1,N,k} = A_{L-1,N,k}X_{N,L,k}$. If we approximate this vector by $[e_{L-1}(k)\, e_{L-1}(k-1) \cdots e_{L-1}(k-N+1)]$, and introduce a stepsize, then we get exactly a stochastic gradient version of an instrumental variable technique. If we furthermore use a second stochastic gradient algorithm to generate the prediction errors $e_{L-1}(.)$, then we get the algorithm proposed [16].

Finally, a non-fast algorithm for the UGWC problem has been provided in [17]. We have proposed a fast alternative.

7.2. Computational Complexity

We propose to exploit a certain shift invariance structure in the problem to arrive at the UGWC and USWC FTF RLS algorithms. These algorithms have a complexity of $O(N) + O(L)$, which is similar to the complexity of some other FTF algorithms, but never exceeds the complexity of the SWC FTF algorithm. The cheapest one appears to be the UGWC algorithm (with the prewindowing approximation). The average complexity can be further reduced by leaving gaps between successive sweeps (data unused). This may lead to an algorithm that has an average complexity that is lower than that of the (N)LMS algorithm, but still converges faster in badly conditioned circumstances [13].

7.3. Tracking Performance

When one compares the RLS and (N)LMS algorithms for a variety of tracking problems, then either one can be the better one, depending on the characteristics of the problem. So the USWC RLS algorithm enlarges the scope of possible algorithms for a better match to the nonstationary characteristics of the tracking problem.

When one demands very high tracking capacity, the prewindowed FTF algorithm with exponential weighting is not satisfactory (numerical stability considerations impose a lower bound on the weighting factor). One has to resort to the SWC FTF algorithm, which is also limited in flexiblity however, due to the rectangular window of constant length. The underdetermined algorithms offer more flexibility: μ can be taken time-varying in the USWC and UGWC2 algorithms (this is not the same as throwing in a stepsize factor in the SWC algorithm to boost the gain, which makes the algorithm lose its LS nature), L can be varied in the UGWC algorithms and one can restart at any desired time. Note that our UGWC algorithms with $\mu > 0$ are different from the GWC algorithm with a soft constraint initialization due to S_k.

In the FNTF algorithm of [3], the convergence dynamics are determined by the spectral variation of the covariance matrix of an order L prewhitened version of the input signal. Hence, with an AR(L) process as input, the convergence is pretty much insensitive to the input spectrum. In the underdetermined FTF algorithms, the deterministic counterpart to the above stochastic consideration holds. If the input signal is persistently exciting of order at most L (sum of L exponentials), then the underdetermined algorithms converge exactly in L steps in the noisless (perfect modeling) case.

An alternative to an algorithm with $\mu > 0$ is to use the algorithm with $\mu = 0$ and use coefficient filtering [18]. The algorithms with $\mu = 0$ provide the fastest tracking characteristics, leading to a substantial estimation noise component in the excess MSE. This can be reduced with a possibly time-varying trade-off between tracking speed and noise averaging.

References

[1] J.M. Cioffi and T. Kailath. "Fast, recursive least squares transversal filters for adaptive filtering". *IEEE Trans. on ASSP*, ASSP-32(2):304–337, April 1984.

[2] G. Carayannis, D. Manolakis, and N. Kalouptsidis. "A fast sequential algorithm for least squares filtering and prediction". *IEEE Trans. ASSP*, ASSP-31(6):1394–1402, 1983.

[3] G.V. Moustakides and S. Theodoridis. "Fast Newton Transversal Filters – A New Class of Adaptive Estimation Algorithms". *IEEE Trans. SP*, SP-39(10):2184–2193, Oct. 1991.

[4] S.H. Leung, B.L. Chang, and S.M. Lau. "On Improving the Convergence Rate of the LMS Algorithm Using Prefiltering". In *Proc. ISSPA90, Signal Processing, Theories, Implementations and Applications*, Gold Coast, Australia, Aug. 27-31 1990.

[5] T. Petillon, A. Gilloire, and S. Theodoridis. "Complexity Reduction in Fast RLS Transversal Adaptive Filters with Application to Acoustic Echo Cancellation". In *Proc. IEEE Int. Conf. ASSP*, San Fransisco, CA, March 1992.

[6] B. Porat. "Second-order equivalence of rectangular and exponential windows in least-squares estimation of Gaussian autoregressive processes". *IEEE Trans. ASSP*, ASSP-33(5):1209–1212, Oct. 1985.

[7] J.M. Cioffi and T. Kailath. "Windowed Fast Transversal Filters Adaptive Algorithms with Normalization". *IEEE Trans. on ASSP*, ASSP-33(3):607–625, June 1985.

[8] D.T.M. Slock and T. Kailath. "A Modular Prewindowing Framework for Covariance FTF RLS Algorithms". *Signal Processing*, 28(1), July 1992.

[9] D.T.M. Slock. "Reconciling Fast RLS Lattice and QR Algorithms". In *Proc. ICASSP 90 Conf.*, pages 1591–1594, Albuquerque, NM, April 3–6 1990.

[10] D.T.M. Slock and T. Kailath. "Fast Transversal Filters with Data Sequence Weighting". *IEEE Trans. Acoust. Speech Sig. Proc.*, ASSP-37(3):346–359, March 1989.

[11] K. Ozeki and T. Umeda. "An Adaptive Algorithm Using Orthogonal Projection to an Affine Subspace and its Properties". *Trans. IECE Japan*, J67-A:126–132, 1984.

[12] T. Furukawa, H. Kubota, and S. Tsuji. "The Orthogonal Projection Algorithm for Block Adaptive Signal Processing". In *Proc. ICASSP 89 Conf.*, pages 1059–1062, Glasgow, Scotland, May 23-26 1989.

[13] T. Furukawa, Y. Kimura, and H. Kubota. "A Fast Adaptive Algorithm Based on Orthogonal Projection onto Multi-Dimensional Subspace". In *Proc. ISSPA90, Signal Processing, Theories, Implementations and Applications*, Gold Coast, Australia, Aug. 27-31 1990.

[14] G.K. Boray and M.D. Srinath. "Conjugate Gradient Techniques for Adaptive Filtering". *IEEE Trans. Circuits and Systems-I: Fund. Theory and Appl.*, 39(1):1–10, 1992.

[15] J. Benesty, S. Li, and P. Duhamel. "A Gradient Based Adaptive Algorithm with Low Complexity, Fast Convergence, and Good Tracking Characteristics". In *Proc. IEEE Int. Conf. ASSP*, San Fransisco, CA, March 1992.

[16] M. Mboup, M. Bonnet, and N. Bershad. "Coupled adaptive Prediction and System Identification: A Statistical Model and Transient Analysis". In *Proc. IEEE Int. Conf. ASSP*, San Fransisco, CA, March 1992.

[17] A. Dembo and J. Salz. "On the Least Squares Tap Adjustment Algorithm in Adaptive Digital Echo Cancellers". *IEEE Trans. Communications*, COM-38(5):622–628, May 1990.

[18] M. Niedzwiecki. "Identification of Time-Varying Systems using Combined Parameter Estimation and Filtering". *IEEE Trans. ASSP*, 38(4):679–686, April 1990.

SIGNAL PROCESSING VI: Theories and Applications
J. Vandewalle, R. Boite, M. Moonen, A. Oosterlinck (eds.)

CLASSIFICATION OF TEXTURE BY AN ASSOCIATION BETWEEN A PERCEPTRON AND A SELF-ORGANIZING FEATURE MAP

Eric MAILLARD*, Benoit ZERR†, Jean MERCKLE*

* I. R. P. , 34 rue Marc SEGUIN, 68067 MULHOUSE-Cedex, FRANCE
† G.E.S.M.A. , 29200 BREST NAVAL, FRANCE

- ABSTRACT - An original pattern classifier is presented. This model introduces the collaboration between supervised and unsupervised neural networks. A Multi Layer Perceptron is trained to classify patterns. A Feature Map is trained on the same data set, then its cells are labellized. Two kinds of uncertainty regions of the parameter space are detected: confusion regions and empty regions. When a pattern from one of these regions is presented, the Feature Map restrains the classification results of the Perceptron. Sigma-pi units are used in order to "mix" the answer of each constituent model to produce the final output of the global model. We show the efficiency of this paradigm through classification of textures.

1. INTRODUCTION

The process of pattern recognition consists roughly in two stages : feature extraction and classification [1]. The first step extracts useful information from raw data by mapping them on a feature space. This process is application dependant. The feature vectors are then classified by assigning them a class. Two ways of improving the efficiency of the classification are, first to banish wrong classification when the previous knowledge allows a right one, second to detect unknown patterns to avoid the risk of assessment error due to a lack of knowledge.
Neural networks are efficiant classifiers [2], they are divided in two categories according to the data they need during learning. Supervised neural nets can form arbitraty association between input and output vectors given a training set of associations (input/output). After training they exhibit generalization capacities, so any input pattern raises an output signal. During unsupervised learning, neural nets cluster the input vectors according to a similarity or distance measure. So they are able to detect novelty when presented a test pattern.
The specific capacities of each approach are added in a new paradigm under name of collaboration. The first aim ot this association is to increase the reliability of pattern classification using supervised neural net. After a description of the two neural nets, the mechanism of collaboration is presented and some simulations on texture classification are repported.

2. DESCRIPTION OF THE MODELS INVOLVED

A Multi-Layer Perceptron (MLP) is trained with the classical algorithm known as the back-propagation of the gradient of the error with addition of a momentum to the updating of the weights [3] [4]. The weights are updated according to:

$$\Delta w_{ji}(t+1) = \eta(\delta_{pj} o_{pi}) + \alpha \Delta w_{ji}(t)$$

Thanks to the nonlinear activity within each cell, the MLP can form arbitrarily complex decision regions (figure 1).

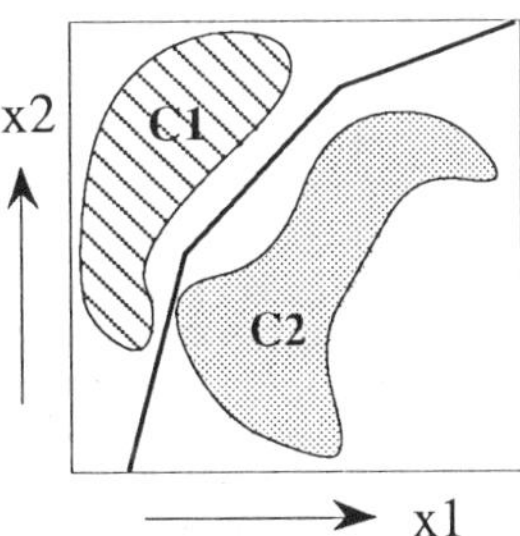

Figure 1: decision regions formed by a MLP.

These regions are formed while trying to correctly classify a set of input-output pairs of

vectors. The internal representation developed by the MLP during this phase allows the network to generalize. Thus the design of the training set and the scheduling of the training phase are crucial.
An adaptive scheduling is introduced. It is performed automatically by varing the time constant g of the activation fonction. The global error decrease is monitored. If the error stands still at a high value, then the time constant of the activation fonction is increased. The activation fonction is given by:

$$o_{pj} = \frac{1}{1 + e^{-net/g}}$$

A high value of the constant g brings the output of the neuron closer to 0.5 (figure 2), this allows the weight change to be maximum.

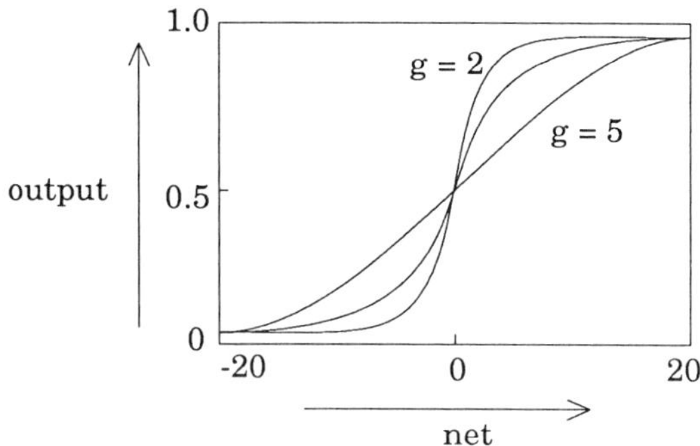

Figure 2: the activation function for various values of g.

Since the network updates its weights after each presentation of a pattern, an augmented change of the weights "shakes" the network. The time constant then slowly decreases toward its previous value, this ensures the global stability of the network.
When training is over, we can expect the MLP to be efficient in its classification task if the input pattern is strongly correlated with one of the training set. But when a new pattern is presented, the classification is deficient. In the better case, it is ambiguous: the activations of the output units are rather low. In the worst case the activation of one output unit is strong while it is low for the others: the pattern lies far from any decision boundary but also far from any known pattern. Whatever the classification, it would be useful to detect the novelty of the input pattern.
Kohonen's self-organizing feature map (SOFM) [5], [6] is a competitive paradigm. The map consists in a grid of interconnected neurons. Each neuron is connected to its eight closest neighbors, and to all the input neurons.

While the MPL is trained with input-output pairs, the SOFM learns without any teacher. The main feature of the map is that it self-organizes in a way that matches the probability density function of the input patterns.
During the training phase an input is presented, the neurons of the grid compete with each other in a winner-take-all manner. Then the winning neuron (i. e. the best matching cell) and the ones in its neighborhood update their weights to get closer to the input vector in the input space. The convergence of the network is forced via the decrease of the learning rate.
As the euclidian distance is used as a proximity measure, we introduce an activation function which allows direct competition between the cells during both training and testing:

$$o_k = \frac{1}{1 + \sqrt{\sum_i (x_i - w_{ki})^2}}$$

As the neuron gets closer to the input pattern, its output signal tends towars 1.0. So during competition, the most active neuron is the one closest to the input pattern. The updating law for the winner and its neighborhood is:

$$\mathbf{w}_i(t+1) = \mathbf{w}_i(t) + \alpha(t)\left[\mathbf{x}(t) - \mathbf{w}_i(t)\right]$$

At the end of the training phase, the grid has "vector quantized" the input space. We use a rectangular grid, and an exponential decrease for both the learning rate and the neighborhood. Once the organization phase is over, the training patterns with addition of noise are presented again to the network and the cells are labbelized by majority voting : they choose the more represented category among the pattern for which they win. If two or more categories are strongly represented, the cell is discarded. The remaining neurons can be viewed as prototypes for their classes (figure 3).

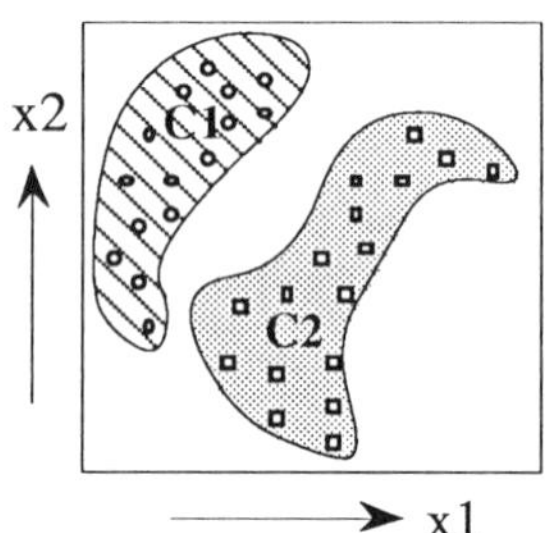

Figure 3: The SOFM matches the probability density function of the training set.

3. COLLABORATION BETWEEN THE MODELS

The aim of this collaboration is to give the MLP classification a confidence rate. The key issue is the possible openess of the decision regions formed by the MLP. The only assessment a MLP neuron can think of is : this pattern is on *that* side of my hyperplane. As the SOFM matches the probality density function of the input pattern, its neurons cluster where most of the training patterns are. The distance between a pattern and the closest neuron is an hint to the *familiarity* of that pattern.
The collaboration of the two neural networks is shown in figure 4.

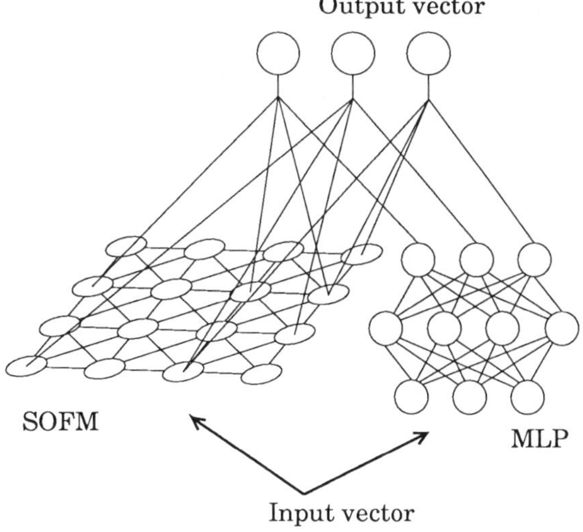

Figure 4: The architecture of the global model.

The output of each network are part of the input to a layer of sigma-pi units. The activation of the output units of the Perceptron are multiplied by the normalized activation of the Feature Map. The dynamic of the model is :
- the MLP classifies the input pattern : the activations of the output neurons are update
- the SOFM neurons within the vicinity of the pattern undergo a competition according to:

$$o_i(t+1) = o_i(t) + \frac{1}{N}\sum_j o_j(t) - \frac{1}{M}\sum_k o_k(t)$$

where N is the number of active cells of the same class and M is the number of other active cells
- the sigmap-pi units update their activations, using the outputs of the MLP and of the SOFM as combined input
- the classification is given by the ouput of the sigma-pi units.

The resulting classification has three different meanings depending on the activation of the sigma-pi units:
- one sigma-pi unit shows a strong activation since the networks agreed on the classification.
- two or more sigma-pi units are activated: the perceptron have classified the pattern while the SOFM decided that the pattern was in an boundary region so the classification is uncertain
- no sigma-pi unit is activate: the SOFM has detected that the pattern is absolutely new so no prior knowledge allows the classification.

4. TEXTURE CLASSIFICATION

There is no universal and precise definition for the word "texture". In fact, on a sea bottom image, an observer detects the limit between sandy surfaces and sand ripple areas as a change of "texture", eventhough both are genuine sand. The texture features used for classification must be translation invariant. So instead of working on pixels grey levels, a preprocessing task extracts basic patterns from homogeneous texture areas from four different texture categories: sand, sand ripples, rocks and pebbles. (see figure 5). The methods used in the preprocessing stage are cooccurrence matrix and sum and difference histograms. The coocurrence matrix is a set of second order spatial statistics for a given distance. Gagalowicz [7] shows the importance of those parameters in human visual system. Each element of the matrix is the proportion of coocurrence of pairs of luminance when their pixel locations differ by a given displacement vector. As it would be difficult to use all matrix elements as features, we compute some expressions that synthesize the coocurrence matrix values configuration : contrast, correlation, variance...[8].
As the forementionned method is time and memory resources consuming, we use an alternative to compute this matrix : the sum and difference histogram proposed by UNSER [9]. Those histograms represent, in a picture, the proportion of occurrence of grey level sum and grey level difference for a given pixel displacement vector. If the pixels pair is the realization of two random and independent variables the cooccurrence matrix can be compute from the sum and difference histograms. The reality is not absolutely so, but the experiment shows the validity of the approximation.
The training set is composed of 645 feature vectors manually labellized in four categories. We extract randomly ten exemplars of each categories and then train the neural nets independently on those 40 patterns. When the networks have converged, a test is carried out

using the 645 patterns. The result is shown in figure 6. When a test pattern is presented to the Feature Map, the network activates the prototypes that lies in the vicinity of the patterns. As shown in figure 6, the size of this vicinity modifies the performance of the final classification. When the size is too small, no prototypes are activate when the test pattern is slightly different from known exemplars. On the opposite, when the size is too big many prototypes from different categories are activated, inducing a confusion in the classification. But when the size is optimum, the performance of the model is promising. The network is able to detect wrongly labellized pattern in the test set.

5. CONCLUSION AND DISCUSSION

We presented here the collaboration of two neural networks as a new model of neural classifier. This model takes advantage of both a supervised and an unsupervised learning algorithm. The results on texture classification shows the efficiency of this new paradigm.
The feature space determination has a key role in the design of a pattern recognition system. Future developments include an adaptive decision-making process with the use of previous performance results as an hint to the feature space discrimination potential.
The discarded neurons of the self-organizing map will be used to address the major issue of texture boundary. Since they got mixed up between different texture patterns, those neurons may be used as boundary detectors.

REFERENCES

[1] Duda, R. O. and Hart, P. E., Pattern classification and scene analysis (John Wiley & sons, New York, 1973).
[2] Lippmann, R. P. and Ng, K., Technical report 894 (MIT, Lexington, 1991).
[3] Rumelhart, D. E., Hinton G. E. and Williams, R. J. "Learning internal representations by error propagation," in Rumelhart, D. E. and McClelland, J. L. (Eds.), Parallel Distributed Processing, vol. 1, (M.I.T. Press, Cambridge, MA, 1986).
[4] Widrow, B. and Lehr, M. A., Proceedings of the IEEE (1990) 78.
[5] Kohonen, T., Biological Cybernetics (1982) 43.
[6] Kohonen, T., Proceedings of the IEEE (1990) 78.
[7] Gagalowicz, A. and Song De Ma, CVGIP (1985) 30.
[8] Castrec, P. and Kernin J-P., Proceedings of the Underwater Defence Technology (1988).
[9] Unser, M., IEEE Pattern Analysis and Machine Intelligence (1986) 8.

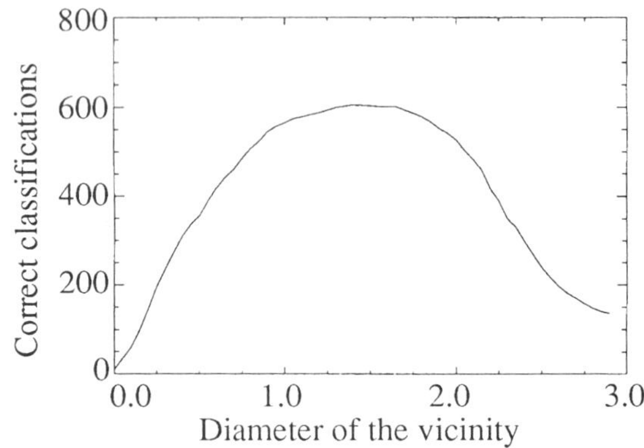

Figure 6: Results of the classification of 645 patterns.

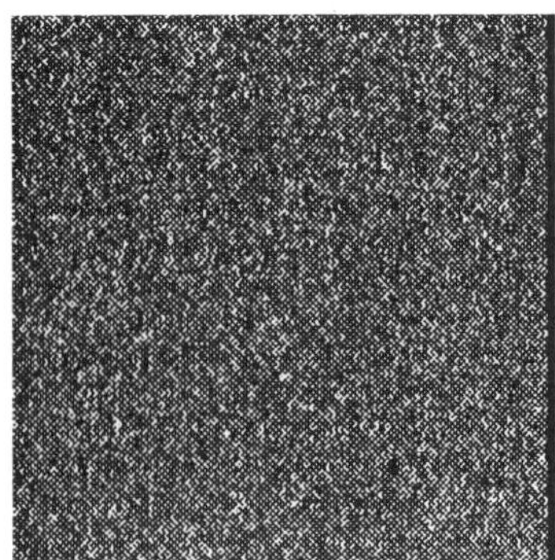

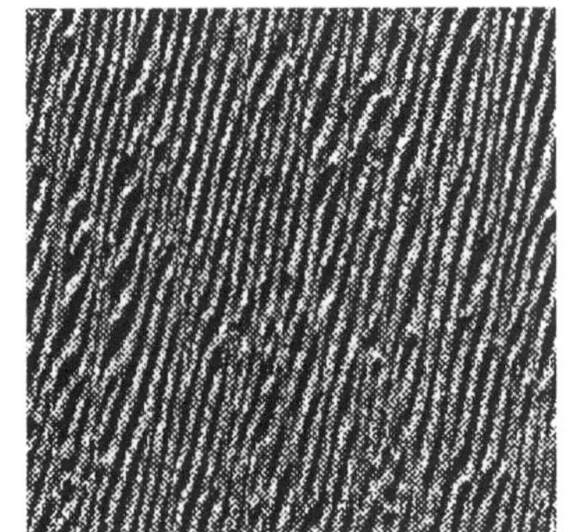

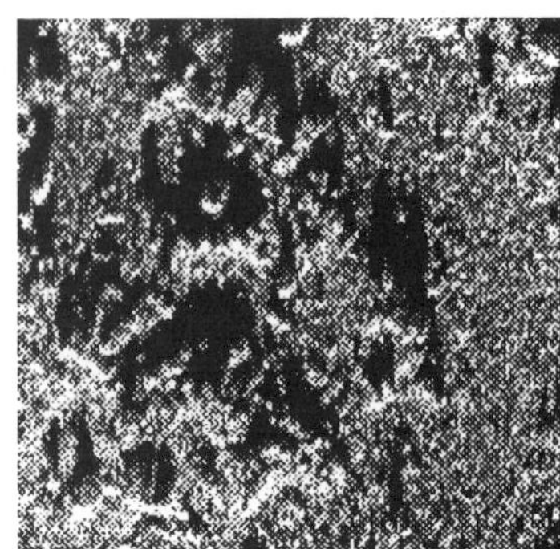

Figure 5: different textures from left to right : sand, sand ripples, rocks.

SIGNAL PROCESSING VI: Theories and Applications
J. Vandewalle, R. Boite, M. Moonen, A. Oosterlinck (eds.)

TWO-LAYER LEARNING VECTOR QUANTIZER FOR COLOR IMAGE QUANTIZATION

C. Kotropoulos E. Augé I. Pitas

Department of Electrical Engineering, University of Thessaloniki,
Thessaloniki 540 06, GREECE

A novel two-layer LVQ architecture is proposed which incorporates second order statistics in its training phase and allows training parallelism by splitting patterns into groups. The learning and recall procedures for the LVQ networks in the first and second layer are described. The proposed algorithm is based on statistical tests on the mean vectors and the dispersion matrices. A simplification based on a proximity test is presented. The application of the two-layer LVQ to color image quantization is also discussed.

1. INTRODUCTION

Neural networks is a rapidly expanding research field which attracted the attention of engineers and scientists in the last decade. One of the most prominent neural networks in the literature is the Learning Vector Quantizer (LVQ) [1,2]. It is an autoassociative nearest-neighbor classifier which classifies arbitrary patterns into p-many classes using an error-correction encoding procedure related to the competitive learning. It has already found extensive application for phoneme recognition [3,4] as well as in image processing, control and combinatorial optimization [2]. The dynamic weighting of input signals and a definition of neighborhood in the LVQ by a minimal spanning tree are proposed in [5]. The modification of the LVQ by using a differential competitive learning algorithm is discussed in [6].

The main contribution of this paper is the design of a novel two-layer LVQ architecture which incorporates second order statistics in its training phase and allows training parallelism by splitting patterns into groups. The proposed two-layer LVQ architecture is shown in Figure 1. It is comprised of N LVQ networks working independently in the first layer and a single LVQ network in the second layer. The training patterns of the first layer LVQ's are input patterns. The training patterns of the second layer LVQ are the weight vectors of the first layer after the convergence of the first layer LVQ's. The second layer classifies the weight vectors provided by the N networks of the first layer. Let us suppose that the N LVQ's of the first layer classify q-dimensional data into p-many classes, then the second layer LVQ has q input nodes and $N \times p$ output nodes at most. Some of them have been trained by patterns extracted from the same population, therefore they must be merged. Some others are reference vectors associated with different populations, therefore they must be preserved. The incorporation of homogeneity and proximity statistical tests based on second-order statistics in the second layer LVQ learning algorithm is proposed in order to group partial results provided by the first-layer LVQ's and considered in the evaluation of the final winner vectors. Thus, the proposed learning algorithm takes into account the presence of outliers and provides more accurate reference vectors for the clusters presented in the input patterns. Furthermore, the proposed two-layer LVQ architecture is easily parallelized and consequently makes faster computationally intensive tasks such as color image quantization and segmentation.

2. TWO-LAYER LVQ ALGORITHM

In the following, the learning and recall procedures of the first and second layer LVQ's are described. The learning procedure of each first layer LVQ is the typical one of a multiple winner LVQ [2,3]. The recall procedure of each LVQ network in the first layer is applied only to the patterns used for the training of this network. It provides the necessary information about the sample mean vector and the sample dispersion matrix of the classes produced at the output of the network. Let $\mathbf{V}_l = (v_{1l}, \ldots, v_{ql})^T \;\; l = 1, \ldots, p$ be the weight vectors for a first layer LVQ. The recall procedure of a network in the first layer has the following steps:

1. Initialize the sample mean vector $\mathbf{m}_j$, the sample dispersion matrix $\mathbf{S}_j$ and the number of patterns n_j associated with each class.

$$\mathbf{m}_j(0) = \mathbf{0}_{q\times 1} \;\; \mathbf{S}_j(0) = \mathbf{0}_{q\times q} \;\; n_j(0) = 0 \;\; j = 1, \ldots, p \tag{1}$$

2. Determine the class C_g represented by the weight vector $\mathbf{V}_g$ to which the training pattern $\mathbf{X}(k)$ is most closely associated with.

$$\mathbf{X}(k) \in C_g \text{ if } \|\mathbf{X}(k) - \mathbf{V}_g\| = \min_{j=1}^{p}\{\|\mathbf{X}(k) - \mathbf{V}_j\|\} \quad (2)$$

3. Increment the number of patterns belonging to C_g by one and update the sample mean vector and the sample dispersion matrix of this class.

$$\begin{aligned} n_g(k) &= n_g(k-1)+1 \\ \mathbf{d}_g(k) &= \mathbf{X}(k) - \mathbf{m}_g(k-1) \\ \mathbf{m}_g(k) &= \mathbf{m}_g(k-1) + \frac{1}{n_g(k)}\mathbf{d}_g(k) \\ \mathbf{Q}_g(k) &= \mathbf{Q}_g(k-1) + \frac{n_g(k-1)}{n_g(k)}\mathbf{d}_g(k)\, \mathbf{d}_g^T(k) \quad (3) \\ \mathbf{S}_g(k) &= \frac{1}{n_g(k)}\mathbf{Q}_g(k) \end{aligned}$$

For the remaining classes $j = 1, \ldots, p\ j \neq g$, the number of patterns, the sample mean and the sample dispersion matrix are not altered:

$$n_j(k) = n_j(k-1) \quad \mathbf{m}_j(k) = \mathbf{m}_j(k-1) \quad \mathbf{S}_j(k) = \mathbf{S}_j(k-1) \quad (4)$$

The LVQ network of the second layer is used to find the input vectors which are candidates for merging or not. The criterion of minimum Euclidean norm used in the LVQ is not sufficient for the above-described task because it does not take into account the presence of outliers. Consequently, additional tests must be implemented in order to test the similarity between the weight vector provided by the first layer LVQ's and the winner vector determined by the second layer LVQ. The following learning algorithm for the second layer LVQ is proposed:

1. Initialize randomly all the weight vectors $\mathbf{W}_l = (w_{1l}, \ldots, w_{ql})^T \; l = 1, \ldots, p'$ where $p \leq p' < Np$.

2. For each weight vector provided by the first layer LVQ's $\mathbf{V}(k) = (v_1(k), \ldots, v_q(k))^T$:

   a. Find the closest weight vector of the second layer LVQ, i.e., the final winner vector $\mathbf{W}_g(k)$ by using:

$$\|\mathbf{V}(k) - \mathbf{W}_g(k)\| = \min_{j=1}^{p'}\{\|\mathbf{V}(k) - \mathbf{W}_j(k)\|\} \quad (5)$$

   b. If there is an output node of the second layer LVQ which has not been activated, i.e., there exists a free class:

   (i) Test the similarity between $\mathbf{W}_g(k)$ and $\mathbf{V}(k)$

   (ii) If $\mathbf{W}_g(k)$ and $\mathbf{V}(k)$ are proved similar, then merge them. The final winner is updated as LVQ suggests:

$$\mathbf{W}_g(k+1) = \mathbf{W}_g(k) + a(t)[\mathbf{V}(k) - \mathbf{W}_g(k)] \quad (6)$$

   Modify the sample mean vector and the sample dispersion matrix. Let $n_g(k), \mathbf{m}_g(k), \mathbf{S}_g(k)$ denote the number of patterns, the sample mean vector and the sample dispersion matrix associated with the class of $\mathbf{W}_g$ respectively. Let also $n_V, \mathbf{m}_V, \mathbf{S}_V$ be the corresponding quantities associated with the class of $\mathbf{V}(k)$. The above-mentioned modifications are given by:

$$\mathbf{m}_g(k) = \frac{n_g(k-1)\mathbf{m}_g(k-1) + n_V\mathbf{m}_V}{n_g(k-1) + n_V} \quad (7)$$

$$\mathbf{S}_g(k) = \frac{n_g(k-1)\mathbf{S}_g(k-1) + n_V\mathbf{S}_V}{n_g(k-1) + n_V} \quad (8)$$

   If $\mathbf{V}(k)$ were previously merged with another class, say C_e, $\mathbf{m}_e(k), \mathbf{S}_e(k)$ would also be modified by replacing addition with subtraction in both the numerator and the denominator of (7),(8).

   (iii) Otherwise, assign $\mathbf{V}(k)$ to a free class, say C_f. Update $\mathbf{W}_f(k)$ by using (6). Set the sample mean vector $\mathbf{S}_f(k)$ and the sample dispersion matrix $\mathbf{m}_f(k)$ of the free class as follows:

$$\mathbf{m}_f(k) = \mathbf{m}_V \quad \mathbf{S}_f(k) = \mathbf{S}_V \quad (9)$$

   c. If there is no free class, merge unconditionally $\mathbf{V}(k)$ and $\mathbf{W}_g(k)$. Update the winner vector and modify the sample mean vector and the sample dispersion matrix associated with the class C_g by using (6)–(8).

3. Repeat step 2 for $t = 1, 2, \ldots$ until convergence is attained.

The recall procedure of the second layer LVQ is used for the classification of input patterns which have either been taken from the training set or not. It is used to determine the class C_g represented by $\mathbf{W}_g$ to which the input pattern $\mathbf{X}(k)$ is most closely associated with.

$$\mathbf{X}(k) \in C_g \text{ if } \|\mathbf{X}(k) - \mathbf{W}_g\| = \min_{j=1}^{p'}\{\|\mathbf{X}(k) - \mathbf{W}_j\|\} \quad (10)$$

The homogeneity of the winner vectors evaluated by the LVQ in the second layer and the input weight vectors provided by the LVQ's of the first layer can be tested (step 2.b.(i)) by employing statistical tests on the mean vectors as well as on the dispersion matrices. Let $\mu_g(k), \Sigma_g(k)$ denote the statistical mean vector and the statistical dispersion matrix associated with the class of $\mathbf{W}_g$ respectively. Let also μ_V, Σ_V be the corresponding quantities associated with the class of $\mathbf{V}(k)$. The homogeneity of the dispersion matrices Σ_g and Σ_V is tested by using the statistic [8] :

$$T_1 = n_g \ln|\mathbf{S}_g^{-1}\mathbf{S}| + n_V \ln|\mathbf{S}_V^{-1}\mathbf{S}| \quad (11)$$

where

$$\mathbf{S} = \frac{1}{n_g + n_V}(n_g\mathbf{S}_g + n_V\mathbf{S}_V) \quad (12)$$

and $|.|$ denotes the determinant of a matrix. The statistic T_1 is distributed as $\chi^2_{\frac{q(q+1)}{2}}$. The following two cases are considered in order to test the homogeneity of the mean vectors:

1. Inhomogeneous dispersion matrices: A test statistic for the hypothesis that μ_g, μ_V are homogeneous is given by [8]:

$$T_2 = (\mathbf{m}_g - \mathbf{m}_V)^T \; (\frac{1}{n_g}\mathbf{S}_g + \frac{1}{n_V}\mathbf{S}_V \;)^{-1} \; (\mathbf{m}_g - \mathbf{m}_V) \leq k_\alpha \quad (13)$$

The threshold k_α in (13) can be approximately evaluated by the procedure described in [8].

2. Homogeneous dispersion matrices: A test statistic for the hypothesis that μ_g, μ_V are homogeneous is the following [7]:

$$T_3 = |\mathbf{I}_{\mathbf{q}\times\mathbf{q}} + \mathbf{S}_s^{-1}\,\mathbf{B}|^{-1} \quad (14)$$

where

$$\begin{aligned} \mathbf{S}_s &= n_g\mathbf{S}_g + n_V\mathbf{S}_V \\ \mathbf{B} &= \sum_{i=g,V} n_i(\mathbf{m}_i - \mathbf{m})\,(\mathbf{m}_i - \mathbf{m})^T \quad (15) \\ \mathbf{m} &= \frac{n_g\mathbf{m}_g + n_V\mathbf{m}_V}{n_g + n_V} \end{aligned}$$

The statistic T_3 is approximately distributed according to the Wilks distribution $\Lambda(q, n_g + n_V - 2, 1)$.

In many practical cases the above-described rigorous procedure is computationally demanding since it requires matrix inversion (although matrix inversion lemma [9] can be invoked) and the calculation of the determinant of a matrix. In addition, the number of matrices to be handled may be extraordinarily large as is the case in color image quantization to be discussed in the section 3. These problems can be alleviated by testing if there is any intersection between the hyperellipsoids associated with the winner vector of the second layer LVQ and the weight vectors provided by the first layer LVQ's. A simple proximity test of the form:

$$\frac{|w_{ig} - v_i(k)|}{\sqrt{S_{g_{ii}}} + \sqrt{S_{V_{ii}}}} \leq 1 \quad (16)$$

where $S_{g_{ii}}, S_{V_{ii}}$ are the i-th diagonal elements of the sample dispersion matrices of the corresponding classes can be used in step 2.b.(i). Inequality (16) implies that the hyperellipsoids are approximated by hyperparallelipipedes and simply tests if there is overlap along the i-th dimension. If such an overlap exists along any dimension, it is inferred that $\mathbf{V}(k)$ and $\mathbf{W}_g$ are similar.

3. EXPERIMENTAL RESULTS

The two-layer LVQ architecture has been applied successfully to color image quantization both in serial and parallel implementations. Figure 2 shows an RGB color image of dimensions 256 × 256 with 24 bits per pixel. Color image quantization aims at encoding each color pixel with one byte instead of three, thus reducing the number of RGB triplets to 256. The major drawback of the typical LVQ algorithm is the excessively large duration of the training phase due to the number of input pixels, i.e., color triplets to be considered. Therefore, parallel LVQ implementation is considered for speed-up. A straightforward parallelization (e.g. a parallel computation of the Euclidean distances) results in an implementation which has heavy communication load. If communication operations are slow, the implementation is very slow as well. On the contrary, if training parallelism is applied by splitting pixels into groups and the above-described novel LVQ architecture is used, a faster training procedure is attained. A two-layer LVQ having 16 networks in the first layer has been used. Each LVQ network of the first layer has 3 input nodes and classifies 1000 randomly selected pixels into 256 classes. The second layer LVQ receives 4096 weight vectors determined after four iterations of the 16 first layer networks. Six iterations have been shown adequate for the training of the second layer LVQ. The result of quantization is shown in Figure 3. The performance of the proposed algorithm has been compared to the one of the following algorithms:

(1) standard LVQ (single-layer) [2,3] having 3 input nodes and producing 256 output classes

(2) a modified LBG vector quantization algorithm which produces a codebook of 256 color vectors in 8 iterations [4]

(3) nonuniform sampling of each color histogram by first equalizing each color histogram separately, uniformly sampling each equalized histogram in a number of predetermined samples and inverse transforming. Seven samples have been chosen for red channel, six for green channel and six for blue one. In total, 252 RGB triplets have been used.

In the comparative study outlined above, we have used as figures of merit the mean-squared-error (MSE) and the signal-to-noise ratio (SNR) measured in dB, given by:

$$\begin{aligned} \text{MSE} &= \sum_{i=1}^{M}\sum_{j=1}^{M} ||\mathbf{X}(i,j) - \tilde{\mathbf{X}}(i,j)||^2 \\ \text{SNR}_{\text{dB}} &= 10\log\frac{\text{MSE}}{\sum_{i=1}^{M}\sum_{j=1}^{M}||\mathbf{X}(i,j)||^2} \quad (17) \end{aligned}$$

where $\mathbf{X}(i,j) = (X_R(i,j), X_G(i,j), X_B(i,j))^T$ represents the (i,j) pixel in the original color image, $\tilde{\mathbf{X}}(i,j)$ is a (3×1) vector which represents the (i,j) pixel in the quantized image and M is the number of rows/columns.

The results of the comparison are summarized in Table 1.

Table 1: FIGURES OF MERIT FOR COLOR IMAGE QUANTIZATION

Method	MSE	SNR
Two-layer LVQ	156	-25.6 √
Single-Layer LVQ	207	-24.3
Non-uniform sampling of each color histogram	297	-22.8
LBG	3777	-11.7

It is seen that the proposed two-layer LVQ is superior than any other method used.

A two-layer LVQ which incorporates the proximity test (16) has also been implemented using 16 transputers T800 working at 20 MHz under HELIOS operating system in farm topology. A speed-up of 3 has been observed between the parallel two-layer LVQ and a single-layer LVQ running in one transputer using the same number of pixels both in the learning and recall phase. A much larger speed-up would have been obtained, if other languages (e.g. OCCAM) supporting much faster communication had been used.

ACKNOWLEDGMENT

The first author is supported by a scholarship from the State Scholarship Foundation of Greece and the Bodosaki Foundation.

References

[1] T.K. Kohonen, *Self-Organization and Associative Memory*, 3rd ed., Berlin, Heidelberg, Germany: Springer-Verlag, 1989.

[2] P.K. Simpson, *Artificial Neural Systems*, Pergamon Press, 1990.

[3] T.K. Kohonen, "The Self-Organizing Map", *Proceedings of the IEEE*, vol. 78, no. 7, pp.1464–1480, September 1990.

[4] S.C. Ahalt, A.K. Krishnamurthy, P. Chen, D.E. Melton, "Competitive Learning Algorithms for Vector Quantization", *Neural Networks*, vol. 3, pp. 277–290, 1990.

[5] J.A. Kangas, T.K. Kohonen, J.T. Laaksonen, "Variants of Self-Organizing Maps", *IEEE Transactions on Neural Networks*, vol. 1, no. 1, pp. 93–99, March 1990.

[6] S.-G. Kong, B. Kosko, "Differential Competitive Learning for Centroid Estimation and Phoneme Recognition", *IEEE Transactions on Neural Networks*, vol. 2, no. 1, pp. 118–124, January 1991.

[7] K.V. Mardia, J.T. Kent, J.M. Bibby, *Multivariate Analysis*, Academic Press, 1979.

[8] G.A.F. Seber, *Multivariate Observations*, J. Wiley, 1984.

[9] A.S. Householder, *The Theory of Matrices in Numerical Analysis*, Blaisdell Waltham, Mass., 1964.

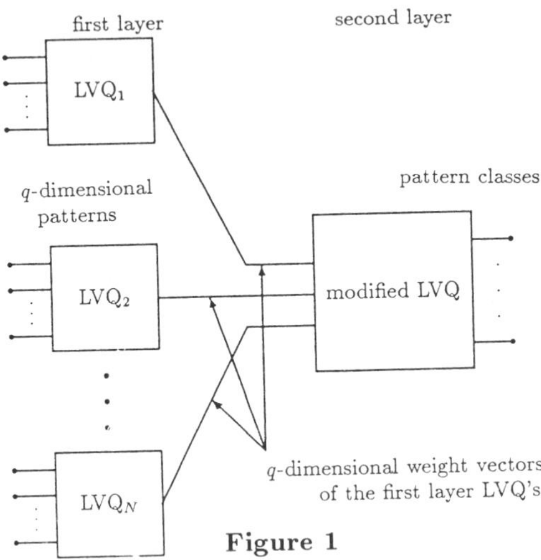

Figure 1

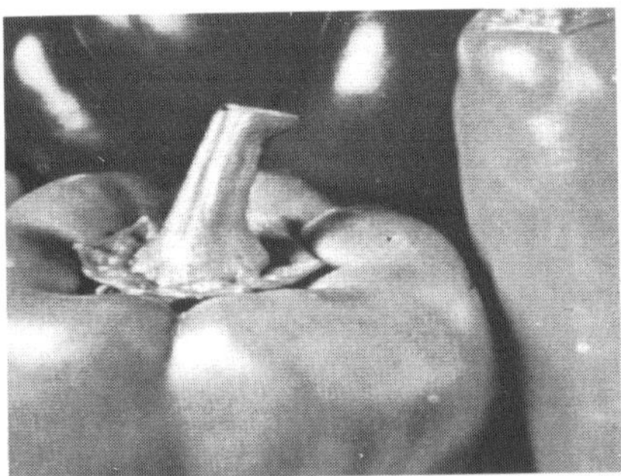

Figure 2

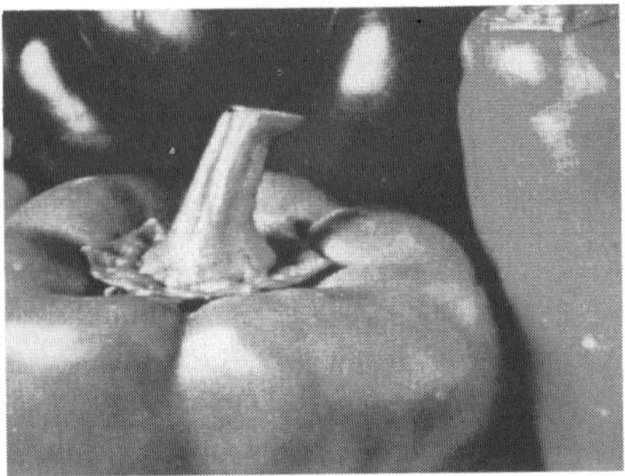

Figure 3

SIGNAL PROCESSING VI: Theories and Applications
J. Vandewalle, R. Boite, M. Moonen, A. Oosterlinck (eds.)

SELECTION OF HIDDEN LAYER NODES IN NEURAL NETWORKS BY STATISTICAL TESTS

Özer Ciftcioglu† and Erdinç Türkcan‡

A statistical methodology for selection of the number of hidden layer nodes in feedforward neural networks is described. The method considers the network as an empirical model for the experimental data set subject to pattern classification so that the selection process becomes a model estimation through parameter identification. The solution is performed for an overdetermined estimation problem for identification using nonlinear least squares minimization technique. The number of the hidden layer nodes is determined as result of hypothesis testing. Accordingly the redundant network structure with respect to the number of parameters is avoided and the classification error being kept to a minimum.

1 Introduction

In the last decade a number of feedforward neural network applications in many diverse areas is reported in the literature. The fast increase of the number of applications and application areas is still on the way and conspicuous. The underlying reasons for this trend can be viewed as the ease in implementations and the rather successful ensuing outcomes. It seems that the particular application in hand does not need to be modelled very much and the solution can be obtained by training from empirical data with little or no prior information about the application. In other words, in order to determine a sufficiently large training sample and an appropriate network architecture, a training algorithm is used to adapt the network parameters for an acceptable solution for implementation. However, it is seldom clearly stated what constitutes a sufficiently large training sample and an appropriate network architecture. The appropriate network structure essentially concerns the number of hidden layer nodes in a feedforward neural network. Viewing the complexity measures in classification problems for pattern recognition in the way as one considers entropy in communication and coding theory, a criterion for optimal number of hidden layer nodes might be of value to study using Shannon entropy [1]. Based on Shannon's entropy some results are reported in literature [2, 3]. In the following, a method for optimal number of hidden layer nodes, based on hypothesis testing is described to form optimal feedforward neural network architecture for a particular application in hand.

2 Pattern classification and parameter identification

Pattern classification takes an important place in the study of neural networks. Let us consider a collection of data of the form

$$(\underline{x}^i, \underline{f}^i) \qquad 1 \leq i \leq N \tag{1}$$

where $\underline{x}^i$ is the input vector forming a pattern; $\underline{f}^i$ is the output vector corresponding to $\underline{x}^i$. As result of pattern classification it is expected that a functional relationship is established so that

$$f(\underline{x}^i) = \underline{f}^i \,. \tag{2}$$

Here, one can easily see that such a problem is similar to classical interpolation and estimation problems, data points being replaced by a data vector as a pattern in function $f(\underline{x})$. The classification is accomplished by means of so-called classifiers of a linear form given by

$$\sum_{j=1}^{N} w_j x_j \leq \theta \tag{3}$$

where w_j and θ are constants to be determined.

On the other hand, a neural network is composed of perceptrons. A perceptron is a simple computational element which is provided with some number of real quantities as input and it performs an affine linear transformation of those inputs of the form

$$\sum_{j=1}^{N} w_j x_j + \theta \tag{4}$$

with a patent computational similarity, referring to inequality (3) and eventual sigmoidal nonlinearity of the form

$$O(x) = \frac{1}{1 + e^{-\sum_{j=1}^{N} w_j x_j + \theta}} \,. \tag{5}$$

Hence, it is clear that a perceptron can act as a classifier in a system for pattern recognition. In a neural network system, perceptrons are arranged in the form of layers so that the output of one layer is taken as the input for the following. The outputs of the last layer are weighted and summed together to produce the output of the network. A network may have multiple outputs providing different weightings and summations at the last layer. In order to surmise the potential performance of a neural network, Cybenko's theorem [4] is essential.

This implies the universality of neural networks with only one hidden layer. However, it makes no commitment about the number of nodes needed. In other words, given an arbitrary collection of input/output data and by using a network with adequate number of hidden nodes in a single hidden

† Istanbul Technical University, Electrical Engineering Faculty, 80191 Teknik Universite, Istanbul, Turkey
‡ Netherlands Energy Research Foundation ECN, P.O. Box 1, 1755 ZG Petten, The Netherlands

layer, the network parameters can be found to achieve any desired modelling error criterion.

Feedforward neural network techniques for classification and recognition applications may be viewed as non-model based. That is, a mathematical model relevant to the classification problem is not required for the neural network approach to the problem. What is needed is a sufficiently large training data and an appropriate network architecture so that network parameters are adapted in a suitable manner for solution. However, the same problem can be considered as a parameter identification problem. Parameter identification involves the estimation of parameters for a model from measured experimental data points. The model may be based on the physical principles underlying a process or it may be based on empirical relationship which is the very relationship between the neural network input/output data, for example. Hence, in this respect, identification of the variables in a network structure turns out to be a model parameter estimation problem where the neural network is a structural model of the problem. The exact structure, i.e. number of hidden layer perceptrons, remaining to be determined.

In model parameter estimation, the estimated parameter values minimize some error functional that measures the difference between the experimental data and the model estimations from these observations. There are several forms of the error functional and a variety of techniques for identifying the optimum parameter set to minimize the functional. The form adapted for use in this work is the weighted least squares error functional. The weighted sum of squares of the deviations provides a measure of the discrepancy between the model estimation and experimental observations.

Let us denote that the experimental data point or 'demand' is y^j and the neural network output is $Y(\underline{x}^j, \underline{w})$ where $\underline{w}$ is the weight parameter vector and $\underline{x}^j$ is the pattern vector of the pattern number j. In the least squares method, the minimization scheme is set up so that the function of 'merit' or 'cost'

$$L(\underline{y}, \underline{w}) = \sum_j [y^j - Y(\underline{x}^j, \underline{w})]^2 \tag{6}$$

is minimized with respect to the parameter vector $\underline{w}$ where $\underline{y}$ is the vector of demands y^j and $Y(\underline{x}^j, \underline{w})$ is the function that estimates the corresponding demand. Assuming that the demands y^j are experimental yields from measurement with statistical uncertainty σ_j in y^j as standard deviation, then for independent measurements, the function

$$\chi^2(\underline{y}, \underline{w}_m) = \sum_{j=1}^{N} \frac{[y^j - Y(\underline{x}^j, \underline{w}_m]^2}{\sigma_i^2} \tag{7}$$

is a random variable with χ^2 statistics, where $\underline{w}_m$ is the weight parameter vector at the minimum; N is the number of patterns introduced.

This statistic reveals whether the solution is consistent with experimental accuracy of the data, that is whether $Y(\underline{x}^j, \underline{w}_m)$ and y^j are equivalent. If all demands y^j are obtained from the same instrumentation system, one can assume that $\sigma_j =$ σ constant and Eq. 7 takes the form

$$\chi^2(\underline{y}, \underline{w}_m) = \frac{1}{\sigma^2} \sum_{j=1}^{N} [y^j - Y(\underline{x}^j, \underline{w}_m]^2 \ . \tag{8}$$

The solution should be consistent within the experimental accuracy which implies that the χ^2 statistics should be less than or equal to N. The most likely value is $\chi = N - p$, where p is the number of parameters being estimated [5]. For the number of parameters p_1 and p_2, the variances are defined as

$$\sigma_{1,2}^2 = \frac{1}{N - p_{1,2}} \sum_{j=1}^{N} [y^j - Y(\underline{x}^j, \underline{w}_m)]^2 \ . \tag{9}$$

Using Eq. 8, we write

$$\chi_{1,2}^2(\underline{y}, \underline{w}_m) = \frac{\sigma_{1,2}^2}{\sigma^2} \nu_{1,2} \tag{10}$$

with the respective degrees of freedom $\nu_1 = N - p_1$ and $\nu_2 = N - p_2$.

Considering that $(N - p_1)\,\sigma_1^2 - (N - p_2)\,\sigma_2^2$ is independent of σ_2^2, as is the case in linear least-squares estimation, we form from Eq. 10

$$\chi_d^2 = \chi_1^2 - \chi_2^2 = \frac{[(N - p_1)\sigma_1^2 - (N - p_2)\sigma_2^2]}{\sigma^2} \ . \tag{11}$$

By the partition theorem for the χ^2-distribution [6], χ_d^2 is independent of the variable χ_2^2 and has a χ^2-distribution with $\nu_d = p_2 - p_1$ degrees of freedom. Hence

$$F = \frac{\frac{\chi_d^2}{(p_2 - p_1)}}{\chi_2^2/(N - p_2)} = \frac{\frac{N-p_1}{N-p_2}\sigma_1^2 - \sigma_2^2}{\sigma_2^2} \frac{N - p_2}{p_2 - p_1} \tag{12}$$

is a F_{ν_1,ν_2}-distributed variable with $\nu_1 = N - p_1$ and $\nu_2 = N - p_2$ degrees of freedom.

For the minimization use is made of a rapidly convergent variable metric method [7, 8, 9] according to Davidon. A brief description is given elsewhere [10].

3 Hypothesis testing

If the function $Y(\underline{x}^j, \underline{w}_m)$ with the number of parameters p_1 is an appropriate representation of the demand y^j then $E[\sigma_{p_1}^2] = E[\sigma_{p_2}^2] = \sigma^2$, where E is the expectation operator. Thus determination of the model order represents a hypothesis testing, namely

$$H_0\colon \sigma_{p_1}^2 = \sigma_{p_2}^2 \tag{13}$$

against the hypothesis

$$H_1\colon \sigma_{p_1}^2 > \sigma_{p_2}^2 \tag{14}$$

where H_0 is the null hypothesis in the statistical terminology. This test cannot be carried out by considering the statistic χ_d^2 from Eq. 11 as it includes the true variance σ^2, which is unknown. Therefore, we have to consider the F statistic given by Eq. 12, which has a $F_{p_1,N-p_2}$ distribution.

For the application of F-test a decision level l must be defined in advance, corresponding to type I error probability in the hypothesis testing, such that the null hypothesis is accepted if $F \leq l$ and rejected if $F > l$. The decision level l corresponding to the type I error α is computed from

$$\int_l^\infty f(F)\,dF = \alpha \tag{15}$$

where $f(F)$ is the probability density function of the F-variable.

In feedforward neural network, the number of parameters p_1, p_2 are related to the number of the hidden layer nodes m, so that hypothesis testing is performed using m and $m+1$ number of nodes.

4 Experimental investigation

Experimental investigation is performed by means of the data from the Borssele nuclear power plant. The Borssele power plant is a two-loop pressurized water reactor with nominal electrical power output of 477 MW. The on-line signal analysis system designed for the multi-level mode operation is capable of monitoring the plant states by tracking 32 DC and 32 AC signals simultaneously since 1983 [11]. In this study two data sets (in wide a range of operation) were selected for the demonstration of universality of the methodology:
Two data sets [3] were collected, one during a $9\frac{1}{2}$ hours of shutdown and another during normal operation with temporary power reduction for a certain duration and sampled with 8 s/s for AC and 1 s/s for DC. Afterwards the DC data are formed with a sampling rate of 1 s/minute to be used in this study. For the first set the corresponding electricity power varied from 375 MWe to 91 MWe and for the second set the power was reduced from 475 MWe to 146 MWe and after a certain lapse of time from 146 MWe to 475 MWe back again. From the 32 DC signals the set of selected signals used in this study are shown in Fig. 1 and indicated in Table 1.

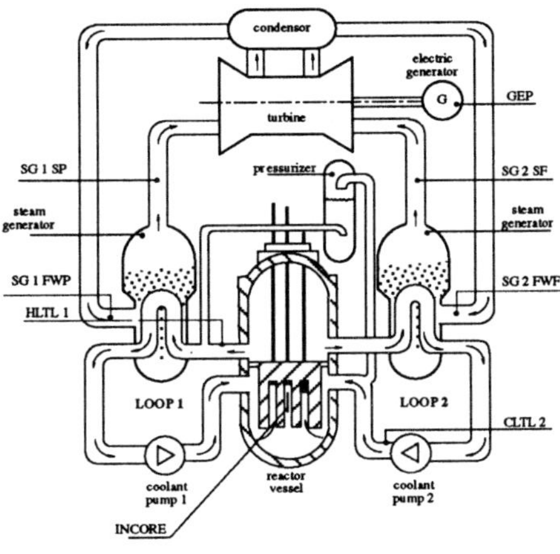

Figure 1: *Schematic representation of the nuclear power plant and the sensors used for the measurements.*

The first data set was obtained during the stretch out followed by normal shutdown. From the 32 available signals the most relevant ones are selected for the test. Table 1 gives the data sets and the selected signals in the neural network application. In the data set 1, a total number of 137 patterns are used out of possible 569 patterns with constant skipping factor of 3 during the training. The remaining 21 patterns are considered for extrapolation.

Signal description	Data set 1 (1 sample/min, 567 minutes data) Sensor selection for the analyses						Data set 2 (1 sample/min, 300 min.)
Number of CH's Number of hidden layers	Identification	Case 1 2 4	Case 2 4 4 4, 32	Case 3 6 4	Case 4 7 7 4, 32	Case 5 10 4	Case 6 4 4 4, 32
INPUTS							
Cold leg temp. (loop 1)	CLTL1	x	x	–	–	x	x
Hot leg temp. (loop 1)	HLTL1	x	x	–	x	x	x
Cold leg temp. (loop 2)	CLTL2	–	x	–	x	x	x
Hot leg temp. (loop 2)	HLTL2	–	x	–	–	x	x
I-core neutron detector	INC	–	–	–	x	x	–
Steam gen. 1 feed water pres.	SG1FWP	–	–	–	x	–	–
Steam gen. 1 feed water flow	SG1FWF	–	–	x	–	–	–
Steam gen. 2 feed water flow	SG2FWF	–	–	x	x	x	–
Steam gen. 1 steam pres.	SG1SP	–	–	x	x	x	–
Steam gen. 2 steam pres.	SG2SP	–	–	x	–	x	–
Steam gen. 1 steam flow	SG1SF	–	–	x	–	x	–
Steam gen. 2 steam flow	SG2SF	–	–	x	x	x	–
OUTPUT							
Generated electric power	GEP	x	x	x	x	x	x
Figures					2		3,4

Table 1: *Selected channels and calculated cases*

The second data set was obtained from a normal operation including considerable reduction of power for a certain time interval. In data set 2, a total number of 100 pattern are used out of possible 300 patterns with constant skipping factor of 2 during the training.
Several cases (Table 1) have been studied using the two different data sets. For data set 1, having 7 different channels information with relatively little correlated signals is used as an input to predict generated electrical power. The F-test described in the preceding section resulted in the number of hidden layer nodes as M=4 for 1000 iterations for training using data set 1. In the statistical terminology, type I and type II errors are two important concepts [5]. In this work, the type I error probability adopted for the F-test is 0.01. The signals involved are shown in Table 1 case 4 (also see Fig. 1) and the results obtained after the recalling procedure are given in Fig. 2. Increasing the number of hidden layer nodes to 32 (case 5) did not improve the prediction error. Increasing the data set up to 10 different channels of information the number of hidden layer nodes is found to be 4 again, as in the previous cases. Since the information is more compared to the preceding case, the prediction error is slightly smaller.

During normal operation there is a power reduction of 70% for a duration of about 2 h as shown in Figs. 3 and 4. The number of hidden layer nodes determined by the F-test is found to be M=4 (Fig. 3). The same experiment is carried out for M=32. Also in this case the differences between the data for training and the predicted values are not improved (Fig. 4), even they are slightly worsened.

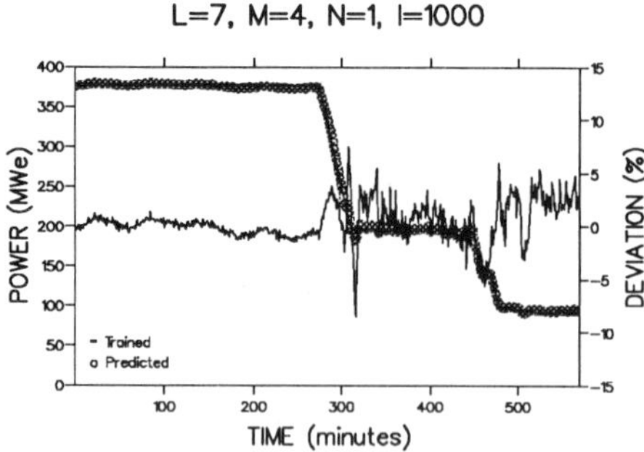

Figure 2: *Extrapolated neural network estimation of generated electric power during the normal shut-down. The fractional error in per cent is indicated at the right, were the number of input nodes L=7; number of output nodes N=1; hidden layer nodes M=4; number of iterations =1000; for data set 1 case 3.*

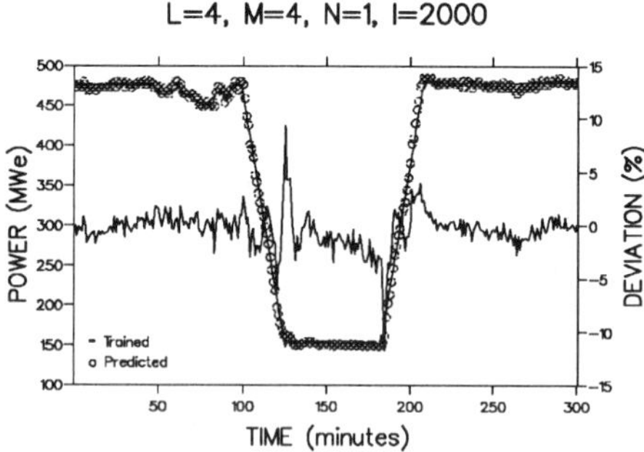

Figure 3: *Extrapolated neural network estimation of electric power during the normal operation including reduction of power (data set 2). Number of iterations = 2000, number of hidden layer nodes M=4.*

5 Discussion and Conclusions

Indeed, the neural networks patently are suitable for using as classifiers. Since the data applied to neural network are essentially for pattern recognition, the network is in a way an appropriate empirical model for the model parameters, i.e. weights subject to identification. In this respect hypothesis testing yields the adequate number of weights through the number of hidden layer nodes so that the true network dimensionality can be identified. The excessive number of weights beyond those required by the true network's modelling structure are simply redundant and they do not add to the network dimensionality at all. This conclusion corroborates with the experimental results which were obtained using the same data but different number of nodes at the hidden layer determined by the hypothesis testing procedure as well as conventional empirical ones, e.g., see [2]. Indeed, as a matter of fact in the overdetermined case, the network performance is slightly better as one should expect with reference to classical parameter estimation problem. In the underdetermined case the error of recognition for pattern recognition is relatively higher, the network performance becomes finally poor in this case.

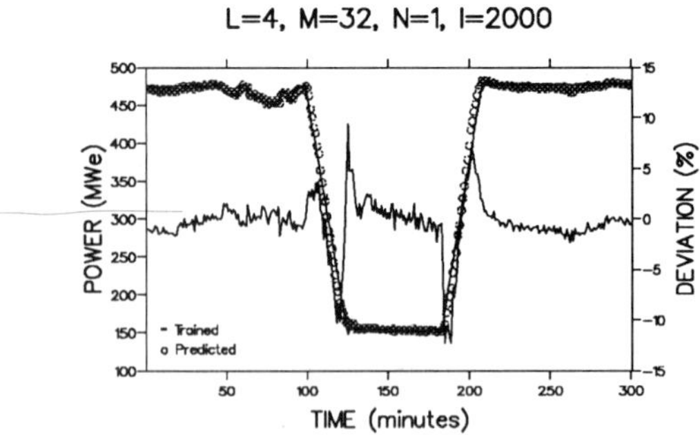

Figure 4: *The same experiment as in Fig. 3, only the number of hidden layers is raised to 32.*

Hence it is to conclude that the hypothesis testing is an important tool to identify the number of hidden layer nodes in a feedforward neural network structure. Additionally, as result of the optimization process, the errors estimates of the network parameters are obtained together with the correlations between the parameters through the inverse second derivative (Hessian) matrix. This additional information is an important aid for further statistical analyses on neural networks.

References

[1] Shannon, C.E.: *A Mathematical Theory of Communication*, Bell System Technical Journal, Vol. 27, pp. 623–656, (1948).

[2] Eryürek, E., Upadhyaya, B.R. and Uhrig, R.E.: *Development and Application of Multi-Layer Neural Networks for Estimation of Power Plant Variables.* DOE/NE/88ER 12824-01, (1991).

[3] Eryürek, E. and Türkcan, E.: *Signal Processing and Neural Network Applications in Pressurized Water Reactors*, ECN-R-91-007, (1991).

[4] Cybenko: *Approximation by Superpositions of a Sigmoidal Function, Mathematics of Control, Signals and Systems*, 2, pp. 303–314, (1989).

[5] Bevington, P.R.: *Data Reduction and Error Analysis for the Physical Sciences*, McGraw-Hill, New York, (1969).

[6] Hald, A.: *Statistical Theory with Engineering Applications*, New York, John Wiley & Sons, (1952).

[7] Garbow, B.S., ANL Z013S, Argonne National Laboratory, Argonne, Illinois (1965).

[8] Davidon, W.C., ANL-3990 (Rev. 2), (Revised February 1966).

[9] Fletcher, R., and Powell, M.J.D., Computer J., Vol. 6, p. 163, (1963).

[10] Ciftcioglu, Ö., and Türkcan, E.: *Statistical Methodology for Selection of Hidden Layer Nodes in Neural Networks*, ECN-R-92-001, (1992).

[11] Türkcan, E. et al.: *Operational Experiences on the Borssele Nuclear Power Plant using Computer Based Surveillance and Diagnostic System On-line*, ECN-RX-91-057 (1991).

SIGNAL PROCESSING VI: Theories and Applications
J. Vandewalle, R. Boite, M. Moonen, A. Oosterlinck (eds.)

CONVERGENCE ANALYSIS OF THE HAMMING NETWORK

K. KOUTROUMBAS and N. KALOUPTSIDIS

UNIVERSITY OF ATHENS, DEPT. OF INFORMATICS

PANEPISTIMIOPOLIS,TYPA BUILDINGS, ATHENS 157 71, GREECE.

ABSTRACT. In this paper results concerning the convergence of the parallel and asynchronous modes of the Hamming Network are presented. Also, an upper bound of the maximum number of steps that are required to reach a stable state is given.

I. Introduction

Neural Networks have drawn considerable attention in the recent years mainly because of their interesting learning abilities. Many Neural Networks architectures have been proposed during these years but all of them can be classified in two major categories: feed-forward Neural Networks and recurrent Neural Networks. The latter category includes the Hamming Network which is the objective of the present work.

One interesting application of the Hamming Network is minimum distance decoding. Also, the Hamming Net may be employed as a lateral inhibition mechanism in competitive learning architectures such as the Kohonen's Self-organizing map and the Counter Propagation Network (see for example [4], [5]).

A typical architecture for the Hamming Network is illustrated in figure 1 (see also [1]). The Hamming Net consists of two subnets. We assume that the patterns are presented as inputs in the lower subnet are bipolar vectors of length N with values -1 and +1. The lower Net has M nodes each one corresponding to one of M exemplar patterns.

The goal of the lower subnet is to compute M matching scores that are equal to $N-d_i$, $i=1,\ldots,M$, where d_i is the Hamming distance (Hamming distance of two patterns is the number of bits the patterns differ) between the input pattern and the $i-th$ exemplar pattern. These scores are integers satisfying:

$$0 \leq N - d_i \leq N, i = 1, \ldots, M.$$

The scores obtained by the lower subnet are then forwarded to the upper Network. The goal of this Net is to pick up the maximum matching score, by forcing all matching scores corresponding to all nodes to zero, except the matching score that corresponds to the node with the maximum initial matching score. The output of the upper set of nodes can then be used for further processing (the complete algorithm can be found in [1]).

The Network will operate until a desired stable state is reached, that is, until one node possesses a positive value and the remaining nodes become zero. It is stated in [1] that the Hamming Net always finds a node with a maximum value when $\varepsilon < \frac{1}{M}$. It turns out that this is not always the case. A more delicate situation rises for example, when there exist two initially equal and maximum matching scores. A complete convergence analysis of the Hamming Net is discussed in the following sections. More specifically, we introduce a deterministic extension of the operation of the Hamming Net which enables us to deal with both parallel and asynchronous modes. We give some results concerning the convergence properties of the generalized Hamming Net. An upper bound for the number of steps required in order to obtain a solution is derived.

II. Basic definitions and theorems

From now on we will refer only to the upper subnet of fig. 1. As can be seen from 1-b step of the Hamming algorithm, the updating at each step of the algorithm is fully parallel ie., all nodes are updated simultaneously. We next define two new modes of operation, the *partially parallel* and the *serial* or *asynchronous* mode of operation. In the *partially parallel* mode of operation only a fraction of the set of nodes is updated simultaneously. In the *serial mode* of operation each node is updated asynchronously with the others. From now on we will refer to the partially parallel mode of operation and we will consider the other modes as special cases of the partially parallel mode of operation. Before we proceed we will define here the *time step* and the *updating cycle*. *Time step* is the time interval in which an updating of a specific group of nodes takes place and *updating cycle* is the number of time steps needed for the completion of one update of all nodes. In more specific terms let us consider fig. 2. The set of nodes is divided in $[\frac{M}{A}]$ groups ($[r]$ denotes the integer part of r) each one containing A nodes, $1 \leq A \leq M$. At each updating cycle we first update the first group, then the second group and so on. With t we shall denote the updating cycle and with $[\frac{k}{A}]+1$ the time steps needed for the update of the $k-th$ node in a given updating cycle. Thus, $\mu_k(t,l)$ denotes the state of the $k-th$ node at the time cycle t and at time step l in the $t-th$ cycle. The term $[\frac{k}{A}]+1$ denotes the time steps that must be spent in a specific time cycle in order the updating of the $k-th$ node takes place.

Example:

Let $A=5$ and $M=20$. Then $\mu_3(2,4)$ denotes the state of node 3 at the time cycle 2 and at time step 4 in the time cycle 2. $\mu_9(2,[\frac{9}{A}]+1)$ or $\mu_9(2,3)$ denotes the state of node 9 at time cycle 2 and at time step 3 in the time cycle 2. At this time step, node 9 will be updated for the time cycle 2. We set:

$$c_k = [\frac{k}{A}]+1, \qquad 0 \leq k \leq M-1.$$

In order to simplify notation we will write $\mu_k(t)$ instead of $\mu_k(t,c_k)$. Note that knowing k we can easily find that at the $[\frac{k}{A}]+1$ time step of updating cycle t the node k will be updated. For example, if $M=20$ and $A=5$, $\mu_9(t)$ is qual to $\mu_9(t,[\frac{9}{A}]+1)$ or $\mu_9(t,2)$. Note that c_k is independent of the time cycle t. To fascilitate the analysis we shall make three simplifications which however can be removed. First, we divide the set of nodes in groups, each one containing the same number of nodes. Second, we assume that A divides M. Finally, the nodes are grouped in their natural ordering. Alternate groupings can be accomodated via permutations of the nodes without affecting subsequent analysis. On the basis of the above discussion, the updating equation of the upper Hamming subnet becomes:

$$\mu_k(t+1) = f[\mu_k(t) - \varepsilon \sum_{m=0}^{[k/A]A-1} \mu_m(t+1)$$

$$-\varepsilon \sum_{m=[k/A]A, m\neq k}^{M-1} \mu_m(t)]$$

with $k=1,2,\ldots,M$ and $t=1,2,\ldots$.

f is defined as:

$$f(x) = \begin{cases} 0, & x<0 \\ x, & 0 \leq x \leq N \\ N, & x<N \end{cases}$$

Eq. (1) states that in order to update node k at time cycle t+1 we use the updated values up to $[k/A]$ node and all the values of the remaining nodes from the previous time cycle t. Note that for the nodes that belong to the same group with k, contribute with their value at time t in the above updating equation. In the sequel we shall assume that the network parameter ε is chosen to satisfy the following bound:

$$\varepsilon < \rho(A,N,1)$$

where

$$\rho(A,a,b) = \begin{cases} \frac{1}{(M-1)}, & A = M \\ \frac{b}{[\frac{M-1}{A}]A(M-1)a}, & A < M \end{cases}$$

Let $\mathcal{G}$ denote the set of sets that initially have at least one node with maximum value. Let $G \in \mathcal{G}$ denote the set that is updated last.

Let T_k be defined as

$$T_k = sup\{t : \mu_k(t) \neq 0\}, \qquad k = 0, 1, \ldots, M-1.$$

Proposition 1: If T_k is finite then $\mu_k(t) = 0$, for $t > T_k$.

Let

$$\Lambda_k = \lim_{t\to\infty} \mu_k(t).$$

Let $i_1 < i_2 < \ldots < i_p$ such that

$\mu_{i_1}(0) = \ldots = \mu_{i_p}(0) = max_i[\mu_i(0,i)]$.

Assuming that G contains i_p and no other maximum, the following hold:

Theorem 1: $\Lambda_k = 0$, for $k = 0, \ldots, M-1$ and $k \neq i_p$ and T_k is finite. Moreover,the limit Λ_{i_p} is attainable after a finite number of steps and satisfies $\Lambda_{i_p} > 0$.

The next Theorem shows that if there are two or more initial maxima in G, then all nodes of the net stabilize to zero after an infinite number of steps.

Assume that $\mu_{i_q}(t), \ldots, \mu_{i_p}(t)$ (with $q < p$) belong to G and initially have the maximum value.

Theorem 2: It holds

$$\Lambda_{i_q} = \ldots = \Lambda_{i_p} = 0.$$

Moreover, these limits are attained in infinite number of steps.

Remark: The asynchronous case results when $A = 1$ whereas the parallel case occurs for $A = M$. In the latter case the bound $\frac{1}{M-1}$ for ε slightly improves the condition $\varepsilon < \frac{1}{M}$ reported in [1].

Theorem 3: An upper bound for the number of steps required to obtain the solution, when only one maximum belongs to G, is:

$$t_b = max\left[\ln(\frac{\mu_{i_p}(0)}{\Delta\mu_{i_p j}(1)})\frac{1}{\ln(1+\varepsilon)} + 2, \right.$$

$$\left. \ln(\frac{\mu_{i_p}(0) - \frac{K}{\varepsilon}}{\Delta\mu_{i_p j}(1) - \frac{K}{\varepsilon}})\frac{1}{\ln(1+\varepsilon)} + 2\right],$$

where

$$K = \varepsilon^2(M-1)N([\frac{M-1}{A}]A+1),$$

It is worth mentioning here that in the fully parallel mode of operation, the Hamming Net will converge after exactly

$$\ln(\frac{\Delta\mu_{i_p j}(t_0-1)}{\Delta\mu_{i_p j}(1)})\frac{1}{\ln(1+\varepsilon)} + 2$$

steps. An upper bound for the number of steps in that case is

$$\ln(\frac{\mu_{i_p}(0)}{\Delta\mu_{i_p j}(1)})\frac{1}{\ln(1+\varepsilon)} + 2.$$

III. Conclusions and remarks

Summarizing the above mentioned facts we have that if $0 < \varepsilon < \rho(A, N, 1)$ and the group of nodes G that is updated in one time step contains only one node with the maximum initial value and updated after all other nodes with maximum initial value belonging to other groups, the Hamming Net will converge in a finite number of steps to a nonzero solution. Furthermore the limiting values of all nodes will be zero exept the limiting value of the node in G with the maximum initial value, which will be positive. An upper bound for the number of steps required has been obtained.

If the group G contains more than one maxima then the values for all nodes asymptotically approach zero.

Further insight into the above results is provided by the following geometric interpretation. We note first that the group of nodes that controls the solution is group G. The values of nodes in G are represented by a point in R^A space and follow a trajectory as time passes. If the initial point is not in one of the hyperplanes that bisect the angles between the reference hyperplanes $\{(x_1, \ldots, x_A) : x_1 = \ldots = x_{i-1} = x_{i+1} = \ldots = x_A = 0\}$, $i = 1, \ldots, A$ then the limit point will be on the axis that corresponds to the node i_p, ie. $\{(x_1, \ldots, x_A) : x_1 = \ldots = x_{i_{p-1}-1} = x_{i_{p+1}+1} = \ldots = x_A = 0, x_{i_p} \neq 0\}$. In the opposite case, the limit vector will be the $(0, 0, ..., 0)$ point. If $A = 1$ we *always* get a solution since group G will cintain only one node with maximmum initial value.

If $A = M$ and there exist two or more nodes with maximum initial value, then the limiting point is zero.

It follows from the above discussion that if we are interested in the speed of convergence, large A are preferable. On the other hand, if we are interested in taking an accurate solution in most of the cases, a small A is better off.

It is also important to note that the above analysis concerns only the upper subnet. The only condition that involves parameters of the lower subnet is eq. (3). Our results become independent of the lower subnet, if eq. (3) is replaced by where b is the maximum possible initial value to the upper subnet and a is the minimum possible difference between the values of two nodes.

References

[1] R.P. Lippmann, *"An Introduction to Computing with Neural Nets"*, IEEE ASSP Magazine, 4-22, April 1987.

[2] K. Koutroumbas and N. Kalouptsidis, *"Qualitative Analysis of the Synchronous and Asynchronous Modes of the Hamming Network"*, under review.

[3] J. Bruck and J. Goodman, *"A Generalized Convergence Theorem for Neural Nets"*, IEEE Trans. Inf. Theory, Vol. 34, No. 5, September 1988.

[4] T. Kohonen, *"The Self-Organizing Map"*, Proc. IEEE, Vol. 78, No.9, September 1990.

[5] R. Hecht-Nielsen, *"Counterpropagation Networks"*, Applied Optics, Vol. 26, No. 23, pp. 4979-4984, December 1987.

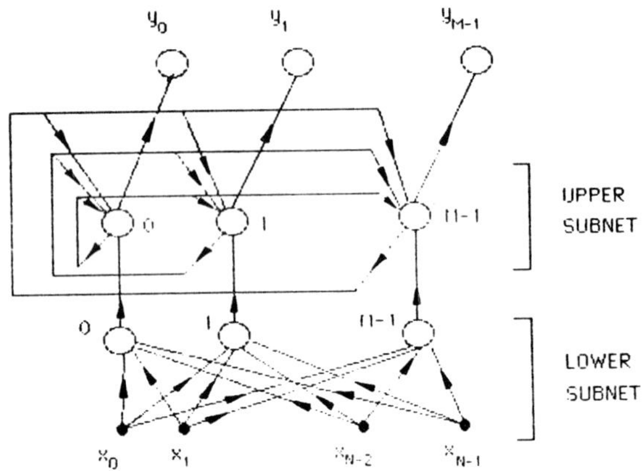

Fig 1: The Hamming Network Architecture

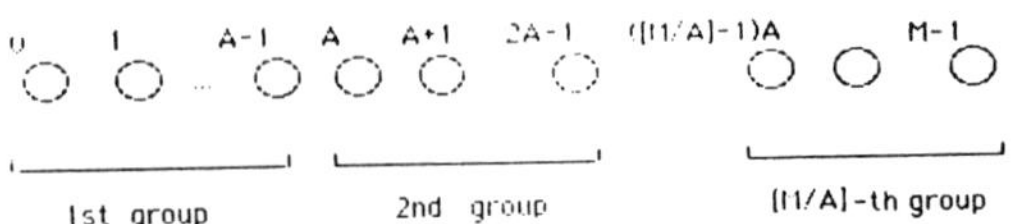

Fig. 2 The grouping of the nodes.

SIGNAL PROCESSING VI: Theories and Applications
J. Vandewalle, R. Boite, M. Moonen, A. Oosterlinck (eds.)

A NOVEL ALGORITHM FOR TRAINING A BIDIRECTIONAL ASSOCIATIVE MEMORY

Marina MEILA PREDOVICIU, Petre STOICA

Department of Automatic Control,Polytechnic Institute of Bucharest
Spl. Independentei 313, 77206 Bucharest, Romania

A training algorithm for discrete-time bidirectional associative memories (BAMs) is presented. It ensures the storage of the desired items as stable equilibrium points with domains of attraction of given radius.The algorithm is proved to work for a broad class of nonlinearities, provided the desired memories lie in the corners of the state-space. Some necessary conditions for the existence of a solution are outlined, and simulations illustrate the algorithm's performance.

1. INTRODUCTION

A key problem in designing neural associative memories is finding the connection weigths matrix which ensures the storage of the desired items as stable equilibrium points of the network.

The outer product method [1,3] accomplishes this task in a direct and easy way. However, as it offers only a storage capacity of about 15 percent of the number of neurons in the net [1,6] and does not offer stability of the stored equilibrium points, it has been replaced by methods which allow the storage of a greater number of items [4,5] and the control of the domain of attraction of each stored state [2,7]. Except for [4], these works deal with Hopfield memories [1], an important special case of associative memory.

This paper focuses on *bidirectional associative memories*. A new algorithm is devised, which improves significantly the storage capacity and ensures the desired minimal radius for the domain of attraction of each stored item. Section 2 introduces the BAM model and the encoding problem, Section 3 presents the training algorithm and Section 4 extends it to networks with neurons having a bounded, nondecreasing activation function. In Section 5, some necessary conditions for the existence of a solution are deduced, while Section 6 shows the results of simulations.

2. THE BAM MODEL

The BAM model used in the following has been introduced by Kosko [3]. The memory is built up of two *layers* containing n_x and n_y *neurons* respectively. Each neuronal output can take only two values, let them be {-1,+1}. The outputs of all the neurons in one layer are connected to the inputs of each neuron in the other layer and viceversa.

If we denote by

$$X(t) = [\ X_1(t)\ X_2(t)\ \dots\ X_{nx}(t)\ ]^T$$
$$Y(t) = [\ Y_1(t)\ Y_2(t)\ \dots\ Y_{ny}(t)\ ]^T$$

the vectors of the neurons outputs for the two layers at time t=0,1,2,.. then the evolution of the network is given by the following equations:

$$X(0) = X0 \tag{1}$$
$$Y(t+1) = \mathrm{Sgn}(W^T X(t)) \tag{2}$$
$$X(t+2) = \mathrm{Sgn}(WY(t+1)) \qquad t=0,1,2,..$$

where W is the $(n_x | n_y)$ connection *weight matrix* and Sgn is the signum function applied to each component of its vector argument.

$$\mathrm{Sgn}(z) = \begin{cases} 1 \text{ if } z > 0 \\ -1 \text{ if } z < 0 \end{cases} \tag{3}$$

If a component of the arguments in (2) is equal to 0, then the corresponding neuron output is not modified. In this context Sgn is called *activation function*.

The pair (X(t), Y(t)) is called the BAM *state* at time t.

It has been shown [3] that from any initial state the network (2) evolves in a finite time to an equilibrium state where no more changes in either X or Y occur. The set of all equilibrium states for a BAM is determined by its weigth matrix W.

Hence, the problem of encoding some

desired items in a BAM can be stated as follows:

Being given the set of pairs $\{(\xi^\mu, \eta^\mu), \mu=1,..,p\}$, $\xi^\mu \in \{-1,+1\}^{nx}$, $\eta^\mu \in \{-1,+1\}^{ny}$, find a matrix $W = (n_x | n_y)$ such that the convergence condition (CC) below holds:

$$\xi^\mu \circ WY > 0$$

$$\eta^\mu \circ W^T X > 0 \qquad \text{(CC)}$$

for each X and Y with

$$d(Y,\eta^\mu) \le f_y$$

$$d(X,\xi^\mu) \le f_x \qquad (4)$$

and $\mu = 1,...,p$.

In (4), d denotes the Hamming distance,

$$d(X,X') = \frac{1}{2}\sum_{i=1}^{nx} | X_i - X'_i | \qquad (5a)$$

$$d(Y,Y') = \frac{1}{2}\sum_{j=1}^{ny} | Y_j - Y'_j | \qquad (5b)$$

and f_x, f_y are two positive integers - the desired radii of the domains of attraction.

It should be noted that along with the pair (ξ^μ, η^μ), the pair $(-\xi^\mu, -\eta^\mu)$ is stored as well. Note also that W in (2) can be chosen to be integer-valued without any restriction. Let us denote by $\mathbf{W}$ the set of $(n_x | n_y)$ matrices W satisfying (CC).

Lemma 1 Whenever the set $\mathbf{W}$ of matrices satisfying (CC) is nonempty, there exists an element W^* of $\mathbf{W}$ such that $[W^*_{ij}]$ are integer-valued.

Proof Obviously, $\mathbf{W}$ is an open set. Suppose it is not empty.Since $\mathbb{Q}^{nx \times ny}$ is dense in $\mathbb{R}^{nx \times ny}$, there exists a $W \in \mathbf{W} \cap \mathbb{Q}^{nx \times ny}$. Note that αW is in $\mathbf{W}$ for every W in $\mathbf{W}$ and for every $\alpha > 0$. Then, by chosing $W^* = N.W$ where N is a common multiple of the denominators of W_{ij} for all i and j, the lemma is proved. ■

3. THE TRAINING ALGORITHM

The following training procedure is an extension of the training with noise procedure introduced by Gardner et al. [2] for Hopfield networks.

TRAINING ALGORITHM 1

For $k = 0$: $W(0) = \sum_{\mu=1}^{p} \xi^\mu \eta^{\mu T}$ (6)

For k = 1,2,.. repeat

1. Chose at random X and Y satisfying (4) for some $\mu=1,..,p$.
2. Compute

$$\varepsilon = \text{diag}\{\varepsilon_i,\ i = 1,...,nx\}$$

$$\varepsilon' = \text{diag}\{\varepsilon'_j,\ j = 1,...,ny\}$$

$$\varepsilon_i = \begin{cases} 0 \text{ if } (\xi^\mu \circ WY)_i > 0 \\ 1 \text{ otherwise} \end{cases}$$

$$\varepsilon'_j = \begin{cases} 0 \text{ if } (\eta^\mu \circ W^T X)_j > 0 \\ 1 \text{ otherwise} \end{cases}$$

3. $W(k) = W(k-1) + \Delta W(k)$

$$\Delta W(k) = \varepsilon \xi^\mu Y^T + X \eta^{\mu T} \varepsilon'$$

until W(k) satisfies (CC).

Theorem 1 If $\mathbf{W}$ is nonempty, then the Training Algorithm 1 converges, regardless of the value of W(0), to a $W^* \in \mathbf{W}$ in a finite number of steps.

Proof Suppose there exists some W^* to satisfy (CC). We will show that all ε , ε' can take nonzero values only for a finite number of steps. Let us define the scalar product and corresponding norm on the set of real $(n_x | n_y)$ matrices:

$$\langle V,W \rangle = \sum_{i=1}^{nx} \sum_{j=1}^{ny} V_{ij} W_{ij}$$

$$\| W \| = \langle W,W \rangle^{\frac{1}{2}}$$

Since there is a finite number of inequalities to be satisfied in (CC), there exist $\alpha > 0$ and $\delta = \alpha / \| W^* \|$ such that

$$\xi^\mu_i . (\sum_{j=1}^{ny} W^*_{ij} Y_j) > \alpha = \delta . \| W^* \| \qquad \forall i=1,..,n_x$$

$$\eta^\mu_j . (\sum_{i=1}^{nx} W^*_{ij} X_i) > \delta . \| W^* \| \qquad \forall j=1,..,n_y$$

Then, at each step k

$$\langle W^*, \Delta W \rangle = \sum_{i=1}^{nx} \varepsilon_i (\xi_i \sum_{j=1}^{ny} W^*_{ij} Y_j) + \sum_{j=1}^{ny} \varepsilon'_j (\eta_j \sum_{i=1}^{nx} W^*_{ij} X_i) \qquad (7)$$

$$> (e+e') . \delta \| W^* \|$$

$$\frac{1}{2} (\| W + \Delta W \|^2 - \| W \|^2) \qquad (8)$$

$$\le \sum_{i,j} W_{ij} \xi_i Y_j \varepsilon_i + \sum_{i,j} W_{ij} \eta_j X_i \varepsilon'_j + \sum_{i,j} \varepsilon_i^2 \xi_i^2 Y_j^2 + \sum_{i,j} \varepsilon'^2_j \eta_j^2 X_i^2$$

$$\le n_x . e + n_y . e'$$

In the above we omitted the superscript μ and we made the notations

$$e = \sum_{i=1}^{nx} \varepsilon_i \qquad e' = \sum_{j=1}^{ny} \varepsilon'_j \quad \text{and} \quad S(N) = \sum_{k=1}^{N} (e(k) + e'(k)).$$

By summing (7) and (8) respectively for k=1,..,N we obtain

$$\langle W^*, W(N) \rangle > \langle W^*, W(0) \rangle + \delta \| W^* \| S(N)$$

$$\| W(N) \|^2 \le \| W(0) \|^2 + 2\max(n_x, n_y) S(N)$$

Then

$$\frac{\langle W^*, W(N) \rangle}{\| W^* \| . \| W(N) \|} > \frac{\delta . S(N) + \text{const}}{[2\max(n_x, n_y) . S(N) + \text{const}.]^{\frac{1}{2}}} \qquad (9)$$

The left term of (9) is obviously less than 1; meanwhile, the right term tends to infinity as $S(N) \to \infty$. Hence S(N) must be bounded, which implies that

$e(k) = e'(k) = 0$ for $k \ge N_0$, q.e.d. ■

Remark that given the initialization in (6), the above algorithm will only produce integer-valued W matrices. Thus, the algorithm will converge to an integer-valued W^* in $\mathbf{W}$ (according to Lemma 1, there always exists such a W^*).

The rule the algorithm uses to adapt the weigths is a Hebb - type rule: to the connection weight between two neurons a positive quantity (equal to 1) is added if the desired states of the two neurons corresponding to the given input and output patterns have the same sign (and thus the two neurons should mutually activate themselves): otherwise, a negative quantity is added.

4. GENERAL ACTIVATION FUNCTION

Let us replace now the Sgn function in (3) by a function $\boldsymbol{\sigma}$ satisfying

$$\boldsymbol{\sigma} \text{ is non-decreasing} \qquad \boldsymbol{\sigma}([-\infty,\infty]) = [-1,1] \tag{10}$$

Then, equations (2) become

$$Y(t+1) = \boldsymbol{\sigma}(W^T X(t)) \tag{11}$$
$$X(t+2) = \boldsymbol{\sigma}(WY(t+1)) \qquad t=0,1,..$$

and the network state space will fill the hypercube $[-1,1]^{nx} \times [-1,1]^{ny}$.

Such a network is generally not globally stable. But we will show that if the pairs (ξ^μ,η^μ) are corners of the state space, i.e.

$$|\xi_i^\mu| = |\eta_j^\mu| = 1 \quad \text{for all } i,j,\mu \tag{12}$$

and the training set is given by

$$\boldsymbol{T}_x = \{ X \in [-1,1]^{nx}, \exists\mu \;\; d(X,\xi^\mu) \le f_x \} \tag{13}$$
$$\boldsymbol{T}_y = \{ Y \in [-1,1]^{ny}, \exists\mu \;\; d(Y,\eta^\mu) \le f_y \}$$

then the training algorithm will create stable equilibrium points with attraction domains defined by f_x and f_y near the given pairs. In (13) and in the following, the distance d defined in (5) generalizes the Hamming distance on the two spaces.

Since $\boldsymbol{\sigma}$ may not reach its extreme values for finite input, we shall admit a tolerance $\Delta>0$ for the accuracy of the retrieval. From (10), there exists $M>0$ such that

$$|\boldsymbol{\sigma}(z)| > 1-\Delta \quad \text{for } |z|>M. \tag{14}$$

The convergence conditions (CC) should be rewritten

$$\xi^\mu \circ WY > M \tag{CCM}$$
$$\eta^\mu \circ W^T X > M \quad \text{for } X \in \boldsymbol{T}_x,\ Y \in \boldsymbol{T}_y$$

The algorithm will be modified as follows:

TRAINING ALGORITHM 2

For $k = 0$: $W(0) = \sum_{\mu=1}^{p} \xi^\mu \eta^{\mu T}$

For $k = 1,2,..$ repeat

1. Chose at random X and Y satisfying (13)
2. Compute

$$\varepsilon = \mathrm{diag}\{\varepsilon_i,\ i = 1,\ldots,nx\}$$
$$\varepsilon' = \mathrm{diag}\{\varepsilon'_j,\ j = 1,\ldots,ny\}$$
$$\varepsilon_i = \begin{cases} 0 \text{ if } (\xi^\mu \circ WY)_i > M \\ 1 \text{ otherwise} \end{cases}$$
$$\varepsilon'_j = \begin{cases} 0 \text{ if } (\eta^\mu \circ W^T X)_j > M \\ 1 \text{ otherwise} \end{cases}$$

3. $W(k) = W(k-1) + \Delta W(k)$

$$\Delta W(k) = \varepsilon \xi^\mu Y^T + X \eta^{\mu T} \varepsilon'$$

until $W(k)$ satisfies (CCM).

Theorem 2 If $\boldsymbol{W}$ is not empty and $\boldsymbol{\sigma}$ satisfies (10) , then the Training Algorithm 2 converges in a finite number of steps, for any value of $W(0)$.

Proof Take $W_1 \in \boldsymbol{W}$, α as in the proof of Theorem 1 and $W^* = W_1 . M/\alpha$. Then W^* satisfies (CCM). Further the proof follows that of Theorem 1.

Theorem 3 If W satisfies (CCM) for every X in $\boldsymbol{T}_x$ and every Y in $\boldsymbol{T}_y$, then W satisfies (CCM) for every X and Y for which $d(X,\xi^\mu) \le f_x$ and $d(Y,\eta^\mu) \le f_y$ for some μ.

Proof Let

$$B(\xi^\mu,f_x) = \{ X \in [-1,1]^{nx},\ d(X,\xi^\mu) \le f_x \}$$
$$B'(\xi^\mu,f_x) = \{ X \in \{-1,1\}^{nx},\ d(X,\xi^\mu) \le f_x \}$$

$B(\xi^\mu,f_x)$ is a simplex and $B'(\xi^\mu,f_x)$ is the set of its extreme points. Since $B'(\xi^\mu,f_x)$ satisfies the linear inequalities (CCM), it follows from the Krein-Milman theorem that all X in $B(\xi^\mu,f_x)$ satisfy (CCM). A similar reasoning for Y completes the proof.See [8] for mathematical background and [9] for the detailed proof.

5. BOUNDS ON THE ATTRACTION DOMAINS

The above algorithms converge to a solution of the associative encoding problem, provided such a solution exists. In the following, some direct necessary existence conditions are given.

A. From the fact that the domains of attraction of each two stored states should be disjoint, it follows that:

$$2f_x+1 \le d(\xi^\mu,\xi^\nu) \le n_x-2f_x-1 \tag{15}$$
$$2f_y+1 \le d(\eta^\mu,\eta^\nu) \le n_y-2f_y-1$$

B. For the compatibilty of system (CC), it is necessary that

$$\frac{f}{n} < \frac{1}{2}\left(1 - \sqrt{1 - d/n} \right) \tag{16}$$

where f,n stay for either f_x, n_x or f_y, n_y and

$$d = \min_{\mu \ne \nu} [\ d(\xi^\mu,\xi^\nu),\ d(\xi^\mu,-\xi^\nu)\ ]$$

or

$$d = \min_{\mu \ne \nu} [\ d(\eta^\mu,\eta^\nu),\ d(\eta^\mu,-\eta^\nu)\ ]$$

respectively. For the demonstration of this condition see [10].

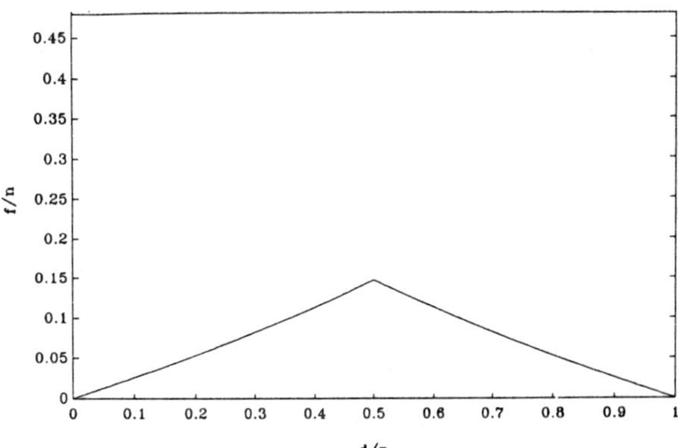

Figure 1 Upper limit of the attraction radius versus minimum distance between patterns (scaled with 1/number of units)

6. SIMULATIONS

Numerical simulations have been performed on a BAM having n_x=182 and n_y=64. For the outer product encoding method of [3], unreliable storing has occured at rather small values of p (average p=5.5), depending on the items stored.

For every training set, the following procedure was used: f_x and f_y were kept equal to 0 till all p equilibrium points were stored. Then both f_x and f_y were increased by 1 and training proceeded by chosing X and Y at distance f_x=f_y from the stored items. After the error frequency dropped under a given value, f_x and f_y were again increased by 1.

Note that the values of f_x and f_y determine the domains of initial states from which the BAM converges to (ξ^μ,η^μ) in no more than two time steps. The real domains of attraction are larger and not necessarily regularly shaped. Fig. 2 shows a case of retrieval from a distance greater than f_x in more than 2 steps. Separate error frequency measurements for initial states randomly chosen around the stored items were performed using the W obtained as explained above. The results summarized in Fig. 3 show that although the training was stopped before convergence, the stored items are attractors even for initial states as far as 11 units.

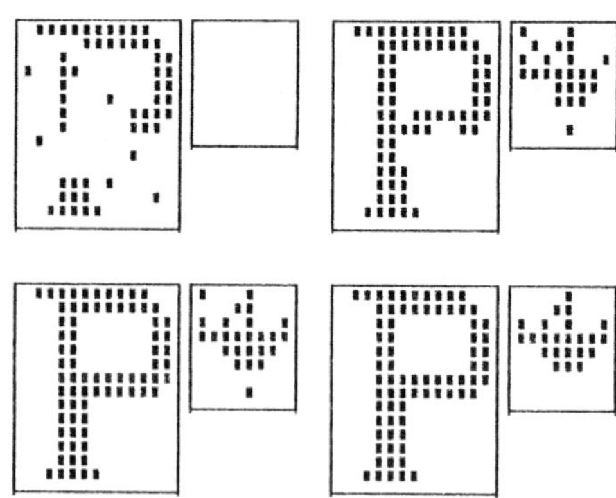

Figure 3. Retrieval in the above memory from an initial state (upper left) at d=28 from the stored item (pair lower right).

It has been also shown that the algorithm can be used for BAMs with general activation functions σ. The closeness of the network state to the desired items in this case can be prescribed if σ is known.

Some necessary conditions for the existence of a solution for a given set of data were outlined, but stronger covergence conditions are still to be found.

7. CONCLUSIONS

The training algorithm introduced in this paper offers very good storage capabilities using simple computations in only fixed point arithmetic. But, as the number of restrictions grows, it might become computationally intensive.

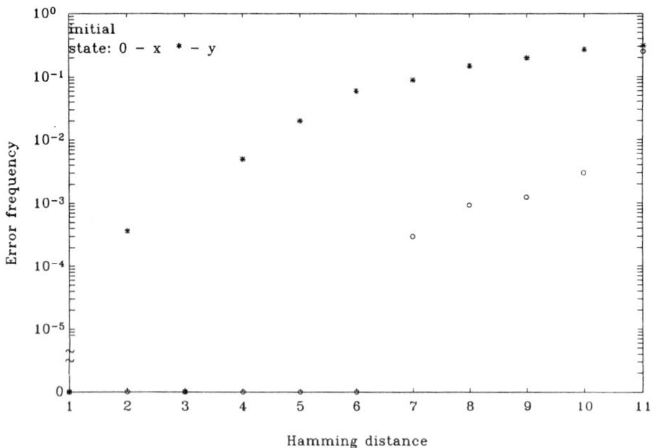

Figure 2 Average retrieval error frequency vs. Hamming distance for p=16 stored items, nx=182, ny=64 and 5025 training cycles with fx=fy=2.

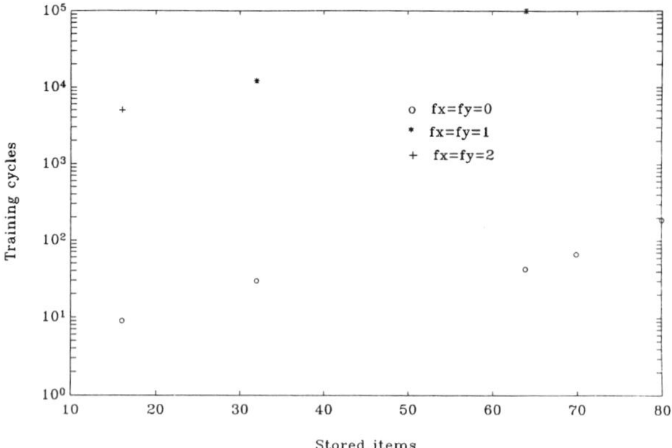

Figure 4 Number of training cycles (1 cycle=2p steps, one for each ξ and η) for different p and f for a 182 x 64 BAM.

ACKNOWLEDGEMENT

The authors are thankful to the colleagues of Elisabeth Gardner with the Physics Laboratory of the University of Edinburgh for their friendly support at the beginning of this work.

REFERENCES

1. Hopfield, J. J., Proc. Nat. Acad. Sci.79 (1982) 2554.
2. Gardner, E., Stroud, N., Wallace, D. J., J. Phys. A 22 (1989) 2019.
3. Kosko, B., IEEE Trans. SMC-18 (1988) 49.
4. Kohonen, T., Associative memory: A system-theoretical approach (Springer Verlag, Berlin, 1978).
5. Personnaz, L., Guyon, I., Dreyfus, G., Phys. Rev. Lett. A 34 (1986) 4217.
6. Amit, D. J., Gutfreund, H., Sompolinsky, H., Phys. Rev. A 32 (1985) 1007.
7. Michel, A., Farrell, J., Sun, H. F., IEEE Trans. CS-37 (1990) 1356.
8. Kantorovitch, L., Akilov, G., Analyse fonctionnelle (Mir, Moscow, 1981)
9. Meila Predoviciu, M., PIB System Control Gr. Rep. 3/92 (1992)

SIGNAL PROCESSING VI: Theories and Applications
J. Vandewalle, R. Boite, M. Moonen, A. Oosterlinck (eds.)

A BANK OF MULTILAYER PERCEPTRONS FOR THE ON-LINE RECOGNITION OF HANDWRITTEN CHARACTERS

Eveline J. Bellegarda†, Jin H. Kim‡, and Jerome R. Bellegarda†

†IBM T.J. Watson Research Center, Yorktown Heights, New York 10598, USA
‡Center for Artificial Intelligence, Kaist University, Seoul, Korea

We describe a neural network appproach to tackle the problem of automatic recognition of handwritten characters. We introduce a novel segment-based approach to handwriting, in which a segment can be viewed as a representative building block of handwriting. The segmentation scheme is used in conjunction with a bank of multilayer feedforward neural networks to perform character recognition. We have evaluated the approach on tasks involving (i) discrimination between similarly-shaped characters and (ii) recognition of discretely written upper-case characters.

I. INTRODUCTION

The automatic recognition of handwritten characters is of paramount importance in applications where handwriting is the desirable input channel, such as personal note taking. As it allows natural interactions between man and machine without the use of a keyboard, this mode of input is increasingly gaining acceptance. Pen-based user interfaces capture handwriting through a special pen operating on an electronic tablet; the handwriting data recorded on the tablet is then sent to a recognizer which outputs a string of typeset characters. Such operation requires the on-line recognition of the handwritten data, which is to be distinguished from optical character recognition (OCR).

So far, the traditional approach to on-line handwriting recognition has been through template-matching algorithms, where character recognition is performed by comparing the unknown character against a set of prototypes constructed during a training phase. Common strategies have focused on stroke-based character recognition, where a stroke is defined as the trajectory of the pen while it remains in contact with the tablet [1]. This approach suffers from inherent difficulties such as (i) a large set of strokes is needed to cover the entire handwriting alphabet, and (ii) stroke lengths and shapes vary greatly for different input modes (discrete, run-on, cursive, or unconstrained handwriting); cf. Fig. 1. Alternatively, a neural network-based strategy was proposed in [2], where a system was designed to perform digits and upper case letter recognition on a touch terminal. This solution is more appealing than template-matching because of its lesser sensitivity to noise; however, it tends to require more training data.

In this paper, we present a novel neural network approach to character recognition. The new algorithm employs a bank of multilayer feedforward neural networks for more efficient use of the available training data. Also, it relies on a preprocessing scheme significantly different from the ones exposed previously, which aims at a more robust description of the different variations (allographs) around each character. More specifically, it is our contention that each handwritten character can be described by a small number of representative segments (building blocks), where a segment corresponds to a finer subdivision of a stroke. These segments are used as inputs to the bank of neural networks.

We define a segment as the amount of information between two consecutive feature points, where a feature point is characterized by a local extremum in the change in the angle of the tangent to the curve at this point. In essence, all points within a segment share somewhat similar characteristics. Since each character is obtained as a collection of distinct segments, various segment-encoding schemes can be investigated to consistently represent each allograph. The encoding parameters are suitably selected to encompass the number of points in a segment and the location of these points within the segment. Once the segments have been isolated, we construct one multilayer neural network for each category of i-segment allographs resulting from this segmental encoding.

The paper is organized as follows. In the next section, we present the feature extraction front-end associated with the segment-based approach. In Section III, we describe the neural network architecture used in conjunction with each segment-based representation. In Section IV, we present experimental results on two different tasks: (i) discrimination between similarly-shaped characters, such as (Z, 2, z), and (ii) discretely-written upper-case character recognition.

II. FEATURE EXTRACTION

1. Pre-processing

Handwriting can be interpreted as a continuous signal $(x(t), y(t))$, where (x, y) are the coordinates of the path followed by the pen and t is the time. The data captured by the tablet are represented as a sequence of points regularly sampled in time. The pre-processing procedure (cf. [3]) is composed of four different steps. In the first step, we aim at reducing variability in the raw data, such as time and scale distortion, by normalizing the sequence of pair coordinates to make all the character sizes equal. In the second step, centerizing is performed, to have the characters fall within the same box. In the third step, an elastic filter is applied to remove some points for smoothing effects. In the fourth step, the variations in writing speed are compensated for by resampling the data to obtain regularly spaced points. Due to this re-sampling, the average number of pair coordinates decreases from approximately 50 to about 20.

2. Segmentation

Segments are obtained by chopping a stroke into appropriate sub-constituants. This is done according to the change in the angle of the tangent to the curve at successive points. For each point P_i of coordinates (x_i, y_i), the angle of the tangent to the pen trajectory at P_i is defined by:

$$\theta_i = \arctan \frac{\Delta y_i}{\Delta x_i}; \tag{1}$$

where the horizontal and vertical incremental changes are computed as: $\Delta x_i = x_i - x_{i-1}$, $\Delta y_i = y_i - y_{i-1}$. The difference in angle between two successive points is thus obtained as: $\Delta\theta_i = \theta_{i+1} - \theta_i$. For each stroke, the chopping procedure is initialized by taking as feature points the starting and ending point. The direction of change $\Delta\theta_i$ is then computed at the midpoint of the stroke, where the arclength is half of the total pen trajectory. If $\Delta\theta_i$ is a local extremum, the corresponding point P_i is selected as a possible feature point candidate, and the procedure is re-applied recursively. Among all feature point candidates, the points for which $||\Delta\theta_i|| > T_\theta$, where T_θ is a pre-set threshold, are kept as final feature points.

Several issues arise regarding the choice of the threshold angle T_θ. If T_θ is "too small," this will result in a large number of segments, which may allow more detailed representation of the stroke but may be more sensitive to noise. On the other hand, "too large" a value of T_θ, resulting in a small number of segments, might induce too much smoothing, and local features might get lost in the process. Thus, several values of the threshold angle have to be experimented with to find a reasonable compromise.

3. Segment Encoding

Among all points belonging to a given segment, we select the x and y coordinates of 7 points, 5 of them equispaced and the last 2 taken at the midpoint of the beginning and the end sections of the segment (see Fig. 2). We content that these extra two points are required to suitably encode the critical information present at the beginning and ending of the segment. In addition, information regarding the connection of one segment to another was taken into account by encoding the direction of change $\Delta\theta$ at the beginning and end of the segment. In summary, each segment was represented by $7 \times 2 + 2 = 16$ numbers.

We have empirically determined (see Section IV) that for the upper-case or lower-case Roman alphabet, each handwritten character can be adequately represented with at most 6 segments. This enables a segment-based recognition implemented using 6 different neural networks. Through the front-end described above, each character is directed toward the proper network according to the category to which it belongs.

III. NEURAL NETWORK ARCHITECTURE

As described above, the training data is partitioned into i−segment data ($i = 1, \cdots, 6$), and one fixed-size neural network is constructed for each category of characters. The i−segment training data also defines the number K_i of output classes for the letters observed. Hence, each neural network differs by its number of input and output units, due to segmental encoding and data-dependent labels, respectively.

For each neural network, a multilayer feed forward architecture is considered; as in [2], the layers successively perform higher level feature extraction and the final classification. We implement 6 feed forward perceptrons with one input layer, two hidden layers, and one output layer. The two hidden layers have 20 and 10 hidden units, respectively. Each neural network has the following number of connection weights: $[(i \times 16) \times 20] + 20 \times 10 + (10 \times K_i)$, where $i = 1, \ldots, 6$.

Each neural network is trained by the standard backpropagation algorithm [4]. A training token is composed of an input pattern and its corresponding output label (one of the K_i labels). In this context, the neural network can be viewed as a mapping from input patterns to target patterns; and the backpropagation algorithm is an iterative version of the mapping function computation. The results of training are the weights associated with all the connections. Before training, the weights are initialized with random values, uniformly distributed between 0 and 1. Subsequently, the weights are modified according to the backpropagation error.

The learning times of the different networks are reported in terms of epochs. An epoch, or iteration step, is one presentation of the entire set of L training patterns. Both local and global updates are investigated. In the case of global update, the weights are updated after the entire set of L training patterns has been seen, i.e., we have one update per epoch. In the case of local update, the weights are updated after seeing each input pattern, i.e., this results in L updates per epoch. In general, local update results in faster learning time but in more oscillations in the sum of the squared errors (SSE) for all the outputs. The learning progression is measured by plotting the curve of SSE vs. the number of epochs. When it falls below some fixed value or when the number of epochs for the current session reaches the maximum number of iterations MI, the learning is completed.

The backpropagation learning parameters of interest are: μ, the learning rate; α, the momentum factor; and r the range of the random initial weights. Selection of those learning parameters is something of a black art, and small differences in these parameters can yield large differences in learning times [4]. The learning parameters can be manipulated to trade off speed and accuracy. It was suggested in [4] that an r value of 1, a μ value of 0.5 to 0.9, and an α value of 0.9 to 0.5 should, in principle, yield faster learning. Of course, those "rules" have to be used as guidelines since the implementation of the network is strongly problem-dependent. A sigmoid prime shift function was considered (see [4]) to eliminate the "flat spot" and speed up the training. Notice that, contrary to [2], our weights were independent, making the present approach more general.

During recognition, each network uses the weights acquired during the training phase to classify the unknown input character against the respective inventory of classes.

IV. EXPERIMENTAL RESULTS

1. Discrimination Task

One major source of errors in character recognition comes from the lack of discrimination between similarly-shaped characters such as, for example, $(2, Z, z)$; $(5, S, s)$; (U, V, u, v); (P, D); $(t, +)$. Template matching algorithms, in particular, tend to discriminate poorly between such characters, which originally provided some motivation to develop the segment-based neural network. To illustrate the potential of the above approach to automatically extract features distinguishing confusable pairs, we consider in this section the problem of discriminating between 2, Z, and z, and compare the results with those obtained with a template-matching algorithm. The performance of the bank of neural networks was evaluated for both writer-dependent and writer-independent recognition tasks. In writer-dependent recognition, the templates or classes obtained during the training of a particular writer are used to recognize or classify handwriting of the same writer, while in writer-independent recognition the writers used for training are totally independent from the writers used for decoding.

The tablet used had a resolution of 0.1 mm and a sampling rate of 70 Hz. We considered a collection of $(2, Z, z)$ from 10 writers. On the average, each of the writers provides 23 2's, 12 Z's, and 12 z's. For writer-dependent recognition tasks, the data from each writer was split into training (80%) and testing (20%); while for writer-independent, the first 8 writers provided the training data set and the 2 remaining writers the test data set.

For the neural network training, the following learning parameters were used: $\mu = 0.5$ for the learning rate, $\alpha = 0.9$ for the momentum rate, and $T_\theta = 90^0$ for the threshold angle for the direction of change. The input layer to the network had $16 \times i$ for $i = 2, 3, 4$ input units and 3 output units, representing the three classes. The two hidden layers had 10 and 4 hidden units, respectively.

For writer-dependent decoding task, the recognition accuracy of the bank of neural networks was comparable to the template-matching algorithm. For writer-independent recognition task, the bank of neural networks achieved a recognition accuracy of 83.4 % while the performance of the template matching algorithm was 79.6 %. Similar results were obtained when discriminating between $(t, +)$; (U, V); (P, D).

2. Upper-Case Character Recognition

We considered the vocabulary of 26 characters in the upper-case Roman alphabet. Training data was collected in boxed-discrete mode from 7 writers over a period of several days. Among the writers, two were left-handed. The training data consisted of 2353 characters, or 330 characters per writer on the average. The classification performance of the networks was measured on data provided by 2 additional (test) writers; who each provided about 350 test characters.

To evaluate the relative contribution of each neural network in the bank, we computed for the training data, the frequency distribution of the number of segments as obtained through the segmental encoding described earlier. The distribution of i-segment upper-case characters $(i = 1, \cdots, 6)$, from the 7 writers considered is displayed in Fig. 3. It can be observed that most of the upper case characters fall in the 3 segments category. Thus, the performance of the 3-segment neural network will be critical to the ultimate recognition accuracy. Also, it was found that the 3 and 4 segments neural networks generated the largest number of classes; thus, it is expected that classification with such networks will be more difficult than with the other networks.

For the training, the following learning parameters were used: $\mu = 0.5$ for the learning rate, $\alpha = 0.9$ for the momentum rate, $T_\theta = 60^0$ for the threshold angle for the direction of change. Local update policy was selected to speed up the training. The parameter SSE was recorded every 10 epoch. When it fell below some fixed value of 10^{-2} or when the number of epochs for the current session reaches the maximum number of iterations (by default set to 10000), training was deemed completed.

Fig. 4 shows the individual performance of (i) the 2-segment neural network, in the 85% recognition rate; (ii) the 3-segment neural network, in the upper 70% recognition rate; and (iii) the 4-segment neural network, also in the upper 70% recognition rate. The individual performance of the 1-segment neural network was in the 90% recognition rate and the 5-segment neural network in the upper 60%. The performance of the 6-segment neural network is not given, since no test character was chopped into 6 segments.

Fig. 5 shows the global recognition rate obtained when all the neural networks are combined, as a function of the number of writers used for training. The overall performance of the bank of neural networks is around 80%. Note that the recognition rate increases noticeably as more writers are added to the training data base but levels off after a critical number of writers is reached. This may indicate that for this particular pool of writers (all from the New York City area) three or four writers are sufficient to capture essentially all the necessary allographs.

Note that the simplest classification scheme was used; i.e., the decision of the network was the one corresponding to the output unit having the highest value. In future studies, it might be worthwhile to investigate a classification scheme which provides a way to automatically rejects ambiguous and/or meaningless patterns. This can be embedded into 2 rules based on the output values of the neural network. The first rule rejects the character when the activation level $L1$ of the highest of the outputs does not exceed a threshold T_{max}, i.e.; $L1 < T_{max}$. The second rule defines the character as ambiguous if the difference between the highest activation level $L1$ and the second highest activation level $L2$ is less than a threshold T_{diff}, i.e., $L1 - L2 < T_{diff}$.

V . CONCLUSIONS

The segment-based approach described in this paper appears to be a viable route to the recognition of handwritten characters. Novel pre-processing and segment encoding schemes were developed, with which an acceptable recognition accuracy was reached. In this respect, we have not explored all possibilities; for example, increasing the number of points used in the encoding scheme may lead to a more accurate description, at a probable expense in computational cost.

Moreover, additional efforts are required to fine tune the different neural networks; slightly different topologies and learning parameters may yield better results. Nonetheless, the bank of neural networks was found to outperform a conventional template-matching approach, which is especially noteworthy for writer independent tasks. In the future, one may envision using such neural networks in the second stage of a template-matching system, for example to discriminate between confusable pairs.

REFERENCES

[1] C.C. Tappert, C.Y. Suen, T. Wakahara, "The State of the Art in On-Line Handwriting Recognition," *IEEE Trans. PAMI,* Vol. 12, No. 8, pp. 787-808, 1990.

[2] I. Guyon, P. Albrecht, Y. Le Cun, J. Denker, and W. Hubbard, "Design of a Neural Network Character Recognizer for a Touch Terminal," *IEEE Trans. Pattern Recognition,* Vol. 24, No. 2, 1991.

[3] C.C. Tappert, "Recognition System for Run-On Handwritten Characters," U.S. Patent No. 4, 731, 857, issued 3/15/88.

[4] S.E. Fahlman, "An Empirical Study of Learning Speed in Back-Propagation Networks," Technical Report, CMU-CS-88-162, June 1988.

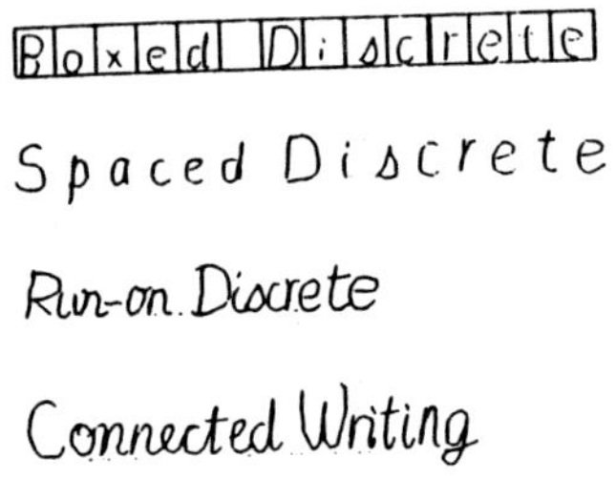

Boxed Discrete

Spaced Discrete

Run-on Discrete

Connected Writing

Fig. 1. Various types of handwriting.

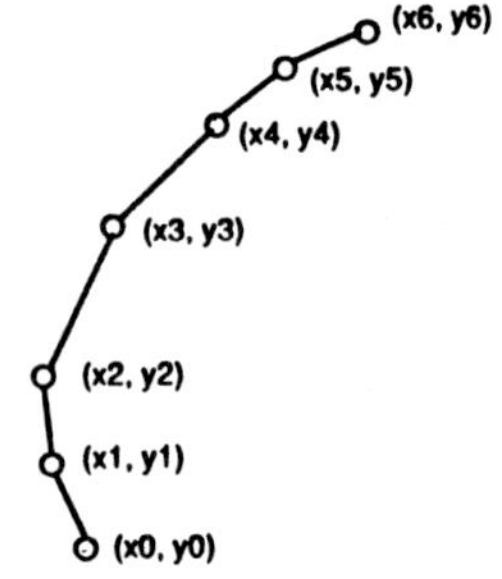

Fig. 2. The 7-point encoding scheme.

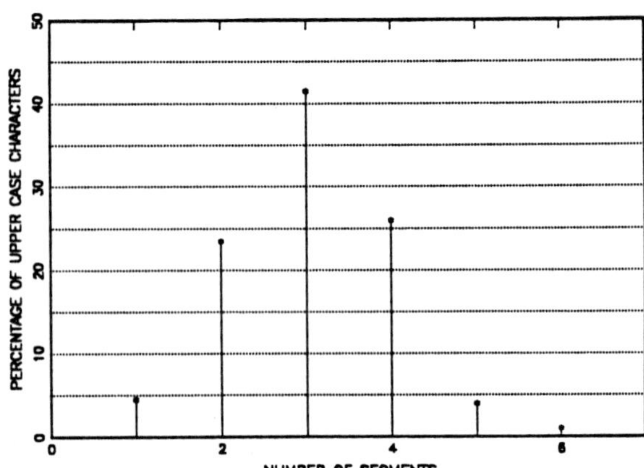

Fig. 3. Distribution of i–segment characters.

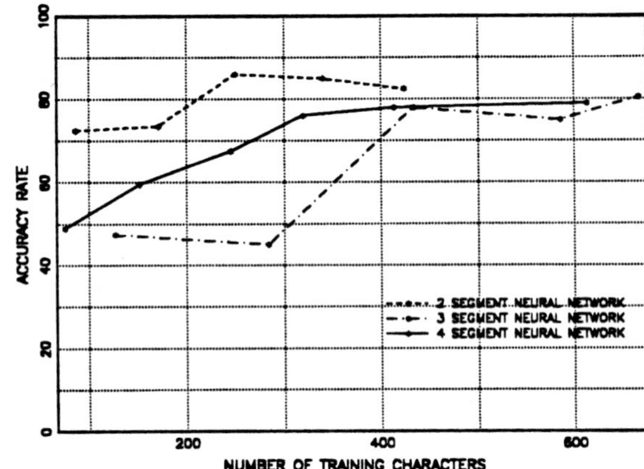

Fig. 4. Recognition rate vs. number of training characters for individual neural networks

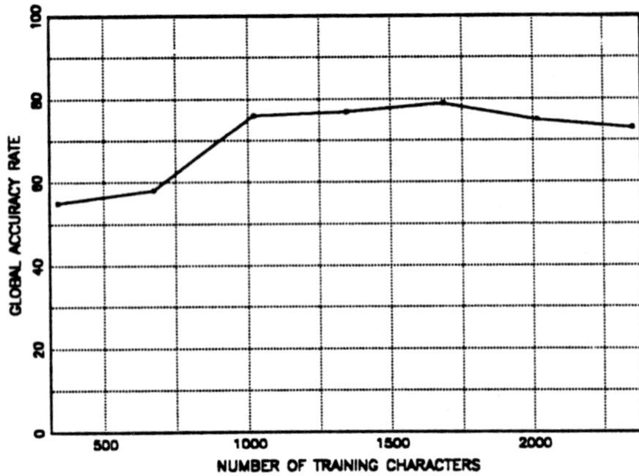

Fig. 5. Global recognition rate vs. number of training characters (all the neural networks combined)

SIGNAL PROCESSING VI: Theories and Applications
J. Vandewalle, R. Boite, M. Moonen, A. Oosterlinck (eds.)

LEFT VENTRICULAR BOUNDARY DETECTION IN ECHOCARDIOGRAPHICAL IMAGES USING ARTIFICIAL NEURAL NETWORKS

Y. Xin P. Boekaerts M. Bister Y. Taeymans J. Cornelis

Vrije Universiteit Brussel
Department of Electronics / Research Group IRIS
Pleinlaan 2, 1050, Brussel, Belgium.
Tel. 0032/2/641,29,82
Fax. 0032/2/641,28,83

INTRODUCTION

Two dimensional real time echocardiography is a noninvasive technique widely used in clinical cardiology. It is especially attractive because of its low cost, the minimal discomfort it causes to the patient, the absence of ionising radiation, and its possible application for patient monitoring through real time processing.

It is however well-known that images produced by sonic methods have poor noise characteristics and low spatial and grey value resolutions [1]. Quantitative analysis and boundary detection on such images is difficult, which explains the large amount of research in this area [2] - [5]. In [2], a temporal tracking algorithm is proposed; [3] describes a radial boundary search algorithm based on simulated annealing; [4] uses optical flow calculation; [5] is a 1-D matched filtering approach. [2], [3] and [4] are computationally heavy and require large amounts of memory. [2] requires important temporal averaging which leads to uncertainty in the location of the ventricular border; [3] has problems to find acceptable heart cavity borders when the valves are open; [5] suffers from the difficulty to define a good model of a "cardiac boundary".

The *aim* of the research behind this paper is to explore a new approach for reliable echographic left ventricular boundary detection using artificial neural networks. The *novelty* resides in the heart cavity delineation — even when the valves are open — without having to define explicitly a model for the cardiac boundary. The *importance* of the work resides in the fact that the intrinsic parallel nature of artificial neural networks allows an implementation bringing the time performances in a range which is acceptable for physicians.

METHODS AND EXPERIMENTAL RESULTS

For the detection of parts of the left ventricular boundary in local windows, one should face the problem that the orientation of the border of the cavity changes for each window position. To reduce this orientation problem, the Cartesian coordinate system of the echographical image is transformed to a polar coordinate system with an origin lying inside the cavity ([3],[5]). *All subsequent processing is applied on polar images (radius, angle).*

1. The NN Boundary Detector

1.1 Method

A Feedforward Neural Network (NN) is used as a local matched filter in order to classify the central pixel of the filter window based on the information of its environment. Let $\mathbf{x}=(x_1,x_2,...,x_n)$ be the vector of grey values of n pixels in a local rectangular window. We define 2 classes of windows (A and B). In the case of left ventricular boundary detection, A is defined as the class of windows for which the central pixel belongs to the boundary and B is the class of windows for which the central pixel does *not* belong to the boundary. We want to design a Feedforward Neural Network $N(\mathbf{x}) \in [0,1]$, which performs the following mapping operation:

$$N(\mathbf{x}) \to 1, \quad \text{if } \mathbf{x} \in A;$$
$$N(\mathbf{x}) \to 0, \quad \text{if } \mathbf{x} \in B. \qquad (1)$$

During training, several windows (i = 1,..., M) are clamped to the input of the neural network and a cost function $\mathbf{J} = \sum J_i$ is minimized. J_i is given by

$$J_i = (N(\mathbf{x}_i) - 1)^2, \quad \text{if } \mathbf{x}_i \in A;$$
$$J_i = N(\mathbf{x}_i)^2, \quad \text{if } \mathbf{x}_i \in B. \qquad (2)$$

The Gradient Backpropagation learning algorithm ([6],[7]) is used to force the neural network to generate a high (or low) activity level $N(\mathbf{x})$ in the output neuron if the central pixel of the window belongs to the boundary (or not) in order to minimize the cost $\mathbf{J}$. The learning set used to train the neural network to recognize the boundary is extracted from the polar image of grey values in which we selected the boundary by hand. It is composed of a set of windows of which we know whether the central pixel belongs to the left ventricular boundary or not.

The trained neural network is applied sequentially to all windows centred around the points lying on a given horizontal line in the polar image. The central pixel of the window with maximum output response N(**x**) in the line is selected as the left ventricular boundary point. This operation is repeated for each horizontal line in the polar image.

1.2 Experiments and Results

The Gradient Backpropagation algorithm was used to train the neural network on 200 samples (the learning set). This learning set contains 100 patterns (i.e. **x** vectors) in windows of class A and 100 patterns of class B.

Although most windows of class A and B are chosen at random from a given polar image, some manual effort has been done to eliminate the so called "ambiguous" patterns. It can be easily understood that, if the boundary lies too closely near the central point of a class B window, it could be easily misclassified by the neural network as belonging to class A. Such patterns can cause local minima during learning and decrease the convergence speed. We have found experimental evidence that even if these patterns are rejected from the learning set, the neural network still classifies them properly afterwards: the output response of the neural network decreases when the boundary slightly shifts from the centre of the window.

Furthermore, to make the neural network immune to the typical speckle noise of echocardiographical images, 10 % of the windows of the learning set were selected to contain important speckle noise.

As a last constraint in the choice of the learning set, we took about 5% of the windows on the mitral valves, when they are open. We found experimental evidence that such choice makes the neural network immune to the configuration of the mitral valve opening during the application phase.

For the evaluation of the performance of the trained neural network, we used two test images which were not used in the learning set. A mean square geometric distance between the manually selected boundary points and the computer selected boundary points is defined as an error measure.

$$\text{error} = \frac{1}{n}\sqrt{(d_{c,i} - d_{m,i})^2} \qquad (3)$$

where d_c is the computer selected distance and d_m is the manually selected distance of the boundary point relative to the origin of the polar image. Fig.1 shows the evolution of the error measure (3), expressed in number of pixels, for a trained Feedforward Neural Network during testing on two images (error1 for PICT 1 and error 2 for PICT 2) as a function of the window size. From this experiment, we can conclude that the error made by the network for the detection of the left ventricular boundary significantly drops when the window size becomes larger than 9 x 9. This is especially true for image PICT 1. This can easily be understood if one realizes that image PICT 1 contains an open mitral valve and also contains more speckle noise then image PICT 2.

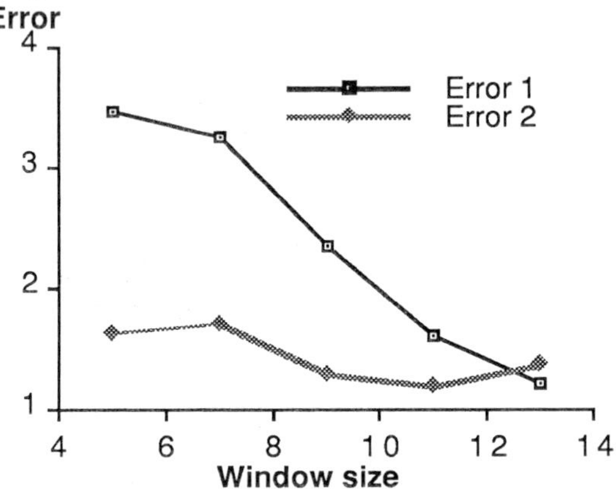

Fig. 1 The evolution of the error expressed in number of pixels, in function of the window size (3 neurons in hidden layer, 200 learning samples). Error 1 corresponds to PICT 1; Error 2 corresponds to PICT 2

The window size affects the convergence time during training in a negative way however. Therefore, one should find a good compromise between the window size of the boundary detector and the learning speed during training. Test and application time (for 1500 window positions and a neural network with 81 input neurons, 3 hidden neurons and 1 output neuron) is typically of the order of 30 seconds for simulations on a SPARC station 1+.

Fig.2 shows the boundary detection accuracy (expressed in number of pixels) of a given trained neural network (200 learning samples, window size 9 x 9 pixels) for both test images as a function of the number of neurons in the hidden layer.

From this kind of experiments, we concluded that the extension of the hidden layer with more than three neurons does not increase the accuracy significantly.

The influence of the number of learning samples on the quality of the boundary detection has been investigated. As could be expected, the accuracy of the network increases if one extends the number of learning samples. Obviously, if we increase the number of learning samples, the classification task becomes more and more well-defined allowing the network to learn a good internal representation of

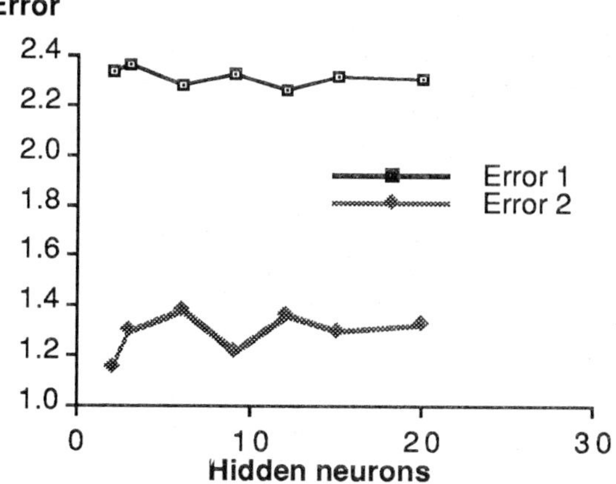

Fig. 2 The evolution of the error expressed in number of pixels, in function of the number of hidden neurons (200 learning samples, window size 9 x 9 pixels).

the input space.

2.The Hybrid boundary detector

2.1 Method

In some of the previous type of experiments, the neural network boundary detector tends to fail to classify the left ventricular boundary properly for echographical images in which the mitral valve is in the open configuration (see PICT1, the boundary is not detected correctly near the open mitral valve). One possible way to solve this problem is to combine the output of the neural network with a smoothness term.

$$s = \frac{(d_i - d_{i-1})^2 + (d_i - d_{i-2})^2}{2d_i^2} \tag{4}$$

The smoothness term s for the current pixel i is equal to the normalized mean squared difference in radius(or distance from the origin) of the two neighbouring boundary points i-1 and i-2. This term is combined with the NN output $N(\mathbf{x})$ in a linear decision function:

$$D = \alpha N(\mathbf{x}) + \beta s \tag{5}$$

or

$$D = \mathbf{w}\mathbf{y} \tag{6}$$

where $\mathbf{w} = [\alpha, \beta]$, $\mathbf{y} = [N(\mathbf{x}), s]^\tau$. The value of the weighting coefficient $\mathbf{w}$ is obtained, during the learning phase, by the Fisher discriminant analysis [8].

$$\mathbf{w} = S^{-1} (\mathbf{m}_A - \mathbf{m}_B) \tag{7}$$

where:

- the vectors $\mathbf{m}_A$ and $\mathbf{m}_B$ are the mean of vectors $\mathbf{y}$ respectively of class A and B.
- S is the variance-covariance matrix.

We can make the decision:

$$\mathbf{x} \in A \quad \text{if } D > \theta$$
$$\mathbf{x} \in B \quad \text{if } D < \theta \tag{8}$$

where θ is the threshold which decides about the attribution of a pixel to class A or B.

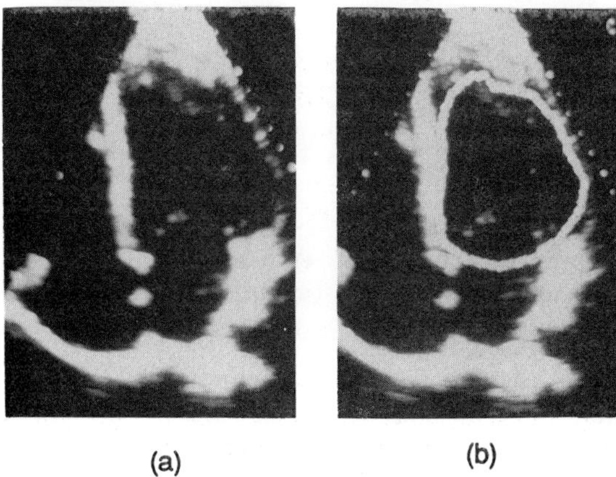

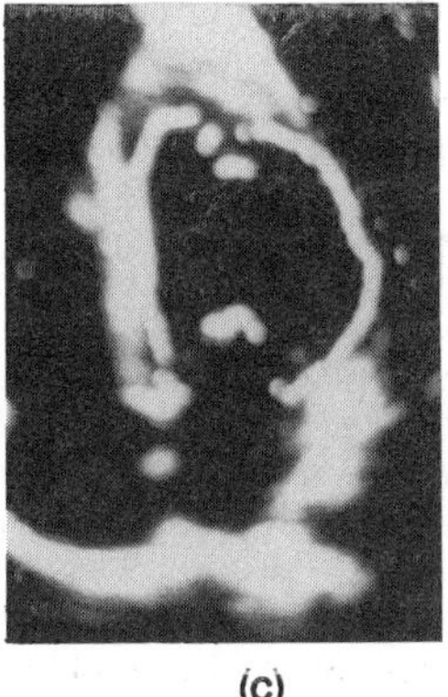

Fig. 3 Comparison of a neural network based boundary detector without smoothness term (c) to a hybrid system with smoothness term (b) for PICT 1 (a).

2.2 Experiments and Results

Experiments showed that the hybrid detector improves the accuracy of boundary detection significantly. Tests on PICT 1, show that the

boundary detection error given in (3) decreases from 2.4 pixel to 1.1 pixel after introducing the smoothness term (for a network with 3 hidden neurons and a window size of 9 x 9 pixels).
Figure 3 gives an idea of the difference in accuracy of boundary detection for a simple neural network boundary detector without smoothness term (b) compared to a hybrid system with smoothness term (c).

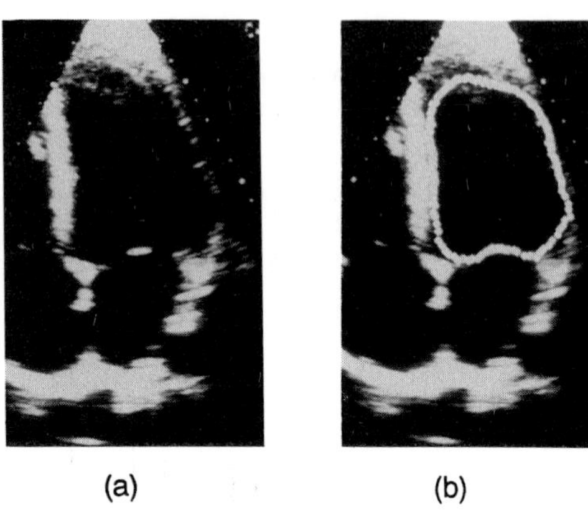

(a) (b)

Fig. 4 Result of the hybrid boundary detector for PICT 2, (a) original image, (b) boundary detection result.

Figure 4 shows boundary detection with the hybrid detector. Experiments on sequences of images from different patients reveal that the accuracy of boundary detection of the hybrid system is equally good during most of the myocardial cycle, except when the heart cavity reaches its minimal size.

DISCUSSION AND CONCLUSION

As a general conclusion, we can state that the neural network based boundary detector, if trained properly with the Gradient Backpropagation learning algorithm, is able to detect the left ventricular boundary in images with a closed mitral valve. Problems related to the detection of the left ventricular boundary in images with an open mitral valve can be solved by combining the response of the neural network with a smoothness term, forming a so called hybrid detector.

In order to apply such boundary detector for clinical use in real time patient heart monitoring, the intrinsic parallel nature of neural networks should be exploited in the implementation to speed up the application time. Remaining problems, such as large contrast changes during the heart cycle and changes in orientation of the boundary within the window, will be investigated in the near future. Large scale testing should be performed before clinical application, mainly on the generalisation properties (from patient to patient) of the trained networks.

REFERENCES

[1] M. Bister, Computer analysis of cardiac MR images, PHD Thesis, Vrije Universiteit Brussel,1990.

[2] D. Adam, O. Hareuveni, and S. Sideman, Semi-automated border tracking of cine echocardiographic ventricular images, IEEE transactions on Medical Imaging. Vol. 6, 1987.

[3] Friedland,N. and Adam, D. Automatic ventricular cavity boundary detection from sequential ultrasound images using simulated annealing, IEEE transactions on Medical Imaging. Vol. 8, No. 4, December 1989.

[4] Mailloux, G. E., Bleau, A., Bertrand, M., Petitclerc, R., Computer analysis of heart motion from two-dimensional echocardiograms, IEEE transactions on Biomedical Engineering, Vol 34, No. 5, May 1987.

[5] Detmer,P. R., *et al.*, Matched filter identification of left-ventricular endocardial borders in transesophageal echocardiograms, IEEE transactions on Medical Imaging. Vol. 9, No. 4. December 1990.

[6] Lipmann, R. P., An introduction to computing with neural nets, IEEE ASSP Magazine, April 1987.

[7] Rumelhart, D. E., McClalland, J. L. and the PDP Research Group, Parallel distributed processing, Vol. 1, The MIT Press, 1988.

[8] Cornelis, A., Nyssen, E., Cornelis, J., Van Denhaute, E., Van Denhaute, F., Van hamme, H., Steenhaut, O., Linear discriminant Analysis applied to various examples in medical imaging, III Simposium international de ingenieria biomedica-Madrid-Espana, October 1987.

SIGNAL PROCESSING VI: Theories and Applications
J. Vandewalle, R. Boite, M. Moonen, A. Oosterlinck (eds.)

A SPECIFIC PROCESSOR FOR THE COMPUTATION OF TDNN ALGORITHMS WITH APPLICATIONS IN PHONETIC CODING

M. Pérez-Castellanos*, V. Rodellar*, V. Peinado#, A. Diaz# and P. Gómez-Vilda*

*Facultad de Informática, Urb. Montepríncipe, Boadilla del Monte, 28660 Madrid-Spain
#Escuela Universitaria de Informática, Km. 7 Carretera de Valencia, 28035 Madrid-Spain

This paper describes the structure of a Specific Processor capable of supporting Time Delay Neural Network algorithms in real time, as well as other NN and digital signal processing algorithms sharing an inner product computational structure. The specific processor has been designed using serial arithmetics. It has the advantadge of presenting a very flexible data path, through the inclusion of a set of multiplexers which allow independent access to the different functional units of the processing elements. The structure described is being used for Phonetic Coding applications.

1. INTRODUCTION

TDNN algorithms have been widely used in different speech problems due to its main characteristic of hybridating the concepts of Digital Filter and Neural Network algorithms [Ham.90], [Unn.91]. We have been using TDNN algorithms in the application of phonetic coding adapted to the Phonetics of Spanish language [Rod.91b]. The particular topology of the algorithm suitable for such a problem is shown in Fig.1, consisting in an input layer containing six nodes, a hidden layer of eight nodes and an output layer of eight nodes also, the digital filter part has order two, and the net is fully connected. The reasons for choosing such a topology may be found in [Rod.91a].

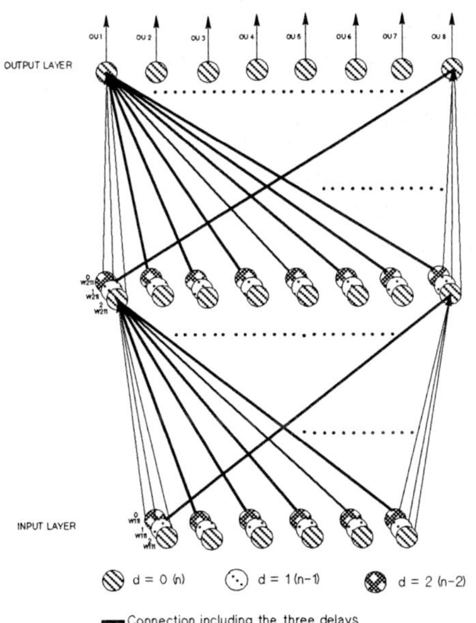

Fig.1 TDNN algorithm topology.

The generic expression of a TDNN node may be written as:

$$S_k(n) = F\left[\sum_{j=1}^{P} \sum_{d=0}^{2} X_j(n-d)\, w^d_{ijk}\right] \quad (1)$$

where "k" indicates the node number under computation inside the hidden or output layers, $1 \le k \le 8$, "n" is the discrete time index, "F" is the non-linear function, "j" is the node number of the source layer, wich contributes to the evaluation of the current node, $1 \le j \le P$; P being the total number of nodes in the source layer, then $1 \le j \le 6$ for the hidden layer and $1 \le j \le 8$ for the output layer, "d" is the delay time index, "X(.)" represents the node inputs, "w" is the set of coeficients obtained during the trainig period and finally "i" distinguishes between the hidden (i=1) and output layers (i=2).

Let's see the data dimensionality and computational structure of the algorithm. The NN will be fed by a vector IN[r] containing three sets of samples of speech spectrum parameters taken in three consecutive time instants, then r=6x3=18. Similarly, the output of the hidden layer will be a vector HI[s] containing three data sets corresponding to the operations performed in three consecutive time instants on the nodes involved in the computation of the current node, hence s=8x3=24. The output of the NN will be a vector OU[v], with v=8. The set of weights, obtained during the trained simulation phase, associated to the hidden layer will be a matrix WH[v,r], being v=8 and r=18, giving a total amount of 144 elements to be stored. The set of weights for the output layer will be another matrix of 192 elements, WO[v,s] with v=8 and s=24. And finally the computation of the non-linear function will be done by means of a look-up table of eight elements LUT[8]. The computation of the algorithm is as follows; first we will evaluate the nodes of the hidden layer, this operation may be seen as a matrix-vector product:

$$WH[v,r]\ IN[r] = Q_n[v] \quad (2)$$

where $Q_n[v]$ is a vector of eight components corresponding to a node of the hidden layer. To complete the evaluation of this layer, the results of this vector must be passed through the look-up table in order to compute the non-linear function F(.):

$$F[Q_n[v]] = H^0 \quad (3)$$

the computation of the output layer may also be written as a matrix-vector product:

$$WO[v,s]\ HI[s] = P_n[v] \quad (4)$$

and after the computation of the non-linear function the final results will be obtained:

$$F[P_n[v]] = OU[v] \quad (5)$$

2. THE SPECIFIC PROCESSOR

The structure of the algorithm described above will be supported by a chip containing four processing units, each PU being locally connected to its neighbours, border processors having connection capability with the external world. The control of the chip is done by means of a PLA, which provides a flexible basis for the reprogrammability of the chip's control. The basic structure of one of these PU's is shown in Fig.2, it consists in an Arithmetic Unit, a double port RAM memory, a ROM memory, a set of multiplexers, two input data lines and two output data lines. The internal data paths of the PU's are bit serial in order to save silicon area.

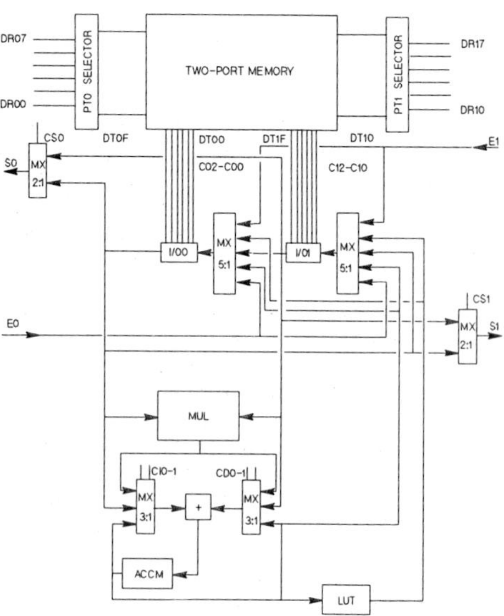

Fig.2 Basic structure of the PU's

The kernel operations involved in the computation of this algorithm are matrix-vector products, then the Arithmetic Unit will consist basically in an Inner Product Unit structure, containing a multiplier of 16 bits, which uses fixed point arithmetics and extended sign for protection against overflows, a 16-bit accumulator and a 16-bit adder. The structure also incorporates two 3:1 multiplexers that allow to route selectively data from the RAM memory, the multiplier or the accumulator to the adder. The double port RAM memory will be devoted to store the weights, input data and auxiliary variables, its capability is 256x16, and the data transfers from and to memory are done in parallel throught the interface registers I/00 and I/01 which provide serial/parallel and parallel/serial conversion. The ROM memory locates the look-up table values for the computation of the non-linear function, its size is 8x16. The structure of the PU's is completed with a set of multiplexers that provides an enormous flexibility to the internal data routing in the rigid schema of a bit serial approach. The multiplexers 5:1 introduce selectively data from the input E1, the result from the look-up table, the contents of the port I/01 or I/00 respectively, the result of the accumulator or the input E0 into the RAM memory registers I/00 or I/01. And finally the 2:1 multiplexers allow routing the contents of both I/O RAM memory registers to the PU's outputs.

The next step is to show how to compute the proposed algorithm in a set of four PU's with a reduced architecture as the one just described. If we take a look to the data dimensionality of the problem, for example keeping our attention in the matrices WH[] and WO[], we notice that none of them can be located in the RAM memory of the same PU, then a segmentation and mapping strategy for the algorithm must be adopted. Different segmentation and mapping strategies may be proposed using as performance parameters the load balance among PU's, memory occupation, and communication simplicity, avoiding waiting times among transfer data. Some possible strategies have been already proposed [Rod.91a].

Although the PU structure has been designed having in mind a given algorithm for a very specific aplication, this PU can also easily support other NN models and Digital Signal Proccessing algorithms with an inner product computational structure. This can be easily accomplished altering the programming of the PLA, for such a purpose a Transfer Register Language is being used.

3. TDNN ARITHMETICS

As it was mentioned before, the Inner

Product Arithmetic Unit of each PU, was designed on a 16-bit fixed point data format, using Serial Arithmetics. At a first glance, it seems that such a format will produce a very degraded behaviour in the arithmetics, due mainly to precision losses and overflows, when compared to wider floating point formats, but this is not necessarily so when some properties of Back Propagation Neural Networks are adequately exploited. For instance, when expression (1) is carefully examined, one can easily see that the basic operation to compute the output of a given node, is a 3xP order Inner Product, involving 3xP multiplications and the same amount of additions. Regularly, the Inner product is carried out by simultaneously addressing each element of the corresponding vectors (V_{1i+1}, V_{2i+1}), multiplying them, and adding this product to the accumulated partial result A_i, by the recursive expression:

$$A_{i+1} = A_i + V_{1i+1} \times V_{2i+1} \qquad (6)$$

When the data involved in these operations are represented using a S+F bits normalized fixed point format, where S bits are used for sign extension, and F bits are used for fraction, it may easily be deducted that after each multiplication, a 2(S+F) bit number will be obtained. In it, the initial S bits will be disregarded, the following S+F bits will be used as the result, and the final F bits will also be rejected, giving a result again in S+F bits. When we assume that both vectors V_1 and V_2 have their elements normalized in the interval [-1, +1], the result would also be a normalized S+F fixed point number. The multiplication would produce a loss of precision if V_{1i+1} or V_{2i+1} or both data are much lower than the unity, because, in that case, the significant bits of the result could appear in the rejected F final bits of the 2(S+F) product. On the contrary, the addition, could produce results higher than the unity, and in this case, an invasion of the S guard bits reserved for the sign extension would results. The worst case would take place if the first sign bit would be invaded, because the sign would be lost. This effect is more critical as the number of additions to be performed increases. As this number is given by 3xP, it would be worse for the output layer (P=8) than for the inner layer (P=6). To prevent these facts, the best solution is to adequately dimension S and F, keeping the constraint that S+F=16. For such, a set of simulations were conducted, using different S+F formats, and contrasting them with standard double precission floating point conventions. Randomly generated numbers were used as elements of the test vectors. A summary of the simulation results is presented in Table 1 for the more restrictive case of P=16.

The Maximum Number of Sign Invasions

S	F	MNSI						MNLB	MAPL	MQE
		#6	#5	#4	#3	#2	#1			
4	12	-	-	7	6	4	1	2	.008	.00012
6	10	8	6	2	1	1	0	3	.049	.00049
6	12	7	5	4	1	1	0	2	.012	.00012

Table 1. Sign Invasions and Precision Losses

(MNSI) is plotted for each sign bit from the least significant (#6) to the most significant (#1). The Maximum Number of Lost Bits (MNLB), Maximum Absolute Precision Loss (MAPL) and the Maximum Quantizing Error (MQR) are also plotted under the corresponding columns. It can be seen that with S=4 we could represent numbers in the interval [-8, +8), whereas with S=6 the interval would be [-32, +32). The Inner Product in (6) would produce a maximum absolute value of 24 for P=8, or 48 for P=16 only when both vectors have their elements equal each to the other with a value of +1 or -1, but these cases are nonsense. Experimental estimations show that the maximum expected absolute value is under one third of these maximum figures, which means 8 for P=8 and 16 for P=16. This means that the format 4+12 is just in the acceptable limit, but the format 6+10 well inside limits. In fact, in this last case from Table 1, one can see that the first bit is never invaded for P=16, which covers the case of P=8. The MAPL keeps reasonably low, about 100 times above the MQE, due to its cumulative effect. In a practical case, with P=8 these effects would be almost neglectable. The results for a format 6+12 are given for a comparison, showing that the gain in precision does not justify the increase in Silicon area required (11%). On the other hand, maintaining the restriction S+F=16 an easy change among different formats could be done just "retailoring" the results after multiplications, as was commented above. Precision losses in Neural Networks are not as critical as they could be in other Signal Processing Applications, because of the nonlinear features of most Neural Networks. In fact, the second operation to be considered in (1) is the nonlinear mapping from the Inner Product Interval, for instance [-32, +32), to the Node State Interval [-1, 1]. This nonlinear mapping is mathematically expressed as a Sigmoid Function. In our case, this fuction is implemented by a Look-Up-Table. This structure may be seen as a Transcoding ROM which receives the S most significant bits of the result in (6), and uses them to address a 16 bit output out of a table of 2^S possible results. Having in mind this effect, a loss of precision in the multiplications does not seem critical, in the sense that it will seldomly affect the bits used to address within the LUT. These facts have been checked using real data, and the proposed format, against a double precision floating point, and the performance degradation is rather acceptable, although the matter is still under study.

4. INTEGRATED CIRCUIT DESIGN

The whole structure is being prepared for its integration under a 1,5 μm CMOS technology, following the floorplan given in Fig. 3., although this floorplan is scaled for a 2μm process, under which different critical parts, as the multiplier, have been casted already.

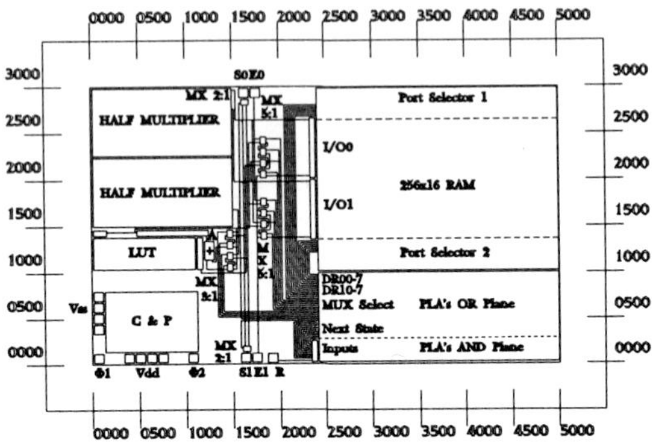

Fig. 3. Floorplan of the supporting IC

The floorplan shows the physical disposition of the structures in Fig. 2 in a real scale. It may be seen that the most area consuming section is the Two Port Memory (34.6%), followed by the Arithmetics (20%), Data Paths (18%), Control (17.3%) and Clocks&Power (10%). These figures give a good justification to assume a fixed-point serial data format, because memory, arithmetics and data paths are very sensitive to data formats in the sense that more than 72% of the integrated circuit area is affected by these aspects. The floorplanning is roughly based in a "Manhattan Skyline" style, most of data and control routing being done using a central highway connecting Memory to Arithmetics. Following CMOS design restrictions, general tri-state buses have been substituted by the massive use of multiplexers. Having in mind that address buses DR00-7 and DR10-7 require parallel routing, the PLA generating these signals have been placed in the same waterfront side than the Two-Port Memory. The data routes are kept under reasonable limits in length. This can't be said of Clock and Power, which must be routed all through the chip. These routes are not shown for the sake of clarity. Vss and φ_1 are routed along the periphery of the chip, and the central corridor is used for Vdd and φ_2. The design have been carried out using customized cells of regular structure in the multiplier, registers and multiplexers, and standard library cells for the PLA, LUT and Memory. The whole chip requires an area of 15 mm^2 in the 2μm process, although a reduction thus allowing the piling of up to 4 of these PU's in a medium size integrated circuit.

5. CONCLUSIONS

A customized design of the Arithmetic Unit has been casted and is currently under test. Other parts of the described structure are also under design, and a whole chip containing one PU will be available in brief. Communication schemes among PU's are being simulated on a four Transputer system. We have also studied different possible segmentation and mapping strategies of the TDNN algorithm in order to obtain the maximun computational efficiency and several solutions that fulfill the requirements for real time proccesing have been found. Actually we are working in the programming of the PLA in order to cast the structure proposed in a 1,5 micron CMOS technology. The applications of the described structure are found in the support of a joint auditory model for frecuency extraction of the Speech trace and of a TDNN. The whole system will be devoted to real time phonetic transcription of Spanish in Computer Aided Speech Learning.

6. ACKNOWLEDGEMENTS

This work is being supported by the grants: Acción concertada UPM A91-0020-02-78, Acciones Especiales de Microelectrónica MIC88-0398-E and MIC90-1219-E, and Plan de Investigación de la CAM C072/90.

7. REFERENCES

[Ham.90] J. B. Hampshire II and A. H. Waibel. "A Novel Objetive Function for Improved Phoneme Recognition Using Time Delay Neural Networks". IEEE Trans. on Neural Networks, Vol.1, No.2 (June 1990), pp. 216-228.

[Rod.91a] V. Rodellar, M. Pérez-Castellanos, P. Gómez-Vilda, V. Peinado and J. Bobadilla. "Segmentation and Mapping Time delay Neural Networks algorithms on VLSI Processors", Proceedings of the 1991 European Simulation Multiconference. CSS, Erik Mosekilde Ed., pp. 254-259.

[Rod.91b] V. Rodellar, F. Naharro, C. Garcia, S. Martin, M. Muñoz and P. Gómez-Vilda, "A Neural Network for the Extraction and Characterization of the Phonetic Features of Speech", Proceedings of Neuro Nimes 91, pp. 203-212.

[Unn.91] K. Unnikrishnan, J. Hopfield and W. Tank. "Connected Speaker-Dependent Speech Recognition Using a Neural Network with Time-Delay Connections". IEEE Trans. on ASSP, Vol.39, No.3, March 1991, pp. 698-713.

SIGNAL PROCESSING VI: Theories and Applications
J. Vandewalle, R. Boite, M. Moonen, A. Oosterlinck (eds.)
1992 Elsevier Science Publishers B.V.

A SAMPLE PROTOCOL FOR THE HOPFIELD NEURAL NETWORK

H.W. Klijn Hesselink*, F.F.G. de Bruin and L.P.J. Veelenturf.

University of Twente, Dep. of Electrical Engineering,
P.O. Box 217, NL-7500 AE Enschede, The Netherlands.

A Hopfield neural network can be used as a pattern recognition system. In this paper a sample-controlled protocol for sequential pattern recognition with the Hopfield network is introduced.

I. INTRODUCTION

The paper by Hopfield of 1982 [1] explored the ability of a system of highly interconnected artificial neurons to have useful collective computational properties. He showed that such a artificial neural network can be used as a Content-Addressable Memory (CAM). An item that is stored in a CAM can be retrieved on basis of a disturbed version of that item. Therefore a CAM can be very well used for pattern recognition problems. In this paper we show that we can retrieve a pattern not only on the basis of a disturbed version of the pattern, but also from a sequence of local samples of the pattern to be recognized. Moreover the number of local samples can be reduced by controlling the sampling process on the basis of the previous responses of the neural network. Results on sample controlled recognition of hand-written numerals are given.

II. HOPFIELD NETWORK

In 1984 Hopfield introduced his time-continuous neural model [2]. An electronic scheme of such a neuron is shown in figure 1. This neuron contains a non-linear amplifier with an input u_i and output

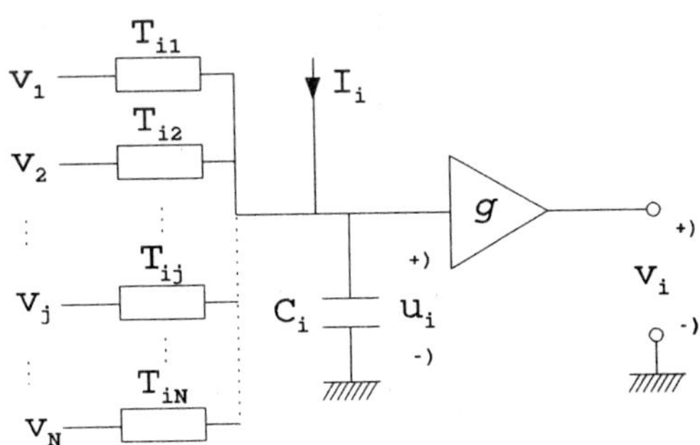

fig. 1: Hopfield neuron model

v_i.
The transfer function $v_i = g(u_i)$ should be monotone and bounded [2].
We will consider the hard-limiter transfer function:

$$v_i = g(u_i) = \begin{cases} +1 & \text{if } u_i \geq 0 \\ -1 & \text{if } u_i < 0 \end{cases} \tag{1}$$

The neural network of Hopfield is totally connected by conductances T_{ij}. The conductance T_{ij} and capacity C_i establishes a time delay between a v_j and state variable u_i. Each output v_i of a

*Currently at University of Groningen, Department of Computing Science, P.O.Box 800, NL-9700 AV Groningen, the Netherlands.
E-mail: klijn@cs.rug.nl.

neuron is an output component of the total system. The output vectors **v** of the total system correspond with state vectors **u**.
Only a few output vectors **v** of a Hopfield neural system with N neurons are stable. An output vector is a stable output vector if and only if:

$$v_i = g(\sum_{j=1}^{N} T_{ij} v_j + I_i) \quad \forall i \tag{2}$$

The initial state of the system corresponds with some selected output vector. For all output vectors which are not stable output vectors the dynamics of the system let these vectors converge to a near(est) stable vector (measured in Hamming distance).
If the network is used as a CAM for pattern recognition, then prototype patterns will be encoded as prototype vectors (PVs), with elements -1 or 1. These PVs will be stored in the neural net as stable states by means of adapting the connection matrix **T** with the use of a learning rule. We used the geometric or outer-product rule [4] to establish the storage of stable states.
This learning rule creates a matrix **T** which projects every vector **v** on the sub-space spanned by the PVs. So all PVs are eigenvectors of the matrix **T** with eigenvalue 1 and are stable according to (2) with $I_i = 0$ for all i. Our experiments revealed that with the geometric rule a good CAM can be constructed, i.e. disturbed patterns (= unstable output vectors) converge to their nearest prototype pattern (output vector), under the conditions:
A. The distortion of the pattern is within certain limits;
B. The number of stored prototype patterns M is not to high ($M_{max} \approx 0.3N$) and
C. The correlation between prototype patterns is not to high [3].
If the diagonal elements of the connection matrix **T** are set to zero, it can be proved that the PVs remain stable and better results with the CAM can be obtained. [3].

III. SAMPLE-CONTROLLED ALGORITHM

If a pattern **w** is offered to a Hopfield network to be recognized the whole pattern **w** can be used to realize the corresponding initial state of the system. The initial state can be imposed by using the inputs I_i of each neuron. The complete pattern **w** to be recognized is then encoded as a input (current) vector $\mathbf{I}=\mu\mathbf{w}$. After initialization of the system the input vector **I** is set to the zero vector and the system will converge to the nearest stable output pattern.
In our approach the initial state of the system corresponds with some randomly chosen prototype vector **v** and an initial input vector **I**(**w**) composed of zero's and one entry μw_k, corresponding with a local pixel sample w_k, is presented to the net. When a stable state is reached a second input vector **I**(**w**) composed of zero's and two entries μw_k and μw_l corresponding with pixel samples w_k and w_l, is presented to the net. And so on.
It turns out that in this case only about 40 % of the pattern needs to be sampled for recognition. The second method has thus the advantage of using less information of a pattern than the first, especially if the sampling process can somehow be controlled on basis of previous responses. A good example of a system which has an efficient sampling process for scanning pictures is the Human Visual System (HVS): The eyes of a human being very rapidly jump (not observable for the human itself) from one "interesting" part of a picture to another. Interesting parts are mostly parts with a sharp luminance difference. The HVS also only needs a small part of a total picture for recognition [5].
We developed a similar method to control the sampling process of patterns with the Hopfield network using information of the internal state during successive stages of the converging dynamical process of the network. The state of the network with N neurons can be described by the N-dimensional vector **u**. An initial state will converge to a stable state for which [6]:

$$u_i = R_i \sum_{j=1}^{N} T_{ij} v_j \quad \forall i$$
$$\text{with } \frac{1}{R_i} = \sum_{j=1}^{N} |T_{ij}| \tag{3}$$

It can be shown that the absolute value of a component u_i of a stable state vector **u** is small the more the corresponding components v_i of all the stable PVs are different from each other. These component v_i correspond to pattern samples which are different for different prototype patterns at the same position in the pattern. One might expect that sampling a pattern mainly on the parts for which it differs from other patterns would yield a better performance. However selecting continually a sample position corresponding with the minimal absolute value of a state variable did not reduce significantly the percentage of necessary samples for recognizing a pattern in comparison to random sampling a pattern.
We can however use also the minimal absolute value of components of unstable states which we encounter during the preceding process of conver-

gence. It turns out that a better performance is obtained by sampling the pattern at those positions (i.e. w_k) where in a preceding dynamical convergence process the absolute value of the state component u_i becomes minimal.
One can consider the initial (randomly chosen) state, corresponding with some prototype vector $\mathbf{v}_1$ as an initial hypothesis about the pattern $\mathbf{w}$ to be recognized. The initial state together with the initial input vector $\mathbf{I}_1(\mathbf{w},\mathbf{u})$ will cause the system to converge to the nearest stable output vector $\mathbf{v}_2$, i.e. the next hypothesis about the pattern $\mathbf{w}$. During the first convergence process unstable vectors $\mathbf{u}$ will be passed. Small values of components u_i of these unstable vectors $\mathbf{u}$ indicate that $\mathbf{w}$ and $\mathbf{v}_1$ differ at position i. Thus it will be profitable to sample the pattern $\mathbf{w}$ at that position. This, together with $\mathbf{I}_1(\mathbf{w},\mathbf{u})$, gives us the second input vector $\mathbf{I}_2(\mathbf{w},\mathbf{u})$ for the second output vector $\mathbf{v}_2$, etc. until $\mathbf{v}$ represents the same pattern as $\mathbf{w}$.

IV Experiments

A network with N neurons was learned with M prototype vectors/patterns.

Table 1

	sample controlled	random samples
N=30 M=5 mHd=5	21.6%	45.2%
N=30 M=5 mHd=10	32.7%	40.6%
N=30 M=10 mHd=10	37.9%	38.0%
N=100 M=15 mHd=15	23.5%	40.6%
N=100 M=30 mHd=15	29.2%	34.2%

First experiment: For every PV as an initial state and for every PV, different from the initial state, as an pattern to be recognized an experiment was performed. We used the sampleprotocol and compare it with random sampling. The mutual minimum Hamming distance (mHd) between PVs could be adjusted, this mHd is a measure for the correlation between prototype patterns. The results show the average percentage of the input vector which has to be sampled before the network converges successfully.

The experiments show that the sample-controlled protocol operates better than the random sample protocol (table 1). Especially if the number of prototypes is low in comparison to the number of neurons, and the minimum Hamming distance between the PVs is low, i.e. the correlation between the PVs must be high (see N=100 M=15 mHd=15 and N=30 M=5 mHd=5). In the other cases the difference is small or negligible.
We note that the performance of a Hopfield network as pattern recognition device in the classical way, thus without the sample controlled protocol, is low in case that the correlation between PVs is high.

Second Experiment: In this experiment pictures of numerals (0-9) where used as patterns. 10 prototype numerals have been encoded to PVs by dividing the picture in 100 pieces and let each piece corresponded with -1/+1 if less/more than a quarter of the pixels of the piece were gray. A network with 100 neurons has been learned with these PVs. Initially the network is in a stable state, representing a prototype numeral. Another prototype numeral is offered for recognition (figure 2 and 3).

Table 2

	random samples	sample controlled
Average	22.4%	34.4%
Maximum	36.0%	65.0%

The results show that the sample-controlled protocol operates better than the random sample protocol (table 2).

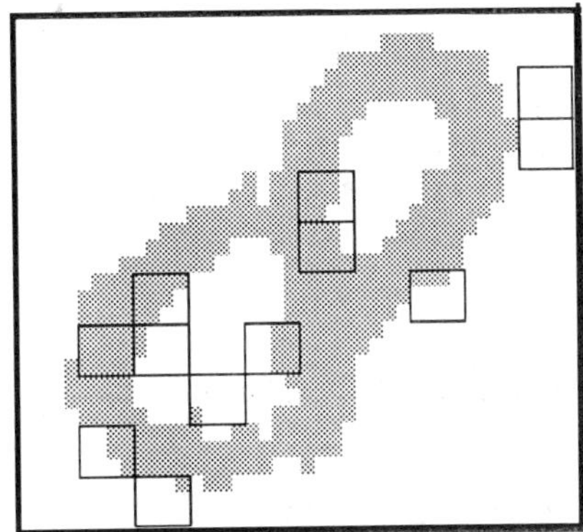

fig. 2: Controlled sampled "8" with initial network state representing a "9". 12% need to be sampled.

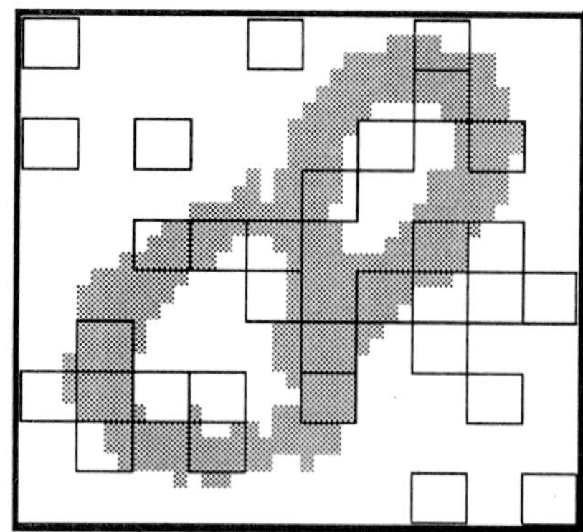

fig. 3: Random sampled "8" with initial network state representing a "9". 30% need to be sampled.

V CONCLUSIONS

Using a Hopfield network for pattern recognition it is not necessary to present the whole pattern. Random sampling the pattern can reduce the required information from the pattern to be recognized with about 60 %.

A sample-controlled sequential input protocol is developed, so that especially networks with a low number of highly correlated prototype vectors will need less samples (2 times less is possible) than a random sample-protocol. This controlled sample-protocol does not use prototype information.

REFERENCES

[1] Hopfield J.J., Neural Networks and physical systems with emergent collective computational abilities; Proc. Natl. Acad. Sci. USA Vol.79 (1982) pp 2554-2558.

[2] Hopfield J.J., Neurons with graded response have collective computational properties like those of two-state neurons; Proc. Natl. Acad. Sci. USA Vol.81 (1984) pp 3088-3092.

[3] Klijn Hesselink H.W., Analysis of the Hopfield network with respect to sample-controlled sequential pattern recognition, Ref.:080-DV-9113 (University of Twente, Dept. of Electrical Engineering: Group ICS, Enschede, the Netherlands 1991).

[4] Denker J.S., Neural network models of learning and adaption; Physica 22D (1986) pp. 216232.

[5] Hubel D.H., Eye, Brain and Vision (Scientific American Library, New York, 1988).

[6] Veelenturf L.P.J., Dynamical Neural Networks, Ref.:080.1181 (University of Twente, Dept. of Electrical Engineering: Group ICS, Enschede, 1988).

SIGNAL PROCESSING VI: Theories and Applications
J. Vandewalle, R. Boite, M. Moonen, A. Oosterlinck (eds.)

A COMPARISON OF ERROR DIFFUSION AND THE HOPFIELD NEURAL NETWORK AS METHODS TO QUANTIZE HOLOGRAM DISTRIBUTIONS

P. VAN DEN BULCK

EE-Department, Eindhoven University of Technology, Eindhoven
P.O. Box 513, NL-5600 MB Eindhoven, The Netherlands

In computer-generated holography, the distribution of a hologram that is used to generate a given reconstruction is calculated by means of a digital computer. Since we use in the production step of such a hologram an output device that is able to generate a binary output only, the calculated distribution has to be quantized to a binary signal. This will lead to quantization noise in the reconstruction plane. With the so-called error diffusion halftoning algorithm it is possible to calculate a binary hologram with low noise power within a chosen region of the reconstruction plane. Using a linear model for the error diffusion process, optimal diffusion coefficients of a first-order causal feedback filter are calculated. The error diffusion concept with causal error feedback filters can be generalized to error diffusion with noncausal feedback filters. The algorithm then describes the dynamics of a discrete-time Hopfield neural network with serial updating. This iterative calculation of the binary hologram is computationally more expensive, but a lower noise power is achieved.

1. INTRODUCTION

Holograms are used in optical systems to realize certain signal processing functions. In computer-generated holography the hologram transmittance is calculated by means of a digital computer and drawn by e.g. a laser-printer or an e-beam writer. When a hologram with transmittance $\phi(x,y)$ is illuminated with a coherent light source, we measure at some distance behind the hologram the wavefront $\Phi(u,v)$. In the case of Fourier holograms, the reconstruction $\Phi(u,v)$ of the hologram and the hologram distribution $\phi(x,y)$ are Fourier transform pairs.

Suppose that we want to generate a certain object $\Psi(u,v)$ in the reconstruction plane of the hologram. With the inverse Fourier transform we calculate the hologram distribution $\psi(x,y)$. In general this will be a complex signal. Since the transmittance of an amplitude hologram is real and limited in dynamic range, we take the real part of $\psi(x,y)$ and perform a scaling $\phi(x,y) = [\Re\psi(x,y)]/C$, with $C = \max[\Re\psi(x,y)]$. For symmetry-reasons we will use a bipolar signal $-1 \leq \phi(x,y) \leq 1$, without considering the extra mapping to a positive transfer function that is necessary for an amplitude hologram. Due to these operations, the hologram will reconstruct $\Phi(u,v) = [\Psi(u,v) + \Psi^*(-u,-v)]/(2C)$, with $\Psi^*(-u,-v)$ called the twin image of the original object. Special care must be taken to avoid overlap of the original and its twin image in the reconstruction plane. Therefore we define a region $\mathcal{R}$ where $\Psi(u,v)$ can be nonzero only [1].

In practice we sample the reconstruction plane and use the inverse discrete Fourier transform to calculate the samples $\phi[n_1,n_2] = \phi(n_1\Delta x, n_2\Delta y)$ of the hologram distribution. In the digital-to-analog conversion (performed by printing the hologram), the calculated hologram samples are interpolated and rectangular cells with a binary transmittance (black or white) are formed. This means we must quantize our calculated hologram samples to binary samples. This operation will always lead to quantization noise in the reconstruction plane. Neglecting interpolation effects due to the finite size of the cells, we will consider two methods to separate the quantization noise from the object in the reconstruction plane.

The error diffusion algorithm [2] is often used as a halftoning-technique for pictures. It shapes the spectral content of the quantization noise to higher spatial frequencies and thus makes the quantization noise more or less invisible to our 'low-pass' eyes. In the reconstruction plane of a Fourier hologram we measure the spectrum of the quantization noise. Thus, when using error diffusion to quantize the hologram distribution [3], we are able to shape the quantization noise in the reconstruction plane and minimize the noise power in the region $\mathcal{R}$.

The error diffusion algorithm processes the two-dimensional hologram distribution in a one-dimensional spatial ordering. To ensure recursive computability, errors are diffused to certain directions only. In a generalization of this concept called the Hopfield neural network, errors are diffused in all directions. The binary hologram distribution is now calculated in an iterative way. This procedure is computationally more expensive than the error diffusion algorithm, but a lower noise power can be expected.

2. ERROR DIFFUSION

Calculating the binary signal $b[n_1,n_2]$ for a discrete signal $\phi[n_1,n_2]$ with the error diffusion algorithm, we use the following equations

$$\begin{aligned} s[n_1,n_2] &= \phi[n_1,n_2] - h[n_1,n_2] ** q[n_1,n_2] \\ b[n_1,n_2] &= \mathcal{Q}s[n_1,n_2] \\ q[n_1,n_2] &= b[n_1,n_2] - s[n_1,n_2]. \end{aligned} \quad (1)$$

In (1) $**$ denotes two-dimensional convolution and $\mathcal{Q}(\phi \geq 0) = 1$, $\mathcal{Q}(\phi < 0) = -1$. Error diffusion can be regarded as processing $\phi[n_1,n_2]$ with a nonlinear two-dimensional filter, as shown in Fig. 1. This filter consists of a

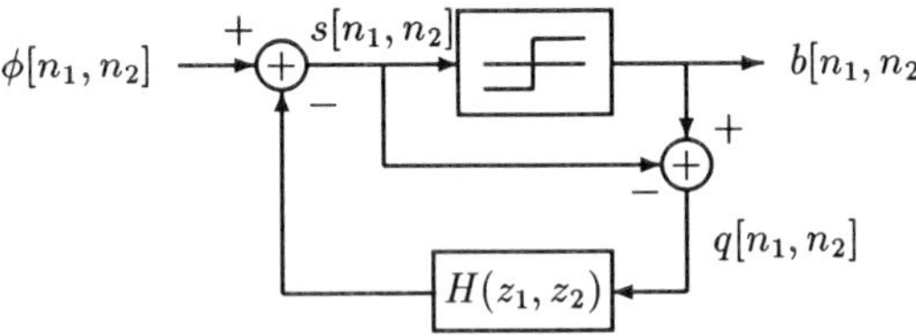

Figure 1: Two-dimensional nonlinear filter.

nonlinear operation, i.e., a quantizer and a filter with a finite impulse response $h[n_1,n_2]$ in the feedback loop. The *quantizer* errors $q[n_1,n_2]$ that have been made in the quantization of 'past' pixels, are weighted and diffused to the present pixel in order to influence the decision of the quantizer. To ensure recursive computability of this algorithm, the impulse response of the error feedback filter $h[n_1,n_2]$ must be causal, and therefore has asymmetrical half-plane support $\mathcal{W}$: $\{1 \leq n_1 \leq M, 0 \leq n_2 \leq M\} \cup \{-M \leq n_1 \leq 0, 1 \leq n_2 \leq M\}$, as is shown in Fig. 2. M denotes the filter-order.

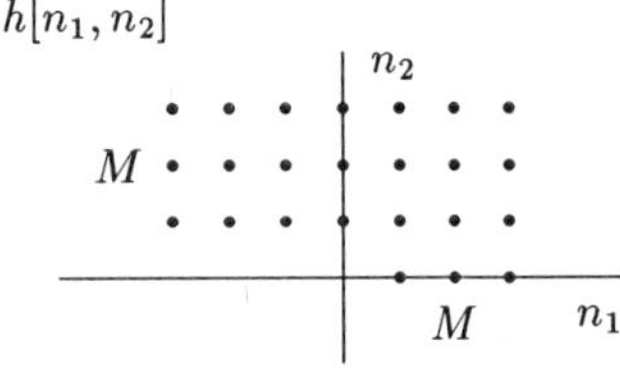

Figure 2: Error-feedback filter with asymmetrical half-plane support.

With the $2(M^2+M)$ filter coefficients $h[n_1,n_2]$ we are able to shape the spectrum of the *quantization* noise $e[n_1,n_2] = b[n_1,n_2] - \phi[n_1,n_2]$. Since we measure this spectrum in the reconstruction plane of the Fourier hologram, we try to find the filter coefficients that cause the smallest noise power in the region $\mathcal{R}$. To this end we introduce a linear model for error diffusion. In this model the quantizer error $q[n_1,n_2]$ is regarded as an independent noise source. The set of equations (1) then simplifies to $s = \phi - h ** q$ and $b = s + q$. The input-output relation of the linear model is

$$B(z_1,z_2) = \Phi(z_1,z_2) + [1 - H(z_1,z_2)]Q(z_1,z_2), \quad (2)$$

with the two-dimensional Z-transform of a signal $x[n_1,n_2]$ defined as

$$X(z_1,z_2) = \sum_{n_1}\sum_{n_2} x[n_1,n_2] z_1^{-n_1} z_2^{-n_2}. \quad (3)$$

Since $E(z_1,z_2) = B(z_1,z_2) - \Phi(z_1,z_2)$ is the two-dimensional Z-transform of the quantization error $e[n_1,n_2]$ we can write

$$E(z_1,z_2) = [1 - H(z_1,z_2)]Q(z_1,z_2). \quad (4)$$

Apparently the *quantizer* error is filtered with a noise shaping filter $1 - H(z_1,z_2)$ to form the *quantization* error. For $z_1 = e^{j\Omega_1}$ and $z_2 = e^{j\Omega_2}$ the Z-transform passes into the Fourier transform and (4) becomes an expression for the noise in the reconstruction plane. If we assume the quantizer noise to be white, with noise power $\mathcal{E}|Q(e^{j\Omega_1},e^{j\Omega_2})|^2 = \sigma_q^2$, the total amount of noise in the region $\mathcal{R}$ of the reconstruction plane will be

$$P = \sigma_q^2 \int\int_{\mathcal{R}} |1 - H(e^{j\Omega_1},e^{j\Omega_2})|^2 d\Omega_1 d\Omega_2. \quad (5)$$

When we construct a coefficient-vector $\underline{h}$ containing the filter-coefficients, we can rewrite (5) in a general quadratic form

$$P = \alpha + \underline{a}^T\underline{h} + \frac{1}{2}\underline{h}^T A \underline{h}. \quad (6)$$

Due to the feedback loop, instability can occur for some sets of filter coefficients. This means that the internal signal $s[n_1,n_2]$ can grow without bound for a bounded input $\phi[n_1,n_2]$. Thus, we call the algorithm stable, if for every bounded input $|\phi[n_1,n_2]| \leq 1$ the internal signal $s[n_1,n_2]$ is bounded (BIBO stability). A sufficient condition to guarantee stability is [4,5]

$$\sum_{(n_1,n_2)\epsilon\mathcal{W}} |h(n_1,n_2)| \leq 1. \quad (7)$$

Finding the optimal filter-coefficients is then a constrained quadratic optimization problem: minimize P under constraint (7).

For the region $\mathcal{R}$: $0 \leq \Omega_1 \leq \frac{\pi}{2}, 0 \leq \Omega_2 \leq \frac{\pi}{2}$ we found as optimal diffusion coefficients for a filter of order $M = 1$: $h[1,0] = 0.3574$, $h[1,1] = -0.1022$, $h[0,1] = 0.0000$ and $h[-1,1] = 0.5404$. Calculation of higher-order filters did not result in a significantly lower

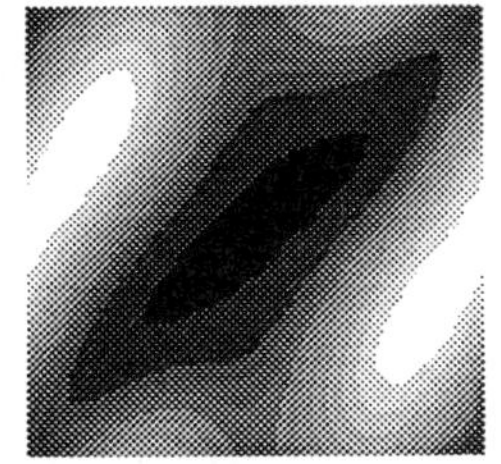
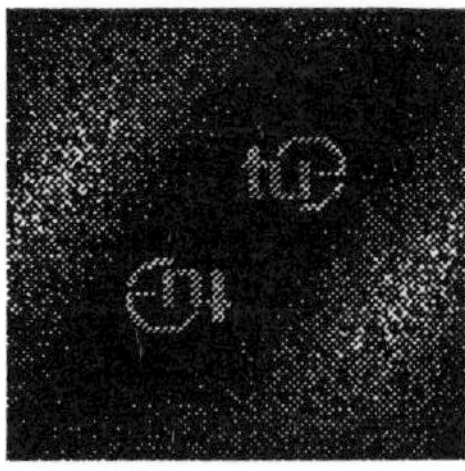

Figure 3: Calculated noise distribution (left figure) and simulated reconstruction (right-hand figure).

noise power. With these error diffusion coefficients, a hologram distribution (size: $N^2 = 128^2$) that generates the EUT-logo in the given region has been quantized. A comparison of the theoretical noise distribution and the simulation result (Fig. 3) shows that the noise distribution is well predicted. To measure the noise power we used an approximation

$$P \approx \frac{1}{N^2} \sum_{k_1,k_2} |B(e^{j\frac{2\pi k_1}{N}}, e^{j\frac{2\pi k_2}{N}}) - \Phi(e^{j\frac{2\pi k_1}{N}}, e^{j\frac{2\pi k_2}{N}})|^2, \quad (8)$$

where the summation extends over samples in the region of interest. A numerical result of about 55.7 was found.

The causal filter $h[n_1, n_2]$ takes only errors in account that have been made in the quantization of previous samples. This makes the algorithm recursive computable, provided that we process the samples in the right spatial ordering. This concept of error diffusion can be generalized, however. In that case, the decision of the quantizer for a given sample is influenced by the quantization errors of all other samples. Of course, it is not possible anymore to calculate the binary output recursively.

To derive an expression for noncausal error diffusion we rewrite (1) into

$$\begin{aligned} s[n_1,n_2] &= \phi[n_1,n_2] - g[n_1,n_2] ** e[n_1,n_2] \\ b[n_1,n_2] &= \mathcal{Q}s[n_1,n_2] \\ e[n_1,n_2] &= b[n_1,n_2] - \phi[n_1,n_2], \end{aligned} \quad (9)$$

which leads to

$$\begin{aligned} b[n_1,n_2] &= \mathcal{Q}\{-g[n_1,n_2] ** b[n_1,n_2] \\ &+ g[n_1,n_2] ** \phi[n_1,n_2] + \phi[n_1,n_2]\} \end{aligned} \quad (10)$$

The coefficients $g[n_1, n_2]$ form the impulse response of a filter with transfer function $G(z_1, z_2)$. This filter is related to the original feedback filter $H(z_1, z_2)$ according to

$$G(z_1,z_2) = \frac{H(z_1,z_2)}{1 - H(z_1,z_2)}. \quad (11)$$

A filter $G(z_1, z_2)$ associated with an error feedback filter $H(z_1, z_2)$ with asymmetrical halfplane support, is also causal and has wedge support. Not restricting ourselves to the causality condition, we can also choose sequences $g[n_1, n_2]$ with four quadrant support. As we will see, (10) then describes the serial updating rule for a discrete-time Hopfield neural network.

3. THE HOPFIELD NEURAL NETWORK

We consider a Hopfield network with N equal processor-elements or neurons. With the interconnection strength between neuron n and m denoted by w_{nm}, the matrix W describes the overall interconnection of the network. We assume that the interconnection is symmetrical and that neurons are connected to other neurons only. Therefore W is a symmetrical matrix with zeros on its main diagonal. Each neuron sums its incoming input signals. If the sum is greater than its threshold, the neuron will generate 1 as output signal, -1 otherwise. The thresholds of all neurons form the threshold-vector $\underline{t}$. In [6] Hopfield has shown that such a network is described with a set of nonlinear differential equations

$$\tau \frac{d\underline{u}}{dt} = W\mathcal{Q}(\underline{u}) - \underline{t} - \underline{u}, \quad (12)$$

with $\underline{u}$ the neuron-state vector and τ a time constant. The hard-clipping operator $\mathcal{Q}$ operates on each component of $\underline{u}$. When the network has converged, it is in a stationary state

$$\underline{u} = Wf(\underline{u}) - \underline{t}. \quad (13)$$

Since we are interested in the binary output of the network, we introduce the binary output vector $\underline{b} = \mathcal{Q}(\underline{u})$. Thus, in equilibrium the binary output vector satisfies

$$\underline{b} = \mathcal{Q}[W\underline{b} - \underline{t}]. \quad (14)$$

In order to simulate the network on a digital computer, we divide the time-axis in intervals Δt. As a consequence we must consider which neurons are updated in each interval. We will consider two extreme cases, namely serial and parallel updating. With serial updating, we select (according to some rule) only one neuron in every interval. To calculate the new binary value of the selected neuron we use

$$b^i_m = \mathcal{Q}\left[\sum_{n=0}^{N-1} w_{mn} b^{i-1}_n - t_m\right] = \mathcal{Q}\left[W\underline{b}^{i-1} - \underline{t}\right]_m, \quad (15)$$

with $b^i_m = b_m(i\Delta t)$. For this updating rule it is known [6] that the 'energy'

$$H(\underline{b}) = -\frac{1}{2}\underline{b}^T W \underline{b} + \underline{t}^T \underline{b} \quad (16)$$

can not increase during updating. Since the energy H has a absolute minimum value, the network will always converge to a stable state.

In order to use a Hopfield network in finding binary hologram distributions with minimal noise power in the region $\mathcal{R}$, we must relate the quantization problem to the energy function H. We are then able to calculate the weight matrix W and the threshold vector $\underline{t}$ and let the Hopfield network search for a solution. In order to maintain a simple notation we will first consider one-dimensional signals $\phi[n]$ and $b[n]$, represented by the signal-vectors $\underline{\phi}$ and $\underline{b}$. The spectra of these signals are approximated with the discrete Fourier transforms $\underline{\Phi} = F\underline{\phi}$ and $\underline{B} = F\underline{b}$, with F the Fourier transform matrix. We now try to find a binary signal $\underline{b}$ that minimizes the distance

$$D(\underline{b}, \underline{\phi}) = ||KF(\underline{b} - \underline{\phi})||^2, \tag{17}$$

with K a real diagonal frequency weighting matrix. This matrix determines the frequencies for which the noise power is minimized. It is possible to show [5] that minimizing the above equation is equivalent to minimizing

$$\tfrac{1}{2}\underline{b}^T A\underline{b} - \underline{\phi}^T A\underline{b}, \tag{18}$$

with $A = F^H K^H K F = F^{-1}K^2 F$ a real circulant matrix. Without loss of generality we can scale the main diagonal elements of A to one, and write $A = G + I$, with I the identity matrix. When we compare the above equation with the energy of a Hopfield network, we can conclude that a Hopfield network with $W = -G$ and $\underline{t} = -A\underline{\phi}$ will solve the quantization problem. For the serial updating rule this gives (in vector-notation)

$$\underline{b}^i_m = \mathcal{Q}\left[-G\underline{b}^{i-1} + (G+I)\underline{\phi}\right]_m . \tag{19}$$

Since the matrix G is also circulant, we can rewrite this equation in the form of a (periodic) convolution (denoted by $\otimes$)

$$b[n] = \mathcal{Q}\{-g[n] \otimes b[n] + g[n] \otimes \phi[n] + \phi[n]\}. \tag{20}$$

When we compare the two-dimensional extension of this equation with (10), we can conclude that the Hopfield network is a generalization of the causal error diffusion concept.

In order to save calculation time we will use parallel updating. With this updating rule, all neurons are updated in each interval. To derive this rule we approximate the derivative in (12) with a first-order backward difference and obtain

$$\tau\frac{\underline{u}^i - \underline{u}^{i-1}}{\Delta t} = W\mathcal{Q}(\underline{u}^{i-1}) - \underline{t} - \underline{u}^{i-1}, \tag{21}$$

with $\underline{u}^i = \underline{u}(t_i) = \underline{u}(i\Delta t)$. When we define $\alpha = \Delta t/\tau$ the parallel updating rule passes into

$$\underline{u}^i = \alpha\left[W\mathcal{Q}(\underline{u}^{i-1}) - \underline{t}\right] + (1-\alpha)\underline{u}^{i-1}. \tag{22}$$

The interval Δt must be smaller than the time-constant τ and therefore $\alpha \ll 1$. Although convergence is not guaranteed for parallel updating, we found that for a small α the Hopfield network always converged to a stable state with (local) minimal energy. Substitution of the calculated interconnection matrix and threshold vector into (22) gives

$$\begin{aligned}\underline{u}^i &= \alpha\left[-F^{-1}K^2F\mathcal{Q}(\underline{u}^{i-1}) + \mathcal{Q}(\underline{u}^{i-1})\right.\\ &\left.+F^{-1}K^2F\underline{\phi}\right] + (1-\alpha)\underline{u}^{i-1}.\end{aligned} \tag{23}$$

This updating rule describes an iteration between space and frequency domain [7]. In the space domain we perform quantization, in the frequency domain filtering. The extension to two-dimensional signals is straightforward. When using the fast Fourier transform, this algorithm is of complexity $O(N^2 \log N)$.

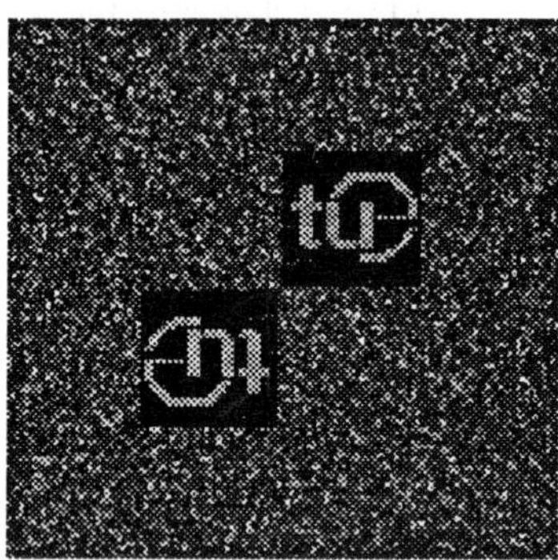

Figure 4: Simulated reconstruction.

We quantized the hologram distribution that generates the TUE-logo with the Hopfield network. In Fig. 4 the reconstruction of the binary hologram after 30 iterations is shown. The Hopfield network indeed tries to minimize the quantization noise in the given region. Since the noise shaping is better matched to the region $\mathcal{R}$, we may expect a lower noise power. Numerical calculation with (8) resulted in a noise power of 40.2, which is indeed smaller than that obtained with the error diffusion algorithm.

REFERENCES

[1] Wyrowski, F., J. Opt. Soc. Am. A 7 (1990) pp. 383-393.

[2] Floyd, R. and Steinberg, L., SID symposium. Digest of Papers (1975) pp. 36-37.

[3] Weissbach, S., Wyrowski, F. and Bryngdahl, O., Opt. Commun. 67 (1988) pp. 167-171.

[4] Broja, M., Eschbach, R. and Bryngdahl, O., Opt. Commun. 60 (1986) pp. 353-358.

[5] Anastassiou, D., IEEE Trans. Circuits and Systems CAS-36 (1989) pp. 1175-1186.

[6] Hopfield, J.J. Proc. Natl. Acad. Sci. USA 81 (1984) pp. 3088-3092.

[7] Just, D. and Ling, D.T., Opt. Commun. 81 (1991) pp. 1-5.

SIGNAL PROCESSING VI: Theories and Applications
J. Vandewalle, R. Boite, M. Moonen, A. Oosterlinck (eds.)

NEURAL PREDICTION IN IMAGE VECTOR QUANTIZATION

F.G.B.DeNatale, G.S.Desoli, S.Fioravanti, and D.D.Giusto

Dept. of Biophysical and Electronic Engineering, University of Genoa, Italy

A prediction scheme, based on a Perceptron neural network, is proposed, which allows one to exploit the residual inter-block correlation after VQ, without loss in the quality of reconstructed images. Main advantages lie in its efficiency (even for images not included in the training set), and in reduced memory requirements.

1. INTRODUCTION

Vector quantization [1] is one of the most widely used coding approaches, thanks to its high performances, in spite of its simplicity. However, it is worth noting that, after the encoding process, a notable amount of redundancy remains between neighbouring vectors, which are highly correlated. This fact is illustrated in Figure 1, where the height of each entry (i,j) of the matrix is proportional to the probability for finding the vector i, given the vector j in the previous position. Considerable efforts have been devoted to exploiting this inter-block correlation.

In a recent paper [2], a prediction scheme was proposed that allows one to obtain good results in terms of an increasing in the compression factor, and that does not cause any further loss in the SNR. But, such a scheme exhibits some drawbacks that make it not so attractive for software and hardware implementation. First of all, it requires the storage of large matrices, called transition matrices, which contain the conditional probabilities of finding the codevector i given the codevector j in the previous position along the principal directions (0, 45, 90, and 135 degrees). Another problem is the computation of the score functions for all the codevectors; this task requires one to access the transition matrices too many times, and is therefore very time-consuming. Moreover, the scheme presents the disadvantages of the storage and re-ordering of the address codebook.

The approach proposed in this paper exhibits the advantage (over the above-mentioned scheme), of an easy hardware realization; furthermore, no additional storage (except for the synaptic weights of the neural network) is required. The approach involves the implementation of a *Perceptron* neural network [3], trained in such a way as to reconstruct an image block on the basis of the neighbouring vectors.

The network is made up of an input layer (whose dimensions depend on the vectors' size and on the predictor's memory), and of an output layer, of the same dimensions as the codevectors. No hidden layers are utilized, thus making it possible to speed up the prediction and to reduce storage requirements (see Figure 2). For the learning phase, the classical back-propagation algorithm [4] has been adopted, which aims to minimize the distance between the vector generated by the network on the output layer and the actual block present in the training set, when the input layer is presented a vector made up of the neighbouring blocks. During the coding phase, every time a vector has to be transmitted, its neighbouring blocks are given as input to the trained network, and the prediction pattern is generated by the output layer. The obtained set of patterns is then used to organize a codebook of smaller dimensions than the original

one, thus requiring that a smaller number of bits be addressed. In the case of a wrong prediction, a recovery is provided.
In the following Sections, the architecture of the proposed network is described, the learning and encoding-decoding phases are detailed, and some experimental results are reported.

2. NETWORK ARCHITECTURE AND TRAINING

Figure 2 shows the scheme of the network architecture. The input layer consists of a number of neurons that is equal to the number of elements of the four vectors **Vi-1,j, Vi-1,j-1, Vi,j-1, and Vi+1,j-1** adjacent to the vector **Vi,j** to be predicted. The output layer has the dimensions of the vector **Vi,j**. No hidden layers are present, and the network is fully connected (e.g., when using 4x4-sized vectors in the quantizer, the number of synaptic connections is 1K).
The learning algorithm adopted is the classical back-propagation one. At each iteration, a set of configurations of blocks (surrounding the one to be predicted and preceding it in a left-to-right, top-down scan) are presented to the network input. The following updating rule is applied to the synaptic weights to minimize the mean square error (MSE) between the vector generated by the network (using the input and the weights computed during the previous iteration) and the target (i.e., the actual vector in the training set):

$$\Delta \mathbf{w}_{t+1}(i,j) = \eta\, \delta_i\, \mathbf{a_j} + \alpha \Delta \mathbf{w}_t(i,j)$$

where η is the gain, α is the momentum, and δ is given for each neuron a_i of the output layer by the following equation:

$$\delta_i = \left(\mathbf{t_i} - \mathbf{a_i}\right)\mathbf{a_i}\left(\mathbf{1} - \mathbf{a_i}\right)$$

For the present specific implementation of the algorithm, the initial weights are set to small random values, while the values of η and α, which strongly affect the speed of convergence, are dynamically updated according to the variations in the MSE.

3. THE ENCODING-DECODING PROCESS

The values of the synaptic weights computed during the learning phase are stored and used by the encoder and the decoder to achieve the prediction. During the encoding process, each block **Vij** is first vector-quantized to obtain the actual associated codevector $\mathbf{C_k}$ ($k = 1,\dots,2^n$), where n stands for the number of bits required to address the overall codebook. Then, the estimated pattern $\hat{\mathbf{V}}_{ij}$, generated by the neural network on the basis of the previously coded vectors $\mathbf{V}_{i-1,j}$, $\mathbf{V}_{i-1,j-1}$, $\mathbf{V}_{i,j-1}$, and $\mathbf{V}_{i+1,j-1}$, is used to organize a codebook of smaller dimensions, in which the codevectors $\mathbf{C'_h}$ ($h=0,\dots,2^{b-1}$, $b<n$), at a minimum distance from $\hat{\mathbf{V}}_{ij}$ are contained. Finally, if the actual codevector $\mathbf{C_k}$ is one of the $\mathbf{C'_h}$, only b bits are required to transmit the address h, otherwise, the b bit configuration left free in the reduced codebook is transmitted as a fault configuration, followed by the k bit address of the codevector $\mathbf{C_k}$ in the overall codebook. In the case of a correct prediction, compression is therefore increased by a factor of $\frac{\mathbf{n}}{\mathbf{b}}$, while, if the prediction is wrong, it is reduced by a factor $\frac{\mathbf{b+n}}{\mathbf{b}}$.
Denoting by $\mathbf{P_{hit}}$ the probability of a correct prediction, the factor **F** that multiplies the compression ratio can be expressed as:

$$\mathbf{F} = \frac{\mathbf{1}}{\mathbf{P_{hit}}\left(\frac{\mathbf{b}}{\mathbf{n}}\right) + \left(\mathbf{1} - \mathbf{P_{hit}}\right)\frac{\mathbf{b+n}}{\mathbf{n}}}$$

In order to obtain an increase in the compression factor, the threshold value $\mathbf{P^T_{hit}}$ for the probability of a correct prediction can be computed by imposing **F** equal to 1 and solving as a function of $\mathbf{P_{hit}}$:

$$\mathbf{P^T_{hit}} = \frac{\mathbf{b}}{\mathbf{n}}$$

The decoder must be synchronous with the encoder, in the sense that it has to accomplish the same sequence of operations, except for quantization, which is replaced by a simple look-up table. For each block, the prediction is calculated in the same way as in the encoder (the decoder contains an exact copy of the neural network), thus generating the reduced codebook. If the b-bit configuration received is a valid address, the vector is selected from the reduced codebook; otherwise (fault condition), the next n bits are used to address the overall codebook directly. The global architecture of the encoding-decoding system is displayed in Figure 3.

4. RESULTS AND CONCLUSIONS

A new algorithm for inter-block redundancy reduction in VQ has been presented, which is based on the use of a neural network as a block predictor. The advantages offered by this approach lie in the reduced memory requirements (both static and dynamic) and in a lower computational complexity (also for a hardware implementation). The proposed scheme seems to be very efficient, thnaks to the predictor structure and to the coding scheme. Several tests were performed on learned and unlearned images, using codebooks of dimensions ranging from 128 to 1024 vectors, and reduced codebooks of dimensions from 7 to 31 vectors. The diagram in Figure 4 shows the results of the tests carried out on the classical image of "Tiffany". The block dimensions were 4*4, and the image was not used for the construction of the codebook, nor for the learning of the predictor. The best results were obtained by using a 256-codevector VQ (n=8) and a reduced codebook of dimensions 7 (b=3). For such parameters, the value of $\mathbf{P}^{\mathbf{T}}_{\mathbf{hit}}$ that ensured a reduction in the bit rate was about 37%, and the $\mathbf{P}_{\mathbf{hit}}$ obtained ranged from 60 to 75% (corresponding to a factor **F** of 1.3 to 1.6). In the encoding phase, the synaptic weights can be further updated to take into account the specific characteristics of the image to be coded. Another possible alternative that was not considered in this paper is to assign variable-length configurations of bits to the vectors in the reduced codebook on the basis of the distances of such vectors from the prediction.

ACKNOWLEDGMENTS

This work has been supported by the National Research Council of Italy (CNR: Progetto Finalizzato Trasporti) and was carried out within the framework of the European Eureka project. Prometheus (ProCom sub-program).

REFERENCES

[1] R.M.Gray, *Vector Quantization*, IEEE ASSP Magazine, pp. 4-29, April 1984.

[2] F.G.B.DeNatale, G.S.Desoli, D.D.Giusto, and G.Vernazza, *A framework for high compression coding of color images*, Visual Communications and Image Processing '89, Proc. SPIE, vol. 1199, pp. 1430-1439, November 1989.

[3] M.Minsky and S.Papert, *Perceptrons*, MIT Press, Cambridge, 1972.

[4] D.E.Rumelhart, G.E.Hinton, and R.J.Williams, *Learning representations by back-propagating errors*, Nature, vol. 323, October 1986.

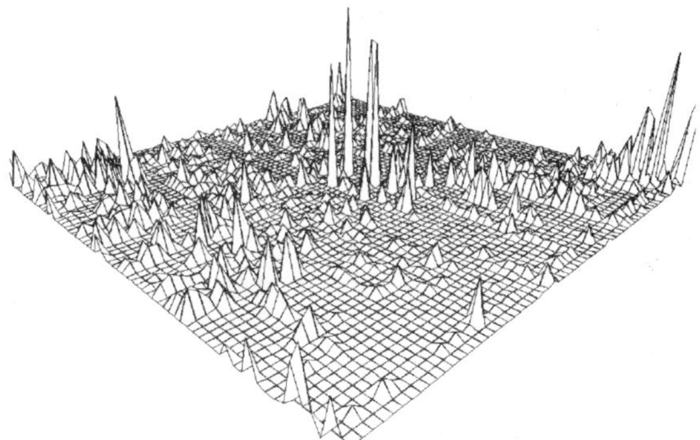

Figure 1. 3D view of a transition probability matrix

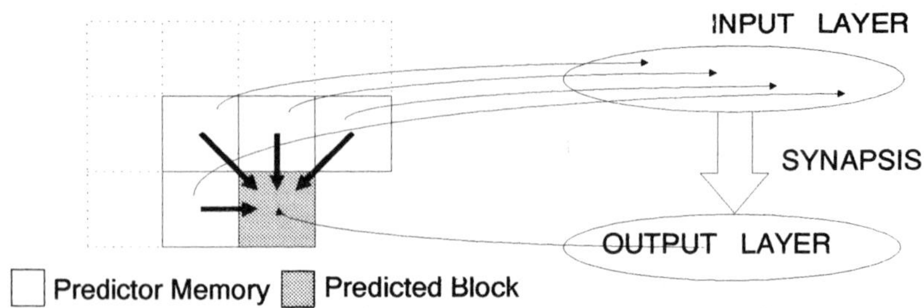

Figure 2. Predictor structure vs neural network architecture

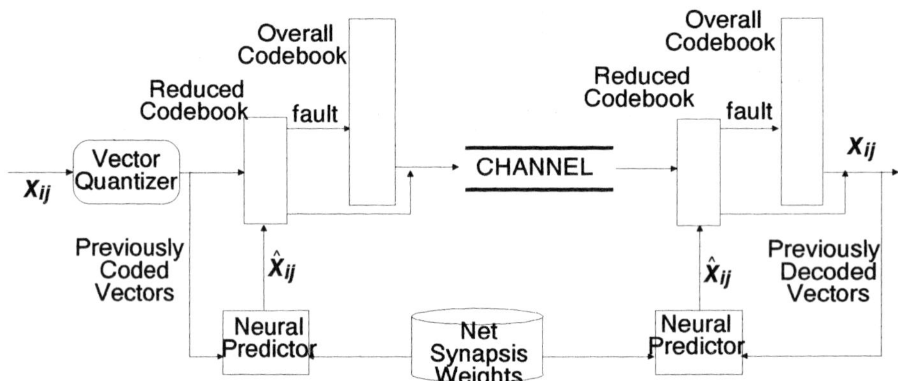

Figure 3. Architecture of the coding system

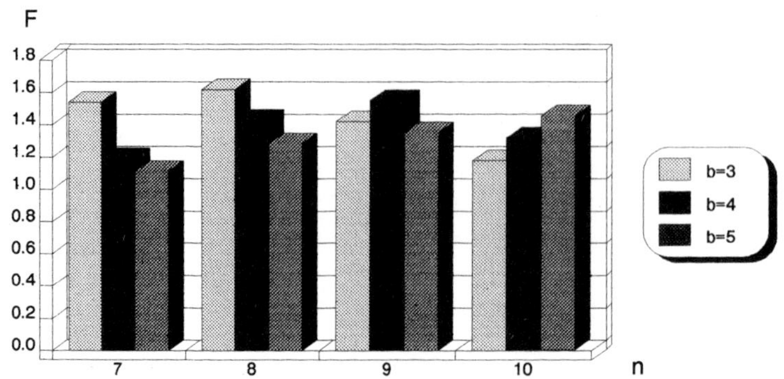

Figure 4. **F** as a function of **b** and **n** for a classical test image (Tiffany).

SIGNAL PROCESSING VI: Theories and Applications
J. Vandewalle, R. Boite, M. Moonen, A. Oosterlinck (eds.)

NEW APPROACH FOR MULTI-PERCEPTRON ARCHITECTURE DESIGN APPLIED TO THE EDGE DETECTION PROBLEM

Ch. Molina, D . Bastard, P. Baylou and M. Najim

Equipe Signal et Image and GDR 134 - CNRS
ENSERB. Université de Bordeaux I
351 cours de la Libération . 33405 Talence cedex. FRANCE

This paper deals with a new method to determine the architecture for a one-hidden-layer perceptron restricted to the convex problems solution in pattern recognition. The main contribution of this method is to train and design the neural network in a polynomial time in opposition to the impredictability of classical methods. This method is derived from the gift-wrapping algorithm which is mainly used in computational geometry for the construction of convex hulls in a *d*-dimensional euclidean space. To demonstrate its efficiency we apply this approach to the edge detection in image processing. We show that the edge detection can be solved by a neural network with only one hidden layer as an alternative to classical methods which employ the local maxima module of the gradient.

INTRODUCTION

The multiperceptron network is, at the present time, one of the most popular among the neuromimetic models. Its design is usually made by choosing *a priori* the neural network architecture. Pattern recognition is one of its privileged fields of application. When the learning phase ends, the neural net demarcates at best the decision regions delivering an exact answer for the learning set. The used algorithm for the learning phase is usually based on the back propagation rule [1].

Although this learning algorithm has proved its efficiency in solving numerous problems, it presents several disadvantages. The most important is the absence of rules allowing to fix the neural net architecture. The consequences can be redhibitory, because of the dependence between the learning capacity of the neural network and its architecture [2]. An under-dimensioned neural network with respect to a posed problem is revealed unable to perform a correct demarcation of the decision regions, leading to a miss-classification of the learning set. Likewise, an over-dimensioned neural network will tend to "memorize" or, in others words, to associate each point to a decision region and will not be efficient to generalize. On the other hand, it has been proved recently that learning a neural network by back propagation is a NP-complete problem [3].

In section one, we present the basis of a new method to design the one hidden layer perceptron architecture restricted to the convex problem solution in pattern recognition. The first idea of this approach is inspired by the paper of Lipmann [4] in which he brought emphasis to the relationship between the geometrical and topological properties of the decision regions of a pattern recognition problem and the neural network structure. The method is derived from the gift-wrapping algorithm [5] which is mainly used in computational geometry for the construction of convex hulls in a *d*-dimensional euclidean space. For a better understanding we only present the two dimensional version: the Jarvis's march algorithm [5].

In section two we show that edge detection can be solved by a neural network with only one hidden layer as an alternative to classical methods like Sobel and Prewitt [6] approaches which employ the local maxima module of the gradient.

Finally, in section three, we provide some numerical results using the geometrical method described above. The performances are very similar to those obtained by a Prewitt mask since the learning sample set employed to train the neural network was an image filtered by this edge operator.

1. USING A GEOMETRICAL ALGORITHM AS LEARNING RULE FOR NEURAL NETS

1.1. Definitions and notations

We present beforehand some geometrical definitions concerning computational geometry.

linear variety

Given k distinct points $p_1,..,p_k$ in E^d ($k<=d$), the linear combination

$a_1p_1 + a_2p_2 +...+ a_{k-1}p_{k-1} + (1 - a_1 - ... - a_{k-1})\,p_k$

($a_j \in \mathbb{R}$, $j = 1,...,k-1$)

is a linear variety of dimension $(k-1)$ in E^d.

Convex set

A domain D in E^d is convex if, for any two points p_1 and p_2 in D, the segment $\overline{p_1p_2}$ is entirely contained in D.

Convex hull and polytope

The convex hull of a set of points S in E^d is the boundary of the smallest convex domain in E^d containing S.

The convex hull of a finite set of points in E^d is a convex polytope and *viceversa*. A *two*-dimension polytope is called a polygon.

A polyhedral set in E^d

A polyhedral set in E^d is the intersection of a finite set of closed half-spaces (a half-space is the portion of E^d lying on one side of a hyperplane).

A polyhedral set is convex, since a half-space is convex and the intersection of convex sets is also convex.

A convex polyhedral set partitions the space E^d in two disjoint regions, the interior bounded and the exterior unbounded that are separated by the polyhedral. The term polyhedral denotes the union of the boundary and of the interior.

1.2. The geometrical algorithm

The Jarvis's march algorithm determines the convex polygon edges containing a set of points in a two-dimensional euclidean space. To initialize the algorithm, the knowledge of at least one of the convex polygon vertices is required. We choose the point having the lowest abscissa. This point called p_1 is certainly a hull vertice. Our purpose is to find the next consecutive vertice p_2 on the convex hull. The point p_2 is the one having the least polar angle superior or equal to zero with respect to p_1. Likewise, the next point p_3 has the least polar angle superior or equal zero with respect to p_2 taken as origin, and so forth. The algorithm stops when the last detected vertice p_n coincides with vertice p_1. If h is the actual number of vertices of the convex hull, the Jarvis's March algorithm runs in $O(hN)$ time, where N is the number of points belonging to the convex set.

In the d-dimensional case, the basic idea of the algorithm remains inchanged and the gift-wrapping algorithm may be used to find the polytope of a finite convex hull. In this case the algorithm complexity becomes $O(N^{[d/2]+1})+O(N^{[d/2]}\log N)$.

We give afterwards the steps for the construction of one hidden layer neural network applied to pattern recognition restricted to convex problems. We want to discriminate between two disjoint pattern classes, C_1 and C_2, characterized by d real parameters $\{e_1, e_2, ..., e_d\}$. The patterns are represented by points in a E^d space. The point coordinates correspond to the pattern classification parameters e_i. The gift-wrapping algorithm gives us the minimal convex polytopes containing each convex set of points. If it exists at least a convex polytope, as defined previously, which contains all the points belonging to only one class, then the two classes can be effectively separated by the convex polytope provided by the algorithm.

In this case it is straightforward to construct a one hidden layer neural network from the facets of the polytope. The input to the neurons of the hidden layer is the pattern. Each polytope facet forms a linear separation $a_1p_1 + a_2p_2 + ... + a_{k-1}p_{k-1} + (1 - a_1 - ... - a_{k-1})\,p_k$ ($a_j \in \mathbb{R}$, $j = 1,...,k-1$) in E^d of the delimited class. The linear separation coefficients a_j are computed from the k-vertices p_j which define the polytope facet. Each linear separation corresponds to a hidden layer neuron [4] and the coefficients a_j correspond to the neuron coefficients. The neurons of the hidden layer are thresholded by a sign function. The single neuron in the output layer is the AND logical function of the binary output from the hidden layer neurons. Thus we obtain a neural net where the architecture and coefficients are fully determined (see figure 1).

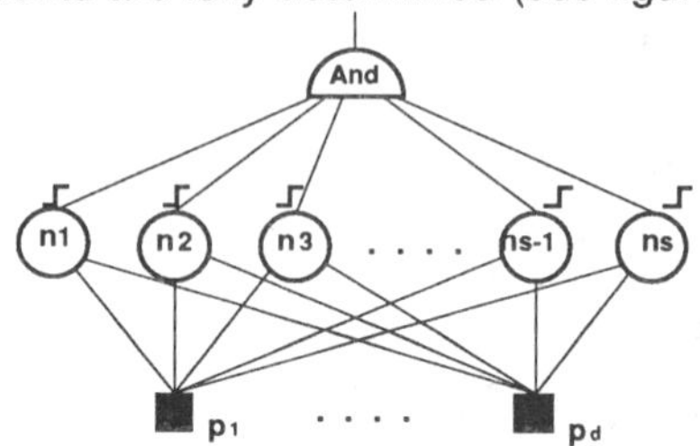

figure 1

Although the two pattern classes discrimination is made without error, in certain cases the problem can be be solved using a polytope with less facets. To overcome this drawback, we suggest a method which eliminates the useless facets of the polytope. A polytope facet (or neuron) is useless if after its elimination the neural net output remains the same for all the learning set patterns. This operation is performed in O(hN) time.

2. EDGE DETECTION USING A NEURAL NETWORK

An image can be considered like a continuous function *f(m,n)* with a gray-level value for each pixel position *(m,n)*. Its gradient denotes the gray-level variation and will be maximum in the direction of the edge. Pixels with a high gray-level variation correspond conventionally to edge points in the image. Let *H* be a *(pxp)* mask. We define, for an arbitrary image *U*, their correlation at location *(m,n)* as

$$< U,H >_{m,n} \triangleq \sum_i \sum_j h\,(i,j)\; u\,(i+m,\, j+n) = U\,(m,n) \otimes h\,(-m,\, -n) \qquad (1)$$

For gradient operators a set of masks $\{H_1, H_2, ..., H_i, ... , H_k\}$ is used to measure the gradient of the image *u(m,n)* in *k* directions. Defining the *i*-directional gradient

$$g_i\,(m,n) \triangleq < U,\, H_i >_{m,n} \qquad (2)$$

the gradient vector magnitude is often calculated as

$$g(m,\, n) \triangleq \; max_{i=1}^{k}\, \{|g_i(m,\, n)|\} \qquad (3)$$

This function agrees with the membership union function $\mu_{\tilde{A}}\,(m,n)$ of the fuzzy logic defined as

$$\mu_{\tilde{A}}(m,n) \triangleq \; max_{i=1}^{k}\; \{\mu_{\tilde{A}_i}(m,n)\} \qquad (4)$$

where $\mu_{\tilde{A}_k}\,(m,n)$ is the membership function of a fuzzy set $\tilde{A}_k$ which corresponds to the function $|g_i\,(m,n)|$ and $\tilde{A}$ is the fuzzy set of all the image points *(m,n)*.

The membership union function *m(m,n)* represents the degree of membership of couple *(m,n)* relative to the class $\tilde{A}$. We suppose that the fuzzy set $\tilde{A}$ is normal, i.e, the range of the values of the membership function is a subset of the nonnegative real numbers with the unity as the least upper bound.

The pixel location *(m,n)* is declared an edge location if $\mu_{\tilde{A}}\,(m,n)$ exceeds some threshold *t*. The locations of edge points constitute an edge map *ε(m,n)*, which is defined as

$$\varepsilon(m,n) \begin{cases} 1, & (m,n) \in Ig \\ 0, & otherwhise \end{cases} \qquad (5)$$

where

$$I_g \triangleq \{(m,n)\;;\; \mu_{\tilde{A}}(m,n) > t\}$$

The function $\mu_{\tilde{A}}\,(m,n)$ can be also defined using the membership intersection function and the membership function of the complement as follows

$$\mu_{\tilde{A}}\,(m,n) = 1 - min_{i=1}^{k}\, \{1 - \mu_{\tilde{A}_i}(m,n)\} \qquad (6)$$

Let us redefine the edge location function *ε(m,n)* as

$$\varepsilon(m,n) \begin{cases} 1, & (m,n) \in Ig \\ 0, & otherwhise \end{cases} \qquad (7)$$

where

$$I_g \triangleq \left\{(m,n);\;\; min_{i=1}^{k}\;\, \{1 - \mu_{\tilde{A}_i}\,(m,n)\} > t\text{-}1\right\}$$

Expressed is this form, it is clear that the edge location function is a convex function due to the convexity of the fuzzy intersection function [7].

Let us consider $\mu_{\tilde{A}}\,(m,n)$ like a couple of hyperplanes ($|g_i\,(m,n)|$ denotes two hyperplanes with the same coefficients but opposite signs) in the E^d euclidean space. Therefore, the edge location function denotes the intersection of a finite half-spaces set resulting in a polyhedral convex set [5]. This half-space set forms an *d*-dimensional polytope which can be transformed to a one-hidden layer neural network as depicted in the precedent section. Points outside of the polyhedral sub-space form the edge pixel class, while others are positioned inside the polyhedral. The polytope constitutes the frontier between the two classes. The frontier position depends on the choice of the threshold *t*. A high value of *t* increases the number of points inside the polyhedral sub-space and so reduce the number of edge points. Nevertheless, the polyhedral form remains the same independently of the threshold value.

3. EXPERIMENTAL RESULTS

We propose to find a new edge operator by a learning procedure. To this end, we use an image and its corresponding edge (see images 1-b and 1-b resp.), resulted from a Prewitt mask, as learning sample set. The images and the Prewitt edge operator have been obtained from the SIMPA library (Signal and IMages softwares PAckages - GDR 134 CNRS) [8]. The sample set for the learning phase contains one hundred samples of edge pixels and the same quantity of points which do not belong to the image edge. The input dimension of the neural network is four, which correspond to the orthogonal neighbors of the pixel to be evaluated.

At the end of the learning phase (which includes the elimination of useless neurons), we obtain a neural network derived from the polyhedral frontier between the two classes of pixels. The number of neurons for the hidden-layer is approximately ten (this number depends on the distribution of the learning samples but remains practically the same after each learning phase). In image 1-c, we illustrate the edge image given from the neural network for the learning image. As one can observe, the results are very similar to those obtained by the Prewitt mask.

4. CONCLUSION

An alternative method to neural network design and train, restricted to the convex problems solution for pattern recognition, has been presented. It has been applied to the edge detection and similar results have been obtained than classical edge operators. We summarize afterwards the inner advantages of the method:

1) The method differentiates the complexity of a problem in three categories: linear problems, convex problems and complex problems.

2) If the problem is convex, the method provides its neural network architecture (number of neurons in the hidden-layer) and the coefficient values.

3) The method insures a bounded learning time, which is polynomial in time and depends on problem dimension.

The main advantages of the method when applied to the edge detection are:

1) The edge detection can be thought to a neural network by sample in a polynomial time. We have chosen a Prewitt mask as a model for the learning phase and have obtained similar results but a more appropriate edge operator or a synthetical edge may be used to this goal.

2) The resulting neural network architecture matchs well with the classical perceptron model and can be easily transformed to a classical edge operator structure.

image 1-a

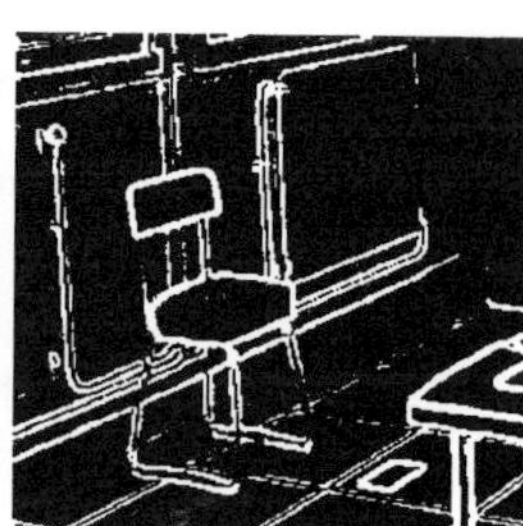

image 1-b

image 1-c

ACKNOWLEDGMENTS

The authors want to thank Dr. J.M. Chassery for his valuable comments on an early unpublished version of this paper. We would also like to thank Mr. Dai Mo and Mr. A. Zolghadri for their scientific contributions.

REFERENCES

[1] D. E. Rumelhart, G. E. Hinton, and R. J. Williams "Learning internal representations by error propagation," in Distributed Processing : Explorations in the Microstructure of Cognition, Vol. 1. Cambridge, MA : M.I.T. Press, 1986.

[2] Gavin, J. Gibson and Colin F. N. Cowan, "On the decision Regions of Multilayer Perceptrons," in Procedings of the IEEE, vol. 78, No. 10, pp. 1590-1594, October 1990.

[3] A. Blum and R. Rivest, "Training a 3-node neural network is NP-complete," in Proc. IEEE Conf. Neural Inform. Processing Syst., Denver, CO, 1988.

[4] R. P. Lipmann, "An introduction to computing with neural nets, "IEEE ASSP mag., pp4-22, April 1987.

[5] F. P. Preparat "Computational Geometry. An introduction, " Springer-Verlag, 1985.

[6] A. K. Jain "Fundamentals of digital image processing", Prentice-Hall, 1989.

[7] Yoh-Han Pao "Adaptive Pattern Recognition and Neural Networks," Addison-Wesley, 1989.

[8] SIMPA - Reference Manual, CNRS - Paris.